Biology of the Plant Bugs
(Hemiptera: Miridae)

Biology of the Plant Bugs
(Hemiptera: Miridae)
Pests, Predators, Opportunists

Alfred G. Wheeler Jr
Department of Entomology
Clemson University

with a foreword by
Sir T. Richard E. Southwood FRS
University of Oxford

COMSTOCK PUBLISHING ASSOCIATES
a division of
CORNELL UNIVERSITY PRESS
Ithaca and London

First published 2001 by Cornell University Press

Printed in Hong Kong

Library of Congress Cataloging-in-Publication Data

Wheeler, Alfred George, 1944–
    Biology of the plant bugs (Hemiptera:Miridae) : pests, predators, opportunists / Alfred G. Wheeler, Jr.
        p. cm.
    Includes bibliographical references (p.).
    ISBN 0-8014-3827-6 (cloth : alk. paper)
    1. Miridae.   I. Tile.
    QL523.M5 W44 2001
    595.7'54—dc21
                                                        00-011505

Cornell University Press strives to use environmentally responsible suppliers and materials to the fullest extent possible in the publishing of its books. Such materials include vegetable-based, low-VOC inks and acid-free papers that are recycled, totally chlorine-free, or partly composed of nonwood fibers. Books that bear the logo of the FSC (Forest Stewardship Council) use paper taken from forests that have been inspected and certified as meeting the highest standards for environmental and social responsibility. For further information, visit our website at www.cornellpress.cornell.edu.

Cloth printing   10 9 8 7 6 5 4 3 2 1

*To
Bertil Kullenberg,
whose basic research on mirid biology
remains unsurpassed in its breadth and detail,
and to
I. M. Kerzhner,
whose vast knowledge of
heteropteran systematics
is so willingly shared*

Contents

Foreword

Mirids are often among the most numerous insects found on vegetation; Dr Wheeler emphasises this in Chapter 1. Mirids are treated prominently in the economic literature both as pests of major crops and as natural enemies of certain key pests. Yet if one searches the indices of texts on insect-plant biology, on ecology, or on evolution one will find little reference to them. Does this matter? It does, because the habits of this group cut across some of the assumptions underlying certain key generalisations in ecology and evolution.

Most models of natural population dynamics are based on predator-prey systems; a feedback is implicit in the underlying Lotka-Volterra equation. Taken to the limits, a large predator population will kill so many prey that they then become rare, causing the predator's numbers to be brought down by starvation. Most simply this can lead to oscillating cycles of relative abundance. Theoretical ecologists have explored how this pattern will differ if the predator is a generalist and has a functional response to the relative density of different prey. If the predator kills increasing numbers of a prey species that is becoming more abundant, this will dampen the oscillations and tend to lead to a more stable ecosystem. But a generalist predator needs some prey, and therefore they may not be present on the plant at the very start of a population buildup. Unchecked, this may lead to an outbreak such as occurs, for example, in many annual crops when a pest species arrives in a virtual "ecological vacuum" (so far as enemies and competitors are concerned) and thus has the potential for the classic logistic rate of increase.

But the versatile and varied feeding habits of mirids so comprehensively reported in this synthesis of basic and applied literature invalidate these simple models for any ecosystem where they occur. Facultative predators may exist in the absence of their prey, but like a fire brigade they may be available to dampen down any outbreak. Few other families of insects exhibit such a range and plasticity of diet. The shifts to predation in lycaenid butterflies are well recognised, but mirids are both more diverse and abundant. Undoubtedly in many ecosystems mirids perform an important role in buffering population swings; the predilection of some species for the reproductive organs of plants can make them an impor-

tant factor in plant, as well as invertebrate, population dynamics.

Mirids also provide novel insights into the evolutionary controversies surrounding speciation. Whilst it is maintained that most speciation is allopatric, with separate geographical ranges, mirids have at least the potential for allochronic speciation. The breeding seasons of mirid populations may become separated in time. Many have very short adult lives. For example, the males of *Harpocera thoracica* occur in the field for little more than a week; during the other 51 weeks of the year gene flow is impossible. This species is restricted to a single host plant, but the development time of any individuals that colonised another species of plant would almost certainly change by more than a week, thus providing a barrier to gene flow. This may have happened, for example, in the species-rich subgenus *Neolygus*, more particularly in those attached to trees. I have suggested that this latter aspect is also a reflection of both the generally lower level of dispersal in arboreal species and the structural features of trees—their large patch size. Of course, as in so many features, mirids are very varied. This hypothesis of allochronic speciation cannot apply to those with relatively long adult lives, for example, *Lygus rugulipennis*, perhaps the most polyphagous and dispersive of species. There are also many examples of several members of the same genus with the same host plant; careful study may show fascinating niche partitioning (e.g., *Orthotylus* on broom).

Thus in these two key areas of biology, ecology and evolution, a familiarity with mirid biology could greatly enrich our understanding. Why have they been so neglected compared with other comparably abundant groups? Undoubtedly they are delicate. They do not take kindly to experimental handling and appear to be difficult to breed in the laboratory. They do not have the macroscopic glory of the adult butterfly or the resistance to gross collection into alcohol of the aphid. However, all these difficulties may be overcome. I believe that when it is more widely appreciated that the mirids provide such abundant and interesting models to investigate these less orthodox ecological and evolutionary processes, they will be more widely studied. This book, with its masterly and comprehensive account of our present knowledge, will surely play a key part in this development. Because of the intrinsic interest of the Miridae, it is not simply for specialists but a work of reference of value to all organismal biologists. It is a book, as well, of much value to agricultural workers and specialists in biological control.

T. R. E. SOUTHWOOD

Oxford

Preface

Although the Miridae deserve more attention from researchers, interest in this largest family of the true bugs is increasing. In fact, recent worldwide activity in mirid biology and systematics inspired this review. Systematists not only are continuing to describe new species and revise genera, but also are reevaluating higher classification, applying cladistic methods to reconstruct phylogenetic relationships, and analyzing distributions in particular geographic areas. From an emphasis on integrated pest management (IPM), applied entomologists are increasingly aware of plant bugs as pests of various crops or as natural enemies of crop pests. They are developing mirid-resistant crop varieties, evaluating plant bugs for the biological control of weeds, and incorporating predacious mirids into IPM programs. Researchers have now specifically identified the female sex pheromone in several mirids, characterized the defensive secretions of the adult metathoracic scent glands, identified various enzymes and other constituents of salivary secretions, analyzed behavioral events and phagostimulants that initiate a feeding response, assessed the factors that terminate egg diapause, and devised artificial media for rearing.

This book's emphasis on trophic habits reflects my interests and specialization. At times I might be guilty of a tell-you-all-I-know approach, believing that numerous examples of a particular feeding habit or symptom of plant injury are needed for adequate coverage of the topic and for access to the literature. Brues (1946), in his *Insect Dietary*, admitted to having included matter peripheral to the main theme of his book. I also include biological material tangential to feeding—defensible, I feel, because feeding has consequences beyond supplying nutrients and energy. In some cases, I provide references for topics covered only superficially or introduced casually in a particular discussion. Without question, this book's tenor is influenced by the philosophy of Schaefer and Mitchell (1983): "If—as we believe—entomologists are interested in insects as insects, and not only as units for the study of evolution, physiology, or genetics, or as pests to be controlled, then a knowledge of what insects eat is interesting in its own right."

Some of the information presented is anecdotal; readily apparent will be the biological aspects needing more rigorous study. Moreover, some of the published notes on natural history that I have chosen to include could be considered trivial, but they assume relevance when placed into context within a synthetic work. If this review stimulates further research on plant bug biology, I shall feel amply rewarded. I would be pleased if evolutionary biologists turn to mirids for studying "explosive speciation" in phytophagous insects, if ecologists assess the attributes that seemingly preadapt certain mirids for colonizing new host plants, or if biologists begin to address some of the suggestions for research presented in the final chapter.

Specialists in crop protection and extension entomology, particularly those working with alfalfa, apple, cocoa, coffee, cotton, ornamentals, small grains, and other major world crops, should benefit from the detailed information on the habits of plant bugs and from the descriptions and photographs of the symptoms associated with mirid feeding. Biological control workers can learn more about the habits of species that are potentially useful in suppressing populations of pest insects or weeds. This review might interest naturalists and biologists studying pollination ecology, food webs, and insect-plant interactions. It also should be useful to plant pathologists because the external symptoms of mirid feeding sometimes closely resemble those induced by various plant-disease agents. I hope this review will serve as a starting point and reference for students, taxonomists, general and applied entomologists, ecologists, pest-management and extension specialists, plant inspectors, and others seeking biological information on plant bugs.

Despite the many gaps in our knowledge of the trophic habits of mirids and related areas of their bionomics, the literature containing information on this subject amounts to many thousands of publications. I hope that from my efforts "new scientific knowledge [will be] made meaningful by sorting and sifting the bits and pieces to provide a larger picture" (Day 1988).

Acknowledgments

This book is the outcome of studies on plant bugs that began officially in the early 1970s but really had much earlier beginnings. My parents encouraged my childhood fascination with natural history, particularly insects, and some excellent teachers in the elementary grades fostered it. For their nurturing of more sophisticated biological pursuits, I express deep appreciation to two outstanding professors I had as an undergraduate student at Grinnell College: Kenneth A. Christiansen and Waldo S. Walker. I am also grateful to James A. Slater of the University of Connecticut for encouraging my interest in terrestrial Heteroptera when I contacted him for advice during my years as a graduate student at Cornell University, and for his continued support of my work.

My attempt to make the literature on mirid biology more accessible has benefited immensely from the cooperation of students and researchers throughout the world. One of the most satisfying aspects of this study has been the opportunity to interact with this international community of biologists. So many people have aided my work that their individual contributions cannot be detailed, nor their professional affiliations listed; yet I hope that my omission of such information will not be construed as depreciating their valued assistance. The many persons who have enhanced this work not only helped directly in various ways, but also helped indirectly by sustaining my efforts when my enthusiasm for the project would wane. For patiently listening to progress reports and encouraging me to persevere during the book's long gestation, I am especially grateful to Peter Adler, Allen Cohen, Karen Hauschild, Thomas Henry, Richard Hoebeke, Gary Miller, Frank Stearns, James Stimmel, and Craig Stoops. I have benefited from the synergistic enthusiasm generated from working on mirids with Tom Henry since the early 1970s and have also enjoyed his companionship during numerous collecting trips in the continental United States, as well as in Puerto Rico, China, Japan, and Taiwan. I have similarly enjoyed trips with Rick Hoebeke to do fieldwork in the eastern United States, Canadian Maritime Provinces, and Puerto Rico. In addition, Tom also identified mirids so that unpublished observations could be used in the book, and scanned images for preparation of the figures, and Rick always responded to requests to check certain references at Cornell University. Gary Miller kindly provided illustrations for Figure 17.1.

Among the many persons who sent prepublication copies of their work, furnished references, lent transparencies or photographs, permitted me to cite unpublished information as personal communications, translated articles published in various languages, answered questions, or otherwise aided this project, I thank P. H. Adler, A. E. Akingbohungbe, R. D. Akre, J. R. Aldrich, W. W. Allen, O. Alomar, S. Appanah, E. Aranda, A. Arzone, B. Aukema, D. A. Avé, E. A. Backus, R. M. Baranowski, L. A. Bariola, D. A. Barstow, D. Bartlett, G. A. C. Beattie, J. W. Begley, J. S. Bentur, A. L. Bishop, R. L. Blinn, G. Bockwinkel, G. Boivin, R. B. Borthwick, N. J. Bostanian, C. C. Bower, D. W. Boyd, S. K. Braman, W. L. Brown Jr., R. L. Bugg, J. R. Byers, R. A. Byers, D. L. Caldwell, E. A. Cameron, D. M. Caron, M. E. Carter, J. C. M. Carvalho, J. F. Cavey, Y.-C. Chang, V. C. Chattergee, A. Chinajariyawong, T. C. Cleveland, A. C. Cohen, J. R. Coulson, P. H. Craig, W. S. Cranshaw, T. W. Culliney, H. Curtis, D. W. Davis, W. H. Day, J. D. DeAngelis, J. W. Debolt, R. F. Denno, S. Devasahayam, K. Dhileepan, J. F. Dill, M. Dissevelt, P. H. van Doesburg, W. R. Dolling, D. Donnelly, G. P. Donnelly, J. E. Eger, M. D. Eubanks, G. Fauvel, P. S. F. Ferreira, R. V. Flanders, S. J. Fleischer, M. L. Flint, D. Folland, J. H. Frank, A. Freidberg, R. C. Froeschner, R. W. Fuester, D. Furtek, D. G. Furth, R. Gabarra, W. C. Gagné, M. J. Gaylor, D. R. Gillespie, R. D. Goeden, L. L. Goering, C. S. Gorsuch, J. C. Guppy, K. J. Hackett, D. G. Hall, D. T. Handley, J. D. Hansen, A. M. Harper, H. Hayashi, B. S. Heming, H. J. Hendricks, T. J. Henry, D. A. Herms, G. B. Hewitt, I. G. Hiremath, E. R. Hoebeke, H. W. Hogmire, J. K. Holopainen, G. A. Hoover, K. Hori, S. C. Hoyt, L. J. Hribar, R. A. Humber, C. G. Jackson, W. T. Johnson, W. A. Jones, N. Jonsson, G. L. Jubb Jr., J. A. Kamm, L. Kassianoff, M. Kawamura, S. Keller, I. M. Kerzhner, P. G. Kevan, K. C. Khoo, J. B. Knight, C. S. Koehler, B. Kullenberg, J. D. Lattin, D. A. Leatherman, T. F. Leigh, B. Lenczewski, J. MacFarlane, R. P. Macfarlane, J.-C. Malausa, R. F. L. Mau, J. R. Mauney, G. C. McGavin, R. McGraw, R. R. McGregor, J. D. McIver, F. J. Messina, J. G. Millar, G. L. Miller, W. C. Miller III, R. F. Mizell, R. Muhamad, M. K. Muliyar, G. M. Mullen, C. A. Mullin, A. J. Musa, J. W. Neal Jr., B. R. Nelson, B. E. Norton, K. F. Nwanze, W. P. Nye, M. S. Okuda, P. F. O'Leary, F.

Önder, P. A. C. Ooi, C. E. Palm, J. M. Palmer, D. J. Parker, M. P. Parrella, S. Passoa, M. Pathak, J. D. Pinto, C. D. Pless, S. A. Podlipaev, S. E. Pohl, D. A. Polhemus, J. T. Polhemus, Y. A. Popov, R. Preston-Mafham, J. F. Price, C. Prior, V. G. Putshkov, E. B. Radcliffe, K. Raman, R. H. Ratcliffe, P. S. Rattan, A. Rauf, R. E. Rice, H. W. Riedl, C. E. Rogers, D. J. Rogers, B. D. Schaber, C. W. Schaefer, L. Schaub, G. Schmitz, H. Schmutterer, D. J. Schotzko, T. D. Schowalter, R. T. Schuh, M. F. Schuster, M. D. Schwartz, D. R. Scott, J. G. Scott, M. K. Sears, J. Šedivý, M. Semel, H. C. Sharma, B. M. Shepard, D. J. Shetlar, E. Show, C. H. Sim, B. C. Simko, T. E. Skelton, J. A. Slater, E. S. C. Smith, J. P. Smith, G. L. Snodgrass, M. G. Solomon, J. J. Soroka, D. B. South, S. M. Spangler, F. G. Stearns, G. J. Steck, J. L. Stehlík, A. G. Stephenson, W. L. Sterling, R. K. Stewart, J. F. Stimmel, C. A. Stoops, P. Štys, D. L. Sudbrink Jr., V. Sudoi, C. A. Sutherland, M. H. Sweet, M. Takai, G. Taksdal, Y. Tanada, H. M. A. Thistlewood, T. C. Tignor, H. Toxopeus, N. P. Tugwell, S. A. Ulenberg, A.-L. Varis, D. J. Voegtlin, N. Waloff, L. F. Wilson, M. R. Wilson, S. W. Wilson, T. Yasunaga, A. Youdeowei, H. J. Young, O. P. Young, and L. Y. Zheng.

For reviewing chapters or parts of chapters, I gratefully acknowledge P. H. Adler, J. R. Aldrich, D. A. Andow, A. Asquith, D. A. Avé, E. A. Backus, L. Bishop, D. W. Boyd, J. H. Cane, R. D. Cave, A. C. Cohen, W. H. Day, J. W. Debolt, R. F. Denno, H. Eichenseer, G. C. Eickwort, S. D. Eigenbrode, A. Erhardt, J. L. Frazier, K. S. Gibb, F. E. Gildow, D. T. Handley, J. D. Hansen, T. J. Henry, D. A. Herms, R. J. Hill, E. R. Hoebeke, K. Hori, G. W. Hudler, L. A. Hull, R. A. Humber, W. A. Jones, G. L. Jubb Jr., P. G. Kevan, K. C. Khoo, I. M. Kerzhner, B. Kullenberg, S. Laurema, R. D. Lehman, T. F. Leigh, R. P. Macfarlane, J. V. McHugh, J. E. McPherson, P. W. Miles, G. L. Miller, C. A. Mullin, L. R. Nault, J. W. Neal Jr., D. A. Polhemus, E. G. Rajotte, R. E. Rice, C. W. Schaefer, G. A. Schaefers, D. J. Schotzko, R. T. Schuh, H. C. Sharma, O. D. V. Sholes, J. A. Slater, T. P. Spira, S. D. Stewart, J. F. Stimmel, M. B. Stoetzel, G. M. Stonedahl, C. A. Tauber, M. J. Tauber, A.-L. Varis, C. Vincent, R. A. Welliver, Q. D. Wheeler, R. N. Wiedenmann, T. K. Wood, K. V. Yeargan, and D. K. Young.

I. M. Kerzhner (Zoological Institute, Academy of Sciences, St. Petersburg, Russia) read and commented on nearly all the chapters, giving invaluable advice on nomenclatural matters and calling my attention to recent papers involving relevant synonymy. Several other reviewers read more than one chapter, provided particularly insightful reviews, or both. Although their substantial input is not distinguished in the above list of names, I hope those who took the time to write exceptionally useful critiques realize how much I appreciated their help. Their additions, corrections, and other comments did much to improve the manuscript, as did those in an anonymous review commissioned by Cornell University Press. All errors of fact or interpretation are my responsibility.

I gratefully acknowledge the financial support received from many sources. Through a research grant in 1985, the American Philosophical Society supported initial work on the book. I am indebted to the Book Fund of Cornell University's College of Agriculture and Life Sciences and to former dean David L. Call, former dean D. B. Lund, and the late Warren T. Johnson, for a grant that substantially subsidized publication; and to the Entomological Society of America for a travel grant that enabled me to attend the XIX International Congress of Entomology in Beijing in 1992, and also allowed me to do fieldwork in Taiwan and Japan following the congress. Dow Agrosciences of Indianapolis also partially subsidized publication of this book.

I thank the Pennsylvania Department of Agriculture, especially E. B. Wallis Jr. and K. Valley, for granting me time for library research at Cornell University. There, department chairs R. A. Morse and M. J. Tauber graciously arranged for the use of an office and access to secretarial help, as well as secured my status as Visiting Fellow during three stays of two months each. I wrote the drafts of several chapters at Cornell in an atmosphere that facilitated my work, and revised all chapters and prepared the back matter while at the Department of Entomology, Clemson University. I sincerely thank former department chair Thomas E. Skelton for approving my appointment as adjunct professor and former chairs Randall P. Griffin and Paul M. Horton, and current chair Joseph D. Culin, for granting the use of departmental facilities and supplies. The camaraderie and congeniality that pervade Clemson's entomology department provided just the right academic climate for finishing the manuscript. To my colleagues at Clemson, Peter Adler in particular, I extend sincere thanks. In addition, Laura Reeves and especially Tammy Morton gave valuable assistance, mainly in helping secure authors' and publishers' permissions to reprint material in the book, doing computer scans of artwork, and producing a final version of the manuscript.

Special thanks are also due the many publishers, authors, and photographers who generously permitted me to reproduce their work. I am especially pleased to receive permission to use several superb transparencies from Premaphotos Wildlife, courtesy of Ken and Rod Preston-Mafham. The sources of the transparencies, black and white photographs, and other material are acknowledged in the figure legends. I give special appreciation to Jim Stimmel of the Pennsylvania Department of Agriculture, for his photographic skills and willingness, often on short notice, to photograph mirids that I would bring in from the field. T. Yasunaga, who was such a gracious host when I visited Kyushu, Japan, kindly arranged for the loan of several excellent images of Japanese mirids. P. Harpootlian's help in improving several transparencies is appreciated. I thank Clemson's D. W. Boyd for helping prepare several figures, K. T. Hathorne for redrawing several illustrations, and C. S. Gorsuch for his help with Figure 17.1.

My efforts to compile the world literature on the feeding habits of plant bugs were substantially aided by the diligent efforts of interlibrary loan and reference librarians at Clemson University, Cornell University, and Pennsylvania State University. This work would have been considerably less comprehensive without their help.

I am especially grateful to Robb Reavill, former science editor at Cornell University Press, for her encouragement, good humor, and patience, and to Helene Maddux, former senior manuscript editor, for patiently answering my many questions about format and style, her unfailing agreeable nature, and her editorial skills that improved the final text. My appreciation also extends to manuscript editor Nancy Winemiller, freelance copy editor Mary Babcock, and freelance indexer David Prout for their excellent work. In addition, I thank Cornell University Press's science editor, Peter Prescott, for patiently awaiting submission of the manuscript and for surmounting obstacles for publishing a manuscript that, when finally submitted, was substantially longer than originally intended; his assistant, Lynn Coryell, for her helpfulness and good cheer; and Richard Rosenbaum, former production director, and designer Bob Tombs, for their help in producing the book.

Now that this project is nearly finished, I can say "thanks" to Maureen Carter and actually mean it. She first suggested that I abandon attempts to prepare a review article on mirid feeding habits and instead write a book on the subject. I also take pleasure in acknowledging the assistance of Susan Pohl, who typed the entire manuscript, provided editorial guidance, and somehow managed to maintain her characteristic good humor through countless manuscript revisions. I much appreciate her superb skills and her exemplary work on this project.

Finally, I am grateful to Sir Richard Southwood for writing the foreword. His early work focused on biology and systematics of the Heteroptera, including mirids, and his later research in community ecology and population dynamics brought him international acclaim as an ecologist. He, therefore, seems ideally suited to introduce a book on mirid biology and ecology.

A. G. WHEELER JR.

Clemson, S. C.

Biology of the Plant Bugs
(Hemiptera: Miridae)

1/ Introducing the Plant Bugs

The last thing we decide in writing a book is what to put first.
— Blaise Pascal 1670 (in Winokur 1986)

Mirids or plant bugs have not inspired the passion devoted to dragonflies, longhorned beetles, tiger beetles, and butterflies—insects that are favored by amateurs. The Miridae admittedly are mostly small and fragile insects, are not often showy, and can be difficult to identify. Ecologists have not emphasized plant bugs, perhaps because these insects can be hard to observe in the field and are not easily reared in the laboratory. But the Miridae exhibit great morphological diversity, display remarkable trophic plasticity, and play a key role in natural systems and agroecosystems—both as herbivores and as predators.

Plant bugs are often exceedingly abundant on annual and perennial plants in both temperate and tropical regions. Yet their great diversity goes unnoticed if collecting is restricted to sweeping field crops or weeds along roadsides and in other disturbed areas. Much of mirid species richness can be appreciated only by tapping branches of trees and shrubs over a beating net or sheet during the relatively limited period when adults are present. Conventional collecting techniques, such as use of the sweepnet, generally yield only the most common and widespread species. The use of blacklight traps, effective in attracting certain species, provides scant ecological information. On the other hand, Malaise and yellow pan traps generally fail to capture mirids.

Ecologists have mostly emphasized chewing herbivores (defoliators), whose effects as consumers of foliage and other plant parts are readily observed (e.g., Thompson and Althoff 1999) or at least are more often apparent than those caused by arthropods with piercing-sucking mouthparts. Attempts to estimate the levels of consumption by sucking herbivores usually involve homopterans such as aphids, leafhoppers, and spittlebugs (e.g., Schowalter et al. 1981) rather than mirids and other heteropterans. Seed-chewing insects perhaps also are better known than are seed-sucking insects (Slansky and Panizzi 1987). North American entomologists and agriculturists tend to be familiar with common, widespread species such as the tarnished plant bug (*Lygus lineolaris*). Familiarity with mirid life histories and habits may even stop with the pestiferous lygus bugs,

which are atypical of most mirids in that they overwinter as adults and have a wide host range.

Knowledge of plant bug habits is generally sparse for all but a few groups that are of agricultural importance. Reviewing the species associated with fruit crops in Canada, Kelton (1983) noted, "Doubt exists whether they are harmful or beneficial, confusion exists in naming them, and little is known of their biology." The problem of inadequate information for most mirids is similar to, but much greater than that mentioned for the reasonably well-studied lycaenid butterflies: "Knowledge of the habits of a given species is usually fragmentary and derived from brief periods of daylight observation made by a single worker—whose findings may never have been confirmed" (Cottrell 1984).

Succeeding chapters, particularly in Parts I and II, provide many of the basic ecological, morphological, physiological, and taxonomic facts readers might want to know about plant bugs. Those needing quicker access to general information on the Miridae should refer to Table 1.1.

Ecological Success

The true bugs or Heteroptera can be considered an ecologically successful group, one of the more successful of the Exopterygota (Goodchild 1966, Schuh and Slater 1995). This suborder's largest family, the Miridae, has persisted at least since the Jurassic (see Chapter 6). As Wilson (1987) pointed out in discussing the ecological success of ants and their persistence through geological time, sheer longevity depends on four general population-level qualities: high species diversity, penetration of unusual or unique adaptive zones, large populations, and wide geographic range. Plant bugs qualify as ecologically successful in terms of their diversity, range, and densities. They have not occupied unusual adaptive zones to the extent that ants and certain other holometabolous groups have, but they are among the few heteropterans that live as commensals in spider webs and that specialize on glandular-hairy and even carnivorous plants. They show greater feeding plasticity than the ecologically successful Auchenorrhyncha and Sternorrhyncha (Homoptera) and other Heteroptera. Indeed, some mostly phytophagous plant bugs, for

Table 1.1. A plant bug primer[a]

- Common names

  Plant bugs, English: from occurrence on vegetation; sometimes capsids, from former name of family (see Chapter 2)

  Blindwanzen, German: referring to absence of ocelli (see Chapter 5)

  Blomstertaeger, Danish: from frequent association with flowers (see Chapter 10)

- Position in animal kingdom

  Phylum Arthropoda: invertebrates with hardened external skeleton, segmented appendages

  Class Insecta: arthropods with wings, three pairs of legs

  Order Hemiptera: exopterygote insects with sucking mouthparts (hinged stylets), labial and maxillary palps lacking

  Suborder Heteroptera: true bugs; adults with forewings (when present) folded flat over body, usually with basally thickened corium, apical membranes overlapping; nymphs with dorsal abdominal scent glands, adults with metathoracic scent glands (see Chapter 5)

  Infraorder Cimicomorpha: ca. 16 families, including the 2 largest in Heteroptera, Miridae and Reduviidae; also Anthocoridae and Nabidae

  Superfamily Miroidea: Miridae and phytophagous families Thaumastocoridae and Tingidae

  Family Miridae: for characteristics, see Recognition features (below) and Chapter 5

- Recognition features

  True bugs (Heteroptera): cuneus in forewing, 1–2 cells in membrane, male terminalia asymmetrical, ocelli absent except in isometopines (Chapter 5)

- Size and appearance

  Length: 1.5–15 mm; mostly 3–7 mm

  Coloration: often cryptically green or brown; some aposematically red or orange and black

  Form: oval to elongate; resemblance to ants (myrmecomorphy) common (see Chapters 5, 6)

- Taxonomic diversity

  Species: ca. 10,000 validly described; actual number, including undescribed species, estimated to be 20,000

  Genera: ca. 1,400; largest genus, *Phytocoris*: >600 species (see Chapter 6)

  Subfamilies: 8 or 9 (see Chapters 3, 17)

- Distribution

  Zoogeography: all major regions

  Altitudinal range: below sea level to ca. 5,400 m (see Chapter 6)

- Antiquity

  Oldest fossils: Upper Jurassic, ca. 150,000,000 years before present (see Chapter 6)

- Habitats

  Terrestrial: widespread in diverse communities, both natural and disturbed, rare or absent in areas devoid of plant cover

  Semiaquatic: some species on aquatic macrophytes (see Chapter 6)

- Host plants and range

  Vascular plants: ferns, gymnosperms, angiosperms (dicots, monocots) (see Chapter 6)

  Mosses: a Japanese bryocorine develops on a moss (see Chapter 6)

  Fungi: some cylapines on pyrenomycetes (see Chapter 13)

  Diet breadth: monophagous to polyphagous; most species host restricted (see Chapter 6)

- Feeding habits

  Mode: lacerate (or macerate) and flush with extraoral digestion; a particulate slurry imbibed from plant or animal tissues (see Chapter 7)

  Range: strict phytophagy (see Chapters 9, 10, 12) to obligate (strict or nearly so) zoophagy (see Chapter 14)

- Development

  Paurometabolous: egg; nymph (sometimes termed *larva*), with typically 5 instars; adult (see Chapters 5, 6)

- Life cycle

  Premating period: 1–7 days; generally 3–6

  Preoviposition period: 1–20 days depending on temperature and species

  Incubation period: 10–16 days for nondiapausing eggs at 20°C; highly temperature dependent

  Nymphal development: 12–35 days at 20–30°C

  Annual generations: often 1, especially in species developing on woody plants; weed feeders of early-successional communities and some predators multivoltine (see Chapter 6)

- Dispersal

  Walking: principal means of locomotion

  Flight: well developed in most species but ranges little known; flight active, or passive on air currents (see Chapter 6)

- Overwintering stages

  Egg: primary stage in temperate regions, particularly among tree and shrub feeders

  Nymph: rare in mirids

  Adult: common in certain multivoltine species (see Chapter 6)

- Natural enemies

  Invertebrate predators: e.g., spiders, mantids, anthocorids, nabids, geocorid lygaeoids, reduviids, other mirids, coccinellids, ants, sphecids

  Vertebrate predators: birds, mammals, reptiles, amphibians

  Egg parasitoids: eulophids, mymarids, scelionids, trichogrammatids

Table 1.1. *Continued*

Nymphal parasitoids: euphorine braconids
Adult parasitoids: tachinids (*Phasia* [= *Alophorella*] spp.)
Parasitic mites: erythraeids, trombidiids
Parasitic nematodes: mermithids
Microbial pathogens: entomophthoraceous fungi, bacteria, microsporidian protozoa (see Chapter 6)
- Defense mechanisms
  Primary: aposematism, crypsis, mimicry
  Secondary: chemical; active, e.g., hopping, running, flying (see Chapter 6)
- Agricultural importance
  Detrimental: key or major pests of alfalfa, apple, cashew, cocoa, cotton, tea, and other crops; limit production of certain tropical crops (see Chapters 9, 10, 12); mostly ineffective vectors of phytopathogens (see Chapter 8); can destroy natural enemies (e.g., phytoseiid mites, parasitic wasps) of crop pests (see Chapter 14)
  Beneficial: naturally occurring predators of crop pests potentially useful in integrated pest management (IPM); omnivory in predators enhances survival during periods of prey scarcity; used successfully in classical biological control of planthoppers; help suppress populations of thrips and whiteflies in greenhouses (see Chapter 14); mostly inefficient pollinators of crop plants (see Chapter 11); possible biocontrol agents of weeds (see Chapters 6, 9)
- Medical importance
  Nuisance "biters": bites of certain species cause pain, itching, chiggerlike welts; effects usually transitory (see Chapter 15)
- Ecological and evolutionary importance
  Food webs: many species omnivorous; cannibalism, intraguild predation, scavenging common (see Chapters 14, 15)
  Food source: important in diet of farmland and grassland birds (see Chapter 6)
  Plant communities: seed predators might affect host population dynamics (see Chapter 10); other phytophages may affect phenology and physiology of native shrubs (see Chapters 9, 10)
  Plant interactions: intimate association with plants invites studies on cospeciation vs. host transfer or colonization, and sympatric speciation
  Herbivory on glandular plants: many dicyphines develop on hosts whose glandular trichomes and exudate deter most other herbivores, exploiting generally unavailable resources and sequestering secondary metabolites for own defense (see Chapters 7, 9, 15)
  Predation: use of extraoral digestion suggests need to reevaluate traditional ideas on functional response kinetics (see Chapter 14); feeding on small prey (e.g., arthropod eggs, mites, scale crawlers) unnoticed by investigators affects studies of fecundity and longevity (see Chapter 6)
- Importance in conservation biology
  Specialized communities: associated with plants of granite outcrops, pitch pine–scrub oak barrens, serpentine barrens, shale barrens, etc.; potentially lower fitness of plants of special concern
  Biodiversity: apparently rare species are threatened by habitat destruction; species of monotypic genera deserve preservation for their phylogenetic uniqueness (see Chapter 6)
  Biomonitoring: susceptibility to certain pesticides and vulnerability to habitat disturbances make mirids potentially useful indicators of ecological changes (see Chapter 6)

Note: Many statements are generalized or simplified; exceptions are covered in later chapters. See Glossary for technical terms.
[a] Cross-references indicate chapters that contain more information on a topic.

---

example, *Lygus hesperus*, can complete their nymphal development while on a plant-only or prey-only diet; in this case, carnivory is substitutable for phytophagy (Naranjo and Gibson 1996). Conversely, in the case of certain mainly predatory mirids, phytophagy seems substitutable for zoophagy (Ghavami 1997). Even though a mirid might be able to complete its development on a particular plant or animal diet, different plant parts and different prey generally are not nutritionally equivalent, and therefore, such variation in food resources can affect life-history attributes and population dynamics (e.g., Eubanks and Denno 1999). Regardless of prey densities, plant feeding is necessary for at least certain omnivorous mirids (Gillespie et al. 1999).

Some of the reasons for the ecological success of mirids relate to their interactions with plants, small size, ability to detoxify secondary metabolites, poikilothermy, and possession of a sclerotized exoskeleton

(e.g., Bernays 1982). Adaptations and modifications, including endophytic oviposition and extraoral digestion (see Chapters 6, 7), enable mirids to deal with the problems confronting all users of green plants: desiccation, attachment, and food, namely, overcoming host defenses and obtaining adequate nourishment from tissues composed largely of carbohydrates. Desiccation was a particular problem for phytophagous and predacious insects that moved from the more humid litter layer or low-growing plants to an exposed, arboreal environment (Southwood 1973, Strong et al. 1984, Norris 1991).

Overlooking Plant Injury and Predation by Mirids

Because mirid injury is frequently disproportionate to the numbers of bugs present—that is, small populations can cause severe injury—growers and applied entomologists tend to underestimate the importance of plant bugs (Putshkov 1966). According to Osborn (1939), "The injury is so widely distributed, so general in character and so obscure in its outward manifestation that it may go unnoticed by the casual observer." Smith and Franklin (1961) suggested that injury from lygus bugs and similar species goes largely unrecognized because it tends to be scattered and usually does not result in the destruction or defoliation of host plants. They also noted that mirid injury often fails to be associated with the proper causal agent. Mirid-induced injury, often inconspicuous initially, can go unrecognized until further plant growth intensifies the feeding symptoms (Putshkov 1966, Becker 1974).

Mirid feeding can be expressed in a variety of overt symptoms—from leaf crinkling to fruit abscission, dieback to witches'-brooms, chlorosis to cankers, shot holing to stunting—that can persist long after the bugs disappear. The subtle and insidious effects of their feeding include wilted and yellowed foliage, thickened leaves and stems, undeveloped flowers, retarded fruit formation, and lagging growth rates. Feeding by mirids on reproductive plant parts can result in injury that is not readily distinguishable from normal flower drop. Feeding also can be expressed cryptically as embryoless seeds and as wheat grains of inferior baking quality, effects unlikely to be ascribed to mirids. Various types of mirid injury might be attributed to chewing insects, mites and other sucking arthropods, plant pathogens, drought, abnormally high temperatures, frost, hail, mechanical injury, insufficient pollination, nutrient deficiency or toxicity, herbicides, or pollution. Additionally, some mirids, even when present in large numbers, leave no discernible trace of their feeding (see Chapter 9).

Mirids also are underappreciated as predators (e.g., Whitman et al. 1994). They tend not to be considered predatory because even obligately carnivorous species are morphologically unspecialized for capturing prey (e.g., Myers 1927b). Despite this perception, members of this ancestrally predacious family (see Chapter 16) feed in a carnivore-like manner, and some possess potent salivary enzymes and venoms (Cohen 1996; see Chapter 14). The family includes species that prey only on arthropod eggs and incapacitated individuals, as well as "true predators" (Miles 1972) whose saliva can rapidly immobilize active prey.

State of Biological Knowledge

Much of what we know about courtship and oviposition behavior, fecundity, longevity, and voltinism for mirids is the result of studies by applied entomologists and others who work with agriculturally important species. This impressive body of knowledge, however useful in assessing the bionomics of the Miridae, has led to unwarranted preconceptions about trophic habits and

ecology. Because only 1% or 2% of all plant bugs are crop pests or important predators, much more of this diverse group needs to be surveyed to characterize the family ecologically.

The life histories of most plant bugs unimportant to agriculture are lacking. Biological data can be gleaned from the ecological and taxonomic literature, but in contrast to many other insect families, there is a dearth of detailed biological information on local faunas. Kullenberg's (1944) study of Swedish mirids is a prominent exception. He examined 92 species in nature and sectioned and stained plant tissues in the laboratory to determine more precisely the sites used for feeding and the nature of the injury. Published in German, this thorough account of mirid biology and ecology is often overlooked and underappreciated. A cadre of devoted and enthusiastic "scholarly amateurs" (Usinger 1960) has helped make the heteropteran faunas of Britain and Japan especially well known. Butler (1923) and Southwood and Leston (1959) summarized considerable biological data on British mirids. Kiritshenko (1951), Putshkov and Putshkova (1956), and Putshkov (1966) reviewed species of the former USSR; the latter provided an especially thorough discussion of bionomics. Ehanno's (1983–1987) review of the distribution and ecology of several hundred French species includes detailed information on habitats and host plants. Handbooks of the Japanese (Kawasawa and Kawamura 1975, Yasunaga et al. 1993) and German (Wachmann 1989) Heteroptera contain biological notes and color photographs of mirid species.

Comparable information on biology is lacking for the faunas of most other regions, including North America, although faunistic data from restricted geographic regions are available in numerous smaller papers. Blatchley's (1926) manual of eastern North American Heteroptera, Knight's (1923, 1941, 1968) taxonomic studies, and Kelton's (1980b) review of mirids of the Canadian prairie provinces contain valuable host lists and discussions of biology. Schuh and Slater (1995) provided a brief but useful summary of mirid biology and discussed faunistic studies on a world basis. The economic importance of mirids as crop pests (Wheeler 2000a) and as natural enemies of injurious arthropods (Wheeler 2000b) has been the subject of recent reviews.

Mirid Uniqueness: An Invitation to Further Study

Within the Hemiptera—that is, Heteroptera + Auchenorrhyncha, Coleorrhyncha, and Sternorrhyncha—the Miridae show unparalleled trophic diversity and plasticity. Among heteropterans, they are the ultimate omnivores. Their haustellate mouthparts and salivary secretions facilitate the exploitation of even dried carrion, feces, and other food sources (Adler and Wheeler 1984). Even within the Insecta, few other families contain species that

- are major crop pests;
- are potential biological control agents of weeds;
- are important predators of crop pests;
- have been used successfully in classical biological control of pest insects; or
- can impair the effectiveness of parasitic wasps and herbivores released for biological control.

Given the proper conditions, a mirid that is normally a key crop pest can become an important predator of other pests in that crop. Some species have been described either as pests requiring control or as useful natural enemies, depending on locality, season of observation, host cultivar, stage of plant growth (especially availability of flowers and fruits), pesticide applications, and availability of prey or alternative food sources such as nectar or pollen (Wheeler 1976b, Wiedenmann and Wilson 1996). In ecological studies, simply assigning plant bugs to a sap-feeding or sucking guild is not only inaccurate but also misleading. Mirids are not true sap suckers (see Table 1.1), and in ecological surveys they usually occupy more than one feeding guild. For example, they potentially can be placed in six of the nine guilds recognized in Kirchner's (1977) study: plant-tissue feeder, pollen and nectar feeder, seed feeder, predator, omnivore, and scavenger.

Many plant bugs move freely along a continuum having strict phytophagy at one end and obligate zoophagy at the other. Studies on facultatively predacious heteropterans showed that a classification of mirids as zoophytophages and phytozoophages is strictly subjective (Wiedenmann and Wilson 1996) and that the pest concept lacks real ecological validity (Alomar and Albajes 1996). Mirids, because of their omnivorous habits and trophic "switching" (Cohen 1996), can blur the definition of a "plant pest," as rendered by regulatory agencies such as departments of agriculture (e.g., Ramsay 1973), as well as complicate attempts to define a "beneficial organism" or "natural enemy." Yet many predatory mirids, as omnivores, have certain advantages in biological control: in the absence of prey, plant feeding can sustain their numbers, and in theory they can stabilize prey populations better than "pure" predators can (Gillespie et al. 1999).

Applied entomologists and agriculturists probably will become more aware of the Miridae as plant bugs associated with native and naturalized plants adapt to introduced crops and transgenic cultivars, as previously innocuous bugs reach outbreak levels when crop acreages increase or when natural enemies are killed by pesticides, and as adventive mirids of possible agricultural importance are detected. Pests long forgotten also may regain pest status under future agricultural practices. Heightened attention to mirids is desirable in the tropics because the introduction of high-yielding crops lacking pest resistance can cause plant bug densities to explode. Moreover, mirid injury should be accurately diagnosed and beneficial plant bugs recognized as such if crop-management programs are to succeed.

If researchers pursue some of the studies suggested in Chapter 18, more references to mirids might begin to appear in texts on insect-plant biology, ecology, or evolution. Such works, as Sir Richard Southwood points out in his Foreword, often fail to mention mirids. Insect ecologists, terrestrial biologists, and entomologists, including graduate students, eventually may turn more and more to mirids as research organisms for illustrating key biological principles and testing hypotheses. Discovery of this intriguing family by more investigators, both applied and basic, might someday give rise to a "mystique" currently associated only with some of the more popular insect groups.

2/ Format and Scope of Review

It takes a long time to write a book; but it takes even longer to decide what is to be included . . . and what is to be left out.

—R. H. Arnett Jr., 1961

A book on mirid biology that stresses trophic behavior and the diversity of foods consumed could use one of several approaches. Such a review could include only the applied literature and be organized by commodities: mirids of alfalfa, apple, cotton, and other major crops. Treatment of the family could be narrow in geographic coverage or global in scope.

I chose a world treatment that integrates information from the applied literature with host data and predation records available mainly in papers of a systematic or an ecological nature. The coverage thus includes studies of both agricultural and natural systems; it also reviews material relevant to the developing field of nutritional ecology. Examples refer to economically important mirid species (albeit some are no longer injurious under modern practices of crop production), as well as innocuous but intrinsically interesting plant bugs. To ignore noneconomic species would give a distorted view of mirid feeding and deny data that may prove useful in understanding the evolution of pestiferous habits in the family and clarifying theoretical issues in ecology. Knowledge of tropical mirids is much less than that of species occurring in temperate regions, but a global coverage helps avoid an overemphasis on temperate species and a neglect of the rich tropical fauna. I do not review chemical control per se but sometimes mention chemicals used to control certain pests and various techniques used in the integrated pest management (IPM) of plant bugs.

Literature Coverage and Use of Unpublished Data

As a step toward a synthesis of mirid ecology, I have evaluated the literature on feeding habits and other aspects of biology, although I do not treat papers published in the Slavic and South Asian languages as thoroughly as those from North America. The cutoff date for literature is 2000. I found information scattered in government publications and "gray" literature, such as theses, that are not widely available and potentially useful data in articles whose titles did not suggest information relating to mirids. I usually present information on species from North America first, followed by that from other regions. This review also includes my original observations (labeled "unpubl. data" or "pers. observ.") and some unpublished observations of other workers, which are cited as personal communications ("pers. comm.").

A detailed analysis and discussion of the origin of mirid-host associations—cospeciation versus host transfer or colonization (e.g., Mitter et al. 1991, Anderson 1993)—is beyond the scope of the present study. I also did not attempt a compilation of mirid hosts similar to that made by C. W. Schaefer and colleagues for several heteropteran taxa, including Berytidae (Wheeler and Schaefer 1982), Coreoidea (Schaefer and Mitchell 1983), and Pyrrhocoroidea (Ahmad and Schaefer 1987), and by Ehanno (1987b) for the French mirid fauna. A partial list of mirid host plants is available in Schuh's (1995) catalog. Moreover, the family's sheer size precludes an exhaustive review of world literature, particularly for the subfamily Bryocorinae, a diverse group containing pests of tropical crops—cocoa, coffee, kola, and tea—that are rich in methyl purines (Leston 1970). The literature on mirid-cocoa associations is particularly extensive. Mirids as cocoa pests are discussed as leaf and stem feeders in Chapter 9 and as fruit (pod) feeders in Chapter 12. In concert with Leston (1970) and Entwistle (1972), I use cocoa to refer to both the crop and the tree, rather than use cocoa for the former and cacao for the latter.

Almost as overwhelming as the cocoa literature is that on plant bugs as pests of cotton. Schmitz's (1958) study of Helopeltis species associated with cotton in Central Africa is a useful source of information and may be consulted for additional references on this important group of pests. Another invaluable source of additional references to cotton-inhabiting mirids and to species associated with other economically important plants is Otten's (1956) fascicle on Heteroptera in Tierische Schädlinge an Nutzpflanzen. Also useful are Cadou's (1993) review of the mirids associated with cotton in Africa and Madagascar and the edited work on cotton arthropods in the United States (King et al. 1996a). Important regional works on crop pests are Kalshoven's (1981) coverage of the Indonesian fauna and King and Saunders's (1984) handbook of pests of annual food crops in Central America. In addition, Wyniger's (1962) *Pests of Crops in Warm Climates and Their Control*, listing mirids injurious to 15 crops or crop groups, summarizes the plant parts fed on and feeding symptoms, the latter usually accompanied by photographs. Ferreira (1999) enumerated species affecting crop and ornamental plants in Brazil. Caswell (1962) and Lavabre (1970) briefly reviewed the mirid species of agricultural importance in the tropics. Libby's (1968) review of Nigerian insect pests discusses mirid injury to 13 crops or crop groups, and mirids are included in a more comprehensive guide to insect pests of Nigerian crops (Federal Ministry of Agriculture and Natural Resources, Nigeria 1996). Schmutterer (1969) reviewed plant bugs as pests of crops in Northeast and Central Africa.

Explanation of Format

This review of mirid biology that emphasizes feeding habits is in five parts. The remaining chapters of Part I introduce names of the principal genera discussed and contain an outline of higher classification (see Chapter 3), which provides the taxonomic framework needed for appreciating feeding trends among tribes and subfamilies, and summarize early ideas about the diet of plant bugs (see Chapter 4). In Part II, Chapter 5 reviews the morphology of adults, nymphs, and eggs and characters useful in specific identification; this chapter is not intended as a detailed treatment of external morphology, and the reader is referred to more thorough descriptions by Knight (1941) and Schuh and Slater (1995). Chapter 6 is an overview of mirid ecology and behavior that brings together a widely scattered literature to emphasize biological attributes other than feeding habits; references providing more detailed information are suggested. Part II also introduces morphology, physiology, and behavior in relation to feeding, including mouthpart structure, feeding strategies, salivary secretions, and plant-wound responses and their possible effects on mirids (see Chapter 7). It concludes with a review of mirids as vectors of plant pathogens (see Chapter 8).

Discussion of feeding habits in Parts III and IV is mainly by food source rather than subfamily or other higher category. In Part III, feeding is considered in relation to leaves and stems (see Chapter 9), inflorescences (including developing seeds) (see Chapter 10), nectar and pollen (see Chapter 11), and fruit (see Chapter 12). Alternative plant-based food sources— fungi, honeydew, and artificial diets—are reviewed in Chapter 13. Additional comments on organization are made at the beginning of most of the chapters in Parts III and IV. Similarity of symptoms induced by plant bugs and pathogens is stressed throughout, and the transmission of plant-disease agents is referred to occasionally in Part III.

Predacious and scavenging tendencies of the family are covered in Part IV. Predators (see Chapter 14) are discussed by prey type (e.g., arthropod eggs, mites) and scavengers (see Chapter 15), by specialized habits (e.g., spider commensalism). Other animal-associated foods (e.g., human blood) are also treated in Chapter 15.

Some mirid taxa are mentioned throughout the book: as feeders on leaves, stems, inflorescences, and fruits of numerous hosts, and as predators of various prey organisms. Feeding trends, therefore, might be obscured, but Part V discusses the origin of feeding habits in the Heteroptera and in the Miridae (see Chapter 16) and attempts to identify trends among higher-level taxa (see Chapter 17). Future research needs are highlighted in the final chapter (see Chapter 18).

Terminology and Use of Insect and Plant Names

Hemiptera has been used as the order name for both heteropterans (mirids and other true bugs) and homopterans: cicadas, leafhoppers and planthoppers, whiteflies, aphids, scale insects, and their relatives. In that classification scheme, Heteroptera and Homoptera are suborders of Hemiptera. The decision to use Heteroptera as a suborder or as an order involves various nomenclatural and phylogenetic considerations (see discussion in Henry and Froeschner [1988] and Sorensen et al. [1995]). Here, I follow Schuh and Slater (1995) in considering Heteroptera a suborder and also retain the use of Homoptera in its traditional sense, even though it has been shown to be paraphyletic and, therefore, to lack taxonomic validity (e.g., von Dohlen and Moran 1995, Schaefer 1996). Hemipterists do not yet agree on a systematic scheme to replace the concept of a classic Homoptera (Schuh 1996).

Higher classification of the Miridae is undergoing revision (e.g., Schuh 1975, 1976, 1984, 1986a; Akingbohungbe 1983; see also Chapter 3). Nomenclature hereinafter generally follows the world catalog of the group (Carvalho 1957–1960) and its revision (Schuh 1995), regardless of authors' original usage. I. M. Kerzhner checked the mirid names in Appendix 1 for taxonomic and nomenclatorial accuracy and recommended modifying several names cited by Schuh (1995) so that they are formed correctly. Some exceptions to the Schuh (1995) catalog are my use of *Creontiades dilutus* as the valid name for *Megacoelum modestum* (Gross and Cassis 1991, Malipatil and Cassis 1997); the spelling *norwegicus*, rather than *norvegicus*, in the genus *Closterotomus* (formerly in *Calocoris*) (Cassis and Gross 1995, Rosenzweig 1997, Kerzhner and Josifov 1999); and the spelling *seriata*, rather than *seriatus*, for the cotton fleahopper in the genus *Pseudatomoscelis* (I. M. Kerzhner, pers. comm.). Name changes involving some mirids of economic importance are summarized in Table 2.1.

Readers who consult the primary literature will find mirid names that differ from those I use, but only in a

Table 2.1. Some name changes involving mirids of economic importance, including predators

Former name	Current name	Common name
Adelphocoris apicalis	*Megacoelum apicale*	
Calocoris biclavatus	*Closterotomus biclavatus*	
Calocoris fulvomaculatus	*Closterotomus fulvomaculatus*	Hop capsid (mirid)
Calocoris norvegicus	*Closterotomus norwegicus*	Potato mirid
Calocoris striatellus & *C. ochromelas*	*Rhabdomiris striatellus*	
Campylomma livida (in part)[a]	*Campylomma liebknechti*	Apple dimpling bug
Campylomma nicolasi	*Campylomma verbasci*	"Mullein bug"
Campylomma subflava	*Campylomma plantarum*	
Carvalhoia arecae	*Mircarvalhoia arecae*	
Creontiades pallidifer	*Creontiades pacificus*	
Cyrtopeltis caesar	*Nesidiocoris caesar*	
Cyrtopeltis modestus	*Engytatus modestus*	Tomato bug, "tomato girdler"
Cyrtopeltis nicotianae	*Engytatus nicotianae*	
Cyrtopeltis tenuis & *C. volucer*	*Nesidiocoris tenuis*	
Deraeocoris fulleborni	*Deraeocoris oculatus*	
Diaphnocoris chlorionis	*Blepharidopterus chlorionis*	Honeylocust plant bug
Diaphnocoris pellucida & *D. provancheri*	*Blepharidopterus provancheri*	
Dicyphus minimus	*Tupiocoris notatus*	Suckfly
Dicyphus rhododendri	*Tupiocoris rhododendri*	
Dionconotus cruentatus	*Dionconotus neglectus*	
Eucerocoris suspectus	*Ragwelellus suspectus*	
Eurystylus immaculatus	*Eurystylus oldi*	
Halticus tibialis	*Halticus minutus*	
Heterocordylus flavipes	*Pseudophylus stundjuki*	
Lamprocapsidea coffeae	*Ruspoliella coffeae*	
Lepidopsallus minusculus	*Phoenicocoris minusculus*	
Letaba bedfordi	*Palomiella bedfordi*	
Lygocoris lucorum	*Apolygus lucorum*	
Lygocoris spinolae	*Apolygus spinolae*	
Lygus desertinus	*Lygus elisus*	Pale legume bug
Lygus disponsi	*Lygus rugulipennis*	European tarnished plant bug
Lygus ravus & *L. varius*	*Lygus shulli*	
Macrolophus caliginosus	*Macrolophus melanotoma*	
Macrolophus nubilus	*Macrolophus pygmaeus*	
Macrolophus rubi	*Macrolophus costalis*	
Megacoelum modestum	*Creontiades dilutus*	Green mirid
Megaloceraea recticornis	*Megaloceroea recticornis*	
Paramixia carmelitana	*Sthenaridea carmelitana*	
Paramixia suturalis	*Sthenaridea suturalis*	
Plesiocoris rugicollis	*Lygocoris rugicollis*	Apple capsid (mirid) bug
Psallus ancorifer	*Lepidargyrus ancorifer*	
Pseudatomoscelis seriatus	*Pseudatomoscelis seriata*	Cotton fleahopper
Pseudodoniella laensis	*Pseudodoniella pacifica*	
Ragmus importunitas	*Moissonia importunitas*	Sunnhemp mirid
Rhodolygus milleri	*Niastama punctaticollis*	Dimpling bug
Sejanus albosignatus	*Sejanus albisignatus*	
Taylorilygus pallidulus	*Taylorilygus apicalis*	
Trigonotylus coelestialium	*Trigonotylus caelestialium*	
Trigonotylus doddi	*Trigonotylus tenuis*	

Note: Nomenclatural changes can involve synonymy, homonymy, generic transfer, misidentification, or misspelling; see Schuh (1995) and Kerzhner and Josifov (1999) for additional information. The first word of common names is capitalized above, but the entomological convention is to lowercase these words except those that are conventionally capitalized: thus, green mirid, potato mirid, but European tarnished plant bug; see also Appendix 2.

[a] The true *C. lividum* (note correct spelling) is restricted to the Oriental region; *C. liebknechti* is an Australian species (Malipatil 1992, Kerzhner and Josifov 1999).

few cases have I added a parenthetical reference to synonymy. Some published records may be based on misidentification, but I accepted determinations at face value unless I am aware of a correction. I also question certain identifications when the site of study is well removed from the known distribution of the species or genus. The North American tarnished plant bug (*Lygus lineolaris*) was misidentified in the early literature as *L. pratensis*, a Palearctic species, and the synonym *L. oblineatus* was also used for *L. lineolaris* (Slater and Davis 1952). In Europe, the name *L. pratensis* has often been misapplied; *L. rugulipennis* is the more important crop pest, one deserving the name European tarnished plant bug (Southwood 1956c, Varis 1995). Many of the older references to *L. pratensis* refer to *L. rugulipennis*, *L. gemellatus*, or some other species of the genus (Linnavuori 1951, Woodroffe 1966, Rácz and Bernáth 1993).

Determination of *Helopeltis* species, bryocorine pests of many crops in warm, humid regions of Africa, Asia, and the Pacific Islands, is also difficult. Intraspecific color variation and similarity of color patterns among species have resulted in many misidentifications (Crowe 1977, Gapud et al. 1993). Stonedahl's (1991) taxonomic review of the Oriental species of *Helopeltis* contains a key to species and discusses known or suspected misidentifications.

The terms *injury* and *damage* often are used interchangeably in the entomological literature (an exception is White and Schneeberger [1981]). Here I generally follow Bardner and Fletcher (1974) by using the word *injury* to refer to the symptoms of mirid feeding, such as chlorosis or shot holing, and reserve the term *damage* for injury that results in measurable loss of yield or reduction in economic value (see also Bos and Parlevliet [1995]).

The common names used are those approved by the Entomological Society of America (Bosik 1997), with unofficial names placed in quotation marks (e.g., "cottondauber"). I also use common names listed in *Invertebrates of Economic Importance in Britain* (Seymour 1979) or in the work by Southwood and Leston (1959), those given to Australian insects (Naumann 1993), or those applied to economically important species in other geographic regions (e.g., sorghum earhead bug in India, boll shedder bug in Africa). The scientific names of all mirid species mentioned are listed in Appendix 1, with their authors (which are omitted in the text) and subfamily designations. A subfamily or tribal name is often used in the text for a particular mirid species, usually as an adjective (e.g., bryocorine, dicyphine, mirine, orthotyline, stenodemine; see also Chapter 3). A list of mirids referred to in the text by common name appears in Appendix 2.

The common name for the Miridae—plant bugs—reflects the occurrence of nearly all species on vegetation and their mainly phytophagous habits (Smith 1896). This designation is somewhat misleading because it depreciates the predacious tendencies of mirids (e.g., Cleveland 1987, Henry 2000, Wheeler 2000b); but it seems preferable to "leaf bugs," a common name sometimes used. This name, based on the assumed dominance of leaf feeding among mirids (Slingerland 1893), or simply their presence on leaves (Surface 1906), is also inaccurate because it neglects inflorescence feeding in the family. Kellogg (1905) is one of the few North American workers to have referred to the family as "flower bugs," which in many ways is an appropriate common name (see also Essig [1915] and Chapter 10). Though North American entomologists generally restrict the name "plant bugs" for the Miridae (e.g., Cranshaw 1992, Schuh 1995), authors in North America and elsewhere have sometimes used this name to refer to members of the squash bug family Coreidae; in a broad sense for plant-feeding heteropterans (e.g., Glover 1876, Fletcher 1985); for all plant-associated heteropterans, including predators (Cherrill et al. 1997); or for all phytophagous hemipterans (e.g., Ramakrishna Ayyar 1940; Miles 1968a, 1968b; Leftwich 1976).

Immature or preimaginal stages of mirids are termed *nymphs*, in keeping with traditional North American usage of reserving the word *larvae* for immatures of insects having complete metamorphosis or holometabolous development (reviewed by China et al. [1958], Davies [1958], Fox and Fox [1964], Stehr [1987]), even though such convention tends to obscure physiological similarities in development of the immature stages (Chapman 1991). *Holarctic* is used to refer to mirids occurring in both the Old and the New World regardless of their zoogeographic status—that is, whether they are indigenous or adventive in either the Nearctic or Palearctic region (e.g., Benson 1962, Wheeler and Henry 1992; cf. Slater 1993). *Omnivory* occasionally is used synonymously with *polyphagy* (e.g., Harborne 1988), but here it means feeding at more than one trophic level, and *omnivore* refers to a mirid species that feeds at more than one trophic level. Either *predatory* or *predacious* is used as an adjective referring to predators; "predaceous" can be considered a less preferred spelling (Frank and McCoy 1989). *Parasitoid* (Frank and McCoy 1989), rather than *parasite* (e.g., Doutt 1959), is used throughout in referring to parasitic Diptera and Hymenoptera. This term distinguishes between these insects and true parasites, recognizing that their effect on prey populations is comparable to that of predators (e.g., Daly et al. 1978, Price 1984, Sabelis 1992; see also Eggleton and Gaston's [1990] review of definitions of *parasitoid*). The noun *parasitoidism* and verb *parasitoidize* (Frank and McCoy 1989, Quicke 1997) are not in general use and are avoided here in favor of *parasitism* and *parasitize*. Technical terms (except chemical compounds, pesticides, and nouns or adjectives based on animal and plant taxa) are defined in the Glossary.

Common names only are used in the text for the numerous plant species mentioned. The naming of economically important plants follows Beetle's (1970) *Recommended Plant Names*, Dirr's (1975) *Manual of Woody Landscape Plants*, Terrell et al. (1986), and Brako et al. (1995). If I am unaware of any common name, some coinage (e.g., sundewlike byblidaceous plant) is used, or the generic name is simply lowercased (e.g., astrotricha, dillwynia). Common plant names appear alphabetically in Appendix 3, which also lists botanical names, authors' names, and families; the use of familial names follows that by Mabberley (1987).

3/ Higher Classification and Principal Genera

The position and classification of Capsidae [Miridae] have engaged the attention of numerous eminent Rhynchotists [hemipterists], and yet, of the several systems proposed, there is not one which can claim the acceptance of all.
—H. Singh-Pruthi 1925

A thorough discussion of higher classification is inappropriate in a review of mirid feeding habits. But for those not versed in mirid systematics, some familiarity with the names of subfamilies and tribes is essential for an appreciation of the phylogenetic context presented in Part V. This chapter also introduces the names of most of the principal genera discussed in the text.

The framework of mirid classification rests largely on the careful analysis by the Finnish hemipterist O. M. Reuter in the late nineteenth and early twentieth century. Building on the classification outlined by Fieber (1861), Reuter (1905, 1910) argued for the recognition of nine subfamilies, mainly on the basis of pretarsal structures, and was the first author to offer a character-based defense of his views of phylogenetic relationships in the family. Acknowledging possible variation in pretarsal structures, Reuter concluded that they were more stable than other available characters. Indeed, it is mainly pretarsal structure (see Schuh [1976] and Chapter 5 for terminology) that provides the basis for the classification used in two works familiar to many nonspecialists: Knight's (1941) *Plant Bugs, or Miridae, of Illinois* and Southwood and Leston's (1959) *Land and Water Bugs of the British Isles*. Other characters that have been used to diagnose subfamilies are the pronotal collar and wing cells (Leston 1961b).

As Carvalho and Leston (1952), Odhiambo (1961), Schuh (1974, 1976), and Cobben (1978) noted, several workers (e.g., Kullenberg 1947b, Wagner 1955) challenged a classification based largely on pretarsal structures, arguing that other characters are less subject to variation. Higher classification was much confused until Carvalho (1952), relying on Reuter's studies, proposed a universal system (Akingbohungbe 1983). Detailed comparative studies of female (Slater 1950) and male genitalia (Kelton 1959) and testis follicle numbers (Leston 1961b), with a reevaluation of pretarsal structures (Schuh 1976), have resulted in several refinements.

The subfamilies and tribes accepted by Schuh (1995), along with their distribution by faunal regions, are

Table 3.1. Subfamilies and tribes currently recognized in the Miridae, and their distribution by faunal regions

Mirid taxon	Faunal region[a]
Bryocorinae	
Bryocorini	C[b]
Dicyphini	C
Eccritotarsini	C[b]
Cylapinae	
Cylapini	C
Deraeocorinae	
Clivinematini	N, NT, P
Deraeocorini	C
Hyaliodini	C
Saturniomirini	A
Surinamellini	AT, N, NT, O
Termatophylini	A, AT, NT, O, P
Isometopinae	
Isometopini	C
Mirinae	
Herdoniini	AT, N, NT, P
Hyalopeplini	A, AT, O, P
Mecistoscelidini	A, O, P
Mirini	C
Restheniini	N, NT
Stenodemini	C
Orthotylinae	
Halticini	C
Nichomachini	AT, P
Orthotylini	C
Phylinae	
Auricillocorini	O
Hallodapini	C
Leucophoropterini	C
Phylini	C
Pilophorini	C
Psallopinae	A, NT, P

Source: Schuh (1995); Cassis (1995) for Deraeocorinae: Termatophylini. The spellings "Clivinematini," "Mecistoscelidini," and "Restheniini" follow those from Steyskal (1973) and Kerzhner and Josifov (1999).

[a] A, Australian (including Oceania); AT, Afrotropical (Ethiopian); N, Nearctic; NT, Neotropical; O, Oriental; P, Palearctic; C, Cosmopolitan.

[b] Tribe is mainly circumtropical.

Table 3.2. Selected mirid genera discussed in text

Subfamily	Tribe	Genus
Bryocorinae	Bryocorini	*Bryocoris, Monalocoris*
	Dicyphini	*Bryocoropsis, Campyloneura, Dicyphus, Distantiella, Engytatus, Helopeltis Lycidocoris, Macrolophus, Mircarvalhoia, Monalonion, Nesidiocoris, Pameridea, Platyngomiriodes, Pseudodoniella, Rayieria, Sahlbergella, Setocoris, Tupiocoris*
	Eccritotarsini	*Caulotops, Halticotoma, Hesperolabops, Mertila, Neoneella, Pycnoderes, Sixeonotus, Tenthecoris*
Cylapinae	Cylapini	*Cylapocoris, Cylapus, Valdasus*
Deraeocorinae	Clivinematini	*Clivinema*
	Deraeocorini	*Alloeotomus, Deraeocoris, Eurychilopterella, Romna*
	Hyaliodini	*Hyaliodes, Paracarnus, Stethoconus*
	Termatophylini	*Termatophylidea*
Isometopinae	Isometopini	*Corticoris, Diphleps, Isometopus, Myiomma, Palomiella*
Mirinae	Herdoniini	*Barberiella*
	Hyalopeplini	*Hyalopeplus*
	Mecistoscelidini	*Mecistoscelis*
	Mirini	*Adelphocoris, Agnocoris, Apolygus, Calocoris, Capsodes, Capsus, Closterotomus, Creontiades, Dagbertus, Dichrooscytus, Dionconotus, Eurystylus, Horistus, Irbisia, Liocoris, Lygocoris, Lygus, Megacoelum, Neurocolpus, Orthops, Phytocoris, Platylygus, Poecilocapsus, Polymerus, Ruspoliella, Sidnia, Stenotus, Taedia, Taylorilygus, Tropidosteptes*
	Restheniini	*Opistheurista, Prepops*
	Stenodemini	*Collaria, Dolichomiris, Leptopterna, Megaloceroea, Notostira, Stenodema, Trigonotylus*
Orthotylinae	Halticini	*Halticus, Labops, Orthocephalus*
	Orthotylini	*Blepharidopterus, Ceratocapsus, Cyrtorhinus, Globiceps, Heterocordylus, Heterotoma, Hyalochloria, Labopidea, Lopidea, Malacocoris, Orthotylus, Paraproba, Parthenicus, Pseudoxenetus, Reuteria, Schaffneria*
Phylinae	Hallodapini	*Coquillettia, Orectoderus, Systellonotus, Trichophthalmocapsus*
	Leucophthoropterini	*Sejanus, Tytthus*
	Phylini	*Amblytylus, Atractotomus, Campylomma, Chlamydatus, Harpocera, Hoplomachus, Lepidargyrus, Megalocoleus, Microphylellus, Moissonia, Plagiognathus, Psallus, Pseudatomoscelis, Ranzovius, Reuteroscopus, Rhinacloa, Rhinocapsus, Spanagonicus*
	Pilophorini	*Pilophorus*

Note: See Appendix 1 and Index for all genera mentioned in this work.

listed in Table 3.1. Most of the principal genera discussed in the text are given in Table 3.2. The names of subfamilies and tribes appear alphabetically because of changing views regarding a phylogenetic arrangement for mirids (see Chapter 17) and, therefore, the unsettled state of internal classification; as scientific hypotheses, classifications are subject to subsequent testing (e.g., Q. D. Wheeler 1995). Schuh's (1995) concept of tribal classification differs slightly from that presented by Schuh and Slater (1995). The current framework is somewhat similar to that outlined by Carvalho (1952) and used in his (1957–1960) catalog, a work that might have had the "single most important modern influence on the classification of the Miridae" (Schuh 1986a). Higher taxa in the family are still defined mainly by pretarsal structure but reflect the use of other characters, especially male genitalia, to resolve artificial groupings within the Orthotylinae and Phylinae (Schuh 1974, 1976). The reorganization of orthotyline and phyline tribes that had been based mainly on a myrmecomorphic habitus has helped refine the higher classification.

Schuh's (1995) classification (see Table 3.1) differs from Carvalho's (1957–1960) arrangement mainly in the recognition of isometopids or jumping tree bugs as mirids (subfamily Isometopinae) (Carayon 1958, Slater and Schuh 1969); placement of the tribe Pilophorini in the Phylinae (Schuh 1974); and erection of Psallopinae, reorganization of tribes within the Bryocorinae, and placement of the Dicyphini in Bryocorinae (Schuh 1976, Štys 1985). Two major changes proposed by Schuh (1976)—placement of Orthotylini as a tribe within Phylinae and Deraeocorini within Mirinae—have not been followed (e.g., Schuh 1984, Henry and Wheeler 1988). Akingbohungbe (1974b), following Knight (1943), considered the hyaliodines deserving of subfamily rank, but this change likewise has not gained widespread acceptance (Schuh 1976, Akingbohungbe 1983). Additionally, Schuh's (1995) subtribe Palaucorina of the bryocorine tribe Eccritotarsini is sometimes considered a subfamily, the Palaucorinae (Carvalho 1984a, Gorczyca 1997).

4/ Mirid Feeding Habits and Host Plants: Historical Sketch

Old work is given space . . . to trace the development of ideas, to place newly acquired facts in a proper perspective, . . . and above all to extricate the thread of continuity which in the final analysis directs the pattern of our efforts.

—V. G. Dethier 1976

I previously traced the evolution of ideas regarding the predacious habits of mirids (Wheeler 1976b). This paper should be consulted for a more complete account of early observations on predation and for references omitted from the historical account that follows.

Early European Literature: Zoophagy and Phytophagy

European workers of the mid-nineteenth century not only were aware that some mirids were partially predacious but even tended to overemphasize the importance of arthropods in the bugs' diet. Comments by Burmeister (1835), Amyot and Serville (1843), Meyer-Dür (1843), and Sahlberg (1848) imply that members of the family are almost exclusively carnivorous (Flor 1860). Clausen (1940) probably had their works in mind when he remarked, "Several authors state that the family as a whole subsists principally upon other insects." Unfortunately, these European authors did not provide specific examples of predator-prey associations.

During the same period, other Europeans (e.g., Blanchard 1840, Westwood 1840) mentioned only the plant-feeding habits of mirids. In their well-known studies of British Heteroptera, Douglas and Scott (1865) and Saunders (1892) referred only to phytophagy, either overlooking earlier references to carnivory in the continental European and British literature (Budgen 1851, Curtis 1860) or regarding as unfounded the possibility of predatory tendencies in a family known as leaf or plant bugs. When Jennings (1903) credited Douglas (1895) with the first specific record of plant bugs as predators, Reuter (1903) pointed out previous references to predation that he had discussed in a brief review of mirid food habits (Reuter 1875). Kirkaldy (1907) added observations published after Reuter's (1903) paper and a few that Reuter overlooked.

Predacious Habits of North American Plant Bugs

Mirids were long considered strictly phytophagous in North America, even though early applied entomologists such as W. Le Baron (1871) and C. V. Riley (1871) had observed several species feeding on insect eggs and soft-bodied arthropods. P. R. Uhler, the first Heteroptera specialist in the United States (Schwarz et al. 1914, Osborn 1931), also referred to predatory habits in the family (e.g., Uhler 1876). In addition to *Hyaliodes vitripennis*, whose predatory habits Uhler had noted, Glover (1876) thought it likely that "many other species of the *Capsides* [Miridae], hitherto considered as plant-feeders, also occasionally vary their diet by sucking out the juices of other insects."

Nonspecialists often overlooked these early references to carnivory, which usually were made in passing. Thus, Weiss (1921) characterized mirids as plant feeders despite acknowledging in his review of heteropteran food habits that predacious tendencies existed in other mainly phytophagous families, such as the Lygaeidae. Until recently, most textbooks of entomology characterized mirids as herbivores, even though H. H. Knight (1921), the first North American worker to concentrate on mirid taxonomy (Slater 1978; Wheeler 1979, 1983b), emphasized that some plant feeders are facultative predators and that the Deraeocorinae are primarily carnivorous.

Study of Hosts and Plant Injury in North America

Despite an emphasis on mirids as plant feeders, the early North American literature seldom mentioned specific hosts. In describing new species, Thomas Say occasionally noted that a particular mirid occurred on pine or oak (e.g., Say 1832). T. W. Harris (1841), in his classic *Report on the Insects of Massachusetts, Injurious to Vegetation*, was perhaps the first to mention damage by a New World mirid. In 1838, he had observed *Phytocoris lineolaris* (actually *Lygus lineolaris*, the tarnished plant bug) "ravaging" potatoes and damaging plants in the flower garden. Henderson (1858) proposed in *Scientific American* the unfounded theory that the tarnished plant bug feeds underground on potato tubers and caused the mysterious potato rot or blight that had engendered the catastrophic Irish famine (detailed by Wheeler [1981c]). Other mirids said to be destructive were the fourlined plant bug (*Poecilocapsus lineatus*) (Harris 1851) and the rapid plant bug (*Adelphocoris rapidus*) (Glover 1856).

By 1870, agricultural entomology had become recognized as a scientific profession in North America, and its practitioners led their European counterparts in the range of investigation and application of their findings (Sorensen 1988). The Hatch Act establishing state agricultural experiment stations was passed in 1888, the gypsy moth (*Lymantria dispar*) was detected in Massachusetts the next year, and the accidentally introduced San Jose scale (*Quadraspidiotus perniciosus*) began to devastate eastern U.S. orchards during the 1890s. These and other events brought greater respect and funding to applied entomology (e.g., Howard 1930, 1933). But during the rise of government-sponsored applied entomology, plant bugs were largely ignored. That mirids were not well known as pests in the mid-nineteenth century is evidenced by Riley's (1870) comment: "None but the species under consideration [tarnished plant bug] have thrust themselves upon public notice by their evil doings." Moreover, Cook (1876), acknowledging that the family is quite extensive, said only one mirid, the tarnished plant bug, inflicts serious injury.

Mainly because of an emphasis on describing new species of the poorly known New World fauna, host records and other ecological information accumulated slowly during the second half of the nineteenth and early twentieth century. Uhler and E. P. Van Duzee, who succeeded Uhler as North America's leading heteropterist (Essig and Usinger 1940), sometimes mentioned host plants in their taxonomic papers (e.g., Uhler 1878, Van Duzee 1887). Gillette and Baker (1895) provided several host associations for species occurring in Colorado. But when asked about the "faunistic value" of mirids, Uhler (1901) admitted the group had not been collected carefully enough for him to assess its role. During this period, the only paper emphasizing mirid-host associations in North America was that by Heidemann (1892) on species of the Washington, D.C., area. Johnston (1928), in compiling the known hosts of North American mirids, remarked that entomologists only recently had begun to pay attention to host plants. Several species that damage fruit, vegetable, and forage crops, however, did receive attention from applied entomologists during the 1890s and first two decades of the twentieth century; many of the papers resulting from this work are referred to in Part III.

The development in North America of crop monocultures, often involving nonindigenous plants, as well as other changing agricultural practices of the late 1800s and early 1900s, almost certainly induced outbreaks of plant bugs. Some mirid species, absent or present only in small numbers on host plants occurring in scattered colonies, attain large densities when their hosts are concentrated (e.g., Moore et al. 1982). Grazing practices

in the Southwest promoted increases in crotons and other preferred weedy hosts of the cotton fleahopper (*Pseudatomoscelis seriata*), allowing the bugs to invade cotton fields (Fletcher 1930, 1940b; Schuster et al. 1969). Lygus bugs, thriving under the conditions of modern agriculture, proved remarkably adaptable to introduced crops and weeds (e.g., Domek and Scott 1985).

Even with the flurry of research on mirids near the end of the nineteenth century, Crosby and Leonard (1914), in their comprehensive review of the tarnished plant bug, could list only 10 additional North American species of Miridae for which substantial information on immature stages was available. Osborn (1918) acknowledged that few mirids had been studied from an economic standpoint. The appearance of Knight's (1923) fascicle on mirids in the Hemiptera of Connecticut, a work amounting essentially to a monograph of the family (Parshley 1923), stimulated not only taxonomic work on North American plant bugs but also studies on their life history and habits.

Advances in the knowledge of North American mirids did not immediately follow the appearance of Knight's (1923) publication, for Watson (1928) apparently was aware of only 12 species to include in a discussion of economically important plant bugs. Today, perhaps 25 or 30 species are known as at least occasional pests of various North American crops and landscape plants.

Mirids as Pests outside North America

Information on mirids as crop pests in other parts of the world was scant at the turn of the century. Quaintance (1913) was aware of fewer than 10 injurious species occurring outside North America, although his literature review was not exhaustive, and Dudgeon (1910) stated that "a large number of extremely injurious species" belong to the genus *Helopeltis*. Otten's (1956) review of economically important Heteroptera contains several pre-1913 references to mirid species not cited by Quaintance (1913). Putshkov (1966, 1975) noted that more than 50 mirids are known to injure field and garden crops within what is now the former USSR.

In England, where entomology was advanced relative to North America yet lagged slightly in the development of its applied aspects (Wheeler 1981c, Barnes 1985), mirids in the early twentieth century were only beginning to be appreciated as pests. When Curtis (1860) discussed the insects of British field crops, the habits of plant bugs were little known and the group consequently received little attention in his book. Carpenter (1912) stated that mirids "rarely attract notice by seriously damaging cultivated plants," but as Horne and Lefroy (1915) pointed out, these bugs deserve much more attention as agricultural pests and cause considerably more damage than is generally realized. Fryer

(1916) commented that mirids as apple pests were much better known in North America. Changes in British agricultural practices were thought responsible for increasing the host range of plant bugs and inducing outbreaks on crops such as apples (Fryer and Petherbridge 1917).

K. M. Smith (1920a, 1920b), working in England, was the first to suggest the possible ways in which mirids could injure plants. He stated that injury to apple crops could be purely mechanical, from laceration of the tissues by the bugs' stylets; pathological, from the inoculation of bacteria with their saliva; or toxic, from the injection of virulent salivary secretions into plant tissues (see also Chapters 7, 8).

5/ Family Characterization and Identification

The eye is bewildered by the excessive variety and number of patterns of ornamentation which are present on every hand; while the mind is delighted with their [mirids'] graceful proportions and light elegance of form.
—P. R. Uhler 1884

At the beginning of Chapter 3, I remarked that an extensive discussion of higher classification is not needed in a book that emphasizes biology of the Miridae. Somewhat similarly, a detailed review of the morphology of mirid life stages is considered unnecessary here. Even so, some appreciation of the structural modifications in this diverse group of insects, as well as the principal characters on which taxonomic keys are based, should benefit those who study various biological aspects of the family.

In the Heteroptera, in contrast to the Homoptera, the forewings or hemelytra usually have a dissimilar texture: the basal part is thickened and leathery and the apical part is thin, membranous, and overlapping. Fully developed hemelytra are unique to the Heteroptera, and much of the bugs' morphology is associated with the operation of hemelytra in flight (Wootton and Betts 1986). The hindwings are membranous and shorter than the forewings. Heteropterans hold their wings folded flat over the abdomen instead of in the rooflike configuration seen in leafhoppers, planthoppers, psyllids, and certain other homopterans. Adults of some species are wingless or have greatly reduced wings. The segmented beaklike labium, typical of hemipterans, encloses the paired, bristlelike mandibular and maxillary stylets, which are fitted for piercing-sucking, allowing fluids and liquefied solids to be withdrawn from plant and animal tissues. The term *rostrum* refers to the ensheathing labium and the stylets. As in all hemipterans, labial and maxillary palps are lacking (mirid stylet structure is detailed in Chapter 7). Metamorphosis is paurometabolous—that is, incomplete with a gradual change in body proportions during development.

According to Schuh and Štys (1991), the following characters (in addition to the two-segmented trochanter) define the Miridae (see also Schuh and Slater [1995]):

Wings: costal fracture long, delimiting cuneus; membrane with a stub on the distal angle of a single cell or pair of cells; membrane with one or two short cells (if two, anterior one attached to cuneus), rarely with a few emanating veinlike structures, and with one posterior free vein.

Abdomen: scent gland absent between abdominal terga 4 and 5 (present between 3 and 4) in nymphs.

Genitalia and insemination: male terminalia asymmetrical, of mirid type, with left paramere generally more strongly developed than the right; spermatheca absent.

Adults

COLLECTING AND PRESERVING MIRIDS

Collectors should kill adult mirids in vials or jars separately from other insects. Specimens should be mounted promptly to prevent damage to antennae, legs, and pubescence, and alcoholic preservation should be avoided. Killing or storing plant bugs in alcohol can deform, bleach, and dehydrate them (Martin 1977, Schuh and Slater 1995), making identification difficult or impossible, especially in the Orthotylinae, Phylinae, and other groups in which pubescence is used extensively in keys (e.g., Carvalho and Leston 1952; Schuh 1984; Schuh and Schwartz 1985, 1988; Stonedahl 1990). In addition, legs easily become detached in alcohol-preserved material. Potassium or sodium cyanide is preferred as a killing agent because ethyl acetate killing jars tend to "sweat" (Schuh and Slater 1995). Point mounting—gluing adults to small triangles punched from heavy archival paper or card stock such as no. 2 Bristol board—is recommended for mirids rather than direct pin mounting (Parshley 1919, Torre-Bueno 1925, Schuh and Slater 1995).

EXTERNAL AND INTERNAL CHARACTERISTICS

Adult mirids vary considerably in size and appearance (Figs. 5.1–5.3). They range from 1.5 mm to as much as 15 mm in some Neotropical Restheniini, have a four-segmented labium that is often long and tapering but is sometimes short and stout, four-segmented antennae, and three-segmented tarsi (Isometopinae, Psallopinae, and some Bryocorinae and Cylapinae have only two). The head is declivent to porrect and is sometimes

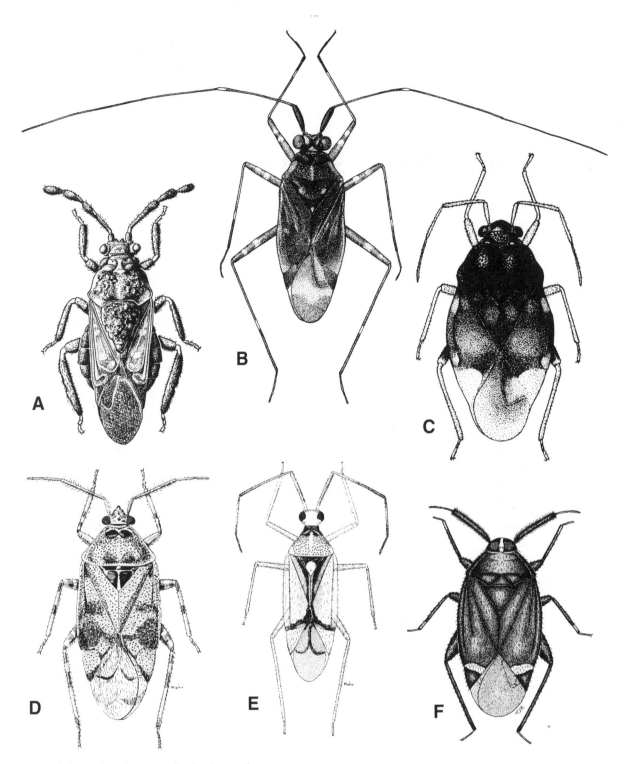

Fig. 5.1. Adult mirids. A. *Distantiella theobroma* (from Entwistle 1972). B. *Cylapus tenuicornis* (from Froeschner 1949). C. *Pycnoderes medius* (from Froeschner 1949). D. *Deraeocoris aphidiphagus* (from Knight 1941). E. *Hyaliodes harti* (from Knight 1941). F. *Myiomma cixiiforme* (from Henry 1979c). (A reprinted by permission of P. F. Entwistle and Addison Wesley Longman Ltd; B, C, by permission of R. C. Froeschner and University of Notre Dame; D, E, by permission of Illinois Natural History Survey; F, by permission of T. J. Henry and the Entomological Society of Washington.)

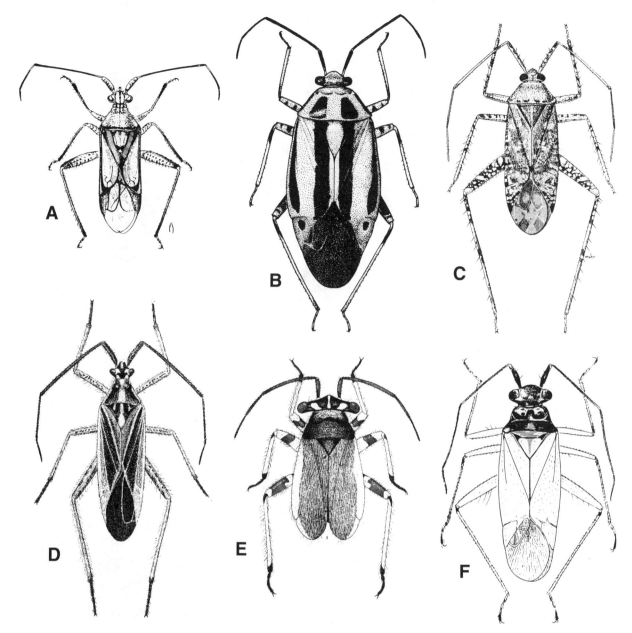

Fig. 5.2. Adult mirids. A. *Hyalopeplus pellucidus* (from Zimmerman 1948a). B. *Poecilocapsus lineatus* (from Froeschner 1949). C. *Phytocoris salicis* (from Knight 1941). D. *Leptopterna dolabrata* (from Knight 1941). E. *Labops hirtus* (from Slater 1954). F. *Cyrtorhinus fulvus* (from Zimmerman 1948a). (A, F reprinted by permission of E. C. Zimmerman; B, by permission of R. C. Froeschner and University of Notre Dame; C, D, by permission of Illinois Natural History Survey; E, by permission of J. A. Slater and the New York Entomological Society.)

inserted into the pronotum up to the eyes (Fig. 5.4). The adults have paired scent glands situated lateroventrally in the metathorax; the placement and shape of the peritreme or opening vary. These glands, which usually have an auricular evaporatory area (Fig. 5.5), are nonfunctional in some adult cocoa mirids (Leston 1973a).

Good general references on the external morphology include the works by Parshley (1915) and Knight (1941). Barbagallo (1970) provided details on *Closterotomus trivialis*. Hoke's (1926) discussion of heteropteran wing venation includes the Miridae and illustrations of five species. Carayon (1977) presented an overview of the

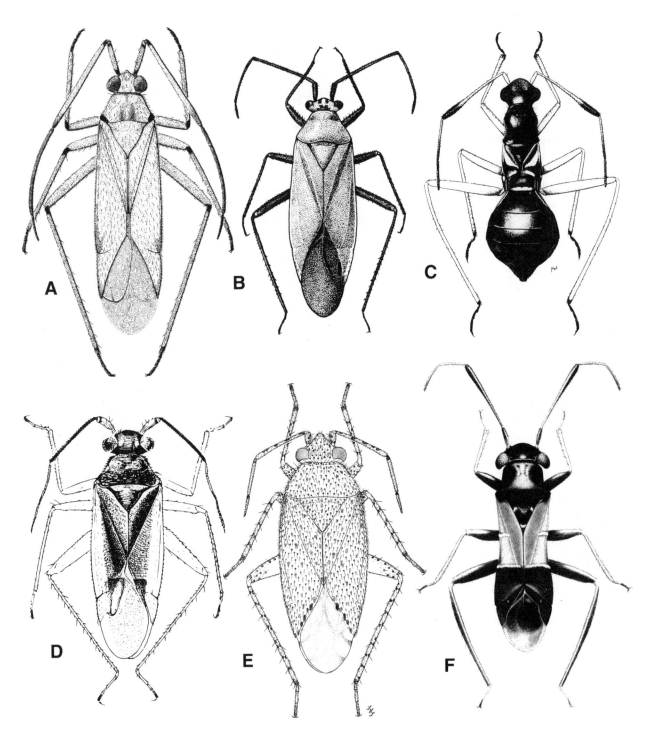

Fig. 5.3. Adult mirids. A. *Blepharidopterus angulatus* (from Kelton 1983). B. *Lopidea media* (from Froeschner 1949). C. *Orectoderus obliquus* (from McIver and Stonedahl 1987b). D. *Tytthus mundulus* (from Zimmerman 1948a). E. *Pseudatomoscelis seriata* (from Henry 1991). F. *Pilophorus amoenus* (from Schuh and Schwartz 1988). (A reprinted by permission of L. A. Kelton and the Minister of Public Works and Government Services Canada, 1997; reproduced from Agriculture and Agri-Food Canada publications "Plant Bugs on Fruit Crops in Canada"); B, by permission of R. C. Froeschner and University of Notre Dame; C, by permission of J. D. McIver and the New York Entomological Society; D, by permission of E. C. Zimmerman; E, by permission of T. J. Henry and New York Entomological Society; F, by permission of R. T. Schuh, courtesy of The American Museum of Natural History.)

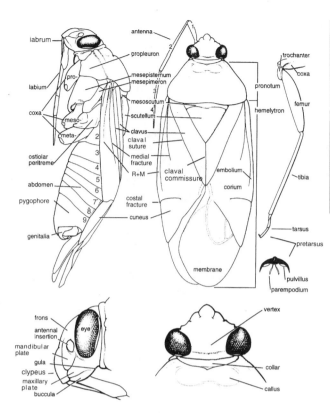

Fig. 5.4. General mirid morphology as typified by a *Lygus* species. (Modified, with permission of M. D. Schwartz and the Minister of Public Works and Government Services Canada, 1997, from M. D. Schwartz and R. G. Foottit 1992b; reproduced from Agriculture and Agri-Food Canada publications "Lygus Bugs on the Prairies".)

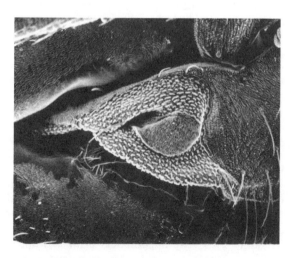

Fig. 5.5. Metathoracic scent efferent system and ostiolar opening of *Cariniocoris geminatus* (×221). (Reprinted, by permission of T. J. Henry and the New York Entomological Society, from Henry 1989.)

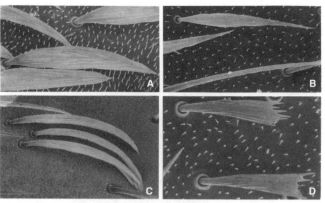

Fig. 5.6. Sericeous setae. A. *Lepidargyrus ancorifer*. B. *Campylomma verbasci*. C. *Pseudatomoscelis seriata*. D. *Atractotomus prosopidis*. (Reprinted, by permission of R. T. Schuh, courtesy of The American Museum of Natural History, from Schuh and Schwartz 1985.)

external and internal morphology of cocoa mirids, and Slater (1982) reviewed the family's external (and internal) morphology. Schuh and Štys (1991) and Schuh and Slater (1995) discussed characters diagnostic for mirids and reviewed heteropteran morphology.

Mirids are mostly fragile, delicate bugs (Butler 1923)—in Blackburn's (1888) words, "among the frailest of all insects." Hemelytral pubescence, consisting mainly of silky, woolly, flattened, scalelike, and setiform hairs or setae (Fig. 5.6; Carvalho and Leston 1952), is easily abraded. After females are killed, decomposing eggs occasionally release oily substances that penetrate the cuticle, forming a film over the pubescence and rendering the hairs nearly invisible (Hussey 1954b).

Plant bugs are often green or brown, as might be expected of an insect group living on plants (Chinery 1993). They are sometimes brightly colored, and even those of drab coloration are hardly unattractive, which is how Tillyard (1926) described the appearance of most Australian and New Zealand species. Mirids are characterized by the absence of ocelli (except in the subfamily Isometopinae), a condition responsible for the German common name for mirids, *Blindwanzen* or "blind bugs," and by the presence (except in shortwinged or brachypterous forms) of a cuneus, a small triangular area at the apex of the corium on the outer wing margin just beyond the embolium (see Fig. 5.4). The membrane contains one or two unequal cells divided by a short longitudinal vein, a character that is apomorphic for mirids. Species of a few mirid groups, especially litter inhabitants, are remarkable in lacking a membrane and having coleopteroid hemelytra (see Fig. 5.7; Slater and Gross 1977, Schuh 1986b). This type of wing modification might be an adaptation for reducing water loss in arid habitats (Schuh and Slater 1995). Some antlike species are dimorphic in color and form, with only the brachypterous females strongly resembling ants. Myrmecomorphic species also display other color,

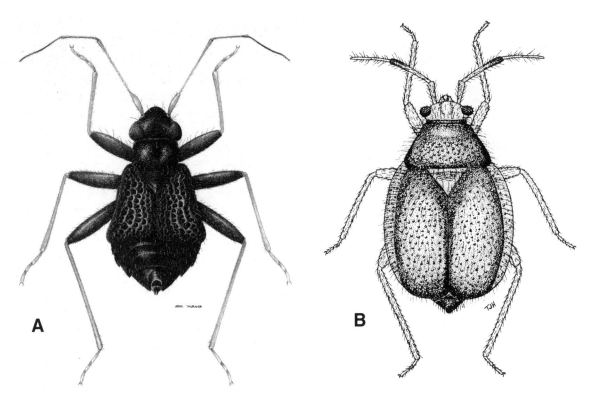

Fig. 5.7. Coleopteroid hemelytra. A. *Carvalhoma taplini* (from Slater and Gross 1977). B. *Bothynotus johnstoni* (from Henry 1979a). (A reprinted by permission of J. A. Slater and the Australian Entomological Society; B, by permission of T. J. Henry and the Florida Entomological Society.)

morphological, and behavioral modifications that create the impression or illusion of ants (Brindley 1935, McIver and Stonedahl 1993; see also Chapter 6). Only eight segments of the essentially nine-segmented abdomen are clearly visible (segments 10 and 11 are greatly reduced).

Mirid antennae are usually linear, with the last two segments (antennomeres) relatively slender and threadlike (Knight 1941). An inflated second segment and segments 3 and 4 held at an angle ("elbowed") are characteristic of certain myrmecomorphic species (McIver and Stonedahl 1993). In some groups (e.g., *Ceratocapsus, Diplozona, Eustictus*), all four segments are of nearly equal thickness. Antennal characters such as a clavate or incrassate second segment are used extensively in keys. Segment 2 is thickened in both sexes of some species (e.g., in *Atractotomus, Teleorhinus*), while in others (e.g., *Criocoris saliens*), this segment is thickened only in the male. Segment 3 is swollen in the genus *Hambletoniola*, whereas in another western North American genus, *Larinocerus*, segments 2 and 3 are inflated (Henry and Schuh 1979). Some taxa show striking modifications. Antennae of the Palearctic *Harpocera thoracica* are sexually dimorphic: segment 2 in males is swollen ventrally and bears an oval pad of 200–250 adhesive setae, which grip the female's pronotum during copulation (Stork 1981). Members of the mainly Neotropical genus *Hyalochloria* have spines or projections on segment 1 or 2, or on both segments (Fig. 5.8; Henry 1978). These projections apparently are used to grasp antennae of the female during copulation (Schuh and Slater 1995). Carvalho's (1955) key to world genera includes illustrations of many other, often bizarre, types of antennal modifications.

Plant bugs usually have a smooth, unarmed scutellum, but in certain tropical bryocorines, mostly odoniellines, the scutellum is lobed or armed with spines, spurs, or tubercular swellings (China 1944). In fact, the Bryocorinae are noted for "peculiar outgrowths and modifications of the external cuticular structure, found nowhere else in the Miridae to such an extent" (Odhiambo 1962). Some herdoniine Mirinae and certain other myrmecomorphic species have similar scutellar spines that resemble those of ants.

Mirids are unique among heteropterans in having specialized sensory hairs, usually 2–8, restricted to the lateral and ventral surfaces of the mesofemur and metafemur (Fig. 5.9; Schuh 1975, McGavin 1982, Schuh and Slater 1995). The hypothetical primitive number might have been six for the mesofemur and seven or eight for the metafemur (Schuh 1975). These hairs can be recognized by being generally longer and more slender than other femoral setae (McGavin 1979) and

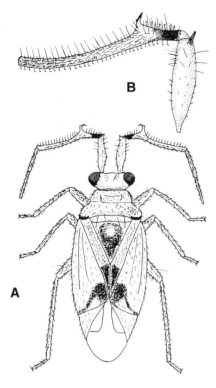

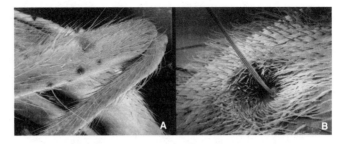

Fig. 5.9. Femoral trichobothria of *Pachymerocista pilosa*. A. ×200. B. ×1,350. (Reprinted, by permission of R. T. Schuh, courtesy of The American Museum of Natural History, from Schuh 1975.)

Fig. 5.8. Antennal modifications of *Hyalochloria longicornis*. A. Adult male. B. Segments 1 and 2. (Reprinted, by permission of T. J. Henry and the American Entomological Society, from Henry 1978.)

are referred to as *trichobothria* in mirids and other insects. Steyskal (1991), however, argued that the term *trichobothrium* should refer only to the cuplike receptacle into which the seta is inserted. In mirids, the seta is often oriented at nearly a 90° angle to the cuticle (rather than an acute angle), and the bothrium, or receptacle, is usually a pitlike depression or a domelike structure (Schuh 1975).

Pretarsal structures vary more in the Miridae than in any other heteropteran family (Fig. 5.10; Schuh and Slater 1995). As discussed in Chapter 3, they are used extensively in mirid higher classification. The pretarsus (Fig. 5.11) includes the claws, which can be simple or toothed. Pretarsal structure usually also includes the parempodia (= arolia of H. H. Knight), paired processes (lamellate or setiform) arising from the unguitractor plate and between the claw bases, as well as pulvilli (= pseudoarolia of Knight), which are bladderlike or padlike appendages associated with the ventrobasal claw surface. Some bryocorines have accessory parem-

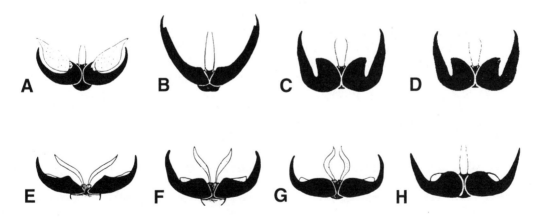

Fig. 5.10. Pretarsal structures. A. *Macrolophus separatus* (Bryocorinae: Dicyphini). B. *Cylapus tenuicornis* (Cylapinae: Cylapini). C. *Deraeocoris ruber* (Deraeocorinae: Deraeocorini). D. *Hyaliodes vitripennis* (Deraeocorinae: Hyaliodini). E. *Barberiella formicoides* (Mirinae: Herdoniini). F. *Lygus vanduzeei* (Mirinae: Mirini). G. *Heterocordylus malinus* (Orthotylinae: Orthotylini). H. *Rhinocapsus vanduzeei* (Phylinae: Phylini). (Reprinted, by permission of the Illinois Natural History Survey, from Knight 1941.)

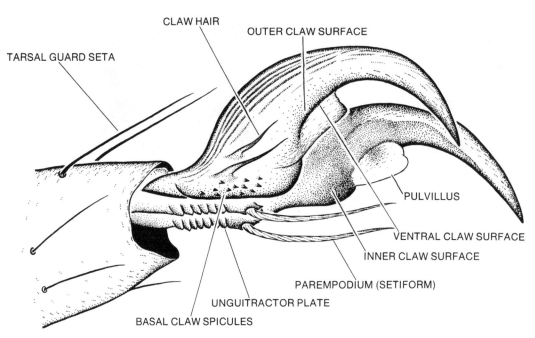

Fig. 5.11. Hypothetical mirid pretarsus, lateroventral view. (Reprinted, by permission of R. T. Schuh, courtesy of The American Museum of Natural History, from Schuh 1976.)

podia (= pseudopulvilli), paired fleshy structures at the base of the claws (Schuh 1976). The pretarsus may help the legs hold onto plant surfaces (Strong et al. 1984: Fig. 2.5; Dolling and Palmer 1991). Cobben (1978) discussed the numerous pretarsal structures that occur in this family and the putative adhesive function of mirid pretarsi. Setiform parempodia might serve as mechanoreceptors, whereas fleshy parempodia possibly have an adhesive function (Schuh 1976; Schuh and Slater 1995).

The claws and pulvilli are used to grip the substrate, for instance, trichomes by dicyphine mirids (Southwood 1986; see Chapter 15). *Ranzovius* species use their claws to grip strands of spider webs; their claws, though not as modified as those of some other spider web–inhabiting heteropterans, are shorter and straighter than those of most other phylines (Henry 1999). Mirids also use their tarsal claws to grip the substrate during oviposition, eclosion, and molting (see Chapter 6). Studies similar to those on oak aphids (Kennedy 1986) or the potato leafhopper (*Empoasca fabae*) (Lee et al. 1986) might reveal possible morphological and behavioral differences in the way mirids grip smooth versus pubescent plant surfaces. Although mirid pretarsal structures are adaptive, much remains to be learned about their functional morphology in relation to the substrate (Cobben 1978). Are the parempodia used to detect the nature of the substrate so that the bugs behave differently on smooth than on rough leaves, and how do the roles of the claws and pulvilli change with the type of substrate?

Genital structure, especially that of the male (ninth abdominal segment) (Fig. 5.12), is used extensively in

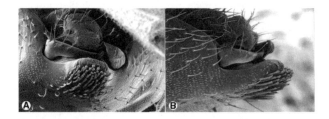

Fig. 5.12. Male genital capsule of *Proboscidotylus carvalhoi*. A. Caudal aspect (×426). B. Lateral aspect (×447). (Reprinted, by permission of T. J. Henry and the Entomological Society of Washington, from Henry 1995.)

mirid classification and identification. The German miridologist E. Wagner was the first to consistently illustrate aedeagal structure in mirids (Schuh and Slater 1995). H. H. Knight was the first North American worker to provide good illustrations of male genitalia (Fig. 5.13) and the first miridologist to emphasize the usefulness of parameres (claspers) in distinguishing closely related species (Fig. 5.13B, C; Knight 1958). Male and female (Fig. 5.14) genitalia have been studied in the family (e.g., Kullenberg 1941b, 1947a, 1947b; Slater 1950; Kelton 1959), those of males more comprehensively than females. The extent of intraspecific variation has been analyzed in taxa such as *Lopidea nigridia* (Asquith 1990), and D. A. Polhemus and J. T. Polhemus (1984) described variability in the male genitalia of the orthotyline *Ephedrodoma multilineata* (Fig. 5.15). Scanning electron microscopy, allowing the three-dimensional structure of the phallus to be more easily

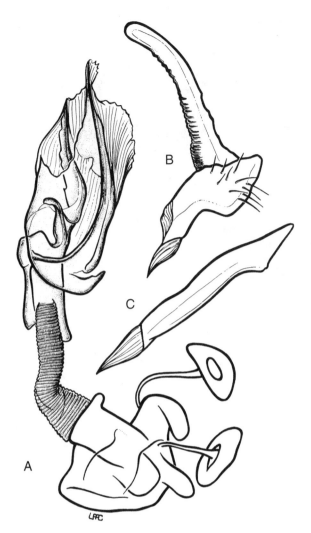

Fig. 5.13. Male genitalia of *Ceratocapsus guaratibanus*. A. Aedeagus. B. Left paramere. C. Right paramere. (Reprinted, by permission of T. J. Henry, from Carvalho et al. 1983.)

interpreted, has revealed setae, spines, and teeth on the membranous vesica of Mirinae (Clayton 1989). Davis (1955) conducted a noteworthy study of the female reproductive system in mirids. De Jong (1934) described the female and male internal reproductive organs of *Helopeltis antonii*, and Youdeowei (1972) and Oppong-Mensah and Kumar (1973) described these organs in several cocoa mirids.

Males of most mirids are more slender and slightly smaller than females, sometimes have slightly larger eyes, and have a more cylindrical abdomen that bears two hooklike, sclerotized parameres near the tip. At rest, the parameres typically lie folded across the end of the pygophore; the right paramere is usually less developed (see Figs. 5.12, 5.13) or rarely vestigial (Kullenberg 1944, Kelton 1959, Drake and Davis 1960, Schuh and Slater 1995). In the female, the cleft in the genital segments is medioventral, with the ovipositor, at rest, lying in a groove (Fig. 5.16A). The long shaftlike, sword-like, or laciniate ovipositor (Fig. 5.16B–D) is used for

inserting eggs in plant tissues or in crevices on the host plant (Drake and Davis 1960). Females of some sexually dimorphic species have a much broadened abdomen. Males vary little in weight throughout the year, whereas the weight of a female can increase as much as twofold when ripe eggs are being formed and the size of the fat body is increasing (Woodward 1949). Stewart et al. (1992) noted that *Lygus hesperus* females are about 1.5 times heavier than males.

Data on chromosome, testis follicle, and ovariole numbers might help elucidate relationships within the family. On the basis of numbers of diploid chromosomes observed in 64 species, ranging from 2n = 24 in orthotylines to 2n = 48 in dicyphines, Leston (1957a) considered 32A + X + Y the basic mirid karyotype. Kumar (1971b) reported on the karyotype of cocoa mirids and the behavior of their chromosomes in meiosis. Akingbohungbe (1974a), who determined the karyotype for 80 additional mirids, suggested that reference to 32A + X + Y as the basic karyotype be avoided until more data are available, and proposed agmatoploidy rather than polyploidy as the most likely mechanism for evolution of the mirid karyotype from an ancestral condition. Thomas (1987) discussed agmatoploidy in the Heteroptera, noting that the available evidence fails to support fragmentation (Akingbohungbe 1974a, Ueshima 1979) as an important mode of chromosome evolution in mirids and in nearly all other heteropteran groups. Instead, fusion, simple aneuploidy, and possibly polyploidy (Thomas 1996) explain most karyotype evolution in heteropterans.

The mirid diploid number ranges from 4 to 80, and m-chromosomes can be present. *Capsus ater* has a chromosomal complement of 2n = 32, including a pair of large autosomes and a fairly large X-chromosome (Nokkala and Nokkala 1986), as well as a pair of tiny m-chromosomes (Ueshima 1979). Its complement can be expressed as 2n = 28A + 2m + XY.

Ueshima (1979) listed cytological data for 73 mirid genera and 167 species, noted that the 80 chromosomes found in several *Lopidea* species is the highest number recorded among heteropterans, discussed trends in chromosome numbers in mirid higher taxa, and provided additional references to cytological studies of the family. He also pointed out that chromosomal polymorphism in the Holarctic *Stenotus binotatus* suggests the possibility of sibling species and warrants further study.

Nokkala (1986) analyzed segregational behavior of autosomal univalent chromosomes during meiosis in *Rhabdomiris striatellus*. Nokkala and Nokkala (1986) provided details of the meiotic behavior of chromosomes in males of four mirid species, involving achiasmatic meiosis of the collochore type. The latter authors suggested that determination of cytological characteristics in all cimicomorphan families might help resolve phylogenetic relationships.

Testis follicle numbers in mirids range from one to eight, with seven considered the ancestral condition

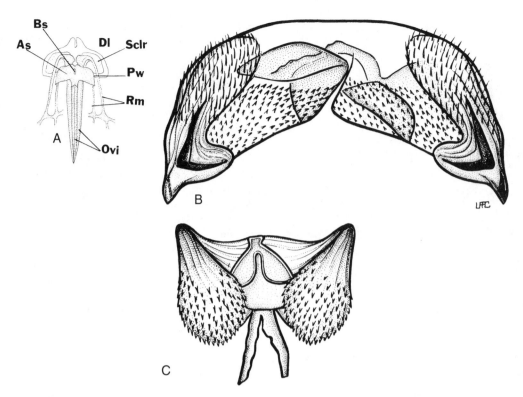

Fig. 5.14. Female genitalia. A. Schematic drawing of a *Neurocolpus* species. Abbreviations: As, A structure; Bs, B structure; Dl, dorsal lobe; Ovi, ovipositor; Pw, posterior wall; Rm, ramae; Sclr, sclerotized rings (from Henry and Kim 1984). B. *Ceratocapsus bahiensis*, sclerotized rings. C. *Ceratocapsus bahiensis*, posterior wall (from Carvalho et al. 1983). (A reprinted by permission of T. J. Henry and the American Entomological Society; B, C, by permission of T. J. Henry.)

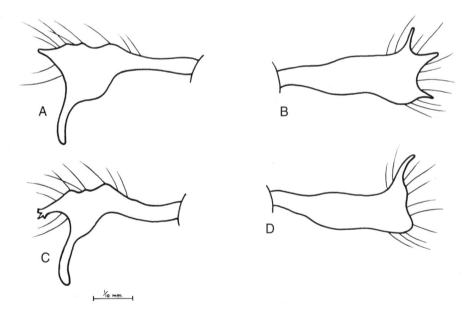

Fig. 5.15. Intraspecific variability in male genitalia of *Ephedrodoma multilineata*. A, C. Left paramere. B, D. Right paramere. (Reprinted, by permission of D. A. Polhemus and the Entomological Society of Washington, from D. A. Polhemus and J. T. Polhemus 1984.)

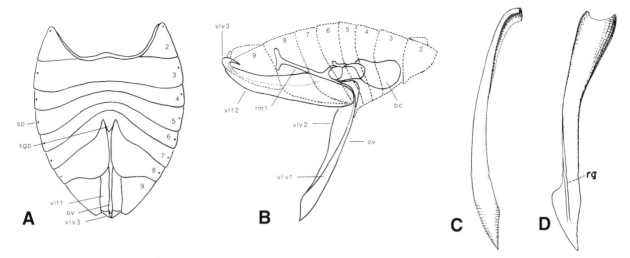

Fig. 5.16. Female abdomen and external genitalia of *Lygus lineolaris*. A. Ventral view of abdomen (from Schuh and Slater 1995). B. Lateral view of abdomen and genitalia. C, D. Lateral view of first valvula from left side of ovipositor (C) and lateral view of left second valvula (D) (from Davis 1955). Abbreviations: bc, bursa copulatrix; ov, ovipositor; rm 1, first ramus; sgp, subgenital plate; sp, spiracle; vlf1, first valvifer; vlf2, second valvifer; vlv1, first valvula; vlv2, second valvula; vlv3, third valvula. (A reprinted from R. T. Schuh and J. A. Slater: *True Bugs of the World [Hemiptera: Heteroptera]: Classification and Natural History*. Copyright © 1995 by Cornell University. Used by permission of the publisher, Cornell University Press; B–D, by permission of N. T. Davis and the Entomological Society of America.)

in the Heteroptera, but lower numbers might be plesiomorphic for the family (Leston 1961a, 1961b; Akingbohungbe 1983). The reduced numbers characteristic of certain species might result from fusion (Matsuda 1976). Jamieson (1987) elucidated the spermatozoan ultrastructure for several heteropteran families, but demonstrated mostly autapomorphies. Jamieson's review did not mention mirid studies, and it is unknown if comparative work on spermatozoa of this family would yield phylogenetically useful information. Studies of mirid sperm are apparently lacking.

Fertilization in the Miridae takes place after ovulation in the pedicels of the ovarioles (Carayon 1954, Hinton 1962). Miyamoto (1957) reported ovariole numbers of seven and eight for 20 mirid taxa. Wightman (1973) described the microstructure of the ovariole in *Lygocoris pabulinus*, analyzed ovarioles of other mirid subfamilies, and compared the ovariole structure of the family with that in other Heteroptera (see also Büning [1994]). Mirids, as well as cimicids, lack apical dividing cells in the germarium, and the trophic nuclei secrete RNA (Wightman 1973). Ma and Ramaswamy (1987) described the ovariole structure in *Lygus lineolaris* (seven ovarioles, of telotrophic type) and the seven-day gonotrophic cycle that consists of three stages: previtellogenic, vitellogenic, and choriogenic.

They also studied the histochemistry of yolk formation in the ovaries of this mirid (Ma and Ramaswamy 1990).

IDENTIFICATION AND VARIABILITY

Generic and specific identifications rely on male genitalic characters and pubescence (vestiture) of the body and appendages. Other characters that can be useful in identifying mirids are the relative lengths of the antennal segments, rostral length, punctation of the dorsum, pronotal structure, scent-gland structure, and pretarsal morphology.

The type of nymphal diet—pollen versus arthropod prey—can affect morphological characters, such as pronotal width, in adult omnivorous mirids (Bartlett 1996). Host quality can affect overall size. Specimens of *Dichrooscytus elegans* developing on cupressaceous hosts growing under arid conditions are smaller than those that develop on these plants under more favorable conditions (Kelton 1972). Host-plant information facilitates the identification of most phytophagous, and even predacious, mirids. The different hosts of polyphagous mirids can affect adult coloration, size and density of dorsal spots, and body length (Kelton 1980a, Henry and Kim 1984, Henry 1991, Barlow et al. 1999). Even females from known hosts cannot be reliably identified in certain groups (e.g., Štys and Kinkorová 1985). Some intraspecific variation in British Phylinae was once suggested to result from the selection of different genotypes on different hosts, with plant environment having a modifying effect (Southwood and Blakith 1960); this proposal, however, was invalidated by the taxonomic study of Schuh et al. (1995). Complicating the identification of different British populations of *Blepharidopterus angulatus* is a male polymorphism in antennal and hind tibial length (Leston 1958). Woodward (1952) discussed differences in body size, proportionate antennal length, and form of the left paramere between the two generations of *Notostira elongata* (as *N. erratica*; see Woodroffe [1977]) in England, suggesting that in both generations there are two different genetic populations comprising two varieties. Because this hypothesis remains untested, it is not known if varieties or sibling species are involved.

Color changes in various body parts within two hours of the imaginal molt were detailed for the cocoa mirid *Distantiella theobroma* by Kumar and Ansari (1974). Adults of many mirids, including *Taylorilygus* species (Taylor 1947b), *Lygus abroniae* (Kelton 1973), *Lygocoris tinctus* (Wheeler and Henry 1976), *Helopeltis clavifer* (Smith 1979), and *Campylomma verbasci* (Smith and Borden 1991), do not develop their typical (darker) coloration for several days after the final molt. Genitalia and other structures of such teneral individuals often become distorted and collapsed, making specific identification of preserved material difficult or impossible.

The common eastern North American mirines *Metriorrhynchomiris dislocatus* and *Taedia scrupea* occur in various color forms, most of which have been named as varieties or subspecies. They apparently have no geographic significance—Blatchley (1926, 1928) referred to them as "spotted dogs"—and Henry and Wheeler (1988) synonymized these names under the respective nominate species. Studies are needed to determine whether the distinct color patterns in these and other mirids are under genetic control and reflect a polymorphism similar to that among populations of the cercopid *Philaenus spumarius* (Stewart and Lees 1988), or whether some of the differences in color morphs are determined environmentally.

Color polymorphism of some mirids might be correlated geographically, a reddish or orange morph (rather than black or brownish) occurring in more southern portions of the range. Examples in North America include the myrmecomorphic *Barberiella formicoides* (Wheeler and Henry 1980a), *Pseudoxenetus regalis* (Henry 1985c, Blinn 1988), and *Schaffneria davisi* (Henry 1994). In the last-named species, the color morphs closely match the color of their presumed ant models (see Chapter 6). Because the genetics of these dimorphisms have not been studied, it is unknown if coloration might be controlled by a simple-locus, two-allele system similar to that in a dimorphic Japanese sawfly (Naito 1983).

Different predator complexes might select for certain color patterns in mirids (McIver and Lattin 1990).

Substantial color variation characterizes species such as *Campylomma liebknechti* (Malipatil 1992), *Eurystylus oldi* (Fig. 5.17; Stonedahl 1995), and *Lopidea nigridea* (Asquith 1990), as well as the previously mentioned *M. dislocatus* and *T. scrupea*. Color can be influenced by temperature and humidity, photoperiod, host plant, and age and sex of the specimen (e.g., Kullenberg 1941a, Boness 1963). Historically, a type of color variation in adults was attributed to the influence of heat generated in their bodies during mating (Uhler 1887). Areas of lower humidity and higher temperatures tend to produce light-colored specimens of *Mecomma* species (Kelton and Knight 1962). Adults of *Lygus lineolaris* and other *Lygus* species darken with age (Stewart and Gaylor 1990; Schwartz and Foottit 1992b, 1998). *Lygus* females containing chorionated eggs can be darker than those with previtellogenic eggs (Gerber and Wise 1995). Color change in several *Lygus* species, at least under constant laboratory conditions, is nonreversible and mediated by temperature rather than photoperiod, thus supporting the hypothesis that progressive pigmentation is the result of age (Wilborn and Ellington 1984).

Adults of *Apolygus nigritulus* developing in Japan during September to November show a seasonal variation in color that results in their blending in with the litter in which they overwinter (Yasunaga 1992a, 1992c). Overwintered adults can change color dramatically after they emerge in spring (Butler 1923). *Notostira elongata* females, for example, change from pinkish-ochreous to green. The color of *N. elongata* changes gradually as overwintered females begin to feed in spring, the green coloration first appearing in the abdomen (China 1925b). *Stenodema calcarata* adults are ochreous before hibernation, but overwintering females become green by February and usually remain so until they die in summer (Ellis 1940, Woodward 1949). Overwintered adults of *L. lineolaris* can overlap with those of the spring generation, the latter being readily distinguishable by their brighter markings (Painter 1929b). Hemelytral color (dark red rather than greenish yellow) can be used to separate overwintered *L. lineolaris* adults from first-generation adults (Kelton 1975, Stewart and Khoury 1976, Schwartz and Foottit 1998).

The use of biochemical characters and electrophoretic techniques can help resolve problems in recognizing sibling species of insects (e.g., Menken and Ulenberg 1987, Claridge 1989, Menken and Raijmann 1996). Sluss et al. (1982) used electrophoretic studies of allozyme variability to help distinguish similar species of lygus bugs, and Saahlan et al. (1986) suggested that peptidase polymorphism, as demonstrated electrophoretically in *Helopeltis theivora*, might prove useful in detecting morphologically similar species in the genus. Sibling species might occur not only in *Stenotus binotatus* (Ueshima 1979) but also in *Campy-*

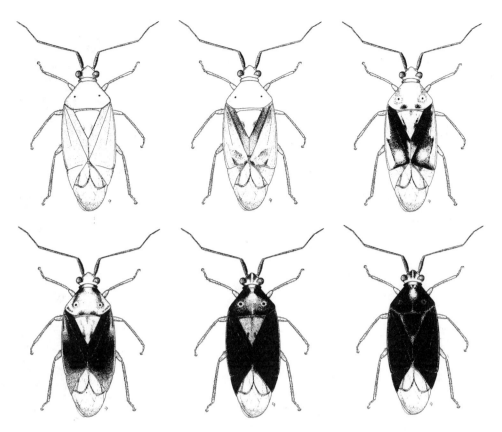

Fig. 5.17. Intraspecific color variation in females of *Eurystylus oldi*. (Reprinted, by permission of G. M. Stonedahl and CAB International, from Stonedahl 1995.)

lomma liebknechti in Australia (Chinajariyawong and Walter 1990).

Teratological specimens occur less frequently in cimicomorphan than in pentatomomorphan Heteroptera (Štusák and Stehlík 1977). Štusák and Stehlík (1978, 1979, 1982) provided examples of anomalous development of mirid antennae, legs, and hemelytra. Antennal oligomery—the number of segments differing on the two sides (Leston 1952a)—is seen occasionally in mirids (e.g., D. A. Polhemus and J. T. Polhemus 1985). Yasunaga (1996a) described an aberrant male of *Lygocoris spinolae*, the specimen appearing to have a pair of right parameres.

Nymphs

Several keys are available for placing an immature heteropteran in the Miridae (Jordan 1951a, Leston and Scudder 1956, DeCoursey 1971, Herring and Ashlock 1971, Slater and Baranowski 1978, Vásárhelyi 1990, Dolling 1991, Lawson and Yonke 1991). The instar of a mirid nymph can be determined by referring to Southwood's (1956a) or Dolling's (1991) keys, and fourth and fifth instars can be keyed to subfamily by referring to Akingbohungbe's (1974b) publication. The appearance

of the different nymphal stages can vary, depending on whether an individual is starved or fully fed; a fully fed nymph, for example, is longer than a starved individual (Petherbridge and Husain 1918, Petherbridge and Thorpe 1928a).

Nymphs share many characteristics with adult mirids, including a four-segmented labium and antenna, but nymphs have a two- rather than three-segmented tarsus and possess a dorsal abdominal scent gland, which nearly always opens between the third and fourth abdominal terga. The paired metathoracic scent glands of the adults are absent in nymphs. Trichobothria are restricted to the mesofemur and metafemur (McGavin 1979). Nymphs also can differ from adults in the types and numbers of antennal sensilla that are involved in olfactory reception (see Chapters 6, 7). The antennae of nymphal tarnished plant bugs (*Lygus lineolaris*) are shorter and have fewer sensilla than those of the adult. Nymphs presumably lack the increased antennal surface area with its associated sense organs found in adults because they have no need to locate mates or oviposition sites (Chinta et al. 1997).

Immatures of paurometabolous insects can be characterized as resembling adults of their species. They are usually small and oval, elongate, or antlike (Fig. 5.18; Yonke 1991). Certainly the difference in appearance of

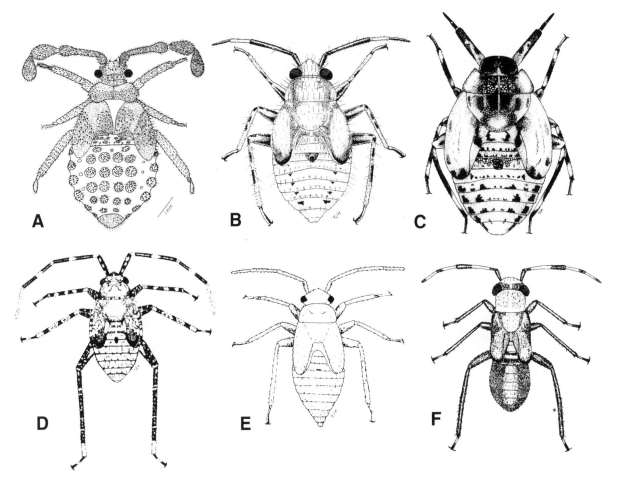

Fig. 5.18. Mirid nymphs. A. *Distantiella theobroma* (from Kumar and Ansari 1974). B. *Deraeocoris nebulosus* (from Wheeler et al. 1975). C. *Myiomma cixiiforme* (from Wheeler and Henry 1978a). D. *Phytocoris breviusculus* (from Wheeler and Henry 1977). E. *Blepharidopterus chlorionis* (from Wheeler and Henry 1976). F. *Pilophorus juniperi* (from Wheeler and Henry 1977). (A reprinted by permission of R. Kumar and Academic Press Ltd.; B, C, E, by permission of the Entomological Society of America; D, F, by permission of the American Entomological Society.)

a mirid nymph and its adult is less than that between the larva and adult of holometabolous species. Yet it is often difficult to visualize the eventual adult form and color from examining even the fifth instar. Butler (1922) said of *Deraeocoris ruber* nymphs, "I very much doubt whether any one who had not actually reared them would guess what they would ultimately become."

The principal changes that occur during nymphal development involve the appearance and expansion of the wing pads (third through fifth instars), as well as differences in the length and width of the head and lengths of antennal segments, in the color and markings, in the relative lengths of the mesothorax and metathorax, in the position of the labial apex relative to the thoracic segments, and in the number and length of the setae.

Mirids, unlike many heteropterans (Schaefer 1975), do not undergo ontogenetic changes in numbers or arrangement of trichobothria (Schuh 1975, McGavin 1979). Pronounced ontogenetic changes in coloration are not as common as in certain other heteropteran families (e.g., Largidae [Booth 1990]), but they do occur in some species (see Chapter 10). Coloration can be variable (Ratnadass et al. 1994), that in predatory mirids sometimes varying with the type of prey consumed (Collyer 1952, Morris 1965; see also Chapter 14). The color of some species blends in with that of host foliage (e.g., Southwood and Leston 1957; Lightfoot and Whitford 1987; Blinn 1988, 1992; Gagné 1997) or inflorescences (Schuh 1974, Schuh and Slater 1995, Yasunaga 1998; see also Chapter 10).

Wing pads or buds usually appear in the third instar except in mirids characterized by a variable number of instars, such as *Cyrtorhinus lividipennis*, in which wing pads appear in the second instar of three-instar populations (Napompeth 1973, Liquido and Nishida 1985c). McGavin (1979) observed that in some sexually dimorphic species, nymphs with small wing pads produce brachypterous adults, whereas nymphs with large wing pads produce macropterous adults.

In certain tropical bryocorines such as *Helopeltis* species, a scutellar spine (Fig. 5.19) appears in the second instar. The visibility of genital structures through the semitransparent cuticle and a narrower abdomen in males usually allow the sex of a fifth-instar nymph to be distinguished (Osborn 1918, Haviland 1945, Lal 1950, Gopalan and Basheer 1966, Napompeth 1973, El-Dessouki et al. 1976, Ambika and Abraham 1979, McIver and Asquith 1989, Ratnadass et al. 1994), except in the youngest fifth instars (McGavin 1979). The female genital opening is sometimes apparent in fourth instars (Cory and McConnell 1927, Hiremath and Viraktamath 1992); a clearing of tissues (e.g., from storage in 70% ethyl alcohol) may be necessary to observe the ovipositor (Betsch 1978). In some species, male nymphs are slightly shorter than female nymphs (Hiremath and Viraktamath 1992). Brachypterous adults, especially in myrmecomorphic species, can be mistaken for nymphs, but close examination reveals fully developed genital structures (Fig. 5.20).

Specific (or even generic) determination of mirid nymphs is often impossible—those of the majority of species have not been described or illustrated. Schwartz

and Foottit (1992b) diagnosed and described all nymphal stages of six *Lygus* species occurring in the Canadian prairie provinces and provided a key to fourth and fifth instars. The only paper treating nymphs of more than a few North American species is that by Akingbohungbe et al. (1973), who provided a generic key to fifth instars and briefly described the fifth instars of 55 species occurring in Wisconsin. McGavin (1979) treated the immature stages of many British mirids. DeCoursey (1971) keyed mirid subfamilies based on nymphs but was unable to differentiate the Orthotylinae and Mirinae based on tarsal characters. Akingbohungbe et al. (1973) and Akingbohungbe (1974b) found that nymphs of these subfamilies were more easily distinguished by the type of abdominal scent-gland opening. McGavin (1979) remarked that the tarsal characters seen in adults—the parempodia are convergent in Orthotylinae but divergent in Mirinae—are not displayed in nymphs. The typical convergent parempodia, though, are readily visible in orthotylines, but for mirine nymphs preserved in alcohol this character can be difficult to interpret.

Mirids possess a single dorsal abdominal gland opening between terga 3 and 4. These glands, although complex histologically, are filled with a fluid that appears to serve no defensive function; the contents of the glands are not discharged when a predator attacks (Aryeetey and Kumar 1973). Carayon (1977), however, remarked that the opening of the dorsal abdominal glands is not sealed with cuticle or occluded, as Aryeetey and Kumar (1973) stated, and that the glands of several bryocorines are highly active (see also Gupta

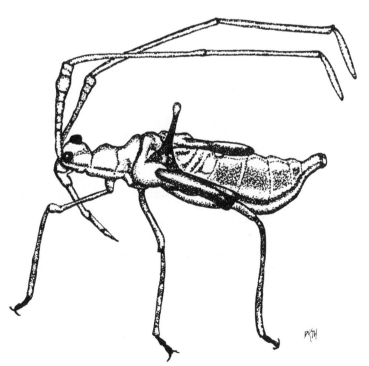

Fig. 5.19. Fifth instar of a *Helopeltis* species showing scutellar projection. (Redrawn from Leefmans 1916.)

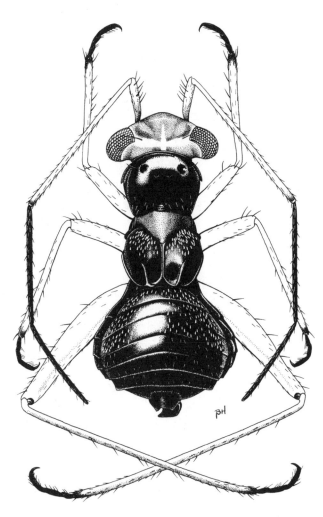

Fig. 5.20. Adult male of *Myrmecophyes oregonensis*. (Reprinted, by permission of R. T. Schuh, courtesy of The American Museum of Natural History, from Schuh and Lattin 1980.)

[1961], Leston [1978a], McGavin [1979], and Chapter 6). Akingbohungbe (1974b) described the generalized and specialized conditions found in mirid abdominal glands, recognizing six categories on the basis of the type of opening. The opening of the dorsal abdominal gland is sometimes absent or very small, as in Nearctic and Palearctic *Dichrooscytus* species (Akingbohungbe et al. 1973, Wheeler and Henry 1977, McGavin 1979).

The host plant can influence nymphal coloration (Palmer and Knight 1924). The nymphs of African *Eurystylus* species vary from red to green depending on the variety of their castor hosts (Boyes 1964) or the color of sorghum grains on which they feed (Ratnadass et al. 1994). The nymphs of congeners, however, can be of a different color, even when they feed on the same parts of a particular host species. For example, nymphs of *Polymerus tinctipes* feeding on reproductive structures of moss phlox are green, whereas those of *P. wheeleri* feeding on generally the same parts of this host are dark red (Wheeler 1995).

The nymphs of a few species show striking morphological modifications. One of the most unusual is that of a Cuban species (suggested to be a *Paracarnus* sp.), which has flattened horns on the vertex and an anchor-like process on the pronotum (China 1931; cf. Bruner 1934). These thoracic horns somewhat resemble those of certain membracids (Hogue 1993).

Eggs

Mirid eggs differ in form and have been described variously as shaped like a banana, bean, cigar, club, flask, paddle, pear, sac, sausage, or test tube. The typical egg is pale, creamy, or white and often glistening (color changes occur before eclosion; see Chapter 6). It has the shape of an elongate cylinder, but is slightly curved and rounded posteriorly and tapered toward the compressed and often truncate anterior pole (Figs. 5.21, 5.22). The length usually varies from 0.6 to 2.0 mm (Kullenberg 1944), but the egg of *Hyalopeplus smaragdinus* is about 2.5 mm long (Roepke 1919). Upon deposition, the eggs can appear flat-sided and collapsed, but they soon become turgid with slight changes in breadth occurring during development (e.g., Neal et al. 1991, Stewart and Gaylor 1993). The eggs increase in length during the maturation period (Stewart and Gaylor 1993). Most eggs are widest near the middle, where they are almost round in cross section (e.g., Johnson 1934). Intraspecific variation in egg width and length can occur. *Lygus rugulipennis* eggs vary in length from 0.83 to 1.14 mm, their length positively correlated with the length of females (Varis 1972).

The chorionic surface of mirid eggs is smooth or sometimes bears faint hexagonal sculpturing; the operculum, "an elliptical domed cap-like continuation of the shell" (Hartley 1965), is typically elaborately sculptured. An asymmetrical operculum, often concealed by the rim of the chorion, appears characteristic of mirids (Southwood 1956b). Chorionic thickness varies from slightly more than 2 μm to almost 19 μm (Kullenberg 1944), and the anterior pole in particular shows great structural diversity. Considered uniquely derived or synapomorphic for the family is the presence of two micropyles, which vary in shape, extent, and position (Hinton 1962, 1981; Cobben 1968). Before Hinton's (1962) study, the presence of micropyles had gone undetected; when micropyles were mentioned (e.g., Johnson 1934), it was the aeropyles that actually were under observation.

An elevated, collarlike prolongation of the rim often occurs around the operculum. This noncellular collar usually appears honeycomb-like or latticelike (Fig. 5.22B). Mirid eggs also possess a subopercular yolk plug (Johnson 1934, Southwood 1956b), which may be extruded soon after oviposition (Morris 1965).

Because the shells of embedded eggs must resist pressure from the growing tissues of host plants, the mirid operculum is remarkably resistant to lateral pressure.

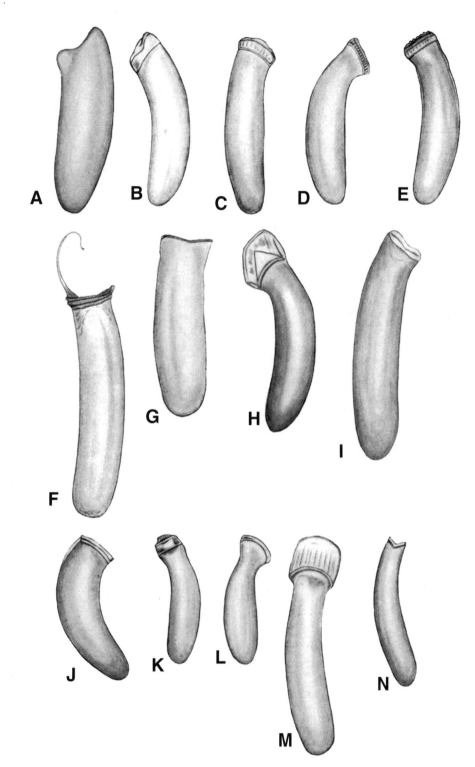

Fig. 5.21. Mirid eggs. A. *Monalocoris filicis*. B. *Adelphocoris lineolatus*. C. *Closterotomus norwegicus*. D. *Capsus ater*. E. *Lygocoris pabulinus*. F. *Megacoelum infusum*. G. *Orthops kalmii*. H. *Leptopterna dolabrata*. I. *Notostira erratica*. J. *Halticus apterus*. K. *Heterotoma merioptera*. L. *Orthotylus marginalis*. M. *Harpocera thoracica*. N. *Pilophorus perplexus*. (Reprinted, by permission of B. Kullenberg and Kungl. Vetenskapakademien, from Kullenberg 1942.)

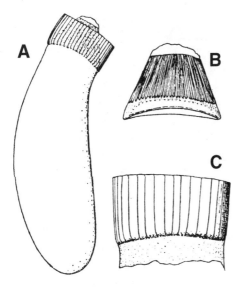

Fig. 5.22. Egg of *Horistus orientalis* (A) showing detail of the collar (B) and operculum (C). (Reprinted, by permission of Société Entomologique de France, from Silvestri 1932.)

The elaborately formed operculum consists mainly of a crest composed of an elliptical cylinder having a series of transverse plates. This appears to prevent the collar region from being constricted by growing plant cells or those undergoing wound repair (Hartley 1965). Opercular thickness tends to be greater in species overwintering in the egg stage than in those that hibernate as adults (Southwood 1956b). Egg shape can change slightly as a result of pressure from the wood of host tissues (e.g., Petherbridge and Husain 1918). Steer (1929) observed that eggs are laterally compressed as they pass down the ovipositor and that lateral pressure from the wood maintains this shape. Any accommodation in shape must take place longitudinally (Johnson 1934, Usinger 1945). The yolk or serosal plug appears to be a mechanism allowing the "egg to swell in one direction and to survive what would otherwise be a fatal bursting due to the inherent tendency of the egg to swell" (Johnson 1934).

Readers interested in egg morphology should consult Kullenberg's (1942, 1943) papers, which contain descriptions and illustrations of the eggs of more than 100 Swedish mirids. Putshkov and Putshkova (1956) illustrated the eggs of economically important mirids in the former Soviet Union. Cobben's (1968) study provides detailed information on egg architecture, embryology, and eclosion (see also Chapter 6).

6/ Overview of Ecology and Behavior

Insects are inextricably tied to the ecological particulars of their nutritional resources.

—D. W. Tallamy 1994

Much of this wide-ranging chapter, although not dealing directly with the trophic habits of mirids, touches on their nutritional ecology. The duration of the behaviors that are mentioned is highly temperature dependent, but temperature and other variables are not always available in the literature. And data obtained from laboratory rearings of mirids often show consistently high variation (Bryan et al. 1976). More studies of the Miridae similar to those on delphacid planthoppers (Denno 1994a) and certain other insect groups (Roff 1992) are needed to place traits such as fecundity, age to first reproduction, and voltinism in the context of life-history theory.

Diversity and Distribution

Mirids belong to the cimicomorphan line of the Heteroptera, a monophyletic, though heterogeneous assemblage that includes several other well-known families (Schuh and Štys 1991, W. C. Wheeler et al. 1993, Schuh and Slater 1995). Cimicomorphan families probably familiar to nonspecialists are the Anthocoridae, minute pirate or flower bugs, a mainly predacious group; Cimicidae, the bed bugs, temporary ectoparasites of birds and mammals; Reduviidae, the ambush, assassin, and threadlegged bugs, a predacious family containing some species that feed on vertebrate blood; Nabidae or damsel bugs, another predacious family; and Tingidae, the lace bugs, an exclusively plant-feeding family. In the most recent analysis of cimicomorphan relationships, the Miridae, Tingidae, and Thaumastocoridae make up the superfamily Miroidea (Schuh and Štys 1991).

SPECIES RICHNESS AND GLOBAL DIVERSITY

Represented by nearly 10,000 described species in about 1,400 genera, the Miridae are the largest heteropteran family (Table 6.1; Schuh 1995). In one genus (*Phytocoris*), about 650 species have been described (Schuh and Slater 1995), and the total number of mirid species eventually might reach 20,000 (Henry and Lattin 1987,

Henry and Wheeler 1988). Yet Froeschner's (1985) comment that the number of mirids approaches the combined total of species in all other heteropteran families is overstated; the Miridae represent less than a third of all described Heteroptera. Butler (1923) had made similar statements about the European fauna: that mirids represent about 60% of all British heteropterans and perhaps half of all terrestrial bugs in the Palearctic region. In addition, the estimate for the Canadian fauna that as many species remain undescribed or unrecorded as are currently known (600 in both cases [Scudder 1979]) might be exaggerated.

As in most other insect groups, our knowledge of mirids and their distributions tends to reflect the activity of collectors and researchers. Schaefer (1990) emphasized the appallingly inadequate knowledge of North American mirids, particularly in the western states (Polhemus 1994), where numerous species remain undescribed. The fauna in many other regions has received even less attention. For example, in South Africa, where only about 60 species had been recorded, Slater (1974) estimated that the fauna actually contained 1,000 species. Since the 1940s, about 2,000 new species from Middle and South America have been described (Carvalho and Froeschner 1987, 1990, 1994). The rich, endemic mirid fauna of the Hawaiian Islands also remains largely undescribed (Asquith 1993b, Gagné 1997).

Global biotic diversity is greatest in the equatorial regions of South America, Africa, and Asia (e.g., Wilson 1992). Maximum species richness in mirids might be predicted for the rainforests of the neotropics, perhaps in Panama, where plant and insect diversity is particularly great (e.g., Erwin 1982), or maybe in Colombia, Peru, or Venezuela. A rich, highly endemic fauna undoubtedly characterizes other New World regions, such as the coastal forests of Ecuador, the Brazilian Amazon (e.g., P. H. Raven 1983, Myers 1988), and southern Mexico (e.g., Schaffner and Ferreira 1995b). Taxonomic studies comparable to those of J. C. M. Carvalho in the neotropics have not been undertaken on mirids of the Old World tropics, but tropical Southeast Asia probably also is a center of species richness. Use of insecticidal fogging techniques in tropical tree canopies (Erwin 1983, 1989) will reveal additional mirid diversity.

Table 6.1. The largest families of Heteroptera worldwide

Family	No. of genera	No. of species
Miridae	1,383	9,805
Reduviidae	930	6,500
Pentatomidae	760	4,100
Lygaeidae	628	4,045[a]
Tingidae	250	1,900
Coreidae	250	1,800
Aradidae	211	1,800

Source: Schuh (1995), Schuh and Slater (1995), Slater and O'Donnell (1995).

[a] Figure is based on the traditional—that is, paraphyletic—concept of the family; see Henry (1997) for proposed reclassification of the Lygaeoidea.

Myrmecomorphic (antlike) mirids, although occurring in temperate regions (especially Phylinae: Hallodapini [Schuh 1974]), proliferate in the tropics. Orthotylines and mirines predominate in the neotropics, and phylines in the Old World tropics. Within the Insecta, myrmecomorphy reaches its greatest diversity in the Miridae, having arisen independently at least 10 times (Schuh 1986a, McIver and Stonedahl 1993). The diurnal activity of most mirids on plants and their vulnerability to visually oriented predators are thought to have favored the evolution of myrmecomorphy (McIver and Stonedahl 1993). The most striking ant mimics, at least in certain phyletic lines of the Orthotylinae and Phylinae, are the more derived taxa (Schuh 1984, 1991; McIver and Stonedahl 1993; Schuh and Slater 1995).

FOSSIL MIRIDAE

Heteroptera began to appear in the fossil record during the Triassic period (Labandeira and Sepkoski 1993), and the Cimicomorpha, including Miroidea, appeared by the Jurassic period (e.g., Popov 1981, Strong et al. 1984). The earliest-occurring mirids are recorded from the Upper Jurassic of Karatau, a locality in Kazakhstan between the Aral Sea and Lake Baikal (Kukalová-Peck 1991). Fossil mirids are also known from the Tertiary (Bekker-Migdisova 1962, Jordan 1972), including Eocene cylapines and a deraeocorine from Baltic amber (e.g., Carvalho and Popov 1984, Herczek and Gorczyca 1991). The first fossil isometopines recently were described from Baltic and Dominican amber (Popov and Herczek 1992, Santiago-Blay and Poinar 1993). Many mirine fossils described from Florissant shale apparently should be assigned to other cimicomorphan families (Carvalho 1959). Spahr (1988) provided a list of fossil mirids known from amber (see also Herczek [1993], Popov and Herczek [1993], Weitschat and Wichard [1998]).

SPECIATION AND RADIATION

Primary speciation in the Miridae perhaps took place during the Upper Cretaceous or Lower Cenozoic (Leston 1979b), following the angiosperms' rise to dominance in the world's vegetation (e.g., Crepet 1979, Crane 1989). Leston (1979b) hypothesized that the main trend in the radiation of phytophagous mirids has been from arboreal to herbaceous hosts, thus paralleling the history of angiosperms. Although the major groups might have been present by the Upper Cretaceous or Lower Cenozoic, speciation almost certainly continued through the Tertiary and, in some Northern Hemisphere groups, it perhaps was associated with Pleistocene events (e.g., Wheeler and Henry 1992).

Important in the early radiation of mirids may have been the development of tarsal structures and adaptations (Schuh 1976, Southwood 1986, Dolling 1991, Dolling and Palmer 1991) that enabled these bugs to overcome the evolutionary hurdle of attaching to plant surfaces (Southwood 1973). As in curculionoid Coleoptera (Anderson 1995), endophytic oviposition in mirids can be considered an adaptation crucial to a life on plants (Chapter 1). Sweet (1964) compared speciation in mirids with that in the Lygaeidae, which, with about 4,000 species (Schuh and Slater 1995, Slater and O'Donnell 1995), is the fourth largest heteropteran family (see Table 6.1; Schuh and Slater 1995). Speciation in rhyparochromid lygaeoids (predominantly litter inhabitants feeding on fallen seeds), as in small mammals, is considered habitat specific rather than host specific. Speciation patterns in plant bugs, by contrast, tend to be host related (Sweet 1964).

Actual modes of speciation in mirids have been little studied. Asquith's (1993a) examination of four monophyletic groups of the orthotyline genus Lopidea (about 60 spp. restricted to the Western Hemisphere) suggests that vicariance accounts for at least 50% of speciation events, that about 25% can be attributed to a sympatric mode, and that only one example fits a peripheral isolation model. Even though these conclusions may be altered when additional data on the distribution of Lopidea species become available and when additional groups of the genus are considered, this approach might be extended to other species-rich genera of the family. In the orthotyline genus Sarona, which is endemic to the Hawaiian Islands, Asquith (1995b) proposed that sympatric speciation, presumably the result of host switches, played a greater evolutionary role than did allopatric colonization of new islands or isolation on mountains or volcanoes (see also Asquith [1997]). Gagné (1997) analyzed insular radiation and speciation in the endemic Hawaiian genus Nesiomiris, including aspects such as interisland founder events, intraisland geographic isolation, intergeneric host-plant transfers, and multiple invasions of the same host species. Polyploidy might be involved in mirid evolution, but its role in speciation in the family is unclear (Thomas 1996).

Mirids are among the most successful of all insect groups (as noted in Chapter 1), not only in numbers of species and abundance of individuals but also in their range, which extends to all zoogeographic regions except the Antarctic. Of all terrestrial heteropterans, a mirid (*Agraptocoris margaretae*) is found in the highest altitude: nearly 5,400 meters in the Karakoram Range of Indian Tiber (Hutchinson 1934, 1965; Mani 1962). Several species occur above the timberline (>3,600 m) in the high tundra grasslands of Colorado, apparently exploiting the short, unpredictable summer seasons (D. A. Polhemus and J. T. Polhemus 1988). Species diversity generally decreases with increasing latitude, and mirids are among the insect groups clearly underrepresented in the Arctic regions of Canada (Danks 1986).

Latitudinal and altitudinal effects on mirids are evident from studies on lygus bugs. The main pest species of *Lygus* in North America dominate at lower altitudes (<500 m), whereas above 1,200 meters, they are replaced by lesser-known members of the genus. Increasing latitude might have similar effects, as the economically important species of *Lygus* are less often noted as pests in Canada than in areas farther south (Scott 1987). But when the economic literature is surveyed and authoritative identifications are considered (Schwartz and Foottit 1998), it is apparent that *Lygus* species economically important in the United States are also pests of Canadian crops.

Major advances in the biogeographic analysis of insects and other groups of organisms have taken place in recent years. Analysis has moved from an emphasis on centers of origin and reliance on dispersal to explain present distributional patterns, to a historical approach that establishes areas of endemism and interprets observed biotic distributions in relation to the disruption of ancient land masses. Historical biogeography seeks congruence between observed distributions and phylogenetic relationships. Cranston and Naumann (1991) present a good overview of biogeographic theory, and Schuh and Slater's (1995) discussion is essential for an understanding of heteropteran biogeography.

The principal mirid subfamilies (Schuh 1995, Schuh and Slater 1995; see Table 3.1) are cosmopolitan, whereas a more limited geographic distribution is sometimes evident at the tribal level, for example, in the deraeocorine tribe Saturniomirini, which is restricted to Australia and New Guinea (Gross and Cassis 1991). Various mirid taxa show distributions that correspond with long-recognized biogeographic regions (e.g., Australian or Neotropical), or they exhibit intercontinental patterns (e.g., Gondwanan or Paleotropical [Schuh and Slater 1995]). Biogeographic studies are beginning to identify centers of endemism in mirids, to analyze their interrelationships, and to determine whether areas of species richness tend to correlate with areas of high endemism. Schuh (1974) discussed patterns of endemism in the South African fauna, and Schuh (1984) and Schuh and Stonedahl (1986) identified taxa endemic to the Indo-Pacific, relating its fauna to that of other regions. A biogeographic analysis of the Pilophorini revealed an apparent tropical Gondwanan origin for this tribe, which later spread into and differentiated in the temperate Northern Hemisphere (Schuh 1991). Wheeler and Henry (1992) reviewed the Miridae that are common to the Old and New World (at that time 98 spp. were known), noting whether they are naturally Holarctic members of the North American fauna or whether they should be considered adventive.

Various anthropogenic influences have altered mirid densities and distributions (e.g., Putshkov 1966, Štys 1974, Dolling 1991, Stehlík 1998; see also Chapter 4). Examples of human influences include the advent of monocultural crop systems, the shipment of woody plants such as fruit and shade trees (Wheeler and Henry 1992, Yasunaga et al. 1999), and even the planting of rock gardens (Cobben 1960).

HABITATS AND HOST PLANTS

As in several other groups of phytophagous insects (Futuyma 1983), the main radiation of host-plant usage in mirids probably followed, rather than accompanied, diversification of the angiosperms. Colonization, with host switching presumably mediated by plant secondary chemistry, has resulted in host associations that exhibit considerable parallelism and convergence (Miller and Wenzel 1995). Evidence that mirid host associations have evolved independently from plant diversification has accrued from several cladistic analyses (e.g., Stonedahl and Schwartz 1986, Schuh 1991).

Plant bugs are abundant in a variety of habitats (e.g., Ehanno 1987b) ranging from mountain tops to salt marshes and coastal dunes (Southwood and Leston 1957, Marples 1966, Denno 1977, Chinery 1993), but they are absent or rare in areas devoid of plant cover (Putshkov 1966). In one British study, plant bugs were the dominant heteropterans occurring in all secondary successional stages surveyed: ruderal, young field, old field, and woodland (Brown 1982a). In the Czech Republic, mirids represented nearly half of the 83 heteropteran species in a floodplain forest ecosystem (Stehlík 1995), and the family represented up to 45% of heteropteran species in a tropical rainforest in Indonesia (Hodkinson and Casson 1991).

Most crop plants, both annuals and perennials, have at least one mirid species among their complement of pests. Agricultural production in the tropics, particularly of cocoa, can be limited by mirids (e.g., Carter 1973), and alfalfa, cotton, and fruit trees such as apple harbor a diverse plant bug fauna. In the case of apple trees in North America and Britain, some of the injurious mirids were originally associated with other rosaceous plants, such as crabapple and hawthorn trees (Knight 1915, Wellhouse 1922), whereas other plant bug colonists of apple trees used native hosts in nonrosaceous families, for example, the Salicaceae (Fryer 1929).

Numerous plant bugs are associated with coniferous gymnosperms and a diverse group of angiosperms, both dicots and monocots. An association with monocots is considered primitive in groups such as halticines, pilophorines, and stenodemines (Schuh 1991). A few mirids develop on fungi (some cylapine spp.) and ferns (*Bryocoris*, *Felisacus*, and *Monalocoris* spp.). Mosses (e.g., Butler 1918), other bryophytes, and lichens until quite recently remained undocumented as hosts. Moss may be the primary habitat of the predatory *Bothynotus pilosus* in Great Britain (Southwood and Leston 1959, Woodroffe 1969), and in Hawaii, the endemic and possibly predacious *Kamehameha lunalilo* is found on mosses and ferns that clothe the trunks and branches of various tree species (Zimmerman 1948a). The primary habitat of the latter bug is actually dead, dry fern litter below banks of fern (D. A. Polhemus, pers. comm.). Species of the orthotyline genus *Pseudoclerada*, which are typically coleopteroid and presumably predacious, are usually collected from moss-covered branches in Hawaii (Asquith 1997). Of the two known specimens of the recently described mirine *Monopharsus annulatus*, one was collected on moss on the ground and the other from mats of a liverwort on a tree in New Zealand (Eyles and Carvalho 1995), but it is doubtful if an actual host relationship exists with these lower plants. The first actual moss-feeding plant bug is the Japanese *Bryophilocapsus tosamontanus*, for which a new genus was described to accommodate this unique bryocorine (Yasunaga 2000).

Diet breadth in mirids ranges from monophagy to polyphagy. The nymphs of the majority of species appear to be oligophagous. Some closely related plants in mixed colonies sometimes support distinct, essentially nonoverlapping mirid faunas (e.g., Pinto and Velten 1986). Herbivores that are monophagous or oligophagous at a particular locality can use a greater number of hosts (regional polyphagy; e.g., Fox and Morrow [1981], Strong et al. [1984], Bernays and Graham [1988], Polis [1991]) throughout their range. In northern areas such as the Yukon, a mirid can be restricted to a single host because only that species of the bug's known hosts occurs there (Scudder 1997). *Pinophylus carneolus*, which is known from New York to Florida, develops mainly on Virginia pine but uses jack pine and pitch pine in more northern areas and sand pine at the southern extreme of its range (Wheeler 1999). In mirids and other heteropterans, hosts tend to remain relatively constant over the entire range of a particular species (Putshkov 1960; pers. observ.), though few plant bugs have been studied carefully throughout their geographic ranges.

Mirids frequently develop on adventitious hosts in arboretums, botanical gardens, and college campuses (Wheeler and Henry 1992). For example, nymphs and adults of *Lopidea robiniae*, which is normally restricted to black locust, were observed in Rochester, N.Y., on false indigo, a native plant on which *Lopidea hesperus* frequently occurs (pers. observ.). The presence of numerous exotic plants in arboretums and botanical gardens favors the adoption of novel host plants. Examples from New York include the native North American mirines *Tropidosteptes amoenus*, *T. cardinalis*, and *T. plagifer* that colonize European ash, and another Nearctic mirine, *Lygocoris vitticollis*, which develops on Amur maple (pers. observ.), a plant that has been introduced from Asia. Similar examples involving coniferous hosts are the development of the native phyline *Psallovius piceicola* on oriental spruce and Nordmann fir in Pennsylvania (pers. observ.) and the use of Asian and North American arborvitae and juniper species by native mirids in the Czech Republic (Stehlík 1998).

Monophagy in a strict sense occurs in the two known members of the South African genus *Pameridea* (Bryocorinae: Dicyphini): each has an apparent obligate relationship with one species of the endemic South African genus *Roridula* (Dolling and Palmer 1991; see also Chapter 15). In the Hawaiian Islands, an endemic dicyphine, *Cyrtopeltis kahakai*, is restricted to a goodeniaceous strand plant (Asquith 1993b), and nearly all *Sarona* species develop on a single host species (Asquith 1995b). Thirty-nine of the 50 described species in the endemic Hawaiian genus *Nesiomiris* are also associated with a single host species (Gagné 1997). Examples of monophagous Mirini include *Neoborella* species on dwarf mistletoes parasitic on conifers in western North America (Stevens and Hawksworth 1970, Kelton and Herring 1978); some conifer-feeding *Bolteria*, *Dichaetocoris*, and *Dichrooscytus* species in the western United States (Polhemus 1988); members of the New World genus *Platylygus* on pines (Kelton and Knight 1970); and the Palearctic *Pinalitus coccineus* and *P. viscicola*, which specialize on mistletoes parasitic on deciduous trees (Southwood and Leston 1959; Bin 1970; Štys 1970b, 1975). In addition, members of the diverse mirid fauna associated with bald cypress in the United States are probably restricted to that tree species (Knight 1941). Other examples of host-restricted mirids include eccritotarsine bryocorines found mainly in the southwestern United States and Mexico: species of *Caulotops* on agaves, *Halticotoma* on yuccas, and *Hesperolabops* on opuntias (Schuh and Slater 1995; pers. observ.).

In contrast, certain polyphagous mirines, such as *Lygus hesperus* (Scott 1977a) and *L. lineolaris* (Young 1986) in North America and *L. rugulipennis* in Eurasia (Hori and Hanada 1970, Holopainen 1989, Holopainen and Varis 1991), are known from more than 100 host plants. Large, mostly phytophagous genera such as *Lopidea* and *Lygus* tend to include both host-restricted and polyphagous members (Asquith 1991, Schwartz and Foottit 1998; pers. observ.). The range of host-restricted mirids is often much less than that of their host plants (e.g., Danks 1979), suggesting that precipitation, edaphic conditions, and other factors help determine their distributions.

Some predominantly carnivorous mirids prefer particular plants (e.g., Slater and Baranowski 1978, Yonke 1991), a phenomenon seen in other predacious heteropterans such as anthocorids, nabids, and reduviids (Cobben 1978). Certain *Deraeocoris* and *Phytocoris* species develop only on oaks, whereas other species occur on pines. Host specialization in the Pilophorini, a mainly predacious tribe of the Phylinae (see Chapter 17), appears strongly influenced by the presence of preferred homopteran prey (Schuh and Schwartz 1988, Schuh 1991) that presumably stimulate reproduction. Partial phytophagy in *Deraeocoris* might allow host recognition in species in which egg hatch must be synchronized with the appearance of specific prey species (Razafimahatratra 1980). Plant material might supply important dietary requirements for mainly carnivorous mirids (Razafimahatratra 1980, Stonedahl 1988). Referring to the apparent restriction of many *Deraeocoris* species to certain plants, Slater and Baranowski (1978) commented, "Interesting problems concerning the evolution of predatory habits certainly await the careful investigator."

Some plants serve as hosts for co-occurring mirids of certain predacious genera. Species that occur simultaneously appear to partition available host resources. For example, the various *Phytocoris* species found on certain conifers in western North America can be associated mainly with the cones, foliage, branches, or trunks of their hosts (Stonedahl 1988).

Predatory heteropterans often have broader host ranges than do phytophagous species. *Blepharidopterus angulatus* (Collyer 1952) and *Deraeocoris nebulosus* (Wheeler et al. 1975) track populations of diverse prey on numerous, unrelated host plants. Mixed feeders, or bugs that feed on plant and animal tissue, tend to have host ranges intermediate between those of predators and plant feeders (Kinkorová and Štys 1989). That mirids show a higher proportion of polyphagous species among phytophagous insects in Britain might be due to the relatively large number of plant bugs that are facultative predators (Ward and Spalding 1993). Overall, though, most phytophagous mirids, as noted earlier, should be considered oligophagous.

HOST AND HABITAT PREFERENCES

Arboreal mirids of temperate regions tend to prefer forest-edge conditions; they are often abundant on isolated or open-grown trees or on trees, shrubs, and the associated vines and understory vegetation of hedgerows (Knight 1941, Pollard 1968, Lewis 1969, Ehanno 1976; pers. observ.). Species such as the mainly predatory *Heterotoma planicornis* and *Phytocoris ulmi* are characteristic of hedgerows (Pollard 1968). Different types of management within a hedgerow, such as the periodic cutting of branches, can increase architectural complexity. Because not all hedgerows are cut at the same time, a shifting pattern of successional stages is created in the landscape, which can lead to changes in the mirid fauna: from mostly predatory species after branches are trimmed to a predominance of plant feeders when branches become overgrown (Burel 1996). The faunal richness of scrub oak (>40 species are associated with this host in the eastern United States) might be partly due to the occurrence of this shade-intolerant plant in communities such as pitch pine–scrub oak barrens that are characterized by a sparse, interrupted canopy (Wheeler 1991b). Open-grown shrubs and trees in urban or suburban landscapes—either under conditions of environmental stress or those promoting more vigorous growth than is typical in native habitats—are especially liable to infestation by mirids.

Some herb-associated mirids are shade-intolerant, their abundance declining with increasing levels of shade (Greatorex-Davies et al. 1994). Some exceptions are *Dicyphus gracilentus*, which lives on an herb in deep, shady woods (Knight 1941); *Orthonotus rufifrons*, which inhabits nettles growing in shade or semishade (Southwood and Scudder 1956, Leston 1961d); and *Bryocoris pteridis* (Kullenberg 1944) and *Monalocoris filicis* (Van Duzee 1887), which develop mainly on ferns in shaded areas (see also Gorczyca [1994]).

Although plant bugs are said to be not as diverse or numerous in dense woods (Knight 1941), many species of temperate regions use coniferous and deciduous hardwood trees as host plants (Knight 1941, 1968; Whittaker 1952; Southwood and Leston 1959; Kelton 1980b; Southwood and Kennedy 1983; Gorczyca 1994). Mirids represented one of the more abundant arthropod groups in samples from forest canopies in West Virginia (Butler et al. 1997). *Deraeocoris* and *Phytocoris* species prey on several important forest pests in western North America (Razafimahatratra 1980, Stonedahl 1988; see Chapter 14), and noble fir in the Pacific Northwest is a host of 11 mirid species (G.M. Cooper 1981).

Even so, mirids of temperate areas seldom are discussed in textbooks of forest entomology (e.g., Anderson 1960), or they are stated to be common on most forest trees but not to cause serious problems in forests (Hoberlandt 1972, Knight and Heikkenen 1980, Coulson and Witter 1984). Because fogging of oak canopy in Britain produced unexpectedly large numbers of the seldom-collected *Psallus albicinctus*, this bug is probably a canopy species likely to be missed by routine collecting (Kirby 1992). This is likely true of other plant bugs in temperate areas. Some tropical tree-associated mirids appear to descend only occasionally from the canopy (e.g., Leefmans 1920), which might also be true in the case of the acacia-feeding *Platycapsus acaciae* in Egypt (Linnavuori 1964). Mirids such as species of the orthotyline genera *Nesiomiris* and *Sarona* are characteristic forest insects in Hawaii, where they are common in the canopies of dominant metrosideros and other trees (Gagné 1979, 1982, 1997; Asquith 1994). Mirids also occur in the canopies of Australian eucalypts (Morrow 1977). Indeed, mirids are characteristic of

Old World forests, including rainforests, as indicated by the faunal richness recorded from the canopies of sites in Australia (Basset 1991), Borneo (Stork 1991), Indonesia (Casson and Hodkinson 1991, Hodkinson and Casson 1991), and New Guinea (Basset et al. 1996).

Some tree- and shrub-associated mirids show age-related host preferences. Knight (1917b) and Drake (1922) commented on the preference of *Tropidosteptes pubescens* for very young white ash plants (seedlings and saplings) in New York. In Nevada, *Bolteria juniperi* is found on juniper bushes 1.5–2.1 meters high rather than on larger junipers (Knight 1968). Some juniper- and pine-associated species of *Phytocoris* and *Pilophorus* in eastern North America also occur more often on juvenile hosts (1–3 m high) than on mature plants (pers. observ.). Nymphs of pine-inhabiting species restricted to developing on microsporangiate strobili occur only on host individuals that have reached a permanent reproductive stage (Wheeler 1999). In the case of predators such as *Phytocoris* and *Pilophorus* species, an apparent preference for juvenile hosts might reflect the presence on those plants of greater numbers of certain prey.

Unusual and Specialized Habitats

Some plant bugs are restricted to specialized habitats. *Cylapus tenuicornis* often occurs on logs, especially those covered with pyrenomycete fungi (Q. D. Wheeler and A. G. Wheeler 1994; see Chapter 13), and *Ranzovius* species inhabit the webs of subsocial spiders (Wheeler and McCaffrey 1984; see Chapter 15). The orthotyline *Schaffneria pilophoroides* has been collected in litter beneath red cedar in Texas and is suspected of feeding on fallen fruits of these trees (Knight 1966); but this bug is not limited to the ground layer, having been collected in or near ant-attended aphid colonies on scrub oak (Wheeler 1991b) and in crowns of bunchgrasses (pers. observ.). Pitfall trapping of *Bothynotus* species (Deraeocorinae), in which females can be micropterous or coleopteroid, suggests ground-dwelling habits (Henry 1979a, Hoffman 1992, Scudder 1995). Some coleopteroid species of the Australian cylapine genus *Schizopteromiris*, which have been collected by sieving litter or in Berlese samples (Schuh 1986b), also appear to represent true ground dwellers, and the saltatorial *Nesidiorchestes hawaiiensis* inhabits leaf litter in Hawaii (Zimmerman 1948a). Mirids, especially several predatory *Phytocoris* species, inhabit bark crevices (Knight 1923, 1941; Kelton 1983) where their dark, mottled color pattern renders them nearly invisible when at rest (Breddin 1896, Stonedahl 1988). Some greenish white and fuscous *Phytocoris* species blend in with the color of lichens on host branches (Breddin 1896, G. M. Cooper 1981, Stonedahl 1988; pers. observ.). In the garden of London's Buckingham Palace, the occurrence of unusually large numbers of a dark-colored species, *P. dimidiatus*, on tree trunks where the lighter-colored *P. tiliae* normally is found might be the result of "blackening of the habitat by smoke" (Southwood 1964; see also Kettlewell [1973] and Sargent et al. [1998] for discussion of industrial melanism). Species of *Eurychilopterella*, *Hesperophylum*, and related deraeocorine genera also are considered a mostly bark-inhabiting group (Wheeler 1991a, Stonedahl et al. 1997).

Although no aquatic mirids are known, some species develop on host plants growing in water (Butler 1923), including grasses and aquatic macrophytes. Such plant bugs might be considered semiaquatic. Woodroffe (1955b) found *Amblytylus nasutus* and *Lopus decolor* in large numbers at the base of rushes and grasses growing in a swamp, even though both species are considered characteristic of dry, grassy areas. An orthotyline, *Zanchius alatanus*, is found on water-lily in Israel's Lake Hula (Linnavuori 1961). Nymphs of the bryocorine *Eccritotarsus catarinensis* feed on the lower leaf surfaces of water-hyacinth; this Brazilian plant bug has been evaluated as a potential biological control agent against this invasive weed of the waterways in South Africa (Hill and Cilliers 1996, Hill et al. 1999) and in Australia (Stanley and Julien 1999). *Eccritotarsus catarinensis* most likely is the unidentified mirid reported as heavily attacking water-hyacinth in an ornamental fish pond at Belém, Brazil (Bennett and Zwölfer 1968). Another bryocorine, *Sixeonotus unicolor*, develops on the foliage of discoic (or few-bracted) beggarticks growing in or at the edge of ponds in Virginia, and *Deraeocoris histrio* can be found on the mats of mild water-pepper in Pennsylvania (pers. observ.). *Cyrtorhinus lividipennis* and certain other specialized predators of rice-inhabiting delphacids and cicadellids (see Chapter 14) also can be considered semiaquatic.

Restricted Habitats and Conservation Biology

Mirids are of interest in insect conservation biology as apparently rare or phylogenetically distinct or unique species needing preservation, and as indicators of the vitality or changes in ecosystems. Among heteropterans, mirids might offer the greatest potential as indicators of ecological changes, mainly because they are susceptible to many insecticides and are vulnerable to habitat disturbances (Fauvel 1999). Brown (1991) listed mirids ("myrids") among taxonomically mature insect taxa that might serve as "indicator" groups in efforts to conserve biodiversity in the neotropics. Plant bugs might even merit attention as host-specific herbivores that could decrease the densities of endangered or threatened plant species.

During the 1990s, an increasing awareness of the importance of insect conservation (e.g., Samways 1994) extended to the Heteroptera, including mirids. Kirby's (1992) review of Hemiptera considered scarce or threatened in Great Britain contains information on 40 mirid species. For each species, he discussed its status in

Britain, threats such as commercial development and other types of habitat destruction, and management considerations. Aukema (1994) discussed the terrestrial Heteroptera regarded as rare in the Netherlands, encouraging the preservation or restoration of unique habitats and communities to conserve bugs of special concern.

In the United States, mirids have not received comparable attention from those interested in preserving biodiversity, largely because the data on plant bug abundance and distribution are not well enough known for the family to warrant treatment in publications devoted to insects categorized as rare, endangered, or threatened. One of the rarest or least often collected eastern U.S. mirids is *Hesperophylum heidemanni*; although it seems genuinely rare, more intensive fieldwork might show that it is more common than the few collection records suggest (Wheeler 1991a). Other North American plant bugs of possible interest to conservation biologists include several seldom-collected species that are characteristic of specialized communities. *Polymerus wheeleri*, found on moss phlox at only 12 of 79 sites in eastern shale barrens and shale outcrops (Wheeler 1995), might be threatened by quarrying, road construction, and other disturbances. Some little-known plant bugs that are characteristic of pitch pine–scrub oak barrens in the northeastern United States include *Largidea davisi* on pitch pine, *Schaffneria* species on scrub oak, and *Pilophorus furvus* on both plant species (Wheeler 1991b, unpubl. data). The presence of these and other Miridae might indicate the health or vitality of a particular pine barren. A mirid collected in only one northeastern pitch pine–scrub oak barren, and apparently rare throughout the Northeast (pers. observ.), is *Hadronema militare*, which develops on wild lupine (and perhaps also on wild indigo). One of the best-known insects in conservation biology is the Karner blue butterfly (*Lycaeides melissa samuelis*), which is restricted to feeding on this same plant—wild lupine—in pine barrens and elsewhere (Lane and Weller 1994). Conservation biologists involved in managing populations of this federally endangered butterfly (U.S. Department of the Interior Fish and Wildlife Service 1992) might also look for the black and reddish plant bug, *H. militare*, during their work with lupine. The bug is easily detected because of its conspicuous color pattern and the chlorotic blotches it causes on the leaves of its host.

Seasonal and Diel Activity

A diverse phytophagous mirid fauna characterizes certain herbs, shrubs, and trees (Wheeler and Henry 1976, Wheeler 1991b). The activity of plant bugs is usually restricted to the period of the hosts' active vegetative growth, flowering, and fruiting. With mirid activity concentrated in a several-month period, some seasonal overlap in the fauna is to be expected, but the species typically show temporal separation (e.g., Waloff and Southwood 1960, Dempster 1964, Pinto 1982, Pinto and Velten 1986). Some competition, however, might occur among the various broom-associated mirids, particularly between *Orthotylus concolor* and *O. virescens* (Waloff 1965). A phenological separation among the seven principal species developing on Canada goldenrod suggests "diffuse competition" in the community, although the factors involved in population control remain unknown (Reid et al. 1976). Interspecific competition can be inferred from Gagné's (1997) discussion of insular evolution and speciation in endemic Hawaiian species of *Nesiomiris*. Suspected interspecific competition between phytophagous insects is not always upheld by rigorous experimentation as, for example, between the grass-feeding stenodemines *Megaloceroea recticornis* and *Notostira elongata* (Wetton and Gibson 1987; see Chapter 10).

The activity of mirid collectors, except for light trapping, is mostly diurnal, although that of the bugs on plants may not be. Species such as *Irbisia cascadia*, *I. sericans*, and *Phytocoris varipes* in the western United States are swept in much larger numbers at night (Schwartz 1984; G. M. Stonedahl and T. J. Henry, pers. comm.), suggesting that nocturnal sweeping and beating of vegetation could be profitable for collecting these and other plant bugs, particularly species known primarily or solely from light-trap collections. Some *Phytocoris* species can be collected easily at night on the same plants that yield few or no specimens earlier in the day (Stonedahl 1988). Predation by some omnivorous plant bugs can take place mainly after dark (Neal et al. 1972).

Wing Reduction, Wing Polymorphism, and Flight

VARIATION IN DISPERSAL CAPACITY

Several types of wing reduction and polymorphism occur in the Miridae. Deviations from the fully winged or macropterous condition in insects are typically grouped under "brachyptery," even though various intermediate conditions occur. Slater (1975) categorized the types of wing modification found in heteropterans. In mirids, males can be completely winged, with most females either entirely winged or with a low proportion brachypterous, or both sexes can be macropterous or brachypterous. Infrequently, both sexes are brachypterous (Carvalho and Southwood 1955, Southwood and Leston 1959, Southwood 1961a, Brinkhurst 1963, Dolling 1991). Males often retain their wings to locate females more effectively (Roff 1986, Denno et al. 1991), and when wing reduction is restricted to one sex, it is always the female (Fig. 6.1A, Plate 1; Carvalho and Southwood 1955, Southwood 1961a, Putshkova 1971, Schmitz and Štys 1973). As a consequence of wing retention, macropterous males probably contribute

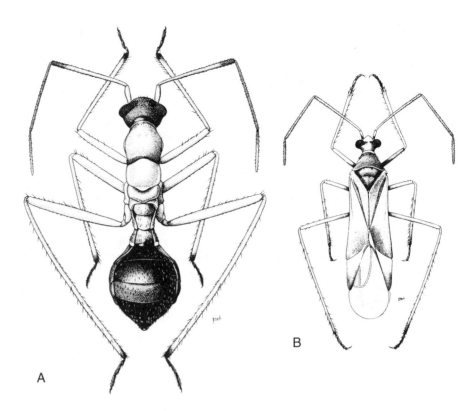

Fig. 6.1. Adult female (A) and male (B) of *Coquillettia insignis*. (Reprinted, by permission of J. D. McIver and the New York Entomological Society, from McIver and Stonedahl 1987a.)

A

B

more to the gene flow among populations than do the brachypterous males (Schuh 1974). Wing dimorphism sometimes differs among congeners: both sexes of *Orthocephalus coriaceus* may be macropterous or brachypterous, whereas only *O. saltator* males are always fully winged (Southwood and Leston 1959, Dolling 1991). Morphs of a wing-dimorphic mirid can look similar, or their external appearance can be completely different (Roff 1986), especially in myrmecomorphic species (see Fig. 6.1; McIver 1987).

Wing reduction in mirids is seen in the Bryocorinae, Cylapinae, and less frequently, Deraeocorinae (e.g., *Bothynotus*); in Mirinae, in which it is common among stenodemines and infrequent in the tribe Mirini; in Orthotylinae, especially halticines; and in phylines such as the Hallodapini (see Fig. 6.1), Leucophoropterini, and Phylini. In females of three *Globiceps* species, Woodroffe (1959a) observed a tendency toward brachyptery with increasing dampness of the habitat. In at least *Bryocoris pteridis*, brachyptery is more common in mountains and colder parts of the range (Kullenberg 1944, Stehlík 1952, Southwood and Leston 1959). More northern populations of *Leptopterna dolabrata* in the eastern United States tend to exhibit a greater proportion of brachypterous to macropterous females (Garman 1926), although more thorough field observations are needed to verify this trend. As in lygaeids, apparent correlations between mirids of northern distribution and higher elevation and the prevalence of brachyptery probably involve slowed succession and increased habitat persistence (Stehlík

1952, Carvalho and Southwood 1955, Sweet 1964, Schuh and Slater 1995; see also Roff [1990] and Denno et al. [1991]).

Wing dimorphism is characteristic of many ground-dwelling or geophilous plant bugs of xeric habitats, particularly myrmecomorphs (Schuh 1974). Geophiles sometimes exhibit a wing modification, known as *coleoptery*, in which the hindwings are lost (Schmitz and Štys 1973). In Australian phylines of the genus *Carvalhoma* (Slater and Gross 1977) and cylapines of the genus *Schizopteromiris* (Fig. 6.2; Schuh 1986b) and in the Nearctic *Bothynotus johnstoni* (see Fig. 5.7B; Henry 1979a), the membrane is lost and the fused coriaceous portions of the hemelytra are fused into beetlelike "elytra" that meet along the midline.

Flightless morphs apparently are unknown in arboreal mirids; all tree-dwelling species in Britain, for example, are fully winged (Southwood and Leston 1959, Dolling 1991). Though typically myrmecomorphic, the largely arboreal Pilophorini do not exhibit wing reduction. Moreover, arboreal members of the large genus *Phytocoris* are generally macropterous, whereas species that feed on grasses or herbs are often brachypterous (Wagner 1970, Stonedahl 1988). In the small, superficially similar genus *Gracilomiris*, whose species are restricted to grasses, females are brachypterous or at least submacropterous (Stonedahl and Henry 1991). Despite the habitat persistence of trees, flight is likely advantageous owing to the architectural complexity or three dimensionality of the host plants (Slater 1977, Waloff 1983, Denno 1994a).

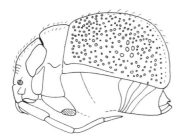

Fig. 6.2. Coleopteroid adult male of *Schizopteromiris carayoni*. (Reprinted, by permission of R. T. Schuh and Société Entomologique de France, from Schuh 1986b.)

As in other insect groups (Roff 1986), some plant bugs have wings but lack flight muscles. This condition has been little investigated in the Miridae, in contrast to lygaeids (e.g., Solbreck 1986). Among mirids, macropterous females of *Chlamydatus pullus* in Greenland show no evidence of longitudinal flight muscles and apparently are incapable of flying (Böcher 1971). Flightlessness resulting from histolysis of wing muscles is unknown in mirids.

The proportion of flightless individuals in populations of wing-dimorphic mirids is seldom mentioned. In Delaware and New Jersey populations of *Halticus bractatus*, sweeping captured a high proportion (about 88%) of brachypterous females (Day 1991). Macropterous females of the halticine *Labops hesperius* comprised only about 4% of a population in Oregon; the bulk of the population consisted of 53% brachypterous females and about 42% macropterous males (Fuxa and Kamm 1976b). Sweepnet samples of a British population of *Leptopterna ferrugata* yielded only 11 macropters among 250 females collected over a full season (Woodward 1949).

A density-dependent variation in the proportion of macropterous versus brachypterous morphs has been demonstrated only for *L. dolabrata* (Braune 1983). Studies in Germany established population density as the primary determinant of wing-morph expression in this species. Braune's (1983) experiments, discussed later, help explain the fluctuating ratios of macropters to brachypters reported previously for *L. dolabrata* in England (Southwood and Leston 1959) and the United States (Garman 1926, Jewett and Townsend 1947).

EVOLUTION OF DISPERSAL STRATEGIES

A discussion of wing polymorphism in delphacid planthoppers can draw on a wealth of data involving temperate and tropical species, including factors that promote the evolution of dispersal and control of wing-form expression (e.g., Denno and Roderick 1990; Denno 1994a, 1994b). Among terrestrial heteropterans such as lygaeids, considerable information is also available on wing polymorphism (e.g., Fujita 1977; Fujisaki 1985,

1986a, 1986b, 1992, 1993a, 1993b; Solbreck 1986; Solbreck et al. 1990); studies on lygaeids have emphasized environmental effects on the histolysis of flight muscles and on flight behavior (Dingle 1966, 1968; Solbreck and Pehrson 1979).

Comparable studies of wing polymorphism in the Miridae have seldom been attempted, perhaps because many species are not as easily reared or manipulated in the laboratory as are delphacids or lygaeids. In certain plant bugs, an obligatory egg diapause hinders laboratory experimentation. Some of the more interesting wing-dimorphic plant bugs are geophilous myrmecomorphs that are rare or at least seldom seen in the field; they cannot be studied easily in their natural environments.

Although wing dimorphism is common in mirids, the extent of wing reduction in the Miridae is not easily perceived; a list of genera exhibiting wing dimorphism has yet to be compiled. Braune (1983) noted that about 16% (48 of 300 species) of German mirids show wing polymorphism in one or both sexes. Future studies on mirids should attempt to correlate flightlessness with temporal and spatial heterogeneity of the environment and with habitat dimensionality—that is, low-profile versus arboreal vegetation (Roff 1990; Denno 1994a, 1994b). Consistent with the hypothesis that flightlessness prevails in persistent habitats is the observation that the highest proportion of flightless lygaeids corresponds with the oldest stable geographic areas (Slater 1977).

Studies of other insect groups have established that flight capability imposes a reproductive penalty and that certain life-history traits are constrained (e.g., Harrison 1980, Roff 1986, Denno et al. 1989, Roff and Fairbairn 1991, Zera and Denno 1997). Wing polymorphism in mirids also can be expected to reflect a balance between the costs and benefits of macroptery and brachyptery. Data documenting trade-offs between dispersal and reproduction, including age at time of first reproduction and fecundity, are apparently lacking for wing-dimorphic mirids, but such trade-offs are known for other heteropterans (Fujisaki 1985, 1986a, 1993a; Solbreck 1986). Osborn (1918) did observe that short-winged females of *Leptopterna dolabrata* appear to produce more eggs than macropters (on average, 60–70 vs. 2). He further noted that the presence of wing polymorphism in this plant bug offers "many interesting biological problems for investigation." Some 80 years later the challenge of placing wing polymorphism in mirids into a context of life-history theory remains.

Apparently the only experimental study on the effects of environmental factors on wing polymorphism in mirids involves *L. dolabrata*, the species Osborn (1918) mentioned as offering possibilities for such research. Braune (1983) showed that population density during nymphal development is critical in determining the proportion of macropterous to brachypterous females. When *L. dolabrata* nymphs are experimentally crowded, a greater proportion of macropters is produced; fourth and fifth instars are the stages most sensitive to crowding and other environmental stimuli. Low temperatures can partially counteract the effects of crowding, possibly because of reduced physical contact between nymphs. Interpopulation differences in the proportion of the two morphs, as in other wing-dimorphic hemipterans, are under genetic influence. The actual genetic basis of wing-morph determination remains to be clarified (Braune 1983).

Dispersal polymorphisms in other mirids undoubtedly also have a genetic basis. Whether their determination involves a single locus system, perhaps with brachyptery dominant, or is polygenic needs to be determined. One can envision that in certain species, a presumed hormonally mediated developmental switch is under genetic control, as it is in the lygaeid *Horvathiolus gibbicollis* (Solbreck 1986, Solbreck et al. 1990). In other mirids, a threshold response to environmental cues might be involved, as in the lygaeid *Cavelerius saccharivorus* (Fujisaki 1986a, 1986b). The mixture of genetic and environmental factors responsible for the proportion of fully winged individuals in populations of geophilous, myrmecomorphic hallodapine mirids need not necessarily be the same as those operating in grass-feeding stenodemines, such as *L. dolabrata*.

In addition to *L. dolabrata*, certain other wing-dimorphic mirids seemingly could be adapted to laboratory studies and to observation in natural and managed environments. *Labops hesperius*, a halticine that injures range grasses (see Chapter 9), seems well suited for investigations on the evolution of dispersal polymorphisms in mirids. It has been studied exten-

sively by applied entomologists, and data on preoviposition period, fecundity, and other life-history traits are available (e.g., Fuxa and Kamm 1976b), but studies might be extended to a comparison of its macropterous and brachypterous morphs. Another candidate for studies on wing-dimorphic mirids is the garden flea-hopper (*Halticus bractatus*) (Fig. 6.3). This species has proved suitable for laboratory studies and attains high densities in nature (see Table 6.7).

The most important point emerging from this preliminary treatment of a complex subject is that additional investigations of wing dimorphism and evolution of dispersal polymorphisms in the Miridae need to be initiated. Such studies should enhance our understanding of life-history evolution in the family, their population dynamics, the genetic structure of their populations, and perhaps factors relevant to managing populations of pest species. Research on dispersal polymorphisms in mirids could include assessment of (1) the extent of wing dimorphism among higher taxa of the family; (2) the incidence of wing dimorphism in relation to habitat persistence, including consideration of altitudinal and latitudinal effects under similar conditions of habitat dimensionality; (3) the trade-offs between reproduction and dispersal, comparing age at first reproduction and fecundity of macropters versus brachypters, and assessing development time and longevity; (4) the correlations between wing morphology and male fitness; (5) the genetic basis of wing dimorphism, that is, a single locus or a polygenic system; and (6) the interplay of genetic and environmental factors in controlling wing-form expression.

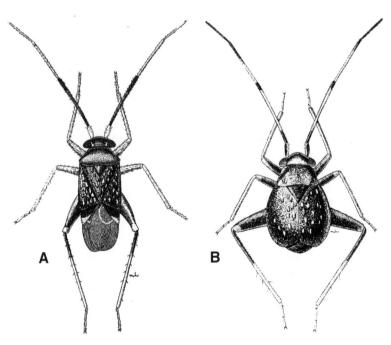

Fig. 6.3. Adult male (A) and female (B) of *Halticus bractatus*. (Reprinted, by permission of the Illinois Natural History Survey, from Knight 1941.)

Walking is the primary means of locomotion for mirids (Fryer 1916, Southwood and Johnson 1957, Southwood 1960b, Waloff and Bakker 1963); it allows nymphs to move among field crops (e.g., Snodgrass 1993). But flight is well developed in the family. In a study of flight activity of mirid specialists on Scotch broom, about 40% of the adult population emigrated from a broom plantation in 1962 (Waloff and Bakker 1963). By radioactively labeling the broom mirid *Orthotylus virescens*, Lewis and Waloff (1964) demonstrated a marked "edge effect" of the habitat on its dispersion. The slight movement of adults released in the middle of the plantation might have resulted from chemical attraction to the host, as well as from dense growth creating a mechanical barrier (Glen and Barlow 1980).

Jordan (1951b) pointed to the Miridae and Corixidae as exceptions to the generally weak flight capabilities of heteropterans. Mirids fly more than any other heteropteran family (Southwood 1960b; cf. Lewis and Taylor 1964) and rank first in spread potential, which is calculated not only from flight activity but also from numerical abundance (Leston 1957b). Plant bugs and other heteropterans of temporary habitats and broad host ranges tend to fly more than do species occupying more permanent habitats and having narrower host ranges (Southwood 1960b). In the case of *Distantiella theobroma*, flight is possible 18–20 hours after the final molt (Leston 1973a).

Leston (1973a) discussed the flight behavior of cocoa mirids, including initiation of flight, path, speed, and range. Šedivý and Honěk (1983) provided detailed information on the daily flight of *Lygus rugulipennis* in relation to peak activity, ovarian development, and meteorological conditions. Betts (1986a, 1986b, 1986c) analyzed the morphology of heteropteran wings (including those of the mirid *Stenodema calcarata*), the functioning of the wings and axillary sclerites during flight, and kinematics in free flight. Škapec and Štys (1980) studied heteropteran wings at rest—that is, whether the right or left forewing is uppermost in an overlapping position. In contrast to an asymmetrical arrangement in some insects, mirids, like most heteropterans, are symmetrical. At any given moment, about half of the resting individuals in a population show a dextral (right wing on top) and half a sinistral (left wing superior) position rather than maintaining a constant dextral or sinistral arrangement.

Trivial versus True Flight

Mirids engage in trivial flight—"flitting" or short everyday movements (Solbreck et al. 1990)—within the habitat, which is temperature dependent, generally continues throughout adult life, and is often associated with mating (Southwood 1960b, Waloff and Bakker 1963, Thistlewood et al. 1989a). Southwood (1960b) sug-gested that short-duration flits are more common in species that overwinter in the egg stage. "True flight" frequently occurs outside the habitat, which leads to dispersal, mainly by sexually immature adults (Johnson 1960; Southwood 1960b, 1962; Waloff and Bakker 1963). Macropterous females of *Labops hesperius*, for example, fly soon after imaginal ecdysis but before maturation of the gonads, conforming to the generally held "oogenesis-flight" syndrome of migratory insects (Fuxa and Kamm 1976b). Macropterous females of *Leptopterna dolabrata* similarly engage in prereproductive dispersal flights, probably owing to mechanical reasons associated with wing loading. Sexually immature females weigh substantially less (about 13 vs. 27 mg) than ovipositing females (Braune 1983). Some polyphagous mirids, however, use dispersal strategies that do not conform to an oogenesis-flight syndrome. For example, *Lygus lineolaris* females that colonize new host patches are mostly older, mated individuals rather than prereproductive ones (Stewart and Gaylor 1991, 1994a). Migration and reproduction, thus, are not always clear-cut alternative physiological states (Rankin et al. 1986, Sappington and Showers 1992, Stewart and Gaylor 1994a).

Mirids of temperate and tropical regions are often the most abundant heteropterans attracted to light (e.g., Thomas 1938; Frost 1952, 1955; Paula and Ferreira 1998). Bark inhabitants, such as certain *Phytocoris* species and certain seldom-collected deraeocorines, are often nocturnal and attracted to lights (Stonedahl et al. 1997). Males of many species are captured in light traps (and in other types of traps) more frequently than are females (Thomas 1938; Hughes 1943; Southwood 1960b, 1961b; Leston and Gibbs 1968; Leston 1973a; Gertsson 1982; Stonedahl 1988; Göllner-Scheiding 1989). Thomas (1938) reported the light trapping of 57 mirid species during 1933–1936, of which only 7% (99 of 1,414) were females. In some species, males show a greater tendency for trivial flight than do females (e.g., Muir 1958, Southwood 1960b, Glen and Barlow 1980; cf. Niemczyk 1967). Several species are crepuscular, flying mainly at dusk, at dawn, or during both periods (e.g., Thomas 1938, Meurer 1956, Southwood 1960b, Lewis and Taylor 1964, Stride 1968, Mueller and Stern 1973a, Liquido and Nishida 1985b). *Lygus hesperus*, for example, tends to fly primarily an hour after sunset and an hour before sunrise (Butler 1972). On the basis of known flight behavior, Southwood (1961b) sug-gested that in diurnal species, males greatly outnumber females in light traps, but where the sex ratio is about equal, the natural flight period is crepuscular or to some extent nocturnal. The flight activity of both sexes is about equal during the day in some species, but males are far more active at night (Southwood 1960b). Flight is generally favored by warm, humid, calm nights (Bech 1965). Šedivý and Honěk (1983) found that especially favorable nights for mirid flight in the Czech Republic are those in which temperatures at 2100 hours are 17.3–25.3°C and wind velocity is low; such nights often follow a period of unfavorable weather. In temperate regions few mirids are trapped when temperatures drop below 15.6°C (Frost 1952), although lygus bugs can be trapped during winter when temperatures exceed 9.4°C (Landis and Fox 1972). The extent of flight can vary greatly between fully winged congeners, for example, *Orthops campestris* and *O. kalmii* (Southwood 1960b).

Dispersal in mirids involves a combination of active flight and passive conveyance by air currents (Southwood 1960b). Some "flitters" become "flyers" when they are caught by currents (Waloff and Bakker 1963), such individuals having been referred to as "vagrants" because their physiological state differs from that of true flyers (Southwood 1962). The potential for long-range dispersal by convective currents depends on the time of day, season, and other factors (Southwood 1960b).

Flight Range and Migration

Flight range probably is limited in most heteropterans (Southwood 1960b), and near the ground, short, discontinuous or hoplike flights are characteristic of some mirids (Waloff and Bakker 1963). The majority of mirid flights probably consist of short-duration flits (Southwood 1960b). Haws and Bohart (1986) were potentially misleading in stating that Fuxa and Kamm (1976b) found migratory flight in *Labops hesperius* to be limited to 2 meters; the latter authors reported that such flights occurred at *altitudes* up to 2 meters.

The actual active flight range of mirids is little known. MacCreary (1965) regarded the capture of several species in traps on a lighthouse nearly 5 km from the nearest shore as evidence of sustained flight, and mirids have even been caught at sea up to 600 km from land (Johnson 1969), though not necessarily the result of active dispersal. The potential flight range of *Distantiella theobroma* is estimated to be 1.1 km for males and 2.3 km for females, with a mean flight speed calculated as 3.1 ± 0.5 m/sec at 23°C (Leston 1973a). *Cyrtorhinus lividipennis* engages in nocturnal migration, which apparently results in flights of six hours or more and in long-distance dispersal (Riley et al. 1987, 1995; Reynolds and Wilson 1989). Migration of *C. lividipennis* enhances the ability of this planthopper predator to suppress prey populations (Döbel and Denno 1994; see also Chapter 14); migratory flights of predator and prey often occur at the same time and altitude (Riley et al. 1995). Capture of a rare macropterous female of *Mecomma dispar* at about 90 meters above the ground suggests that "these forms take to flight and are scattered far and wide to form new colonies" (W. E. China to J. A. Freeman in Freeman 1945).

Mirids, as noted below, can be captured above the flight boundary layer, which is defined as the zone of air near ground level at which an insect's air speed exceeds wind speed; outside this layer, they have limited control over their movements relative to the ground, and move downwind (Taylor 1974, Drake and Farrow 1988). Species caught at such heights are involved in migration rather than trivial flight. Persistent active flight is necessary to maintain an insect at altitudes above the boundary layer (Reynolds and Wilson 1989). Continuous flight is necessary for nocturnal migrants to maintain their altitude because convective lift is absent at night during undisturbed weather (Riley et al. 1995).

Pseudatomoscelis seriata has been captured by day at heights up to 1,500 meters in Louisiana (Coad 1931; see also Gaines and Ewing 1938). *Lygus lineolaris* in the United States has been taken at a maximum of more than 1,500 meters above ground (Glick 1939, 1957), and *L. rugulipennis* in England occasionally has been captured at a height of more than 900 meters (Johnson and Southwood 1949). Because *L. rugulipennis* was the most common heteropteran caught in nets during a two-season study, Johnson and Southwood (1949) suggested it possesses some peculiarity of behavior—climbing to the tips of leaves, or a flight periodicity coinciding with a time of maximum upward convection—that makes it especially likely to attain great heights.

Plant bugs, though, often fly at much lower heights (e.g., Wipfli et al. 1991) and orient downwind (Pruess and Pruess 1966). More than 90% of *L. lineolaris* caught in sticky traps during late July to early August in New York were about 2 meters above the ground; only 7 of 323 adults were trapped at heights exceeding 2.7 meters (Ridgway and Gyrisco 1960b). Most captures (74%) of *L. lineolaris* in a Quebec study were below 1 meter (Boivin and Stewart 1984), and 88% of adults caught in sticky traps in Alabama were within 1.8 meters of the ground (Stewart and Gaylor 1991). In a study of insects attracted to light traps placed at different heights in Pennsylvania, mirids were more numerous at lower levels (1.5–1.9 m) than at the middle (2.1–3.2 m) and upper levels (3.3–4.4 m) (Frost 1958). Mirids in a black walnut plantation in North Carolina flew at a height of 1–7 meters (McPherson et al. 1983). Although few data are available on mirid flight in the tropics, Penny and Arias (1982) reported that plant bugs were 10 times more numerous in light traps in tree canopies than in traps at ground level.

Colonization Ability

Schuh and Stonedahl (1986) considered the Miridae desirable for studies of historical biogeography because of the group's apparent limited powers of flight and, consequently, limited active dispersal. They ascribed to plant bugs "only the most grudging dispersal and colonizing ability" (Slater 1993). Despite minimizing the flight capabilities of mirids, Schuh and Stonedahl (1986) appropriately invoked vicariance to explain the present distribution of many taxa.

An apparent good power of dispersal, coupled with the ability of many species to accept plant or animal food (Becker 1975), and thus to persist in a new environment, should make mirids good colonizers of islands. These attributes could help offset a tendency toward host specialization in the family (Schuh and Stonedahl 1986), which might tend to inhibit their establishment. Although most species of the endemic Hawaiian genus *Nesiomiris* are mainly sedentary, those associated with a summer-deciduous host tree apparently are active dispersers (Gagné 1997). In addition, with the advent of international trade, mirids were predisposed to become good colonists because their oviposition habits facilitate transport in commerce, particularly in shipments of nursery stock (Wheeler and Henry 1992).

As evidence of their colonizing ability, mirids were relatively well represented in an analysis of heteropteran faunas on four groups of islands: the Madeira and Canary Islands, the Galápagos Islands, and the California Channel Islands (Becker 1992). In addition, several mirid species were among arthropods immigrant to a newly emerged volcanic island in the North Atlantic (Surtsey, Iceland) (Lindroth et al. 1973), trapped as transoceanic migrants on Aldabra atoll in the Indian Ocean (Frith 1979) and on Willis island in the Coral Sea (Farrow 1984), and caught in insect drift over the North Sea (Hardy and Cheng 1986). An unidentified mirid species was among the airborne arthropod fallout trapped on lava flows in Indonesia's Anak Krakatau (Thornton et al. 1988). Records of mirids stranded on snow at high altitudes (Heidemann 1903, Van Dyke 1919) lend further support to the high spread potential of the family (Scudder 1963). Mirids were also among arthropods that rapidly (<6 months) recolonized defaunated small spartina islands off the northwest coast of Florida (Rey 1981). As Leston (1961b) noted, a plasticity in food habits helps the Miridae to exploit new localities. Experimental data, however, are needed to demonstrate that mirids are good island colonists—that is, good dispersers and persisters (Simberloff 1981). Moreover, any consideration of the heteropteran fauna of oceanic islands should include not only the bugs' dispersal capabilities but also the movement of land masses to explain current distributions (Schuh and Slater 1995).

Mirids also can locate "habitat islands." Four specialists on Scotch broom colonized rapidly, invading the plant during the first year of plot establishment (Waloff and Bakker 1963, Waloff and Richards 1977). One specialist and three generalist plant bugs were among good colonizers of isolated patches of stinging nettle in England (Davis 1975).

Chemical and Acoustical Communication

Mirid females attract males by releasing pheromones. This type of long-range mate location contrasts with that of other heteropterans, in which an evolutionary emphasis has been placed on males for attracting females by chemical means (Aldrich et al. 1976; Aldrich 1988, 1995). Aldrich (1988) regarded this behavior in mirids as specialized displacement that facilitates immigration. Several species of Bryocorinae, Mirinae, and Phylinae use sex pheromones, mainly butyrate esters. Although most mirid pheromonal codes have yet to be broken (Aldrich et al. 1988, Smith et al. 1991, Aldrich 1996), considerably more information is now available on the family than is found in Fletcher and Bellas's (1988) review of heteropteran pheromones.

Recent research on *Lygus lineolaris* showed that apparent olfactory sensilla, which are not present in the last-stage nymph, develop during the final molt. Higher levels of an antennal-specific protein, which associates with olfactory sensilla, and greater numbers of sensilla in adult males appear to be involved in detecting the female-produced sex pheromone (Dickens et al. 1995, Dickens and Callahan 1996, Chinta et al. 1997, Dickens 1997).

SEX PHEROMONES IN MIRIDS

Kullenberg (1944) suggested that males search for females by using antennal sensory receptors before using visual discrimination to determine their exact position. Two decades earlier, Cotterell (1926) had observed males of *Distantiella theobroma* aggregating late in the evening around cages containing mature females. Graham (1988) demonstrated that sexual attraction in *Lygus hesperus* is indeed an olfactory response. Removal of the entire antennal flagellum in males eliminated their response to pheromones, but removal of the last two segments did not substantially alter their ability to locate females. Pheromone receptors in the Miridae apparently are concentrated on the second antennal segment (Graham 1988), the third and fourth segments tending to be diminutive (Aldrich 1996). That dead females of *L. lineolaris* attract males from downwind indicates the attraction is not acoustical (Aldrich et al. 1988).

The presence of a female sex attractant is known for the mirines *L. lineolaris* (Scales 1968, Blumenthal 1978), *L. hesperus* (Strong et al. 1970, Graham 1988), *Lygocoris communis* (Boivin and Stewart 1982a), *L. pabulinus* (Blommers et al. 1988; Groot et al. 1996, 1998), *Phytocoris relativus* (Millar et al. 1997), and *P. californicus* (Millar and Rice 1998). Graham (1987) reported interspecific attraction of *Lygus elisus* and *L. lineolaris* males to traps baited with virgin females of these species. Through olfactometer studies he showed that males of *L. hesperus* respond only to conspecific females. Traps baited with virgin females of *L. lineolaris* also attract *Adelphocoris lineolatus* and *Stenotus binotatus* males, indicating that these mirines might use a similar pheromonal system (Slaymaker and Tugwell 1984). Field and laboratory observations suggest that *Neurocolpus nubilus* females emit a sex pheromone (Lipsey 1970b), and *Trigonotylus caelestialium* females also are suspected to have a sex pheromone (Kakizaki and Sugie 1997).

Within the Bryocorinae, a female sex attractant is known in *D. theobroma* (King 1973, Kumar and Ansari 1974) and *Helopeltis clavifer* (Smith 1977a). The release of sex pheromones by *H. clavifer* is controlled by a change in light intensity or humidity. Males are attracted three days after the females appear, with the attraction of males greatest in the first few hours after sunset (Smith 1977a). In New Guinea, the field observation of *Ragwelellus horvathi* males accompanying mating pairs suggests that females of this bryocorine employ a sex pheromone (Smith 1977b).

Sex pheromones are known in two species of Phylinae. *Campylomma verbasci* females move to the tops of mullein plants, assuming an abdomen-elevated calling pose. Males are attracted to traps baited with females, and *n*-butyl butyrate produced mainly in the head and thorax of females may be a female-specific pheromone (Thistlewood et al. 1989a). A female-produced sex pheromone is also present in *Atractotomus mali* (Smith et al. 1994).

Males can be attracted by pheromones during the day (*D. theobroma*, *Lygus* spp.) or night (*C. verbasci*, *H. clavifer*). In at least some species possessing a female sex attractant (*Lygocoris communis*, *Lygus lineolaris*), males exhibit greater flight activity (Boivin and Stewart 1982a). The flitting of *C. verbasci* chiefly involves males and sexually immature females; the period of male flitting corresponds with the crepuscular calling period of females (Thistlewood et al. 1989a).

Although the evidence is meager, the type of chemical communication system found in mirids possibly accounts for the greater numbers of males of some species that are taken in light traps (as discussed earlier in this chapter) and might help explain why in certain other heteropteran groups females predominate in light-trap catches. Thomas (1938) observed diminished flight activity among males after they had copulated. In *C. verbasci*, which is known to employ a sex pheromone, flitting is typically restricted to males and immature females (Thistlewood et al. 1989a). But disparate flight activity among males of apparently closely related species (Southwood 1960b) does not support the hypothesis that the type of chemical communication used by mirids explains why more males are attracted to lights compared to males of other heteropteran families, and other explanations are possible. Hughes (1943) observed that gravid females of *Adelphocoris lineolatus* seem to have difficulty flying from one alfalfa plant to another. To account for the dominance of males taken at light traps in Minnesota from late June to early September (363 males vs. 46 females), Hughes suggested that females are less able to fly, simply because they are heavier. Another possibility is that *A. lineolatus* females are less likely to be caught in light traps owing to some behavioral difference between the sexes. In some species, the higher number of males engaged in flitting might simply reflect their greater sensitivity to temperature compared to females (Waloff and Bakker 1963). Not all mirids in which a sex phero-mone has been demonstrated, for example, *D. theobroma*, are even attracted to light (Gibbs et al. 1968). Finally, among species in which mostly males come to light, the presence of a sex pheromone is unknown.

ATTRACTANT COMPOUNDS

Several volatile constituents have been identified in *Lygus lineolaris*, including the ester (*E*)-2-hexenyl butyrate as a possible component of the attractant system (this compound is more abundant in males than in females), but the particular compound or compounds making up the sex pheromone have not been determined (Gueldner and Parrott 1978). The release of hexyl butyrate is not sex specific in *L. lineolaris*; it is unattractive to males and actually reduces the number of males caught when it is applied to cages baited with females (Blumenthal 1978).

Aldrich et al. (1988) summarized the research characterizing mirid pheromonal systems and reviewed attempts to identify specific attractant compounds. In several species, a female elevates her abdomen to assume a calling position and releases a pheromone; attraction occurs in a circadian pattern, and insemination interferes with the further release of pheromones. Mating eliminates calling in *Distantiella theobroma*; females of *L. hesperus* lose their attractiveness to males after mating but are again attractive after five days (Strong et al. 1970, King 1973, Aldrich et al. 1988). The loss of attraction after mating probably also is temporary in *Campylomma verbasci* (Thistlewood et al. 1989a).

In the case of *Lygus* species and two predatory orthotylines, hydrolysis of esters in the scent-gland reservoir is retarded, with the esters remaining the dominant secretory components (Knight et al. 1984). In *Leptopterna dolabrata*, (*E*)-2-octenol, an aldehyde, predominates in scent glands (Collins and Drake 1965), and only terpenoids are produced in the scent glands of *Harpocera thoracica* (Hanssen and Jacob 1982). Several of these scent substances are discussed as defensive mechanisms later in this chapter.

Butyrate ester concentrations in *Lygus lineolaris* and *L. elisus* differ in the scent glands of males and females and in airborne-trapped volatiles, whereas concentrations are not sexually dimorphic in *L. hesperus*. Compounds identified from mirid metathoracic scent glands also differ from those in airborne extracts. These findings suggest that compounds other than butyrate esters are involved in the pheromonal system of *L. hesperus* (Aldrich et al. 1988).

Attractant pheromones in Heteroptera have been identified and artificially mimicked only for certain pentatomids (Aldrich 1988, Aldrich et al. 1988) and the mirids *C. verbasci* (Smith et al. 1991), *Phytocoris relativus* (Millar et al. 1997), and *P. californicus* (Millar and Rice 1998). Sex pheromones from mirid females might not always originate in the metathoracic glands, and glands associated with the reproductive system are possibly involved (Aldrich et al. 1988). For example, sealing the ovipositor of *L. hesperus* with nail polish reduces the attractiveness to males (Graham 1988). The spermatheca, which in mirids no longer functions as an organ for storing sperm (Carayon 1954, Davis 1955, Drake and Davis 1960, Strong et al. 1970), might be associated with pheromone production. That evaporative areas around scent-gland openings in *Distantiella* and *Helopeltis* species are much reduced suggests that their attractant pheromones originate elsewhere. Whether butyrate esters from the metathoracic gland act synergistically with compounds from unidentified glands awaits determination.

Smith et al. (1991) were the first to provide the specific identity of a mirid sex pheromone and to develop a synthetic lure that traps males. They demonstrated that the pheromonal system in *C. verbasci* is composed of butyl butyrate and (*E*)-crotyl butyrate acting synergistically (the latter trace chemical is inactive alone). The two components occur naturally in a 16:1 ratio. Used in this ratio, synthetic butyl butyrate and (*E*)-crotyl butyrate rivals the attractiveness of live females. Because airborne and thoracic (but not abdominal) extracts of females attract males in the field, the metathoracic scent gland might be the pheromone source (Thistlewood et al. 1989a). Both principal components of the sex pheromone are typical defensive compounds produced in the metathoracic scent glands, illustrating the sexual use of defensive secretions and thus, the semiochemical parsimony that is prevalent among arthropods (Blum 1996).

The only other sex-attractant pheromones actually identified in the Miridae involve members of the largest genus, *Phytocoris*. Millar et al. (1997) determined that the pheromone of *P. relativus* is a 2:1 blend of hexyl acetate, a saturated ester produced by both sexes, and *E*2-octenyl butyrate, an unsaturated ester produced only by the female. In field trials, males were attracted by different doses and blend ratios of the pheromone. The pheromone chemistry of *P. californicus* is similar, consisting of a 2:1 blend of hexyl acetate and *E*2-octenyl acetate; the second component apparently maintains reproductive isolation in these often sympatric species (Millar and Rice 1998).

To summarize, the release of attractant pheromones is known for several mirid species. As in most Lepidoptera, calling females, with abdomen elevated, attract males flying upwind. No characteristic calling position, however, has been observed in *Lygocoris pabulinus* females (Groot et al. 1998). In several species, butyrate esters are sexually dimorphic volatile constituents of the metathoracic glands. Mirid scent-gland esters tend to remain intact instead of being converted to acids, alcohols, and aldehydes as in many other, mostly phytophagous heteropterans (Aldrich et al. 1988, Aldrich 1996). The pheromone source in most cases appears to be the metathoracic scent gland (involvement of the reproductive system is also possible), which suggests

that a sexual function has been superimposed on this gland's original role in defense (Aldrich 1996). Another point needing clarification is whether seemingly redundant compounds produced by females are involved in preventing the attraction of sympatric congeners, as suggested by Millar et al. (1997). Further work on chemical communication in mirids will undoubtedly establish the presence of sex pheromones in additional species, including those in the large and diverse subfamily Orthotylinae, and promises to elucidate the evolution of pheromonal systems in the family. Precise identification of additional sex pheromones might allow synthetic attractants to be developed for use in managing populations of injurious plant bugs.

ACOUSTICAL COMMUNICATION

A few Phylinae possess a stridulatory device involving the lateral costal margin of the corium, which is serrate and forms the stridulitrum or file, and the metafemur, which is roughened on its inner surface to form a plectrum or scraper (Fig. 6.4; Schuh 1974, 1984; D. A.

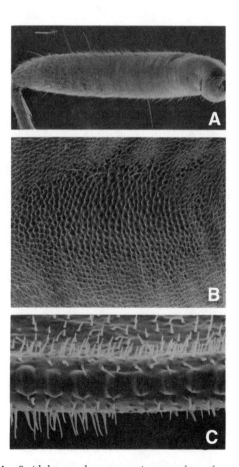

Fig. 6.4. Stridulatory plectrum on inner surface of metafemur of *Hallodapus albofasciatus* (A) with detail of plectrum (B) and wing-edge stridulitrum (C). (Reprinted, by permission of R. T. Schuh, courtesy of The American Museum of Natural History, from Schuh 1984.)

Polhemus and J. T. Polhemus 1985). The existence of probable stridulatory devices in mirids also includes the deraeocorine tribe Hyaliodini (Akingbohungbe 1979), the Orthotylini in the Orthotylinae (Schaffner and Ferreira 1995a), and the cylapine tribe Fulviini (Gorczyca 1998). Acoustic emission has not been studied in mirids (Gogala 1984), and it is unknown if a metafemoral-corial margin structure is used in courtship, as a premating isolating mechanism, or possibly in defense.

Reproductive Behavior

Certain aspects of mirid reproduction, such as chemical communication, already have been treated in this chapter, whereas others, such as egg laying, will be taken up later in the chapter. Here, reproductive events—premating, courtship, and mating—are considered, as well as atypical modes of reproduction, such as parthenogenesis and paedogenesis.

PREMATING PERIOD

A short premating period—the duration between the last molt and the first copulation—is characteristic of most mirids (Kullenberg 1944, Leston 1961b). This period varies with temperature (e.g., Groot et al. 1998), but two to seven days is perhaps typical. *Distantiella theobroma* females usually mate within four to five days after the final molt (King 1973), and the premating period in *Nesidiocoris tenuis* is two to three days in summer and three to four days in winter (El-Dessouki et al. 1976). The premating period is seven days in *Labops hesperius* (Coombs 1985), six days in *Lygocoris pabulinus* (Groot et al. 1996, Blommers et al. 1997), three to six days in *Horcias nobilellus* (Sauer 1942), and five days in *Lygus hesperus* (Strong et al. 1970, Strong 1971). *Helopeltis schoutedeni* males become sexually mature within 48 hours of the final molt; females, within 24 hours (Schmitz 1958). *Halticus bractatus*, however, mates within a few hours or even within five minutes after the imago appears (Beyer 1921), and copulation of *Psallus ambiguus* reportedly takes place immediately after the female's teneral period of about two hours (Morris 1965). Copulation of *Helopeltis antonii* similarly is said to occur immediately after adult emergence (Puttarudriah 1952) or one day later (Ambika and Abraham 1979).

COURTSHIP AND MATING

Observations of mirids mating within minutes after emergence probably require verification; it is questionable whether the female would be receptive that soon and whether sexual aggressiveness in the male would be developed. Sexual aggressiveness in the male *Lygus hesperus* begins within about five days after the final molt and depends on development of the accessory glands rather than on development of the testes; females become sexually receptive when the first eggs mature, again usually after five days (Strong et al. 1970). *Lygus lineolaris* does not become sexually mature and mate until adults are at least four days old (Scales 1968, Bariola 1969). Yet observations by Palmer and Pullen (1998) indicate that some mirids do mate shortly after the final molt. They reported that adult males of *Falconia intermedia* straddle fifth-instar females, with mating occurring soon after the female molts.

Most observations on mirid mating behavior are made during the day, but some species mate at night (e.g., Sauer 1942). Smith and Borden (1991) reared *Campylomma verbasci* in the laboratory from January through September in successive years without ever seeing mating pairs. Mating of *Pseudatomoscelis seriata* is rarely observed in the daytime, either in the field or in the laboratory (Reinhard 1926). And, as already noted, females of some mirids release sex pheromones at night. *Lygus elisus* has no distinct mating period, mating during both photophase and scotophase (Graham et al. 1987).

Elaborate courtship behavior can be either present or absent in mirids. Too few observations on close-range courtship are available to discern behavioral trends among higher taxa; detailed, comparative studies of species in all subfamilies are needed.

An apparently simple courtship behavior occurs in *Lygus* species. A male *L. hesperus* walks slowly toward the female from in front, touches her with the antennae, and if she is receptive, jerks his abdomen before attempting intromission (Strong et al. 1970). A male *L. lineolaris* approaches a female from behind and to the right and jabs or touches the female's posterior with his abdomen thrust forward and under; a receptive female slowly raises her abdomen, lowering the ovipositor to expose the genital opening (Bariola 1969).

A male *Cyrtorhinus lividipennis*, with wings flapping, walks around the female, which employs similar wing-flapping behavior. The male grasps the female's abdomen with his forelegs, and with wings flapping, slips back, turns around to position his abdomen below the female's, and inserts the aedeagus. The pair's coordinated movements leading to copulation last only 10 seconds (Liquido and Nishida 1985b).

Mating behavior in the cocoa mirid *Sahlbergella singularis* involves a short burst of flight between 1,700 and 1,800 hours. A male remains in one spot, opens his wings, and uses the metatarsus to stroke the tip of his abdomen. The male approaches a female, and after about 25 seconds of erratic movements followed by short bursts of flight, alights on the female's dorsum facing in the same direction. With the female motionless, the male moves backward so that the tip of his abdomen contacts the female's; after intromission, the pair face in opposite directions. Little sexual display occurs before copulation (Youdeowei 1970, 1973).

Nesidiocoris tenuis males search for females on host

plants, mating nearly as soon as they meet. The male approaches, faces the female, calms her by stroking her abdomen with his antennae, and jumps onto her dorsum, using his forelegs to hold her (El-Dessouki et al. 1976). A male *N. caesar* approaches the female's side and touches her abdomen several times with the antennae; if receptive, the female becomes immobile and the male mounts, holding her with his forelegs (Chatterjee 1984a).

A male *Helopeltis antonii* becomes alert ("agitated") when a female approaches, and lightly probes her dorsal surface with his antennae; receptive females respond passively. The male mounts the female from the posterior region, stroking her dorsum with the labium. At 27.8–33.4°C, the arousal period lasts 10–30 seconds; mounting, 25–60 seconds (Devasahayam 1988).

A male *Calocoris angustatus* aggressively pursues a female, sometimes being repelled by her hind legs, approaches dorsolaterally, and grasps her. Copulation lasts 30–120 seconds (Hiremath and Viraktamath 1992).

Males of *Macrolophus tenuicornis* (Wheeler et al. 1979), *Harpocera thoracica* (Stork 1981), and *Trigonotylus caelestialium* (Kakizaki and Sugie 1997) simply approach the female before jumping onto her. Males of *H. thoracica* possess adhesive setae (Fig. 6.5) on the

second and third antennomeres (similar to adhesive tarsal setae in certain beetles [Stork 1981]) that are used for gripping the female's pronotum (Fig. 6.6) during mating. Stork noted that the male is displaced to the female's right during copulation and that the roughly parallel arrangement of the sexes, as described by Kullenberg (1944), is likely in error. Similarly, the spined antennae of male *Hyalochloria* species (see Fig. 5.8) are used to hold the female during copulation (Beingolea 1959b).

The premating signals used by other mirids appear to involve more ritualistic behavior, although behavioral descriptions by different workers cannot be considered equivalent and should be compared with caution. An antennal "patting" among certain species of the endemic Hawaiian genus *Nesiomiris* might be involved in recognizing a conspecific individual and avoiding hybridization (Gagné 1997). Precopulatory behavior in *Lygocoris pabulinus* includes antennation involving both sexes and vibration of the abdomen by males. Contact with a female is unnecessary to induce male vibration. Courtship lasts an average of 10 minutes at 25°C (Groot et al. 1998). A male *Moissonia importunitas*, having approached from behind, strokes the female's hind legs, wings, and head with his antennae; the

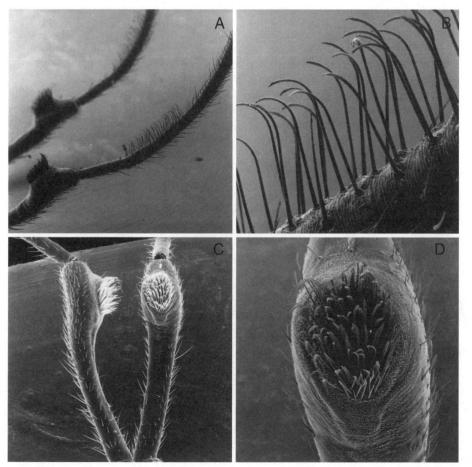

Fig. 6.5. Adhesive setae on male antenna of *Harpocera thoracica*. A. Lateral view of distal and proximal regions of second and third segments. B. Lateral view of adhesive setae on third segment. C. Lateral and ventral views of second segment. D. Ventral view of adhesive pad. (Reprinted, by permission of N. E. Stork and Taylor & Francis Ltd., from Stork 1981.)

process can be repeated and last for more than 10 minutes. The female indicates acceptance by raising her abdomen after the male has mounted. After unsuccessful attempts, males sometimes feed on a leaf before additional attempts at courtship (Gopalan and Basheer 1966).

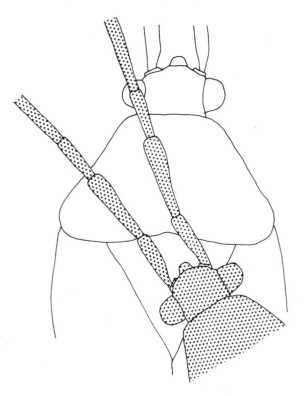

Fig. 6.6. Position of the male (stippled) and female of *Harpocera thoracica* during pairing, with second antennal segments of the male attached to the female's pronotum. (Reprinted, by permission of N. E. Stork and Taylor & Francis Ltd., from Stork 1981.)

Lipsey (1970b) described the courtship behavior of *Neurocolpus nubilus* in captivity. After flying in the rearing dish, the male settles near the female, approaches with antennae pointing at her, then stops. The female assumes a calling position, hind legs spread and abdomen raised. This position initiates antennal scissoring in the male: alternate raising and lowering of each antenna over the female's abdomen. The female backs into the male as he spreads his forelegs and middle legs and continues to scissor her dorsum. Copulation occurs suddenly and lasts only seven seconds. Nonreceptive females may kick a male away with their hind legs.

Mating Positions

Male genitalia show an asymmetrical configuration in all mirid groups, including isometopines (cf. Drake and Davis 1960). The left paramere characteristically is larger than the right and is more complicated in structure (Fig. 6.7; Singh-Pruthi 1925, Roberts 1930, Michalk 1933, Schuh and Štys 1991, Schuh and Slater 1995). Pairing, therefore, usually involves displacement of the male to the right of the female (Fig. 6.8; Kullenberg 1944, Drake and Davis 1960). A *Lygus hesperus* male places his left hind leg over the female's hemelytra and rotates his body about 45° to the right of hers (Strong et al. 1970). Plant bugs often face in opposite directions after initial coupling (Fig. 6.9). Puttarudriah (1952) observed a *Helopeltis antonii* female, facing opposite the male, continuing to feed on the host plant during mating; Devasahayam (1988), however, reported that copulatory pairs in this species remain stationary, showing little body movement unless disturbed, and do not feed. Mating pairs of *L. hesperus* show similar motionless behavior (Strong et al. 1970). Males of

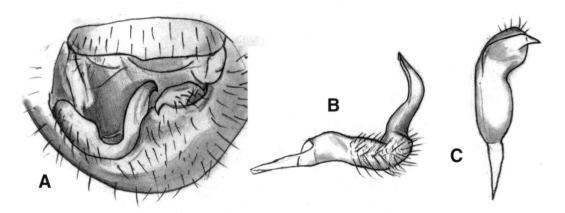

Fig. 6.7. Male terminalia of *Stenodema laevigata*. A. Genital capsule, posterior view showing left and right parameres. B. Left paramere. C. Right paramere. (Reprinted, by permission of B. Kullenberg and Zoologiska institutionen, Uppsala Universitet, from Kullenberg 1947a.)

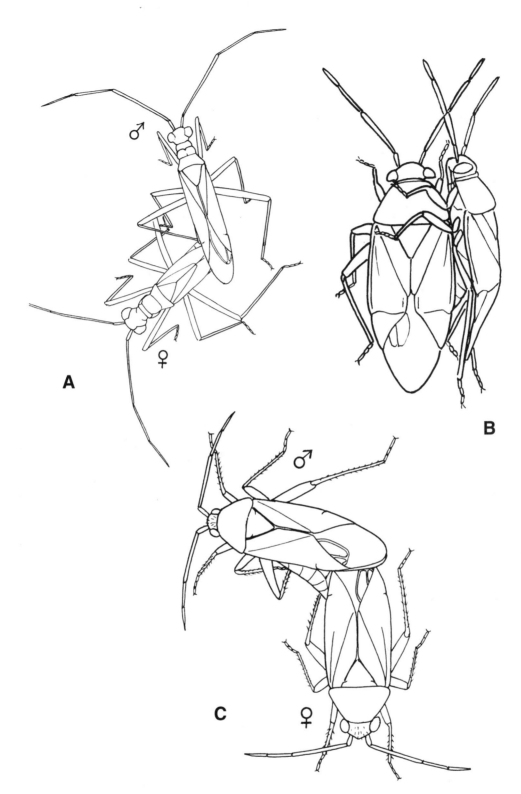

Fig. 6.8. Position of mating pairs. A. *Dicyphus constrictus*.
B. *Orthops campestris*. C. *Plagiognathus arbustorum*.
(Reprinted, by permission of B. Kullenberg and Zoologiska
institutionen, Uppsala Universitet, from Kullenberg
1944.)

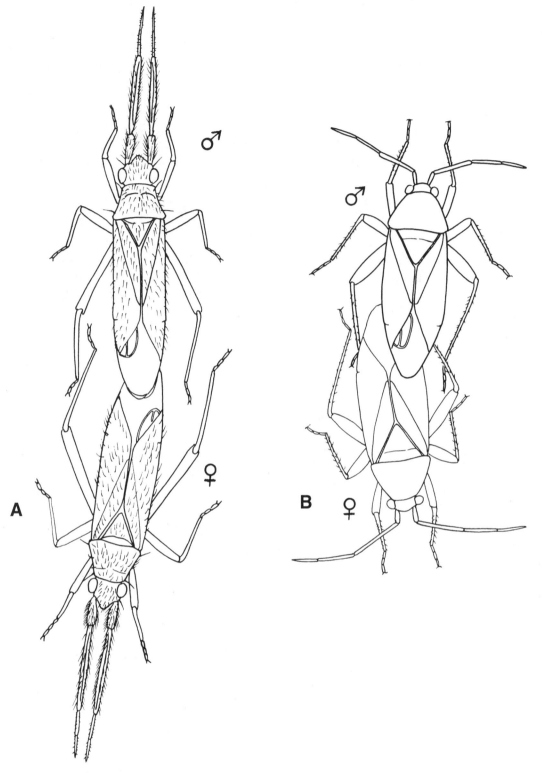

Fig. 6.9.   Position of mating pairs. A. *Heterotoma merioptera*. B. *Psallus betuleti*. (Reprinted, by permission of B. Kullenberg and Zoologiska institutionen, Uppsala Universitet, from Kullenberg 1944.)

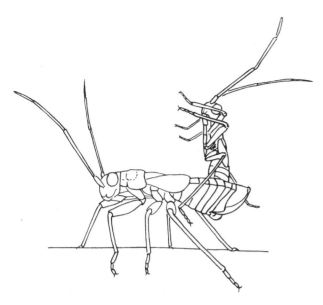

Fig. 6.10. Mating pair of *Pithanus maerkelii*, with male showing cataleptic behavior. (Reprinted, by permission of B. Kullenberg and Zoologiska institutionen, Uppsala Universitet, from Kullenberg 1944.)

several species become cataleptic during pairing (Fig. 6.10; Kullenberg 1944), and when disturbed, a female *Psallus ambiguus* will drag the inactive male behind her (Morris 1965). Kullenberg (1944, 1947a) described and illustrated a variety of positions assumed by mating pairs. The latter work includes a detailed discussion of the stability of the copulation substrate, strength of the copulatory position, duration of copulation, and complexity of male genital structures.

Duration of Copulation

Depending on the species and temperature, pairs copulate for a few minutes—about 1.5 minutes in *Lygus hesperus* (Strong et al. 1970), one to two minutes in *Lygocoris pabulinus* (Groot et al. 1998), and even less in *Calocoris angustatus* and *Neurocolpus nubilus*, as noted above—or remain in copula for several hours. The Indian bryocorine *Pachypeltis maesarum* copulates in the laboratory for as long as nine hours (Trehan and Phatak 1946). Some of this time might represent a postcopulatory riding phase, as identified in other heteropterans (e.g., Heming-van Battum and Heming 1986). In heteropterans (in which females store sperm), prolonged copulation is considered adaptively significant as a male strategy to prevent sperm displacement by subsequent matings (e.g., Sillén-Tullberg 1981, McClain 1989, Carroll and Loye 1990, Alcock 1994). Similarly, prolonged copulation could represent male behavior to ensure sperm priority. Mirid sperm are received by a median unpaired sac, the seminal depository, on the anterior genital chamber. This sac, rather than the spermatheca, apparently serves a sperm-storage function in mirids (Carayon 1954, Davis 1955, Strong et al. 1970;

see also later discussion of extragenital insemination), and a fully inflated seminal sac indicates a recent mating (Strong et al. 1970, Stewart and Gaylor 1991). Postinsemination associations in the Miridae are not well understood, and any adaptive value of such behavior has not been critically evaluated.

Disengagement of the genitalia is often abrupt (e.g., Roberts 1930). In some cases only a slight disturbance causes the pairs to separate (e.g., Reyes and Gabriel 1975), but mating pairs of *Lygus hesperus* can be picked up and handled without copulation being interrupted (Strong et al. 1970). A male *L. hesperus* always initiates the interruption of copulation by stroking the tip of the female's abdomen with his right hind leg (Strong et al. 1970). A postcopulatory genital cleaning, using the proboscis or hind legs, has been observed in both sexes of *Nesidiocoris caesar* (Chatterjee 1984a), *N. tenuis* (El-Dessouki et al. 1976), and *N. volucer* (*sensu* Kerzhner and Josifov 1999) (Roberts 1930). Genital cleaning (no details provided) is also known to occur in *Helopeltis antonii* (Devasahayam 1988).

Multiple Matings

Multiple matings to fertilize continuously maturing eggs in the ovaries might seem unnecessary (Connell 1970a), but multiple matings appear to be common in mirids, at least under laboratory conditions. The effects of mating frequency on fecundity, fertility, and longevity in the Miridae have received little attention.

Laboratory observations led Smith (1977) to suggest that *Helopeltis clavifer* females need to mate multiple times to maintain deposition of fertile eggs for several weeks. In *H. antonii*, males mate with 8–10 females (Jeevaratnam and Rajapakse 1981a), although in the laboratory a female *H. bergrothi* that mated only once laid eggs for 78 days (Kirkpatrick 1941). Reinhard (1926) considered a single mating sufficient to fertilize the typical number of eggs laid by a female of *Pseudatomoscelis seriata*. A single mating of *Lygus hesperus* is sufficient for the production of viable eggs over a female's egg-laying life. Males of this species, however, can mate a maximum of seven times; females, three times (Strong et al. 1970, Strong 1971). Mating only once allows *Lygocoris pabulinus* females to maintain egg production for 2–3 weeks at 20°C (Blommers et al. 1997) but might be insufficient for deposition of the entire complement of fertilized eggs. Copulations in this species can occur on consecutive days (Groot et al. 1998). Under laboratory conditions, *Nesidiocoris caesar* females occasionally kill their mates following copulation (Chatterjee 1984a), but in nature such behavior apparently is unknown and probably rare.

Extragenital Insemination

Carayon (1984) reported details of mirid reproduction that previous workers overlooked. In a few species, such as *Acetropis gimmerthalii*, insemination consists of an extragenital, traumatic (or hemocoelic) type similar to that of cimicids (Carayon 1966) and certain other cimicoid families (Carayon 1984). Insemination results in copulatory scars made by the male's parameres on the female's abdominal integument; deposition of spermatozoa in an anterior diverticulum of the vagina, the seminal sac; and migration of spermatozoa to the oocytes. The process in mirids, however, is not as well developed as in cimicids, differing in several important aspects, and requires evolutionary interpretation (Carayon 1984). It might be considered a precursor to true extragenital insemination.

ATYPICAL MODES OF REPRODUCTION

Parthenogenesis

Parthenogenesis evidently occurs in a few mirid species in which males are rare. In Sicilian populations of *Campyloneura virgula*, the male genitalia appear to be nonfunctional (Wagner 1958). By breeding females isolated after the imaginal molt, Carayon (1989) was able to demonstrate a constant parthenogenesis in European populations of this species. That populations of *C. virgula* are typically parthenogenetic might have been known to Herrich-Schaeffer (1836) when he described this species; the specific epithet *virgula* means "virgin" (I. M. Kerzhner, pers. comm.). Populations in North Africa (the subspecies *C. virgula marita*) have a male-female ratio of 1:2 and reproduce sexually (Wagner 1968).

Van Doesburg (1964) considered the absence of *Termatophylidea opaca* males suggestive of parthenogenesis but noted that more detailed study is needed to confirm this supposition. After examining more than 800 specimens of *Chlamydatus pullus* from Greenland, all of which were females, Böcher (1971) suggested that Arctic populations of this phyline reproduce parthenogenetically, a habit that is relatively common among other Arctic insects (Danks 1981). Fowler (1980) collected a series of *Halticotoma valida* from New Mexico that consisted only of adult females, but he was unable to determine whether males had been present earlier in the season (before November) or if the population was strictly parthenogenetic. I consider the latter suggestion improbable because of his limited late-season observations (female-biased sex ratios are discussed later in this chapter). In addition, males of this species occur in the eastern states (Wheeler 1976a), and even if Fowler's (1980) species should prove not to be conspecific with *H. valida*, males are present in all other western species of the genus. Parthenogenesis is possible in the rarely collected *Hesperophylum heidemanni* (Wheeler 1991a), although the recent report of a *H. arizonae* male renders the possibility of parthenogenesis in *Hesperophylum* species less likely (Stonedahl et al. 1997). *Campylomma verbasci* has mistakenly been referred to as parthenogenetic (Hagen et al. 1999).

Paedogenesis

Only one mirid, the bryocorine "*Hemisphaerodella mirabilis*," has been said to reproduce by paedogenesis (Maldonado Capriles 1969). Henry and Carvalho (1987), however, discovered that *H. mirabilis* is merely the nymphal stage of *Cyrtocapsus caligineus*, characterized by its peculiar round form and beetlelike hemelytra. They synonymized the former name under *C. caligineus*. Paedogenesis thus is unknown in the Miridae.

Oviposition Behavior

Many workers have observed mirid oviposition behavior, noted various sites used for egg deposition, and commented on the visibility of eggs in host tissue and on injury to hosts during egg laying. Here, I summarize the fragmented literature relating to oviposition in the Miridae.

PREOVIPOSITION

The preoviposition period refers to the length of time from emergence of the adult female to deposition of the first egg (e.g., Neff and Berg 1966). A short preoviposition period is associated with reproduction earlier in life. The preoviposition period in *Sahlbergella singularis* is shorter in paired than in unpaired females, suggesting a stimulatory effect of males that accelerates the rate of egg maturation (Eguagie 1977). Data on preoviposition times, as well as on fecundity, longevity, and other parameters, eventually need to be analyzed for trends among mirid higher taxa and to be placed within the context of population dynamics and life-history theory.

The preoviposition period of mirids, varying from less than a day to about 20 days, is usually highly temperature dependent. For example, this period in *Lygus hesperus* is 27 days at a constant temperature of 12.8°C, 17 days at 15.6°C, and 7 days at 26.7°C (Strong and Sheldahl 1970). The preoviposition period in *Pseudatomoscelis seriata* reared at constant temperatures varies from 4.6 days at 26.7°C to 6.7 days at 35°C, although these differences are not statistically significant (Gaylor and Sterling 1975a). This period is slightly longer in winter than in summer for *Nesidiocoris tenuis* (El-Dessouki et al. 1976). Preoviposition lasting 2–4 days is characteristic of *Calocoris angustatus* in the rainy season, but afterward this period lasts 5–8 days (Sharma and Lopez 1990a). Hiremath and Viraktamath (1992) determined that the preoviposition period for *C. angustatus* under field conditions during a 12-month study ranged only from 1.2 to 3.0 days. Preoviposition times for *L. rugulipennis* vary according to generations (Hori and Hanada 1970) but are not statistically different on different host plants (Hori and Kuramochi 1984). Preoviposition times for *L. hesperus* are longer on cotton than alfalfa, which is a more suitable reproductive host (Cave and Gutierrez 1983).

Halticus bractatus (Cagle and Jackson 1947; cf. Beyer 1921) and *Creontiades pacificus* (Lal 1950) apparently begin laying eggs within a day after emergence, although the preoviposition period in the multivoltine *H. bractatus* varies greatly among generations. The preoviposition period is only 1 or 2 days for *Moissonia importunitas* (Gopalan and Basheer 1966) and averages 2.6 days at 25°C for *Sthenaridea carmelitana* (Matrangolo and Waquil 1991). For *Helopeltis antonii*, Devasahayam (1988) reported a mean preoviposition period of 3.6 days at 37.8–38.4°C. *Helopeltis schoutedeni* females begin to oviposit 3–15 days after the final molt (Schmitz 1958). The preoviposition period is about 8 days in *L. lineolaris* (Bariola 1969, Stewart and Gaylor 1994a), 10–14 days in *Plagiognathus chrysanthemi* (Guppy 1963), 12–16 days in *Neurocolpus nubilus* (Lipsey 1970b), and 14–18 days in *Adelphocoris lineolatus* (Craig 1963).

Psallus ambiguus females supplied with excess animal food in the laboratory lay their first eggs sooner than they do in nature (8–9 vs. 14–15 days) (Niemczyk 1967). The length of the preoviposition period for other predatory species also can depend on prey availability, as in the mainly predacious *Hyalochloria denticornis* and the facultative predator *Nesidiocoris tenuis* (Beingolea 1960, Torreno and Magallona 1994). The discrepancy among observations on *N. tenuis* by different workers—oviposition on day of emergence to a mean of 3.2 days—can be attributed not only to diet but also to different rearing conditions and other factors (Torreno and Magallona 1994).

OVIPOSITION

The selection of oviposition sites by plant bugs probably involves the factors governing the preferences of homopterans that insert their eggs in plant tissues (Denno and Roderick 1990, Denno 1994a). These factors could include competition for available sites on the host plant, which might assume considerable importance in phytophagous mirids that use high-quality resources on woody hosts; searching behavior of hymenopteran egg parasitoids; physical characteristics of the host such as hardness and thickness of the substrate, which might affect the potential for water uptake by eggs; and chemical compounds in the host plant that stimulate oviposition (Constant et al. 1996a). Oviposition rates in polyphagous species can vary substantially on different host plants (e.g., Purcell and Welter 1990b). The hypothesized inverse density-independent egg laying by *Horistus infuscatus* females on asphodel clones of varying density (Izhaki et al. 1996) invites research on the oviposition behavior of this species. In nontemperate species that breed nearly continuously, preferred oviposition sites might be those that will provide optimal nutrition during the succeeding month: an approximate one-week incubation period plus three weeks for nymphal devel-

opment (Muhamad and Way 1995b). In the case of *Lygus lineolaris* and *L. hesperus*, plant species and varieties and sites chosen for oviposition are not those that show maximum viability or that favor nymphal development (Taksdal 1963, Curtis and McCoy 1964, Barlow et al. 1999). Gerber (1997), however, stressed the complexity of oviposition responses shown by *L. lineolaris*. Because its preferences for genera in the same plant families can change over time—for example, during days 1–6 versus 7–13—researchers using test periods of a week or less would draw erroneous conclusions regarding oviposition preferences of *L. lineolaris* and perhaps other polyphagous plant bugs.

Selection of Sites

The search for oviposition sites often involves several minutes of touching the substrate with the antennae and probing the surface with the labium. Sensilla at the labial tip presumably play a role in site selection (Cobben 1978), but stimuli received during feeding and plant chemicals that elicit or deter oviposition are little known in mirids (e.g., Curtis and McCoy 1964). An ovipositional nonpreference is involved in crop resistance to certain pest mirids, such as lygus bugs (Alvarado-Rodriguez et al. 1986b).

Under experimental conditions, oviposition sometimes can be obtained only when the preferred site on host plants is made available (e.g., Johnson 1934). Females of certain plant bugs will accept an artificial substrate for oviposition (e.g., Shimizu and Hagen 1967, Debolt 1982, Constant et al. 1996a), which can result in reduced fertility (Albajes et al. 1996). Some species oviposit mainly at night or in early morning (e.g., Beyer 1921, Lean 1926, Reinhard 1926, Puttarudriah 1952, Jeevaratnam and Rajapakse 1981a, Devasahayam 1988). Sites selected for oviposition involve lenticels, flowers and flower buds, bud scales, seeds, fruits, bark crevices or wounds, wood of a particular age (including dead wood and fenceposts), axils of stems, midribs, petioles, and leaf sheaths of grasses. Species such as *Adelphocoris superbus* oviposit only in stems with a narrow range of widths (Lilly and Hobbs 1956). Evidence suggests that females of the photonegative *Sahlbergella singularis* become temporarily photopositive during oviposition and are attracted to cocoa foliage growing in the light (Madge 1968). Some kairomones produced by flowering plants appear to trigger oviposition by the polyphagous *Lygus hesperus* (Scott 1983). Oviposition by *L. rugulipennis* increases on pine seedlings grown under conditions of a higher nitrogen supply; a preliminary experiment suggested that more eggs are laid on seedlings with significantly higher levels of alanine, glutamic acid, and proline (Holopainen 1990b). The predatory *Cyrtorhinus fulvus* deposits more eggs on plants harboring eggs of its delphacid prey, with oviposition apparently stimulated by the presence of prey eggs (Matsumoto and Nishida 1966). The presence of delphacid prey apparently also induces oviposition by another egg predator, *C. lividipennis* (Bae and Pathak 1966).

Some species show a distinct preference in their oviposition sites—for example, *Brachynotocoris puncticornis* for lenticels of European ash (Wheeler and Henry 1980b). Others, such as *Pseudatomoscelis seriata* (Reinhard 1926), *L. lineolaris* (Painter 1927, Fleischer and Gaylor 1988), and *L. rugulipennis* (Varis 1972), deposit eggs on nearly all aerial plant parts, which can be expected to result in differential rates of parasitism of eggs in different sites on the host. Partitioning of available sites can occur within the mirid fauna of a particular host, perhaps reflecting competition for oviposition sites. Accordingly, the various species associated with Scotch broom (Waloff and Southwood 1960, Waloff 1965), apple (Sanford 1964a), honeylocust (Wheeler and Henry 1976), and oak (Ehanno 1987b) show little overlap in their oviposition sites.

Oviposition preferences in some multivoltine species vary between generations, the selected sites being related to overwintering strategies. Eggs of some species probably overwinter in nonhost plants, as is known in delphacid planthoppers (Denno and Roderick 1990) and other hemipterans with endophytic oviposition. In the case of *Adelphocoris lineolatus*, eggs are deposited on alfalfa stems 0.3 meter or more above the ground in early season, but in late summer they are laid in less succulent growth near the stem base, such sites presumably offering greater protection during the winter (Hughes 1943). Eggs of the overwintering generation in *Campylomma verbasci* are generally more deeply embedded than those of the summer generation (Collyer 1953a). First-generation females of *Blepharidopterus provancheri*, a mainly predatory bug found on various woody plants, oviposit in apple leaves during mid-June to mid-July. In contrast, second-generation females deposit eggs in the bark from mid-September to mid-October (Steiner 1938). A similar oviposition pattern prevails in *Malacocoris chlorizans* (Geier and Baggiolini 1952). Bark tissues obviously give greater protection to overwintering eggs than do the leaves of deciduous trees such as apple. The predatory and multivoltine *Deraeocoris nebulosus*, which uses apple trees as a host and overwinters in the adult (rather than egg) stage, lays eggs in the foliage (McCaffrey and Horsburgh 1980).

The lace bug predator *Stethoconus japonicus* deposits its first-generation eggs in the midribs of young azalea leaves, whereas those of the second generation are placed in stems. Eggs laid in foliage are surrounded by vascular fluid and usually hatch within two weeks. Those inserted in stems are not in direct contact with the vascular system and do not hatch under laboratory conditions, but in nature they do hatch the following spring. This switch of oviposition sites on azalea coincides with an increase in leaf age and decline in leaf moisture, suggesting that water uptake in the host plant and by the eggs, rather than an accumulation of degree-days alone, is needed for the hatching of overwintered eggs (Neal et al. 1991, Neal and Haldemann 1992). *Stethoconus japonicus* females apparently can perceive changes in the moisture content of host tissues. Even after a site of oviposition is switched from the midribs of young leaves to the stems, oviposition can be redirected to the midrib following substantial summer rainfall before overwintering eggs are laid in the stems (Neal and Haldemann 1992).

Egg Deposition

Details of oviposition behavior are available for relatively few mirids. A notable exception is the study by Ferran et al. (1996) on *Macrolophus melanotoma* (as *M. caliginosus*), in which a modified video camera was used to analyze the position of the female's body, rostrum, and ovipositor throughout oviposition. Rostral probing of the plant surface, which appears to involve "locating, recognizing and marking of the oviposition site," is followed by a forward and downward thrusting of the ovipositor. A successively deeper probing before oviposition suggests that the ovipositor is used to test physical and chemical properties of the host. The antennae remain motionless throughout the sequence, suggesting that at least in this species they play no role in the search for a suitable oviposition site (Ferran et al. 1996).

Observations on oviposition behavior of several North American mirids inhabiting apple trees (Knight 1915) probably are typical of species that lay eggs in woody tissue. In the following composite account, observations by other workers are incorporated; some of their observations involve herbaceous hosts.

The female explores the substrate with her antennae and touches it periodically with the tip of her stylets. When a female locates a suitable site, she often uses her stylets to initiate a hole, the process lasting 10 minutes or more. The stylets just pierce the plant surface or can be inserted in woody tissues to their full extent (Steer 1929). The differences in the degree of stylet insertion perhaps reflect differences in the oviposition substrate, for example, living plant tissue versus dead wood. The stylets can be inserted two or three times before oviposition begins (Lupo 1946). With her abdomen arched, she stands as high as possible (see Fig. 6.11A), unsheaths the ovipositor, and with help from the stylets, locates the hole previously prepared. This is accomplished by bending forward to bring the ovipositor over the site where the stylets were inserted, turning her head under with the stylets in the prepared hole, and inserting the ovipositor (Plate 1) as soon as the stylets are withdrawn (Knight 1915, Chatterjee 1984a, Rauf et al. 1984b). If a female fails to locate the hole, she moves a short distance and continues stylet probing (Thontadarya and Channa Basavanna 1962). The tarsi are used to grip the substrate (Osborn 1918, Kullenberg 1944, Devasahayam 1988), and the female often faces the base of the shoot during oviposition (Petherbridge and Husain 1918; Petherbridge and Thorpe 1928b; Austin 1930b, 1931b).

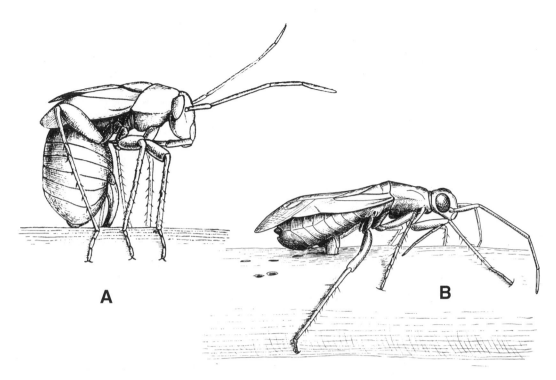

Fig. 6.11. Oviposition behavior. A. *Halticus apterus.*
B. *Cyrtorhinus caricis.* (Reprinted, by permission of
B. Kullenberg and Zoologiska institutionen, Uppsala
Universitet, from Kullenberg 1944.)

She sometimes makes several attempts before the
ovipositor can be inserted, the abdomen wobbling side
to side, and she may have to return to probing with the
stylets to enlarge the initial opening (Knight 1915,
Osborn 1918, Roberts 1930, Gopalan and Basheer 1966,
Haws and Thompson 1978). The ovipositor is embedded (see Fig. 6.11B) by a rocking or rapid jerky motion
of the abdomen (Knight 1915, Steer 1929), through a
series of sawlike motions (Massee and Steer 1928,
Austin 1931a), or by a flexing of the abdomen along its
longitudinal axis (McIver and Stonedahl 1987a). Butler
(1923) likened the sawlike ovipositor (see Fig. 6.11B; see
also Fig. 5.16), which is composed of four blades (Slater
1950, Davis 1955), to that of sawflies (Hymenoptera:
Symphyta). A female removes the ovipositor from
woody tissue by moving her body side to side, up
and down, or both (Petherbridge and Thorpe 1928a,
Austin 1930). Typically one to five minutes passes
before another egg is deposited (e.g., Knight 1915,
Austin 1930b).

Thontadarya and Channa Basavanna (1962) reported
oviposition behavior similar to this general pattern in
Helopeltis antonii on cashew trees, emphasizing the use
of stylets. The female probes the plant surface with
the labial tip and on finding a suitable site, inserts
the stylets deep into tissues. If the stylet puncture
is not found immediately with the ovipositor, the *H.
antonii* female moves slightly and reinitiates stylet
probing.

A female of the egg predator *Tytthus mundulus* first
sucks or merely punctures a planthopper egg; the ovipositor appears unable to penetrate unbroken plant tissues.
She bends her ovipositor to the site of the sampled egg
before alternately using her mouthparts and ovipositor
to prepare the site for oviposition. Preparatory behavior
can last several minutes, whereas egg deposition is
completed in 45 seconds (Verma 1955a).

When deposited, mirid eggs are usually coated with
an egg cement (Kullenberg 1947a). Davis (1955) was
unable to confirm Kullenberg's (1947a) supposition that
glandular tissue at the base of the ovipositor's second
valvulae secretes the cement. The spermatheca, which
in mirids does not receive and store sperm, might
produce the cement (Davis 1955).

After an egg is deposited and the ovipositor withdrawn, a sequence frequently requiring several minutes,
the proboscis often touches the oviposition site
(Gossard 1918, Osborn 1918) and apparently is used to
seal the wound. A *Halticus bractatus* female, after the
ovipositor is withdrawn, will apply a drop of clear fluid
to the exposed end of the egg (Beyer 1921). *Heterocordylus genistae* appears to use salivary secretions to
moisten the exposed portion of eggs immediately after
oviposition (Brown 1924). In the case of *Halticotoma
valida*, the operculum is often surrounded by an "irregular mass of light colored material" (Haviland 1945).
Females of *Mertila malayensis* secrete a white fluid
from the anus and spread it over the exposed opercula.

Eggs of this eccritotarsine, which are usually laid in groups of two to eight, can be recognized by the pasty covering or "egg spots" (Capco 1941). Similarly, oviposition sites of an Australian species of *Rayieria* are covered with a waxy material through which the aeropyles protrude (Plate 1; Donnelly 1986). Rostral exploration of the oviposition site, which characterizes the behavior of the predatory *Macrolophus melanotoma*, might deter other females from laying eggs in the same area (Ferran et al. 1996).

In some cases, egg deposition requires only 10–15 seconds. This might typify mirids, such as *Halticus bractatus* (Beyer 1921), that oviposit in the leaves of herbaceous plants, or *Trigonotylus caelestialium* (Wheeler and Henry 1985) and *Calocoris angustatus* (Hiremath and Viraktamath 1992) that oviposit in leaf sheaths and panicles of grasses. Oviposition by the grass-feeding *Labops hesperius* requires only four seconds (Haws and Thompson 1978), and *Helopeltis theivora* lays eggs in the pericarp of cocoa pods in 8–10 seconds (Tan 1974a).

Eggs frequently are inserted singly but they can be laid in groups of two to five (Fig. 6.12) or sometimes 20 or more. Eggs tend to be deposited singly on woody plants at sites offering resistance to the ovipositor, whereas two or more eggs are laid in more accessible sites, such as cankers, ends of broken twigs, or pruning scars (Horsburgh 1969). *Blepharidopterus angulatus* females lay eggs in batches of up to 12, then feed intensively for a few days before depositing more eggs (Collyer 1952). As in other insects, depositing eggs in small groups might spread the "risk" in space and increase survival (e.g., Denno and Roderick 1990, Denno 1994a, Tallamy and Schaefer 1997). Those of *H. antonii* are laid singly or in groups of two to four in single or double rows (Devasahayam 1988). More than 300 eggs of *Labops hesperius* have been found in a grass stem, representing oviposition by more than one female (Haws and Thompson 1978). It has even been suggested that an ovipositing *Labops* female deposits a pheromone that attracts other females to the same stem (Paraqueima 1977). Several eggs sometimes are placed in a single incision in the plant (e.g., Shull 1933, Craig 1963, Coombs 1985, Chung et al. 1991). Even when eggs are laid together, with their opercula contiguous, each lies in a distinct cavity (Collinge 1912, Austin 1929, Collyer 1952), with small amounts of plant tissue between the opercula (Coombs 1985). Further details on mirid oviposition behavior can be found in the publication by Kullenberg (1944).

An iteroparous mode of reproduction (e.g., Fritz et al. 1982) characterizes the Miridae. Egg laying is usually spread over several days, as in *Adelphocoris lineolatus* (Craig 1963) and *Nesidiocoris tenuis* (El-Dessouki et al. 1976), or over a several-week period punctuated by bouts separated by intervals of several days (e.g., Blommers et al. 1997). In the case of cocoa mirids and other plant bugs, oviposition occurs more or less over the life of the female (Kumar and Ansari 1974, Niemczyk 1978). The tingid predator *Stethoconus praefectus* lays most of its eggs within a week after fertilization (Mathen and Kurian 1972), and oviposition in the facultatively predacious *Campylomma verbasci* declines steadily after day 16 (Smith and Borden 1991). Egg production in another predator, *Deraeocoris nebulosus*, begins on day 4 when females are provided a constant source of whitefly prey, and ranges from about 10 to 14 eggs per day until day 22, when the oviposition rate decreases with increasing age. A female that lived 58 days laid her last eggs on day 57 (Jones and Snodgrass 1998). The availability of arthropod prey can extend the oviposition period in predatory mirids (e.g., Torreno and Magallona 1994). Discontinuous oviposition sometimes extends for two months, as in *Lygus lineolaris* (Painter 1929b) and *H. schoutedeni* (Schmitz 1958). The length of the oviposition period for *L. hesperus* averages 36 days but lasts up to 128 days (Jackson 1987), and *L. rugulipennis* sometimes lays eggs for more than 10 weeks (Boness 1963).

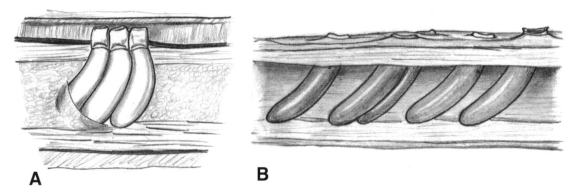

A   **B**

Fig. 6.12. Eggs inserted in plants. A. *Orthocephalus coriaceus* (×15). B. *Adelphocoris lineolatus* (×15). (Reprinted, by permission of B. Kullenberg and Zoologiska institutionen, Uppsala Universitet, from Kullenberg 1944.)

The average duration of oviposition for *H. bergrothi* was 75 days, with one female continuing to lay eggs for 125 days (Kirkpatrick 1941). The egg-laying period often declines as temperatures increase (e.g., Khattat and Stewart 1977), and the oviposition rate can decrease on low-quality hosts (e.g., Fleischer and Gaylor 1988).

Discontinuous oviposition is especially characteristic of many tropical Miridae. For example, ovarial development in *Helopeltis* species (probably *H. bradyi*; see Stonedahl [1991]) takes place only under favorable conditions in Java (Roepke 1916). An abrupt termination of oviposition and resumption of egg laying might explain some of the erratic nature of *Helopeltis* populations: sudden outbreaks followed by cessation of attacks (Giesberger [1983]).

Egg Placement and Oviposition Types

Endophytic oviposition (e.g., Leston 1961b) is practiced by heteropterans such as anthocorids, mirids, and nabids, and by homopterans such as many leafhoppers, planthoppers, psyllids, and treehoppers. Endophytic oviposition is an adaptation for reducing water loss in eggs, a stage particularly vulnerable to desiccation because of the high surface-volume ratio (Southwood 1973, Byrne et al. 1990). As in membracids (Wood 1993) and planthoppers (Denno 1994a), endophytic oviposition helps minimize winter mortality. This habit, intimately associated with the evolution of univoltinism in temperate regions, helps correlate the hatching of overwintered eggs with host phenology in spring. The insertion of eggs into host tissues also offers a degree of protection from natural enemies (Kumar 1971a, Tallamy and Schaefer 1997). An early hatching of eggs, a synchronous development of nymphs, and a relatively short life cycle might represent a strategy to minimize predation, as suggested by Jonsson (1985), or to lessen parasitism (Myers 1981). Late-stage mirid nymphs and adults are generally more difficult for predators to catch than are those at earlier stages. Mirids that oviposit in tissues of viscid-pubescent hosts might gain further protection, the plants' glandular trichomes serving as a physical barrier to some egg parasitoids (Torreno and Magallona 1994).

Placement of eggs

How mirids place their eggs depends partly on how readily the ovipositor can penetrate plant tissues. Different species vary in the extent to which cutting teeth on the ovipositor blade are developed (Davis 1955). On the stems of certain grasses, a *Leptopterna dolabrata* female, unable to insert her ovipositor, lays eggs on the surface (Garman 1926). *Adelphocoris superbus* will not oviposit in heavily lignified stems of alfalfa (Lilly and Hobbs 1956). A plant bug female occasionally dies when the ovipositor cannot be withdrawn from a hard stem (Smith and Franklin 1961) or she falls victim to some predator while ovipositing (Massee and Steer 1928).

Eggs of *Helopeltis antonii* are typically embedded in tender shoots, but when tissues of the host plant harden, they are laid between the bark and wood (Ramachandra Rao 1915). When offered cocoa chupons categorized as soft, medium, and hard, *Distantiella theobroma* females more often oviposit in soft chupons (Pickett 1968). Eggs of *Monalonion annulipes*, under laboratory conditions, are laid in the soft tissues of young cocoa shoots rather than in the pods or seedlings (Villacorta 1977a). *Lygus hesperus* and *L. lineolaris* also deposit their eggs in tender plant tissues (Alvarado-Rodriguez et al. 1986a), the oviposition sites on a particular plant species being influenced by host maturity (Graham and Jackson 1982). The stems of grain amaranth become less succulent as the plant matures and less attractive to *L. lineolaris*; oviposition then switches to leaves and eventually to inflorescences (Wilson and Olson 1990), suggesting the importance of water absorption in egg hatch of this species. These observations contrast with those of Haseman (1913), who reported that eggs of *L. lineolaris* are laid only in flowers and noted that the ovipositor seems unable to penetrate even soft tissues of weeds. Although this species often deposits its eggs in the flowers of composites and other weeds, as Haseman (1913) reported, it also inserts them into the tissues of various plant parts (e.g., Painter 1927). Females of the predacious *Tytthus mundulus*, apparently unable to penetrate the plant surface, oviposit in cavities that once contained eggs of their prey (Verma 1955a). In another specialized egg predator, *Cyrtorhinus lividipennis*, oviposition occurs either in emptied egg shells of its delphacid prey or in cavities made by the bug's ovipositor (Napompeth 1973).

In other mirids, tissues must be a certain hardness for oviposition to occur. For example, increasing the hardness of certain artificial substrates results in greater oviposition by *Macrolophus melanotoma* females (Constant et al. 1996a). Females of this species are apparently able to detect variation in the hardness of the host tissues and perhaps also in the thickness of the substrate. The interaction of host hardness, substrate thickness, and allelochemicals from the host requires clarification (Constant et al. 1996b). In addition, green chupons of cocoa apparently are less suitable as an oviposition substrate for *Sahlbergella singularis* than are more mature, harder chupons (Eguagie 1977). Similarly, *Deraeocoris serenus* oviposited in lignified stems of eggplant under laboratory conditions but did not accept tender stems for egg laying (Fauvel 1999).

Dense plant pubescence sometimes deters oviposition (e.g., Miller 1941). The cotton fleahopper (*Pseudatomoscelis seriata*), however, deposits more eggs on pubescent than on glabrous varieties of cotton (Holtzer and Sterling 1980), as does *L. hesperus* (Benedict et al. 1983). *Tupiocoris notatus* females deposit significantly more eggs on glandular than on nonglandular plants of downy thornapple, which is polymorphic in trichome morphology (van Dam and Hare 1998).

Types of oviposition

Eggs laid in stems can be placed in various positions between parallel and 90° to the long axis. Often only the operculum and a portion of the egg are exposed (Plate 1), or the egg can be flush with the surface or slightly below it (Figs. 6.12, 6.13E,F; Kullenberg 1944). In some mirids that develop on herbs, the female pushes her egg through the stem so that it projects from the side opposite to where her ovipositor is inserted, with the operculum embedded in the far side of the stem (Michalk 1935).

These principal types of oviposition—deeply embedded, partially implanted, and inserted through a stem—are termed *Profund-implantiert*, *Parum-implantiert*, and *Tradukt* (Michalk 1935; see also Gäbler [1937]). The last-named type of oviposition would seem to be uncommon or accidental, and it apparently has been observed only once in a North American mirid: an egg of *Lygus lineolaris* was pushed entirely through an alfalfa stem (Smith and Franklin 1961). Kullenberg (1944) challenged the accuracy of Michalk's (1935) figures, particularly the illustration of *Tradukt* oviposi-

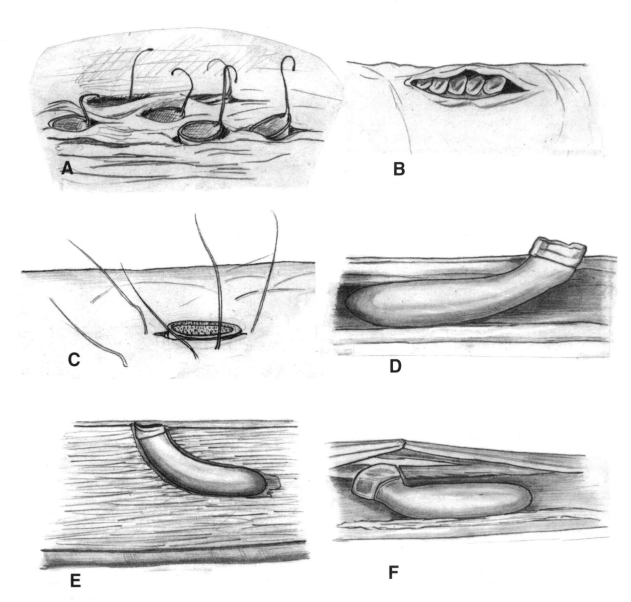

Fig. 6.13. Eggs inserted in plants. A. *Megacoelum infusum* (×40). B. *Cyllecoris histrionius* (×30). C. *Liocoris tripustulatus* (×60). D. *Calocoris roseomaculatus* (×30). E. *Calocoris roseomaculatus* (×15). F. *Leptopterna ferrugata* (×30). (Reprinted, by permission of B. Kullenberg and Zoologiska institutionen, Uppsala Universitet, from Kullenberg 1944.)

tion, but the ovipositor of *L. rugulipennis* can indeed be inserted entirely through a stem (Varis 1972).

Oviposition patterns do not always correspond well to Michalk's (1935) categories, which probably reflect only different degrees of a similar method of oviposition (Leston 1953). To replace Michalk's scheme, Southwood (1956b) proposed four categories for egg deposition by terrestrial Heteroptera based on relationship to the environment. Three of Southwood's categories pertain to mirids: (1) semiexposed, as in axils of stems; (2) embedded in plant tissues that are dead or will die before egg hatch; and (3) embedded in or otherwise intimately associated with living tissues of host plants.

Examples of variations in mirid oviposition habits include differences in the angle and depth of insertion by *Lygocoris rugicollis* females, and consequently, in the exposure of the operculum (Austin 1929). Painter (1927) commented on the extreme variability of egg insertion by *Lygus lineolaris*: "In some cases the eggs are buried for their full length in the tissue; in others the egg may protrude for practically its entire length, or only a small tip may remain exposed; and occasionally they may even be found lying loose on a fold in the leaf." Painter's (1927) observation of eggs lying loose in leaf folds might be explained by the subsequent opening of leaf buds in which eggs had been deposited. Eggs of *Deraeocoris nebulosus* typically are inserted so that only the micropylar process protrudes (McCaffrey and Horsburgh 1980), but at least under laboratory conditions they can protrude conspicuously from the plant surface or fail to be embedded (Jones and Snodgrass 1998). Eggs of *Adelphocoris rapidus* can be inserted loosely in alfalfa flowers or flower-bud clusters (Smith and Franklin 1961), and those of fulviines generally are deposited in crevices rather than inserted in plant tissues (Schmitz and Štys 1973). The Palearctic *Malacocoris chlorizans* also shows an atypical habit: eggs are superficially placed (exophytic oviposition [Leston 1961b]) and sometimes exposed on the surface (Collyer 1952) rather than inserted in host wood. At least in rearing cages, *Halticus bractatus* females with damaged ovipositors lay eggs only on host surfaces rather than insert them in plant tissues (Cagle and Jackson 1947). *Rhinacloa callicrates*, which typically inserts its eggs under the epidermis of a leaf rachis on host plants, under experimental conditions laid an egg on the leaf surface of a novel host (Donnelly 2000).

Despite the unusual oviposition habits of *M. chlorizans* and fulviines, and the atypical behavior observed in *H. bractatus*, the statement that mirid eggs are simply laid on leaves or stems of host plants (Jacobs 1985) fails to characterize the family's oviposition habits. Similarly misleading is Miller's (1971) comment that oviposition on the plant is the usual method rather than deposition of eggs into softer plant parts. In addition, the statement that *Dionconotus neglectus* (as *D. cruentatus*) oviposits in the ground (Avidov and Harpaz 1969) is erroneous (Bodenheimer 1951), and the observation that *Apolygus lucorum* sometimes deposits eggs

in the soil (Chu and Meng 1958) needs verification. But eggs of species that oviposit in nonwoody plants do sometimes drop to the ground, and the stem tips of rushes that harbor *Tytthus pygmaeus* eggs sometimes decay, allowing eggs to fall to the litter layer (Rothschild 1963).

Egg Visibility

Mirid eggs can be invisible on plant surfaces (Crosby 1911, Crawford 1916, Fryer 1916, Cotton 1917, Petherbridge and Husain 1918, Butler 1923, Johnson 1934, van Turnhout and van der Laan 1958, Guppy 1963, Niemczyk 1967, Smith 1991, Smith and Borden 1991), though the operculum is sometimes visible to the unaided eye (Figs. 6.13, 6.14, Plate 1), and they can usually be observed with a stereoscopic microscope. As noted earlier, sometimes eggs are incompletely embedded in host tissues (see Fig. 6.14, Plate 1), or they are laid in semiexposed sites such as the axils of stems (Kullenberg 1944, Southwood 1956b). Sites of egg deposition can be obscured by plant hairs (e.g., Collinge 1912, Dickerson and Weiss 1916, Austin 1929, Roberts 1930, Collyer 1952, Gopalan and Basheer 1966). They eventually can become visible as a small but distinct darkened, discolored, or bruised area (e.g., Dumbleton 1938, Waloff and Southwood 1960, Guppy 1963, Wheeler et al. 1979, Chatterjee 1984a) or as scars that appear as elongate or irregular slits (Usinger 1945); or they can be marked by a blister, bump, pouch, or swelling on woody stems (e.g., Brittain 1919, Collyer 1952, Sanford 1964a, Bin 1970, Ciampolini and Servadei 1973, Wheeler and Henry 1976, Gagné 1997, Donnelly 2000). Oviposition sites of *Lygidea mendax* on apple trees can be located by the reddish brown appearance of the normally light-colored lenticels (Knight 1915). After a few months, the growth

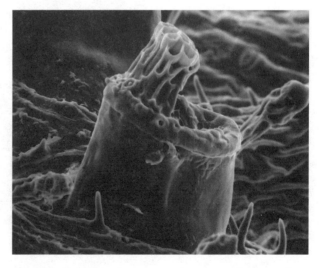

Fig. 6.14. Egg of *Stethoconus japonicus* partially inserted in midrib of azalea leaf (×400). (Reprinted, by permission of J. W. Neal Jr. and the Entomological Society of America, from Neal and Haldemann 1992.)

of algae and accumulation of dust can again make it impossible to detect eggs (Petherbridge and Husain 1918), and when flaps of plant tissue surrounding egg scars begin to dry, they conceal the inconspicuous chorionic rims (Waloff and Southwood 1960). Sanford (1964b) stated that eggs of *Atractotomus mali* laid between a leaf stem and twig near the abscission layer can become visible when the leaf drops, but bark often grows over the once-exposed opercula. Eggs inserted flush with the plant surface can protrude noticeably just before hatching (Roberts 1930, Geering 1953, Chatterjee 1984a).

Detection of mirid eggs is a time-consuming part of life-history studies (Guppy 1963; pers. observ.). Because eggs often are so difficult to find, oviposition periods and fecundity have been calculated from the time when first instars emerge (Haviland 1945, Al-Ghamdi et al. 1995, Gerber 1995). For *Campylomma verbasci*, the distribution of overwintered eggs on apple trees was estimated from the abundance of first instars on different types of foliage (Thistlewood and Smith 1996).

But heteropteran and homopteran eggs that are embedded in leaves usually can be detected by boiling host material and using an indicator dye to stain the eggs (e.g., Curtis 1942). Other staining techniques (Khattat and Stewart 1980, Benedict et al. 1983, Alvarado-Rodriguez et al. 1986a, Khan and Saxena 1986, Backus et al. 1988, Ferran et al. 1996) also can reveal mirid eggs in the leaves or stems of herbaceous plants. Staining techniques sometimes fail to detect mirid eggs in the stems of woody plants (Tonhasca 1987). Eggs deposited in woody stems, or in some herbaceous plants, will often stick to the inner surface when the bark or epidermis is peeled off (e.g., Petherbridge and Husain 1918, Roberts 1930). The eggs typically remain in place where only the outer cork area is peeled away (Petherbridge and Thorpe 1928a).

At certain times of the year, the greenish yellow egg caps of *Lygocoris pabulinus* are prominent on the stems of woody plants (Petherbridge and Thorpe 1928b). The egg caps of *Adelphocoris lineolatus* are visible as tiny specks on alfalfa stems (Hughes 1943), and those of *Labops hesperius* are visible on grass stems (Haws and Thompson 1978). Certain tropical bryocorines have two chorionic processes (Dudgeon 1895), which are two- to three-fifths an egg's length (Fig. 6.15; see also Fig. 6.17, Plate 1; De Silva 1961, Smith 1979), but even the white filamentous processes of these eggs are not easily seen unless the plant part harboring them is held against a dark background (Miller 1941). If the plant is in the "right light," eggs are just visible to the unaided eye (Conway 1971).

The oviposition habits of most plant bugs do not allow their eggs to be easily detected by plant inspectors of regulatory agencies. This has aided the establishment of numerous Old World species in North America, mainly as a result of their introduction with shipments of nursery stock (e.g., Henry and Wheeler 1973; Wheeler and Henry 1973, 1992). Movement of shade trees such as honeylocust and other nursery stock also results in establishment of indigenous mirids in areas of a country where they are not native (Wheeler and Henry 1992, 1994; Polhemus 1994; Yasunaga et al. 1999). In one case, an endemic South African mirid (*Zanchius buddleiae*) belonging to a genus otherwise restricted to the Southwest Cape or Karoo became established in another region of the country when its host plant was transported as an ornamental (Schuh 1974). Moreover, *Closterotomus fulvomaculatus*, which sometimes oviposits in hop poles (Theobald 1895, 1896, 1929; Massee 1937, 1942a), is believed to

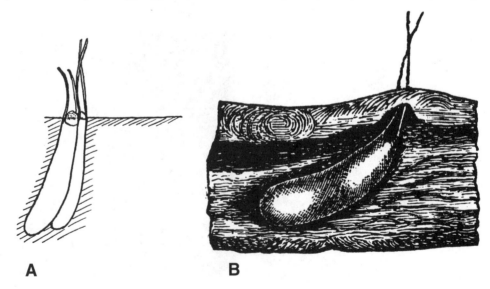

A   **B**

Fig. 6.15. Chorionic processes of bryocorine eggs. A. *Helopeltis schoutedeni* (from Schmitz 1958.) B. *Helopeltis theivora* (from Dudgeon 1895). (A reprinted, by permission of G. Schmitz.)

have been moved with such poles to a previously uninfested English hop garden (Theobald 1928, Steer 1929). At times *Lygocoris pabulinus* might have been introduced into England with egg-infested crabapples imported from Holland (Petherbridge and Thorpe 1928b), although this mirid probably is native to Britain. Mirid injury to top-grafted apple trees in England has been attributed to the introduction of *L. rugicollis* as eggs in grafted wood (Hey 1935).

Noting that mirids (Bryocorinae) associated with cocoa appear not to have been spread by commerce, Entwistle (1977) remarked, "The nature of mirid biology precludes the likelihood of their carriage over any appreciable distance." His statement applies to the active stages of most species, but a more appropriate biological characterization of the family would be that endophytic oviposition predisposes the bugs to long-distance transport in commerce (Wheeler and Henry 1992).

Oviposition Injury and Plant-Wound Responses

Mirid injury to host plants, as in psyllids (Waloff and Richards 1977) and other homopterans that practice endophytic oviposition, can result from both feeding and egg laying (Taylor 1908, Hammer 1939, Huber and Burbutis 1967, Nuzzaci 1977; see also Chapter 9). Mirid oviposition involves the penetration of plant tissues by the ovipositor, with the number of punctures made generally exceeding the number of eggs deposited (Smee 1928b, Putshkov 1966). Oviposition in plant bugs thus includes considerable piercing with the stylets (e.g., Gopalan and Basheer 1966), behavior similar to that of females of certain psyllid species whose "trial incisions" intensify scarring on the surface of host stems (Watmough 1968). Scars made by *Orthotylus ramus* can be identified on pecan wood at least eight years after egg deposition (Chung et al. 1991). Insertion of mirid eggs need not adversely affect the host; oviposition sites can be indicated merely by drops of plant sap (Smith and Franklin 1961). Crop yield (Jewett and Townsend 1947) and plant health (Cech 1989, Torreno and Magallona 1994) sometimes remain unaffected. But egg deposition, involving a disorganization of plant tissues through mechanical injury rather than cellular breakdown (Fig. 6.16; e.g., Smith 1925, 1926; Reid 1968), sometimes results in conspicuous or even economic injury on herbaceous and woody plants (Plate 1, Table 6.2). Its exact nature depends on the sensitivity of the host and the plant parts involved. On apple trees, the severity of injury to the fruit caused by *Lygus lineolaris* depends on the host's cultivar and the weather (Howitt 1993). Several eggs inserted near each other can cause stems to split (Silvestri 1932, Sorenson 1939, Hughes 1943, Sorenson and Cutler 1954).

On Scotch broom, eggs of *O. concolor* and *Heterocordylus tibialis* penetrate the xylem cylinder (Waloff and Southwood 1960). *Lygocoris pabulinus* eggs can be

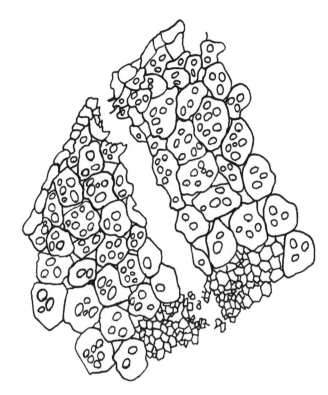

Fig. 6.16. Cellular destruction in potato stem from oviposition by *Closterotomus norwegicus*. (Redrawn, by permission of the Association of Applied Biologists, from Smith 1926.)

embedded in the phloem and penetrate the xylem of currant (Petherbridge and Thorpe 1928a); those of the apple capsid (*L. rugicollis*) penetrate the bark, phloem, and often the xylem, and on small young shoots, will reach the pith (Austin 1929). Eggs of *L. pabulinus* deposited on apple trees can be surrounded by corky material of unknown significance (Wightman 1968), and a nonspecific plant-wound response would seem to be involved. Usinger (1945) reported dead corky tissue around the eggs of *Tropidosteptes illitus* in California, and in Britain, hardened tissues almost encase the egg of *Blepharidopterus angulatus* (Collyer 1952). Apparently similar plant-wound responses are known in eggs of the tingid *Stephanitis rhododendri*, which are surrounded by hardened, corky tissue (Crosby and Hadley 1915). Oviposition wounds from membracids of the *Enchenopa binotata* species complex can induce masses of corky tissue in black walnut stems, resulting in the formation of wound periderm and callus tissue (Armstrong et al. 1979). In membracids and probably also in mirids, eggs inserted deep in woody stems are sometimes forced out by plant-defensive responses, rendering the eggs more vulnerable to predation (Tallamy and Schaefer 1997). In addition, eggs of the mirid *Macrolophus melanotoma* on plants less favorable for oviposition desiccate from an antibiosis reaction: a cankering of tissues around the eggs (Constant et al. 1996a).

Table 6.2. Examples of oviposition injury by mirids

Mirid	Host plant	Locality	Remarks	Reference
Bryocorinae				
Engytatus modestus	Tomato	Hawaii, USA	Rings of scars on fruit	Illingworth 1937a
Helopeltis theivora	Tea	India	Cracking of bark, followed by tissue dieback and callusing over by adjoining tissue	Das 1984
Mertila malayensis	Moth orchid	Philippines		Capco 1941
Mirinae				
Adelphocoris lineolatus	Alfalfa	Minnesota, USA	Splitting of stems	Hughes 1943
Adelphocoris lineolatus, A. rapidus	Birdsfoot trefoil	New York, USA	Wilting, blighting of terminals	Neunzig and Gyrisco 1955
Closterotomus norwegicus	Asparagus	New Zealand	Dieback of soft stem tissue, leaving brown area dotted by insertion scars	Townsend and Watson 1982
Creontiades pallidus	Cotton	Africa		Soyer 1942
Dichrooscytus elegans	Juniper	Ohio, USA	Tip dieback similar to that associated with fungal pathogens	Shetlar et al. 1992
Eurystylus oldi	Sorghum	Mali	Oviposition possibly more injurious than the bug's feeding on earheads; see Stonedahl (1995) for nomenclature	Sharma 1985c
	Sorghum	Niger	Developing kernels can deteriorate following entry by fungi	Steck et al. 1989
Leptopterna dolabrata	Rye, wheat	Finland	Stems injured	Vappula 1965
Lygus hesperus	Douglas fir	Oregon, USA	Seedling terminals injured	Schowalter et al. 1986
Lygus lineolaris	Alfalfa	Kansas, USA	Foliar wilting, death of leaf buds; injured buds can give rise to shot-holed leaves, allowing disease organisms to enter	Smith and Franklin 1961
	Apple	Missouri, New York, USA	Oviposition in young fruit causes depressions ("dimples") that can distort and downgrade the fruit; injury resembles that of weevils or other insect pests of apple	Taylor 1908, Felt 1910, Hammer 1939
	Celery	New York, USA	Leaf stalks injured	Hill 1932
	Lima bean	California, USA	Possible crop losses	Elmore 1955
	Green pepper	Delaware, USA	Possible crop losses	Huber and Burbutis 1967
Lygus rugulipennis	Apple	England	Oviposition in young fruit causes injury resembling that from apple scab	Collinge 1912
	Apple	Germany	Injury to young shoots	Abraham 1937b
	Groundsel	England	Flower buds blackened internally; possible abnormal development	Austin 1931b
	Sugar beet	Finland	Oviposition in hypocotyl can kill seedlings	Varis 1972
Orthops kalmii	Fennel	Italy	Oviposition punctures contribute to malformation of vascular bundles, leading to blockage of sap flow to leaves	Nuzzaci 1977
Orthops scutellatus	Parsnip	Nova Scotia, Canada	Oviposition in stalks can cause drooping of flower heads	Brittain 1919
Tropidosteptes amoenus	Ash	New Jersey, USA	Foliar curling, distortion	Dickerson and Weiss 1916, Weiss 1918

Note: See text for additional examples and discussion.

Oviposition by the egg predator *Tytthus mundulus*, a phyline mirid important in the biological control of planthoppers (see Chapter 14), can cause minor disruption of host tissues. Surfaces of sugarcane leaves that harbor its eggs are sometimes whitened by local decay (Kirkaldy 1908).

Fecundity and Longevity

The cost of reproduction is the most prominent trade-off in the evolution of life-history traits (e.g., Stearns 1989); trade-offs between reproduction and dispersal were discussed earlier in this chapter in relation to wing polymorphisms in mirids. Longevity in insects generally depends on the oviposition rate—that is, the number of eggs laid per day (e.g., Strong and Sheldahl 1970). The relationship between reproduction and length of adult life in plant bugs would be a fruitful area for additional study; a phenotypic trade-off between the two attributes seems characteristic of most mirids (Schmitz 1958, Hori and Hanada 1970, Eguagie 1977, Jackson 1987, Fleischer and Gaylor 1988, Stewart and Gaylor 1994b). For *Cyrtorhinus lividipennis*, however, the feeding regime (30 planthopper eggs/day) that produced maximum fecundity did not result in a statistically significant decrease in longevity (Chua and Mikil 1989).

Data on mirid fecundity and longevity too often come from studies that are not conducted under prescribed or well-monitored environmental conditions. Temperature, as well as relative humidity and photoperiod, is frequently not stated in papers discussing these traits. The number of matings, though often unspecified, can affect insect fecundity and longevity (Ridley 1988, Wang and Millar 1997), as can nymphal or larval diet (which might not be described in detail) and the availability of water to adult females (e.g., Zavodchikova 1974, Landolt 1997, Riudavets and Castañé 1998); access to water might not affect mirid fecundity if fresh host material is available. The importance of the oviposition substrate was discussed earlier in this chapter. Oviposition in heteropterans also can be limited by above-optimum temperatures, and low temperatures can adversely affect fecundity because a female's feeding period is extended at the expense of time devoted to oviposition (e.g., Nadgauda and Pitre 1986). Even a moderate level of stress during nymphal development ultimately might reduce mirid fecundity (Cross 1971).

Investigators have conducted several mirid studies under controlled conditions, for example, that by Neal et al. (1991) on the predacious *Stethoconus japonicus* (see also Strong and Sheldahl [1970], Gaylor and Sterling [1976a, 1976b]). Data accruing from well-controlled studies are not directly comparable to those from less precise studies. Various assessments of fecundity in a given mirid species can produce widely disparate results, as is the case with *C. lividipennis* (Table 6.3).

FECUNDITY

Mated females of many mirids deposit 30–100 eggs, although both phytophagous and predacious species can produce several hundred eggs (e.g., Hargreaves [1926?], Abraham 1958, Boness 1963, Leigh 1963, Le Pelley 1968, Hori and Hanada 1970, Fauvel et al. 1987, Neal et al. 1991, Gerber 1995, Blommers et al. 1997, Jones and Snodgrass 1998). Endophytic oviposition, requiring energy to insert eggs in plant tissue, probably involves a trade-off with fecundity (Denno and Dingle 1981). Whether a trade-off between egg size and fecundity might occur in the family needs to be determined. As Tallamy and Schaefer (1997) emphasized, the development of a piercing ovipositor was not without its costs. Eggs of hemipterans that practice endophytic oviposition are typically two to three times smaller in volume than those of hemipterans in which the ovipositor is short and flattened.

Unmated females of several mirid species deposit their normal complement of eggs, which, however, are sterile (Kirkpatrick 1941, Verma 1955a, Beingolea 1960, Niemczyk 1967; cf. Jeevaratnam and Rajapakse 1981a). In contrast, Vanderzant (1967) found that unmated females of *Lygus lineolaris* deposit fewer eggs than do mated females, which is characteristic of many other insects (e.g., Ridley 1988). Fecundity of mated *Sahlbergella singularis* females is less than that of unmated females (Eguagie 1977). *Lygus hesperus*

Table 6.3. Fecundity ($\bar{x} \pm$ SE) of *Cyrtorhinus lividipennis* reared on different prey species and under different experimental conditions

Study area	Experimental conditions	Prey species	No. of prey eggs	Fecundity	Reference
Hawaii, USA	Laboratory: 24 ± 2°C, 70 ± 5% RH	*Peregrinus maidis*	25 daily	65.7 ± 0.3	Liquido and Nishida 1985a
India	Greenhouse: 30 ± 5°C	*Nilaparvata lugens*	Unlimited	147.0	Pophaly et al. 1978
Malaysia	Laboratory	*Nilaparvata lugens*	30 daily	34.0 ± 5.2	Chua and Mikil 1989
Philippines	Laboratory	*Nephotettix virescens*	20 every other day	13.5	Reyes and Gabriel 1975
Thailand	Laboratory	*Nilaparvata lugens*	Not specified	7.6 ± 1.6	Tanangsnakool 1975

Source: Chua and Mikil (1989).
Note: SE, standard error; RH, relative humidity.

females begin to oviposit about nine days after the last molt, even if they have not mated; unmated females produce only sterile eggs (Strong et al. 1970).

Fecundity among congeners can vary significantly. *Lygus hesperus*, for example, laid a mean of 117 eggs at 20°C and 162 eggs at 26.7°C, whereas *L. elisus* laid only 38 and 48 eggs at the same temperatures (Mueller and Stern 1973b). In determining age-specific fecundity of *L. hesperus*, Strong and Sheldahl (1970) found that a constant temperature of 32.2°C or alternating 26.7°C with 32.2°C gives the maximal potential rate of population increase. Their paper provides further details of the influence of temperature on fecundity and longevity.

Examples of Reduced Fecundity

The number of eggs laid by *Polymerus cognatus* females is said to be only two (Kiritshenko 1951), a figure seemingly in need of corroboration. In some cases, data on fecundity might reflect the number of eggs observed while dissecting only a few field-collected females. It also is likely that females often do not survive in nature to deposit their full potential of eggs.

Ant-mimetic mirids appear to lay fewer eggs (12–15) than sympatric nonmimetic plant bugs of similar size (20–30 eggs) (McIver and Stonedahl 1987a, 1987b, 1993). McIver and Stonedahl (1993) pointed out that life-table studies of both groups are needed to determine whether the lower fecundity in myrmecomorphic species actually reflects reduced mortality from predation.

Fecundity under laboratory conditions is often less than 15, for example, in *Campylomma verbasci* (Niemczyk 1978), *Cyrtorhinus lividipennis* (Tanangsnakool 1975; see Table 6.3), *Campylomma liebknechti* (Chinajariyawong and Walter 1990), *Moissonia importunitas* (Goot 1927; cf. Gopalan and Basheer 1966), and *Trigonotylus caelestialium* (Blinn and Yonke 1986). Because longevity has an important influence on fecundity (e.g., Leather 1988), some reports of low fecundity in mirids might reflect a decreased longevity under less than ideal laboratory conditions, the unavailability of pollen or prey, and water stress or other effects of rearing procedures (e.g., Whitcomb 1953, Cross 1972, Smith 1973, Eguagie 1977, Gerber 1995, Groot et al. 1998). The rearing of *Sahlbergella singularis* and other cocoa mirids is particularly hampered by low fecundities (Youdeowei 1973, Eguagie 1977).

Todd and Kamm (1974) suggested that differences in fecundity between two populations of *Labops hesperius* were the result of food shortages at a site where the bug density was twice that of another site. Adult movement or migratory behavior seems necessary for the maximum reproduction of *Adelphocoris lineolatus*; the number of eggs laid in confinement by laboratory-reared females or females that developed in field cages is substantially less than the number deposited by confined females that attained sexual maturity in the field (Craig 1963). Gibson (1976) noted that measurement of fecundity in the grass-feeding *Notostira elongata* under controlled laboratory conditions often gives "a very false picture of real events in the field due to changes in the nutrient status and growth form of grasses brought into the laboratory." In addition, fecundity of field-collected *Lygus lineolaris* is more than twice that reported for females in laboratory colonies of this species (Gerber 1995).

Intergenerational Differences

The mean number of eggs produced by *Adelphocoris lineolatus* is much greater in the first generation than in the second (Romankow 1959). From *Helopeltis schoutedeni* reared on castor, Schmitz (1958) obtained more eggs from first-generation females ($\bar{x} = 147$) than from the second- ($\bar{x} = 62$) or third-generation females ($\bar{x} = 45$). The fecundity of first-generation females of *Trigonotylus caelestialium* is similarly higher than that of second-generation females, even though females of the first generation are generally smaller (Mikhaïlova 1979). For certain cocoa mirids, fecundity is reduced during the dry season when the host plant is under water stress (e.g., Cross 1971, Collingwood 1977a, Eguagie 1977, Gibbs 1977), and *Calocoris angustatus* females lay fewer eggs after the rainy season than during it (Sharma and Lopez 1990a). Reports of differences in fecundity between generations suggest the value of studying plant bugs of different life histories and habitats, with the possibility of discerning patterns of compromise between reproductive allocation, growth rate, and future survivorship.

Host-Plant Effects

The amount and rate of consumption of different plant parts, which differ in their nitrogen concentrations (Southwood 1973, Thompson 1983, Hagen et al. 1984, Slansky and Scriber 1985), potentially affect fecundity and other components of fitness (e.g., Saharia 1982, Slansky 1982). Host-influenced effects on fecundity, including different hosts as well as different parts (e.g., foliage vs. fruits) of the same plant species, are known for *Helopeltis antonii* (Sundararaju and Sundara Babu 1998), *Lygocoris pabulinus* (Groot et al. 1998), *Lygus lineolaris* (Khattat and Stewart 1977, ·Stewart and Gaylor 1994b), *L. rugulipennis* (Varis 1972), and *Notostira elongata* (Bockwinkel 1990). Gaylor and Sterling (1976b) determined that different hosts and temperatures affect the net reproductive rates per generation of *Pseudatomoscelis seriata*. *Lygus lineolaris* shows a higher total fecundity on cotton than on horseweed, although net fecundity on the latter plant is higher (Fleischer and Gaylor 1988). The fecundity of certain cocoa mirids varies depending on whether the bugs are reared on host shoots or pods (Houillier 1964a). McNeill (1973) found that average fecundity of *Leptopterna dolabrata* decreases when the bugs are switching from feeding on leaves to seeds (see Chapter 10), a period when high-nitrogen sites are limited and fewer females can fulfill their nitrogen requirements (see also discussion by Price [1984]). De Jong (1938) studied the effects of food quality—tea leaves differing in carbohydrate level from plucked versus unplucked bushes—on egg production by *H. theivora*, and noted that fecundity on the same host species might differ from site to site, owing to differences in the quality of food available to the bugs. Schmitz (1958) provided fecundity and longevity data for *H. schoutedeni* on several host plants. Fertilization of salt marsh grasses results in larger, more fecund *Trigonotylus* females compared to those developing on nitrogen-poor grasses (Vince et al. 1981). Holopainen et al. (1995) reported a nearly linear relationship between the oviposition rate of *Lygus rugulipennis* and increases in nitrogen fertilization of pine seedings.

Effects of an Animal Diet

Because many mirids are facultative predators, availability and type of prey (e.g., relatively low-nitrogen aphids vs. higher-nitrogen arthropod eggs) can affect total fecundity. Researchers assessing mirid fecundity should consider the possibility that the presence of prey, including organisms too small to be easily noticed, potentially affects their results. More eggs may be laid when bugs have access to both plant and prey or are fed a mixed prey diet rather than a single prey species (e.g., Chinajariyawong and Walter 1990, Torreno and Magallona 1994, Albajes et al. 1996, Riudavets and Castañé 1998).

Prey are needed for egg production by *Campylomma liebknechti* (Chinajariyawong and Walter 1990) and *Psallus ambiguus* (Niemczyk 1968, 1978). The fecundity of *Spanagonicus albofasciatus* (Musa and Butler 1967) and *Engytatus modestus* (Parrella and Bethke 1982) increases significantly when aphids or lepidopteran eggs or larvae are added to a plant diet. Similarly, the addition of neonate lepidopteran larvae results in a dramatic increase in mean fecundity of *Nesidiocoris tenuis*, from 5.3 when only the host plant (tobacco) is provided, to 66.6 (Torreno and Magallona 1994), and fecundity in the predatory *Macrolophus melanotoma* is substantially greater on whitefly-infested than on clean tomato leaves (Schelt et al. 1996). Egg production in this species varies substantially when it feeds on different arthropods; fecundity is highest when the bugs feed on lepidopteran eggs and is lowest on a mite diet (Fauvel et al. 1987). *Deraeocoris punctulatus* females do not oviposit when fed a plant-only diet but do so when reared on a diet that includes arthropods; fecundity is greater on an aphid diet than on a diet of either spider mites or thrips (Zavodchikova 1974).

Variable-Instar Species

Fecundity has been determined for different-instar groups of only one mirid that exhibits a variable number of nymphal instars. Liquido and Nishida (1985c) hypothesized that the increased fecundity of four-instar females versus five-instar females in *Cyrtorhinus lividipennis* represents a strategy helping this egg predator maintain a stable interaction with its more fecund delphacid prey.

LONGEVITY

Relationship between Gender and Longevity

Males of both univoltine and multivoltine species usually appear before females and do not live as long, either in nature (e.g., Crawford 1916; Beyer 1921; Haviland 1945; Lal 1950; Collyer 1952; Schmitz 1958; Gopalan and Basheer 1966; Niemczyk 1967; Knight 1941, 1968; Cmoluchowa 1982; Hansen 1988) or in laboratory cultures (e.g., Beyer 1921, Austin 1931a, Sauer 1942, Leigh 1963, Napompeth 1973, El-Dessouki et al. 1976, Khattat and Stewart 1977, Al-Munshi et al. 1982). At least for some species this statement tends to oversimplify the relationship between gender and longevity. Females of the egg predator *Cyrtorhinus lividipennis* outlive males only when reared from the first instar on a diet of 30 planthopper eggs per day. At other feeding rates (5, 10, 15, or 20 eggs/day), male longevity is greater, though not statistically different (Chua and Mikil 1989). As with other life-history parameters, availability of prey needs to be considered; the addition of animal matter can show greater effects on one sex than the other (Torreno and Magallona 1994). Therefore, even though males of the strongly predatory *Spanagonicus albofasciatus* (Butler 1965) live slightly longer than females when the species is reared in the laboratory on pods of green beans (Butler and Stoner 1965), one wonders how an animal diet might affect longevity of males versus females.

Female mirids typically live 25–40 days and often die shortly after oviposition is completed (e.g., Craig 1963, Reyes and Gabriel 1975, Blommers et al. 1997). *Lygus rugulipennis* females live an average of eight days after depositing the last egg (Varis 1972). Those of *Halticus bractatus* can live nearly 100 days (Cagle and Jackson 1947). Second-generation (overwintered) females of *Stenodema laevigata* are notable for living a maximum of 12 months (Woodward 1952). Longevity apparently is only nine days for *Tupiocoris notatus* in Brazil (Moreira 1923), about a week for males of the British *Harpocera thoracica* (Southwood and Leston 1959, Groves 1968), and five days for males of *Creontiades pacificus* in India (Butani 1979). Some of these figures might be based on observations made in the laboratory where longevity can be less than that in field cages or in nature (e.g., Pang 1981). Schmitz (1958) found that *Helopeltis schoutedeni* males feeding on certain plants do not live as long in the laboratory as in the field, but on some hosts, laboratory longevity is greater than that in the field (temperature, relative humidity, and number of observations were not stated). For at least *Horcias nobilellus*, both mated males and females do not live as long as unmated adults (Sauer 1942).

Sex Ratios

Females of some cocoa mirids live three or four times longer than males (Johnson 1962). Collection records of some cocoa mirids, however, do not reflect a high female-male ratio in a population, owing to an apparent greater mobility of females upon disturbance (Collingwood 1977a). Females of the univoltine *Tytthus pygmaeus* can be found in England a month and a half after nearly all males die (Rothschild 1963). The oldest males of *Bryocoris pteridis* in Britain sometimes die by the time all nymphs become adults (Satchell and Southwood 1963).

Even though the sexes of phytophagous mirids are usually produced in nearly equal numbers, early-season collections of temperate-zone species that overwinter in the egg stage tend to contain a high proportion of males to females, whereas late-season collections are composed mostly, or entirely, of females (e.g., Brittain 1916a, Brown 1924, Butler 1924, Fulmek 1930, Collyer 1952, Niemczyk 1967, Austreng and Sømme 1980, Pinto 1982, Schwartz 1984, Coombs 1985, Haws and Bohart 1986, Hansen 1988, Gagné 1997). Males of the univoltine *Labops hesperius* in rangelands of the western United States can predominate (1.5:1) early in the season, but represent only 5% of the population near season's end (Haws and Bohart 1986). Undoubtedly, observations of female-biased populations in late season misled Fitch (1870) into thinking females of the univoltine fourlined plant bug (*Poecilocapsus lineatus*) are much more numerous than males. Females that overwinter as adults can actually dominate in early season, owing to the greater proportion of females entering hibernation, or perhaps a higher mortality among males during the winter (Roberts 1930, Varis 1972, Khattat and Stewart 1980). In tropical species such as *Helopeltis antonii* in India, females consistently predominate in field populations: 1.00:0.49 to 1.00:0.62 over a one-year period (Ambika and Abraham 1979). Males of the predatory *Blepharidopterus angulatus* emigrate before females when prey are plentiful, resulting in female-dominated sex ratios (Glen and Barlow 1980). Inexplicably, females of *Lygocoris pabulinus* in England significantly outnumber males in both summer and autumn broods (Petherbridge and Thorpe 1928a). In assessing the sex ratios of mirids, researchers should keep in mind that the proportion of males to females might differ significantly depending on the sampling techniques used (Stewart and Gaylor 1991).

Sex ratios, as well as longevity (e.g., De Jong 1936, Pankanin 1972, Cave and Gutierrez 1983, Hori and Kuramochi 1984, Fleischer and Gaylor 1988), can be affected by the host. For example, Gaylor and Sterling (1975a, 1976a) obtained equal numbers of males and females of *Pseudatomoscelis seriata* when the species was reared on pieces of green bean or potato, but females predominated (3:1) when the bugs were maintained on certain flowering plants. Fleischer and Gaylor (1988), however, reared *Lygus lineolaris* on various hosts without encountering significant effects on the sex ratio.

Little is known about the fluctuation of sex ratios in obligately predacious mirids contrasted with plant-feeding species. In Hawaii, the sex ratios of *Cyrtorhinus fulvus*, a predator of planthopper eggs, fluctuate widely during the season. Trends are inconsistent, except that females predominate when populations are low, and only females are found in May when mirid numbers are lowest. Prey populations are at their lowest levels during February to June (Matsumoto and Nishida 1966). Another delphacid egg predator, *T. pygmaeus*, can have a sex ratio of 1 male to 2.9 females (Rothschild 1963). Studies are needed to determine if sex ratios tend to fluctuate more in predatory than in phytophagous mirids and whether ratios correlate with population densities.

Egg Physiology

RESPIRATORY SYSTEM AND WATER ABSORPTION

Endophytic eggs need to maintain contact with the ambient air. Connection is effected by means of aeropyles, which vary in number and position on the shell's anterior pole or on the operculum. The aeropylar system in nearly all species is separate from the micropyles, but at least two species have a combinated aeropylar-micropylar system (Cobben 1968). Respiratory horns containing aeropyles have evolved in several subfamilies, their number, structure, and sometimes function varying among species.

Some tropical bryocorines have a unique chorionic respiratory system characterized by the presence of dorsal and ventral horns containing aeropyles (Fig. 6.17; see also Fig. 6.15). The horns might have selective value in permitting atmospheric respiration by projecting above the film of water that flows over the eggs during heavy rains (Hinton 1962). Furthermore, the permanent layer of air trapped between the opercular rim and its projections when an egg of *Lygocoris pabulinus* is submerged in water might function as a weak plastron (Wightman 1972).

Eggs inserted deep in plant tissues also must have a means of absorbing water. In demonstrating the uptake of water by psyllid eggs, White (1968) noted that little work on water relationships had been conducted on insects that use endophytic oviposition. In the case of a Neotropical bryocorine, *Monalonion annulipes*, humidity greater than 90% is needed to maintain egg viability (Villacorta 1977a).

Water enters *Deraeocoris ruber* eggs by permeating the chorion at the posterior pole, flooding the air layer in this region, and being absorbed through subchorionic membranes. If the stem harboring an egg should die, any

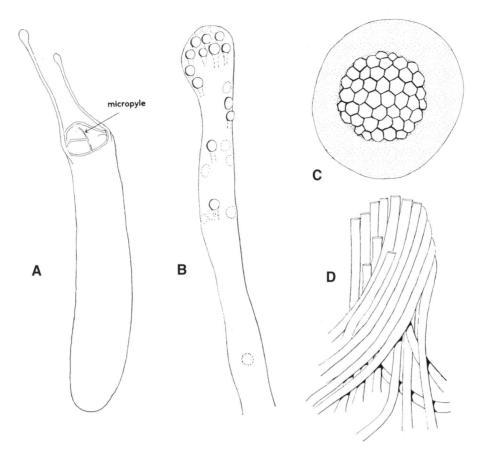

Fig. 6.17. Egg of *Helopeltis schoutedeni*. A. Left side of egg showing one of the two micropyles. B. Apex of dorsal respiratory horn showing aeropyles. C. Section through middle of ventral respiratory horn; note that the diameter of the central aeropyles is greater than that of the peripheral ones. D. Proximal ends of aeropyles of ventral respiratory horn from dorsal side. (Reprinted, by permission of Oxford University Press, from Hinton 1962.)

available external water is sometimes taken up through tubes in the operculum (Hartley 1965). The serosal cuticle apparently is associated with water uptake (Johnson 1937), and in the Miridae (and fulgoroid family Delphacidae) its progressive stretching "may be governed by the selective advantage of sinking the eggs deeper into plant tissue" (Cobben 1968). As eggs absorb water from the host, they can increase slightly in length, with the yolk plug extruded beyond the egg's apex (Usinger 1945). The length, diameter (Fig. 6.18), wet weight, and volume of the *Macrolophus melanotoma* egg increase during embryonic development. Exchanges between egg and host plant of this species mainly involve the uptake of water, with the amino acid content remaining stable during embryogenesis (Constant et al. 1994). Eggs of *Stethoconus japonicus* gain about 60% in weight from day 1 to day 10 (Fig. 6.19; Neal and Haldeman 1992).

SUSCEPTIBILITY TO INSECTICIDES

Mirid eggs resist the action of some insecticides (e.g., Theobald 1911a; Fryer 1914, 1929; Crosby 1915; Abraham 1937b; Collyer 1953d). The projecting filaments or chorionic processes of the bryocorine *Helopeltis theivora* apparently can withdraw, by capillary action, insecticides that contact the egg surface (Andrews [1923]). But the application of ovicides affects some mirid eggs. For example, delayed dormant oil sprays applied after budbreak killed a large percentage of *Lygidea mendax* eggs in Pennsylvanian apple orchards (Frost 1925), and substantial control of this pest was obtained with various oil emulsions and emulsible oils (Chapman et al. 1941). Winter washes of various petroleum types, which once provided partial control of phytophagous mirids on apple trees in England (Austin et al. 1932, Miles 1932, Schoen 1932,

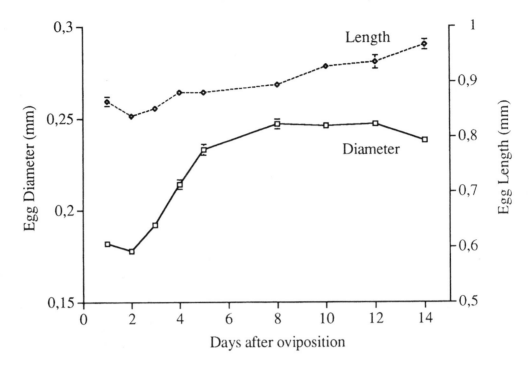

Fig. 6.18. Variation of egg length and diameter during embryonic development of *Macrolophus melanotoma*. (Reprinted, by permission of S. Grenier and the International Organization for Biological Control [IOBC]), from Constant et al. 1994.)

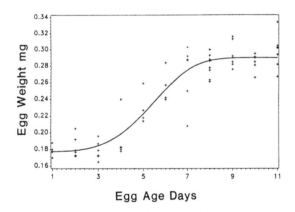

Fig. 6.19. Sigmoidal curve showing the gain in weight of 10 *Stethoconus japonicus* eggs ($R^2 = 0.88$). (Reprinted, by permission of J. W. Neal Jr. and the Entomological Society of America, from Neal and Haldemann 1992.)

Austin 1933, Massee 1937), continued to be recommended a half-century later (Alford and Gwynne 1983). Ash tree–inhabiting plant bugs in California were successfully controlled using dormant oils as ovicides (Usinger 1945).

Insecticides, particularly oil sprays, kill eggs of some beneficial predacious mirids (Steiner 1938, Collyer 1953d, Massee 1955; cf. Horsburgh 1969). Winter sprays

of lime sulfur did not affect the eggs of *Malacocoris chlorizans* in Italian orchards, but dormant oil sprays, as well as oil used as summer sprays, killed all or nearly all eggs of this mite predator (Foschi and Carlotti 1956). Eggs of predatory mirids also can be killed inadvertently when soil is drenched with systemic insecticides (Horsburgh 1969).

Eggs only partially embedded in host tissues can be especially susceptible to chemical treatment. Fryer (1929) and Schoevers (1930) invoked the depth to which eggs are inserted in wood to explain the discrepancy between success and failure to control mirids on apple trees: eggs protruding above the stem surface proved vulnerable to insecticides. Insecticides directed against the eggs of *Lygus lineolaris* appear less effective on hosts in which the eggs are more deeply inserted (Bailey and Cathey 1985).

APPEARANCE OF DIAPAUSING EGGS

Diapausing eggs can look different from nondiapausing eggs. *Pseudatomoscelis seriata* eggs dissected intact from stems remain opaque and yellow when placed in water, in contrast to the nondiapausing egg, which becomes clear and usually reveals the developing embryo (Gaylor and Sterling 1977; see also Danks [1987]). These differences might result from morphological changes or relate to physiological changes such as increases in fat or glycogen levels (e.g., Braune 1976).

MacPhee (1979) measured the hardiness of overwintering eggs of *Blepharidopterus angulatus* to cold temperatures and found that the mean freezing point (–34°C) was sufficient to withstand the lowest temperatures in apple-growing regions of Nova Scotia, where this Palearctic plant bug has become established (Wheeler and Henry 1992). But he suggested that considerable mortality of this predator might occur in parts of New Brunswick and Quebec during cold winters.

Incubation

Typical incubation periods for nondiapausing eggs of temperate-region mirids, reared at a constant temperature of 20°C, are 10–16 days (e.g., Cagle and Jackson 1947, Mols 1990). The incubation period for *Ragwelellus horvathi*, however, is about 21 days, based on bugs caged on potted cardamoms placed outside in the shade in Papua New Guinea (Smith 1977b). Developmental times decrease as temperatures increase (until a developmental maximum is reached) so that an incubation of 6–8 days might be typical for species reared at 30°C (e.g., Ridgway and Gyrisco 1960a, Butler 1970). Incubation times for nondiapausing eggs of *Campylomma verbasci* held at four constant temperatures have a linear relationship with temperature (Fig. 6.20; Smith and Borden 1991). Over a wide range of temperatures, development of mirid eggs is nonlinear because of inhibition at the lower and upper ends of the scale (e.g., Ridgway and Gyrisco 1960a, Ting 1963a, Gaylor and Sterling 1975a, Judd and McBrien 1994). For example, *Lygus hesperus* eggs do not develop at 10° or 40°C (Table 6.4; Champlain and Butler 1967). Although in some insects development under changing temperatures is different from that under constant temperatures, developmental rates of *L. hesperus* eggs do not differ significantly at constant and fluctuating temperatures (Champlain and Butler 1967).

Some temperate, multivoltine species, such as *L. lineolaris* in North America and *L. rugulipennis* in Japan, have a considerably shorter incubation period in the second generation than in the first generation (Painter 1929b, Hori and Hanada 1970). This disparity might relate to differences in physiological condition of the eggs (Hori and Hanada 1970) and possibly also humidity. Laboratory observations of *Spanagonicus albofasciatus* suggest that increased relative humidity shortens the incubation period (Musa and Butler 1967). Rainy weather will shorten the normal incubation period for *Helopeltis schoutedeni* (Schmitz 1958). In areas where reproduction occurs year-round, summer eggs have a shorter incubation than winter eggs (El-Dessouki et al. 1976).

Elevation influences temperature regimes, thereby affecting egg hatch (e.g., Leefmans 1916). Hatching of *Lopidea nigridia*, which is widely distributed between

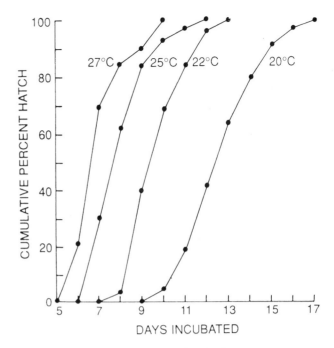

Fig. 6.20. Cumulative percentage hatch of summer-generation (nondiapausing) eggs of *Campylomma verbasci* at four constant temperatures, 16-hour photoperiod. (Reprinted, by permission of R. F. Smith and the Entomological Society of Canada, from Smith and Borden 1991.)

Table 6.4. Mean number of days (± SD) for eggs of *Lygus hesperus* to hatch at constant temperature; 10 replications/test

Temp. °C	Hatching time
10	NA
15	21.0 ± 1.20
20	12.3 ± 0.71
25	8.5 ± 0.59
30	5.9 ± 0.33
35	5.6 ± 0.56
40	NA

Source: Champlain and Butler (1967).
Note: NA, not applicable (no eggs hatched); SD, standard deviation.

1,300 and 2,250 meters in southeastern Oregon, begins about three weeks earlier at lower elevations. Although populations at higher sites start later, egg development occurs faster, presumably because mean ambient temperatures are higher by the time of hatching (McIver and Asquith 1989).

Hatching

The eclosion fracture in mirids is ringlike except in *Bryocoris pteridis*, which has a pseudoperculum, the egg opening through a slit extending over the anterior pole. Apparently all other mirids possess a true operculum, with the nymph hatching (sometimes referred to as *eclosing*) through a roughly circular opening (Cobben 1968). In *Notostira elongata* eggs, a small black ring appears beneath the micropylar region after three or four days of development (Bockwinkel 1990). As development proceeds, a mirid embryo often becomes yellow, orange, or red, and eye spots are visible. Shortly before eclosion, a dark V-shaped streak may be seen on the concave side of an egg, which marks the position of the folded legs and antennae (Mathen and Kurian 1972). As the embryo absorbs water, an egg can appear slightly swollen before hatching (Beyer 1921). The opercular cap, once flush with or slightly below the surface, begins to protrude above the plant surface (Roberts 1930, Geering 1953, Chatterjee 1984a). With incubation, the embryo in field-collected, diapausing eggs of *Labops hesperius* revolves so that the head faces the operculum (Fuxa and Kamm 1976a).

Changes in the egg of *Lygus lineolaris* are well documented. At 27°C, a partial evagination of the operculum is associated with stretching of the serosal cuticle (at age 60 hours), which increases the egg's visibility on the plant surface. Other changes include the appearance of red eye spots (108 hours), an orange dorsal abdominal scent-gland opening on the developing nymphs (132 hours), nymphal antennae as orange stripes (132 hours), and a general green tint (156 hours). These and other changes follow a predictable pattern, allowing the age of the eggs to be estimated accurately (Stewart and Gaylor 1993).

An egg burster is absent in mirids. Instead, the operculum is elevated primarily by the serosal cuticle (Hinton 1962). Air is taken into the gut at the time of hatching, and body fluids pumped into the head cause it to swell (Johnson 1934, Southwood 1956b). When eclosing (Fig. 6.21), a nymph pushes out on the operculum and by a continuous peristalsis of its body, lifts the operculum with its head and emerges head first (Plate 2). The emerging insect sometimes forms a semicircle with the egg, the arched dorsum corresponding to the egg's convex side (Massee and Steer 1929). When the labium and legs are appressed to the body, an emerging nymph undergoing peristaltic movements looks somewhat wormlike (Geering 1953, Smith and Franklin 1961). A nymph sometimes grips the host surface with its foretarsi to help free itself from the embryonic cuticle (Gopalan and Basheer 1966) and swings its body side to side and forward and back (Khristova et al. 1975). These movements (Fig. 6.21E) free the labium and the first two pairs of legs, which are used to free the antennae and metathoracic legs (Beyer 1921).

Under conditions of low relative humidity, a nymph sometimes cannot free itself from the embryonic cuticle (Bech 1969). Mortality from a nymph's inability to free its legs from the egg shell occurs in nature and in laboratory colonies (Horsburgh 1969).

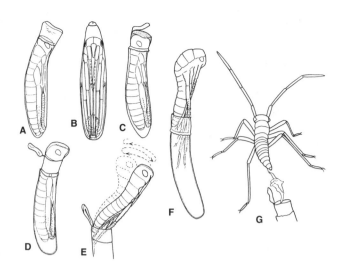

Fig. 6.21. Sequence of events (A–G) during eclosion of *Notostira erratica*. (Reprinted, by permission of B. Kullenberg and Zoologiska institutionen, Uppsala Universitet, from Kullenberg 1944.)

The sequence of withdrawal of different appendages and the duration of the hatching process, often requiring 10–20 minutes (e.g., Hughes 1943), vary among species and with temperature. A *L. lineolaris* nymph can free itself within one or two minutes (Smith and Franklin 1961). *Taylorilygus vosseleri* completes the process within three minutes (Geering 1953), whereas a 45- to 60-minute hatching time typifies *Halticus bractatus* and *Pseudatomoscelis seriata* (Beyer 1921, Reinhard 1926). Johnson (1934) detailed the hatching process of *N. elongata* (as *N. erratica*). In at least one species, *P. seriata*, eclosion occurs mainly at night (Breene et al. 1989b).

In the laboratory, eggs can fail to hatch if adult mirids feed on and severely injure host tissues surrounding the oviposition sites or if saprophytic fungi grow over the sites of oviposition (Geering 1953, Eguagie 1977). Khattat and Stewart (1977) observed that bacterial soft rot infections of celery stalks and other oviposition media that were provided for *L. lineolaris* females affect egg hatch more than do fungal infections of the same substrates.

Nymphal Feeding and Development

FEEDING AND DISPERSAL OF NEONATES

The newly hatched, pale or white nymph moves about the host plant and begins to feed almost immediately (Cotterell 1926, da Silva Barbosa 1959, Smith and Franklin 1961, Smith 1973, Samal and Misra 1977, Haws 1982). Or the nymph remains near the eggshell for several minutes (e.g., Kirkpatrick 1941)—about 5 minutes for *Blepharidopterus angulatus* (Collyer 1952), 10–15 minutes for *Calocoris angustatus* (Hiremath and Viraktamath 1992), and 10–20 minutes for *Lygocoris rugicollis* (Massee 1928, Austin 1931b)—

before searching for feeding sites. Posthatching inactivity can last five minutes to a few hours in the egg predator *Cyrtorhinus lividipennis* (Napompeth 1973). The first instars of *Plagiognathus chrysanthemi* do not feed until three or four hours after hatching (Guppy 1963); those of *Ragwelellus horvathi* feed within a few hours (Smith 1977b). Generally, though, the delicate first instars need to feed soon after hatching to avoid death from desiccation (Geering 1953, Smith 1973), and those of *L. rugicollis* and undoubtedly other species frequently conceal themselves between developing leaf and flower buds (Fryer 1916). Mirids are quite sensitive to changes in humidity (e.g., Kumar 1971a, Nwana and Youdeowei 1976) and appear more sensitive to desiccation than do insects of many other groups (Knight 1941, Madge 1968, Youdeowei 1977; see also Cohen [1982]). To help overcome this problem, they will obtain moisture from dew or raindrops (Kullenberg 1944). Late instars survive better at high temperatures than do the early instars, suggesting that larger nymphs are more resistant to desiccation (Khattat and Stewart 1977). In at least some tropical species, nymphal development cannot be completed unless relative humidity is above 70% (Nwana and Youdeowei 1976).

The first instars of most plant bugs eventually disperse over the host in search of tender tissues. Swaine (1959), however, reported feeding aggregations of the first instars of the African *Helopeltis* species and little dispersion until the third stage, and nymphs of all stages of the yucca plant bug (*Halticotoma valida*) and certain other eccritotarsine mirids aggregate on host foliage (Haviland 1945, Wheeler 1976a, Kalshoven 1981). Nymphs of the phyline *Hoplomachus affiguratus* cluster and feed together on larkspur leaves before moving to reproductive parts when they become available (Jones et al. 1998).

MOLTING

As in nearly all true bugs, nymphs molt five times before the adult stage is attained—that is, five nymphal instars are typical of the family. The report of six nymphal instars in an Indian *Deraeocoris* species, in which the second through fourth instars each last only about one day (Kapadia and Puri 1991), needs verification. A few species show variation in the number of instars, from three to five, four to six, or five to six (Matsumoto and Nishida 1966, Napompeth 1973, Readshaw 1975, Foley and Pyke 1985, Liquido and Nishida 1985c, Sivapragasam and Asma 1985, Geetha et al. 1992). Although four instars is characteristic of certain heteropterans having strongly reduced wings (Dolling 1991), a reduced number of nymphal instars is not necessarily associated with abbreviated wings in mirids. Deviation from the typical five instars in *Cyrtorhinus lividipennis* reportedly depends on food quality (Sivapragasam and Asma 1985). Liquido and Nishida (1985c), however, noted that the inconstancy in the

number of instars in this species is intrinsic to a population because variation occurs in males and females reared under the same environmental and nutritional conditions. They discussed the relationships between number of instars and adult longevity and fecundity, and speculated on the significance of different reproductive strategies among geographic populations. *Cyrtorhinus lividipennis* might be unique among heteropterans in having either a constant or variable number of instars and in exhibiting geographic variation in this number (Štys and Davidová-Vilímová 1989).

Before molting, nymphs of both phytophagous and zoophagous species become sluggish and stop or substantially reduce feeding (Petherbridge and Husain 1918, Austin 1931a, Jeevaratnam and Rajapakse 1981a, Jonsson 1987). The fore region of the midgut is often distended with air bubbles during this process (Goodchild 1952). Ecdysis is sometimes preceded by a throbbing in the region of the pronotal callosities (Petherbridge and Husain 1918). In the case of *Nesiomiris sinuatus*, a propleural stripe characteristic of the adult, but otherwise absent in the last-stage nymph, becomes visible through the overlying nymphal exuviae just before ecdysis (Gagné 1997).

Molting in *Helopeltis schoutedeni* (as *H. bergrothi*) (Lean 1926), *Hyaliodes vitripennis* (Horsburgh 1969), *Lopidea media* (Cory and McConnell 1927), and *Psallus ambiguus* (Morris 1965) occurs mainly on lower leaf surfaces; the process in *Moissonia importunitas* takes place near the leaf margin or depression formed by veins (Gopalan and Basheer 1966). During molting, a nymph extends its legs and grips the host plant with its tarsal claws (Lal 1950). The molting nymph sometimes hangs from the host by only its hind tarsi (Plate 2). Molting in *C. lividipennis* shows a bimodal pattern, its frequency associated with changing light conditions at dawn and dusk (Liquido and Nishida 1985b).

The nymphal cuticle splits along the middorsal line of the thorax during molting. Usually the first to emerge is part of the head, followed by the thorax and abdomen (Fig. 6.22); the antennae, legs, and finally the proboscis are withdrawn. In *Helopeltis antonii*, the abdomen is the last part to be freed from the old cuticle (Jeevaratnam and Rajapakse 1981a). Occasionally the last nymphal skin remains attached to the transparent, wet wings for an hour or more before being shed with help from the hind legs (El-Dessouki et al. 1976). Lintner (1882) provided details of the final molt of *Poecilocapsus lineatus*. Cast skins can be white, gray, or nearly colorless, with dark markings and spots retained on the head, thorax, and legs (Brown 1924, Butler 1924, Lal 1950), or exuviae retain the pigmentation and coloration of the associated instar (Mathen and Kurian 1972).

Under laboratory conditions, considerable mortality can occur during molting (pers. observ.), perhaps because of desiccation or an absence of animal matter in the diet of omnivorous plant bugs. Molting individuals are also more vulnerable to predation (e.g., Austin 1932, Stewart

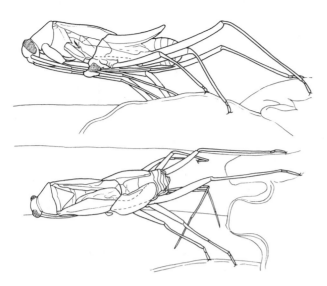

Fig. 6.22. Final molt in *Phytocoris longipennis*. (Reprinted, by permission of B. Kullenberg and Zoologiska institutionen, Uppsala Universitet, from Kullenberg 1944.)

1969a; see also Chapter 14). Mortality during molting is greater on resistant hosts of *Lygus hesperus*, suggesting that antibiotically weakened individuals are unable to complete this energy-consuming process (Alvarado-Rodriguez et al. 1987).

GROWTH RATIOS AND EFFICIENCY

The increase in size occurring with each molt conforms to that of most hemimetabolous insects. The overall mean growth ratios of 1.29 for *Sahlbergella singularis*, 1.31 for *Distantiella theobroma* (Kumar and Ansari 1974), and 1.25 for *Eurystylus oldi* (Ratnadass et al. 1994) are perhaps typical for the family. These figures are similar to the median growth ratio of 1.27 that Cole (1980) calculated for 50 hemimetabolous insects, including the mirid *Plagiognathus chrysanthemi* (based on Guppy's [1963] data). Cole (1980) noted that the growth ratios for holometabolous insects, contrasted with hemimetabolous species, are significantly larger, with a median growth ratio of 1.52 calculated for 55 species. Males of the predatory *Blepharidopterus angulatus* show a lower growth efficiency than do females (Gange and Llewellyn 1989). In the case of *Lygus hesperus*, mean dry weight approximately doubles from the final weight of one nymphal instar to the next (Stewart et al. 1992).

DEVELOPMENTAL TIMES

The first instar of *Megaloceroea recticornis* lasts little more than a day (Gibson 1980). In some other species, the first (Tan 1974a) or second instar is the shortest (Hori and Hanada 1970). Perhaps because of the greater developmental changes and lipid storage before eclosion (Slansky and Panizzi 1987), the fifth instar usually is the

longest by one or two days. Examples include *Blepharidopterus angulatus* (Glen 1973), *Campylomma verbasci* (Smith and Borden 1991), *Coquillettia insignis* (McIver and Stonedahl 1987a), *Creontiades dilutus* (Foley and Pyke 1985, Hori and Miles 1993), *Helopeltis clavifer* (Smith 1973), *Lygocoris pabulinus* (Mols 1990, Blommers et al. 1997), *Lygus elisus* (Shull 1933), *L. hesperus* (Butler and Wardecker 1971), *L. lineolaris* (Ridgway and Gyrisco 1960a), *Moissonia importunitas* (Gopalan and Basheer 1966), *Nesidiocoris tenuis* (El-Dessouki et al. 1976), *Platyngomiriodes apiformis* (Pang 1981), *Pseudatomoscelis seriata* (Eddy 1928), and *Tytthus mundulus* (Verma 1955a). Duration of the fifth instar of *Orthops scutellatus* is about twice that of any of the other four instars (Whitcomb 1953). In *B. angulatus* fed different numbers of lime aphids, Glen (1973) found that instar duration does not increase when food is scarce.

In species having a variable number of instars, the period of nymphal development decreases in the third- and fourth-instar individuals. That period in *Cyrtorhinus lividipennis* is 14.6, 18.6, and 20.2 days for the three-, four-, and five-instar groups, respectively (Napompeth 1973).

Development in Relation to Temperature

The period of nymphal development is generally inversely proportional to temperature (e.g., Betrem 1953, Ting 1963a, El-Dessouki et al. 1976, Chatterjee 1983b, Foley and Pyke 1985, Mols 1990), with developmental rates outside low and high temperature extremes linear in relation to temperature (Fig. 6.23; e.g., Ridgway and Gyrisco 1960a, McNeill 1971, Gaylor and Sterling 1975a, Jeevaratnam and Rajapakse 1981a, Higley et al. 1986, Fauvel et al. 1987, Fleischer and Gaylor 1988). Males, which in many species are smaller, can develop more rapidly than females (e.g., Tanangsnakool 1975, Hiremath 1986, Chua and Mikil 1989), or there can be no statistically significant difference in developmental times of the sexes (Neal et al. 1991, Jones and Snodgrass 1998). Nymphal development of *Helopeltis* populations in Java at an altitude of about 250 meters (25°C) averages 13 days, but 19 days are required at 1,200 meters (19.5°C) (Leefmans 1916). In the laboratory, nymphal development of *H. bradyi* requires 32–34 days at 20°C but only 15 days at 25°C (Betrem 1953). *Boxiopsis madagascariensis* nymphs develop in 23 days at 21°C and 16 days at 25°C (Decazy 1977). The average total developmental times for nymphs of different generations of *Lygus rugulipennis* (as *L. disponsi*) in Japan show the influence of temperature: 23 days in the first generation (May, 19°C), 15 days in the second (July, 24°C), and 10–12 days in the third (August, 28°C) (Hori and Hanada 1970). Whether more rapid development at higher temperatures is associated with reduced survivorship in mirids needs to be determined.

Nymphal development of temperate species typically lasts 12–35 days at temperatures of 20–30°C

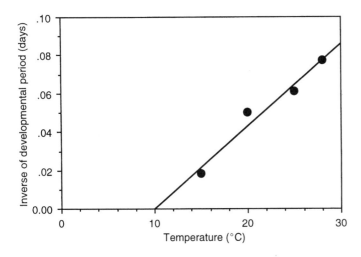

Fig. 6.23. Relationship between developmental time and temperature in *Leptopterna dolabrata*, showing that developmental zero is about 10°C. (Redrawn, by permission of S. McNeill and Blackwell Science Ltd., from McNeill 1971.)

(Ridgway and Gyrisco 1960a, Champlain and Butler 1967, Butler 1970, Butler and Wardecker 1971, McNeill 1971, Gaylor and Sterling 1975a, Blommers et al. 1997, Jones and Snodgrass 1998). At any given temperature, the time required for nymphal development of *Lygus hesperus* is about twice that for eggs (Champlain and Butler 1967, Butler and Henneberry 1976), and this relationship also holds for *Rhinacloa forticornis* and *Spanagonicus albofasciatus* reared at different temperatures (Butler 1970). In small Nearctic species such as *Pseudatomoscelis seriata* and *Halticus bractatus*, nymphs can develop during the warmest part of the season in only 11–12 days (Reinhard 1926, Cagle and Jackson 1947). *Spanagonicus albofasciatus* completes nymphal development in slightly less than 10 days at 30°C in the laboratory (Butler 1970). At 29.4°C, some nymphs of *L. hesperus* develop in 7 days (Stitt 1940). Nymphs of the early-season *Labops hesperius*, whose early instars feed mainly at night, sometimes require five weeks to develop in rangelands of the western United States (Todd and Kamm 1974).

The duration of life stages is difficult to predict under fluctuating temperatures in nature; development under changing temperatures can differ from that under constant temperatures (Higley et al. 1986). Butler and Watson (1974) prepared a computer program for *Lygus hesperus*, in which data derived from laboratory rearing at constant temperatures were used to determine the duration of stages subjected to almost all combinations of fluctuating temperatures. Their model predicts developmental rates more accurately than relying on the average mean temperatures that prevail during each life stage. Taylor (1981) estimated parameter values for development-rate curves of *L. elisus*, *L. hesperus*, and 52 other insect species. A model relating development of *Labops hesperius* to growing degree-hours allows the appearance of the bug's stages in the field to be predicted (Haws and Bohart 1986). Schaub and Baumgärtner (1989) developed a phenology model for *Orthotylus marginalis*, using degree-days above developmental thresholds. Phenological prediction of *Lygocoris pabulinus* in apple orchards also is possible (Mols 1990).

Effects of Diet and Host Plant

As important as the effects of temperature are on nymphal development, the role of an animal diet should not be overlooked. In facultative predators, access to arthropod prey, including eggs, can substantially decrease the time needed for nymphs to develop (Bryan et al. 1976). The development of *Nesidiocoris tenuis* nymphs takes about 14 days when they are provided newly hatched noctuid caterpillars, or about 7 days less than when reared on tobacco alone (Torreno and Magallona 1994; see also Libutan and Bernardo [1995]). In addition, Jonsson (1987) demonstrated that at the same temperature (15°C) the development of *Atractotomus mali* nymphs takes longer on a diet of aphids (28.8 days) than on a psyllid diet (22.6 days). Slightly different developmental times are possible for *Dicyphus tamaninii*, depending on whether thrips or whiteflies are used for rearing (Albajes et al. 1996). For *Dicyphus hesperus*, the development time is shorter (and adult body size larger) when the nymphs are reared on whiteflies rather than on spider mites (McGregor et al. 1999). Similarly, nymphs of another dicyphine, *Macrolophus pygmaeus*, develop move rapidly on whiteflies than on a diet of aphids or mites (Perdikis and Lykouressis 2000).

Nymphal development can be affected not only by different plant species (e.g., Pankanin 1972, Gaylor and Sterling 1976a, Hori and Kuramochi 1984, Hiremath 1986, Perdikis and Lykouressis 2000) but also by host cultivar. Genotypes differing in nymphal development and survival times have been identified in various studies of crop resistance (e.g., Tingey et al. 1975a, 1975b; Alvarado-Rodriquez et al. 1986b). For instance, the growth rate of *Lygus hesperus* on different cultivars of cotton is affected by trichome density. Although the oviposition rate is greater on pilose than on glabrous leaves, the more pubescent cotton phenotypes are sub-

Table 6.5. Mean number of days spent in each nymphal stage of *Pseudatomoscelis seriata* when reared at 26.7°C on different host plants[a]

Host plant	Instar					
	I	II	III	IV	V	Total
Preflowering cutleaf primrose	3.12b	2.64b	3.89d	3.14b	3.00cd	16.00c
Preflowering cotton	3.00b	2.13ab	3.38cd	3.38b	ND	ND
Preflowering spotted beebalm	3.87c	3.57c	3.10c	3.21b	4.17e	15.00c
Preflowering croton	3.36b	2.36b	2.58c	2.10a	3.00bcd	12.25b
Flowering spotted beebalm	2.00a	2.26b	1.95b	2.21a	3.08d	11.55ab
Flowering croton	2.13a	1.91a	1.34a	1.84a	2.41ac	9.63a

Source: Gaylor and Sterling (1976a).

Note: ND, not determined.

[a] Means within columns followed by the same letter are not significantly different at the 0.05 level.

optimal for nymphal development of this plant bug (Benedict et al. 1983). Characteristics such as glandlessness in cotton (see Chapter 10) also influence the growth rates of *L. hesperus* nymphs. In rearing the polyphagous *L. lineolaris* on various plants in the laboratory, Fleischer and Gaylor (1988) determined that the percentage of time spent in different instars is influenced by the host (Table 6.5; see also Gaylor and Sterling [1976a]). Development of *Adelphocoris suturalis* nymphs varies on different parts of alfalfa—leaf, flower bud, flower, and seed pod—with nymphs that feed on flowers developing the fastest. Nymphs that feed on leaves fail to develop beyond the fourth instar (Hori and Kishino 1992). Somewhat similarly, *Helopeltis antonii* nymphs develop optimally when feeding on the tender branches of cashew trees, whereas development is delayed 2–3 days when the diet is young fruits; nymphs fail to develop on a diet of the ripened fruits of cashew (Jeevaratnam and Rajapakse 1981a).

Another factor influencing nymphal development is host water status. Nymphs of *Distantiella theobroma*, for example, take longer to develop on cocoa seedlings maintained on an intermediate-water regime than on seedlings growing under high-water conditions (most nymphs die under a low-water regime). Nymphs presumably imbibe nutrients at a reduced rate on the intermediate-water seedlings (Gibbs and Pickett 1966).

The growth rates of mirid nymphs can be affected by nitrogen availability. *Lygus rugulipennis* nymphs show an increased growth rate on nitrogen-fertilized pine seedlings, possibly because of an increase in concentrations of free amino acids (Holopainen et al. 1995).

Life-History Patterns

Mirid life histories, as those of all other insects, reflect genotypic constraints that are modifiable through an individual's interaction with its environment. Some life-history patterns emerge at a generic or even tribal

level, or when certain food resources are considered (see Chapter 17).

In insects, tolerance to cold temperatures is generally concentrated in a single stage (e.g., Berg et al. 1982). Mirids most often overwinter in the egg stage, especially among phytophagous species associated with woody plants in temperate regions; insects associated with late-successional stages tend to overwinter as eggs (Brown 1990). Early workers, perhaps extrapolating from information on the overwintering habits of other insect groups, incorrectly assumed that the adults of some common mirids, such as the North American *Poecilocapsus lineatus* (Fitch 1870), survive the winter. The assumption that the adults of many common plant bugs overwinter (e.g., Forbes 1900) also might have been based on early work that emphasized the tarnished plant bug (*Lygus lineolaris*), adults of which do overwinter.

Seventy-eight of 92 Swedish mirids studied by Kullenberg (1944) overwinter as eggs; the remaining 14 species overwinter as adults. In the Netherlands, 171 of the approximately 200 known species hibernate in the egg stage, the remainder hibernating as adults, except for one species (*Dicyphus pallicornis*) that overwinters in all stages and another (*Macrolophus pygmaeus*) that overwinters as a fifth instar (Cobben 1968). Nymphs of another dicyphine, *D. tamaninii*, overwinter with adults in warm coastal areas in the Mediterranean basin (Albajes et al. 1996). Late instars of the Nearctic *L. lineolaris* reportedly survive the winter under mullein leaves (Forbes 1884a), which might be possible in some years (Stedman 1899), but this is atypical. Cooley (1900) was incorrect in thinking that immatures of this species normally overwinter under any convenient shelter. The nymphs of an undetermined mirid genus suspected of overwintering within dipteran galls on sagebrush in Wyoming (Fronk et al. 1964) perhaps belong to some other heteropteran family.

In the extreme southern United States, adults of many species are found throughout the winter

(Blatchley 1926, 1934; Sweet 1930). Similarly, there is no actual overwintering period for *Nesidiocoris tenuis* in Egypt (El-Dessouki et al. 1976). The nymphs and adults of *Nesiomiris* species are present throughout the year in Hawaii (Gagné 1997). The nearly continuous reproduction of tropical mirids and their population fluctuations are discussed later in this chapter.

Macrolophus melanotoma exhibits an unusual life history. From November to January in parts of France, the females contain ripe eggs and abundant fat reserves. Oviposition takes place in midwinter, which is an exceptional behavior among heteropterans (Carayon 1986).

OVERWINTERING IN THE ADULT STAGE

Although not the most common overwintering stage, the adult represents the primary overwintering stage for several mirids (e.g., Butler 1923, Kullenberg 1944, Ehanno 1987b), particularly species associated with early-successional communities. Among groups overwintering as adults in temperate climates are members of the economically important mirine genus *Lygus*; other mirine genera such as *Agnocoris*, *Charagochilus*, *Liocoris*, *Orthops*, *Salignus* (Kelton 1955, Cobben 1958, Southwood and Leston 1959, Korcz 1977, Dolling 1991), and *Megacoelum* (Wagner 1967); grass-feeding Stenodemini such as *Notostira* and *Stenodema* (Kullenberg 1944, Woodward 1952, Buczek 1956); several species of the large carnivorous genus *Deraeocoris*; and the bryocorine *Monalocoris filicis* (Kullenberg 1944, Southwood and Leston 1959). Adults of some Dicyphini (Bryocorinae) overwinter (e.g., Knight 1927a, Cobben 1953, Downes 1957, Önder et al. 1983, Dolling 1991), as do those of *Halticus bractatus* (Orthotylinae: Halticini), the garden fleahopper, in South Carolina (Beyer 1921); farther north, however, the latter species overwinters as eggs (Cagle and Jackson 1947). The Palearctic *Chlamydatus evanescens* also overwinters in the adult stage (Kullenberg 1944), an unusual habit among members of the Phylinae. Although Knight (1941) listed *Lygocoris pabulinus* among plant bugs overwintering as adults, eggs of this species overwinter in Europe (Kullenberg 1944) and North America (pers. observ.). Adults of species in several genera, including *Calocoris*, *Creontiades*, *Dionconotus*, *Horistus* (Mirinae), *Nanopsallus* (Phylinae), and *Pachytomella* (Orthotylinae), overwinter in Turkey (Önder et al. 1983).

Overwintering Mortality

Many of the groups in which adults overwinter are multivoltine and suffer a high mortality during winter (Haseman 1918, Painter 1929b, Austin 1931b, Fox-Wilson 1938, Sorenson 1939, Bech 1969, Varis 1972). Warm, dry conditions when lygus bugs move to hibernation quarters favor successful overwintering (Varis 1995). Survival for certain other mirines during winter

Table 6.6. Overwintering of *Lygus rugulipennis* males and females in Finland

Year	Sex	Total no. of bugs	Overwintered bugs (%)
1967–1968	♂	564	2
	♀	540	7
1968–1969	♂	300	7
	♀	360	11

Source: Varis (1972).

is favored by adequate atmospheric humidity (Woodward 1952). Females of *Lygus rugulipennis* often show lower winter mortality than do males (Table 6.6; Varis 1972), although males consistently emerge earlier in spring than do females (Stewart 1969a). The significantly greater numbers of *L. lineolaris* females than males collected early in the season in sticky traps suggest that females of this species also overwinter more successfully than males, or perhaps simply reflect greater flight activity of females in spring (Ridgway and Gyrisco 1960b).

Overwintering Sites and Sex Ratios

Adult plant bugs overwinter in various protected places: among fallen leaves, under leaves of mullein rosettes, in moss, in hollow stems of herbaceous plants, and under bark (e.g., Forbes 1884a, Crosby and Leonard 1914, Haseman 1918, McAtee 1924, Holmquist 1926, Knight and McAtee 1929, Fox-Wilson 1938, Ellis 1940, Froeschner 1949, Tischler 1951, Bech 1969, Varis 1972, Wheeler 1981c, Fye 1982a, Klausnitzer 1988, Wheeler and Stimmel 1988). Mirids also commonly overwinter on pines and other conifers in North America and Europe (Plate 2; Reuter 1909, McAtee 1915, Hofmänner 1925, Kullenberg 1944, Bech 1969, Varis 1972). Females of some species will even oviposit on conifers used for overwintering (Bech 1969). Lygus bugs, which often fly considerable distances to hibernate, more often seek a distant woodland for overwintering than a hedgerow surrounding a crop field (Kullenberg 1944, Stewart 1969a) and are found in sites protected from the prevailing winds (Kullenberg 1944, Varis 1972).

Overwintering populations usually consist of both sexes, with males and females about equally abundant in *Lygus lineolaris* in New York (Crosby and Leonard 1914, Hill 1941), although Smith and Franklin (1961) stated that only the females overwinter. An overwintering female-male ratio of 2 : 1 has been found for *Deraeocoris brevis* in Oregon (Westigard 1973), though in adjacent Washington State only about 60% of nearly 700 adults consisted of females (Horton et al. 1998). Only females of *Notostira erratica* hibernate in Sweden (Kullenberg 1944), although farther south this species overwinters mainly as eggs (Buczek 1956).

Reproductive Diapause in Species Overwintering as Adults

Reproductive patterns differ among mirids that overwinter as adults, particularly in the degree of diapause in ovarian development. Kamm and Ritcher (1972) provided a rapid dissection technique for determining such development in insects, including *Labops hesperius*.

Woodward's (1952) study of two grass-feeding stenodemines of the British fauna, *Notostira elongata* (as *N. erratica*; see Woodroffe [1977]) and *Stenodema laevigata*, illustrates the disparate types of reproductive cycles in mirids. Adults of both sexes of the multivoltine *S. laevigata* hibernate. Females show a complete suppression of reproductive development; no egg rudiments develop before or during hibernation. A large fat body is laid down in the fall, accompanied by reduced water content and increased body weight. Sexual maturation in males is merely slowed, with the testes and accessory glands maturing slowly during fall and winter. Fertilization takes place in May; copulation apparently is triggered by warm conditions in spring. Factors responsible for initiating the reproductive diapause remain undetermined. Woodward's (1952) suggestion that adult diapause is broken by fertilization and exposure to low temperatures requires more experimental evidence in mirids because both cases are uncommon in insects (Tauber et al. 1986).

In contrast, males of *N. elongata* fertilize females in autumn, then die. Females of this bivoltine species do not accumulate large fat reserves. Egg development, which begins in fall, is arrested by low temperatures; reproductive development can resume at any time. Ripe eggs can be formed within 15 days of exposure to 25°C. For *S. laevigata*, a univoltine life cycle ensures that immatures do not develop when conditions are unfavorable, whereas the opportunistic *N. elongata* exhibits what Woodward (1952) termed a more labile life cycle.

Photoperiodic control of diapause appears to allow phenological plasticity in some mirids that overwinter as adults (Dolling 1973). Dolling discussed the ecological consequences of photoperiodic induction and termination of diapause.

Dolling (1973) envisioned that in cool years late-maturing first-generation females of *N. elongata* encounter short days sufficient to trigger diapause, thereby preventing the production of a second generation that might be killed by frost. Two generations would be produced in a warmer year before short days induce diapause. The previous photoperiodic experience of overwintered females does not determine the onset of ovarian development in spring. Rather, females are synchronized so that all begin to oviposit when diapause is terminated. According to Dolling, this not only could allow gene exchange to occur between descendants of different overwintering populations, but also enables females to lay eggs in early spring to take advantage of unseasonably warm periods. Such a "head start" on oviposition would not be possible if ovarian development in overwintered females had to be initiated by a photoperiod equivalent to that experienced by autumn females at the time of their maturation.

Dolling's (1973) work shows that short days induce diapause in *N. elongata*, that photoperiod has a role in determining color phase in this species, and that phase of the parent is involved in determining phase of the reared female. There is, however, no conclusive evidence that photoperiod plays a role in terminating hibernal diapause in *N. elongata*.

Reproductive diapause in North American and European *Lygus* species is photoperiodically induced, with nymphs (especially late instars) responding to short photophase (Fig. 6.24). In southern California, diapausing females of *L. hesperus* (showing atrophied ovaries and well-developed fat bodies) appear in mid-September, most enter diapause by early October, and reproductive individuals reappear by mid-November (Leigh 1966). Reproductive diapause is similarly under photoperiodic induction in the predatory *Deraeocoris brevis*, with light-dark ratios (in hours) between 16:8 and 15:9 being critical (Horton et al. 1998). Long-day conditions, prolonged exposure to low temperatures, or exposure to high temperatures under short-day conditions will terminate diapause in *Lygus* species and other mirines (Boness 1963, Beards and Strong 1966, Wightman 1969a).

OVERWINTERING IN THE EGG STAGE

Univoltinism in an insect family can be associated with diapause in several stages (e.g., Berg et al. 1982). Univoltine mirids of temperate and tropical regions typically overwinter or pass the harshest environmental conditions in the egg stage (Fig. 6.25). According to Dolling (1991), all British orthotylines, many of which are arboreal, overwinter as eggs.

Arboreal, univoltine mirids, like their phytophagous counterparts in other insect groups of northern temperate regions, generally have narrow host ranges (e.g., Southwood 1978a). Plant bugs that specialize on the inflorescences of grasses also tend to be univoltine. Thus, in mirids, the relationship between voltinism and host specificity conforms to that identified in certain other insects: restricted seasonal availability of food resources imposes univoltinism, and some mechanism of egg dormancy conveying such a phenological pattern has evolved in oligophagous or monophagous species (Branson and Krysan 1981, Saulich and Musolin 1996).

Diapause in most mirids occurs at the early germ-band stage of embryogenesis—that is, before protocormic buds are formed (Cobben 1968). Diapausing embryos of *Labops hesperius*, though, show greater advancement, with appendages and eyes developed beyond the germ-band stage (Fuxa and Kamm 1976a).

Because most mirid eggs are embedded in plant tissues and maintain intimate contact with the host plant, egg dormancy and synchronized postembryonic

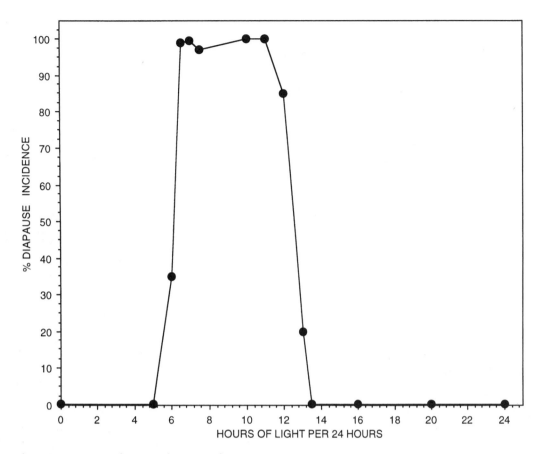

Fig. 6.24. Photoresponse curve for *Lygus hesperus*, showing the relationship between photophase and diapause incidence among 44-day-old females reared in the laboratory at 27°C. (Redrawn, by permission of G. W. Beards and the University of California Agricultural Experiment Station, from Beards and Strong 1966.)

MIRID SPECIES		JAN	FEB	MAR	APR	MAY	JUN	JUL	AUG	SEP	OCT	NOV	DEC
Polymerus tinctipes	**Egg**												
	Nymph												
	Adult												
Polymerus wheeleri	**Egg**												
	Nymph												
	Adult												

Fig. 6.25. Generalized univoltine life cycles of *Polymerus tinctipes* and *P. wheeleri* on moss phlox in mid-Appalachian shale barrens. As in most mirids of temperate regions, overwintering occurs as eggs. (Reprinted, by permission of the Entomological Society of Washington, from Wheeler 1995.)

development allow hatching to coincide with host phenology. In many phytophagous mirids, synchronization with budbreak can be achieved, in contrast to aphids that oviposit into bark crevices and cannot synchronize development with host phenology because their eggs are not able to monitor the hormonal changes in buds (Dixon 1976).

An egg diapause, in mirids often lasting 9–10 months or slightly longer (see Fig. 6.25), allows egg hatch in spring to be synchronized with plant development

(Braune 1976). This phenological pattern is an adaptation to a short period of feeding on resources of limited seasonal availability, such as the flush of growth in spring and buds of woody plants, particularly those producing male flowers (Cobben 1968; see also Chapter 10); high nutrient levels are often associated with budbreak (e.g., Sutton 1984).

Zeh et al. (1989) proposed that attributes of the egg stage are important in diversification of the Insecta. Endophytic oviposition undoubtedly contributed to the evolutionary success of mirids and is particularly important because it facilitates water uptake by eggs, enabling early instars of many species to exploit inflorescences and flushes of new growth at times of peak nitrogen availability (Slansky 1974, McNeill and Southwood 1978). Speciation in arboreal mirids, as in membracids (Wood and Keese 1990, Wood et al. 1990) and probably other phytophagous groups that insert eggs into plant tissues, might involve shifts to novel hosts having different phenologies. This could lead to different host-mediated life histories and reproductive isolation.

Period of Egg Hatch

Plant bugs that develop on woody hosts in temperate regions are often present in their active stages only for a small portion of the growing season. The eggs of *Harpocera thoracica*, an inflorescence feeder on British oaks, hatch in early spring. This species spends nearly 11 months of the year in the egg stage (Southwood and Leston 1959). *Phoenicocoris claricornis* and *Pinophylus carneolus* adults can be collected in Pennsylvania only during a two- to three-week period in May (Wheeler 1999). Adults of the univoltine, oak-associated *Saundersiella moerens* are present only for about 10–12 days in early spring (Josifov 1978). The typical life cycle of mirids that develop on trees and shrubs, even though they are not phloem specialists, is similar to that of many treehoppers (Membracidae) developing on similar woody hosts (Ball 1920). Egg hatch in these membracids occurs relatively early in the season, and feeding takes place during the spring flush of growth on trees such as oaks, when water and nitrogen levels in phloem sap are highest (Keese and Wood 1991).

On a given host at the same locality, the associated mirid species tend to hatch sequentially. Examples of mirids showing sequential phenological patterns include those occurring on honeylocust (Wheeler and Henry 1976) and juniper (Wheeler and Henry 1977) in the eastern United States and on species of ceanothus (Pinto 1982), adenostoma (Pinto and Velten 1986), and juniper (Polhemus 1984) in the western states. Eggs of many chiefly predacious mirids of temperate regions tend to hatch slightly later than those of phytophagous mirids on the same hosts (Jonsson 1985, Schuh and Slater 1995; pers. observ.). Among predatory mirids, an often less synchronous development compared to that of phytophagous species, and the use of different micro-

habitats by different instars and a longer life span, suggest a strategy for maximizing use of the food supply (Jonsson 1985).

Overwintering eggs of most mirids of temperate regions hatch in spring, but several phytophagous plant bugs (*Halticus* and *Pantilius* spp.) in the former USSR are known to hatch in mid to late summer (Putshkov 1966). In some extreme southern regions of the United States—for instance the chaparral community of southern California (Pinto 1982, Pinto and Velten 1986)—the summer and autumn are spent in the egg stage, with nymphs and adults occurring during the period of lowest temperatures and highest rainfall. Eggs of *Horistus infuscatus* under the harsh conditions of Israel's Negev desert region pass the summer in diapause, hatching in the cool, rainy winters (Ayal and Izhaki 1993). Studies of life-history adaptations among the diverse mirid fauna of deserts might show that certain species even undergo a multiannual egg diapause.

Initiation of egg hatch within mirid populations at a given locality can vary by about 10 days, depending on the local temperatures (Sanford 1964b), but is keyed to the growth stage of the woody host. In fact, the beginning of egg hatch can be so attuned to plant development that even in extremely early or late seasons, plant bug activity coincides with host phenology (Petherbridge and Husain 1918, Frost 1922, Wheeler and Henry 1976; see also Schaub and Baumgärtner [1989]). Over the entire range of a mirid species in temperate regions, seasonal history is asynchronous because of varying elevations and latitudes. Further, egg hatch on any one tree may not be synchronized with that on nearby conspecifics; in some aphids, hatching coincides with the average time of host budbreak in the area (Dixon 1987b). As a plant-defense strategy (other explanations for delays in budbreak are discussed by Quiring and McKinnon [1999]), budbreak on oaks and other trees typically occurs over a 10-day period so that a herbivore cannot be synchronized precisely with each tree in the population (McNeill and Southwood 1978, Faeth and Rooney 1993). On English oak, Crawley and Akhteruzzaman (1988) observed a 25-day variation in the date of budbreak between the earliest and latest individuals of a population, with the same trees leafing out first from year to year.

Eggs of the mainly predacious *Orthotylus marginalis* hatch over a relatively short period (Schaub and Baumgärtner 1989), but in populations of some other mirid species, the overwintering eggs do not complete diapause at the same rate (Judd and McBrien 1994). Eggs of *Lygocoris rugicollis* on apple trees can hatch over a three- to four-week period in England (Austin 1931a), with hatching more likely to be prolonged over several weeks if cold weather intervenes soon after hatch begins (Petherbridge and Kent 1926). A similar weather-delayed period in hatching was observed for *Lopidea robiniae* (Leonard 1916b). In the case of *Campylomma verbasci* in British Columbia, the beginning of egg hatch generally coincides with the blooming of apple or pear trees. In some years, 50% of the overwintered eggs of *C. verbasci* hatch within 3–4 days, but an additional week may be needed for 100% to hatch. As in *L. rugicollis*, an intervening cool period prolongs hatching over a several-week period (Fig. 6.26; Thistlewood and Smith 1996). The proportion of eggs hatched relative to tree phenology varies not only between years but also between orchards of similar growing conditions. Hatching of *C. verbasci* eggs laid in the woody tissues (mainly fruit spurs) of apple trees occurs at about the same rate from all compass quadrants. This has important implications for the validity of random sampling and the monitoring of this bug's populations in pest-management programs (Smith 1991, Thistlewood and Smith 1996).

Eggs of the predatory *Blepharidopterus angulatus* hatch from late May to July or August in southern England (Collyer 1952), the process inexplicably beginning about a month later in sprayed orchards (Collyer and Massee 1958). A more restricted hatching period is characteristic of this species in Scotland, probably because of the shorter summer (Glen and Barlow 1980). Eggs of the predacious *Tytthus pygmaeus* hatch over such an extended period that new adults overlap with first instars (Rothschild 1963), and overwintered eggs of *Pseudatomoscelis seriata*, a multivoltine polyphage on many nonwoody hosts, can continue to hatch for two to three months (Reinhard 1928). At least for *Adelphocoris lineolatus*, the sporadic hatching of eggs well after the main period of hatch is ascribed to unfavorable conditions, such as placement of the eggs in alfalfa stems at the bottom of debris piles (Craig 1963).

Induction of Egg Dormancy

The mechanisms inducing egg dormancy are known for relatively few mirid species. For *Adelphocoris lineola-*

tus, a reduction of photoperiod results in egg diapause that is apparently induced by neuroendocrine activity in the females of univoltine Saskatchewan populations (Ewen 1966). Ewen (1966) did not observe similar photoperiodic effects in bivoltine Minnesota females. Egg diapause in *Trigonotylus caelestialium* is induced in Japan by short days, the critical photoperiod of 14 hours occurring in August; 10 days' exposure to a short photoperiod is sufficient to induce diapause. Diapause, however, varies seasonally within Japanese populations of this plant bug. At latitude 39.7°N, a few diapausing eggs appear as early as mid-June (rather than in August at 43.7°N). Deterioration of host grasses or genetic variation among populations might cause the differences in seasonal occurrence of egg diapause (Kudô and Kurihara 1988). The proportion of diapause eggs decreases with increasing age of *T. caelestialium* females transferred from a long to a short photoperiod, which suggests that the sensitivity to a short photoperiod is lost or diminishes with age (Kudô and Kurihara 1989).

Termination of Egg Diapause

The relationship between water absorption and egg diapause has been discussed for many insects, particularly those with chewing mouthparts (e.g., Ando 1972) and for aquatic and semiaquatic heteropterans (e.g., Mori 1986). In contrast, the importance of water uptake by host plants in spring has been appreciated only recently for hemipterans that insert their eggs in woody plant tissues. In membracids of the *Enchenopa binotata* complex, low water levels in host branches during fall, and perhaps prolonged cold periods, can dehydrate the eggs. Eggs are hydrated in spring with the ascent of sap in host trees; the water uptake rather than auxins or nutrients in sap is responsible for terminating diapause. This adaptation promotes synchronous egg hatch and a uniform age structure that facilitates synchronous dispersal, mating, and oviposition (Wood 1987, Wood et al. 1990).

Experimental work comparable to that on membracids has not been conducted on mirids, but the water relations of host plants are likely involved in synchronizing the egg hatch of most phytophagous, univoltine species that develop on woody plants. A role of water uptake in woody hosts is suggested by the egg hatch of *Lygocoris pabulinus* varying from 201 to 325 degree-days (above 4°C, an approximate threshold for development of winter eggs) over a 15-year period (Blommers et al. 1997). A host-mediated termination of egg diapause would be an adaptation for helping to circumvent the unpredictable budbreak of host trees.

Water uptake

Several workers, in fact, have alluded to the significance of water in the hatching of mirid eggs. In addition to Johnson's (1934) discussion of *Notostira elongata*, Crosby (1915) stated that eggs of the red bugs *Hetero-*

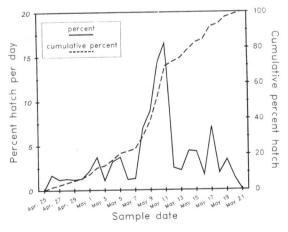

Fig. 6.26. Hatching of *Campylomma verbasci* eggs in an apple orchard in British Columbia. A sharp peak in egg hatch occurs in early May and hatching extends over several weeks. (Reprinted, by permission of H. M. A. Thistlewood and the Entomological Society of America, from Thistlewood and Smith 1996.)

cordylus malinus and *Lygidea mendax* will hatch if apple twigs are brought indoors after March 1 in New York and placed in water to force out the buds. The hatching of *Nesidiocoris volucer* (*sensu* Kerzhner and Josifov 1999) eggs requires moisture (Roberts 1930), water uptake is needed for the eggs of *Taylorilygus vosseleri* to develop (Geering 1953), and the eggs of *Tropidosteptes illitus* swell by absorbing water during the winter following a relatively undeveloped state in summer and fall (Usinger 1945). Alternate drying and wetting apparently trigger egg hatch in *Pseudatomoscelis seriata* (Breene et al. 1989b). The eggs of *Macrolophus melanotoma* develop normally when they are embedded deeply enough in plant tissues to allow water exchange to take place between plant and egg (Constant et al. 1994, 1996b). Reinhard (1928), Puttarudriah (1952), and Schuster et al. (1969) noted correlations between mirid egg hatch and rainfall. Others, such as Austin (1931a) and Collyer (1952), observed that eggs exposed on stem surfaces or partially embedded in stems can fail to hatch; in *Malacocoris chlorizans*, which often deposits its eggs in a more exposed position (Geier and Baggiolini 1952, Collyer 1953b), hatching presumably occurs normally. When twigs harboring *Deraeocoris ruber* eggs are placed in water, leaves appear and hatching occurs, but eggs do not hatch in twigs that are kept dry (Hartley 1965). Egg development in an *Irbisia* species began when dead stems filled with eggs were placed in water and then incubated (Schwartz 1984). It is not known if *Irbisia* eggs will hatch if they are incubated but not placed in water.

Water may or may not play a critical role in species whose eggs are inserted in dead wood, fenceposts, or hop poles (Massee and Steer 1928, Steer 1929). Vanderzant's (1967) observation that mirid eggs inserted in a dry cork fail to hatch suggests that water does play a role, though in nature the species observed (*Lygus hesperus*) oviposits in live plant tissues. As pointed out earlier in this chapter, mirid eggs inserted in host plants sometimes drop to the litter layer when host material decays. Whether such eggs, and those inserted in plant parts that become detached from the host due to wind or other factors, remain viable is not known. If viability is unaffected, what factors trigger egg hatch?

Because membracid eggs in branches that are kept on moist toweling do not hatch, Wood (1987) suggested that egg development depends on water imbibed from the plant. In *Lygus* species, however, eggs maintained in moistened filter paper rolls do hatch (Vanderzant 1967), suggesting that at least in some multivoltine weed feeders, eggs do not always need to imbibe water from the host. In membracids, host-water relations rather than temperature or photoperiod seem critical in mediating egg hatch (Wood 1987, Wood et al. 1990). The role of temperature and photoperiod on egg hatch has been studied in several mirid species (e.g., Fuxa and Kamm 1976a), but the possible interaction of these factors with water uptake requires clarification.

Temperature and photoperiod

Several workers have studied the effects of exposure to low temperatures on dormancy in mirids, but the direct effect of cold on diapause termination per se generally has not been investigated. Tauber et al. (1986) discussed this deficiency in studies of insect dormancy.

Low temperature enhances the hatching of *Psallus ambiguus* eggs (Morris 1965). Eggs of *Blepharidopterus angulatus* hatch after exposure to 4.4 or 7.2°C for 14–16 weeks (Muir 1966a). The hatching of *Lygocoris pabulinus* eggs occurs after they are placed at 3–4°C for six weeks (Wightman 1969a). Eggs of *Labops hesperius* collected in the fall will hatch after about a week's exposure to room temperature when they are held for a month at approximately 4.5°C (Haws et al. 1973). Fuxa and Kamm (1976a) studied the effects of temperature and photoperiod on egg diapause of this early-season species and noted that egg hatch appears to be regulated by a combination of increasing temperature and day length in early spring.

The course of egg development in *Leptopterna dolabrata*, a grass feeder that overwinters in the egg stage, is not influenced by photoperiod (Braune 1971). Dormancy during embryogenesis begins in nature during late July. The four phases delimited in *L. dolabrata*—prediapause, mesodiapause, metadiapause, and postdiapause—react differently to temperature treatment. Braune's (1971) paper gives more detailed information on the effects of temperature on egg diapause and development in this species. Braune (1976) discussed the rates of oxygen consumption during morphogenesis and diapause in *L. dolabrata*, noting low levels of oxygen uptake during diapause and increased levels during prediapause and postdiapause development. Diapause termination is related to exposure to low temperatures, which accelerates the rate of oxygen uptake.

Embryonic development of *L. dolabrata* is arrested by a partial anaerobiosis inside the egg and is maintained by the serosal cuticle, which acts as a barrier to oxygen. This barrier exists until sufficient exposure to low temperatures breaks diapause by restoring permeability of the egg membranes (Braune 1980). Lord (1971) subjected the eggs of *Atractotomus mali* to different photoperiods and also concluded that light is unimportant in diapause termination. Morphological development of *Pseudatomoscelis seriata* eggs in an intense diapause resumes after eggs are subjected to temperatures below 10°C (Gaylor and Sterling 1977).

VOLTINISM AND SEASONALITY

Univoltinism, as previously noted in this discussion of life-history patterns, characterizes most tree- and shrub-feeding mirids of temperate regions. An exception is the mirine *Dichrooscytus repletus*, whose populations on juniper in the eastern United States apparently are polymorphic with respect to egg diapause. A small second generation sometimes is present in late summer or early fall on plants that harbored much larger numbers in spring. Egg hatch on some plants continues until October (Wheeler and Henry 1977). Studies are needed to determine the nature of this intrapopulation variation and to confirm a limited bivoltinism in this plant bug. If such a pattern occurs consistently, what selective forces led to this type of voltinism? Particularly valuable would be comparative studies on the termination of egg diapause in *D. repletus* and the sympatric mirine *Bolteria luteifrons*, a univoltine specialist on juniper and other cupressaceous plants.

Number of Annual Generations

Females of certain species with multiple, overlapping generations have a long postreproductive life (Southwood 1956c); they can overlap adults of the succeeding generation (Kelton 1975), which makes the number of annual generations difficult to determine. *Lygus* species in the western United States might complete six or seven generations (e.g., McGregor 1927, Faulkner 1952). Salt (1945), however, discussed the number of generations that *L. elisus* and *L. hesperus* (perhaps actually *L. keltoni*) produce in Alberta, cautioning that the calculation of their generations from length of the growing season and length of a single generation, as well as pooling different plant bug stages, leads to inaccuracies. Previously, Ball (1920) emphasized the fallacy of such calculations in relation to homopteran life cycles. Even Salt's (1945) data on the number of annual generations of mirids are inaccurate, for they are based on mixed populations of several *Lygus* species (Kelton 1975). Careful, season-long sampling and observations, including determination of preoviposition periods and dissections of females to determine reproductive status, are needed to establish the actual number of generations that multivoltine mirids produce (e.g., Boness 1963, Stewart 1969a). A model study of voltinism in *L. lineolaris* involved not only fluctuations in adult and nymphal densities in Manitoba, but also determination of the reproductive status of adult females throughout the growing season. Females were dissected and their eggs classified as previtellogenic, vitellogenic, or chorionated (Gerber and Wise 1995).

Multiple, often overlapping generations are produced by several polyphagous species associated with field crops and herbaceous weeds, some oligophagous species that feed on legumes and grasses, certain largely predacious mirids, and a few other plant bug groups. Multivoltinism, widespread among insects, characterizes those plant bugs associated with temporary habitats (Southwood 1977, Brown 1982a), allowing high potential rates of population increase (r) and enabling reproduction to be delayed to compensate for a cold, wet spring (e.g., Berg et al. 1982). As in many butterflies (e.g., Shapiro 1975) and other insects, some multivoltine mirids are r-strategists that show high vagility and are associated with disturbed habitats. The tarnished plant bug (*L. lineolaris*), a wide-ranging, early-successional colonizer of numerous weedy herbs, combines attributes of both r—and K-selected species (see Chapter 12).

Fertilization of salt marsh grasses, allowing the plants to remain green longer into the season, might trigger the hatching of *Trigonotylus* eggs that would normally overwinter. This apparently leads to the production of late-season, partial generations of *Trigonotylus* species in high-nitrogen grass plots (Vince et al. 1981). In certain multivoltine species, an additional, partial generation might also be produced in response to an unusually warm summer (e.g., Blommers et al. 1997).

As in many other temperate-zone insects (e.g., Wolda 1988, Denno 1994a), the number of generations in multivoltine mirids tends to increase with decreasing latitude. The multivoltine, polyphagous *Halticus bractatus* produces five or six generations in South Carolina (Beyer 1921) and five in Virginia (Cagle and Jackson 1947). Some mirids are bivoltine in more southern parts of their range but univoltine farther north. For example, *Chlamydatus pullus* is bivoltine in Europe but univoltine in Greenland (Böcher 1971), and *L. rugulipennis* is bivoltine in England and most of continental Europe but is univoltine in Finland, Scotland, and Sweden (Kullenberg 1942, Stewart 1969a, Varis 1972). *Lygocoris pabulinus*, generally bivoltine in temperate Europe, has only a single generation in Sweden (Kullenberg 1944, Blommers et al. 1997). The number of generations of several pest species varies from one in northern regions of the former USSR to two to five in the southern end of their range (Putshkov 1966, 1975). Saskatchewan *Lygus* species are bivoltine on alfalfa at a latitude of 50°N, but at 53°N populations are univoltine, or at most a small second generation is produced; many late instars of this generation are killed by cold weather before attaining the adult stage (Craig 1983). *Adelphocoris lineolatus*, which overwinters in the egg stage, has two generations in southern Minnesota but only one in northern Saskatchewan. A limited bivoltine strain of this species persists in northern Saskatchewan (Craig 1963). In addition, certain plant bugs might produce several generations at low altitudes but only one at higher elevations (Roshko 1976).

Nesidiocoris tenuis produces two winter and six summer and autumn generations in Egypt (El-Dessouki et al. 1976). In cocoa mirids, eight annual generations can be expected in the Old World tropics (Johnson 1962), and *Calocoris angustatus* produced 16 generations during 12 months' continuous rearing under field conditions in southern India (Hiremath and Viraktamath 1992). Ten generations of *Tytthus mundulus* (Williams 1931), an egg predator used in the successful biological control of homopteran pests, are thought possible (see Chapter 14).

Seasonal Variation in Tropical Species

Reproduction in some tropical mirids is initiated with the first rains, and the bugs are less numerous in succeeding generations if the rainy season is short (Delattre 1947). Development of *Helopeltis theivora* on tea plants in southeast Asia is most rapid during the summer monsoon, the bugs becoming rare with the onset of the dry season (Hanson 1963). East African *Helopeltis* populations on tea and guava are also smaller during dry weather (Smee 1928b, Puttarudriah 1952). The time needed for cocoa mirids to complete a generation often lengthens or the bugs are nearly absent during the dry season (e.g., Williams 1954, Taylor 1955, Bahana 1976, Collingwood 1977a). Tropical insects, though, often show complex patterns of seasonal abundance in relation to rainfall and food availability (e.g., Wolda 1978). Although population crashes in tropical plant bugs tend to coincide with the dry season, an extended dry season favors population outbreaks of *H. clavifer* and *Ragwelellus horvathi* in New Guinea (Smith 1972, 1977b). Leston and Gibbs (1971) emphasized the complexity of population changes in cocoa mirids (see also Bruneau de Miré [1977]), cautioning that fluctuations in their numbers should be considered not merely in relation to wet versus dry seasons, but relative to six seasons: dry, sunny; first wet, sunny; first wet, dull; dry, dull; second wet, dull; and second wet, sunny. The correlation between low mirid numbers and the dry season might be related to nutritional stress in host plants and the rapid reduction in humidity. Water stress might adversely affect feeding behavior, increasing nymphal mortality (with water stress above −10 atm) and possibly reducing fecundity and even egg viability (Gibbs and Pickett 1966, Cross 1971, Kumar and Ansari 1974).

Crowe (1977) said that no evidence of diapause or other resting stage is available for *Helopeltis* species in the tropics. These bugs breed almost continuously but often occur at low densities during cool periods and resume population buildup when the host trees bloom or produce new growth. They can also respond to localized climatic events, declining in numbers during periods of heavy rain, high winds, or low relative humidity (Stonedahl 1991). The erratic and dramatic nature of *Helopeltis* attacks—at times, their populations seemingly vanish (Ramachandra Rao 1915)—apparently results from ovarial development occurring only under favorable conditions, leading to abrupt interruption and resumption of oviposition (Roepke 1916, Giesberger [1983]). As Tauber and Tauber (1981) discussed, detection of diapause in species of tropical regions is more difficult than in temperate species, but dormancy (diapause and quiescence) is common in tropical insects (Denlinger 1986). Cessation of ovarial development when conditions are unfavorable or an egg diapause might, therefore, help explain some of the foregoing observations on reproduction and population fluctuations in *Helopeltis* species (and other tropical bryocorines) in relation to the dry and monsoon seasons. Cocoa, coffee, and other important tropical crops on which mirids develop show seasonal cycles in growth (Greenwood and Posnette 1950, Bigger 1993).

Seasonal variation also seems to characterize many tropical mirids of economic importance. Because most species have not been studied carefully over a several-year period and their population differences between years have not been analyzed, the mechanisms and environmental cues responsible for observed patterns in seasonality remain largely unknown.

Population Density, Dispersion, and Spatial Analysis

Examples are given here of various measurements that have been used to estimate mirid population densities, as well as indices of aggregation that have been used to evaluate their numeric (frequency) distribution. Information on the distribution of a species in a habitat and its dispersion is needed to establish effective sampling procedures for basic research and pest management. Dispersion patterns reflect both the biological characteristics of a species and the environmental effects of the habitat (e.g., Sevacherian and Stern 1972a).

Insect spatial relationships usually are inferred from methods using frequency distribution or mean-variance associations. Indices of aggregation are useful for evaluating sampling methods, but they have limited value in assessing spatial distribution. Geostatistics, a direct measure and analysis of spatial dependence, is being used to analyze mirid spatial distributions (Schotzko and O'Keeffe 1989a).

A comprehensive treatment of the considerable, and sometimes controversial, literature on insect population sampling and spatial patterns is not provided here. Readers interested in additional or background information should refer to works by Southwood (1978b), Kogan and Herzog (1980), Taylor (1984), and Kuno (1991).

POPULATION DENSITY

An extended discussion of mirid population dynamics will not be attempted here. Plant bug abundance obviously changes from year to year (e.g., Varis 1995), but various factors involved in such fluctuations are not reviewed, except for some abiotic and biotic mortality factors, which are discussed later in this chapter. Treated earlier in this chapter in a discussion of voltinism and seasonality were annual fluctuations in populations of some tropical mirids in relation to changes in weather and food quality.

Examples of High Densities

Some mirids are low-density pests, but others develop quite large populations (Table 6.7). Many of the species mentioned as crop pests in Chapters 9, 10, and 12 attain outbreak densities, perhaps in part because insecticide applications destroy their natural enemies, alter host-plant nutrition, increase host attractiveness, or change

Table 6.7. Examples of mirids reported to develop large populations

Mirid	Density	Host plant	Locality	Reference
Bryocorinae				
Halticotoma valida	700/plant	Yucca	Maryland, USA	Haviland 1945
Mirinae				
Closterotomus norwegicus	10+/sweep	Alfalfa	New Zealand	Wightman and Macfarlane 1982
Leptopterna dolabrata	15–20/sweep	Bluegrass	Kentucky, USA	Jewett and Townsend 1947, Jewett et al. 1954
Lygus elisus, L. hesperus	120,000/hectare	Alfalfa	Utah, USA	Sorenson 1939
	22/sweep	Alfalfa	Arizona, USA	Stitt 1949
Lygus lineolaris	500–600/sweep	Strawberry	Missouri, USA	Enns 1947
	17/sweep	Alfalfa	Kansas, USA	Smith and Franklin 1961
Taylorilygus vosseleri	1,100/head	Sorghum	Tropical Africa	Ingram 1970b
Trigonotylus caelestialium	15/sweep	Small grains	Russia; Pennsylvania, USA	Mikhaĭlova 1980, Wheeler and Henry 1985
Orthotylinae				
Blepharidopterus chlorionis	Several 100/terminal branch	Honeylocust	Pennsylvania, USA	Wheeler and Henry 1976
Halticus bractatus	100–200/sweep	Red clover	Maryland, USA	Anonymous 1970
Labops hesperius	1,000/clump; >10,000/m²	Range grasses	Western USA	Haws et al. 1973, Haws and Bohart 1986
Orthotylus ramus	66.5 million eggs/hectare	Pecan	Texas, USA	Chung et al. 1991
Phylinae				
Amblytylus nasutus	15–20/sweep	Bluegrass	Kentucky, USA	Jewett and Townsend 1947
Hoplomachus affiguratus	10,000/plant	Tall larkspur	Utah, USA	Ralphs et al. 1997
Pseudatomoscelis seriata	10/sweep	Cotton	Texas, USA	Fletcher 1940b
	680,000 eggs hatching/hectare/1 week	Croton	Texas, USA	Reinhard 1928

behavior of the pest mirids. Some of the examples of high densities involve adventive plant bugs that become unusually abundant in the absence of their normal complement of natural enemies. As in other insect groups (e.g., Price 1991), a series of favorable seasons, particularly those with warmer-than-normal summers, might allow mirid numbers to attain outbreak levels. In the case of mirids associated with hops in the Czech Republic, outbreaks have been correlated with years of above-average temperatures from October to May coupled with below-average precipitation (Šedivý and Fric 1999). Also involved in large plant bug populations might be the introduction of high-yielding crop varieties, application of nitrogen fertilizers, use of irrigation, and various other factors that can affect their population dynamics. Maximum population levels in some pest species are positively correlated with stable temperatures, for example, *Nesidiocoris caesar* in India (Chatterjee 1983b), or with periods of high humidity, as for *Calocoris angustatus* in India (Leuschner et al. 1985). An outbreak of *Helopeltis clavifer* on tea in Papua New Guinea was attributed to excessive rainfall, fertilization, and other factors promoting greater-than-normal flush growth on host trees (Smith et al. 1985).

Factors Affecting Population Estimates

The responses of plant bugs to various abiotic and biotic factors should be considered in assessing their population densities on herbaceous and woody hosts. Population estimates that are obtained at different times during a 24-hour period, owing to flight periodicities (e.g., Bodnaruk 1992), and under different weather conditions can vary substantially. The following examples of mirid responses to weather suggest the importance of considering the bugs' behavior in any attempts to estimate their densities.

In nature, the delicate and presumably more vulnerable early instars (Gibbs and Pickett 1966) of many plant bugs—for example, *Campylomma verbasci*, *Heterocordylus malinus*, *Lygidea mendax*, and *Lygocoris communis*—tend to live under bud scales or within blossoms, flower clusters, and curled leaves (sometimes curled from their own feeding [Gossard 1918]) rather than on exposed plant parts, which is typical of later instars (Crosby 1915, Dustan 1924, Jonsson 1985). Mirids that are pests of apple can also be found in leaves rolled by lepidopteran larvae and in tangled plant parts used previously by noctuid and tortricid larvae (Crawford 1916, Fryer and Petherbridge 1917, Thistlewood and McMullen 1989). This behavior should be considered when beating or limb tapping are used to assess the densities of economically important arboreal species; early instars often are underrepresented in such samples. Moreover, many mirid species are sedentary under cloudy or windy conditions (Dempster 1960). Sheltered areas on host plants are sought during cold, wet periods (e.g., E. H. Smith 1940, Terauds 1971, Gaylor and Sterling 1975b), and species frequenting low-growing or prostrate herbs—for instance *Chlamydatus* species—will seek crevices or hide under stones during cloudy conditions (Woodroffe 1955a). During adverse weather, nymphs of *Polymerus cognatus* can be found in cracks in the ground or under lumps of soil (Dekhtiarev 1927). *Closterotomus fulvomaculatus* hides under loose bark on grape vines during the heat of the day (Tominić 1951), a behavior that might help prevent desiccation.

Many tropical mirids avoid direct sunlight and can be found on trees having a well-developed canopy (Cotterell 1926, Leach and Smee 1933, Lal 1950, Puttarudriah 1952, Tan 1974a, Jeevaratnam and Rajapaske 1981a). These bugs feed mainly during early morning, evening, or night (e.g., Smee 1928b, Cross and King 1973). *Lygocoris viridanus* remains within the ground cover or in protected areas on Sri Lankan tea bushes, emerging to feed during dull weather (Calnaido 1959). *Eurystylus* species associated with castor in South Africa are found on host flowers in early morning but on shady portions of their hosts during the heat of the day (Boyes 1964).

That different mirid stages do not always live in the same part of their hosts can also affect assessments of plant bug populations. Nymphs of economically important mirids often occupy the same parts of their hosts as adults, but younger nymphs of *Helopeltis* species feed more in the center or lower parts of tea plants (Smee 1928b, Lever 1949), and older nymphs are found more often on the green shoots of tea than are adults or younger nymphs (Leach and Smee 1933). Early instars of *Lygus rugulipennis* are more difficult to collect than late instars and adults because they frequent the more humid lower layers of weedy vegetation (Stewart 1969a). Snodgrass (1998) discussed the differences between within-plant distributions of *L. lineolaris* nymphs versus adults and the potential effects such differences can have in assessing populations of this pest in cotton. *Halticus bractatus* nymphs of both sexes generally feed lower on crops such as alfalfa than do the adult females (Day 1991).

DISPERSION

Microclimate is sometimes important in determining mirid distributions within fields. *Lygus rugulipennis* is often more abundant in low-lying, sheltered areas of oat fields in Scotland than in exposed sites on high ground (Stewart 1969a). In Finland, this bug was most numerous in spring wheat in warm, sunny areas such as southern exposures or areas of relief in fields (Varis 1974).

Populations of the egg predator *Cyrtorhinus fulvus* are more numerous along the periphery of taro patches. In this case, the bug's spatial distribution correlates with that of its planthopper prey (Matsumoto and Nishida 1966).

Spatial Distribution

Nymphs of most mirid species are more aggregated than adults, owing to the clustering of eggs and greater adult mobility (e.g., Sevacherian and Stern 1972a, Pieters and Sterling 1974). An aggregated distribution of nymphs might also be related to the availability of plant nitrogen, prey, or shelter (e.g., Thistlewood and Smith 1996). Feeding aggregations like those of *Halticotoma* species on yucca foliage (Plate 2; Haviland 1945; pers. observ.), *Mertila malayensis* on orchids (Capco 1941), and *Dicyphus pallicornis* on digitalis (Cobben 1978) are not as common in mirids (e.g., Hansen et al. 1985b) as in many tingids, some Coreoidea, and homopterans such as aphids. Distribution and spatial dispersion, however, have been studied in relatively few mirids. Studies are needed to evaluate the potential benefits of gregariousness on food consumption, developmental rate, and weight gain.

In celery fields of eastern Canada, *Lygus lineolaris* populations are contagiously distributed, and nymphs are more aggregated than adults (Boivin et al. 1991). The spatial distribution patterns of *L. elisus* and *L. hesperus* in Californian cotton fields show a good fit to the negative binomial frequency distribution, with nymphs more clustered (showing a lower k-value) than adults (Sevacherian and Stern 1972a). In Idaho and Washington, *L. hesperus* shows a negative binomial distribution on lentils (i.e., means are less than the variance), nymphs are more aggregated (have a lower k-value) than adults, and changes in the distribution of the adult population vary seasonally and with density (Schotzko and O'Keeffe 1989a). For details of the aggregation patterns in *L. lineolaris*, in which aggregation increases gradually through the fourth instar and decreases through the adult stage, and development of sampling techniques for estimating population density, the reader is referred to the publication by Mukerji (1973).

Pieters and Sterling (1974) found that nymphs of the cotton fleahopper (*Pseudatomoscelis seriata*) are more aggregated (have a lower k-value) than adults. Dupnik and Wolfenbarger (1978) and Young and Willson (1987) presented additional data on the frequency distribution in this species; Young and Young (1988) demonstrated that insecticide applications disrupt its usual frequency distribution (geometric distribution, i.e., negative binomial with $k = 1$).

On apple trees in Quebec, Boivin and Stewart (1983b) analyzed the within-tree and within-orchard distributions and the spatial dispersion of four mirid species. Nymphs and adults of all species show a random distribution on trees—densities are not significantly different between quadrants or between upper and lower strata—indicating samples can be taken anywhere on the hosts to monitor mirid populations. Young nymphs of all species are contagiously distributed (see Fig. 6.27), with aggregates of individuals the basic components of their population, probably because of clumped oviposition patterns. Later instars of *Lygocoris communis* and *Campylomma verbasci* (see Fig. 6.27) show a similar pattern, remaining grouped throughout their development. *Campylomma verbasci* populations exist as groups of nymphs or single adults in British Columbia (Thistlewood 1989). Nymphs are distributed contagiously, but less so than what Boivin and Stewart (1983b) reported for this species (see Fig. 6.27) in Quebec.

The lupine-feeding *Lopidea nigridia* is patchily distributed among sites examined in Oregon because it oviposits on the same hosts on which development occurs and males and females tend to fly only short distances (McIver and Asquith 1989). Within a site, an extreme aggregation of individuals sometimes occurs, which apparently is related to aposematism (McIver and Lattin 1990). Ting (1965) analyzed the frequency distribution of plant bugs, including *Apolygus lucorum*, in Chinese cotton fields. Adults are distributed according to a Poisson series, although at low densities any clustering would be impossible to detect (Sevacherian and Stern 1972a). Nymphs on cotton are distributed randomly at low densities, but at levels exceeding 0.28 per

plant they show negative binomial distributions (Ting 1965). Aggregation in cocoa mirids is also well known (Youdeowei 1965, Entwistle 1972, Lotodé 1977, Wills 1986). Based on data collected from Australian sugarcane fields, Allsopp and Bull (1990) discussed mathematical relationships for describing the distribution of *Tytthus* species (mainly *T. mundulus*).

Analysis of Spatial Relationships

Geostatistical procedures have been used to analyze the spatial distribution of *Lygus hesperus*. Schotzko and O'Keeffe (1989b) determined that the spatial distribution of adults and nymphs in Idaho lentil fields varies seasonally and with changing population density. Adults show an aggregated distribution after immigrating into lentils when the crop begins to bloom. They are clumped at low densities during the midpoint of the growing season but are uniformly or randomly distributed at higher densities; by season's end, newly emerged adults again show a clumped distribution. Variable, high-density distributions might relate to the bug's reproductive biology: early-season aggregation indicating mating, and midseason reduction in spatial dependence reflecting dispersal for oviposition. Nymphs are uniformly distributed until they become clumped as populations increase in late season. A hexagonal sampling pattern best estimates the spatial structure of *L. hesperus*, whereas random sampling patterns give the poorest estimates of such structure (Schotzko and O'Keeffe 1990).

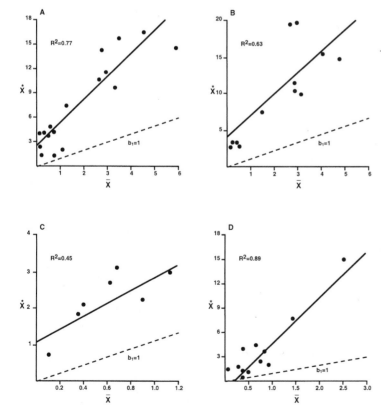

Fig. 6.27. Regression of mean crowding $(\overset{*}{x})$ on mean density $(\bar{x})$ for total captures of nymphs (A), young nymphs (B), old nymphs (C), and adults (D) of *Campylomma verbasci* in Quebec. Both nymphs and adults show a contagious distribution. (Reprinted, by permission of G. Boivin and the Entomological Society of America, from Boivin and Stewart 1983b.)

Mortality Factors

High mirid densities sometimes develop, as noted earlier in this chapter, but various mortality factors usually keep populations below such levels. Mortality results from various abiotic, density-independent factors such as rain and wind, and from a complex of natural enemies, both generalists and specialists. Because some mirids are commonly attacked by predators, the presence of plant bugs such as the cotton fleahopper (*Pseudatomoscelis seriata*) might be beneficial at times by attracting and maintaining generalist predators in cotton (Sterling et al. 1989b).

WEATHER, POLLUTION, AND HOST DESTRUCTION

Weather

Rainfall combined with high winds is a major factor in aphid mortality (e.g., Walker et al. 1984). Wind and rain also inflict heavy losses on mirids (Haseman 1918), particularly the early instars (Phillips and De Ronde 1966), although these bugs are difficult to dislodge from their hosts when they seek shelter beneath and in the axils of the buds (Fox-Wilson 1938). Nymphs that become dislodged from host trees sometimes reach maturity by feeding on understory plants (see Chapter 10). McNeill (1973) reported that heavy rainfall drowns early-instar *Leptopterna dolabrata* on grass blades, and Chatterjee (1983b) observed that *Nesidiocoris caesar* adults are killed by torrential rains. Under simulated conditions of rainfall and wind, death of the cotton fleahopper is greater on glabrous than on pilose strains of cotton (Gaylor and Sterling 1976b; see also Beirne [1970]). The cotton fleahopper sustains some mortality in sprinkler-irrigated cotton (Sterling et al. 1989b). The effect of heavy rainfall on fleahopper dynamics is nearly comparable to that of an insecticide application (Breene et al. 1989c). There is one observation of hail as a mortality factor: a 15-minute storm reduced a population of *Helopeltis orophila* by 39% in the Democratic Republic of the Congo (Lefèvre 1942).

Although Kullenberg (1944) remarked that plant bugs are generally unseasoned against wind, he pointed out that the slender body and narrow wings of certain mirines are adaptations to a life on grass blades and spikelets (Fig. 6.28). Mirids also possess an eversible rectal organ (Fig. 6.29), which apparently is characteristic of the family. It helps nymphs maintain contact with host plants under adverse conditions (Leston 1979b, Wheeler 1980c). Even so, rain, especially when accompanied by wind, results in considerable mortality of nymphs and adults. Muir (1920) reported an apparent exception to this rule: in sugarcane districts of Australia where cyclonic storms are frequent, populations of the delphacid egg predator *Tytthus mundulus* remained unaffected by several weeks of rain and flooding.

Fig. 6.28. Resting position of *Notostira erratica* on grass stem. The narrow wings and elongate body are apparent adaptations for living on grasses. (Reprinted, by permission of B. Kullenberg and Zoologiska institutionen, Uppsala Universitet, from Kullenberg 1944.)

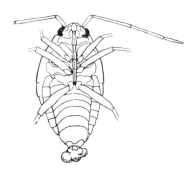

Fig. 6.29. Rectal organ of *Apolygus spinolae*. (Reprinted, by permission of Blackwell Wissenschafts-Verlag GmbH, from Fulmek 1930.)

Pollution

Environmental pollution can induce both increases and decreases in insect numbers and can either retard or enhance their growth and reproduction. The effects of pollution on heteropterans are little known, and most of the studies that have been conducted involve aquatic rather than terrestrial communities (Heliövaara and Väisänen 1993). In an assessment of the impact of air pollution from industrial wastes in Russia, heteropteran biomass and numbers actually increased in the immediate vicinity of an industrial operation (aluminum plant) responsible for fluoride contamination. In that study, heteropterans were represented mainly by mirids and plataspids (Katayev et al. 1983). In three plant communities in Poland, Lis (1992) found that heteropteran densities and species richness (51 of the 115 species captured were mirids) declined with increasing environmental pollution. Results of this study suggest that the stenodemine mirid *Trigonotylus caelestialium* would be a good bioindicator species.

Destruction of Host Material

Removal of the host plant or plant parts harboring mirid eggs, whether related to storms or practiced as a tillage or control technique, causes considerable mortality. As a means of controlling the fourlined plant bug (*Poecilocapsus lineatus*), Slingerland (1893) recommended pruning currant and gooseberry bushes to destroy overwintering eggs, and when *Helopeltis* species severely infest East African tea, the bushes can be pruned and the prunings burned to destroy the eggs (Smee 1928b; see also discussion by Carter [1973]). Similar recommendations have been made for *Helopeltis*-infested tea plants in Sri Lanka (Light 1930), India (Bamber 1893, Das 1984), and Papua New Guinea (Smith et al. 1985). In Italy, *Closterotomus norwegicus* oviposits in stakes used to support bean plants, and in winter, Lupo (1946) recommended immersing stakes for three days in 10% mineral oil to kill eggs of this plant bug.

Harvesting field crops, mowing forages, and plowing under sod can be used to eliminate the eggs of plant bugs (e.g., Osborn 1939, Sorenson 1939, Ridgway and Gyrisco 1960b, Pruess 1974, Radcliffe et al. 1976). The final alfalfa harvest of the season is considered the main reason for the failure of European populations of *Adelphocoris lineolatus* to attain high early-season densities the following year (Carl 1982). In addition to destroying eggs, cutting alfalfa alters the crop microenvironment so that early-instar mirids die from increased temperatures and decreased humidity (Sorenson 1939, Stern and Mueller 1968, Butler et al. 1971, Godfrey and Leigh 1994). Strip cropping of alfalfa leads to a similar mortality of lygus bugs: adults move from cut to uncut strips; they lay eggs in the half-grown hay, which will be harvested in about two weeks; and the resulting nymphs die from exposure to unfavorable temperatures and humidity when the strips are cut (Stern et al. 1964). Harvesting crested wheatgrass fields reduces the densities of the black grass bug (*Labops hesperius*) by destroying its overwintering eggs (Hagen 1982), and the cutting of bermudagrass eliminates substantial numbers of *Trigonotylus tenuis* (Buntin 1988). Disking, mowing, raking, and rototilling reduce hibernating adults of *Lygus* species (Fye 1983b).

Burning to destroy eggs helps suppress populations of the black grass bug in western rangelands (Todd and Kamm 1974, Haws and Bohart 1986), and spring burning reduces plant bug numbers in seed alfalfa fields (e.g., Hughes 1943; Bolton and Peck 1946; Lilly and Hobbs 1962; Craig 1963; Schaber and Entz 1988, 1994; Soroka 1991) and in birdsfoot trefoil (MacCollom 1967, Wipfli et al. 1990b). A spring burning of mulch was once practiced by strawberry growers in Nova Scotia to reduce populations of *C. norwegicus* (Pickett et al. 1944). Osborn (1939) stressed the burning of waste grass strips along fences to help control the meadow plant bug (*Leptopterna dolabrata*) (see also Hardison [1976], Kamm [1979]). For a discussion of the disparate effects that burning can have on heteropteran populations, the reader should consult the publications by Cancelado and Yonke (1970), Nagel (1973), Duffey et al. (1974), Morris (1975), and Warren et al. (1987).

The effects of prescribed burns and fire regimes on mirid populations in pitch pine–scrub oak barrens, remnant prairies and savannas, and other fire-maintained communities are little known. Siemann et al. (1997) included the Miridae, with the tarnished plant bug (*Lygus lineolaris*) as its most numerous representative, in an analysis of the effects of prescribed burning on oak savanna arthropods in Minnesota. They determined that the responses of community members vary widely but often are weak and not significant. One might expect that mirids associated with fire-maintained communities would be generally fire adapted (e.g., Anderson et al. 1989), but the short- and long-term effects of fire on individual species should be assessed. Such effects might differ substantially between multivoltine plant bugs that overwinter as adults and univoltine ones that overwinter as eggs.

Grazing by vertebrate animals, such as cattle and sheep, also affects mirid densities. Nymphs of *Leptopterna dolabrata* are more abundant on ungrazed plots than on grazed chalk grassland (Morris 1967). Grazing can be used to help reduce populations of *Labops hesperius* (Haws and Bohart 1986). In some cases, however, overgrazing creates conditions favorable to population increases of multivoltine, polyphagous mirids that develop on herbaceous weeds (C. C. Smith 1940).

Morris (1979) emphasized the different effects that cutting, as a single catastrophic event whose timing is of critical importance, has on grassland mirids and other heteropterans as contrasted with the effects of grazing, which is a continuous activity. Bivoltine mirids such as *Notostira elongata* are less susceptible to cutting than are univoltine species (Duffey et al. 1974), and the effects of cutting can vary depending on whether a mirid is an inflorescence or a foliage feeder (Morris 1979).

Destruction of host material also kills beneficial mirids. The regular pruning of apple trees, for example, removes eggs of predacious mirids from British orchards (Collyer 1953c). Winter pruning destroys large numbers of eggs of the predacious *Heterotoma merioptera* in France (Herard 1986). The effects of both natural and unnatural phenomena on predatory mirids, though largely unstudied, potentially influence the persistence and stability of predator-prey interactions.

VERTEBRATE PREDATORS

Vertebrates such as lizards, frogs, toads (e.g., Heikertinger 1922, Knowlton 1942, Knowlton et al. 1946), and birds (e.g., Crosby and Leonard 1914; McGregor 1927; Middleton and Chitty 1937; Ford et al. 1938; Knowlton and Maddock 1943; Kullenberg 1944; Knowlton and Harmston 1946; Burghardt et al. 1975; Krištín 1984, 1986) sometimes feed on mirid nymphs or adults. Members of the large orthotyline genus *Nesio-miris* are white, green, or greenish white and match the color of the undersides of host foliage. This suggests that visual predators such as birds have been a selective force in the evolution of this species-rich genus in Hawaii (Gagné 1982, 1997). Despite comments that birds avoid mirids because of their offensive odor (e.g., Smith and Franklin 1961), the scent-gland secretions of even coreids and pentatomids do not always protect heteropterans from bird predation (Schlee 1986, 1992). Mirids, in fact, can be considered beneficial in providing an important food source for farmland birds (Moreby et al. 1997). Plant bugs actually represent a preferred food item for partridge chicks (Vickerman and O'Bryan 1979, Potts 1986, Sotherton 1991). Robel et al. (1995) assessed the nutrient and energetic characteristics (e.g., fat content, crude protein, calcium, and phosphorus levels, and gross energy contents) of mirids and other potential invertebrate prey of grassland birds.

INVERTEBRATE PREDATORS

Plant bugs of all stages are subject to attack by invertebrates. The diverse group of mirid predators includes hunting and web-building spiders, as well as various insects, both those with chewing and those with sucking mouthparts. Field experiments in which the introduction of *Mantis religiosa* into old fields and pastures led to large increases in plant bug numbers suggest the role of generalist predators in limiting mirid densities. Mantid predation on geocorids and nabids, which are natural enemies of Miridae, presumably results in higher mirid survival rates (Fagan and Hurd 1994).

Predation on Nymphs and Adults

Generalist predators such as spiders (Plate 2), nabids, minute pirate or flower bugs (anthocorids), and bigeyed bugs (*Geocoris* spp.) feed on mirid nymphs (Plate 2) and adults (Bilsing 1920; McGregor 1942; Kullenberg 1944; Whitcomb and Bell 1964; Clancy and Pierce 1966; Dempster 1966; van den Bosch and Hagen 1966; Edgar 1970; Perkins and Watson 1972; Varis 1972; Wheeler 1977; LeSar and Unzicker 1978; Tamaki et al. 1978; Whalon and Parker 1978; Crocker and Whitcomb 1980; Hiremath and Thontadarya 1983; Young and Lockley 1985, 1986; Cao 1986; Nadgauda and Pitre 1986; Dean et al. 1987; Nyffeler et al. 1987; Siddique and Chapman 1987; Araya and Haws 1988, 1991; Breene et al. 1988, 1989a; Arnoldi et al. 1991; Sterling et al. 1992). Mirids do not represent the favored prey of some spiders and can be rejected (Bristowe 1941), or small spiders will accept mirid nymphs but not adults (Chant 1956). Daddylonglegs or harvestmen (Opiliones) occasionally feed on mirids (Dempster 1966), as do young scorpions (Polis 1979). Other common polyphagous predators, such as the North American reduviid *Sinea diadema* (Plate 2) and pentatomid *Podisus maculiventris*, feed on mirids

(Hawley 1917, Readio 1924, Balduf 1943, Balduf and Slater 1943, McPherson 1982), and in the Old World tropics reduviids, mantids, longhorned grasshoppers (cf. Marchart 1968), and crickets prey on cocoa mirids (e.g., Ghesquière 1922, Squire 1947, Williams 1954, Johnson 1962, Leston 1970, Kumar 1971a, Entwistle 1972, Collingwood 1977b, Commonwealth Institute of Biological Control 1983). Sundararaju (1984) reported five reduviid species as predators of *Helopeltis antonii* on cashew trees in India. Cockroaches and earwigs attack the eggs and nymphs of *Mircarvalhoia arecae* on palms (Kurian and Ponnamma 1983).

Coccinellids feed on plant bugs (e.g., Everly 1938, Yamamuro and Hoshino 1940, Whitcomb 1953, Matsumoto and Nishida 1966, Breene et al. 1989a), but they generally prefer aphids or other prey (e.g., Fullerton 1961, Varis 1972). In laboratory feeding trials, coccinellids fed on mirid nymphs but not on adults (Tyndall 1958). Adult soldier beetles (Cantharidae), tiger beetles (Carabidae), and melyrids (*Collops* spp., Melyridae) sometimes include mirids among their prey (Wene and Sheets 1962, Haws 1978a, Dolling 1991). Chrysopids and syrphids, generally unimportant enemies of plant bugs (Leigh and Gonzalez 1976), feed on mirid nymphs (e.g., Heidemann 1910, Lean 1926, Isely 1927, Cotterell 1928, Goodman 1953, Tyndall 1958, da Silva Barbosa 1959, Varis 1972, Hedlund 1987), or adults (Ingram 1980). Also among neuropteran enemies of plant bugs are ascalaphid larvae, which at least once have been observed to prey on nymphs of cocoa mirids (Gerard 1966). Robber flies (Asilidae) and shore flies (*Ochthera mantis*; Ephydridae) also prey on mirids (e.g., Kullenberg 1944, Szent-Ivany 1961, Simpson 1975, Scarbrough and Sraver 1979, Scarbrough 1981). Sphecoid wasps provision their nests with plant bugs (e.g., Bohart and Villegas 1976, Carvalho 1976), some species almost exclusively so (Kurczewski 1968, Evans 1969, Kurczewski and Peckham 1970). Carabid and staphylinid beetles sometimes feed on overwintering adults of *Lygus lineolaris* in the eastern United States (Patch 1907). Desiccation and unfavorable temperatures are often responsible for higher winter mortality than are natural enemies among mirid species (especially lygus bugs) that overwinter as adults (see Table 6.6; Painter 1929b, Austin 1931b, Newcomer 1932, Fox-Wilson 1938, Whitcomb 1953, Varis 1972).

Stitt (1940) found that ants are the only predators that noticeably reduce the numbers of lygus bug nymphs in the southwestern United States, and aphid-tending ants may prey on mirids colonizing the same plants (Domek and Scott 1985). Ants can be important predators of cocoa mirids in the tropics (Squire 1947; Marchart 1968; Bruneau de Miré 1969; Leston 1970, 1973b; see Chapter 9) and of similar bryocorines on cashew trees (Ambika and Abraham 1979, Peng et al. 1995). Ants also can reduce the populations of *Monalocoris filicis* on bracken (Heads 1986). One case of an apparent ant-mirid mutualism is known: the arboreal monaloniine *Chamopsis conradti* is an ant-attended species in Ghana whose association with ants might be mediated by trichomelike structures on its antennae (Leston 1980a).

Predation on Eggs

Predators having sucking mouthparts, such as *Geocoris* species, feed on mirid eggs in the laboratory (Clancy and Pierce 1966, Dunbar and Bacon 1972, Cohen and Debolt 1983) and in field cages (Leigh and Gonzalez 1976). Egg predation is difficult to observe but also occurs in nature (Kullenberg 1944, Strawiński 1964a, Ehler 1977). Nabids are particularly common predators of mirid eggs (Kullenberg 1944). The recent development of species- and stage-specific monoclonal antibodies as diagnostic probes for analyzing the gut contents of potential predators should allow natural predation to be detected immunologically (Hagler et al. 1991; see also McIver and Tempelis [1993], Naranjo and Hagler [1998]). When bean pods containing *Lygus hesperus* eggs were placed in cotton fields in California, 55% of the eggs were destroyed within 24 hours (20–60% in another experiment, number of observations unspecified), probably by bigeyed bugs, anthocorids, and nabids (Ehler 1977). Insertion of eggs into plant tissues, however, offers some protection from natural enemies (Williams 1954, Morris 1965), especially from predators with chewing mouthparts (Hagler et al. 1991), and might preclude egg predation from being a significant mortality factor in many mirids. Among congeneric psyllid species occurring on Scotch broom, the species that embeds its eggs in stems is considered less vulnerable to predation than the species having an exposed, superficial egg (Watmough 1968). Plant bug eggs, as noted earlier in this discussion of mortality factors, also sustain incidental losses from grazing herbivores such as cattle and sheep.

Effect of Invertebrate Predators on Mirid Populations

With few exceptions, the impact of polyphagous predators on plant bug populations has not been quantified. In an attempt to assess the effect of mirid natural enemies, Breene and Sterling (1988) irradiated *Pseudatomoscelis seriata* with phosphorus-32 (^{32}P) and suggested ^{32}P labeling as a means of identifying and evaluating its predators. None of this bug's generalist predators—various spiders and an ant—showed a numerical response to prey numbers. As possible "lie-in-wait" predators, they might, however, prevent severe outbreaks of this pest (Breene et al. 1990).

PARASITOIDS OF EGGS

Plant bug eggs are parasitized by wasps of several families: principally mymarids but also eulophids (Fig. 6.30), scelionids, and trichogrammatids. Examples of mymarids that parasitize mirid eggs are *Anaphes iole* (Plate 2; Romney and Cassidy 1945, Jackson 1987,

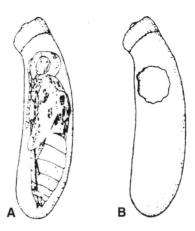

Fig. 6.30. *Tetrastichus miridivorus*, a eulophid parasitoid of mirid eggs. A. Pupa in egg of *Closterotomus trivialis*. B. Emergence hole of the parasitoid. (Reprinted, by permission of S. Barbagallo and Istituto di Entomologia Agraria (Milan), from Barbagallo 1969.)

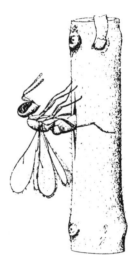

Fig. 6.31. *Telenomus* species (Scelionidae) ovipositing in an egg of *Horistus orientalis*. (Reprinted, by permission of Société Entomologique de France, from Silvestri 1932.)

Huber and Rajakulendran 1988), *Polynema pratensiphagum* (Painter 1929b, Sohati et al. 1989), *Erythmelus miridiphagus* (Dozier 1937), *E. psallidis* (Ewing and Crawford 1939, Rajakulendran and Cate 1986), and *E. helopeltidis* (Gahan 1949, Ibrahim 1989). Species of the scelionid genus *Telenomus* (Fig. 6.31) parasitize mirid eggs (e.g., Silvestri 1932, Connell 1970b, Chang 1982, Commonwealth Institute of Biological Control 1983, Coulson 1987, Ibrahim 1989, Sohati et al. 1989, Al-Ghamdi et al. 1995), as do trichogrammatids such as *Chaetostricha miridiphaga* (Connell 1970b, Viggiani 1971a), *C. thanatophora* (Pinto 1990), *Paracentrobia pulchella* (Claridge 1959), *P. nympha*, *P. subflava* (Burks 1979), and *Ufensia minuta* (Viggiani 1989). Larvae of the eulophid *Cirrospilus ovisugosus* tunnel through the

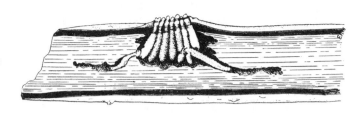

Fig. 6.32. Larvae of the eulophid *Cirrospilus ovisugosus* that have tunneled through a stem to feed on eggs of *Poecilocapsus lineatus*. (From Crosby and Matheson 1915.)

pith of stems to devour the eggs of *Poecilocapsus lineatus* (Fig. 6.32; Crosby and Matheson 1915), and the eulophid *Tetrastichus miridivorus* parasitizes the eggs of the European mirines *Closterotomus trivialis* and *Horistus orientalis* (as *Capsodes lineolatus*) (see Fig. 6.30; Barbagallo 1969).

Egg parasitoids of mirids have been studied mainly in agroecosystems. A natural parasitism rate of about 50% is sometimes maintained in alfalfa fields, but management practices usually limit parasitoid effectiveness. Augmentative releases in North America of mymarids, such as the indigenous *A. iole*, can substantially reduce losses from lygus bugs in seed alfalfa (Graham et al. 1986, Jones and Jackson 1990) and show potential for reducing fruit damage in commercial strawberry fields (Norton and Welter 1996). Foreign exploration for natural enemies of pest mirids in the United States has included the collection of Old World egg parasitoids, but the actual importation of hymenopteran parasitoids has been limited to species that attack mirid nymphs (Hedlund 1987, Jackson et al. 1995).

Because mirid eggs are generally concealed, females of most hymenopteran egg parasitoids must show specialized searching behavior (Conti et al. 1997). Endophytic oviposition affords some protection from parasitoids (e.g., Kumar 1971a), with certain sites on a host plant providing more protection from parasitism than do others (Udayagiri and Welter 2000). Stoner and Surber (1969) found that as eggs of *Lygus hesperus* aged, a smaller percentage was parasitized by the mymarid *A. iole*. They suggested that the operculum of lygus bug eggs might become more resistant to penetration, or that older eggs simply are less attractive to the female wasp. Among tropical bryocorine pests of cocoa, the partially exposed eggs of *Bryocoropsis* and *Helopeltis* species appear more susceptible to parasitism than the more concealed eggs of *Distantiella theobroma* or *Sahlbergella singularis* (Collingwood 1977b). In the case of a mymarid parasitoid of *H. theivora*, the female wasp, during oviposition, holds onto the chorionic processes or projections of the host egg (see Fig. 6.15; Ibrahim 1989).

PARASITOIDS OF NYMPHS

Euphorine Braconidae

Wasps of the braconid genera *Leiophron* and *Peristenus* are specialized parasitoids of plant bug nymphs in North America, and mirids serve as their principal hosts

in Europe (Loan 1974, Loan and Shaw 1987). The life cycles of these euphorine parasitoids are synchronized with those of their hosts. Pupation occurs in the soil, and in temperate climates the pupal stage lasts 8–10 months (Bilewicz-Pawińska 1982, Varis and Bilewicz-Pawińska 1992). Diapausing adults emerge from overwintered cocoons shortly after the host eggs hatch in spring. Diapause is critical in the host specificity of euphorines because it prevents temporal separation from the mirid host after overwintering (Loan and Shaw 1987).

Female euphorines oviposit through the intersegmental region into the hemocoel of early-instar mirids, which walk away almost immediately after being attacked (Loan 1965, 1974; Lim and Stewart 1976b; Glen 1977a; Loan and Shaw 1987) or are sometimes stunned or paralyzed for several minutes (Waloff 1967). Debolt (1981) determined that *L. uniformis* females attempt to oviposit in all nymphal instars of *Lygus hesperus*, but parasitoids could not be reared from attacked fifth instars. Second- and third-instar *L. hesperus* yield more cocoons per female parasitoid than do first or fourth instars. Lim and Stewart (1976b) observed that euphorine females sometimes seize and attempt to oviposit in the exuviae of *L. lineolaris* (see also Condit and Cate [1982]). This led Aldrich (1988, 1995) to postulate that secretions (presumably allomones) of the bug's dorsal abdominal scent gland, whose contents are shed with the cast skin, are used as a host-finding cue or kairomone.

Euphorines mostly develop in the nymphs of their hosts, the last-stage larva emerging from the abdomen of a fourth- or fifth-instar mirid (Loan 1974, Snodgrass et al. 1990, Lattin and Stanton 1999). In some parasitoid species, only a small proportion of larvae emerge from adult bugs (e.g., Loan and Craig 1976, Glen 1977a, Hedlund 1987, Loan and Shaw 1987), but in other species most larvae emerge from teneral adults (Loan 1965), or they develop mainly in the adult (Loan 1966). *Leiophron uniformis* occasionally emerges from adult *Lygus hesperus* when fourth instars are attacked (Debolt 1981). In the British fauna, Leston (1959, 1961c) suggested that parasitoids emerge from adults mainly in early-season, arboreal plant bugs. Among mirids associated with lodgepole pine in the western United States, only the early-occurring species are parasitized by euphorine braconids (Lattin and Stanton 1999).

The gut contents of euphorines can be orange, green, or bluish green, matching the color of the host's fat body or testis sheaths; or they are white like the expanded teratocytes (sac cells) in the host (e.g., Leston 1961c, Clancy and Pierce 1966). Late-instar mirids parasitized by euphorines often stand out by having a distended, shiny abdomen (Plate 3; Butler 1923, Menzel 1928, Hey 1933a, Glen 1977a, Cmoluchowa 1982, Dolling 1991), whereas nymphs (those that do not die) from which the parasitoid larva has emerged have a shrunken, decurved abdomen (Plate 3) and a hesitant gait (Loan 1965, 1974). In addition to external changes in the parasitized nymph, parasitization affects the hemolymph chemistry of the host. The hemolymph of mirids parasitized by euphorines contains teratocytes, which are dissociated trophamnion cells of the parasitoid egg, composed of various amino acids and fatty acids (Cohen and Debolt 1984, Debolt and Cohen 1984).

Some euphorine braconids are bivoltine, whereas others are univoltine. Both generations of *L. rugulipennis* in Poland are attacked by the bivoltine *Peristenus digoneutis* and *P. stygicus*; the univoltine *P. rubricollis* parasitizes only the first generation of this host (Craig and Loan 1981). Sequential parasitism of *L. lineolaris* by two univoltine species once was attributed to a single parasitoid species having two generations (Day et al. 1990).

Parasitoid larvae feed on expanded teratocytes rather than on their hosts' vital organs (Leston 1961c). From an immediate control standpoint, this feeding habit of euphorines allows parasitized mirids to continue to feed for the several weeks needed for parasitoid development (Loan 1965). In the case of *Helopeltis bergrothi*, a female nymph parasitized as a fifth instar can become an adult before the euphorine larva emerges and can live long enough to copulate and oviposit (Kirkpatrick 1941). Because parasitized nymphs take longer to develop, the percentage of parasitism by euphorines will be overestimated if assessments are based on the nymphs (stragglers) remaining when a population consists mostly of adults (Leston 1959, 1961c; Wheeler and Loan 1984; see also Van Driesche [1983]). For measuring mortality from parasitism, dissection of mirid hosts is more accurate than rearing methods (Day 1994).

Effects on mirid populations
Some euphorines appear to function in a density-dependent manner (Bilewicz-Pawińska 1982), causing population crashes of pest mirids (Wheeler and Loan 1984). In a two-year study of the multivoltine *Halticus bractatus*, heavy parasitism by *Leiophron uniformis* (an average 50% of nymphs in all generations) depressed populations significantly (Day and Saunders 1990).

But nymphal parasitism does not seem to play a major role in regulating the numbers of three co-occurring *Orthotylus* species on broom in England, although parasitoids are important in checking the populations of *Asciodema obsoleta* and *Heterocordylus tibialis* on the same host (Waloff 1967, 1968). *Orthotylus concolor*, the least abundant of the five mirids that colonize broom in England (Dempster 1960), is an immigrant species in California, where its densities exceed

those of all plant bugs found on broom in England. The apparent freedom of *O. concolor* from nymphal parasitism in an alien environment (Waloff 1965) appears partly responsible for this population explosion. In the case of a *Peristenus* species and its host, *Blepharidopterus angulatus* in England, a negative correlation exists between percentage of parasitism and host density (Fig. 6.33; Glen 1977a). King (1971) reviewed parasitism of the cocoa mirids *Distantiella theobroma* and *Sahlbergella singularis*, and the Commonwealth Institute of Biological Control (1983) listed parasitoids of *Helopeltis* species.

Factors affecting parasitism
Parasitism of economically important mirids is potentially affected by the presence of exotic plants in the habitat. Native parasitoids that have evolved with their mirid hosts appear to have developed search preferences for particular plant species. Pestiferous mirids of a broad host range, such as certain North American lygus bugs, might escape attack on nonindigenous weeds when parasitoids cue in on the bugs' native hosts (Taksdal 1961, Scott 1987).

Euphorine parasitism of *Lygus lineolaris* is higher in undisturbed weedy fields than on the same wild hosts growing in areas disturbed by agricultural production. Parasitism of *L. lineolaris* by *Leiophron uniformis*, even in undisturbed habitats, does not reach the levels found for *Lygus hesperus* (Snodgrass and Fayad 1991). Southwestern U.S. populations of *L. lineolaris* are

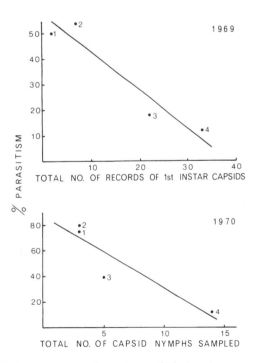

Fig. 6.33. Percentage of parasitism of *Blepharidopterus angulatus* nymphs by the euphorine braconid *Peristenus* species; parasitism was negatively correlated with host density in both years of study. (Reprinted, by permission of D. M. Glen and Blackwell Science Ltd., from Glen 1977a.)

able to encapsulate *Leiophron uniformis* eggs (Debolt 1989a). Encapsulation ability increases with nymphal age, reaching about 70% in second instars 72–96 hours old (Debolt 1991). Prior to Debolt's (1989a) study, such a potent defense mechanism was unknown in the Heteroptera. Parasitoid females readily oviposit in *L. lineolaris* but lay fewer eggs than in other *Lygus* species, suggesting *Leiophron uniformis* uses internal chemical cues to avoid ovipositing in an unsuitable host (Debolt 1989a). The level of immune response varies with the strain and age of *L. lineolaris*; parasitoid females adjust oviposition in an inverse relationship to the level of host immune response. Strains of *Leiophron uniformis* completely resistant to the immune response of *Lygus lineolaris* have been identified (Debolt 1989b).

Certain mirid populations differ in their ability to encapsulate parasitoid eggs and thus, in their susceptibility to euphorine parasitism (Graham and Debolt 1986). In fact, the disparate levels of parasitism between New Jersey and Arizona-Delaware colonies of *L. lineolaris* suggest that considerable reproductive isolation occurs in this species (Graham and Debolt 1986). On the basis of a limited western distribution and an apparent preference for arid habitats, in contrast to the general occurrence of *L. lineolaris* in the East, Clancy (1968) proposed that this mirid has a "western race" or biotype.

The primary mirid parasitoids themselves are attacked by mesochorine ichneumonids (e.g., Nixon 1946; Clancy and Pierce 1966; Waloff 1967; Stewart 1969a; King 1971; Bilewicz-Pawińska 1973, 1976; Day 1987; Dolling 1991) that reduce their effectiveness. *Mesochorus curvulus* is a hyperparasitoid of braconid parasitoids of *L. lineolaris* (e.g., Lim and Stewart 1976a; Carlson 1979). *Mesochorus* species are secondary parasitoids of *Peristenus* species that parasitize *L. rugulipennis* in Europe (Stewart 1969a; Bilewicz-Pawińska 1970, 1973), and *M. melanothorax* attacks primary parasitoids of cocoa mirids in West Africa (Wilkinson 1927, King 1971).

Strepsiptera

Strepsipterans of the family Corioxenidae also parasitize mirids (Kathirithamby 1992). The effects of stylopization (parasitism by a strepsipteran) on plant bug hosts apparently are unknown.

PARASITOIDS OF ADULTS

Parasitic insects rarely attack adult mirids. Tachinid flies of the genus *Phasia* (formerly *Alophorella*), however, specialize on heteropterans; some species parasitize adult plant bugs in the Old and New Worlds (e.g., Leonard 1916c, Painter 1929b, Medler 1961, Dupuis 1963, Clancy and Pierce 1966, Scales 1973, Arnaud 1978, Graham et al. 1986). The apparently bivoltine *P. robertsonii* is an uncommon natural enemy of grass- and legume-feeding mirines in eastern North America; only 54 adults were reared from nearly 12,000 adult

mirids during an 11-year study (Day 1995). Two sarcophagid species apparently are endoparasitoids of adult and fifth-instar *Helopeltis orophila* in the Democratic Republic of the Congo (Lefèvre 1942).

PARASITIC MITES

Nymphal and adult mirids are sometimes collected with mites attached to their bodies (Plate 3; Weber 1930, Kullenberg 1944). Welbourn (1983) summarized much of the scattered literature pertaining to trombidioid and erythraeoid mites associated with mirids and other insects; his mirid records and a few additional ones are given in Table 6.8.

The erythraeid and trombidiid species (see Table 6.8) are considered parasitic on their hosts; they belong to the category "protelean parasites" as used by Eickwort (1983). Postlarval instars are free-living predators, but it is not known if these stages attack mirids in the field. *Lasioerythraeus johnstoni* does so under laboratory conditions (Young and Welbourn 1987), and adults of an undetermined erythraeid apparently attack *Calocoris angustatus* in nature (Hiremath 1989). Species of *Balaustium*, *Leptus*, and *Trombidium* probably are polyphagous ectoparasites not restricted to plant bugs or even heteropterans. *Parathrombium megalochirum* might be restricted to phytophagous Heteroptera as larval hosts, although species of this genus are known from other host orders (Robaux 1974, Welbourn 1983). Da Silva Barbosa (1959) reported a laelapid-mirid association, but without additional information, it cannot be placed in the categories Eickwort (1983) recognized for relationships between parasitic mites and their insect hosts.

Nearly all the scant information available on mites as parasites of mirids relates to polyphagous erythraeids or trombidiids that have unknown effects on their hosts. Such species were not mentioned in Eickwort's (1983) review of mites as potential biological control agents of insect pests that feed on plant surfaces. Yet, as La Munyon and Eisner (1990) emphasized, even transient infestation by an ectoparasitic mite is potentially detrimental to a host.

Young and Welbourn (1987) provided the first detailed information on a mite-mirid association. They determined that the erythraeid *L. johnstoni* can complete its life cycle on *Lygus lineolaris* nymphs, the parasitic larvae characteristically attaching to ecdysial lines on the bugs' head and thorax (Fig. 6.34). As in other erythraeids, the deutonymphs and adults are free-living predators. Deutonymphs feed on first-instar tarnished plant bugs in the laboratory; adults, on first through fourth instars. Development from the larval to the adult stage requires 21 days under laboratory conditions (26°C). Larvae collected in old-field habitats kill their hosts and detach within two days. *Lasioerythraeus johnstoni* holds promise as a biocontrol agent of the tarnished plant bug and related pests. Nearly all its known hosts belong to the

Table 6.8. Examples of mites parasitic or predacious on mirids

Mite	Mirid	Locality	Remarks	Reference
Erythraeidae				
Balaustium sp.	*Lygus lineolaris*	Mississippi, USA		Young and Welbourn 1987
"*Bochartia*" sp.	*Pseudatomoscelis seriata*	Texas, USA	Record might refer to *L. johnstoni*, below; see Welbourn and Young (1987), Young and Welbourn (1987)	Reinhard 1926
Lasioerythraeus johnstoni	*Lygus lineolaris, Polymerus basalis, Pseudatomoscelis seriata, Taylorilygus apicalis*	Mississippi, USA	See text for discussion	Young and Welbourn 1987
	Trigonotylus tenuis	Mississippi, USA		Young and Welbourn 1988
Leptus sp.	*Helopeltis* sp.	Democratic Republic of the Congo		Squire 1947
	Sahlbergella singularis	Ghana, Nigeria		Entwistle and Youdeowei 1965, Collingwood 1977b
Undetermined genus	*Calocoris angustatus, Campylomma lividum*	India	Widespread, nymphs and adults on third–fifth instars and adults of *C. angustatus*	Hiremath and Thontadarya 1983, Hiremath 1989
Undetermined genus	*Halticus bractatus*	South Carolina, USA	Larvae attack early-stage nymphs	Beyer 1921
Trombidiidae				
Parathrombium megalochirum	*Capsus ater, Stenodema calcarata*	France		Robaux 1974
Trombidium holocericeum	*Dicyphus globulifer*	England	Larvae on each side of dorsum, forcing wings up	Leston 1961d
Trombidium parasiticus	*Notostira erratica*	Netherlands		Oudemans 1912
Trombidium teres	*Orthops montanus, Stenodema calcarata*	France		André 1928, 1929
Trombidium sp.	*Taedia hawleyi*	New York, USA		Hawley 1917, 1918
Undetermined genus	*Polymerus cognatus*	Ukraine	Listed among important natural enemies	Vernigor 1928
Laelapidae				
Undetermined genus	*Taylorilygus vosseleri*	Mozambique	9 mites feeding on adult	da Silva Barbosa 1959

Note: See text for additional examples and discussion; see also Kullenberg (1944).

Heteroptera or Homoptera (Young and Welbourn 1987), including the stenodemine *Trigonotylus tenuis* (Young and Welbourn 1988). *Lasioerythraeus johnstoni* (then undescribed) possibly is the "red mite" Isely (1927) observed in Arkansas cotton on nymphs of *Adelphocoris rapidus, Lygus lineolaris,* and *Pseudatomoscelis seriata.*

PARASITIC NEMATODES

Immature stages of mermithid nematodes attack nymphal and adult mirids, but little biological information is available on nematode–plant bug relationships (e.g., La Rivers 1949). Most nematode infections of heteropterans probably are accidental (Poinar 1975). Information on species associated with plant bugs often consists merely of a rearing record. Examples include a mermithid from *Lygus pratensis* and *L. rugulipennis* in Germany (Boness 1963) and a *Hexamermis* species from lygus bugs in France (Nickle 1978); an unidentified nematode from *L. rugulipennis* in Scotland (Stewart 1969a) and from *Calocoris angustatus* in India (Sharma 1985b); an unidentified mermithid from adult *Helopeltis schoutedeni* in Africa (Schmitz 1958) and adult *L. lineolaris* in Mississippi (Scales 1973); a nematode tentatively identified as *Hexamermis arvalis* from an adult *L. lineolaris* in New York (Poinar and Gyrisco 1962); a *Paramermis* species from this mirid in Quebec (Stewart and Khoury 1976); and *Agamermis decaudata* from adult females of *Labops hesperius* in Utah (Coombs 1985). Poinar (1975), Collingwood (1977b), and the Commonwealth Institute of Biological Control (1983) listed additional nematode-mirid associations.

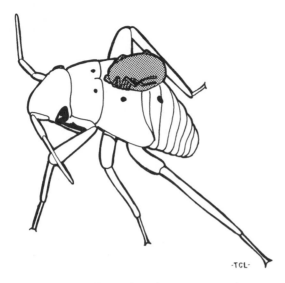

Fig. 6.34. Larva of the erythraeid mite, *Lasioerythraeus johnstoni*, attached to the thorax of a second-instar *Lygus lineolaris*. (Reprinted, by permission of O. P. Young and the Entomological Society of America, from Young and Welbourn 1987.)

Nematode infections of mirids in tropical and temperate regions occur mainly during wet periods (e.g., Painter 1929a, Bilewicz-Pawińska and Kamionek 1973, Collingwood 1977b). The nematodes typically live within the hemocoel, only one occurring per host (Painter 1929a, Reid 1974), but Stewart and Khoury (1976) observed an overwintered female of *Lygus lineolaris* parasitized by six larvae of a *Paramermis* species. Nymphs do not always show external evidence of parasitism (Reid 1974), although infected adults of *L. lineolaris* can have distended, drooping abdomens and a conspicuous integument between the tergites and sternites (Painter 1929a). Rates of parasitism are generally low (Day 1987): 2–3% for *Helopeltis* species on tea plants (Menzel 1922), 0.1% for *H. schoutedeni* on cotton (Schmitz 1958), a maximum of 4% for the cocoa mirids *Distantiella theobroma* and *Sahlbergella singularis* (King 1971), 5–10% for cocoa-infesting *Helopeltis* species under favorable conditions (Giesberger [1983]), and 8–12% among *Slaterocoris* species on goldenrod (Reid 1974).

The ovaries and testes of infected mirids fail to develop (Painter 1929a, Reid 1974), but mermithids apparently are not among the principal biotic agents that limit mirid populations. Moreover, the necessity of having to mass-produce mermithids in vivo limits their potential use in biological control (Kaya and Stock 1997).

MICROBIAL PATHOGENS

Mirids are more susceptible to infection by entomopathogenic fungi than by bacterial or protozoan pathogens. Viral diseases of mirids are unknown (Dolling 1991; see also Martignoni and Iwai [1977], Tinsley and Harrap [1978], Adams and Bonami [1991],

Kurstak [1991]); most viruses infect members of the Lepidoptera, Diptera, Hymenoptera, and Coleoptera (Evans and Entwistle 1987).

Fungi

Entomophthoraceous fungi cause epizootics in several mirine species. An example is *Entomophthora erupta*, which is pathogenic on *Lygocoris communis* (Dustan 1923, 1924; Evans 1989), *Irbisia solani* (Hall 1959), and *Adelphocoris lineolatus* (Wheeler 1972). Additional examples are *E. helvetica*, which is pathogenic on *Notostira elongata* (Geoffroy) (Plate 3; Keller 1981, Ben-Ze'ev et al. 1985), and *Zoophthora anglica*, a pathogen of *N. elongata* (Keller 1982). In addition, Dustan (1924) observed a second host of *E. erupta* on apple trees in Nova Scotia, a small green mirid identified as *Plagiognathus* species (this record likely refers to another phyline, *Campylomma verbasci*). *Lygocoris pabulinus* has been reported as a host of *E. muscae* in Britain (Petch 1948, Leatherdale 1970), but this record might represent a misidentification of some other entomophthoraceous species; an undetermined species of *Entomophthora* has been reported from this plant bug in the Netherlands (Blommers et al. 1997). An unidentified entomophthoraceous fungus is known from *Leptopterna dolabrata* (Osborn 1918), and Kullenberg (1944) recorded entomophthoraceous infections of *Closterotomus norwegicus*, *Fieberocapsus flaveolus*, *N. erratica*, and *Pantilius tunicatus*.

Zoophthora radicans infects the eccritotarsine *Pycnoderes quadrimaculatus* in Hawaii (Zimmerman 1948a; see also Holdaway and Look [1942]). The common hyphomycetes *Beauveria bassiana* and *Paecilomyces fumosoroseus* infect *A. lineolatus* and *Lygus rugulipennis* (Bajan and Bilewicz-Pawińska 1971, Bilewicz-Pawińska 1976, Riba et al. 1986). *Beauveria bassiana* also infects *L. hesperus* (Dunn and Mechalas 1963) and *L. lineolaris* (Steinkraus and Tugwell 1997) and is highly pathogenic on *Helopeltis theivora*, adults dying within 1–3 days after treatment with fungal spores (Lim et al. 1989). Two strains of *B. bassiana* are pathogenic to *Pseudatomoscelis seriata* in laboratory bioassays (Wright and Chandler 1991) and have been evaluated as pathogens of the New World cocoa pest *Monalonion dissimulatum* (Montealegre and Rodriguez 1989). Insecticides containing *B. bassiana* have been tested against *L. lineolaris* (e.g., Brown et al. 1997, Steinkraus and Tugwell 1997). Another hyphomycete, *Acremonium* species (cited as a *Cephalosporium* sp.), infects the sorghum earhead bug (*Calocoris angustatus*) in India (Hiremath 1989), and *Aspergillus candidus* infects *Mircarvalhoia arecae* in India (Plate 3; Dhileepan et al. 1990).

The high mortality among overwintering adults of *L. rugulipennis* in Scotland might be associated with fungal infections (Stewart 1969a). Species of *Fusarium* and *Penicillium* are known from *H. clavifer* in Papua New Guinea, but the nature of the mirid-fungus relationship is unknown (Shaw 1984; see also Grisham et al. [1987]). An unidentified fungal pathogen is associated with a cocoa-infesting *Helopeltis* species in Ghana (Brew 1992). Collingwood (1977b) reviewed the literature on fungal pathogens of cocoa mirids, noting that their incidence was low (<1% of diseased mirids) and that they have not been used as alternatives to chemical control.

Bacteria

Bolton (1973) reported a bacterial pathogen that causes massive septicemia in various cocoa mirids, its transmission occurring when the bugs ingest spores at a previously infected site. Spores develop into a rod-shaped motile form in the bugs' midgut, penetrate the hemocoel, spread to all organ systems, and eventually kill the host. *Distantiella theobroma*, *Helopeltis bergrothi*, and *Sahlbergella singularis* are susceptible to the bacterium (>80% mortality in petri dishes), but disease symptoms fail to develop in *Bryocoropsis laticollis*. At times, mirids reared in the laboratory are subject to bacterial infections (e.g., Piart 1970, Collingwood 1977b). An unidentified bacterial pathogen has been found on *Calocoris angustatus* (Hiremath and Thontadarya 1983).

Protozoa

Outbreaks of microsporidian protozoa sometimes occur in laboratory cultures of plant bugs (Parrott et al. 1975). Flagellate protozoa can be found in the mirid alimentary canal. Known parasite-host associations include *Blastocrithidia miridarum* and *Adelphocoris quadripunctatus* (Frolov and Skarlato 1988), *Proteomonas inconstans* and *Grypocoris sexguttatus* (Podlipaev et al. 1990), and *Leptomonas mycophilus* and a *Phytocoris* species (Frolov and Skarlato 1991). *Blastocrithidia miridarum* is also known from species of *Deraeocoris*, *Lygocoris*, *Notostira*, and *Stenodema* (Podlipaev and Frolov 1987; see also Camargo and Wallace [1994]).

Generally referred to as parasites, these trypanosomatids might be only harmless commensals in their hosts. Most Trypanosomatidae occurring in the gut of insects are considered nonpathogenic (Wallace 1979) or "subpathogenic stressors" (Undeen and Vávra 1997). But infection of certain Heteroptera by flagellates—for example, triatomine reduviids by *B. triatomae* (Schaub and Breger 1988) and the gerrid *Gerris odontogaster* by *B. gerridis* (Arnqvist and Mäki 1990)—results in pathogenic effects such as reduced host vigor associated with food stress. Further study is needed to determine the potential effects of trypanosomatid gut parasites on mirid populations.

LIFE TABLES

Changes in recruitment and mortality cause insect population densities to change from one generation to the

next. Because recruitment takes place over a relatively short period in animals with discrete generations, changes in density are often interpreted in terms of changing mortality, which operates continuously. Life table data have been prepared for few mirids, and seldom is key-factor analysis (Varley and Gradwell 1960) used to estimate the contribution of separate mortalities to overall generation mortality (key-factor analyses sometimes are flawed when the role of natality is ignored; see Hawkins et al. [1999]). Several studies, discussed below, represent notable contributions to the causes of mortality in mirids.

As determined by McNeill (1973), the braconid parasitoid *Peristenus pallipes* is unimportant in regulating populations of *Leptopterna dolabrata* in England. More important are density-dependent losses of third to fifth instars from competition for high-nitrogen feeding sites on grasses (i.e., seeds), and to annual fluctuations in fecundity, another consequence of competition for feeding sites. A density-independent factor—intensity of rainfall—sometimes kills large numbers of early instars (McNeill 1973). McNeill stated that no key factor was obvious, which Podoler and Rogers (1975) confirmed by using an additional method to identify key factors.

In other British studies, Glen and Barlow (1980) monitored the numbers of the carnivorous *Blepharidopterus angulatus* during 1965–1974 and assessed mortality, using *k*-values: parasitism, disease, eggs not laid, and other factors responsible for losses of immature stages and adults. The key factor affecting this bug's population dynamics is death or emigration of females, particularly the loss of prereproductive individuals, which is unrelated to prey numbers at the time of peak migration. Gange and Llewellyn (1989) prepared life tables for *B. angulatus* living on alder windbreaks and determined that emigration of adult females, with the resulting loss of eggs, is the key mortality factor (Fig. 6.35).

Less detailed studies include Williams's (1954) assessment of mortality—mainly from parasitism, predation, and changes in the food supply—in all life stages of the cocoa mirids *Distantiella theobroma* and *Sahlbergella singularis* in West Africa. Additionally, Wightman (1968) followed a first-generation population of *Lygocoris pabulinus* for 10 weeks in England. In that study, the greatest mortality appeared to occur in the egg stage and among early instars as a result of various density-dependent and -independent factors. Dempster (1966) used serological analysis (precipitin test, reviewed by Frank [1979]) to identify arthropod predators of broom-inhabiting mirids in England, and presented circumstantial evidence for the effects of predation on nymphal mortality.

Defense

Because many mirids are low-density pests, it might be expected that these insects would be capable of active

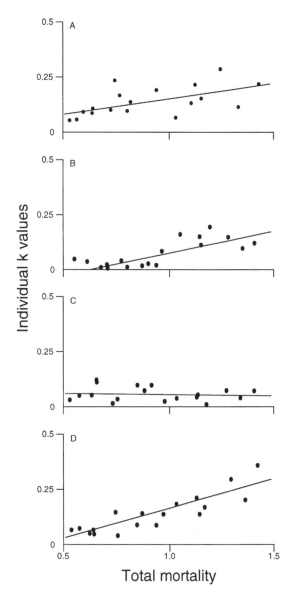

Fig. 6.35. Analysis of key factors affecting mortality of *Blepharidopterus angulatus*. A. k_1, loss of eggs and young nymphs: $y = 0.24x + 0.06$; $r = 0.52$, $d.f. = 16$, $P < 0.05$. B. k_4, loss of old nymphs: $y = 0.35x - 0.19$; $r = 0.759$, $d.f. = 16$, $P < 0.01$. C. k_5, early migration of males: $y = -0.04x + 0.17$; $r = 0.146$, $d.f. = 16$, $P > 0.05$. D. k_6, loss of adult females: $y = 0.518x - 0.19$; $r = 0.815$, $d.f. = 16$, $P < 0.001$. The key factor, by definition, is k_6, with the largest regression coefficient (0.518). (Reprinted, by permission of A. C. Gange and the Association of Applied Biologists, from Gange and Llewellyn 1989.)

defense (Kumar 1971a). Yet Kullenberg (1944) remarked that mirids are rather defenseless animals, but, as Dolling (1991) pointed out, most aspects of a bug's form, physiology, and behavior affect its susceptibility to attack by natural enemies. Some tropical species, such as *Distantiella theobroma*, feed mainly at night, which, in part, may represent predator-avoidance behavior (Cross and King 1973). Kumar (1971a) and Aryeetey and

Kumar (1973) referred to the use of a passive defense mechanism—camouflage—when certain cocoa mirids drop from host trees to avoid ants and blend in with the background to make detection by ground predators more difficult. Nymphs of many plant bugs are more cryptically colored than the adults (Kullenberg 1944). Mirids thus combine primary defensive mechanisms—those that operate regardless of whether a predator is nearby and lessen the chances of an encounter—with secondary mechanisms, such as flight, that operate during an encounter with a predator (Edmunds 1974).

PRIMARY DEFENSE

Anachoresis

Living in crevices or holes is a primary defensive mechanism called *anachoresis* (Edmunds 1974). As temporary anachoretes, mirids are able to avoid some predators as well as be protected from harsh environmental conditions. Living in crevices is also a means of reducing water loss in plant bugs (e.g., Nwana and Youdeowei 1976).

Crypsis

Nymphs of many mirids, especially those of inflorescence specialists (see Chapter 10), blend in with their host plants and, as noted in Chapter 5, some grass-feeding stenodemines are green in early season but later become brown like the color of their mature host plants (Plate 3; Schuh and Slater 1995). Adults of *Orthotylus flavosparsus* bear patches of silvery setae that help camouflage the bugs on chenopodiaceous hosts having a mealy appearance (Southwood and Leston 1959). Hawaiian *Nesiomiris pallasatus* exhibit a hemelytral "salt-and-pepper" pattern that renders the bug cryptic against the fecal spotting it causes on the lower surfaces of host leaves (Gagné 1997). Well-camouflaged or cryptic plant bugs are less likely to be detected by natural enemies. Cryptic coloration and behavior that help mirids escape detection by predators can be called *procrypsis* (Norris 1991).

Aposematism

Contrasted with the cryptically colored *Distantiella theobroma* and *Sahlbergella singularis* on cocoa, the brightly colored (orange and black) *Helopeltis* species lack furtive habits (Squire 1947) and rest mainly on exposed parts of cocoa trees (Youdeowei 1977). Aposematism, another category of primary defense, involves the advertising of dangerous or unpleasant attributes by striking coloration (usually red and black or yellow and black) so that predators avoid such animals (Edmunds 1974; cf. Sillén-Tullberg et al. 1982). In experiments using avian predators, aposematic heteropterans are eaten less frequently than nonaposematic bugs (Schlee

1986). Unpalatability in *Helopeltis* species has not been demonstrated but is suspected (Kumar 1971a). McLain (1984) observed that the mirid *Lopidea instabilis* and the lygaeid *Neacoryphus bicrucis* share both the same habitat (old fields) and a somewhat similar red and black pattern. He determined that both become unpalatable to a vertebrate predator (lizard) when they sequester pyrrolizidine alkaloids from a composite host plant (a senecio), and speculated that the bugs function as Müllerian mimics. The lizards tend not to attack the mirid or lygaeid after an encounter with an individual of either species that has fed on the composite. Müllerian mimicry is a type of defensive mimicry in which several aposematic species evolve similar color patterns that a predator rejects through a single learned avoidance response (Edmunds 1974).

The red and black *L. nigridia* is distasteful to some co-occurring arthropod predators on an alkaloid-producing lupine in Oregon, both in laboratory no-choice feeding experiments (McIver 1989) and in nature, based on serological analysis (McIver and Tempelis 1993). To demonstrate aposematism, McIver (1989) noted that predators must perceive the color pattern of *L. nigridia* and reject them because of an association with an unpleasant attack. Arthropods have not been shown to perceive long-wavelength visible light (McIver 1989), but the bugs might gain an advantage by being distasteful, even if predators do not associate taste with appearance. That *L. nigridia* is gregarious and shows less mobility relative to nonaposematic mirids could enhance the value of its aposematism as a defensive adaptation (McIver and Lattin 1990). By focusing on the nature of color vision and learning ability in predacious arthropods, these researchers are attempting to determine whether aposematism in this mirid has demonstrable effects on survivorship in nature. Similar studies on other plant bugs that show aposematic coloration, such as most restheniines and *Hadronema* species (Schuh and Slater 1995), are needed.

Mimicry

Webster (1897b) remarked that *Pilophorus amoenus* on new growth of pine resembled certain cerambycids, although he was not aware of a co-occurring member of this beetle family. Since then, several mirids have been said to "mimic" beetles occurring on the same hosts: *Campyloneura virgula* and the cantharid *Malthinus flaveolus* (Massee 1959), *Cyllecoris histrionius* and members of the cantharid genera *Malthodes* and *Malthinus* (Butler 1923, McGavin 1979), and *Lopidea robiniae* and the chrysomelid *Odontota dorsalis* (Balsbaugh and Hays 1972). Some of these examples might involve mimicry that affords protection to the bugs by causing errors in choice of prey, but none of these plant bugs has been subjected to experimental study. It is, therefore, not known if any of these cases represents Batesian mimicry, in which a predator avoids a palatable plant bug (mimic) because it resembles a distasteful beetle

(model), or involves a Müllerian system. Mirids in some instances might be involved in complex com-binations of Batesian and Müllerian mimicry (Schuh and Slater 1995; see also previous discussion of aposematism).

Resemblance to wasps
Some mirids, for example, *Helopeltis* species, bear a superficial resemblance to parasitic Hymenoptera such as braconids and ichneumonids (e.g., Ramachandra Rao 1915, Kumar 1971a, Gross and Cassis 1991). The orthotyline *Mecomma angustatum* is said to mimic the ichneumonid *Gelis* species (Carvalho and Southwood 1955); indeed, a synonym of this plant bug is *M. mimeticum*. It is unknown if a resemblance to wasps conveys some selective advantage to the mirids. In addition to a morphological resemblance to wasps, certain mirids (e.g., *Cyllecoris* spp. in Japan) behave like wasps, at least when they are captured in an insect net (Yasunaga 1999a).

Myrmecomorphy
That plant bugs of several subfamilies—especially Orthotylinae, Phylinae, and Mirinae—exhibit a remarkable antlike habitus is well known (Fig. 6.36; see also Fig. 5.3C, Plate 3) (e.g., Poppius 1921). This resemblance to ants can involve wing reduction, enlargement of the abdomen, and modification in shape of the head; in some species, white patches on a dark forewing or streaks of silvery hairs that create the impression of a three-lobed ant enhance the effect (Schuh and Slater 1995). Behavior such as an antlike gait and frequent antennal waving accentuates the resemblance to ants (Jackson and Drummond 1974). An example from the northeastern United States involves orthotylines of the genus *Schaffneria*. On scrub oak, *S. davisi* and *S. pilophoroides* resemble the ant *Dolichoderus taschenbergi* and seem to occur only in or near ant-attended aphid colonies on the host plant. These aphid predators (see Chapter 14) are patchily distributed in pitch pine–scrub oak barrens, occurring on the few scrub oaks that harbor aphids. The scattered nests of *D. taschenbergi* remain in the same positions for several years, and the bugs can be found on the same trees in successive seasons (Wheeler 1991b). The extent of aphid predation in nature is unknown, but behavioral modifications in this mirid might represent aggressive mimicry (McIver and Stonedahl 1993) that facilitates the approach to aphid colonies. Experimental evidence for aggressive mimicry is lacking, however, and the relationship might simply be Batesian.

A remarkable color polymorphism occurs in *S. davisi*. Populations associated with an all-black ant (*D. taschenbergi*) in northeastern pine barrens are entirely black, whereas farther south a red and black morph of the bug occurs in association with a similarly bicolored ant (*D. mariae*) in serpentine barrens (Henry 1994). In California, *Dacerla mediospinosa* generally is dark brown or black at higher elevations (up to ca.

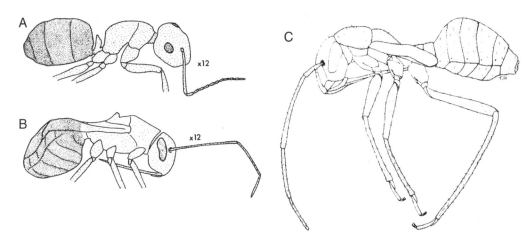

Fig. 6.36. Myrmecomorphic mirids (mimics) and co-occurring ants (models). A. The ant *Camponotus planatus*. B. The antlike *Barberiella* species, a presumed mimic of *C. planatus*; both are involved in a Batesian mimicry complex in Belize. C. Fifth-instar *B. formicoides*, which resembles the co-occurring ants *C. nearcticus* and *Formica subsericea* in the eastern United States. (A, B reprinted, by permission of J. F. Jackson and the University of Notre Dame, from Jackson and Drummond 1974; C, by permission of the Entomological Society of Washington, from Wheeler and Henry 1980a.)

2,600 m) but pale in the foothills or in the coast ranges; a similar trend prevails in the color of associated ant species (Carvalho and Usinger 1957).

A morphological and behavioral resemblance to ants—myrmecomorphy—is often termed *ant mimicry*, but Osborn (1898) advocated caution in using mimicry to refer to species that resemble ants unless biological data support true mimicry. He stated, "Whether the abortion of the wings and elytra is merely the result of such mimicry or connected with advantages of an entirely different nature we are not prepared to guess. I have used the term mimicry in a general way to cover this feature of resemblance but I would dissent from the use of this term in such a loose manner if a better one were available."

Fulton (1918) acknowledged the striking resemblance of *Pilophorus perplexus* to co-occurring ants but concluded that antlike appearance could be only accidental. He overlooked the possibility that myrmecomorphy could benefit a bug by deceiving potential vertebrate predators (Myers and Salt 1926, Donisthorpe 1927). Earlier, Webster (1897b) had considered the antlike behavior of *P. amoenus* to be unnecessary for protection against predators, but stated, "Yet, it seems to me, that the careful investigator would not be justified in dismissing the whole matter as a mere coincidence, but rather in searching elsewhere for the causes of a phenomenon of which the effects only are here perceivable." As untenable as Fulton's (1918) conclusion now seems—Kullenberg (1944) held a similar view—experimental evidence for the selective advantage of myrmecomorphy, until recently, has been lacking for heteropterans (e.g., Oliveira 1985, Cobben 1986).

McIver and Stonedahl (1987a, 1987b) studied the ecology of two myrmecomorphic phylines, *Coquillettia insignis* and *Orectoderus obliquus*. They identified potential selective agents ("operators" or sensitive signal receivers; see Vane-Wright [1976])—vertebrate and invertebrate predators—that perceive these supposed mimics and confuse them with various ant species (models) occurring on the bugs' host plants. Their data suggest that visual arthropod predators, such as nabids, reduviids, and salticid spiders, are part of a Batesian mimicry system with the mirid mimics and their ant models. Under controlled conditions, the reduviid *Sinea diadema* can associate an antlike habitus with an unpleasant encounter with a model. This predator acts as a selective agent by translating such experiences from the model to a mimic (*C. insignis* females) but not to nonmimetic plant bugs offered as alternative prey (McIver 1987).

McIver (1989) summarized results from field observations on *C. insignis* as a member of the lupine fauna in southeastern Oregon and from laboratory feeding trials involving visual arthropod predators that live on lupine. Evidence suggests that this plant bug is a classic Batesian mimic that gains protection from its resemblance to ants and is protected by an ant complex, with different ant species associated with the bug's various developmental stages. Such transformational mimicry (Mathew 1935) might be characteristic of systems in which the mimic—for instance, some heteropteran— undergoes gradual metamorphosis (McIver 1989). A possibly similar transformational mimicry occurs in the myrmecomorphic mirine *Barberiella formicoides* (Fig. 6.36C; Wheeler and Henry 1980a), although no experimental work has been conducted on its relationship to ant models or on co-occurring predatory arthropods.

SECONDARY DEFENSE

Mirids respond to predators or other stimuli by using secondary or active defensive mechanisms. Some species characteristically move opposite the source of disturbance on their hosts (Fitch 1870, Howard 1892, Slingerland 1893, Stedman 1899, Petherbridge and Husain 1918, Böcher 1971, Gagné 1997). Crosby (1911) observed this behavior in *Heterocordylus malinus* and *Lygidea mendax* on apple trees: "As the nymphs grow older they become more active, and when disturbed retreat to the twig where they adroitly dodge to the opposite side like a squirrel." To avoid attack by a euphorine parasitoid, *Blepharidopterus angulatus* nymphs attempt to outrun the wasp and disappear over the edge of a leaf (Glen 1977a). When disturbed, certain mirids walk briskly over the host plant and use their hind legs to skip 50 mm or more, with such behavior often repeated (Fitch 1870). *Halticotoma* individuals roll down or run to the bases of yucca leaves when disturbed (Haviland 1945), and those in the genus *Caulotops* behave similarly on agave (D. A. Polhemus, pers. comm.). This behavior might involve the use of an alarm pheromone, the presence of which has been observed in *Lygus lineolaris* (Wiygul and McKibben 1997). To escape, adult plant bugs can also make short, downward curving or zigzag flights from the host plant or drop to foliage beneath (e.g., Theobald 1895, 1911b; Brittain 1916a; Hawley 1917; Andrews [1923]; Smee 1928b; Leach and Smee 1933; Gilliatt 1935; Bodenheimer 1951). In addition to camouflage, cocoa mirids use a "fall-and-fly" escape as a defense against predators (Leston 1973a). A restheniine, *Platytylus bicolor*, employs similar behavior (Müller and Costa 1964). According to Leston (1973a), such behavior characterizes plant bugs that must disengage their forewings from the scutellar margin and couple the forewings and hindwings; falling and flying provides "a quicker launching than flying out from the substrate when disturbed." Dolling (1991) noted that flight alone is an uncommon means of escape for the Hemiptera except for mirids and whiteflies.

Saltation

Species of *Halticus* (Orthotylinae) not only resemble flea beetles (alticine chrysomelids), but their enlarged hind femora allow a similar jumping ability (e.g., Riley [1931?]). Saltatorial mirids, such as *Nesidiorchestes hawaiiensis* (Fig. 6.37; Zimmerman 1948a), *Chlamydatus pullus* (Böcher 1971), and *Myrmecophyes oregonensis* (Schuh and Lattin 1980), can jump 5–10 cm or more when disturbed. Some species of *Sarona* combine jumping with short, erratic flights (Asquith 1994). McGavin (1979) speculated that *H. apterus* and *H.*

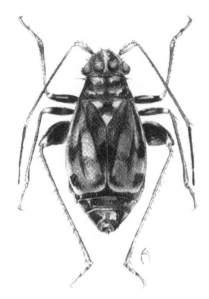

Fig. 6.37. *Nesidiorchestes hawaiiensis* showing enlarged (saltatorial) hind femora. (Reprinted, by permission of E. C. Zimmerman, from Zimmerman 1948a.)

saltator use "an energy storage system" similar to that in Orthoptera, which is related to their saltatorial habits. He suggested that the mechanism involves the metathoracic tibial-femoral articulation; resilin, although not mentioned, might be involved.

Thanatosis and Autotomy

Edmunds (1974) discussed feigning death, or thanatosis, and loss (autotomy) of nonessential body parts as other examples of secondary defense in animals. Plant bugs sometimes enter a cataleptic state lasting several minutes, or suddenly lose a leg when seized, at least under artificial conditions (Larsen 1941, Böcher 1971, Aryeetey and Kumar 1973, Dolling 1991). When legs are shed, the fracture is often between the trochanteral podomeres (a two-segmented trochanter is characteristic of mirids [Goel 1972]), the distal one being shed with the femur and tibia. The concomitant loss of campaniform sensilla possibly raises the threshold for leg shedding on the same side (McGavin 1979).

Leston (1980a) described a behavior more typical of sternorrhynchans than mirids: nymphs of the bryocorine *Prodromus thaliae*, when disturbed, remain on the underside of a leaf against the midrib, their legs and antennae outstretched and flattened against the substrate. When disturbed, some cryptically colored *Nesiomiris* species freeze on the underside of host leaves (Gagné 1997).

Chemical Defenses

Many animals use chemical defenses to repel predators, and it has long been known that certain mirids impart a disagreeable odor to ripe raspberries (Anonymous 1847). Moreover, *Lygus lineolaris* gives off an offensive smell when handled (Saunders 1883); *Leptopterna dolabrata* possesses an intensely strong, disgusting odor (Butler 1923); and some *Calocoris* (or other genera of the *Calocoris* complex), *Helopeltis*, and *Phytocoris* species give off a powerful smell (Brindley 1930, Ho et al. 1995; see also Kullenberg [1944]). In California, workers collecting large numbers of *Lygus hesperus* for experimental use perceived a noxious odor from the thousands of bugs and were afflicted with vertigo and nausea (Blumenthal 1978). Domek and Scott (1985) noted a distinct odor emanating from an insect net containing several hundred lygus bugs. Some of the chemicals produced by male and female mirids, which are not part of the female sex pheromone, probably serve a defensive function; both sexes release volatiles when disturbed (Millar and Rice 1998).

Effectiveness against predators

Forbes (1884a) determined that birds seldom feed on *Lygus lineolaris*. Although it is assumed that the characteristic odors some adult mirids emit from metathoracic scent glands help deter predators (Leston 1973a), the metathoracic secretions in several heteropteran families are ineffective in deterring predation by birds (Schlee 1986, 1992). Birds as natural enemies of mirids were discussed earlier in this chapter. Some heteropterans apparently move their mesotarsus or metatarsus across the orifice of the scent-gland outlet to moisten it with secretion before brushing it on an aggressor (Remold 1963). Blumenthal (1978) described and illustrated the paired, leaflike metathoracic glands in *L. lineolaris*, noting that the reservoir lacks the accessory gland tissue characteristic of most heteropterans. He suggested that the principal gland chemical, the ester hexyl butyrate, acts as a repellent.

Nymphs of the bryocorine *Platyngomiriodes apiformis* are covered with large red tubercles that exude clear, liquid droplets (Pang 1981), but any defensive function of the secretions is unknown. The results of Remold's (1962) study involving nymphs of 10 mirid species suggest a nondefensive function of the dorsal abdominal glands. In testing the hypothesis that secretions of the dorsal abdominal scent gland of bryocorines serve a repugnatorial function, Aryeetey and Kumar (1973) were unable to show that gland secretions afforded protection from ants. They observed, however, that nymphs of *Bryocoropsis*, *Distantiella*, and *Sahlbergella* exude fluid from their dorsal abdominal glands, and Aldrich (1988) speculated that the bugs sequester host alkaloids in the gland and "deliver them when needed via the blood or an epidermal syncitium [*sic*] (as in milkweed bugs) to setae or cuticular weak points." Mirid nymphal secretions presumably are allomones, but they are chemically unknown (Aldrich 1988). Carayon (1971) and Staddon (1979, 1986) reviewed scent-gland structure and biology in the Heteroptera; Pavis (1987) discussed the types of secretions known in heteropterans and their role as allomones, kairomones, and pheromones; and Aldrich (1988) reviewed heteropteran chemical ecology.

Volatile constituents of scent glands

In addition to the need to evaluate the effectiveness of mirid odors against potential predators, much remains to be learned about heteropteran scent-gland structure and function (Weatherston and Percy 1978, Aldrich 1988), particularly in mirids. They differ from certain other families of true bugs in the major components of their scent substances (Staddon 1979, Knight et al. 1984). Several, mainly unbranched, aliphatic substances that occur as principal constituents of volatile materials in mirids are thought to originate in the adult metathoracic scent glands. Such substances include aldehydes [(*E*)-2-octenal] in *Leptopterna dolabrata* (Collins and Drake 1965); mixtures of esters in *Lygus lineolaris* (Gueldner and Parrott 1978; cf. Blumenthal 1978), *Blepharidopterus angulatus*, and *Pilophorus perplexus* (Knight et al. 1984); and monoterpenes in *Harpocera thoracica* (Hanssen and Jacob 1982). The main constituents in the metathoracic scent gland of *Helopeltis fasciaticollis* are acetic acid, 2-octenal, and octyl acetate (Ho et al. 1995).

That esters are the main components of volatile materials in phytophagous and zoophagous mirids suggests that the bugs manufacture their own scent oils and that diet has little influence on the composition of gland substances (Knight et al. 1984). Moreover, it appears that reactions yielding toxic end substances occur outside the gland epithelial cells in most heteropterans (Staddon 1979) but that such extracellular biochemical reactions might not be typical of mirids (Knight et al. 1984).

7/ Morphology, Physiology, and Behavior in Relation to Feeding

> The details of diet reveal the enormous complexity of the feeding processes and diverse and precise specifications to which the physiological and biochemical machinery must be designed.
>
> —V. G. Dethier 1976

Although the morphology of stylets is similar in the Auchenorrhyncha, Sternorrhyncha, and Heteroptera, the structure of the head capsule in these groups differs significantly (see Fig. 7.5). In true bugs this capsule is closed ventrally by a cuticular bridge, the gula (see Fig. 5.4), which moves the rostrum forward where it can be used to explore the plant surface through labial tapping. The possession of a gula permits greater rostral mobility and allows heteropterans to have greater dietary breadth than homopterans (Cobben 1978, Sweet 1979, Dolling 1991, Stonedahl and Dolling 1991, McGavin 1993).

The term *feeding* appears throughout this chapter (and in other chapters), as do other terms relating to the feeding behaviors, strategies, and tactics of hemipterans. I use these terms in their more traditional senses but note that Backus (2000) clarified much of the terminology pertaining to hemipteran feeding.

Heteropterans often feed on plant parts rich in nutrients and high in energy. They are more exploitative or disruptive in their feeding than homopterans, disturbing plant hormone systems (which is also true of homopterans to some extent), sometimes causing violent necrosis of tissues, and triggering various wound responses. Phytophagous Heteroptera, especially flower-, fruit-, and seed-feeding species that can kill individual plants, could be termed *plant predators* (e.g., Southwood 1973); the use of *predator* in this context might be considered inappropriate because it implies that herbivory is necessarily deleterious to plants (Owen 1990). The destructive feeding behavior of most phytophagous mirids, as cell-content feeders, contrasts with that of the generally more plant-parasitic sternorrhynchans, such as aphids, scale insects, and whiteflies, which have delicate stylets that tap directly into the vascular system of their hosts (Pollard 1955, 1973; Kennedy and Fosbrooke 1973; Southwood 1973; Cobben 1978; Strong et al. 1984; Tonkyn and Whitcomb 1987) to conserve, rather than destroy, their food resources (Miles 1972, 1989a; Tjallingii and Hogen

Esch 1993). The often minimal laceration of host cells caused by many homopterans is associated with their stylet sheath-feeding mode (Miles 1972) and their ability to vector plant viruses (Nault 1997; see also Chapter 8). Some aphids, however, ingest from mesophyll parenchyma and are more destructive than are phloem-feeding aphids (e.g., Saxena and Chada 1971), and some homopterans do inflict severe injury, impairing host fitness (e.g., Dixon 1971a) and even killing their host plants within a few days (van Lenteren and Noldus 1990). Chloroplasts of the palisade parenchyma can be destroyed as stylets locate and penetrate phloem sieve elements. Such destruction represents an indirect effect of homopteran feeding (Cockfield et al. 1987).

Homopterans, in contrast to heteropterans, typically compensate for low nitrogen concentrations in plant sap by increasing the "throughput" or consumption (Kennedy and Fosbrooke 1973, Southwood 1973, Tonkyn and Whitcomb 1987). Although feeding behavior varies somewhat within the Homoptera, this is an apt distinction between homopterans and heteropterans. In particular, the relationship between aphids and their hosts has evolved toward parasitic specialization (Shaposhnikov 1987), in which large numbers of individuals often have little impact on their hosts (Southwood 1973, Miles 1989a). This generalization, however, might oversimplify the nature of plant responses to aphid feeding (Miles 1989b; see also Labandeira and Phillips [1996]).

Price (1980) presented a different view, arguing that heteropterans are also plant parasites if the definition of a parasite (as given in *Webster's Third International Dictionary*) "is applied objectively without taxonomic or disciplinary constraints." Some mirids are similar to plant-parasitic nematodes that inject secretions into plant tissue, causing cellular responses or sometimes inducing an imbalance of growth from a systemic reaction by the host (Yeates 1971, Hussey 1989). Among phytophagous heteropterans, Price's (1980) terminology seems most appropriate for leaf- and stem-feeding pentatomomorphans that are stylet sheath feeders—their feeding may minimize the physiological disturbance of their hosts (Miles 1987b). It is also possible to classify all arthropods that feed on plant surfaces as ectoparasites (Janzen 1985, P. W. Price 1997).

Semantic considerations can be extended by noting that some mirids and certain other insects are termed *plant pathogens*. Norris (1979) regarded mirids that cause local-lesion and systemic phytotoxemias as pathogens (causing noninfectious diseases) and referred to the evolution of terminology that has led to the designation of plant-infesting nematodes as pathogens rather than pests (also discussed by Wallace [1973:5–6]). The terminology relating to specialization and evolution among phytophages continues to evolve. Thompson (1994) discussed herbivores, all of which might once have been termed *plant predators*, as parasites, grazers, or predators, depending largely on the selection pressures they exert on their hosts. Certain mirids might be placed in any one of Thompson's (1994) categories, but an evaluation of this terminology's applicability to plant bugs must await a more thorough biological knowledge of the family and the bugs' effects on their host plants.

After lacerate and flush feeding is contrasted with stylet sheath feeding in heteropterans and homopterans, the remainder of this chapter focuses on mirids. The studies of F. Flemion, A. J. P. Goodchild, K. Hori, R. Kumar, S. Laurema, P. W. Miles, P. Nuorteva, F. E. Strong, A.-L. Varis, and others, which are cited, are invaluable in characterizing stylet structure and insertion and in reviewing the physiological aspects of mirid feeding, including host finding and acceptance, salivary secretions, and plant-wound responses. I borrow heavily from Miles (1968b, 1972), Tingey and Pillemer (1977), Cobben (1978), and Backus (1988a). In the absence of extensive behavioral and electrophysiological studies of mirids (some preliminary electronic monitoring of *Distantiella theobroma* [Cross and King 1973] and *Lygus hesperus* [Sevacherian 1975, Leigh 1976] has been done), I use Backus's (1988a) paper to infer the presence and function of various sensory systems, as well as the behaviors that mediate mirid feeding. Painter (1930) provided histological details of the alimentary canal, mouthpart structure, and salivary apparatus in *Pseudatomoscelis seriata*, the cotton fleahopper; and Goodchild (1952, 1966) and Miyamoto (1961) supplied additional information on the mirid alimentary canal. Smith's (1985) review of insect feeding mechanisms includes useful information on heteropterans.

Despite a voluminous literature relating to the feeding habits of mirids and their injury to crops, there are large gaps in our knowledge of mirid-plant interactions. Even such basic questions as the amount of ingestion from phloem and the significance of vascular tissue feeding in the Miridae need clarification.

Stylet Morphology and Insertion

MOUTHPART STRUCTURE

Heteropteran and homopteran mouthparts, the most specialized among insects (Snodgrass 1944, Goodchild

1966), are suited for gaining access to food enclosed within bark, skin, cuticle, or other protective layers (Smith 1985). The highly modified mouthparts of mirids and other heteropterans consist of a slender rostrum, with the generally four-segmented labium making up most of the visible "beak" or proboscis (Fig. 7.1). Within a groove along the dorsal (anterior) surface of the labium lie flexible, chitinous, threadlike or needlelike stylets: an outer pair of mandibles with a pair of maxillae within (Fig. 7.2). The mandibles are equipped with sharp, apical teeth (Fig. 7.3A, B), an adaptation for holding onto tissue beneath the host's integument; the maxillae are usually barbed, the right one more strongly so (Fig.

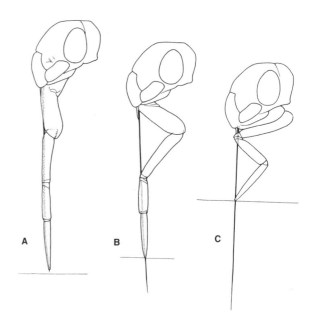

Fig. 7.1. Mirid feeding behavior. A. Labium placed perpendicular to host surface. B, C. Bending of the first and second labial segments as host tissues are penetrated. (Reprinted, by permission of B. Kullenberg and Zoologiska institutionen, Uppsala Universitet, from Kullenberg 1944.)

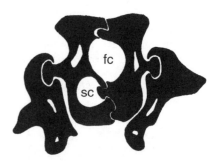

Fig. 7.2. Cross section of the labium of *Notostira erratica* showing the food canal (fc) and salivary canal (sc). (Reprinted, by permission of B. Kullenberg and Zoologiska institutionen, Uppsala Universitet, from Kullenberg 1944.)

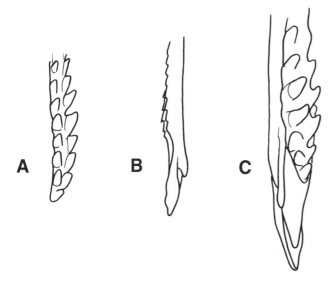

Fig. 7.3. Mouthparts of *Notostira erratica.* A. Apex of rostrum. B. Apex of mandibular stylets. C. Apex of maxillary stylets. (Reprinted, by permission of B. Kullenberg and Zoologiska institutionen, Uppsala Universitet, from Kullenberg 1944.)

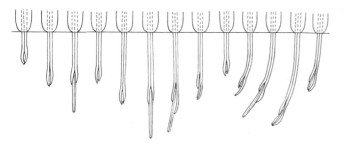

Fig. 7.4. Stylet insertion in mirids showing generalized sequential movement (left to right) of the mandibular (outer) and maxillary (inner) stylets. (Reprinted, by permission of B. Kullenberg and Zoologiska institutionen, Uppsala Universitet, from Kullenberg 1944.)

7.3C). In all hemipterans, the rostral tip does not penetrate the membrane covering a food source (see Fig. 7.1; Awati 1914, Butler 1918, Smith 1926, Cobben 1978, Miles 1987a). Hemipteran stylets were once considered nonliving structures (Pollard 1973).

During feeding, the interlocked maxillary stylets move together (Fig. 7.4), functioning much like a hypodermic needle of an injection syringe, and they are grooved so that their inner surfaces are appressed, forming a dorsal food and a ventral salivary canal (see Fig. 7.2). The food canal is connected with the pharynx; the salivary canal, with the salivary duct by way of the hypopharynx (Snodgrass 1935, Qadri 1959, Cobben 1978).

Even though heteropteran mouthparts have a prognathous orientation (Fig. 7.5A), whereas those of auchenorrhynchans and sternorrhynchans are hypognathous (Fig. 7.5B, C), the general stylet structure in these groups is similar (Smith 1926, Spooner 1938, Goodchild 1952, Pollard 1973, Tonkyn and Whitcomb 1987, Backus 1988a). As Pollard (1973) noted, however, detailed examination reveals certain differences in stylet structure. A major difference occurs between the Heteroptera + Auchenorrhyncha and the Sternorrhyncha. In the first two groups, mandibles are mirror images of one another, with the salivary canal found predominantly in the right maxillary stylet. Mandibles are not mirror images in sternorrhynchans, and their smaller salivary canal is contained mostly within the left maxillary stylet (Cobben 1978).

Horne and Lefroy (1915) compared the similar symptoms induced by the feeding of mirids and leafhoppers, noting the outer stylets of the latter are smooth; after insertion, they can be withdrawn without further tissue laceration. In contrast, the mandibular stylets of mirids have serrate edges (Cobben 1978). The mandibular stylets can be significantly shorter than the paired maxillae in some homopterans and in carnivorous heteropterans. Furthermore, in some taxa there are slight differences in thickness between the mandibles and the maxillae and in the position of the paired mandibles as they surround the maxillary stylets (Pollard 1969, Backus 1985). Other notable differences in mouthpart structure among heteropterans include the atypical maxillary barbs of corixine Corixidae, the elongated stylets of mycophagous Aradidae (Cobben 1978), and the long stylets of certain phloem-feeding Coreidae and Plataspidae (Maschwitz et al. 1987). Faucheux (1975) discussed and illustrated the variations in mandibular and maxillary ultrastructure of heteropterans, relating the differences to phytophagous, phytozoophagous, hematophagous, or predacious habits.

According to Cobben (1978), the possession of strongly barbed maxillae is plesiomorphic in the Heteroptera. Mirids, belonging to a family composed of phytophagous and zoophagous species, show an intermediate condition: maxillae roughened only on the inner surfaces. This contrasts with the moderately barbed maxillae of the mainly predacious Anthocoridae and Nabidae and the smooth maxillae of the Tingidae, a strictly phytophagous group. Several characters to be mentioned in this overview of stylet structure and insertion in mirids are considered derived states: the interlocking of mandibular and maxillary stylets, the deep penetration by the mandibulars, and the folding of the rostral segments (Cobben 1978).

The gross structure of the stylets in plant-feeding and predatory mirids is similar. Their salivary and food canals are about equal in diameter regardless of food habits (Cobben 1978, Cohen 1996). Differences, at least superficially, appear less pronounced than among the predatory asopine pentatomids and geocorid lygaeoids compared to phytophagous members of their respective taxa (Cobben 1978). Stylet morphology, therefore, is not an infallible indicator of trophic habits in the

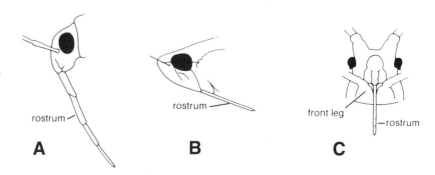

Fig. 7.5. Rostral positions in the Hemiptera. A. Heteroptera, in which the rostrum arises from the front of the head. B. Auchenorrhyncha, in which the rostrum arises from the rear of the head. C. Sternorrhyncha, in which the rostrum arises nearly between the front legs. (Reprinted, by permission of G. C. McGavin and Cassell, London, from McGavin 1993.)

Miridae (Cohen 1996). In Thysanoptera, the range of food sources substantially exceeds the range of variation in their mouthpart structure (Heming 1993), and such is also the case in mirids. Even so, the mandibular stylets of phytophagous mirids do vary, as for example with *Pseudatomoscelis seriata*, in which the mandibular stylets are shorter than the maxillaries (Painter 1930, Pollard 1969). Cobben's (1978) comprehensive discussion of heteropteran mouthparts provides a more detailed treatment of morphology and numerous scanning electron micrographs of stylet structure in mirids.

STYLET INSERTION

When a suitable site is found, feeding begins by a process Miles (1968b, 1972) termed "lacerate and flush." A mirid usually elevates the front of its body and places the labium perpendicular to the host surface (see Fig. 7.1A; e.g., Painter 1930). With the head lowered for penetration into host tissues, the first and second labial segments form a V having about a 30° angle (see Fig. 7.1B, C; Painter 1930). Contraction of the muscles in the proximal segment causes the remaining segments to bend during feeding (Weber 1930).

McGavin (1979) aptly described this characteristic position of the rostral segments during feeding (see Fig. 7.1): "The stylet bundle passes through the first rostral segment and is held firm; the second and third segments are folded back from the bundle during feeding and the stylets are again firmly held by the fourth segment, which also holds the stylet bundle on the desired spot prior to penetration." He (1979) contrasted the flexibility of stylets during feeding by mirids, tingids, and phytophagous pentatomomorphans (cf. Elson 1937) with the apparent rigid posture of stylets in some predatory bugs (e.g., anthocorids, nabids, and aquatic groups) as they pierce their prey (but, as noted below, stylets of predators show great flexibility as they ream prey contents). He also (1979) speculated that this flexibility relates to the possession of campaniform sensilla on the rostrum of plant-feeding taxa. In addition, Miles (1968b) observed that phytophagous heteropterans generally possess flexible mandibles curved at the tips rather than

the relatively straight, stout mandibles of carnivorous taxa (Pollard 1969, Cobben 1978).

Weber (1930) outlined the process of stylet insertion in Hemiptera, though parts of his often-cited model do not apply to auchenorrhynchans (which probe with maxillary stylets ahead) and probably most other hemipterans (Miles 1958a, 1968b; Backus 1988a). According to Weber (1930), one mandible is inserted a short distance, followed by the opposite one to the same depth. Using fixation techniques, Kullenberg (1944) determined that certain phytophagous mirids lower their maxillary stylets, positioning them between the mandibles. Muscles at the base of the mandibles apparently allow a path to be torn out for the maxillae (Heriot 1934). A "groove-ridge" interlocks the mandibular and maxillary stylets, and displacement of the apices of the latter is slight during feeding. Mirids differ somewhat from most other heteropterans in having mandibular stylets that act as "guiding rods" for the maxillae (Kullenberg 1944, Cobben 1978); sternorrhynchans have a similar mechanism (Backus 1988a).

Phytophagous heteropterans penetrate a substrate by using a reciprocating, "sawing" action of the stylets (Miles 1987a). Stylet insertion in mirids is not cautious and slow, as Kullenberg (1944) stated, but consists of a series of rapid thrusts into a small area of tissue (Flemion et al. 1954). Often localized, penetration sometimes is only a millimeter deep (e.g., Tanada and Holdaway 1954) or only 0.3 mm for *Helopeltis clavifer* (Miles 1987a). But it reaches a depth of more than 2 mm in some mirids (Kumar and Ansari 1974). As depth of penetration increases, the bug pushes its head closer to the substrate (Leach and Smee 1933), a pumping action of the head and proboscis becomes more noticeable (Latson et al. 1977), and the bend between the first and second labial segments is more pronounced (Flemion et al. 1954). During deep penetration, the rostrum may show two elbow folds and assume a W shape in side view (Goodchild 1952), particularly in pentatomomorphans (Rathore 1961, Cohen 1996) but apparently also in some mirids (Myers 1929, Goodchild 1952). The stylets do not elongate (Flemion et al. 1954), but deep penetration is made possible by longitudinal sutures on either side of the stylet groove. These sutures allow the rostrum to be shortened ("elbow" folded or telescoped; see Fig. 7.1B, C) and the stylets to be lifted from the labial groove (Cobben 1978). From the angle formed and its configuration relative to the labrum, the depth to which the stylets have penetrated can be estimated (Fig. 7.6; Miles 1987a).

Mirid stylets pierce most plant tissues with ease, those of *Lygus lineolaris* entering celery stalks as easily as one pushes a pin into a cushion (Davis 1893). Stylet penetration results mainly from pressure exerted by the bug. It is both intercellular and intracellular (e.g., Ledbetter and Flemion 1954) but usually becomes strictly intracellular (Fig. 7.7) because cell walls offer little resistance (Smith 1926, Bech 1967). Stylets of *L. lineolaris* enter the epidermis of bean pods between the guard cell of a stoma and the adjacent cell (Fig. 7.8; Flemion et al. 1954). Histological observations indicate that *L. rugulipennis* pierces sugar-beet tissue intercellularly or intracellularly (Hori 1971a) but follows a strictly intercellular path when feeding on pumpkin fruit (Hori et al. 1987). In aphids, the use of electron microscopy helps clarify the path taken by the stylets, although in some cases the distinction between intercellular and intracellular penetration is blurred (Montllor 1991).

Aphids often puncture intercellularly (stylet entry in aphids is detailed by Pollard [1973], Montllor [1991], and Tjallingii and Hogen Esch [1993]) and create a sharply defined feeding channel (e.g., Pollard 1973). Mirids usually form a broader, more irregular feeding track (Wagner and Ehrhardt 1961, Bech 1967). That mirids usually fail to produce a definite stylet track (Flemion

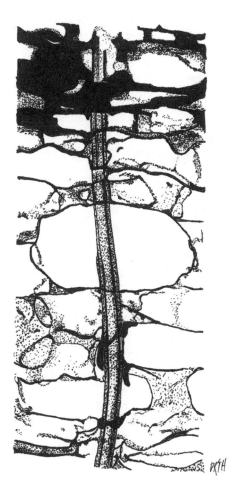

Fig. 7.7. Intracellular path of stylets of *Lygus lineolaris* in bean pod (×200). (Redrawn, by permission of the Boyce Thompson Institute for Plant Research, from Flemion et al. 1954.)

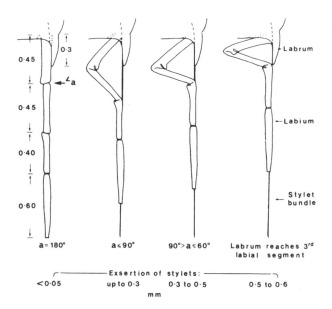

Fig. 7.6. Estimating stylet penetration (exsertion) in *Helopeltis clavifer*. On the basis of typical lengths of the labial segments and labrum (in millimeters), and from the angle of the elbow and its conformation relative to the labrum, it is possible to estimate, within ± 0.05 mm, how far the stylets have been inserted. (Reprinted, by permission of P. W. Miles and Birkhäuser Verlag, from Miles 1987a.)

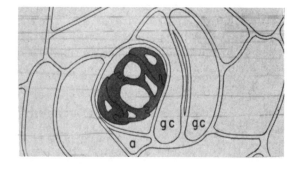

Fig. 7.8. Stylets (shaded) of *Lygus lineolaris* in bean tissue between guard cell (gc) of stoma and adjacent cell (a). (Reprinted, by permission of the Boyce Thompson Institute for Plant Research, from Flemion et al. 1954.)

et al. 1954, Cobben 1978) is evident in material sectioned soon after stylet penetration (King and Cook 1932)—that is, before tissues turn brown (Hori 1971a). Instead of a clearly defined stylet track, a brown discoloration or discolored streak of collapsed cells may be evident (Leach and Smee 1933, Tanada and Holdaway 1954, Hori 1971a). Tanada and Holdaway (1954) observed that the track of *Engytatus modestus* superficially resembles the stylet track of aphids and certain other homopterans (see Fig. 9.9). The stylet tracks of whiteflies, in contrast to those of aphids, are only poorly developed and stain faintly (Pollard 1955). The feeding track of *L. rugulipennis* leaves microgranular substances that might represent salivary material secreted during withdrawal of the stylets or formed by the reaction of saliva with plant substances (Hori 1971a). The stylet path of *Helopeltis clavifer* contains some solid deposits but is characterized by the dark remains of damaged cells (Miles 1987a).

Heteropteran stylets are partly composed of flexible elastic chitin (the rest is cross-linked proteins), allowing them to bend in any direction (Hori 1968a). Although mirid stylets can enter at right angles to the plant surface, they are capable of moving rapidly in different directions, following an irregular pathway (Flemion et al. 1954, Putshkov 1956, Flemion 1958, Hori 1971a). Mirid stylets presumably can orient parallel to leaf surfaces, as Pollard (1955) described for a whitefly and (1959) for a tingid. The stylets can twist considerably after entering a stem (Leach and Smee 1933), take a curved path in the cortex (Squire 1947), or turn up sharply (Taylor 1954). This twisting is readily apparent in the laboratory when plant bugs probe and suck juices from heat-killed or crushed caterpillars (pers. observ.; see also Cheng [1965] and Cohen [1990] for discussion of extreme stylet flexibility in predacious heteropterans).

Closterotomus norwegicus maintains directional control of the stylets. When its mandibles are inserted into polyporus (a strip of polypore fungus), they follow a curved path and the individual maxillaries penetrate in a straight line (Pollard 1969). Phytophagous heteropterans probably are able to control the direction their stylets take through extrusion of the inwardly curved mandibles (Miles 1958a, Pollard 1969). Stylet flexibility allows a mirid to explore an unexpectedly large area of plant surface without having to move (China 1925a, Painter 1930); stylets can be reinserted several times while the tip of the rostrum remains stationary (Awati 1914, Flemion et al. 1954, Strong 1970, Varis 1972). Features of the host plant influence stylet penetration and directional control. Plant maturity and texture affect the stylet behavior of *L. lineolaris* (Flemion et al. 1954, Pollard 1969). Lateral movement of the stylets in *Notostira elongata* depends on how solid the epidermis and parenchyma cells are, with greater stylet flexibility apparent in host leaves having a weak, rather than firm, epidermis (Bockwinkel 1990).

Feeding Strategies of Heteroptera and Homoptera

Backus (1988a) discussed heteropteran and homopteran feeding strategies, comparing sensory systems and behaviors that mediate their feeding and emphasizing differences among four lineages (infraorders) containing plant feeders: auchenorrhynchans, sternorrhynchans, pentatomomorphans, and cimicomorphans. The distinct nutritional patterns that have evolved are correlated with differences in the sensory systems (antennal, labial, stylet, and precibarial sensilla) involved in host orientation and acceptance, as well as in stylet control during feeding. As Backus (1988a) pointed out, the strategies enabling Heteroptera and Homoptera to specialize on particular host tissues are lacerate (or macerate [Miles 1987b]) and flush feeding, and stylet or salivary sheath feeding (Miles 1968b, 1972; Cobben 1978).

LACERATE AND FLUSH FEEDING

Many heteropterans, including mirids and tingids among the Cimicomorpha, and a few homopterans (mainly typhlocybine leafhoppers) feed by a lacerate and flush method, rather than producing a continuous stylet sheath. Stylet sheath feeding (Miles 1968b, 1972, 1987b; Backus 1988a), characteristic of phloem and xylem feeders, is considered the more highly evolved strategy (Goodchild 1966, Cobben 1978). Tonkyn and Whitcomb (1987) pointed out that lacerate and flush feeding is less discriminating than is a stylet sheath-feeding mode, but it affords higher rates of food intake. Although Cobben (1978) assumed that cimicomorphans have lost the ability to produce a stylet sheath or flange, such products of feeding occur in the Reduviidae (Friend and Smith 1971) and in the Anthocoridae and Nabidae (Cohen 1990, 1996). Details of the formation and function of the salivary sheath and of stylet movement during feeding are available in reviews by Miles (1972, 1987b), Backus (1985, 1988a), and Smith (1985).

Baptist (1941) and Nuorteva (1954) suggested that extraoral digestion plays only a minor role in mirids because of their short feeding times and low enzyme activities. But for several reasons (discussed by Cohen [1993]) previous researchers failed to appreciate the importance of a strategy that allows relatively small heteropteran predators to use larger prey; that integrates morphological, behavioral, and biochemical adaptations; and that results in a high ingestion efficiency of nutrient-rich food (Cohen 1989, 1993, 1995, 1998a). Researchers once thought that salivary enzymes injected into plant tissue and later sucked up with sap were involved mainly with digestion in the alimentary canal. Polygalacturonase in *Lygus rugulipennis*, however, functions extraorally to degrade plant cell walls. Amylases play a major role in the extraoral hydrolysis of plant starch (Chippendale 1978), and proteinase also shows activity sufficiently high to be involved in extraoral digestion (Laurema et al. 1985). These major enzymes—amylase, polygalacturonase, and proteinase—are discussed in more detail later in the chapter.

To test host suitability, hemipterans secrete copious amounts of saliva containing digestive enzymes, which are dabbed on the plant surface and sucked up to internal sensory organs (Miles 1968b, 1972). Insertion of the stylets during test probing (testing plant constituents for suitability with short-distance, brief probes) and exploratory probing (penetrating deeper to locate preferred feeding sites) tends to be of short duration and the probes rapidly repeated (Backus 1988a). During feeding, the stylets are inserted intercellularly or intracellularly, lacerating mesophyll cells during active movement. Despite the long-held misconception that mirids feed solely on liquid food (discussed by Putshkov [1966]), digestive enzymes, with the aid of barbed stylets, help create a soupy mixture or "particulate slurry" (Labandeira and Phillips 1996) that can be imbibed. In other words, plant bugs "dilute the tissue to a suckable brew" (Carver et al. 1991). Ingestion of this mixture of saliva and particulate matter typically results in an emptied food pocket or cavity (Fig. 7.9; Hori 1971a, Labandeira and Phillips 1996). Mirids are often lumped with all other phytophagous hemipterans as "suckers," or are designated as "sap feeders" (e.g., Cassis and Gross 1995). But they do not specialize on sap in the sense of aphids and other phloem feeders that tap sieve elements, but instead ingest liquefied by-products, using both cytoplasm, with its included nucleoplasm and organelles, and sap (Southwood 1973, Tonkyn and Whitcomb 1987, Campbell et al. 1994). The term *laceration* is inappropriate when referring to certain heteropterans, including the mirid *Helopeltis clavifer*, that suck out a mass of cells without actually penetrating them. Use of salivary polygalacturonase to macerate a pocket of tissues chemically is best termed "macerate and flush" feeding (Miles 1987a, Miles and Taylor 1994).

The presence of salivary polygalacturonase in mirids apparently relates to their lacerate-flush feeding strategy (Strong 1970, Hori 1975a). Many aphids, which evidently are phloem feeders, also have salivary polygalacturonase but often cause minimal injury to their hosts. Moreover, phloem-feeding coreids cause symptoms similar to those of mirids, yet pectinase is lacking in the coreid species that have been studied (Taylor and Miles 1994). Lacerate and flush feeders mainly attack mesophyll, but they also ingest from secondary sites. This behavior is typical of lygaeids and pyrrhocorids among pentatomomorphans, tingids and phytophagous mirids among cimicomorphans, some typhlocybine leafhoppers among auchenorrhynchans, and some scale insects and adelgids in the Sternorrhyncha (Cobben 1978, Tonkyn and Whitcomb 1987, Backus 1988a).

During host penetration, heteropteran lacerate and flush feeders lead with their mandibular stylets, and the maxillaries follow slightly behind. Backus (1988a) suggested that feeding with the mandibular stylets ahead represents a derived character in cimicomorphans. The relatively few homopteran lacerate and flush feeders insert their mandibular stylets shallowly into plant tissue, with the maxillaries pushed through and past the mandibulars during probing (Smith 1985, Backus 1988a). Many homopteran mesophyll feeders ingest the contents one cell at a time (Tonkyn and Whitcomb 1987).

Phytophagous mirids are mainly lacerate and flush feeders on the mesophyll of leaves and on the ground tissues of stems, inflorescences, fruits, and developing seeds. Mesophyll is a richer, better-balanced nutrient source than phloem or xylem (Tonkyn and Whitcomb 1987), with chloroplasts of the palisade parenchyma containing large amounts of protein and photosynthate (Sadof and Neal 1993). Unlike most mirids, lygus bugs can feed on nearly mature seeds (Putshkov 1975). But plant bugs feed mainly on growing tissues (Hori 1971a) and fruits rather than on dry seeds as do many lygaeoids and pyrrhocorids.

USE OF VASCULAR TISSUES IN MIRIDS

Mirids have been classified as phloem feeders, apparently on the basis of casual field observations (e.g., Dhileepan et al. 1990) or perhaps just the assumption that this habit is common in mirids and other phytophagous heteropterans (e.g., Berenbaum 1990, Lavigne et al. 1991, Becker 1992, Blommers et al. 1997). For most mirids the evidence does not justify this classification.

Lygus rugulipennis, a mirid whose feeding habits have been extensively studied, is stated not to feed from phloem (Hori 1972c), and in the case of the mesophyll-feeding *Labops hesperius*, transmission electron microscopy fails to reveal damage to the xylem or phloem (Campbell and Brewer 1978, Brewer et al. 1979). Some histological studies do show mirid stylet tracks that terminate in the phloem and xylem, in addition

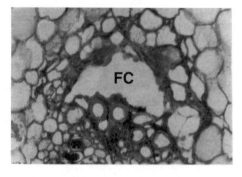

Fig. 7.9. Feeding cavity (FC) in the intrafascicular cambium of a sugar-beet petiole fed on by *Lygus rugulipennis* (×800). (Reprinted, by permission of K. Hori and the Japanese Society of Applied Entomology and Zoology, from Hori 1971a.)

to the mesophyll and other tissues (e.g., Smith 1926, Painter 1930, King and Cook 1932, Leach and Smee 1933, Bech 1967, Hori 1971a, El-Dessouki et al. 1976; see also Kullenberg [1944] and Chapter 9). Smith (1926) noted that mirid stylets take an intracellular rather than intercellular path to phloem. Some plant bugs might use xylem tissue only as a source of water during dry periods, as certain phloem-feeding lygaeoids do (Painter 1928). The significance of xylem feeding by mirids, as in aphids (Montllor 1991), has not been critically studied.

When stylets of a tea-feeding *Helopeltis* species were severed during feeding, they were always found to have touched a vascular bundle (Cohen Stuart 1922). In addition, *Engytatus modestus* feeds mainly on the vascular bundles of tomato plants (Tanada and Holdaway 1954), *Nesidiocoris tenuis* is nearly restricted to feeding in phloem tissues of tomato stems and petioles (Raman and Sanjayan 1984b, Raman et al. 1984), and *Dichrooscytus* species might feed on the phloem sap of junipers (Polhemus 1988). The actual amount of ingestion involved in these observations is uncertain. Miles (1987a, 1987b), noting that mirids appear unable to secrete a continuous, tubular stylet sheath, concluded that hemipterans cannot feed on vascular tissue other than through a complete sheath (excepting the results of Saxena's [1963] study of a pyrrhocorid). Heteropterans are not mentioned in J. A. Raven's (1983) review of xylem and phloem feeders.

In commenting on the tissue preferences of hemipterans, Carter (1973) cautioned that it should not be assumed that hemipteran feeding is limited to the objective tissue, particularly in the case of intracellular feeders. The lygaeoid *Blissus leucopterus* pierces the epidermis, mesophyll, sclerenchyma, bundle parenchyma tissues, protoxylem, and xylem, but the objective of its stylets, and this insect's food source, is the phloem (Painter 1928). An analysis of the composition of fluid excreted by mirids, similar to that conducted on xylem- and phloem-feeding Homoptera (Cheung and Marshall 1973, J. A. Raven 1983), might supply evidence for or against significant ingestion of xylem or phloem sap. Although Cheung and Marshall (1973) determined that cicada urine contained the same ions as xylem fluid, and in the same proportions, this is not true of all Homoptera (e.g., Cochran 1975) and might not be the case in mirids that feed on vascular tissues.

In some mirid studies using light microscopy, the resolution likely was insufficient to determine whether stylet tracks actually terminated in vascular tissue (e.g., Montllor 1991). Plant bugs might be able to change their feeding behavior depending on the plant part attacked, for example, from a mesophyll feeder on leaves to a phloem feeder when using stems, or to alter their behavior depending on the host plant (e.g., Hori 1971a). Such behavioral plasticity occurs in the potato leafhopper (*Empoasca fabae*) (Backus and Hunter 1989). Yet the primary tissues used by certain other leafhoppers

usually do not vary as they switch their feeding behavior (Naito 1976, 1977b; Tonkyn and Whitcomb 1987). An exception is *Nephotettix virescens*, which can switch its ingestion from phloem on susceptible cultivars of rice to xylem on resistant cultivars (Khan and Saxena 1985). In some Miridae, tissue preferences might change with age, as in the pentatomid *Eurydema rugosa*, which alters its feeding behavior on crucifers from vascular tissues in early instars to mesophyll in later stages (Hori 1968a).

Even homopterans that specialize on particular cell types within plant tissue can show plasticity in their feeding behavior. The presumed phloem specialist *Cacopsylla pyricola* ingests from all leaf cell types (Ullman and McLean 1988), and certain psyllids associated with Australian eucalypts feed not only from the phloem tissue of small vascular bundles but also from cells of the sheathing parenchyma (Woodburn and Lewis 1973). Predominantly mesophyll-feeding mirids, therefore, might also ingest from cells of the vascular bundle. They sometimes ingest from stems, petioles, and midribs, which potentially involves some phloem feeding. Or they might become accidental or facultative phloem feeders, ingesting occasionally from phloem after the host loses enough turgor to reduce the stem girth, which could allow deeper penetration by the stylets. Certain leafhoppers appear to display such behavior (Günthardt and Wanner 1981, Hunter and Backus 1989). Mirids, in fact, would be more likely to engage in chance or incidental feeding from vascular cells than would homopterans that tend to ingest cellular contents one cell at a time (Tonkyn and Whitcomb 1987).

Phloem feeding by mirids might consist principally of ingestion from various kinds of parenchyma cells, and thus be similar to feeding by aphids that ingest from phloem parenchyma when placed on nonhost plants but feed from sieve elements on their normal hosts (e.g., McLean and Kinsey 1968). Whitefly stylets similarly can end in the parenchyma (Pollard 1955).

Plant bugs do not tap phloem directly by inserting their stylets into sieve tube elements as do many aphids (Mittler 1957, Tonkyn and Whitcomb 1987) and certain other homopterans (J. A. Raven 1983, Tonkyn and Whitcomb 1987), or adopt the sedentary habits of phloem feeders (Tonkyn 1985). Mirids also penetrate mostly intracellularly, rather than intercellularly as do many phloem-feeding Homoptera (Pollard 1955, Tonkyn and Whitcomb 1987). Mirids clearly do not show the high throughput of sap and honeydew excretion characteristic of aphids and whiteflies. Marchart (1968) characterized cocoa mirids as destroyers of phloem tissue but also noted that the bugs do not secrete honeydew. Mirids and other heteropterans lack the specialized filter chamber that homopterans use to dispose of excess water. Moreover, they lack gastric caeca, which possibly serve as water-excreting organs in some strictly sap-feeding Heteroptera (Goodchild 1963a, 1963b). Mirids typically do not feed in the aggregations that characterize phloem-feeding homopterans and coreids (Goodchild 1977), and they do not produce the large amounts of excrement characteristic of xylem-feeding homopterans (Horsfield 1978) or the copious sugary fluid excreted by a few attended coreids and plataspids (Maschwitz et al. 1987). The black excrement of many mirids is characteristic of mesophyll feeders (see Chapter 9). Some tropical bryocorines, which induce the formation of large lesions into which plant sap exudes, excrete a transparent fluid that is not, however, considered indicative of phloem-feeding habits (Goodchild 1963b). The feeding strategies of most mirids—using developing seeds and ovules and rapidly growing tissues such as new shoots and young leaves, and lacerating or macerating their mesophyll cells—do not create the excess of plant sap or the problem of osmotic regulation that typifies the habits of most Homoptera and some Heteroptera (Goodchild 1963a, 1963b). As Goodchild (1963b) stated, "There is no clear evidence that sap alone constitutes a major portion of the diet" in mirids or tingids.

The salivary secretions of mirids—amylases, proteinases, and other enzymes not possessed by xylem and phloem specialists (Tonkyn and Whitcomb 1987, Cohen 1996)—and their general inability to transmit xylem- and phloem-limited pathogens (see Chapter 8) provide additional evidence that they are not typical vascular feeders. Mirids seem not to possess the behavioral, morphological, or physiological adaptations characteristic of vascular tissue feeders. Although our knowledge of mirid ingestion from vascular tissues is inadequate (ultrastructural studies of paths taken by the stylets are needed), similar imprecision exists for the aphids, the best-studied group of sap-feeding insects. In aphids, the extent and significance of nonphloem ingestion, as well as certain other aspects of aphid-plant interactions, remain unresolved (Montllor 1991).

ENDOSYMBIONTS

Endosymbionts generally are less important to the nutritional needs of heteropterans than of homopterans. Next to the Coleoptera, the Homoptera most frequently harbor endosymbiotic microorganisms (Buchner 1965, Dasch et al. 1984). Transovarially inherited endosymbionts, often found within specialized cells called *mycetocytes*, are almost universally associated with homopterans; endosymbionts of aphids, leafhoppers, and scale insects are among the best studied (Buchner 1965, Houk and Griffiths 1980, Dasch et al. 1984). These homopterans are highly dependent on their endosymbionts, or "biochemical brokers" as Southwood (1985) termed them. Endosymbionts likely supply nutritional requirements such as key amino acids, sterols, and vitamins that are lacking in a diet of plant sap, but investigators have tended to disagree on the exact role the bacteria play in host nutrition (e.g., Houk and Griffiths 1980, J. A. Raven 1983, Dasch et al. 1984, Houk 1987). Molecular studies of aphid-microbial symbioses strongly suggest that endosymbionts do provide aphids with essential amino acids and vitamins (e.g., Baumann et al. 1997, Douglas 1998, Nakabachi and Ishikawa 1999).

In the Heteroptera, the presence of endosymbionts (some might simply represent gut microbes) is not as uniform as in homopterans. When present, endosymbionts most often are found in gastric caeca (Forbes 1892, Glasgow 1914, Buchner 1965). Inheritance of these extracellular symbionts generally is by superficial contamination of the eggshell (Dasch et al. 1984).

The presence of endosymbionts tends to correspond with diet in heteropterans (Buchner 1965). They are absent in aquatic and semiaquatic bugs (Koch 1967, Dasch et al. 1984), although Cobben (1968) reported their presence in a mesoveliid. Goodchild (1966) emphasized the correlation between the presence of caeca and a preference for feeding on soft plant tissues. Gastric caeca are found in phytophagous pentatomoid groups, most coreoids, and many seed-feeding lygaeoids (Southwood 1973). It has long been known, however, that mirids lack these caeca (Dufour 1833; Forbes 1892; Goodchild 1952, 1963b) and do not harbor the endosymbionts that most sap feeders possess (Buchner 1965, Southwood 1973).

One of the few papers treating an association between microorganisms and mirids is that by Chang and Musgrave (1970). From the midgut epithelium of *Stenotus binotatus* collected in Ontario, they reported intracellular rickettsia-like microorganisms that are Gram-negative, somewhat pleomorphic rods (0.3 × 2.5 μm). These microbes do not seem to produce pathological effects, and the possibility of a mutualistic relationship with the mirid host was raised but not investigated.

Coccus forms of bacteria are found in the midgut of *Pseudatomoscelis seriata* (Painter 1930), but the nature of their association with this mirid is unknown. Steinhaus (1941) reported the presence of bacteria in the alimentary tract of *Lygus lineolaris*. Goodchild (1952) recorded a *Micrococcus* species and rod-shaped bacteria from the gut of cocoa mirids, noting that one of the groups might be pathogenic, and that if symbiotes are present to provide supplemental nutrition, they would have to reside in a gut that lacks caeca.

That bacteria occur in the gut of mirids is not surprising. Bacteria could be acquired easily from plant surfaces during labial tapping when watery saliva is secreted and almost immediately sucked up to test host suitability. The discovery in *Stenodema calcarata* of a duct connecting the accessory salivary gland to the fat body suggested that symbiotic bacteria or yeasts found in the fat body are the source of indoleacetic acid in the bugs' saliva. Experiments, however, showed that it was "exceedingly improbable" that plant-growth substances could be transported from the fat body by way of the accessory salivary gland to salivary secretions (Nuorteva 1956a).

Other mirids potentially harbor gut microbes—the family has not been adequately surveyed for the presence of endosymbionts (Chang and Musgrave 1970). As Dasch et al. (1984) pointed out, most workers have examined the gut of heteropterans for obvious caeca, perhaps overlooking intracellular microbes in the midgut epithelium or in abdominal mycetomes.

Viruses and other phytopathogenic microorganisms can affect the chemical composition of plants (e.g., the amount of soluble nitrogen) and, thereby, the development of their vectors or other insects on these plants (Macias and Mink 1969, Gildow 1983). A type of symbiosis might exist between microbes and their insect vectors, including associations that improve plant nutrition. It is not known if this type of symbiosis has any significance for mirids.

Host Finding and Acceptance

Before a herbivore begins to feed, it must locate its host in both time and space. Behavioral events associated with finding and accepting a host have been little studied in mirids, and even morphological, electron microscopic studies of their sensilla have been few. Intraspecific variation in sensory organs, for example, has not received attention. Detailed behavioral and electrophysiological studies on plant bug sensory systems and various aspects of their feeding behavior are needed. Particularly useful would be a comparative approach using phytophagous species that show extremes in diet breadth, specialize on various plant parts, prefer different tissues, produce different feeding symptoms, and display disparate tendencies toward carnivory. Research may establish that phytophagous heteropterans show more plasticity in the chemosensory aspects of finding and accepting hosts and in their feeding behavior than do homopterans (Cobben 1978). The results of studies using representatives from all mirid subfamilies and tribes might modify the current ideas regarding feeding strategies within this diverse family, the tissues accepted by different species, and possible changes in their feeding behavior on different hosts. Also needed is additional research on the chemical basis of host acceptance—for example, the role of secondary plant substances (whether phagostimulatory, deterrent, or toxic) and nutrients in the selection process, as well as the effect of physical defenses such as trichomes (e.g., Bodnaryk 1996) on mirid feeding. The stylet flexibility of mirids discussed earlier in this chapter may allow them to avoid plant parts rich in toxins (see Chapter 10).

The steps associated with hemipterans' finding, examining, and consuming a host—host orientation while in flight, rostral tapping and surface exploration, reingestative tasting of watery saliva, stylet probing and ingestion, and cessation of feeding and stylet withdrawal—involve sensory stimulation. In the case of *Lygus lineolaris*, the behavioral components of accepting and feeding on a host involve relatively stereotyped fixed-action patterns. Hatfield et al. (1983) proposed a model for the sequence of behavioral events, indicating appropriate sensory inputs (Fig. 7.10; Frazier 1986). The associated behaviors can be considered in terms of input (sensory stimuli) and output (insect responses), the two being connected by the central nervous system (Backus 1985). Miller and Strickler (1984) provided an overview of host finding and acceptance by insects and clarification of the terminology; Städler (1984, 1992) reviewed contact chemoreception in insects. Miles's (1958b) study of contact chemoreception in some lygaeids, pentatomids, and pyrrhocorids is helpful in understanding chemoreception in heteropterans, and Pollard's (1973) review of plant penetration by aphids contains material relevant to this process in mirids.

HOST FINDING

Visual Stimuli

Before plant-feeding insects orient toward potential hosts, they show random movement independent of plant cues (e.g., Miller and Strickler 1984, Frazier 1986). Host orientation by visual and olfactory cues is little known in the Hemiptera (Backus 1988a), and even fewer studies have been conducted on mirids than on aphids or leafhoppers. Initial orientation to a host in the tarnished plant bug (*Lygus lineolaris*) involves olfactory and visual stimuli operating at relatively long distances from the plant (Avé et al. 1978, Prokopy et al. 1979). This polyphagous mirid perhaps is a visual generalist (Prokopy and Owens 1978). In contrast to monophagous or oligophagous herbivores that develop on apple, *L. lineolaris* adults do not orient specifically to the hue or form of host structures (Prokopy and Owens 1978). Although this species may show no significant color response (Capinera and Walmsley 1978), significant differences in the numbers of *L. lineolaris* captured on or in traps of certain colors suggest that visual stimuli play some role in this bug's host orientation (Prokopy et al. 1979). The visual response of *Heterocordylus malinus*, *Lygidea mendax*, and other mirids restricted to developing on apple trees and related rosaceous genera might be evaluated as a further test of the hypothesis that monophagous or oligophagous species are more specific than polyphages in orienting to visual characteristics of their hosts.

Experiments demonstrating that *Lygus lineolaris* shows no obvious preference for flowers of a particular color (Painter 1929b) support Prokopy and Owens's (1978) conclusion regarding this bug's nonspecific visual responses. Boness (1963) and Bech (1965) showed that the European tarnished plant bug (*L. rugulipennis*) and other *Lygus* species are attracted to yellow-flowered crucifers. In the laboratory, however, the bugs are also attracted to other ranges of the light spectrum (Bech 1965). The use of yellow pans (Moericke traps) has

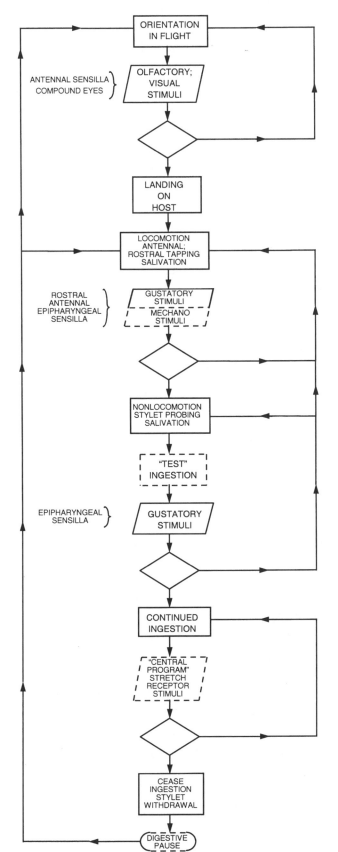

proved unsuccessful in attracting *L. rugulipennis* (Bech 1965, Varis 1972), but significantly more *L. elisus* and *L. hesperus* can be collected from light orange-yellow and deep chrome yellow pan traps in eastern Washington than from traps of other colors (Landis and Fox 1972).

Host Volatiles

As a generalist herbivore, the tarnished plant bug might not differentiate among the volatile constituents of its hosts. Gueldner and Parrott (1981) found little similarity among compounds occurring in volatile extracts of croton, goldenrod, and mustard. They speculated that *Lygus lineolaris* responds to a complex mixture of volatiles associated with its host plants rather than to a few specific compounds, which is true of phytophagous insects in general (e.g., Städler 1992). Phenylacetaldehyde, a volatile component of corn silk and other plants, attracts tarnished plant bugs but is more attractive when combined with other volatiles (Cantelo and Jacobson 1979). Several volatile compounds from red clover also proved attractive to these bugs in preliminary field tests (Buttery et al. 1984).

Electroantennogram techniques demonstrate that *L. lineolaris* has receptors for a wide range of host-plant odors (Chinta et al. 1994). It behaves similarly when orienting to different host plants, but the frequency and duration of the behavioral activities associated with feeding vary with the host (Hatfield et al. 1983). Moreover, as Zaugg and Nielson (1974) cautioned, based on studies with *L. hesperus*, olfactory responses of laboratory-reared and field-collected populations can differ significantly between specimens.

How monophagous and oligophagous mirids might use volatiles to discriminate hosts from nonhosts needs investigation, as well as the importance of antennal sensilla in long-range orientation to host plants. Plant volatiles might play a more prominent role in host orientation in mirids than in some orthopterans (Berenbaum and Isman 1989) and other mainly phytophagous hemimetabolous insects that do not oviposit on their host plants. The orthotyline *Sarona mokihana* frequently is found in methyl eugenol–baited traps used for the Oriental fruit fly (*Batrocera dorsalis*) in Hawaii. Methyl eugenol is a constituent of this plant bug's only known host (Asquith 1994), suggesting the role of this chemical in orientation by *S. mokihana*.

Fig. 7.10. Model of host selection and feeding by *Lygus lineolaris*. Rectangle indicates process; parallelogram, input; diamond, decision; oval, terminal. Dashed lines indicate that function has not been determined experimentally. (Redrawn, by permission of L. D. Hatfield and the Entomological Society of America, from Hatfield et al. 1983.)

Locomotion, Labial Tapping, and Test Probing

After arriving on a potential host, mirids begin exploring the surface, periodically tapping with the labium as they walk slowly over leaves and other structures. This dabbing behavior or touching of the surface precedes actual insertion of or probing by the stylets.

The role of volatile chemicals on the plant surface in eliciting labial tapping appears unstudied in mirids. There is also no conclusive evidence that tarsal sensilla serve a chemosensory function in plant bugs, although the foretarsi and midtarsi of certain other families of phytophagous heteropterans are external contact chemoreceptors (Miles 1958b).

When a mirid walks over plant surfaces, its tarsal sensilla might be sensitive to mechanical cues, enabling the bug to sense the degree of pubescence and to locate major veins, as Backus (1985) suggested for leafhoppers. Although the antennae might not be involved in the selection of feeding sites by cocoa mirids (Youdeowei 1977), the antennal tips of other heteropterans serve as chemoreceptors (Miles 1958b). It seems likely that antennae also play a role in the exploration of plant surfaces by mirids (e.g., Ferreira 1979), but their function in cimicomorphans remains largely unstudied. Heteropteran antennae, because of their greater numbers of sensilla, might have more sensitivity to mechanical and chemical cues than those of sternorrhynchans or of auchenorrhynchans such as fulgoroids (Backus 1988a). Studies on adult *Lygus lineolaris* reveal antennal sensilla with surface pits that appear to represent pores and to house olfactory receptor neurons (Fig. 7.11). The bugs respond electrophysiologically to the plant odorant 1-hexanol, and an antennal-specific protein is known. This protein, which might be an odorant-binding protein, is likely involved in the detection of host-plant odors (Dickens et al. 1995, Dickens and Callahan 1996).

Labial tapping or dabbing is critical in conveying initial information on host suitability for *L. lineolaris* (Ferreira 1979) and probably most other mirids and other phytophagous heteropterans (Miles 1958b). Ferreira (1979) found that stylet probing is not related to duration of feeding in the tarnished plant bug. This contrasts with host discrimination in aphids, which apparently have only mechanosensilla at the labial tip and must rely heavily on stylet probing (i.e., brief, short-distance, generally nonfeeding probes) (McLean and Kinsey 1968, Backus 1988a, Montllor 1991). Some aphids, however, employ rostral tapping of the leaf surface before probing (Klingauf 1987). The duration of rostral tapping by *L. lineolaris* and the mean number of taps per bug (and time spent in grooming and locomotion) are inversely related to the time spent feeding and consequently indicate host suitability (Latson et al. 1977, Ferreira 1979, Hatfield et al. 1983). A similar inverse relationship between rostral tapping and feeding is seen in *Dicyphus tamaninii* on vegetable crops (Gessé Solé 1992).

Watery saliva is often secreted during labial tapping. The amount *L. lineolaris* extrudes at the rostral tip is about $0.1\,\mu l$ (Strong 1970). Saliva is almost immediately sucked up in an apparent tasting by gustatory receptors (Miles 1968b, 1972). Chatterjee (1984b) described and illustrated the gustatory receptor organ of the dicyphine *Nesidiocoris caesar*. Chemosensilla presumably involved in tasting internal plant constituents and locating preferred feeding sites are present in the precibarium, a narrow canal lying within the head. These sensilla in mirids (and pentatomomorphans) occur mainly in regular, double rows (Cobben 1978). Precibarial sensilla are often referred to as the *epipharyngeal organ* (Backus 1985, 1988a).

The labial tip, as in all heteropterans (Cobben 1978), is equipped with sensory sensilla (Fig. 7.12), which in *L. lineolaris* (and no doubt in most other mirids) act as contact chemoreceptors (Flemion et al. 1954, Latson et al. 1977, Avé et al. 1978, Hatfield and Frazier 1980). Two sensory fields are present on the rostral tip of the tarnished plant bug, each with 11 sensilla basiconica. These sensilla enable the bugs to distinguish between susceptible and resistant cotton varieties (Avé et al. 1978). Sensilla trichodea or "trichodeal hairs" on the labial tip appear to function as mechanoreceptors (Avé et al. 1978, Backus 1988a). Sensory hairs at the labial tip of *L. lineolaris* probe the surface carefully before the stylets are inserted in bean pod tissue (Flemion et al. 1954). Setae at the labial tip of the grass-feeding *Notostira elongata* are used similarly to test host suitability (China 1925a).

Apparently the stylets of all heteropterans are innervated, but only in triatomine reduviids are they innervated throughout (Cobben 1978). The mandibles of predacious and phytophagous taxa contain dendrites. Like sternorrhynchans, some pentatomomorphans lack a maxillary nerve supply, and throughout the Heteroptera, the nerve supply to the maxillae is less extensive than that of the mandibles. Six mandibular and five maxillary dendrites characterize *L. rugulipennis* and *N. elongata* (Cobben 1978). Cobben emphasized that in phytophagous heteropterans the role of stylet innervation in mechanoreception and chemoreception needs elucidation.

Test or trial probing might convey additional information on a potential host—for instance, the suitability of internal constituents. McIver and Stonedahl (1987a) used the term "probe search" to refer to probing behavior that precedes feeding, in which the proboscis is used to "briefly sample potential resource patches," with the antennae "gently swaying alternately."

Obviously, the nature of the plant surface—whether hairy, glandular, or toxic—and its covering of epicuticular lipids affects host exploration, acceptance, and location of feeding sites (Juniper and Southwood 1986, Eigenbrode and Espelie 1995). *Closterotomus norwegicus* carefully selects a feeding site on potato plants, sometimes probing leaf surfaces for five to six minutes before penetrating tissues (Smith 1926). On the densely

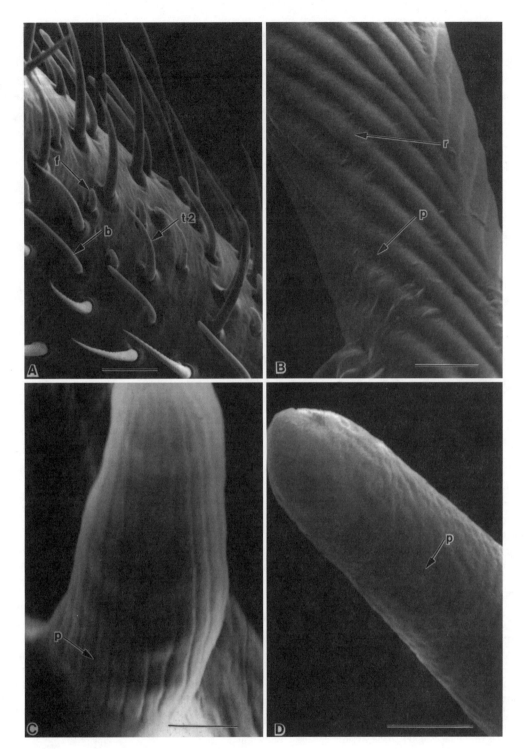

Fig. 7.11. Sensilla on the second antennal segment of a
Lygus lineolaris male. A. Cuticular structures near middle of
second antennal segment. B. Higher magnification of a
trichoid sensillum. C. Higher magnification of fluted
sensillum. D. Higher magnification of tip of basiconic
sensillum. Abbreviations: b, long basiconic sensillum; f,
fluted sensillum; p, pore; r, diagonal ridges; t2, trichoid
sensillum. Bar = 1 μm. (Reprinted from *J. Insect Physiol.* 41,
J. C. Dickens et al., "Olfaction in a Hemimetabolous Insect:
Antennal-specific Protein in Adult *Lygus lineolaris*
[Heteroptera: Miridae]," pp. 857–867, 1995, by permission of
J. C. Dickens and with kind permission from Elsevier
Science, Ltd., The Boulevard, Langford Lane, Kidlington OX5
1GB, UK.)

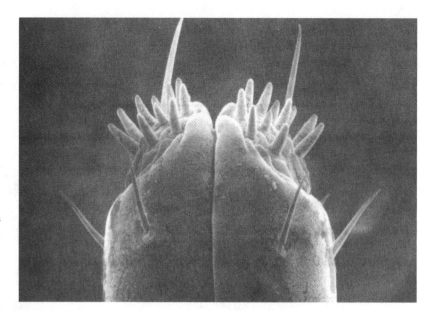

Fig. 7.12. Rostral tip of *Lygus lineolaris*, dorsal view, showing two fields of sensilla basiconica (middle), two long sensilla trichodea (top), and ring of subapical trichodea (bottom). (Reprinted, by permission of D. A. Avé and Kluwer Academic Publishers Group, from Avé et al. 1978.)

pubescent, viscid surfaces of tobacco plants, *Nesidiocoris volucer* makes test probes at 10 or more sites before settling down to feed, a process that can last seven minutes (Roberts 1930). Cocoa mirids, however, insert their stylets in host pods or stems with little hesitation or prolonged exploration (Goodchild 1952). Peak probing of *L. hesperus* males tends to occur in the laboratory (photoperiod = 14 hours) before and after periods of intensive movement, which, under natural conditions, corresponds to crepuscular periods (Sevacherian 1975). The younger the host plant, the greater the number of trial punctures by *L. rugulipennis* (Varis 1972).

Setae that are usually present along the sides of heteropteran (and auchenorrhynchan) labia probably serve as mechanosensilla. As the stylets move through plant tissues during probing, proprioceptive mechanosensilla within the stylet canals detect twisting and turning movements and possibly the degree of labial telescoping or bending during probing. Both the mandibular and maxillary stylets are innervated in cimicomorphan Heteroptera and in Auchenorrhyncha. In contrast, the pentatomomorphan and sternorrhynchan species that have been examined show only a single innervation. The former condition—*dual innervation*—might represent an ancestral character within the Hemiptera (Backus 1988a).

Feeding Stimulants, Including Secondary Metabolites

To this point, chemicals that provide external excitatory input have been scarcely mentioned in this overview of host acceptance in mirids. Information on feeding stimulants is available mainly for the polyphagous tarnished plant bug (*Lygus lineolaris*) and for some European lygus bugs. An amino acid mixture acts as a phagostimulant for *L. hesperus* (Strong and

Kruitwagen 1970), and sucrose is the strongest phagostimulant for *L. lineolaris* (Hsiao 1985), but glucose, methionine, and phenylalanine also are effective phagostimulants (Hatfield et al. 1982). Hatfield et al. noted that the concentrations of sucrose used by Strong and Kruitwagen (1970), which failed to stimulate ingestion, might have been higher than those that gave optimum stimulation in their own study. In addition, the relative concentration of free amino acids in potato varieties affects their attractiveness to the Palearctic *L. pratensis* and *L. rugulipennis* (Turka 1985).

Data obtained from generalist lygus bugs almost certainly are not applicable to cocoa mirids and other Bryocorinae intimately associated with host plants rich in methyl purines, or to other mirids with a narrow host range. Such specialist herbivores presumably use phytochemicals as feeding stimulants; for example, *Dichrooscytus* species are influenced by the terpenoid chemistry of host junipers (Polhemus 1988). Radioactive tracer studies have shown that a specific substance in the cocoa plant stimulates salivation in the bryocorine *Distantiella theobroma* (Cross 1970). Leston (1970) commented on the ability of cocoa mirids to exploit hosts otherwise protected by methyl purines. He noted that it would "be of value to elucidate the role of methyl purines as attractants (unlikely because of chemical structure, but if true, opening up a possible method of control through baiting), phagostimulants or essential substances for bryocorines." Plant secondary metabolites that stimulate feeding by specialist mirids most likely deter feeding by generalist plant bugs, as is the case with other insects (e.g., Dethier 1947, Fraenkel 1959, Ehrlich and Raven 1964, Feeny 1976, Rodriguez and Levin 1976, Rosenthal and Janzen 1979, Carroll and Hoffman 1980, Berenbaum 1983, Mullin 1986, Bell 1987, Wink 1988, Tallamy and Krischik 1989).

Secondary Metabolites as Feeding Deterrents

Some heteropterans perceive secondary plant substances (Schoonhoven and Derksen-Koppers 1973), but the effects of these compounds and the possible metabolic costs of dealing with them are largely unexplored. Contrasted with caterpillars and other chewing insects, relatively little work has been done on compounds that deter sucking insects; the data available pertain mainly to stylet sheath-feeding homopterans rather than to lacerate and flush feeders such as mirids. As cell-content feeders, mesophyll-feeding heteropterans appear more likely to contact and ingest plant secondary compounds than are vascular tissue feeders (Tonkyn and Whitcomb 1987, Rosenheim et al. 1996). But certain mirids develop on hosts in nutrient-rich habitats and with high levels of leaf nitrogen, high growth rates, and predictably low levels of antiherbivore secondary compounds (McNeill and Southwood 1978, McNeill and Prestidge 1982). Mullin's (1986) review is invaluable for understanding the disparate exposure to plant allelochemicals of sucking herbivores versus chewers, and the different strategies that have evolved for dealing with defensive chemicals.

Many mirids feed on actively growing tissues. This feeding mode may carry some liability, for actively growing tissues can contain high concentrations of some secondary metabolites (and auxins) that are potentially toxic when consumed. Concentrations of secondary metabolites are often highest at the time of budbreak and decline with leaf expansion (e.g., Herms and Mattson 1992). Auxin concentrations, though, are apparently so low that hardly any toxic effects can be expected. In addition, *Lygus rugulipennis* can detoxify some ingested indole auxins by converting them to excretable conjugates, by combining them with amino acids or sugars (Hori 1979a, 1979b, 1980). Secondary compounds are produced in or translocated to reproductive tissues, seeds, and young leaves, but they also occur in substantial concentrations in mature leaves and stems (McKey 1979). Secondary compounds are often stored in specialized organelles, in epidermal cells, or in the cytoplasm and cell walls. It is difficult to generalize about their location (e.g., Montllor 1991), but defensive chemicals are stored mainly in intracellular vacuoles (McKey 1979). That lygus bugs and many other mirids feed preferentially at actively growing sites suggests these tissues are actually less toxic and more nutritive than most others on the plant.

How North American mirids and other sucking insects associated with creosotebush in the desert Southwest (Hurd and Linsley 1975) might be affected by host defensive chemistry is unknown. Species that feed on phloem and xylem might be able to avoid much of the nordihydroguaiaretic acid in the surface resin of leaves and terminal stems (Lightfoot and Whitford 1987). But, as discussed earlier, mirids generally are not vascular tissue feeders.

The polyphagous habits of the tarnished plant bug (*L. lineolaris*) are attributed to a lack of response to inhibitory substances in their secondary host plants (Curtis and McCoy 1964). Cotton tannins and other secondary metabolites do inhibit feeding by *L. lineolaris* (Ferreira 1979, Hatfield et al. 1982; see Chapter 10 and review by Frazier [1986]). Glucosinolates also can deter *L. lineolaris* (Hatfield et al. 1982, Bodnaryk 1996), but oilseed rape cultivars with reduced glucosinolate levels (canola crops) are not more attractive to lygus bugs. The suitability of oilseed rape for *L. elisus* is not affected by glucosinolate levels (Butts and Lamb 1990b). Because of the wide host range of *L. rugulipennis*, secondary compounds might not represent important barriers to feeding (Holopainen and Varis 1991), although a high concentration of total glycoalkaloids might be involved in host resistance to European lygus bugs (Turka 1985).

To summarize, leaf chewers, as somewhat indiscriminate feeders on external plant parts, encounter chemicals that are compartmentalized in specialized organelles or tissues and released when cells are broken, whereas phloem-feeding homopterans and heteropterans are able to avoid sites of secondary metabolite concentration. Secondary metabolites tend not to be concentrated in vascular tissues, but some compounds are mobile in the phloem (e.g., MacLeod and Pridham 1965, Wink et al. 1982, J. A. Raven 1983, Dreyer et al. 1985). The role of phloem-derived metabolites in deterring aphids, other homopterans, and heteropterans is unsettled. Whether compounds are present in concentrations sufficient to deter feeding is unknown (Dixon 1985, Mullin 1986, Dreyer and Campbell 1987, Wink 1988).

The extent to which mesophyll-feeding mirids encounter defensive phytochemicals is imprecisely known. A group of secondary phytochemicals they almost certainly encounter is quinones. It is not known, however, if plant bugs that are incidental phloem feeders are affected by secondary metabolites. If mirids should receive lower exposure to secondary metabolites than most chewing insects (exposure varies among mandibulate groups [Berenbaum and Isman 1989]), they also might show decreased detoxification enzyme activity. But there is little or no experimental evidence demonstrating a lower exposure to secondary metabolites by mirids, an associated lower detoxification activity, or both. Indeed, the exposure of mirids and other cell-content feeders might actually be similar to that of chewing herbivores (Rosenheim et al. 1996). Furthermore, few studies have examined the possibility that mirids sequester secondary metabolites, a strategy common in other insect groups. The dicyphine *Tupiocoris notatus* ingests droplets of exudate from the glands on the sticky leaves of downy thorn apple or angel's trumpet. Because the exudate contains acylsugars that deter feeding by other herbivores, this mirid might sequester these secondary metabolites for its defense (van Dam and Hare 1998). Much remains to be learned about the exposure of mirids to defensive plant chemicals, as well as the ways in which they deal with these compounds.

Ingestion and Extended Feeding in Phytophagous Mirids

INGESTION AND DURATION OF FEEDING

Slater and Baranowski (1978) succinctly characterized the rather complex process of fluid ingestion in heteropterans: "In feeding, salivary fluid is pumped down the salivary duct and liquified food material is pumped up the food canal. The pumping action is accomplished by a complex arrangement of plates attached to the stylets that are continuous with a sucking pump in the head called a cibarium." Ingestion in phytophagous heteropterans involves a combination of turgor pressure, capillary action, and suction pressure (Rathore 1961, Dhiman 1985). The flow of fluids through the food canal is governed by Poiseuille's principle, which predicts that the rate of flow depends on the viscosity of the fluid, length of the tube, and especially its diameter. Volumetric flow rate is inversely related to fluid viscosity and depends mainly on the radius of the food canal (e.g., Mittler 1967; Kingsolver and Daniel 1993; Cohen 1995, 1998b; Loudon 1995). Dhiman's (1985) study of the berytid *Metacanthus pulchellus* contains a detailed description of a feeding mechanism apparently similar to that in mirids.

Duration of mirid feeding varies considerably depending on the species, host plant, temperature, and other factors that affect a bug's level of satiation. An adult *Lygus rugulipennis* spends slightly more than a tenth of its total time feeding (Varis 1972). Behavioral observations of three other mirids (third to fifth instars and adults) and construction of time budgets indicate that a considerable amount of active time is devoted to feeding: 39% for *Coquillettia insignis*, 33% for *Lopidea nigridia*, and 30% for *Orectoderus obliquus* (McIver 1987, McIver and Stonedahl 1987a). Similarly, an adult *Lygus lineolaris* spends 34.4% of its time feeding on horseweed, a preferred host; on cotton, it spends only about 10% of its time on feeding (Latson et al. 1977).

Once feeding begins for *Helopeltis clavifer*, there is little movement of the head capsule and stylets (Miles 1987a). During extended feeding, *C. insignis* and *O. obliquus* are relatively unresponsive to external visual stimuli (McIver and Stonedahl 1987a, 1987b). The length of time some cocoa mirids spend during a feeding bout is inversely proportional to temperature: 30 minutes at 24°C, but 10 minutes or less at 30°C (Goodchild 1952). *Helopeltis* species feed at one spot only a few minutes (Leach and Smee 1933), $7\frac{1}{2}$ to 10 minutes (Lever 1949), or as much as 20 minutes (Jeevaratnam and Rajapakse 1981a). An adult *Adelphocoris superbus* can feed on an alfalfa seed pod for about 40 minutes without removing its stylets (Sorenson and

Cutler 1954), and *H. clavifer* can feed at one site on cocoa for an hour or more (Miles 1987a). *Lygus lineolaris*, however, might not feed during the entire period of stylet insertion (Flemion et al. 1952, Flemion 1958), as is the case with the pentatomid *Eurydema rugosa* (Hori 1968a). The feeding period of *L. rugulipennis* is shorter on young, small plants, and the percentage of time it spends feeding varies from host to host (Varis 1972). *Engytatus modestus* may feed in the same region for one or two days, making lesions that encircle the stems or petioles of tomato (Tanada and Holdaway 1954).

A plant bug often interrupts feeding to clean its rostrum by drawing it through the foretarsi or foretibiae (Fig. 7.13; Beyer 1921, Brown 1924, China 1925a, Austin 1931a, Gilliatt 1935); the latter bear a comblike structure at the distal end (McGavin 1979). *Lygus lineolaris*, when grooming, holds the forelegs in an X position and pulls them numerous times across the antennae and proboscis (Latson et al. 1977). McIver and Stonedahl (1987a, 1987b) discussed the grooming behavior of the Nearctic myrmecomorphs *C. insignis* and *O. obliquus*, noting that the foretarsi (rather than the foretibiae) are used to groom head-associated parts. Grooming probably helps maintain the sensitivities of the antennae (Plate 3) and labia by keeping them free of debris. In the case of the dicyphine *Tupiocoris notatus*, the forelegs are used on the body and stylets to remove exudate acquired from downy thorn apple, its glandular, solanaceous host plant (van Dam and Hare 1998).

Hlavac's (1975) review of insect grooming systems includes comments on the lygaeid *Oncopeltis fasciatus*, in which the arrangement and angling of antennal setae approximate mirror images of cleaning setae on the foretibiae. Whether a similar arrangement occurs in the Miridae is not known.

A behavior of uncertain function occurs in tea mosquito bugs (*Helopeltis* spp.). While feeding, they raise their hind legs onto the forewings, which are stroked or patted (Lever 1949).

QUANTITY OF FOOD CONSUMED AND ECONOMIC INJURY

Following the final ecdysis and a short teneral period, intense feeding usually occurs (Petherbridge and Husain 1918, Kullenberg 1944). Few data are available on the quantities of fluid imbibed by phytophagous mirids or the consumption rates of adult females and males. Studies also are needed to determine if consumption generally increases in response to a decrease in food quality. In the case of *Helopeltis theivora* on cocoa, the number of feeding lesions decreases as pods increase in age (Muhamad and Way 1995b).

The amount of fluid that *Lygus lineolaris* imbibes ranges from 0.2 to 2.0 mg (Flemion 1958), and the amount for *L. rugulipennis* also varies with duration of feeding (Varis 1972). Older nymphs and adults of *L. hesperus* ingest greater amounts of ^{14}C-labeled diet per

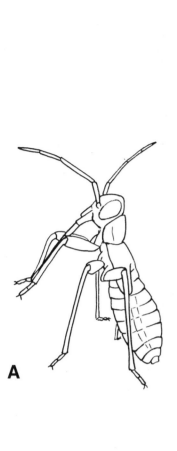

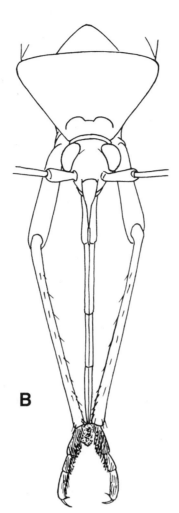

Fig. 7.13. Use of the forelegs to groom the rostrum. A. *Globiceps fulvicollis.* B. *Calocoris roseomaculatus.* (Reprinted, by permission of B. Kullenberg and Zoologiska institutionen, Uppsala Universitet, from Kullenberg 1944.)

feeding bout than do early instars (Stewart et al. 1992). The quantity of food consumed by *L. rugulipennis* increases gradually from 2 mg/day in the first instar to 7.5 mg/day in the fifth. Feeding on rape pods by *L. rugulipennis* adults results in the loss of as much as 13 mg of sap per day; females consume slightly more food per day than do males (Hori 1971b). Similarly, *L. hesperus* females ingest more fluid than males, their consumption of bean juice through a Parafilm membrane equaling 169% of body weight in 24 hours (Strong and Landes 1965; see also Stewart et al. [1992]). But consumption by *Calocoris angustatus* females feeding on sorghum panicles (16.73 mg/day on a dry weight basis) is nearly the same as that of males (16.15 mg/day) (Natarajan and Sundara Babu 1988c).

The frequency of feeding on host plants can be higher in mirid females than in males: for instance, 15% compared to 7% in *L. rugulipennis* (Varis 1972). *Helopeltis* females make many more punctures on tea than do males (Lever 1949), and those of *H. antonii* cause an average of 97 lesions on cashew trees in 24 hours compared to only 25 by males (Sathiamma 1977). Additionally, females can cause more host injury than males: 20–50% greater mortality of seedling alfalfa (unifoliate

stage) by *L. hesperus* (Nielson et al. 1974), greater necrosis and about twice as much damage on cotton terminals by *L. hesperus* (Gutierrez et al. 1977, Mauney and Henneberry 1979), and significantly greater bud abscission on apple trees by *L. lineolaris* (Prokopy and Hubbell 1981). Reid (1968), however, found that *L. lineolaris* females do not differ significantly from males in their ability to injure cotton.

Feeding rates depend on temperature, but little experimental work on the effects of temperature has been conducted. Feeding frequency in *L. rugulipennis* increases with rising temperatures (Varis 1972), and feeding by *L. elisus* increases sharply between 20 and 30°C, departing from a linear relationship (Fye 1982c). Lettuce sustains greater damage from *L. lineolaris* at 25°C than at 21.1°C (Kageyama 1974). The mean number of ingestion bouts ("feeds") per *Leptopterna dolabrata* individual increases with temperature regardless of its size: 2.07 feeds per insect per hour at 15°C versus 3.61 per hour at 25°C (McNeill 1971).

Feeding on plants suitable but not favored for reproduction can be less than on preferred hosts. *Taylorilygus vosseleri* adults feed on cotton (a relatively unacceptable plant for adults) only to maintain a

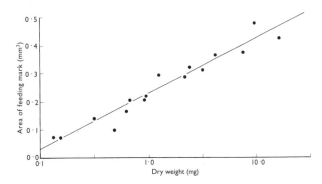

Fig. 7.14. Relationship between size of a feeding mark and dry weight (size) of a *Leptopterna dolabrata* individual. (Reprinted, by permission of S. McNeill and Blackwell Science Ltd., from McNeill 1971.)

Table 7.1. Area of feeding puncture and stylet diameter of *Lygus rugulipennis*

Stage	Diameter of stylets (µm)	Area of feeding puncture (µm)
Adult	21.5	362.9
Instar V	15.6	191.0
Instar IV	11.7	107.5
Instar III	9.8	75.4
Instar II	7.8	47.8
Instar I	6.0	28.3

Source: Hori (1971b).

certain body weight, that is, "just sufficient to replace water lost by evaporation," with feeding increasing as humidity drops, temperatures rise, or both (Stride 1968).

The potential for economic injury varies among mirid nymphal instars (e.g., Tamaki and Hagel 1978). Early instars of some species might be able to obtain sufficient food from only one or two parenchymatous cells beneath the epidermis, as Miles (1969) suggested for some Cimicomorpha, and they probably cannot penetrate auxin-producing organs far enough to disrupt hormonal balance (Strong 1970, Leigh 1976). The stylets of third-instar *Lygus lineolaris* appear incapable of piercing the ovules of green beans (Khattat and Stewart 1975). The mean stylet diameter of *H. theivora* first instars and the depth of lesions they cause on cocoa are less than those of the other nymphal instars (Muhamad and Way 1995b). Older nymphs of *Labops hesperius* cause injury to grass leaves that is orders of a magnitude greater than that of early instars (Ling et al. 1985). In addition, first instars of *Engytatus modestus* produce fewer lesions on tomato stems than do later instars and adults (Tanada and Holdaway 1954), although early-instar *Lygus hesperus* make more feeding punctures through a membrane covering an artificial diet than do older nymphs and adults (Stewart et al. 1992).

Temperature does not necessarily affect the size (surface area) of individual ingestion bouts. In the case of *Leptopterna dolabrata*, the surface area fed on is closely correlated with the size (dry weight) of an individual, owing to increasing stylet length (Fig. 7.14). As *L. dolabrata* increases in size, the longer rostrum allows more cells to be reached during one puncture of its host grasses (McNeill 1971).

Relative to host impact, the diameter of an adult stylet can be three times that of a first instar—and the area of puncture can be more than 10 times greater in the adult (Table 7.1; Squire 1947, Hori 1971b). De Silva (1961) compared the area of feeding punctures produced

by the five nymphal stages and the adult of *H. bradyi* (as *H. ceylonensis*).

Nymphs, however, often cause more injury than adults, for example, *Creontiades pallidus* on cotton in Africa (Soyer 1942) and *Horistus infuscatus* on an asphodel in Israel (Ayal and Izhaki 1993). *Lygus hesperus* nymphs cause about twice as much damage to carrot seed than do adults (Carlson 1956), and the fifth instar of *Helopeltis antonii* can produce 114 lesions on cashew trees during 24 hours compared to 97 for adult females and 25 for males (Sathiamma 1977).

CESSATION OF FEEDING

Sensory adaptation to feeding stimulants is responsible for the cessation of feeding by the lygaeid *Oncopeltis fasciatus*, as apparently are the depletion of salivary secretions, dehydration, and volume saturation (Feir and Beck 1963). Factors leading to the termination of feeding and stylet withdrawal by phytophagous mirids have received scant attention. The quantity of plant juice ingested by *Lygus lineolaris* is governed by precibarial (epipharyngeal) chemosensilla. Accordingly, cessation of feeding in this species "may involve sensory adaptation, central programming, and input from gut stretch receptors" (Hatfield et al. 1983). In West African cocoa mirids, feeding ceases and the bugs become quiescent when food and secretions fill both the first and the second midgut. Digestion of a meal (emptying of the intestine) takes about two hours, although feeding sometimes resumes in less time. The feeding cycle in cocoa mirids, therefore, depends on changes in the midgut (Goodchild 1952).

Feeding Behavior of Predacious Mirids

SEARCHING BEHAVIOR AND PREY CAPTURE

Less is known about the behavioral events leading to predation (see also Chapter 14) than about those associated with phytophagy. The mainly carnivorous orthotyline *Blepharidopterus provancheri* searches methodically for mites over leaf surfaces, moving the beak "rapidly from side to side between the front pair of legs, completely covering a portion of leaf about 1/16-inch [ca. 1.6 mm] in width. By proceeding slowly in this manner in an irregular course over the leaf they located the various stages of the mites. The beak is an extremely sensitive organ by which all food is located" (Gilliatt 1935). While feeding, *B. provancheri* uses its forelegs to hold a mite nymph or an adult female (males are seldom captured because they are more active) and quickly inserts its stylets into the prey.

Nymphs of *B. angulatus* also move rapidly over apple leaves "feeling with the tip of the proboscis, which is apparently the only sensitive region, no response being shown when the tarsi contact suitable food" (Collyer 1952). According to Collyer, the bugs move quickly backward and forward, searching unsystematically, sometimes covering the same ground many times. They spend considerable time searching upper leaf surfaces, petioles, and twigs where no prey are found. They apparently perceive prey by touching them with the antennae or the rostral tip (Glen 1975b). When *B. angulatus* perceives a prey with its antennae, it moves forward, sometimes seeming to pounce on the prey. Searching and capturing prey can be hindered by host plants with pubescent leaf veins (Glen 1975b). Papers by Glen (1975b) and Glen and Brain (1978) contain additional information on the rate of searching (speed of searching is significantly lower at night), prey handling time, interference between predators, and other aspects of its searching and prey-capture behavior.

McGavin (1979) briefly described the behavior of the mainly predatory (Collyer 1953b) *Heterotoma merioptera* in probing leaf surfaces: "Antennal segments I and II are held out horizontally, segments III and IV being bent at right angles, with the apex of the IVth segment touching the leaf surface." *Deraeocoris pallens* nymphs probe the substrate in an apparent random search for prey, which are perceived only upon physical contact (Susman 1988).

Mirids on Scotch broom often wander past potential prey several times before stopping to feed, which prevents very active prey from being accepted (Dempster 1964). Movement by the prey about to be captured seems less important to mirids compared to its role in certain other carnivorous heteropterans such as reduviids. For instance, nymphs of the facultative predator *Psallus ambiguus* readily accept anesthetized drosophilid flies (Morris 1965; see also discussion of scavenging in Chapter 15).

Nymphs of the obligate tingid predator *Stethoconus japonicus* approach fifth-instar lace bugs with their antennae recurved and usually from the side or rear. They attack without hesitation, and prey do not respond to their approach. Within a minute of stylet insertion by *Stethoconus japonicus*, nymphs of the azalea lace bug (*Stephanitis pyrioides*) die or become paralyzed. Although nymphs of *Stethoconus japonicus* feed on their prey in situ, adults use their forelegs to invert and hold their prey, feeding on a lace bug's venter (see Fig. 14.10; Neal et al. 1991).

Smith's (1920a) observations on two apple-inhabiting species might provide the only comparison of plant-feeding behavior between a phytophagous and a mainly predacious mirid on the same host. Before feeding, the predatory *P. ambiguus* holds its proboscis upward, secretes saliva that accumulates at the tip, deposits saliva on an apple leaf, and inserts its stylets through the salivary droplet. In contrast to the behavior of *P. ambiguus*, the phytophagous apple capsid (*Lygocoris rugicollis*) first inserts its stylets and then pumps in saliva. An explanation for the behavioral differences in these species perhaps relates to their different food habits and methods used to test the plant surface for suitability. Alternatively, such behavior by *P. ambiguus* might merely represent the use of saliva to excrete excess water (Goodchild 1952).

EXTRAORAL DIGESTION AND FEEDING EFFICIENCY

Mirids and other predatory heteropterans use extraoral digestion, injecting their prey with hydrolytic enzymes to reduce fluid viscosity, and then ingest the concentrated nutrients (Cohen 1993, 1995). The availability of water seems critical for prey feeding in mirids (Gillespie et al. 1999). Preliminary studies suggest that cimicomorphan predators, including some mirids, do not probe as deeply into their prey as do pentatomomorphan predators, and rely more on salivary constituents to liquefy their prey than on mechanical action of their stylets (Cohen 1996). Heteropteran predators typically show a high efficiency of ingestion and protein consumption, owing to extraoral digestion and mechanical disruption of prey tissues (Cohen 1984, 1989, 1995, 1998a). Mirids feed with efficiency even though they lack a stylet flange or collar (Cohen 1990) that would provide leverage during feeding (Cohen 1996). Glen (1973) studied the efficiency with which the mirid *Blepharidopterus angulatus* converts aphid food into body tissues. Early instars are able to compensate for a scarcity of prey through increased efficiency of converting ingested food into body weight, whereas later instars tend to waste prey tissue. Comparable studies on growth efficiency in other predatory mirids are needed.

PREY CONSUMPTION

Prey consumption throughout an instar is often uneven. In the case of *Blepharidopterus angulatus*, it is usually higher in the middle than at the beginning or end of an instar, particularly in late-stage nymphs (Glen 1973, 1975b).

Temperature affects the feeding behavior of predacious as well as phytophagous mirids. Prey consumption by *Atractotomus mali* is significantly higher at 20°C than at 15°C (Jonsson 1987), but the nymphs of

Table 7.2. Numbers of aphids killed per 12 hours by instars II–V of four mirid species[a]

Mirid predator	Instars			
	II	III	IV	V
Blepharidopterus provancheri	1.5 ± 1.0 a	2.1 ± 2.3 ab	3.3 ± 2.5 a	7.9 ± 3.7 a
Deraeocoris fasciolus	2.0 ± 1.2 a	2.4 ± 1.2 bc	7.7 ± 3.7 a	14.8 ± 6.8 b
Phoenicocoris minusculus	3.3 ± 1.2 b	2.3 ± 2.1 ab	3.7 ± 3.7 a	8.3 ± 4.4 a
Phytocoris canadensis	2.5 ± 1.4 ab	4.4 ± 2.7 c	13.0 ± 6.1 c	18.9 ± 8.9 c

Source: Bouchard et al. (1988).

[a] Means followed by the same letter are not significantly different ($P = 0.05$).

Macrolophus costalis consume fewer whitefly eggs as temperatures increase from 22.0° to 25.5°C (Khristova et al. 1975).

Females of several predatory mirids consume more prey than males. This is the case for *Rhinacloa forticornis* (Herrera 1965), *Psallus ambiguus* (Niemczyk 1966a, 1968), *Tytthus mundulus* (Stephens 1975), *Cyrtorhinus lividipennis* (Tanangsnakool 1975; Chua and Mikil 1986, 1989), and *Stethoconus japonicus* (Neal et al. 1991). Likewise, *C. lividipennis* females show significantly higher attack rates than do the males (Heong et al. 1990b). When *B. angulatus* is reared on an aphid diet, males are smaller than females and show a lower growth efficiency (increase in dry weight/dry weight of aphids eaten ×100) (Glen 1973, Gange and Llewellyn 1989).

Parasitized insects sometimes exhibit altered feeding behavior; for instance, parasitized pea aphids (*Acyrthosiphon pisum*) feed at a higher rate than unparasitized individuals (Cloutier and Mackauer 1977). A similar example in a predatory mirid involves *B. angulatus*. Glen (1977a) observed that individuals parasitized by the braconid *Peristenus malatus* weigh 60% more at maturity than unparasitized *B. angulatus*, and determined that parasitized fifth instars consume 2.1 times more in dry weight of aphid prey. He discussed the ecological conditions necessary for the parasitoid's having evolved a feeding mode that induces its host to eat more prey.

First instars of many predatory pentatomids (Asopinae) do not feed, or feed only on unhatched eggs of their species (e.g., Oetting and Yonke 1971, 1975; Waddill and Shepard 1974; Ruberson et al. 1986; Simmons and Yeargan 1988). The observation that first and second instars of some predacious mirids on apple trees feed sparingly or not at all during laboratory trials using mite eggs (Lord 1971) might be the only report of similar behavior in predatory plant bugs. Viggiani (1971b) observed that first instars of *Deraeocoris ruber* apparently are able to subsist on a plant diet. He did not state whether feeding actually occurred, but Cobben (1978) assumed that first-instar mirids do require food. That neither animal nor plant food is required by early instars of predatory mirids, therefore, needs verification.

Fifth instars usually consume more prey than do mirids of any other nymphal instar (Table 7.2; Bouchard et al. 1988, Torreno 1994). In the case of *S. praefectus*, however, the number of lace bugs consumed by fifth instars is less than that taken by third or fourth instars (Mathen and Kurian 1972). Studies on a congener, *S. japonicus*, yielded more typical results: fifth instars of both sexes kill more tingid nymphs (fifth instars) than do fourth instars (Neal et al. 1991).

PREY REMOVAL AND CESSATION OF FEEDING

When mirids prey on aphids or even larger arthropods, the comblike protibial structure used to clean the antennae pushes the prey off the end of the rostrum (McGavin 1979). After *Deraeocoris ruber* removes its prey, the combs are pushed repeatedly down the rostrum and cleaned by a rapid flicking of the forelegs; antennae and mesothoracic tibiae and tarsi are cleaned by drawing them through the combs; and finally the metathoracic tibiae and tarsi are cleaned with the metathoracic claws. Between bouts of predation, *D. ruber* nymphs eject fluids from the rostrum onto a surface and suck it in and out several times. McGavin (1979) thought this behavior washes the sensory papillae at the rostral tip and cleans the food canal. Similar antennal and rostral preening occurs in other species of the genus (Knight 1921), including *D. brevis* during bouts of predation on psyllids (Knowlton and Allen 1936), as well as in the mite predator *Hyaliodes vitripennis* (Gilliatt 1935).

Cessation of feeding has been little studied in predacious mirids. The nymphs of *Blepharidopterus angulatus* show some degree of satiation after an aphid meal, becoming less active and sometimes remaining still in a "digestive pause." The effects of satiation (reduced activity) persist into the next day (Glen 1975b).

Salivary Glands and Secretions

The clear, colorless, slightly alkaline to neutral salivary secretions (Goodchild 1952; Miles 1964, 1972; Hori 1973b; Miles and Slowiak 1976) of Hemiptera are complex. Saliva secreted outside the plant is dilute, consisting mainly of water and lacking hydrolyzing enzymes. The saliva not only contains hydrolytic enzymes, moistens the food, and adjusts pH and ionic content, but also facilitates stylet penetration and moistens and helps clean the mouthparts between feeding bouts (Strong and Kruitwagen 1968, Miles 1972, Takanona and Hori 1974). As noted later in the chapter, excess water in the diet of cocoa mirids can be excreted in the saliva, such regurgitation being characteristic of their feeding behavior (Goodchild 1952).

The following discussion of salivary glands and secretions should be supplemented by Miles's (1972) review of hemipteran salivary composition and function. K. Hori's extensive work (much of it cited here) on the feeding habits of *Lygus rugulipennis* and comparative studies on other plant bugs provide useful information on the salivary glands, digestive enzymes occurring in these glands and in the gut, enzyme activators, variation in enzyme activity, activity of enzymes in relation to pH and temperature, use of carbohydrate, and biochemical changes induced in damaged host tissues.

Many aspects of hemipteran salivary physiology remain unresolved (e.g., Taylor and Miles 1994). The salivary enzymes of additional mirid species need to be determined and their activity characterized in relation to pH, temperature, enzyme and substrate concentration, and activators. It has also been suggested that amino acids facilitate the action of salivary enzymes and help protect them from inactivation by plant-defensive mechanisms (Laurema and Varis 1991). Although the control of salivary discharge is incompletely known, salivary gland function appears to be under control of the nervous system (Miles 1972, Laurema et al. 1985).

SALIVARY GLANDS

Mirids, like all terrestrial heteropterans, have two pairs of salivary glands (Fig. 7.15). The generally bilobed principal glands (four-lobed in some cocoa capsids [Squire 1947]) lie on either side at the front of the thorax. A posterior displacement of these glands—from the thorax in nymphs to the abdomen in adults—is associated with their extraordinary development in some bryocorines (Squire 1947). In at least *Stenodema laevigata*, gland

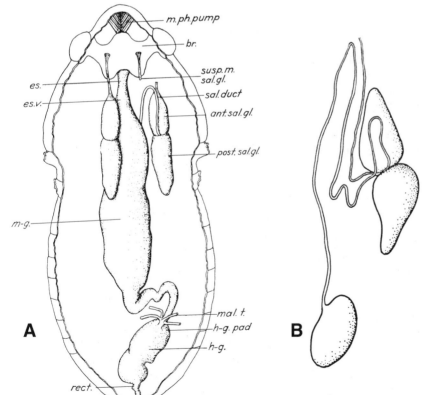

Fig. 7.15. Alimentary canal and salivary glands. A. Alimentary canal of *Pseudatomoscelis seriata*. Abbreviations: ant. sal. gl., anterior salivary gland; br., brain; es., esophagus; es.v., esophageal valve.; h-g., hindgut; h-g. pad, hindgut pad; m-g., midgut; m. ph. pump, muscles of salivary pump; mal. t., Malpighian tubule; post. sal. gl., posterior salivary gland; rect., rectum; sal. duct, salivary duct.; susp. m. sal. gl., suspensory muscle of salivary gland. B. Salivary glands of *Deraeocoris lutescens*. (A from Painter 1930; B reprinted, by permission of Sir Richard Southwood and Nederlandse Entomologische Vereniging, from Southwood 1955.)

structure is similar in the fifth instar and the adult, except for shorter ducts in the nymph (Southwood 1955). In contrast to those in the Homoptera, mirid principal glands have a distinct lumen that holds secretory products originating in the gland walls (Schuh and Slater 1995). In some cases, accessory glands, which are associated with each principal gland (Baptist 1941, Goodchild 1952, Southwood 1955, Nishijima and Sogawa 1963, Hori 1968b), account for only about 10% of total gland volume (Laurema and Varis 1991).

Accessory glands in cimicomorphans are vesicular rather than tubular as in pentatomomorphans; in mirids (except some bryocorines), the ducts are looped over the anterior lobe of the principal gland (Southwood 1955, Nishijima and Sogawa 1963). The duct of each principal gland is joined by the accessory gland duct, and the principal ducts unite to form a common salivary duct that enters the salivary pump (Southwood 1955, Miles 1972). The accessory gland vesicles apparently help maintain salivary circulation during the flushing out of plant-cell contents and produce a copious flow of watery saliva used in dabbing the plant surface to test host suitability. The accessory glands function as a vehicle for transferring hydrolyzing enzymes and salivary amino acids and are the source of polyphenol oxidases (Goodchild 1952, 1966; Miles 1969, 1972; Hori 1975b; Miles and Slowiak 1976; Laurema et al. 1985). An excretory role of the vesicles is likely (Goodchild 1952, 1963a, 1963b, 1966), and hemipteran phytopathogenicity might depend principally on the excretory function of the salivary glands (Hori 1992).

Nuorteva (1956a) called attention to a narrow duct in *S. calcarata* that ostensibly connects the accessory salivary gland to the fat body. Previous workers apparently overlooked this duct because of its thinness and transparency. Even though Nuorteva stated that the duct is not difficult to locate in fresh material, the existence of such a connection between the fat body and accessory salivary gland in mirids needs confirmation.

Salivary gland structure shows little correspondence with external characters at the specific and generic levels but has value in the higher classification of the Heteroptera (Southwood 1955, Miyamoto 1961, Nishijima and Sogawa 1963). The form of heteropteran salivary glands apparently is unrelated to the type of secretions or feeding habits (Smith 1920a), but in species that can produce different kinds of saliva—for example, those used in stylet sheath formation—a more complex salivary gland structure can be expected (Miles 1972). The salivary glands of phytophagous mirids with limited predatory tendencies, such as stenodemines, are similar to those of nearly obligate carnivores, such as deraeocorines (Smith 1920a, Southwood 1955).

The heteropteran alimentary canal, unlike salivary gland structure, varies with feeding habits, and enzyme complements correspond with the type of food consumed (Baptist 1941, Southwood 1955). Hori (1975a) noted that the kinds of enzymes present in the salivary gland and midgut of different heteropteran families vary depending on the nature of the injury produced and the digestive physiology. Predatory mirids that occasionally feed on plant sap do not produce the array of injury symptoms characteristic of most phytophagous species (Putshkov 1956), presumably because they have different enzymes (see also Chapter 9). The presence of an enzyme in an insect's digestive system suggests that the substrate of this enzyme is used in nature, especially if the enzyme is abundant (Hori 1973b).

TYPES AND FUNCTIONS OF SALIVARY ENZYMES

Enzyme complements in mesophyll-feeding leafhoppers differ from those of presumed phloem feeders (Saxena 1954; see also Tonkyn and Whitcomb [1987]). The salivary proteases and amylases of the the former group are absent in phloem-feeding cicadellids. These enzymes are unnecessary in species that feed on phloem "because, from the insect's point of view, the food is already digested" (Nuorteva 1958). Putshkov (1966) distinguished between the action of saliva in plant bugs that feed on vegetative plant parts (mainly leaves) and those feeding on meristematic tissue or, especially, generative organs. In the former group, saliva acts mainly to destroy the cell walls bordering the site of stylet insertion; in members of the latter trophic group, saliva promotes an increase in the flow of plant fluids to the area of penetration by making them more soluble.

Principal Secretable Enzymes

Relatively few important secretable enzymes occur in mirid salivary glands. Those that are most important in mirids are polygalacturonase, amylase, and protease. Plant bugs also have various enzymes that appear to form minor components of their saliva or are nonsecretable constituents of glandular cells.

Polygalacturonase (pectinase) originates in the posterior lobe of the principal salivary glands (Laurema et al. 1985). This enzyme is closely associated with mirids' lacerate and flush method of feeding and apparently is characteristic of the family (Takanona and Hori 1974, Hori 1975a). It is responsible for solubilizing pectin in the middle lamella (Squire 1947, Strong and Kruitwagen 1968, Strong 1970, Tingey and Pillemer 1977, Laurema et al. 1985) by hydrolyzing pectin's α-1,4-polygalacturonic acid residues into oligogalacturonides (Bateman and Basham 1976). Polygalacturonase likely is instrumental in mirid-induced phytotoxemias, allowing much larger feeding lesions to develop compared to those that result from attacks by other sap (cell-content)-feeding heteropterans (Strong 1970, Norris 1979). Only an exopectinase occurs in the salivary gland complex of the lesion-causing fourlined plant bug (*Poecilocapsus lineatus*) (Cohen and Wheeler 1998). Elsewhere in the Hemiptera, pectinase is found mainly in aphids (Adams and McAllan 1958, Laurema and Nuorteva 1961, Strong and Kruitwagen 1968, Strong 1970, Miles 1972, Hori 1975a, Laurema et al. 1985, Dreyer and Campbell 1987, Ma et al. 1990). The activity of salivary polygalacturonase in *Lygus hesperus*, *L. lineolaris*, and *L. rugulipennis* is optimal at an acidic pH (Laurema et al. 1985, Agblor et al. 1994). The polygalacturonase of *L. lineolaris*, although showing maximum activity at pH 5.0, contains an isoenzyme with a pH optimum at pH 6.0 (Agblor 1992). Secretion of polygalacturonase by *Pseudatomoscelis seriata* as it feeds on cotton apparently is involved in the plant's production of stress ethylene (Martin et al. 1988b; see also Chapter 10). A salivary pectinase in *Creontiades dilutus* is considered responsible for this mirine's effects (see Chapter 10) on developing ovules and seeds of alfalfa (Miles and Hori 1977, Hori and Miles 1993).

Salivary pectinase systems in mirids also might be expected to consist of pectinmethylesterase (see Dreyer and Campbell [1987]), although this enzyme is not known in the family. If present, it would demethylate pectin so that polygalacturonase could act more efficiently to depolymerize the main chain of the pectin polymer. Among hemipterans, pectinmethylesterase is known only from aphids (Dreyer and Campbell 1987) and the coreid *Leptoglossus occidentalis* (Campbell and Shea 1990).

Amylase is another important secretable enzyme in mirid salivary glands (e.g., Baptist 1941; Goodchild 1952; Nuorteva 1954; Reid 1965; Hori 1970a, 1973b, 1975a; Takanona and Hori 1974; Varis et al. 1983; Laurema et al. 1985; Agblor et al. 1994; Cohen 1996). The α-amylases solubilize and digest starch and related carbohydrates (Agblor 1992). In the case of *Lygus rugulipennis*, the optimum reaction temperature of this enzyme in vitro is 37°C (Hori 1970b). Starch digestion in the midgut might depend on amylase derived from the salivary glands, where amylase activity is substantially higher than in the midgut (Hori 1971d, 1973b; Takanona and Hori 1974). The anions Cl^- and NO_3^- strongly activate salivary amylase in *Adelphocoris suturalis* and *L. rugulipennis* (Hori 1969a, 1969b, 1970b). These ions, abundant in the reproductive organs of the latter's cruciferous hosts, might activate amylase injected into flower buds or amylase in the midgut when ingested. The ions would promote conditions favorable for ingesting host nutrients (Hori 1970b). This type of activation varies among heteropteran families but only slightly in the mirid species tested (Hori 1972b). In addition, Hori (1972c) reported on the digestibility of raw starches by amylases in the digestive system of *L. rugulipennis* (1973b) and gave comparative values for activation of salivary amylase by Cl^- and NO_3^- (using pH 7.0 at 37°C for 24 hours) for 15 mirid species in 9 genera. Miles (1987a) did not detect salivary amylase in *Helopeltis clavifer*.

Proteases that break down polypeptides into amino acids represent another class of major secretable enzymes in mirids (e.g., Nuorteva 1954, 1956b; Reid 1965; Kumar 1970; Takanona and Hori 1974; Varis et al. 1983; Cohen 1996). They are present in hemipterans that specialize on mesophyll (Hori 1992) and presumably occur in zoophages. The optimum pH for proteinase in *Leptopterna dolabrata* nymphs (nymphs <6 mm long lack proteinase) is slightly alkaline. The proteases restricted to adult females of *L. dolabrata* might facilitate the uptake of proteins needed for egg production (Nuorteva 1956b). Nuorteva's (1954) opinion that *Lygus rugulipennis* adults lack protease is misleading because the activity of salivary proteinase can vary considerably (e.g., Hori 1970a, Kumar 1970, Varis et al. 1983). Gel chromatography and isoelectric focusing have revealed two types of proteolytic enzymes, alkaline and acid proteinase, in *L. rugulipennis*. The former facilitates the use of "food proteins and possibly also the penetration of plant tissues by solubilization of cell-wall bound proteins" (Laurema et al. 1985). According to Hori (1970c), salivary proteinase in *L. rugulipennis* is five times more active than the gut proteinase. Proteolytic enzymes in the salivary glands of the tropical bryocorines *Bryocoropsis laticollis* and *H. corbisieri* also are of two types, one active in the acid range and another in the alkaline. The absence of proteases in the first through fourth instars of both species and the presence in the fifth instars and adults might reflect an increased need for amino acids in the late stages of gonadal development in males and females (Kumar 1970).

Minor Secretable and Nonsecretable Enzymes

Cellulases, present in the saliva of some aphids (Miles 1972, 1987b), are not generally found in plant bugs. *Lygus rugulipennis*, for example, appears unable to use cellulose (Varis et al. 1983), and no salivary cellulase was detected in *Helopeltis clavifer* (Miles 1987a). Cellulase, however, was found in salivary glands of the Oriental phyline *Moissonia importunitas*, apparently the first time this enzyme group has been identified in the family (Gopalan 1976a). Carbohydrases such as invertase and trehalase are minor salivary components or constituents of mirid glandular cells (Laurema et al. 1985). Reid (1965) found that salivary invertase has the strongest and most consistent activity between pH 6.2 and 6.8. Hori (1971c) discussed the activity of invertase in the gut of *L. rugulipennis*; Miles (1987a) determined that salivary glands of *H. clavifer* lack invertase.

Phenol oxidases are salivary enzymes that might help mirids circumvent a plant's wound responses by oxidizing phenolic compounds to nontoxic end products (Miles 1964, 1969, 1972; Strong 1970). Because polyphenol oxidase is hardly detectable in the saliva of *Lygus* species, they apparently have other means of detoxifying harmful phenols (Laurema and Varis 1991).

Phosphatase in *L. rugulipennis* and other *Lygus* species is a nonsecretable lysosomal enzyme in gland tissue (Varis et al. 1983). Lipases are present in the salivary glands of some heteropterans (e.g., Baptist 1941), including cocoa mirids (Goodchild 1952), but they have not been detected in some studies of *Lygus* species (e.g., Nuorteva 1954, Agustí and Cohen 2000) and certain other mirids (Gopalan 1976a). But lower lipid contents and higher acid numbers in lipids have been observed in sunflower seeds injured by *L. rugulipennis* (and the pentatomid *Dolycoris baccarum*) (Shindrova and Ivanov 1982; see also Laurema et al. [1985]). Salivary phospholipase is present in *L. hesperus*, indicating that this pestiferous plant bug is also biochemically adapted for predation (Cohen 1996).

Variation in Enzymatic Activity and Constituents

Miles (1972) noted that saliva can vary in "chemical composition and physical consistency from one moment to the next." He further stated, "There is no reason to believe that the saliva of any one species in the Homoptera (or Heteroptera) necessarily has an invariable composition" (Miles 1987b). This statement refers mainly to the threefold origin of saliva in the Pentatomomorpha: (1) sheath material from the anterior lobe and somewhat different components from the lateral and median lobes (if present); (2) hydrolytic enzymes and amino acids derived from the posterior lobe, which contribute to the watery saliva; and (3) diluent of watery saliva, a dilute ultrafiltrate of the hemolymph, which might also contain polyphenol oxidase and peroxidase and is contributed by the accessory gland. Thus, from moment to moment, there exists a capacity for switching from one secretion to another, or to a different mix of secretions (P. W. Miles, pers. comm.). The pentatomid *Eurygaster integriceps*, for instance, changes feeding sites on its host plants as the season progresses, with corresponding changes in the enzymatic composition of its saliva (McNeill and Southwood 1978).

Even though mirids do not produce sheath material similar to that of pentatomomorphans, they produce some solidifying material in their saliva (Hori 1971a) and have anterior and posterior salivary gland lobes and a separate accessory gland. Mirids, therefore, have a similar potential for differing mixes in their saliva.

The enzymatic content and activity of mirid saliva vary seasonally, with developmental stage and physiological state, and between pairs of salivary glands in a single bug (Nuorteva 1954; Hori 1970a, 1973b; Varis et al. 1983). Variations in composition of the hemolymph, for example, can affect salivary amino acids (Laurema and Varis 1991). Variation in enzymatic activity between *Lygus rugulipennis* individuals of the same stage, collected at the same time and locality, also exists (Hori 1970a). Enzymatic activity is significantly higher in females of *L. rugulipennis* than in males (Varis et al. 1983), and similar differences are seen in salivary amino acids (Laurema and Varis 1991). Salivary glands in *L. rugulipennis* females are 1.33 times larger than those of males (Laurema and Varis 1991). According to Hori

(1970a), amylase and proteinase activities do not differ significantly in the salivary glands of males and females of this bug (cited as *L. disponsi*).

Varis et al. (1983) determined that amylase, proteinase, and polygalacturonase activities vary among individuals of *L. rugulipennis*, but this variation is not observed when the bugs feed on different host plants or on nutrient solutions. This result was unexpected because diet strongly influences enzymatic activity in other Heteroptera, including certain mirids (e.g., Nuorteva and Laurema 1961, Reid 1968). Although active feeding depletes the amount of salivary enzymes in insects of several orders, feeding by *L. rugulipennis* for one-half to one hour does not significantly affect enzyme activity in its salivary glands (Varis et al. 1983). *Lygus lineolaris* secreted 0.05–0.25 µl of saliva during 20–108 minutes of feeding (Flemion et al. 1952), which corresponds to the total volume of *Lygus* salivary glands (Laurema and Varis 1991).

Nonenzymatic Constituents Originating in Food

Mirid saliva also contains nonenzymatic constituents that originate in the food source and are transported after ingestion from the gut to the salivary glands. Dietary amino acids, for example, can pass through the hemolymph, enter the salivary glands, and be introduced into host tissues (Nuorteva 1955, 1956a; Nuorteva and Laurema 1961; Hori 1973c; Tingey and Pillemer 1977). Predominant among free amino acids and related compounds in the salivary glands of *Lygus* species are arginine, glutamic acid, glycerophosphoethanolamine, leucine, lycine, methionine sulfoxide, and proline. Salivary amino acids are similar to those in hemolymph of the whole insect, but they often vary considerably among different populations of a *Lygus* species. No effects of host plants on salivary amino acids of lygus bugs have been detected (Laurema and Varis 1991). The previous host, however, can influence the tissue damage that lygus bugs cause (Reid 1968).

Enzymatic Constituents of Salivary Glands
Versus the Midgut

The enzymatic complements of salivary glands can differ from those of the midgut, where digestion and absorption take place (Takanona and Hori 1974). Digestion in cocoa mirids occurs in the first and second midgut, whereas absorption of degraded products occurs in the third (Goodchild 1952). Takanona and Hori (1974) suggested that in the case of *Stenotus binotatus*, starch digestion takes place in the first and second midguts, with sucrose digested in all three midgut regions.

Effects of Salivary Secretions on Plants

Weaver's (1978) comment regarding the effects of sucking insects on plants—that the role of these insects in causing the plant to produce quinones, the factors in their saliva that promote auxin synthesis, and the importance of other salivary constituents are under dispute—still seems appropriate. Chapters 9 and 12 discuss the numerous injury symptoms involving interactions of stylet penetration and laceration, withdrawal of liquefied contents, introduction of enzymes (especially polygalacturonase) and other biologically active substances, plant-wound responses, and changes in plant growth and metabolism.

Mirids cause mechanical injury to plants by lacerating cells (Taylor 1945; Flemion et al. 1954; Tanada and Holdaway 1954; Dale and Coaker 1958; Flemion 1958; Varis 1972; Boivin and Stewart 1982b; Holopainen 1990a, 1990b), yet these bugs cause such violent effects on plants that more than mere trauma is undoubtedly involved (Carter 1952). In fact, applied entomologists long suspected that mirid-induced injury to plants entails more than a puncturing of tissues and an acceleration of cell division in wounded tissues near the sites of penetration. Harris (1841) and Riley (1870) speculated that the tarnished plant bug (*Lygus lineolaris*) injects a poison or toxin into plants, but Forbes (1884a) thought that injection of a poisonous principle would be "contrary to the order of nature" and stated it is impossible to show that the plant bug benefits from any such supposed "poisoning of its own food." Riley (1885) suggested that the "buttoning" of strawberry fruit (see Chapter 12) is due to a "poisonous and withering influence" rather than a "single, innocuous puncture." Salivary toxins might have been involved in Riley's observations, but mechanical destruction of young meristem can result in growth deformities and sometimes is erroneously attributed to a toxic action of saliva. Use of the word *toxic* to describe the histolytic properties of mirid saliva, although common in the literature, is often inappropriate (e.g., Squire 1947, Entwistle 1972). Hemipteran salivary toxins might not be intrinsically toxic (Taylor and Miles 1994). *Toxin*, as used by those working with lygus bugs and many other mirids (and also used in the present work), probably refers to polygalacturonase (Laurema and Nuorteva 1961, Strong and Kruitwagen 1968).

PHYSIOLOGICAL EFFECTS OF MIRID FEEDING

Smith (1920a, 1926) attributed much of the injury inflicted by the mirines *Lygocoris pabulinus* and *L. rugicollis* to the phytotoxic properties of their saliva. He compared a track made by a mirid stylet with that made by an ovipositor. Whereas the stylet track was accompanied by the pathological effects of saliva, the path of the ovipositor showed only a physical laceration of cells identical to that produced by a sterile needle (Smith 1926).

The diffusible irritant secreted by a *Helopeltis* species on tea plants remains active long after the bugs feed; the cambium is stimulated to produce gallwood weeks or even months later. This bug's effects are evident as much as 12 mm from the area of tissue laceration (Leach and Smee 1933). Circular lesions as large as 4 mm in diameter produced by cocoa mirids are "beyond the possibility of being merely mechanical damage by the stylets" (Goodchild 1952).

Phytophagous heteropterans are potential gall formers (Miles 1968a, 1969), and mirids are indeed capable of inducing several types of necrosis and other phytotoxemias, but any cecidogenic influence of their saliva still is in need of experimental demonstration. Consequently, lygus bugs, despite their production of large amounts of salivary amino acids and an ability to disturb growth and metabolism in their hosts, are not actually cecidogenic (Laurema and Varis 1991). Citing Leach and Smee (1933), Mani (1964) referred to "gnarled galls" made by *H. schoutedeni* on tea bushes, and Wilson and Moore (1985) used the term "ovoid gall" to describe large, swollen lesions made by *Lygus lineolaris* on the stems of poplar. But the lesions and cankers that these bugs induce should not be considered true galls.

Although Flemion et al. (1954) attached more significance to strictly mechanical injury than to enzymatic destruction of tissues, it is generally accepted that plant bugs can inject toxic substances into host tissues (e.g., Putshkov 1956, Carter 1973). Norris (1979) suggested that phytotoxins and phytohormones be termed *phytoallactins*, that is, plant-altering chemicals.

Plant tissues punctured by various species of *Lygus* and related mirine genera deteriorate at different rates, as evidenced by color changes in the damaged tissues. Therefore, the toxicities of the salivary secretions from plant bugs appear to differ (Bech 1967). Enzymes can vary among presumed closely related species. Salivary phytotoxicity within the family ranges from violent to little or no phytotoxic action (Smith 1920a, Hori 1971a).

Cells considerably removed from the stylet tracks of lygus bugs and other mirines are affected by salivary secretions (Putshkov 1956, Reid 1968). Such cells lose their natural shape, collapse, and die. When *Orthops campestris* feeds on dill stems, cell division increases near the site of stylet penetration, leading to the formation of gall-like protuberances; parallel cell walls, with the cells arranged in rows, are evident in meristematic tissue (Bech 1967). Hori (1971a) did not observe cell disorganization along the stylet track made by *L. rugulipennis* on sugar beet, but on pumpkin fruit, cells surrounding the stylet track continued to swell for six days, eventually becoming enlarged and arranged in rows along the track and the feeding cavity that was created. The loss of starch in cells adjoining the cavity might have been due to starch consumption during cell hypertrophy (Hori et al. 1987).

PLANT-GROWTH DISTURBANCES

Many plant bugs, especially mirines, feed preferentially on actively growing tissues, perhaps with limited feeding in the xylem or phloem. Injury inflicted by these bugs might involve the destruction of meristematic tissues by salivary polygalacturonase, coupled with mechanical damage caused by probing with the stylets (Strong and Kruitwagen 1968, Strong 1970). Loss of auxin-producing meristematic tissues of terminal buds and other structures can create a hormonal imbalance at the feeding site. An auxin imbalance could be involved in the various symptoms associated with mirid feeding, including altered vegetative growth, fruit deformation, and abscission of flowers and fruits (Tingey and Pillemer 1977). Reid's (1968) observations that the effects of Lygus lineolaris on cotton tissues depend on whether the bugs have fed previously on a legume such as clover, or on cotton or oats, also suggest the involvement of biological growth regulators in mirid feeding (Scott 1976a).

Role of Plant-Growth Regulators

The petioles of sugar-beet leaves fed on by Lygus rugulipennis show increased levels of amino acids, oxidative enzymes, and phenolic compounds (Hori 1973a, 1973b, 1975c). Secretion of nonenzymatic constituents is implicated in the distortion and crinkling (see Chapter 9) of sugar-beet foliage (Hori 1967, 1971a). Such symptoms might involve an imbalance of growth-regulating substances in the leaf and involve a "continuous hypersensitive reaction brought about by quinones produced in the plant tissue as the result of the damage" (Hori 1973c). According to Hori (1974b, 1975b), quinones produced in the injured part of sugar-beet leaves destroy indoleacetic acid (IAA) or other auxins and inhibit growth. In the surrounding tissues, substances that promote the activity of IAA, or inhibit the activity of IAA-oxidase, spread to promote growth. This imbalance of growth might cause various malformations of the sugar-beet leaf. After additional research, Hori (1975c) altered his thoughts on the causes of leaf malformation, stating that destruction of IAA and inhibition of growth in both tissues (petioles and veins of damaged leaves and in surrounding tissues) result from quinone production.

The salivary glands of adult and nymphal L. rugulipennis contain an IAA synergist that promotes IAA activity (Hori 1974b, 1975d; Hori et al. 1987). It still is not known if this synergist is present in the bug's saliva, and its identity remains undetermined (Hori 1992).

The role of specific plant-growth promoters—IAA, inhibitors of IAA-oxidase, or nutrient metabolites such as free amino acids—in causing host injury is complicated (see Hori's [1992] review) and remains unresolved for mirids, particularly the species that feed on apical meristems. Contrasted with homopterans, few heteropterans secrete salivary IAA (e.g., Hori et al. 1979). In the mirid Stenodema calcarata, Nuorteva (1956a) showed that IAA, which stimulates plant growth and induces phytotoxic reactions, can be transferred from the bug's diet (a synthetic nutrient medium) to its saliva. Scott (1970) suggested that lygus bugs produce IAA in their saliva and that its injection into carrot seeds results in phytostimulation or increased growth of seedlings (see Chapter 10). Strong (1970), however, found no evidence for IAA synthesis in L. hesperus, and in other mirids it has not been substantiated that IAA is transferred to their saliva (Nuorteva 1956a, Miles and Hori 1977, Dixon 1983, Laurema and Varis 1991). In fact, mirids appear not to discharge significant amounts of IAA into their host plants. The amounts of IAA found in the saliva of Creontiades dilutus are at most two orders of a magnitude less than the concentrations ingested (Miles and Hori 1977), which tends to confirm Strong's (1970) conclusion that pectin polygalacturonase, rather than salivary auxins, is the main cause of phytotoxic reactions in plants. Concentrations of plant-growth substances, such as IAA in heteropteran saliva, are small compared to those in the host plant, and effects of their saliva on plants might be minimal (Miles 1987b).

In vitro experiments with extracts of heteropteran salivary glands, including mirids, have mostly failed to demonstrate direct effects on the growth of plant tissues (Hori 1976). When macerated salivary glands of L. rugulipennis are applied to the growing point of sugar-beet seedlings, the gland material can cause small dark spots in the area of application, darkening of the growing point as in plants injured by lygus bug feeding, and eventual malformation of the leaves (Varis 1972). Although such salivary extracts appear to synergize the effects of IAA (Hori 1976, Hori and Miles 1977), they have not yet been identified chemically or demonstrated to occur in the saliva itself (Miles 1978).

The varying results obtained from studies on saliva as mediators of plant responses might reflect the failure of the crude preparations of salivary glands or their parts to show activity similar to the component substances tested alone (Hori and Miles 1977). Madhusudhan et al. (1994) discussed the problem of distinguishing the contents of the insect body, the head, or the salivary glands from the secretions that are ejected—the "pure" saliva. The increased amino acid levels in tissues fed on by some mirids might result from a disruption of translocation in vascular elements of damaged tissue (Tingey and Pillemer 1977). In some instances, the presence of IAA in the salivary glands might be due to its formation from tryptophan by phenol oxidase activity (e.g., Miles 1968a, Hori and Endo 1977). Analysis of mirid saliva and its effects on plants is further complicated by the hormonal imbalance created in tissues when mirids feed on auxin-rich apical meristems. In addition, microorganisms introduced during feeding can affect the levels of plant-growth regulators (Miles 1972, Tingey and Pillemer 1977).

Salivary Toxins

As discussed earlier for cocoa- and tea-feeding bry-ocorines, the injury inflicted by mesophyll feeders might involve salivary toxins. Goodchild (1952) likened mirid feeding on plant cells to attacks by carnivorous heteropterans on the internal tissues of their prey. He hypothesized that the toxic effects of cocoa mirids on plants occur when saliva injected under high pressure fills the intercellular spaces, causing water-soaked lesions to appear. The small diameter of the mirid stylet channel would impart a hydraulic advantage that allows saliva to spread out against the turgor pressure of the plant cells. A strong acidity is produced, possibly because of esterase activity, which has a toxic effect (Cross 1971). When cells die, their walls become permeable and soluble contents are leached out; salivary amylase penetrates and dissolves insoluble carbohydrates (Goodchild 1952). It is unknown if the bug pumps out a lesion or is helped by a "back pressure," with the turgor of unattacked cells pushing out the fluid (Cross 1971).

So far, the only identified "toxins" are salivary enzymes such as polygalacturonase (Strong 1970), but enzymes sometimes are excluded from the definition of *toxin* (e.g., Scheffer 1983). Salivary enzymes also can have indirect toxic effects, such as ethylene production by *Pseudatomoscelis seriata* (Martin et al. 1988b).

Plant-Wound Responses and Effects on Mirid Feeding

In addition to the feeding process itself, the plant-wound responses that characterize the injury inflicted by phytophagous mirids need to be analyzed (Tingey and Pillemer 1977, Raman et al. 1984). The subject of wound responses, like that of the effects of hemipteran enzymes and plant-growth regulators and inhibitors on their hosts, is complicated. Much remains to be discovered. Plant tissues can be sensitive to the release of their own breakdown products (Miles 1987b), which may be produced by a disruption of cells or by exogenous pectinase (e.g., Martin et al. 1988b). Mirid feeding on plants potentially results in reduced nutrition in the damaged tissues as well as in increased nutrition in stressed plants.

PHENOLIC METABOLISM AND WOUND REPAIR

The brown discoloration of mirid lesions, which can be induced by mechanical wounds, should not be attributed to direct salivary action (Goodchild 1952). Nonspecific wounding of plant tissues stimulates responses involving phenolic metabolism, including oxidation of endogenous phenols or de novo production of phenolic compounds (Rhodes and Wooltorton 1978, Jones 1984, Bostock and Stermer 1989, Chessin and Zipf

1990). Phenolics released from damaged tissues can be oxidized to quinones by phenolase, laccase, and peroxidase enzymes, and further oxidized to nontoxic polymers that produce the brown discoloration characteristic of wounded tissue or a hypersensitive (defense) reaction of plants (Rubin and Artsikhovskaya 1964, Goodman et al. 1967, Lipetz 1970, Ishaaya and Sternlicht 1971, Levin 1971, Hori 1973a, Tingey and Pillemer 1977, Rhodes and Wooltorton 1978, Miles 1987b, Bostock and Stermer 1989).

Levin (1971), based on Miles (1968b), asked why an insect or a microorganism might contribute to the oxidation of phenol and the production of toxic quinones (Rodriguez and Levin 1976, Hatfield et al. 1982) by adding to the pool of polyphenol oxidase. That is, why would an organism add to a system supposedly triggered in plants as a defense against pathogens and phytophages? The production of quinones by the host triggers an interaction between the plant's phenolase-quinone content and the attacker's polyphenol oxidase content. Attack is successful if the invader's oxidizing system dominates, oxidizing the plant's quinones. If host quinones prevail, the plant resists attack (Miles 1968b; see also discussion by Norris [1979]).

The precise role of phenols in wound repair is unresolved (Rhodes and Wooltorton 1978) (Bostock and Stermer [1989] reviewed phenol metabolism in relation to wounding). Whether injury inflicted by mirids is accompanied by the oxidation of phenolic compounds to quinones apparently also requires verification (Levin 1976). Polyphenol oxidase enzymes of arthropods might be able to neutralize plant phenolics (e.g., Miles 1968b, 1969; Ishaaya and Sternlicht 1971; Hori 1973a, 1975c). In the case of *Lygus* species, salivary phenolases are hardly detectable (Hori 1974a, Laurema and Varis 1991).

REDUCED NUTRITION IN DAMAGED TISSUES

In addition to the formation of toxic quinones, reduced levels of nutritional factors such as proteins and sugars resulting from an insect attack might deter further plant feeding (Ishaaya 1986). K. Hori's work with *Lygus rugulipennis* was cited by Ishaaya (1986) as an example of a reduction in nutrition. Hori (1973c) actually reported that the concentrations of all 13 amino acids are increased in injured compared to uninjured sugar-beet tissues. All sugars are more or equally abundant in injured tissue. Somewhat different results were obtained with Chinese cabbage, although a similar increase in fructose was observed (Hori and Atalay 1980). Different plants respond differently to infestation by the same insect, which further complicates the analysis of biochemical changes in plants following herbivore attack.

OTHER PLANT DEFENSES

Plants respond to stress agents ("incitants"), including insects, in various ways. A common response to attack, including feeding by mirids, is an increase in respiratory rate (e.g., Gopalan and Subramaniam 1978). Interactions between a herbivore and a plant, marvelously intricate from an evolutionary perspective, are bewilderingly complex when the effects (deterrent as well as potentially beneficial ones) of mirid feeding on plants and the impact of wound responses on the bugs are considered. Many of the comments that follow are speculative and perhaps only marginally relevant, owing to fragmentary data on mirid feeding habits and an imprecise knowledge of plant allelochemicals and wound responses.

A review of rapidly induced plant responses and apparently nonmetabolic effects appearing in tissues remote from the sites of injury (Ryan 1983, Davies 1987, Chessin and Zipf 1990) seems unnecessary for the present discussion of mirids. Information on this topic is available in the works by Fowler and Lawton (1985), Edwards and Wratten (1985, 1987), Tallamy (1986), Karban and Myers (1989), and Haukioja (1990). Contrasted with the generally intercellular penetration of plant tissues by aphids—exploitation by "stealth" (Tallamy 1986)—the lacerate-flush feeding hemipterans, particularly mirid species that feed intracellularly, do not minimize the mechanical disruption to cells (cf. Miles 1987a). To what extent this strategy might trigger systemic plant responses to feeding is unknown for mirids. Tallamy (1986) suggested that phytophagous hemipterans and thrips might not stimulate inducible plant defenses as readily as do mandibulate herbivores (certain other sucking arthropods, such as spider mites, can induce host defenses [e.g., Bruin et al. 1992]). In addition, many mirid hosts might rely on rapid regrowth as a response to attack (e.g., Coley et al. 1985, Price 1991).

Research on wound repair in plants has emphasized the processes elicited by attacks on the bark of woody plants; an overview of this complex subject is provided by Hudler (1984). Mullick (1977) summarized the results of studies on the nonspecific defense mechanisms of bark, emphasizing production of an impermeable layer involving nonsuberized impervious tissue, which blocks further feeding by the balsam woolly adelgid (*Adelges piceae*). In susceptible fir trees, this process might be impeded by the adelgid's salivary secretions. This defense mechanism, described for a sternorrhynchan in which most stages are sessile, likely does not apply to mirids, which tend to be more active and to change their feeding sites more frequently, and to feed differently from adelgids.

PLANT STRESS AND METABOLIC MOBILIZATION

In addition to the putative defensive roles discussed, plant-wound responses appear also to have positive effects on some plant bugs, resulting in improved host quality. Mirids frequently return to injured hosts or to previous sites of feeding; the latter behavior could be interpreted as returning to feed from previously digested

mesophyll. Cocoa mirids are strongly attracted to injured trees (Squire 1947), and *Nesidiocoris tenuis* is attracted to fresh wounds on tomato plants and returns to previous feeding sites (El-Dessouki et al. 1976). The total protein content of tissues injured by the feeding of *N. tenuis* is about 34% higher than that in healthy tissues (Raman et al. 1984). Tanada and Holdaway (1954) emphasized the behavior of *Engytatus modestus* in returning repeatedly to former feeding sites (lesions encircling stems) on tomato plants. The bugs are attracted to fresh wounds in the laboratory (wounded tissues are thought to be more easily pierced), but in the field they are drawn to mature feeding lesions or girdles, suggesting that the physiological condition of the lesions is actually important. Protein quantity is greater in lesions than in control tissues (tissues unaffected by the bugs' feeding), with little difference in protein content above and below the lesions (Tanada and Holdaway 1954). *Engytatus modestus* appears not to feed above lesions on tomato plants; this contrasts with the behavior of the membracid *Spissistilus festinus*, which feeds above girdles made on legumes (Mitchell and Newsom 1984). Feeding by *E. modestus* might not disrupt the vascular tissues, block the flow of translocates, and create a nutrient sink above girdles in a manner comparable to the membracid (Mitchell and Newsom 1984; see also Tonkyn [1985] and Chapter 9). Plant bugs, however, appear capable of inducing physiological changes at feeding sites, including some accumulation of photosynthates above severed vascular bundles (Tingey and Pillemer 1977).

Importance of Plant Nitrogen

Plant bugs might simply orient to volatile chemicals emanating from damaged host tissues, or they might cue in on stress-induced changes in some phagostimulant or other phytochemical. A key to interpreting some of these observations might be nitrogen availability. McNeill and Southwood (1978), White (1978, 1984), Mattson (1980), McClure (1980), and others stressed the significance of nitrogen in the growth of all organisms and in determining animal abundance (see also Felton's [1996] discussion of the importance of protein quality to herbivores). Changes in nitrogen availability affect fecundity, generation times, and survival (Brodbeck and Strong 1987). This critical element is often in short supply in animals, which have much higher nitrogen requirements than plants and use it less efficiently (Mattson 1980). Looking at total nitrogen levels in plants, though, can be misleading (McNeill and Southwood 1978, Montllor 1991) because only a few amino acids (e.g., methionine) can be in short supply (Lindig et al. 1981). If the level of an essential amino acid is inadequate, development does not take place, regardless of the available total nitrogen (Brodbeck and Strong 1987).

Many phytophagous insects select the plant parts highest in nitrogen content and have mechanisms—

facultative predation, scavenging, host switching, possession of endosymbionts, and gall formation—for coping with low and seasonally variable levels of nitrogen. Plants in turn evolve strategies to minimize nitrogen availability to herbivores (McNeill and Southwood 1978, Mattson 1980, Mattson and Scriber 1987).

The importance of nitrogen in mirids is particularly well demonstrated for *Leptopterna dolabrata*. When leaf nitrogen declines, the nymphs move from German velvetgrass to feed on nutrient-rich inflorescences of a congeneric grass species (McNeill 1971, 1973). Competition for scarce high-nitrogen sites leads to a density-dependent mortality of late instars and decreased fecundity among surviving females (McNeill and Southwood 1978; see also Chapter 10). In addition, the degree of injury caused by several cotton-associated mirids is positively correlated with the total nitrogen content of the host plants (Ting 1963b). Moreover, *Distantiella theobroma* nymphs develop faster on cocoa seedlings growing under a high-nitrogen regime than on seedlings under normal- or low-nitrogen conditions (D. G. Gibbs 1969). The numbers of *Trigonotylus* species increase substantially in fertilized salt marsh grasses, which can be attributed to the production of larger, more fecund females and an increased survivorship in the nitrogen-enriched plots (Vince et al. 1981). The significance of plant nitrogen for mirids is further supported by Lightfoot and Whitford's (1987) study of insects inhabiting creosotebush. They found that *Phytocoris vanduzeei* increases disproportionately on fertilized plants, suggesting that this herbivore better exploits variable foliage quality than can less mobile chewing species. Plant bugs are among sucking arthropods that show increased densities on creosotebush growing in nutrient-rich sites (Lightfoot and Whitford 1989). The greater densities of mirids and other sucking herbivores in grain sorghum receiving an inorganic source of nitrogen versus an organic source is attributed to the higher nitrogen concentrations of the inorganic source (Blumberg et al. 1997).

Reallocation of Plant Resources and
Effects on Mirids

Far from responding passively to herbivory, plants respond directly through compensatory growth and reallocation of resources (e.g., McNaughton 1979, 1983; Mattson 1980; Verkaar 1988). Insect feeding, as well as water stress and other disturbances, can trigger the redistribution of reserves and nutrients to meristematic tissues (Mattson 1980, McNaughton 1983). For example, aphids feeding on phloem can secrete substances (e.g., IAA) that affect the host, sometimes to the aphids' advantage. Physiological changes near the feeding site might not produce external signs of disturbance but result in larger, more fecund aphids (Dixon 1985). Improvement of hosts as nutrient sources is characteristic of some gall-inducing aphids (Packham

1982, Miles 1989a), but not of all gall-forming insects (Larsson 1989), and of some planthoppers (Cagampang et al. 1974). Insects of different feeding habits respond differently to stressed plants. In a ranking of insect feeding guilds by response to stressed trees, Larsson (1989) placed sucking insects next to cambium feeders as the most sensitive to stress-induced changes in food quality. His sucking insects category presumably included only aphids and other phloem feeders. Homopterans and heteropterans that empty cell contents from nonvascular tissues might not elicit plant-compensatory responses (Tonkyn 1986). This phenomenon, however, has received little attention in mirids.

Mobilization of carbohydrates and amino acids can result in stressed plants becoming richer nitrogen sources than "healthy" plants (Klein 1952; White 1969, 1974, 1978, 1984; Kennedy and Fosbrooke 1973; Osborne 1973; Schaefer 1981, 1997). As White (1978) stated, "When plants are stressed (by whatever means) their tissues become a richer source of available nitrogen for young animals feeding on those tissues." That is, free amino acids, released by the hydrolysis of proteins, tend to increase in concentration relative to the protein content and are more available than proteins (White 1984, Cockfield 1988). The so-called plant stress hypothesis has received much attention in explaining patterns of herbivory (e.g., Rhoades 1983, White 1984, Mattson and Haack 1987, Cockfield 1988). Although stressed plants might be more susceptible and palatable to some herbivores than healthy plants and show decreased ability to synthesize defensive chemicals (Rhoades 1979), available evidence tends not to support this relationship as a general pattern (Miles et al. 1982, Larsson 1989, Lightfoot and Whitford 1989). Because the proportion of the various amino acids changes after hydrolysis, stressed tissues can actually become less suitable nutritionally for certain herbivores (Cockfield 1988).

Herbivore-induced stress can also have long-term deleterious effects. Feeding by a certain diaspine scale, for example, reduces the concentration of foliar nitrogen available to nymphs the following season (McClure 1980).

Examples of mirid-induced changes in plant nitrogen are relatively few. The immediate and most significant response to feeding by the orthotyline *Labops utahensis* on crested wheatgrass is an increase in nitrogen concentration in the shoots, an effect that is short-lived (Norton and Smith 1975). Gopalan and Subramaniam (1978) suggested that the increase in respiration in plants injured by *Moissonia importunitas* results from the mobilization of metabolites: infested hosts show increased levels of soluble sugars and free amino acids. Similarly, the percentage of total protein is higher in tomato parts infested by *Nesidiocoris tenuis* than in uninfested parts of the host plant (El-Dessouki et al. 1976).

The hypothesis that young animals often die from a nitrogen shortage in their diet was formulated for sap suckers, specifically phloem-feeding psyllids. "Essential nitrogenous nutrients in the food plants are too dispersed—too dilute—in relation to the amount of non-nutritious material that the young insects must 'process' in order to extract sufficient nitrogen to provide for their very rapid growth" (White 1974). The concentration of individual amino acids, though, can increase and decrease without a common pattern so that related psyllids might respond differently to plants under the same amount of stress if they had different amino acid requirements. Stress can affect the composition and concentration of free amino acids and those in protein form. As Brodbeck and Strong (1987) pointed out, White's (1974) hypothesis depends on whether herbivores benefit from such changes in amino acids. The essential amino acids Brodbeck and Strong (1987) mentioned as targets for stress-induced effects on insects are methionine, tryptophan, and sometimes histidine.

Feeding aggregations are not characteristic of the Miridae, although they are observed among cocoa mirids and other bryocorines such as *Halticotoma valida* and *Mertila malayensis* (see Chapter 6). Adaptations for group feeding are well known in aphids and certain other phloem-feeding sternorrhynchans, and feeding aggregations also are relatively common among coreoid Heteroptera (Tonkyn 1986, Slansky and Panizzi 1987, Wheeler and Miller 1990). Whether aggregations of mesophyll-feeding mirids result in a stress-induced enrichment of feeding sites on leaves has not been thoroughly investigated. In the laboratory, *Nesidiocoris tenuis* nymphs and adults aggregate at sites of injury on tomato plants, feeding at necrotic areas on the stems and petioles (Raman and Sanjayan 1984b). Any stress-induced increases in the concentrations of soluble nitrogen might have less effect on these mesophyll-feeding bugs than on phloem feeders, and as Gibson (1976) noted, leaf feeders tend to be more tolerant of low nitrogen levels than flower feeders. Because mirids induce phytotoxemias that result in the breakdown and release of nutrients in host tissues, they are potentially able to exploit any resulting increases in nutritional quality. The possibility, then, exists for mirids to benefit from group feeding.

Certain bryocorine mirids are characterized as having an especially nutrient-poor diet, although it has been termed "well balanced"; that is, the insects have "no urgent need either to conserve or to excrete water" (Goodchild 1963b). Goodchild (1952) commented that "no diet could be more restricted than that of the cacao mirids." He further noted (1966) that some members of this subfamily, which have a diet high in water content and an alimentary canal lacking in water-absorbing structures, represent the "nearest approach to sap suckers of all the Cimicomorpha." Producing rapidly appearing water-soaked lesions might represent a feeding strategy that allows some tropical bryocorines to benefit from increases in host nitrogen. Such increases result mainly from amino acids translocated in the phloem (White 1974), and tropical bryocorines might feed to some extent in phloem as well as in other tissues (e.g., Leach and Smee 1933). There is, however, disagreement concerning the importance of phloem to cocoa mirids, as discussed earlier in this chapter. They are said to feed on superficial parenchyma (Crowdy 1947, Leston 1970), but they might feed on phloem parenchyma (Cross 1971). As Gibbs et al. (1968) stated, factors influencing "the nutrient composition of phloem sap may or may not influence the tissues exploited by capsids [mirids], but the functional behaviour of these tissues is not well known to plant physiologists and no predictions can be made."

McNaughton (1979) cautioned against straightforward generalizations regarding the immediate effects of herbivores on plant growth and resource allocation. In some cases and in certain insects, stress can increase nitrogen and carbohydrate availability, which enhances the phagostimulatory characteristics of the host. Stress also can be associated with effects detrimental to a herbivore—that is, decreased food supplies or increased concentrations of allelochemicals (e.g., Trichilo et al. 1990).

The hypothesis that mirids frequently alter the food quality of their host plants and benefit from stress-induced increases in nitrogen (e.g., Schaefer 1981) remains to be tested. Does this putative strategy apply more to certain strictly phytophagous bryocorines than to other mirids? Or do some bryocorines compensate for possible low plant nitrogen by increasing their "throughput" compared to other mirids, and are they even limited by nitrogen? Is the mesophyll tissue fed on by bryocorines and many other mesophyll-feeding mirids nutritionally limiting? Mesophyll represents a more nitrogen-rich and balanced food source than either phloem or xylem tissue (Mattson 1980, Tonkyn and Whitcomb 1987). Usinger (1934), referring to the distinction between cytoplasm and cell sap, stated that the former "contains a high concentration of protein, lipoids and asparagine as well as starches, sucrose, and certain reducing sugars among the carbohydrates." Is the rate of nutrient removal ever limiting in these mesophyll feeders? Do mirids generally show compensatory changes in food consumption when faced with declining nutrient availability and concentration? The role of phloem in the nutrition of all mirids, not just bryocorines, also needs clarification, for as Brodbeck and Strong (1987) emphasized, amino acid concentrations in phloem are more likely to be improved by stress than those in leaf tissue.

Some mirid hosts, or tissues of those hosts, might have such high nitrogen levels that the bugs do not respond to increases in organic nitrogen, which appears to be true for certain aphids (Miles et al. 1982, Holtzer et al. 1988). After all, many plant-feeding mirids feed preferentially on meristems, buds, and young leaves, resources that represent high-quality food.

Dealing with Low Dietary Nitrogen

Creating higher-than-normal nitrogen concentrations in host tissues would be only one strategy by which plant bugs deal with the problem of low dietary nitrogen (McNeill and Southwood 1978). As mobile insects, mirids use various behaviors to avoid the effects of protein that is of suboptimal quality. Some polyphagous, multivoltine mirids track a succession of weedy herbs, feeding on nitrogen-rich meristems, buds, or flowers. In others, a univoltinism has evolved that allows specialization on staminate catkins of trees and on other temporally restricted plant parts. Some plant bugs change their feeding sites on a particular host species as the season progresses. The total nitrogen content of leaves often declines with age, but it varies and can have several maxima during a season (Kennedy 1958, McNeill and Southwood 1978, Mattson 1980, Raupp and Denno 1983, Mattson and Scriber 1987). Kraft and Denno (1982) suggested that seasonality in the orthotyline mirid *Slaterocoris pallipes*, a univoltine, mesophyll feeder on young leaves of groundsel baccharis or sea myrtle (Wheeler 1981a), is an adaptation that enables this bug to exploit a higher early-season nitro-

gen content of leaves and to avoid low foliar nitrogen in late summer. In addition, the grass-feeding *Leptopterna dolabrata* shows a compensatory increase in feeding rate on low nitrogen food (McNeill and Southwood 1978; see Chapter 10). Finally, mesophyll feeding by mirids—using barbed stylets to flush out the contents of numerous cells at one time and ingesting cell contents at rates higher than those for homopteran mesophyll feeders (Tonkyn and Whitcomb 1987)—can itself be considered an adaptation for obtaining adequate nutrition.

The family Miridae, with its diverse feeding habits, seems well suited for studies aimed at better understanding plant-herbivore interactions. Plant defense and compensatory growth or regrowth can be considered alternative responses to herbivory (van der Meijden et al. 1988). Strategies of herbivory in mirids range from attacks on stressed plants to those on young and vigorous plants. Plant feeding by mirids might be viewed as a continuum (*sensu* Price 1991). Some species create higher-than-normal nitrogen levels on their host plants or otherwise use physiologically stressed hosts, whereas others develop consistently on the most vigorous hosts or rapidly growing tissues.

8/ Mirids and Plant Diseases

Many early claims of [virus] transmission by unusual vectors, or involving many different orders of insects, still await substantiation.

—M. A. Watson 1973

Mirids often are toxicogenic, introducing salivary secretions during feeding that affect host metabolism and induce symptoms of plant disease (e.g., Allen 1951, Carter 1973). Because the symptoms can be indistinguishable from those induced by plant-pathogenic microbes (e.g., Christenson and Smith 1952), the entomological and phytopathological literature contains examples of initial uncertainty about the cause of a particular disease: a mirid or some fungus (e.g., Schmidt 1932, Leach 1935), bacterium (e.g., Steyaert and Vrydagh 1933), or virus (e.g., Laubert 1927, Gadd 1937, Müller and Costa 1964). Before the fungus *Phytophthora infestans* was demonstrated to cause potato late blight, a mirid (*Lygus lineolaris*) actually had attracted considerable attention as a possible causal agent of the "potato rot" responsible for the Irish famine of the 1840s (Wheeler 1981c). Preliminary studies on the relationship of the cotton fleahopper (*Pseudatomoscelis seriata*) to a mysterious disorder of cotton suggested that the bugs were vectors of a virus that caused the actual crop damage (see Chapters 9, 10; Hunter 1926, Painter 1930). Chapters 9 and 12 provide additional examples of the difficulty in determining whether a mirid, a pathogen, or both are involved in a plant-disease problem.

Plant bugs might seem well adapted for transmitting pathogenic microorganisms. Mirids are highly mobile, piercing-sucking insects that are often abundant, and species such as the Nearctic *L. lineolaris* and Palearctic *L. rugulipennis* have exceedingly broad host ranges. Mirids thus have ample opportunity to interact with pathogens and host populations. In some cases it is implied that the role of mirids in transmitting phytopathogens is considerable (e.g., Kiritshenko 1951; Jacobs 1974; Kelton 1975, 1983; Chu and Cutkomp 1992).

Many mirid species certainly are economically significant, but it is misleading to equate the direct injury they inflict as lacerate-flush feeders on crop plants with their indirect importance as vectors of disease organisms. Other sucking insects, such as aphids, leafhop-

pers, and planthoppers, are far more important vectors. With few exceptions, heteropterans are ineffective agents for dispersing plant pathogens and do not transmit plant-pathogenic microbes while they feed. Most mirids associated with infectious plant pathogens show a nonspecific, accidental relationship with fungi or bacteria, serving only to facilitate the pathogen's dissemination and penetration of host tissue. They often merely provide an infection court for the causal organism—that is, facilitate infection—although a few mirid species do transmit plant pathogens during stylet penetration and feeding. In some cases, their oviposition punctures allow fungi or other plant pathogens to invade (e.g., Cech 1989, Whitwell 1993).

Since Leach's (1940) general treatment, the literature involving mirid–phytopathogenic microbe associations has been compiled for bacteria (Harrison et al. 1980) and viruses (Heinze 1951, 1959). Reviewing the literature on disease transmission by mirids, Gibb and Randles (1991) stated that there are only three known examples of mirids as vectors of plant pathogens, but the literature actually contains numerous such reports. Critical evaluation of this literature, however, reveals examples based solely on speculation (e.g., Santoro [1960] for *Pycnoderes quadrimaculatus*) or poor experimental design; and especially for viruses, there is need to confirm reports of suspected transmission (e.g., see Hameed et al. 1975; see also Table 8.2). Mirids are considered relatively unimportant in the spread of plant diseases and have not received much attention from those who study interactions between insects and plant pathogens.

Dissemination of Fungal Pathogens

Although some notable plant pathogenic fungi are intimately associated with insects, the role of insects as vectors of lesser-known pathogens is not well studied. Relationships between insects and plant-pathogenic fungi have often been assumed or inferred from casual observations—experimental proof of transmission is lacking (Agrios 1980). Even less is known about the role of mirids in transmitting plant-pathogenic fungi than for most other insect groups that are potential vectors.

Mirids, like numerous other agents, influence fungal diseases of plants by wounding hosts and allowing the pathogen to enter (Plate 4; e.g., Thorold 1975, Chung and Wood 1989, Lim and Khoo 1990, Varma and Balasundaran 1990). They can also serve as carriers of fungal pathogens by helping disseminate spores. For most of the known mirid-fungal associations, no data are available on the transmission efficiency under field conditions.

CALONECTRIA INFECTION OF COCOA

A well-known example of a mirid-fungal relationship involves the bryocorines *Distantiella theobroma* and *Sahlbergella singularis*. Lesions produced by these cocoa pests provide a common means of entry for the weakly parasitic fungus *Calonectria rigidiuscula*, which causes a dieback of cocoa shoots. Dieback can be rapid on trees weakened by repeated mirid feeding (Crowdy 1947, Chatt 1953, Lavabre 1954, Usher 1962, Tinsley 1964, Thorold 1975). The percentage of lesions infected by *Calonectria* in West Africa can be as high as 80–95% (Entwistle 1972). Other points of entry are mechanical wounds and dead wood. The combination of mirid lesions and resulting deep-seated *Calonectria* infections can be devastating to cocoa (Johnson 1962). *Calonectria rigidiuscula* can also remain viable in old lesions in the xylem for 10 years or more (Owen 1956), giving rise to further infection when conditions are favorable (Booth and Waterston 1964). This fungus enters mirid-induced and other wounds and spreads in the xylem more rapidly than other fungi associated with mirid lesions on cocoa. A close mirid-fungus relationship appears not to exist: fungal spores have not been found on the bugs' mouthparts and have been recovered only occasionally from their body surfaces (Kay 1961). Thorold (1975) provides a more thorough discussion of dieback of cocoa in relation to mirid feeding and fungal infection.

ANTHRACNOSE OF ANNATTO

Helopeltis schoutedeni is associated with an asco-mycete that causes an anthracnose of a dye plant, annatto, in Malawi. As the bryocorine feeds on the succulent annatto pods, its stylet punctures afford entry by *Glomerella cingulata* (anamorph *Colletotrichum* sp.). Although the wind may be the principal means of spore dissemination, mirid feeding appears to be important in infection (Peregrine 1970, 1991). The evidence, however, is strictly circumstantial. Earlier, a *Helopeltis* species was observed to carry numerous spores of *G. cingulata* on tea bushes in India, but infection did not appear to be initiated through the bug's feeding wounds (Tunstall 1928).

VERTICILLIUM WILT OF ALFALFA

Another mirid-fungus relationship involves *Adelphocoris lineolatus* and a *Lygus* species, which have been implicated as dispersal agents of the fungal pathogen causing verticillium wilt of alfalfa. Bugs collected from an alfalfa field in Alberta showing a severe incidence of the disease were contaminated with spores of *Verticillium albo-atrum*, and it was supposed that the fungus could enter the host through feeding wounds. Contaminated bugs, however, did not cause the disease when they were caged on healthy alfalfa plants, and their effectiveness as vectors of *Verticillium* species remains undetermined (Harper and Huang 1984).

YEAST SPOT DISEASES

Mirids are among various insects that have been studied as potential vectors of the *Nematospora* species responsible for yeast spot diseases. *Lygus elisus* has been suspected of transmitting a nematosporaceous fungus while it feeds on cotton (McGregor 1927, Fawcett 1929). In addition, this bug's feeding (and that by *L. hesperus*) in California induces necrotic spots on lima bean seed that resemble the necrotic pitting caused by *N. coryli*. Because the pathogen could not be isolated from lygus bug–induced lesions, it was concluded that injury actually resulted from the bugs' direct feeding and injection of toxicogenic saliva (Baker et al. 1946). On soybean, *L. lineolaris* appears incapable of transmitting the yeast spot fungus *N. coryli* (Daugherty 1967). *Lygus* species (mainly *L. borealis* and *L. elisus*) collected from Saskatchewan mustard fields were infected internally with *N. coryli*, suggesting their potential as transmitters of the yeast to mustard crops. Yet infected adults failed to transmit the yeast to seeds in healthy mustard pods (Burgess et al. 1983). Other mirids associated with nematosporaceous fungi are *Helopeltis* species, which might vector *N. coryli* and *Ashbya gossypii* on cashew trees in East Africa (Batra 1973).

BOLL ROTS OF COTTON

In the southern United States, the tarnished plant bug (*Lygus lineolaris*) is associated with boll rot of cotton, caused by *Alternaria* and *Fusarium* species (Bagga and Laster 1968). The fungus *Aspergillus flavus*, which also incites boll rot of cotton, has been isolated externally and internally from tarnished plant bugs collected on cotton bolls in Arizona. Lygus bugs and other insects that attack bolls and squares can carry the fungus to injured tissues, where infection develops in the feeding wounds and exit holes made by lepidopteran larvae (Stephenson and Russell 1974, Widstrom 1979). *Creontiades pallidus* is associated with boll or lint rot of cotton in Africa. Spores of *Rhizopus stolonifer* (as *R. nigricans*; see Holliday [1980]), which are carried on a bug's rostrum, can enter bolls through wounds inflicted by mirids and other insects (Kirkpatrick 1925, Soyer 1942). *Horcias nobilellus* is associated with boll rot in Brazil (Moreira et al. 1994).

OTHER FUNGAL DISEASES

Lygus species were among several arthropods shown to harbor propagules of various wood-rot fungi that attack pome and stone fruit trees in Idaho orchards. The role of lygus bugs in disseminating the fungal pathogens has yet to be established (Helton et al. 1988).

An additional association of mirids with fungal pathogens involves *Orthops campestris* and the imperfect fungus *Stemphylium radicinum*, which causes black rot of carrot. Bech (1967) demonstrated that this mirine is able to transmit the fungus with its mouthparts.

Transmission of Prokaryotes: Bacteria and Mollicutes

Insects play a more prominent role in the spread of bacterial than fungal pathogens (Eastham 1915, Caesar 1919, Carter 1973), often helping disseminate bacteria from plant to plant or different parts of the same plant, or inoculating hosts with bacterial pathogens. The older literature contains frequent references to the possible role of insects in spreading bacteria-induced plant diseases (e.g., Ballard 1921), sometimes without mention of the particular insect species involved (Leach 1940). Among the various associations of mirids with phytopathogenic bacteria (see Table 8.1; Harrison et al. 1980), the best known involves *Lygus* species and fire blight of pome fruits, a disease caused by *Erwinia amylovora* (van der Zwet and Keil 1979). Van der Zwet and Beer (1991) inadvertently omitted lygus bugs from their discussion of fire blight, as their practical guide to disease management includes other mirids that play a lesser role in transmission than that played by lygus bugs.

BACTERIA

Fire Blight of Apple and Pear

Forbes (1884a, 1884b) implicated the tarnished plant bug (*Lygus lineolaris*) in spreading the fire blight bac-

Table 8.1. Some mirids implicated in transmitting bacterial pathogens

Pathogen	Disease	Locality	Mirid species	Reference
Clavibacter michiganensis subsp. *sepedonicus*	Ring rot of potato	Quebec, Canada	*Lygus lineolaris*	Duncan and Généreux 1960
Erwinia amylovora	Fire blight of apple, pear	New York, USA	*Adelphocoris rapidus, Campylomma verbasci, Heterocordylus malinus, Lygidea mendax, Lygocoris communis, Plagiognathus politus, Polymerus basalis, Taedia colon*	Stewart and Leonard 1915, 1916
		England	*Lygocoris pabulinus*	Emmett and Baker 1971
		Colorado, USA	*Lygus elisus*	Stahl and Luepschen 1977
		Illinois, USA	*Lygus lineolaris*	Forbes 1884b
		New York, USA	*Lygus lineolaris*	Stewart 1913a, 1913b; Stewart and Leonard 1915; Jones 1965
		California, USA	*Lygus lineolaris*	Thomas and Ark 1934
		Colorado, USA	*Lygus lineolaris*	Stahl and Luepschen 1977
		Ontario, Canada	*Neurocolpus nubilus*	Crawford 1916
		England	*Orthotylus marginalis*	Emmett and Baker 1971
		Denmark	*Phytocoris ulmi*	Thygesen et al. 1973
Erwinia carotovora subsp. *carotovora*	Heart rot of celery	USA	*Lygus lineolaris*	Hill 1932, Richardson 1938
Erwinia sp. (as *Bacillus gossypina*)	Cotton boll rot	Southern USA	*Adelphocoris rapidus, Pseudatomoscelis seriata*	Morrill 1910
Pseudomonas syringae pv. *aptata*	Beet bacterial disease	Poland	*Lygus rugulipennis, Orthotylus flavosparsus*	Bilewicz-Pawińska 1967
Xanthomonas campestris pv. *malvacearum*	Angular leaf spot, blackarm of cotton	Nigeria, Uganda	*Helopeltis* spp., *Taylorilygus vosseleri*	Hayward 1967, Logan and Coaker 1960

terium in Illinois pear orchards, but it was not until Stewart's (1913a, 1913b) work in New York that a causal relationship with *Erwinia amylovora* was demonstrated. He observed that peak numbers of mirids on apple trees tended to coincide with, or just precede, new fire blight infections. He further showed that the bugs visited blighted tissues and became smeared with the bacterial ooze or exudate, and that contaminated individuals caged on healthy shoots could transmit the bacterium, initiating infection. Because of its abundance in New York nurseries, *L. lineolaris* seemed to be the most important mirid vector, although other mirids were able to disseminate and inoculate the blight pathogen (Stewart and Leonard 1915, 1916; see also Rand and Pierce [1920] and Table 8.1). Plant bugs considered capable of spreading blossom blight in nurseries included the apple pest *Campylomma verbasci*, as well as species that develop on herbaceous plants and would not likely occur on apple: *Adelphocoris rapidus*, *Orthotylus flavosparsus*, and *Polymerus basalis*. Although experimental data were lacking, other mirids were implicated as disseminators of fire blight in apple and pear orchards: *Heterocordylus malinus*, *Lygidea mendax*, *Lygocoris communis*, and *Taedia colon*. Miridae generally were considered important in spreading the disease, especially when rainfall was abundant (Stewart and Leonard 1915).

Stahl and Luepschen (1977) found that *Lygus elisus* and *L. lineolaris*, under experimental conditions in Colorado, feed on pear fruits rather than on leaves or shoots. This agreed with observations that *L. lineolaris* feeds on pear flowers and blossom buds in New York (Jones 1965). Contaminated lygus bugs can infect pear fruits, but noncontaminated adults fail to initiate infection. The bugs' feeding creates wounds on the fruit surface that serve as infection courts for the fire blight bacterium (Stahl and Luepschen 1977).

Fire blight, indigenous to North America, was spread to England during the late 1940s or early 1950s, probably through shipments of contaminated fruit, propagating wood, or nursery stock (van der Zwet and Keil 1979). Several mirids associated with pome fruits in Britain were tested for their ability to transmit the blight bacterium, and feeding by *Lygocoris pabulinus* and *Orthotylus marginalis* resulted in an enhanced bacterial infection of pear shoots but not apple (Emmett and Baker 1971). *Phytocoris ulmi* has been implicated in the transmission of *E. amylovora* in Denmark (Thygesen et al. 1973).

Lygus bugs have been termed direct inoculators that introduce the fire blight bacterium into the wounds they create (Du Porte 1919); the bugs can become contaminated both internally and externally with the pathogen (Stahl 1976). They apparently also provide wounds that allow epiphytic populations of *E. amylovora* disseminated from some other source to enter healthy plant tissues (Harrison et al. 1980). Although mirids sometimes are important in initiating primary infections on blossoms or fruit, their role in

spreading infection to leaves and shoots needs further study.

In assessing the role of leafhoppers in spreading the fire blight bacterium, Pfeiffer et al. (1999) observed transmission by a species that feeds on the vascular tissues of growing shoot tips, whereas a species associated with older leaves was not involved in transmission. Because mirids often feed on young, succulent tissues of their hosts (though not necessarily on vascular tissues), they might be expected to play a role in the spread of the fire blight bacterium—if not as actual vectors, then as facilitators for the entry of epiphytic bacteria into wounds (see Pfeiffer et al. 1999).

Angular Leaf Spot of Cotton

In addition to inoculating cotton bolls with rot-inducing fungi, mirids can transmit the bacterial pathogens responsible for boll rot. Morrill (1910) included *Adelphocoris rapidus* and *Pseudatomoscelis seriata* among heteropterans that may cause bacterial boll rot in the southern United States; the pathogen was cited as *Bacillus gossypinus*, now considered a rejected name (Bradbury 1986). Morrill's (1910) observations actually refer to the boll rot phase of bacterial blight caused by *Xanthomonas campestris* pv. *malvacearum* (Pinckard et al. 1981, Hillocks 1992).

Other mirid species are associated with the foliar phase of *Xanthomonas* bacterial blight of cotton, a stage known as angular leaf spot. *Helopeltis* species are associated with this disease in Nigeria (Hayward 1967), and the role of *Taylorilygus vosseleri* in transmitting the bacterium has been studied in Uganda (Logan and Coaker 1960). Lesions resulting from mirid feeding are so similar to bacterial lesions that determination of a causal agent can be difficult, and observers in that East African country suspected that *T. vosseleri* vectors the blight bacterium. Under controlled conditions, Logan and Coaker (1960) demonstrated that contaminated *T. vosseleri* can inoculate healthy cotton plants with the bacterium, causing lesions from which the bacteria could be isolated. Infection, however, must be initiated within a few hours of feeding because injured hosts quickly form impervious tissue that prevents the bacterium from entering the wounds. The bacterium is carried mainly on the mirids' legs and head, but whether *T. vosseleri* inoculates the causal organism during feeding remains to be determined. Moreover, bugs that feed on artificially infected leaves fail to transmit the bacterium to healthy plants caged outside. Additional work is needed to clarify the importance of this mirid as a vector of angular leaf spot under crop conditions (Verma 1986).

Cotton fleahoppers (*P. seriata*) that are artificially infested with *X. campestris* pv. *malvacearum* can transmit the bacterium to healthy cotton plants and induce the typical symptoms of bacterial leaf blight. But this mirid does not appear to be an important vector under field conditions in Texas (Martin et al. 1988a).

Heart Rot of Celery

The relationship of *Lygus lineolaris* to heart rot of celery further illustrates the uncertainty of defining the role of a mirid and a bacterium in causing plant disease. The effects of injury from the tarnished plant bug and those associated with a soft rot caused by *Erwinia carotovora* subsp. *carotovora*, combined with physiological "blackheart" (calcium deficiency), make it difficult or impossible to pinpoint the initial cause of the problem (Hearst 1918, Hill 1932, Richardson 1938). The mirid's role in spreading the heart rot bacterium apparently is minimal, but experimental work is needed to clarify its relationship to the pathogen (Leach 1940, Harrison et al. 1980).

Other Bacterial Diseases

Some additional examples of mirids known to disperse plant-pathogenic bacteria or facilitate their penetration of host tissue are listed in Table 8.1. But not all mirids tested as possible vectors are involved in the transmission of bacterial pathogens. *Lygus lineolaris* reared on diseased tomato plants and transferred to healthy plants failed to induce bacterial canker caused by *Clavibacter michiganensis* subsp. *michiganensis*. The bacterium also could not be isolated externally or internally from bugs that had fed on diseased tomato plants (Ark 1944). Neither *Adelphocoris rapidus* nor *L. lineolaris* appear capable of transmitting *Xanthomonas campestris* pv. *phaseoli*, which causes common blight in beans (Hawley 1922). Similarly, the tarnished plant bug, as well as several other mirids common in the eastern United States, are not involved in vectoring *Erwinia stewartii*, which causes Stewart's wilt of corn (Poos and Elliott 1936, Elliott and Poos 1940).

FASTIDIOUS VASCULAR BACTERIA

Mirids are not known to transmit fastidious vascular bacteria, prokaryotic plant pathogens once referred to as rickettsia-like bacteria or organisms (e.g., Raju and Wells 1986). The failure of mirids to serve as vectors would be predicted on the basis of their feeding habits (see Chapter 7), although mirids have been little studied in relation to these pathogens. *Lygus hesperus* failed to transmit alfalfa dwarf (caused by a fastidious xylem-limited bacterium) in an early study in California when a virus was thought to be the causal agent (Weimer 1937).

MOLLICUTES

Phytoplasmas

Prokaryotes lacking cell walls are classified as *mollicutes*, a group that includes the phytoplasmas (previously called mycoplasma-like organisms or MLOs) and spiroplasmas. Phytoplasmas, once thought to be viruses, are transmitted mainly by homopterans such as leafhoppers, planthoppers, and psyllids (Tsai 1979, Ploaie 1981). Few associations of phytoplasmas with mirids or other heteropterans are known, although vector relationships in this group of mollicutes have been relatively little studied (e.g., McCoy et al. 1989). Because recent literature states that all known mollicute vectors are phloem-feeding homopterans (e.g., Fletcher et al. 1998), records of mirids as vectors of phytoplasmas perhaps need verification.

Mirids are presumptive transmitters of stolbur (big bud) among solanaceous plants. In the former Soviet Union, *Lygus rugulipennis* and several less common species of the genus transmit the stolbur phytoplasma from diseased tomato and field bindweed to healthy tomato and pepper plants (Neklyudova and Dikii 1973). The bugs' role in the epidemiology of the disease remains unknown. A dicyphine, *Nesidiocoris tenuis*, is suspected of transmitting a witches'-broom–inducing phytoplasma on paulownia trees in China (Jin et al. 1981, Guozhong and Raychaudhuri 1996).

Halticus minutus (as *H. tibialis*) mechanically transmitted a pathogen responsible in Papua New Guinea for witches'-broom of sweetpotato, a disease reported as "little leaf" and erroneously said to be viral induced. Other insects tested as possible vectors (a mealybug, a whitefly, and an aphid) failed to transmit the pathogen from infected sweetpotato to an indicator plant (Van Velsen 1967). The disease is now known to be caused by a phytoplasma and, in other areas of Southeast Asia and Oceania, to be transmitted mainly by leafhoppers (Jackson and Zettler 1983).

In studies on aster yellows in New York, *L. lineolaris* proved incapable of transmitting the disease pathogen (Kunkel 1924, 1926). The tarnished plant bug also failed to transmit the phytoplasmas responsible for little peach, peach yellows (Manns 1942), peach rosette (McClintock 1931), and purple top wilt of onion (Bonde and Schultz 1953).

Spiroplasmas

Mirids, as well, have not been implicated as vectors of spiroplasmas (e.g., Chiykowski 1987), which are helical mollicutes that cause plant diseases characterized by stunting and abnormal growth and development. Several serologically distinct spiroplasmas have been isolated from mirids, including the green leaf bug spiroplasma from *Trigonotylus ruficornis*. The pathogenic significance, if any, to the bugs or to their host plants has not been determined (Lei et al. 1979, Clark 1982; see also Hackett and Clark [1989], Williamson et al. [1989]). The mirid possibly is a vector in a complex natural cycle or a host in a symbiotic or parasitic relationship (Markham and Oldfield 1983). In most cases, the relationship between a spiroplasma and its insect host probably is benign (Fletcher et al. 1998).

Transmission of Viruses and Viroids

Reviews of insects as vectors of plant viruses often omit a discussion of mirids (e.g., Storey 1939, Freitag 1950, Smith and Brierley 1956, Ossiannilsson 1966, Heathcote 1976, Harris and Maramorosch 1980). Black (1954) declined to treat the family because so little was known of its role in viral transmission, yet many reports of putative mirid-transmitted viruses can be found in the literature (Table 8.2; see also Smith 1931a). North American and European species of *Lygus* and other mirine genera were used in the majority of these studies. In North America, lygus bugs apparently also are involved in the transmission of a viruslike disorder in carrots. Their feeding on developing seeds results in foliar symptoms; plants grown from injured seed are stunted and exhibit red- or yellow-mottled leaves (Scott 1972, 1983).

The omission of mirids by nearly all reviewers of insect-transmitted viruses reflects the unverified status of most early reports (e.g., Watson 1973). This early work was conducted during a period when methods for detecting and identifying plant viruses and knowledge of vector specificity were limited. Results from studies using generally unreliable techniques for virus identification, outmoded concepts of transmission, and in some cases, inadequate experimental design should not be accepted uncritically and should be considered suspect. The transmission of potyviruses by lygus bugs seems improbable, although transmission might occasionally occur when insects not belonging to typical vector groups are tested. Almost certainly erroneous are reports of transmission of luteoviruses, a phloem-limited plant virus group characterized by well-developed vector specificity and whose transmission by insects is limited to aphids (e.g., Kassanis 1952; Rochow and Israel 1977; Gildow 1987, 1991, 1993; Casper 1988; Waterhouse et al. 1988; Hull 1994; Halbert and Voegtlin 1995). Moreover, Smith (1931a) doubted the validity of most records of mirids as virus vectors, suggesting that their direct feeding on plants induced symptoms that were mistaken as evidence of viral diseases, and later

Table 8.2. Reports of putative transmission of plant viruses by mirids

Virus	Locality	Mirid species	Remarks	Reference
Bean common mosaic potyvirus	Georgia, USA	*Lygus lineolaris*	Possible transmission	Jenkins 1940
Beet mosaic potyvirus	Ukraine	*Orthotylus flavosparsus, Polymerus cognatus*		Novinenko 1928
Beet mosaic potyvirus	Poland	*Polymerus cognatus*	Important vector	Bilewicz-Pawińska 1967
Potato leafroll luteovirus + other virus diseases of potato	Ireland	*Closterotomus norwegicus*		Murphy 1923a, 1923b; Murphy and McKay 1926, 1929
Potato leafroll luteovirus + other virus diseases of potato	Germany	*Lygocoris pabulinus*		Blümke 1937
	Netherlands	*Lygus pratensis*	Putative vector might have been *L. rugulipennis*	Elze 1927
"Rape savoy virus"	Germany	*Lygus pratensis*	Putative vector might have been *L. rugulipennis*; record perpetuated by A. Gibbs (1969) but questioned by Carter (1973)	Pape 1935, Kaufmann 1936, Neumann, 1955
Sowbane mosaic sobemovirus	California, USA	*Halticus bractatus*		Bennett and Costa 1961
"spinach blight"	Virginia, USA	*Lygus lineolaris*	Causal agent has not been characterized (Watson 1973)	McClintock and Smith 1918
Tobacco mosaic tabamovirus	Brazil	*Engytatus modestus*	Tobacco to tobacco transmission noted; based on work by Costa and Carvalho	Gibbs and Harrison 1976
Turnip mosaic potyvirus	Germany	*Lygus rugulipennis*		Bech 1967

Note: Lygus bugs and other putative mirid vectors of plant viruses are typically excluded in discussions of insect transmission in the CMI/AAB "Descriptions of Plant Viruses."

noted that none of the records had actually been confirmed (Smith 1958). *Lygus lineolaris* was included (as *L. pratensis*, misplaced under leafhoppers) in a list of insect vectors in Bawden's (1943) *Plant Viruses and Virus Diseases*, but this plant bug was excluded from a similar list of insect vectors of plant viruses in the revised edition of the book (Bawden 1950). Flemion (1958) stated that *L. lineolaris* does not transmit viruses, and Harris (1981) considered transmission by lygus bugs improbable except by purely mechanical means. Since then, plant virologists have revised the concept of mechanical transmission (Ammar 1994). That vectors remain unknown for few, if any, economically important viruses—at least in temperate regions—weighs against the possibility that mirids are significant in the transmission of plant viruses. In addition, vector groups have been identified for most of the common virus groups, indicating that viruses generally have only one type of vector (a single genus or family).

Evidence for lygus bugs and other mirids as vectors of plant viruses is often limited to single reports that usually are not cited in the descriptions of plant viruses published by the Commonwealth Mycological Institute and Association of Applied Biologists (CMI/AAB). Furthermore, the early literature provides many examples of the failure of mirids to transmit viruses (Table 8.3).

Table 8.3. Examples of the failure of mirids to transmit plant viruses

Virus	Locality	Mirid species	Remarks	Reference
Bean common mosaic potyvirus	Wisconsin, USA	*Lygus lineolaris*		Fajardo 1930
	New York, USA	*Lygus lineolaris*		Hawley 1922, Harrison 1935
Bean yellow mosaic potyvirus	Ohio, USA	*Lygus lineolaris*		McLean 1941
Beet leaf curl rhabdovirus	Germany	*Lygocoris pabulinus*		Proeseler 1964
Beet mosaic potyvirus	England	*Lygocoris pabulinus*		Smith 1934
Beet yellows closterovirus	Netherlands	*Lygus pratensis*	Vector tested might have been *L. rugulipennis*	Roland 1936
Black currant reversion	England	*Lygocoris pabulinus*	Causal agent remains to be clarified (Wood 1991)	Massee 1952, Smith 1962
Celery mosaic potyvirus	California, USA	*Lygus lineolaris*		Severin and Freitag 1938
Cucumber mosaic cucumovirus	Wisconsin, USA	*Lygus lineolaris*		Doolittle 1920
	Philippines	*Nesidiocoris tenuis*		Torreno and Magallona 1994
Dahlia mosaic caulimovirus	New York, USA	*Lygus lineolaris*		Brierley 1933
Onion yellow dwarf potyvirus	Iowa, USA	*Labopidea* spp., *Lygus lineolaris*		Tate 1940
Potato leafroll luteovirus + other virus diseases of potato	New York, USA	*Lygus lineolaris*		Black 1937
	New York, USA	*Lygus lineolaris*		Dykstra and Whitaker 1938
	Maine, USA	*Lygus lineolaris*		Folsom 1942
	Netherlands	*Lygus pratensis*	Vector tested might have been *L. rugulipennis*	Elze 1927
	England	*Closterotomus norwegicus*, *Lygocoris pabulinus*		Smith 1927, 1929
Red clover mottle comovirus	Sweden	*Halticus apterus*, *Lygus* spp.		Gerhardson and Pettersson 1974
Tobacco mosaic tobamovirus	Zimbabwe	*Nesidiocoris tenuis*		Roberts 1930
	Philippines	*Nesidiocoris tenuis*		Torreno and Magallona 1994
Tobacco streak ilarvirus	Brazil	*Engytatus modestus*		Costa and Carvalho 1961
Turnip yellow mosaic tymovirus	England	*Closterotomus norwegicus*, *Lygocoris pabulinus*, *Lygus pratensis*		Markham and Smith 1949
Wheat streak mosaic potyvirus	Kansas, Nebraska, USA	*Trigonotylus* sp.		Connin and Staples 1957

Like early work reporting mirids as virus vectors, the credibility of research demonstrating a lack of transmission by mirids is open to question, and in at least some cases, experimental studies to reexamine the possibility of mirid transmission seem desirable.

Lists of viruses and their purported mirid vectors are most useful only for historical purposes. These compilations are beset by problems similar to those Mink (1993) emphasized in reviewing pollen- and seed-transmitted viruses: frequent uses of local disease names and names now considered synonyms, resulting in redundancies because a virus or disease is listed under several synonyms. In these lists some of the diseases reported as viral induced are caused by phytoplasmas (mycoplasma-like organisms).

Perhaps the first to review mirid-virus relationships was Smith (1931a), who listed only one credible example: the tarnished plant bug (*L. lineolaris*) as a vector of spinach blight (this bug was also noted to transmit potato spindle tuber viroid); four presumptive mirid-transmitted viruses were said to need confirmation (see also Smith [1958]). Other early reviews were those by Cook (1935), Leach (1940), Jensen (1946), and Heinze (1951, 1959). The reviews by Heinze are the most comprehensive, but they suffer from problems of redundancy and nomenclatural confusion of viruses, as well as the use of outdated names for the mirids. There clearly is need for a comprehensive assessment of viral transmission by plant bugs, including an updating of viral names and identification of records that are obviously erroneous or need corroboration. Here, I emphasize recent (1990s) experimental studies on mirids as vectors of plant viruses and discuss previous work on mirids in relation to plant viroids.

PLANT VIROIDS

An association of mirids with potato spindle tuber viroid dates from Goss's (1930, 1931) work in Nebraska with the tarnished plant bug. This mirid was one of several insects reported to transmit the causal agent (then considered a virus) of spindle tuber, both type and unmottled curly dwarf strains. This mirid-viroid relationship is cited in the CMI/AAB's description of the viroid (Diener and Raymer 1971) and is mentioned in reviews of potato diseases (e.g., Hodgson et al. 1973, O'Brien and Rich 1979, Beemster and de Bokx 1987). Diener and Raymer (1971) and Diener (1979), however, cautioned that new transmission studies were needed to confirm various insect species as actual vectors because the viroid can be spread easily by machinery and through contact with foliage.

Schumann et al. (1980) evaluated six insect pests of potato and found that insects serve as infrequent vectors of the viroid. Spindle tuber was detected in only 2 of 183 test plants, both of which had been infested with *Lygus lineolaris*. These results suggested that insects play only an insignificant role in the spread of the viroid in potatoes.

Lygus pratensis (but possibly *L. rugulipennis*) is included among insect vectors of potato spindle tuber viroid in Poland and the former USSR (e.g., Leont'eva 1962, Werner-Solska 1983). This plant bug is also said to transmit the potato carlaviruses S and M (Fomina and Lebedeva 1975, Lebedeva and Fomina 1977, Fomina et al. 1979). Because these viruses are transmitted by aphids and are easily transmitted mechanically (Wetter 1971, 1972), the ability of mirids to serve as vectors should be substantiated.

PLANT VIRUSES

Velvet Tobacco Mottle Virus

The most critical experiments on virus transmission by mirids are those by K. S. Gibb and J. W. Randles, whose studies involve velvet tobacco, velvet tobacco mottle sobemovirus (VTMoV), and *Engytatus nicotianae*, an Australian-Indo Pacific dicyphine, as the vector. Their terminology relating to viral transmission is retained here, although the reader may wish to consult Nault's (1997) system that combines terminology based on persistence and on mechanism or mode of transmission. Developed from a knowledge of homopteran-transmitted viruses, that system may or may not be appropriate for viruses that are transmitted by mirids or other heteropterans.

Salivary secretions released by *E. nicotianae* during feeding induce a silvering of leaf surfaces and sometimes foliar distortion on velvet tobacco (K. S. Gibb, pers. comm.). The bug also transmits the related solanum nodiflorum mottle, southern bean mosaic, and sowbane mosaic sobemoviruses (Greber and Randles 1986; Gibb and Randles 1988, 1991). The association of the mirid and VTMoV can be classed as both circulative and noncirculative—it defies categorization using traditional characteristics based on studies of aphid and leafhopper vectors.

Randles et al. (1981) first noted that *E. nicotianae* could transmit VTMoV. Transmission was found to involve a short acquisition period (minimum <60 seconds) and no detectable latent period, characteristics of a noncirculative, nonpersistent virus-vector relationship. The transmission rate increased with increasing acquisition time up to about 30 minutes, and then increased only slightly (Fig. 8.1), which characterizes either a noncirculative, semipersistent or a circulative association. Other characteristics of semipersistent or circulative transmission were a minimum inoculation time of one to two hours and persistence of infectivity for up to 10 days after an acquisition access period of 24 hours. Nymphs also acquired infectivity by feeding, but they lost the ability to transmit VTMoV within five to nine days (Table 8.4). The bugs' ability to transmit transstadially (after a molt) characterizes a circulative relationship. Transstadial transmission did not result from the probing of shed cuticles or the bugs' excrement on the surfaces of the rearing containers (Gibb and Randles 1988, 1989).

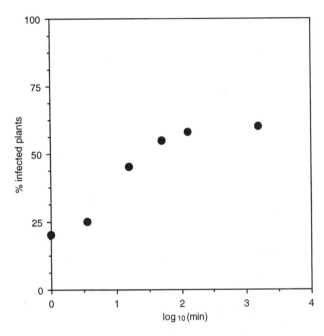

Fig. 8.1. Relationship between acquisition period for *Engytatus nicotianae* ($\log_{10}$ min) and percentage of tobacco plants inoculated with velvet tobacco mottle virus. Each point represents the percentage of plants infected by the mirid at each acquisition time. (Redrawn, by permission of K. S. Gibb and the Association of Applied Biologists, from Gibb and Randles 1988.)

Engytatus nicotianae nymphs also became infective when VTMoV was injected into the hemocoel. Following acquisition feeding on infected plants, the virus was detected in the gut, hemolymph, and feces of infective bugs, indicating that sap accumulates in the gut during feeding and that VTMoV moves from the gut to the hemolymph. The ability of the mirid to transmit the virus is independent of the presence of the virus in the hemolymph or the gut (Table 8.5). Although VTMoV appears to circulate in the plant bug, the mirid's salivary glands probably are not involved (virus could not be detected in salivary glands) and the virus is not propagative. Transstadial transmission, transmission following injection, and detection of virus in the bugs' hemolymph were considered evidence for virus translocation rather than a true circulative transmission in which the virus is associated with the vector's salivary system. Studies to detect low levels of the virus in sali-

Table 8.5. Relationship between site of detection of velvet tobacco mottle virus and the ability of *Engytatus nicotianae* to transmit the virus[a]

Site of virus	No. mirid nymphs	No. nymphs transmitting
Hemolymph only	4	3
Hemolymph and gut	7	5
Gut only	12	6
Neither	4	3

Source: Gibb and Randles (1990).
[a] Samples of hemolymph were taken after acquisition, and bugs were transferred to test plants for 2 days before gut samples were assayed for virus antigen.

Table 8.4. Loss of virus antigen and infectivity in groups of 10 *Engytatus nicotianae* after acquisition for 48 hours

Days after acquisition completed	No. positive[a] by ELISA	Mean (± SE) virus contact of mirids positive by ELISA (ng)	No. transmitting[b]
1	10/10	1,760 ± 500	8/10
2	9/10	2,440 ± 900	1/10
3	3/10	630 ± 160	1/10
4	2/10	1,030 ± 30	0/10
5	4/10	720 ± 180	1/10
6	3/10	490 ± 150	1/10
7	2/10	470 ± 320	1/10
8	1/10	330	NT
9	0/10	—	NT
10	0/10	—	NT

Source: Gibb and Randles (1989).
Note: ELISA, enzyme-linked immunosorbent assay; SE, standard error; NT, not tested.
[a] Minimum amount of virus that could be detected by ELISA = 12 ng.
[b] Numerator is numbers of plants infected; denominator is numbers of plants used.

vary glands are needed (Gibb and Randles 1989, 1990, 1991).

Gibb and Randles (1990, 1991) suggested an ingestion-egestion mechanism that does not involve the bug's salivary glands, to explain certain characteristics of noncirculative transmission by the mirid: short acquisition threshold and increase in transmission rate with increasing acquisition and inoculation times. An ingestion-defecation model, essentially a new mode of viral transmission by insects (Ammar 1994), was proposed as a mechanism of persistent, but nonpropagative, transmission. Virus ingested during feeding accumulates in the gut but does not propagate, and it is retained for as long as nine days. Virus also enters the hemocoel, and virus in both the gut and the hemocoel is eliminated slowly, serving as a reservoir for long-term transmission. The bugs can defecate infective plant sap up to six days after ingestion. According to the model, virus could be inoculated when the bugs ingest infectious feces from leaf surfaces and regurgitate them, or the virus could be transmitted through contamination of the mouthparts during probing (Gibb and Randles 1990, 1991).

Mirids as Unimportant Vectors of Plant Viruses

Why have so few other examples of definite mirid-virus relationships come to light? As Harrap (1973) noted, "In theory . . . any type of organism feeding on . . . an infected plant could act as a vector and transmit a virus to another healthy plant but not all do." Gibb and Randles (1991), though, acknowledged that the Miridae would be expected to be among the more important vector groups. Plant bugs are widely distributed, highly mobile sucking insects whose salivary glands, as in aphids, may serve an excretory function (Goodchild 1952, 1963a, 1963b, 1966; see also Chapter 7). Their mouthparts and basic feeding behavior resemble those of most phytophagous Hemiptera, seemingly making them well suited for virus transmission (e.g., Smith 1977). That they are not important vectors could accurately reflect the limited role of plant bugs in viral transmission, or it could merely be the result of inadequate attention to mirids by virologists (Gibb and Randles 1991).

Explanations for the paucity of vectors known in the Miridae (Smith 1926, 1931a, 1951; Leach 1940; Roberts and Boothroyd 1984) center on two ideas: (1) mirid feeding kills cells, making infection by obligate parasites difficult; (2) mirid saliva directly inactivates viruses. Regarding the second point, Carter (1973) suggested that in view of the temporary changes in either the quality or the quantity of mirids' oral secretions, the bugs' capacity for transmitting plant viruses might also vary. Even though tarnished plant bugs can ingest tobacco mosaic tobamovirus from their hosts, they fail to transmit the virus (as do other insects). The bugs' saliva inhibits the virus in vitro, but whether infectivity is inhibited in vivo is unknown (Orlob 1963).

Maramorosch (1963), citing Day and Irzykiewicz (1954), also referred to tobacco mosaic virus being inhibited by the saliva of the tarnished plant bug. But the heteropteran whose saliva Day and Irzykiewicz (1954) showed to be inhibitory was actually the pentatomid *Nezara viridula*.

Prospects of Discovering Additional Mirid Vectors

Most of the limited attention to mirids as virus vectors has concentrated on *Lygus* species and those of other mirine genera such as *Calocoris* and *Lygocoris*. The work generally was done during the early years of virus research, and as already noted, nearly all reported cases of transmission by mirids require confirmation. Studies using lygus bugs mainly involve the widespread, polyphagous *L. lineolaris* in North America and *L. rugulipennis* in Europe. Monophagous or oligophagous species of the genus apparently have not been tested as possible vectors. Studies on more host-restricted lygus bugs, though, would seem desirable: among allied insect species, one may fail to transmit a virus but another may serve as a vector (e.g., Butler and Jones 1949). Even lygus bugs that have failed to transmit certain viruses should be reevaluated in light of technological advances in virus detection and identification.

Gibb and Randles's work involved a species of Bryocorinae (rather than Mirinae), one of limited geographic and host-plant range. Bryocorines suspected of transmitting viruses of crop plants, such as *Helopeltis* species on cashew trees (Ramakrishna Ayyar 1942) and *Tupiocoris notatus* on tobacco plants (Moreira 1923), should be further examined as potential vectors.

Ideas regarding the importance of various insect groups as virus vectors, as well as mites, nematodes, and fungi (Smith 1965), have changed as more sophisticated research techniques and equipment have been used. At one time nearly all insect vectors of plant viruses were thought to be homopterans, and in that group the leafhoppers were once relegated to a minor role as vectors (Smith 1958). Sucking insects now known to transmit viruses, such as whiteflies and thrips, as well as mandibulate groups, were then unappreciated as vectors.

Matthews (1991), summarizing Gibb and Randles's (1988) paper on *Engytatus nicotianae*, raised the possibility of discovering additional vector species in the economically important family Miridae. Probably most species to be tested will prove unable to transmit plant viruses, a result that might be predicted on the basis of their feeding behavior. Mirids are lacerate-flush feeders that destroy cells and tend to induce violent reactions in host tissues and to trigger plant-wound responses (see Chapter 7), thus inhibiting some viruses. In contrast, aphids create minimal irritation to and disorganization of cells (Tjallingii and Hogen Esch 1993) and would be expected to be more successful vectors (e.g., Smith 1926). Mesophyll-feeding hemipterans in general are not vectors of plant viruses (Tonkyn and Whitcomb 1987). Moreover, plant bugs that are mesophyll specialists would not be expected to vector phloem-limited viruses. Yet *E. nicotianae* has now been shown to exhibit a somewhat complex relationship with plant viruses (Gibb and Randles 1988, 1989, 1990). In all probability, additional vectors await discovery in this diverse plant-associated family.

9/ Leaf and Stem Feeding

If the symptoms [of insect feeding] as described in the literature are assembled, it is possible to group them in certain broad categories, necessarily with ill-defined limits, which serve nevertheless to "classify" them on a basis of complexity.

—W. Carter 1973

The classification of leaf- and stem-feeding mirids according to the feeding symptoms they produce is for convenience only. This scheme does not necessarily correspond with the cell and tissue types used for feeding or with the enzymatic complements of the species involved. Furthermore, the symptoms described are not as clear-cut and mutually exclusive as the categories I have chosen might suggest. Expression of injury often varies according to how long the bugs have fed, and can be affected by the bugs' previous host and the physiological condition of the host, including its water status. Some plant bugs induce several types of macrosymptoms, such as chlorosis, blasting of new growth, dieback, and leaf tattering. A symptom is often transitory, soon leading to another in a series of easily visualized expressions of plant responses to feeding. The succession of symptom types sometimes corresponds to the various stages of the host's maturity (e.g., Fox 1993). Even in mirids of economic importance, the precise manner of their feeding (i.e., the cell and tissue types used), salivary constituents, and effects on plant metabolism are unknown. In some cases I have extrapolated from data on other insect groups and speculated on the etiology of certain plant responses to feeding.

Discussed first are instances in which feeding by plant bugs does not cause visible injury, including a few observations that pertain to reproductive structures in addition to leaves and stems. The remaining coverage pertains to most of the types of leaf and stem injuries that are apparent to a field observer. This discussion of leaf- and stem-feeding Miridae includes species that have been evaluated as possible agents for the biological control of weeds (e.g., Julien 1992; Woods 1992; Balciunas et al. 1994; Cassis and Gross 1995; Donnelly 1995, 2000; Hill and Cilliers 1996; Palmer and Pullen 1998; Hill et al. 1999; Stanley and Julien 1999).

The etiology of particular symptoms is not always equivalent—the bugs might use different enzymes or exploit different cell types or tissues—but the end result is the same. Mirid injury can include local lesions, such as chlorosis and similar local-lesion phytotoxemias, as well as various systemic effects such as growth and differentiation disorders. Some mirids induce a range of phytotoxemias that includes a combination of local and systemic symptoms (Norris 1979). Such injury involves the secretion of enzymes of various phytotoxicity, the secretion of nonenzymatic substances that create hormonal imbalances in affected tissues, and plant-wound responses (see Chapter 7).

Not all plant parts are suitable feeding sites because they can differ in nutritional value or presence of secondary metabolites (e.g., Thompson 1983). As in many herbivores (White 1978), mirids generally feed on tissues with the highest content of available nitrogen. Many of the species to be discussed ingest from the mesophyll of leaves and young shoots and less frequently from stems (Smith 1926, Putshkov 1956). Early-instar *Tropidosteptes illitus* feed only on the new leaves of ash, whereas later instars feed on stems in addition to leaves, seeds, and male flowers (Usinger 1945). *Lygus rugulipennis* spends less time feeding on the stems of sugar-beet seedlings than on the growing point or foliage of seedlings (Varis 1972). *Campylomma verbasci* feeds on the flowers and fruits of its hosts but will puncture the stems of mullein, leaving corky warts on the surface (Thistlewood and Smith 1996). Certain other mirids are either clear-cut leaf or stem feeders. On Scotch broom, four species feed on young growing stems, whereas a species with a shorter rostrum is mainly a leaf feeder (Dempster 1964). Use of stem tissue is particularly important among sap-feeding bryocorines on cocoa (Entwistle 1972, 1985a), the dicyphines *Engytatus modestus* (Tanada and Holdaway 1954) and *Nesidiocoris tenuis* on tomato (Raman and Sanjayan 1984b), and the grass-feeding *Capsus ater* (Kullenberg 1944). Putshkov (1966) noted that mirids also commonly feed on stems below the cotyledons. Certain types of stem feeding—that involving pedicels and peduncles—result in the abscission of buds and flowers, a symptom that is indistinguishable from feeding on the actual generative organs (discussed in Chapters 10 and 12).

Mirids feed to some extent on phloem (see Chapter 7), but their feeding behavior has not been extensively studied using an electronic monitoring system, as has been done for certain sternorrhynchans and auchenor-rhynchans (Backus 1994). It is, therefore, unknown if any mirids show the feeding plasticity characteristic of the potato leafhopper (*Empoasca fabae*): using different tactics (vascular and nonvascular) when feeding on different parts of the same host, or showing different feeding behaviors on the same parts of different hosts (Backus and Hunter 1989, Hunter and Backus 1989).

Mirids feed on mixed buds that contain leaves and flowers, and certain species associated with fruit trees and grasses prefer reproductive structures but must feed on the leaves and stems before fruit set or the "heading" of grasses. For example, *Lygocoris communis* feeds extensively on the tender, succulent twigs of apple trees before the fruits are available (Brittain 1916a). The adventitious shoots arising from the base and trunk of fruit trees are sometimes used by inflorescence feeders, whereas mature foliage is avoided (e.g., Smith 1907). Mature leaves presumably are more resistant to penetration by stylets because of the accumulation of lignin and other polysaccharides in the cell walls (e.g., Bateman and Millar 1966, Agblor 1992). *Hoplomachus affiguratus* feeds on the leaves of larkspurs before the plants' reproductive parts are available (Jones et al. 1998). In addition, lygus bugs feed to some extent on stems when developing seeds are unavailable on their host plants (e.g., Romney et al. 1945). These generally inflorescence-associated species that only occasionally feed on leaves and stems are discussed mainly in Chapter 10.

Injury Unapparent

Leaf-feeding insects, even chewers, often "apparently do little more than reduce the total leaf area and thereby lower the effective photosynthetic capacity of the plant" (Osborne 1973). But the effects of chewers or defoliators are more obvious than those of most suckers (e.g., Brown 1982b). The effects of sap-feeding arthropods on plants are often subtle and unapparent until considerable pressure is exerted, and the magnitude of their consequent physiological effects is frequently unappreciated (Dixon 1971b, Shure 1973). Feeding by aphids can cause negligible symptoms or no immediately discernible injury, but the resulting carbohydrate drain can depress growth and crop yield (Miles 1989b).

Phytophagous and predacious mirids, as well as other heteropterans, commonly pierce plant structures without giving rise to the discernible symptoms of sap (actually cell-content) consumption (Putshkov 1956)—the hosts remain symptomless, or symptoms go unnoticed because they occur at the cellular level. Species of the orthotyline genus *Sarona*, for example, fail to cause foliar chlorosis, even in high numbers (Asquith 1994), although some members of the sister group, *Slaterocoris* + *Scalponotatus*, do so (see Table 9.2). *Scalponotatus albibasis*, however, does not cause chlorosis on the foliage of antelope bitterbrush in Arizona, even though black excrement typical of mesophyll feeders is present on the lower leaf surfaces (pers. observ.). Biochemical and other internal changes in plant tissues fed on by mirids can differ among plant species. On bean pods, lygus bugs cause more internal breakdown of the tissues than damage at the surface (Flemion et al. 1954). But even high infestations of *Lygus rugulipennis* on Chinese cabbage (150 bugs/plant for 5 days) do not lead to the feeding symptoms seen on sugar-beet leaves; instead, the bugs produce only small brownish spots around the feeding punctures (Hori and Atalay 1980). Large sugar-beet leaves fed on by this bug are less likely to show obvious injury than small leaves (Varis 1972). Smith (1931b) stated that stinging nettle seems immune to mirid salivary secretions, the feeding of *L. rugulipennis* and two other species failing to result in symptoms. Feeding by various mirid species on cotton stems does not always result in external symptoms, although asymptomatic stems, when sectioned near the point of stylet entry, do reveal internal injury (Ewing 1929). Feeding by certain leafhoppers apparently does not produce even internal injury in host tissues (Medler 1941).

Empoasca fabae is a lacerate-and-flush-feeding leafhopper that probes both phloem and mesophyll tissues. It does not cause stippling when feeding on leaves, as do most other members of its subfamily. The absence of typical stippling symptoms results from short-duration probes that do not completely empty the cell contents (Hunter and Backus 1989). Some laceration-feeding mirids might show similar feeding behavior.

In some cases, the absence of external symptoms of mirid feeding might be related to incidental feeding on vascular tissues (see Chapter 7). Hori (1968a) discussed the pentatomid *Eurydema rugosa* as a mesophyll and vascular tissue feeder, noting the absence of external symptoms when vascular elements are pierced. In particular, stylet penetration by mirids belonging to Putshkov's (1956) trophic group IV tends not to produce distinct symptoms, although a whitish or brownish spot can appear at the site of penetration. Sometimes only a dried droplet of plant juice, invisible without magnification, remains (Putshkov 1966). Even when injury near the point of stylet insertion is unapparent, mirid feeding can lead to yellowing, wilting, retardation of growth, abnormal growth and differentiation, reduction in yield, or alteration of chemical composition in the roots—effects that are only later expressed or detectable. Allen (1947) suggested that insects can affect the activity of or response to plant-growth substances by injecting or removing materials, which indicated to Carter (1973) "that the term 'mere withdrawal' (of cell sap) may be an empty phrase." The effects of leaf-feeding aphids probably depend not only on what is removed but also on what is put into the leaves (Dixon 1971a). Even limited foliar injury by sucking insects such as typhlocybine leafhoppers can be important for its effect on water loss (Southwood 1985, Whittaker 1992). Sap or cell-content feeding that does not result in external symptoms can be associated with physiological changes in the host that also are symptoms of attack.

CONIFEROUS HOSTS

The leaves of many hosts more readily show feeding injury than do the stems, but slight foliage feeding does not always produce symptoms (Table 9.1). Because plants and tissues differ in their sensitivity to attack, even heavy feeding does not result in symptoms on certain hosts. This appears especially true for mirids developing on fir, pine, and spruce trees in North America (pers. observ.), although some conifer-feeding mirids do cause noticeable injury (e.g., Krausse 1923, Francke-Grosmann 1962, Wheeler and Henry 1973).

Large numbers of the phytophagous mirines *Bolteria luteifrons*, *Dichrooscytus elegans*, and *D. repletus* fail to produce visible injury to the foliage of ornamental junipers—even large populations are seemingly innocuous (Wheeler and Henry 1977). In the western United States, juniper-associated *Dichrooscytus* species, which tend to insert their stylets between the scales of the scalelike leaves (Plate 4), also do not cause conspicuous injury (D. A. Polhemus, pers. comm.). Although the immediate impact of feeding by these bugs is not apparent, their possible effects on plant growth should be studied.

Similar examples involving conifer feeders include the phyline *Parapsallus vitellinus*, which feeds on new needles of spruce, larch, and Douglas-fir in the eastern United States (Henry and Wheeler 1973) and on Siberian larch in Siberia (Rozhkov 1970). Similarly, the phyline *Phoenicocoris australis* does not appear to

injure the male flower buds or bud stems of slash pine in Florida (Ebel 1963).

LEAF VEINS

Taylor (1945) observed that the mechanical or phytotoxic effects of leaf feeding by mirids does not penetrate the veins. Extensive feeding on the actual midrib and major lateral veins, however, can trigger various physiological and biochemical changes in the veins and surrounding tissues and can result in the expression of symptoms (e.g., Molz 1917, Blattný and Starý 1942, Hori 1975c). But mirids that occasionally pierce midribs and lateral veins do not always leave external traces of their feeding. Examples of such behavior include Forbes's (1885a) observation of *Lygus lineolaris* piercing the midrib of a strawberry leaf in Illinois, as well as several Swedish species that feed on leaf veins (Kullenberg 1944). The feeding by the stenodemine *Collaria oleosa* on the leaf veins of yams in the West Indies is of "negligible importance" (Fennah 1947).

Whether mirid feeding on leaf veins involves significant use of vascular tissues, especially phloem, needs to be determined. Mesophyll represents high-quality food that contains a balanced mixture of proteins, carbohydrates, lipids, nucleic acids, vitamins, and minerals (Tonkyn and Whitcomb 1987; see also Chapter 7). Mirids might feed mostly on the mesophyll of leaf veins but obtain supplemental nutrition from the phloem or imbibe water from the xylem during dry periods.

PUBESCENT LEAVES AND STEMS

Pubescence, the collective trichome cover of the plant cuticle (also called *indumentum*), affects host explo-

Table 9.1. Examples of mirids whose feeding may not induce symptoms on host plants

Mirid	Host plant	Locality	Remarks	Reference
Bryocorinae				
Dicyphus errans	Potato	England		Wightman 1974
Macrolophus tenuicornis	Hayscented fern	Eastern USA		Wheeler et al. 1979
Mirinae				
Coccobaphes frontifer	Red maple	Eastern USA	Cited as *Coccobaphes sanguinarius*; see Henry (1985b) for nomenclatural clarification	Wheeler 1982a
Lygocoris aesculi	Ohio buckeye	Missouri, USA	Symptoms absent on foliage in laboratory	Blinn 1992
Lygus elisus	Prostrate kochia	Utah, USA		Moore et al. 1982
Lygus rugulipennis	Sugar beet	Finland	Symptoms not always produced	Varis 1972
Taylorilygus vosseleri	Cotton	Mozambique	No feeding symptoms apparent on old foliage	da Silva Barbosa 1959
Orthotylinae				
Orthotylus aesculicola	Ohio buckeye	Missouri, USA	Feeding failed to produce symptoms on foliage in laboratory	Blinn 1992

Note: See text for additional examples and discussion.

ration, acceptance, and choice of feeding sites by mirids. In leafhoppers, trichomes can limit access to the preferred oviposition and feeding sites, interfere with mobility and attachment to the surface, and when hooked, physically entrap nymphs. Pubescence is associated with varietal resistance to leafhoppers; the duration of nymphal feeding is reduced and injury is significantly less on resistant crop lines (Tingey 1985).

Some mirids prefer pubescent over glabrous hosts (see also Chapter 15). The cotton fleahopper (*Pseudatomoscelis seriata*) is usually more abundant on pubescent cotton strains (see Chapter 10), but such plants may be more tolerant to mirid feeding than glabrous strains (Schuster and Frazier 1977, Meredith and Schuster 1979). Some pubescent native or naturalized plants also sustain little, if any, noticeable injury from mirid feeding.

The several hundred individuals of *Campylomma verbasci* on the densely woolly foliage of mullein in Nova Scotia produce no evidence of feeding symptoms (Pickett 1938). Similarly, the fourlined plant bug (*Poecilocapsus lineatus*), whose distinctive water-soaked foliar lesions are discussed later in this chapter, does not induce the typical feeding symptoms on plants having strongly pubescent leaves (Wheeler and Miller 1981). Whitish nymphs of the phyline *Larinocerus balius*, which blend in with the leaves and stems of bladder sage in the deserts of the western United States (Henry and Schuh 1979), might not cause noticeable injury on the pale greenish-white, gray-pubescent foliage and stems. This also seems likely for mirids that feed on the lower, white-tomentose surfaces of hoaryleaf ceanothus foliage in California (Pinto 1982); for *Keltonia pallida* on the pubescent, pale greenish-white leaves and stems of a stemodia in Texas (Henry 1991); and for the bluish green nymphs of *Phytocoris tillandsiae* that blend in with and feed on the filiform leaves of Spanish moss in the southeastern United States (pers. observ.).

GLANDS, EXTRAFLORAL NECTARIES, AND RESIN

Use of glandular structures and their secretions as a food source usually does not result in external symptoms of feeding, although when glands of cotton are punctured, the point of penetration can turn black (Putshkov 1966). Mirids associated with plants having glandular leaves and stems (e.g., European filbert, hempnettle, stinging nettle, wood woundwort) often feed on viscid plant parts. In the case of *Dicyphus constrictus*, feeding occurs on the apices of glandular hairs on labiates (Kullenberg 1944). Another dicyphine, *Tupiocoris notatus*, feeds on the exudate from glands on the sticky leaves of its solanaceous hosts (van Dam and Hare 1998).

Many insects feed at extrafloral nectaries of native and crop plants in temperate and tropical regions (e.g., Hagen 1986). These include *Lygus lineolaris* and other mirids on cotton in the southern United States (Schuster et al. 1976b, Rogers 1985; see also Chapter 10).

Robertson (1928) reported this species from extrafloral nectaries of a wild bean ("tangle mealybean") in Illinois. *Lygus* nymphs and adults feed at stipular extrafloral nectaries of hybrid vetches in Georgia (Bugg et al. 1990a, 1990b), and *L. elisus* feeds in nectaries of a thelesperma in Colorado (Lavigne et al. 1991).

Similar feeding behavior is known in European mirids—for example, *Apolygus lucorum* and *Lygocoris pabulinus* on extrafloral nectaries of thistle (Rammner 1942) and *Liocoris tripustulatus* on floral nectaries of tall buttercup (Kullenberg 1944). Hespenheide (1985) recorded mirids as infrequent visitors to foliar nectaries of a sterculiaceous liana in Costa Rica.

Kullenberg (1944) mentioned another habit not likely to induce visual evidence of feeding: the sucking of resin exuding from the stems and other structures of conifers. He suggested that resin feeding might account for some of the records of mirids preserved in amber.

PLANT FEEDING BY PREDATORS

Some mainly predacious mirids—*Atractotomus mali*, *Deraeocoris brevis*, *Orthotylus marginatus*, and *Psallus ambiguus* (see Chapter 14)—also feed on plant tissues without causing noticeable injury. Plant feeding by predacious species can be induced by a scarcity of prey (e.g., Chinajariyawong and Harris 1987). Such behavior frequently occurs on apple (Petherbridge and Husain 1918; Smith 1920a; Rostrup and Thomsen 1923; Collyer 1952; Niemczyk 1963, 1968, 1978) and pear trees (McMullen and Jong 1967).

Chlorosis: Bleaching, Spotting, and Stippling

Chlorosis can be defined as the blanching or yellowing of plant parts from a failure of chlorophyll to develop. At a molecular or biochemical level, a decrease in chlorophyll content can result from an absence of light, deficiency of various nutrients, drought, and temperature-related stress. Plant pathologists and entomologists, in contrast to plant physiologists, sometimes use this term to refer to the breakdown or removal of chlorophyll. Chlorotic symptoms can result from the prevention of chloroplast development in emerging leaves or the degradation of fully developed chloroplasts in expanded foliage (Culver et al. 1991). Some workers use *chlorosis* to refer only to the yellowing of plant parts, whereas others broaden the definition to include fading of the normal green coloration to light green, yellow, or white.

The loss of chlorophyll and the resulting chlorosis are expressed in numerous ways, depending on the causal agent, type of plant, age of leaf tissue, and length of the feeding period (e.g., Tedders 1978). Affected leaves can appear uniformly blanched or yellowish, can appear mottled, or can show a mosaic pattern. Various other terms have been used to describe various types of chlorosis: banding, bleaching, bronzing, flecking, glazing, silvering, splotching, spotting, and stippling. Symptoms such as flecking and stippling can also involve necrosis or tissue death (Treshow 1970).

In the present discussion, *chlorosis* is used in a general sense to refer to the removal or destruction of chlorophyll by plant bugs, the term applicable more at a tissue than a molecular level. Emphasis is on the external appearance of affected foliage rather than etiology; in nearly all the examples given, chlorosis is assumed to result from mesophyll rather than phloem feeding.

Mirids often cause a stippled, spotted, or bleached appearance of foliage typical of certain other sap-feeding arthropods. Whitish stippling or spotting on the adaxial surface resembles that produced by many typhlocybine leafhoppers (Smith and Poos 1931, Günthardt and Wanner 1981, Whittaker 1984), certain aphids (Stace-Smith 1954), and some tetranychid (Baker and Connell 1963) and eriophyoid mites (Wilson and Cochran 1952, Gilmer and McEwen 1958). Stippling results from the removal of parenchyma and the appearance of air-filled spaces beneath the epidermis (Horne and Lefroy 1915, Kullenberg 1944, Putshkov 1966). Air that rushes in to fill the emptied cells simply leaves a clear spot or stippled area on the leaf (Tonkyn and Whitcomb 1987). Although stippling is mainly due to air replacing emptied palisade cells, some discoloration can result from the destruction of chlorophyll by diffusing saliva. Chloroplasts sometimes also degenerate in cells that are pierced but not emptied (Pollard 1968, Whittaker 1984). Some of the effects of mirid feeding, as in thrips (Lewis 1987), might be indirect, resulting from a desiccation of pierced and surrounding cells.

Pollard (1959) distinguished chlorotic spots caused by whiteflies, scale insects, and many heteropterans when they destroy chloroplasts, from those caused by lace bugs, which result mainly from the extraction of cell contents. As lacerate-flush feeders, mirids and tingids cause chlorosis differing from that of scale insects and other homopterans that indirectly destroy chloroplasts as their stylets probe for phloem sieve elements. Certain tingids always insert their stylets through stomata on the lower leaf surface (Ishihara and Kawai 1981, Buntin et al. 1996), but such behavior may or may not be typical of mirids. Cockfield et al. (1987) referred to the chlorotic spots that arise on foliage when armored scale insects destroy chloroplasts in the palisade layer as their stylets penetrate phloem sieve elements. Such feeding can reduce leaf chlorophyll and net carbon dioxide assimilation. Mesophyll feeding by certain leafhoppers produces greater physiological effects (reduced rate of photosynthesis, increased conductance of water vapor) than the small extent of foliar stippling might suggest (Whittaker 1984). Welter (1989) reviewed the effects of insect-induced chlorosis on photosynthesis and water relations in the host plant; apparently no studies of mirids were available for his

discussion of the effects of mesophyll feeders. Mirids, like other mesophyll-feeding arthropods (Welter 1989), can be assumed to reduce the photosynthetic rates of their hosts. But studies assessing the extent of such reduction and the effects on transpiration rate are needed.

Typical mesophyll-feeding mirids excrete dark-colored material that often coats the lower surface of infested leaves, potentially impairing gas exchange in the host plant by clogging the stomates (Putshkov 1966). Similar brown or black excrement is produced by chlorosis-causing lace bugs (e.g., Bailey 1951, Silverman and Goeden 1979). Dark or viscous fecal material is also characteristic of mesophyll-feeding leafhoppers. In contrast, phloem feeders produce clear excrement that must be stained with chromatographic dye to be seen (Smith 1940, Saxena 1954, Backus 1989). The significance of the yellow to green excrement of lygus bugs and other mirines (Parrott and Hodgkiss 1913, Willcocks 1922, Kirkpatrick 1923, Flemion et al. 1951, Stam 1987) in relation to their feeding habits and metabolism of ingested indole-3-acetic acid (e.g., Hori 1979b) and other compounds warrants further study.

Assuming mirid-induced chlorosis is similar to that inflicted by typhlocybine leafhoppers, injury to spongy and palisade parenchyma is mostly confined to the cells actually penetrated by the stylets and is discrete and permanent. The extent of chlorosis is related to the duration of feeding, and it does not enlarge with time. In typhlocybine leafhoppers, typical chlorosis is produced when stylets are inserted perpendicular to the leaf surface, whereas an oblique insertion is associated with the production of chlorotic areas having dark centers (Smith and Poos 1931, Quisenberry and Yonke 1981).

Leafhopper-induced stippling varies somewhat among species, some producing a finely peppered adaxial surface and others a more coarse pattern. Some feed on palisade and spongy mesophyll, whereas others seem to prefer, or be limited to, the upper leaf surface (Günthardt and Wanner 1981). Mirid-induced stippling also takes slightly different forms on the upper surface, depending on the species and probably also the host plant. Comparative studies of symptomatology of mesophyll-feeding mirids is needed, especially those that co-occur on a particular host.

The flexibility of mirid stylets (see Chapter 7) allows them to be redirected following insertion so that a single puncture can result in the removal of contents from several mesophyll cells. Withdrawal of cell contents during numerous probes, which radiate from a central point of stylet entry, seems to determine the characteristic shape of the stippling (Day and Saunders 1990). Stippling is often localized, but the entire adaxial surface can be affected.

Stippling and chlorosis fall within the category of "local lesions at the insect's feeding point" in Carter's (1973) classification of phytotoxemias. Yet it is difficult to show conclusively that injury from sap- and cell-content–feeding arthropods is merely local or that it involves a systemic or persistent salivary toxin. As Jeppson et al. (1975) noted for mites, "It is often difficult to determine whether the unhealthy plant symptoms or plant abnormality caused by mites result entirely from physical means or whether some local toxin is secreted into the plant during feeding." Some insects, including the mirines *Adelphocoris lineolatus* and *Lygus lineolaris* (Carter 1973, Norris 1979), include a combination of local lesions and systemic symptoms. The amino acids in heteropteran salivary glands that induce simple chlorosis can differ markedly from those causing tissue malformation. Salivary amino acids of species associated with the latter type of plant injury disturb the metabolic processes in their hosts and are involved in the altered amino acid pattern seen in injured tissue (Hori 1975b; see also Chapter 7).

The following review of mirids that cause chlorosis probably includes a few instances of incidental chlorosis. Some phloem-feeding heteropterans and homopterans—for example, the piesmatid *Piesma quadrata* and certain leafhoppers (Putman 1941, Naito 1976)—only incidentally touch and empty the contents of mesophyll cells and produce a bleaching of foliage (Cobben 1978:62). As discussed in Chapter 7, phloem feeding appears to be infrequent in mirids, and specialized vascular feeders are unknown in the family.

BRYOCORINAE

Bryocorines are discussed later in this chapter mainly as pests that produce stem lesions and cankers or that injure reproductive structures (see Chapter 12). This diverse subfamily, however, contains many species, particularly eccritotarsines, that feed on foliage and induce stippling rather than water-soaked lesions (Table 9.2). Chlorosis-producing bryocorines are leaf feeders that are not generally associated with feeding on the buds or flowers of their hosts.

Eccritotarsini

The yucca plant bug (*Halticotoma valida*), a pest of ornamental yuccas, feeds gregariously, causing irregular pale spots on the foliage and sometimes severe yellowing or browning of the plants (Plate 4; Haviland 1945, Wheeler 1976a). Fowler (1980) described chlorosis and locally severe injury caused by *H. valida* on native datil yucca or Spanish bayonet in New Mexico. The Neotropical *Caulotops distanti*, detected on potted cuttings of yucca in Florida and traced to plant material imported from Costa Rica, causes similar injury to ornamental yuccas (Henry 1985a). *Hesperolabops gelastops* causes chlorosis on the joints of cactus in the western United States (Mann 1969; pers. observ.).

Feeding by the Neotropical *Tenthecoris orchidearum* produces similar symptoms on orchid foliage; severely injured plants lose their leaves, fail to flower, and sometimes die within a few years (Reuter 1907a; Gimingham 1928; Lepage 1941, 1942; Occhioni 1944; Ossiannilsson 1946). In the early twentieth century, *T. orchidearum* occasionally became established in New Jersey greenhouses (Weiss 1917), and at one time this and other orchid-feeding mirids were often intercepted at U.S. ports of entry on shipments of cattleyas

Table 9.2. Examples of mirids that cause foliar chlorosis

Mirid	Host plant	Locality	Reference
Bryocorinae			
Bromeliaemiris bicolor	Orchids	Indonesia	Schumacher 1919b
Cyrtocapsus caligineus	Morning-glory	Florida, USA	MacGowan 1988
Dicyphus errans	Zonal geranium	Germany	Schewket Bey 1930
Eccritotarsus catarinensis	Water-hyacinth	Brazil	Hill and Cilliers 1996
Macrolophus separatus	False foxglove	Arkansas, South Carolina, USA	Pers. observ.
Pachypeltis polita	Eggplant	India	Gubbaiah et al. 1976
Pachypeltis vittiscutis	Villebrunea	Indonesia	Leefmans 1916
Pachypeltis sp.	Bell pepper	Indonesia	Kalshoven 1981
Pycnoderes dilatatus	Wide-leaved spiderwort	North Carolina, USA	Pers. observ.
Pycnoderes medius	Dayflower	Pennsylvania, USA	Pers. observ.
Ragwelellus suspectus	Paper-bark tree	Australia	Cassis and Gross 1995
Sixeonotus albicornis	Canadian wild lettuce	Florida, South Carolina, USA	Pers. observ.
Sixeonotus areolatus	Gaillardia	Texas, USA	Fletcher 1940a
Sixeonotus unicolor	Garden coreopsis	Michigan, USA	Pers. observ.
	Discoic (few-bracted) beggarticks	Virginia, USA	Pers. observ.
Mirinae			
Collaria columbiensis	Pasture grasses	Colombia	Barreto and Martinez 1996
Collaria meilleurii	Downy wildrye, panic grasses	Kentucky, USA	Pers. observ.
Collaria oleosa	Corn, rice, wheat	Central America	King and Saunders 1984
	Rice	West Indies	Schmutterer 1990b

Table 9.2. *Continued*

Mirid	Host plant	Locality	Reference
	Rice	Brazil	Silva et al. 1994
Collaria scenica	Panic grasses	Brazil	Costa Lima 1940, Kalvelage 1988
Litomiris debilis	Wheat	Utah, USA	Knowlton 1956
Prepops latipennis	Beans, potato	Central America	King and Saunders 1984
Orthotylinae			
Anapus spp.	Grasses, herbs	Former USSR	Putshkov 1956
Brachynotocoris puncticornis	European ash	New York, USA	Wheeler and Henry 1980b
Carvalhoisca jacquiniae, *C. michoacana*	Jacquinias	Mexico	Schaffner and Ferreira 1995a
Euryopicoris nitidus	Legumes	Former USSR	Putshkov 1956
Falconia intermedia	Lantana	Mexico	Palmer and Pullen 1995, 1998
Hadronema militare	Wild lupine	New York, USA	Pers. observ.
Halticus apterus	Dandelion, other hosts	Sweden	Kullenberg 1944
	Red clover	Nova Scotia, Canada	Pers. observ.
Halticus chrysolepis	Pasture grasses	Hawaii, USA	Gagné 1975a
Halticus intermedius	Canadian anemone	Illinois, USA	Pers. observ.
Halticus minutus	Cucurbits, legumes, sweetpotato	China, Indonesia	Maki 1918, van der Goot 1929, Tong and Wang 1987
	Peanut	Papua New Guinea	G. R. Young 1984
Halticus saltator	Bean, phlox, pot-marigold, potato	England, Germany	Butler 1925, Abraham 1937a
Halticus sp.	Muskmelon	India	Viswanath and Viraktamath 1969
Ilnacora malina	Giant ragweed	Arkansas, Pennsylvania, USA	Pers. observ.
	Thinleaf sunflower	Virginia, USA	Pers. observ.
Lopidea dakota	Caragana (Siberian peashrub)	Alberta, Canada	Strickland 1953, see also Ives and Wong 1988
Lopidea heidemanni	Carolina vetch	Pennsylvania, USA	Pers. observ.
	Hairy vetch	West Virginia, USA	Wheeler 1995; pers. observ.
Lopidea hesperus	False indigo	Eastern USA	Pers. observ.
Lopidea media	Bushy seaoxeye	Georgia, USA	Pers. observ.
Lopidea robiniae	Black locust	New York, Delaware, USA	Leonard 1916b; pers. observ.
Lopidea staphyleae	Bladdernut	Pennsylvania, Virginia, USA	Pers. observ.
Nesiomiris sp.	Araliaceous tree	Hawaii, USA	Bianchi 1966
Orthocephalus coriaceus	Oxeye daisy, composites in herb gardens	Eastern USA	Knight 1918c, Stear 1923; pers. observ.
	Tansy	Nova Scotia, Canada	Pers. observ.
Orthocephalus funestus	Artemisia	Japan	Yasunaga et al. 1993; see Plate 5
Orthocephalus saltator	Chicory[a]	Pennsylvania, Virginia, USA	Wheeler 1985, Henry and Kelton 1986, Wheeler and Henry 1992
Orthotylus flavosparsus	Lambsquarters, sugar beet	Eastern USA	Forbes 1900; pers. observ.
	Orache	Prince Edward Island, Canada	Pers. observ.
Orthotylus iolani	Wilwilli tree	Hawaii, USA	Beardsley 1958
Pseudopsallus viridicans	Biennial gaura	Texas, USA	Pers. observ.
Slaterocoris atritibialis	Mugwort	Pennsylvania, USA	Pers. observ.
Slaterocoris hirtus	Cup rosinweed	Arkansas, USA	Pers. observ.
Slaterocoris pallipes	Groundsel baccharis	North Carolina, USA	Wheeler 1981a
Slaterocoris stygicus	Giant ragweed	Pennsylvania, USA	Pers. observ.
Phylinae			
Europiella decolor	Artemisias	Kansas, USA	Pers. observ.
Hoplomachus affiguratus	Tall larkspur	Western USA	Ralphs et al. 1998
Macrotylus amoenus	Aromatic aster	Virginia, USA	Pers. observ.
Macrotylus sexguttatus	Clasping heart-leaf aster	Maine, USA	Pers. observ.

Note: See text for additional examples and discussion.

[a] Records of *Orthocephalus coriaceus* from chicory in Wheeler (1985) pertain to *O. saltator*.

from Mexico, Colombia, Venezuela, and Brazil (Weigel and Sasscer 1935, Hsiao and Sailer 1947). Beginning in 1876 and continuing until 1886, *T. orchidearum*, *T. bicolor*, and *T. colombiensis* (see Carvalho [1951] for nomenclatural clarification) were introduced with orchids imported to British glasshouses and conservatories (Westwood 1877, Carvalho and Leston 1952: fn. 9). These bugs caused a foliar blistering at feeding sites (Groves 1968), but they are no longer considered orchid pests in England (Hussey et al. 1969). *Tenthecoris* species were also imported into France from Brazil, resulting in serious injury to orchids in the plant houses of Lyons (Denis 1908).

Mertila malayensis causes similar injury to orchids in Indonesia, including premature leaf drop (Roepke 1918, Schumacher 1919b, Dammerman 1929); in the Philippines, gregarious nymphs of this species cause a yellow spotting and sometimes, death of the leaves (Capco 1941, Otanes 1948). Swezey (1945) reviewed the records of *M. malayensis* and other orchid-associated plant bugs that have been intercepted in quarantine in Hawaii and elsewhere.

Pycnoderes quadrimaculatus, the bean capsid, was once extremely numerous and injurious to cucumber plants in Arizona (Cockerell 1899). This bryocorine feeds on the lower leaf surfaces, producing a characteristic mottling on the upper leaf surfaces of beans and cucurbit crops (Wehrle 1935). Referred to as the "humped-back melon bug" on Mexico's west coast, this "ever-present menace" causes injury similar to that inflicted by the mirid *Halticus bractatus* on tomatoes (Morrill 1925, 1927b). In hot, dry regions of Hawaii, as many as 45 bean capsids per leaflet have been recorded; black excrement on the abaxial surfaces accompanies stippling on the adaxial surfaces (Holdaway and Look 1942). Leaves of squash in Puerto Rico become speckled before they wither and die (Cotton 1918, Wolcott 1933). The bugs hide under the leaves by day, particularly where vines trail along the ground, where they can attain high densities without attracting attention (Fennah 1947). *Pycnoderes quadrimaculatus* once ruined several rows of Mexican-grown cucumbers (Morrill 1927a) and injured gourds (Lebert 1935) and cantaloupe (Johnston 1940) in the southwestern United States. In Central America, this species feeds on the undersides of christophine (chayote) leaves, causing mottling, premature senescence, and premature fruit drop on this cucurbitaceous vegetable (King and Saunders 1984). *Pycnoderes monticulifer* is a pest of cucurbit crops in the Caribbean (Schmutterer 1990b).

Dicyphini

Members of the bryocorine tribe Dicyphini less often cause the stippling so characteristic of eccritotarsine feeding. Dicyphines can induce foliar crinkling and stem lesions, as discussed later in this chapter, or feed on inflorescences and seeds of tobacco and tomato (see Chapter 10). On tobacco in North Carolina, the suckfly (*Tupiocoris notatus*) once caused leaves to become spotted and pale and to lose body (Leiby 1927). This bug causes chlorosis on the foliage of desert tobacco in Arizona (pers. observ.) and of downy thorn apple or angel's trumpet in California (van Dam and Hare 1998).

MIRINAE

Mirini

Lygus bugs and their relatives typically feed on unfolding leaves, floral buds, and fruits but occasionally feed on expanded foliage. On some hosts, they produce stippling and chlorosis in addition to necrosis, but they belong to the trophic group associated with feeding on young vegetative or reproductive plant parts rather than leaves (Putshkov 1956, 1966). According to Putshkov (1966), bugs of trophic group IV, in contrast to the leaf feeders of group III, do not destroy cell walls or do so only slightly. When they do feed on the laminae of leaves, the injury usually occurs only as small, spreading whitish spots. The white marks appearing on foliage of castor at sites of penetration by *Polymerus cognatus* (Barteneva 1986) probably should not be considered typical of mirid-induced chlorosis.

Examples of mirines that produce chlorosis include *Lygocoris pabulinus* (Austin 1931a), *Lygus rugulipennis* (Southwood and Leston 1959), *Orthops basalis* (Ulenberg et al. 1986), and *O. campestris* (Kullenberg 1944) in Britain and on the European continent, and *Proba fraudulenta* in Chile (Olalquiaga 1955). *Lygus rugulipennis* can cause yellowing, as well as crinkling and death, of sugar-beet leaves (Bilewicz-Pawińska 1967).

The stippling and necrotic spots that species of *Lygus*, *Lygocoris* (Plate 4), and presumed related genera produce are usually not as pronounced as those of certain other mirines. An exception is *Lygus bradleyi*, large numbers of which caused extensive chlorotic blotches on the upper surfaces of yellow bush lupine in Washington State (pers. observ.). Stippling and spotting, though, are generally limited to cells near the stylet track and might reflect different feeding behavior, different salivary constituents, or both. Backus (1988b) demonstrated that the potato leafhopper (*Empoasca fabae*) can feed on both mesophyll and phloem, yet it feeds differently from typical mesophyll feeders and causes no stippling of host foliage (Hunter and Backus 1989). Leafhoppers feeding on shoots and the vascular tissues of leaves usually do not cause obvious injury under field conditions, but high densities can produce chlorotic discoloration and leaf curl (Varty 1963). Some mainly phloem-feeding aphids cause slight chlorosis when they feed on parenchyma (Miles 1989b). Whether the chlorotic symptoms induced by species of *Lygus* and similar mirines involve typical mesophyll feeding or reflect incidental chlorosis associated with some phloem feeding is uncertain (see Chapter 7). In the case of *L. bradleyi* nymphs on yellow bush lupine (pers. observ.), the copious dark excrement associated with their feeding resembles that produced by typical mesophyll feeders.

Lygus bugs also induce another type of chlorosis. Feeding by *L. lineolaris* on potato stems for two weeks in the greenhouse results in a foliar wilting that resembles fusarium wilt and is preceded by marked chlorosis (Leach and Decker 1938). This type of chlorosis, arising from feeding on stems below the affected leaves, differs from chlorosis due to direct feeding on foliage. The yellowing of midribs on citrus caused by a mostly stem-feeding restheniine, *Platytylus bicolor*, is considered a secondary symptom possibly due to a partial girdling of the host's conducting tissues (Müller and Costa 1964). Chlorosis associated with stem feeding can be categorized as a systemic phytotoxemia involving limited translocation of a toxic entity (Carter 1973). Another chlorosis different from that caused by typical mesophyll feeding is the localized yellowing (e.g., Whitney and Duffus 1986: pl. 140) that results mainly from feeding on leaf veins, injury similar to that induced by vein-feeding aphids (Tedders 1978); this yellowing contrasts with the whitish stippling of foliage caused by typical chlorosis-causing mirids.

Nearly all *Tropidosteptes* species develop on native and ornamental ash (Knight 1917b) and produce distinct spotting and bleaching of foliage (Plate 4). The ash plant bug (*T. amoenus*) feeds on the lower leaf surfaces of green and white ash in the eastern United States, causing pale spots on the upper surface (Dickerson and Weiss 1916). In heavy infestations involving ornamental plantings, leaves can turn brown, curl, and drop prematurely (Wheeler 1982b). Infested trees may also grow more slowly, show dieback of small branches, and become susceptible to injury by borers (Pellitteri and Koval 1981). Other species of the genus, such as *T. adustus*, discolor the foliage of ornamental ash trees in the East (Henry 1980). Plant bug–infested ash trees occasionally need control measures under landscape conditions, although the trees seem to tolerate the injury (Dreistadt 1994). The western equivalent of *T. amoenus* is *T. illitus*, an important pest of ornamental ash in California. The first instars feed on opening buds, and the later instars prefer new foliage and stems. Leaves eventually curl and branches wilt; complete defoliation occurs when plant bug populations are large. Like *T. amoenus*, *T. illitus* produces unsightly black excrement on the undersides of ash foliage (Usinger 1945). Other pests of ornamental ashes in the western United States are *T. pacificus* and *T. vittifrons* (Knight 1968, Dreistadt 1994). *Tropidosteptes chapingoensis* is a pest of ash in suburban and urban areas of Mexico (Cibrián Tovar et al. 1995).

Although many grass-feeding mirines attack the culm (stem) and produce a condition known as *silver-top*, or show a preference for seeds and other reproductive parts, species of several mirine and stenodemine genera feed on young or fully developed leaves. Some species feed on terminal foliage and on inflorescences (Kullenberg 1944; see also Chapter 10). Both leaves and inflorescences represent high-quality food sources.

Feeding by *Irbisia californica*, *I. pacifica*, and *I. solani* in the western United States produces a yellow spotting and sometimes a curling of foliage on barley, oats, and wheat (Essig 1915, Shull 1941; Schwartz [1984] corrected several identifications of the mirid species involved). *Irbisia pacifica*, a species of valleys and lowlands, causes yellowing and death of apical growth on grasses (Knowlton 1951). This bug feeds preferentially on grasses of semiarid rangeland, with Great Basin wildrye the most heavily damaged and crested wheatgrasses (which are favored by the orthotyline *Labops hesperius*) the least preferred (Hansen 1986). *Irbisia* species, in contrast to those of *Labops*, usually cause greater injury at lower elevations (Knowlton 1966). Feeding results in a reduced green leaf area and can also decrease the production of new foliage (Hansen 1988). Great Basin wildrye fed on by *I. pacifica* is particularly susceptible to drought stress. In contrast to intermediate wheatgrass, this wildrye grass shows greater reduction in green leaf area when subjected to plant bugs and drought. The bugs indirectly affect the plants by disrupting the acquisition or allocation of resources (Hansen and Nowak 1988).

Another species, *I. sericans*, feeds on the leaf blades of bluejoint reedgrass in Alaska, producing chlorotic spots and, in heavy infestations, dwarfing the plants. Injured leaves show decreased weight, cell contents, and nonstructural carbohydrates (McKendrick and Bleicher 1980). Schwartz (1984) noted that all species of the genus feed on grass blades, the injury consisting of pale white or yellow chlorotic spots. *Irbisia* species might depend more on foliage than *Labops* species; all *I. brachycera* and *I. pacifica* died after being caged for two days on growing seeds of crested wheatgrass, whereas *L. hesperius* survived (Haws and Bohart 1986). In addition, *I. brachycera* populations reach maximum abundance in Utah rangeland during the leaf stages of crested wheatgrass but rapidly decline as the boot stage is reached (Spangler and MacMahon 1990).

Irbisia species extensively damage giant wildrye in Utah (Haws et al. 1973), but the amount of biomass removed from crested wheatgrass by *I. pacifica* is small, and the amount of livestock forage lost is negligible even when bug populations are high (Hansen and Nowak 1985). The chlorophyll content in host plants is higher than would be predicted by the intensity of leaf injury. Visual ratings developed to estimate injury indicate the intensity of feeding rather than the actual chemical or physiological changes in the host plant. Hansen and Nowak (1985) suggested that this anomaly might be explained by a compensatory production of chlorophyll in undamaged leaf cells, or by the bugs' damaging only epidermal and palisade mesophyll immediately below the epidermis, with the dead and discolored cells masking underlying green cells. *Irbisia pacifica* feeds preferentially on the second and third youngest leaves, rather than on the youngest and oldest leaves, regardless of grass species, and shows an intraleaf preference for the tips of all grasses studied (feeding was least intense at leaf bases). Hansen (1987) hypothesized that a preference for middle-aged leaves (Fig. 9.1) was associated with biochemical changes in the host grasses, leaves intermediate in age providing a more balanced diet of carbon and nitrogen. The observed feeding patterns might be associated with accessibility of plant parts or olfactory cues that the plants produce, in addition to nutritional quality of the foliage (Hansen 1987, 1988).

Numerous other mirines produce chlorotic or mottled foliage (see Table 9.2), presumably from a lacerate-flush feeding on mesophyll with little local diffusion or systemic action, although few histological and physiological studies on the nature of injury have been conducted. Feeding by *Opistheurista clandestina* results in extensive white spots on the upper leaf surfaces of pole beans, the injury in Louisiana described as resembling that inflicted by mites or thrips (Jones 1921). This plant bug causes large, irregular blotches on the upper leaf surfaces of maypop passion-flower (Plate 4) in Arkansas and coats the lower surface with black excrement (pers. observ.). King and Saunders (1984) described a somewhat similar injury by this species on bean, cowpea, and sweetpotato plants in Central America: white flecking or mottling of foliage, sometimes accompanied by necrosis. *Lygocoris* species are mostly inflorescence feeders, but a few, such as *L. viburni*, produce a chlorosis and curling of foliage (Knight 1941); this

plant bug also causes small necrotic spots on newly emerged leaves and severe tattering of expanded foliage (Johnson and Lyon 1988; pers. observ.).

Species of *Capsodes* and *Horistus*, once placed in the small tribe Horistini that now has been merged with the Mirini, also induce foliar chlorosis. Silvestri (1932) discussed chlorosis caused by *H. orientalis* (as *C. lineolatus*) on grasses in Italy. Önder and Karsavuran (1986) reported similar injury to an asphodel in Turkey (see also Moursi and Hegazi [1983]).

Stenodemini

The stenodemine *Trigonotylus caelestialium* produces chlorosis on the blades of barley, oats, and various wild grasses (Wheeler and Henry 1985), and *T. tenuis* causes foliar chlorosis of bermudagrass in Georgia (Buntin 1988). In Russia, *T. caelestialium* and *T. ruficornis* cause yellowing or white spots on winter cereal crops (Kiritshenko 1951, Sannikova and Garbar 1981); the latter bug is most destructive to cereals in the forest-steppe zones and on irrigated land (Putshkov 1975). Because Putshkov noted that this bug feeds mainly on flowers, ovaries, and maturing seed, any chlorosis produced might be incidental to injury on the reproductive structures (see Chapter 10). A *Trigonotylus* species sucks sap from the leaves and stems of rice in Fiji (Hinckley 1965a), but any symptoms resulting from its feeding have not been noted.

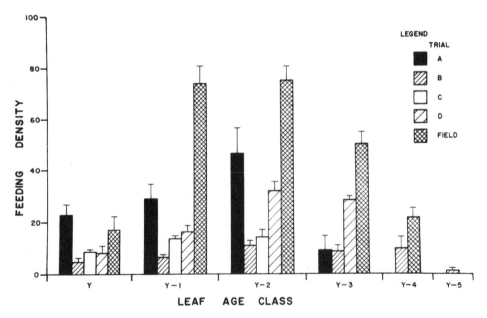

Fig. 9.1. Feeding density ($\bar{x} \pm$ SE) of *Irbisia pacifica* based on leaf area (Trials A, B, and Field) or leaf length (Trials C, D) among leaf-age classes. Trials A, C: crested wheatgrass; Trial B, quackgrass-bluebunch wheatgrass; Trial D, Great Basin wildrye; Field Trial, intermediate wheatgrass. Among leaf-age classes, Y represents the youngest leaves; Y-1, the next youngest; etc. (Reprinted, by permission of J. D. Hansen and the Kansas Entomological Society, from Hansen 1987.)

Leptopterna dolabrata, the meadow plant bug, occasionally invades wheat fields, feeding on inflorescences (see Chapter 10) and on leaves before the plants have headed. Large numbers of bugs once injured wheat along the edge of an Ohio field adjacent to a bluegrass pasture. Their feeding resulted in brownish yellow leaves and total loss of about 0.4 hectare of wheat nearest the pasture (Osborn 1939). Meadow plant bugs similarly invaded wheat during May 1992 in Pennsylvania after an adjacent alfalfa-timothy pasture was cut. The timothy served as host for the nymphs, which then moved into the wheat field, causing foliar blotching (Plate 4) on plants along the periphery. Injury extended 3–4 meters into the field (pers. observ.).

Notostira erratica feeds preferentially on the leaf petioles and blades of grasses in Europe (Kullenberg 1944), producing a white spotting of blades (China 1925a, Lehmann 1932a). Other stenodemines cause chlorosis before the nymphs move to the heads of grasses and cereals (Buczek 1956). The morphologically similar *N. elongata* is also a characteristic leaf feeder (Plate 4; Putshkov 1975, Bockwinkel 1990), although in England second-generation bugs feed on flowers when they seek new hosts as a result of extremely low nitrogen levels in hosts fed on during the first generation (Gibson 1976). Grass species differ in their suitability as hosts of *N. elongata*, with different grasses varying in nitrogen content between sites and seasonally at the same site. Nitrogen content can even vary "between different races or species of grass at the same site and time" (Gibson 1980). Gibson also pointed out that plant secondary chemistry, in addition to nitrogen content, helps determine the host preferences of *N. elongata*.

Bockwinkel (1990) demonstrated the preference of *N. elongata* for quackgrass among four grasses that were tested with the flowers absent. The proportion of feeding injury, fecundity, and adult survivorship were greatest and nymphal mortality was lowest on quackgrass, contrasted with the three other grasses (Fig. 9.2). Bockwinkel's (1990) work is particularly important for showing how plant structural characteristics profoundly influence feeding habits and help determine host suitability. The weak epidermis of quackgrass, the host that is used most efficiently under laboratory conditions, allows the stylets of *N. elongata* to penetrate easily; the bugs move their stylets back and forth to ingest the contents of parenchyma cells. On the least suitable host, orchardgrass, the firmer epidermis and more solid parenchyma cells necessitate repeated piercing. Feeding marks on this grass are more numerous and smaller (hardly wider than the stylets) than on quackgrass, where the area of injury can be as large as 4 × 6 mm (Bockwinkel 1990).

The stenodemine *Megaloceroea recticornis* produces diffuse feeding marks on the foliage of its hosts and is sometimes characterized as a leaf feeder on grasses (Gibson 1976, 1980; Gibson and Visser 1982; Bockwinkel 1990), but it feeds more extensively on inflorescences (see Chapter 10). Table 9.2 gives examples of several other stenodemines that produce chlorosis.

Mecistoscelidini

Feeding by an Oriental mirine, *Mecistoscelis scirtetoides*, which belongs to the small tribe Mecistoscelidini, results in characteristic symptoms (Plate 4) on its graminaceous hosts. White, tetragonal spots are produced on the leaves of bamboo seedlings (Chang 1981, 1982).

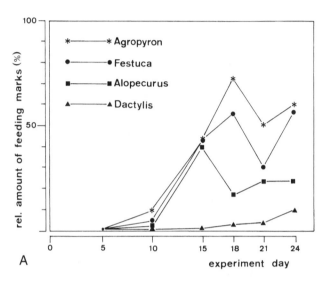

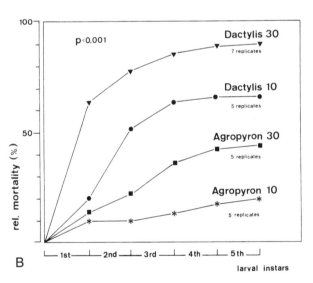

Fig. 9.2. Feeding behavior of *Notostira elongata* nymphs on different grasses, and nymphal mortality at different densities on two grass species. A. Preference for foliage of four grass species expressed as relative proportion of feeding marks made by first-generation nymphs. B. Mortality of neonate nymphs of first generation at densities of 10 and 30 per cage on quackgrass (*Agropyron*) and orchardgrass (*Dactylus*). (Reprinted, by permission of G. Bockwinkel and E. Schweitzerbart'sche Verlagsbuchhandlung, from Bockwinkel 1990.)

Numerous species of this subfamily (see Table 9.2) produce chlorotic symptoms on herbs, shrubs, and trees, and some are important crop pests. Feeding by several members of the tribe Halticini (Plate 5) results in the chlorosis of foliage typical of active, mesophyll-feeding heteropterans that extract sap (cell contents) for a considerable time from a relatively small area (Putshkov 1956).

Halticini

The garden fleahopper (*Halticus bractatus*) feeds mostly on the upper surfaces of numerous hosts (Plate 5), including legumes, vegetables, and flower garden and greenhouse plants (Weiss 1916, McDaniel 1931, Day and Saunders 1990). A few pale spots appear, followed by an overall foliar bleaching and a pale sickly color (Chittenden 1899). Feeding can stunt the growth of alfalfa and clover, the plants sometimes dying from severe injury (Webster 1897a, Beyer 1921). Webster (1897b) believed that the discoloration of red clover leaves by the garden fleahopper affords the nymphs protection because the "changed color more nearly harmonizes with that of their bodies." *Halticus bractatus* once destroyed several thousand acres of tomatoes on the western coast of Mexico (Morrill 1925) and was such a serious pest that expensive aerial applications of insecticides to old fields were necessary to protect young fields nearby (Morrill 1927b). Its accidental introduction into Hawaii resulted in extensive foliar chlorosis on several crop plants (Davis et al. 1985). Intensive feeding pressure from this bug once resulted in defoliation of some South Carolina colonies of kudzu (Nettles 1963), a pestiferous vining legume. Injury to this weed can be considered beneficial.

All *Halticus* species, including the grass fleahopper (*H. chrysolepis*) (Gagné 1975a), have similar feeding habits. The Nearctic *H. intermedius* injures foliage of native and ornamental clematis (Knight 1927a; pers. observ.), and *H. minutus* is an occasional pest of vegetables in the Philippines (Deang 1969). Further examples of plants injured by *Halticus* species are given in Table 9.2.

Another halticine, *Labops hesperius*, affects the productivity and palatability of rangelands (Plate 5) in the western United States and Canada, especially at elevations between 1,800 and 2,800 meters (Haws 1978b, Haws and Bohart 1986). The revegetation of native rangeland with exotic grasses, particularly wheatgrasses introduced to increase the carrying capacity of rangelands, might have triggered outbreaks of this native, previously innocuous insect. Its densities usually are low in native grasses (Todd and Kamm 1974). Problems began to develop during the mid-1940s to the early 1950s (Knowlton 1945; Denning 1948; Bohning and Currier 1967; Watts et al. 1982, 1989; Lattin et al. 1994, 1995). Losses of range grasses now ascribed to *L. hesperius* were once blamed on drought or early freezing (Haws et al. 1973).

Damage by *L. hesperius* occurs from the Yukon, Alberta, and the Pacific Northwest south to Arizona and New Mexico (Bohning and Currier 1967). In one field in Utah, this mirid consumed more animal units per month than did livestock (Haws and Thompson 1978). Annual losses in potential gain in cattle weight have been estimated at $5.1 million in Utah's intermountain region (Brewer et al. 1979). Cattle may not eat heavily infested grasses, possibly because of the bugs' fecal deposits or their injection of distasteful substances (Wiebe et al. 1978, Haws 1982), although reduced palatability might result from feeding-induced increases in fiber content or concentrations of secondary metabolites. Studies designed to evaluate free-choice feeding in cattle are needed to determine whether livestock actually reject bug-infested grasses (the incidence of larkspur toxicosis in livestock is lessened because cattle in western U.S. rangelands avoid eating plants injured by the phyline *Hoplomachus affiuratus* [Ralphs et al. 1997, 1998; Jones et al. 1998]). The real importance of feeding damage by *L. hesperius* lies in the loss of potential forage production (Hansen and Nowak 1988).

Although populations of *L. hesperius* are mostly higher on introduced wheatgrasses than on native range grasses, and higher yet on fertilized grasses (Higgins et al. 1977), not all introduced grasses show an increased susceptibility. Great Basin wildrye and western wheatgrass, both native species, are more susceptible to *L. hesperius* than are the introduced reed canarygrass or orchardgrass (Hansen et al. 1985a). Vegetational diversification of rangelands through reseeding (establishment of polycultures) might help lower the densities of grass bugs (Spangler and MacMahon 1990, Lattin et al. 1994). Decreased numbers of predators in monocultures, contrasted with those in mixtures of seeded and native grasses, might be partly responsible for the increased incidence of grass bugs (Haws et al. 1973, Vallentine 1989, Lattin et al. 1995).

Feeding by *L. hesperius* produces small white spots on the blades of wheatgrass, followed by a coalescence of spots and general yellowing of foliage (Todd and Kamm 1974, Malechek et al. 1977). Two or three nymphs can yellow a blade of crested wheatgrass in 4–6 days (Watts et al. 1989). In an otherwise healthy stand of grasses, circular discolored patches up to 6 meters in diameter can develop (Bohning and Currier 1967). Histological examination of tissues fed on by *L. hesperius* indicates that injury is generally confined to mesophyll cells, with the xylem and phloem remaining unaffected. Cell destruction, which can occur within 30 minutes of feeding (Brewer and Campbell 1977), involves the disappearance of cytoplastic organelles, including dictyosomes, endoplasmic reticulum, and mitochondria (Brewer et al. 1979). The bugs' feeding decreases leaf length, seedhead height, and carbohydrate reserves in the root crown of crested wheatgrass, consequently reducing plant vigor (Ansley and McKell 1982). Feeding by this mirid reduced seed number, seed weight, and germination rate of wheatgrass in a field nursery (Hewitt 1980).

Because the majority of *L. hesperius* individuals in a population are brachypterous, this halticine spreads slowly. Its colonization of uninfested rangeland depends mainly on the dispersal powers of the macropterous females, which at one study site in Oregon constituted only 4% of the population (Fuxa and Kamm 1976b). Only rarely does *L. hesperius* invade nearby wheat fields in numbers sufficient to cause foliar chlorosis (e.g., Mills 1939, 1941; Pepper et al. 1956).

Suppression of *L. hesperius* has emphasized cultural and, especially, chemical control. Insecticidal treatment has shown varying degrees of success (Glover 1978). Osman and Brindley (1981) discovered that *Labops* females are more tolerant than males to carbaryl (a somewhat similar insecticide susceptibility in *Tytthus mundulus* males is known [Verma 1955b]) and differ in their potential to detoxify this insecticide by monooxygenases. Losses from *L. hesperius* and other grass bugs, however, do not often warrant insecticide applications (Vallentine 1989). Economic models are needed to assess the cost-benefit relationships of control strategies (Haws 1978b).

Selected grass clones, which show a high degree of tolerance to the feeding of *L. hesperius*, have the ability to outgrow injury from the bugs' feeding (Hansen et al. 1985b; J. D. Hansen, pers. comm.). Research suggests that genetic diversity in range grasses is sufficient to develop resistant cultivars with a high density of relatively large trichomes that could inhibit the bugs' movement and ability to feed (Campbell et al. 1984, Ling et al. 1985). In addition, biochemical differences among hosts varying in their resistance to grass bug injury suggest that phospholipids and proteins in the chloroplasts play a "kairomonic (= attractive)" role in feeding by *L. hesperius*, whereas phenolics serve as repellents (Windig et al. 1983).

Other *Labops* species, which have not been as thoroughly studied as *L. hesperius*, induce similar symptoms on host grasses. In parts of Utah, the damage *L. hirtus* causes in rangeland is as severe as that of *L. hesperius*, and *L. utahensis* produces similar injury (Haws and Thompson 1978). Knowlton (1945) observed *L. utahensis* producing whitish dots on range grasses in Utah, the discolored plants being visible from a quarter mile. In Siberia (Yakutia), the Holarctic *L. burmeisteri* reportedly decreases the productivity of hay fields and pastures, although confirmation of this damage is lacking (Vinokurov 1988).

Orthotylini

In addition to the Halticini, species of numerous genera in the tribe Orthotylini cause foliar chlorosis. The Nearctic *Lopidea davisi*, which often infests ornamental phlox, produces a chlorosis from a breakdown of the parenchymal layer; spaces are left where cells collapse (Cory and McConnell 1927). The phlox plant bug (*L. davisi*) is likely the mirid that once forced some Arkansas gardeners to abandon phlox culture, rather

than *L. media* as originally reported (Becker 1918, Knight 1927a). Other *Lopidea* species that cause similar symptoms are listed in Table 9.2.

Another orthotyline causing similar chlorosis (see also Table 9.2) is the hollyhock plant bug (*Brooksetta althaeae*) (Plate 5; Cook 1891, Hussey 1922b). The onion plant bug (*Labopidea allii*), which once severely damaged cultivated onions in the Midwest (Henry 1982), discolors leaves and can cause the apical portion to turn brown and wilt (Bryson 1937). The Holarctic *Orthotylus virescens* extracts the contents of palisade and spongy parenchyma on Scotch broom, giving rise to extensive white blotches (Dempster 1964). This mirid, accidentally introduced into British Columbia and California with Scotch broom imported from Europe (Scudder 1960, Waloff 1966, Wheeler and Henry 1992), injures broom in the Pacific Northwest (Downes 1957).

PHYLINAE

Relatively few North American phylines—tribes Phylini and Pilophorini—cause foliar chlorosis. The reddish or dark necrotic spots caused by certain species might involve a type of feeding that differs from that of typical mesophyll feeders.

Spanagonicus albofasciatus, the whitemarked fleahopper, can discolor the blades of bentgrass on golf greens. In an outbreak of this bug in Missouri, growth of bentgrass was only 10% of normal over a several-week period (Knight 1927a, Satterthwait 1944). Knowlton (1954) reported serious injury to grasses from feeding by *Conostethus americanus* in Utah but did not describe symptoms; it can be inferred from the brief reports of Mills (1939, 1941) that this bug causes chlorosis on the leaves of wheat. This phyline, which is common on wheatgrass in Utah (Spangler and MacMahon 1990), might cause some of the leaf spotting observed on this grass, although the wheatgrass plots used for study contained a mixture of *C. americanus* and *Irbisia brachycera* (S. Spangler, pers. comm.); the latter bug, as already discussed in this chapter, belongs to a genus whose members definitely cause foliar chlorosis on grasses. The slight chlorosis caused by the Holarctic *Europiella decolor* in the western United States is mostly obscured on sericeous or tomentose leaves of host artemisias (pers. observ.; see also Table 9.2).

Rhinacloa cardini is a pest of elephant's ear or earpod in Cuba. Foliage of this shade tree shows yellow dots during June, then turns pale yellow; complete defoliation can occur by early autumn (Barber and Bruner 1946). For a congener, *R. callicrates*, Donnelly (2000) observed that its feeding marks on the foliage of Mexican palo-verde can take the form of round white spots. Feeding by the sunnhemp mirid (*Moissonia importunitas*) causes foliar chlorosis of sunnhemp in India (Ramakrishna Ayyar 1940, Reddy 1958, Banerjee and Kakoti 1969, Gopalan 1976b, Gopalan and Subramaniam 1978). High densities of this bug cause plants to become pale, appear sickly, and even die (Atwal 1976, Agarwal and Gupta 1983). In Pakistan, where *M. importunitas* is generally a minor pest, the bugs sometimes swarm over sunnhemp, causing the foliage to curl and turn pale yellow (Shabbir and Choudhry 1984). The yellowing might have an etiology different from typical chlorosis and involve a disruption of functions other than photosynthesis. The sunnhemp mirid also feeds on shoot apices and buds, resulting in phytotoxemic effects and a loss of pods and seeds (van der Goot 1927, Gopalan and Basheer 1966, Gopalan and Subramaniam 1978). The yellowing of aloe foliage by *Aloea australis* in South Africa (Jacobs 1985) is an example of plant injury by a phyline of the largely predacious tribe Pilophorini.

Blasting and Wilting of New Growth

Other mirids, instead of producing typical chlorotic symptoms that persist on expanded foliage, severely injure the buds and expanding leaves. Although feeding takes place on mesophyll so that stippling is apparent initially, the rapid distortion, wilting, and drying of affected leaves can obscure any chlorosis stage of injury. The distinction between the symptoms of leaf spotting and those of blasting is not always clear-cut. Indeed, the manner of feeding might be similar for plant bugs that induce both types of symptoms, although some of the species to be discussed are placed in another trophic group—those preferring the reproductive structures of their hosts (Putshkov 1956). Species causing chlorosis, as well as foliar lesions, can induce a curling or drying of particularly vulnerable growing points of their hosts and can kill seedlings.

Applied entomologists use the term *blasting* to refer to the destruction of tender, growing or fruiting portions of host plants (Metcalf et al. 1962). Plant pathologists sometimes use this term (or *blast*) for the sudden death of buds, flowers, or fruit (Shurtleff 1966). Here, the word *blasting* is used to encompass severe injury to new growth, especially buds, shoot tips, and new leaves. Included are symptoms that can be described as wilting and dieback. In some cases they represent the later or final expression of symptoms that first appear as chlorosis.

HONEYLOCUST PLANT BUG

A blasting of terminal growth is caused by the honeylocust plant bug (*Blepharidopterus chlorionis*) feeding on ornamental honeylocust in eastern North America. Hatching of the overwintered eggs coincides with vegetative budbreak on host trees in early spring. Newly hatched bugs move to the leaflets for protection and begin feeding within the unfolding buds. As leaflets expand, stippling and curling become apparent (Plate 5), but populations are often so large that they exert severe feeding pressure: leaflets are rapidly distorted, fail to expand, and abscise (Wheeler and Henry 1976, Betsch 1978). Leaflet distortion (Plate 5), sometimes

accompanied by a twisting of petioles, can easily be mistaken for injury by 2,4-D (2,4-dichlorophenoxy acetic acid) and other phenoxy herbicides (Feucht 1987, 1988). Premature leaf fall and complete defoliation can occur (Goble 1969, 1970; Wheeler and Henry 1976). In Ohio, Herms et al. (1987) observed defoliation on 'Moraine' honeylocust by mid-June during peak adult densities and determined that reduced starch levels correspond with the period of midsummer refoliation. Refoliated trees are able to compensate so that starch levels are replenished by the end of the growing season. Defoliation and subsequent refoliation, however, can predispose honeylocust trees to colonization by wood borers (Herms 1986).

A PLANT BUG PEST OF MESQUITE

In the southwestern United States, another green orthotyline causes injury somewhat similar to that of *Blepharidopterus chlorionis*. Jones (1932) reported an abnormal condition of mesquite that initially involves a discoloring of developing foliage, followed by a distortion of leaves, which later dry and drop prematurely. Feeding by a species of *Apachemiris*, probably *A. vigilax* (Henry and Wheeler 1979), results in the plants' failure "to develop leaves and flowers at the proper time." Before recognition of the mirid as the causal agent, this abnormal appearance of mesquite bushes was attributed to winter injury, spring frosts, drought, and even industrial pollution (Jones 1932).

Later in the 1930s, Arizona cattlemen complained about mesquite foliage drying up and dying, as though affected by a blight (Cassidy 1937). They were concerned because cattle feed on mesquite before the summer rains begin. The insect believed responsible for the injury was identified tentatively as *Orthotylus mimus*, but it, too, probably was *A. vigilax*. This plant bug is often abundant on new growth of mesquite in Arizona, causing discolored spots before distorting the leaflets (pers. observ.).

OTHER MIRIDS THAT FEED ON NEW GROWTH

Tropidosteptes species, which typically cause chlorosis and sometimes leaflet distortion on ash, also feed on emerging leaflets. In New York and Pennsylvania, *T. cardinalis* causes a dieback or blasting of terminal growth on white ash (Plate 5; pers. observ.).

Feeding by the polyphagous fourlined plant bug (*Poecilocapsus lineatus*) produces water-soaked lesions on foliage, often leading to symptoms of shot holing and tattering. If feeding is severe, new growth is killed and the plant appears scorched (McDaniel 1931). Blackened, dead terminals are conspicuous on certain hosts (pers.observ.).

In the eastern United States, feeding by the phyline *Criocoris saliens* causes terminals of various bedstraws to wither (Plate 5; unpubl. data). Another plant bug that causes similar symptoms is the mirine *Polymerus venaticus*, which prefers unfolding leaves at the top of goldenrod plants (Messina 1978). Feeding by several nymphs on a terminal can kill new growth (pers. observ.). *Polymerus fulvipes*, a characteristic insect of pitch pine–scrub oak barrens, injures the growing tips of whorled loosestrife in the northeastern United States (pers. observ.), and *P. venustus* causes similar injury to the new growth of lizard's-tail in Arkansas (pers. observ.).

Polymerus cognatus, a Holarctic polyphagous species most often associated with chenopods (Wheeler and Henry 1992), is a pest of sugar beet in the former USSR (Putshkov 1975). Its feeding causes the tips of plants to darken, curl, and wither; stems can be covered with dark spots of excrement. In first-year plantings, foliage is sometimes sucked dry, the tips and margins becoming shriveled and twisted, and the petioles darkened (Kirit-shenko 1951). Several mirine species injure sugar beet during its cotyledon stage in England (Dunning 1957) and in the western United States (Tamaki and Hagel 1978). Mirines similarly injure many other plants (Table 9.3).

The feeding habits of mirids that injure nitrogen-rich meristematic tissues or shoot apices, such as *P.*

Table 9.3. Examples of plant bugs that cause blasting and wilting of new growth

Mirid	Host plant	Locality	Remarks/symptoms	Reference
Bryocorinae				
Pachypeltis polita	Ornamental cuphea	Sri Lanka	Wilting, disfiguration of young shoots	Green 1901
Mirinae				
Adelphocoris lineolatus	Asparagus	Michigan, USA	Tip dieback of newly emerging spears	Grafius and Morrow 1982
Adelphocoris rapidus	Potato	Eastern North America	Wilting of growing tips	Webster and Stoner 1914, Beirne 1972
Apolygus lucorum (+ other mirines)	Hops	Czech Republic	Drying of terminals	Šedivý and Fric 1999
Apolygus lucorum	Mugwort	Japan	Distortion, dark spots on new growth	Pers. observ.
Closterotomus norwegicus	Asparagus	New Zealand	Twisting, withering of fall-harvested spears; injury resembles that from *Fusarium* infection; slight browning of unopened bracts of spring-harvested crop	Findlay 1975, Watson and Townsend 1981, Townsend and Watson 1982
Creontiades dilutus	Cotton	Australia	Tip wilting of seedlings	Chinajariyawong 1988
Creontiades pacificus	Potato	India	Wilting, drying of growing tips, young leaves	Lal 1950
	Melon	India	Wilting, drying of foliage	Butani 1979
Lygocoris pabulinus	Potato	Ireland	Wilting, death of foliage	Carpenter 1920
Lygus lineolaris	Potato	Eastern North America	Wilting, death of tips	Fenton 1921, Folsom et al. 1949
	Celery	Eastern North America	Adults feed on outer leaves, causing yellowing, wilting of tips; nymphs feed on tender heart leaves; injury, once attributed to sun scald (Davis 1893), resembles that of bacterial rot and blackheart (see also Chapter 8)	Hill 1932, 1941; MacLeod 1933, Maltais 1938, Whitcomb 1953, Boivin et al. 1991, Stevenson 1994
	Asparagus	Ontario, Canada	Shrivelling, darkening of spears soon after emergence; spears desiccate or rot from secondary bacteria or *Botrytis*; infestations do not always result in sub-stantial injury (see Drake and Harris 1932)	Wukasch and Sears 1982
Lygus pratensis	Cruciferous vegetables	Russia	Withering, loss of cotyledonous leaf	Znamenskaya 1962
Phylinae				
Spanagonicus albofasciatus	Beet, carrot, chard	Hawaii, USA	Blasting, death of cotyledons	Holdaway 1944

Note: See text for additional examples and discussion.

cognatus, more closely resemble those of species belonging to group IV of Putshkov's (1956) scheme of trophic relationships in the Heteroptera. They feed on young vegetative parts and generative organs (see Chapters 10, 12). These mirids often feed on shoots below the cotyledons, causing a wilting and withering of their hosts (Putshkov 1966).

Leaf Crinkling and Crumpling

Mirid-induced leaf crinkling, crumpling, and puckering are associated with feeding on young foliage of cotton, hop, sugar beet, tobacco, and tomato plants, as well as trees and shrubs (Table 9.4). The bugs' feeding on main veins can produce a shriveling that looks like a drawn thread, causing the leaf to pucker (Taylor 1978). Symptoms resemble those caused by aphids (e.g., Stace-Smith 1954), plant viruses (Fulmek 1930, Roberts 1930), or root problems (Roberts 1930). Many of the plant bugs that induce foliar crinkling and similar symptoms are mirines—species of *Adelphocoris*, *Closterotomus*, *Lygocoris*, and *Lygus*—whose feeding results in systemic phytotoxemias. When *C. norwegicus* and *Lygocoris pabulinus* feed on the midribs or lateral veins of sugar-beet leaves, they cause a puckering of leaf lamina, followed by distortion and yellowing of the apical portion (Bovien and Wagn 1951). The yellowing closely resembles that caused by virus infection (Jones and Dunning 1972, Dunning and Byford 1982) and should be distinguished from the whitish or yellowish spots caused by typical mesophyll-feeding plant bugs. Crinkling, however, can be preceded by reddish or dark necrotic spots.

Foliar crinkling was formerly attributed to an injury of tissues near mirid feeding sites, followed by collapse of the affected areas, which causes contraction of the leaf surface (Roberts 1930, El-Dessouki et al. 1976). The findings of later studies on the physiological basis of sugar-beet injury caused by *Lygus rugulipennis* (Plate 5) suggest that leaf crinkling results not only from possible mechanical injury associated with stylet penetration and enzymatic destruction of meristematic tissue, but also from the secretion of amino acids and phenolic compounds. The precise physiological and biochemical activities involved in mirid injury have yet to be elucidated. K. Hori's extensive work on *L. rugulipennis* (see Chapter 7) led to the following conclusions regarding leaf crinkle on sugar-beet plants, as summarized by Tingey and Pillemer (1977): "At the feeding site, localized secretion of amino acids coupled with mechanical injury lead to decreased growth-promoting activity through increased levels of oxidative enzymes and phenolic compounds. At the same time, growth-promoting factors secreted during feeding diffuse away from the feeding site, causing increased growth. Thus, the simultaneous promotion and inhibition of growth in adjacent tissues leads to crinkling and distortion of leaves."

Leaf Tattering and Shot Holing

Mirid injury expressed as wrinkled or crumpled leaves can lead to fraying or shot holing. Shot holing, a symptom well known to plant pathologists (e.g., Samuel 1927, Cunningham 1928; see also Cook [1924]), can be produced by various abiotic and biotic factors. Mirids often feed on young, still-folded leaves (Coaker 1957). Injured foliage can assume a shot-holed or tattered appearance (Fig. 9.3) as tissue at a feeding site becomes thinner and sinks below the leaf surface, and dead tissue drops from areas of stylet penetration (e.g., Petherbridge and Husain 1918), or as leaves tear or holes enlarge during expansion (e.g., Reichle et al. 1973). More severely injured leaves often become tattered or ragged as they unfold, whereas those slightly fed on develop only small holes (Coaker 1957). In the case of *Taylori-*

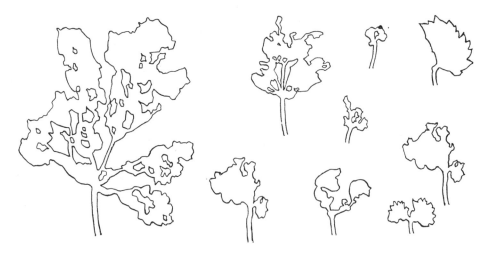

Fig. 9.3. Grape leaves crinkled and tattered by the feeding of *Apolygus spinolae*. (Reprinted, by permission of Blackwell Wissenschafts-Verlag GmbH, from Fulmek 1930.)

Table 9.4. Examples of plant bugs that cause foliar crinkling and similar symptoms

Mirid	Host plant	Locality	Remarks/symptoms	Reference
Bryocorinae				
Dicyphus errans	Zonal geranium	Germany	Crinkling	Schewket-Bey 1930
Helopeltis anacardii	Cashew	East Africa	Crumpling	Swaine 1959
Helopeltis clavifer	Annatto	Papua New Guinea	Crinkling	Smith 1978
Helopeltis schoutedeni	Cotton	East Africa	Crumpling	Harris 1937; see also Taylor 1978
	Mango	Malawi	Puckering	Leach 1935
Helopeltis theivora	Cashew	Malaysia	Crinkling	Khoo et al. 1982
Nesidiocoris tenuis	Tomato	Egypt	Rolling, puckering, and unevenness of leaves	El-Dessouki et al. 1976
Nesidiocoris volucer	Tobacco	Zimbabwe	Crinkling; injury once erroneously attributed to poorly functioning root system	Roberts 1930
Mirinae				
Adelphocoris lineolatus	Birdsfoot trefoil	Wisconsin, USA	Cupping, puckering	Wipfli et al. 1990a
Apolygus spinolae	Hops	Austria	Crinkling	Fulmek 1930
Closterotomus norwegicus	Sugar beet	England, northern Europe	Puckering of leaf lamina, distortion of tip	Jones and Dunning 1972, Dunning and Byford 1982
	Linseed	England	Crinkling, curling	Ferguson et al. 1997
Lygidea mendax	Apple	Pennsylvania, USA	Occasional crinkling	Hodgkiss and Frost 1921
Lygocoris pabulinus	Artichoke	England	Crinkling	Smith 1920a
	Bean	Ireland	Crinkling	Carpenter 1920
	Potato	England	Wrinkling around growing points of leaflets	Wightman 1974
	Sugar beet	England, northern Europe	Puckering of leaf lamina, distortion of tip	Jones and Dunning 1972, Dunning and Byford 1982
Lygocoris viridanus	Tea	Sri Lanka	Inrolled leaf margins appear wavy or "corroded"; cited as *L. pabulinus* (see Wheeler and Henry 1992); injury once erroneously attributed to *Cercosporella* disease	Webster 1954, Calnaido 1959
Lygus lineolaris	Birdsfoot trefoil	Wisconsin, USA	Cupping, puckering	Wipfli et al. 1990a
	Sugar beet	New York, USA	Cupping, kinking from feeding on midrib	Crosby and Leonard 1914; see also Hill 1941
Lygus rugulipennis	Chrysanthemum, herbaceous ornamentals	England	Puckering	Penna et al. 1983, Alford 1995
	Cucumber	Finland	Wrinkling	Varis 1972
	Sugar beet	Austria, Japan	Crinkling, wrinkling	Schreier and Faber 1954; Hori 1967, 1971a
Polymerus cognatus	Sugar beet	Ukraine	Wrinkling	Dekhtiarev 1927
Taylorilygus vosseleri	Cotton	Mozambique	Crinkling	da Silva Barbosa 1959
Tropidosteptes cardinalis	Ash	New York, USA	Crumpling if feeding is severe	Leonard 1916a
Orthotylinae				
Heterocordylus malinus	Apple	Pennsylvania, USA	Occasional crinkling	Hodgkiss and Frost 1921
Lopidea robiniae	Black locust	New York, USA	Occasional crumpling	Leonard 1916b
Phylinae				
Plagiognathus chrysanthemi	Birdsfoot trefoil	Wisconsin, USA	Cupping, puckering	Wipfli et al. 1990a
Sthenarus rotermundi	White poplar	Germany	Crumpling, wrinkling	Schumacher 1917d

Note: See text for discussion.

lygus vosseleri, Taylor (1945) suggested that the water content of cotton foliage affects the severity of shot holing and tattering, with leaves during dry weather exhibiting small spots where the bugs feed and only minimal holing or tearing. "Succulent" cotton plants thus show more leaf injury than do "drier" individuals. A leaf can be fed on while still in the bud stage, which gives rise to small necrotic areas that cannot grow as the leaf expands and are left as holes (e.g., Wolfe et al. 1946, Stride 1968). These symptoms are induced by leaf-feeding mirids or by characteristic fruit and inflorescence feeders whose early instars develop on new growth. Injury can also result from tiny, purplish brown or red, necrotic, pinholelike spots on leaves, such as those caused by *Plagiognathus punctatipes* on ninebark (Wheeler and Hoebeke 1985) and by *Lygocoris communis* (Brittain 1916a), *L. pabulinus* (Fig. 9.4; Smith 1920a, Rostrup and Thomsen 1923), *L. rugicollis* (Petherbridge and Husain 1918), and *Lygidea mendax* (Gossard 1918, Hodgkiss and Frost 1921) on apple, as well as from larger foliar lesions. Spots can be inconspicuous on more mature foliage, but they become apparent when leaves are viewed by transmitted light (Petherbridge and Thorpe 1928a). Shot holing is not a symptom generally associated with the typical chlorosis-inducing mirids— that is, white-spotted or stippled leaves usually do not become shot-holed or frayed. This injury can be followed by some of the abnormal growth symptoms that are discussed later in this chapter.

Expression of shot holing and similar symptoms varies according to the host plant. Petherbridge and Thorpe (1928a) observed severe feeding injury caused by the common green capsid (*Lygocoris pabulinus*) on bindweed, currant, and potato plants, whereas groundsel and stinging nettle were little injured. The dead tissue of black currant resulting from feeding by the mirine *L. rugicollis* drops out sooner than that of apple trees (Fig. 9.5) because of currant's thinner foliage (Petherbridge and Husain 1918). These observations contrast with those of Smith (1920a), who noted he was unaware of any mirid harmful to one plant species but not another. Smith (1931b) later reported that stinging nettle, unlike other hosts of *L. pabulinus*, shows no symptoms from this bug's feeding.

A ragged or shot-holed effect of foliage makes determination of the causal agent difficult (Plate 6; Hancock 1935, Wheeler and Valley 1981); it is an expression of arthropod feeding seldom ascribed to species with sucking mouthparts (e.g., Koehler 1987). Other possible causes of this injury include chewing by flea beetles (halticine chrysomelids) (e.g., Isely 1920, Jewett 1929), feeding by sucking arthropods such as eriophyoid mites (Wilson and Cochran 1952) or some scale insects (Westcott 1973: 417), oviposition by thrips (Hely et al. 1982), or feeding punctures (pinholes) made by females of agromyzid flies (LaBonte and Lipovsky 1967). Possible abiotic causes of foliar ragging and shot holing are hail or mechanical damage, application of arsenical sprays (Wheeler and Valley 1981) or certain other pesticides during hot weather (Partyka 1982), or injury to leaf buds from low temperatures in early spring (Wilson and Ellett 1980).

Mirid-induced shot holing can be confused with typical flea beetle feeding. For example, injury to tomato leaves caused by *Lygocoris pabulinus* and *Lygus pratensis* (perhaps actually *L. rugulipennis*) is nearly indistinguishable from that caused by the flea beetle *Psylliodes affinis*, except the severity of the

Fig. 9.4. Grape leaf, fed on for four days by *Lygocoris pabulinus*, showing necrotic, pinholelike spots. (Reprinted, by permission of Danmarks Jordbrugsforskning, from Rostrup and Thomsen 1923.)

Fig. 9.5. Foliar distortion and shot holing on apple crops caused by the feeding of *Lygocoris rugicollis*. (Reprinted, by permission of Danmarks Jordbrugsforskning, from Rostrup and Thomsen 1923.)

Fig. 9.6. Similarity in symptoms caused by the feeding of a sucking herbivore and a chewing herbivore. A. Shot holing of a hop leaf by a mirid, *Closterotomus fulvomaculatus*. B. Similar injury to hop by a chrysomelid, *Psylliodes attenuata*. (Redrawn from Massee 1937.)

chrysomelid injury is greater (Pape 1925). Foliar injury from feeding by the mirine *Closterotomus fulvomaculatus*, known as the hop capsid in Britain, and the flea beetle *P. attenuata* on hops is also similar (Fig. 9.6; Massee 1937). Irregular holes 1–10 mm in diameter on cucumber foliage resemble those made by the striped flea beetle (*Acalymma vittatum*) but can actually be traced to feeding by *Lygus lineolaris* (Harcourt 1953). Not all chrysomelid-induced shot holing resembles that caused by co-occurring plant bugs. Holes made by the leaf beetle *Neochlamisus platani* on sycamore (Neal 1989) are considerably larger than those produced by the "sycamore plant bug" (*Plagiognathus albatus*) (Plate 6; Wheeler 1980b, Baxendale and Johnson 1990).

Plant pathologists have erroneously attributed irregular holes in cotton leaves to pathogens when mirids were the causal agent (Chu and Meng 1958). Moreover, a "disease" of salvias characterized by perforated leaves was initially attributed to a virus but was eventually shown to be caused by mirid feeding (Gadd 1937). Fungi, bacteria, phytoplasmas, and viruses, however, can cause several diseases that involve shot-holing and leaf-tattering symptoms (Dunegan 1932, Gilmer and Blodgett 1976, Nyland et al. 1976, Sinclair et al. 1987).

THE FOURLINED PLANT BUG
(*POECILOCAPSUS LINEATUS*)

This polyphagous mirine provides the most dramatic example of shot-hole injury caused by a mirid. It feeds on growing tips and expanded foliage of more than 250 plant species in North America, mainly those of the families Lamiaceae, Solanaceae, and Asteraceae (Wheeler and Miller 1981). Hosts growing in the open tend to be more susceptible to infestation than those in shaded situations (Fitch 1870).

Early instars of *P. lineatus* feed on unfolding leaves, whereas late instars and adults often feed on expanded or partially expanded foliage. Terminal buds and growth

can be destroyed or blasted before expansion (Hight 1990), but conspicuous injury also appears on fully expanded leaves. Each leaf can be punctured several hundred times (Fitch 1870). The immediate plant response to stylet insertion by the fourlined plant bug and its injection of saliva is readily apparent on expanded foliage; lesions appear as soon as feeding commences (Wheeler and Miller 1981).

Lesions caused by *P. lineatus* appear water soaked (Plate 6) as the bug withdraws the pulpy or soupy mixture, darken (Plate 6), and on some hosts, become semitransparent. The spots, or feeding scars as McDaniel (1931) called them, are roughly circular with a diameter of about 2 mm, but the size and shape vary depending on leaf texture, pubescence, and venation of the host plant. Foliar lesions typically persist throughout the growing season (Plate 6). The injury is sometimes similar to fungal or bacterial spotting (Fitch 1870, Daughtrey and Semel 1987, Pirone et al. 1988). Old lesions can resemble the dark foliar lesions caused by toxins of *Alternaria* species (e.g., University of California 1990:75) or black leaf spot of delphinium caused by *Pseudomonas syringae* pv. *delphinii* (Pirone 1978). Lesions made by *P. lineatus* on red clover soon darken and appear as black spots (Plate 6). On plants with rather thin foliage—for example, bittersweet nightshade, forsythia, or hosta—membranous tissue drops from the spots or lesions, imparting a shot-holed effect or severe leaf tattering (Plate 6; Wheeler and Miller 1981; pers. observ.). Fitch (1870) noted that the leaf edges sometimes appear notched. In severe infestations, injury can include foliar wilting due to a loss of water from damaged tissues, as well as leaf abscission (Daughtrey and Semel 1987).

As Slingerland (1893) noted, *P. lineatus* "shows no impartiality, notwithstanding the juices of the leaves may be acrid, bitter, aromatic, mucilaginous, bland, or sweet, and their surfaces be rough or smooth." This bug's salivary secretions might enable various plant defenses to be circumvented by causing the immediate formation of water-soaked lesions from which the soupy contents can be imbibed. This feeding strategy perhaps employs a "phenol-phenolase" system (Miles 1969) to overcome plant phenolics or other toxins.

The violent histolysis caused by *P. lineatus*, the most rapid known for any plant bug, was initially thought to involve a potent lipid enzyme (Wheeler and Miller 1981). According to Carter (1952), the extreme rapidity with which mirid lesions can develop is evidence for their toxic secretions. The example he provided of "immediate reaction," however, was the dark spots that appear on cotton leaves and other structures three hours after attack by *Taylorilygus vosseleri* (Taylor 1945). Other examples of rapidly appearing symptoms include transparent spots that appear on tea leaves within 10–15 seconds of attack by *Helopeltis* species (Lever 1949); circular, water-soaked lesions that develop within "a few minutes" after *H. bradyi* begins to feed on cocoa pods (De Silva 1961); and the discolored areas appearing on

Areca palm as soon as *Mircarvalhoia arecae* pierces a leaf (Nair and Mohan Das 1962). In addition, *Labops hesperius* produces yellowish spots on grass blades immediately or a few seconds after stylet insertion (Campbell and Brewer 1978, Haws 1982).

The immediate and spectacular damage caused by *P. lineatus*, involving a complete loss of cellular integrity and removal of the palisade cell layer (Cohen and Wheeler 1998), might be unique among mirids. As typical mesophyll feeders, mirids are not otherwise known to specialize on palisade parenchyma. Recent attempts to elucidate the mechanism of this bug's feeding damage revealed a salivary gland complex proportionately larger than is known in other mirids: 13–21% of fresh body weight. This contrasts with the 4–7% of total body mass for the salivary gland complex of *Lygus hesperus* (Cohen and Wheeler 1998) and 1.5% for *L. rugulipennis* (Laurema et al. 1985). An abundance of rough endoplasmic reticulum suggests a high potential for producing salivary proteins. Tests for various enzymes—amylase, cellulase, lipase, pectinase, and proteinase—were positive only for exopectinase activity (Cohen and Wheeler 1998). Additional work is needed to clarify the biochemical and physiological basis of the fourlined plant bug's damage to host tissues.

OTHER PLANT BUGS THAT CAUSE SHOT-HOLING SYMPTOMS

Lygus lineolaris, the tarnished plant bug, feeds on the bud leaves of tobacco in Tennessee, with spots 1.5 mm to more than 1 cm in diameter soon forming. Sloughing of tissue can occur within three days (Plate 7), with the holes enlarging as the leaves expand; foliage shows conspicuous holes having ragged necrotic margins. Holes in mature leaves become smooth and calloused (Plate 7) so that injury appears as though it is inflicted by chewing insects, such as the tobacco budworm (*Heliothis virescens*). Severe injury occurs only on burley breeding lines lacking normal trichome exudates; plants with secreting trichomes are not damaged (Pless and Miller 1986). Of the plants tested, the only other that reacted like the sensitive tobacco lines was redroot (rough) pigweed (C. D. Pless, pers. comm.).

A pest of maples, principally red maple in nurseries and landscape plantings, is *Lygocoris vitticollis*. Although this bug's injury symptoms were known to Forbes (1885b), its importance as a red maple pest was not reported for almost 100 years (Wheeler and Valley 1981, Wheeler 1982a). The overwintered eggs hatch soon after leaf flush on host trees, generally in early May in Pennsylvania. The young nymphs feed on the unfolding leaves, causing thin, round, windowlike spots (Plate 7) to appear within 7–10 days. Nymphs feed mostly from the lower leaf surfaces and move from expanded leaves to colonize the next pair of leaves to unfold at the shoot apex. Tissue at the feeding sites tears as leaves expand, resulting in shot-holed, tattered, and often dis-

torted foliage (Plate 7). Injury can be so severe that leaves on every terminal branch are ragged and misshapen and nursery-grown trees become unsalable. The symptoms, which persist throughout the growing season, can resemble those of the pear thrips (*Taeniothrips inconsequens*) on maple (Wheeler 1993).

Hop culture in the eastern United States was once affected by the hop plant bug (*Taedia hawleyi*), which riddles the foliage with holes (Hawley 1917, 1918). *Closterotomus fulvomaculatus*, a European bug now known from Alaska and western Canada (Kelton 1980b, Wheeler and Henry 1992), injures hops similarly in England. A British common name for this mirine is the hop capsid. Although Theobald (1895, 1896) attributed foliar injury on hops to an anthocorid bug, *Anthocoris nemorum*, he eventually altered that opinion, noting that the anthocorid is a beneficial insect and that feeding by the mirid induces a ragged appearance of foliage (Theobald 1911b). Injured leaves tend to split at sites punctured by the bug, especially along the main vein (Moreton 1964).

Feeding by mirines (and plant bugs of other subfamilies) causes similar shot-holing, tattering, and raggededge symptoms on the foliage of other field crops, as well as on trees and shrubs (Table 9.5, Plate 7). These symptoms are expressed more often than I assumed when I discussed leaf tattering of London plane and sycamore trees caused by the North American phyline *Plagiognathus albatus* in the eastern United States (Wheeler 1980b). *Occidentodema polhemusi* causes similar leaf tattering and shot holing of Arizona sycamore in Arizona (pers. observ.). Shot holing, leaf tattering, and similar types of injury, therefore, should be considered typical symptoms associated with mirid feeding.

Silvertop

Silvertop or "white ear" of grasses and cereals is a condition describing silvery white heads that appear mature but have largely sterile flowers and lack viable seeds (Plate 7). The cause of this distinct injury long puzzled entomologists. Suggested causal agents of this "elusive mystery" (Peterson and Vea 1969) included fungi, thrips, mites, chloropid flies, and leafhoppers, in addition to plant bugs (Hardison 1959, Starks and Thurston 1962, Arnott and Bergis 1967, Peterson and Vea 1971, Gagné et al. 1984). Although the causal agents differ from region to region, mirids are often the most important.

Mirids causing silvertop pierce the leaf sheath of grasses and cereals to feed on the stem (culm). The injection of toxic salivary secretions can cause a shriveling and constriction of the culm at the feeding site that leads to blockage of the phloem (Wagner and Ehrhardt 1961; Peterson and Vea 1969, 1971). As Putshkov (1966) emphasized, the effects of stem

Table 9.5. Examples of plant bugs that cause foliar ragging, shot holing, and tattering

Mirid	Host plant	Locality	Remarks/symptoms	Reference
Bryocorinae				
Dicyphus hesperus	Cucumber	British Columbia, Canada	*Lygus*-like tattering under greenhouse conditions	Gillespie et al. 1999
Engytatus modestus	Tomato	Hawaii, USA		Tanada and Holdaway 1954
Helopeltis schoutedeni	Avocado	East Africa	Angular spots, resembling those caused by bacteria, lead to shot holing and "hen's foot" appearance of leaves	Harris 1937, McDermid 1956, Benjamin 1968
	Quinine	Africa		Gerin 1956
	Tea	East Africa		Benjamin 1968
Helopeltis theivora	Indian long pepper	India		Abraham 1991
	Tea	India	Predicted to become a major pest (Peal 1873), it soon caused "enormous damage" (Bamber 1893)	Huque [1970], Eden 1976
Helopeltis spp.	Cotton	Africa	Leaves peppered with holes resemble injury caused by mandibulate insects	de Pury 1968
Macrolophus costalis	Tobacco	Bulgaria		Reh 1929
Mircarvalhoia arecae	Areca palm (betel nut), oil palm	India	Tissue darkens, foliage shredded; a major pest of oil palm; see Plate 7	Nair and Mohan Das 1962, Dhileepan 1991
Nesidiocoris tenuis	Tobacco	Indonesia	Tearing of dry leaves lowers value of plants used for cigar wrappers	Kalshoven 1981
Nesidiocoris volucer	Tobacco	Zimbabwe		Roberts 1930
Pachypeltis maesarum	Betel pepper	India	Shot holing may cause total crop loss	Trehan and Phatak 1946, Jagdale et al. 1986
Ragwelellus horvathi	Cardamom	Papua New Guinea	Water-soaked lesions lead to shot holing and tattering of older leaves	Smith 1977b
Tupiocoris notatus	Tobacco	Florida, USA	Hundreds of bugs may occur on a leaf, coating lower surface with black excrement; heavily infested leaves are difficult or impossible to cure properly	Quaintance 1898, Howard 1899
Mirinae				
Apolygus spinolae	Grape	Europe	Severe tattering	Paoli 1923, 1924; Fulmek 1930; Russ 1959; Caccia et al. 1980
Closterotomus fulvomaculatus?	Rhubarb, rose	Finland	Holes in older leaves	Vappula 1965
Closterotomus norwegicus	Bean	Italy	Irregularly shaped holes 1–12 mm in diameter	Lupo 1946
	Potato	England, New Zealand	One of first plant bugs observed to injure plants; long known as a pest in British potato gardens (Balkwill 1846, Curtis 1860)	Smith 1926, 1931b; Chapman 1976
	Sugar beet	England		Jones and Dunning 1972, Jones and Jones 1984
Lygocoris pabulinus	Fuchsia, prunus (ornamental)	England	Foliar spotting, tattering, and shot holing	Smith et al. 1983, Alford 1995
	Raspberry	Scotland	Severe perforation of foliage	Hill 1952b, 1952c
	Rhubarb, rose	Finland	Holes in older leaves	Vappula 1965
	Rose	England	Young leaves can be destroyed	Deacon 1948

Table 9.5. *Continued*

Mirid	Host plant	Locality	Remarks/symptoms	Reference
Lygocoris viburni	Downy viburnum, nannyberry	New York, USA	Foliar curling, ragging, and tattering	Johnson and Lyon 1988
	European cranberry-bush	Pennsylvania, USA	Tiny, dark necrotic spots, followed by larger necrotic areas and tattering on expanded leaves; dark fecal spots present; see Plate 7	Pers. observ.
Lygocoris viridanus	Tea	Sri Lanka	Cited as *L. pabulinus* (see Wheeler and Henry 1992)	Webster 1954, Calnaido 1959
Lygus hesperus	Magnolia	California, USA	Tattered, distorted foliage; sooty moldlike fungus may colonize sap flowing from injured leaves; see Plate 6	Koehler 1963
"*Lygus*" *muiri*	Cotton	Fiji	Tattered foliage; species is given status of *incertae sedis* by Schuh (1995)	Swaine 1971
Lygus pratensis	Dahlia	Germany	Shot holing, tattering	Bech 1967
Lygus rugulipennis	Potato, umbelliferous crops	Europe	Tattering	Molz 1917, Ellis and Hardman 1992
Taylorilygus vosseleri	Cotton	East Africa	Ragged edges, shot holing; succulent leaves more heavily fed on than "dry" ones	Hancock 1935, Taylor 1954, Coaker 1957, Buyckx 1962, Stride 1968
Phylinae				
Campylomma verbasci	Cotton	North Africa	Cited as *C. nicolasi*; leaf margins malformed; foliage perforated	Cowland 1934, Bedford 1940, Pearson 1958, Ripper and George 1965
Chlamydatus associatus	Cotton	California, USA	Irregular holes in leaves, curled or ragged edges; injury indistinguishable from that caused by the western flower thrips, *Frankliniella occidentalis*	Smith 1942
Macrotylus nigricornis	Fava bean	Middle East	Necrotic spots collapse, forming holes	Bardner 1983
Plagiognathus punctatipes	Black walnut	Eastern USA	Distortion with some shot holing	Johnson and Lyon 1988
Pseudatomoscelis seriata	Cotton	Southern USA	Tattering	Parencia 1978
Rhinacloa clavicornis	Avocado	Florida, USA	Possibly causes shot holing	Wolfe et al. 1946

Note: See text for additional examples and discussion.

feeding (in this case, silvertop) can be indistinguishable from those resulting from attacks on the reproductive structures themselves (see Chapters 10, 12). Plant bugs implicated in causing silvertop belong to the subfamily Mirinae, especially its tribe Stenodemini (Table 9.6). Most of the species are common to the Palearctic and Nearctic regions (Wheeler and Henry 1992). Certain mirines that are otherwise injurious, such as lygus bugs, fail to cause silvertop symptoms under experimental conditions (Arnott and Bergis 1967).

Mirid oviposition in grass stems can also result in silvertop symptoms. The eggs of *Notostira erratica* can

interrupt the flow of substances in the phloem, which leads to complete or partial blanching of the heads (Kiritshenko 1951).

Lesions, Cankers, and Other Secondary Symptoms

Not all mirid feeding results in the diffusion of toxic salivary secretions from the immediate area of stylet penetration and causes profound effects on plant metabolism. For example, of four species that feed similarly

Table 9.6. Plant bugs (subfamily Mirinae) implicated in causing silvertop of grasses

Mirid	Host plant	Locality	Remarks	Reference
Mirini				
Capsus cinctus	Kentucky bluegrass	Minnesota, USA	Single bug feeding 10 min on culm can induce silvertop	Peterson and Vea 1971; see also Hewitt 1980
Stenodemini				
Leptopterna dolabrata	Grasses	Czech Republic, Germany		Wagner 1960, Wagner and Ehrhardt 1961, Rotrekl et al. 1985
	Kentucky bluegrass	Kentucky, USA	May not be important cause of silvertop but causes blasting of seed heads	Jewett and Spencer 1944, Starks and Thurston 1962
Leptopterna ferrugata	Fine fescues, Kentucky bluegrass	Oregon, USA	Single probe may destroy all seeds of panicle	Kamm 1979
Litomiris debilis, *Megaloceroea recticornis*, *Stenodema* spp., *Trigonotylus ruficornis*	Merion bluegrass, other grasses	British Columbia, Canada	Suspected causal agents; *M. recticornis* injures developing seeds but appears to be unimportant in causing silvertop	Arnott and Bergis 1967
Unspecified species	Timothy	Quebec, Canada	Suspected causal agents	Gagné et al. 1985

on apple trees in England, only the apple capsid (*Lygocoris rugicollis*) causes injury (Smith 1920a). Feeding by *Polymerus cuneatus* on the flower stems of bean plants in the West Indies sometimes leads to the shedding of flowers (Fennah 1947), but this might result from severe, localized attacks and a weakening of stems rather than any systemic action of the bugs' salivary secretions.

But some stem feeding by mirids does involve a translocation of toxic secretions (see Chapter 7), which diffuse through the plant's conducting system in "rising and falling currents" (Putshkov 1956). Such feeding can also lead to secondary symptoms such as cankers and split lesions. Some secondary symptoms are difficult to separate from those produced by plant viruses and are analogous to those induced by plant growth hormones (Carter 1973). Mirid lesions are sometimes also similar to those caused by bacteria, fungi, or plant-parasitic nematodes (e.g., Lee 1971, Crowe 1977, Lehman and Miller 1988; see also Chapter 8). Subsequent infection by fungi and bacteria can substantially alter the character of a local lesion caused by a plant bug.

Mirids are perhaps the most important insects whose feeding produces local lesions with development of secondary symptoms (Carter 1973). Although some tropical bryocorines rapidly induce local lesions when they feed only for a short time, their feeding can result in secondary injury that is not apparent for several weeks and can persist for years (Squire 1947). Bryocorines often limit production in tropical ecosystems. Recurrent attacks by *Helopeltis* species on tree crops such as cashew and cocoa are severe "because the tree never has a chance to grow into its natural size and shape" (Crowe 1977). Plant bugs of temperate regions appear to be less important than those in the tropics in causing severe stem cankers and lesions.

LYGUS LINEOLARIS AND LESIONS OF HYBRID POPLAR

The well-studied Nearctic mirid *L. lineolaris* was not known to cause stem lesions on woody plants until Sapio et al. (1982) determined that its feeding produces split-stem lesions on young hybrid poplar in Michigan (see also Wilson and Moore [1985], Ostry et al. [1988]). A fully formed lesion ranges from less than 0.5 to 5.0 cm long and appears as an irregular split having a swollen, flared area of necrotic bark and xylem tissue around the stem (Plate 8). Experimental work showed that lesions caused by the tarnished plant bug are not the result of its oviposition injury, mechanical stimuli, or activity by mirid-transmitted pathogens, although the role of bacteria in lesion formation requires further research (Juzwik and Hubbes 1984, 1986). The presence of horseweed, a favored host of *L. lineolaris*, reduces the incidence of lesions on poplars (Sapio et al. 1982).

The bacterial canker of hybrid poplar reported by Mosseler and Hubbes (1983) in Ontario is actually caused by *L. lineolaris* (Juzwik and Hubbes 1986). Bacteria can also be isolated from adult tarnished plant bugs collected on poplar and weeds; various bacteria, mainly facultatively anaerobic species, can be isolated from discolored areas of the xylem and phloem of stem lesions. These bugs might introduce bacteria when they puncture stems, or bacteria might be introduced immediately after such wounding.

LESIONS ON COTTON

Few North American mirids have attracted more attention than the cotton fleahopper (*Pseudatomoscelis seriata*), a Nearctic mirid first implicated as a cotton pest in the late nineteenth century (Howard 1898).

This bug is known for the stem swellings and split lesions it produces (see also Chapter 10 for effects on reproductive parts of cotton). The apparent systemic condition of its feeding, however, might result merely from numerous punctures, the effect of this insect's secretions remaining localized and not transmitted appreciably from the point of penetration (Painter 1930, King and Cook 1932).

An outbreak of the cotton fleahopper in 1926 created widespread popular alarm in the South (Isely 1927), causing an estimated $12 million in losses to cotton growers (Hyslop 1938). More recently, yield reduction was estimated to be 34% without control measures and 12% with control (Schwartz 1983); in the United States, 1983 losses were about 50,000 bales (Sterling et al. 1989b). In parts of Texas during 1987, this pest was considered responsible for a greater yield reduction than the boll weevil or bollworms and budworms (King et al. 1988).

In the 1960s, insecticides used to control this key pest of cotton (and the boll weevil) triggered outbreaks of budworms and bollworms by killing their natural enemies (Adkisson 1973a, Almand et al. 1976, Adkisson et al. 1982). Classical biological control of the cotton fleahopper and other plant bug pests of cotton is not practiced (Cate et al. 1990), but foreign exploration for parasitoids of presumed related mirid taxa is under way. Exotic natural enemies might help suppress the numbers of fleahoppers on herbaceous weeds that serve as alternative hosts (Cate 1985). The impact of natural enemies (mostly native predators) on fleahopper abundance in Texas has been modeled, and growers are able to monitor and conserve predators in conjunction with the Texas Cotton Insect Model, known as TEXCIM (Sterling et al. 1989b, Breene et al. 1989c).

Much has been written on the cotton fleahopper, and the bibliography of Sterling and Dean (1977) should be consulted for papers detailing its feeding habits and host plants. The following account is taken largely from work by Hunter (1926), Painter (1930), and King and Cook (1932).

Pseudatomoscelis seriata develops on numerous weeds, especially croton (see Plate 11), evening primrose, and horsemint (Knight 1926a, Reinhard 1926, Eddy 1927, Fletcher 1940b, Hixson 1941, Almand et al. 1976, Bohmfalk 1982, Henry 1991). More eggs are laid on plants beginning to flower than on plants in earlier or later growth stages (Holtzer and Sterling 1980). An early buildup of fleahopper populations on weeds in the lower Rio Grande Valley of Texas can be precipitated by the migration of adults from Mexico (Schuster and Boling 1974). As their weed hosts become less succulent, the bugs invade cotton fields. Indeed, the massive numbers emigrating from weeds make this otherwise relatively innocuous bug such an important cotton pest (Sterling et al. 1989b).

Although fleahoppers will migrate into presquaring cotton, populations generally do not build substantially until squaring begins (Almand et al. 1976, Snodgrass et al.

1984a). Densities in Texas tend to be higher in cotton planted in late June than in early (late April) plantings (Slosser et al. 1994). The bugs move from maturing to more succulent cotton plants. Populations often increase after periods of rainfall (Glick 1983). Irrigated plants, especially highly fertilized ones, attract greater numbers than nonirrigated plants (Adkisson 1957). Cotton grown on the heavier soils of the Texas Blacklands is often more severely injured than cotton on lighter soils because the bugs' more frequently used weedy hosts are generally scarce. Consequently, the bugs usually remain longer in these larger and less weedy fields (Thomas and Owen 1937, Little and Martin 1942). In 1929, extensive outbreaks caused injury on the Brazos River floodplain soils of Texas but little injury to the cotton on hill-section soils (Sterling et al. 1989b).

Fleahopper feeding often results in the shedding of cotton squares (see Chapter 10), but the main stem and lateral branches are also involved, in many cases leading to typical symptoms of "hopper damage." When lesions are mildly severe, injury is confined to the cortex, or cortex and collenchyma. Cortical cells enlarge and their numbers can increase. With severe injury, more than half of the stem is affected and the epidermis and collenchyma are forced out, forming external swellings of various sizes and shapes; under pressure, the swellings split lengthwise (Fig. 9.7). Some lesions can involve the vascular cylinder. As a result of cell destruction, the position of the xylem in a vascular bundle is altered, and cell structure may change. In areas where cells have been destroyed or in lesions that have split and been exposed to the air, cell walls become so devoid of cellulose that only a pectinlike material remains. The amount of injury to cotton is often disproportionate to the number of insects present (Reinhard 1926).

Under experimental conditions, *Lygus lineolaris* and several other Nearctic mirids (but not leafhoppers) can cause lesions similar to those of the cotton fleahopper (Ewing 1929, King and Cook 1932, Ewing and McGarr 1933, McGarr 1933). The split lesions that *L. lineolaris* causes are severe wounds a centimeter or more in length (King 1929). Reid (1965) correlated a wilting of terminal leaves of cotton with stem swellings or lesions that were attributed to feeding by *L. lineolaris*. Lesion-inducing species on cotton also include the mirines *Adelphocoris rapidus*, *Creontiades debilis*, *Polymerus basalis*, *Proba vittiscutis*, and *Taylorilygus apicalis*; the orthotyline *Orthotylus leviculus*; and the phylines *Lineatopsallus biguttulatus*, *Megalopsallus atriplicis*, *M. latifrons*, *Psallus pictipes*, and *Reuteroscopus ornatus*. Several species, including *A. lineolatus*, cause fleahopper-like symptoms on cotton in the former USSR. The gradual growth of tissues adjacent to the sites of stylet penetration results in external swellings that can burst and discharge resin (Putshkov 1956).

The African bryocorine *Helopeltis schoutedeni* can invade cotton as soon as the cotyledons open (McDermid

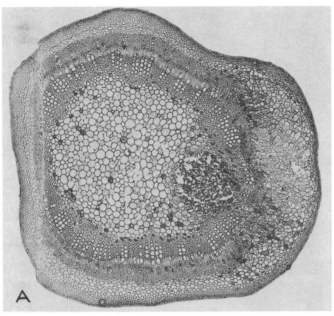
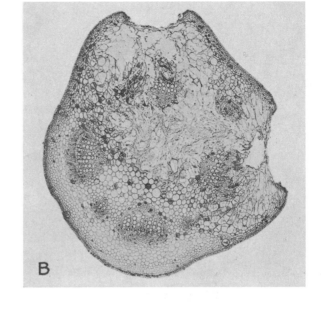
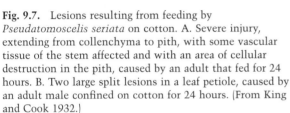

Fig. 9.7. Lesions resulting from feeding by *Pseudatomoscelis seriata* on cotton. A. Severe injury, extending from collenchyma to pith, with some vascular tissue of the stem affected and with an area of cellular destruction in the pith, caused by an adult that fed for 24 hours. B. Two large split lesions in a leaf petiole, caused by an adult male confined on cotton for 24 hours. (From King and Cook 1932.)

1956). Nymphs puncture the petioles and tender branches of young plants (adults mostly injure the bolls [Schmutterer 1969]), which in southeastern Africa induces distortion (Plate 8) and plant death. In Mozambique, fields once appeared "as if a flame thrower had been used . . . burning down every single plant" (da Silva Barbosa 1959). In northern Africa, its feeding results in lesions resembling those associated with a bacterial blight (blackarm disease), except that spots turn light brown rather than dark brown or black (Ripper and George 1965). Vrijdagh's (1936) studies in the Democratic Republic of the Congo showed conclusively that stem cankers (see Cadou [1993:67] for a color photograph of the injury) on cotton are produced mainly by *Helopeltis* feeding rather than by any phytopathogen. Diagnosis of the problem can be difficult because bacterial blight and *Helopeltis* injury occur together, and where injury by the mirine *T. vosseleri* is also apparent, a confusing array of symptoms is presented (Prentice 1972; see also Chapter 8).

In first noting the presence of *Helopeltis* species on cotton in the Ivory Coast, Delattre (1947) suggested that the bugs might have been present for many years but went undetected because their injury is similar to that caused by the bacterial pathogen responsible for cotton blight. Most of the earlier reports of *H. bergrothi* on cotton actually refer to *H. schoutedeni* (Carayon and Delattre 1948, McDermid 1956). Schmitz (1968) and

Leston (1973c) noted that "*H. bergrothi*" might represent a complex of several species.

LESIONS ON TOMATO

The dicyphine *Engytatus modestus* ranges from the southern United States through Mexico to South America and occurs in the West Indies and Hawaii (Henry and Wheeler 1988). Detected in Hawaii in 1924, *E. modestus* assumed agricultural importance in the early 1940s when the tomato variety 'Bounty', which is suited to low elevations, was introduced. This allowed Hawaiian tomato production to expand into lowland areas favorable to the plant bug (Holdaway 1945).

This pest, known as the tomato bug but sometimes also called the "tomato girdler" (Fullaway and Krauss 1945), causes lesions or "feeding rings" that encircle the tomato stems and petioles. Bugs are significantly more abundant on plants of high nitrogen content (Fig. 9.8); they also return to mature lesions, which have a higher protein content than nearby undamaged tissues (Tanada and Holdaway 1954). Girdling produced by *E. modestus* is similar to that inflicted by a membracid, the three-cornered alfalfa hopper (*Spissistilus festinus*), and can be confused with disease symptoms (Reynard 1943). Both the mirid (Tanada and Holdaway 1954) and membracid (Mitchell and Newsom 1984) feed primarily on vascular

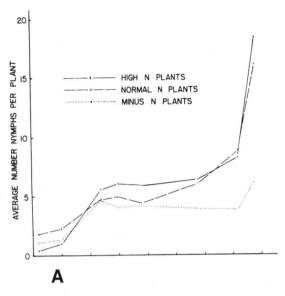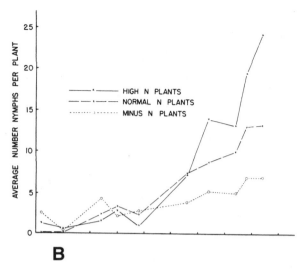

Fig. 9.8. Densities of *Engytatus modestus* in relation to different nitrogen (N) levels on tomato. A. Numbers of nymphs on plants grown outdoors. B. Numbers of nymphs on plants in greenhouse. (From Tanada and Holdaway 1954.)

tissues. Although physiological changes are induced in mirid lesions, it is uncertain if *E. modestus* creates a nutrient sink in tomato stems like that induced by the membracid on leguminous hosts (Mitchell and Newsom 1984). In the case of another dicyphine mirid that causes similar injury on tomato stems, Raman and Sanjayan (1984b) proposed that the "accumulation of peroxidases, polyphenol oxidases, acid phosphatases, proteins and tannins occurring in higher gradients at the feeding spot and the stylet path . . . and the gradual decline in the concentration in the cells of the neighbouring areas indicate that the insect activates the cells at the feeding spot enabling [it] to obtain a continuous nutrition suggesting a 'sink-source' relationship."

On stems fed on by *E. modestus*, punctures are about 1 mm deep; dead cortical cells collapse, the epidermis may sink below the surface, and tumorlike growths may develop. As Tanada and Holdaway (1954) described, the "cells bordering the puncture enlarge and divide in a plane parallel to the stylet path. The pressure exerted by enlarging and dividing cells presses the walls of dead cells together, forming a reddish-brown streak of cell walls and dead protoplasm extending inward from the epidermis." This streak superficially resembles the stylet track of homopterans (Fig. 9.9). Most of the stylet tracks end in vascular tissue, pith, and internal phloem, although xylem is also penetrated. Wound parenchyma replacing elements of the vascular bundle often becomes so brittle that the stems break easily (Plate 8; Tanada and Holdaway 1954, MacCreary and Detjen [1947?]). Much of this bug's injury is attributed to the mechanical effects of stylet insertion and the withdrawal of fluids rather than to a violently phytotoxic salivary secretion (Tanada and Holdaway 1954). But *E. modestus* sometimes also causes a secondary wilting of organs some distance from the sites of penetration when vascular bundles are injured and produces bushy plants showing swollen nodes and short internodes.

Nesidiocoris tenuis causes similar injury to tomato plants in Ghana (Van Halteren 1969) and Egypt (El-Dessouki et al. 1976). It not only develops on solanaceous crops such as tobacco but also on cucurbits in India (Sreeramulu et al. 1975). On tomato plants, it feeds mainly on vascular bundles, especially the phloem. Water-soaked brownish spots on the petioles and stems coalesce and, within 48–72 hours, give rise to a brown ring encircling the stem (Plate 8). The secretion of various substances and mechanical injury apparently are responsible for the necrosis at the feeding site. The wound response in tomato is characterized by increases in phenolic compounds and oxidative enzymes, which can result in decreased growth-promoting activity (Raman et al. 1984, Raman and Sanjayan 1984b; see Chapter 7). In the Philippines, *N. tenuis* causes only slight feeding injury on tobacco and instead is considered an important predator of lepidopteran pests on this crop (Torreno 1994, Torreno and Magallona 1994; see also Chapter 14). *Nesidiocoris tenuis* is an adventive species in the southern United States (Wheeler and Henry 1992), and although it has not injured tomato or other solanaceous plants, it causes some yellowing and necrosis on the young leaves of head lettuce in New Mexico (C. A. Sutherland, pers. comm.).

The Nearctic dicyphine *Dicyphus hesperus* can cause foliar lesions on tomato plants. When its numbers are high under greenhouse conditions, the bugs produce necrotic lesions on the oldest leaves of seedlings (Gillespie et al. 1999).

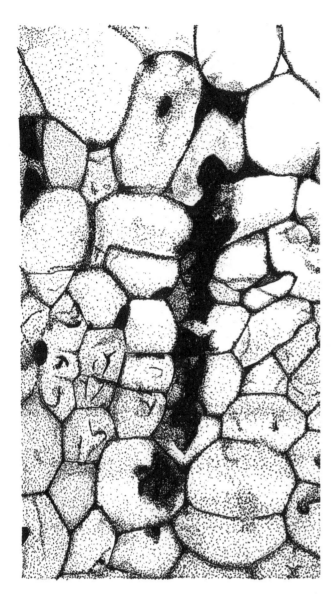

Fig. 9.9. Cross section of tomato stem fed on by *Engytatus modestus* and stained with safranin and fast green. The bug's stylet path is indicated by the collapsed cell contents within the row of cells of the vascular bundle. (Redrawn from Tanada and Holdaway 1954.)

LESIONS OF CITRUS

Platytylus bicolor (Mirinae: Restheniini) is a pest of grapefruit, orange, and other citrus crops in Brazil. The syndrome of citrus, referred to as *false exanthema* or *variola*, once was thought to be caused by a nutrient deficiency or virus. Müller and Costa (1964), however, demonstrated that the mirid's direct feeding causes the symptoms observed on nearly all Brazilian citrus—that is, gum-filled blisters on stems, gum exudation, and split-stem lesions. The bugs mostly feed by day, causing more damage on vigorous than poor hosts. Feeding seems most prevalent on young twigs in shaded portions of citrus trees. A nymph or adult of *P. bicolor* can

Fig. 9.10. Tea shoot, injured by *Helopeltis schoutedeni*, showing elongated, diamond-shaped marks caused by the bugs' feeding on or near the leaf midrib. (Redrawn from Smee 1928b.)

produce about 10 lesions in 24 hours. Severely damaged trees can become weakened and show decline (Müller and Costa 1964).

LESIONS AND CANKERS OF TEA

Helopeltis schoutedeni, the "mosquito bug," produces water-soaked areas on the stems of tea plants in East Africa, the lesions appearing while a bug still feeds (Leach and Smee 1933). The youngest stems are preferred, but *H. schoutedeni* and other species of the genus also feed on shoots, young leaves, and buds (Plate 8; Leach and Smee 1933, Lever 1949, Banerjee 1983b). *Helopeltis* species developing on tea belong to a sap-feeding guild associated mainly with the soft branches and succulent leaves of young plants (Banerjee 1983a, Rattan 1992). When feeding takes place on the midrib, diamond-shaped lesions appear on the foliage (Fig. 9.10; Smee 1928b). Stem cankers are apt to form when stylets penetrate pericycle parenchyma rather than xylem or phloem. Sunken areas mark the position of these cankers, which are usually 15–18 mm long and 1.5–2.0 mm wide. Stems can be girdled from severe cankering, and the development of gall wood around cankers imparts a gnarled appearance to the stem. Saprophytic or weakly parasitic bacteria and fungi sometimes invade, resulting in collapse of the cortex (Leach and Smee 1933, Eden 1953).

Stem canker of tea in East Africa was once attributed to a fungal pathogen rather than an insect. This supposition was reinforced when applications of a fungicide, Bourdeaux mixture, and pruning of affected shoots helped alleviate the condition (Smee and Leach 1932). Leach and Smee (1933) demonstrated that fungi play only a secondary role in the etiology of this syndrome.

Helopeltis schoutedeni is still a major tea pest in East Africa—one capable of causing severe crop losses (Rattan 1992) and requiring chemical control. In Malawi, the application of pesticides is based on the monitoring of *Helopeltis* populations and economic damage thresholds (Rattan 1992). The use of annatto as a trap crop in tea estates is another means of reducing the dependence on insecticides. Mosquito bugs are attracted to annatto and feed on its succulent pods when this crop (the seed coat coloring is used in cosmetics and food products) is grown adjacent to tea fields (Peregrine 1991). In addition, different tea clones show varying degrees of susceptibility to *Helopeltis*, with the more tolerant clones appearing to possess more acidic sap (Rattan 1992).

Feeding by *Helopeltis* species causes small brown spots on the foliage of tea plants in southern Vietnam. The leaves shrivel and dry when the punctures are numerous, and plants sometimes are devastated when the entire shoot is ravaged. In addition to the immediate loss of harvest, long-term yields can be diminished. Damage is most severe in weedy plantations and those suffering from poor nutrition (Hanson 1963; see also Smee [1928b], Light [1930]).

Helopeltis species occurring on tea in Indonesia penetrate the vascular bundles (Cohen Stuart 1922), producing dark, necrotic spots and malformation or death of the twigs, and allowing pathogenic fungi to enter. Severely damaged plantations appear blackened, and up to 80% losses in yield are common. Weakened plants are more rapidly colonized than vigorous hosts (Kalshoven 1981). The thinning of shade trees for the control of a fungal blight of tea promotes *Helopeltis* injury (Vollema 1951). Use of persistent insecticides helped alleviate the mirid problem in Indonesia, although populations of *Helopeltis* species have become increasingly resistant to chemicals (Kalshoven 1981). Carter (1973) summarized the early attempts to control *Helopeltis* populations on tea using techniques that involve the bugs' responses to environmental conditions.

Tea growers in India began to notice injury from *H. theivora*, the so-called mosquito bug or blight, about 1865 (Peal 1873, Westwood 1874; see Stonedahl [1991] for nomenclatorial explanation). Watt and Mann (1903) summarized early research on this "most alarming of all Indian tea pests." *Helopeltis theivora* was a serious tea pest in India before the 1950s, with crop losses sometimes reaching 100% (Muraleedharan 1987, 1992) and the very existence of tea estates in parts of Travancore being threatened (Rao 1970). DDT later provided effective control (Das 1963, Cranham 1966), but when DDT's use was no longer permitted, this insect reestablished itself as a pest in discontinuous patches of northeastern India (Banerjee 1983b, Das 1984).

LESIONS AND CANKERS OF CASHEW

The effects of stem feeding by *Helopeltis ancardii* on cashew trees in east central Africa were summarized by Carter (1973), apparently based on work by Swaine (1959): "Hyperplasia originates, after a week or 10 days, from unaffected cells in or near the border of the affected area. It develops in the parenchyma, pushing out the cortex to form a slight swelling. This cortex may become necrotic due to the invasion of secondary microorganisms. . . . Hyperplasia of the pith often occurs; that of cambium rarely, but when it does, the canker may form an open wound with the development of wound wood." Wheatley (1961) discussed *Helopeltis* injury to cashew trees in Kenya, including the long, black stem lesions caused by late instars.

Helopeltis antonii, the most serious pest of cashew trees in southern India, induces dense, necrotic cankers that cause a drying of terminal shoots (Ambika and Abraham 1979, Devasahayam and Radhakrishnan Nair 1986). Stem lesions are slightly more than 9 mm long and nearly 2 mm wide (Sathiamma 1977). Biochemical changes in the young shoots and leaves that are infested for 24 hours include a decrease in sugar, chlorophyll, and carotenoid content and an increase in total phenols. Certain biochemical changes can be detected within six hours of this bug's infestation of cashew leaves and shoots (Nagaraja et al. 1994). Shoot dieback occurs only when the pathogenic fungus *Botryodiplodia theobromae* invades mirid-induced lesions (Varma and Balasundaran 1990). This mirid also affects cashew production in Sri Lanka. Development of the bug is more rapid and fecundity greater when the nymphs feed on tender branches rather than fruits (Jeevaratnam and Rajapakse 1981a). *Helopeltis theivora* and an undetermined member of the genus (perhaps *H. bradyi*; see Stonedahl [1991]) sometimes co-occur with *H. antonii* on cashew trees in India; under laboratory conditions they cause necrotic streaks on shoots and truncated foliar lesions (Ambika and Abraham 1983). *Pachypeltis maesarum*, another bryocorine, causes similar injury to cashew in India: necrotic lesions around feeding punctures on tender stems and young leaves (Remamony and Abraham 1977).

West Malaysian cashew plantations are subject to injury by *H. theivora*. Its feeding sometimes kills foliar shoots (Khoo et al. 1982).

LESIONS AND CANKERS OF COCOA

Cocoa mirids are *K*-selected specialists of forest trees that have adapted to cocoa and other plants rich in methyl purines, such as caffeine and theobromine (Leston 1978a). This family "towers far above" other insect groups that damage cocoa (Braudeau 1974). The feeding habits of native bryocorine mirids appear to have changed relatively quickly, their association with cocoa in West Africa having arisen during the twentieth century (Southwood 1960a). At least 35 mirid species are now associated with cocoa grown in the Old and New World tropics (Entwistle and Youdeowei 1965, Leston 1970), and bryocorines might be further adapting to this crop (Entwistle 1985a).

The propensity of bryocorines to shift to introduced cocoa from local faunas (Strong 1974) suggests that the bugs are preadapted to this crop plant (Conway 1969) because of biochemical similarity between their previous hosts and cocoa, as well as their own metabolism (Southwood and Kennedy 1983; Southwood 1973, 1985). Some species feed on cocoa only as adults. Even those species that reproduce on the tree sometimes vary geographically in their preference for cocoa (e.g., Collingwood 1977a).

Bryocorines are consistently devastating to cocoa, their injury in Sri Lanka having been reported as early as 1863 (Entwistle 1972). The extensive work on the so-called cocoa capsids is largely buried in departmental publications and internal reports, but Entwistle (1972) provided a useful evaluation and summary of previous research (about 400 references). The interactions of cocoa mirids with their tropical forest environment are exceedingly complex, especially their association with ants (Leston 1970, Majer 1972, Taylor 1977, Bigger 1981, Jackson 1984). During rainy periods that follow the dry season, mirid-damaged cocoa trees tend to regenerate weak shoots that are susceptible to infestation by shoot tip–feeding homopterans and lepidopterans (Collingwood 1977b). Cocoa mirids should also be considered in relation to swollen shoot virus (Thresh 1960, Thresh et al. 1988) and black pod (formerly *Phytophthora* pod rot [Wood and Lass 1985]) because "cocoa degeneration mirrors the complexity of the tropical forest environment" (Leston 1978a; see also Clark et al. [1967]). Moreover, the literature discussing the amount of light entering the canopy, the relationship of the prevalence of injury to plant growth, and the movement and buildup of mirid populations "is diffuse and full of apparent contradictions: it is not an easy matter to simplify it" (Johnson 1962). Leston and Gibbs (1971) referred to the complexity of analyzing mirid population fluctuations, emphasizing that consideration of the bugs' changes in density should not be dissociated from that of cocoa growth and cropping.

Readers wanting more detailed information on mirids as cocoa pests should consult Johnson (1962), Entwistle and Youdeowei (1965), Cochrane and Entwistle's (1964) bibliography of cocoa mirids, Leston (1970), Lavabre's (1977) *Les Mirides du cacaoyer* (a work Leston [1978a] attacked as mostly repetitive and valueless), and especially Entwistle's (1972) *Pests of Cocoa*. Giesberger's ([1983]) review of the biological control of *Helopeltis* species in Java is a valuable English summary of studies by J. G. Betrem, P. van der Goot, S. Leefmans, W. Roepke, and others that were published in Dutch. Carter (1973) summarized the research on cocoa mirids in his *Insects in Relation to Plant Disease*. Johnson's (1962) summary of the growth and architecture of cocoa trees and Entwistle's (1972) review are helpful in understanding the development and form of mirid injury, while Tox-

opeus (1985), Wood (1986), and Young (1994) provide a general account of the tree.

Mirids as Key Pests in West Africa

Problems with cocoa mirids was one reason for the crop's abandonment in Uganda in the early 1920s (Ingram 1970a). One could travel hundreds of miles along West African roads without losing sight of damage by cocoa mirids (Carter 1973). Ghanaian farmers once referred to these bugs as *sankonuabe*, meaning "go back to your palms" (Squire 1947); palm oil was the country's leading agricultural product before cocoa was established (Cotterell 1926). West African farms abandoned because of problems with cocoa mirids sometimes recovered when growth of the bush (secondary growth) "restored conditions more like those of the natural forest" (Galletti et al. 1972). In Ghana during the early 1950s, about 30% of dry bean yield was lost each year to insects, mainly to cocoa mirids (Kay 1961, Entwistle 1965, Collingwood 1972). By the mid-1950s, millions of trees in Ghana alone had been killed and many more millions had been severely retarded. Unsprayed trees might not bear fruit until they are 10 years old, whereas trees sprayed to control cocoa mirids begin to bear fruit after four years (Johnson 1962).

A cautious optimism about cocoa mirids prevailed in Africa during the 1960s as synthetic organic insecticides such as hexachlorocyclohexane (HCH) (= benzenehexachloride or BHC) and DDT successfully controlled these pests (Raw 1959b, Johnson 1962). Because HCH was relatively inexpensive, efficient, and safe, it gained widespread use in West Africa (Collingwood 1977c). The effectiveness of HCH might relate to its fumigant action, which helps compensate for imperfect spray coverage (Entwistle 1972). In Ghana during the late 1960s, well-timed insecticide applications against cocoa mirids resulted in next-year increases in yield of about 250 kg/ha (Collingwood 1972). The potential production of a plantation could be doubled by chemically controlling cocoa mirids over three years (Kumar and Youdeowei 1983).

Heavy pesticide use (Entwistle 1985b), however, placed intense selection pressure on the bugs. Even in the early 1960s, resistance to HCH and other cyclodiene chlorinated hydrocarbons was reported in some Ghanaian populations of *Distantiella theobroma* (Dunn 1963), which had supplanted another bryocorine, *Sahlbergella singularis*, as the most important cocoa pest in Ghana. This bug generally appears to respond more quickly to insecticide pressure than does *S. singularis* (Collingwood 1977c). But *S. singularis* also began to show resistance, beginning in Nigeria in the early 1960s (Entwistle 1966, 1985b; Booker 1969; Collingwood 1972). Resistant populations of both species continued to spread through West Africa during the 1970s (Collingwood 1977c, Entwistle 1985b). Alternating insecticides belonging to different classes is recommended to combat increasing insecticide tolerance (Bruneau de Miré 1985). Such resistance, or at least a differential sensitivity to chemicals indicating an impending problem, also occurs in populations of *Helopeltis theivora* in Malaysia (Dzolkhifli et al. 1986). Alternative insecticides such as synthetic pyrethroids are being evaluated in an attempt to counter the problem of resistance to commonly used cyclodiene and organochloride compounds (Owusu-Manu 1985, Idowu 1988).

Habits, Injury, and Economic Importance

Although pod feeding by cocoa mirids can injure the trees (see Chapter 12), their vegetative feeding is more destructive—just a few lesions can result in death of an apical shoot (Leston 1970). The tendency toward stem feeding is stronger in odonielline Bryocorinae (species of *Distantiella*, *Pseudodoniella*, and *Sahlbergella*) than in monaloniines (*Helopeltis* and *Monalonion* spp.) (Entwistle 1985a). Although Entwistle (1985a) also included *Platyngomiriodes* among largely stem-feeding odoniellines, the only species of this monotypic genus, *P. apiformis*, feeds extensively on cocoa pods (Pang 1981). Some bryocorines contain histolysis-inducing pectinases (see Chapter 7) that seem to have an affinity for the cambium of young shoots. Because of their feeding mode and possession of toxic saliva, these and certain other bryocorines belong to a group of heteropterans termed "low density pests" (Conway 1969, Southwood 1973). Their injury is often disproportionate to the numbers of bugs present (e.g., Wester 1915, Cotterell 1926, Harris 1937, Squire 1947, Lever 1949, Williams 1954). Injury is much more apparent than the pest itself (Khoo 1987). A "dangerously high" level on cocoa is estimated to be only 400 individuals/ha, which would contain 200–240 trees (Kay 1961, Entwistle 1965). Marchart (1968) and Leston (1970) pointed out that absolute population size in cocoa mirids is often underestimated.

Losses from mirids are difficult to determine, but in West Africa they can be at least 20% (Entwistle 1985b). Leston (1970) summarized information on the three main types of injury (Plate 8): "blast—small discolored lesions on young green shoots which later wilt, the dead leaves remaining on the trees; staghead—a thin crown with many leafless branches; [and] capsid pockets—trees reduced to bare trunks with numerous lateral shoots." Voelcker and West (1940), Williams (1953), Taylor (1954), Kay (1961), and Johnson (1962) described injuries further.

Description of foliar and stem lesions
Feeding by cocoa mirids mostly involves superficial parenchyma (Crowdy 1947); their relationship to phloem is uncertain. On woody stems, the cortex and vascular bundles are penetrated. Whether changes in the nutrient composition of phloem sap influence the feeding sites selected is unknown (Gibbs et al. 1968).

Miles (1987a) found that the dimensions of the lesions made by *Helopeltis clavifer* feeding on cocoa are not determined by the maximum reach of the bugs'

stylets. Bugs are able to evacuate the contents of parenchyma cells lying 3 mm or more from the point of stylet insertion; these cells can be drained without mechanical damage within an hour after feeding commences. Miles (1987a) proposed that saliva rapidly infiltrates the intercellular regions, possibly aided by the action of polygalacturonase on cell walls. The release of hydrolytic enzymes containing osmotically active substances causes an outflow of secretions that might be periodically sucked back by the bugs. Lesions produced by *H. clavifer* can be melanized and visible externally, or be occluded, lying between or below vascular tissue several millimeters below the surface (Miles 1987a).

Helopeltis species and other bryocorines of cocoa, such as *Boxiopsis madagascariensis* in Madagascar (Decazy 1974) and *Pseudodoniella pacifica* in Papua New Guinea (Szent-Ivany 1961), cause foliar and stem lesions, branch cankers, and dieback (Plate 8) similar to those discussed in connection with other tropical crops. But the larger and apparently more phytotoxic mirids, *Sahlbergella singularis* and *Distantiella theobroma*, are even more important cocoa pests. Youdeowei (1973) reviewed information on their life history, including mating and oviposition behavior, incubation period, nymphal development, and fecundity. Leston (1973) noted that the term *cocoa capsid* is most commonly applied to these two species, which belong to his "Group B": mirids that infest pods and vegetative structures. They can be characterized as regenerative tissue feeders (Gibbs and Leston 1970). *Sahlbergella singularis* and *D. theobroma* build in numbers on the main crop of pods (July or August to October) and transfer to vegetative tissues after peak harvest. Populations living on vegetative tissues can also develop independently of those on pods (Gibbs et al. 1968; see Chapter 12).

Within 2–5 minutes after *D. theobroma* begins to feed, there appear water-soaked areas up to 200 mm long; in just one night a fifth instar can produce 20 such stem lesions (Kumar and Ansari 1974). *Helopeltis bradyi* in Sri Lanka produces an average of 180–225 lesions during nymphal development (De Silva 1961). Elliptical lesions are generally most conspicuous on unhardened shoots, where a blackening of lesions usually takes place within a few hours. The scarring and cankering of bark that eventually results from mirid attack provides feeding sites and shelter for mealybug vectors of swollen shoot virus (Strickland 1951).

Lesion size decreases with increasing turgor pressure of the stem; at a high level of water stress (−14 atm) only small brown spots, rather than large, diffuse lesions, may be present at probing sites on chupon tissue (Cross 1971). Late instars and adults produce larger lesions than do early instars. Johnson (1962) described in more detail the lesions caused by *D. theobroma* and *S. singularis*.

Damage in relation to light
Cocoa generally is grown under the shade of other trees, with the amount of shading affecting pest populations. Host trees growing in light gaps of the canopy would

grow more rapidly and might be preferred by cocoa mirids. As noted earlier, the literature relating to this point is equivocal, but mirid damage often begins in areas where the canopy is broken (e.g., Entwistle 1972, Collingwood 1977a). A blast type of mirid injury is associated with an absence of overhead shade; pockets, with injury localized beneath gaps in shade cover (Entwistle 1985a). Injury is often patchy because of the bugs' tendency to aggregate (Kay 1961; Entwistle 1972, 1985b; Wills 1986). The index of aggregation is constant and characteristic for each species, but the basis for such clumping is unknown (Youdeowei 1977). Several workers have observed that most cocoa mirids are photophobic and feed mainly at night (e.g., Madge 1968; see also Entwistle [1972]). Preference for feeding in darkness is most pronounced in *Helopeltis* species, the least so in *Distantiella* (Youdeowei 1977). All nymphal stages of *S. singularis* are strongly photonegative at constant temperature and relative humidity; fourth instars, the stage used for laboratory experiments, prefer darkness to various low light intensities (Madge 1968).

In Papua New Guinea, the widespread planting of cocoa under leucaena, a host of *H. clavifer* (Smith 1978), provides less shade and promotes pest problems (Smith 1985). The relationship of mirid population dynamics to light, however, is complex. The amount of light interacts with factors such as humidity and also affects cocoa physiology (Entwistle 1972, Khoo 1987, Wood and Chung 1989). Readers wanting more detailed information on the influence of light intensity on cocoa mirids should consult the publication by Entwistle (1972).

Cocoa Mirid Distribution and Ants

The distribution of *Distantiella theobroma* and *Sahlbergella singularis* in plantations depends substantially on the fauna of dominant ants (Majer 1972, Leston 1973b). Leston's ant mosaic theory, briefly stated, holds that ants represent the numerically preponderant insects inhabiting tree crops and other permanent vegetation in the humid tropics and that each dominant ant species is antagonistic to nearly all other dominant ants. Because of mutual exclusion, a mosaic distribution is the only one possible. The actual two- or three-dimensional mosaic is determined by the ants' specific preferences for nest sites, availability of material for nest construction, and presence or absence of preferred homopteran species tended by the ants. The composition of a cocoa farm's mosaic is influenced by the shade regime and nature of the overhead and surrounding vegetation (Leston 1973b, 1975; see also Hölldobler and Wilson [1990], Majer [1993]).

Each dominant ant appears "positively or negatively associated with an assemblage of pest species" (Leston 1973b). *Helopeltis* species apparently are chemically protected from attacks by the red tree ant (*Oecophylla longinoda*) (Leston 1973b, 1973c). The mirids are usually avoided, and if a worker ant seizes a nymph or adult, it immediately lets go (Aryeetey and Kumar 1973). But *O. longinoda* and the felt nest ant (*Macromischoides aculeatus*) are important predators of *D. theobroma* and *S. singularis*. Mirids are usually absent from territories occupied by the ants. The ants' potential for natural control in Ghana is considerable, for their territories can cover 60–70% of the trees in a plantation. Pesticides should not be applied to cocoa trees harboring nests of either ant species (Leston 1973b). Leston (1978a) stressed that problems from cocoa mirids are likely to intensify when management practices change the ant mosaic from one dominated by *Oecophylla* and *Macromischoides* species to one in which *Crematogaster* and *Camponotus* species predominate. Species of the latter genera are not active predators of cocoa mirids, and in the case of *Crematogaster*, the species tend mealybugs that transmit swollen shoot virus.

The practicality of an integrated control program that relies heavily on *Oecophylla* predation (Leston 1971) has been questioned (Squire 1947, Marchart 1971). Attempts to manipulate ant distribution by releasing red tree ants in Ghana have been judged unsuccessful (Kumar and Youdeowei 1983). Entwistle (1985b) summarized ant-mirid relationships in Ghana, Nigeria, Cameroon, and Papua New Guinea and commented on the economic implications of these ant mosaics (see also Bennett et al. [1976], Collingwood [1977b]; see Chapter 12). Ooi (1992) and Way and Khoo (1992) reviewed the role of ants in cocoa pest management.

LESIONS AND CANKERS OF OTHER SUBTROPICAL AND TROPICAL PLANTS

A *Rayieria* species, a monaloniine bryocorine, was imported from Australia into South Africa for evaluation as a biological control agent against invasive acacias or wattles. The bug attacks phyllodes and leaves of its hosts (Plate 9). In the laboratory, clear watery lesions 2–4 mm in diameter form in less than an hour after adults begin to feed. The severely spotted (Plate 9), nearly leafless condition of acacias in Australia is attributed to this bug. Because of its potential threat to Australian acacias used commercially in South Africa (e.g., blackwood and black wattle), it was not recommended for South African release (Donnelly 1986, Dennill and Donnelly 1991).

Helopeltis species and other bryocorines feed on various tropical crops, causing lesions and cankers similar to those described for cashew, cocoa, and tea plants (Table 9.7). Injured hosts sometimes appear as though scorched by fire.

Abnormal Growth and Differentiation

Mirids are responsible for causing various plant abnormalities—including "grotesque" growth forms (Leigh

Table 9.7. Examples of mirids of the subfamily Bryocorinae that cause lesions and cankers

Mirid	Host plant	Locality	Remarks/symptoms	Reference
Dicyphini: Monaloniina				
Felisacus elegantulus	Fern	Australia	"Papery patches" on young tissue of fronds	Jones and Elliot 1986
Helopeltis antonii	Allspice, black pepper	India	Elongated necrotic lesions on young shoots	Devasahayam et al. 1986
	Guava	India		Ramakrishna Ayyar 1940
	Horseradish tree	India	Dieback of terminals, foliar abscission	Sasidharan Pillai et al. 1980
	Neem	India	Tender shoots appear scorched; affected twigs show dark patches, oozing gum, and eventually dry up	Ramakrishna Ayyar 1940; see also Schmutterer 1990a
Helopeltis orophila	Quinine	Democratic Republic of the Congo	Necrotic lesions along secondary veins, cankers on shoots; young plantations may be destroyed; for discussion of species injuring quinine in Africa, see Carayon (1949), Schmitz (1968), Leston (1973c)	Lefèvre 1942, Steyaert 1946
Helopeltis schoutedeni	Avocado, guava, mango, other crops	East Africa	Cankerlike growths on stems resemble bacterial lesions	Harris 1937
	Mango	Malawi	Water-soaked lesions on stems; lesions may crack, resulting in open wounds	Leach 1935
Helopeltis theivora	Guava	Malaysia	Lesions and cankers may kill young terminals, distal parts of lateral shoots	Lim and Khoo 1990
"*Helopeltis*" spp.	Rubber-tree	Bolivia	Necrotic patches on stems of saplings develop into open cankers; causal agent remains uncertain because the genus *Helopeltis* is restricted to the Old World	Squire 1972
Mansoniella nitida	Formosan sweetgum	Taiwan	Foliar lesions, necrosis	Pers. observ.
Pararculanus piperis	Pepper	East Africa	*Helopeltis*-like lesions are produced on leaves	Poppius 1912
Poppiusia leroyi	Combretum	Ghana	Nymphs are concealed by the large brown lesions they produce	Leston 1980a
Ragwelellus suspectus	Paper-bark tree	Australia	"Papery patches," foliar spotting	Jones and Elliot 1986, Balciunas et al. 1994, Cassis and Gross 1995
Dicyphini: Odoniellina				
Chamopsis tuberculata	Guava	Sudan	Large foliar lesions on young plants; severely injured leaves dry up and abscise; considered a minor pest	Schmutterer 1969
Distantiella collarti	Grapefruit, orange, other citrus	Democratic Republic of the Congo	Nymphs can puncture stems 10–20 times/night, resulting in lesions, breakage of stems, and invasion by fungi	Decelle 1955, Buyckx 1962
Lycidocoris mimeticus	Coffee	East Africa	Foliar spots or lesions resemble those caused by *Helopeltis*, the pentatomid *Antestiopsis orbitalis*, and *Colletotrichum* fungi	Hargreaves [1926?], Le Pelley 1968; see also Hancock 1926
Pseudodoniella chinensis	Chinese cinnamon	China	Cracks and lesions on stems	Zheng 1992

Note: See text for additional examples and discussion.

Fig. 9.11. Aster plant severely dwarfed by the feeding of *Lygus lineolaris*. The three branches that escaped injury are growing normally. (Redrawn from Crosby and Leonard 1914.)

1976)—that are often associated with phytohormones which alter the amount of, or the balance among, growth-regulating plant hormones. These abnormalities, which include growth and differentiation disorders (Putshkov 1956, Norris 1979), reflect the stress that mirids can have on their host plants—the bugs can prevent plant growth from attaining optimal levels (Fig. 9.11). The preference of many mirids for actively growing tissues increases the possibility of such injury. Insects that feed near meristematic tissue are considered "more likely *a priori* to affect subsequent growth of the food plant than those that feed on mature or senescent parts" (Miles 1989a). At least in lygus bugs, injury apparently results more from the destruction of cells controlling growth than from injection of any particular substance by the bugs (Strong 1970, Laurema and Varis 1991). When the bugs feed on seedlings having only the first pair of true leaves, they destroy the terminal enzymatically, the injury similar to that of mechanical removal. Lygus bug injury to older seedlings cannot be mimicked mechanically and results in patterns of abnormal regrowth (Strong 1970). Strictly mechanical damage to the main stem apex of cotton by mirid feeding increases cell division in the stem tips, causing the plant to grow taller (Dale and Coaker 1958). When terminal buds and shoots are involved, proliferation of secondary or even tertiary shoots can occur, resulting in deformed and stunted plants (Smith 1978).

The cause of physiological disorders such as stunting, whiplike growth, bushiness, multiple leaders, thickened leaves and stems, and excessive branching (witches'-brooming) often is not readily apparent. A mirid responsible for a particular symptom can appear suddenly on a plant, as with adult *Lygus lineolaris* on fruit trees, and feed for a short time before migrating to other hosts. Because symptoms might not appear for a week or more after feeding (e.g., Varis 1972), an observer assessing injury several weeks later would find it difficult to establish the causative agent. Moreover, some of these mirid-induced problems are not perceived as deviations from normal growth or, if recognized as abnormal, are misdiagnosed. Mirid injury might be considered genetic in origin; thought to involve poor soil or weather-related stress; or attributed to herbicide injury, nematodes or other plant pathogens, or other sap-feeding arthropods. For example, some of the plant abnormalities induced by eriophyoid mites (Keifer et al. 1982) closely resemble those caused by mirids.

The various manifestations of mirid injury involve feeding on seedlings and on the leaves and stems of older plants. But in some cases, especially with cotton and alfalfa, they reflect cumulative, secondary effects of feeding on flower buds and inflorescences. These particular plant-growth disorders are discussed here (see also Tables 9.8, 9.9, and Plate 9) rather than in Chapter 10, which covers inflorescence feeding.

FIELD AND FORAGE CROPS

Some of the best-known examples of mirid-induced stunting and other abnormalities involve alfalfa and other legumes, cotton, and sugar beet (see also Table 9.8).

Cotton

Lygus bugs are most destructive to cotton after squares form (see Chapter 10), but they will feed on presquaring cotton. On seedling cotton in the southern United States, *Lygus* species sometimes feed on succulent cotyledons, retarding plant growth as much as 40%. Plants can exhibit delayed development of true leaves, forked growing points, and a witches'-brooming that results from growing points having small leaves. Feeding at the seedling stage can also produce plants with large, thick, leathery leaves on elongated petioles and multiple stems (Wene and Sheets 1964). When the western tarnished plant bug (*L. hesperus*) feeds on the terminals of older cotton plants (4–8 nodes present), a candelabra type of regrowth results (Strong 1970). Injury to still older plants, principally by *L. hesperus*, results in a malformation because of excessive branching from adventitious buds following the destruction of terminal buds (Plate 9; Cassidy and Barber 1938). When an adult *L. hesperus* is allowed to feed on presquaring cotton for 72 hours, injury consists of aborted terminals, deformed leaves, swollen nodes, shortened internodes, and witches'-brooming of the main stem (Hanny and Cleveland 1976, Hanny et al. 1977). The swelling of nodes and shortening of internodes are associated with internal necrosis of the nodal tissues. A diversion of cotton nutrients into secondary branches, as well as tissue destruction, might be involved in the shortening of

Table 9.8. Examples of mirids that cause abnormal growth of field and vegetable crops

Mirid	Host plant	Locality	Remarks	Reference
Bryocorinae				
Engytatus modestus	Squash	California, USA	Plant deformity, stunting, slow growth	Cecil 1940
	Tomato	Hawaii, USA	Swollen nodes, shortened internodes	Holdaway and Look 1940, Tanada and Holdaway 1954
Helopeltis antonii	Black pepper	Malaysia	Leaf deformation	Blacklock 1954
Mirinae				
Adelphocoris lineolatus	Asparagus	Ontario, Canada	Witches'-brooming of fern branches; *Lygus lineolaris* causes similar injury	Wukasch and Sears 1982
Closterotomus fulvomaculatus	Hops	England	Forked stems (bines), checked growth; dense growth of lateral shoots	Smith 1931b, Massee 1944
Closterotomus norwegicus	Flax (fiber)	Ireland	Arrested growth, bushiness	Lafferty et al. 1922
Creontiades dilutus	Sunflower	Australia	Deformed leaves, shortened internodes of seedlings	Miles and Hori 1977
Lygocoris pabulinus	Blackberry, loganberry	England	Excessive branching, distortion and shortening of shoots; absence of pith	Hey 1937
	Chrysanthemum	Germany	Stunted shoots, deformed flowers; *Lygus rugulipennis*, *Orthops campestris* also involved in injury	Bodenheimer 1921, Bech 1964
Lygus elisus	Potato	California, western USA	Distorted growth; *L. hesperus* causes similar injury	Flint 1992
Lygus hesperus	Safflower	California, USA	Reduction in diameter of seed heads	Carlson 1966
Lygus lineolaris	Aster	Eastern USA	Imperfect flowers, stunting	Arnold 1913, Crosby and Leonard 1914, Weiss 1920
	Bean	New York, USA	Malformed leaves, stunting	Taksdal 1963
	Celery	Florida, USA	Stunting	Quaintance 1896
	Chrysanthemum	Ohio, USA	Bushy, multiflowered plants	Weaver and Olson 1952; see also Crosby and Leonard 1914
	Daffodil	Massachusetts, USA	Deformed flowers	Russell 1947
	Dahlia	Eastern USA	Shortened stems, distorted growth; deformed blossoms	Patch 1906, Hitchings 1908, Brierley 1933; see also Crosby and Leonard 1914, Weiss 1920
	Lettuce	New York, USA	Stunting, irregular leaf arrangement, asymmetrical heads	Kageyama 1974
	Potato	Maine, USA	Shoot proliferation, modification of plant growth	Folsom et al. 1949, 1955
"Lygus" muiri	Cotton, eggplant	Fiji	Malformation, stunting; species is given status of *incertae sedis* by Schuh (1995)	Lever 1941, 1942; Hinckley 1965b; Swaine 1971
Lygus pratensis	Crucifers	Russia	Retarded growth	Znamenskaya 1962
Lygus rugulipennis	Chrysanthemum	England, Finland	Stunted, malformed shoots; misshapen blossoms	Austin 1932, Fox-Wilson 1938, Vappula 1965, Becker 1974
	Cucumber	Finland	Distorted shoots, shortened internodes, retarded growth of plants in greenhouse	Varis 1972, 1978
	Oats	Sweden	Retarded growth	Sömermaa 1961
Orthops campestris	Dahlia	England	Deformed buds, flowers	Woodroffe 1954
Polymerus cognatus	Kenaf	Dagestan (former USSR)	Excessive branching, thickened stems	Nevinnykh and Riabov 1931
Proba fraudulenta	Artichoke	Chile	Size of heads reduced	Olalquiaga 1955
Proba sallei	Guayule	California, USA	Retarded growth, witches'-brooming; *Lygus hesperus* also involved in injury	Lange 1944, Romney et al. 1945, Addicott and Romney 1950

Table 9.8. *Continued*

Mirid	Host plant	Locality	Remarks	Reference
Taedia hawleyi	Hops	New York, USA	Deformed, stunted vines	Hawley 1917, 1918
Taylorilygus apicalis	Chrysanthemum	Florida, USA	Twisted, blind-tipped stems; lateral shoot growth; deformed flowers (Plate 9)	Poe 1972
	Chrysanthemum	Japan	Bending of stems, arrested growth	Yasuda 1993
Trigonotylus caelestialium	Oats	Pennsylvania, USA	Retarded growth	Wheeler and Henry 1985
Trigonotylus ruficornis	Small grains	Former USSR	Retarded growth; *Notostira erratica* can cause similar injury	Kiritshenko 1951
Trigonotylus tenuis	Bermudagrass	Arizona, California, Georgia, USA	Distortion of unfolding leaves, stunting	Buntin 1988, Rethwisch et al. 1995
Orthotylinae				
Labops hesperius	Wheatgrass, other grasses	Colorado, Utah, USA	Stunting	Bohning and Currier 1967, Haws et al. 1973
Labops utahensis	Great Basin wildrye	Idaho, USA	Stunting	Youtie et al. 1987
Phylinae				
Moissonia importunitas	Crotalaria, sunnhemp	India	Stunting; twisted stems	Gopalan and Basheer 1966, Banerjee and Kakoti 1969, Gopalan 1976b van der Goot 1927
	Crotalaria	Indonesia	Stunting	

Note: See text for additional examples and discussion; table includes a few floral crops.

Table 9.9. Some mirids of the subfamily Mirinae responsible for growth disorders of fruit crops

Mirid	Host plant	Locality	Remarks/symptoms	Reference
Adelphocoris lineolatus, Closterotomus norwegicus	Peach	Italy	Stunting	Cravedi and Carli 1988
Creontiades dilutus	Passionfruit	Australia	Feeding on terminal shoots may arrest growth so severely that plants do not reach top of trellis and wires before winter	Hely et al. 1982
Lygocoris communis	Apple	Nova Scotia, Canada	Stunting	Dustan 1924
Lygocoris pabulinus	Gooseberry	England	Side-shooting	Petherbridge and Thorpe 1928a
	Raspberry	Scotland	Branching of canes caused by 1st generation; formation of crown of small branches near apex of nearly full-grown canes by 2nd-generation bugs	Hill 1952b
	Red currant	England	Side-shooting followed by stunting of terminals	Petherbridge and Thorpe 1928a Austin 1931a; see also Massee 1949
Lygocoris rugicollis	Apple	England	Excessive production of short lateral twigs, or side-shooting, giving trees a characteristic appearance	Fryer 1916, Smith 1920b, Austin 1931a
	Pear	England	Shortening of new growth, forking of tips	Massee 1956
Lygus rugulipennis	Peach	Italy	Stunting	Cravedi and Carli 1988

Note: See text for additional examples and discussion.

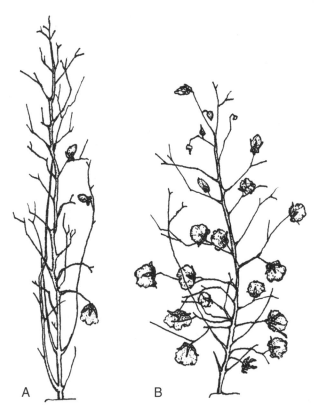

Fig. 9.12. Effects of feeding by *Pseudatomoscelis seriata* on cotton. A. Abnormally tall plants showing increased vegetative growth and poor boll production. B. Normal mature cotton plant. (Reprinted, by permission of A. D. Telford, from Telford 1957.)

internodes (Leigh 1976). Nymphal feeding by *L. hesperus* causes increased growth of the main stem and vegetative branches (Jubb and Carruth 1971). Tall, spindly, unfruitful cotton results from the removal of large numbers of squares (see Chapter 10), diverting more of the plants' energy resources to vegetative growth (University of California 1984). *Lygus lineolaris* is apparently also capable of causing a witches'-brooming of cotton (Reid 1965). The disorder of cotton characterized by abnormal branching in the upper parts of the plant and subsequent sterility is sometimes referred to as "crazy top" or acromania (Cook 1924).

Pseudatomoscelis seriata, the cotton fleahopper (discussed earlier in this chapter for the split-stem lesions it causes on cotton), induces abnormally tall or whip-like growth of the main stems, usually with widely spaced nodes, and sometimes an increased number of vegetative branches (Fig. 9.12). These responses to fleahopper injury—whiplike growth and rank vegetative growth (Hixson 1941)—are associated with this bug's feeding on the growing tips of cotton and the resulting loss of small squares (see also Chapter 10). When the terminal growing point is destroyed, an umbrella-like growth without a central stalk can result (Reinhard 1926, Brett et al. 1946, Faulkner

1952). The cotton fleahopper can also retard growth, shorten internodes, and induce excessive branching of seedling cotton (Eddy 1927, 1928). Some of the effects of stunting, however, can be outgrown (Powell 1974).

Feeding by the mirine *Taylorilygus vosseleri* in east central Africa results in cotton plants of a columnar shape because of shortened branches having bunched leaves and numerous weak secondary, tertiary, and axillary branches (Taylor 1945). Decreased yield is associated with this abnormal vegetative branching or candelabra effect (Matthews 1989), and plants can be stunted when infestations of this bug are severe (Dale and Coaker 1958, Stride 1968). Plants injured by *T. vosseleri* in Mozambique have an upright rather than conical shape (da Silva Barbosa 1959). Symptoms of prolonged feeding on cotton involve the entire plant and are analogous to those associated with hormonal growth-inhibiting substances (Carter 1973).

Feeding by *Helopeltis* species in Ghana retards bud production on cotton or, in severe cases, results in stunting (Cotterell 1928). Attacks by the bryocorine *H. schoutedeni* on young cotton tissues in northeastern Africa arrest vegetative growth (Steyaert 1946, Ripper and George 1965). Bedford (1938, 1940) implicated the mirine *Creontiades pallidus* and phyline *Campylomma verbasci* (as *C. nicolasi*) as the cause of crazy top in cotton in northeastern Africa (see also Gentry [1965]). The mirine *Horcias nobilellus* causes abnormally tall growth of Brazilian cotton (Hambleton 1938), and *"Lygus" muiri* (this name is considered *incertae sedis* in *Lygus* [Schuh 1995]) causes a stunting of cotton in Fiji (Swaine 1971).

Alfalfa and Other Legumes

In the western United States, *Lygus* species can reduce green and dry weights and stunt vegetative growth in alfalfa seed fields (Shull et al. 1934, MacLeod and Jeppson 1942, Stitt 1948), in addition to injuring buds, florets, and seeds (see Chapter 10). Stems may be unusually short, thickened, and terminated by rosettes of small, distorted racemes of buds. The reduction in stem length is nearly proportional to the intensity of infestation. Leaves sometimes are smaller than normal and distorted. This retarded growth can be misdiagnosed as soil deficiency, high temperatures, or the plants' need for a dormant period (Stitt 1948). Extended infestations cause proliferation of alfalfa, leading to excessive branching, leafiness, and a stringy appearance of plants in seed fields (Carlson 1940, Davis et al. 1976).

Despite their importance as pests of alfalfa grown for seed, lygus bugs usually cause little or no damage to alfalfa hay in California (Stern 1966). Mirids, however, do injure alfalfa grown for forage in the eastern United States to the extent that insecticidal control has been recommended (Medler and Fisher 1953). Plant regrowth can be delayed (Newton and Hill 1970). Chemical suppression of lygus bugs can increase hay yields, total amount of digestible nutrients, and yield of protein in first-crop alfalfa (Walstrom 1983). *Lygus lineolaris* has been dismissed as a pest of prebloom alfalfa in Ontario (Smith and Ellis 1983), but in other studies it and another mirine, *Adelphocoris lineolatus*, stunted plant growth, reduced stem length, and decreased dry weight (O'Neal and Peterson 1971, Jensen et al. 1991). Loss of yield is possible, although no losses from mirid feeding were detected in a two-year study in Wisconsin (Jensen et al. 1991).

Adelphocoris superbus retards and decreases vegetative growth of alfalfa in Utah (Sorenson and Cutler 1954). The morphologically similar *A. lineolatus*, or alfalfa plant bug, induces local lesions and systemic symptoms (Norris 1979). It has been implicated in *hopperburn* of alfalfa, a term describing symptoms induced by certain homopterans, particularly the potato leafhopper (*Empoasca fabae*). Plants affected by the leafhopper show yellowed or reddened foliage and are stunted (Monteith and Hollowell 1929). The alfalfa plant bug causes similar injury in Minnesota. Symptoms of feeding by the potato leafhopper and the mirid are so similar that they could be confused (Plate 10; Radcliffe and Barnes 1970).

The alfalfa plant bug's salivary secretions might disrupt phloem transport, as postulated for a Japanese leafhopper (*E. sakaii*) that also produces reddened foliage on alfalfa (Naito 1977a). *Empoasca fabae* is mainly a lacerate-and-flush feeder on alfalfa, performing different behaviors when feeding on stems contrasted with the interveinal area of leaves (Backus 1988b, Hunter and Backus 1989). These differences in feeding tactics probably explain why stem feeding, but not leaf feeding, results in hopperburn symptoms (Kabrick and Backus 1990, Ecale and Backus 1995, Ecale Zhou and Backus 1999). Blockage of phloem resulting in hopperburn may involve a secretion of watery saliva near the phloem; the enzymes responsible for cellular change and, indirectly, for vascular blockage are under investigation (Backus 1988b). *Adelphocoris lineolatus* might induce hopperburn in a manner similar to that suggested for *E. fabae*, although the mirid feeds primarily on reproductive structures rather than vegetative parts of alfalfa.

When the potato mirid (*Closterotomus norwegicus*) feeds on lotus seed crops at an early stage of plant development, it causes a proliferation of secondary stems, which produces greater numbers of flower heads. In New Zealand, the bug's feeding can actually result in increased yields if flower heads are later protected with insecticide (Clifford et al. 1983, Chapman 1984). Harris (1974) discussed the phenomenon of increased yield following insect feeding, including that by sucking species (for examples involving mirids, see Chapter 10). Crawley (1987) and Whitham et al. (1991) addressed the controversy involving the question of whether herbivory can be good for plants.

Sugar Beet

Feeding by *Lygus lineolaris* in the western United States imparts a bushiness to sugar beet when plants initiate growth in the axils of the outer leaves (Maxson 1920). Multiple crowns form when lygus bugs kill heart leaves and injure the crown region of the sugar beet (Yun 1986). On this crop in England, the common green capsid (*Lygocoris pabulinus*) feeds on the new leaves of young plants, retarding growth and causing an abnormal extension of cotyledons, which become thickened and brittle. This plant bug apparently can invade sugar beets from nearby hedgerows where bindweed and brambles serve as hosts. Beets in the seedling stage also suffer injury from the bugs' feeding on stems and cotyledons (Walton and Staniland 1929). Feeding by the European tarnished plant bug (*Lygus rugulipennis*) sometimes causes heart tissue to swell and cotyledons to enlarge and thicken (Dunning 1957, Jones and Dunning 1972, Varis 1972, Dunning and Byford 1982; see also Cooke [1992]). A summary of arthropod injury to sugar beet in England and Wales from 1947 to 1974, including that by mirids, is found in the publication by Dunning (1975).

Sugar-beet seedlings fed on by *L. rugulipennis* in Finland often exhibit retarded growth, multiple crowns, larger numbers of leaves, and smaller roots. Plants with multiple crowns have a larger leaf area for photosynthesis, and after the growing point is destroyed by lygus feeding, sugar-beet plants are able to grow faster than plants having a single leaf rosette. Roots of multiple-crowned plants can be slightly heavier than those of injured plants with single crowns, but both types of injured plants have smaller roots than healthy plants (Varis 1972). Injury is more severe in the inner portions of fields than within 10 meters of the edge (Varis 1974). Because only several minutes of feeding by *L. rugulipennis* on newly emerged seedlings can injure or destroy the growing point, control of this pest is a problem for the grower. Application of insecticides, however, can increase yields (Varis and Rautapää 1976). Another mirine, *Closterotomus norwegicus*, causes multiple-crowned beets in Denmark (Bovien and Wagn 1951).

VEGETABLE CROPS

Vegetable seedlings fed on by overwintered *Lygus hesperus* adults in the western United States sometimes become distorted (Fye 1984). *Lygus rugulipennis* adults feed on young cruciferous vegetables and other plants in Norway. Injured cotyledons of rutabaga and forage rape do not wilt as quickly in response to water deficits because they become abnormally thick (Plate 10) and dark green until new shoots are initiated at the destroyed growing point. Later, poor stems may develop, with a reduction in plant growth, quality, and yield. Injured cabbage shoots produce small, loose heads of no value unless the grower removes the excessive shoots. Cauliflower plants fed on by *L. rugulipennis* fail to form the terminal, partially developed flower clusters or "curd" (G. Taksdal, pers. comm.; see also Table 9.8).

FRUIT CROPS

Mirid injury to peach nursery stock in North America attracted much attention in the late nineteenth and early twentieth century. The well-defined symptoms —crooked ("doglegged") and stunted trees—characterized a syndrome referred to as "stop-back," "bush-head," or "bunch-head." Symptoms were often attributed to the work of mites, thrips, plant pathogens, or soil problems (Johnson 1898; Quaintance 1912; Back and Price 1912; Crosby and Leonard 1914; Haseman 1913, 1918) rather than to feeding by plant bugs. Slingerland (1895) in New York and Webster and Mally (1899) in Ohio concluded that *Lygus lineolaris* was actually responsible for the problem, but Quaintance (1912) considered plant bugs too scarce on peach trees to be responsible for the damage. Experimental evidence for this mirid as the causal agent was not forthcoming until Back and Price's (1912) work in Virginia. They determined that when tarnished plant bugs "sting" the tender terminal buds of a peach leader (Plate 10), the "upward growth . . . is checked, the nourishment is thrown into the laterals, and a short bushy tree which has lost much or all of its market value is the result" (Leonard 1915).

Even after the cause of losses to peach nursery stock was established, alleviation of the problem proved difficult (Back and Price 1912). Before the advent of chlorinated hydrocarbon insecticides (Zia-ud-Din 1951), satisfactory control of tarnished plant bugs could not be achieved with feeding deterrents or available insecticides (Crosby and Leonard 1914). As Haseman (1918) noted, the key to minimizing damage lay in clean culture and other preventive measures.

Lygus lineolaris continued to damage peach nursery stock in Ontario into the 1930s, often appearing in destructive numbers during late June to early July (Ross and Putman 1934). More recently, Missouri nurseries experienced damage to first-year growth of peach scion stock, the problem being most severe during mid-June to mid-July, which coincided with the occurrence of late first-generation and early second-generation adults (Caldwell 1981). The tarnished plant bug remains a significant problem because its feeding tends to stunt plants, which sometimes must be lowered in grade (stock is sold according to grades determined by height); creates a bushy plant more difficult to store, wrap, and ship; and produces unaesthetic, misshapen plants having less sales appeal in retail outlets. Injury to understock plants can also reduce the survival rate of grafted buds (Caldwell 1981).

Mirine plant bugs other than *L. lineolaris* cause similar injury to fruit crops (Fig. 9.13). Growth disorders

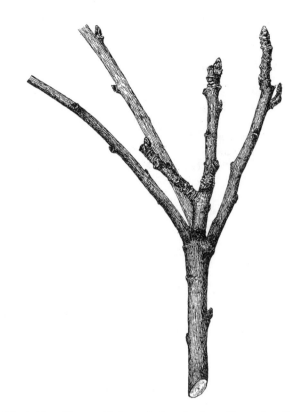

Fig. 9.13. Abnormally branched apple shoot ("crow-foot") caused by the feeding of a mirid, probably *Lygocoris pabulinus*. (Reprinted, by permission of Danmarks Jordbrugsforskning, from Rostrup and Thomsen 1923.)

of brambles, tree fruits, and passionfruit, a woody climber, are summarized in Table 9.9.

TREES AND SHRUBS

Meristem feeding by *Lygus lineolaris* shortens the internodes and stunts the shoots of hybrid poplar in Ontario. These symptoms involve inoculation of microorganisms into the plant and production of ethylene (Juzwik and Hubbes 1986). An arrested growth of rose shoots has been attributed in Ireland to feeding by *Lygocoris pabulinus* (Carpenter 1912).

Plant bugs also induce abnormalities of subtropical and tropical trees and shrubs. Hargreaves ([1926?]) described the effects of feeding by the bryocorine *Lycidocoris mimeticus* on the foliage and terminal buds of Ugandan coffee: brown patches on leaves, destruction of young shoots, and induction of secondary growth, which in turn may be attacked, giving rise to dense, bunchy plants. Growth sometimes became so badly distorted that no coffee crop could be produced and affected trees had to be "stumped immediately" (Hancock 1926). *Helopeltis ancardii* causes a witches'-brooming, stunting, and general malformation of cashew trees in east central Africa (Swaine 1959, Wheatley 1961); trees fed on by *Helopeltis* species sometimes have a short main stem that divides into heavy branches (Ohler 1979). Chronic infestations of *H. theivora* on cocoa lead to repeated growth and dieback that causes witches'-brooming similar to that induced by the fungus *Crinipellis perniciosa* (Khoo 1987). Long-standing infestations by *H. orophila* on quinine in the Democratic Republic of the Congo arrest the growth of terminal buds and cause a proliferation of secondary (and sometimes tertiary) lateral growth (Lefèvre 1942). Continued attacks by the bryocorine *Mircarvalhoia arecae* can lead to arrest of the crown, foliar stunting, and retarded leaf production of areca-nut (betel-nut) in India (Kurian and Ponnamma 1983). Feeding by the mirine *Lygocoris viridanus* on young succulent tissues of tea bushes in Sri Lanka destroys "flushing points," stunting the plants and causing severe economic loss (Calnaido 1959). Harris (1937) reviewed the types of injuries that *Helopeltis* species inflict on avocado and other economic plants in East Africa, including a bunching type of growth and growth retardation.

Froghopper refers to the injury that sucking insects inflict on young wattle or acacia trees in South Africa. This injury, consisting mainly of a witches'-brooming and bushiness or excessive branching, is so named because a leafhopper (*Iassomorphus cedarnus*) was once thought to be the main causative agent. It produces spittle resembling the frothy, anal secretions of nymphal cercopids, the spittlebugs, or froghoppers (Smit 1964).

A mirine plant bug, *Lygidolon laevigatum*, has now been established as the sole cause of wattle injury, but the misnomer *froghopper* persists. Some growers regard this mirid as the most serious insect pest of the plantation crop black wattle (Connell 1970a, 1974). The brown wattle mirid feeds on vegetative buds and foliage, its apparently phytotoxic saliva producing necrotic areas on foliage, distorted leaflets, and premature leaf fall. This bug's feeding on terminal buds results in witches'-brooms on branch tips and a bushy appearance instead of a single, dominant leading shoot. Because the leader does not follow the line of the original trunk, crooked timber results. The growth of trees can be retarded by six months, with trees that are weakened by drought or other stresses being more severely affected. Some trees never outgrow the effects of froghopper injury. Timber yields of black wattle, as well as bark extracts for tanning, resins, and adhesives, can be reduced (Ingham et al. 1998). Fertilization, intercropping, and other cultural practices help protect young wattle trees or lessen the severity of plant bug injury (Ripley 1927, 1929; Connell 1970a, 1974; Jacobs 1985). The development of more effective monitoring methods might enable growers to be alerted to impending outbreaks of this pest (Ingham et al. 1995, 1998).

Abnormal growth also is a symptom associated with the feeding of mirids that have been used in the biological control of weeds. *Rhinacloa callicrates*, a phyline introduced to Australia to help suppress populations of the adventive Mexican palo-verde, has distorted and stunted the growth of palo-verde trees in Arizona where this plant bug is native (Woods 1992).

CONIFER SEEDLINGS

Forsslund (1936) noted the potential for the mirine *Lygocoris pabulinus* to injure conifer seedlings in Sweden. In this often-overlooked paper, he reported that several individuals sometimes feed on stems and cotyledons, causing the latter to fade and lose their ability to shed the testa, and the stems to wither. Small yellowish or brownish spots appear on the stems in the areas of stylet insertion.

Since the early 1980s, growers in North America, Europe, and New Zealand have been concerned about a syndrome of conifer seedlings characterized by the production of multiple (bushy-topped or forked) leaders (Fig. 9.14, Plate 10; South 1991b). This injury, occurring mainly on first-year (1–0; one year in a seedbed, none in a transplant bed) seedlings, can involve older seedlings and transplants. Split-stem lesions below damaged terminals, as well as distorted needles, can also occur (Schowalter et al. 1986, Overhulser and Kanaskie 1989). Herbicide application, fertilizer burn, air pollution, nutrient imbalance, frost, plant pathogens, and genetics were considered possible causes of the problem (South 1986, 1991a, 1991b; Holopainen 1986, 1990a, 1990b). In widely separated regions of the United States and in British Columbia and Finland, lygus bugs are now known as important causal agents, and they have been implicated as the cause of multiple leadering in conifer seedlings in Great Britain. As reviewed by South

Fig. 9.14. Bushy-topped or multiple-leadered conifer seedling caused by the feeding of *Lygus lineolaris*. (Reprinted, by permission of W. N. Dixon and the Florida Department of Agriculture and Consumer Services, Division of Plant Industry, from Dixon 1989.)

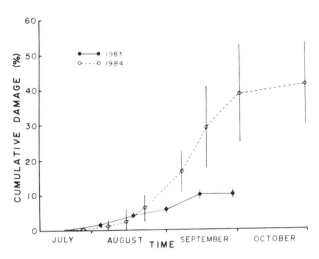

Fig. 9.15. Injury to conifer seedlings by *Lygus hesperus*. Accumulation of bud-abortion symptoms during rapid shoot elongation of first-year Douglas-fir seedlings (70/m²) at an Oregon nursery in 1983 and 1984. (Reprinted, by permission of T. D. Schowalter and Kluwer Academic Publishers Group, from Schowalter et al. 1986.)

(1991b), the injury has increased in recent years, perhaps because of the declining use of mineral spirits to control the weed hosts of lygus bugs in conifer nurseries. This aromatic petroleum oil might have minimized or prevented lygus bug injury through its repellent properties rather than from any direct, insecticidal effects.

Another hypothesis to account for the recent status of these generalist bugs as conifer pests involves the carbon-nutrient balance of the plants (Holopainen et al. 1995). They suggested that nitrogen fertilization in nurseries triggers the colonization of conifers, a plant group that lygus bugs do not otherwise prefer as hosts. An increased susceptibility of nitrogen-fertilized pine seedlings to *Lygus* might result from greater concentrations of free amino acids in host needles along with a reduction in host defenses via carbon-based secondary metabolites such as phenolics (Holopainen et al. 1995).

Lygus hesperus, the western tarnished plant bug, causes bud abortion of Douglas-fir in Oregon, mainly on first-year seedlings in forest nurseries. The bug's feeding results in a loss of apical dominance and a multiple-topped seedling. Bud abortion appears in late July; the damage increases steadily from early August to late September and ceases to accumulate once bud set occurs in late September (Fig. 9.15; Schowalter et al. 1986). The provenance of Douglas-fir seedlings and their proximity to alfalfa fields affect their susceptibility to injury (Schowalter and Stein 1987). Schowalter (1987) discussed the influence of sampling date, seedling type,

and cultural practices (top-pruning, fertilization, and irrigation) on the abundance and distribution of *L. hesperus* on pine and Douglas-fir seedlings.

The tarnished plant bug (*L. lineolaris*) causes similar injury to loblolly and other pine seedlings in Alabama (South 1986, 1991a). Timely application of insecticides reduced the percentage of injured seedlings in one nursery from 17% to less than 1% (South 1991a). Feeding by lygus bugs on the terminals of conifers, particularly pines, causes multiple-leadered, stunted seedlings in British Columbia; terminal needles can become thicker and stouter (Shrimpton 1985, Sutherland et al. 1989). In Florida, severely injured pine seedlings usually do not survive the growing season. Preventive insecticide applications in several Florida forest nurseries reduced the amount of injury from *L. lineolaris* and increased seedling survival rates compared to untreated plots (Dixon 1989).

Feeding by nymphal and adult *L. rugulipennis* on Scotch pine (Plate 10) in Finland leads to death of the apical meristem, to opening of the lateral buds, and eventually, to multiple leadering of seedlings (Holopainen 1986). Poteri et al. (1987) reported similar growth disturbances in one-year-old Scotch pine in Finland—bushy topped and multiple-shooted seedlings. Transmission of viral pathogens appears not to be involved in the symptoms. Development of multiple leadering is seen on first-year (1–0) pine seedlings, whereas on second-year (2–0; two years in a seedbed, none in a transplant bed) and older seedlings, growth-disturbance symptoms usually do not appear until the shoots elongate the next season. Lygus bugs can reduce the mean shoot length by 30–46%. A significant positive correlation exists between peak bug densities and the proportion of Scotch pine seedlings that show growth disturbances (Holopainen 1990b).

Multiple-leadered pine seedlings (Plate 10) often must be culled in nurseries, and the Finnish Ministry of Agriculture and Forestry prohibits their sale. The performance of such seedlings under natural conditions, however, needs further evaluation (Holopainen 1990a); multiple leadering might not actually affect growth in the field (South 1991b). Various herbaceous weeds in conifer nurseries—groundsel, for instance—are preferred by *L. rugulipennis* for oviposition and development. The weeds are potentially useful as trap plants that decrease lygus bug–induced injury to conifer seedlings (Holopainen 1989).

Taylorilygus apicalis might also injure conifer seedlings. This mirine is found in pine nurseries in the United States, Guatemala, and South Africa. When caged on pines, it can cause "bushy-top" symptoms identical to those caused by lygus bugs (South et al. 1993).

10/ Exploitation of Inflorescences

Few of the existing records made by Hemipterists specify the particular part of the plant on which the bugs [heteropterans] occur. . . . But, in my own experience, when the insect occurs on the upper parts, it is almost always on or under the leaves, except in the case of the *Umbelliferae*, when the flowers are preferred.
—E. A. Butler 1923

The extent of feeding on inflorescences, a relatively energy-rich food source (Pellmyr and Thien 1986), seems underappreciated for North American mirids. It is well known that pest species develop on the buds and flowers of alfalfa, cotton, and other crops, and that mirids such as lygus bugs (*Lygus* spp.) and the cotton fleahopper (*Pseudatomoscelis seriata*) frequently occur on the flowers of herbaceous wild hosts. But the comprehensive taxonomic works of Knight (1923, 1941, 1968) and Kelton (1980b) did not mention plant bugs as pollen feeders (see Chapter 11), and among species restricted to inconspicuous insect- or wind-pollinated flowers of trees, they mentioned only the association of *Orthotylus ramus* with male catkins of hickory and pecan trees (Knight 1941). In recording pollen feeding by *L. hesperus*, Rosenthal and Lipps (1973) remarked that they were unaware of any other hemipteran known to feed on pollen.

Even for the better-known British fauna, Butler (1918, 1923) seemed unaware of the extent to which mirids and certain other heteropterans feed on flowers. Moreover, Barth (1985) did not discuss heteropterans in his *Insects and Flowers: The Biology of a Partnership*, dismissing hemimetabolous insects as unimportant flower visitors.

Kullenberg (1944), however, emphasized the importance of pollen sacs and pollen as mirid food, providing numerous examples of the preferential use of these resources. He suggested the common name *Blumenwanzen*, or "flower bugs," for the family; indeed, this is a Danish name (*Blomstertaeger*) for these insects (Gaun 1974). Southwood and Leston (1959) referred to many of Kullenberg's (1944) observations on flower feeding in their discussion of the habits of British mirids. Some European mirids, although present on their hosts throughout the vegetative period, increase considerably in density once anthesis occurs (e.g., Shindrova 1979).

Putshkov (1966) noted that lygus bugs are among the relatively few plant bugs that can feed on mature seeds. The effects of mirid feeding on the seed production of certain native hosts might even regulate plant densities and affect host population dynamics, although long-term studies are needed to assess the actual effects of their herbivory (e.g., Ayal and Izhaki 1993, Izhaki et al. 1996). In some cases, feeding by plant bugs might affect the phenology and physiology of native shrubs (Haws et al. 1990, Nelson 1994).

Certain mirids feed preferentially on male rather than female flowers (Kullenberg 1944). The nectar and pollen rewards associated with pistillate and staminate flowers of monoecious and dioecious plants can be unequal (Baker 1983), but the phenomenon of sex-related feeding on plant parts, until recently (e.g., Krischik and Denno 1990), has received relatively little attention. Male plants of dioecious species often experience greater herbivory than do females (e.g., Willson 1991, Herms and Mattson 1992, Muenchow and Delesalle 1992).

The only experimental study of mirids in relation to male versus female hosts involves foliage rather than inflorescence feeders. *Dichrooscytus* species occurring on Rocky Mountain juniper show greater diversity and higher densities on male trees, owing to lower terpenoid levels compared to female trees (Polhemus 1988). Examples of sex-related feeding by mirids on apetalous inflorescences of several tree species are discussed later in this chapter.

The present chapter is not organized by injury symptoms (as was done in Chapter 9 for leaf and stem feeders) because mirids feeding on inflorescences and immature seeds often leave only slight evidence of their activity (Putshkov 1956). Feeding on mature yet not fully hardened seeds, as by lygus bugs, sometimes results only in "stings"—that is, dark spots on the surface (Strong 1970). This chapter instead is organized by plant groupings: herbaceous weeds and other herbs, wild grasses and cereal or grain crops, legumes used for forage and seed, cotton and other fiber crops, vegetable crops, shrubs, and trees. Castor, variable in growth habit, its ancestral type thought to be a tree, now tends to be an annual of herbaceous habit (Singh 1976); here it is considered a field crop. Plants that grow as a tree, shrubby tree, or shrub, depending on conditions, could fit in either of the last-named categories; one such problem group, mesquites, is discussed as a shrub.

Inflorescence feeding is broadly interpreted here to involve all floral structures, in addition to nectar and pollen. Deferred to the following chapter are a discussion of mirids as pollinators, a proposed mechanism allowing access by sucking insects to the nutrients within pollen grains, and comments on nectar as a food source (feeding at extrafloral nectaries is covered in Chapter 9). The present chapter uses the word *fruit* in the botanical sense of any ripened ovary of a seed-bearing plant; in the literature on cotton, *fruit* is used loosely to encompass squares (buds), flowers, and bolls (Stewart and Sterling 1989a). Feeding on fruits that can be classified as edible forms of ripened ovaries—mainly sweet fruits eaten out of hand—is covered in Chapter 12; an exception is elderberry, which has edible fruits but is treated here.

Lygus bugs and their relatives dominate the North American literature dealing with mirids that feed on floral buds, unripe fruits, and seeds of alfalfa, cotton, and other crops. The principal pests in *Lygus* are the tarnished plant bug (*L. lineolaris*) in eastern North America and western tarnished plant bug (*L. hesperus*) and pale legume bug (*L. elisus*) in the West. Other injurious species are of more limited importance: *L. atriflavus*, *L. borealis*, *L. robustus*, *L. shulli*, and *L. unctuosus* (Scott 1987, Schwartz and Foottit 1998). The names *L. desertus* and *L. desertinus* now refer to *L. elisus* (for nomenclatural clarification, see Lattin et al. [1992]), and *L. varius* is a synonym of *L. shulli* (Schwartz and Foottit 1998). *Lygus* species are difficult to identify because they show an inordinate degree of intraspecific variation, sexual dimorphism, and seasonal differences in color (Kelton 1975; Schwartz and Foottit 1992b, 1998; see also Chapter 5). Misidentifications, therefore, abound in the literature. Similar taxonomic confusion exists in Europe, with many records of injury by *L. pratensis* actually referring to the European tarnished plant bug (*L. rugulipennis*) (Southwood 1956c, Woodroffe 1966, Stewart 1969a). An important pest on the European continent, this species is less injurious in Great Britain (Stewart 1969a) and is not found in Scotland (Stewart 1969b).

I have selected examples of lygus bug injury from an extensive literature, which was summarized by Tingey and Pillemer (1977). Additional references are available in bibliographies of *Lygus* (Scott 1980) and the *Lygus* complex (Graham et al. 1984), the annotated host list for *L. hesperus* (Scott 1977a), a bibliography of *L. lineolaris* on apple (Parker and Hauschild 1975) and a list of its more than 300 hosts (O. P. Young 1986), and Crosby and Leonard's (1914) review of early economic literature on *L. lineolaris*. Holopainen and Varis (1991) recorded the extensive hosts (>400 species) of the Palearctic *L. rugulipennis*. Sterling and Dean's (1977) bibliography gives additional references to damage by the cotton fleahopper (*P. seriata*).

Herbaceous Weeds and Other Herbs

Various life forms of plants (Brown and Southwood 1983) are considered here. Represented are dicotyledonous annuals, especially *r*-selected opportunists of mainly ephemeral floras; biennials and perennials exclusive of trees and shrubs, including *K*-selected species characteristic of climax or near-climax vegetation; monocots except grasses and their relatives, which are covered later in this chapter; and a few pteridophytes. Lygus bugs sometimes destroy the flower spikes of ornamental gladiolus (Davis 1927, Bourne 1931), and several other cultivated forms, such as aster, chrysanthemum, dahlia, and daylily, are briefly mentioned in the following discussion.

LYGUS BUGS ON WEEDY AND CULTIVATED HERBS IN NORTH AMERICA

Mirids developing on the flower buds and inflorescences of composites and other herbaceous weeds in the eastern United States belong mainly to the subfamilies Mirinae and Phylinae. The tarnished plant bug (*Lygus lineolaris*) feeds in the flowers of daisy fleabane, apparently "pushing apart the disc flowers and sinking its proboscis into the tender base portion" (Smith and Franklin 1961). On horseweed (Plate 11), another composite, 60% of its feeding and nearly 70% of total feeding time take place on the receptacles (Latson et al. 1977). Goldenrods are important hosts of *L. lineolaris* (e.g., Patch 1906, Pree 1985, Maddox and Root 1987, Young and Lockley 1990), its feeding taking place "either between or on top of flower heads" (Sholes 1984). The bugs also probe florets with their stylets, possibly taking nectar from the plants (Sholes 1984; see also Chapter 11). Adult tarnished plant bugs sometimes disperse to daylily flowers in landscape plantings and feed on the petals and stamens (pers. observ.).

Polyphagous, multivoltine species such as *L. lineolaris* and certain other *Lygus* species (e.g., *L. borealis* [Kelton 1955]) track a succession of flowering hosts (Fig. 10.1), which remain uncolonized until they produce floral buds or flowers (Fleischer and Gaylor 1987). Pigweed, which is attractive at all stages of growth, is an exception (Woodside 1947). Clearcut logging, road building, and similar disturbances that perpetuate early-successional communities and promote the growth of herbaceous plants create favorable lygus bug habitats (Domek and Scott 1985). These highly mobile pests exploit various herbs concurrently, a strategy that helps offset changes in the relative abundance of herbaceous plants from year to year and place to place (Dixon 1987a). The abundance of *L. lineolaris* on Appalachian groundsel in Georgia, in contrast to a monophagous lygaeid and an oligophagous rhopalid occurring on this composite, is independent of bud density (McLain 1981). These findings suggest that bud density is less critical to generalist herbivores than to stenophagous species.

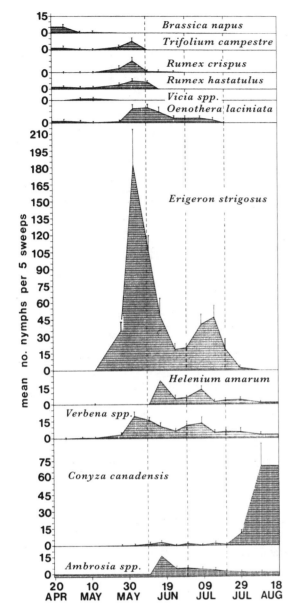

Fig. 10.1. Seasonal abundance of *Lygus lineolaris* nymphs on roadside hosts near cotton fields in the coastal plain of Alabama, 1983. The mirid colonizes host plants only when they enter a reproductive growth stage. Dashed lines indicate the early-square stage of early-planted (area to left) vs. late-planted cotton (area to right). Lack of a shaded area or standard error bar indicates the host was not sampled on that date. (Reprinted, by permission of S. J. Fleischer and the Entomological Society of America, from Fleischer and Gaylor 1987.)

Populations of *L. lineolaris* usually peak on herbs after most of the flowers open, remain stable until the seeds mature, and then collapse (Woodside 1947). In the Southwest, *Lygus* species sometimes remain and even develop on beets and turnips when these plants are going to seed (Stitt 1940). When weeds are cut along roadsides or damaged in cropland, growth is delayed,

maturity is prolonged, and their availability to lygus bugs is extended (Cleveland 1982).

The sequential occurrence of *L. lineolaris* on wild hosts in the Mississippi Delta region remains relatively constant from year to year (Cleveland 1982). Annual fleabane and horseweed are consistently used as hosts in the eastern states (Fenton et al. 1945, Woodside 1947, Phillips 1958, Tugwell et al. 1976, Latson et al. 1977, Cleveland 1982, Sapio et al. 1982, Snodgrass et al. 1984b, Young 1985, Fleischer and Gaylor 1987). Annual fleabane attracts adults from cotton and retains the bugs even in the presence of squaring cotton (Fleischer et al. 1988). Large numbers of *L. lineolaris* and other lygus bugs can colonize horseweed in the western United States, where this plant bug is usually much less abundant than other species of the genus (Stitt 1940, 1949; Domek and Scott 1985; Graham et al. 1986). *Lygus elisus* and *L. hesperus* can also be found on horseweed in the western states (Fye 1982b). In Great Britain and continental Europe, the semicosmopolitan horseweed (Fernald 1950) is a host of the European tarnished plant bug (*L. rugulipennis*) and other *Lygus* species (e.g., Fox-Wilson 1938, Kullenberg 1944, Bilewicz-Pawińska 1970, Holopainen and Varis 1991).

Hairy aster is another important late-season host of *L. lineolaris* in the southeastern United States (Young 1989). Even a generalist herbivore such as the tarnished plant bug does not use all available plants for development, and some common weeds usually go uncolonized (Womack and Schuster 1987).

Many weedy chenopods, composites, and crucifers are such important hosts of western lygus bugs that the presence of these reservoir plants should be considered in programs designed to manage mirid populations in forage seed and vegetable crops (Fye 1982b). A chenopod, nettleleaf goosefoot, and a crucifer, Londonrocket hedgemustard, are the primary winter and spring hosts of *L. elisus* and *L. hesperus* in California. Feeding occurs mainly on the blossoms and fruits (Clancy 1968). The influx and population buildup of lygus bugs in crucifers and legumes may preclude some members of these plant families (Brassicaceae and Fabaceae) from being used successfully as cover crops for enhancing the biological control of pest arthropods in orchards (Fye 1983c, Bugg and Waddington 1994, Prokopy 1994).

Lygus lineolaris shows different rates of reproduction on different hosts (Fig. 10.2; Curtis and McCoy 1964; see also Stewart and Gaylor [1994b]), and a "host switching" to exploit inflorescences can affect this bug's survival. Al-Munshi et al. (1982) evaluated the effects of different plants on the fecundity and longevity of *L. hesperus*, suggesting that interhost movements might be important if detrimental effects on life history are carried into the next generation. Their studies indicated that the overall success of species changes when individuals switch hosts. Similar studies on Indian populations of the dicyphine *Nesidiocoris* (formerly *Cyrtopeltis*) *tenuis* showed that reproductive rates differ on different hosts and that host switching affects

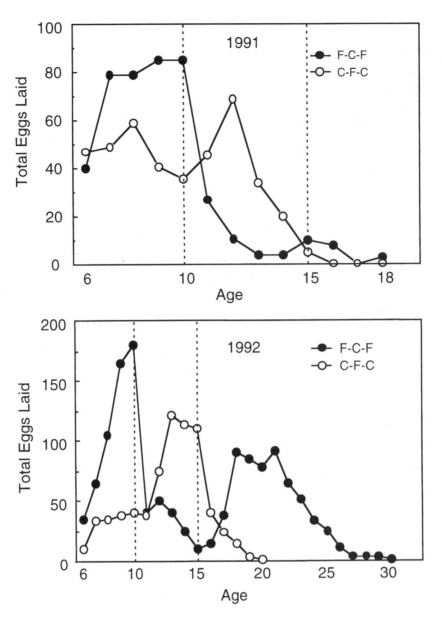

Fig. 10.2. Daily total of eggs laid by *Lygus lineolaris* when switched between cotton (C) and fleabane (F) in the sequences indicated. Oviposition immediately declined when bugs were switched from fleabane to cotton. Days of host switching are indicated by dotted vertical lines. (Reprinted, by permission of S. D. Stewart and the Georgia Entomological Society, from Stewart and Gaylor 1994b.)

longevity. Survival of *N. tenuis* "was not only affected by the host plant on which the ovipositing female feeds, but also by the interaction of nutrients provided by the host on which the species subsequently develops" (Raman and Sanjayan 1984a). In studies on *L. rugulipennis* in Japan, different first-generation hosts affected the survival rate, developmental period, and fecundity of first-generation bugs, suggesting that food plants affect population density in the next generation (Hori and Kuramochi 1984). Different first-generation hosts,

however, did not influence the growth of second-generation nymphs or the longevity of second-generation adults on alfalfa, wheat, and a cruciferous weed. Hori and Kuramochi (1984) assumed that the effects of host switching by *L. rugulipennis* do not carry over into the next generation.

Another consequence of the wild host spectrum of *L. lineolaris* involves interactions with its host plants and natural enemies. Parasitism by the euphorine braconid *Peristenus pseudopallipes* is high when the bug occurs on fleabanes, but partial protection (Hassell and Southwood 1978) is afforded to its populations on other host plants. Parasitoid females, attracted to fleabane flowers, apparently search them in preference to less favored hosts (Streams et al. 1968). Gravid parasitoid females also respond to odors from fleabane plants, suggesting that this braconid uses olfactory and visual cues to locate inflorescences (Shahjahan and Streams 1973). The wasp might also be attracted to the high-quality nectar of fleabane flowers because this potential food source is preferred to the flowers of several other common hosts of *L. lineolaris* (Shahjahan 1974).

OTHER WEED-ASSOCIATED MIRIDS IN
THE NEW WORLD

In the southeastern United States, many of the composites used by *Lygus lineolaris* also serve as hosts of the mirines *Polymerus basalis* and *Taylorilygus apicalis* (Snodgrass et al. 1984c, Young and Lockley 1990, Fontes et al. 1994). Tall goldenrod and white heath aster are important late-season hosts of both species (Young and Lockley 1990). Among members of the phyline genus *Plagiognathus*, which are common on goldenrod in eastern North America (Reid et al. 1976), the nymphs and adults of *P. cuneatus* and *P. politus* feed almost exclusively on goldenrod flowers, frequently occurring with the mirines *L. lineolaris* and *Adelphocoris rapidus* (Messina 1978). The phylines *Chlamydatus associatus*, *P. politus*, and *Reuteroscopus ornatus* feed on the mature heads of another composite, common ragweed (Shure 1970, 1973; pers. observ.); the first-named species is known as the ragweed plant bug. Croton (Plate 11) is a common wild host of *Pseudatomoscelis seriata*, the cotton fleahopper. Flowering cutleaf primrose and spotted beebalm are better hosts of this bug (i.e., survival and developmental rates are higher) than are preflowering individuals of these plants (Gaylor and Sterling 1976a).

Some additional records of mirids associated with the inflorescences of weeds and other herbs in North America are listed in Table 10.1. The numerous papers of R. D. Goeden and colleagues (e.g., Goeden and Ricker 1974, 1986a, 1986b; Goeden and Teerink 1993) on phytophagous insect faunas of weeds, mainly composites, occurring in southern California provide additional records of inflorescence-associated mirids. Some of the species they list actually develop on the plants, whereas others probably are present only for adult feeding.

HERBS AS HOSTS OF OLD WORLD PLANT BUGS

Palearctic mirids that specialize on the inflorescences of herbs include the phyline *Orthonotus rufifrons* and mirines *Apolygus spinolae*, *Grypocoris sexguttatus*, and *Liocoris tripustulatus*, which feed on the buds and fruits of stinging nettle. The phyline *Megalocoleus*

tanaceti feeds on the flower buds, inflorescences, and unripe fruits of tansy, and *Oncotylus viridiflavus* feeds mainly on the flower heads of black knapweed, another composite (Kullenberg 1944, Southwood and Scudder 1956, Southwood and Leston 1959, Groves 1968).

Species of the *Calocoris* complex (Mirinae) are active flower visitors that fly between flower heads (Plate 11; Betts 1986c). The phyline *Placochilus seladonicus* sometimes swarms on the heads of field scabious (blue buttons) in England (Hawkins 1989). *Lygus pratensis* (possibly *L. rugulipennis*) feeds at the base of chrysanthemum flowers in British greenhouses, resulting in a distorted bloom (see Chapter 9); the bugs also feed on blossoms, disfiguring the petals (Austin 1932). Asters, chrysanthemums, and dahlias grown outdoors in Finland are subject to similar injury from lygus bugs (Vappula 1965). *Orthops campestris*, a mirine specialist on umbellifers, feeds on the ovaries and immature fruits of various wild members of this family in Sweden (Kullenberg 1944). Woodroffe (1954) reported an unusual habit for this species: intensive feeding on the buds and flowers of cultivated dahlias, causing a drying and browning of inflorescences.

Some additional Old World mirids associated with inflorescences are the mirines *Taylorilygus vosseleri*, which shows an obligatory relationship with the flowering stage of a woody leguminous herb (a pseudarthria) in East Africa (Stride 1968), and *Adelphocoris suturalis*, a flower bud feeder on red sorrel in Japan (Hori 1969a). Japanese populations of *Lygus rugulipennis* feed on the receptacle, filament, and ovarial wall of the flower bud, as well as the shell, septum, and developing seeds of cruciferous pods (Hori 1973b). The phyline *Chlamydatus pullus* feeds on the reproductive parts of cinquefoil in Siberia (Yakutia) (Vinokurov 1988).

CAMOUFLAGE OF NYMPHS ON INFLORESCENCES

Nymphs of inflorescence feeders on herbs, for example, *Polymerus basalis* on red sorrel (Plate 11), as well as on other plant types, often blend with the color of their hosts (Table 10.2). Similarly, many leaf- and stem-feeding mirids are cryptic on host structures (Plate 11; Breddin 1896, Southwood and Leston 1957, Henry 1991, Gagné 1997; pers. observ.; see also Chapter 9 and Lightfoot and Whitford [1987]). Crypsis among anthophilous insects as a presumed means of protection from predators has received scant attention from biologists (e.g., Kevan 1978b, 1983).

Haseman (1918) commented on the relationship of *Lygus lineolaris* to horseweed: "The nymph really takes on gorgeous markings resembling... the maturing seed clusters." Although the green nymphs of this variably colored species blend well with the seed clusters of horseweed, the red nymphs stand out against the green involucres (Shahjahan and Streams 1973). That the rate of parasitism of the red morph can be three times that

Table 10.1. Some mirids associated with inflorescences of herbaceous weeds and other herbs in the New World

Mirid	Host plant	Locality	Reference
Mirinae			
Adelphocoris lineolatus	Red sorrel	New York, USA	Pers. observ.
Lygus abroniae, L. oregonae	Beach ragweed	California, Washington, USA	Schwartz and Foottit 1998
Polymerus basalis	Red sorrel	Connecticut, Pennsylvania, USA	Pers. observ.
Polymerus testaceipes	Ragweed parthenium	Puerto Rico	Pers. observ.
Proba vittiscutis	Ragweed parthenium	Florida, USA	Henry and Wheeler 1982; pers. observ.
Taylorilygus apicalis	Horseweed	Southeastern USA	Pers. observ.
	Ragweed parthenium	Florida, USA	Henry and Wheeler 1982; pers. observ.
Orthotylinae			
Lopidea teton	Astragalus	Colorado, USA	Lavigne et al. 1991
	Silky loco	Colorado, USA	Asquith 1991
Orthotylus catulus	Pussy-toes	North Carolina, USA	Pers. observ.
Pseudopsallus puberus	Desert evening-primrose	Nevada, USA	Knight 1968
Phylinae			
Lepidargyrus ancorifer	Black knapweed, tansy ragwort, other composites	Oregon, USA	Frick and Hawkes 1970
Plagiognathus albifacies	Pale-flowered leafcup	Tennessee, USA	Pers. observ.
Plagiognathus brevirostris	Tall meadowrue	Virginia, USA	Pers. observ.
Plagiognathus chrysanthemi	Sheep sorrel	New York, USA	Pers. observ.
Plagiognathus nigronitens	American feverfew	Virginia, USA	Pers. observ.
Plagiognathus obscurus	Oxeye daisy	Western USA	Uhler 1872
	Stinging nettle	Eastern USA	Pers. observ.
Reuteroscopus ornatus	Ragweed parthenium	Florida, USA	Pers. observ.
Rhinacloa basalis, R. clavicornis	Hairy beggarticks	Florida, USA	Needham 1948
Rhinacloa forticornis	Hairy golden aster	Colorado, USA	Kumar et al. 1976

Note: See text for additional examples and discussion.

Table 10.2. Some examples of mirids whose nymphs resemble flowers and unripe or ripe fruits of herbaceous hosts

Mirid	Host plant	Locality	Reference
Mirinae			
Calocoris roseomaculatus	Salad burnet	England	Southwood and Leston 1959
Polymerus basalis	Red sorrel	Eastern USA	Pers. observ.
Tinginotum sp.	Laportea (male flowers)	Ghana	Leston 1980a
Phylinae			
Megalocoleus molliculus	Yarrow	Sweden, eastern USA	Kullenberg 1944, Henry and Wheeler 1979
Oncotylus viridiflavus	Black knapweed	England	Leston 1952b
Semium hirtum	Spotted spurge	Eastern USA	Wheeler 1981b

Note: See text for additional examples and discussion.

of the cryptically colored green morph raises several questions. Does color dimorphism have a genetic basis, and if so, how is the red morph maintained in the population?

RESOURCE USE ON HOST INFLORESCENCES

Few mirids (except *Lygus* spp.) that develop on inflorescences of herbs have been associated with particular plant parts. In addition to the exceptions discussed earlier, nymphs of *Lygocoris pabulinus* in England feed on the central pistil of an arum lily (Wightman 1968). Pollen is important in mirid nutrition (see Chapter 11), and radionuclide tracer studies have shown that on ragweed in eastern North America, *Chlamydatus associatus*, *Plagiognathus politus*, and *L. lineolaris* derive most of their food from pollen tissue (Shure 1970, 1973). Four European phylines—*Hoplomachus thunbergii*,

Megalocoleus molliculus, *M. tanaceti*, and *Oncotylus punctipes*—are pollen feeders on their asteraceous or composite hosts (Reuter 1907b, Schumacher 1917b, Kullenberg 1944). The mirines *Adelphocoris lineolatus* and *Calocoris roseomaculatus* show similar habits (Kullenberg 1944). In addition, two bryocorines, the bracken bug (*Monalocoris filicis*) and the fern bug (*Bryocoris pteridis*), feed on the sporangia of ferns (Plate 11; Kullenberg 1944, Elton 1966, Ottosson and Anderson 1983), mainly the developing green sporangia (Srivastava et al. 1997).

DISPERSAL TO HERBS FOR ADULT FEEDING

Adults of mirids that use host plants of various growth habits, including trees and shrubs, sometimes disperse to herbs (Tables 10.3, 10.4) to obtain pollen, nectar, or both (see Chapter 11). In some cases, these supplemental food sources might supply the nitrogen needed for egg production (Schwartz 1984). Plants used solely for adult feeding can be called *food plants*, to distinguish them from true hosts on which development ("breeding") takes place (e.g., Taylor 1947b). Kinkorová and Štys (1989) discussed the different types of heteropteran "hosts" (also reviewed by Putshkov [1960]). A word of caution, though, is needed because mirids can develop on adventitious hosts (see Chapter 6). Records from such plants might be assumed to involve only adult bugs were it not for observations to the contrary. For example, Kirby's (1991) record of *Liocoris tripustulatus* on the flowers of cultivated mullein in England might be dismissed as referring only to adults—this mirine

generally develops on stinging nettle—except that nymphs were present for three consecutive seasons. Yet certain other records of mirid nymphs from plants that would represent atypical hosts are dubious. While adults of the grass-feeding *Leptopterna dolabrata* might well visit the flowers of oxeye daisy, as suggested by Heidemann's (1899) observations, the presence of nymphs on this plant should be attributed to a misidentification of the plant bug involved or to the possibility that nymphs simply were observed in an insect net (Heidemann noted that he collected Heteroptera from oxeye daisy by "sweeping" fields) and assumed to have been on the daisy rather than on grasses in the same field.

Umbellifers

Plants of the Apiaceae (= Umbelliferae) have open floral systems that do not restrict access to their nectar or pollen (Lindsey 1984). They consequently attract numerous insects (e.g., Bohart and Nye 1960, Bugg and Wilson 1989), including Diptera (Grensted 1946, Downes and Dahlem 1987), ichneumonids (Townes 1958, Kevan 1973) and other parasitic wasps (Leius 1960), as well as plant bugs (Plate 11; Rammner 1942; see also Table 10.3). Curtis and McCoy (1964) reported that a cultivated umbellifer, celery, furnished good oviposition sites and food for *Lygus lineolaris* nymphs in the laboratory, but they were puzzled by the rather large numbers of adults occurring but reproducing little on wild Apiaceae. Many of the individuals they observed in the field might

Table 10.3. Some mirids attracted to inflorescences of umbellifers

Mirid	Food plant	Locality	Reference
Mirinae			
Adelphocoris rapidus	Rattlesnake master	Illinois, USA	Adams 1915
Closterotomus norwegicus	Toothpick ammi	California, USA	Bugg and Wilson 1989
Lygocoris belfragii	Poison hemlock	Eastern USA	Knight 1917a
Lygocoris pabulinus	Cowparsnip	Alberta, Canada	Kelton 1955
Lygus lineolaris	Water hemlock	Iowa, USA	Hendrickson 1930
	Water hemlock	Pennsylvania, USA	Pers. observ.
Lygus spp.	Toothpick amni	California, USA	Bugg and Wilson 1989
Mermitelocerus annulipes	Umbellifers	Japan	Yasunaga 1993
Metriorrhyncomiris fallax	Wild parsnip	Pennsylvania, USA	Pers. observ.
Neurocolpus pumilus	Water hemlock	Pennsylvania, USA	Pers. observ.
Phytocoris spp.	Umbellifers	Illinois, USA	Robertson 1928
Stenotus binotatus	Wild parsnip	Utah, USA	Knowlton 1956
Orthotylinae			
Lopidea spp.	Umbellifers	Illinois, USA	Robertson 1928
Slaterocoris spp.	Poison hemlock	Pennsylvania, USA	Pers. observ.
Phylinae			
Plagiognathus brevirostris	Water hemlock	Pennsylvania, USA	Pers. observ.
Plagiognathus chrysanthemi	Water hemlock	Pennsylvania, USA	Pers. observ.
Plagiognathus cornicola	Water hemlock	Pennsylvania, USA	Pers. observ.
Plagiognathus spp.	Wild parsnip	Pennsylvania, USA	Pers. observ.

Note: See text for additional examples and discussion.

Table 10.4. Some North American and European mirids recorded from flowers of herbs and weeds that may not serve as breeding hosts

Mirid	Food plant	Locality	Reference
Bryocorinae			
Dicyphus errans	Hawkweed	England	Payne 1984
Deraeocorinae			
Deraeocoris ruber	Canada goldenrod	Austria	Schuler 1982
Deraeocoris schwarzii	Yarrow	Wyoming, USA	Stephens 1982
Mirinae			
Adelphocoris lineolatus	Grassleaf goldenrod	Pennsylvania, USA	Pers. observ.
Adelphocoris rapidus	Yarrow	Wyoming, USA	Stephens 1982
Capsodes flavomarginatus	Spurge	England	Woodroffe 1959b
Capsodes gothicus	Devilsbit	England	Groves 1976
Capsus sp.	Tansy ragwort	England	Harper and Wood 1957
Closterotomus norwegicus	Corn chrysanthemum	Ireland	Lafferty et al. 1922
Irbisia spp.	Lupine	Western USA	Schwartz 1984
Lygocoris pabulinus	Goldenrod	Austria	Schuler 1982
	Hawkweed	England	Payne 1984
Lygus sp.	Common knotweed	California, USA	Bugg et al. 1987
Metriorrhynchomiris dislocatus	Revolute meadowrue (male flowers)	Pennsylvania, USA	Pers. observ.
Phytocoris longipennis	Canada goldenrod	England	Groves 1976
Polymerus wheeleri	Branching draba	Virginia, USA	Wheeler 1995
Prepops eremicola	Yarrow	Wyoming, USA	Stephens 1982
Prepops insignis	Boneset	Ontario, Canada	Judd 1969
Orthotylinae			
Lopidea heidemanni	Appalachian groundsel	Virginia, USA	Wheeler 1995; pers. observ.
Lopidea minor	Oxeye daisy	Virginia, USA	Wheeler 1995; pers. observ.
Lopidea nigridia	Malacothrix	California, USA	Miller and Davis 1985
Phylinae			
Plagiognathus arbustorum	Helleborine	Belgium	Verbeke and Verschueren 1984
Plagiognathus chrysanthemi	Musk mallow	Ontario, Canada	Judd 1974
Plagiognathus politus	Teasel	Ontario, Canada	Judd 1984

Note: See text for additional examples and discussion; Table 10.3 lists mirids attracted to inflorescences of umbellifers.

have dispersed to obtain nectar or pollen from food plants—species that were not actually used for reproduction.

Among British heteropterans, mirids are particularly attracted to the inflorescences of umbellifers (Budgen 1851, Butler 1923). A synonym of *L. pratensis* is *L. umbellatarum*, the latter epithet perhaps referring to this bug's propensity for visiting umbelliferous flowers (Curtis 1849, 1860). Parfitt (1884) recorded several British mirids from inflorescences of Apiaceae, and Blathwayt (1889) observed *Grypocoris sexguttatus* swarming on the flowers of English umbellifers. Willis and Burkill (1903) reported the orthotyline *Heterocordylus tibialis* from flowers of cow-parsnip, Imms's (1971) *Insect Natural History* contains a color photograph of three adult *Leptopterna dolabrata* feeding on flowers of cowparsnip, and J. M. Price (1997) recorded *Phytocoris varipes* as numerous on seed heads of umbellifers. Plant bugs visit European umbellifers not only for floral nectar but also apparently to feed at extrafloral nectaries (Rammner 1942; see also Chapter 9). The crop mirid (*Sidnia kinbergi*) can be found on the flowers of wild umbellifers in New Zealand (Eyles 1999).

Additional Plant Families

Adults of other mirids on the flowers of plants that do not serve as breeding hosts include the mirine *Pinalitus approximatus*, a conifer feeder that is frequently attracted to largeleaved goldenrod in the eastern United States (Knight 1923). In South Carolina, *Pseudatomoscelis seriata* visits the flowers of aster, goldenrod, and ragweed, plants that appear not to support nymphs of this otherwise polyphagous bug (Eddy 1927). Robertson (1928) and Hendrickson (1930) provided records of numerous other mirids collected from the the flowers of North American herbs, and Willis and Burkill (1895, 1903, 1908), Knuth (1909), Kullenberg (1944), Porsch (1957, 1966), and Southwood and Leston (1959) recorded European mirids visiting flowers (see also Table 10.4). Kawasawa and Kawamura's (1975) book on Japanese Heteroptera, and the revision by Yasunaga et al. (1993), contain several color photographs of mirids on the inflorescences of herbs and other plants.

Wild Grasses and Cereal or Grain Crops

Many grass- and cereal-feeding plant bugs occur mainly on host inflorescences (Table 10.5, Plate 11) where they pierce the ovaries and unripe fruit, often destroying the endosperm (Kullenberg 1944). Such feeding preferences are common in mirines of the tribe Stenodemini, including the European *Stenodema calcarata*, *S. holsata*, *S. laevigata*, *S. trispinosa*, and *S. virens*. *Stenodema laevigata* and the meadow plant bug (*Leptopterna dolabrata*) require developing flowers and seed heads to complete their life cycle (McNeill 1973, Gibson 1976). The stenodemine *Megaloceroea recticornis*, once thought to feed mainly on grass leaves and to compete for this resource with *Notostira elongata* (Gibson 1980, Gibson and Visser 1982; see Chapter 9), actually requires grass flowers to complete its development (Wetton and Gibson 1987). *Notostira erratica*, mainly a leaf feeder, also feeds on the inflorescences of grasses (Kullenberg 1944). *Trigonotylus* species feed extensively on vegetative plant parts but often feed on the heads of their grass and cereal hosts. Their feeding on leaves and stems before developing seeds are available can reduce seed yields just as if they had fed directly on the seeds (Rethwisch et al. 1995; see Chapter 9). *Pithanus maerkelii* feeds on grass blades, but on moor matgrass in Sweden it sometimes sucks the endosperm of unripe seeds. Some European grass-associated mirids also feed on the inflorescences of rushes and sedges (Kullenberg 1944).

Most grass-feeding leafhoppers in Britain tend to be generalists in host range but specialists in terms of their nitrogen requirements (McNeill and Prestidge 1982, Prestidge and McNeill 1983); these highly mobile auchenorrhynchans track the most suitable hosts. A similar pattern might be characteristic of grass-associated plant bugs, although some species, like certain British grassland leafhoppers, probably are host specialists and nitrogen generalists (Gibson 1976).

Table 10.5. Numbers of *Stenotus binotatus* on different parts of orchardgrass in Japan, mid-June to early August

	Plant part		
Stage of mirid	Head	Stem	Leaf
Instars I–III	332	0	0
Instars IV–V	200	0	0
Adult	108	8	0

Source: Hori et al. (1985).

There are few reports of mirids developing on corn, and even *Lygus rugulipennis* and other pestiferous species of the genus failed to damage corn during a 10-year study in Hungary (Rácz and Bernáth 1993). The widely distributed *Trigonotylus ruficornis* sometimes feeds on corn in the former USSR, the injury it causes resulting in partially filled ears (Putshkov 1975). Nymphs and adults of the mirine *Creontiades pallidus* feed on pistillate inflorescences (silks) in Egypt (Willcocks 1925), and this species can be reared on the staminate inflorescences or tassels (Soyer 1942). In India, the sorghum earhead bug (*Calocoris angustatus*) also feeds on the tassels (Cherian et al. 1941), and *Creontiades pacificus*, which feeds on the earheads of maize, is occasionally a serious corn pest in India (Ullah 1940). The phyline *Campylomma verbasci* (as *C. nicolasi*) develops on maize silks in Egypt (Willcocks 1922, 1925; Pearson 1958). The nymphs of *Stenotus transvaalensis* are found on corn in Ghana, and as many as 30 well-camouflaged adults can occur on a head of male flowers; they are typically absent from female flowers (Leston 1968, 1980a). In Central America, King and Saunders (1984) included corn among plants on which a *Lygus* species feeds on ovaries, fruits, and developing grain but did not state whether this mirid develops on corn; the species could be either *L. lineolaris* or *L. mexicanus*, both of which occur as far south as Guatemala (Schwartz and Foottit 1998). The nymphs and adults of *Sthenaridea carmelitana* feed on the reproductive parts of corn (Matrangolo and Waquil 1991), and the availability of staminate inflorescences favors an increase in numbers of this pilophorine on corn in Peru (Marín and Sarmiento 1979).

Mirids using other host plants will disperse to corn. In the 1870s, corn in Michigan was severely injured by *L. lineolaris* (Cook 1876), and in Illinois, the bug sucked sap from tassels (Forbes 1883). Blackened holes in the kernels are symptomatic of this bug's feeding (Flood et al. 1995). *Plagiognathus obscurus* and *Adelphocoris rapidus* fed on kernels at the tips of ears in Illinois (Forbes 1905, Neiswander 1931). Tullgren (1919) reported that *Leptopterna dolabrata* sometimes migrates from meadows and grasses to attack the periphery of Swedish corn fields, and in Uzbekistan, the mainly predatory *Deraeocoris punctulatus* feeds on the tassels and other parts of corn (Popova 1959).

Corn pollen, which often accumulates along the depression of midribs or at leaf axils, is an especially rich source of carbohydrate (Todd and Bretherick 1942). Several arthropods feed on shed pollen of corn; these include phytoseiid mites (McMurtry and Johnson 1965), adults of some chrysopids (Sheldon and MacLeod 1971), coccinellids and other Coleoptera (Everly 1938), and the anthocorid *Orius insidiosus* (Dicke and Jarvis 1962). Ashmead (1887a) reported an undetermined mirid species feeding on corn pollen in Florida. Pollen that accumulates on corn leaves might serve as a supplementary food source for other plant bugs (see Chapter 11).

RICE

Mirids generally are not important rice pests (Kiritshenko 1951), although their relationship to the plant remains undetermined in some areas—for example, Thailand (Japan International Cooperation Agency 1981). In Central America, the mirine *Garganus albidivittis* sucks sap from developing seeds and leaves (King and Saunders 1984). Other records of injury to rice in the New World involve mirids that cause foliar chlorosis (see Chapter 9) rather than injury to the reproductive structures: an undetermined reddish species in Haiti (Myers 1931b) and the stenodemine *Collaria oleosa* in Costa Rica (Ballou 1933) and the Caribbean (Schmutterer 1990b).

Trigonotylus caelestialium, known as the rice leaf bug in Japan, is responsible for black rot and severe deterioration of rice grains (Hachiya 1985, Dale 1994). Another pest of Japanese rice is *Stenotus rubrovittatus* (Plate 12), a multivoltine mirine that develops on meadow grasses and invades paddies when rice begins to head (Hayashi and Nakazawa 1988). Hayashi (1989) identified five types of injury to kernels ("pecky rice") caused by this bug (Plate 12) and correlated different symptoms with the period of host flowering. The numbers of pecky grains caused by third instars and adults (per day per head) of *S. rubrovittatus* were 0.27–0.36 at full heading, 0.72–1.69 at the milk-ripe stage, 2.00–2.50 at the dough-ripe stage, and 0.10–0.40 at the yellow-ripe stage.

SORGHUM

Several plant bug species pierce the inflorescences and grain of sorghum in different parts of the world. They are major pests in Africa and Asia but are considered unimportant in Australia and in North, Central, and South America. Panicle-feeding pests are particularly injurious because they affect crop development at a late stage and impair grain yields (Harris 1995). Teetes (1985) and Seshu Reddy (1988) discussed the methods of assessing reduced yields from feeding by mirids and other sorghum head bugs.

The Sorghum Earhead Bug (*Calocoris angustatus*)

This mirine attracted attention from economic entomologists in South India as early as 1893 (Pradhan 1969), and it continues to be a serious pest in the states of Andhra Pradesh, Karnataka, and Tamil Nadu (Jotwani 1983, Gahukar 1991, Hiremath 1995). Problems from this bug might move north as high-yielding varieties of sorghum are introduced (Teetes et al. 1979).

Calocoris angustatus feeds on sorghum panicles (heads) as they emerge from the flag leaf, inhibiting grain set and decreasing yields (Cherian et al. 1941). Adults in nearby crops are attracted to the preflowering stage of sorghum (Natarajan et al. 1988), which is preferred for oviposition (Hiremath and Thontadarya 1984a); no eggs are laid in panicles following grain set (Hiremath and Viraktamath 1992). As many as 350 nymphs can be found on a panicle (Cherian et al. 1941). The sorghum earhead bug causes grains to become chaffy, shriveled, and distorted (Plate 12). If nymphs feed on the earhead before emergence, the ear may dry up and produce no grain (Ballard 1916, Cherian et al. 1941, Atwal 1976, Mote and Jadhav 1990). Hiremath et al. (1983) discussed histochemical changes in infested grain, including decreased starch content of the endosperm. The bug's feeding increases amylase activity, which affects grain quality both physically and biochemically. During the rainy season, grain is also rendered susceptible to infection by molds (Sharma and Lopez 1989, 1990a).

Daily food consumption by fourth instars on panicles is greater than for second, third, or fifth instars (first instars were not studied), and adult consumption is about three times greater than that of fourth instars (Natarajan and Sundara Babu 1988c). For females ovipositing at the half-anthesis stage, the economic injury level is less than one *C. angustatus* per panicle. Higher levels can be tolerated for nymphs and feeding adults, with the actual density depending on the insect stage and insecticide used for control (Natarajan and Sundara Babu 1988b). Economic injury levels range from 0.2 to 1.4 bugs per panicle, depending on the cultivar (Sharma and Lopez 1989).

Coexistence of *C. angustatus* with a pod borer, the noctuid *Helicoverpa armigera*, increases the damage potential. Natarajan and Sundara Babu (1990) found that grain loss is higher than that caused by either the noctuid or the mirid alone.

At least two generations of *C. angustatus* develop on sorghum when the ripening of panicles is staggered (Teetes et al. 1983). The incidence of this pest is lower in normal years when the crop life averages about 91 days than in unfavorable seasons when sorghum has a mean life of 105 days (Balasubramanian and Janakiraman 1966).

Infestations of *C. angustatus* in South India are greater in irrigated (April–June) than rain-fed sorghum (August–January), and on cultivars with compact rather than loose panicles (Cherian et al. 1941). Later work showed that damage is more severe on short-duration cultivars that mature during periods of moderate temperatures and high humidity when the numbers of bugs are highest. Early-season buildup of plant bug populations also subjects late-maturing sorghums to greater damage (Leuschner et al. 1985). Sharma (1985a) determined that temperature, sunshine, relative humidity, and rainfall do not influence population increase (highest densities generally occur during the second half of the rainy season). In other studies, relative humidity and rainfall generally exerted a positive influence, and maximum temperature and hours of sunshine, a negative effect on incidence (Natarajan et al. 1988; see also Sharma and Lopez [1990a], Gahukar [1991], Hiremath [1995]).

Calocoris angustatus prefers the milk stage of the panicle to the preflowering, complete-anthesis, or ripening stages, and its populations decline rapidly when grain matures (Hiremath and Thontadarya 1984b, Prabhakar et al. 1986). Populations in South India are greater during the dough stage, regardless of the planting date, than during the milk stage, except in summer. The numbers of nymphs and adults are always lowest at the preflowering stage (Natarajan et al. 1989). Crucial in preventing losses from sorghum earhead bugs is protection during the initial stages of grain development, particularly the time of complete anthesis (Sharma and Lopez 1989).

The sorghum earhead bug is a polyphagous pest, but grain sorghum is the preferred host: developmental time is shorter and adult size and weight greater on this plant than on other grasses and grains (Hiremath 1986). The introduction of high-yielding sorghum varieties in India has increased the bug's incidence and severity of infestation (Young and Teetes 1977, Steck et al. 1989). Cultivars showing effective resistance have yet to be developed (Sharma 1985b), although some of the nearly 15,000 germplasm lines that have been screened possess a moderate level of resistance (Sharma 1985a, Sharma and Lopez 1992, Sharma et al. 1995). An important component of this resistance is cultivar preference—chemical and morphological characteristics of sorghum that affect this bug's host selection and oviposition. Another is antibiosis, which involves prolonged nymphal development, reduced survival, and less efficient food consumption and use on certain genotypes (Sharma and Lopez 1990b, Ramesh 1994, Sharma et al. 1995). Losses from *C. angustatus* should be minimal on cultivars with loose or semicompact panicles (Sharma and Lopez 1989). Sharma and Lopez (1991) found that no cultivar is stable relative to grain damage at higher infestation levels and concluded that damage is strongly influenced by environmental conditions.

Although *C. angustatus* is closely associated with sorghum panicles and apparently restricted to this crop's reproductive phase (Sharma and Lopez 1990a), it can be reared successfully on nonreproductive parts. Natarajan and Sundara Babu (1988a) used sprouted sorghum seedlings as an easy means of culturing bugs for use in evaluating host resistance.

Other Plant Bugs of Sorghum

The mirine *Eurystylus oldi* (Plate 12) is the most abundant and injurious of the more than 30 heteropteran species found on sorghum panicles in Africa (Steck et al. 1989, Sharma et al. 1992, Ratnadass et al. 1994, 1995a). Much confusion, though, has surrounded the identity of *Eurystylus* species on sorghum. In reevaluating African species of the genus, Stonedahl (1995)

established that *E. oldi* varies considerably in coloration and size (see Fig. 5.17). He proposed *E. immaculatus* as a synonym of *E. oldi* and showed that previous identifications of *E. marginatus* on sorghum represent misidentifications of *E. oldi*. *Eurystylus oldi* is the principal sorghum pest in West Africa; a minor pest is *E. bellevoyei* (see Stonedahl [1995] for nomenclatural and taxonomic details).

High-yielding sorghum cultivars that flower in mid-season are most susceptible to damage by *E. oldi*. The bugs, mostly females, colonize the crop in late August. Two succeeding generations are produced; the populations in Niger peak in September at about 80 bugs per panicle (Steck et al. 1989). In addition to causing oviposition injury (discolored, floury endosperm), *E. oldi* reduces yields and grain quality (Steck et al. 1989, Ratnadass et al. 1995a). The incidence of grain mold can also be greater on panicles heavily damaged by the mirid (Sharma et al. 1992). Measurable damage can occur with as few as 10 bugs per panicle. Maximum densities and damage are associated with panicles infested at the complete-anthesis stage of crop development (Steck et al. 1989, Sharma et al. 1992).

Gahukar et al. (1989) observed large numbers of *E. oldi* on late-flowering cultivars in Mali. The nymphs and adults feed on maturing grain, which results in brownish spots on the grains not covered by glumes. When feeding is intensive, the grains shrivel and their quality deteriorates. Studies on host resistance to *E. oldi* are under way in West Africa (Doumbia et al. 1995, Ratnadass et al. 1995b). Management of *E. oldi* populations on sorghum in West Africa might also include the disruption of its life cycle on castor, a common alternative host during the dry season (Ratnadass et al. 1997). Other plant bugs that feed on sorghum panicles are listed in Table 10.6 (see also Bowden [1965], Steck et al. [1989], Harris [1995], Stonedahl [1995]).

TIMOTHY AND OTHER GRASSES

In New York, Howard (1892) observed numerous two-spotted grass bugs (*Stenotus binotatus*) on the heads of timothy, which became spotted with excrement. The 6–15 bugs on nearly every head were suspected of destroying immature seeds. As an adventive species in New Zealand, this plant bug occurs in huge numbers on the heads of various grasses (Myers 1922). Feeding by *S. binotatus* on the heads of orchardgrass affects seed production in Japan (see Table 10.5; Hori et al. 1985). *Stenotus rubrovittatus* (Plate 12) feeds on the inflorescences of Sudan grass in Japan, reducing seed set by mechanically or physiologically preventing pollination (Yamada and Watanabe 1952). In West Africa, *S. gestroi* and *S. transvaalensis* are seed feeders on grasses (Gibbs and Leston 1970), and the stenodemine *Dolichomiris linearis* sometimes feeds at night on the seeds of tall grasses in Paraguay (Carvalho and Hussey 1954).

Some of the grass-associated plant bugs that produce silvertop symptoms when they feed on culms (see

Table 10.6. Some mirids recorded from sorghum panicles

Mirid	Locality	Remarks/symptoms	Reference
Mirinae			
"Adelphocoris apicalis"	Africa	Identity of mirid is in question: perhaps *Megacoelum apicale*, but not *Adelphocoris seticornis* as suggested by Harris (1995); the latter species is not known from Africa (Carvalho 1959, Schuh 1995)	Seshu Reddy 1991, Ratnadass et al. 1997
Collaria oleosa	Cuba	Possible reduced yields	Ryder et al. 1968
Creontiades pacificus	India		Ullah 1940
Creontiades pallidus	Northern Africa		Willcocks 1922, 1925; Sharma 1985a; see also Seshu Reddy 1991
Creontiades rubrinervis	Central America		King and Saunders 1984
Eurystylus bellevoyei	India, Africa		Odhiambo 1958, Sharma 1985a, Sharma and Leuschner 1987, Sharma and Lopez 1990a; see also Seshu Reddy 1991
Polymerus basalis, *P. cuneatus*	Cuba	Possible reduced yields	Ryder et al. 1968
Taylorilygus ricini	East Africa		Taylor 1947b, Boyes 1964
Taylorilygus virens	East Africa		Taylor 1947b
Taylorilygus vosseleri	Tropical Africa	Maximum densities coincide with "green seed" stage; >1,000 individuals may occur on a panicle	McKinlay and Geering 1957, Stride 1968, Ingram 1970b
Phylinae			
Campylomma angustius	Nigeria, West Africa		Nwanze 1985, MacFarlane 1989; see also Seshu Reddy 1991
Campylomma plantarum	Nigeria, West Africa		Nwanze 1985, MacFarlane 1989; see also Seshu Reddy 1991
Campylomma spp.	India		Sharma and Lopez 1990a
Sthenaridea suturalis	Nigeria, West Africa		Nwanze 1985, MacFarlane 1989; see also Seshu Reddy 1991

Note: See text for additional examples and discussion.

Table 10.7. Effects of mirids on seed germination of fine fescue from caging late instars and adults on 2 newly emerged panicles at different densities; each density was replicated 6 times

Density of bugs	% Germination (n = 200)	
	Leptopterna ferrugata	*Megaloceroea recticornis*
0	85	87
2	22	0
4	7	1
6	1	0
8	0	0

Source: Kamm (1979).

Chapter 9) also feed on inflorescences. For example, *Leptopterna ferrugata* and *Megaloceroea recticornis*, Palearctic bugs that are now established in North America (Wheeler and Henry 1992), feed on the developing seeds of bluegrass and fine fescue. Both species can destroy endosperm and cause inviability, in addition to causing silvertop (Table 10.7; Kamm 1979).

The early instars of several mirids feed on grass blades before moving to seed heads when they become available. Examples include two Old World bugs now common on Kentucky bluegrass in the eastern United States: *L. dolabrata*, the meadow plant bug (Osborn 1918, 1919), and the phyline *Amblytylus nasutus* (Jewett and Spencer 1944, Jewett and Townsend 1947). Osborn (1918) observed a female meadow plant bug piercing the glumes of witchgrass in Ohio and inserting the proboscis "between and down into the anthers, penetrating them and causing them to burst, and probably sucking juices from the ovules." Injury to Kentucky bluegrass is more pronounced in dry years when infested plants grow less, stems are shortened, and seed heads are smaller and less compact (Jewett and Spencer 1944). *Leptopterna dolabrata* also infests the heads of sheep fescue grown as an ornamental in Pennsylvania (Wheeler and Henry 1992).

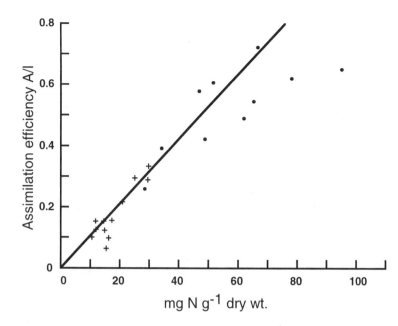

Fig. 10.3. Energetic digestive efficiency (A/I) relative to total nitrogen (N) levels in the grass-feeding *Leptopterna dolabrata*. As is the case in many insects, a compensatory increase in feeding rates on low-nitrogen food is accompanied by a decrease in assimilation efficiency. (Reprinted, by permission of S. McNeill and Academic Press, London, from McNeill and Southwood 1978.)

Leptopterna dolabrata, like the leafhoppers that colonize German velvetgrass (creeping softgrass), shows a compensatory increase in its feeding rate on low-nitrogen food sources, coupled with a decrease in its assimilation efficiency (Fig. 10.3; McNeill and Southwood 1978, McNeill and Prestidge 1982). The highest assimilation efficiencies of *L. dolabrata* occur when late instars move from German velvetgrass, on which the early-stage nymphs feed on mesophyll cells of blades, to the seeds of an earlier-flowering velvetgrass, Yorkshire fog. A decrease in leaf nitrogen in the early host triggers a switch to high-nitrogen feeding sites on the new host, which eventually affects the bug's population dynamics (McNeill 1971). Because of competition for the scarce seeds of velvetgrass, late instars suffer density-dependent mortality; surviving females show a similar density-related reduction in fecundity (McNeill 1973, McNeill and Southwood 1978, Price 1984). McNeill's (1973) research is important for documenting the interactive influence of nutrition, competition, and weather on population regulation in insects (P. W. Price 1997).

Energy transfer in mirids has received little attention, making the studies on *L. dolabrata* particularly important. McNeill (1971) calculated production (P), assimilation (A), and ingestion (I) ratios: A/I = 31.5, P/A = +55.1, and P/I = 17.3 (see also discussion by Wiegert and Petersen [1983]). The energy ratio A/I (assimilation/ingestion) indicates a rather low nitrogen content of food resources ingested by *L. dolabrata*. Its P/A ratio, which is lower than the mean calculated for various aphid species, indicates higher respiration costs, possibly because of slow growth resulting from low nitrogen levels (Llewellyn 1982).

Chorosomella species share the biological feature of feeding in the tops of cereal grasses. This Old World genus, formerly placed in the Stenodemini, subfamily Mirinae, belongs to the orthotyline tribe Halticini (Kerzhner 1962). Among Palearctic species of the mainly predacious genus *Phytocoris* (see Chapter 14), *P. varipes* feeds on unripe fruits and flowers of grass heads (Kullenberg 1944, Southwood and Leston 1959). In Oregon, where this mirid is adventive (Stonedahl 1983a), it ascends grass stems at night to feed on seed heads (Stonedahl 1988; T. J. Henry and G. M. Stonedahl, pers. comm.). The Nearctic *Deraeocoris nigrifrons*, another member of a predominantly carnivorous group, occasionally probes the florets of timothy (Stephens 1982).

WHEAT AND OTHER GRAIN CROPS

Plant bugs are not important pests of wheat in North America (Hatchett et al. 1987), although a Michigan farmer once experienced a 10% loss of crop from the tarnished plant bug (*Lygus lineolaris*) (Cook 1876). Seed damage in wheat fields of western Canada consists of irregular bleached areas that extend into the endosperm (Wise et al. 2000). Adults are sometimes attracted to wheat in early spring (Haseman 1918), and nymphs and adults can puncture the kernels, extract the milk, and cause them to shrivel and dry up or become moldy and discolored (Webster 1884). In more recent work with this pest in New York, beans intercropped with winter wheat harbored greater bug densities than did beans grown in monoculture plots. Dispersal from the maturing or drying inflorescences of wheat might have increased the tarnished plant bug numbers in the intercropped plots (Tingey and Lamont 1988).

Adelphocoris rapidus once injured spring and fall wheat in the midwestern United States, the adults apparently causing a withering of heads and abortion of kernels (Webster 1886). Lugger (1900), noting that the native North American stenodemine *Stenodema vicina* is abundant on the heads of wheat and other grains, predicted this bug would cause damage if it became more numerous. This species sometimes develops on the seed heads of annual bluegrass in Pennsylvania (pers. observ.), but small grains have yet to sustain damage. The meadow plant bug (*Leptopterna dolabrata*) can invade midwestern wheat when grass surrounding the fields dries up. This stenodemine's feeding on newly formed heads sometimes causes severe damage, which is usually limited to a few fields or a small area of one field (Cleveland 1925, Drake 1928). Foliar injury to cereals by the meadow plant bug was discussed in Chapter 9.

In Central America, King and Saunders (1984) stated that the mirine *Proba sallei* feeds on developing grain of wheat. They also listed a *Lygus* species as a grain feeder; only *L. lineolaris* and *L. mexicanus* are known from Central America (Schwartz and Foottit 1998).

Mirids are thought to have replaced pentatomoids as the major heteropteran pests of European cereals, owing to insecticide applications and changes in agricultural practices (Bilewicz-Pawińska 1982). In regions of the former USSR where *Eurygaster* species (Scutelleridae) do not cause severe damage, lygus and other plant bugs adversely affect the production of wheat, barley, millet, and oats (Putshkov 1975). Shek and Evdokimov (1981) noted increases in populations of *Trigonotylus caelestialium* in the wheat fields of Kazakhstan, and Krăsteva and Apostolov (1990) recorded *L. rugulipennis* and *T. caelestialium* among the more abundant heteropteran species collected in the wheat fields of Bulgaria, and discussed their seasonal occurrence in relation to crop development. Because the stylets of both *Lygus* and *Trigonotylus* species are not heavily chitinized, these bugs are not as well adapted to feeding on later developmental stages of wheat as are *Eurygaster* species (Vilkova 1976).

Lygus rugulipennis, the most abundant of several lygus bug species occurring in wheat fields in Finland, occasionally reaches outbreak levels on this crop (Vappula 1965, Varis 1995). Unusually large populations can occur in wheat and rye fields in Russia (Shapiro 1956). By labeling bugs with C^{14}, Nuorteva and Reinius (1953) determined that *L. rugulipennis* transfers oral secretions from the sites of penetration into tissues surrounding the wheat kernels. Its feeding, however, does not impair baking quality. Adults lack the proteases capable of destroying gluten in wheat dough. Nymphs possess these enzymes but generally are present on wheat before the heads emerge (Nuorteva 1954, Nuorteva and Veijola 1954). Amylase occurs in the saliva of nymphs and adults, but the bugs do not affect the yield and starch content of infested grain (Rautapää 1969).

Varis (1991) evaluated the effects of feeding by *L. rugulipennis* on developing wheat grains and on plants grown from injured grain. Two or five bugs feeding on ears from the milk stage through ripening cause the grains to shrivel; ear and grain weight are significantly reduced. Germination of injured grains is reduced 4–5%, and plants grown from such seed are stunted and produce smaller ears and grain than those grown from uninjured grain. Feeding thus does not have a phytostimulatory effect, as occurs in bean and carrot plants grown from seed damaged by lygus bugs (see Chapter 9). Even though the infestation levels of *L. rugulipennis* used in the laboratory were greater than those usually found in wheat fields, yield losses are possible in years of high bug populations (Varis 1991).

The feeding behavior of *L. rugulipennis* on wheat differs from that of *Leptopterna dolabrata*. Whereas the former inserts its stylets between the husks, the latter enters a kernel from above. When sites of penetration are mapped for several hundred kernels, differences in the distribution of visible spotting are apparent. Sites pierced by *L. rugulipennis* are concentrated on the sides; those of *Leptopterna dolabrata*, mainly at the tops (Nuorteva 1956c). Although feeding by *L. dolabrata* on wheat decreases the yield per head and per plant, it does not affect starch or gluten quality (Rautapää 1970).

In New Zealand, the adventive *Stenotus binotatus* can cause "slimy gluten" or "sticky dough" of wheat during fermentation (Morrison 1938, Swallow and Cressey 1987). Dough made from wheat fed on by *S. binotatus* and certain lygaeoids behaves abnormally (Cottier 1956). Meredith (1970) offered a more graphic description of the problem: The bug's salivary secretions break down the dough structure rapidly so that "in the bakery, usually during the night, the dough pieces suddenly turn runny and sticky while passing through the machines." This creates a "sticky mess . . . quite unusable in the baking trade and . . . embarrassing even as a waste product." Damaged wheat kernels exhibit collapsed sides, usually at the brush ends, and the cavities that appear in the endosperm reduce the weight of the kernels by an average 30% (Every et al. 1992).

Adult mirines sometimes invade wheat fields from their reproductive or breeding hosts. *Closterotomus* species in Italy will disperse from peach trees to feed on the flowers of wheat and other plants (Pegazzano 1958). In addition, *C. norwegicus* adults sometimes feed on developing seeds in the heads of barley and rye (Lupo 1946). While not considered injurious to winter wheat in England, this species is a dominant heteropteran of the cereal ecosystem, occurring in particularly high densities in weedy fields (Moreby 1996). Nymphs can also develop on wheat in Europe (e.g., Afscharpour 1960). Vagrant individuals of *C. norwegicus* occasionally are found in New Zealand wheat fields. When caged on wheat at the watery ripe stage of grain development, the bugs cause pale circular spots (1.0–1.5 mm in diameter) on the grain. Although capable of puncturing and marring wheat grains, this mirine does not reduce the baking quality (Cressey et al. 1987).

In New Zealand, *C. norwegicus* and another mirine, the crop mirid (*Sidnia kinbergi*), are capable of severely injuring starch granules of "Rangolea" wheat, but injury is confined to a small area around the site of penetration (Lorenz and Meredith 1988). Kernels typically show a pale area on the surface, the bugs' point of entry occasionally indicated by a dark spot (Every et al. 1992). Injury does not adversely affect the physiochemical properties of starch—minor effects on kernel weight are evident—or baking potential. Insect injury to "Aotea" wheat can affect baking quality, but the species involved were not identified (Lorenz and Meredith 1988). The work by Lorenz and Meredith (1988) suggested that the causal agents of impaired baking quality of wheat—lygaeids, mirids, and pentatomids—need to be reevaluated. More recent studies verified that certain species of all three families cause visible injury to kernels and a loss of weight, but only *Stenotus binotatus* and the lygaeid *Nysius huttoni* can cause proteolysis of gluten sufficient to impair the baking quality of bread. All recent (1980s) commercially important damage is attributed to the lygaeid (Every et al. 1992; see also Cressey et al. [1987]).

Several mirines, including *Adelphocoris ticinensis, A. variabilis,* and *S. rubrovittatus,* cause seed sterility or reduce the seed quality in Japanese cereal crops (Hori et al. 1985). The leaf-feeding (see Chapter 9) *Trigonotylus caelestialium* also feeds extensively on the reproductive structures of small grains. In Russia, it injures grain in a manner similar to that of the well-known scutellerid pest of wheat, *Eurygaster integriceps* (Mikhaïlova 1979, Sannikova and Garbar 1981, Bilewicz-Pawińska and Varis 1985). *Trigonotylus ruficornis,* which occurs with *T. caelestialium* in parts of the former USSR, injures the flowers, ovaries, pistils, and maturing seeds of nearly all cultivated cereals. It can significantly reduce the seed production of summer wheat, barley, and oats (Putshkov 1975).

Legumes for Forage and Seed

Lygus bugs are important pests of alfalfa and other legumes, especially when these plants are grown for seed. The smaller seeds of alfalfa and clover are more severely damaged by plant bugs than are the larger seeds of legumes such as sainfoin (Putshkov 1966). The bugs prefer meristematic tissue, particularly developing reproductive organs of their hosts (e.g., Jeppson and MacLeod 1946, Strong 1970, Varis 1972, Leigh 1976, Gupta et al. 1980). Mature fruits, as well as leaves and stems, are usually avoided (Taylor 1945, Strong 1968). Plant taxonomists once mistakenly interpreted mirid-induced injury to alfalfa—small aborted racemes on long lateral branches—as a new kind of alfalfa (Haws et al. 1990).

Alfalfa serves as a reservoir for lygus buildup, and development of resistant cultivars and use of biological control (Hedlund and Graham 1987, King et al. 1996b) would minimize movement of the bugs from alfalfa into cotton and other crops (Nielson et al. 1974, Yeargan 1985). Although differential resistance to lygus bugs has long been known (e.g., Aamodt and Carlson 1938)—consisting of tolerance among alfalfa cultivars in greenhouse tests and nonpreference in field tests—no actual resistant cultivars have been developed (Sorensen et al. 1988). Successful classical or augmentative biological control through releases of parasitoids of lygus bug eggs and nymphs is difficult to achieve because of the bugs' migratory behavior and sequential colonization of hosts (Norton et al. 1992, DeGrandi-Hoffman et al. 1994; cf. Norton and Welter 1996). The use of degree-day accumulations, allowing egg hatch to be predicted in spring, may have potential for managing lygus bug populations. For example, degree-day models might enable sampling and treatment schedules to be optimized during the prebloom of alfalfa (e.g., Simko and Kreigh 1998). Berberet and Hutchison (1994) reviewed sampling methods for these key pests of alfalfa seed production in relation to integrated pest management (IPM).

LYGUS BUGS AS ALFALFA PESTS

Seed Crops in the Western United States

Numerous workers have described lygus bug injury to alfalfa, especially by the western tarnished plant bug (*Lygus hesperus*) and the pale legume bug (*L. elisus*) to seed crops in the western United States. Growers sometimes refer to the former species as "brown lygus" and the latter as "green lygus" (Johansen and Retan 1975, Gupta et al. 1980). Injury symptoms usually include premature abscission of flower buds ("stripping") and pods when feeding takes place on young ovaries, as well as small holes or depressions in seeds and shriveled seeds (Plate 13; Sorenson 1932b, 1939; Shull et al. 1934; Carlson 1940; Stitt 1940; Jeppson and MacLeod 1946; Smith and Michelbacher 1946; Gupta et al. 1980). Lygus-damaged seeds do not germinate (Berberet and Hutchison 1994). Lygus bugs not only cause direct injury to seeds but also so extensively injure the blossoms that honey bees avoid or restrict their visits to alfalfa fields (Linsley and MacSwain 1947, Melton et al. 1971). Lygus bugs destroy only a portion of the flowers when they initiate feeding after the lower flowers of a raceme begin to bloom (Stitt 1940). Carlson (1940) further treated the effects of these bugs on alfalfa flowers, and Strong (1970) discussed the physiology of injury to alfalfa seed.

Lygus bugs caused losses to alfalfa seed crops in the United States of an estimated $16 million in 1944 (Haeussler 1952). A too-frequent and probably often unnecessary use of DDT against lygus bugs in western U.S. alfalfa fields promoted resistance to this insecticide

(Michelbacher 1954, Andres et al. 1955). The lygus bug threat to alfalfa was indirectly responsible for honey-crop failures in western Utah in 1946. Researchers recommended a dusting with DDT during the bud stage, but growers who had difficulty estimating bug populations also treated alfalfa in full bloom, which caused honey bee poisoning (Todd and McGregor 1952). The potential for loss of pollinators is now considered in recommendations for pest management on alfalfa seed crops. Some chemicals effective against lygus bugs are highly toxic to an important pollinator, the alkali bee (*Nomia melanderi*). The alfalfa leafcutting bee (*Megachile rotundata*) also needs to be safeguarded, especially the insecticide-susceptible three- to five-week-old bees (Johansen 1981). Certain organophosphorous insecticides, such as trichlorfon, show minimal effects on pollinators (Johansen and Eves 1972).

Trichlorfon, which once controlled lygus bugs in the alfalfa seed fields of the Pacific Northwest, shows diminishing effectiveness because of the bugs' development of insecticide resistance. Esterases in *L. hesperus*, specifically carboxylesterases, play an important role in this resistance. Esterase activity occurs in head, thoracic, and abdominal regions (Zhu and Brindley 1990b). A reduction in the sensitivity of acetylcholinesterase to inhibition by organophosphorous and carbamate insecticides might also be involved in insecticide resistance (Zhu and Brindley 1990a). Substitution of the β-exotoxin of *Bacillus thuringiensis*, a highly selective pesticide toxic to lygus bugs, would protect alfalfa pollinators in seed IPM programs (Tanigoshi et al. 1990).

The cutting of alfalfa hay or sugar beet grown for seed can result in a heavy migration of lygus bugs into alfalfa seed fields (Stitt 1949, Gupta et al. 1980, Schaber et al. 1990). When feeding pressure is severe, buds are blasted (Plate 13) and nearly all can drop within 2–5 days of attack, the field assuming a grayish cast (Davis et al. 1976). This injury to alfalfa is known as "white-top" (Bolton and Peck 1946) or "lygus blast" (Knowlton et al. 1951). Feeding by *L. hesperus* occurs mainly on basal flower parts (Sorenson 1932b), but also on ovaries and other succulent floral parts, which results in bare rachises or "stripped racemes" (Plate 13; Gupta et al. 1980). In Arizona and southern California, *L. hesperus* probably causes the greatest damage to seed crops, *L. elisus* the least, with *L. lineolaris* intermediate in its potential for limiting alfalfa seed production (Stitt 1944). During 1939–1940, the impact of *L. hesperus* on seed production in Arizona was so severe that a cooperative cultural control program was established. As a management strategy, all seed growers in a district of Yuma County cleaned up alfalfa and weeds to eliminate reservoirs of the bugs, planted seed crops within a designated period, and followed prescribed cropping practices (Stitt 1941). Another management technique tested in alfalfa seed fields involved uncut trap strips

attractive to lygus bugs, which could then be sprayed without having to treat the entire field with insecticide (Scholl and Medler 1947).

Population buildup of *L. hesperus* in Arizona increases at a relatively uniform rate in spring and decreases at a similar rate in late summer and fall; a similar amplitude occurs between years (Butler and Wardecker 1970). Lygus injury to alfalfa in Alberta is low during drought conditions or when rainfall is excessive; seed loss is highest during periods of normal rainfall (Butts 1984).

The proportion of *L. elisus* to *L. hesperus* occurring in alfalfa in Washington is usually equal in spring, but the harvesting of alfalfa hay and mint induces a late-summer or fall migration into seed fields, resulting in 75–100% of the population being *L. hesperus* (Johansen and Retan 1975). *Lygus hesperus* shows such a strong preference for alfalfa that volunteer plants near seed fields should be destroyed (Fye 1982b). The predominance of *L. hesperus* over *L. elisus* in alfalfa also involves the former's greater insecticide tolerance and the late-season preference of the latter species for alkali weed (Johansen and Retan 1975). Feeding by lygus bugs on carbohydrate-rich young pods and seeds of alfalfa might contribute to an insecticide tolerance not observed in bugs in hay fields (Bacon et al. 1964).

Abscission of alfalfa pods and the fruit of other *Lygus* hosts are likely related to the destruction of auxin-producing structures (apical meristems, ovules, and pollen grains) and the creation of hormonal imbalances (see Chapter 7). Auxin is primarily responsible for preventing premature abscission (Strong 1970, Tingey and Pillemer 1977). Early-instar lygus bugs appear incapable of penetrating floral buds to a depth sufficient to destroy auxin-producing organs, and at least on cotton, the late-stage nymphs and adults cause premature shedding of young buds (Leigh 1976). The application of plant hormones can counteract abscission induced by lygus feeding (Fisher et al. 1946, Allen 1951).

Lygus bugs thus can destroy their preferred food source—developing reproductive tissues. Strong (1968) realized the paradox posed by the elimination of preferred feeding sites and the bugs' inability to survive and reproduce on the vegetative organs of alfalfa. He postulated a selective advantage accruing to these insects when they blast floral buds. *Lygus hesperus* and *L. lineolaris* appear to have evolved a mechanism for ensuring a continuous food supply: pruning of host plants such as alfalfa results in the production of new floral buds. According to Strong (1968), "Plants fed upon by a population of lygus which causes minimal blasting (or pruning) would produce fewer total buds over the growing season, forcing such bugs to feed on suboptimal plant material, resulting in their ultimate decline." Coley (1983) and Coley et al. (1985) emphasized the capacity of plants to regrow after herbivore attack.

A too-severe pruning by lygus bugs can cause alfalfa to revert to vegetative growth, and feeding on plants that develop new fruits under the influence of pruning is only one strategy used by generalist herbivores (Scott 1983). Another strategy is to track a succession of flowering hosts and to use plants such as carrot that produce new fruiting structures throughout the growing season.

Seed Crops in Europe

Lygus bugs also damage alfalfa seed crops in Europe (e.g., Blattný et al. 1948, Šedivý 1972, Bournoville 1975, Holopainen and Varis 1991). The damage can consist of shriveled seeds as well as a decrease in seed germination and yields (Šedivý 1972).

OTHER MIRIDS OF ALFALFA SEED CROPS

In western North America, other mirines—the alfalfa plant bug (*Adelphocoris lineolatus*) (Plate 13), the rapid plant bug (*A. rapidus*), and the superb plant bug (*A. superbus*)—damage alfalfa seed crops by blasting buds and causing flower abscission, poor bloom (Plate 13), and shriveling of seeds (Sorenson 1932a, 1932b; Dean and Smith 1935; Hughes 1943; Lilly and Hobbs 1956; Smith and Franklin 1961; Beirne 1972; Manglitz and Ratcliffe 1988; Soroka and Murrell 1993). Before the alfalfa plant bug was identified as an important pest in Minnesota's alfalfa seed fields, growers during the 1930s attributed declining yields to problems with plant density, soil, weather, scarcity of bees, and grasshopper and thrips damage (Hughes 1943). Hughes determined that the alfalfa plant bug feeds mainly at the base of a flower, particularly near the ovary. If *A. lineolatus* males are deprived of alfalfa flowers by being fed vegetative structures of the host or an experimental diet of 10% glucose solution, their spermatic cells degenerate, resulting in diminished fertility. Females denied access to flowers show a delay in the morphogenesis of prefollicular tissue and suppression of oocyte development at previtellogeny (Masner 1965).

Adelphocoris lineolatus, an Old World species adventive in North America (Knight 1930, Wheeler and Henry 1992), is an alfalfa pest in Poland (Romankow 1959), Hungary (Benedek and Jázai 1968, Benedek et al. 1970), France (Bournoville 1975), and elsewhere in Europe. In Tadzhikistan and elsewhere in the former USSR, it feeds on inflorescences, including petals, calyxes, and peduncles, and causes premature petal fall. Seed crops can be reduced by 50% or more (Karpova 1945, Kiritshenko 1951). In the Czech Republic, this bug's feeding increases the number of shriveled seeds and decreases the absolute weight of seeds (Rotrekl 1973). Other *Adelphocoris* species—*A. seticornis* and *A. vandalicus*—similarly injure legume seed crops in

the former USSR (Putshkov 1975), as does *A. suturalis* in Japan (Hori and Kishino 1992).

A phyline with similar feeding habits on alfalfa in eastern North America is the adventive *Plagiognathus chrysanthemi* (Guppy 1963, Wheeler 1974a), which also injures the flowers of this crop in Europe (Blattný et al. 1948, Putshkov 1975). In western Canada, *P. medicagus* causes blasting of buds, flower abscission, and damage to developing seeds, reducing the crop yield (Beirne 1972).

During warm years, the Holarctic mirine *Polymerus vulneratus* is an important pest of alfalfa seed crops in Hungary (Benedek et al. 1970, Erdélyi et al. 1994), and in Russia the Holarctic *P. cognatus* is a significant pest of alfalfa grown for seed (e.g., Bochkareva and Vdovichenko 1974). The green mirid (*Creontiades dilutus*) feeds on the developing seeds of alfalfa in Australia, causing a yellowing, browning, and withering of the pods. Nymphs feed preferentially on the developing ovules, their stylets usually entering a flower through the sepals (Fig. 10.4); they cannot complete their devel-

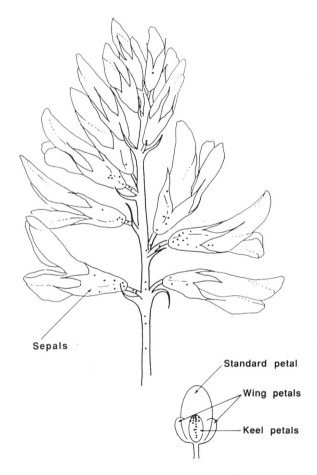

Fig. 10.4. Composite diagram showing sites (black dots) on alfalfa flowers pierced by four fifth instars of *Creontiades dilutus* during five hours' observation. (Reprinted, by permission of K. Hori and CSIRO Publishing Australia, from Hori and Miles 1993.)

opment on the vegetative parts of alfalfa (Miles and Hori 1977, Hori and Miles 1993). Injury to alfalfa seed in New Zealand is attributed to a grass-feeding Holarctic mirine, *Stenotus binotatus*, which feeds during the early bud stage and can destroy a seed crop. Its feeding on young racemes can cause alfalfa to revert to vegetative growth (Iversen and White 1959). Mirines such as the crop mirid (*Sidnia kinbergi*) and potato mirid (*Closterotomus norwegicus*) affect alfalfa and lotus seed crops in New Zealand by blasting buds, causing podless racemes, and shriveling the seeds. The crop mirid, a native species, is favored by warm, dry conditions, whereas the adventive *C. norwegicus* prefers cool, moist conditions (Macfarlane et al. 1981, Wightman and Macfarlane 1982, Wightman and Whitford 1982). The mirine *Lincolnia*

lucernina feeds on developing flower buds and is a potential alfalfa pest in New Zealand (Eyles and Carvalho 1988b). As part of integrated control strategies for legume seed pests in New Zealand, insecticides used to suppress plant bug populations are evaluated for possible toxicity to honey bees, other pollinators, and predators (Wightman and Whitford 1982).

MIRIDS OF OTHER LEGUMES

Plant bugs feed on the flowers and pods of forage legumes other than alfalfa (Table 10.8), some of which are grown for fodder and as vegetable crops. Some similar examples of mirid injury are given later in this chapter in a discussion of vegetable crops.

Table 10.8. Examples of mirids that feed on flowers and seed pods of forage legumes

Mirid	Host plant	Locality	Remarks	Reference
Mirinae				
Adelphocoris lineolatus	Sainfoin	Montana, USA	Significant negative correlation between bug densities and quality of seed produced	Morrill et al. 1984
	Lupine	Former USSR	Bugs feed on flowers	Kiritshenko 1951
Adelphocoris rapidus	Red clover	USA	Injurious but actual effects difficult to evaluate	Osborn 1939; see also Osborn 1888
Closterotomus norwegicus	White clover	New Zealand	Important pest of seed crops	Schroeder et al. 1998
Creontiades rubrinervis	Cowpea	Brazil	Bugs feed on flower buds, young pods	Singh et al. 1990
Creontiades sp.	Pigeon pea	India	Flower buds abort	Lateef and Reed 1990
Eurystylus sp.	Pigeon pea	India	Flower buds abort	Lateef and Reed 1990
Horciasinus signoreti	Cowpea	Brazil	Bugs feed on flower buds, young pods	Singh et al. 1990
Hyalopeplus similis	Black gram	East Africa	Bugs feed on flowers; cited as *H. rama*, but apparently *H. similis* (see Miller 1971)	Smee 1928a, 1928b
Lygocoris pabulinus	Lupine	Former USSR	Bugs feed on flowers	Kiritshenko 1951
Lygus borealis	Sainfoin	Montana, USA	Significant negative correlation between bug densities and quality of seed produced	Morrill et al. 1984
Lygus elisus, L. hesperus	Ladino clover, red clover	California, USA	Important pests in seed fields	Rincker and Rampton 1985
Lygus hesperus	Red clover	Oregon, USA	Bugs feed on florets and small seed pods, causing seed loss; injury difficult to detect on pollinated florets, easily confused with lack of pollination	Kamm 1987
Sidnia kinbergi	White clover	New Zealand	Feeding reduces seed yield under experimental conditions; effects on seed production in field are unknown	Pearson 1991
Phylinae				
Campylomma sp.	Pigeon pea	India	Flower buds abort	Lateef and Reed 1990
Lepidargyrus ancorifer	Red clover	Oregon, USA	Bugs may induce flower distortion, possible seed loss	Capizzi and Penrose 1978
Plagiognathus chrysanthemi, P. politus	Crownvetch	Pennsylvania, USA	Nymphs, adults feed on flowers	Wheeler 1974a

Note: See text for additional examples and discussion.

Soybean

Henry and Lattin (1987) questioned why mirids, including the polyphagous tarnished plant bug (*Lygus lineolaris*), feed so little on soybean, a major world crop. This species is sometimes numerous in the soybean fields of the midwestern United States and southern Canada (Broersma and Luckmann 1970) and will migrate into soybeans after alfalfa is cut (Poston and Pedigo 1975, Whitfield and Ellis 1976). When caged over soybean plants, tarnished plant bugs can reduce the numbers of seeds per pod and the average weight of seeds (Broersma and Luckmann 1970). Only small numbers of nymphs are usually found on soybeans in Arkansas (Freeman and Mueller 1989), although densities of more than 4,500 per hectare have been observed during late August in Mississippi (Lambert and Snodgrass 1989). Reproduction on soybean is limited to the period when flowers and small pods are available (Lambert and Snodgrass 1989), and this crop is generally not a favored host of *L. lineolaris* (Freeman and Mueller 1989, Boyd and Thomas 1994).

Several other mirids occur in U.S. soybean fields (e.g., Tugwell et al. 1973, Dietz et al. 1980), but most are probably associated with weeds within the fields or are otherwise accidental on soybean. *Adelphocoris rapidus* once developed in large numbers on soybean in Arkansas (Isely 1927), and *A. lineolatus* can migrate from alfalfa into soybean (Poston and Pedigo 1975).

Birdsfoot Trefoil

On birdsfoot trefoil in New York, *Lygus lineolaris* feeds mainly on newly formed and partially mature floret buds (Neunzig and Gyrisco 1955), causing a blasting of immature buds but little injury to older buds or blossoms. Shriveled seed results from feeding on green pods later in the season (Neunzig and Gyrisco 1955). A blasting of floral buds and blossoms of birdsfoot trefoil is also attributed to *Adelphocoris lineolatus* and *A. rapidus* in New York (Neunzig and Gyrisco 1955). MacCollom (1967) reported a negative correlation between seed yield and numbers of tarnished plant bugs and other mirids present on birdsfoot trefoil in Vermont. *Adelphocoris lineolatus* and *L. lineolaris* cause flower and pod abscission of this crop in Michigan as well as a loss of plant vigor (Copeland et al. 1984).

These species and the trefoil plant bug (*Plagiognathus chrysanthemi*) are abundant in birdsfoot trefoil fields in Wisconsin, feeding on buds and young florets. On a per-insect basis, *A. lineolatus* causes the most damage, but *P. chrysanthemi* is usually twice as numerous and might pose a relatively greater threat to seed production. During June and July, damage inflicted by *A. lineolatus* and *P. chrysanthemi* is so severe that plants do not bloom (Wipfli et al. 1989, 1990a). When weighted counts of the three mirids in birdsfoot trefoil are combined (based on a weight of 1.00 for *A. lineolatus*) to give plant bug equivalents, seed yield is significantly reduced at a level of 153 per 20 sweeps. Plant bug densities below 127 do not reduce the yield measurably, but in the absence of control measures, levels above 127 per 20 sweeps result in delayed blossoming of birdsfoot trefoil because of bud abortion (Peterson et al. 1992).

Cotton and Other Fiber Crops

A diverse plant bug fauna is associated with cotton. Because of the role of mirids as pests of this plant (Table 10.9), and cotton's importance as a world crop and its contribution to pesticidal pollution—more than 30 million hectares are grown in some 80 countries (Bottrell and Adkisson 1977)—the mirid fauna of cotton (upland cotton) is treated in greater detail than is that for other field crops. This more extensive coverage is further justified because of the controversies surrounding the pest status of lygus bugs on cotton and the effects of early-season loss of squares on yields (e.g., Bacheler et al. 1990, Gutierrez 1995, Mann et al. 1997).

Lygus and other plant bugs are usually attracted to cotton only after squares (flower buds) form. Early-instar lygus bugs feed on tender vegetative tissues, but later-stage nymphs generally use the reproductive structures when they become available (University of California 1984). Mirids feed on cotton terminals, resulting in tattered foliage, loss of apical dominance and development of abnormal growth (see Chapter 9), and flat squares. They also feed on anthers of larger squares, which can cause small bolls to be shed, and sometimes, generally in late season, they feed on developing seeds of bolls (Falcon et al. 1971, Tugwell et al. 1976, Schuster 1977, Leigh and Matthews 1994).

Table 10.9. Economic effects of mirids on U.S. cotton production, 1992

Mirid impact	West	Southwest	Midsouth	Southeast	Entire U.S.
Estimated % crop loss	1.7	0.5	1.0	0.2	1.0
Estimated no. of insecticide applications	0.5	0.5	0.8	0.1	0.6

Source: Luttrell (1994).

Lygus bugs and other mirids occasionally reduce cotton yields when populations are high throughout the growing season, and they are notable for producing various defects that cannot be measured by yield. The type of injury and its severity depend not only on the timing of plant bug infestations but also on the yield potential of the plants (Tugwell et al. 1976). In the opinion of Gutierrez (1995), misperceptions regarding yield cause the threat of lygus bugs as cotton pests to be overemphasized, which results in the needless use of insecticides. Early-season applications of insecticides for lygus bug control often kill natural enemies of other cotton pests, such as bollworms and budworms, which can then multiply later in the season (Falcon et al. 1968, 1971; Stern 1969, 1976; Adkisson 1971, 1973a; van den Bosch 1971a; van den Bosch et al. 1971; Ehler et al. 1973; Eveleens et al. 1973; Reynolds et al. 1975; Bottrell and Adkisson 1977). Control of secondary pests such as spider mites and various lepidopterans is costly, and their damage can prove greater than that which uncontrolled lygus bugs would cause (Beck et al. 1975). The excessive, prophylactic use of insecticides against cotton pests (van den Bosch 1971b) led to a tolerance in the western tarnished plant bug (*L. hesperus*) and the tarnished bug (*L. lineolaris*) that encompasses chlorinated hydrocarbons, carbamates, and organophosphates (Leigh et al. 1977, Cleveland 1985, Snodgrass and Scott 1988, Snodgrass and Elzen 1995, Snodgrass 1996, Hollingsworth et al. 1997). In some populations of *L. lineolaris* in the Midsouth, resistance to pyrethroids is sufficiently high to cause control failures in the field (Snodgrass and Stadelbacher 1994, Leonard 1995). Resistance levels of lygus bugs to various insecticides, however, can change substantially from one season to the next (Russell et al. 1997).

In at least some areas of the Cotton Belt, the importance of plant bugs as pests has increased with an emphasis on early crop maturity and short-season production. The bugs can affect production by delaying crop maturity (e.g., Isely 1927, Gilliland 1972), which can result in unfavorable conditions at harvest and decreased harvest efficiency (Scott et al. 1986, Luttrell 1994). Delayed maturity also results in cotton's having to be protected longer from late-season attacks by bollworms (Noctuidae), which increases production costs (Schuster 1977, Lidell et al. 1986).

The composition of the mirid complex of cotton and the importance of plant bugs as pests vary with the management and production practices in different regions, proximity to other crops (e.g., alfalfa and safflower) and to weeds harboring large numbers of bugs that can migrate into cotton, and the time of migration relative to the presence of pests such as the boll weevil (*Anthonomus grandis*) (Adkisson 1973a, Schuster 1977, Schuster and Frazier 1977, Sterling et al. 1989b). Usually no more than three generations of multivoltine, migrant plant bugs are completed on cotton (University of California 1984).

LYGUS BUGS AS COTTON PESTS

Long regarded as pests in most cotton-growing regions of the United States, lygus bugs are sometimes called "cotton daubers" because of the yellowish daubs or globules of excrement they leave on plant surfaces (McGregor 1927, Faulkner 1952). Their piercing of squares and bolls frequently leads to abscission (McGregor 1927). Heavy, season-long infestations of *Lygus lineolaris*, the tarnished plant bug, can reduce yields in Mississippi as much as one bale per acre (Scales and Furr 1968). Heteropterans (mainly pentatomids and mirids) were once responsible for an average loss of $5 per bale in Arizona (Cassidy and Barber 1938). At times, lygus bugs in the western states were as detrimental to cotton production as the boll weevil was in the Midsouth (Newsom and Brazzel 1968). Some researchers consider *L. hesperus* a key pest of cotton in California—annual losses, including control costs, amount to $20–30 million (Stern 1969, 1973; cf. Gutierrez 1995). The lygus bug problem intensified in the late 1970s and early 1980s, perhaps because of changed production practices: decreased early-season use of pesticides, allowing the bugs to build up; later planting that results in delayed fruiting; herbicide use; excessively thick stands; and excessive nitrogen fertilization, which increases the crop's vulnerability to injury later in the season (Gilliland 1981). As will be discussed later, the pest status of the tarnished plant bug is particularly controversial, and its threat to cotton in the Midsouth and Southeast might be exaggerated (Bacheler et al. 1990). But increases in populations of the tarnished plant bug and other mirids are anticipated with the more widespread adoption of genetically modified or transgenic cottons expressing α-endotoxins by altered genes from *Bacillus thuringiensis* var. *kurstaki*—so-called Bt cotton. The use of Bt cotton will reduce the number of mid- to late-season insecticide applications against bollworms and budworms, which inadvertently provide control of the plant bugs. An anticipated eradication of the boll weevil from the South, which will further reduce insecticide use, might also lead to increased problems with the tarnished plant bug (Layton 1996, Hardee and Bryan 1997, Mensah and Khan 1997, Gantz 1998).

SOURCES AND DISPERSAL OF *LYGUS* IN COTTON

As inhabitants of ephemeral, unpredictable habitats, lygus bugs show a strong tendency to disperse (e.g., Cleveland 1982, Fleischer and Gaylor 1987, Stewart and Gaylor 1991, Raulston et al. 1996). Crops such as alfalfa, grain sorghum, and sugar beet, especially when in bloom, are sources of plant bug infestations in cotton. Migration into cotton is affected by the proportion of land planted with crops serving as alternative hosts, their proximity to cotton and time of harvest, the density of weeds and other surrounding vegetation, and rainfall (Cassidy and Barber 1940, Stern and Mueller 1968, Sevacherian and Stern 1975, Stern 1976). Faulkner (1952) looked at the influence of desert vegetation on the migration of mirids into cotton, concluding that in New Mexico's Mesilla Valley, marginal vegetation is as important a reservoir as alfalfa for *Lygus* species and other plant bugs. Pigweed can be an important source of *Lygus* in southern Arizona during July and August (Fye 1975). Outbreaks in cotton in the San Joaquin Valley of California are correlated with winter rains and, especially, spring rains, which favor the growth of native vegetation on pasture lands (Leigh 1987). Millions of tarnished plant bugs sometimes spend the night in dense weeds bordering cotton fields before invading the crop at sunrise (Folsom 1932). *Lygus lineolaris* can reduce early boll set in central Texas, especially when decreased rainfall produces an early maturation of favored wild hosts, leading to earlier-than-normal dispersal into cotton (Anderson and Schuster 1983). Acknowledging the tarnished plant bug as only an occasional cotton pest, Frisbie et al. (1983) noted that its effects on pest-management programs can be devastating when conditions favor its dispersal.

The movement of adult *L. lineolaris* from weedy hosts into fields can be monitored with a physiological marker, rubidium, which is taken up from treated plants and retained for at least several days (Fleischer et al. 1986). Fleischer et al. (1988) studied the dispersal of tarnished plant bugs following the destruction of plots of annual fleabane used as a nursery crop in a cotton field. Some adults move through cotton during dispersal to other available hosts, whereas others feed and reproduce in cotton before emigrating, or the bugs remain in cotton in a spatial pattern resembling diffusion from fleabane plots.

In the southeastern United States, where large tarnished plant bug populations generally do not develop on other crops, area-wide management by manipulating preferred flowering wild hosts, such as annual and rough fleabane, hairy aster, and tall goldenrod, might help reduce migration into cotton fields (Fleischer and Gaylor 1987, Young and Lockley 1990). An early-season host in the Mississippi Delta is cutleaf geranium, a winter annual of roadsides and field margins. Two generations of *L. lineolaris* can develop on cutleaf geranium, but a single application of insecticide or mowing reduces its numbers, and consequently the threat of migration into cotton, when this weed dies in late May or early June (Stadelbacher 1987).

Not only weeds but also cultivated plants affect the incidence of lygus bugs in cotton. Safflower and alfalfa are examples of what Kennedy and Margolies (1985) termed the *nursery-crop phenomenon*.

In California's San Joaquin Valley, early-season problems with *Lygus* species in cotton can often be traced to a dispersal from maturing safflower (Mueller and Stern 1974). Mirids are safflower pests that bronze, blast, and distort the young buds (Carlson 1964, DePew 1967), although injury to safflower can actually be considered trivial (Gutierrez 1995). Insecticide applications to this crop, when properly timed, can significantly reduce lygus bug numbers, a practice most

effective when large plantings are treated at the same time. An organized treatment of about 28,000 hectares of safflower enabled the number of annual insecticide applications on cotton to be reduced from the usual four or five to two or three. Organized spraying during the early 1970s produced additional savings in cost per hectare by minimizing disruption to the complex of predators and parasites occurring in cotton (Mueller and Stern 1974). Lygus bug populations in safflower tended to be lower in the 1980s, perhaps because of the introduction of new varieties of this crop (University of California 1984).

Cotton is not particularly favored by *L. hesperus* and *L. elisus* (e.g., Sevacherian and Stern 1975); the former species prefers alfalfa (Cave and Gutierrez 1983, Godfrey and Leigh 1994), the most extensively grown crop in the Central Valley of California (van den Bosch and Stern 1969). *Lygus elisus* shows some preference for cruciferous plants (e.g., Getzin 1983) and reproduces little on cotton (T. F. Leigh, pers. comm.), although both lygus bug species have a wide host range. Laboratory and field studies comparing the fecundity, longevity, and survivorship of *L. hesperus* reared on alfalfa and on cotton further support this insect's preference for the former plant (Cave and Gutierrez 1983).

Lygus elisus was, however, one of the early pests of cotton in the Imperial Valley of California after this crop was introduced commercially in 1909 (Morrill 1917; McGregor 1927, 1961), and it and *L. hesperus* were first noted as cotton pests in Arizona in 1914 (Morrill 1918). With the harvesting of alfalfa, which generally must be cut every 30 days (Southier 1974), millions of lygus bugs (Stern 1969) can migrate into adjacent crops such as cotton (Fig. 10.5) when "overnight a dense and lush plant cover many acres in extent is obliterated" (van den Bosch and Stern 1969). Shortening the intervals between the cutting of alfalfa for hay is sometimes recommended in Arizona to lessen the migration of *L. hesperus* into cotton (Butler et al. 1971). Monitoring lygus populations in alfalfa gives an idea of the numbers of bugs that can be expected to invade cotton if measures are not taken to limit their dispersal (Fye 1975).

To minimize ecological change and stabilize the alfalfa arthropod community, strip cutting (or harvesting half of a field where open irrigation ditches make strip cutting difficult) and border harvesting can be practiced to protect cotton, as lygus bugs remain in the uncut portions of the alfalfa fields (Stern and Mueller 1968; van den Bosch and Stern 1969; Stern 1973, 1976; Summers 1976; Flint and Roberts 1988). Adults, but not nymphs, move into half-grown alfalfa borders when mature alfalfa is cut (Rakickas and Watson 1974). Alfalfa also can be interplanted in cotton to serve as a trap crop, attracting the bugs when safflower dries up or alfalfa is cut (Stern 1969, 1976; Sevacherian and Stern 1974, 1975). Interplanted alfalfa must be cut periodically to prevent lygus bugs from attaining excessively high densities; a frequent cutting schedule, every 14 or

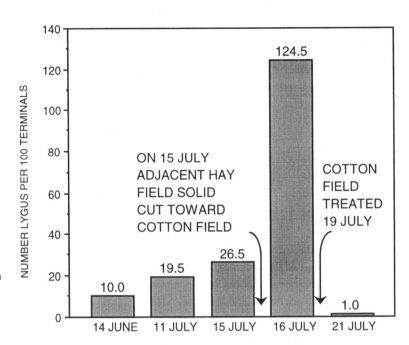

Fig. 10.5. Movement of lygus bugs from alfalfa into cotton when an entire alfalfa field is cut. (Redrawn, by permission of V. M. Stern and CRC Press, from Stern 1981, *CRC Handbook of Pest Management in Agriculture*, Vol. I. CRC Press, Boca Raton, Fla. ©1981.)

28 days, hinders growers from adopting this management technique (Godfrey and Leigh 1994). Studies on safflower and alfalfa in California provide good examples of modern tactics of insect control. Knipling (1979) and Stern (1981) summarized the successes realized from this intensive field research.

NATURE OF *LYGUS* DAMAGE TO COTTON

Feeding by lygus bugs in plant terminals and on "pinhead" cotton squares (about 2 mm long) often leads to shedding of the buds (Plate 13), but the extent of injury and effect on yield are difficult to estimate; plant pathogens and various physiological stresses induce abscission that also affects fruit set (Gilliland 1972, Guinn 1982, Mauney and Henneberry 1984, Johnson et al. 1996). After early to peak bloom, most square abscission might be due to nutritional and carbon stress that results from a heavy fruit load (Leigh et al. 1988). Other insects also affect squares, and in some Arizona plantings the early-season loss of squares can be the result of thrips rather than lygus injury (Terry and Barstow 1988). Lygus bugs should be considered "only *one* cause of the loss of early fruit set"; yields "are determined by how much fruit is retained, *not* how much is lost" (Oakman 1981). Williams et al. (1987) described a field technique allowing growers and scouts to determine whether freshly shed pinhead squares are the result of tarnished plant bug or bollworm feeding or physiological stress. The effects of the bugs are most critical in California's San Joaquin Valley when excessive loss of squares has its maximum impact on yield: the beginning of the third week of squaring to the end of the sixth week (University of California 1984).

Buds of at least 8 mm are too large for *L. hesperus* to penetrate deeply enough to injure auxin-releasing sites (Strong and Villanueva-Barradas 1968; see also Chapter 7). Studies on lygus bugs by Pack and Tugwell (1976), Tugwell et al. (1976), and Mauney and Henneberry (1979) showed that developing pollen sacs are the principal feeding sites on squares 3 mm in diameter or larger. Damaged squares appear normal, but when the calyx and corolla are removed, necrotic anthers are evident (Plate 13); anthers darken within 24 hours of feeding. Williams and Tugwell (2000) provided a histological description of the feeding by *L. lineolaris* on small squares, and the degree of injury to anthers can be used to estimate feeding intensity by *L. lineolaris* (Maredia et al. 1993). Pollen sacs often appear shrunken and can lose their turgidity from the emptying of contents. Bracts, calyx, and corolla usually show little injury when anthers are pierced. Injured squares can be shed within 1–4 days after attack, though the majority remain on the plant. Shriveled squares are often trapped in young foliage, a scar remaining at the point of abscission (University of California 1984).

Lygus bugs can also be found within the blooms—nearly 300 adult *L. hesperus* were recorded during a study of insect visitors to cotton flowers in Arizona (Moffett et al. 1976). *Lygus* species sometimes cause warty growths on flower petals and brown spots on pistils and stamens (Telford 1957).

The most detailed work on *Lygus* feeding behavior has been conducted on nymphs because adults, with their flighty habits, are more difficult to study. Adults are used mainly in the greenhouse or in cages. In greenhouse studies the distribution of plant bugs on cotton probably is biased because of unnatural plant-growth patterns. Field studies show that the majority of adults occur on bolls, whereas in the laboratory there is a bias toward the terminals, with most adults found on leaves. Boll feeding might occur more often in the field than once thought (Wilson et al. 1984; see also Snodgrass [1998]).

Symptoms of feeding on young bolls appear as dull, dark spots 1–2 mm in diameter (Plate 13), often with a shiny black spot at the center. These spots might result from a toxic effect of excrement (University of California 1984); dark, necrotic spots are not observed where excrement is left on bracts, calyx cups, or foliage. Feeding injury is revealed by opening the bolls, which often show brown necrosis and callouslike growths on the inner boll wall (T. F. Leigh, pers. comm.). Lygus feeding does not always produce internal damage in young bolls, but nearly mature bolls sometimes show a staining of lint, which lowers the grade. Bolls remain vulnerable to lygus feeding until their walls are too tough to be penetrated—usually about 10 days after flowering (University of California 1984). Some of the injury to bolls that is attributed to stink bugs (Pentatomidae) might be due to feeding by *L. lineolaris*. Tarnished plant bug adults caged on cotton produce injury to young bolls indistinguishable from that caused by late instars of the southern green stink bug (*Nezara viridula*) (Greene et al. 1999).

Bolls injured by mirids and other heteropterans are harder to pick. Lint sticks to proliferations within damaged bolls, preventing them from opening fully (Plate 13), and several "pulls" might be required to pick them. The work of migrant pickers once was so hindered in Arizona, and their pay thus reduced, that widespread dissatisfaction led them to move from camp to camp in search of "clean" fields (Cassidy and Barber 1940).

Feeding by *L. lineolaris* can induce boll rot by inoculating bolls with the fungi (species of *Alternaria* and *Fusarium*) responsible for this disease (Bagga and Laster 1968). *Lygus hesperus* can also play a role in disseminating inoculum of the fungus *Aspergillus flavus*, which incites boll rot (Stephenson and Russell 1974). This nonaggressive fungus generally invades insect-damaged plant tissues and produces the metabolite aflatoxin that contaminates cotton and other agricultural products (Widstrom 1979, Lee et al. 1987; see also Chapter 8).

EFFECTS OF MANAGEMENT AND PRODUCTION PRACTICES ON *LYGUS*

Because water availability and fertility affect the vegetative-fruiting balance of cotton plants (Leigh et al. 1969), it follows that lygus bugs, as specialists on reproductive tissues, would also be affected. In Arizona and California, frequently irrigated cotton, and plants receiving greater nitrogen fertilization, attract larger numbers of *L. hesperus* than do fields receiving less water and lower nitrogen levels (Leigh et al. 1970, 1974; Flint et al. 1996). A similar positive relationship between irrigation and mirid densities holds for the cotton fleahopper (*Pseudatomoscelis seriata*) in Oklahoma (Brett et al. 1946).

Higher plant densities (narrow-row plantings) also favor large numbers of *L. hesperus*, and factorial separation of both variables—water availability and plant density—suggests that the effects of spacing override those of irrigation schedules. Dense plant canopies associated with regular irrigation, high nitrogen, and high crop densities tend to produce more favorable insect habitats (Grimes [1986?]).

The use of drip rather than furrow irrigation in southwestern cotton fields also tends to promote lygus injury. Drip-irrigated cotton can lead to earlier establishment of the bugs and development of larger populations throughout the season (Zwick and Huber 1983). Guinn and Eidenbock (1982) tested the hypothesis that water deficits increase catechin and tannin levels in cotton, which would contribute to insect resistance. Their study produced no evidence for increased concentrations, but partial defruiting resulted in increases of catechin and tannin in bolls and leaves. Fruit loss might promote certain biochemical changes in the host that convey antibiotic activity (Guinn and Eidenbock 1982).

The use of certain herbicides on cotton can result in increased numbers of lygus bugs (e.g., Stam et al. 1978). Certain conservation tillage systems, such as the use of rye as a winter cover crop, seem not to increase damage by lygus bugs or to affect cotton yields (Gaylor et al. 1984). In general, more research is needed to evaluate yields in relation to the complex interactions among herbicide application, nitrogen fertilization, planting dates, and other production practices (Gaylor et al. 1983).

SAMPLING *LYGUS* AND ECONOMIC THRESHOLDS

Of all cotton pests in the United States, lygus bugs (and the cotton fleahopper) are labeled the most controversial. Workers not only disagree on how much plant bug feeding affects crop yield but also on the significance of damage thresholds and the accuracy of sampling techniques (Lincoln 1974; Parvin et al. 1989; Pendergrass 1989; Sterling et al. 1989b; Bacheler et al. 1990; Baker et al. 1993; Snodgrass 1993; Gutierrez 1995, 1999). The most accurate sampling methods are often too time-consuming and expensive to be used by growers and crop consultants (e.g., Young and Tugwell 1975, Snodgrass and Scott 1997). Clarification of damage to cotton by lygus bugs depends on an accurate determination of nymphal and adult feeding rates (including early vs. late instars and adult males vs. females) and whether the amount of ingestion tends to be more critical than the number of feeding punctures (Stewart et al. 1992). Currently used economic thresholds for cotton tend to be static and nominal, having been based mainly on the field experiences of research and extension personnel.

Cotton is produced in the United States in the irrigated deserts of the West, the semiarid Southwest, and the humid Midsouth and Southeast, regions characterized by distinct agroecosystems (Adkisson 1973b, Suguiyama and Osteen 1996). Within the Midsouth and Southeast, differences in rainfall patterns result in harvest delays having more economic impact on the Midsouth because of greater late-season (October) rainfall at the time of harvest (Mann et al. 1997). Consequently, realistic thresholds for lygus bugs can be expected to vary according to the different lengths of the growing season, rainfall patterns, planting times, and plant densities. Control recommendations should consider (1) plant bug densities, (2) predator populations, (3) fruiting rate, and (4) budworm and bollworm populations (Oakman 1981). Cotton's ability to compensate for herbivory (e.g., Rosenheim et al. 1997) should also be considered in any assessment of lygus bugs and other mirids as cotton pests. Parvin and Smith (1996) analyzed the complex interactions between cotton phenology and management of insect pests.

The research on interfield and intrafield movement of lygus bugs, which I summarized earlier, and the determination of economic thresholds in California led to a revision of the generally accepted level of 10 bugs per 50 sweeps (1 nymph = 2 adults) that had been established in the early 1950s (Falcon et al. 1971; Stern 1973, 1976). Control measures began to be recommended only when this level was sustained in consecutive samples taken at four- to five-day intervals during prebloom to two weeks following onset of flowering. After the early fruiting period, a larger population (20–30 bugs/50 sweeps) could be tolerated without significant effects on yield or quality (Smith and Falcon 1973, Stern 1973). Sevacherian and Stern (1972b) and Sevacherian (1976) gave the details of sequential sampling plans for Californian cotton fields, in which the clumped distribution of lygus bugs is considered (see Chapter 6). Byerly et al. (1978) compared three methods of sampling *L. hesperus*—use of the sweepnet, suction machine (D-vac), and whole-plant-bag samples; all methods underestimated densities and gave similar population trends but were generally inadequate for assessing bug populations (Gutierrez 1995). The sweepnet similarly underestimates the numbers of *L. lineolaris* adults compared with visual sampling of cotton, but multiplying by three the mean number of adults collected by sweeping can be used as a correction factor (Snodgrass and Scott 1997).

Cotton exhibits indeterminate growth, and more than 80% of its fruit is shed as squares and small bolls regardless of the presence of insects; plants, therefore, are able to compensate for some insect-induced abscission. Because plant bugs typically feed on only young fruits, the cotton plant expends little time and energy to replace them (Falcon et al. 1968; Gutierrez et al. 1977; Gaylor et al. 1983; Stewart and Sterling 1989a, 1989b; Gutierrez 1995). The pruning of excess

bolls can result in larger bolls and even increased yields (van den Bosch 1978). To determine the loss of fruit caused by feeding by *L. hesperus*, Gutierrez et al. (1977) developed a simulation model for the phenology and population dynamics of this plant bug. *Lygus hesperus* apparently is a cotton pest only when it is extremely abundant or when plants are under physiological stress. An optimization study on the impact of this mirid indicated that fluctuations in cotton yield can be explained by weather patterns and that this mirid is not generally a serious pest (Gutierrez et al. 1979; cf. Baker et al. 1993). Considering a threshold of 10 bugs per 50 sweeps "grossly obsolete," especially the counting of nymphs twice, Gutierrez et al. (1977) suggested that the threshold be raised (see also Wilson [1986?], Stewart et al. [1992]). The simulation model of Gutierrez et al. (1977) gave a lower feeding rate than the bioassay of Mauney and Henneberry (1979), who determined that each bug damaged a mean of 3.9 squares per day at 30°C (or 2.83 if adjustment is not made for nymphs that die). Mauney and Henneberry's highest rate of injury (8 squares/insect/day) was about nine times that calculated by Gutierrez et al. (1977).

Using field observations, Mauney and Henneberry (1984) further calculated the rate of feeding by lygus bugs (mostly *L. hesperus*); the results proved intermediate between those of Gutierrez et al. (1977) and their own previous work, which had been based on laboratory and field-caging studies (Mauney and Henneberry 1979). Noting the difficulty of assessing the various factors responsible for square abscission, Mauney and Henneberry (1984) stated that "the regression equation relating plant bug numbers to square damage is not yet precise." Leigh et al. (1988) calculated still different rates of square abscission by *L. hesperus*, stressing that because their methods differed greatly from those used in other studies, similar results should not have been expected.

Extrapolating laboratory or greenhouse results to the field can bias the data on the distribution of lygus bugs on the cotton plant and possibly give an inaccurate picture of their feeding behavior (Wilson et al. 1984). In contrast to the microsimulation model of Gutierrez et al. (1977), Mangel et al. (1985) presented an analytical model of *L. hesperus*–cotton interactions involving this mirid's effect on square and boll development; losses in the field could be forecast from data on the numbers of squares per area and on the numbers of bugs per 50 sweeps. The model might allow growers to predict yields and to determine optimal strategies for spraying.

The monitoring of lygus bugs in California's San Joaquin Valley emphasizes control thresholds that are not fixed, relying instead on a field's fruit load. The need for control measures using IPM depends on a *Lygus*-square ratio (Frisbie et al. 1989). Sterling et al. (1989b) summarized the economic thresholds for *Lygus* species in cotton and other economic considerations relating to their injury. The need for chemical control of lygus bugs in the San Joaquin Valley is no longer considered routine (Frisbie et al. 1989). Goodell et al. (1990) described how California's integrated expert system for cotton production and management—CALEX/cotton—helps growers decide if chemical control of lygus bugs is needed.

Because it is often the first insect to require control measures during the growing season, the tarnished plant bug is considered a key pest of cotton in Mississippi (Craig et al. 1997). *Lygus lineolaris*, however, can be considered only a potential cotton pest in North Carolina. According to Bacheler et al. (1990), automatically applying insecticides because of the bug's presence—that is, preventive spraying—is unjustified (see also Gilliland [1972]). This practice can disrupt natural enemies, leading to mid- and late-season outbreaks of other pests and possibly inducing the development of insecticide-resistant pest strains. Published information on the tarnished plant bug as a cotton pest in the Midsouth and Southeast tends to overestimate its importance, and the economic thresholds established in some states might be too conservative. In addition, it appears that late-harvest penalties are exaggerated, yield loss estimates are questionable, and research on the effect of preventive treatments is lacking (Bacheler et al. 1990; cf. Baker et al. 1993).

DEVELOPMENT OF RESISTANCE TO *LYGUS* AND OTHER PLANT BUGS

The breeding of cotton resistant to lygus bugs and other pests has received considerable attention (Painter 1951, Niles 1980, Benedict [1986?], Milam et al. 1989, Watson 1989, Gannaway 1994, Jenkins and Wilson 1996), but the search for resistance to plant bugs might deserve even greater emphasis (Frisbie et al. 1992). Efforts to develop strains less susceptible to *Taylorilygus vosseleri* in East Africa emphasized cotton's physical characteristics, although plant chemistry, particularly feeding deterrents (see Chapter 7), was suggested to be a likely source of resistance (Reed 1974). The type and density of trichomes affect cotton's susceptibility to lygus and other plant bugs (e.g., Wilson and George 1986). Here, only three morphological characters in cotton—nectarilessness, frego bract, and glandlessness—are discussed. Additional references and discussion of other characteristics evaluated for their effects on cotton pests are provided by Schuster and Frazier (1977), Tingey and Pillemer (1977), Niles (1980), Tugwell (1983) Milam et al. (1989), El-Zik and Thaxton (1989), Smith (1992), Gannaway (1994), and Jenkins and Wilson (1996).

Alvarado-Rodriguez et al. (1987) discussed the use of life-table data—fecundity, longevity, and survivorship to adulthood—to evaluate host-plant resistance to *L. hesperus*. In evaluating the effectiveness of the characters associated with resistance to the cotton fleahopper

(glabrous, pilose, and nectariless traits), Lidell et al. (1986) stressed the value of looking at various genetic backgrounds and using different infestation levels.

Nectarilessness

Nishida (1958) discussed the use of extrafloral secretions by insects, remarking that extrafloral glands are well known to botanists but generally little appreciated by entomologists. But it was cotton entomologists who took the lead in assessing the agricultural significance of extrafloral nectars to insects (Rogers 1985).

Trelease (1879) realized that certain insects feed on the secretions produced by the cotton plant's foliar nectaries (Plate 14), which are usually present on the lower surface of each leaf and can be found below, between, and inside the bracts (Tyler 1908, Meyer and Meyer 1961). These have been termed *leaf, circumbracteal,* and *subbracteal nectaries* (Butler et al. 1972), but all such nectaries on vegetative structures, as well as the floral nectaries not involved in pollination, can be lumped as *extrafloral* (Baker et al. 1978, Elias 1983, Rogers 1985). Nectar is secreted in varying amounts during a 24-hour period (Butler et al. 1972). Whitefly and leafhopper infestations can result in increased secretory activity by cotton, but thrips-infested plants do not produce more nectar, possibly because thrips' feeding habits differ from those of the homopterans (Mound 1962). The possible effects of mirid feeding on nectar production in cotton apparently have not been evaluated.

Clark and Lukefahr (1956) identified several carbohydrates in the extrafloral nectar of cotton and suggested that a nectariless (sometimes termed *nectarless*, e.g., Maxwell [1977]) plant might suppress populations of certain cotton pests. Cotton nectar consists of various amino acids, carbohydrates, and lipids (Stone et al. 1985). Although 24 amino acids are present, the four considered essential for insect growth and reproduction are missing (Hanny and Elmore 1974, Hagen 1986). Feeding at extrafloral nectaries is similar to the use of floral nectar as a food source (see Chapter 11), though the amino acid complements of the two types of nectar differ, even on the same plant (Baker and Baker 1986).

Some plant bugs feed on extrafloral nectar (see Chapters 9 and 13), which is an important, though sporadically available (see Chapter 11) food source for *Lygus* nymphs and adults (Plate 14; Latson et al. 1977, Parencia 1978, Benedict et al. 1981, Bailey et al. 1984, Wilson et al. 1984; cf. Ferreira 1979). After Meyer and Meyer (1961) elucidated the inheritance of nectarilessness—this trait is controlled by two pairs of recessive genes transferred to upland cotton from Hawaiian cotton, which lacks extrafloral nectaries (Meyer and Meyer 1961, Rhyne 1966, Niles 1980)—the breeding of nectariless cotton could be pursued. Initially these lines were agronomically inferior, but those with yields and fiber quality nearly equal to nectaried strains were soon developed (Meredith et al. 1973, Meredith 1980).

Nectariless cotton tends to reduce populations of *L. lineolaris* (e.g., Meredith et al. 1973, Schuster and Maxwell 1974, Maxwell et al. 1976, Schuster et al. 1976b, Henneberry et al. 1977, Maxwell 1977, Scott et al. 1986, Gannaway 1994), the trait imparting resistance through antibiosis and nonpreference owing to a nutritional deficiency (Schuster and Frazier 1977, Bailey et al. 1984). Nectariless cotton more consistently reduces the numbers of adults than nymphs of the tarnished plant bug in Mississippi (Scott et al. 1988). Nectarilessness also conveys an early crop maturity, probably because early-season feeding by the bug is lessened (Meredith 1976). Although the presence of extrafloral nectaries can increase the survival rate of lygus nymphs (Butler et al. 1972), nectariless plants evaluated initially in California did not show substantial resistance to *L. hesperus* (Tingey et al. 1975a; see also Flint et al. [1992]). Later work indicated that the nectariless trait, in some genetic backgrounds, does reduce the survival rate of *L. hesperus* and that plants also possess antibiosis to this pest (Benedict et al. 1981, 1982). Assessments of nectariless cottons, however, do not consider that some of the apparent "resistance" might involve reduced numbers of prey, such as neonate lepidopteran larvae, available to this omnivorous plant bug (Naranjo and Gibson 1996).

Cotton lacking extrafloral nectaries yielded 68 more pounds of lint per acre than did nectaried isolines in one analysis in the Midsouth (Jenkins 1986). Research on the effects of large acreages of nectariless varieties on lygus bugs is needed, particularly the evaluation of plant performance under current production practices (Scott et al. 1986). Increased field size can diminish the effectiveness of nectariless cotton (Scott et al. 1988).

Because some predacious and parasitic insects feed on extrafloral nectar, beneficial arthropods might be adversely affected by the unavailability of extrafloral nectaries (Schuster et al. 1976a, Yokoyama 1978, Agnew et al. 1982, Rogers 1985, Treacy et al. 1987, Scott et al. 1988). In a four-year study in Queensland, Australia, beneficial arthropods were significantly reduced in nectariless cotton. Numbers of a predatory mirid, the brown smudge bug (*Deraeocoris signatus*), were reduced 51% compared to 28% for the pest *Creontiades dilutus* (=*Megacoelum modestum*) and 37% for *Campylomma liebknechti* (Adjei-Maafo and Wilson 1983). Burton (1978), however, regarded the advantage of *Lygus* suppression alone as sufficient reason to consider this trait for inclusion in cotton pest-management programs in the United States. Decreased insecticide use might more than offset any reduction in predators resulting from the use of nectariless cultivars (Scott et al. 1988). An alternative hypothesis is that the presence of ants on nectaried cultivars offsets the perceived advantages of nectariless cotton (Hagen 1986). In this "protectionist" school of thought (Rogers 1985, South-wood 1986, Ganeshaiah and Veena 1988, Koptur 1989), the role of extrafloral nectaries is to attract ants, which feed on the nectar and help protect plants from herbivory.

Frego Bract

Frego bract or "rolled bract" cottons are characterized as "an Upland mutant inherited as a monofactorial recessive" trait (Niles 1980). This trait produces a more exposed boll, imparts resistance to the boll weevil and boll rot, and enhances insecticide coverage. It also increases susceptibility to *L. hesperus*, *L. lineolaris*, and sometimes other plant bugs such as the clouded plant bug (*Neurocolpus nubilus*) and cotton fleahopper (*Pseudatomoscelis seriata*) (e.g., Lincoln et al. 1971, Leigh et al. 1972, Newsom 1974, Scales and Hacskaylo 1974, Schuster et al. 1976a, Maxwell 1977, Maredia et al. 1993). Frego bract's vulnerability to attack is associated with sensory input from the bugs' receptors; a preference for frego disappears when the rostral tip and associated antennal sensilla are removed (Avé et al. 1978). Yet the attractiveness of frego appears not to depend on significant qualitative differences in its sugars (Holder et al. 1975). Field tests in Arizona when *Lygus* populations were low showed that frego bract strains are no more susceptible to the bugs than normal cotton (Wilson and Wilson 1978). Tingey et al. (1975a) reported greater nymphal survival of *L. hesperus* on frego cotton than on a commercial cultivar used as a control. Frego strains showing resistance to tarnished plant bugs have been developed in Mississippi (Milam et al. 1985), although no cultivars released for production in the United States have frego bracts (Smith 1992).

Glandlessness

Dark pigment glands occurring as spots on all aerial parts of cotton contain various terpenoid aldehydes, principally gossypol. This polyphenolic yellow pigment is toxic to monogastric (nonruminant) animals, and poultry and swine are especially susceptible (Bottger et al. 1964, Tingey et al. 1975b, Abou-Donia 1976).

Cottonseed, a by-product from the removal of fiber from the seed coat, is a rich source of edible vegetable oil and protein meal. Terpenoids, however, must be removed from cottonseed oil through refining, and the meal remaining after oil extraction can be fed to monogastric animals only if gossypol and related terpenoids have been inactivated. The special procedures required are expensive and reduce the protein content of the meal (Tingey et al. 1975b, Bell and Stipanovic 1977, Kromer 1978, Leigh et al. 1985). Discovery of glandless cotton—at first a glandless-stem type useful only as a genetic marker (McMichael 1954) but eventually genotypes having glandless seeds (McMichael 1960)—stimulated interest in breeding for this trait and suggested an increased value and market potential for cottonseed.

Glandless cotton, however, is susceptible to certain insect pests (Almeida 1980), including lygus bugs (Bottger et al. 1964). Studies on *L. hesperus* showed that growth and nymphal survival rates are higher on glandless compared to glanded cottons (Tingey et al. 1975b, Benedict et al. 1977). Gossypol presumably has evolved in cotton as a defensive mechanism, and its removal renders the plant more vulnerable to insect herbivory (Lukefahr et al. 1966, Maxwell 1977; cf. Fryxell [1978]).

Bottger et al. (1964) reported 90% mortality for *L. hesperus* nymphs within three days after plants were sprayed with an acetic acid solution containing 5% gossypol. In the case of *L. lineolaris*, gossypol also deters the uptake of sucrose, a feeding stimulant. Deterrent properties of gossypol and other allelochemicals apparently are due to their detection by precibarial chemosensilla internal to the food canal (see Chapter 7), which alters the normal feeding behavior (Ferreira 1979, Hatfield et al. 1982). Because *L. hesperus* avoids puncturing the scattered pigment glands, factors other than gossypol might also be involved in plant bug resistance of glanded cottons (Leigh et al. 1985).

Given the insect susceptibility of available glandless lines, Benedict et al. (1977) suggested that if these strains are grown commercially, they be used in more pest-free areas of the cotton belt—for example, the Texas High Plains (Metzer 1975, Maxwell 1977). Further research revealed genetic variability among glandless cottons and the potential for developing lines that are no more pest susceptible than isogenic glanded plants (Meredith et al. 1979, Leigh et al. 1985). Another possibility is the eventual development of genotypes with glanded leaves and flower buds but glandless seeds (Ridgway and Bailey 1978).

OTHER MIRIDS OF COTTON AND RELATED PLANTS

Whether the ecological conditions are arid or humid, plant bugs seemingly are pests wherever cotton is grown. Hargreaves (1948), Pearson (1958), and Cadou (1993) listed and discussed the numerous species known from this crop, many of which feed on squares, blossoms, or bolls (Table 10.10).

The Rapid Plant Bug (*Adelphocoris rapidus*)

Perhaps the first reference to plant bugs as cotton pests in North America was that of Glover (1856), who referred to and illustrated a sucking insect that punctured the flower buds and bolls. The bug was later said

Table 10.10. Some mirids that feed on reproductive parts of cotton

Mirid	Locality	Remarks	Reference
Bryocorinae			
Helopeltis schoutedeni	Africa	Bolls open prematurely or abscise, predisposing them to bacterial and fungal pathogens; single bug makes 80 punctures/day	Pearson 1949, Schmitz 1958, Ripper and George 1965
Mirinae			
Adelphocoris lineolatus	Uzbekistan	Fruit abscises; problem once attributed to insufficient irrigation	Demidov 1940
Adelphocoris lineolatus, A. taeniophorus	China	Reproductive parts are shed; early plantings with dense growth and fields near green manure crops are most vulnerable; treatment recommended when 2 bugs/100 plants are found or when 3% of plants show injury	Chu and Meng 1958, Chiang 1977
Apolygus lucorum	China	See *Adelphocoris* spp. above	Chu and Meng 1958, Chiang 1977
Horcias nobilellus	Brazil	Excessive shedding of squares and young bolls is induced; crop losses intensified by delayed planting; still a major pest but data on crop losses are unavailable	Hambleton 1938, Sauer 1942, Ramalho 1994
Phytocoris obscuratus	Turkey	Implicated in injury to cotton	Nizamlioglu 1962
Phylinae			
Campylomma diversicorne, C. verbasci (as *C. nicolasi*)	Turkey	Considered pests in western Turkey	Karman and Akşit 1961
Campylomma liebknechti	Australia	May cause square abscission, though cotton is an inadequate host; pest status is uncertain	Bishop 1980, Chinajariyawong and Walter 1990
Campylomma plantarum, C. verbasci (as *C. nicolasi*)	Africa	May injure reproductive parts	Pearson 1958, Odhiambo 1959, Deeming 1981

Note: See text for additional examples and discussion.

Fig. 10.6. Cotton boll showing feeding punctures of *Adelphocoris rapidus*. (From Sanderson 1906.)

to be *Adelphocoris rapidus* (Glover 1878), a mirine now called the rapid plant bug. This species often feeds between the involucres and bolls, leaving a small black dot at the point of puncture, a mark often attributed to the bollworm (*Helicoverpa zea*). Injury nearly always causes the boll to flare and drop, or stains the tuft of cotton in that section of the boll (Mally 1893). Ashmead (1895) stated that rapid plant bug nymphs and adults feed on cotton blossoms, petals, and the corolla. Sanderson (1906) noted that the black spots appearing around each feeding puncture (Fig. 10.6) resemble the early stages of anthracnose on bolls. *Adelphocoris rapidus*, referred to as the "cotton leaf-bug" in early literature (Sanderson 1906, Morrill 1910), is no longer considered an important pest of cotton. It causes problems only when local conditions favor its dispersal into cotton fields (Brown and Ware 1958, Schuster 1977).

The Cotton Fleahopper (*Pseudatomoscelis seriata*)

Discussed in Chapter 9 for the stem lesions and abnormal growth it produces on cotton, the cotton fleahopper consistently causes serious injury to squares, especially primordial ones (Walker and Niles 1984). Severe outbreaks of this key pest have occurred, such as one nearly throughout the cotton belt in 1926 (Coad 1929, King 1929) and in the Lower Rio Grande Valley of Texas in 1970 (Adkisson 1973a). Early-season treatments of *P. seriata* can trigger outbreaks of the tobacco budworm in cotton (Adkisson 1971; van den Bosch 1971b, 1978). In a study of fruit abscission in Texas, the cotton fleahopper was the most abundant cotton pest encountered (Stewart and Sterling 1989b).

Cotton, as already noted, shows an inherent ability to compensate for some removal of squares by mirids (Hamner 1941, Coaker 1957, Guinn 1982, Mauney and Henneberry 1984, Sterling et al. 1989b). Plant compensation makes it difficult to distinguish crop injury from actual economic damage (Sterling 1984). After *P. seriata* inflicts early-season fruit mortality, cotton compensates

by retaining more of the surviving fruit within a cohort, defined as fruiting sites that initiate squares at about the same time (Stewart and Sterling 1988). The reader may wish to consult Stewart and Sterling's (1989a) discussion of cotton's compensation for fruit lost to pests in relation to maximizing plant fitness and minimizing risk of excess resource loss, as well as the publication by Sterling et al. (1989b) that discusses action levels for this mirid in relation to pest management. In addition, Ring et al. (1993) calculated economic injury levels (EILs) from fleahopper density–cotton yield response functions and discussed EILs in relation to changes in market values and management costs. Sterling et al. (1996) discussed the benefits of control in relation to varying fleahopper densities.

Square abscission in cotton is related to the production of ethylene, an air-pollutant gas that is the product of aging flowers, fruits, and vegetables; it regulates natural fruit abscission and is produced in greater-than-normal amounts when the plant is stressed (Lipe and Morgan 1972, Yang and Pratt 1978, Davies 1987, Abeles et al. 1992). Abscission of cotton flower buds is also regulated by another phytohormone, indoleacetic acid (IAA). Ethylene production is enhanced by increases or decreases in IAA concentration (Burden et al. 1989). Duffey and Powell (1979) investigated the possibility that some of the cotton fleahopper's effects on cotton—square abscission and stunting—involve the production of stress ethylene. Their results suggest that the bugs inoculate host buds with microorganisms that promote ethylene synthesis (Fig. 10.7), cause plant tissue to produce ethylene, or both.

Pseudatomoscelis seriata is usually contaminated with pathogenic and nonpathogenic fungi and bacteria. *Penicillium* species are common contaminants of its salivary glands. Species of the fungal genera *Alternaria*, *Fusarium*, and *Penicillium* and the bacterial genus *Xanthomonas* are contaminants of field-collected nymphs and those hatching in the laboratory. That these microorganisms promote ethylene synthesis in cotton further supports the hypothesis that the cotton fleahopper inoculates its host with bacteria and fungi, inducing ethylene production and leading to square abscission (Grisham et al. 1987, Martin et al. 1987). Under experimental conditions favoring disease transmission, *P. seriata* can acquire microorganisms from cotton, transmit them to healthy plants, and produce symptoms typical of bacterial leaf blight in cotton (Martin et al. 1988a; see also Chapter 8).

The means by which cotton fleahoppers become contaminated with microorganisms—external contamination of eggs or contact with pathogenic or secondary species during feeding and movement over host plants—is uncertain. In addition, how the bugs transmit inoculum, whether through their salivary glands, fecal material, externally on body parts, or at feeding punctures, is unknown. Because polygalacturonase from salivary gland extracts of *P. seriata* appears to be involved

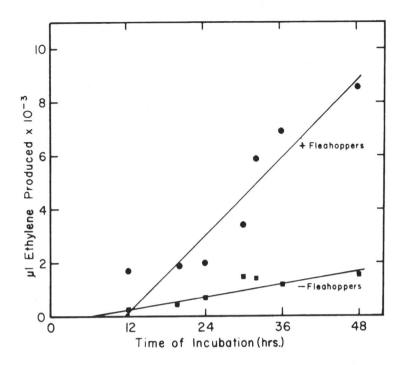

Fig. 10.7. Production of ethylene by excised cotton buds. Lines were obtained by linear regression with $r = 0.57$ for the line without the cotton fleahopper (*Pseudatomoscelis seriata*) and $r = 0.82$ for the line with the fleahopper. (Reprinted, by permission of the Entomological Society of America, from Duffey and Powell 1979.)

in producing stress ethylene in injured buds, some cotton injury might be enzymatic in nature (Martin et al. 1988b). Mirids other than the cotton fleahopper, for example, lygus bugs, might also be involved in the production of stress ethylene in cotton and in additional indeterminant-flowering plants sensitive to ethylene-induced abscission of fruit (Grisham et al. 1987, Martin et al. 1987).

In addition to IAA, the ethylene precursor 1-aminocyclopropane-1-carboxylic acid (ACC) is found in cotton fleahoppers reared in the laboratory and in nymphs collected from wild hosts. Because cotton fleahopper eggs do not contain ACC, the bugs' diet is considered the most likely source of this phytohormone. Injury to cotton by *P. seriata* and the induction of ethylene production in its hosts might involve some mechanical damage, introduction of microorganisms and hydrolytic enzymes, and alteration of phytohormone balance by ingestion or injection of IAA and ACC (Burden et al. 1989).

Much remains to be learned about this insect-microbe-plant interaction. Although perhaps not an example of a phytopathogenic symbiotic complex, this plant bug–microbe system points to an inadequate knowledge of interactions between phytopathogenic insects and their microbial associates (Norris 1979). Moreover, differential concentrations of a bacterium (*Bacillus* sp.) in leaf and square tissues possibly affect the levels of resistance to *P. seriata* in different cotton cultivars (Bird et al. 1979).

Feeding by *P. seriata* also results in delayed fruiting, which runs counter to the management practices recommended for Texas cotton production, and in a loss of

yield. Early fruiting minimizes damage by the boll weevil, whereas a later-maturing crop is more susceptible to bollworm injury (Walker et al. 1974).

Plant resistance to the cotton fleahopper has been evaluated since the 1960s. Its populations are significantly lower on glabrous than on hirsute strains of cotton regardless of the presence of pigment glands or nectaries (Fig. 10.8; Lukefahr et al. 1968, 1970; Tingey and Pillemer 1977; Mussett et al. 1979; Niles 1980; Sterling et al. 1989b), although sensitivity to injury can be greater on glabrous than on hairy cottons (Walker et al. 1974; cf. Lukefahr et al. 1976). Cotton fleahopper densities are often reduced in nectariless cotton (Schuster and Maxwell 1974, Schuster et al. 1976b; cf. Agnew et al. 1982). Certain cultivars with high trichome densities tolerate high fleahopper densities (Walker et al. 1974, Ring et al. 1993). High gossypol strains also provide resistance (Lukefahr and Houghtaling 1975). The Texas Cotton Insect Model (TEXCIM), designed to help farmers make crop-management decisions, allows fleahopper densities and fruit injury to be predicted with some reliability (Legaspi et al. 1989; see also Sterling et al. [1989b], Wagner et al. [1996], and Chapter 9). Higher bug densities can be tolerated on plants that have set squares if the soil moisture is adequate (Sterling et al. 1989b).

Creontiades Species

Creontiades species (Mirinae) are characteristic feeders on cotton squares and bolls in North America and elsewhere. In the southern United States, *C. debilis* has

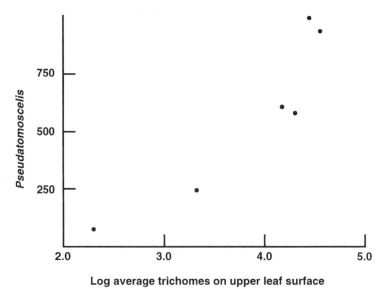

Fig. 10.8. Abundance of *Pseudatomoscelis seriata* on different cotton varieties based on data from Lukefahr et al. (1970). Cotton fleahopper density is positively correlated with trichome densities of its host. (Reprinted, with permission of Sir Richard Southwood and Cambridge University Press, from Southwood 1986.)

been implicated in "hopper damage" (Ewing 1929). This species once caused significant damage in the Yaqui Valley of Mexico; the bugs fed on buds, injuring anthers and spotting the petals, and reduced yields by about 50% in a 160-hectare field and 15% over 1,200 hectares (Morrill 1928). *Creontiades rubrinervis* shows similar habits in the southern states (Cassidy and Barber 1938, Schuster and Boling 1974), in Mexico (Garcia 1974), and in Puerto Rico (Fife 1939). Cassidy and Barber (1940) reported greater numbers of this species in Arizona than the combined populations of *Lygus hesperus* and *L. lineolaris*, noting that insect for insect it was more destructive than lygus bugs. They observed that *C. rubrinervis* can maintain itself in cotton and multiply throughout the season. During the 1990s, a species of *Creontiades* (probably *C. rubrinervis*) has assumed greater importance as a cotton pest in the Lower Rio Grande Valley of Texas, where pesticides have been needed to help suppress the bug's populations during the second half of the season (Norman and Sparks 1998).

Feeding by the green mirid (*C. dilutus*) on cotton squares and young fruits in Australia can cause severe early-season losses of squares (Mensah and Khan 1997, Murray and Lloyd 1997, Sadras and Fitt 1997). But different densities of this plant bug do not affect the yield of lint. Because plants apparently compensate for damage to squares at all population levels of this pest, economic injury levels or thresholds have not been determined. Its feeding, however, can delay boll maturation by 6–7 days, which increases the risk that lint will be downgraded by adverse weather conditions (Chinajariyawong 1988). Insecticides used against other cotton pests normally suppress plant bug populations, but the reduction in the use of chemical sprays associated with the use of transgenic cottons may intensify the problems with *C. dilutus* and

other mirids found in Australian cotton (Fitt 1994, Mensah and Khan 1997). Interplanting alfalfa as a trap crop in commercial fields might reduce damage from the green mirid in Australian cotton (Mensah and Khan 1997).

Creontiades pallidus, a mirine known in northern Africa as the boll shedder bug, feeds on flower buds, especially the anthers, and young bolls. Infested buds and bolls are usually covered by bright yellow excrement; bolls are often pierced through the resin glands, causing an irregular spotting on the carpel walls (Willcocks 1922, Kirkpatrick 1923, Stam 1987). The black discoloration on green bolls might be associated with a discharge of resins from the boll wall (Stam 1987, Cadou 1993). The bug's feeding results in bud and boll abscission and can delay maturation of the crop (Adair 1918, Soyer 1942, Samy 1958, Buyckx 1962, Ripper and George 1965, Schmutterer 1969). Severely injured older bolls can dry up, split prematurely, be invaded by fungi, and produce no sound lint (Kirkpatrick 1923). Although *C. pallidus* is sometimes labeled a major cotton pest in northeastern Africa, it might not remove more squares than those that are typically shed as a result of physiological factors (Goodman 1955). Studies in Syria established that yields can be reduced by about 50% and that chemical treatment of *C. pallidus* is needed when populations reach seven individuals per 50 sweeps during June and July, even though the use of insecticides can induce outbreaks of bollworms (Stam 1987). This species is also considered a cotton pest in Madagascar (Delattre 1958) and western Turkey (Karman and Akşit 1961).

Additional New and Old World Mirid Pests of Cotton

Other mirids associated with cotton inflorescences in North America include the mirines *Proba vittiscutis*

(McGarr 1933), *Adelphocoris superbus* (Eyer and Medler 1942), and *Neurocolpus nubilus.* The last-named species, the clouded plant bug, is a sporadic pest in Arkansas and Mississippi; it typically feeds on the anthers of squares and can cause a loss of lint, boll abscission, and reduced seed quality (Bibby 1946, Pack and Tugwell 1976, Tugwell et al. 1976, Tonhasca and Luttrell 1991). An orthotyline mirid of cotton, *Partheni-cus aureosquamis,* was suggested to damage small squares in Arizona (Bibby 1961). In addition to the cotton fleahopper (*Pseudatomoscelis seriata*), phylines that feed on cotton include *Keltonia tuckeri* (as *K. sulphurea*), *Megalopsallus atriplicis,* and *Reuteroscopus ornatus* (Painter 1930, McGarr 1933, Brett et al. 1946).

Taylorilygus vosseleri (Hancock 1935, Taylor 1945, Dale and Coaker 1958, Stride 1968, Cadou 1993) and *Megacoelum apicale* (Pearson 1958, Cadou 1993) adversely affect cotton production in Africa. The former, referred to as *Lygus vosseleri* and *L. simonyi* in earlier literature (Taylor 1947a), feeds on flowers and terminal buds in addition to young leaves (see Chapter 9). Its feeding on buds sometimes results in a loss of yield. Several workers, however, stated that because of cotton's ability to compensate for a considerable loss of squares, plants can withstand injury from *T. vosseleri* without sustaining decreased yields (Coaker 1957, McKinlay and Geering 1957). Bowden and Ingram (1958) and Davies and Kasule (1964) demonstrated that various bollworm species (Noctuidae) are responsible for some of the damage attributed to *T. vosseleri,* at least in drier areas of Uganda (Bowden 1970). Reed (1974) reviewed the search for resistant cotton varieties in Uganda, noting that lines least damaged by *T. vosseleri* also give the poorest yields. Stride (1968) did not consider *T. vosseleri* a true "cotton insect," referring to its high nymphal mortality and a low oviposition rate on this host. In contrast to the habits of *T. vosseleri, M. apicale* prefers young bolls (leaves and squares are not fed on), which exhibit irregularly shaped dark areas that develop split and broken scars (da Silva Barbosa 1959). In Nigeria at least, the application of nitrogen fertilizers does not increase cotton's attractiveness to *Adelphocoris* species, *T. vosseleri,* or other plant bugs (Hayward 1972).

In the moist tropics of Africa, *Helopeltis schoutedeni* feeds not only on the young stems, shoots, and leaves of cotton (see Chapter 9), but also on green bolls. This bug shows two population peaks, the first (mid-August in the Sudan) tending to coincide with the main period of vegetative growth of cotton and the second (September–October) with the period when immature bolls undergo their most rapid development (McDermid 1956). The circular, shallow lesions produced by its feeding can leave the boll surface looking like a "lunar landscape." When lesions are numerous, the boll wall is converted to woody, cankerous tissue that cannot expand. Because growth of the boll's contents can no longer be accommodated, the boll sometimes splits pre-

maturely, especially when secondary bacteria and fungi invade and cause a rotting of the boll contents (Pearson 1958).

Injury to a related cotton, Sea Island, in the West Indies, is ascribed to a phyline, the western plant bug (*Rhinacloa forticornis*). Its feeding on reproductive parts causes small squares to turn brown and blacken, the injury (blasting) resembling a "spent match-head." Larger squares show ragged petals and darkened anthers. Although premature shedding of fruit can delay crop formation, plants usually offset this loss with compensatory growth (Ingram 1980).

MIRIDS ASSOCIATED WITH KENAF

Kenaf, a shrubby herb belonging to the same family as cotton (Malvaceae), also sustains plant bug injury. In Dagestan in the former USSR, *Polymerus cognatus* and *Lygus pratensis* feed on the small buds and ovaries of this fiber crop, causing both structures to abscise (Nevinnykh and Riabov 1931). An increased cultivation of kenaf in other areas might be expected to produce additional problems from plant bugs. In the southern United States, kenaf is a potentially important host of *L. lineolaris* (Craig et al. 1997).

Sugar Beet, Tobacco, and Other Field Crops

Solanaceous crops such as tomato and tobacco are preferred hosts of several dicyphine mirids. These species are more important for their leaf and stem injury (see Chapter 9); their feeding lesions can encircle a seed-stalk, severing the head before seeds have matured (Thomas 1945). But dicyphines also feed on host inflorescences and developing seeds. Leonard (1930) reported the tomato bug (*Engytatus modestus*) as swarming on the flower heads of tobacco in Puerto Rico, reducing seed production on affected plants. *Tupiocoris notatus* sometimes feeds on the ovaries of tobacco and tomato plants in Central America, which results in flower abscission (King and Saunders 1984). *Nesidiocoris tenuis* occurs more often on flower buds or inflorescences than on other parts of tobacco in India; when the bugs feed on flowers, the calyx is preferred (Krishna Prasad et al. 1979, Patel 1980). *Nesidiocoris caesar,* which also feeds on tobacco and other Solanaceae, has become a pest of bottle gourd, a cucurbitaceous crop in India. Its feeding on inflorescences prevents fruit set, and that on young fruit causes abscission (Chatterjee 1983a).

Hills (1941) evaluated the effects of lygus bugs on sugar beet (table or garden beet is discussed later in this chapter) grown for seed in Arizona. When introduced at the blossom or late-blossom stage and allowed to remain on plants until harvest, the bugs (a mixture of *Lygus hesperus* and *L. lineolaris*) reduce seed viability. They do not significantly affect the quantity of seed produced or decrease the size of seed balls. Yun (1986) noted that lygus bugs feed only on developing seeds of sugar beet, which causes collapse of the embryos.

Oilseed rape is a cruciferous field crop grown extensively in western Canada. High-yielding cultivars with low levels of erucic acid and glucosinolates are referred to as canola (Butts and Lamb 1990a, 1990b). *Lygus lineolaris* and other *Lygus* species (Schwartz and Foottit [1992a] discussed mirid species richness and diversity on this crop in the Canadian prairie provinces) generally invade rape fields when buds become available, the first peak in plant bug numbers coinciding with the early flowering stage (Butts and Lamb 1991a). Thus, the stage of plant development, rather than seeding date, determines canola's attractiveness to colonizing lygus adults (Leferink and Gerber 1997). Their feeding causes the buds and flowers to abscise, but the plants compensate for flower loss so that the number of pods is not reduced. Lygus bugs, however, cause seeds to collapse and reduce the weight of seeds produced per pod, which can affect seed yields (Butts and Lamb 1990a, Turnock et al. 1995). Field studies in Alberta demonstrated that the percentage of seeds injured increases and yield decreases with increasing bug densities. Insecticide applied at the pod (but not at the bud or flower) stage results in substantial yield increases. These increases ranged from 11% to 35% but were significantly different in only two of the five tests (Butts and Lamb 1991b).

Lygus bugs also colonize rapeseed in other areas of North America. In reporting that *L. lineolaris* nymphs and adults feed on the buds, flowers, and pods of rapeseed in Tennessee, Boyd and Lentz (1999) suggested that rapeseed could serve as an early-season host of the tarnished plant bug, adults of which might migrate into cotton. *Lygus rugulipennis* is a minor pest of winter-planted oilseed rape in Germany, the overwintering adults generally restricted to areas of the field nearest woods. In Germany, *Closterotomus norwegicus* is found almost exclusively on summer rape (Hossfeld 1963).

Phillips et al. (1973) reported *Calocoris barberi* (as *Hercias* [sic] *sexmaculatus*) and *L. lineolaris* congregating on and damaging the heads of cultivated sunflowers in Texas. Lygus bugs—mainly *L. rugulipennis* but also *L. gemellatus* and *L. pratensis*—are pests of sunflower in Vojvodina in Yugoslavia (Čamprag et al. 1986a, 1986b, 1986c). The bugs, present only in small numbers when heads begin to form, become abundant during flowering and seed ripening. Populations are higher when alfalfa fields are nearby (Čamprag et al. 1986a). Feeding by lygus bugs can reduce the oil content of seeds by 17% in heavy infestations and affect germination as much as 60% (Čamprag et al. 1986c). Seed damage to sunflower during 1975–1985 averaged 10%, with yields lowered by 3% (Čamprag et al. 1986c). Feeding by *L. rugulipennis* reduced the oil and protein content of sunflower seeds in Bulgaria (Shindrova and Ivanov 1982). *Adelphocoris lineolatus*, a mirine that typically develops in and damages legume seed crops, is an occasional pest of sunflower seed plants in the former USSR (Putshkov 1975).

Mirids are associated with flax grown both for fiber and for seed (linseed), although they appear to be more injurious when the plant is grown for fiber (Ferguson et al. 1997; see also Tables 9.4, 9.8). On linseed in southeastern England, Ferguson et al. (1997) found that *Closterotomus norwegicus* (61%) and *L. rugulipennis* (33%) are more abundant than *Lygocoris pabulinus* (4%) and *Grypocoris sexguttatus* (2%). Species of the mirid complex found on linseed feed on the flower buds and flowers, causing mostly insignificant losses in yield near field boundaries. In Italy, *C. norwegicus* feeds on the flowers and developing seeds of hemp (Goidanich 1929), another herb that yields fiber.

Plant bugs are the most destructive insect pests of castor flowers and fruit. Mirid damage in Kenya and Tanzania is intensified when this crop (sometimes called castorbean) is grown near large fields of corn or sorghum (Weiss 1971). Conversely, in West Africa, castor serves as an alternative host for sorghum pests such as *Eurystylus oldi* that can migrate into and damage sorghum (Ratnadass et al. 1997). Introduced dwarf varieties of castor are particularly susceptible to injury by mirids and other heteropterans (Bohlen 1973). In South Africa, *Eurystylus* species feed on male and female flowers, causing them to shrivel and die; in some cases, the entire stem dies. Nymphs of these mirines depend on the inflorescences of castor. When a flower becomes unsuitable for *Taylorilygus ricini*, the nymphs search for another flower, a young leaf, or growing points or switch to another castor plant. *Eurystylus* nymphs, however, are capable of migrating only to another inflorescence on the same plant (Boyes 1964). March (1972), in discussing a *Eurystylus* species and *T. ricini* as important pests of male flower buds in Mozambique, listed the mirid species collected on castor in other geographic regions.

Vegetable Crops

This discussion of mirids as vegetable pests includes some feeding on reproductive plant parts other than inflorescences—for example, bean pods and developing carrot seeds. Tomato and green (bell) pepper are treated as vegetables, although botanically they are fruits; observations involving potato flowers are also included. Quinoa, a tropical American chenopod cultivated as a grain (it once may have been used as a vegetable in Mexico [Simmonds 1976]), is discussed with garden beet, another chenopod, and grain amaranth.

Because lygus bug injury is generally similar to that described earlier in this chapter for the legumes used for

forage and seed, only a few examples involving leguminous vegetables will be discussed. Additional information on lygus bugs as vegetable pests is available in the
publications by Bohart and Koerber (1972) and Scott
(1987), bibliographies by Scott (1980) and Graham et al.
(1984), and reviews of injury to vegetable and forage
seedlings by Fye (1982c, 1984).

BEANS AND OTHER EDIBLE LEGUMES

Lygus bugs (and other mirines) have long been known
as pests of bean crops (Table 10.11; e.g., Davis 1897,
Hawley 1922), the nature of their injury depending
partly on plant age. Feeding during the early bud and
flowering stages can reduce yields, whereas feeding
on developing seeds causes pitting and blemishes
("dimples") on table-market beans and reduces germination in seed beans (Burton 1991). Other mirines, such
as *Adelphocoris rapidus*, can cause similar injury to
bean seeds (Hawley 1920).

Lygus bugs can be important pests of the lima bean
in the United States. Damage in California was
observed as early as the mid-1940s when lima bean
crops adjacent to alfalfa fields could not be harvested
or showed poor yields (Stone and Foley 1959). Later
studies documented that an early-season application of
insecticides results in increased yields, as measured by
the total number of marketable beans per plant (Shorey
et al. 1965). Schwartz and Klassen (1981), however,
stated that crop pests not directly reducing yields or
causing adverse changes in quality need to be examined
more critically. The bean–lygus bug relationship
was one they suggested as needing further study to
determine whether control measures are actually
needed.

Lygus elisus and *L. hesperus* cause several types
of injury to lima bean crops in California: a shedding
of blossoms and small pods, shriveling of seed, and
necrotic pitting of seed, a symptom similar to that
associated with yeast-spot disease except for the
absence of any pathogen. Pods fed on by lygus bugs
generally lack external symptoms, but intumescences
or swellings appear on the inner surface of pod walls.
The shedding of young pods occasionally causes total
crop loss in small areas, and seed pitting results in a
downgrading of beans and the added cost of sorting
damaged beans by hand. Damage is most severe in fields
adjacent to alfalfa and seed beet (Baker et al. 1946).
Populations of *L. hesperus* exceeding 0.5 bug per sweep
are capable of reducing yields. Failure to control the
bugs results in vegetative plants that produce few pods
(Bushing et al. 1974). Similarly, *L. lineolaris* can limit
yields of lima beans in New York (McEwen and Hervey
1960).

Middlekauff and Stevenson (1952) and Middlekauff
(1956) described injury resulting from feeding by *L.
elisus* and *L. hesperus* on young pods and developing
seeds of cowpea (black-eyed pea) crops in California.
Seed pitting caused mainly by *L. hesperus* reduces crop
quality and yield (Bosque-Perez et al. 1987). *Lygus*
species (mainly *L. elisus* and *L. hesperus*) disperse to
fields of green or garden beans when alfalfa is cut, when
seed alfalfa matures, or when annual weeds dry up.
Feeding by lygus bugs during blooming and development not only decreases yields in the Pacific Northwest
but also causes pitting or scarring that can lower the
market value of bean seed (Shull 1933, Hagel 1978).
Bean varieties least susceptible to injury generally have
thick-walled pods that might preclude the bugs' stylets
from reaching the developing seeds (Shull and Wakeland
1931). Even the most susceptible varieties can be
immune to lygus bug feeding once the pods become less
succulent (Shull 1933). Khattat and Stewart (1980) suggested that *L. lineolaris* does not develop on green bean
crops in Quebec, the late instars immigrating to the
bean plant from adjacent host plant species. Adults can
also migrate into beans, but they do not always invade
in large enough numbers or sufficiently early in the
season to be considered important pests of beans (Stoltz
and McNeal 1982).

Lygus hesperus is also a major pest of lentil
(Summerfield et al. 1982). In Idaho, its feeding produces

Table 10.11. Some mirine plant bugs associated with flowers of beans and other edible legumes

Mirid	Host plant	Locality	Remarks/symptoms	Reference
Closterotomus norwegicus	Green bean	Italy	Flowers and young pods are injured	Lupo 1946
Creontiades dilutus	Bean	Australia	Feeding on axillary buds inhibits formation of flower-bearing branchlets ("blind eye"); complete crop failure may result	Anonymous 1940, Hely et al. 1982
Creontiades pacificus	Azuki bean	Papua New Guinea	Feeds on flowers	G. R. Young 1984
Creontiades rubrinervis	Cowpeas	Brazil	Pods shrivel and abscise	Lukefahr 1981
Neurocolpus mexicanus	Hyacinth-bean	Central America	Feeds on flowers and ovaries	King and Saunders 1984

Note: See text for additional examples (mainly *Lygus* spp.) and discussion of lygus bugs as bean pests.

depressed, pitted areas on the seed coat (Plate 14), which often is accompanied by a chalky appearance of cotyledons. Seed viability can be reduced (Schotzko and O'Keeffe 1989a). On white lupine in Washington State, where this legume is grown experimentally as a replacement for lentils and peas, *L. hesperus* and *L. elisus* significantly reduce yields by feeding on young pods (Tanigoshi and Babcock 1989).

Mirids are consistently associated with groundnut (peanut) wherever this crop is grown, although they are not usually economically important pests. A *Creontiades* species reduced flower production in a greenhouse experiment in Australia, and mirids are considered potential pests of groundnut in India (Wightman and Ranga Rao 1994).

SOLANACEOUS VEGETABLES: GREEN PEPPER, POTATO, AND TOMATO

Huber and Burbutis (1967) reported damage by *Lygus lineolaris* to green pepper in Delaware. Nymphs and adults cause abscission of the buds and blossoms and reduce the numbers of fruit produced. In the Netherlands, feeding by another mirine, *Liocoris tripustulatus*, causes necrotic blotches on the fruit of green peppers (Plate 14) to be dried for paprika (Simonse 1985, Ulenberg et al. 1986).

Lygus bugs sometimes leave their preferred breeding hosts to feed on tomato crops in California, their feeding causing the fruit surface to dry out and crack (Toscano et al. 1990). Preliminary studies in New York suggested that tomato yields can be reduced by *L. lineolaris* (Davis et al. 1963; see also Pitblado [1994]).

Injury to potato by *L. lineolaris* in Idaho includes wilted blossoms (Wakeland 1931), and in Canada, this bug destroys potato flowers (Hodgson et al. 1973) in addition to feeding on flower stems and growing tips (Gorham 1938). The green mirid (*Creontiades dilutus*) invades coastal potatoes in Australia (New South Wales) where it feeds in the blossoms. The bugs develop on this crop and can suppress flower production, but they apparently do not affect the tubers (Hely et al. 1982).

Solanaceous plants are the preferred hosts of several dicyphine mirids, which are more important for their stem and leaf injury (see Chapter 9), but they also feed on inflorescences and developing seeds. In Texas, Johnston (1930) attributed the abscission of tomato blossoms to attacks by a dicyphine known as the suckfly (*Tupiocoris notatus*); a few years earlier, infestations of this plant bug made it nearly impossible to grow tomatoes in the Austin area (High 1924). In Hawaii, the tomato bug (*Engytatus modestus*) feeds on tomato flowers, particularly the developing ovaries, which results in abscission of the blossoms and reduction (or prevention) of fruit set (Illingworth 1929). Feeding at the base of stamens causes discoloration and in severe cases, shriv-

eling of a group of stamens. Serial sections of damaged stamens showed that the bugs' stylets penetrate near vascular traces, the path sometimes continuing through a stamen. Cells around vascular elements become disorganized masses of protoplasm and cells walls (Tanada and Holdaway 1954). Tanada and Holdaway reviewed additional economic literature pertaining to *E. modestus*, including reports of blossom drop.

The dicyphine *Macrolophus pygmaeus* occasionally injures tomato flowers in Spain and the Canary Islands (Gomez-Menor 1955), and the facultatively predacious *Dicyphus tamaninii* feeds on tomato fruit in Spain, particularly when whitefly populations are low (see Chapter 14); fruit injury is not closely correlated with the mirid's density. This bug's feeding causes discolored areas surrounded by a red halo, which can result in a tanned area on the fruit surface (Albajes et al. 1988, Gabarra et al. 1988). Feeding by the Old World *D. errans* on tomato fruit causes similar injury (Plate 14). When *D. melanotoma*, which is not native to Great Britain, is introduced to help manage whitefly populations in tomato crops, it can damage both the fruits and the foliage (Sampson and Jacobson 1999). The Nearctic *D. hesperus* will feed on tomato fruit only if tomato leaves are unavailable (Gillespie et al. 1999).

CARROT AND OTHER UMBELLIFERS

Injury to Carrot by Lygus Bugs

Lygus bugs readily feed on the reproductive structures of umbellifers, inducing symptoms characteristic of mirids. *Lygus hesperus* causes flower or seed abortion in California when it feeds on the blossoms or developing seeds of carrot, as well as a blasting of mature seeds (Carlson 1956, 1959). *Lygus elisus* and *L. hesperus* significantly reduce the seed yields (20% or more with one bug per umbel) of Idaho carrots; they are the most destructive pests of carrot seed production (Scott et al. 1966, Scott 1977b). But feeding by plant bugs on umbelliferous crops can also produce atypical effects—accelerated growth or phytostimulation, as well as embryolessness.

Phytostimulation and Embryolessness

Researchers in North America and Europe have devoted considerable attention to phytostimulation and embryolessness in carrot and related plants. Developing carrot seeds subjected to lygus bug feeding are smaller, germinate more slowly than protected seeds, and can exhibit premature bolting. Damaged seeds show a ninefold increase in IAA and a 40% increase in a protein inhibitor–inducing factor (Scott 1983). In the field, however, plants originating from damaged seeds produce heavier roots and surpass the growth of plants grown from larger, undamaged seeds (Scott 1969). Scott

(1970) discussed the increased growth and yield of carrots in response to lygus bugs' feeding on seed, and reported phytostimulation in bean seeds. Secretion of a growth-promoting factor in the developing seeds is considered the most likely explanation for this phenomenon. Phytostimulation might be related to auxins, either directly through a secretion of auxin precursors that change auxin levels in seeds, or indirectly when auxins increase locally in response to wounding (Tingey and Pillemer 1977).

Discovering other crops in which increased yields result from lygus bug attacks on seeds is potentially important, particularly if the effects of phytostimulation can be realized without the accompanying economic loss caused by feeding on the developing seed crop (Scott 1976a). The lygus bug–carrot association also involves a viruslike disorder that apparently is transmitted during its feeding on developing seeds (Scott 1972; see also Chapter 8). Scott (1983) reviewed the intricate relationships between *Lygus hesperus* and carrot, including the accelerated growth from damaged seeds. Because increased yields are correlated with taller shoots, Belsky (1986) suggested that the carrot plant is able to overcompensate in total biomass for lost tissues.

Erratic germination and embryoless seed of Apiaceae (=Umbelliferae) are common in widely separated regions of the Old and New World. When lygus bugs feed on carrot, celery, dill, fennel, parsley, and parsnip seeds, the embryo is often destroyed, whereas the endosperm remains normal in appearance. Seed yields are not greatly affected, but the worthless, embryoless seeds cannot be segregated by external appearance, size, or weight (Flemion and Henrickson 1949). Although the condition occurs in other plant families (Robinson 1954), it is most common in the Apiaceae (Flemion et al. 1949, Flemion 1958).

Flemion and MacNear (1951) noted that lygus bugs select an embryo for feeding even though this plant stage represents only a small part of the umbellifers exposed to the bugs. Stylet flexibility can be considered a behavioral adaptation that allows lygus bugs to avoid external oil ducts on the seed coat of these plants. The ducts contain secondary substances such as furanocoumarins and toxic essential oils. *Lygus lineolaris* can feed on umbellifer seeds without contacting the furanocoumarin-containing vittae (Weis and Berenbaum 1989, Berenbaum and Zangerl 1992).

When *L. lineolaris* feeds on umbellifers at or near the flowering stage rather than at postflowering, the ovaries and flowers are destroyed and seed yields are substantially reduced (Flemion and MacNear 1951). Robinson (1954) reviewed studies undertaken to assess the factors potentially responsible for the nearly universal problem of embryolessness—climate, soil type, fertilization, genetic influence, varietal differences, plant spacing, seed size, and position of seed on the plant—and experiments that established lygus bugs as the main

causative agents. Other mirines, such as *Orthops campestris* and *O. kalmii*, also cause embryolessness in carrot seed (Wagn 1954).

Injury to Umbellifers by *Orthops* Species

The seed yields of carrots in British Columbia were once severely reduced by another mirine, *Orthops scutellatus* (Wagner and Slater [1952] discussed the taxonomic confusion with the Palearctic *O. campestris*). This species, abundant on the flower heads of seed carrots, is rarely observed on plants grown for roots (Handford 1949). Carrot seed crops in British Columbia sustained little injury from this plant bug in the years following the outbreaks of 1947–1948 (Arnott 1956). *Orthops scutellatus* once also damaged the heads of cultivated parsnips in Nova Scotia (Brittain 1919).

Overwintered adults of *Orthops* species sometimes feed on trees and shrubs in the former USSR before colonizing umbellifers. These bugs are capable of blighting the reproductive structures of host plants grown as vegetables or for medicinal purposes. *Orthops basalis* and *O. campestris* prefer fields, gardens, and other open areas, whereas *O. kalmii* is attracted to more shaded sites—for example, forest clearings and banks of waterways—and seldom damages cultivated umbellifers (Putshkov 1975).

Orthops species can injure nonreproductive parts of umbellifers in Europe (e.g., Nuzzaci 1977, Ulenberg et al. 1986). When these bugs feed on a pedicel of a developing umbel, all buds on the umbel can be destroyed (Putshkov 1966). *Orthops campestris* and *O. kalmii* sometimes co-occur on the flower heads of European umbellifers, with the former bug usually more abundant (Korcz 1977). Both species can be pests of celery grown for seed in Poland (Anasiewicz and Winiarska 1995). Feeding by these species on carrot ovaries and fruits results in aborted endosperm and embryos, reduced germination percentage of seeds because of an embryoless condition, and decreased yields ("seedless umbels") (Wagn 1954, Kho and Braak 1956, van Turnhout and van der Laan 1958, Taksdal 1959). Insecticides applied against *Orthops* species increased seed yields in the Netherlands from 53% to either 65% or 89% (van Turnhout and van der Laan 1958). In Denmark, Wagn (1954) observed that the bugs always insert their stylets into seeds "at a point on the ridge between the two spine-bearing secondary ribs, (i.e. in or beside the central rib) one third or one quarter of the way down the seed, reckoning from the base of the pistils." Germination could be increased 10% with insecticidal control of plant bugs in Danish seed fields.

The Green Mirid (*Creontiades dilutus*)

An occasional pest of carrots and parsnips in New South Wales, Australia, is the green mirid. The bugs suck sap from buds and flowers, causing their abscission and a failure of the heads to set seed (Hely et al. 1982).

Lygus bug injury to seeds of garden beet, a chenopodiaceous vegetable, is not as great as with certain other vegetables, such as carrot. Feeding by *L. hesperus*, however, on beet seeds in California can result in decreased yields and reduced weight of seeds (Carlson 1961).

On another chenopod, quinoa, a *Lygus* species (possibly *L. shulli*) feeds on developing heads, causing seed abortion and "light seeds" (Cranshaw et al. 1990). They noted that the phyline *Atomoscelis modesta* also feeds on quinoa heads in Colorado. An annual herb native to the South American Andes, quinoa is grown mainly for its high-protein seeds.

Lygus lineolaris is the most important insect pest of grain amaranth in North America. Injury to this alternative grain crop results from the bug's feeding on floral buds, immature blossoms, and developing embryos, which causes immature seeds to shrivel and discolor. Plant bug numbers increase as amaranth heads develop (Clark et al. 1995). Field studies have shown that the tarnished plant bug can reduce the seed production of grain amaranth. Infestations during the first two weeks of head development are not critical, but seed weight is significantly reduced when the heads are fed on from the third through fifth weeks (Olson and Wilson 1990, Wilson and Olson 1992).

USE OF VEGETABLES FOR ADULT FEEDING

Mirid dispersal to the inflorescences of food (nonhost) plants to exploit various resources was discussed earlier in this chapter and will be treated more thoroughly in the next two sections on shrubs and trees. Here, it is noted that the phyline *Lepidargyrus ancorifer* can invade onion fields in Oregon to feed on flower and seed heads. Infestations can result in a blighting of inflorescences, the injury most severe at field margins. Densities of 25–75 bugs per head and losses of 5–50% have occurred (Thompson 1945). *Lygus lineolaris* sometimes also congregates on onion flowers, causing a reduction in seed yields (Tate 1940). Cabbage seed crops bordered by overwintering hosts of *L. elisus* are vulnerable to invasion by adults in western Washington (nymphs are uncommon in cabbage seed fields). Feeding occurs on various floral structures, but caged adults injure differentiating cabbage terminals more than mature buds; the bugs cause little or no injury to flowers or immature siliques (Getzin 1983). The tendency of a congener, *L. lineolaris*, to congregate on cabbage flowers was noted in the nineteenth century (Glover 1876). *Closterotomus norwegicus* sometimes feeds on cabbage and turnip pods in Italy (Lupo 1946).

Shrubs

Plant bugs develop on shrubs of diverse inflorescence types (Table 10.12). Perhaps because mirids often do not produce obvious symptoms on the reproductive structures of commercially grown shrubs (they can severely injure the foliage; see Chapter 9), flower-feeding species associated with shrubs are not particularly well known. An especially rich mirid fauna is associated with flowering shrubs in the chaparral and desert regions of western North America (e.g., Knight 1968, Pinto and Velten 1986; pers. observ.).

SHRUBS AS BREEDING HOSTS

Several mirids that develop exclusively on the flowers of shrubs in eastern North America occur on plants with a cymose inflorescence (see also Table 10.12). *Lygocoris atrinotatus* develops on the inflorescences of smooth hydrangea (Wheeler and Henry 1983). Species of *Lygocoris*, subgenus *Neolygus*, overwinter as eggs, their hatching synchronized with the development of host flower buds. Early instars of *L. atrinotatus* are present in Pennsylvania when hydrangea buds are small, and adults begin to appear during the early-bloom stage (Wheeler and Henry 1983).

The nymphs of two *Cariniocoris* species (Phylinae) develop mainly, and perhaps exclusively, on the staminate flower buds and inflorescences of holly. *Cariniocoris ilicis*, for example, feeds on winterberry in New York, and farther south (Pennsylvania to Florida and Texas), *C. geminatus* is associated with the male flowers of American holly, inkberry, possumhaw, and yaupon (Henry 1989; pers. observ.). Nymphs of the clouded plant bug (*Neurocolpus nubilus*) develop on the staminate flowers of American holly in Alabama and Virginia and on inkberry in Florida and Pennsylvania (pers. observ.).

The inflorescences of ericaceous shrubs harbor a diverse mirid fauna (see also Table 10.12). *Lygocoris laureae* nymphs develop on the unfolding flower clusters of mountain laurel in eastern North America (Plate 14; Knight 1917a), feeding mainly on the ovaries and discs rather than on the petals, stamens, or pollen (Clayton 1982). *Rhinocapsus vanduzeei*, a phyline known as the azalea plant bug, feeds on the flowers of ornamental azaleas, often piercing the filaments and stamens (Plate 14); after dehiscence, the bugs will probe pollen adhering to the plant surfaces (Wheeler and Herring 1979).

Three species of the bryocorine genus *Neoneella* (mainly *N. bosqi*) occur on philodendron plants that are shrubby or viny. Observations of these bugs living on an undetermined philodendron species that probably is not a shrub—the plant grew as an epiphyte on the larger branches of forest trees in Paraguay—are nevertheless included here. According to Carvalho and Hussey (1954), the bryocorines "occurred principally on the inner side of the spathe rather than on the spadix of the

Table 10.12. Some New and Old World mirids that develop on inflorescences of shrubs

Mirid	Host plant	Locality	Reference
Mirinae			
Dagbertus fasciatus	Bayberry	Connecticut, New York, USA	Pers. observ.
	Groundsel-baccharis	Maryland, USA	Pers. observ.
	Yellowbells (yellow elder)	Cuba	Maldonado Capriles 1986
Eocalocoris albicerus	Japanese clethra	Japan	Yasunaga and Takai 1994, Yasunaga 1995
Lampethusa anatina	Spiny hackberry	Texas, USA	Pers. observ.
Lampethusa collaris	Lantana	Mexico	Palmer and Pullen 1995
Lygocoris belfragii	Dogwood	Eastern North America	Knight 1917a
	Viburnum	Pennsylvania, USA	Wheeler 1980a; pers. observ.
Lygocoris knighti	Viburnum	New York, Pennsylvania, USA	Wheeler 1980a; pers. observ.
Lygocoris neglectus	Virginia sweetspire	Georgia, USA	Pers. observ.
Lygocoris spp.	Panicled hydrangea	Japan	Yasunaga 1991a, 1992a, 1992b
Neurocolpus arizonae	Mesquite	Western USA	Knight 1968
Neurocolpus flavescens	Redbay	Florida, USA	Pers. observ.
Neurocolpus jessiae	Elderberry	Illinois, Missouri, USA	Knight 1934, Froeschner 1949; see also Henry and Kim 1984
Neurocolpus nubilus	Buttonbush	Arkansas, Illinois, USA	Needham 1903, Lipsey 1970a
	Chinese sophora, slender deutzia	Pennsylvania, USA	Pers. observ.
	Virginia sweetspire	Georgia, USA	Pers. observ.
Platytylus bicinctus	Desert hackberry	Texas, USA	Pers. observ.
Taedia deletica	Sand sagebrush	Nebraska, USA	Pers. observ.
Taedia externa	Yaupon	Florida, USA	Pers. observ.
Taedia hawleyi	Red chokeberry	Pennsylvania, USA	Wheeler 1991b
Taedia heidemanni	Shrubby St. John's-wort	Virginia, West Virginia, USA	McAtee 1916; pers. observ.
Taedia scrupea	Common hoptree	Pennsylvania, USA	Pers. observ.
Taedia virgulata	Lime pricklyash	Texas, USA	Pers. observ.
Orthotylinae			
Heterocordylus genistae	Dyer's greenwood	England	Brown 1924
Orthotylus ericetorum	Heath, heather	Europe	Kullenberg 1944, Southwood and Leston 1959
Phylinae			
Atractotomus prosopidis	Mesquite	Western USA	Knight 1968
Eumecotarsus breviceps, E. kiritshenkoi	False tamarisk	Former USSR	Kerzhner 1962
Glaucopterum atraphaxius	Atraphaxis	Turkmenia	Putshkov 1977
Moissonia schefflerae	Schefflera	New Guinea	Schuh 1984
Plagiognathus cornicola	Dogwood	Eastern North America	Knight 1923; pers. observ.
Plagiognathus luteus	Fremont's grape holly	Arizona, USA	Knight 1929a
Plagiognathus punctatipes	Ninebark	New York, Pennsylvania, USA	Wheeler and Hoebeke 1985
Psallus physocarpi	Ninebark	New York, Pennsylvania, USA	Wheeler and Hoebeke 1985
Pseudatomoscelis flora	Wislizenia	Mexico	Van Duzee 1923

Note: See text for additional examples and discussion.

flower, and nymphs of older stages were found together with the adults. The spadix and the inner surface of the spathe were thickly covered with sticky drops of a yellowish exudate, possibly resulting from the punctures in the plant tissues made by the multitude of bugs, but these insects seemed not to become entangled in it. Sometimes melolonthine beetles were also attracted to those flowers and worked their way into the deepest part of the corolla, becoming thoroughly smeared with the sticky exudate; when these beetles were present the *Neoneella* were usually found only in the upper part of the flower and on its outer surface."

DISPERSAL TO NONHOST SHRUBS

Mirids developing on plants of various growth habits disperse to feed on the flowers of shrubs (Table 10.13; Porsch 1957, Schwartz 1984, Schwartz and Foottit 1998). According to Knight (1941), "Even among species which always breed on a single host plant, a general dispersal of individuals usually takes place. Following the time of emergence and mating, individuals of *Tropidosteptes cardinalis* Uhler and *Lopidea staphyleae* Knight and others have been observed to migrate from their host plant to shrubbery in the general vicinity."

Table 10.13. Some North American and European mirids that disperse to flowers of shrubs that appear not to be used as breeding hosts

Mirid	Food plant	Locality	Reference
Deraeocorinae			
Deraeocoris schwarzii	Shrubby cinquefoil	Wyoming, USA	Stephens 1982
Mirinae			
Adelphocoris lineolatus	Elderberry	Pennsylvania, USA	Frost 1979
Adelphocoris rapidus	Broadleaf meadowsweet	New York, USA	Blackman 1918
	New Jersey tea	Illinois, Virginia, USA	Banks 1912, Robertson 1928
Agnocoris pulverulentus	Common privet	Pennsylvania, USA	Uhler 1893
Apolygus nigritulus	Panicled hydrangea	Japan	Yasunaga 1992c
Lygocoris belfragii	American hydrangea	North Carolina, USA	Pers. observ.
Lygocoris caryae	American hydrangea	North Carolina, USA	Pers. observ.
Lygocoris communis	Silky dogwood	Ontario, Canada	Judd 1975
Lygocoris fagi	American hydrangea	North Carolina, USA	Pers. observ.
Lygocoris omnivagus	American hydrangea	North Carolina, USA	Pers. observ.
	Virginia sweetspire	Pennsylvania, USA	Pers. observ.
Lygocoris ostryae	Hawthorn	Iowa, USA	Johnston 1928
Lygocoris pabulinus	Elderberry	New York, USA	E. R. Hoebeke, pers. comm.
Lygocoris quercalbae	Mountain maple	Virginia, USA	Pers. observ.
	Virginia sweetspire	Pennsylvania, USA	Pers. observ.
Lygocoris semivittatus	American holly	Florida, USA	Pers. observ.
	Photinia	South Carolina, USA	Pers. observ.
Lygocoris tiliae	Cockspur hawthorn	Minnesota, USA	Pers. observ.
Lygus elisus	White rubber rabbitbrush	Utah, USA	Moore et al. 1982
Lygus rugulipennis	Bumald spirea, common snowberry, Morrow's honeysuckle, ninebark, shrubby cinquefoil	Norway	Taksdal 1965
	Bluebeard	Germany	Schaarschmidt 1983
Lygus vanduzeei	Broadleaf meadowsweet	Maryland, New York, USA	Blackman 1918, Drake 1922, Knight and McAtee 1929
	Mountain maple	Virginia, USA	Pers. observ.
Neurocolpus nubilus	New Jersey tea	Virginia, USA	Banks 1912
	Silky dogwood	Ontario, Canada	Judd 1975
Pinalitus approximatus	Elderberry	Pennsylvania, USA	Frost 1979
Taedia scrupea	Silky dogwood	Ontario, Canada	Judd 1975
Taylorilygus apicalis	Groundsel baccharis	Texas, USA	Palmer 1987
Phylinae			
Atractotomus miniatus	American holly	Georgia, USA	Pers. observ.
Chlamydatus associatus	Hawthorn	Indiana, USA	Blatchley 1926
Microphylellus modestus	Arrow-wood viburnum	District of Columbia, USA	Pers. observ.
Microphylellus tsugae	Oakleaf hydrangea	Tennessee, USA	Pers. observ.
Plagiognathus arbustorum	European meadowsweet	England	Parfitt 1884
Plagiognathus brevirostris	Meadowsweet	Pennsylvania, USA	Pers. observ.
Plagiognathus caryae	Oakleaf hydrangea	Kentucky, USA	Pers. observ.
Plagiognathus cornicola	Meadowsweet	Pennsylvania, USA	Pers. observ.
Plagiognathus guttulosus	American holly	Georgia, USA	Pers. observ.
	Herculesclub pricklyash, honey mesquite	Texas, USA	Pers. observ.
Plagiognathus politus	New Jersey tea	Virginia, USA	Banks 1912
	Oakleaf hydrangea	Kentucky, USA	Pers. observ.
Plagiognathus punctatipes	New Jersey tea	North Carolina, USA	Pers. observ.
Psallus physocarpi	New Jersey tea	North Carolina, USA	Pers. observ.
Psallus variabilis	Bayberry, downy serviceberry, hawthorn, viburnum	New York, USA	Wheeler and Hoebeke 1982; pers. observ.
Rhinocapsus vanduzeei	American hydrangea, New Jersey tea	South Carolina, USA	Pers. observ.

Note: See text for additional examples and discussion.

Knight's (1941) observations, while generally true, do not apply to all host-restricted mirids. Some species are mostly confined to their hosts (e.g., Wheeler 1995) except during periods of host plant deterioration. In addition, the gap in host usage between adults and nymphs is generally less in mirids than in other heteropterans, such as coreids and lygaeids (Putshkov 1956, 1960). Other mirids that readily disperse to nearby plants are various mirines: species of *Dagbertus* (Knight 1923), *Lygocoris* (Knight 1917a, Clayton 1982), and *Neurocolpus* (Henry and Kim 1984). Members of the large orthotyline genus *Lopidea* often behave similarly (Knight 1917c, Asquith 1991). Dispersing mirids can injure plants used only for adult feeding—for example, rose (Vosler 1913, Brittain 1929) and buddleja (Burke 1923).

Dispersal to Sumac

Among native shrubs of eastern North America, staghorn sumac is especially attractive to mirids (pers. observ.). In addition to *Neurocolpus nubilus*, whose nymphs develop on sumacs (Henry and Kim 1984, Tonhasca and Luttrell 1991), adults of 14 other species have been collected from the terminal panicles of this plant in New York and Pennsylvania (Wheeler 1980a, Wheeler and Hoebeke 1985; pers. observ.): the mirines *Lygocoris belfragii*, *L. knighti*, *L. semivittatus*, *Lygus lineolaris*, *Neurocolpus jessiae*, *Taedia scrupea*, and *Trigonotylus caelestialium*; the orthotylines *Lopidea heidemanni* and *L. media*; and the phylines *Plagiognathus chrysanthemi*, *P. politus*, *P. punctatipes*, *Psallus physocarpi*, and *Rhinocapsus vanduzeei*. Many individuals were found deep within inflorescences where they appeared to feed on nectar, pollen, or both (pers. observ.). Female flowers of staghorn sumac produce more nectar than do the male flowers, but I did not look at possible sex-related differences in resource use on this dioecious plant. Knight (1917a) recorded two additional mirines, *Lygocoris caryae* and *L. tiliae*, from sumac flowers in New York (the latter species from smooth sumac), and *L. communis* occurs on sumac inflorescences in Quebec (Boivin et al. 1982). Arnoldi et al. (1992) observed the phylines *Campylomma verbasci* and *Plagiognathus obscurus* on the male inflorescences of staghorn sumac in Quebec. Knight and McAtee (1929), commenting on a record of the phyline *Americodema nigrolineatum* (as *Plagiognathus nigrolineatus*) on smooth sumac, surmised that the adult was attracted to the flowers.

Dispersal to Poison Ivy and Other Plants

Poison ivy is also attractive to mirids. Mirines such as oak-feeding *Lygocoris* species and the phyline *Atractotomus miniatus* disperse to the inflorescences of this high-climbing shrub or vine. Large numbers of these and other mirid species occur on poison ivy inflores-cences during late May to early June in New York, Pennsylvania, and Virginia (unpubl. data). Flowers of American hydrangea and New Jersey tea also attract numerous mirids (see Table 10.13). Among mainly predacious plant bugs, *Deraeocoris schwarzii* feeds in the inflorescences of shrubby cinquefoil in Wyoming (Stephens 1982). In England, *Plagiognathus arbustorum*, a facultative predator (Southwood and Leston 1959), visits spirea flowers (Parfitt 1884); the mainly predacious *D. lutescens* disperses to the flowers of English ivy (Verdcourt 1949), a climbing shrub whose flowers attract numerous other insect visitors (Elton 1966). The facultatively predacious phyline, *Atractotomus mali*, has been found dusted with pollen on a bramble flower in Britain (Groves 1969). Porsch (1957) recorded additional European mirids from flowers of various shrubs (see also Table 10.13).

Trees

Studies on insect-flower interactions often emphasize "typical" pollinator species (see also Chapter 11) and until recently have tended to stress extreme coadaptive mechanisms of pollination. Nonpollinators and inefficient pollinators that visit the flowers of trees for nectar and pollen, and obligate feeders on tree pollen and other floral structures, are overlooked or ignored in many ecological studies of inflorescences. The inconspicuous flowers of wind-pollinated or anemophilous trees are especially neglected as potential food sources for insects. As pointed out by Ogle (1878), "Their flowers have inconspicuous petals, or none; exhale no sweet odour; secrete no sweet nectar. They have no object in attracting insects, and consequently appeal to none of their senses." In his "Windpollen und Blumeninsekt," Porsch (1956) did not include mirids or other heteropterans among insects that visit anemophilous flowers. Yet mirids exploit the apetalous and inconspicuously petalous flowers of trees, whether they are pollinated by insects or the wind.

As used below, "apetalous" and "petalous" inflorescences are somewhat arbitrary categories recognized on the basis of perianth types. The former category, loosely interpreted botanically, includes several tree species whose flowers have inconspicuous petals—for example, ash, black gum, honeylocust, and the mimisoid flowers of acacias.

APETALOUS INFLORESCENCES

Presented first among examples of mirids associated with apetalous flowers of trees are species that develop on and disperse to the inflorescences of deciduous hosts, followed by those that occur on conifer "flowers." Consideration of resource switching, which is sometimes necessary for specialized bugs to complete their development, concludes this discussion.

Deciduous Trees

Few references to feeding on trees bearing catkins or other types of inconspicuous flowers are found in the North American literature on plant bugs. Knight (1941) remarked that *Orthotylus ramus* feeds on the staminate catkins of hickories and pecan but only sparingly on the pistillate catkins. Nymphs of *O. ramus*, *Plagiognathus caryae*, and *Lygocoris caryae* complete their development just before or as male catkins of pecan wither and drop from the trees (pers. observ.). *Orthotylus ramus* and *P. repletus* (probably a misidentification of *P. caryae*) feed on the staminate catkins of pecan trees in Georgia, in addition to the pistillate flowers, developing nuts, terminal growth, and young leaves (Tedders 1965). *Orthotylus juglandis* occurs on the staminate flowers of walnut in Texas (Henry 1979b).

Four plant bugs—the phylines *P. delicatus* and *P. gleditsiae* and mirines *Lygocoris tinctus* and *Taedia gleditsiae*—specialize on staminate inflorescences of honeylocust in the eastern United States. Their nymphs are camouflaged on the greenish yellow racemes of native and ornamental honeylocusts (Plate 14; Wheeler and Henry 1976).

Inflorescences of other North American trees, notably black gum and willow (Table 10.14), support a rich mirid fauna, the nymphs developing mostly on male flowers. An exception is the mirine *Agnocoris pulverulentus*, which feeds on the female catkins of willow and cottonwood (Plate 15; pers. observ.); the European *A. rubicundus* apparently has similar habits on willow (Gulde 1921). In addition, nymphs of *O. dorsalis* and *O. modestus* feed on the female catkins of sandbar willow in the eastern United States (pers. observ.). In eastern North America, the nymphs and adults of the adventive phyline *Sthenarus rotermundi*, which feed on the seeds of big-tooth and quaking aspen and on white poplar, are concealed among the white pubescent capsules of their hosts (Henry and Wheeler 1979, Wheeler and Henry 1992; T. J. Henry, pers. comm.).

Among arboreal mirids, species diversity in North America probably is greatest on oaks (e.g., Wheeler 1991b, Polhemus 1994), and many of the associated species are catkin feeders. Included in the rich plant bug fauna of scrub or bear oak in the eastern states are several species that develop mainly on the staminate catkins: the mirines *Lygocoris semivittatus* (Plate 15) and *Phytocoris olseni*; the phylines *Plagiognathus guttulosus*, *Psallus variabilis*, and *Teleorhinus tephrosicola*; and the orthotyline *Pseudoxenetus regalis*. Overwintered eggs of *L. semivittatus* and *P. regalis* hatch when staminate catkins are developing but before vegetative budbreak (Wheeler 1991b). Early instars of the latter species feed on the staminate catkins (Plate 15) and expanding foliage of white oak in Missouri (Blinn 1988).

Several mirines and phylines feed predominantly or exclusively on the staminate catkins of other eastern oaks, such as post oak and pin oak. Overwintered eggs begin to hatch when dormant buds containing inflorescence primordia are beginning to expand: mid to late March on post oak in North Carolina and about 4–6 weeks later on pin oak in Pennsylvania (unpubl. data). The nymphs of catkin feeders, such as the mirine *Neo-*

Table 10.14. Some New and Old World mirids that develop on "apetalous" inflorescences of trees

Mirid	Host plant	Locality	Reference
Mirinae			
Castanopsides hasegawai	Castanopsis	Japan	Yasunaga 1992d
Lygocorides rubronasutus	Daimyo oak	Japan	Yasunaga 1996b
Lygocoris kyushuensis	Castanopsis	Japan	Yasunaga 1991b
Lygocoris nyssae	Black gum	Eastern USA	Pers. observ.
Neurocolpus nubilus	Paper mulberry	Mississippi, USA	Tonhasca and Luttrell 1991
	Black gum	Eastern USA	Pers. observ.
Pinalitus conspurcatus	A Mediterranean sumac	Israel	Furth 1985
Salignus duplicatus,	Willow	Asia, Canada, western USA	Schwartz 1994
S. tahoensis			
Taedia johnstoni	Live oak	Texas, USA	Pers. observ.
Taedia scrupea	Black gum	Eastern USA	Pers. observ.
	White mulberry	New York, USA	Pers. observ.
Taedia sp. nr. *evonymi*	Osage-orange	Maryland, New York, USA	Pers. observ.
Orthotylinae			
Pseudoloxops takaii	Castanopsis, oak	Japan	Yasunaga 1997
Phylinae			
Atractotomus miniatus[a]	Black gum	Eastern USA	Henry 1989; pers. observ
Cariniocoris nyssae	Black gum	Maryland, USA	Henry 1989
Plagiognathus sp.	Black gum	Eastern USA	Pers. observ.

Note: See text for additional examples and discussion.
[a] *Lepidopsallus nyssae* is a synonym; see Stonedahl (1990).

capsus leviscutatus on post oak, resemble the color of host inflorescences, and a striking ontogenetic color change occurs in at least one species. Whereas the greenish yellow early instars of *Lygocoris quercalbae* blend in with the new catkins (Plate 15), late instars, yellow and distinctly mottled with brown, are camouflaged on the darker withered catkins (Plate 15).

Numerous catkin-feeding plant bugs occur on western North American oaks (Polhemus 1994; pers. observ.), but their seasonal histories and habits remain largely unstudied. Members of the *Atractotomus miniatus* group, of which seven of the eight species are restricted to the western states or Mexico, reach their greatest densities during the flowering period of host oaks. They are possible pollen feeders (Stonedahl 1990).

Mirids are also inflorescence specialists on European trees, feeding on the staminate catkins of alder, aspen, birch, hazel (Plate 16), oak, and willow (Kullenberg 1944). The diverse plant bug fauna of oak (e.g., Kullenberg 1944; Leston 1951; Ehanno 1965, 1987b; Taksdal 1965; Goula 1986) includes several early-season species. Catkin feeders such as the phyline *Harpocera thoracica* require only about two weeks for nymphal development (Southwood and Leston 1959). A mirine known as the catkin bug (*Pantilius tunicatus*), though sometimes stated to be a leaf feeder on trees (Ellis 1940, Massee 1956), feeds on the catkins of wild Betulaceae and on the male catkins of cultivated European filbert in Italy (Plate 16), causing abscission and limiting nut production (Ciampolini and Servadei 1973, Arzone 1983). *Dryophilocoris miyamotoi*, an orthotyline known only from Japan, occurs mainly on the flowers of oaks (Yasunaga 1999a). A Japanese mirine, *Castanopsides falkovitshi*, might feed mainly on pollen in the inflorescences of its juglandaceous hosts (Yasunaga 1998).

This discussion has so far emphasized mirids that develop on the inflorescences of trees, but some species also feed on the fruits and developing seeds. The nymphs of *Psallus lepidus*, a Palearctic phyline adventive in eastern North America, feed on the fruits (samaras) of European ash and other Old World species of *Fraxinus*. At the Arnold Arboretum near Boston, this bug discolors the fruits and spots them with dark excrement (Plate 16; Wheeler and Hoebeke 1990). An undetermined Australian mirid causes the young seed pods of a wattle or acacia tree to wilt and turn brown (Van den Berg 1982).

Tree flowers, like those of other plant types covered in this chapter, attract adults of mirids that breed elsewhere. Adults of *Lygus lineolaris* and *Metriorrhynchomiris fallax* disperse to the staminate catkins of white oak in Pennsylvania (pers. observ.), and overwintered adults of the former mirid are attracted in early spring to willow catkins in Nova Scotia (Brittain and Saunders 1918; see also Robertson [1928]). *Lygocoris caryae*, *L. fagi*, *L. quercalbae*, *L. semivittatus*, and *L. tiliae*, as well as the phyline *Plagiognathus guttulosus*,

that develop on other deciduous trees can be collected in the eastern states on the staminate flowers of black gum (pers. observ.). Some of the species collected on the inflorescences of trees that do not serve as nymphal hosts are merely seeking shelter or are of accidental occurrence, that is, vagrants (e.g., Southwood et al. 1982). More often, though, they likely feed on the nectar or pollen (see Chapter 11). Certain mirids of phytophagous and predacious habits can even use such trees as occasional breeding hosts (e.g., Štys and Kinkorová 1985).

Conifers

Mirids also specialize on the "flowers" of coniferous trees. The nymphs and adults of the phylines *Phoenococoris claricornis* and *Pinophylus carneolus* feed on the male cones (microsporangiate strobili) of Virginia pine in Pennsylvania (Plate 16; Wheeler 1999). These plant bugs, which are most abundant on Virginia pines bearing large numbers of male cones, seem less common on native table mountain pine (Wheeler 1999; pers. observ.). The free amino acid content of pine pollen, noted for its generally poor nutritive value to honey bees, differs significantly among species (Stanley and Linskens 1974). Could the apparent scarceness or absence of *Ph. clavicornis* and *Pi. carneolus* from populations of table mountain pine be related to lower levels of key amino acids in this tree, or does it simply reflect this tree's more limited range compared to Virginia pine? In addition, *Pi. carneolus* pierces the staminate strobili of jack pine in conifer seed orchards in Wisconsin (Rauf et al. 1985). Rauf et al. reported similar habits for another phyline, *Microphylellus flavipes*, but this record is based on a misidentification of males of the sexually dimorphic *Pi. carneolus* (Wheeler 1999). Ebel (1963) collected nymphs and adults of *Phoenococoris australis* from the male strobili of slash pine in the southeastern United States, and Mattson (1975) noted the presence of mirids (undetermined taxa) in the male strobili of red pine in Minnesota.

Adults of the pine conelet bug (*Platylygus luridus*) feed on the conelets of jack pine (nymphs develop mainly on the developing shoots and needles), causing abortion and reducing seed yields. In cages, the intensity of conelet feeding appears to increase as the number of available conelets decreases (Table 10.15). Under experimental conditions, abortion occurs 2–10 weeks after the introduction of adults. In one Wisconsin orchard, the conelet abortion rate averaged about 75% in three of four years. Aborted conelets can remain on branches for a year or more (Plate 16). Although the precise mechanism of abortion is unknown, stylet penetration into ovules (Plate 16) can destroy nucellar walls (Rauf et al. 1984a, 1984b). Other mirids occurring on jack pine, including *Dichrooscytus suspectus*, *Phytocoris michiganae*, and *Pilophorus amoenus*, do not cause conelet abortion. *Phytocoris michiganae* inserts its

Table 10.15. Effect of feeding by *Platylygus luridus* adults (4) on conelet abortion of jack pine under different durations of exposure and varying numbers of conelets

	Average % conelet abortion at exposure duration (days)[a]	
No. of conelets/cage	6	12
2–4	52.8	100.0
5–7	41.7	83.9
≥8	25.3	77.9

Source: Rauf et al. (1984a).
[a] Abortion was nil in cages without *P. luridus*.

stylets deep into the cones and might feed on seeds (Rauf et al. 1984a, 1985).

In addition to the use of pollen by mirids that actually develop on pine, species breeding on noniferous hosts occasionally disperse to pine to feed on pollen. Schwartz (1984) gave circumstantial evidence for *Irbisia bliveni*, a member of a mainly grass-feeding genus (see Chapter 9), migrating to pines to exploit pollen; several pollen-covered individuals were collected from pines in the mixed conifer forest of California's Sierra Nevadas.

PETALOUS INFLORESCENCES

Fewer tree-feeding mirids in North America develop on petalous than on apetalous flowers. An early observation was that of the phyline *Plesiodema sericea* on the inflorescences of American basswood or linden. In the Washington, D.C., area, Heidemann (1892) noted that this mirid "likes to hide in the withered blossoms, and evidently punctures the forming fruit; it is easily overlooked, as its straw-yellow color does not offer the slightest contrast with the faded blossoms." Nymphs of *P. sericea* feed on the developing buds of basswood flower clusters in Minnesota, adults appearing as the flowers enter full bloom (Knight 1926b). *Neurocolpus tiliae* develops on American basswood and littleleaf linden flowers in the eastern United States (pers. observ.). Adults of *Lygocoris communis* and another mirine, *N. nubilus*, and the phylines *Microphylellus modestus* and *Plagiognathus politus* are attracted to basswood flowers in Ontario (Judd 1980). Flowers of basswood (lime) trees in Europe are fed on by the mirines *Lygocoris viridis*, *Pinalitus cervinus*, and *P. viscicola* (Southwood and Leston 1959).

Nymphs of *N. nubilus* sometimes develop in the large, showy flowers of tuliptree. In Pennsylvania, their feeding results in a brown stippling on filaments and a premature dropping of stamens (Henry and Kim 1984).

Tuliptree flowers also attract the adults of mirids that breed on other hosts. The most common plant bugs collected on tuliptree in New York, *Lygocoris caryae* and *L. omnivagus*, are more abundant on young than

old flowers. Three additional *Lygocoris* species, the mirine *Taedia* species, and phyline *Monosynamma bohemanni* also can be found in tuliptree flowers, and all seven species feed mostly on the lower portions of the carpels. Because only 11 of 20 individuals marked with fluorescent dust remained for a day on the same flower, mirids were thought to "move enough to be able to respond rapidly to factors such as changes in flower quality and availability" (Andow 1982). In Missouri, Blinn and Yonke (1985) collected *L. hirticulus* from a tuliptree flower, and *L. hirticulus* and *L. fagi* sometimes disperse to tuliptree flowers in Pennsylvania (pers. observ.).

Each season during a five-year study of northern catalpa in the University of Michigan's Matthei Botanical Garden, the nymphs and adults of *Taedia scrupea* fed on flower buds, causing them to blacken, to shrivel before opening, or in some cases, to abscise. At times injury was so severe that only one or two flowers of a panicle opened (A. G. Stephenson, pers. comm.). Adults of *Lygocoris* species, *Lygus lineolaris*, *Neurocolpus tiliae* (Mirinae), *Atractotomus magnicornis* (Phylinae), and other mirids in the eastern United States disperse to the flowers of northern and southern catalpa (pers. observ.).

SPECIALIZATION AND RESOURCE
SWITCHING ON TREE INFLORESCENCES

Specialization in the feeding habits of a herbivore allows its life cycle to be linked closely to host phenology (e.g., Feeny 1975). The small size of mirids and their generally rapid development allow short-lived resources to be used (e.g., Bernays 1982). But the exploitation of ephemeral resources such as catkins carries obvious risks (e.g., Southwood 1977, 1978a). In the absence of the preferred resource, an inflorescence specialist must use other options: move to flowers of a conspecific individual or of a different plant species, use a different part of the usual host, or enter dormancy until the preferred resource becomes available (e.g., Jensen 1951, Sholes 1980; see also Simpson and Simpson [1990]). Seed-feeding heteropterans sometimes must use other plant tissues when seeds are unavailable (Slansky and Panizzi 1987).

Adults of most catkin-feeding mirids on oak appear about the time pollen is shed or catkins begin to wither. Yet in some years or on certain trees, catkins drop before mirid feeding is completed, forcing nymphs to feed on vegetative tissues or other resources (perhaps animal matter) to complete their development. In the laboratory, late instars of *Lygocoris quercalbae* probe dehisced anthers and pollen-coated areas on leaves and stems (see Chapter 11), and in the absence of reproductive parts, they feed on young foliage, particularly midribs and lateral veins, and on crushed caterpillars (pers. observ.).

Nymphs of arboreal mirids are sometimes blown onto understory plants (e.g., Brown 1982a). I have frequently collected nymphs of tree-feeding *Lygocoris* species on understory vegetation, high winds or some other factor having dislodged them from host trees. In Pennsylvania, I observed late instars of an oak-associated *Lygocoris* species apparently feeding on the heads of wheat growing under white oaks that bordered one edge of the field (pers. observ.). In a New York arboretum, fifth instars and teneral adults of *L. geneseensis* (or similar species) fed on the unfolding, terminal leaves of a composite (perhaps an aster) growing beneath white oak. That nymphs can survive on this nonhost plant was indicated by the small, discolored spots on the foliage (pers. observ.). Brittain (1916a) made similar observations on the pear plant bug (*Lygocoris communis*) in Nova Scotian orchards. When nymphs were dislodged from fruit trees by rain, wind, or pesticide spray, they fed on various herbs or grass. According to Brittain (1916a), "Even when forced to feed on these plants early in the nymphal life the insects seemed to be able to complete their transformations, but once they had obtained wings, they invariably sought the fruit of the apple or pear."

In Europe, when *L. pabulinus* nymphs are dislodged from apple trees by wind or rain, they sometimes find suitable nourishment on herbaceous plants beneath (Rostrup and Thomsen 1923). In the case of this mirid, herbs can also serve as summer hosts, this bivoltine species undergoing what Rostrup and Thomsen termed a "primitive migration" between woody, first-generation hosts and herbaceous, second-generation hosts.

11/ Nectar and Pollen Feeding and Pollination

Although pollen is thought to have been the original food sought by the most primitive anthophiles, nectar is now the reward most commonly sought.

—P. G. Kevan and H. G. Baker 1983

Even with the concept of a definite relationship in mind, it is not always easy to draw the line between pollinators and accidental visitors.

—K. Faegri and L. van der Pijl 1979

In the previous chapter, I showed that many mirids are flower specialists that develop exclusively on inflorescences. The family also contains species with various trophic habits that disperse to feed on the nectar and pollen of nearby nonhost plants. This distinction between true or breeding hosts and plants that are used solely for adult feeding was made in Chapter 10. Entomologists and ecologists often assume that mirids do not feed on flowers, dismissing plant bugs as transients (e.g., Travis 1984; see also Chapter 10). Yet inflorescence visits and the use of sugar-rich resources in flowers can be expected to improve mirid fecundity and thus their reproductive potential.

Heteropterans, generally considered insignificant pollen consumers and pollinators (e.g., Crepet 1983, Primack 1983), are, nevertheless, conspicuous and common nonholometabolous flower visitors (Kevan and Baker 1983, 1984). Little information is available on the importance of nectar and pollen in their diet or their importance as pollinators, but the ecological significance of their occurrence on flowers should not be disparaged automatically (e.g., Torre-Bueno 1929). Also, the possible use of floral rewards other than pollen and nectar, such as stigmatic exudates (Simpson and Neff 1983), has not been investigated. In this chapter, nectar as a nutrient source for mirids is reviewed, their pollen-feeding habits are examined, and the literature on their role as pollinators is compiled and evaluated.

Nectar as a Food Source

Nectar is considered an important floral reward (rather than a primary attractant; e.g., Meeuse 1972) for insects, especially Lepidoptera, Diptera, and Hymenoptera (e.g., Kevan and Baker 1983, 1984; Simpson and Neff 1983; Barth 1985). The monosaccharides and disaccharides of nectar represent the principal energy source for flower visitors. Different plants have different sucrose-hexose ratios. Certain families are characterized by sucrose-rich or hexose-rich nectars, whereas others show diversity. Taxonomically unrelated plants of the same pollinator type can show similar sugar and amino acid ratios (Baker and Baker 1983b), but this relationship does not always hold (Gottsberger et al. 1984, 1989). Closely related plant taxa with different pollinators can also show dissimilar sugar ratios (Baker and Baker 1983b).

Secretion and concentration of nectar vary with plant age, weather, microclimate, genetics, physiological and nutritional stage, diel rhythms, and other factors (e.g., Boggs 1987). Nectar secretion, a complex physiological process, involves more than the production of an aqueous sugar solution. Nectar-gland cells actively remove and add certain substances from phloem sap. Among other nectar constituents are amino acids, including the 20 commonly found in proteins; lipids; proteins; vitamins; organic acids; and toxic chemicals such as alkaloids, glycosides, and phenolics, the last named perhaps affecting only the taste of nectar (e.g., Beutler 1953; Baker et al. 1978; Baker 1983; Baker and Baker 1983a, 1983b, 1986; Kevan and Baker 1983, 1984; Barth 1985; Roubik 1989). Floral nectars do not constitute a nutritionally complete diet for insects because they rarely contain all 10 amino acids essential for growth and reproduction (Hagen 1986), but nectar feeders could obtain all essential amino acids by visiting a variety of flowers. Pollen-contaminated nectar contains more amino acids than plain nectar because of passive diffusion of free pollen amino acids into a sucrose medium (Hagen 1976, Erhardt and Baker 1990).

Hocking (1953) pointed out that a large proportion of adult insects depend on nectar as a carbohydrate fuel source. Nectar also can serve as a source of moisture (Kevan 1973). Heteropterans and many other hemimetabolous insects are underappreciated as nectarivores. Mirids and other heteropterans are not mentioned in Elton's (1966) discussion of the "concourse" of British insects that imbibe floral nectar. But berytids, largids, lygaeids, pyrrhocorids, and other heteropterans are at times nectarivorous (e.g., van Doesburg 1968, Wheeler and Henry 1981, Opler 1983, Wheeler 1983a); in the case of pyrrhocorids, their nectar sources have been referred to quaintly as "adult playhouses" to distinguish such plants from true (breeding) hosts (Myers 1927b).

Müller (1883) wrote that plant bugs seek nectar on the flowers of composites, umbellifers, and willows. Earlier, Kerner (1878) had observed an adult of the orthotyline *Globiceps flavomaculatus* adhering to the viscid flower stems of nodding silene in Austria, one of many insect species believed to have visited the plant for its nectar. Rammner (1942) discussed nectar feeding by several mirid species, and Kullenberg (1944) emphasized that the habit is well developed in the family. A lack of dietary sugar, which may be obtained from floral nectar or honey, results in decreased fertility in *Adelphocoris lineolatus* males and the arrest of oocyte development at previtellogeny in females (Masner 1965). Adult tarnished plant bugs (*Lygus lineolaris*) feed on the nectar of goldenrods (Sholes 1984). Flower-visiting mirids (unspecified taxa) seek nectar on an Australian astrotricha and a dillwynia (Armstrong 1979), *Plagiognathus* species imbibe nectar from geranium flowers in Germany (Schaarschmidt 1982), and *Nesidiocoris tenuis* feeds on the nectar of bottle gourd flowers in India (Shrivastava 1991).

Mirids often imbibe the nectar of their breeding hosts, and many other species disperse to nonhost plants to take nectar. An example of a mirid that takes nectar from a plant not used for nymphal development is *Metriorrhyncomiris dislocatus* on wood lily flowers (Barrows 1979). Whether adult mirids seek particular kinds of plants for feeding is not known, but they have been recorded most frequently from composites and umbellifers (Rammner 1942, Porsch 1957, Sidlyarevitsch 1982). The floral nectar of the shallow flowers in these families is easily accessible (Kevan and Baker 1983). Sidlyarevitsch (1982) pointed out that umbellifers such as dill and mustard (Apiaceae), planted to serve as a carbohydrate source for parasitoids that attack agricultural pests, could also attract predators such as the mirid *Campylomma verbasci*. Flowers of herbaceous plants might be visited more often than those of trees and shrubs; the amino acid concentrations in the nectar of herbs are generally higher than those in woody plants of even the same family (Baker and Baker 1986). The presence of some nonresident mirids (those breeding on other plant species) recorded from Canada thistle (Maw 1976), goldenrod (Reid et al. 1976, Messina 1978), and other composites (e.g., Graenicher 1909, Robertson 1928) might represent more than "sitting" records.

Because mirids frequently feed on nectar and pollen, entomologists and ecologists should not routinely disregard mirids observed on any plants—including those that seem unlikely to be used even for adult feeding. The bugs' relationship to the plants may be more than one of chance occurrence. Even so, they still would qualify as "tourists" (Moran and Southwood 1982), for most of their nutrition is obtained from other plant species.

Pollen Feeding by Mirids

Many adult insects exploit pollen, a nitrogenous food source rich in free amino acids and protein (e.g., Kevan and Kevan 1970, Gilbert 1972, Stanley and Linskens 1974, McLellan 1977, Bell et al. 1983, Rabie et al. 1983, Simpson and Neff 1983, Hagen 1986). Pollen feeding and carnivory appear closely correlated in an evolutionary sense in several insect groups (Kevan and Baker 1983); the addition of both pollen and animal food to the diet of mirids can enhance life-history traits (e.g., Musa and Butler 1967, Rosenthal and Lipps 1973, Perdikis and Lykouressis 2000; see also Chapter 6). As an often concentrated, high-quality food (Barth 1985), pollen is needed by some species for egg production, development of fat body, and growth. In others, it increases fecundity and longevity. Although many arthropod groups feed on pollen, few authors have described or even proposed the actual method of getting access to and digesting the contents. Stanley and Linskens's (1974) brief review of pollen digestion in insects deals mainly with the honey bee (*Apis mellifera*).

Pollen grains represent a somewhat "difficult diet" for insects (Kearns and Inouye 1993). Pollen feeding is easier to visualize for mandibulate insects than for those with mouthparts adapted for piercing and sucking, particularly in groups that cannot ingest the grains and presumably must rely on extraoral digestion. The extraordinarily resistant pollen wall is almost indestructible (e.g., Heslop-Harrison 1971, Faegri and Iversen 1975, Barth 1985); the outer shell or exine is difficult to break by grinding and other mechanical means (e.g., Vinson 1927, Barker and Lehner 1972). As Crowson (1981) pointed out, pollen's contents may not be readily accessible even to beetles (see also Grinfeld [1975]).

The rather specialized pollen-feeding mechanisms of selected lapping or sucking arthropods, including the heteropteran family Anthocoridae, are summarized in Table 11.1 to highlight the morphological and behavioral adaptations that facilitate the use of pollen. Sometimes authors have casually stated that phytoseiid mites (e.g., Chant 1959, Chant and Fleschner 1960, Osakabe et al. 1986, McMurtry and Rodriguez 1987), anthocorids, and other small sucking arthropods feed on pollen, often without supporting evidence or regard to the presumed difficulty of obtaining nutrients once the exine is formed during microgametogenesis. For arthropod groups omitted from this overview, the reader should consult reviews by Baker and Hurd (1968), Proctor and Yeo (1972), Faegri and van der Pijl (1979), and Kevan and Baker (1983, 1984).

The nymphs and adults of numerous mirid species pierce the immature anthers of their hosts to feed on pollen sacs (microsporangia) containing nutritive cells and young pollen grains (Kullenberg 1944, Pack and Tugwell 1976; pers. observ.; see also Chapter 10). By radiolabeling ragweed plants, Shure (1973) determined that three mirid species associated with this host obtain

Table 11.1. Pollen-feeding habits in selected groups of insects with lapping or sucking mouthparts

Species	Pollen-feeding behavior and adaptations	Reference
Acarina		
Laelapidae		
Pneumolaelaps longanalis	Feeds on nectar coating bumblebee-collected grains and on pollenkitt, apparently using salivary secretions; certain pollen types also may be ruptured, with mites possibly feeding on nutrients of pollen core	Royce and Krantz 1989
Phytoseiidae		
Amblydromella caudiglans	Pierces individual grains and empties the contents	Putman 1962
Amblyseius similoides	Picks up pollen grains individually with chelicerae; ruptures exine using alternate movements of chelicerae and sucks grain contents	Flechtmann and McMurtry 1992
Euseius stipulatus	See remarks under *Amblyseius similoides*	Flechtmann and McMurtry 1992
Euseius tularensis (as *E. hibisci*)	Gut contents turn yellow following exposure to pollen, indicating grains are consumed; population growth is correlated with fallout of wind-borne pollen	McMurtry and Johnson 1965, Kennett et al. 1979
Typhlodromus pyri	Grasps a grain with chelicerae, sucks out contents, and discards exine	Chant 1959
Araneae		
Araneidae		
Araneus diadematus	Grains caught on adhesive spirals of web provide food for spiderlings, doubling their life expectancy; an "exinase" might be involved because grains are too large to pass through cuticular platelets of pharynx	Smith and Mommsen 1984
Heteroptera		
Anthocoridae		
Orius insidiosus	Bugs pierce individual grains and suck out contents; nymphal development can be completed on pollen diet; fecundity and longevity are increased	Dicke and Jarvis 1962, Kiman and Yeargan 1985
Orius pallidicornis	Specializes on pollen of a wild cucumber, feeding in male flowers on grains that have fallen on petals; stylets are inserted in a grain and are alternately and rhythmically protracted	Carayon and Steffan 1959
Orius vicinus	Grain size and surface sculpturing determine plant species used for pollen feeding	Fauvel 1974
Paratriphleps laeviusculus	Feeds on pollen of a sapodilla; nymphs complete development on piece of flower with pollen-bearing filament	Bacheler and Baranowski 1975
Thysanoptera		
Aeolothripidae, Phlaeothripidae, and Thripidae		
Various spp.	Suck out pollen contents one grain at a time, leaving the pollen wall; mouthcone is placed on grain without manipulation or use of palps and forelegs; size of grain and presence of spines on exine do not affect feeding	Kirk 1984, 1985; see also Grinfeld 1959, Lewis 1973
Diptera		
Ceratopogonidae		
Atrichopogon pollinivorus	Rostral tip is applied to dehisced anther and a fresh pollen grain located by probing; grains begin to collapse within 3 sec and totally collapse in 3 more sec	Downes 1955
Syrphidae		
Various spp.	Grasp filament with foretarsi and pull anther to mouth; pollen identified in guts of individuals observed collecting pollen in field	McAlpine 1965, Gilbert 1981
	Sparsely pubescent species with unbranched body hairs and short proboscis collect pollen directly from anthers; larger, more pubescent species with elongate mouthparts and palyonophilic hairs use front and hind tarsi to comb pollen trapped among body hairs; pollen is ingested from bristles on front tarsi	Holloway 1976

Table 11.1. *Continued*

Species	Pollen-feeding behavior and adaptations	Reference
	Nutrients likely are obtained from pollen through diffusion; pollen remains intact in esophagus and crop but yields contents in midgut; nectar in gut may trigger exudation of grain contents	Gilbert 1981, Haslett 1983
Lepidoptera Nymphalidae *Heliconius* spp.	Dry pollen mass forms on ventral side of proboscis near head; clear liquid (probably nectar) exudes from proboscis tip and is mixed with pollen; wet pollen load is agitated for several hours through coiling and uncoiling of proboscis	Gilbert 1972; see also Dunlap-Pianka et al. 1977, Gilbert 1980
	Clear fluid in which pollen begins to germinate is saliva, not regurgitated nectar	Boggs 1987
	Passive diffusion, rather than active release of free pollen amino acids, is used to obtain nutrients	Erhardt and Baker 1990
Battus, Parides spp.	Pollen use similar to that by *Heliconius* is suggested, although pollen feeding might not represent typical behavior	DeVries 1979; see also Boggs 1987
Hymenoptera Ichneumonidae *Scambus buolianae*	Adults do not consume dry pollen but swallow grains with nectar; pollen breakdown occurs in midgut	Leius 1963
Apoidea: Apidae *Apis mellifera*	Adults use enzymes to penetrate membrane and pores of exine; digestion takes place in midgut, and grains are devoid of contents by time hindgut is reached	Whitcomb and Wilson 1929
	Microbial digestion of pollen may occur	Stanley and Linskens 1974; cf. V. Turner 1984
	Digestion in alimentary tract may be preceded by osmotic shock to grains, which burst due to differences in osmotic pressure as they pass between honey sac and ventriculus; thick-walled dandelion pollen does not burst immediately; differences in osmotic pressure between crop and midgut cause swelling of germination pores; pollen wall ruptures and collapses, leading to gradual extrusion of protoplasmic contents	Kroon et al. 1974, Peng et al. 1985
	Starches, sugars, pectic acids, proteins, and nucleic acids released from pollen are digested and absorbed; some complex proteins, sporopollenin, and waxes cannot be digested	Klungness and Peng 1984

most of their food from pollen tissue when that resource is available. Mirids also appear to feed on mature pollen grains (multicellular male gametophytes) on the anthers or, following dehiscence, on pollen that adheres to plant surfaces. Plants often intercept some of the vast quantities of pollen produced by wind-pollinated taxa, the grains accumulating or being trapped on leaves and stems (e.g., Whitehead 1983, Niklas 1985).

OBTAINING NUTRIENTS FROM POLLEN

Accessing pollen's protoplasmic contents would seem to present a problem for sucking arthropods once exine formation begins after meiosis and sporopollenin is deposited on the outer exine layer (sexine) upon separation of the tetrad. Or maybe access to the protoplasm is

not necessary—that is, the bugs might derive energy-rich nutrients from the sporopollenin content of the exine and from other wall layers and the pollen coat or pollenkitt.

During pollination, mirids such as *Adelphocoris lineolatus* and *Calocoris roseomaculatus* suck up pollen from newly opened anthers or from stigmas. Realizing pollen grains might be too large to be sucked up through the food canal, Kullenberg (1944) suggested that mirid salivary enzymes dissolve the pollen wall, permitting access to the contents. He observed that the food canal sometimes becomes so filled with a mushy pollen-saliva mixture that it filters out between the maxillary stylets. Few comments on pollen feeding in mirids have been made since Kullenberg's (1944) studies. Schwartz (1984) thought *Irbisia* species feed on pollen and said dissection of the gut might verify Kullenberg's (1944) hypothesis that mirids suck up pollen grains after mixing them with copious amounts of saliva.

Kullenberg's (1944) hypothesis has not been experimentally tested. In preliminary work with several eastern North American mirids restricted to developing on inflorescences and suspected of pollen feeding, the use of pollen stains did not reveal the presence of exines in the gut (unpubl. data). Pollen grains of most species are 20–40 μm (length and width); the mean diameter of most wind-pollinated taxa ranges from 17 to 58 μm. Extremes of 3–5 μm or more than 200 μm are recorded, the smaller grains usually associated with specialized pollination systems and the larger grains mostly transported by animals (Wodehouse 1935, Stanley and Linskens 1974, Faegri and Iversen 1975, Muller 1979, Barth 1985). Probably all but the smallest grains are too large to be ingested (Fig. 11.1)—thus, pollen feeding in mirids is not analogous to sucking beads through a straw. Coreids, thyreocorids, and other heteropterans (unspecified families) visiting the flowers (cyathia) of euphorbias in the desert Southwest did not contain pollen in their guts, whereas the guts of most Coleoptera, Diptera, and Hymenoptera contained pollen (Ehrenfeld 1979).

Also relevant to this discussion on the use of pollen by mirids is feeding by the mirine *Lygocoris communis* on conspecific individuals infected with an entomophthoraceous fungus. The possibility that conidia (15–18 × 17–23 μm) could be sucked up into the alimentary canal was dismissed because these structures are too large (Dustan 1924). Yet some thrips are able to ingest chloroplasts, which otherwise would be too large to be sucked through the maxillary tube, apparently by distorting them into a sausage shape (Lewis 1987). Pollen's rigidity would seem to prevent such a change in shape.

Another possibility is that mirid stylets pierce the pollen wall, a capability inferable for anthocorids (Carayon and Steffan 1959, Dicke and Jarvis 1962, Kiman and Yeargan 1985). Because mirid mouthparts do not seem as well adapted for piercing pollen as those of Thysanoptera (see Table 11.1), the toughness of the wall

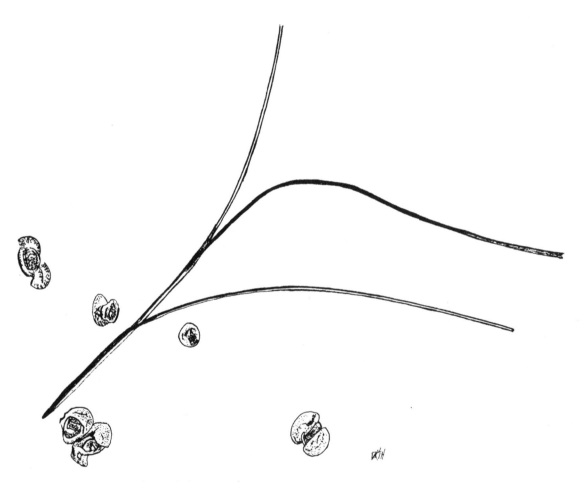

Fig. 11.1. Size of the stylet tip of *Pinophylus carneolus* relative to that of pollen grains from its host, Virginia pine (×250). (Redrawn from a photomicrograph.)

might preclude penetration of thicker areas. Compared to larger apertures of the exine, however, the mirid labial tip is slender enough to probe the furrows and pores and extract the inner contents. The size of the labial tip probably is not as important as the circumstances that would allow pollen to be punctured—that is, a grain would need to be held fast so that the stylets could obtain enough leverage to pierce the pollen wall. This method might allow a rich supplemental food source to be tapped when other sources are inadequate, or be important in some species, but feeding on individual grains might be a relatively inefficient means of obtaining protein and energy. The observation that nymphs of *Pseudatomoscelis seriata* and *Campylomma verbasci* insert their stylets into pollen grains (Burden et al. 1989, Bartlett 1996) suggests that some pollen-feeding mirids exhibit this behavior, but it does not contradict the diffusion theory to be presented later in this chapter.

Possible also is that mirids and anthocorids have evolved salivary enzymes capable of destroying exines. Some Collembola may possess a midgut "exinase" that helps digest the outer pollen wall (Scott and Stojanovich 1963). Although the presence of such an enzyme in collembolans has been accepted (e.g., V. Turner 1984), it apparently has not been demonstrated. Honey bees were once thought to degrade pollen enzymatically, the enzymes originating in the bee or perhaps in the microflora or the pollen (Barker and Lehner 1972). The results of a study demonstrating that spiderlings of the araneid *Araneus diadematus* feed on pollen caught on their adhesive webs suggested the presence of an exinase (Smith and Mommsen 1984).

That insects have enzyme systems for digesting pollen exines has not been verified (Faegri 1971). Enzymes might facilitate the extraction of nutrients from pollen. At least certain substances and microorganisms can move through the wall apertures (Faegri 1971, Rowley 1971, Stanley and Linskens 1974), and such openings could allow mirid saliva to invade thinner areas of the wall. Direct enzymatic action through germinal apertures in the wall is believed to allow pollen digestion in vertebrates—for instance, in certain marsupials (Richardson et al. 1986) and birds (Wooller et al. 1988). The chemistry of the process remains uninvestigated (Richardson et al. 1986) and might involve other mechanisms, such as pollen tube formation (V. Turner 1984). An enzymatic breakdown of pollen by mirids is possible and should be investigated using appropriate chemical techniques.

POLLEN DIGESTION BY NUTRIENT DIFFUSION

More likely is that mirids gain access to pollen nutrients not from direct enzymatic breakdown but from diffusion—or a complementary diffusion-enzymatic extraction of nutrients. Digestion of pollen by nutrient diffusion is not a new idea. Kevan and Kevan (1970) credited Müller (1873), who had observed undamaged pollen grains in the guts of syrphid flies, with the thought that diffusion allows access to nutrients held in the protoplast. Kevan and Kevan (1970) suggested that diffusion occurs in honey bees and in pollenivorous Collembola and some calyptrate Diptera (Muscidae).

Erhardt and Baker (1990) emphasized diffusion as a means of removing nitrogenous compounds from pollen. In providing the first report of butterflies as pollen feeders, Gilbert (1972) had pointed out earlier that pollens release free amino acids and protein when incubated in a sucrose solution; at 25°C, about 50% of the free amino acids are released within the first minute. Amino acids, however, probably diffuse passively out of pollen grains rather than being actively released into the medium, as Gilbert (1972, 1980) believed (see Erhardt and Baker [1990]). *Heliconius* butterflies mix pollen with saliva extruded from the tip of the proboscis, which apparently triggers the release of nutrients without the need for a complement of gut enzymes. Only sucrose was reported to release proteins and amino acids (Gilbert 1972), but other sugars or osmoeffectors might act similarly. By strict physico-chemical considerations, water having minimal or no solutes (e.g., dew) could drive a solvent-solute movement and elicit a stronger or faster reaction. Such a mechanism would be an energy-efficient means of exploiting pollen.

Gilbert's (1972) much-cited paper has sometimes been misinterpreted. First, diffusion was suggested not as a mechanism allowing pollen to germinate in the gut of butterflies, as is sometimes stated (e.g., Haslett 1983, Peng et al. 1985), but was assumed to involve active release, as suggested by the work of Stanley and Linskens (1965) and Linskens and Schrauwen (1969) on pollen physiology. Pollen was thought to be digested extraorally and then imbibed. *Heliconius* species do not ingest pollen grains. Dunlap-Pianka et al. (1977:490, note 9) stated that pollen is mixed with liquid, which induces it to pregerminate and release amino acids into solution.

Second, diffusion permits access to nutrients contained in the protoplast, even though V. Turner (1984) implied that this mode of digestion, as proposed by Gilbert (1972), involves wall-borne proteins rather than the protoplast. Turner noted that most nitrogenous compounds occur in the protoplast and that in the two Australian marsupials studied, removal of pollen contents required the intine to be broken. Drawing on Linskens and Mulleneers's (1967) paper on the formation of instant pollen tubes, V. Turner (1984) proposed that instant or natural pollen tubes are formed in the gut of the pollen-feeding honey possum (*Tarsipes rostratus*) and that disruption of such tubes through mechanical activity in the gut could expose the protoplast to digestive enzymes. Later studies on certain pollen-feeding marsupials and birds discounted the role of pollen tube initiation and invoked direct enzymatic action to explain pollen digestion (Richardson et al. 1986, Wooller et al. 1988). Nectar-induced germination of pollen in the crop of Darwin's finches (*Geospiza* spp.) may facilitate their digestion of pollen (Grant 1996).

Strengthening a diffusion explanation for extraction of pollen nutrients in insects is the presence in the hindgut of entire grains devoid of contents (e.g., Whitcomb and Wilson 1929, Haslett 1983, Peng et al. 1985). The studies of Kroon et al. (1974) and Peng et al. (1985) on pollen digestion in honey bees suggested a sugar-initiated exudation of protoplasmic contents (see Table 11.1). A gradual rupturing of grains, probably beginning with formation of a germination tube, also seems to occur in syrphid flies as a means of access to pollen nutrients (Haslett 1983). In certain bombyliid flies, the liquid (salivary secretions or perhaps material regurgitated from the gut) that appears to be mixed with pollen as it is transported along the proboscis (Deyrup 1988) might serve a similar function. Nectar and urea in the stomach of glossophagine bats initiates pollen germination and access to cellular contents (Howell 1974).

A MECHANISM OF POLLEN DIGESTION IN MIRIDS

Pending empirical verification of pollen-feeding mechanisms in mirids, I propose that saliva is applied to lumps of pollen or to numerous grains coating a small area of leaf or other plant surface. Osmotic effectors, probably sugars or minerals obtained from plant sap, trigger pollen germination and a gradual release of protoplasmic contents. The soupy pollen-saliva mixture is then sucked up through the food canal.

A diffusion explanation of access to the contents of pollen grains would support Kullenberg's (1944) observation of a mushy pollen-saliva mixture filtering out between the bugs' maxillary stylets. Proteolytic enzymes in the saliva might enable mirids to extract more of the nitrogen content from pollen than would be possible from passive diffusion alone. Erhardt and Baker (1990) suggested a role for salivary enzymes in allowing *Heliconius* butterflies to obtain increased levels of amino acids. Plant bugs also appear capable of using dry pollen, which contrasts with the feeding capabilities of certain mites and thrips that apparently must have fresh grains (Putman 1962, 1965).

The mechanism of pollen digestion in mirids, which is only conjectural, is offered to stimulate experimental work. Invertebrate (Simpson and Neff 1983) and vertebrate (e.g., certain birds [Wooller et al. 1988]) pollen feeders do not necessarily have specialized digestive systems for handling pollen. Specializations of the alimentary tract are known (e.g., Richardson et al. 1986, Richardson and Wooller 1990), but the mechanisms of extracting nutrients from within the indigestible exine (McLellan 1977, Simpson and Neff 1983, Brice et al. 1989) are poorly understood or controversial for most

groups. As I noted earlier, the method of pollen use in the well-studied honey bee has been debated. Among mammals that ingest pollen (e.g., certain Australian marsupials), the mode of digestion is uncertain: direct enzymatic action (Richardson et al. 1986) and initiation of pollen tubes (V. Turner 1984) are possibilities. The means by which bats digest pollen also has not been fully resolved (Howell 1974, Rasweiler 1977, V. Turner 1984).

In proposing diffusion as the key to pollen digestion in mirids, I suggest a model based on Gilbert's (1972) studies on *Heliconius* butterflies but invoke diffusion rather than active release as the mechanism allowing nitrogenous compounds to be extracted. Gilbert (1972) suggested that other animals are able to use pollen in a similar manner. Indeed, diffusion would seem important not only in *Heliconius* species but also in certain beetles, and in mites, heteropterans, and other nonmandibulate arthropods incapable of ingesting pollen. Kevan and Baker (1983) said most pollenivorous insects extract nutrients by diffusion.

A diffusion theory of pollen digestion, which may or may not involve pregermination, seems compatible with most observations on the behavior of pollen in the gut of insects and some vertebrates. Diffusion would allow nitrogenous compounds in the protoplasm and those associated with the pollen wall to be ingested. Experimental work will determine whether diffusion is the key to nutrient removal by mirids and whether a similar mode of extraction operates in other insects. Biochemical studies are needed to elucidate the differences and similarities of pollen use among the various groups of pollenivorous invertebrates and vertebrates, as well as their ability to use pollen of different plant species.

Because pollen can be incorporated into nectar during pollination (Baker and Baker 1986, Erhardt and Baker 1990) and is sometimes abundant in nectar (Todd and Vansell 1942, Beutler 1953), mirids might also obtain nutrients from pollen when they imbibe nectar. Similarly, deer flies and horse flies (Tabanidae) ingest pollen with nectar (Magnarelli et al. 1979), and diffusion of nutrients from pollen may enrich the nectar fed on by some collembolans (Kevan 1978a).

Mirids as Pollinators

Many plant bugs are intimately associated with flowers (see Chapter 10). Their nymphs develop not only on buds but also on open flowers, feeding on the nectar, pollen, and floral tissues. But, as in other insects (Faegri and van der Pijl 1979), only the adults merit consideration as potentially important pollinators. When flower-associated resources are no longer available on their breeding hosts, adults of numerous Miridae visit flowers of other species (food or nonhost plants). Some leaf-feeding mirids also leave their hosts to visit flowers of other plants. Do adults pollinate their breeding hosts, and are their visits to nonhost plants potentially pollinating? Are mirids ever effective pollinators? To begin answering these questions, I review the attributes that make these bugs potential pollinators of host and nonhost flowers, review specific instances in which they have been mentioned in relation to pollination, and assess their role in pollination. Mainly on the basis of floral structure, I also consider the types of plants that mirids seem most likely to pollinate.

POLLINATION POTENTIAL

Müller (1883), Knuth (1906, 1909), Robertson (1928), Porsch (1966), McMullen (1993), and many others have reported the occurrence of mirids on flowers. Some of the plants are true hosts of the species cited, whereas others probably were serving only as adult food plants (the distinction between food plants and true ["breeding"] hosts on which development occurs was made in Chapter 10). The extensive work of Kullenberg (1944), records compiled by Southwood and Leston (1959), and observations reported in this chapter and in Chapter 10 document mirids as common anthophiles. This association with inflorescences is the most basic requisite for a pollinator.

Although adults that disperse to inflorescences of food plants are seldom numerous (exceptions are noted in Chapter 10), mirids do develop large populations on their breeding hosts. Species of the bryocorine genus *Neoneella* can be abundant in and on the flowers of an epiphytic philodendron in Paraguay (Carvalho and Hussey 1954). Some of the examples of high mirid densities given in Chapter 6 (see Table 6.7) involve inflorescence feeders. An abundance on flowers might help compensate for the small size of mirids relative to most known pollinators (e.g., bats, hummingbirds, butterflies, moths, beetles, and flies). But even less frequent visitors can be effective in regularly transferring pollen to conspecific stigmas; a pollinator's importance is the product of its effectiveness and abundance (e.g., Young 1988, Herrera 1989).

Mirids can also become dusted with pollen (e.g., Butler 1923, Groves 1969, Anderson 1976, Andow 1982). Müller (1883) observed several species carrying pollen and acknowledged that they could be active pollinators, but said mirids seem unspecialized morphologically for pollination. He noted that the long proboscis is found in allied groups that do not visit flowers and that no flowers seem especially adapted to the visits of mirids or other hemipterans.

The ability to carry pollen is not enough to characterize an organism as a pollinator (e.g., Kevan 1972, Primack 1978, Gess 1996). Individuals must pick up pollen during flower visits and transfer it to receptive stigmas within the period of viability of the grains being carried. Some heteropterans carry pollen but fail to touch anthers or stigmas (Williams 1977). Morphologically and behaviorally specialized insects that consis-

tently visit flowers of the same plant species and do not waste pollen on other species tend to be more effective pollinators than those that only sporadically visit a given flower species. Mirids, despite fulfilling several general criteria for pollination, are not mentioned or are quickly dismissed in the reviews of crop pollination by Free (1970) (but see Free [1993]), McGregor (1976), and Crane and Walker (1984). Thomson et al. (1982) considered all potentially effective insect taxa as pollinators of spiny aralia but excluded hemipterans, ants, and various tiny species. In contrast, Petanidou and Ellis (1993) included not only well-known groups such as bees but also heteropterans, including mirids, as pollinators (albeit of limited importance) in an East Mediterranean ecosystem (see also Webb [1989]).

Pollinators of particular plants and their pollination effectiveness are often predicted on a priori assumptions regarding insect behavior and floral morphology (O'Brien 1980). A few plants might show a one-to-one, tightly coevolved relationship with a pollinator; most, however, are pollinated by a diversity of species (e.g., Morse and Fritz 1983, Waser et al. 1996). Generalist pollinators may be assumed to be less effective than specialists, but this is not necessarily the case (e.g., Motten et al. 1981, Kearns 1992). In certain situations, such as plants relying on a diversity of visitors for cross-pollination or using an opportunist system, minor pollinators or visitors considered irrelevant (Baker et al. 1971) can function as important pollen vectors. They may also ensure pollination during years in which environmental fluctuations affect the abundance of major pollinator species, or at certain times when pollinator diversity and numbers fluctuate seasonally. Thrips, though often dismissed as pollinators, might be important as "widespread low-level pollinators" (Kirk 1988). For the Miridae, seldom have their behavior (including flight patterns) and pollen loads been monitored and evaluated. It, therefore, seems desirable to review examples in which mirids have been discussed as pollinators, or at least considered potential pollinators, and to assess their role in self- and cross-pollination, rather than to conclude that they are unimportant from only a cursory review of the literature.

FLOWER VISITATION AND
POLLINATION EFFICIENCY

Members of most insect orders have been observed on flowers or cited as potential pollinators (McGregor 1976). As pollinators of crop plants, bees have been studied more than other groups and most research has emphasized honey bees (Bohart 1966, Erickson 1983, Kevan 1987). The pollinating ability of mirids has not been examined critically.

Among flower visitors, mirids are usually considered unimportant or inefficient pollinators (Torre-Bueno 1929; Simão and Maranhão 1959; H. J. Young 1986, 1990; Table 11.2). Occasionally, though, they can be the most abundant insect visitors to flowers of a

Table 11.2. Some mirids collected during studies on pollination ecology

Mirid	Plant	Locality	Comments	Reference
Mirinae				
Adelphocoris lineolatus	Onion	New York, USA	3 individuals collected at 1 site	Caron et al. 1974
Adelphocoris spp.	Sunflower	Italy	Ineffective as pollinators, but anthocorids suggested as important in pollination	Bagnoli 1975
Closterotomus norwegicus	Buttercup	New Zealand		Primack 1983
Grypocoris sexguttatus	Giant hogweed	Great Britain	2 individuals carrying a trace of pollen	Grace and Nelson 1981
Irbisia brachycera	Bigflower cinquefoil, cliff-bush, marsh-marigold, pineywoods geranium, scarlet globemallow	Colorado, USA		Clements and Long 1923
Lycocoris sp.	Green-gentian	Colorado, USA		Beattie et al. 1973
Lygus elisus	Carrot	Utah, USA	Considered an inefficient pollinator	Bohart and Nye 1960
Lygus lineolaris	Onion	New York, USA	57 individuals collected at 3 sites	Caron et al. 1974
	Green-gentian	Colorado, USA		Beattie et al. 1973
	American lotus	Wisconsin, USA	8 individuals collected; an unimportant pollinator	Sohmer and Sefton 1978
Lygus sp.	Continental lady's-tresses	Colorado, USA		Clements and Long 1923
Orthops campestris	Giant hogweed, hogweed	Great Britain	49 individuals carrying a trace of pollen; 26 individuals carrying slightly more than a trace	Grace and Nelson 1981
Orthops scutellatus	Carrot	Utah, USA	An inefficient pollinator	Bohart and Nye 1960
Stenotus binotatus	Hebe	New Zealand		Heine 1937
Orthotylinae				
Labops hesperius	Green-gentian	Colorado, USA		Beattie et al. 1973
Phylinae				
Chlamydatus associatus	Carrot	Utah, USA	An inefficient pollinator	Bohart and Nye 1960
Microphyllelus sp.	Carrot	Utah, USA	An inefficient pollinator	Bohart and Nye 1960
Plagiognathus arbustorum	Hogweed	Great Britain		Grace and Nelson 1981
	Helleborine	Belgium	8 individuals collected with pollen	Verbeke and Verschueren 1984
	Indian balsam	Germany	8 individuals collected	Schmitz 1994
Psallus aethiops	Baneberry	Sweden		Pellmyr 1984
Psallus perrisi	Baneberry	Sweden	8 individuals collected with pollen	Pellmyr 1984

Note: See text for additional examples and discussion.

particular plant. Their visits to radish flowers in British Columbia made up 34% of the total insect visits during a two-month period; small Diptera accounted for 33% and honey bees 22% of all visits to the crop (Treherne 1923). In New York, several mirid species (mainly *Lygocoris caryae* and *L. omnivagus*) readily dispersed from their breeding hosts to tuliptree flowers; individuals became dusted with pollen, suggesting that they pollinate the plant (Andow 1982). Pollen also adhered to the legs or abdomens of the few mirids collected from littleleaf linden flowers in Connecticut (Anderson 1976).

The numbers of pollen grains that mirids carry have been recorded only for guayule in California (grain deposition on stigmas has not been studied in mirids). *Lygus hesperus* individuals carried an average of 500 grains, a figure higher than that recorded for anthomyiid flies and coccinellids collected on guayule (Gardner 1947) but much less than that carried by honey bees and other specialized pollinators (e.g., Free 1970, Kendall and Solomon 1973, Macfarlane and Ferguson 1984).

The pollinating ability of mirids has rarely been assessed by caging them on flowers. Moreover, data from such studies are difficult to relate to natural conditions (Primack and Silander 1975). On caged safflower in Arizona, *L. hesperus* moved little pollen between the flower heads during the three to four weeks the flowers were open. The percentage of crossing (28%) was greater than that for two beetle species but less than that for the Hymenoptera tested (35–56%), and the number of seeds produced per head did not differ significantly from the number that unvisited control plants produced (Levin et al. 1967). These results agree with those G. E. Bohart and colleagues obtained in Utah. Using size of insect, amount of pollen on the body, and activity on flowers of several crop plants, some investigators rated mirids as particularly inefficient pollinators (Bohart and Nye 1960, Bohart et al. 1970, Nye and Anderson 1974). Other investigators have come to similar conclusions regarding the pollination efficiency of heteropterans (Lindsey 1984), including mirids (Caron et al. 1974).

It seems apparent that visits by mirids to food plants generally do not coincide with the period of pollen viability or receptivity of stigmas. Plant bugs tend to exploit flowers of nearby plants when resources on their breeding hosts deteriorate. Their use of food plants (adult hosts) is unpredictable, occurring, for example, when the harvesting of host crops forces migration to other plants.

Because of their small size, mirids also might not contact both stamens and stigmas. Compared to bees, they also have a narrow "tongue" (labium) relative to the flower entrance, and individuals that feed on nectar may fail to contact stamens. The slender (usually hairless) labium is unlikely to transfer much pollen, and the remainder of the body may not contact pollen. Williams (1977) found that heteropterans (three species of Coreidae and one each in the families Pentatomidae, Pyrrhocoridae, and Reduviidae) carried the pollen of pigeon pea but did not touch the anthers or stigmas and, therefore, contributed little to cross-pollination.

Most mirids, as unselective flower visitors carrying small amounts of pollen, perhaps have a more parasitic than mutualistic relationship to adult hosts, as do certain butterflies on their nectar plants (Wiklund et al. 1979). Other butterfly species have been termed *facultative mutualists* on flowers (Murphy 1984). In some cases, mirids could be viewed as nectar thieves: animals such as certain lepidopterans that do not make holes in the corolla or destroy floral tissue (as nectar robbers do) but take nectar without contacting stamens and stigmas or effecting pollination (Inouye 1980, 1983; Kevan and Baker 1983, 1984). In at least one instance, a mirid, *Metriorrhynchomiris dislocatus*, was termed a nectar robber—that is, it takes nectar from a flower without pollinating it (Barrows 1979). Mirids could also be regarded as commensals that receive a benefit but provide none in return (Schemske and Horvitz 1984). In some instances, the removal of small amounts of nectar

might even be considered beneficial, for legitimate pollinators must then spend more time on the "robbed" flowers, which contributes to their effectiveness (e.g., Meeuse 1972).

Mirids also do not show behavioral adaptations for flower visitation, such as the specialized buzz collecting of pollen by certain bees (Roubik 1989); unlike bees, they apparently do not learn the location of plants or how to manipulate the flowers. They also are unlikely to exhibit regularity in their flower-positioning behavior. Although foraging rates are unknown for any plant bugs, they do not forage from flower to flower (pers. observ.). On any particular plant, they do not visit a number of flowers in sequence, and interplant movement is infrequent or at least substantially less than that of insects considered efficient pollinators. In addition, any pollination benefit from mirids is potentially negated by their injury to flower parts and fruit (Kullenberg 1944), although in most cases, plant bugs probably have an insignificant negative effect on plant reproductive success. They might reduce plant fitness or yields by feeding on the pollen on anthers, or possibly by destroying the pollen on stigmas before fertilization. Similarly, thrips can adversely affect flowers (Kirk 1987, 1988), although they can also be important pollinators (Appanah and Chan 1981).

Scott (1976b) pointed out that pollination is an aspect of insect-plant interactions that should not be overlooked, especially in the case of lygus bugs and other mirids that spend much of their time on flowers. In most settings, mirids will not be pollinators, but exceptions might yet exist within the family.

POSSIBLE MIRID-POLLINATED PLANTS

Müller (1883) realized that mirids are potential pollinators of composites, crucifers, umbellifers, and willows—plants characterized by a floral structure allowing easy access to nectar or pollen. Their blossoms are categorized (without phylogenetic consideration) as dish or bowl shaped and head or brush shaped (Faegri and van der Pijl 1979). A more recent comment is that mirids may play some role in pollinating the flowers of composites and umbellifers (Willemstein 1987). In this chapter (and in Chapter 10), the numerous references to plant bugs on the inflorescences of herbs pertain mostly to the Asteraceae, Brassicaceae, and Apiaceae (see also Robertson [1928]). The only mirid Müller (1881) reported from his study of pollination in European alpine communities was associated with a composite. Several references to mirids as crop pollinators also involve the Asteraceae (guayule, safflower, and sunflower), the Apiaceae (carrot), and the Brassicaceae (radish).

Mirids possibly contribute to the general pollinator assemblage associated with these plant groups, and experimental work to assess their importance as polli-

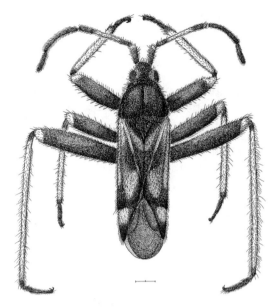

Fig. 11.2. *Calocoris roseomaculatus* feeding on pollen of a composite (Asteraceae). (Reprinted, by permission of B. Kullenberg and Zoologiska institutionen, Uppsala Universitet, from Kullenberg 1944.)

Fig. 11.3. *Pameridea roridulae*, adult habitus. This strongly pubescent bug, once considered specially adapted for pollinating its sole byblidaceous host, might be only a minor pollinator. Scale bar = 1 mm. (Reprinted, by permission of W. R. Dolling, J. M. Palmer, and Blackwell Science Ltd., from Dolling and Palmer 1991.)

nators of wild or crop plants might begin with composites. Candidates for study are plant taxa fed on by the densely pubescent phylines *Hoplomachus thunbergii*, *Megalocoleus molliculus*, *M. tanaceti*, and *Oncotylus punctipes* (Kullenberg 1944). Pollen often coats the venter and extremities of these bugs. Less pubescent species that Kullenberg (1944) observed as pollen carriers on composites, including the common mirines *Adelphocoris lineolatus*, *Calocoris roseomaculatus* (Fig. 11.2), *Closterotomus norwegicus*, and *Lygus pratensis*, also could be evaluated.

Experimental methods of pollination biology should be used to explore the relationship between South African dicyphines of the genus *Pameridea* and their byblidaceous hosts. *Pameridea marlothi* and *P. roridulae* were once believed to pollinate glandular-pilose roridulas, each plant bug species restricted to a herbaceous or shrubby host species (Marloth 1903, 1910; Reuter 1907b; Schumacher 1919a). Reuter suggested that pubescence (Fig. 11.3) on the bugs' extremities, antennae, and apical segments of the rostrum was particularly suited for pollen collection. Marloth noted that the bugs move freely over sticky leaf surfaces without becoming entrapped and appear to feed on plant tissues. The mirids were stated to probe the bases of the anthers, which effects the release of pollen on the bugs. Evidence indicates that *Pameridea* species are scavengers or predators of insects caught in plant secretions (see also Chapter 15) and that they are only minor pollinators of roridulas (Dolling and Palmer 1991) rather than "specially adapted" pollinators of their hosts (Marloth 1903).

Mirids also might be involved in pollinating plants of the Chenopodiaceae, Polygonaceae, and Rosaceae, families in which the flowers tend not to exclude unspecialized pollinators. Mirids should be considered potential pollinators of grasses and sedges and wind-pollinated trees, especially under conditions in which wind movement of pollen is limited. Soderstrom and Calderón (1971) discussed the possible role of insects in pollinating grasses in the rainforest understory where there is no wind to effect pollination. Plant bugs in temperate regions are potential pollinators of sporadically flowering perennials, such as green-gentian (monument weed) (Beattie et al. 1973), that depend on a diversity of generalist insect visitors for cross-pollination. They might also be considered in relation to other erratically flowering perennials, such as rainforest trees that bloom profusely every 3–10 years in the aseasonal tropics (Kavanagh 1979), especially dipterocarps (Appanah 1981, 1987; Ashton et al. 1988). In western Malesia, mirids of the phyline genus *Decomia* pollinate mass-flowering dipterocarps of the genus *Shorea*, section Brachypterae (Appanah 1987, Ashton 1988, Ashton et al. 1988). These lowland tropical rainforest trees have small flowers that do not open fully and undergo nocturnal anthesis (Appanah 1987). Although the adult bugs feed at the base of the style, they apparently do not cause substantial flower injury (Appanah 1990). Mirids also might be among the generally small and unspecialized insects that visit the small flowers of dioecious and other trees in tropical forests (Bawa and Opler 1975, Bawa et al. 1985).

Plant bugs might be studied in harsh (e.g., alpine, boreal) environments and in insular situations such as oceanic islands and remote mountaintops where bees and other specialist pollinators often are scarce; in these areas, plants sometimes must rely on generalized pollinators (Primack 1978, Kevan and Baker 1983, Kearns 1992, Kevan et al. 1993). Mirids possibly help adventive plants become established, although their flower-visiting behavior contrasts with that of the generalized pollinators considered important in fertilizing island taxa (e.g., Woodell 1979). As pollinators of crop plants, mirids warrant consideration in New Zealand or other areas where effective pollinating Hymenoptera are scarce or absent (Primack 1978, Faegri and van der Pijl 1979, Godley 1979, Kevan and Baker 1983, Donovan 1990).

The selective effects of potential pollinators on plants can vary significantly, which can lead to an evolutionary specialization of mutualism (Schemske and Horvitz 1984). Mirids might represent an early evolutionary stage of primarily parasitic anthophiles that could become "engaged" by their host plants as pollen vectors. In fact, Bell (1971) noted a relatively high percentage of "bugs (Hemiptera)" visiting flowers of an umbellifer, black sanicle. Most Apiaceae are characterized by promiscuous pollination by numerous unspecialized species. Bell (1971), however, suggested that black sanicle is developing a weakly specialized interaction with short-tongued bees, and possibly another line of specialization with bugs.

ASSESSING THE POTENTIAL OF
MIRIDS AS POLLINATORS

The Miridae, although potential pollinators of many plants, tend to be flower visitors rather than effective pollinators. They visit flowers only for their own immediate needs (rather than forage to obtain food for their young, as bees do), are thought to be behaviorally and morphologically unspecialized for pollination, and show little, if any, flower constancy. They may contribute to the generalized pollination systems of certain taxa, and pollination biologists should at least be aware of the large densities associated with the inflorescences of some hosts, as well as the possibility that damage to the ovaries could offset the benefits from pollination.

Plant bugs developing on inflorescences, rather than using them strictly for adult feeding, have not been distinguished in traditional lists of flower visitors, and they are not usually segregated in current studies of pollination biology. Adults of inflorescence feeders, though, would seem more likely to injure the reproductive structures of their hosts than adults of species that seek nectar or pollen on nonhost or food plants. Quantitative studies of the effects of mirids on seed set of host plants are needed.

Even as minor or chance pollinators, mirids warrant additional work. Foraging behavior, nocturnal activity, number of pollen grains carried, location of pollen on the integument, the numbers and identities of grains delivered or transferred, and the directionality of transfer require study. Do some anthophilous species visit certain flowers for nectar and others for pollen, or is their occurrence on flowers determined merely by proximity of other plants to their breeding hosts? Do flowers of a certain color, odor, or maturation pattern (dichogamy) tend to attract plant bugs? Robertson (1928) listed a few mirid visitors to the flowers of dioecious taxa, including willows (see Chapter 10), but the efficacy of prospective insect pollinators on willow has not been assessed (Meeuse 1978). Does the movement of plant bugs between male and female individuals of dioecious or monoecious plants ever contribute significantly to pollination?

A recent study of insect pollination in cucurbits suggests that mirids moving between male and female flowers of the monoecious, night-blooming bottle gourd effect pollination (Shrivastava 1991). According to Shrivastava, males of *Nesidiocoris tenuis* typically feed on bottle gourd shoots, but after anthesis, the bugs are said to orient to the staminate flowers, having been attracted by their odor. After alighting on the petals, males crawl down into the perianth cup, where their bodies become dusted with pollen. They fly to the pistillate flowers of bottle gourd, where the "wingless" *N. tenuis* females (possibly nymphs; I am not aware of aptery or brachyptery in this species) feed on the exposed nectaries. As a pollen-loaded male bug reaches the nectary of a female flower, it is contacted by the capitate stigma whose sticky surface traps the bug's pollen. The mirid, which can occur in large numbers on bottle gourd where it infests male and female flowers in addition to apical shoots (Sreeramulu et al. 1975, Sinha et al. 1981, Chatterjee 1983b), is considered a key pollinator of this crop plant (Shrivastava 1991). Although intrafloral behavior of these abundant bugs makes them candidates for pollination, data on pollen transfer and fruit set are needed. Because other gourds are pollinated by bees (Free 1993), the existence of a highly coordinated system, as suggested by Shrivastava's (1991) observations, seems improbable.

Despite the putative close relationship reported for a mirid and bottle gourd, biologists have not recognized any "mirid flowers"—that is, a pollination syndrome or suite of characters relating floral adaptations to distinct pollinator types (e.g., Baker 1961, Proctor and Yeo 1972, Faegri and van der Pijl 1979, Wyatt 1983, Bertin 1989, Waser et al. 1996)—in this case, plant bugs. Plant-insect relationships in several insect groups have evolved from one of chance or incidental pollination to a symbiotic or mutual interdependence (Feinsinger 1983, Pellmyr 1992). Such a specialized, or at least coevolved, relationship probably does not exist in the Miridae. Even so, the pollination syndrome approach, with its emphasis on a single pollination group for each plant, can result in widespread but minor pollinators such as thrips (Kirk 1988) being ignored. In recent years, the status of beetles as pollinators has been revised as specialized relationships with flowers have been discovered (H. J. Young 1986, Irvine and Armstrong 1990). A comparable change in ideas regarding mirids as pollinators is unlikely, but it would not be surprising if plant bugs were found to contribute more to pollination than previously suspected.

12/ Fruit Feeding

In little more than a quarter-century, the redbugs [two mirid pests of apple] rose from obscurity, devastated crops, and sank back into the status of a minor pest.
—C. W. Schaefer 1974

Mirids that feed on fruit can be referred to as *direct pests* (e.g., Roitberg and Angerilli 1986), in contrast to *indirect pests* that feed on other (nonmarketable) plant parts and generally cause less severe injury. Mirid-induced foliar and stem lesions and cankers, however, can be devastating (see Chapter 9). Lygus bugs are among the most insidious pests of fruits and vegetables (Scott 1987). Because plant bugs suck juices from the fruits without effecting seed dispersal, they might be considered nectar thieves (Owen 1980). Although mirid feeding on fruitlets typically affects cell division, which leads to visible symptoms, their feeding on more mature fruits might not lead to detectable injury because fruit growth at this later stage occurs through cell expansion (e.g., Reding and Beers 1996).

In this chapter, feeding by mirids is mostly restricted to their use of edible fruits and seeds (injury to tomato fruit was mentioned in Chapter 10). The coverage, which includes beverage, confectionary, and nut crops, is thus broader than fruit in the horticultural sense of "something . . . eaten fresh and out of hand" (Samson 1980). Crops are discussed alphabetically rather than by economic importance or geographic area. Some feeding on the inflorescences of subtropical and tropical crops is considered here rather than in Chapter 10. This has been done not only for convenience but also because mirids often feed on both inflorescences and fruit; in addition, their attacks on flowers sometimes involve the failure to set fruit or lead to the abscission of young fruit. Some of the mirids cited in older literature as fruit pests were associated with cultivars or production practices that are no longer used; these bugs usually are not pests under current management practices. In a few cases, the record of a plant bug as a pest of fruit trees is based on the incidental collection of adults of an herb-feeding species on trees (Putshkov 1960); presumed incidental species (e.g., Niemczyk 1963), though not necessarily injurious on fruit trees, might feed on nectar or pollen (see Chapter 11).

Omitted from further consideration are a few fruit crops infested by mirids but not to the extent to warrant a section of this chapter. Examples include red currant, whose berries are punctured by *Lygocoris pabulinus* in England (Austin 1930a), and mulberry, which is subject to fruit feeding by *L. viridis* in Korea (Yamamuro and Hoshino 1940). Pumpkin is another fruit mentioned only briefly in the literature as sustaining mirid injury. Hori et al. (1987) discussed the formation of tubercles (Plate 17) on pumpkin fruit fed on by *Lygus rugulipennis* in Japan, noting that such swellings are atypical symptoms of heteropteran feeding. Mirids are usually unimportant to banana crops (e.g., Anitha and Rajamony 1991), but in 1936 the polyphagous *Helopeltis westwoodi* injured the fruit in western Africa (Carayon and Delattre 1948).

Apple

The diverse fauna of apple, a pome fruit, includes phytophagous mirids as well as predatory species, which are discussed in Chapter 14. Among the phytophagous species, lygus bugs are discussed first, followed by the North American red bugs *Heterocordylus malinus* and *Lygidea mendax*, and by other mirids recorded as apple pests. A summary of the biological information on phytophagous mirids associated with apple crops in North America can be found in the publication by Braimah et al. (1982). Boivin and Stewart (1982b) characterized plant bug injury to apple crops in Quebec.

LYGUS BUGS

Crosby and Leonard (1914) summarized the early reports of injury to apples by the tarnished plant bug (*Lygus lineolaris*), dating from 1860. Parker and Hauschild's (1975) bibliography refers to later studies on this bug as an apple pest. Nymphs are infrequent on apples (e.g., Prokopy et al. 1982), and only the adults are considered injurious (Beers et al. 1994).

Injury and Economic Importance of *Lygus lineolaris*

Overwintered adults of *Lygus lineolaris* move into orchards with warm temperatures in spring, usually April in the northeastern United States (Weires 1983). Feeding on buds usually begins soon after the delayed dormant bud stage. Injury can be more severe when warm weather occurs before plants of the ground cover have begun to develop (Howitt 1993). This bug's pest potential is enhanced by its capabilities of dispersal. The tarnished plant bug combines features common to good colonizing species (r-strategists) and good competitors (K-strategists) (Stearns [1976] and Southwood [1981] reviewed r- and K-selection in relation to life-history tactics). An aggregate KKr rating was assigned to this species (Croft and Hull 1983), based on its reproductive potential (K), survivorship (K), and dispersal (r) attributes (see also Fleischer and Gaylor [1988]).

On 'McIntosh' and 'Red Delicious' trees in Massachusetts, adults induce abscission of the flower buds when they feed during the silver tip through the tight cluster stages. This injury is of no or slight economic importance in productive commercial orchards because of the small number of blossoms needed to set a crop. Fruit malformation (dimples and scabs; Plate 17) results from feeding during the early pink stage to two weeks after petal fall (Prokopy and Hubbell 1981, Prokopy et al. 1982). Because most of the feeding takes place on the calyx tube near the bases of the sepals and petals, injury to mature fruit often appears at the calyx end (Howitt 1993, Hull et al. 1995). Tarnished plant bug injury to the fruit resembles that inflicted by the apple seed chalcid (Torymus varians) and apple curculio (Anthonomus quadrigibbus) (Hammer 1939). Vincent and Hanley (1997) measured how well apple pest-management specialists agree when asked to identify injury by pest insects. The average interobserver agreement for 22 species on apple trees was the second lowest (83.2%) for the tarnished plant bug.

Integrated Management of *Lygus lineolaris*

Lygus lineolaris is sometimes considered the principal insect pest damaging apple fruit in the Northeast (Prokopy et al. 1982), and among all insect pests, plant bugs (mainly *L. lineolaris*) can be the most difficult to control at commercially acceptable levels in the integrated pest-management (IPM) program used in Quebec apple orchards (Bostanian and Coulombe 1986, Bostanian et al. 1989). Losses resulting from its feeding, however, can be relatively minor compared to factors such as size, color, and hail (Weires 1983). In some years the tarnished plant bug has been responsible for only 0.16% of apples culled and 0.08% of those downgraded in eastern New York. Injury observed in the field can have a significant positive relationship to loss from cullage and downgrading in the packinghouse (Weires et al. 1985).

Lygus lineolaris is one of four major apple pests in Massachusetts orchards for which effective alternatives to insecticidal control are unavailable (Prokopy et al. 1990). Synthetic pyrethroid insecticides usually provide better control of tarnished plant bugs than do other materials (Weires 1983). Emphasis in New Hampshire and elsewhere is placed on integrating insecticide and

fungicide applications. A combined fungicide, insecticide, and oil treatment is applied before the fruit buds are in the tight cluster to the pink stages of development, to control tarnished plant bugs and other pests (Gadoury et al. 1989).

Use of sticky traps placed near the periphery in apple orchards allows the bugs to be detected as soon as they invade (Prokopy et al. 1979, Boivin et al. 1982). Sticky-coated traps are significantly more effective in detecting adults than is sweeping the ground cover, jarring limbs over a drop cloth, or making visual counts of adults on apple trees. The number of adults caught cumulatively on sticky traps through the tight cluster stage is positively correlated with fruit injury (Prokopy et al. 1980). An action threshold of 3.0 adults accumulated per trap through the tight cluster stage, or 4.4 through the late pink stage, can be used to determine the need for insecticide applications (Prokopy et al. 1982).

Economic thresholds established for *L. lineolaris* should be revised to reflect changes in the costs of insecticide applications and monitoring practices. In Michigan, for example, the threshold of three adults per 100 fruit clusters in IPM orchards (Hoyt et al. 1983) was raised to five adults per 100-leaf sample (Whalon and Croft 1984). Information on control costs, the market price of apples, and the timing of pest monitoring and spray applications is not always used to establish economic thresholds (Michaud et al. 1989a). To relate the level of injury at the time of harvest to the density of *L. lineolaris* in the spring, Michaud et al. developed economic injury levels and thresholds for this pest in Quebec. Regressions of the percentage of fruit damaged on the number of adult bugs caught during three seasons (April–October) showed a significant positive linear relationship when insect number was expressed as mean cumulative captures per sticky trap or 50 limb taps. Estimated economic injury levels ranged from 0.5 to 2.8 cumulative captures of adults per sticky trap and 0.8 to 4.5 nymphs and adults per 50 limb taps. The economic thresholds calculated were based on the costs of the insecticides and their application and allowed 24 hours for initiating controls. Whalon and Croft (1984) and Michaud et al. (1989a) provided thresholds for *L. lineolaris* in other states and provinces; Beers et al. (1994) reviewed the relationship between sampling methods, economic thresholds, and IPM. In New York, treatment is generally not considered economically profitable, and monitoring for this pest is not currently recommended in orchards (Agnello et al. 1993).

Lygus Species in the Western United States

Lygus elisus and *L. hesperus* occur in all apple-growing districts of California. Feeding by lygus bugs on flower buds in early spring is usually unimportant, but mid-season infestations on the fruit result in round pits, and late-season feeding causes irregularly shaped pits that resemble stink bug injury. Problems are most frequent in orchards with a permanent cover crop or those adjacent to crop fields and vegetation harboring large densities of lygus bugs. *Lygus* species can migrate into orchards at any time during the growing season, often feeding first on trees in the outside rows. Because injury is often patchily distributed, each orchard block should be monitored thoroughly for the presence and abundance of these bugs. Injury to one apple in 100 necessitates further sampling and evaluation of the need for pesticides (Pickel and Bethell 1990).

APPLE RED BUGS

The red bugs *Heterocordylus malinus* and *Lygidea mendax*, whose original hosts probably were native crabapple (Plate 17) and hawthorn (Wellhouse 1922, Knight 1941, Schaefer 1974), first attracted the attention of New York orchardists in 1896. Crosby (1911) briefly summarized the life history of these mirids and referred to fruit deformation that results from their feeding punctures. Apple varieties react differently to feeding, and Knight (1918a, 1922) provided descriptions and photographs of injuries to the varieties then commonly grown in New York. The early instars of both mirids feed on tender, young leaves before moving to immature fruits. Growers were advised to use the appearance of foliar spotting to detect the bugs' presence and to initiate immediate control measures if red bugs were found (Hodgkiss and Frost 1921). Because overwintered eggs of *H. malinus* hatch 7–10 days before those of *L. mendax*, nymphs of the former mirid feed mainly on foliage rather than on developing fruit. *Lygidea mendax* thus has greater potential as a fruit pest (Knight 1918a, 1922). Both species, however, can be reared on apple foliage (Crosby 1911).

Apple red bugs cause two main types of fruit injury (Plate 17): deep, sunken pits in young fruit resulting from penetration to the core, and rusty-brown wounds or russet scars that form on fruit that is too large for the bugs' stylets to penetrate the core. Scars are sometimes enlarged by the rosy apple aphid (*Dysaphis plantaginea*) (Knight 1918a, 1922). By tagging and photographing apples fed on by mirids, Knight was able to distinguish plant bug injury from that caused by other apple pests and by agents such as frost, pesticides, and mechanical injuries. Once almost devastating to the apple culture in the northeastern United States, red bugs have been eliminated from orchards by modern spray practices (e.g., Boivin and Stewart 1983a) or, at worst, have been reduced to the status of unimportant pests (Cutright 1963, Schaefer 1974).

OTHER PLANT BUGS OF APPLE

Former or Occasional Pests in North America

Although nymphs of the clouded plant bug (*Neurocolpus nubilus*) infested apples in Ontario orchards in the early 1900s (e.g., Crawford 1916, Henry and Kim 1984), the damage was never widespread. *Taedia pallidula* was formerly a pest of apple fruit in Ontario (Caesar 1913). In Canada, Braimah et al. (1982) and Kelton (1983) noted that adults of *Capsus ater* and *Stenotus binotatus*, grass-feeding mirines, sometimes disperse to apple trees to feed on the fruit when grass is cut or the season is particularly dry.

The Pear Plant Bug

An important pest in Nova Scotian orchards in the early 1900s was the pear plant bug (*Lygocoris communis*), which causes malformation and abscission of the young fruit (Brittain 1916a). Its feeding on blossoms produces symptoms that can be mistaken for fire blight or a frost problem (Brittain 1916b). Sometimes 8–10 nymphs feed in a flower, causing death and curtailing or eliminating fruit production. It was once common to see entire orchards devoid of apples, and the amount of damage was immense (Dustan 1923, 1924). Because this pest was not amenable to the routine control methods used in the early twentieth century, exceptionally careful spray procedures had to be followed to curtail its injury (Sanders and Brittain 1916). The entomogenous fungus *Entomophthora erupta* helped eliminate *L. communis* from the orchards in Nova Scotia. Dissemination of the fungus (Dustan 1924) marked the first attempt to control a native apple pest with an apparent native pathogen (LeRoux 1971). After populations of *L. communis* peaked in 1919–1920, this plant bug was seldom abundant. In 1928, however, nymphs substantially affected apple production when they concentrated on the fruit of an unusually light crop (Brittain 1929).

Interest in *L. communis* as an apple pest was rekindled in the 1980s. This mirine continues to injure fruit in unsprayed orchards in Quebec (Boivin and Stewart 1982b). Feeding by young nymphs (first through third instars) results in badly deformed 'McIntosh' apples at harvest, whereas fruits fed on by late instars show scars or depressions of less importance. Each nymph caged on fruit at the time of petal fall damaged an average of 2.1 apples (Michaud and Stewart 1990). Michaud et al. (1989b) developed economic injury levels and thresholds for first through third instars. Economic injury levels ranged from 0.4 to 5.8 young nymphs per sample unit, depending on the sampling method (number per fruit cluster, leaf cluster, or 50 limb taps) and the insecticide used. Economic thresholds were calculated on the basis of control costs and the market price of apples, and assumed a 24-hour period necessary to begin control measures and a 100% insecticide efficiency.

Atractotomus mali and *Campylomma verbasci*

North America

Two Holarctic phylines—*Atractotomus mali*, sometimes known as the "apple brown bug," and *Campylomma verbasci*, the "mullein bug"—can be pests of apple fruit in North America. Both species also feed on small arthropods and their eggs (see Chapter 14), and early instars of the mullein bug apparently need arthropod prey or pollen for survival (Niemczyk 1978, Smith 1991, Smith and Borden 1991, Bartlett 1996). MacPhee and MacLellan (1972) cited periodic outbreaks of *A. mali* in their discussion of new pest problems that have developed in dynamic orchard ecosystems since the development of integrated pest control in Nova Scotia. The mullein bug has been only a sporadic pest in British Columbia (Madsen et al. 1975), but in more recent years its occurrence on apples has been more consistent than intermittent (Thistlewood et al. 1990). This plant bug has been a widespread, though sporadic, pest in Washington since the late 1980s (Reding and Beers 1996). Damage from *C. verbasci* often follows the collapse of aphid or mite populations in an orchard (Thistlewood and Smith 1996).

Both *A. mali* and *C. verbasci* cause disfiguration (Plate 17) and development of callus tissue sufficient to render the apples salable only for juice (MacPhee 1976). The bugs' feeding is usually most injurious to developing fruits when it occurs during the sensitive period between the bloom stage and two weeks after petal fall (Thistlewood and Smith 1996). Fruit can be injured when the bugs feed directly on fruitlets and apparently also when they feed on the flower receptacle during the bloom stage (Kain et al. 1997).

MacLellan (1979) reviewed the population levels of *A. mali* and *C. verbasci* and the percentage of fruit injured in integrated control orchards of Nova Scotia from 1953 to 1977. Application of pyrethroid insecticides has since drastically reduced the populations of both bugs in Nova Scotia (Hardman et al. 1988); the application of a sterol-inhibiting fungicide, flusilazole, used in apple pest management in Ontario sometimes is associated with increased numbers of *C. verbasci* (Biggs and Hagley 1988). Damage by *C. verbasci* exceeded the 1% economic injury level of culled apples in 17 of 40 commercial orchards sampled in British Columbia's Okanagan Valley, which ranked the bug among the top three apple pests in western Canada. A positive linear relationship was found for mirid numbers and damage at harvest (Fig. 12.1; Thistlewood et al. 1989b). Their studies suggest that an economic injury level of one nymph per tap be established for the susceptible cultivar 'Golden Delicious' and four per tap for 'Red Delicious'. Densities of first-generation nymphs can be predicted directly by limb-tap sampling in spring (Thistlewood and McMullen 1989). A nonlinear developmental model allows egg hatch to be predicted more accurately and the timing of insecticide applications to be improved (Judd and McBrien 1994).

A sex pheromone–based monitoring system, using the number of *C. verbasci* males caught during the fall to predict nymphal populations, might enable economic injury levels to be anticipated 6–8 months before overwintering eggs hatch the following spring. A positive and significant relationship exists between the numbers

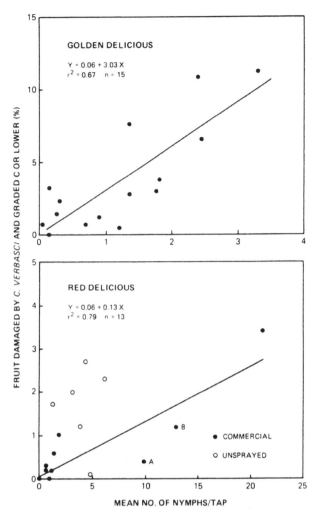

Fig. 12.1. Linear regressions of serious damage at harvest on peak first-generation numbers of *Campylomma verbasci* for "Golden Delicious" and "Red Delicious" apples in a commercial orchard in British Columbia; data from unsprayed sites are included for comparison. Points A and B are from an orchard in which damaged fruit was removed by hand thinning and culling at harvest. (Reprinted, by permission of H. M. A. Thistlewood and the Entomological Society of Canada, from Thistlewood et al. 1989b.)

of *C. verbasci* trapped in the fall and the numbers of nymphs per limb the following spring. A similar, though weaker, relationship is shown for trees adjacent to those in which traps are placed (Fig. 12.2; Smith and Borden 1990, McBrien et al. 1994). The commercial availability of a sex pheromone for managing *C. verbasci* might reduce a reliance on labor-intensive limb-tapping methods for sampling this bug's populations (Hardman 1992). Additionally, the atmospheric permeation with a synthetic sex pheromone, using a natural 94:6 blend of butyl butyrate and (*E*)-crotyl butyrate (Judd et al. 1995), can disrupt the mating of *C. verbasci* in apple orchards and help reduce population densities (McBrien et al. 1996, 1997).

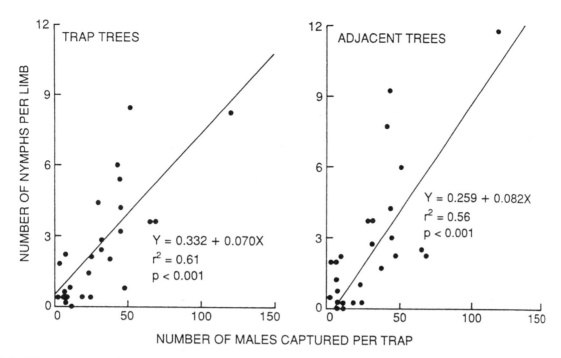

Fig. 12.2. The positive, significant relationships between catches of male *Campylomma verbasci* attracted to sex pheromone in female-baited traps, mid-September to mid-October 1987, in British Columbia, and numbers of nymphs per limb observed the following spring. (Reprinted, by permission of R. F. Smith and the Entomological Society of America, from Smith and Borden 1990.)

Damage by nymphs of *A. mali* and *C. verbasci* can depend on the apple cultivar (Pickett 1938), availability of prey or pollen for these facultative predators (Lord 1971, Sanford 1964b, Bartlett 1996), crop load, and other factors (Thistlewood et al. 1989b). Thus, *C. verbasci* is known in North America as an important enemy of the European red mite and as a pest capable of severe damage (e.g., Pickett 1938, Hagley and Hikichi 1973, MacPhee 1976, Boivin and Stewart 1982b, Smith 1991, Thistlewood and Smith 1996).

Europe

Campylomma verbasci is sometimes considered a beneficial species on apple trees in England (Collyer 1953a). It is a recent pest of apple crops in the Netherlands, where the corky scars on the fruit are sometimes referred to as "dimple disease." In Belgium, the corky spots were initially attributed to hail damage (Stigter 1996).

Theobald (1913) noted the role of *Atractotomus mali* as an apple pest in England, but Fryer (1916), Fryer and Petherbridge (1917), and most subsequent British workers considered this species to be of little economic consequence. Sanford (1964b) summarized the European literature in which *A. mali* is reported in a beneficial and a destructive role on apples. Jonsson (1987) reported that few *A. mali* can be reared to adulthood on apple foliage in the laboratory and concluded that some

arthropod food (see Chapter 14) is needed for normal development and survival.

The Apple Capsid Bug and Other European Mirids

Fryer (1914, 1916), Petherbridge and Husain (1918), Massee (1937, 1956), and others studied the phytophagous mirid fauna of apple in England. Fryer's (1914) study represented the first extensive work on mirids as apple pests in Britain. He noted that some symptoms of mirid injury—fruit distortion, severe cracking of the skin, discoloration or russetting, and surface pimpling—had once been attributed to cold winds or excessive moisture. Injury by plant bugs had also been confused with physiological defects of the trees or that caused by other insects, for example, thrips (Theobald 1911a, Fryer 1916). The mirid species responsible for damaging apples had not yet been established, but the mirine *Lygocoris rugicollis* and an orthotyline, the dark-green apple capsid (*Orthotylus marginalis*), were considered more likely causal agents than were *A. mali*, *Lygus pratensis* (probably *L. rugulipennis*), and *Psallus ambiguus* (Fryer 1914, 1916).

The British mirid that proved to be most important as an apple pest was *Lygocoris rugicollis*, the apple capsid bug. Once limited to feeding on native willows in England, *L. rugicollis* apparently adopted apple as a host between 1900 and 1910. Its life cycle on apple trees appears to have become advanced by about a month compared to that on willow (Fryer 1929). Fruit damage by *L. rugicollis* nymphs soon became so great that the trees were crippled—adults severely damage terminal shoots (Fryer 1916, Petherbridge and Husain 1918)—and crops were rendered unsalable. Severely injured young apples looked more like poorly shaped potatoes than proper fruits (Fryer 1929). Malformation resulted from the restriction of tissue growth near feeding sites; as the fruit grew, some parts developed more rapidly, causing distortion (Carter 1973). Injury sometimes also resembled the symptoms of apple scab on fruit (Smith 1920b).

Lygocoris rugicollis remained a serious threat to apple crops until 1946 when DDT sprays were introduced in Britain (Massee 1956, Southwood and Leston 1959). This bug has not since reinfested orchards (Dicker 1967). Before the use of synthetic insecticides, the apple capsid bug was more dreaded than any other British apple pest (Massee 1937).

The mirine *L. pabulinus*, the common green capsid, now occurs commonly on apple trees in England. Because Jones and Jones (1984) thought early workers failed to mention this bug as a pest in English orchards, they suggested that its habits had changed. Injury to apple fruit (Plate 18) was observed in Irish orchards as early as 1916–1917 (Carpenter 1920), and although Fryer (1914) and Petherbridge and Husain (1918) did not encounter this species on apples in England, it was found on apple trees there in 1926 (Petherbridge and Thorpe 1928a).

Lygocoris pabulinus occurs on the shoot tips in Dutch apple orchards before it begins to feed on developing fruits after petal fall. Analysis of beating samples is inefficient for monitoring its populations because of frequent poor weather during sampling and the clumped distribution of bugs within orchards. Even lowering recommended thresholds from four to one or two nymphs per 100 branches does not ensure an injury-free crop (Bus et al. 1985). The use of sex pheromone traps may allow the presence or even density of *L. pabulinus* to be monitored in orchards (Blommers et al. 1988); development of a phenological model also enables its populations to be forecast (Mols 1990). Contrasted with the visual monitoring trap developed for *Lygus lineolaris* in eastern North America (Prokopy et al. 1979), a possible disadvantage of a sex pheromone trap for *Lygocoris pabulinus* would be the attraction of only one sex (male) and individuals from outside the orchard (Prokopy et al. 1982). An undergrowth of wildflowers (various herbaceous plants serve as summer hosts of this bug) can trigger outbreaks that require treatment with white oil sprays in integrated management programs for apples in the Netherlands (Gruys 1975, 1982).

Among important early studies on mirids injurious to apple crops on the European continent were those by Schøyen (1916), Rostrup and Thomsen (1923), Schoevers (1930), Lehmann (1932b), Speyer (1933, 1934), and Abraham (1935, 1936). Their work dealt mainly with *A.*

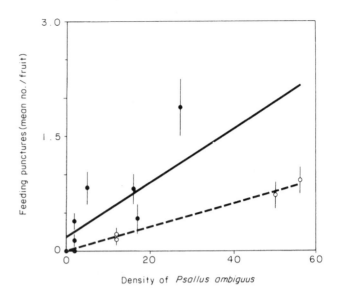

Fig. 12.3. Mean numbers of feeding punctures per fruit as a function of density of *Psallus ambiguus* in 1983 and 1984 in Switzerland. The confidence interval is defined with standard errors. Note: The original source should be consulted for probable explanation for the disparity in regression coefficients between years. (Reprinted from *Crop Prot. 7*, L. Schaub et al., "Elements for assessing mirid [Heteroptera: Miridae] damage thresholds on apple fruits," pp. 118–124, 1988, by permission of L. Schaub and with kind permission from Elsevier Science Ltd., The Boulevard, Langford Lane, Kidlington OX5 1GB, UK.)

mali, *L. pabulinus*, and *L. rugicollis*, species already discussed in this chapter, but also with *O. marginalis* and *P. ambiguus*. Even though sometimes abundant on apples (Schaub et al. 1987), *O. marginalis* is now considered unimportant in causing fruit injury (e.g., Asgari 1966, Schaub et al. 1988). In contrast, Schaub et al. (1988) determined that *P. ambiguus* is a key pest in Swiss orchards; injury occurs from the time of flower initiation until "stem hollow formation" (i.e., fruit about one-third developed). Wildbolz et al. (1955) and Wildbolz and Henauer (1957) illustrated and discussed the fruit malformation caused by this phyline. The bug's densities significantly affect the mean number of punctures per fruit on 'Glockenapfel', the injury to fruits occurring in a negative binomial distribution. Although the number of fruits damaged by *P. ambiguus* is positively related to its densities (Fig. 12.3), an inability to assess plant bug density accurately has prevented establishment of a reliable action threshold for this species (Schaub et al. 1988).

Old World Mirid Pests of Apple outside Europe

Helopeltis antonii causes a scabbing on young apple fruit in India (Puttarudriah and Appanna 1955). In Japan, *Pseudophylus stundjuki* (as *Heterocordylus flavipes*) once caused fruit deformation (Nitobe 1906, Crosby 1911), though it no longer occurs in Japanese apple orchards (Yasunaga et al. 1996). Adults of *Campylomma liebknechti*, a phyline called the apple dimpling bug, invade orchards in New South Wales, Australia, from the late pink stage to about petal fall, their migrations generally coinciding with warm periods in September or October. The bugs feed on floral parts (Plate 18) and developing fruits up to 10 days after petal fall, causing raised, scabby areas that become dimpled on larger fruits. This damage is similar to that of internal cork due to boron deficiency. All apple varieties can be damaged, but 'Delicious' and 'Granny Smith' are particularly susceptible. Nymphs of *C. liebknechti* are usually killed by the insecticides used against the codling moth, and the remaining individuals that develop on the apples do not injure the fruit or foliage (Lloyd 1969, Hassan 1977, Swaine et al. 1991).

The mirine *Niastama punctaticollis* induces a similar injury in Australia (Hassan 1977). Called the dimpling bug in Tasmania, this univoltine species invades apple orchards from its apparent sole breeding host, Monterey cypress, hedges of which have been planted as windbreaks near orchards. Large numbers of adults leave the cypress in September and October to feed on other plants. The bug's dispersal into orchards typically coincides with the pink bud stage of apple development. Feeding by *N. punctaticollis* on developing flowers can distort the fruitlets. Fruit malformation, which was especially severe on 'Granny Smith' in northern Tasmania during the 1960s, generally decreases with increasing distance from the cypress (Terauds 1970, 1971). Feeding by another mirine, the green mirid (*Creontiades dilutus*), deforms apples in Victoria, Australia, producing fruit that shows small, sunken, discolored areas at feeding sites (Victorian Plant Research Institute 1971).

Avocado

The mirine *Dagbertus fasciatus* and phyline *Rhinacloa clavicornis* feed on the opening buds and flower clusters of avocado crops in Florida, causing abscission of the young fruits and fruit malformation (Wolfe et al. 1946). Their threat to avocado production, however, might be overrated (Ebeling 1959), for Wolfenbarger (1963) considered these plant bugs to be of little economic importance. A more recent comment is that flower abscission and fruit malformation resulting from mirid feeding on avocado can occasionally be severe (Leston 1979a).

Dagbertus olivaceus is known from avocado flowers in Cuba (Alayo 1974). Nymphs of a mirid cited as a *Lygus* species., but probably belonging to another mirine genus (see Schwartz and Foottit [1998]), are sometimes found on avocado flowers in Fiji (Swaine 1971). In Malawi, injury to avocado fruit by *Helopeltis* (probably *H. schoutedeni*) resembles symptoms from mango scab. Its feeding results in a white crystalline exudation from scabs, a problem one grower attributed to a fungal pathogen and tried to control with

Bordeaux mixture (Leach 1935). *Helopeltis antonii* sometimes feeds on avocado in the Philippines (Quayle 1938).

Blackberry and Raspberry

Lygus lineolaris, the tarnished plant bug, infests bramble crops such as blackberry and raspberry, which are aggregate fruits consisting of a cluster of small drupelets. In contrast to strawberry, another aggregate fruit, the mirids associated with brambles have received little attention.

Lygus bugs have long been known to occur on brambles (Wier 1875, Stedman 1899, Frank 1920, Enns 1947), but Mundinger and Slate (1952) might have been the first to describe specific feeding injury to the fruit: abnormal development of blackberries in New York. Previously, Wier (1875) referred to a "winter killing" of stems, and Stedman (1899) stated that small fruits, including blackberry and raspberry, suffer considerable injury from *L. lineolaris*. The tarnished plant bug is now included in guides to bramble production and management for the eastern United States (e.g., Goulart et al. 1989, Pritts and Handley 1989). Its feeding on buds, blossoms, and developing berries can deform the fruit and apparently decrease yields (Spangler and Agnello 1989, Schaefers 1991). Deformed berries resulting from plant bug feeding (rather than from poor pollination, infertility, or viral problems) are characterized by an unusually small number of fully formed drupelets, the injured drupelets appearing shriveled and seedlike (Goulart et al. 1989). This bug's feeding on mature fruits results in a whitening of drupelets (Schaefers 1991). Other causes of white drupelets in raspberries, such as solar injury or sunscald, however, are possible (Renquist et al. 1987).

Several research papers also discuss mirids of bramble crops. Studies in Quebec showed that adults of *L. lineolaris* are present on raspberry bushes throughout the growing season, especially when the fruit ripens (mid to late July). Because only small numbers of nymphs occur on canes (shoots), this bug might use raspberry bushes for adult feeding (mainly on fruits) rather than as a breeding host (Mailloux et al. 1979, Boivin et al. 1981). But *L. lineolaris* nymphs occur in greater numbers on brambles than once thought. Nymphs and adults are common in New York from the late bloom through the green fruit stages, and adults show a second peak during fruit ripening on summer-bearing blackberry and raspberry bushes. In addition, *Plagiognathus politus* and *P. obscurus* are abundant enough to be considered potential pests of brambles (Spangler et al. 1993).

The caragana plant bug (*Lopidea dakota*) is an occasional raspberry pest in western Canada (Arnason et al. 1939, Kelton 1982a). It once destroyed crops in Alberta (Jacobson 1939) and Saskatchewan (MacNay 1952). None of the mirid species Hill (1952a) reported from cultivated raspberry crops in Scotland were observed to injure the fruit. The comment by Gordon et al. (1997) that, according to Hill (1952c), *Lygocoris pabulinus* was once an important pest of "cane fruit" could be interpreted to mean the berries are injured. Most likely Gordon et al. were referring generally to bramble crops because Hill (1952b, 1952c) stated that raspberry fruits in Scotland are unaffected by this plant bug, the injury consisting only of foliar perforations and branching of canes (see Chapter 9).

Cashew

A mirine, *Orthops palus*, develops on cashew trees in East Africa, apparently feeding on the inflorescences (Taylor 1947b). Bryocorines of the genus *Helopeltis* are important insect pests of cashew crops in Africa and India. *Helopeltis antonii* is considered the most severe cashew pest on India's west coast (Hari Babu et al. 1983). Because *Helopeltis* species are polyphagous, hosts such as cotton should not be used for intercropping in plantations (Ohler 1979).

Helopeltis antonii feeds on the inflorescences of cashew trees in India, occurring principally on the main axis around the nodes but also on the developing nuts (Abraham 1958, Pillai and Abraham 1975). It produces necrotic lesions that allow secondary fungi to invade and give rise to inflorescence blight (Nambiar et al. 1973, Thankamma Pillai and Pillai 1975, Ambika and Abraham 1979). Lesions appear on the main rachis and secondary rachises within 5–6 hours after inflorescences are exposed to *H. antonii*. The lesions become pinkish brown within 24 hours and, after 2–3 days, enlarge to a maximum of 13 mm and become "scabby." With the coalescing of lesions, affected inflorescences develop a scorched appearance (Nambiar et al. 1973). The biochemical changes resulting from this bug's feeding on leaves and shoots were described in Chapter 9.

Helopeltis antonii is responsible for the shedding of cashew fruit in various stages of development: early (mustard), peanut, and later stages (Thankamma Pillai and Pillai 1975). This pest is active mainly from October to May, building in numbers with the emergence of new flushes and panicles after the monsoon rains and peaking in January when trees are in full bloom. The bugs are typically absent during the monsoon season, but on young trees that produce flushes nearly continuously, they can be present almost year-round (Sathiamma 1977, Hari Babu et al. 1983, Sundararaju 1984, Devasahayam 1985). Trees established by vegetative propagation are particularly susceptible because of their prolonged flowering period (Devasahayam and Radhakrishnan Nair 1986). *Helopeltis* populations may not attain injurious levels in years of low rainfall (Ohler 1979).

Losses in yield of 30% or more can result from infestations by *H. antonii* (Nair 1975, Pillai et al. 1976, Devasahayam and Radhakrishnan Nair 1986). Damage

to cashew trees may be more severe on shaded sides of the canopy, with outbreaks altering the general pattern of fruit production and abscission. In 1980, a heavy loss of immature fruits fed on by *H. antonii* resulted in compensatory increases in the numbers of fruits set (Subbaiah 1983). Usually three insecticide applications are needed for control: at emergence of new flushes, at panicle development, and at fruit set (Pillai 1987; see also Sundararaju and Sundara Babu [1998]). This mirid also is the primary causal agent of inflorescence blight (rather than various fungi) in Sri Lanka, where it is considered important in limiting cashew production (Jeevaratnam and Rajapakse 1981a, 1981b).

In East Africa, nymphs and adults of *H. anacardii*, *H. schoutedeni*, and perhaps other species of the genus feed on developing cashew nuts, causing them to shrivel and abscise. The presence of lesions can result in lowered prices for the nuts. Older nuts are reduced in size and show surface pocking. Nuts that dry out and crack prematurely are invaded by fungi and insects (Drosophilidae and Nitidulidae), promoting decay (Swaine 1959, Wheatley 1961, Martin et al. 1997). The use of predatory ants in cashew plantations and clones showing decreased susceptibility to *Helopeltis* and other sucking pests is being evaluated in Tanzania (Martin et al. 1997). Evaristo and Pais (1970) discussed the identity of *Helopeltis* species injurious to cashew crops in Mozambique.

Cashew fruit in Papua New Guinea is sometimes malformed and shows sunken areas from the effects of *H. clavifer* (Plate 18; Smith 1978). Since the late 1980s, *H. pernicialis* has become the main insect pest of cashew in Australia's Northern Territory, where its feeding on inflorescences can prevent nut set. In addition to feeding on new growth and foliage, this bug sometimes injures nuts before the shell is hardened (Stonedahl et al. 1995). Insecticides generally must be applied to ensure a worthwhile nut harvest in tropical Australia. Use of the green ant (*Oecophylla smaragdina*) might prevent mirid damage from exceeding 6–10%, a level that can be considered a control threshold (Peng et al. 1997).

Citrus

Mirids are occasional pests of grapefruit, orange, and other citrus crops, feeding on the young shoots and twigs (see Chapter 9) as well as the flowers. The mirine *Closterotomus trivialis* punctures the flowers of citrus trees in the Mediterranean region, inducing blossom drop (Barbagallo 1970). *Dionconotus neglectus*, the orange blossom bug, is a polyphagous mirine that develops on grasses and herbaceous weeds in Mediterranean countries and migrates to orange trees. In Turkey, it develops on plants of numerous families during February and March; adults first appear in mid-March and live until the last half of May, and injury to citrus flowers is restricted to the Aegean and Mediterranean regions

(Demirdere 1956). Feeding by *D. neglectus* causes the flower buds and blossoms to drop, and sweet sap oozing from the wounds attracts various adult Diptera and Hymenoptera (Bodenheimer 1951, Talhouk 1969). Blossoms produced in response to flower abortion result in leathery, poor-quality fruit (U.S. Department of Agriculture 1968). An Australian mirine, *Austropeplus* species, feeds on the small flower shoots of coastal citrus in New South Wales. Its effect on trees that will bloom profusely appears to be minimal, even when numerous blossom shoots are destroyed, but on trees that blossom lightly, injury can be severe. Flower production is checked, thereby reducing fruit set (Gellatley 1967, Hely et al. 1982). In Bolivia, a mirid reported as a *Helopeltis* species (but misidentified because this genus is strictly Old World) causes a black spotting of orange crops (Squire 1972).

Cocoa

On cocoa trees, mirids often feed either on vegetative or on reproductive structures, and reports of the same species feeding on pods and on foliage or stems might involve misidentifications (Kumar and Ansari 1974). Populations of some cocoa mirids on vegetative tissues perhaps develop more or less independently from those on pods (Gibbs et al. 1968). The cocoa tissues fed on can be affected by the water status of the host and by the stage of a particular mirid species, with pods perhaps providing better nutrition than shoots because of a higher water tension (Entwistle 1972). Species preferring pods may switch to vegetative tissues with the harvesting of fruit. Yet the role of pods versus climate in population fluctuations of cocoa capsids is not fully understood (Williams 1954, Tan 1974a), especially for species unspecialized in their feeding preferences (Entwistle 1972).

Pod feeding by cocoa mirids mainly involves the parenchymatous husk tissues outside the sclerotic layer (Plate 18; Entwistle 1972). Although damage to the pods by mirids can be significant (e.g., from *Helopeltis theivora* in Malaysia [Tan 1974b; K. C. Khoo, pers. comm.]), it is usually not as detrimental to production as the bugs' feeding on shoots and branches (see Chapter 9). For the direct effects of mirid feeding to be harmful, it usually must be exceptionally heavy on young pods (Toxopeus and Gerard 1968).

COCOA MIRIDS OF THE OLD WORLD TROPICS

Helopeltis Species

Helopeltis bradyi tends to favor cocoa pods in Sri Lankan plantations. Leaf and stem feeding is rather rare and generally associated with poor shade over trees (De Silva 1957). Some workers contend that invasion of pathogenic fungi through mirid-induced lesions or cracks in the husk is more important than the bugs'

direct feeding, but it is uncertain that fruit feeding predisposes the pods to infection by *Phytophthora* fungi (Entwistle 1965, 1972, 1985a).

The rapid and spectacular injury caused by *H. bradyi* in Sri Lanka consists of water-soaked lesions (3–6 mm in diameter, as many as 60 produced in 24 hours) at the point of stylet entry; injured areas sink and blacken. Young pods can fail to develop or drop prematurely. Older pods, which often show concentrated spotting, can be malformed, grow more slowly than normal, and attract larvae of a pod-boring lepidopteran (Fernando and Manickavasagar 1956, De Silva 1961).

Helopeltis theivora is a key cocoa pest in Peninsular Malaysia, feeding on cherelles and pods (Plate 18) in addition to new flush shoots (e.g., Khoo 1987; Wood and Chung 1989, 1992; Khoo et al. 1991; Muhamad and Way 1995a). Even though 100 punctures can be produced in 24 hours, the seeds remain unaffected (Miller 1941). Tan (1974b), however, reported that small pods (<5 cm long) wilt following infestation by *H. theivora* and that the weight of large pods and the dry weight of beans are reduced, even though they appear normal. Failure to apply insecticides results in a substantial loss of yield. The mirid can reduce the number of cherelles developing into pods and cause a long-term decline in plant vigor (Chung and Wood 1989); young cherelles are especially susceptible to mirid feeding (Muhamad 1994). The magnitude of "natural" cherelle wilt complicates an assessment of the additive or complementary effects from a direct pest such as *H. theivora*. Crop losses vary substantially, depending on the time and intensity of mirid feeding on cherelles and pods (Muhamad and Way 1995a). Pods injured by *H. theivora* can also attract arthropods characterized as secondary invaders—for instance, the coffee bean weevil (*Araecerus fasciculatus*) (Tan 1974b).

Pods are critical to the ecology of *H. theivora* in Malaysia. Females caged on shoots in the field do not produce offspring, and adults live more than three times longer on pods than on shoots. Third through fifth instars show higher rates of development on pods than on shoots (Azhar 1986, Awang et al. 1988). The availability of pods, which are produced in two annual peaks, plays an important role in this bug's reproductive success and interacts with weather to influence its population dynamics. Malaysian populations of *H. theivora* increase not only with increasing numbers of pods but also with precipitation. Populations can drop with decreasing precipitation even though pod numbers remain high (Tan 1974a, Wills 1986).

Van der Goot (1917) established interrelationships among Indonesian populations of *Helopeltis* (probably *H. theivora*), ants, and mealybugs. The black cocoa ant (*Dolichoderus thoracicus*) is intimately associated with the mealybug *Planococcus lilacinus*, which dies out when the ant is absent but thrives on ant-infested trees. Ants obtain honeydew from mealybug colonies, their protection resulting in lower mortality and somewhat accelerated development of *P. lilacinus*. In addition, the dense masses of ants might deter oviposition and pod feeding by *Helopeltis* (van der Goot 1917; see also Chin et al. [1988], Khoo and Chung [1989], Ooi [1992], Way and Khoo [1992]). Mealybug-infested pods apparently are also less attractive to the mirids, as determined by P. Levert (cited in Giesberger [1983]). Because the mealybug is considered innocuous, growers were encouraged to introduce the ant into their cocoa plantations (Kalshoven 1981).

For many years, growers practiced this ant-mealybug system of biological control in the cocoa estates of Java, but certain underlying principles of this technique were challenged during the 1930s, and other explanations (e.g., physiological condition of trees and sap composition) for the lower incidence of *Helopeltis* on trees harboring ants and mealybugs were advanced. Some researchers even said the ant did not control *Helopeltis* effectively and that the system should be rejected (Giesberger [1983]). J. K. De Jong suggested that trees favoring the mealybug are unfavorable to *Helopeltis*, and therefore, it is pointless to introduce the black cocoa ant to reduce mirid injury (references in Giesberger [1983]; see also Khoo [1987]).

Giesberger ([1983]), summarizing numerous investigations in Java, reported that the black ant-mealybug system is no longer being used against *Helopeltis*. He hoped, though, that growers and scientists might rediscover this essentially self-perpetuating means of biological control. More recent studies on ant-*Helopeltis* relationships in Malaysia showed that the dominant ant species, especially the weaver ant (*Oecophylla smaragdina*) (Azhar 1986, Chin et al. 1988), can provide almost complete protection from injury by *H. theivora* (Way and Khoo 1989, 1992). Plantation workers, however, might find the presence of this aggressive ant unacceptable (Khoo and Chung 1989, Way and Khoo 1991).

Cocoa monocultures tend to favor outbreaks of *H. clavifer* in Papua New Guinea. Extended dry seasons and cultivation under low shade might promote population increases; the effects of natural enemies and the production of pods and vegetative flush seem minimal (Smith 1972). Like *H. theivora* (Muhamad and Khoo 1983), this pod-feeding bryocorine (Plate 19) can be reared in the laboratory on fresh cocoa pods (Smith 1973).

Distantiella theobroma and *Sahlbergella singularis*

Even though the African odoniellines *Distantiella theobroma* and *Sahlbergella singularis* generally feed more on stems than do monaloniine Bryocorinae (*Helopeltis* spp.) associated with cocoa, they feed extensively on the pods (Plate 19; Cotterell 1926). In fact, unripe pods can be used for rearing both species in the laboratory (Raw 1959a, Prins 1965). Females having access to pods attached to trees lay more eggs than do those provided detached pods or those reared on chupons or seedlings. The availability of pods improves fecundity more in *D. theobroma* than in *S. singularis* (Houillier 1964a).

Lavabre (1969) reviewed attempts to rear cocoa mirids, emphasizing conditions that seem to favor successful breeding: a nutrient-rich diet, high humidity without moisture condensing on containers, darkness or near darkness, ample space, and minimal disturbance of the bugs.

The feeding-induced lesions of these mirids are concentrated on the peduncular ends of the pods (Houillier 1964b, Youdeowei 1977). The sites of penetration can be detected by the plug of blackened tissues left on the pod surface, but this injury is considered unimportant because the wound is superficial (Williams 1953). In one study in Nigeria, however, *S. singularis* reduced the weight, but not the size, of the pods, the number of beans per pod remaining unaffected. Because feeding occurs mainly on the husks, the loss of weight of the beans might result from changes in the pod physiology (Akingbohungbe 1969). Pods injured by this species can be shorter and contain slightly more woody seeds than undamaged pods; both pod and husk weights are reduced when more than half the pod surface is covered with lesions. Pod length and width, bean weight, and total number of beans per pod are not significantly affected (Ojo 1981, 1985). Other studies on the economic importance of pod feeding by *S. singularis* and other bryocorines suggest that only the husk is affected, rather than the weight of the peeled beans, even when more than half the surface of a mature pod is blackened by lesions. Mirid feeding generally does not affect yields after the cherelle stage is reached (Toxopeus and Gerard 1968, Marchart 1972).

Populations of *D. theobroma* and *S. singularis* undergo characteristic fluctuations, often disappearing near the end of the dry season (March–April) in West Africa and increasing during the fruiting season (November–December). Williams (1954) stated that the buildup in numbers should not be attributed solely to the presence of pods (see also Bruneau de Miré [1970]), although he acknowledged that they provide the mirids additional nourishment for their development.

Gibbs et al. (1968) proposed that seasonal cycles of cocoa mirids in Ghana depend not only on the availability of pods but also on many other factors: intermittent flushes of vegetative growth (4–5/yr), to which the bugs transfer after peak harvest; nutritional quality of external parenchymatous tissues; structure of the canopy; weather, especially relative to harmattan, a seasonal southward movement of dry air from the desert of West Africa; and other superimposed variables. *Sahlbergella singularis* declines during harmattan because of the combined effects of desiccation and the bugs' inability to feed when plant water stress exceeds –12 atmospheres (Nwana and Youdeowei 1976). In the case of mirid species restricted to or at least showing a preference for pods, the presence of fruit can allow populations to increase, but in species of less pronounced feeding habits, the role of pods in population fluctuations is unclear. Entwistle (1972) discussed these fluctuations in more detail.

Additional Bryocorine Pests of Cocoa

Other bryocorines, such as *Bryocoropsis laticollis*, feed mainly on developing green pods in Ghana, causing scarring and some distortion but little economic loss (Williams 1953, Kumar 1971a). In Madagascar, the odonielline *Boxiopsis madagascariensis* feeds on pods in addition to stems and branches. Infestations are more common on pod-bearing trees, and the density of this bug is proportional to the number of pods available (Decazy 1974).

Mirids are among the most damaging cocoa pests in all areas of Papua New Guinea except the North Solomons Province (Moxon 1992, O'Donohue 1992). In addition to *Helopeltis clavifer*, previously discussed, injury to cocoa pods in New Guinea involves *Pseudodoniella pacifica* and *P. typica* (Dun 1954). These bugs might have developed on fig fruits in the rainforest of New Guinea before invading cocoa plantations after the felling of host trees in their natural habitat (Szent-Ivany 1965). During feeding, they appear to inject a toxin that induces the death of tissue near the point of stylet entry (Henderson 1954). Each individual can produce 40–80 lesions in 24 hours, mainly at the base of fruit. Feeding scars are invaded by insect larvae and by a fungus, a *Gloeosporium* species, which causes raised pustules to form on the surface (Dun 1954, Smee 1963). *Pseudodoniella pacifica* and *H. clavifer* are categorized as pod-sucking pests in Papua New Guinea (Szent-Ivany 1961, Room and Smith 1975).

A Malaysian pest of cocoa is *Platyngomiriodes apiformis*, a bryocorine of beelike habitus (Plate 19) whose common name is the bee bug (Khoo et al. 1991). In Sabah, the gregarious nymphs and adults feed at the base of the pods and inflict more severe damage than does *H. clavifer*, a co-occurring species that also prefers pods to shoots (Conway 1971). Infestations of *P. apiformis* tend to be localized and restricted to dense, shaded pods and those on older trees (Pang and Syed 1972, Pang 1981). Its feeding lesions are larger (up to 7 mm in diameter) and more sunken than those of *H. clavifer* (Plate 19). Conway (1971) described the symptoms: "After a while cracks appear in the lesions, particularly around the circumference, so that a black plug-like body is produced, covered in a powdery white bloom. This then becomes raised ... as a scab, the bloom being replaced by a buff powdery mycelium." If insecticides are not applied, cocoa yields can be affected (Pang 1981).

NEOTROPICAL MIRIDS OF COCOA

Plant bugs associated with cocoa in the neotropics have not been as thoroughly studied as those infesting this tree in the Old World tropics. Even though devastating dieback fungi do not invade cankers induced by Neotropical mirids and intensify the initial injury, plant bugs are among the most serious cocoa pests in Central and South America (Posnette and Smith 1985, Abreu et al. 1989).

Monalonion Species

Species of the New World bryocorine genus *Monalonion* take the place occupied in the Old World by *Helopeltis* species (Schmutterer 1977). The occurrence of *Monalonion* species on cocoa was first noted in 1909 in Brazil, although their injury might have occurred earlier (Entwistle 1965). These bugs can inflict serious injury: desiccation of young pods and deformation of larger pods. Husks sometimes break open, rendering the beans worthless. Pod feeding by *M. dissimulatum* reduced bean production from 7.0 to 2.7 tons per hectare during the 15 years after its detection in Venezuelan cocoa estates (Entwistle 1972). Estimated losses as great as 75% of harvest are possible (Hernandez et al. 1953).

Monalonion dissimulatum, known locally as the "mosquilla" or mosquito bug, was once considered the most important insect pest of cocoa in Peru. Its feeding is not restricted to pods, but injury can be severe on young green fruits. Green pods develop pustules and dark, warty spots; injured pods can eventually wilt and die, resulting in a complete loss of crop. This pest thrives under conditions of high temperature, high humidity, and heavy shade (Wille 1944, 1952). Bondar (1937, 1939), Knight (1939), and Costa Lima (1940) described a similar injury to Brazilian cocoa pods by *M. bondari*, *M. schaefferi*, and other species of the genus. A species of *Monalonion* scars and disfigures pods in Bolivia but is considered a minor pest (Squire 1972). In the laboratory, *M. annulipes*, a common pest of cocoa in Central America (Saunders 1981), feeds more on mature pods than on green pods (Villacorta 1977a). Its numbers seem greatest on unshaded than on shaded trees, the largest populations occurring in Costa Rica during October to November (Villacorta 1977b). Abreu (1977) reviewed the *Monalonion* species associated with cocoa in the New World.

Ant Mosaic

An ant mosaic has been described for cocoa in the New World tropics. Although its composition differs from that of the Old World (see Chapter 9)—dolichoderine and ponerine ants predominate in the neotropics—the mosaic probably is similarly important in determining the distribution of mirids and other pests on host trees (Leston 1978b).

Coffee

The mirine *Ruspoliella coffeae* sometimes limits coffee production in Africa. It feeds on flower buds and especially on anthers; petals and stamens usually blacken, and flowers can abort and fail to set fruit. Withered buds remain on the trees as "black caps" (Stoffels 1941, Fiedler 1951, Wyniger 1962). Injury from feeding by *R. coffeae* usually can be distinguished from that by pentatomids (*Antestiopsis* spp.) in that mirids leave the style uninjured (LePelley 1942). One plantation in Kenya suffered a nearly complete loss of crop in 1930, and about half the crop on more than 1,000 hectares was lost to *R. coffeae* (LePelley 1932, 1968). Its feeding can be considered beneficial because excess flowers are pruned, which helps prevent overbearing at lower altitudes (de Pury 1968). Mirines causing similar injury to the reproductive structures of coffee in Africa are *Taylorilygus ghesquierei*, *Volumnus obscurus*, and *Xenetomorpha carpenteri* (Ghesquière 1939, Buyckx 1962, LePelley 1968, Coste 1992).

Grape

NEW WORLD PESTS

Several plant bugs are associated with the inflorescences and berries of grape vines in the eastern United States. Overwintered adults of *Lygus lineolaris* sometimes invade vineyards in early spring, their feeding not only blackening the edges of tender leaves (Murtfeldt 1902) but also injuring the opening buds and blossoms (Bruner 1895). Fruit injury can be inferred from Stedman's (1899) comments on the economic importance of the tarnished plant bug in Missouri.

The mirine *Taedia scrupea*, which feeds on the blossom clusters and tender foliage of wild grape (Parrott and Hodgkiss 1913), once attracted attention in New York vineyards. Felt (1915, 1916) reported an irregular blasting of blossoms, with the bugs producing black spots on young fruit. Referred to as the banded grape bug, this mirid was thought responsible for the abscission of young grapes, and one grower estimated that it destroyed half his crop; some of the damage, however, might have been caused by poor pollination (Felt 1916). In 1979, chemical treatments were needed to reduce sporadic infestations of *T. scrupea* in three Pennsylvania vineyards (Jubb 1979). Recent studies in New York showed that feeding on 'Concord' grape clusters and shoots at the prebloom stage can significantly reduce the number of berries per cluster and the average berry weight. The total crop weight can be reduced by nearly 70%. Feeding by *T. scrupea* at the bloom or postbloom stage does not affect cluster weight. This low-density pest can cause economic crop losses at levels of 0.5 or 1 nymph per vine during a three-week period that corresponds to the time of rapid shoot elongation and cluster formation (Martinson et al. 1998).

Although *T. scrupea* occurs on cultivated muscadine grape vines in North Carolina without causing apparent injury (McGiffen and Neunzig 1985), its feeding elsewhere in the Southeast (in association with *L. lineolaris*) on buds and blossoms injures emerging leaves and blossom clusters. The incidence of injury is low, and its potential for damaging muscadine grape is rated only as moderate (Dutcher et al. 1988). At least three other species of the genus develop on the inflorescences of wild grape in eastern North America: *T. casta*, *T. multisignata* (Knight 1918b, Froeschner 1949), and *T. floridana* (pers. observ.).

The nymphs and adults of the mirine *Lygocoris inconspicuus* feed on grape flowers (Knight 1918b). Fox, frost (winter), and other wild grapes serve as hosts in Pennsylvania (pers. observ.). This plant bug probably is the one Uhler (1878) observed on grape inflorescences in Maryland and that Parrott and Hodgkiss (1913) reported (as *Lygus invitus*) from grape plants in New York. Early instars feed on the unfolding leaves, move to the developing flower buds, and sometimes puncture the peduncles and pedicels of the unopened blossoms and young fruit, resulting in imperfect clusters of grapes (Parrott 1913, Parrott and Hodgkiss 1913).

Other mirids that develop on grape flowers include the phyline *Sthenarus viticola* on fox grape in Pennsylvania and on mustang grape in Texas, and the mirine *Dagbertus fasciatus* in Florida (pers. observ.). *Sthenarus mcateei*, collected from grape flowers in Arkansas (pers. observ.), might also be an inflorescence feeder. The nymphs and adults of the restheniine *Prepops insitivus* develop on wild muscadine grape in Florida, and an adult fed on a flower bud (pers. observ.). But the habits of *Prepops* species are so little known that it is uncertain if these bugs are typical inflorescence and fruit feeders. In the western United States, *Labopidea simplex*, an orthotyline probably associated with composites (Kelton 1980b), occurs on grape plants when they are in bloom (Uhler 1877).

OLD WORLD PESTS

Closterotomus fulvomaculatus has been called the grape-vine flower bug in Europe. Infestations in the vineyards of Croatia once led to a complete loss of crop (Tominić 1951). Another mirine, *Capsodes sulcatus*, can be a serious problem on grape vines in France, its pest status dating from 1890 (Kiritshenko 1951). It feeds on weeds in the vineyards but will move to the grape plants when the inflorescences appear. This mirid's feeding discolors the pedicels and can affect yields (Stellwaag 1928). In Spain, *C. sulcatus* feeds on the flower buds, which turn brown and dry up (Garcia-Tejero 1993). Feeding by *Helopeltis antonii* on developing fruit in India gives rise to circular depressions that sometimes cover the entire surface. Affected grapes can dry up, rot, or drop at the slightest touch (Puttarudriah and Appanna 1955).

Guava

HYALOPEPLUS PELLICIDUS

The flower buds of guava are necessary for the development of the transparentwinged plant bug (*Hyalopep-*

lus pellucidus) (Plate 19) in Hawaii. The anthers of the buds fed on by the nymphs (Plate 19) show a necrotic blackening (Plate 19). Damage to the anthers, which are important auxin-producing sites, apparently causes flower bud abscission. The current practice of dividing orchards into separate production units, so that fruiting occurs sequentially, provides the bugs a nearly continuous food supply. This production practice might allow *H. pellucidus* to become an even more important pest in Hawaiian guava orchards (Mau and Nishijima 1989).

HELOPELTIS SPECIES AND OTHER BRYOCORINES

Guava growers in West Malaysia must contend with *Helopeltis theivora*, a serious pest that can go unrecognized because it occurs at low densities and causes injury that is sometimes attributed to fungi. Its feeding results in necrotic lesions (Plate 19) that lower the market value or render the fruit unmarketable and can lead to cracking as the fruit expands. In addition, scabby fruit canker, caused by the secondary fungus *Pestalotiopsis psidii*, begins as water-soaked spots resulting from mirid punctures (Lim and Khoo 1990).

Feeding by *H. antonii* in India produces ugly warts, blisters, or scales on guava fruit, the injury often attributed to a fungal pathogen (Puttarudriah 1952, Nair 1975). This bryocorine pest reduces the market value of fruits (Sudhakar 1975) and causes a premature dropping of younger fruits (Puttarudriah 1952). The incidence of *H. antonii* on Indian guava ranged from about 35% to more than 80% among 11 varieties, and the effects tended to be more severe on younger trees. This preference might be due to increased shade provided by young trees, or possibly results from nutritional differences between young and old trees (Gopalan and Perumal 1973). Sundararaju and Sundara Babu (1998) determined the net reproductive rate of *H. antonii* on guava under laboratory conditions.

Helopeltis schoutedeni causes similar injury to guava in West Africa (Gerin 1956). *Helopeltis clavifer* (Smith 1978) and another bryocorine, *Ragwelellus festivus*, cause a scabbing of fruit in Papua New Guinea, in addition to distorting and producing side-shooting of the flush tissue (Greve and Ismay 1983).

Mango

A *Helopeltis* species (probably *H. schoutedeni*) severely injures mango fruit in southeastern Africa (Malawi). If the bugs' stylets penetrate only to the middle layer of skin on unripe fruit, round, water-soaked lesions 2–5 mm in diameter soon form, sink and darken within 24 hours, and eventually become black scabs. Heavy scabbing leads to fruit abscission. When the inner skin is pierced, a fruit rot develops and spreads in a manner that suggests fungal or bacterial infection (Leach 1935). Cooperation by plant pathologists and entomologists in

diagnosing this and similar types of injury is desirable. Otherwise, it is possible that "symptoms such as these may start investigations along wrong lines, leading to the waste of much time and energy in studying the aetiology of disease and may possibly lead to unnecessary expense in attempting to combat the diseases" (Leach 1935).

Additional mirids associated with mango are inflorescence-feeding *Dagbertus* and *Rhinacloa* species in Florida (Peña 1993) and in Dominica (Whitwell 1993). The effects of these plant bugs on mango are unknown, but in Florida their feeding may cause flower abscission (Peña et al. 1996). *Orthops palus*, and probably also *Taylorilygus virens*, develop on mango inflorescences in East Africa (Taylor 1947b). The nymphs and adults of a *Lygus* species (but probably belonging to another mirine genus) suck mango blossoms in Fiji without causing significant injury (Swaine 1971), and *Campylomma austrinum* and *C. liebknechti* can occur on mango flowers in Australia (Malipatil 1992).

Olive

Closterotomus trivialis is a pest of olive crops in the Mediterranean region of Italy. The nymphs and adults feed on young shoots, buds, and blossoms, which results in the premature dropping of flowers. Trees can tolerate substantial flower abscission, but chemical treatments may be needed when the injury is severe (Barbagallo 1970, Monaco 1975). *Closterotomus trivialis* shows similar habits in western Turkey. When the densities of this mirine are high, 4–8 flowers of every cluster (containing 30–40 flowers) can be destroyed. Because natural abscission from physiological factors is about 95%, such an injury level is considered unimportant. This plant bug, therefore, is considered a secondary pest of olive (Kaya 1979). The olive pest in Turkey that was recorded under the name *Calocoris rubrinervis* (Gentry 1965) apparently also refers to *Closterotomus trivialis*. This mirid is not considered economically important on olive trees in Greece (Drosopoulos 1993).

Several workers have recorded *Lygus pratensis* as an olive pest in Turkey. Önder (1972), however, was unable to collect this species from olive trees, or from plants under olive trees, during his studies on Turkish Mirinae.

Peach and Nectarine

Several mirid species injure the fleshy fruit of peach, a stone fruit classified as a drupe. Much of the literature on plant bugs as pests of peach trees concerns North American species of *Lygus*. Plant bugs should also be important pests of nectarine crops, but there are few references to fruit injury by mirids on this variety of peach. California's pest-management guidelines, however, specifically mention lygus bugs as pests of nectarine fruit (Barnett and Rice 1989, Barnett et al. 1990), as does the orchard monitoring guide for the mid-Atlantic region (Polk et al. 1995). Because lygus bugs do not reproduce on fruit trees, only adults are found on peach and nectarine crops (Polk et al. 1995).

LYGUS BUGS

Overwintered and first-generation adults of *Lygus lineolaris* feed on the tender terminal buds of peach nursery stock, resulting in a loss of apical dominance and a dwarfed or stunted appearance known as "stop back" (Back and Price 1912; see also Chapter 9). Adults can also cause the blasting of newly opened blossoms (Taylor 1908), but their feeding on fruit is more serious. Since the early 1890s, when injury by *L. lineolaris* was termed a rare occurrence (Hall and Lowe 1900), numerous workers, from southern Canada (Roberts and Pree 1983) to Florida (Zak 1986), have recorded the tarnished plant bug as an important pest of peaches. This bug causes many of the young fruits to drop prematurely, and injured fruits that continue to develop show a blemishing usually referred to as *catfacing* (Plate 20; Moore and Fox 1941). This condition is characterized by scarred, sunken areas that typically become corky and hard, grow much slower than the surrounding tissues, and lack fuzz (Plate 20; Porter et al. 1928, Howitt 1993, Polk et al. 1995). Strong (1970) explained that typical catfacing involves "asymmetrical destruction of auxin-producing areas" on fruits 8–20 mm in diameter.

In the early 1940s, orchardists in Oklahoma drew attention to the catfacing of peach fruits, including 18% damage to the peaches in one orchard in early season 1943. Because the tarnished plant bug was prevalent in affected orchards, it was assumed to be the causal agent. Fenton et al. (1945) showed that *L. lineolaris* adults when they are caged can produce similar catfacing symptoms on young fruits. In another study, caged tarnished plant bugs caused catfacing of newly set fruit within seven days (Snapp [1947b]). Bobb (1970) termed catfacing a major problem for peach growers in Virginia, noting that the tarnished plant bug causes more fruit injury than any other insect. In West Virginia, *L. lineolaris* is the most abundant of the insect species responsible for catfacing and related types of injuries to peach (Hogmire and Custer 1982).

Catfacing by lygus bugs is indistinguishable from that inflicted by stink bugs (Pentatomidae) and generally occupies a larger area than injury caused by the plum curculio (*Conotrachelus nenuphar*) (Chandler 1955). In addition, Rice ([1938?]) commented on the similarity between catfacing and bacterial spot of peach, noting that damage to the fruit by tarnished plant bugs in Delaware once was attributed to a pathogen.

Attracted by swelling blossom buds and opening flowers, overwintered *L. lineolaris* adults usually begin migrating into the orchards during the delayed-dormant bud stage (Hammer 1939) to the pink stage and leave shortly after petal fall (Porter 1926; Woodside 1947, 1950). In South Carolina, the largest number of overwintered adults are found from the time of full bloom until petal fall (Snapp 1947a). In Delaware, first-generation (rather than overwintered) adults are more common on peaches (Rice [1938?]), and first-generation adults usually cause the greatest injury to peaches in the Niagara Peninsula of Ontario; their appearance in orchards tends to coincide with the blooming of alfalfa (Phillips 1958, Phillips and DeRonde 1966).

Damage is sometimes more severe in orchards planted with a permanent cover crop of an alternative host such as alfalfa or red clover (Venables and Waddell 1943) and is often influenced by environmental conditions. Hot, dry weather and drought in May and June in Ontario favor the migration of tarnished plant bugs from once-succulent weed hosts into orchards, where they generally feed on the fruit of trees near the periphery (Roberts and Pree 1983, Pree 1985). Weed management in orchards can minimize catfacing by tarnished plant bugs (Fogle et al. 1974, Killian and Meyer 1984, Meagher and Meyer 1990).

In addition to catfacing the fruits, lygus bugs also produce bleeding or gummosis (Plate 20) (sometimes spelled "gumosis," e.g., Stearns [1956]), an exudation of a clear, gummy substance (Rings 1958, Roberts and Pree 1983, Polk et al. 1995). Gummosis usually occurs within a few hours after feeding, and injured fruit can continue to exude gum for several weeks (Phillips 1958). Although often associated with catfacing, gummosis can occur alone, particularly in late season (Howitt 1993).

Lygus elisus and *L. hesperus* are implicated in causing a similar injury to maturing peach fruit in Utah (Sorenson and Gunnell 1936). The former species and *L. shulli* sometimes cause catfacing in the Pacific Northwest (Moore and Fox 1941), and *L. elisus* and *L. hesperus* are occasional pests of peach crops in British Columbia (Buckell 1939, Twinn 1939). Damage to peach and nectarine fruit by *L. elisus* and *L. hesperus* is sporadic in California but in some years results in severe economic losses (Barnett and Rice 1989, Barnett et al. 1990).

LYGOCORIS SPECIES

Howard (1901a) might have been the first to report a species of the mirine genus *Lygocoris* as a pest of peaches, although the earlier reported injury by undescribed species of *Lygus* (Riley 1893b) and that which Hall and Lowe (1900) attributed to *Lygus lineolaris* might actually refer to a species of *Lygocoris* undescribed at the time (Rings 1958). At least five *Lygocoris* species are now known to leave their breeding hosts, mainly oak and hickory trees, to infest peach fruit in eastern North America (Rings 1958, Howitt 1993). Injury is usually concentrated in orchard rows adjacent to woodland (Thatcher 1923, Rings 1959).

Lygocoris quercalbae once seriously damaged peach crops in Ontario, its feeding so severely scarring fruit that 90% of the fruit in half of a 2.4-hectare orchard was unmarketable (Caesar 1921). Ross and Caesar (1922) mentioned *L. omnivagus* as a pest of peach trees in the same Canadian province. Because of recurrent problems with *Lygocoris*, growers in the Niagara fruit belt of Ontario were once advised to avoid growing peaches near oak and hickory trees (Ross and Putman 1934).

The hickory plant bug (*L. caryae*) is generally a more important pest in New York than *L. quercalbae* and other species of the genus (E. H. Smith 1940), its feeding causing more severe injury to peaches. At one time, growers confused *Lygocoris* and *Lygus* injury with that of the plum curculio (Smith 1950). Although stink bugs are more important than plant bugs in causing catfacing in some areas, only *Lygocoris* species and *Lygus lineolaris* are implicated in this problem in Connecticut (Garman et al. 1953). A few *L. omnivagus* and *L. quercalbae* adults can be collected from peach trees in West Virginia (Custer 1981), but no catfacing injury is attributed to these species (Hogmire and Custer 1982). Catfacing of peaches in Ontario, initially attributed to *Lygus lineolaris* (as *L. pratensis*), has been caused by *Lygocoris caryae* adults that migrated from black walnut trees (Ross 1939, Twinn 1939).

OTHER MIRID PESTS OF PEACH IN
NORTH AMERICA

When the spring rains end in California, *Irbisia* species will invade various cultivated plants (e.g., Duncan and Pickwell 1939). For example, a black grass bug, *Irbisia solani*, once migrated into orchards from grasses and weeds, punctured the skin of the peaches, and caused sap to ooze from the wounds (Vosler 1913). Lockwood (1933) presented circumstantial evidence for this species as a cause of catfacing in Californian peach orchards. The injury is sometimes similar to that inflicted by lygus bugs but at other times takes the form of narrow creases on the fruit. Damage occurs in orchards adjacent to grassy and weedy slopes after the bugs' host plants, various grasses, senesce (Lockwood 1933). Lockwood and Gammon (1949) observed light to heavy catfacing of peaches, cherries, and plums by *I. solani* (cited as *Irbisia* sp. but identified by Schwartz [1984]) in California.

Lyle (1936) reported an unusual habit for *Lopidea robiniae*, an orthotyline nearly restricted to developing on black locust trees. Adults once dispersed in mid-June from their breeding hosts and severely injured peach fruit in Mississippi.

OLD WORLD PESTS OF PEACH

Closterotomus fulvomaculatus, *C. norwegicus*, and *C. trivialis* cause cracks and gummy secretions on fruit in Italy, particularly on trees planted among forage crops (Pegazzano 1958). In France, *Adelphocoris lineolatus*, *C. norwegicus*, and *Lygus rugulipennis* are associated with leguminous crops or herbaceous weeds, but

they can invade peach orchards in spring when alfalfa is cut or wild hosts go to seed. They cause malformation and gummosis of young fruit and sometimes dark spots and soft pulp on mature fruit (Cravedi and Carli 1988). In Italy, where *L. rugulipennis* has been a pest in IPM orchards since 1990, weed management is reducing the need for chemical controls against this plant bug. The mowing of alternate rows in orchards is particularly useful in preventing the bugs from moving into peach trees (Tavella et al. 1996).

In Australia, deep-seated puncturing of the flesh of peaches by the green mirid (*Creontiades dilutus*) causes gummosis and dark spots on the skin, with corky areas occurring beneath. This bug's punctures are deeper and more numerous than those made by the rutherglen bug (*Nysius vinitor*, Lygaeidae). Damage to 'Elberta' peaches by *C. dilutus* can be so severe that every fruit on a tree is affected (Pescott 1940), and exudations from its feeding punctures can ruin the fruit (Victorian Plant Research Institute 1971). This mirid produces a pitting of skin and malformation of peach fruit (also nectarines and prunes) in New South Wales (Anonymous 1940). Its feeding on green fruit results in the exudation of long, persistent columns of gummy exudate (Hely et al. 1982). In New Zealand, Cunningham (1950) mentioned catfacing by an unidentified mirid species.

Pear

MIRIDS AS PESTS OF PEAR IN NORTH AMERICA

Lygus Bugs

Walsh (1860) wrote that adults of *Lygus lineolaris* puncture and sometimes kill blossom buds of pear. Wier (1875) reported the destruction of pear flowers by this bug in Illinois and Knaus (1889), in Kansas. In Missouri, overwintered adults once invaded pear trees in early spring, and their feeding injuries resulted in trees that appeared scorched from fire (Haseman 1918). Crosby and Leonard (1914) reviewed other early reports of damage to pear by the tarnished plant bug.

Lygus bug injury to pear buds in Washington State consists of a blasting and brownish globules exuding from the wounds (Webster and Spuler 1931). The sap oozes for several hours and, when bugs are numerous, it can nearly wet an entire tree (Rolfs [1932?]). *Lygus elisus*, *L. hesperus*, and *L. lineolaris* once severely damaged pear fruit on young trees in British Columbian orchards with a heavy growth of alfalfa (Buckell 1939, Twinn 1939). In Oregon, Lovett and Fulton (1920) called attention to the injury caused by mirids (unspecified taxa) on the mature fruit of pear and apple trees: funnel-shaped pits or surface dimples, often with an irregular russetted spot in the depression.

Lygus elisus and *L. hesperus* were recognized only as minor pests of pear crops in California during the late 1960s. Since then, however, these plant bugs have become important causes of cull fruit (Bethell and Barnett 1978). Increased mirid injury is attributed to the development of pesticide resistance in *Lygus*, the greater number of lygus host plants adjacent to pear orchards, and the planting of cover crops that serve as hosts (Barnett et al. 1976, Bethell and Barnett 1978; see also Chapter 10). Consultants also are better able to recognize fruit damage inflicted by these bugs.

Feeding by lygus bugs on pear fruit in California produces pustules and small depressions with swollen centers. The latter symptom resembles the pitting caused by boron deficiency. Late-season injury can be confused with that caused by the stink bug *Euschistus conspersus*. Lygus injury appears as a hard mass of cells (stone cells) beneath the skin, whereas stink bugs cause white pithy areas to form at the stem end. Growers in California are instructed to do additional monitoring of lygus bug populations and to consider control measures if sampling reveals one damaged pear in 100 (Bethell and Barnett 1978, Flint 1991).

Lygocoris Species

Parrott and Hodgkiss (1913) described injury to pears by the pear plant bug (*Lygocoris communis*) in New York: scarring and deformation of young fruit (Fig. 12.4) with development of hard, granular areas forming cores in the flesh, and cracking of skin. The piercing of immature pears is often accompanied by a considerable flow of sap from the wounds (Fig. 12.5). Symptoms that might seem inconsequential on young fruits generally intensify as the pears mature (Fig. 12.6). Riley's (1893b) brief mention of an undetermined species of "*Lygus*" injuring young pears in New York also refers to *Lygocoris communis*, as do records of pear injury in New York dating from the mid-1880s (Parrott and Hodgkiss 1913). Brittain (1916a) reported a similar injury—woody pears—in Nova Scotia, noting that adults sometimes pierced the fruit until the bugs became covered with sap oozing from the punctures. He felt that no other pest of pear crops compared with *L. communis* in the amount of damage inflicted or in the difficulty of eradication. This pest sometimes migrated from apple into pear orchards in Nova Scotia as soon as the adult stage was reached, rendering the latter crop unfit for market within a few days (Dustan 1924).

Other Pests of Pear in North America

Occasional pests of pear include *Campylomma verbasci* and *Taedia pallidula*, which once punctured young fruit soon after blossom drop in a New York orchard (Parrott

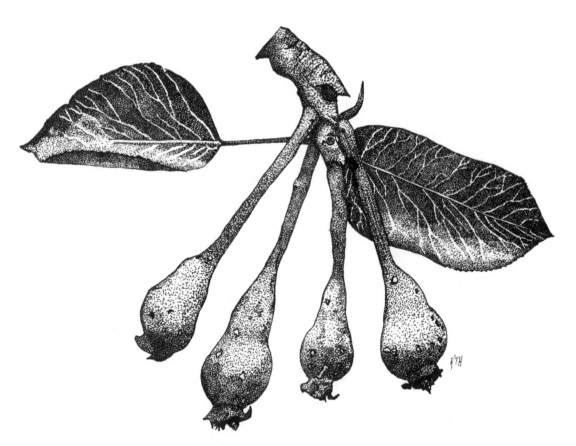

Fig. 12.4. Scarring of young pears resulting from feeding by *Lygocoris communis*. (Redrawn from Parrott and Hodgkiss 1913.)

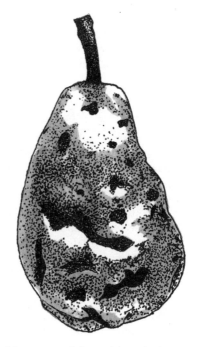

Fig. 12.5. Sap exuding from young pear fed on by *Lygocoris communis*. (Redrawn from Parrott and Hodgkiss 1913.)

Fig. 12.6. Mature pear deformed from feeding by *Lygocoris communis*. (Redrawn from Parrott and Hodgkiss 1913.)

1913). Whereas *C. verbasci* is often injurious on apple trees, it is seldom so on pear trees, where it can be an important predator of psyllids (McMullen and Jong 1970, Thistlewood and Smith 1996; see also Chapter 14). This bug is an intermittent pest of apple crops, but on pears, its feeding sites usually remain unaffected, so that the fruit undergoes normal growth and retains its shape (Thistlewood and Smith 1996).

MIRIDS AS PESTS OF PEAR IN EUROPE

In Switzerland, pear has been injured by *Closterotomus biclavatus* and *Lygocoris pabulinus* and by the partially or mostly predacious *Atractotomus mali*, *Orthotylus marginalis*, and *Pilophorus clavatus*. At times, injury by the predatory species might be due to a scarcity of aphids and caterpillars on the bugs' host trees (Zschokke 1922). Feeding on newly set fruit by *C. fulvomaculatus* in France once resulted in mature pears showing sunken, stony areas; as much as half the crop was rendered unsalable (Trouvelot 1926a, 1926b). Studies in the Loire Valley showed that *C. fulvomaculatus* is not the only mirid responsible for stony pears in French orchards. In addition to several pentatomids, the mirines *Lygus pratensis* (probably *L. rugulipennis*) and *Orthops campestris* (presumably only adults of this umbellifer-feeding species) cause similar symptoms (Coutin et al. 1984).

The common green capsid (*Lygocoris pabulinus*) has been a pear pest in England since at least 1914 (Petherbridge and Thorpe 1928a). Massee (1942b) reported an outbreak of *L. pabulinus* in a British orchard, the bugs migrating from black currant growing in the orchard to feed on developing pears. Young fruits show

irregular pitting, whereas older pears are malformed and sometimes develop corky or warty patches (Plate 20). Vanwetswinkel and Paternotte (1968) discussed the habits of *L. pabulinus* as a pest of pears in Belgium. Certain common insecticides cannot be used against *L. pabulinus* in Dutch pear orchards without disrupting the natural control of injurious psyllid species (Trapman and Blommers 1992). Several mirids injure pear crops in Norway. *Lygocoris pabulinus*, which causes stony tissue to develop at the fruit surface (Plate 20), was an important pest in the late 1960s (Taksdal 1970). The heavy use of organophosphorous insecticides to control *L. pabulinus* and other mirids on pear trees is largely responsible for the development of insecticide resistance in Norwegian populations of the pear psylla (*Cacopsylla pyri*) (Edland 1997). *Orthotylus marginalis* and *Psallus ambiguus* can also be responsible for stony pits in pears (Taksdal 1983). Other plant bugs capable of causing stony-pit symptoms in Norway are *Lygus rugulipennis*, *Miris striatus*, and *Plagiognathus chrysanthemi* (Sørum 1977). The management of the bugs' alternative hosts, including frequent mowing within orchards and removal of nearby weeds and shrubs, is more effective in limiting their injury in Norwegian pear orchards than the application of pesticides at the time of petal fall (Hesjedal and Vangdal 1986).

Pistachio

The seeds of pistachio, a drupe, yield the edible nuts. Growers in California's Sacramento and San Joaquin valleys are plagued by lesions in the epicarp (Plate 21). The problem, which can involve epicarp, mesocarp, and endocarp tissues, was once attributed to a physiological disorder; mirids are now known to be important causal agents.

CLOSTEROTOMUS NORWEGICUS AND LYGUS HESPERUS

The mirines *C. norwegicus* and *L. hesperus* are among the insects responsible for epicarp lesions of pistachio fruits (Rice et al. 1985). *Closterotomus norwegicus*, the main causal agent in some orchards (Purcell and Welter 1990b), can substantially lower yields (Purcell and Welter 1991). Damage to pistachio crops occurs when adults disperse from nearby alfalfa fields or weeds that serve as host plants. A significant positive correlation exists between the incidence of epicarp lesions and the proximity to alfalfa (Fig. 12.7; Purcell and Welter 1990b).

Within four hours of mirid feeding on immature fruits, the shell tissue at the sites of penetration becomes liquefied. Cavities in the shell are formed but are covered by the internal cuticular membrane. This structure ruptures within 24–48 hours, exposing a cavity 2.0–2.5 mm in diameter and 1 mm deep (Plate 21). The kernel necrosis and other internal symptoms that typify injury by the coreid *Leptoglossus clypealis* or pentatomids do not occur (Rice et al. 1985).

Rice et al. (1985) further contrasted mirid feeding behavior and the internal and external symptoms they induce on pistachio fruits with those of other insects that cause epicarp lesions. Mirids generally feed on the developing fruit as long as they can penetrate the soft epicarp and mesocarp. Near the end of their feeding period on pistachio (mid to late May), the hardened pericarp tissues preclude stylet penetration, and the bugs switch their feeding to the fruit base, a soft area of about 10 mm². Coreids and pentatomids can puncture the partially hardened endocarp, although they also feed more at the stem end as pericarp firmness increases (Michailides 1989).

Closterotomus norwegicus and *L. hesperus* feed on young pistachio fruits in orchards, and when caged on healthy clusters, they produce necrotic lesions on large fruits and a blackening and shriveling of small fruits (Uyemoto et al. 1986). In northern California, pistachio fruit is susceptible to injury by *C. norwegicus* between the time of nut set and shell hardening—that is, early April to early June (Purcell and Welter 1991).

Uyemoto et al. (1986) thought pistachio injury results from direct wounding by mirids rather than their salivary secretions because mechanical injury causes similar symptoms. Additional study verified that the epicarp lesion is not directly involved with heteropteran salivary enzymes. The disorder apparently results from wound-induced peroxidase activity. Further research is needed to test the hypothesis that ethylene, or some other hormone generated by wounded epicarp cells, induces peroxidase activity in nearby cells (Bostock et al. 1987).

Pistachio fruits should be protected from plant bugs for 6–8 weeks after fruit set (Uyemoto et al. 1986). Effective weed control in and around pistachio orchards reduces or eliminates the need for insecticide sprays against *C. norwegicus*. Dispersal of *L. hesperus* from nearby alfalfa or safflower fields needs to be detected early so that control measures can be implemented before damage occurs (Rice et al. 1988, Purcell and Welter 1990b). In addition, Purcell and Welter's (1990a) degree-day model allows the development of *C. norwegicus* populations to be predicted so that insecticides can be timed to control late-instar bugs in the ground cover.

OTHER MIRIDS AS PISTACHIO PESTS

Another mirid pest of pistachio is the phyline *Psallus vaccinicola*, which sometimes migrates into orchards in early April to feed on small fruits (4 mm in diameter) before *Closterotomus norwegicus* or *Lygus hesperus* cause damage (Plate 21). Other mirids capable of causing epicarp lesions on pistachios in California are the phyline *Lepidargyrus ancorifer* and the mirines *Neurocolpus longirostris* (Plate 21) and *Phytocoris relativus* (Plate 21; Michailides et al. 1987). More information on the symptomatology and habits of these bugs is available in the publications by Michailides et al. (1987) and Rice et al. (1988).

Strawberry

Certain mirid species injure strawberry fruit, whereas others feed mainly on the blossoms. Rather than split

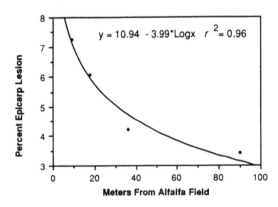

Fig. 12.7. Correlation of epicarp lesion caused by *Closterotomus norwegicus* on pistachio trees in California with distance from the border of an adjoining alfalfa field. (Reprinted, by permission of M. Purcell and the Entomological Society of America, from Purcell and Welter 1990b.)

observations on this crop between two chapters (here and in Chapter 10 on inflorescences), references to mirids as strawberry pests are grouped in this discussion of fruit feeding. Mirids are important pests of strawberry plants because even low densities can inflict cosmetic damage that renders the fruit unsalable on the fresh market (Norton et al. 1992). Such fruit must be sorted from berries of fresh-market quality, increasing a grower's harvest cost. In addition, insecticides used to suppress plant bugs can trigger outbreaks of other insect pests of strawberry crops (Zalom et al. 1993, Pickel et al. 1994).

The aggregate fruit of strawberry (the "berry") consists of an edible, fleshy receptacle bearing numerous achenes, the true fruit, on its surface. For the receptacle to grow beyond the blossom stage, the achenes or "seeds" must be pollinated and fertilized. If adjacent achenes are damaged or destroyed, growth regulators (mainly auxin) are not released into the receptacle; tissue in the area of injury does not grow and the resulting berry is misshapen (Handley and Pollard 1993). Malformed strawberries in which the apical end of the receptacle fails to develop show a concentration of achenes, a condition referred to as *apical seediness* (Plate 22). Such deformed fruits are called *buttons*, and the injury is termed *buttoning, catfacing*, or *nubbining* (Schaefers 1966). Way (1968), noting that some fruit malformation is accepted as normal in Britain, discussed the role of frost, genetic variation, and incomplete pollination or inadequate fertilization as causes of the problem (see also Darrow [1966]). Mirid feeding probably is the main cause of underdeveloped, somewhat woody berries with fully developed, but apically grouped achenes. Poor pollination results in small fruit that differs from that injured by mirids in showing hollow, hairlike projections where the achenes normally would be (Goulart et al. 1989). Hollow, straw-brown achenes are a good indication of lygus bug injury (Allen 1959).

LYGUS HESPERUS AND L. LINEOLARIS

Walsh and Riley (1869) thought that the tarnished plant bug (*L. lineolaris*) might be responsible for the blackened, diseased appearance of strawberry plants in Illinois. Forbes (1884a) was the first to present actual evidence for mirid injury to the fruit: that feeding by *L. lineolaris* before expansion of the receptacle causes a hardening and deformation. Misshapen berries are distinguished from those produced by lack of fertilization by having well-formed, rather than blighted, achenes. In Florida, similar symptoms were attributed to mirid feeding: deformed fruit or, in severe cases, blackening and death of the berries (Quaintance 1897).

Damage by lygus bugs is now known to range from slight to total loss of crop—berries 100% unsalable—and the economic threshold can be as low as one or two nymphs per plant (Schaefers 1981, Maas 1984). An action threshold generally recommended is 0.25 nymph per flower cluster (Mailloux and Bostanian 1988, 1989).

Lygus lineolaris: Injury and Economic Importance

Overwintered adults of *L. lineolaris* colonize strawberry fields when flower buds become available (e.g., Mailloux et al. 1979). Weedy conditions within or adjacent to fields favor the buildup of overwintered tarnished plant bugs in strawberries (Mailloux et al. 1979, Vincent et al. 1990). When the bugs invaded fields of southwestern Missouri in May 1946, as many as 500–600 adults could be caught with a single stroke of an insect net (Enns 1947).

Tarnished plant bug nymphs, rather than adults, are usually responsible for fruit malformation in certain seasons and in some areas (e.g., Cooley et al. 1996). Haseman (1928) described severe losses in Missouri strawberries from feeding by nymphs on berries of a crop that appeared late because of spring frosts; some fields suffered nearly 100% loss of crop. First-generation nymphs of *L. lineolaris* tend to be most abundant when June-bearing strawberries in New York are most susceptible to injury. A significant positive correlation exists between nymphal densities and the percentages of injured fruit (Fig. 12.8; Schaefers 1980). Nymphal populations begin to increase at the beginning of petal fall and reach maximum density in Quebec during the second half of June when fruits are at the green stage (Mailloux and Bostanian 1989, 1991). In Iowa, Rose (1996) found that the maximum numbers of early-instar tarnished plant bugs are present at or near peak bloom of June-bearing strawberries, with late instars most abundant during the green-fruit stage.

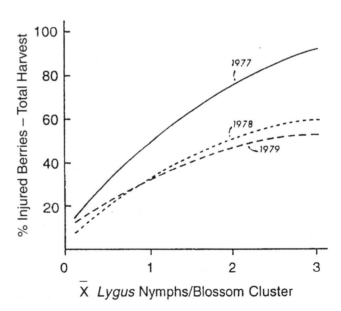

Fig. 12.8. Relationships between numbers of *Lygus lineolaris* nymphs per blossom cluster and percentage of damaged strawberries in total harvest, Geneva, N.Y., 1977–1979. (Reprinted, by permission of G. A. Schaefers and the Entomological Society of America, from Schaefers 1980.)

Investigations in New York verified that *L. lineolaris* is the primary cause of apical seediness (Schaefers 1966). Seediness caused by its feeding differs from that induced by several plant pathogens in that the former shows localized patches of seeds (Schaefers 1987). Schaefers (1966) observed as many as five bugs per blossom and found that seediness increased in each of the five pickings of the first fruiting period (mid-June to early July in 1962). Injury is most severe at field margins, especially where plantings border alfalfa or hedgerows. Tarnished plant bugs not only feed on achenes but also penetrate the flower bud between rudimentary achenes (Schaefers 1979).

Economic injury levels

Schaefers (1980) investigated the relationship between the numbers of *Lygus lineolaris* nymphs per blossom cluster and the percentage of damaged fruit at harvest. In an unusually warm season (1977), 67% damage to the crop resulted in a 30% reduction in mean berry weight; 36% and 31% damage produced 14% and 11% reductions in mean berry weight the following two years. Densities of one nymph per inflorescence can result in 20–30% of the fruit damaged at harvest (Schaefers 1972). The earliest fruits of June-bearing varieties may nearly escape injury in New York, but the later berries can be severely damaged. The appearance of late-season fruit of everbearing varieties often coincides with high lygus bug numbers, and the berries consequently suffer severe damage (Schaefers 1981).

The relationship between density of *L. lineolaris* nymphs per blossom cluster and its effect on strawberry production has been determined in Quebec. On 'Redcoat', a cultivar with a carrying capacity of 0.90 nymph per cluster, the economic injury level is estimated to be 0.95–0.99 nymph per cluster when fruit weight is considered. At this density, 13% of fruit at harvest shows injury, an unacceptable level for fresh-market consumption. When percentage of damaged fruit is considered, an action threshold of 0.26 nymph per blossom cluster produces only slight damage involving 3.5% of berries (Mailloux and Bostanian 1988). Mailloux and Bostanian (1989) developed a binomial or presence-absence sequential sampling plan that can be used in the field to classify infestation levels and to assess the need for control measures. Additional information on threshold levels and sequential sampling plans for this pest in Quebec strawberries is available in the publication by Bostanian and Mailloux (1990).

Degree-day models

The seasonal increase of *Lygus lineolaris* in Quebec can be related to degree-days (accumulated air thermal units). Nymphs are predicted to attain maximum density when 173 degree-days are accumulated from April 1, using 12.4°C as a lower threshold of development and 33°C as an upper threshold (Bostanian et al. 1990). A slight modification of this model relates increases in the nymphal population to a temperature-based phenological model that describes strawberry growth (Mailloux and Bostanian 1991).

Determining the cause of fruit malformation

Blossom feeding by *Lygus lineolaris* (Plate 22) can produce fruit malformation as a result of enzymatic and mechanical injury. Because auxin treatments induce some tolerance to the tarnished plant bug, Handley and Pollard (1989) proposed that its feeding interferes with auxin synthesis in achenes or the transport of this growth regulator to receptacle tissue. They related malformation to the length of exposure to feeding. The resulting injury depends not only on the length of exposure under greenhouse conditions but also on plant development at the time of mirid feeding and on strawberry cultivar (Handley and Pollard 1990). Certain cultivars show consistently lower levels of fruit damage and higher marketable yields (Handley 1991).

Only recently has the cause of characteristic injury to strawberry by *L. lineolaris* been studied in detail. Previously, this mirid was observed puncturing individual achenes (Howitt et al. 1965), and it was hypothesized that its feeding on achenes blocks growth stimulation of the receptacle, leading to fruit malformation (Allen and Gaede 1963, Schaefers 1980). Handley and Pollard (1993) used light and scanning electron microscopy to examine the feeding behavior of the tarnished plant bug and to corroborate some of their previous hypotheses concerning fruit malformation (Handley and Pollard 1989, 1990). They determined that the achenes serve as primary feeding sites for nymphs and adults when strawberry flowers are between the stages of anthesis and petal fall (Fig. 12.9). At anthesis, the bugs' stylets enter the side of an achene, which exhibits small holes that correspond to the size of the stylets (Figs. 12.10, 12.11); achenes soon become discolored.

Handley and Pollard's (1993) research further suggests that destruction of the endosperm within achenes inhibits the synthesis of indoleacetic acid and its translocation to the receptacle. Receptacle tissue fails to enlarge without this hormone, leading to fruit malformation. As fruit develops, the achenes enlarge and become increasingly lignified, which makes stylet penetration by the bugs more difficult. *Lygus lineolaris* then switches to receptacle tissue, usually feeding near an achene (Figs. 12.12, 12.13). Such sites appear to be more attractive than others on the receptacle because of the vasculature and nutrient supply associated with the developing embryo in each achene. In contrast to the effects on the achenes during early fruit development, feeding on receptacle tissue of more mature fruit results only in localized creases and indentations (Handley and Pollard 1993).

Injury and Economic Importance of *Lygus hesperus*

The principal mirid pest of strawberry in the western United States is another lygus bug, *L. hesperus*. Typical fruit injury in California can be produced by caging the

Fig. 12.9. *Lygus lineolaris* feeding on achenes of a strawberry flower. A. Anthesis. B. Petal fall. (Reprinted, by permission of D. T. Handley and the Entomological Society of America, from Handley and Pollard 1993.)

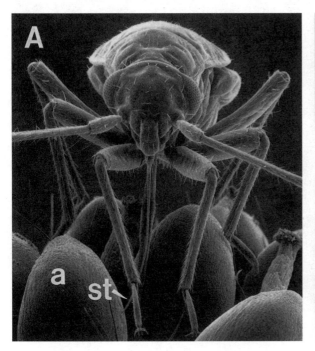

Fig. 12.10. *Lygus lineolaris* feeding on strawberry, its stylets entering the side of an achene. A. ×32. B. ×100. Abbreviations: a, achene; m, rostrum; st, stylets. (Reprinted, by permission of D. T. Handley and the Entomological Society of America, from Handley and Pollard 1993.)

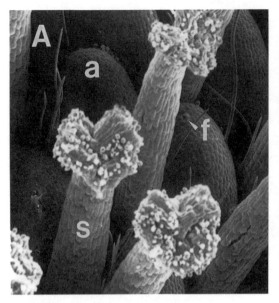
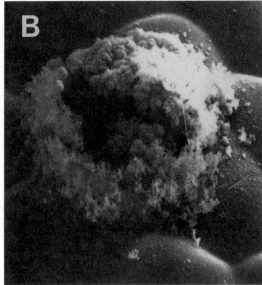

Fig. 12.11. Feeding holes made by *Lygus lineolaris* in strawberry achenes, about 48 hours after anthesis. A. ×60. B. ×1,000. Abbreviations: a, achene; f, feeding hole; s, style. (Reprinted, by permission of D. T. Handley and the Entomological Society of America, from Handley and Pollard 1993.)

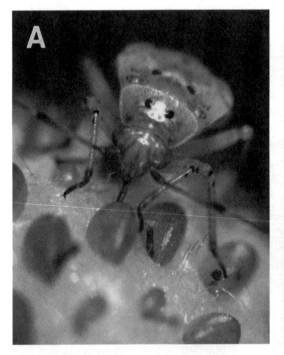

Fig. 12.12. *Lygus lineolaris* feeding on strawberry receptacle tissue near achenes. A. Nymph feeding at achene-separation stage. B. Adult feeding on mature tissue in depressed "well" of receptacle. (Reprinted, by permission of D. T. Handley and the Entomological Society of America, from Handley and Pollard 1993.)

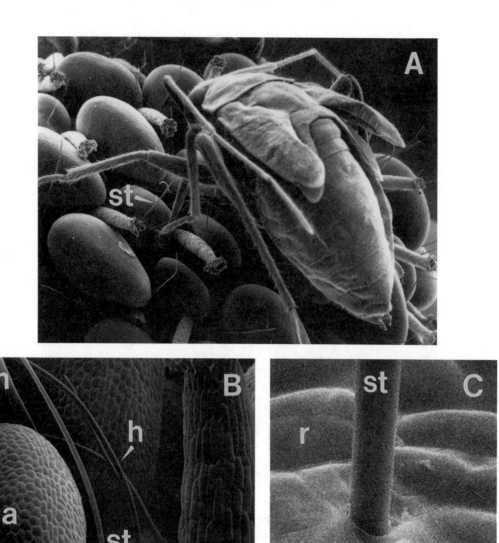

Fig. 12.13. *Lygus lineolaris* feeding on a strawberry, its stylets in receptacle tissue. A. ×22. B. ×100. C. ×1,000. Abbreviations: a, achene; h, epidermal hair; m, rostrum; r, receptacle; st, stylets. (Reprinted, by permission of D. T. Handley and the Entomological Society of America, from Handley and Pollard 1993.)

bugs on plants in the greenhouse. Allowing an adult western tarnished plant bug to feed for 24 hours on the buds, flowers, and berries leads to fruit deformity, with the injury more pronounced when feeding takes place immediately after petal fall. Nymphs of all stages can induce catfacing, the distortion restricted to punctured areas on the fruit. Injury might result from the piercing of individual achenes, which become hollow and sometimes brown. Direct feeding on the receptacle does not produce misshapen fruit (Allen and Gaede 1963). An infestation level of 0.5 nymph or adult per plant leads to economic damage (Fig. 12.14; Zalom et al. 1990).

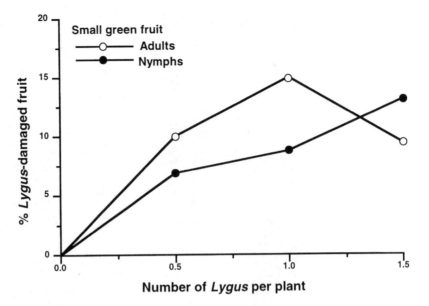

Fig. 12.14. Average percentage of strawberries damaged in California by *Lygus hesperus* nymphs or adults in field cages at three infestation levels 14 days after initial infestation. Berries were at the small green fruit stage (<1.2-cm diameter). The study was conducted three times with adults and three times with nymphs, and replicated three times at each infestation level. (Reprinted, by permission of F. G. Zalom and Intercept, Andover, UK, from Zalom et al. 1990.)

Riggs (1990) observed *L. hesperus* feeding directly on achenes, but the holes he reported in achenes apparently are larger and less defined in shape than those made by *L. lineolaris* (Handley and Pollard 1993). Auxin production is stopped, either by mechanical damage alone or by the removal of cellular material. Feeding on day-neutral strawberry varieties during the early stages of achene development (within 1–2 days of pollination) results in fruit deformity and decreases fruit weight in the Pacific Northwest (Riggs 1990).

Lygus injury to strawberries in California usually occurs only after weedy hosts dry up in late spring (Allen 1959, Maas 1984), but lygus bugs are considered key pests of the fruit (Zalom et al. 1990, 1993). Injury is particularly important in the central and south coast growing areas and is more severe in areas where fruit production continues through the summer and fall (Pickel et al. 1991). A suction machine—hydraulically driven vacuum fans and suction hoses mounted on a tractor (Street 1989)—can be used to suck lygus bugs from strawberry plants. This mechanical control device, used with other IPM practices in Californian strawberries, reduces bug populations while protecting beneficial arthropods that feed on spider mites, thrips, and other insect pests of strawberry crops (Grossman 1989). Vacuum devices reduce the numbers of lygus nymphs and adults compared to untreated plots, but fruit damage remains unacceptably high (Welch et al. 1990, Pickel et al. 1994), and the cost of the machines is considerable. The overall crop damage is reduced only 10%, and the use of suction machines may increase problems with fungal diseases (Pickel et al. 1991; see also Vincent and Lachance [1993]). Augmentative releases of the egg parasitoid *Anaphes iole* might reduce *L. hesperus* densities and fruit damage in commercial strawberry fields (Norton and Welter 1996). For additional information on various nonchemical methods of

managing lygus bug populations in California strawberries and an evaluation of sampling methods, the reader should consult the works by Flint (1990) and Zalom et al. (1990, 1993).

OTHER MIRID PESTS OF STRAWBERRY IN NORTH AMERICA

Forbes (1884a) implicated *Adelphocoris rapidus* as a cause of deformed strawberry fruit in Illinois. Another mirine, *Closterotomus norwegicus*, is also an occasional pest of this crop. Nova Scotian strawberries sometimes fail to develop or produce malformed fruit. Once attributed to a viral disease, drought, winter injury, incomplete pollination, infertility, or root rot, this disorder actually results from adults of *C. norwegicus* feeding on developing fruit (Pickett et al. 1944, Andison 1956). *Lygus shulli* (as *L. varius*) can produce fruit deformity in British Columbia (Kelton 1982a), and an orthotyline, the caragana plant bug (*Lopidea dakota*), once injured strawberry fruit in western Canada (Arnason et al. 1939).

OLD WORLD PESTS OF STRAWBERRY

Lygus rugulipennis causes fruit deformity of strawberries in continental Europe (Plate 22; Taksdal and Sørum 1971), and in the United Kingdom its feeding is considered the main cause of malformed fruit in late-season strawberry crops (Easterbrook 2000). In Britain, an increased use of continuously flowering and fruiting (late June–October) day-neutral cultivars has led to late-season damage by this pest (Easterbrook 1996, 1997). Strawberry crops also suffer from infestations of the phyline *Plagiognathus arbustorum* in Norway and Sweden. A congener, *P. chrysanthemi*, is a less important-tant pest. The injury from *Plagiognathus* species resembles that inflicted by North American lygus bugs. Feeding on flowers prevents achenes from developing in the affected areas, a problem sometimes blamed on lack of fertilization, whereas penetration of fruit gives rise to deformities (Plate 22; Sørum and Taksdal 1970; Taksdal and Sørum 1971; Gertsson 1979, 1980). Severe fruit malformation occurred in Norway during 1992, the injury resembling that inflicted by *P. arbustorum* but attributable to feeding by *Closterotomus norwegicus* (G. Taksdal, pers. comm.). *Lygocoris pabulinus* once destroyed strawberry crops in Latvia (Ozols and Zirnits 1927, Carl 1965), and another mirine, the crop mirid (*Sidnia kinbergi*), can seriously injure strawberry flowers and fruits in New Zealand, resulting in considerable crop losses (Baker 1978).

13/ Other Plant-Associated Foods and Artificial Diets

Predation often provides a source of protein, while alternative foods such as nectar, honeydew or plant sap can provide a subsistence diet.
—J. S. Edwards 1963

The development of a simple method of rearing a number of species under conditions suitable for comparative studies should greatly enhance the usefulness of these insects [Hemiptera] as tools for biological research.
—R. I. Sailer 1952

Included here is a summary of the known relationships between mirids and fungi. Because of their growth habit and plantlike vegetative body, fungi were traditionally placed in the plant kingdom. They are now classified in their own kingdom, although the classification and evolutionary relationships of eukaryotes continue to be debated (e.g., Whittaker 1969; Lipscomb 1985, 1991). Also covered is a review of artificial diets, a subject included because the various techniques developed for rearing mirids in the laboratory depend largely on plant parts or their constituents. This chapter includes the use of honeydew—essentially unchanged plant sap—as well as the occasional use of other nonfloral sources of sugars.

Fungi

Mycophagy or mycetophagy (some authors use the term *fungivory* equivalently) remains one of the least understood aspects of plant bug biology. Even for well-known insect groups such as the Lepidoptera (Rawlins 1984) and some families of Coleoptera (Lawrence 1989), knowledge of the precise feeding habits and confirmation of mycophagy are often lacking.

Evidence supporting the existence of mycophagy in mirids is equivocal. The six plant bug species observed on several types of fungi in Switzerland were thought to be using them as a food source rather than merely for shelter (Simonet 1955). Even casual mycophagy, however, seems improbable for any of the species Simonet listed; these typical inhabitants of vascular plants would seem merely accidental on fungi, or at most, to use them for shelter. Once considered mycophagous, the Isometopinae have now been shown to be predators (see Chapter 14). That leaves the Cylap-

inae as the only mirid group for which observations suggest actual mycophagous habits.

Fungi generally occur in habitats occupied by cylapine mirids. In auchenorrhynchans of the fulgoroid families Achilidae and Derbidae, the presumption that nymphs are mycophagous is based on their collection under bark or in cavities in logs (O'Brien and Wilson 1985), habitats in which fungi flourish. The assumption that members of these families are mycophagous is logical because carnivory does not occur in homopterans. In mirids that occur in somewhat similar habitats, both mycophagy and carnivory are possible.

The presence of potential prey organisms living on and within many fungi complicates the determination of dietary habits for fungus-associated mirids. Typical of the uncertainty regarding trophic relationships of plant bugs collected on fungi is Leston's (1980a) comment that he was unable to determine if the Ghanaian *Rhinomiridius ogoouensis*, a cylapine having a long rostrum and green gut contents, is a carnivore or a fungivore. Here, I review the evidence for and against the occurrence of mycophagy in mirids.

ISOMETOPINAE

That isometopines can be beaten from dead, fungus-covered branches of trees probably led to the implication that the group is mycophagous (e.g., McAtee and Malloch 1924). But observations in South Africa (Hesse 1947) and in eastern North America (Wheeler and Henry 1978a) established that several isometopines are specialized predators of scale insects. This mirid subfamily is not discussed further here, but details of predation on scale insects by isometopines and additional references are provided in Chapter 14.

CYLAPINAE

Reviewing the Evidence for Mycophagy

Cylapus tenuicornis, the only North American member of a predominantly Neotropical genus, is a widely distributed but seldom-collected cylapine. It is usually encountered on the bark of dead trees or on fungi covering dead wood (Heidemann 1891, Uhler 1891, Banks 1893). Knight and McAtee (1929) remarked that this species "frequents fallen limbs and trunks especially those having velvety fungus growths." Leston (1961b), Schuh (1974), and Cobben (1978), among others, stated, without supporting evidence, that cylapines are predacious. Schuh (1976), reconsidering his earlier opinion, suggested that evidence leans toward the Cylapinae as mycophagous. In support of a fungus-feeding hypothesis, Schuh (1976) cited several older papers and more recent ones by China and Carvalho (1951b) and Carvalho (1954b), none of which contain substantial evidence for mycophagy. China and Carvalho (1951b) stated merely that *Xenocylapus nervosus* was collected on a fallen log in South America, probably during oviposition. Carvalho (1954b) described *Cylapocoris pilosus* and *C. tiquiensis* as new species from Brazil, noting they occur in association with *Fulvius quadristillatus* and feed on fungi (Auricularia) "growing on rotten trees in the forest. Since nymphs were taken it is probable that the species feed and complete their life cycle on this fungus."

Schuh (1976) gave additional observations supporting partial or total mycophagy in the Cylapinae. He observed *Cylapus ruficeps* feeding on pyrenomycete fungi in Brazil, and in Peru he collected *C. citus*, a *Valdasus* species, and a *Xenocylapus* species from a log covered with pyrenomycetes, and two species of *Cylapocoris* from "soft mushroom-like fungi on rotting logs." He also noted that two species of *Fulvius* occur under bark in larger numbers than might be expected for a predator and that fungi are always present (Schuh 1976). In reviewing the cylapine fauna of Papua New Guinea, Carvalho and Lorenzato (1978) remarked that many species feed on fungi.

Little doubt exists that most cylapines live under bark, on rotting logs, and in other habitats where fungi occur (Table 13.1). The available evidence, however, suggests that some cylapines (as classified by Schuh [1995]) are predacious rather than mycophagous. At least two tropical species of *Fulvius* are predators of beetle eggs (see Chapter 14). In addition, Woodroffe and Halstead (1959) found *F. brevicornis* living in stored Brazil nuts and suggested that the mirid feeds on insects and mites inhabiting the nuts. They characterized

cylapines as predators or partial predators living on tree trunks or under bark. Herring (1976) described the new cylapine *Trynocoris lawrencei* from Panama, noting that the coleopterist John Lawrence considered it a predator of beetle larvae living within fungi growing on trees. Kelton (1985) recorded *F. imbecilis* from a nearly dry pile of poplar logs in Manitoba. Nymphs and adults were hiding under loose bark and were observed "feeding on dipterous and small coleopterous larvae and on other soft-bodied arthropods found in damp areas under the bark or in fungi." The habits of *F. imbecilis* in a pile of black locust logs in Pennsylvania also suggest carnivory rather than mycophagy. These quick-moving bugs can be found beneath loose bark in moist areas of the wood pile in company with fly puparia, isopods, and collembolans; predacious forms such as pseudoscorpions, anthocorids, and emesine reduviids are often present (pers. observ.). No fungal parts were detected in the guts of specimens of a North American species of *Fulvius* (Q. D. Wheeler and A. G. Wheeler 1994).

Among mirids, *Carvalhofulvius gigantochloae* has the unique habit of living within the internodes of giant bamboo shoots. Entering shoots through cracks in the internode wall or through holes made by other insects, this Malaysian bamboo specialist develops in partly water-filled reservoirs of the internodes, a microhabitat used by various other arthropods. Observations of two bugs inserting their stylets in moist debris on the inner wall of a bamboo shoot suggest that the mirid is mycophagous and that it uses the fungus in this substrate as a staple food source (Stonedahl and Kovac 1995).

Another possible mycophagous cylapine is *Punctifulvius kerzhneri*. In the Russian Far East, I. M. Kerzhner (pers. comm.) observed large numbers of nymphs and adults on a polyporaceous fungus that appeared too young to have been infested by other arthropods (see also Yasunaga et al. [1999]).

The most compelling evidence for mycophagous habits in mirids involves a temperate and a tropical species of *Cylapus*. Two individuals of *C. tenuicornis* collected from the xylariaceous fungus *Hypoxylon frag-*

Table 13.1. *Fulvius* species (Cylapinae) recorded from fungi or habitats where fungi abound

Mirid	Habitat	Locality	Reference
Fulvius angustatus	Under bark, in rotten log, under fallen fruit, among dead leaves	Guam	Usinger 1946
Fulvius imbecilis	Under bark of fallen tree, in pile of cut wood	Michigan, USA	Hussey 1922a
	On dry fungi at base of stumps	Eastern USA	Blatchley 1926
	Under bark of dead elm	Eastern USA	Hoffmann 1942
	On pig carrion	South Carolina, USA	Payne et al. 1968
Fulvius slateri	In baited malt trap	Florida, USA	Hussey 1954a
	Associated with and apparently feeding on wood-rotting higher fungi	Wisconsin, USA	Ackerman and Shenefelt 1973
Fulvius sp.	Reared from conks of *Fomes* sp.	Wisconsin, USA	Akingbohungbe et al. 1972

Note: See text for discussion.

iforme in New York were found to have ascospores in their guts. One of the specimens showed large numbers of densely packed spores. These could have been ingested during predation on fungus-associated arthropods, but the presence of dense spores in the gut renders that possibility unlikely. Dissection of a *Cylapus* species collected from a pyrenomycete in lowland Amazonian rainforest in Peru did not reveal spores, but the gut contained hyphal fragments and ovoidal structures about 5 μm long that appeared to be conidial stages of a xylariaceous fungus (Q. D. Wheeler and A. G. Wheeler 1994).

The probing behavior of *C. tenuicornis*, previously thought to involve a search for arthropod prey living within pyrenomycete fungi (T. J. Henry and A. G. Wheeler, pers. observ.; see also Q. D. Wheeler and A. G. Wheeler [1994]), might actually reflect the bugs' attempts to locate ostioles on the surface of fruiting bodies. The bugs might insert their stylets through the ostioles to gain access to asci within perithecia. The feeding habits of mirids might preclude their ability to ingest most fungal spores (Dustan 1924, Madelin 1966). Although the size of the food canal in *C. tenuicornis* is unknown, it ranges from about 5 to 10 μm at the widest point in other mirids (see Cobben 1978: figs. 136, 154, 157–158, 160). *Cylapus tenuicornis* can ingest the spores of *H. fragiforme*, which are about 5 μm wide and 10–12 μm long and are flexible before their ejection from fruiting bodies. Mirids, however, appear unable to ingest pollen grains, which typically range from 20 to 40 μm in diameter (see Chapter 11).

Need for Empirical Studies on Mycophagy

A more precise knowledge of the relationships between mirids and fungi awaits experimental studies on the diverse cylapine fauna of the neotropics—the information on food habits presented here should be considered provisional. Observations of egg predation do not establish *Fulvius* species as mainly predacious. Even primarily phytophagous species of most mirid tribes and subfamilies are facultative predators (see Chapter 14). Similar doubts concerning the food habits of fungal inhabitants exist for other groups of insects. For instance, members of the coleopteran genus *Lordithon* (Staphylinidae) are now known to prey on dipteran larvae living in mushrooms, but the larvae were once considered at least partially mycophagous (Campbell 1982; see also Ashe [1993]).

In the Miridae, mycophagy possibly evolved from predation on the arthropods inhabiting fungi. A switch to feeding on the host fungus, for example, is thought to have occurred in certain staphylinid beetles (Newton 1984). Despite chemical differences between fungal and plant tissues (Martin 1979, Kukor and Martin 1987), the nutritive content of most fungi is similar to that of herbaceous plants, and access to their nutrients does not present a problem for many general herbivores (Rawlins 1984). Although it can be questioned whether a mostly predatory heteropteran would become a phytophage (or mycophage) with lower dietary nitrogen than animal tissues (Schaefer 1997), unaggressive predators do evolve into phytophagous lineages (see Chapter 16).

A thorough biological investigation of fungus-associated mirids might show that certain species—perhaps those in the genus *Fulvius*—are mainly carnivorous, whereas other Cylapini might be chiefly mycophagous. Just as many mirids living on plants include both plant and animal matter in their diet, a given mirid occurring on fungi might feed on the fungus as well as its arthropod inhabitants. Further work on the feeding habits of presumed mycophagous microarthropods often reveals that they feed at several trophic levels: for example, on algae, bacteria, bryophytes, and arthropods in addition to fungi (Walter 1987).

The consistent occurrence of certain cylapines on pyrenomycete fungi, combined with the presence of spores and other fungal structures in the guts of a temperate and a tropical species, suggests a widespread, perhaps ancestral, association with fungi (Q. D. Wheeler and A. G. Wheeler 1994). Additional research might attempt to assess the extent of mycophagy in the Cylapinae, especially their ingestion of spores, as well as possible use of mycelia and context tissues making up the sporocarp body of macrofungi.

Honeydew and Other Alternative Sugar Sources

Honeydew is a waste secretion of aphids, scale insects, psyllids, treehoppers, and a few other homopterans. It is a nutritious food source sought by ants, bees, flies, and other insects (Imms 1971, Hagen 1986, Roubik 1989). As Elton (1966) noted, honeydew is only a slight modification of plant sap (see also Mittler [1958]); it can differ from sap, nectar, and secretions from extrafloral nectaries mainly by the presence of the disaccharide trehalulose, the trisaccharides glucosucrose and melezitose, and the tetrasaccharide stachyose (Byrne and Miller 1990, Dolling 1991).

Like nectar, honeydew is an aqueous solution rich in amino acids and carbohydrates; it also contains various lipids, minerals, organic acids, and vitamins (Way 1963, Johnson and Stafford 1985). Honeydew composition varies with the homopteran species that excretes it and with the host plant and its physiological condition (Hagen et al. 1970, Hagen 1986, Hendrix et al. 1992). Unlike nectar (see Chapter 11), honeydew is generally devoid of alkaloids, glycosides, and other potentially toxic compounds, although melezitose can adversely affect certain insects (Zoebelein 1956).

Butler (1968) stated that few insects are known to feed on honeydew, omitting Elton's (1966) discussion and the work of Zoebelein (1956, 1957). Zoebelein (1956) referred to the use of honeydew by numerous insects, but no heteropterans were included in a list of honeydew-feeding species belonging to six orders.

A

B

C

D

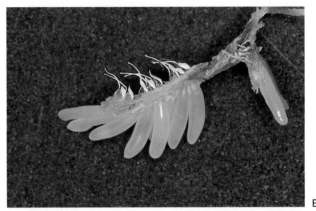

E

F

Plate 1. Morphology and biology. A. Brachypterous female of *Orthonotus rufifrons*. B. Eggs of the fourlined plant bug (*Poecilocapsus lineatus*) in ninebark stem. C. *Helopeltis clavifer* ovipositing in cocoa pod. D. Waxy secretions covering oviposition scars of a *Rayieria* species on acacia or wattle stem. E. Aeropyles of *Rayieria* eggs dissected from acacia or wattle shoot. F. Oviposition scars of *Horistus infuscatus* on asphodel stem. (Photo A courtesy of G. C. McGavin; B courtesy of J. F. Stimmel; C photo by C. Prior, courtesy of E. S. C. Smith; D, E courtesy of D. Donnelly; F courtesy of F. Önder.)

Plate 2. Biology and ecology. A. Nymph of the honeylocust plant bug (*Blepharidopterus chlorionis*) hatching from egg in honeylocust twig. B. Molting fifth instar of the western tarnished plant bug (*Lygus hesperus*) hanging from lentil stem. C. European tarnished plant bug (*L. rugulipennis*) adult overwintering on Norway spruce. D. Nymphs and adults of the yucca plant bug (*Halticotoma valida*) aggregating on yucca leaf; note the dark spots of excrement. E. Crab spider (*Misumenoides formoceipes*) preying on adult cotton fleahopper (*Pseudatomoscelis seriata*). F. Western bigeyed bug (*Geocoris pallens*) preying on *Lygus* nymph. G. Spined assassin bug (*Sinea diadema*) nymph preying on *Lopidea nigridia* adult. H. Mymarid (*Anaphes iole*) ovipositing in *Lygus hesperus* egg. (Photo A courtesy of J. F. Stimmel; B courtesy of D. J. Schotzko; C courtesy of B. Kullenberg; D obtained from Department of Entomology, Cornell University; E courtesy of W. L. Sterling; F courtesy of R. D. Akre; G photo by T. Steen, courtesy of J. D. McIver; H courtesy of C. G. Jackson.)

Plate 3. Natural enemies and behavior. A. Distended abdomen of a fifth-instar honeylocust plant bug (*Blepharidopterus chlorionis*) parasitized by a euphorine braconid (*Peristenus henryi*). B. Shrunken abdomen of a fifth-instar *Taedia scrupea* from which a larva of a euphorine parasitoid has emerged. C. *Pycnoderes medius* adult, its wings displaced by a parasitic (trombidiid) mite. D. *Notostira elongata* nymph infected by the fungus *Entomophthora helvetica*. E. *Mircarvalhoia arecae* infected by the fungus *Aspergillus candidus*. F. *Stenodema holsata* adult camouflaged on a grass inflorescence. G. Myrmecomorphic nymph of *Teleorhinus tephrosicola* feeding on a dead caterpillar. H. Myrmecomorphic adult female of *Paradacerla formicina* using forelegs to clean the left antenna. (Photos A–C, G courtesy of J. F. Stimmel; D courtesy of S. Keller; E courtesy of K. Dhileepan and the Incorporated Society of Planters (Malaysia); F courtesy of B. Kullenberg; H courtesy of J. D. McIver.)

Plate 4. Feeding and feeding injury. A. Lesions on cocoa pod fed on by *Helopeltis clavifer*, showing subsequent invasion by fungi. B. Feeding by a *Dichrooscytus* species on juniper; conifer-feeding plant bugs often do not cause noticeable symptoms on their hosts. C. Severe chlorosis on ornamental yuccas caused by the yucca plant bug (*Halticotoma valida*). D. Stippling of apple leaf fed on by *Lygocoris communis*. E. Green ash foliage showing typical injury (chlorosis) from the ash plant bug (*Tropidosteptes amoenus*). F. Chlorosis on maypop passion-flower foliage caused by *Opistheurista clandestina*. G. Wheat blades showing chlorosis from feeding by the meadow plant bug (*Leptopterna dolabrata*). H. Quackgrass leaf showing chlorosis from feeding by *Notostira elongata* (x20); black spots are the bugs' excrement. I. Chlorotic blotches on bamboo caused by *Mecistoscelis scirtetoides*. (Photo A courtesy of E. S. C. Smith; B courtesy of D. A. Polhemus; C courtesy of E. R. Hoebeke; D courtesy of R. K. Stewart; E courtesy of J. F. Stimmel; F courtesy of T. J. Henry; G courtesy of P. H. Craig; H courtesy of G. Bockwinkel; I courtesy of Y.-C. Chang.)

Plate 5. Feeding injury. A. *Orthocephalus funestus* and chlorotic blotches it causes on artemisia foliage. B. Garden fleahopper (*Halticus bractatus*) adults and their chlorosis on morning-glory leaves. C. Intermediate wheatgrass damaged by *Labops hesperius*; bugs were killed with insecticide in dark-green areas. D. Hollyhock showing chlorosis from feeding by *Brooksetta althaeae.* E. Stippling and curling of honeylocust leaflet fed on by early instars of the honeylocust plant bug (*Blepharidopterus chlorionis*). F. Leaflet distortion and premature drop of honeylocust leaflets caused by *B. chlorionis*. G. Withering of bedstraw shoot apices fed on by *Criocoris saliens*. H. *Tropidosteptes cardinalis* adult and blasting of terminal of white ash. I. Sugar-beet leaf crinkled from the feeding of the European tarnished plant bug (*Lygus rugulipennis*). (Photo A by M. Takai, courtesy of T. Yasunaga; B, D courtesy of T. J. Henry; C courtesy of J. A. Kamm; E–H courtesy of J. F. Stimmel; I courtesy of K. Hori.)

Plate 6. Feeding injury. A. Ragged and shot-holed chrysanthemum leaves fed on by *Taylorilygus apicalis*. B. Magnolia leaf showing necrosis and shot holing from feeding by the western tarnished plant bug (*Lygus hesperus*). C. *Plagiognathus albatus* adult and initial stage of injury—slight discoloration—on a London plane leaf. D. London plane leaves tattered by *P. albatus*. E. Close-up of newly formed, water-soaked lesions caused by the fourlined plant bug (*Poecilocapsus lineatus*) on bittersweet nightshade. F. Older, darkened lesions caused by *P. lineatus* on bittersweet nightshade. G. Dark lesions on red clover foliage fed on by *P. lineatus*. H. Darkened lesions of *P. lineatus* on leaf of an ornamental dogwood. I. Lesions and shot holing of hosta leaf fed on by *P. lineatus*. (Photo A courtesy of J. F. Price; B courtesy of C. S. Koehler; C, H courtesy of D. L. Caldwell; D photo by D. L. Caldwell, courtesy of J. F. Stimmel; E–G, I courtesy of J. F. Stimmel.)

Plate 7. Feeding injury. A. Early symptoms on tobacco from feeding by the tarnished plant bug (*Lygus lineolaris*) on bud leaves. B. Tobacco foliage showing sloughing of tissue from feeding by *L. lineolaris*. C. Expanding red maple leaf showing two small areas of thin tissue where *Lygocoris vitticollis* nymphs have fed. D. Expanded red maple leaf tattered by the feeding of *L. vitticollis*. E. Viburnum foliage tattered by *L. viburni*. F. Areca palm showing dead, tattered leaves fed on by *Mircarvalhoia arecae*; the foliage, initially discolored, turns yellow and brown and becomes shredded when necrotic tissue drops out. G. Silvertop of bluegrass caused by a shriveling of the stem when the meadow plant bug (*Leptopterna dolabrata*) feeds above uppermost node. H. Heads showing mirid-induced symptoms of silvertop scattered in field of smooth brome. (Photos A, B courtesy of C. D. Pless; C–E courtesy of J. F. Stimmel; F courtesy of K. Dhileepan; G courtesy of J. A. Kamm; H courtesy of M. S. Okuda.)

Plate 8. Feeding injury. A. Split-stem lesion on hybrid poplar caused by the tarnished plant bug (*Lygus lineolaris*); lesions once were thought to involve attack by some plant pathogen. B. Cotton foliage and stems distorted by *Helopeltis schoutedeni*. C. Tomato stem broken at feeding rings made by the tomato bug (*Engytatus modestus*). D. Feeding rings (necrotic areas) on tomato stem fed on by *Nesidiocoris tenuis*. E. Lesions on tea foliage made by *H. schoutedeni*. F. Shoot dieback of cocoa caused by *H. theivora*. G. Advanced stages ("blast," "staghead") of mirid damage on cocoa trees. (Photo A courtesy of L. F. Wilson and USDA Forest Service; B courtesy of H. Schmutterer; C courtesy of M. P. Parrella; D courtesy of J.-C. Malausa; E courtesy of V. Sudoi; F courtesy of K. C. Khoo; G courtesy of A. Youdeowei.)

Plate 9. Feeding injury. A. *Rayieria* adult and the lesions this plant bug causes on acacia or wattle. B. Acacia leaves injured by *Rayieria* (left) and leaves of an uninfested branch. C. Chrysanthemum stem with witches'-brooming from feeding by *Taylorilygus apicalis* and an unaffected stem (right). D. Chrysanthemum stem fed on by *T. apicalis* with witches'-broom effect and late-forming flowers (left); affected stem resulting in a single, malformed flower (middle); and normal stem (right). E. Normal chrysanthemum flowers (left) and flowers with an incomplete array of florets caused by *T. apicalis*. F. Aborted cotton terminal and secondary branching from feeding by the western tarnished plant bug (*Lygus hesperus*). (Photos A, B courtesy of D. Donnelly; C–E courtesy of J. F. Price; F courtesy of T. F. Leigh.)

Plate 10. Feeding injury. A. Alfalfa yellowed by the alfalfa plant bug (*Adelphocoris lineolatus*). B. Thickened cotyledons of rutabaga caused by the European tarnished plant bug (*Lygus rugulipennis*). C. Terminal shoot of peach injured by the tarnished plant bug (*L. lineolaris*). D. Bushy-topped loblolly pine seedlings caused by *L. lineolaris*. E. *Lygus* *rugulipennis* adult feeding on pine seedling. F. Multiple-leadered Norway spruce seedling caused by *L. rugulipennis*. (Photo A courtesy of E. B. Radcliffe; B courtesy of G. Taksdal; C courtesy of D. L. Caldwell; D courtesy of T. C. Tigner; E, F courtesy of J. K. Holopainen.)

Plate 11. Ecology and behavior. A. Tarnished plant bug (*Lygus lineolaris*) adult feeding on flower bud of horseweed. B. Cotton fleahopper (*Pseudatomoscelis seriata*) adult feeding on croton buds. C. *Calocoris roseomaculatus* adult feeding in flower of knapweed. D. Nymph of *Polymerus basalis* camouflaged and feeding on flower buds of red sorrel. E. *Creontiades debilis* adult camouflaged on saltwort. F. *Monalocoris filicis* on fern; sporangia serve as the bug's main food source. G. Common green capsid (*Lygocoris pabulinus*) on inflorescence of an umbellifer. H. Meadow plant bug (*Leptopterna dolabrata*) on quackgrass inflorescence. (Photos A, B, D, E courtesy of J. F. Stimmel; C, H courtesy of B. Kullenberg; F photo by M. Takai, courtesy of T. Yasunaga; G reprinted, by permission of T. Yasunaga and Zenkoku Noson Kyoiku Kyokai, Tokyo, from Yasunaga et al. 1993.)

Plate 12. Feeding habits and injury. A. *Stenotus rubrovittatus* on rice. B. Aborted rice grains fed on by *S. rubrovittatus*. C. Rice grains injured by *S. rubrovittatus*. D. Sorghum panicles showing grains injured by the sorghum earhead bug (*Calocoris angustatus*). E. *Eurystylus oldi* adult on sorghum panicle. (Photo A courtesy of T. Yasunaga; B, C courtesy of H. Hayashi; D courtesy of K. F. Nwanze; E courtesy of G. J. Steck.)

Plate 13. Injury to alfalfa and cotton. A. Alfalfa seeds damaged by lygus bugs (left) and normal seeds (right). B. Alfalfa flower buds blasted by lygus feeding. C. "Stripped racemes" of alfalfa caused by mirid feeding. D. Poor bloom of alfalfa (left) caused by the alfalfa plant bug (*Adelphocoris lineolatus*) and normal blooming (right). E. *Adelphocoris lineolatus* adult feeding on alfalfa flower buds. F. Cotton square aborted from feeding by the western tarnished plant bug (*Lygus hesperus*). G. Two cotton squares injured by lygus bugs. H. Necrotic anthers of cotton fed on by *L. hesperus*. I. Injury by the tarnished plant bug (*L. lineolaris*) to young cotton bolls and (far right) a boll that did not open fully. (Photo A by W. P. Nye, courtesy of D. W. Davis; B courtesy of B. D. Schaber; C–E courtesy of J. J. Soroka; F, H courtesy of T. F. Leigh; G courtesy of J. R. Mauney; I courtesy of G. L. Snodgrass.)

Plate 14. Lygus bugs and alternative food sources. A. Extrafloral nectary of cotton. B. Western tarnished plant bug (*Lygus hesperus*) feeding at extrafloral nectary on cotton. Feeding injury and habits. C. Injury to lentil seeds by *L. hesperus*. D. Injury to pepper fruit by *Liocoris tripustulatus*. E. Tomato fruit injured by *Dicyphus errans* (right) and uninjured fruit (left). F. *Lygocoris laureae* nymph feeding on flower buds of mountain laurel. G. Azalea plant bug (*Rhinocapsus vanduzeei*) adult feeding on azalea stamen. H. *Plagiognathus gleditsiae* nymph on buds of staminate flowers of honeylocust. (Photo A courtesy of C. E. Rogers; B courtesy of T. F. Leigh; C courtesy of D. J. Schotzko; D photo by A. van Frankenhuyzen, courtesy of S. A. Ulenberg; E courtesy of J.-C. Malausa; F–H courtesy of J. F. Stimmel.)

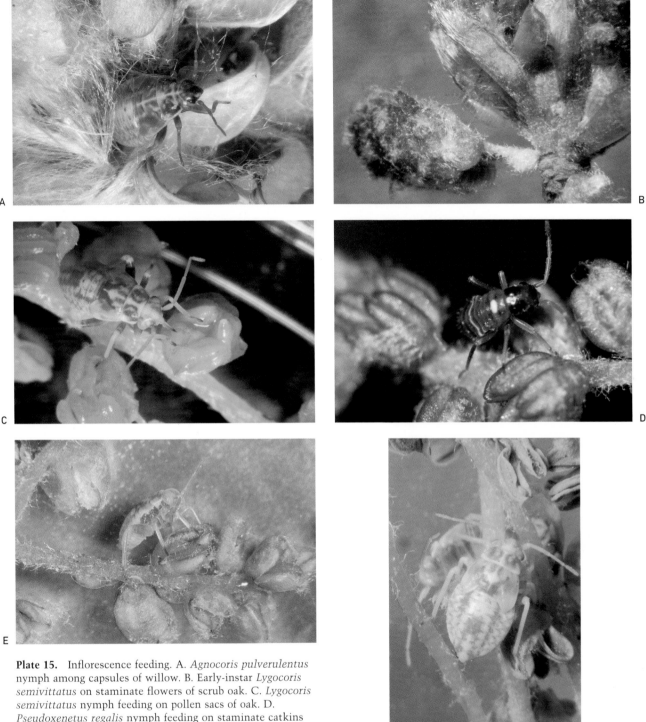

Plate 15. Inflorescence feeding. A. *Agnocoris pulverulentus* nymph among capsules of willow. B. Early-instar *Lygocoris semivittatus* on staminate flowers of scrub oak. C. *Lygocoris semivittatus* nymph feeding on pollen sacs of oak. D. *Pseudoxenetus regalis* nymph feeding on staminate catkins of oak. E. Early-instar *L. quercalbae* feeding on developing staminate catkins of oak. F. Fifth-instar *L. quercalbae* on withered oak catkins. (Photos A–F courtesy of J. F. Stimmel.)

Plate 16. Fruit and inflorescence feeding. A. *Pantilius tunicatus* on filbert catkin. B. Discolored samaras of European ash fed on by *Psallus lepidus*. C. Early-instar *Phoenicocoris claricornis* on microsporangiate stobilus of Virginia pine. D. Fifth-instar *P. claricornis* on microsporangiate cone of Virginia pine. E. *Pinophylus carneolus* adult feeding on microsporangiate cone of Virginia pine. F. Conelets of jack pine aborted from feeding by the pine conelet bug (*Platylygus luridus*). G. Jack pine ovule showing penetration by stylets of *P. luridus*. (Photo A courtesy of A. Arzone and the Institute of Agricultural Entomology, University of Turin (Italy); B–E courtesy of J. F. Stimmel; F, G courtesy of A. Rauf.)

Plate 17. Fruit and inflorescence feeding. A. Tubercles on pumpkin caused by the European tarnished plant bug (*Lygus rugulipennis*). B. Dimples on apple from feeding by the tarnished plant bug (*L. lineolaris*). C. Lygus bug injury to immature 'Golden Delicious' apples. D. *Heterocordylus malinus* adult feeding on crabapple flower bud. E. Scarring and pitting of apple by *Lygidea mendax*. F. Scarring of 'Red Delicious' apple by *Campylomma verbasci*. (Photo A courtesy of K. Hori; B courtesy of H. W. Hogmire; C courtesy of S. C. Hoyt; D courtesy of J. F. Stimmel; E courtesy of R. K. Stewart; F photo by E. H. Beers, courtesy of S. C. Hoyt.)

A

B

C

D

E

F

Plate 18. Fruit and inflorescence feeding. A. Dimpling injury to apple by the common green capsid (*Lygocoris pabulinus*). B. Scarring of apple by *L. pabulinus*. C. Apple dimpling bug (*Campylomma liebknechti*) feeding on apple flower. D. Lesions on developing cashew nuts caused by *Helopeltis clavifer*. E. *Helopeltis* injury to cocoa pods. F. Close-up of *H. theivora* and its lesions on a cocoa pod. (Photos A, B courtesy of M. G. Solomon; C photo by C. C. Bower, courtesy of G. A. C. Beattie; D courtesy of E. S. C. Smith; E courtesy of H. Schmutterer; F courtesy of K. C. Khoo.)

Plate 19. Fruit feeding. A. *Helopeltis clavifer* nymphs and the lesions they cause on cocoa. B. Nymphs of *Sahlbergella singularis* and their lesions on a cocoa pod. C. Bee bug (*Platyngomiriodes apiformis*) feeding on cocoa pod. D. Adult *P. apiformis* and its feeding lesions on cocoa pod. E. Transparentwinged plant bug (*Hyalopeplus pellucidus*) adult feeding on immature guava fruit. F. Nymph of *H. pellucidus* feeding on guava flower bud. G. Necrotic anthers of guava buds fed on by *H. pellucidus*. H. Injury to guava by *Helopeltis theivora*. (Photo A courtesy of E. S. C. Smith; B courtesy of H. Schmutterer; C, H courtesy of K. C. Khoo; D courtesy of P. A. C. Ooi; E–G reprinted, by permission of R. F. L. Mau and the Hawaiian Entomological Society, from Mau and Nishijima 1989.)

Plate 20. Fruit feeding. A. Catfacing of peach by lygus bugs.
B. Scarring of peach by the tarnished plant bug (*Lygus
lineolaris*). C. Large scar on peach caused by *L. lineolaris*. D.
Gummosis on peach from feeding by *L. lineolaris*. E. Warts
and corky patches on pear from feeding by the common green
capsid (*Lygocoris pabulinus*). F. Stony pits of pear caused by
L. pabulinus. (Photo A by E. H. Beers, courtesy of S. C. Hoyt;
B–D courtesy of H. W. Hogmire; E, F courtesy of G. Taksdal.)

Plate 21. Injury to pistachio. A, B. External epicarp lesion caused by plant bugs. C. Soft shell pitting caused by the western tarnished plant bug (*Lygus hesperus*). D. Pits at base of funiculus caused by *Psallus vaccinicola*. E. Pits on hardening shell caused by *Neurocolpus longirostris*. F. Pitting of soft shell from feeding by a *Phytocoris* species (Photos A–F courtesy of R. E. Rice.)

A

B

C

D

Plate 22. Injury to strawberry. A. Apical seediness caused by lygus bugs. B. Adult tarnished plant bug (*Lygus lineolaris*) feeding on flower. C. Fruit malformation caused by the European tarnished plant bug (*L. rugulipennis*).

D. Deformation of fruit from feeding by *Plagiognathus arbustorum*. (Photos A, B courtesy of J. F. Dill; C, D courtesy of G. Taksdal.)

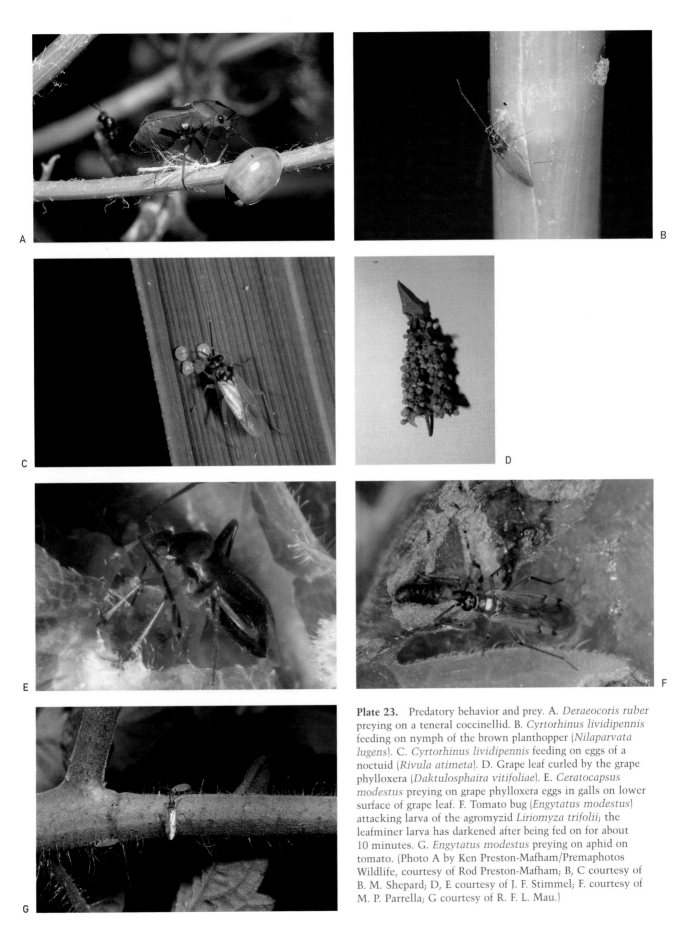

Plate 23. Predatory behavior and prey. A. *Deraeocoris ruber* preying on a teneral coccinellid. B. *Cyrtorhinus lividipennis* feeding on nymph of the brown planthopper (*Nilaparvata lugens*). C. *Cyrtorhinus lividipennis* feeding on eggs of a noctuid (*Rivula atimeta*). D. Grape leaf curled by the grape phylloxera (*Daktulosphaira vitifoliae*). E. *Ceratocapsus modestus* preying on grape phylloxera eggs in galls on lower surface of grape leaf. F. Tomato bug (*Engytatus modestus*) attacking larva of the agromyzid *Liriomyza trifolii*; the leafminer larva has darkened after being fed on for about 10 minutes. G. *Engytatus modestus* preying on aphid on tomato. (Photo A by Ken Preston-Mafham/Premaphotos Wildlife, courtesy of Rod Preston-Mafham; B, C courtesy of B. M. Shepard; D, E courtesy of J. F. Stimmel; F. courtesy of M. P. Parrella; G courtesy of R. F. L. Mau.)

Plate 24. Predation, scavenging, and use of alternative animal food. A. *Myiomma cixiiforme* nymph and its prey, the obscure scale (*Melanaspis obscura*), on trunk of pin oak. B. Adult *M. cixiiforme* feeding on a female obscure scale with cover removed. C. Western tarnished plant bug (*Lygus hesperus*) adult feeding on conspecific adult trapped in spider web. D. *Deraeocoris quercicola* nymph feeding on exuviae of the periodical cicada (*Magicicada septendecim*). E. *Ranzovius clavicornis* adult in web of a subsocial theridiid spider (*Anelosimus studiosus*). F. *Aoplonema uhleri* adult feeding on a meloid beetle (*Lytta crotchi*). G. Chiggerlike welt on human skin resulting from the "bite" of the azalea plant bug (*Rhinocapsus vanduzeei*). (Photos A, B, E courtesy of J. F. Stimmel; C courtesy of D. J. Schotzko; D courtesy of T. J. Henry; F reprinted by permission of J. D. Pinto and the Pacific Coast Entomological Society, from Pinto 1978; G courtesy of D. G. Hall.)

Several records of the use of honeydew by mirids are available. Plant bugs noted in passing to use homopteran honeydew include *Pilophorus perplexus* (Fulton 1918; see also Brindley [1935]), *Campylomma verbasci, Microphylellus modestus* (Knight 1923), *Deraeocoris incertus* (Razafimahatratra 1980), and *Irbisia* species (Schwartz 1984) in North America and *Atractotomus mali* in Europe (Finţescu 1914). Kullenberg (1944) also observed honeydew feeding in several Swedish mirids, including *Miris striatus, Myrmecoris gracilis,* and *Systellonotus triguttatus,* and Southwood and Leston (1959) noted that *Campyloneura virgula* feeds on honeydew in England. In Hawaii, the egg predator *Tytthus mundulus* imbibes honeydew on corn (Williams 1931), and another predator of homopteran eggs, *Cyrtorhinus fulvus,* can survive at low prey densities partly by feeding on honeydew (Matsumoto 1964; see also Chapter 14). Additionally, Tiensuu (1936) speculated that nearly all the insects, including seven mirid species, occurring on aphid-infested willows in Finland feed on honeydew. Sidlyarevitsch (1982) stated that *A. mali* and *P. perplexus* do not visit flowering plants for nectar, like certain other predacious mirids on apple trees (see Chapter 11), but instead obtain carbohydrates from aphid honeydew. In England, Glen (1973) reported that honeydew from lime aphids prolongs survival and allows modest weight gains in *Blepharidopterus angulatus* nymphs (Table 13.2), but he noted they could not be reared to maturity on a honeydew diet. This bug's use of honeydew might explain why so much time is seemingly wasted in searching upper leaf surfaces where its prey do not occur (see Chapter 7); honeydew is more likely to be found on the adaxial (upper) surfaces of host foliage (Glen 1975b).

Because *B. angulatus* and *P. perplexus* commonly prey on aphids (see Chapter 14), the use of honeydew as a food source might be a secondary phenomenon, the bugs originally having used the exudate to discover aphids. A similar idea was proposed for a chrysopid (Neuroptera) in which the larva feeds on aphids and also obtains a sugar meal from its prey (Downes 1974).

Honeydew-feeding habits of plant bugs are best known for *Lygus* species and other mirines. A small quantity of honey or aphids is needed for the laboratory rearing of *Adelphocoris lineolatus, A. taeniophorus,* and *Apolygus lucorum* (Chu and Meng 1958). Butler (1968) could not maintain *Lygus hesperus* nymphs on alfalfa until he noticed they were able to survive on a mealybug-infested plant. Suspecting that honeydew was responsible for their survival, he showed that 93% of nymphs (number of individuals not stated) having access to mealybug honeydew lived a week compared to only 13% surviving that long on alfalfa stems without mealybugs. Survival was also increased by the availability of honeydew from pea aphids and spotted alfalfa aphids. Survival of *L. hesperus* was enhanced when the bugs fed on alfalfa stems dipped in different concentrations of melezitose. Honeydew might serve as the main source of sugars for lygus bugs before alfalfa begins to flower (Butler 1968). Also possible is that the increased nymphal survival that Butler attributed to the use of aphid or mealybug honeydew could have been aided by predation on the aphids (Bryan et al. 1976). Another possibility is that bacteria, sooty molds, and other contaminants of honeydew play a nutritional role in mirid diets. In the case of *L. rugulipennis,* use of aphid honeydew allows a generation to be produced on the vegetative parts of wheat before the heads of this crop are available (Hori and Hanada 1970).

Lindquist and Sorenson (1970) reported a higher survival rate for *L. lineolaris* caged on alfalfa sprayed with a 10% sucrose solution than on untreated plants, and they observed an attraction to sucrose-treated plants. The use of artificial honeydews might actually have a detrimental effect on crops if lygus bug populations in crop fields increase following the application of sucrose and honey solutions. But mirid numbers did not increase when a protein hydrolysate food spray (Feed Wheast), plus honey and glycerine, was applied to cotton (Butler and Ritchie 1971). Because potato plants sprayed with several artificial honeydews are attractive to various insects, including *L. hesperus* and several predatory species, this attraction might be used to help control this pest. According to Ben Saad and Bishop (1976), either the bug's natural enemies could be attracted to the crop or its olfactory responses could be manipulated.

The nectariless trait of cotton—plants without extrafloral nectaries—is associated with reduced lygus bug numbers because an important food source is elim-

Table 13.2. Survival and increase in weight of *Blepharidopterus angulatus* at 14°C on a honeydew diet

Instar	No. of individuals	Survival rate (%) to next instar	Length of instar (days)	Weight increase during instar (μg)	Weight increase as % of normal
I	13	70	20 ± 1.1	33 ± 11	39
III	17	100	14.5 ± 0.95	141 ± 15	49
V	10	70	33 ± 1.4	220 ± 28	35
Adult female	6	NA[a]	35 +	172 ± 16	23

Source: Glen (1973).
[a] NA, not applicable.

inated (see Chapter 10). Schuster (1980) speculated that nectariless cotton's effectiveness might be neutralized by the presence of aphid or whitefly honeydew on the plants.

Plant bugs also exploit other sugar sources. *Atractotomus mali* will imbibe juices emanating from tunnels made in apples by larvae of the codling moth *(Cydia pomonella)* (Rostrup and Thomsen 1923). Another phyline, *Sejanus albisignatus*, is attracted to fermenting brown sugar (Dumbleton 1938), and the mainly carnivorous *Blepharidopterus angulatus* sometimes feeds on young apples without blemishing the surface, but this bug is said to be "partial to rotten fruits both on the tree and on the ground" (Collyer 1952). A *Lygus* species considered a chance visitor to slime flux on a tree (Fox-Wilson 1926) might have been exploiting sugar sap exudate in a similar manner.

IMPORTANCE OF SUGAR SOURCES TO MIRIDS

No mirid is known to visit extrafloral nectaries, honeydew, or other sugar sources to the exclusion of flowers. These supplemental sources of energy are characterized by an absence or low concentrations of essential amino acids and a seasonal unpredictability. At times, however, they might sustain longevity in plant bugs, enhance egg production, and in the case of predatory species, help stabilize predator-prey interactions.

Artificial Diets and Other Rearing Methods

Experimental studies on insects are easily affected by the source, quantity, and quality of test subjects available. For example, individuals of the western tarnished plant bug *(Lygus hesperus)* reared on green beans in the laboratory respond differently to chemicals in alfalfa than do field-collected bugs. Therefore, use of individuals from laboratory colonies in behavioral or resistance studies in alfalfa can lead to erroneous results (Zaugg and Nielson 1974). To facilitate biological and control studies on economically important mirids, researchers need large numbers of individuals of uniform age and nutritional background, and of minimal genetic variability, without having to maintain the bugs on their natural hosts.

The rearing of phytophagous hemipterans is often hampered by problems involving the food supply, humidity, and sanitation (Sailer 1952, Lavabre 1969, Kumar and Ansari 1974). Growth of saprophytic fungi on the oviposition substrate, which prevents egg hatch and leads to high nymphal mortality, is a particular problem (e.g., Geering 1953). Development of practical diets for other phytophagous insects during the 1950s and 1960s stimulated an interest in rearing mirids on artificial diets. Studies on aphid nutrition led to the dis-

covery that the dietary requirements of plant-sucking insects, contrary to some earlier statements, are similar to those of other insects. In the development of chemically defined diets, the need for dietary sugar—sucrose or glucose—is often critical (Auclair 1969).

EARLY ATTEMPTS TO REAR LYGUS BUGS

Following attempts to rear the tarnished plant bug *(Lygus lineolaris)* and other mirines on green beans (Waters 1943, Sailer 1952), Beards and Leigh (1960) succeeded in rearing *L. hesperus* on fresh green beans (see also Patana [1969]), and a similar method was used to rear *L. elisus* and *L. lineolaris* (Bottger 1966). The similarity in amino acid profiles of tarnished plant bug nymphs and green bean pods indicates that a green bean diet generally fulfills the bug's amino acid requirements (Lindig et al. 1981). Landes and Strong (1965) determined that *L. hesperus* will feed on bean juice through a Parafilm membrane and observed a survival rate similar to that obtained by using fresh beans; the bugs lived significantly longer on a sucrose solution than on alfalfa juice. Auclair and Raulston (1966) found that *L. hesperus* grew substantially and lived a relatively long time (maximum survival of 30 days) on a holidic or chemically defined diet enclosed in a membrane. Their diet was an aqueous solution of dry matter consisting of amino acids, water-soluble vitamins, sucrose, mineral salts, and cholesterol benzoate. *Lygus hesperus* and *L. lineolaris* can be reared from egg to adult on artificial diets containing alfalfa meal. Incubation, nymphal duration, and preoviposition compare favorably with values obtained by using natural food; longevity, however, is only half that obtained from culturing the insects on beans (Vanderzant 1967).

ATTEMPTS TO DEVELOP MERIDIC DIETS

Further work on artificial diets for *Lygus hesperus* (Raulston and Auclair 1968) and *L. rugulipennis* (Hori 1972a) showed that the bugs cannot be reared from egg to adult, and adults reared from the third instar fail to reproduce. Strong and Kruitwagen (1969) reported limited success in rearing *L. hesperus* on a meridic diet, but, as discussed in Chapter 14, the faster maturation, greater fecundity, and increased longevity observed among the small percentage of survivors were illusory owing to cannibalism in the rearing cages (Strong and Kruitwagen 1970). The failure to rear *L. rugulipennis* on a chemically defined diet containing casamino acids and, thus, mainly peptides, suggests that amino acids rather than peptides are essential for growth (Hori 1977). Attempts to rear another mirine, *Lygocoris pabulinus*, on a natural diet of bean pods, carrots, and pieces of apple and potato, and on an artificial diet of potato dextrose agar, sucrose, casein extract, and yeast extract failed because females would not oviposit. Prevention of this bug's migratory flight might have broken the sequence of behavioral and physiological events required for reproduction (Wightman 1969b).

Because of problems associated with rearing lygus bugs on various artificial diets (Debolt 1982), North American workers continue to rear lygus bugs on green beans or on beans to which heat-killed lepidopteran larvae are added (Bryan et al. 1976, Cleveland 1987, Snodgrass and McWilliams 1992). Green beans have also been used in Australia to rear the green mirid (*Creontiades dilutus*) (Foley and Pyke 1985), and bean seedlings have been used in Europe to rear *L. rugulipennis* continuously for five generations (Kodys 1971).

Although improved techniques facilitate the mass rearing of lygus bugs on green beans (e.g., Parrott et al. 1975), beans tend to deteriorate rapidly under rearing conditions (pesticide-contaminated beans from a commercial source can also affect the results of studies; Kiman and Yeargan [1985], Braman and Yeargan [1988]). To circumvent this problem, sprouted potatoes can be used as an efficient means of mass rearing *L. lineolaris*. Early instars feed on the tender portions of sprouts, whereas older nymphs and adults feed on the sprouts and tubers; females use the sprouts for oviposition. Development is normal on this food source—that is, various developmental indices are similar to those obtained by other workers (Slaymaker and Tugwell 1982).

Debolt (1982) was unable to rear *L. hesperus* on Vanderzant's artificial diet, but by adding and deleting various ingredients to a diet based on dry lima beans, wheat germ, sucrose, and hen's eggs, he cultured the mirid for 13 generations. This represented the first continuous rearing of a mirid on a meridic diet. In this method, the bugs pierce a Parafilm membrane to feed on a liquid diet contained in glass shell vials or glass petri dish covers. The fecundity and rates of development and survival for *L. hesperus* reared individually to prevent cannibalism, and reared in a group, are comparable or even superior to values obtained when green beans are used. Development of a heat-sealed disposable diet packet for the feeding and oviposition of *L. hesperus* allows the Debolt diet to be used for mass rearing and eliminates the need for glassware (Patana 1982). *Lygus lineolaris* can also be reared on this diet (Debolt 1987).

PROSPECTS FOR IMPROVED REARING TECHNIQUES

Several techniques might lead to improvements in devising artificial diets for lygus bugs and other mirids. An analysis of the nymphal carcass of *Lygus lineolaris* for essential and total amino acids indicates that the relative abundance of these acids is similar to that present in the host plant (Lindig et al. 1981). Lindig et al. (1981) suggested that such a procedure might help in selecting an artificial diet when the amino acid content of the insect's natural feeding site cannot be defined. Identification of the concentrations of various trace minerals in field-collected *L. hesperus* might aid in preparing a defined diet for this bug (Cohen et al. 1985). A recently developed undefined (oligidic) diet, using chicken eggs and consisting of a "semisolid slurry," promises to reduce the costs of rearing *Lygus* species while maintaining the bugs' biological fitness (Cohen 2000).

Artificial diets have not been developed for the cotton fleahopper (*Pseudatomoscelis seriata*), but green beans and potato slices can be used as a dependable diet for laboratory rearing (Breene et al. 1989b). *Macrolophus melanotoma* (as *M. caliginosus*), a predator of whiteflies and other insects, can be reared on an artificial medium and on the medium with a geranium leaf added (Grenier et al. 1989).

Singh (1977) provided additional references and a summary of the composition and preparation of mirid diets, as well as performance or development rates for plant bugs reared on the different diets that have been used. Debolt and Patana (1985) and Patana and Debolt (1985) summarized the rearing of *L. hesperus*, including a review of the necessary facilities and equipment, diet preparation and composition, colony maintenance and rearing schedule, and monitoring of insect quality.

14/ Predation

Feeding specialization in predatory insects has, by and large, escaped scrutiny.

—C. M. Bristow 1988

Once thought to feed exclusively on plant material (e.g., Kirby 1892, Riley [1931?]; see also Chapter 4), mirids are now known as frequent users of animal food. Probably most species exhibit both phytophagous and zoophagous feeding strategies (Southwood 1996), with adults and late-stage nymphs of omnivorous species tending more toward predation than the early instars (Kullenberg 1944, Herrera 1965, Libutan and Bernardo 1995). Such "life-history" omnivory (e.g., Polis and Yamashita 1991)—differing trophic use among life stages and age classes—is common in mirids. Dependence on animal matter ranges from occasional scavenging and opportunistic predation in primarily phytophagous species to specialized, obligate carnivory in other taxa.

Heteropteran families, especially the Miridae, in which both phytophagy and zoophagy are well developed show what Cobben (1978) referred to as unbalanced feeding types. Kullenberg (1944) considered 25 of 100 mirid species in Sweden to be facultatively predacious. For the British fauna, a similar percentage of species—25.6% (52 of 203)—was mentioned as being at least partly predacious (Southwood and Leston 1959), but this figure is conservative because predation was not noted for several species whose predatory habits are well established. Of 58 mirid species determined to be predacious in Poland, Strawiński (1964a) classified 33 as zoophytophages, 24 as phytozoophages, and 1 as a zoophage.

The predacious mirid fauna of apple orchards is particularly well documented. Madsen and Madsen (1982) found that mirids dominated the predatory heteropteran fauna of apple trees in British Columbia. References to and information on the habits of predatory mirids in Canadian apple orchards are found in the works by Bouchard et al. (1982), Braimah et al. (1982), and Arnoldi et al. (1992). Kelton's (1983) review of plant bugs of Canadian fruit crops lists several mirids as predators of mites, aphids, psyllids, whiteflies, and other arthropods, usually without mentioning particular prey species. Fifty-seven of the 81 species occurring on fruit crops in Canada are at least partially predacious (Kelton 1983).

Collyer (1953a, 1953b, 1953c) included numerous predatory mirids in her studies on the natural enemies of the European red mite (*Panonychus ulmi*) in British apple orchards. Mirids and anthocorids are generally considered the most important predatory insects attacking arthropod pests of apple trees in Britain (Solomon 1987). Predatory plant bugs are often absent from orchards characterized by intensive management practices (Schaub et al. 1987), and they are not among the key natural enemies emphasized in European integrated pest management (IPM) programs (Blommers 1994). In Denmark, Rostrup and Thomsen (1923) recorded three mirid species as aphid predators in apple orchards. Among other works on predatory mirids of European apple trees are those by Zeletzki and Rinnhofer (1966) for Germany, Korcz (1967) for Poland, Arčanin and Balarin (1972) for Croatia, Austreng and Sømme (1980) and Jonsson (1983b) for Norway, Rácz (1986) for Hungary, Schaub et al. (1987) for Switzerland, and Tolstova and Atanov (1982) for Russia and Ukraine. A particularly useful account of mirids as beneficial insects in European apple orchards is that by Fauvel (1976; see also Fauvel [1999]). Other regional studies of the European fauna, containing information on both phytophagous and predacious mirids of apple crops, are mentioned in Chapter 12.

Until recently, few mirids, whether specialist or generalist predators, had been the subject of experiments to determine their developmental requirements, to assess their potential for pest suppression, or to clarify factors that affect their activity. The laboratory evaluations that have been conducted should be interpreted with caution because predatory behavior can be distorted under experimental conditions (e.g., van den Berg et al. 1992). In the field, plant bugs have been identified as predators mainly from direct observations rather than from serological techniques such as analysis of pest-specific monoclonal antibodies (Hagler and Naranjo 1994, Agustí et al. 1999a) or from DNA markers (Agustí et al. 1999b).

Determining predator-prey relationships is time-consuming, and the effects of natural enemies on prey densities are difficult to evaluate (e.g., Kiritani and Dempster 1973, Greenstone 1989, Hagler and Naranjo 1994). The role of predacious mirids in helping to suppress pest populations is often underestimated (e.g.,

Carayon 1961). Predation rates are difficult to derive for small, often secretive species with sucking mouthparts. The effects of predation sometimes must be inferred by subtracting from the total measured mortality those deaths that are definitely attributable to other causes (Greenstone and Morgan 1989). Egg mortality due to mirids often gets assigned to a "sucking predator" category. A predacious mirid also can remove evidence of its activity, the bug walking over the plant with an egg impaled on its stylets (Nuessly and Sterling 1994) and later dropping the egg shell. Yet the most important contribution of mirids and other predatory heteropterans to natural biological control might be their feeding on the eggs (and early instars) of pest arthropods (Yeargan 1998).

Mirids might be neglected as biological control agents partly because life-history studies have been conducted on relatively few species in this diverse family. Many workers also are unaware of the extent of cannibalism and the use of pollen, honeydew, and other food resources by predatory mirids. The omnivorous tendencies of predatory mirids probably increase their survival during periods of prey scarcity, helping prevent desiccation during dry weather (Glen 1973) and contributing to the persistence of their populations. This trophic flexibility should enhance their value as biological control agents (e.g., Ehler 1990, Naranjo and Gibson 1996, Gillespie et al. 1999; cf. Carayon 1961). Greater attention to generalist predators (e.g., Murdoch et al. 1985, Ehler 1998) and an increased use of invertebrate predators in biological control (e.g., Döbel and Denno 1994) are helping to clarify the role of mirids in agroecosystems (e.g., Alomar and Wiedenmann 1996).

Facultative predation is frequently noted in passing in articles treating other biological aspects of the Miridae. In their catalogue of predators of insect pests, Thompson and Simmonds (1965) listed many of the mirid-prey associations that are scattered in the literature of applied entomology. Many more records of predation by mirids are available, particularly for aphids and for the eggs and larvae of various insects, but I have not compiled all the published records. Spiders and opilionids, for example, occasionally serve as prey (Rothschild 1963, Niemczyk 1968, Wheeler 1974b) but are not treated further here. Mirids that feed on a wide range of arthropods are mentioned under more than one prey category. In general, a mirid that feeds on the eggs of a particular arthropod but preys mainly on it in other life stages is discussed under the taxonomic group to which the arthropod prey belongs; for example, a species attacking psyllids of all stages but feeding more often on the nymphs or adults is covered in a discussion of psyllids rather than arthropod eggs.

In the following review of predation by mirids, I first discuss species that are more or less specialized as predators of a particular category of prey. Only a fraction of the extensive literature on Cyrtorhinus lividipennis as a predator of leafhopper and planthopper eggs is mentioned, and the reader is encouraged to consult the work by Döbel and Denno (1994), who reviewed factors that influence this mirid's stability and persistence as a planthopper predator. Likewise, the rapidly accumulating literature on various dicyphine mirids—mainly species of Dicyphus and Macrolophus—as predators of thrips, whiteflies, and other arthropod pests of outdoor and protected crops in Europe is not reviewed exhaustively. After the specialist predators of any group are presented, species that feed occasionally or facultatively on the same prey group are discussed. The coverage of unspecialized carnivory includes opportunistic feeding by mirids that are plant pests.

Most of the records presented here are strictly qualitative, and I include anecdotal information based on single observations to give a more complete picture of prey range, and perhaps to stimulate research on neglected species. A few of the mirids discussed, particularly certain specialized egg predators, have been used in classical biological control (DeBach and Rosen 1991, Rosen and DeBach 1992, Waage and Mills 1992)—that is, as exotic natural enemies, inoculatively released and permanently established to help suppress a nonindigenous pest. Often included in the definition of biological control is the use of natural enemies against indigenous pests (Caruthers and Onsager 1993, Van Driesche and Bellows 1993; cf. Lockwood 1993). Most of the records cited in this chapter represent "natural" (sometimes termed "passive") or naturally occurring biological control by native species (Thompson 1951, van den Bosch and Messenger 1973, Luff 1983; see also Lockwood [1993]).

Predation involves the capture of live animals (e.g., Edwards 1963), but the distinction between predation and scavenging (treated in Chapter 15) often is obscured. As Cobben (1978) pointed out, it is not easy to determine whether a particular heteropteran "prey" was killed or was already dead. In analyses of feeding guilds in other invertebrates (e.g., polychaetes [Fauchald and Jumars 1979]), carrion feeding is treated as a subset of carnivory.

Records of feeding on prey normally too large for capture by a mirid represent opportunistic predation on temporarily vulnerable individuals (or scavenging)—for example, Lopidea media feeding on a carabid beetle (Knight and McAtee 1929), Macrolophus praeclarus attacking a disabled ant (Bruner 1934), Lygus borealis feeding on a molting individual of the spined assassin bug (Sinea diadema) (Lilly 1958), and Deraeocoris ruber feeding on a coccinellid (Plate 23). Some mirids, as users of extraoral digestion, might be expected to attack prey heavier than themselves (e.g., Cohen and Tang 1997, Cohen 1998a). Although Miles (1972) noted that mirids tend to lack the salivary enzymes that carnivorous heteropterans such as reduviids and members of some aquatic families use to immobilize their prey, Stethoconus japonicus causes almost immediate prey paralysis or death of lace bugs. Even the mainly phytophagous lygus bugs can quickly incapacitate small prey such as aphids (Cohen 1996; see Chapter 16). Here I treat all feeding records as examples of predation unless an author specifically mentioned scavenging (see Chapter 15).

Among works that should be consulted for additional records of predation by mirids are those by Reuter (1909), Gulde (1921), Butler (1923), Kullenberg (1944), Southwood and Leston (1959), Putshkov (1961a, 1961b), and Strawiński (1964a, 1964b). I previously reviewed facultative predation in the genus *Lygus* (Wheeler 1976b). Hagen et al. (1999) briefly reviewed the Miridae in relation to classical and natural biological control (see also Wheeler [2000b]).

Arthropod Eggs

The exposed, easily pierced and nitrogen-rich eggs of many plant-inhabiting arthropods are preyed on by various heteropterans (Richards and Waloff 1961, Cohen and Debolt 1983). Mirids that feed on insect and mite eggs include obligate or chiefly predatory species of *Deraeocoris* and *Phytocoris*, as well as lygus bugs and other important pests (e.g., Putshkov 1966, Wheeler 1976b). Few are specialized predators of arthropod eggs, but those few provide outstanding examples of the use of insects to achieve biological control of crop pests. Mirids that are specialized egg predators (most also feed on young nymphs) are associated mainly with auchenorrhynchans—especially delphacids and cicadellids—and belong to the genera *Cyrtorhinus* (Orthotylinae) and *Tytthus* (Phylinae). Several of the "*Cyrtorhinus*" species reviewed by Usinger (1939) are now placed in *Tytthus* (Carvalho and Southwood 1955). In natural systems, *Tytthus* species prey on the eggs of salt marsh–inhabiting delphacids (Döbel and Denno

1994). Species such as *C. lividipennis* and *T. mundulus* are emphasized as natural enemies of pest planthoppers, for egg mortality is often a key factor in delphacid population dynamics. In both natural and agricultural systems, a significant negative relationship between predator density and increase of prey populations (e.g., Cook and Perfect 1985) characterizes mirid-planthopper interactions, as well as temporal and spatial synchrony in their life histories. Mirids tend to show the strongest and most consistent numerical responses to prey densities of all the major delphacid predators (Fig. 14.1; Döbel and Denno 1994). The role of mirids and other major predators in suppressing populations of agriculturally

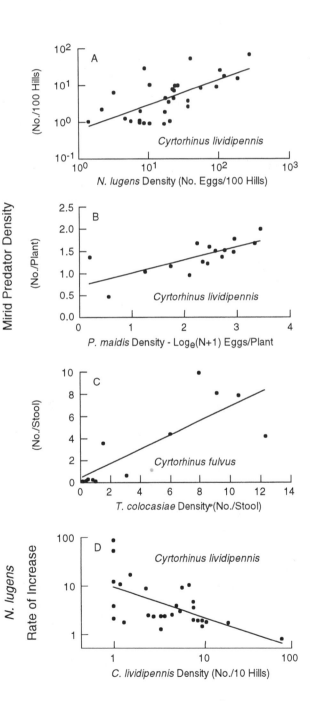

Fig. 14.1. Numerical responses of *Cyrtorhinus* species (Orthotylinae), specialized predators of auchenorrhynchan eggs, to increasing densities of delphacid planthoppers. A. Relationship between the densities of *C. lividipennis* and *Nilaparvata lugens* (no./100 hills) in rice in Malaysia; $Y_{pred} = 2.29 + 0.34X_{prey}$ for nontransformed data, $R^2 = 0.58$, $P < 0.01$ (from Ooi 1988). B. Relationship between densities of *C. lividipennis* and egg densities of *Peregrinus maidis* [$log_e (N + 1)$/plant] in corn in Hawaii; $Y_{pred} = 0.80 + 0.29X_{prey}$ for log_e-transformed data, $R^2 = 0.55$, $P < 0.01$ (from Napompeth 1973). C. Relationship between densities of *C. fulvus* and *Tarophagus colocasiae* (no./stool) in taro in Hawaii; $Y_{pred} = -0.44 + 0.64X_{prey}$, $R^2 = 0.64$, $P < 0.01$ (from Matsumoto and Nishida 1966). D. Relationship between rate of population increase of *N. lugens* (generational growth from one rice-growing period to the next, 1 = no growth) and densities of *C. lividipennis* (no./10 hills) in Malaysia; $Y_{prey\ increase} = 0.99 - 0.67X_{pred}$ for log_{10}-transformed data, $R^2 = 0.37$, $P < 0.01$ (from Ooi 1988). (A,D reprinted, by permission of P. A. C. Ooi; B, by permission of B. Napompeth; C, modified, by permission of B. M. Matsumoto and the University of Hawaii Agricultural Experiment Station.)

important planthoppers exceeds that of parasitoids (Napompeth 1973, Benrey and Lamp 1994, Döbel and Denno 1994). Mirids also prey on the stalked eggs of chrysopids (Samways 1979) and on unparasitized and parasitized auchenorrhynchan eggs; feeding on both the predator and the parasitoids is an example of intraguild predation.

SUGARCANE DELPHACID

The use of *Tytthus mundulus* against the sugarcane delphacid ("leafhopper" in older literature), *Perkinsiella saccharicida*, is recognized as one of the most successful examples of classical biological control (Hagen and Franz 1973, van den Bosch and Messenger 1973, Rosen 1985). This story, combining good basic science and luck (DeBach and Rosen 1991), illustrates the use of an introduced natural enemy that provides long-standing, effective control of a pest (Huffaker et al. 1971). The mirid-delphacid relationship can be used (Dent 1991) to support the hypothesis that biological control agents should be sought from the same area as the target pest (old associations; Rosen and DeBach [1992], Bellows and Legner [1993]). Others contend that agents should be selected from outside the area of endemism or should not have a close evolutionary history with the pest (new associations theory of Hokkanen and Pimentel [1984], Waage and Greathead [1988]; see also Waage [1990a]).

An adventive species detected in Hawaii in 1900, the sugarcane delphacid soon threatened the islands' sugarcane industry, prompting an extensive search for natural enemies. Beginning in 1904, the importation and release of several species of parasitic Hymenoptera, aided by native parasitoids and predators, gave substantial but unsatisfactory relief from this pest. A few plantations still experienced considerable damage. Even the most efficient natural enemy, a mymarid egg parasitoid imported from Australia, was ineffective in wetter areas of plantations and additional exploration for natural enemies was needed (Howard 1916; Imms 1926; Swezey 1928, 1929).

In 1920, Frederick Muir discovered that *T. mundulus* suppressed populations of *P. saccharicida* in sugarcane fields in Queensland, Australia. Nymphs and adults pierce and at least partially empty the contents of delphacid eggs, as well as kill parasitic wasps when they attack parasitized hosts (Muir 1920, Swezey 1936). Muir had observed the mirid on sugarcane in Fiji during 1905–1906, but not realizing it was an egg predator, he had experimented only with planthopper nymphs and adults as possible prey (Muir 1920). Once Muir realized that the mirid preys mainly on eggs, he decided to use it in biological control. The decision to import the bug to help suppress the sugarcane delphacid was considered somewhat risky, for predacious heteropterans had not yet been used in biological control and this member of a mainly plant-feeding family might injure sugarcane

(Imms 1926, DeBach and Rosen 1991). Zimmerman (1948b), however, stated the mirid was not released until its beneficial habits were "fully ascertained."

Tytthus mundulus was soon shipped from Australia to Hawaii, studied, and released in a sugarcane plantation, but the small shipment did not result in immediate establishment (Bianchi 1977). Later in 1920, Cyril Pemberton was sent to Fiji to collect additional material. His shipments resulted in the rapid establishment of *T. mundulus* on Hawaii, and the predator soon spread on its own power, or passively with air currents, to Maui and Molakai (Swezey 1936).

Within three years, *T. mundulus*, which under favorable conditions can produce 10 generations each year compared to three or four for its prey (Williams 1931, Swezey 1936), brought the delphacid under control and saved the Hawaiian sugarcane industry (Usinger 1955, DeBach 1974). It is now known that a significant negative relationship exists between the population increase of *P. saccharicida* and the density of *T. mundulus*, suggesting that this mirid's action on prey populations is density dependent (Southwood 1969, Bull 1981, Döbel and Denno 1994). The mirid's shorter generation time helps to offset the planthopper's higher fecundity. Furthermore, the lower fecundity of the sugarcane delphacid compared to that of the wing-dimorphic corn delphacid (*Peregrinus maidis*) in Hawaii (see discussion below) allows the mirid to track its prey reproductively and to suppress populations on sugarcane (Döbel and Denno 1994). The impact that the introduction and establishment of *T. mundulus* had on the islands' economy was so great that the savings could hardly be expressed in millions of dollars no matter how high the count was carried (DeBach 1974). The story of the sugarcane delphacid's eventual suppression in Hawaii by natural enemies is summarized by Sweetman (1958) and DeBach and Rosen (1991) and is detailed by Muir (1920), Imms (1926), Timberlake (1927), Williams (1931), Swezey (1936), and Pemberton (1948).

Tytthus parviceps is also an egg predator that shows a strong numerical response to its prey. It, too, can suppress populations of the sugarcane delphacid, but it accounts for only about 20% of the *Tytthus* population in Australian sugarcane. The composition of predator populations can be partly explained by food availability. The sugarcane delphacid lays most of its eggs in the third to sixth open leaf, and *T. mundulus* also prefers this upper or crown portion of the plant; in contrast, *T. parviceps* prefers the region slightly below the mature leaf axils (Bull 1981).

Although *T. mundulus* was released in Florida against the sugarcane delphacid after this pest was detected there in 1982 (Sosa 1985), this mirid did not become established in sugarcane fields (Frank and McCoy 1993). Instead, the adventive *T. parviceps* was found in Florida sugarcane infested with this planthopper (Sosa 1985, Hall 1988, Bennett et al. 1990, Wheeler and Henry 1992, Hall and Bennett 1994).

In addition to being a key predator of the sugarcane delphacid, *Tytthus mundulus* preys on the eggs of other delphacids (Usinger 1939, Chiu 1979) and occasionally leafhoppers (Van Zwaluwenburg and Rosa 1951). It apparently can survive on delphacid nymphs in the absence of eggs (Verma 1955a). It was hoped *T. mundulus* would effectively suppress populations of the corn delphacid in Hawaii, even though Williams (1931) had noted the bug was not attracted to cornfields.

Verma's (1955a) studies, and those by Napompeth (1973), revealed biological characteristics of the corn delphacid–mirid relationship that help explain the failure of *T. mundulus* to suppress this delphacid. The mirid prefers mature tissues for oviposition, usually depositing its eggs in old egg cavities of its prey; when females are forced to deposit eggs in young corn tissues, the eggs fail to hatch. *Peregrinus maidis*, however, oviposits in young, soft tissues, frequently in the roots of young plants where ants deter predation by the mirid. Deposition of the eggs in corn roots, where they are less susceptible to predation than eggs laid in higher strata of host plants, provides a spatial refuge for the planthopper (Verma 1955a, Döbel and Denno 1994). *Tytthus mundulus* shows a delayed numerical response, migrating to cornfields later than *P. maidis* because hardened tissues in which to oviposit are necessary for the predatory mirid's survival. Corn is less than ideal for the mirid because this short-term crop dies when the bug's populations are increasing. The shorter plants and relatively small plantings also do not offer as much protection from wind as does sugarcane. Moreover, the corn delphacid's life-history traits—reproductive and dispersal capabilities—render it less likely than the sugarcane delphacid to be suppressed effectively by *T. mundulus*. The corn delphacid is substantially more fecund than the mirid and exhibits wing dimorphism, which promote escape from predators and population outbreaks (Döbel and Denno 1994).

A mirid introduced from Guam, *Cyrtorhinus lividipennis*, eventually proved successful in suppressing infestations of the corn delphacid in Hawaii (Napompeth 1973; Liquido and Nishida 1985a, 1985b, 1985c). This predator, however, fails to provide effective control during years in which its populations are not well synchronized with those of its prey; inoculative releases of the bug at the beginning of a new corn crop might help ensure adequate suppression of the delphacid (Napompeth 1973). Native to the Indo-Pacific, *C. lividipennis* oviposits on corn in sites similar to those of its prey: midribs and leaf sheaths (Liquido and Nishida 1985b). It also shows a strong numerical response to prey density (see Fig. 14.1), sometimes colonizing cornfields early because of the presence of cicadellid eggs as alternative prey (Napompeth 1973, Benrey and Lamp 1994, Döbel and Denno 1994). This mirid, once falsely accused of damaging rice in Southeast Asia (Horváth 1906) before carnivory in mirids was well known, is discussed below as a predator of rice planthopper eggs.

Tytthus mundulus has been used against other delphacid pests (Clausen 1978) and has been imported from Mauritius to southern Africa for evaluation in the biological control of an introduced tropiduchid planthopper (*Numicia viridis*) on sugarcane. Although the mirid feeds on fulgoroid eggs in the laboratory, it cannot be reared over several generations (Anonymous 1969, Simmonds 1969). The ability to culture *T. mundulus* on eggs of the house fly (*Musca domestica*) (Stephens 1975) and oriental fruit fly (*Bactrocera dorsalis*) (Takara and Nishida 1981) might facilitate mass rearing and shipment, thereby enhancing its usefulness against other exotic pests.

TARO DELPHACID

Another successful project in classical biological control involves the taro delphacid (*Tarophagus colocasiae*) and its predator, *Cyrtorhinus fulvus*, in Hawaii. The fulgoroid, discovered in the islands in 1930, can be devastating to young plants; this pest resisted attempts at eradication by flaming and insecticidal treatments. By 1937, a search for natural enemies was needed to protect the culture of taro, an important food crop in the Pacific Islands (Fullaway 1940, Fatuesi et al. 1991).

Cyrtorhinus fulvus apparently is a specific predator of *T. colocasiae* and lives only on taro (Fullaway 1940, Matsumoto and Nishida 1966). It was obtained from the Philippines with several species of hymenopteran parasitoids and released in Hawaiian taro plantings in 1938 (Fullaway 1940, Fullaway and Krauss 1945). The results were dramatic: "The increase and spread of the predaceous bug in the field was really astonishing, and it soon became evident that *Cyrtorhinus fulvus* alone would keep the leaf-hoppers down to such insignificant numbers . . . that no further fears need be entertained about the decline of taro production" (Fullaway 1940).

That *C. fulvus* alone can suppress Hawaiian populations of *T. colocasiae* (hymenopteran parasitoids imported to help control this delphacid either failed to become established or were recovered from taro many years later [Clausen 1978]) is supported by the studies of Matsumoto and Nishida (1966). They demonstrated this predator's strong numerical response. The mirid can maintain pest numbers at subeconomic levels because fluctuations in predator populations are highly correlated with those of the prey, and spatial distributions of predator and prey overlap. Age distribution of the delphacid shows a low proportion of early instars, and hand removal of *C. fulvus* produces an increase in prey numbers (Fig. 14.2). Adults of *C. fulvus* are stronger fliers than those of the mainly brachypterous *T. colocasiae*, and their oviposition in taro petioles is stimulated by the presence of prey eggs (Matsumoto and

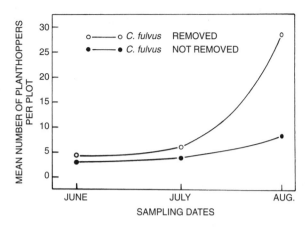

Fig. 14.2. Effect of *Cyrtorhinus fulvus* on populations of the taro delphacid (*Tarophagus colocasiae*) in Hawaii. (Modified, by permission of B. M. Matsumoto and the University of Hawaii Agricultural Experiment Station, from Matsumoto and Nishida 1966.)

Nishida 1966). The dispersal advantage of the mirid over its prey compensates for its reproductive disadvantage, allowing this predator to track spatial changes in prey populations (Döbel and Denno 1994). In addition, *C. fulvus* can survive at low prey densities by feeding on honeydew or its own eggs (Matsumoto 1964).

Cyrtorhinus fulvus, intentionally introduced to Guam in 1947 (Pemberton 1954), similarly suppressed the taro delphacid on that island (Matsumoto and Nishida 1966). In the Pacific Islands, the delphacid is still generally held in check by this obligate egg predator (Mitchell and Maddison 1983) except when pesticide use reduces its effectiveness (Waterhouse and Norris 1987). Although *T. mundulus* co-occurs with *C. fulvus* in some Hawaiian taro fields, the populations of this predacious mirid are small (<1%) compared to those of the latter species (Liquido and Nishida 1983). The value of *C. fulvus* tends to be unappreciated by growers, who often do not recognize this small bug of cryptic habits (Fatuesi et al. 1991). This predator's apparent lack of success in suppressing *Tarophagus* on taro in some areas might reflect sympatry among taro-feeding members of the genus and the failure of workers to distinguish the three species (all were referred to as *T. proserpina*) and, thus, the taxonomic identity of the target pest (Asche and Wilson 1989). They further speculated that variation in populations of *C. fulvus* might also be involved in the disparate results of biocontrol releases of this predator.

RICE PLANTHOPPERS

Cyrtorhinus lividipennis plays an important role in the population dynamics of the brown planthopper (*Nilaparvata lugens*) and whitebacked planthopper (*Sogatella furcifera*) in rice fields, even though these wing-dimorphic pests have polymorphic life histories

that promote escape from predators (Döbel and Denno 1994). Both the predator and the brown planthopper oviposit in the leaf sheaths of rice (Liquido and Nishida 1985b). The mirid feeds mainly on planthopper eggs but attacks nymphs (Plate 23) and adults more readily than do other mirids that are specialized egg predators, such as *C. fulvus* and *T. mundulus* (Bae and Pathak 1966, Pophaly et al. 1978, Chiu 1979, Misra 1980, Sivapragasam and Asma 1985). First- and second-instar *C. lividipennis*, though, die when fed planthopper nymphs rather than eggs. Only 80% of third instars provided nymphs are able to reach adulthood, and individuals reared on planthopper nymphs show decreased fecundity and longevity (Chua and Mikil 1986, 1989). Olfactometer responses of *C. lividipennis* support this predator's preference for delphacid eggs over nymphs. Rice treatments with brown planthopper eggs attract more *C. lividipennis* than do plant treatments with hopper nymphs (Rapusas et al. 1996). *Cyrtorhinus lividipennis* prefers brown planthopper eggs to those of other planthoppers or of leafhoppers such as *Nephotettix virescens* (Pophaly et al. 1978, International Rice Research Institute 1981, Sivapragasam and Asma 1985, Manti 1989, Heong et al. 1990b; cf. Dyck and Orlido 1977, Greathead 1983).

Under some conditions, *C. lividipennis* does not suppress rice planthopper densities as effectively as spiders and other generalist predators that feed on prey nymphs (Holdom et al. 1989, Fowler et al. 1991). Spider predation on the mirid can limit the bug's value as a natural enemy (Holdom et al. 1989, Heong et al. 1991, Döbel and Denno 1994). In addition, *N. lugens* shows a decided reproductive advantage over the mirid, which is not substantially offset by the plant bug's shorter generation time (Döbel and Denno 1994). Yet planthopper populations have been suppressed effectively (Hinckley 1963, Stapley 1976, Otake 1977, Cook and Perfect 1985, Manti 1989, Döbel and Denno 1994), and the mirid's conservation in rice pest-management programs is recommended (Manjunath et al. 1978, Pophaly et al. 1978, Ooi 1979, Kalode 1983). This predator also severely limits brown planthopper densities in greenhouses (van Vreden and Ahmadzabidi 1986).

Cyrtorhinus lividipennis and Suppression of Rice Hoppers

The literature on the role of *C. lividipennis* as a predator of rice planthoppers is potentially confusing. For example, it is considered both effective (Manti 1989) and ineffective (Dyck and Orlido 1977, Chandra 1978) in the Philippines, where predation tends to be highest in late season and lower in dry than in wet seasons (Cook and Perfect 1989). At times, the mirid might be an opportunist, exploiting prey populations only during outbreaks (Kenmore 1980, Baltazar 1981, Greathead 1983). Because not all reports of successes and failures of the mirid to suppress hoppers in rice fields are placed in an ecological context, a reader cannot easily appreciate its life-history traits, as well as those of its prey, and the environmental factors that either help stabilize or tend to uncouple predator-prey interactions. Döbel and Denno (1994), however, summarized this literature, emphasizing the functional and numerical responses of planthopper predators, including *C. lividipennis*, and the life-history attributes of both predator and prey that affect prey suppression and population stability.

Cyrtorhinus lividipennis shows a relatively high daily capture and consumption rate involving high instantaneous search rates and short handling times (Sivapragasam and Asma 1985, Manti 1989, Döbel and Denno 1994). Like many other predacious mirids (see Chapter 7), *C. lividipennis* females consume significantly more prey than do males—in this case, mainly planthopper eggs (Chua and Mikil 1989, Heong et al. 1990b; cf. Reyes and Gabriel 1975). Both sexes show a Type II functional response (Sivapragasam and Asma 1985, Manti 1989, Heong et al. 1990b), except at higher temperatures (35°C) (Song and Heong 1997), which is likely to result in inverse density-dependent predation (Döbel and Denno 1994). This mirid's ability to respond numerically (see Fig. 14.1) to increases in planthopper populations in rice and other crops helps offset the destabilizing effects of its functional response. It is able to aggregate in areas of high prey density and to increase its reproduction in response to increasing prey densities (Napompeth 1973, Chang and Oka 1984, Kuno and Dyck 1985, Balasubramanian et al. 1988, Qingcai and Jervis 1988, Heong et al. 1990a, Ghosh et al. 1992, Döbel and Denno 1994).

Considerable evidence indicates that *C. lividipennis* can track and suppress populations of its planthopper prey, especially when predator-prey populations are temporally synchronized. In parts of the tropics, the mirid often migrates at the same time as its prey (Riley et al. 1995). Effective suppression can occur in the field when predator-prey ratios are 1:2 to 1:20 (Stapley 1976, Otake 1977, Pophaly et al. 1978, Chiu 1984, Manti 1989); at times, ratios of 1:1 are attained (Balasubramanian et al. 1988). Correlations between predator and prey populations in rice are more likely when reservoirs of the mirid occur nearby. *Cyrtorhinus lividipennis* sometimes invades rice fields from nearby grassy areas that harbor alternative prey (Stapley 1975, Otake 1977), and it can colonize rice from weeds growing in fields during intercropping periods (Stapley 1976).

As reviewed by Benrey and Lamp (1994) and Döbel and Denno (1994), data on the effectiveness of *C. lividipennis* are available from predator-removal experiments, as well as from studies that demonstrate a negative relationship between mirid density and rate of increase in planthopper populations. Most failures of the bug to suppress the brown planthopper can be attributed to late colonization of rice and the mirid's delayed numerical response (Dyck and Orlido 1977, Bentur and Kalode 1987, Manti 1989). Manti showed that under experimental conditions, late arrival by *C. lividipennis* results in prey outbreaks.

The mirid's ability to migrate (e.g., Kisimoto 1979; Ooi 1979; Riley et al. 1987, 1995; Reynolds and Wilson 1989; Rutter et al. 1998) allows colonization of distant areas. Although this predator cannot overwinter in Japan, it migrates there annually from mainland China (Cheng et al. 1979, Kisimoto 1981). But it colonizes much later than its wing-dimorphic planthopper prey, leading to asynchrony between predator and prey populations. Mainly because this predator is unable to track prey populations effectively in temperate Asia, it plays a less important role in suppressing the brown planthopper in Japan than it does in tropical Asia and the Pacific region (Kuno and Hokyo 1970, Kiritani 1979, Benrey and Lamp 1994, Döbel and Denno 1994).

Despite these limitations to the bug's effectiveness, the dispersal and reproductive capabilities of *C. lividipennis* facilitate its suppression of rice planthoppers. Although various factors can disrupt interactions with its planthopper prey, the bug's persistence is enhanced by an ability to survive on planthopper nymphs when its preferred prey, planthopper eggs, are scarce, or to switch to leafhopper eggs (Hinckley 1963, Napompeth 1973, Chua and Mikil 1989, Heong et al. 1990b). In India, when rice planthopper populations are low during November to January, the mirid feeds on the whitestriated planthopper (*Nisia nervosa*) on a weedy cyperus (Bentur and Kalode 1987). Eggs of noctuids (Plate 23; van den Berg et al. 1988, 1992) and pyralid leaffolders (Ooi and Shepard 1994) can serve as alternative prey in rice fields, and eggs of the Asiatic rice borer (*Chilo suppressalis*) are attacked in the laboratory (Shepard and Arida 1986). *Cyrtorhinus lividipennis* will cannibalize (O'Connor 1952, Hinckley 1963) when preferred prey are scarce and will also feed to a limited extent on the rice plant (Sivapragasam and Asma 1985; cf. Liquido and Nishida 1985b). The abilities to mass propagate *C. lividipennis* on eggs of the Mediterranean fruit fly (*Ceratitis capitata*) (Liquido and Nishida 1985a) and to rear it on eggs of the rice moth (*Corcyra cephalonica*) (Bentur and Kalode 1985a, Geetha et al. 1992) increase the possibility of using augmentative releases of this predator to help manage rice planthopper populations.

Uncoupling *Cyrtorhinus*-Prey Interactions

Interactions between *C. lividipennis* and its planthopper prey can be disrupted by insecticide applications. The mirid is sensitive to pyrethroids and other types of insecticides applied to rice fields (Stapley 1976, Dyck and Orlido 1977, Reissig et al. 1982a, Kalode 1983, Fabellar and Heinrichs 1984, Cheng 1985, Ooi and Waage 1994, Schoenly et al. 1996), generally more so than are other predators (Reissig et al. 1982b). Even though pyrethroids kill the mirid, its populations can increase in sprayed plots because of rapid immigration and reproduction in response to resurgent planthopper populations. Mirid numbers can also be retained through replacement by nymphs that hatch after insecticides are applied (Holdom et al. 1989). In some cases, the resurgence of *N. lugens* following chemical treatments is attributed to a stimulating effect on prey fecundity and improved plant growth rather than to an insecticide-induced decrease in mirid numbers (Chelliah and Heinrichs 1980, Heinrichs et al. 1982, Reissig et al. 1982b; see also Hardin et al. [1995]).

Despite the potential for increases in *C. lividipennis* populations following application of pyrethroids (Holdom et al. 1989), use of insecticides tends to uncouple predator-planthopper interactions and induce outbreaks of its prey (Döbel and Denno 1994). Development of more selective insecticides is imperative for the effective integrated management of rice planthoppers and leafhoppers (Fabellar and Heinrichs 1986).

Combining Cultivar Resistance with Use of *Cyrtorhinus*

The use of resistant rice cultivars can help conserve *Cyrtorhinus lividipennis* and other predators by decreasing insecticide usage (Heinrichs 1994). Plants resistant to herbivores can either adversely affect or enhance the efficiency of a pest's natural enemies (e.g., Price et al. 1980, Rapusas et al. 1996), but cultivar resistance and biological control can often be combined to provide efficient insect suppression (e.g., Smith 1994). Several studies indicated that *C. lividipennis* can be combined with the use of resistant rice cultivars, helping offset the brown planthopper's reproductive advantage (Döbel and Denno 1994, Cuong et al. 1997). Certain cultivars developed for planthopper resistance, however, are less attractive to *C. lividipennis* than are more susceptible cultivars, as measured by this predator's olfactory responses to various rice genotypes (Rapusas et al. 1996). Mortality of the brown planthopper, whitebacked planthopper, and green leafhopper from plant resistance and predators can be simply additive (i.e., with no clear interactions), or it can be synergistic (Kartohardjono and Heinrichs 1984, Cheng 1985, Salim and Heinrichs 1986, Senguttuvan and Gopalan 1990, Döbel and Denno 1994, Heinrichs 1994, Smith 1994). Kartohardjono and Heinrichs (1984) and Senguttuvan and Gopalan (1990) suggested that the mirid's effectiveness is enhanced on resistant cultivars because its planthopper prey move more in search of feeding sites ("restless" behavior). In addition, the role of volatile phytochemicals in allowing the bug to orient to different rice genotypes (see Rapusas et al. [1996]) needs to be considered in evaluating the effects of resistance on natural enemies, as well as an assessment of two-versus three-trophic-level effects (e.g., Pfannenstiel and Yeargan 1998) of resistant rice plants on *C. lividipennis*. Hare (1994) and Rapusas et al. (1996) stressed the need for long-term studies to evaluate the compatibility of resistant rice cultivars and biological control by predators such as *C. lividipennis*.

Other Mirid Predators of Rice Hoppers

Additional mirids that prey on the eggs of the brown planthopper and other homopteran pests of rice are *Tytthus chinensis*, *T. mundulus*, *T. parviceps*, and the stenodemine *Trigonotylus tenuis* (O'Connor 1952, Hinckley 1963, Chiu 1979, Bentur and Kalode 1985b; see also Döbel and Denno [1994]). The last-named species is a facultative rather than a specialized egg predator.

A RUSH-INHABITING DELPHACID

Rothschild (1963) studied the habits of another specialized egg predator of delphacids, *Tytthus pygmaeus*. Closely associated with *Conomelus anceps* on rushes in England, this univoltine plant bug usually oviposits in stems containing planthopper eggs. Adults and nymphs of all stages search for delphacid eggs; first instars can feed on 2 eggs a day, whereas adults consume a daily average of 3.7 eggs. Ten to fifteen minutes are required to empty the contents of an egg, after which the predator continues to destroy all the eggs remaining in a batch. Eggs fed on by mirids can be distinguished from those attacked by hymenopteran parasitoids, or that hatch normally, by the intact operculum and absence of projecting embryonic membranes. *Tytthus pygmaeus* does not feed on other prey offered (mites, collembolans, aphids, and immature spiders) and in the laboratory feeds on nymphs of *C. anceps* only in the absence of eggs. The mirid is also able to survive for up to 10 days on the stems of host rushes in the absence of animal food (Rothschild 1963).

Although *T. pygmaeus* is one of the more important predators of *Conomelus* eggs, sometimes killing more eggs than do parasitoids, it inflicts relatively low mortality: 7–15% in autumn (spring predation could not be assessed owing to loss of antiserum stocks). Mirid nymphs are most abundant only after the main hatch of the delphacid in spring. In late summer when prey eggs are again available, adult mirid populations are low. Adults prey on *C. anceps* eggs as soon as the delphacid begins to oviposit, but the percentage of predation declines as more planthopper females reach maturity (Rothschild 1966). Rothschild (1966) also noted that extensive predation by *T. pygmaeus* almost certainly occurs on first- and second-instar *Conomelus* nymphs in the field, and that a second mirid species, *Cyrtorhinus caricis*, preys on delphacids of almost all stages except the eggs.

PREY OF UNSPECIALIZED EGG PREDATORS

Mirids as facultative and generalized egg predators include species of mainly predatory genera—for example, *Ceratocapsus*, *Deraeocoris*, and *Phytocoris*—as well as mixed feeders belonging to phyline genera such as *Rhinacloa* and *Spanagonicus*. Egg predation is also common in mirids that are important plant pests, including lygus bugs (Wheeler 1976b, Cleveland 1987, Hagler and Naranjo 1994), the garden fleahopper (*Halticus bractatus*) (Buschman et al. 1977), and the cotton fleahopper (*Pseudatomoscelis seriata*) (McDaniel and Sterling 1979, 1982; Agnew et al. 1982; Sterling 1982; Gravena and Sterling 1983; Johnson et al. 1986; Sterling et al. 1989a; Nuessly and Sterling 1994). Most records of egg predation by plant bugs refer to holometabolous prey (Hinton 1981), including beneficial insects (Table 14.1) such as chrysopids and syrphids as well as weevils imported for the biological control of weeds (Goeden and Louda 1976, Goeden and Kirkland 1981). Mirids also prey on the paurometabolous Heteroptera. Early instars of an unspecified plant bug, for instance, feed on the unguarded eggs of a presocial pentatomid (*Elasmucha grisea*) (Melber et al. 1980), and *Miris striatus* feeds on pentatomid eggs (Kullenberg 1944). Whether predatory mirids show a preference for unparasitized eggs or readily feed on eggs containing parasitoid larvae or pupae is apparently unknown (see Ruberson and Kring [1991]). As noted earlier in this chapter, however, mirids that prey on planthopper eggs feed indiscriminately on unparasitized and parasitized eggs of their prey. Because many of the facultative egg predators presented here also feed on small larvae of the same species, the reader is referred to the discussion of insect larvae and pupae as mirid prey later in this chapter.

Several species of *Deraeocoris* feed on arthropod eggs, including those of important homopteran, coleopteran, and lepidopteran pests (Table 14.1; see also Viggiani [1971b], Chinajariyawong and Harris [1987]). *Phytocoris* species in the New and Old World feed on a wide range of prey, including eggs of mites (Gilliatt 1935, Braimah et al. 1982), codling moth (*Cydia pomonella*) (e.g., MacLellan 1962, 1963, 1972), larch casebearer (*Coleophora laricella*) (Ryan 1985), and numerous other arthropods (e.g., Strawiński 1964b). MacLellan (1962) discussed other mirids as predators of codling moth eggs in Nova Scotia. In addition to *P. tiliae*, the orthotyline *Blepharidopterus angulatus* feeds on codling moth eggs in Swedish (Subinprasert and Svensson 1988) and in British apple orchards. The intensity of predation appears unrelated to egg density in Britain (Glen 1977b). Glen and Brain (1978) described a model of predation on codling moth eggs that included mirids and other heteropterans. They showed that the number of eggs destroyed depends on the number and size of the predators (three mirid species were most important) and their searching capacity, which is influenced by the density of spider mites (their principal prey), interference from other predators, and weather (amount of sunshine). The importance of mirid predation on codling moth eggs appears to be greater in England than in Canada or the United States (Glen 1975a), although the disparity might in part reflect differing usage of insecticides: unsprayed orchards were studied in England; integrated control orchards, in

Canada. The phyline *Sejanus albisignatus* feeds on eggs (and small larvae) of the codling moth in New Zealand (Collyer and van Geldermalsen 1975).

The orthotyline *Ceratocapsus modestus* feeds on the eggs of the grape phylloxera (*Daktulosphaira vitifoliae*) in Pennsylvania. The nymphs and adults inhabit foliage heavily infested with phylloxeran galls, the leaf edges curled to form a tube (Plate 23). *Ceratocapsus modestus* inserts its stylets into gall openings on the upper leaf surface to feed on phylloxeran eggs (Plate 23; Wheeler and Henry 1978b). Gravena and Sterling (1983) reported a *Ceratocapsus* species feeding on the eggs of the cotton leafworm (*Alabama argillacea*) in a Texas cotton field. The *Ceratocapsus* species of Gravena and Sterling probably is *C. puntulatus*, which was recorded as a predator of bollworm (*Helicoverpa zea*) eggs in Texas (Nuessly and Sterling 1994). This bug showed the highest rate of egg predation in 24 hours among 11 arthropods that fed on radiolabeled eggs of the noctuid. *Ceratocapsus dispersus* and *C. mariliensis* prey on cotton leafworm eggs in Brazil (Gravena and Pazetto 1987); the former bug can also be reared in the laboratory on eggs of the pink bollworm (*Pectinophora gossypiella*) and the cotton leafperforator (*Bucculatrix thurberiella*) (Encalada and Viñas 1989).

Various dicyphine mirids (see also Table 14.1) feed on the eggs of lepidopteran pests in tobacco fields of the southern United States (e.g., Rosewall and Smith 1930, Thomas 1945, Madden and Chamberlin 1954) and Cuba (Ayala Sifontes et al. 1982). As Thomas (1945) observed, predation on hornworm eggs occurs soon after they are deposited—that is, before the chorion fully hardens. In South Africa, *Nesidiocoris callani* and *N. volucer* (sensu Kerzhner and Josifov 1999) prey on the eggs of the chrysomelid *Lema bilineata* on tobacco. Odhiambo (1961) summarized the observations of E. McC. Callan on *N. volucer*: "In the laboratory I have observed it inserting its stylets through the delicate chorion of the eggs and sucking them dry. These mirids were very numerous in tobacco fields near Grahamstown in March 1959, and practically every *Lema* egg mass I found had been destroyed by them."

The dicyphines *Dicyphus tamaninii* and *Macrolophus melanotoma* are important predators of noctuids in the Mediterranean Basin. At times, they may be responsible for as much as an 80% egg mortality of *Helicoverpa armigera* in the field (Agustí et al. 1999a).

Following Wille's (1951) listing of the orthotyline *Hyalochloria denticornis* as a predator of *Heliothis* eggs in Peru, Beingolea (1959b, 1960) determined that the eggs of the lesser cotton leafworm (*Anomis texana*) represent its main summer prey. *Hyalochloria denticornis*, as well as *Rhinacloa* species (see also Table 14.1), are predators included in the guidelines for the integrated control of Peruvian cotton pests. Differential dosages of systemic insecticides are recommended against aphids so that *Rhinacloa* species are conserved. Populations of *H. denticornis* and other biological control agents are

Table 14.1. Examples of mirids as facultative egg predators

Mirid	Prey	Locality	Reference
Bryocorinae			
Campyloneuropsis cincticornis	*Gratiana spadicea* (Coleoptera: Chrysomelidae)	Brazil	Becker and Frieiro-Costa 1988
Engytatus modestus	"Cabbage butterfly" (Lepidoptera: Pieridae)	Hawaii, USA	Illingworth 1937b
Cylapinae			
Fulvius nigricornis	Weevil (Coleoptera: Curculionidae)	Malaysia	China 1935
Fulvius variegatus	*Rhabdoscelis* sp. (Coleoptera: Curculionidae)	Hawaii, USA	Williams 1931, Zimmerman 1948a
Deraeocorinae			
Deraeocoris annulipes	*Zeiraphera diniana* (Lepidoptera: Tortricidae)	Switzerland	Graf 1974, Delucchi et al. 1975
Deraeocoris brevis	*Cacopsylla pyricola* (Homoptera: Psyllidae)	Oregon, USA	Westigard 1973
	Pristiphora erichsonii (Hymenoptera: Tenthredinidae)	Manitoba, Canada	Ives 1967
Deraeocoris diveni	*Coleophora laricella* (Lepidoptera: Coleophoridae)	Oregon, USA	Ryan 1985
Deraeocoris flavilinea	*Palomena prasina* (Heteroptera: Pentatomidae)	Sicily	Boselli 1932
Deraeocoris laricicola	*Pristiphora erichsonii* (Hymenoptera: Tenthredinidae)	Manitoba, Canada	Ives 1967
Deraeocoris nebulosus	*Corythucha ciliata* (Heteroptera: Tingidae)	North Carolina, USA (laboratory)	Horn et al. 1983b
Deraeocoris nubilus	*Adelges cooleyi* (Homoptera: Adelgidae)	Pennsylvania, USA	Stinner 1975
	Coleophora laricella (Lepidoptera: Coleophoridae)	Wisconsin, USA	Webb 1953
Deraeocoris signatus	*Heliothis* sp. (Lepidoptera: Noctuidae)	Australia	Donaldson and Ironside 1982; see also Zalucki et al. 1986, Chinajariyawong and Harris 1987
Deraeocoris spp.	*Pyrrhalta luteola* (Coleoptera: Chrysomelidae)	District of Columbia, USA	Riley 1893a, Howard 1901b
Hyaliodes beckeri	*Chrysopa externa* (Neuroptera: Chrysopidae)	Brazil	Samways 1979
	Erinnyis ello (Lepidoptera: Sphingidae)	Brazil	Samways 1993
Hyaliodes harti	*Spilonota ocellana* (Lepidoptera: Tortricidae)	Nova Scotia, Canada	Stultz 1955
Mirinae			
Closterotomus biclavatus	*Phaedon cochleariae* (Coleoptera: Chrysomelidae)	Finland	Kanervo 1946
Lygocoris rugicollis	Leaf beetles (Coleoptera: Chrysomelidae)	Finland	Kanervo 1946
Orthotylinae			
Heterotoma planicornis	Butterflies (Lepidoptera)	England	Merrifield 1906
	Phytodecta olivacea (Coleoptera: Chrysomelidae)	England	Dempster 1960
Lopidea sp.	*Rhagoletis suavis* (Diptera: Tephritidae)	Eastern USA	Brooks 1921
Malacocoris chlorizans	*Holocacista rivillei* (Lepidoptera: Heliozelidae)	Italy	Alma 1995
Orthotylus marginalis	*Galerucella lineola* (Coleoptera: Chrysomelidae)	Finland	Kanervo 1946
Orthotylus spp.	*Phytodecta olivacea* (Coleoptera: Chrysomelidae)	England	Dempster 1960
Sericophanes obscuricornis	*Listronotus bonariensis* (Coleoptera: Curculionidae)	Argentina	Lloyd and Ahmad 1971, Dymock 1989
Slaterocoris atritibialis	*Ophraella sexvittata* (Coleoptera: Chrysomelidae)	New York, USA	Messina 1978

Table 14.1. *Continued*

Mirid	Prey	Locality	Reference
Phylinae			
Asciodema obsoleta	*Phytodecta olivacea* (Coleoptera: Chrysomelidae)	England	Dempster 1960
Atractotomus mali	*Eurhodope advenella* (Lepidoptera: Pyralidae)	Germany	Dreyer 1984
	Spilonota ocellana (Lepidoptera: Tortricidae)	Nova Scotia, Canada	Sanford 1964b
Campylomma diversicorne	*Heliothis* spp. (Lepidoptera: Noctuidae)	Australia	Room 1979
Chlamydatus associatus	*Leptinotarsa decemlineata* (Coleoptera: Chrysomelidae)	Ontario, Canada	Harcourt 1964
Microphylellus modestus	*Pyrrhalta luteola* (Coleoptera: Chrysomelidae)	Eastern USA	Knight 1923
Plagiognathus laricicola	*Pristiphora erichsonii* (Hymenoptera: Tenthredinidae)	Manitoba, Canada	Ives 1967
Psallus ambiguus	Syrphids (Diptera: Syrphidae)	England	Petherbridge and Husain 1918
Rhinacloa luridipennis	Lepidopteran pests of cotton (Lepidoptera)	Mexico	Herrera 1987
Rhinacloa, Sthenaridea spp.	Lepidopteran pests of cotton (Lepidoptera)	Argentina, Peru	Wille 1958, Beingolea 1959a, Vergara and Raven 1988, Carpintero and Carvalho 1993; see Schuh and Schwartz (1988) for nomenclatural discussion
Spanagonicus albofasciatus	*Anticarsia gemmatalis* (Lepidoptera: Noctuidae)	Southern USA	Butler 1965, Neal et al. 1972, Irwin et al. 1974, Buschman et al. 1977, Godfrey et al. 1989; see also Musa and Butler 1967
	Microlarinus lareynii, M. lypriformis (Coleoptera: Curculionidae)	California, USA	Goeden and Louda 1976, Goeden and Kirkland 1981

Note: See text for additional examples and discussion.

continually monitored, and chemicals are applied only if the numbers of natural enemies are low and injury is apparent (Frisbie 1983).

European *Lygus* species are sometimes classified as zoophytophages in agroecosystems (e.g., Trojan 1989). At times, these often injurious bugs might be important in limiting populations of the Colorado potato beetle (*Leptinotarsa decemlineata*) (Wheeler 1976b). In my review of predation by lygus bugs, I omitted Trojan's (1968) paper in which *Lygus* species in Poland are implicated as predators of Colorado potato beetle eggs when aphid densities are low. Sterling et al. (1979) listed *Lygus* nymphs and adults among entomophagous arthropods present in a cotton agroecosystem in Texas. In addition, Bisabri-Ershadi and Ehler (1981) observed the western tarnished plant bug (*L. hesperus*) as a frequent predator on the eggs of the western yellowstriped armyworm (*Spodoptera praefica*) in Californian hay alfalfa. When offered only armyworm eggs in the laboratory, an adult plant bug consumed as many as an adult *Nabis* species ($\bar{x} = 104$ eggs/24 hours). Cleveland (1987) observed *L. lineolaris* feeding on the eggs of the tobacco budworm (*Heliothis virescens*) under laboratory conditions (Fig. 14.3), and Bugg et al. (1987) and Bugg and Wilson (1989)

implicated a *Lygus* species as a predator of beet armyworm eggs (*S. exigua*) in California. *Lygus lineolaris* feeds on the egg masses of the fall armyworm (*S. frugiperda*) in Georgia (Bugg et al. 1991), and nymphs and adults of *L. hesperus* feed on the eggs of the pink bollworm (*P. gossypiella*) and sweetpotato whitefly (*Bemisia tabaci*) in Arizona cotton (Hagler and Naranjo 1994, 1996).

Insect Larvae and Pupae

Numerous published records are available for plant bugs as facultative predators of insect larvae and pupae, although the prey sometimes are unspecified. Predation on insect larvae is known for the Palearctic deraeocorines *Deraeocoris lutescens*, *D. punctulatus*, and *D. trifasciatus*; the mirines *Closterotomus fulvomaculatus*, *Megacoelum infusum*, *Miris striatus*, and *Rhabdomiris striatellus*; and the orthotyline *Dryophilocoris flavoquadrimaculatus* (Putshkov 1961a, 1961b; Strawiński 1964a; see also Braimah et al. [1982]). Even when prey organisms are identified, the information provided is often unsubstantial. Prey mostly involve

Fig. 14.3. *Lygus lineolaris*, fifth instar, feeding on egg of *Heliothis virescens* (Lepidoptera: Noctuidae). (Reprinted, by permission of T. C. Cleveland and the Entomological Society of America, from Cleveland 1987.)

plant-surface feeders in the four largest holometabolous orders: Coleoptera, Lepidoptera, Diptera, and Hymenoptera. Mirids sometimes attack gall-forming and leafmining larvae and feed on the larvae and pupae of beneficial groups such as coccinellids, syrphids, and coniopterygids (e.g., Collyer 1952). The latter type of interaction—a predacious mirid eating an insect predator that is a potential competitor—can be categorized as intraguild predation (Polis et al. 1989; Polis and Holt 1992; Rosenheim et al. 1993, 1995). Plant bugs are frequently overlooked as predators of mummified aphids, causing death of the aphidiid parasitoids within (discussed below under Hymenoptera).

The tendency of predatory mirids to feed on various natural enemies and other organisms during times when their preferred prey are scarce may keep them from dying or emigrating (Collyer 1953c), thus promoting their local persistence. Because mirids feed as predators of, and scavengers on, various caterpillars, they might disseminate baculoviruses of lepidopteran larvae, although this has not been documented for the Miridae. Certain other predatory heteropterans—nabids, pentatomids, and reduviids—can serve as baculovirus vectors after they feed on infected caterpillars (D. J. Cooper 1981, Moscardi and Correa Ferreira 1985, Young and Yearian 1987).

Dempster (1960) studied mirid predation on the broom-feeding chrysomelid *Phytodecta olivacea*, using the precipitin test (see Frank [1979]) on more than 7,000 individuals. The adults and older nymphs of five of the six mirid species living on broom—four orthotylines and a phyline—attacked young beetle larvae, although alternative prey formed a higher proportion of their diet. The predation rates varied among the broom mirids, and availability of early-stage larvae differed for the various plant bug species, which hatch in succession and therefore show different patterns of seasonality.

Other records of chrysomelid predation by mirids include *Neurocolpus nubilus*, which attacked a *Chrysomela* pupa on willow in Missouri (Froeschner 1949) and *Lopidea robiniae*, which feeds on leafmining larvae on black locust in Pennsylvania (Wheeler and Snook 1986). On a sycamore tree in Maryland, Neal (1989) observed attempted predation by an unidentified mirid on a second- or third-instar chlamisine chrysomelid (*Neochlamisus platani*), which tilted its case toward the bug in response to the threat. *Adelphocoris lineolatus* was observed feeding on the larvae of the asparagus beetle (*Crioceris asparagi*) in France (Lucas 1888), and Kanervo (1946) recorded chrysomelid predation by other European mirid species (see also Jolivet [1950]). *Lygus pratensis* feeds on larvae of the Colorado potato beetle (*Leptinotarsa decemlineata*) in Poland (Kaczmarek 1955). *Deraeocoris erythromelas* preys on the larvae of *Chrysomela vigintipunctata* in Japan, as does *D. brachialis* on the larvae of the same species and *Gonioctena japonica* (Yasunaga and Nakatani 1998).

Three mirid species occasionally attack the larvae or pupae of the alfalfa weevil (*Hypera postica*) in New York alfalfa fields: *Adelphocoris lineolatus*, *Lygus lineolaris*, and *Plagiognathus politus* (Wheeler 1974b). *Deraeocoris pallens* is an important predator of alfalfa weevil larvae in Iran (Vojdani and Doftari 1963). Schewket Bey (1930) illustrated a nymph and an adult of the Palearctic *Dicyphus errans* feeding on a bark beetle (?scolytid) larva. In Panama, the cylapine *Trynocoris lawrencei* might feed on the larvae of ciid beetles inhabiting shelf fungi (Herring 1976; see also Chapter 13).

In English apple orchards, *Blepharidopterus angulatus* feeds on coccinellid larvae when preferred prey are scarce (Collyer 1952). *Phytocoris tiliae* feeds on coccinellid pupae in Britain (Southwood and Leston 1959, Groves 1976), and *Deraeocoris ruber* attacks the pupae of the coccinellids *Adalia bipunctata* and *Coccinella septempunctata* (Disney et al. 1994). D. J. Schotzko (pers. comm.) observed numerous *Lygus hesperus* adults preying on the pupae of a *Hippodamia* species (probably *H. convergens*) in Idaho. Niemczyk (1966a) noted that the larvae of the weevil *Anthonomus pomorum* occasionally serve as prey of *Psallus ambiguus* in Polish apple orchards.

Mirids are common but often underappreciated predators of neonate caterpillars associated with cotton and other field crops and with apple and other fruit trees. At times, predation by plant bugs can result in considerable mortality, as evidenced by the elimination of entire groups of hatchlings of a nymphalid butterfly, *Euphydryas gilletti*, in Colorado (Ehrlich 1984). But the effects of predacious mirids on lepidopteran population dynamics are largely unknown.

Campylomma liebknechti sometimes attacks the larvae of heliothentine (often spelled "heliothine") noctuids in Australian cotton fields (Room 1979, Malipatil 1992) and the young larvae of several lepidopteran pests on grain legumes in Australia (Shepard et al. 1983). *Hyalochloria denticornis* preys on the tobacco budworm (*Heliothis virescens*) in Peruvian cotton (Wille 1951), as do *Ceratocapsus dispersus* and *C. mariliensis* on the same species (Carvalho et al. 1983) and on neonate larvae of the cotton leafworm (*Alabama argillacea*) in Brazil (Gravena and Pazetto 1987, Gravena and da Cunha 1991). *Rhinacloa forticornis*, a phyline, preys on leafmining larvae of the lyonetiid *Bucculatrix thurberiella* in Peruvian cotton (Herrera and Alvarez 1979), and *C. dispersus* can be reared in the laboratory on *B. thurberiella* larvae (Encalada and Viñas 1989). Corn planted in cotton fields in Peru supports populations of predators, such as the pilophorine *Sthenaridea carmelitana*, which often move into cotton fields to feed on lepidopteran eggs and larvae (Wille 1958). In the Philippines, *Nesidiocoris tenuis* feeds preferentially on neonate larvae (<8 hours old) of the cutworm *Spodoptera litura* in tobacco, usually keeping this pest's populations below damaging levels. Predatory efficiency of the bug is not significantly affected by the availability of other prey organisms (Torreno 1994). Nymphs of all stages and adults of *N. tenuis* prey on neonate larvae of the noctuid *Helicoverpa armigera* on tomato plants in the Philippines (Libutan and Bernardo 1995).

The North American *Lygus lineolaris* feeds on the first three instars of *Heliothis virescens* under test conditions (Cleveland 1987). In deriving relative efficiency indices for arthropod predators of heliothentine larvae (and eggs), Ables et al. (1983) included mirids among various entomophagous heteropterans.

Mirids are common predators of lepidopteran larvae on apple trees (Braimah et al. 1982; Table 14.2). *Deraeocoris nebulosus*, for example, feeds on small codling moth larvae in Nova Scotia (Stultz 1955). Early-instar codling moths are attacked by other predatory plant bugs (MacLellan 1962, 1972), and *D. fasciolus*, *Hyaliodes harti*, and *Plagiognathus obscurus* greatly reduce its numbers in apple orchards in eastern Canada (Shteynberg 1962). Lord (1956) and MacLellan (1963) reported *D. nebulosus* and *H. harti* feeding on eyespotted bud moth (*Spilonota ocellana*) larvae in Nova Scotia. *Deraeocoris nebulosus* feeds on the leafroller

(tortricoid) complex associated with apple trees in Virginia (Parrella et al. 1981). In apple orchards in Virginia and Quebec, *H. harti*, *H. vitripennis*, and a *Lygus* species (possibly a species of *Lygocoris*) prey on the larvae and pupae of the gracilariid leafminers *Lithocolletis blancardella* and *L. crataegella* (Beckham et al. 1950, LeRoux 1960, Pottinger and LeRoux 1971).

Phytocoris tiliae frequently preys on the larvae of the winter moth (*Operophtera brumata*) in German apple orchards (Speyer 1933), and several British mirids feed on the winter moth on unspecified hosts (Southwood and Leston 1959). In Australia, precipitin tests showed that *Campylomma liebknechti*, the apple dimpling bug, and *Orthotylus australianus* feed on the larvae of the light apple brown moth (*Epiphyas postvittana*) (MacLellan 1973; cf. Danthanarayana 1983). Collyer and van Geldermalsen (1975), in studies on integrated control of New Zealand apple pests, reported *Sejanus albisignatus* feeding on young codling moth larvae in the laboratory; some predation on larvae also occurs in the field.

In Europe, *Atractotomus mali* is a well-known predator of the larvae and pupae of the yponomeutid webworms *Yponomeuta malinella* and *Y. padella* (Giard 1900; Pommerol 1900, 1901; Finţescu 1914; Junnikkala 1960; Zeletzki and Rinnhofer 1966). This predator also invades the webs of another yponomeutid, *Paraswammerdamia lutarea* (Dreyer 1984). *Deraeocoris olivaceus* similarly invades webworm nests to attack the larvae (Strawiński 1964a). Other Palearctic mirids sometimes also feed on webworm larvae, or at least are found in their webs (Gulde 1921). Mirids are uncommon inhabitants of the colonial webs of *Hyphantria cunea* in Canada (Morris 1972).

In addition to feeding on the larvae of species associated with fruit and hardwood trees, mirids also attack the larvae of conifer-feeding lepidopterans. Thorpe (1933) suspected *Deraeocoris ruber* of preying on larch casebearer (*Coleophora laricella*) larvae or pupae in England. In the western United States, *Phytocoris* species attack larch casebearer larvae and those of the Douglas-fir tussock moth (*Orgyia pseudotsugata*) (Stonedahl 1983b, 1988). The adults and nymphs of *P. calli* and *P. nigrifrons* feed on the young larvae of the tussock moth in white fir forests of northern California. *Phytocoris nigrifrons* adults tested positive for tussock moth predation in an enzyme-linked immunosorbent assay (ELISA), even at low prey densities (Stonedahl 1988).

DIPTERA

Arčanin and Balarin (1972) included the phyline *Psallus variabilis* among natural enemies that attack dipteran adults and larvae in the apple orchards of Croatia. The additional examples of mirids as predators of Diptera that are discussed here involve larvae or pupae; predation on adults is treated later in this chapter.

Plagiognathus politus feeds on the larvae and pupae of the agromyzid leafminer *Liriomyza trifoliearum* on

Table 14.2. Examples of mirids as facultative predators of lepidopteran larvae or pupae

Mirid species	Prey	Locality	Reference
Bryocorinae			
Engytatus modestus	Heliothentines and hornworms (Noctuidae, Sphingidae)	Hawaii, USA, and elsewhere	Tanada and Holdaway 1954
Nesidiocoris tenuis	Noctuid pests of tobacco (Noctuidae)	Indonesia	Schweizer 1939, Kalshoven 1981
Deraeocorinae			
Deraeocoris albigulus	*Choristoneura pinus* (Tortricidae)	Michigan, USA	Hussey 1954b, Allen et al. 1970
Deraeocoris barberi	*Choristoneura lambertiana* (Tortricidae)	Northwestern USA	McGregor 1970
Deraeocoris fasciolus	*Argyrotaenia mariana* (Tortricidae)	Nova Scotia, Canada	Gilliatt 1937
Deraeocoris sp.	*Coleophora laricella* (Coleophoridae)	Western USA	Denton 1979
Kundakimuka queenslandica	*Xylorycta luteodactella* (Oecophoridae)	Australia	Cassis 1995
Romna capsoides	"Small green caterpillar"	New Zealand	Myers 1926, Eyles and Carvalho 1988a
Mirinae			
Lygocoris communis	Tussock moth, green fruitworm (Lymantriidae, Noctuidae)	Nova Scotia, Canada	Brittain 1916a
Miris striatus	*Tortrix viridana* (Tortricidae)	Germany	Gulde 1921
Phytocoris americanus	*Lambdina f. fiscellaria* (Geometridae)	Ontario, Canada	Schedl 1931
Rhabdomiris striatellus	*Tortrix viridana* (Tortricidae)	Austria, Czechoslovakia	Jahn 1944
	Forest lepidopterans	Europe	Starý et al. 1988
Taedia hawleyi	*Biston betularia cognataria* (Geometridae), *Hypena humuli* (Geometridae), *Malacosoma americanum* (Lasiocampidae), *Nematocampa filamentaria* (Geometridae)	New York, USA	Hawley 1917
Taedia scrupea	Caterpillars on grape	New York, USA	Martinson et al. 1998
Orthotylinae			
Heterocordylus malinus	*Paleacrita vernata* (Geometridae)	Illinois, USA	Balduf 1943
Malacocoris chlorizans	*Holocacista rivillei* (Heliozelidae)	Italy	Alma 1995
Phylinae			
Psallus perrisi	*Archips crataeganus* (Tortricidae)	Czechoslovakia	Hochmut 1964
Psallus sp.	*Homoeosoma vagella* (Phycitidae)	Australia	Ironside 1970
Salicarus roseri	Forest lepidopterans	Europe	Starý et al. 1988

Note: See text for additional examples and discussion.

alfalfa in New York (Wheeler 1974b), as does *Lygus lineolaris* on the larvae of the alfalfa blotch leafminer (*Agromyza frontella*) (Wheeler 1976b). *Adelphocoris lineolatus* and *L. lineolaris* prey on all instars of the alfalfa blotch leafminer in eastern Canada, with facultative predation accounting for 7–30% of total larval mortality in study plots in Quebec (Harcourt et al. 1987).

Engytatus modestus attacks leafminer larvae on tomato plants in California. Late instars and adults of this dicyphine feed on the larvae of a serpentine leafminer (*Liriomyza trifolii*) within the mesophyll of chrysanthemum or tomato leaves (Plate 23; Parrella et al. 1982). In studies on the biocontrol potential of *E. modestus*, Parrella and Bethke (1982) determined that nymphs having access only to tomato cuttings do not complete development, whereas those given a cutting and eggs of the pink bollworm develop in 15 days. Development requires only 12 days when first instars

are provided a tomato leaflet containing about seven larvae of the agromyzid *L. sativae*. Nymphs of all stages pierce the leaf cuticle to feed on the larvae within the mines. The mortality was nearly 90% for *L. sativae* in treatment leaflets—that is, exposed to the predacious mirid—but only 6% for those in control leaflets. The addition of lepidopteran eggs also increases the longevity and fecundity of the plant bug. Although the low nitrogen content of tomato leaflets (about 2%) might adversely affect nymphal development, longevity, and fecundity, Parrella and Bethke (1982) suggested that *E. modestus* has stronger carnivorous tendencies than previously thought and that its predation on tomato pests might outweigh any injury to the host stems. Preliminary data indicate that *E. modestus* does not induce stem lesions (see Chapter 9) on chrysanthemum. The nymphs and adults of another dicyphine, *Nesidiocoris tenuis*, readily attack the larvae of the serpentine leafminer (*L. trifolii*) on greenhouse-grown gerbera in Sicily (Nucifora and Calabretta 1986). *Dicyphus tamaninii* also accepts *L. trifolii* larvae under laboratory conditions (Salamero et al. 1987).

During the 1920s, the presence in New Zealand of a pear midge (*Dasyneura pyri*), which had been accidentally introduced from the Palearctic region, prompted a search for European natural enemies that could be released in New Zealand for the biological control of this pest. In France, Myers (1927a) observed *Pilophorus perplexus* piercing rolled leaves to feed on the midge larvae. The mirid's impact on European populations of the cecidomyiid was considered greater than all other mortality factors combined. On the basis of Myers's (1927a) observations, *P. perplexus* was supposed to have received further evaluation as a potential biological control agent of the midge in New Zealand (Muggeridge 1929). No additional studies were conducted (Barnes 1948, Berry and Walker 1989) because it was realized that the mirid apparently had already been unintentionally introduced into New Zealand with consignments of hymenopteran parasitoids of the midge that were shipped from England and France (Miller 1927).

The native phyline *Sejanus albisignatus* may also be a natural enemy of the pear midge in New Zealand. This plant bug will feed on larvae removed from a curled pear leaf, but it is not known if predation occurs in nature (Dumbleton 1938).

Gagné (1989) included mirids (unspecified taxa) among the predators that attack plant-feeding Cecidomyiidae in North America, noting that they feed on midge pupae breaking out of their galls. On grape vines in Pennsylvania, *Taedia scrupea* adults probe and feed on the fleshy galls caused by the cecidomyiid *Lasioptera vitis* or a related species, and in the laboratory, they attack larvae removed from the galls (A. J. Musa and G. L. Jubb Jr., pers. comm.). Herrera (1965) and Vergara and Raven (1988) recorded predation on cecidomyiid larvae by *Rhinacloa* species in Peru.

The phorids *Phalacrotophora berolinensis* and *P. fasciata* parasitize the pupae of coccinellids such as *Adalia bipunctata* and *Coccinella septempunctata*. In England, when the nymphs and adults of *Deraeocoris ruber* prey on the coccinellid pupae, they kill not only the host coccinellids but also the phorid larvae developing within parasitized coccinellid pupae (Disney et al. 1994). *Deraeocoris maoricus* possibly preys on the larvae of the phorid *Megaselia impariseta* (also coleopteran and lepidopteran larvae) living in bumble bee colonies in New Zealand (R. P. Macfarlane, pers. comm.). This would represent an unusual record of mirid predation, although several heteropterans, mainly reduviids, are known from honey bee hives (Caron 1978).

HYMENOPTERA

Predators kill early-stage hymenopteran parasitoid larvae when they feed on the still-active aphids. Parasitized aphids eventually become sluggish as the late-stage larvae of parasitic wasps (Aphidiidae) consume host contents; the host dies when the fourth-instar aphidiid larva attacks the vital organs. The last-stage larva of *Aphidius* species cuts a hole in the lower side of the host skin and, with salivary secretions, attaches it to the leaf surface (e.g., Starý 1962, 1970, 1988). All that remains of the dead aphid is the hardened, parchmentlike outer shell or exoskeleton, the so-called mummy.

Starý (1988) noted that although predators generally do not distinguish healthy aphids from live, parasitized individuals, sucking predators seem to ignore mummified aphids. This is not always so. Some heteropterans, especially mirids, feed opportunistically on the last-instar larvae, prepupae, or pupae of aphidiid wasps within mummified aphids; this interaction is a type of intraguild predation (Ehler 1996).

Morley (1916), perhaps the first to report a true bug attempting to puncture an aphid mummy, observed an unsuccessful feeding attempt by an anthocorid. Successful predation by anthocorids on parasitized aphids in Great Britain is now known (Smith 1966, Dixon and Russel 1972). Mirids can also penetrate aphid mummies to feed on late-stage primary parasitoids (and probably also hyperparasitoids of various stages), functioning as ecological equivalents of secondary parasitoids (Wheeler 1974b), but presumably lacking the complex interactions occurring between some hyperparasitoids and aphids (e.g., Boenisch et al. 1997). Predatory plant bugs also lack the specificity and synchrony typical of associations of a hyperparasitoid with its primary parasitoid host.

Five mirids fed on the pea aphid parasitoids *Aphidius* and *Praon* species in New York alfalfa fields: *Adelphocoris lineolatus* (9 observations), *Chlamydatus associatus* (1), *Lopidea instabilis* (1), *Lygus lineolaris* (14), and *Plagiognathus politus* (21) (Wheeler et al. 1968, Wheeler 1974b). A mirid nymph or adult typically inserted its stylets several times into a mummy, apparently probing for the parasitoid before beginning to feed for an extended period. Confirmation of feeding on pea aphid parasitoids, similar to that obtained for a berytid that preys on braconid prepupae and pupae (Kester and Jackson 1996), is needed.

Glen (1973) reported that second instars of *Blepharidopterus angulatus* develop to the fifth instar when fed a diet of mummified lime aphids (*Eucallipterus tiliae*). The mirine *Garganus fusiformis* fed on the mummy of an aphid on corn leaves (probably the corn leaf aphid [*Rhopalosiphum maidis*] in Pennsylvania) (pers. observ.). Khattat and Stewart (1977) credited Lindquist and Sorensen (1970) with the observation that *L. lineolaris* preys on mummified aphids, but the latter paper reports the attack on live aphids only.

Mirids, therefore, can be included among the mortality agents of aphid parasitoids. Various secondary parasitoids or hyperparasitoids have been assessed as natural enemies of aphidiid wasps (Luck et al. 1981, Horn 1989). The impact of mirids on the population dynamics of primary parasitoids warrants similar studies, although in the case of chrysopid predation on mummified mealy plum aphids (*Hyalopterus pruni*), the effects on primary aphid parasitoids were scant because the larvae of the neuropteran predator fed mostly on mummies containing hyperparasitoids (Al-Rawy et al. 1969). The effect of hyperparasitoids on primary parasitoid populations is usually considered detrimental, but at times secondary parasitism might be beneficial by dampening population fluctuations of a primary parasitoid and maintaining balance in a parasitoid-host population (Sullivan 1988, Boenisch et al. 1997).

Mirids rarely are reported to feed on the larvae or pupae of other Nearctic hymenopterans. Except for published records of predation on aphid parasitoids, there apparently is only the observation by Culliney et al. (1986) in New York of *Lygus lineolaris* nymphs and adults attacking the cocoons of the braconid *Apanteles glomeratus*, a parasitoid of the imported cabbageworm (*Pieris rapae*). This is an additional example of direct intraguild predation (*sensu* Kester and Jackson 1996) in the Miridae. An unpublished laboratory observation is that of *Lygus* species preying on their parasitoids, *Peristenus* species. Last-stage larvae of these euphorine braconids are attacked in rearing cages after they drop from their lygus bug hosts to pupate (R. W. Fuester, pers. comm.). Clancy and Pierce (1966) anticipated the possibility of lygus bugs preying on their braconid parasitoids, for their rearing methods allowed mature parasitoid larvae to drop through a wire mesh screen "where they were protected from attack of *Lygus* spp."

Deraeocoris ruber sometimes feeds on sawfly larvae in Europe (Strawiński 1964a); its prey include the oak-feeding tenthredinid *Caliroa varipes* (Mijušković 1966). In England, the mainly phytophagous *Notostira elongata* was captured with the larva of a gooseberry sawfly (*Pteronidea ribesii*) impaled on its stylets (Groves 1977). Murray and Solomon (1978) used electrophoretic techniques to identify the larvae of the European alder leafminer (*Fenusa dohrnii*) as prey of *Blepharidopterus angulatus* in England. The Palearctic orthotyline *Cyllecoris histrionius* feeds on larvae in oak galls (Kullenberg 1944, Southwood and Leston 1959), whose inhabitants likely were cynipid wasps. *Hallodapus montandoni*, a myrmecomorphic, ground-dwelling phyline, apparently preys on ant pupae in Europe (Southwood and Leston 1959). Gulde (1921) considered the stenodemine *Myrmecoris gracilis* to be myrmecophagous in Germany, but ant predation by this species is doubtful (Cobben 1986).

Mites

Mites, especially spider mites (Tetranychidae), are consumed by numerous mirids. Although some species appear to subsist mainly on mites, others (e.g., *Deraeoc-*

oris pallens, *D. punctulatus*) cannot complete their life cycle on a mite diet (Zavodtshikova 1974, Ghavami et al. 1998). Nymphs, particularly early instars, that consume mites can sometimes be recognized by the reddish brown patch on their abdomen, which represents the red body fluids of their tetranychid prey. In contrast, aphid-feeding nymphs often have a green abdomen (Collyer 1952, Morris 1965). In addition to spider mites, mirids also consume eriophyoids and other phytophagous mites, as well as predatory mites such as phytoseiids.

EUROPEAN RED MITE AND OTHER TETRANYCHIDS ON APPLE

As mite predators, mirids have been studied most intensively as enemies of the European red mite (*Panonychus ulmi*). Heteropterans generally are considered less efficient natural enemies of spider mite pests of fruit trees than are certain predatory mites, species of the coccinellid genus *Stethorus*, spiders, or thrips (Croft and Brown 1975). Mirids, though, are sometimes cited as the most important insect predators of tetranychids (Hussey and Huffaker 1976).

At least 17 mirid species prey on *P. ulmi* of various stages in England (Collyer 1953a, 1953b; Massee 1956; see also reviews by Berker [1958] and Chazeau [1985]). Ten mirid species feed on the European red mite in Finland (Kanervo 1962), and eight species are enemies of this mite in Belarus (Sidlyarevitsch 1965, 1982). Numerous field and laboratory studies, including paper chromatography to detect predation (Putman 1965), have been conducted on the mirid fauna of Canadian apple orchards (McMurtry et al. 1970). Lord (1971) determined the consumption rates of six mirids that attack the eggs of the European red mite in Nova Scotia (see also Braimah et al. [1982], Kelton [1983], Arnoldi et al. [1992]). Among regional surveys of the predacious mirid fauna of apple orchards in the United States are those by Holdsworth (1968, 1972) for Ohio, Childers and Enns (1975) for Missouri, and Parrella et al. (1981) for Virginia.

Heteropterans, including predacious mirids in orchards, were highly sensitive to certain pesticides formerly used against key apple pests (e.g., Clancy and Pollard 1952; MacPhee and Sanford 1954, 1956, 1961; Clancy and McAlister 1956; Reed 1959; Neilson et al. 1970; Sanford and Herbert 1970; Weires and Smith 1979; Bulyginskaya and Kalinkin 1985). In the 1950s, the populations of predatory mirids increased dramatically in a Nova Scotian orchard when a selective insecticide, ryania, replaced DDT (Pickett 1959). More recently, average toxicity ratings computed for the 15 most common families of natural enemies of agricultural pests indicated that mirids are especially tolerant to the major classes of pesticides: insecticides, acaricides, fungicides, and herbicides (Theiling and Croft 1988; see also Croft [1990], Mizell and Sconyers [1992]). Some synthetic pyrethroids have a low toxicity for predacious mirids such as *Deraeocoris nebulosus*, *Hyaliodes harti*, and *H. vitripennis* (Croft and Whalon 1982, Theiling and Croft 1988, Croft 1990), and the use of these pesticides appears to have favored increases of *H. vitripennis* in Pennsylvania orchards (Croft 1982). Similarly, the use of selective insecticides and fungicides in IPM programs in English apple orchards helps conserve predacious mirids (Easterbrook et al. 1985). The increasing use of certain synthetic insecticides to control apple pests might allow these predators to be exploited better in IPM. No predatory mirid species can be expected to keep phytophagous mites at consistently low levels, but the action of a complex of species throughout the growing season might provide reliable suppression of tetranychids (Arnoldi et al. 1992).

The phyline *Plagiognathus politus* can respond to increases in mite populations on apple trees in Ohio and is a potentially important enemy of the European red mite (Holdsworth 1968). But mirids, including *P. politus*, proved uncommon in subsequent surveys of the predator complex in Ohio apple orchards, possibly because of changes in pesticide use since the 1960s (Welty 1995). The Holarctic phylines *Atractotomus mali* and *Campylomma verbasci* attack this mite but also can injure apple fruit (see Chapter 12). Thistlewood and McMullen (1989) noted that *C. verbasci* nymphs are sometimes associated with large populations of the European red mite in British Columbia, where this mite serves as common prey of the bugs. In some years, *H. harti* suppresses the European red mite and brown mite (*Bryobia rubrioculus*) in Nova Scotian apple orchards, but in other years it allows mite populations to fluctuate widely (Sanford and Lord 1962). Densities of *H. harti* in orchards are strongly influenced by abundance of the European red mite (Lord 1971). Nymphs of this mirid (cited as *H. vitripennis*; see Kelton [1983]) are more persistent predators than the adults (Gilliatt 1935).

Hyaliodes harti was imported from Nova Scotia to New Zealand to help control the European red mite. Although the females oviposited, eggs did not hatch the following season, perhaps because of warm winter temperatures (Walker et al. 1989).

Horsburgh (1969) evaluated the effectiveness of *H. vitripennis* against the European red mite in Pennsylvania, noting that the bugs aggregate in areas of apple trees where mites are most abundant. Water sprouts or adventitious shoots should be left on apple trees until mid-June to allow the eggs of *H. vitripennis* to hatch (Asquith and Horsburgh 1969, Hull et al. 1983). Studies in Quebec suggest that this predator significantly suppresses populations of the European red mite in apple orchards (Bostanian et al. 2000). *Deraeocoris brevis* might substantially reduce the numbers of another spider mite pest of apple, *Tetranychus mcdanieli*, in the Pacific Northwest (Hoyt 1969); in Canada, this mite is also attacked by *C. verbasci* (Thistlewood and Smith 1996). *Campylomma liebknechti* is a voracious predator of the twospotted spider mite (*T. urticae*) and the brown mite in Australian apple orchards (Readshaw 1975).

Gilliatt (1935) regarded the orthotyline *Blepharidopterus provancheri* as one of the most important European red mite predators in Nova Scotia. This mirid was also collected from tetranychid-infested apple leaves in Manitoba, and the nymphs were reared on spider mites (Robinson 1952). The nymphs and adults of *Paraproba capitata* feed on European red mites in Nova Scotia, but this predatory bug is not considered one of this pest's more important natural enemies (Gilliatt 1935).

The Holarctic *Blepharidopterus angulatus*, an orthotyline known as the black-kneed capsid or mirid in Britain, has received the most attention among mirids that prey on mites. It attracted notice in the mid-1940s when some spray trials for control of European red mites in England had to be discontinued because of heavy predation by this plant bug (Austin and Massee 1947). The European red mite was generally insignificant in English apple orchards until the 1940s when DDT and other broad-spectrum organochlorine insecticides were introduced and eliminated *B. angulatus* (Davies 1988). Its role in limiting mite populations in England has been thoroughly investigated (Muir 1966b; see also summaries by Moreton [1958], Swan [1964], and Huffaker et al. [1970]). During a seven-year study, this mirid was the principal biological factor responsible for changes in mite densities. Fluctuations in mite numbers are related to mirid densities, and populations of the predator respond to changes in prey numbers. Females of *B. angulatus* tend to lay more eggs when mite numbers are high (Muir 1965a, 1965b).

In other studies, *B. angulatus* proved to be an important, though unreliable, mite predator that allows prey densities to fluctuate widely from year to year (Collyer 1964b). In 1947 and 1948, for instance, it alone so reduced mite populations that other predators did not become numerous, but effective long-term suppression did not occur (Collyer 1953c). Lime sulfur and other fungicides are toxic to the mirid, and the full spray programs used in British apple orchards during the 1950s often eliminated predator populations, as did parathion and other highly toxic insecticides (Collyer and Massee 1958, Collyer and Kirby 1959, Reed 1959, Muir 1965a). This voracious feeder is one of the first predators to colonize orchards sprayed with selective pesticides (Solomon 1982).

To determine the total number of European red mites consumed under favorable conditions by an individual of *B. angulatus*, Collyer (1952) reared the bugs from the first instar and continued her observations through the adult's normal life span. On average, each female consumed more than 3,200 adult female European red mites in the summer of 1947 and more than 4,200 in similar laboratory studies conducted in 1948 (see also Jeppson et al. 1975). Because overwintered eggs of the predator hatch 4–5 weeks later than those of its prey, predation pressure is most effective on second-generation mites; mite injury often occurs before the mirid exerts control. This plant bug can migrate into British apple orchards harboring only low levels of European red mites (e.g., 19/leaf), and a potential for rapid colonization helps compensate for its inability to respond immediately to changing prey densities due to univoltinism (Collyer 1953c, Muir 1965a, Solomon 1975). *Blepharidopterus angulatus* is often abundant on alder, and the planting of alder windbreaks near orchards can aid its colonization of apple trees when mite densities are as low as two per leaf (Solomon 1981, 1982). By carefully pruning the alders serving as windbreaks, this predator's colonization of orchards can be manipulated in programs of integrated mite management (Gange and Llewellyn 1989).

Another well-studied mite predator is the Palearctic phyline *Psallus ambiguus*, which feeds on *Panonychus ulmi* of all stages (Collyer 1953b). Laboratory studies demonstrating that this predator will not reproduce on a mite diet suggest that mites serve only as alternative prey (Niemczyk 1966a, 1968).

The orthotyline *Malacocoris chlorizans*, the delicate apple capsid, was once the most important predator of the European red mite and other phytophagous mites on fruit trees in Switzerland (Geier and Baggiolini 1952). Foschi and Carlotti (1956) discussed its predation on European red mite eggs in Italy, and Niemczyk (1963) said it was the most abundant heteropteran predator in Polish apple orchards. In England, it is less able to survive at low prey densities than is *B. angulatus* (Solomon 1982), and its effectiveness is impaired by a sensitivity to insecticides in Norwegian apple orchards (Hesjedal 1986). Fauvel (1999) provides additional references to mite predation by *M. chlorizans*.

Campylomma verbasci, whose injury to apple crops was discussed earlier (see Chapter 12), was first noticed on English fruit trees during the 1950s (Collyer and Massee 1958). Applications of parathion during June and July killed populations of this facultatively predacious phyline in a commercial apple orchard, but it recolonized late in the season after European red mite populations had built up and caused injury (Collyer 1953c). A single nymph can destroy an average of 580 female mites during its development (Niemczyk 1978). This bug, however, is often low rated as a mite and aphid predator (Thistlewood and Smith 1996).

Other plant bugs that feed on the European red mite of various stages include *Deraeocoris lutescens* in British and French apple orchards. This mirid hibernates as an adult and attacks winter eggs of the mite after the leaves fall (Collyer 1953c, Fauvel and Atger 1981). Fauvel and Atger thought *Orthotylus nassatus* preys on the European red mite in France, mainly on the basis of its spatial distribution in an orchard and on population dynamics curves. In Belarus, a fifth-instar *Pilophorus perplexus* consumed 50 European red mites per day and an adult female fed on 80 mites per day (Sidlyarevitsch 1982). Another mite predator is *Sejanus albisignatus*, whose nymphs and adults feed on European red mites and *Bryobia* species in New Zealand fruit trees, but this phyline is found only occasionally in orchards (Collyer 1964a). Under an integrated spray program, it became one of the more abundant predators on apple (Collyer and van Geldermalsen 1975, Collyer 1976).

OTHER TETRANYCHID MITES AS PREY

Additional examples of mirid-tetranychid associations include a second-instar *Phytocoris neglectus*, collected from Hinoki false cypress in Pennsylvania, that was reared in the laboratory on the spruce spider mite (*Oligonychus ununguis*) (Wheeler and Henry 1977). The phyline *Atractotomus magnicornis* feeds on spruce spider mites in Europe (Schneider 1962, Thalenhorst 1962), and circumstantial evidence suggests that this adventive plant bug (Wheeler and Henry 1992) preys on spruce spider mites in eastern North America. Its numbers are consistently higher on mite-infested Norway and white spruce, and the bugs can be detected by locating mite-discolored areas of host conifers and tapping chlorotic foliage over a net or a white sheet of paper (pers. observ.).

Rhinacloa forticornis was observed feeding on twospotted spider mites in a cotton field in South Carolina; one individual consumed an average of 60 mites per day during eight days' observation in the laboratory (McGregor and McDunough 1917). *Campylomma diversicorne* and *C. verbasci* feed on spider mites in the cotton fields of Turkmenia. Each species can consume more than 100 prey daily in the laboratory (Sugonyaev and Kamalov 1976). *Campylomma agalegae* preys on a *Tetranychus* species in the Agalega Islands (Miller 1956), and *C. liebknechti* appears to be a major predator of *Bryobia rubrioculus* in Australian apple orchards (Readshaw 1971).

Foglar et al. (1990) studied the functional response of the Palearctic dicyphine *Macrolophus melanotoma* (as *M. caliginosus*) to the twospotted spider mite (Fig. 14.4) and the green peach aphid (*Myzus persicae*). Females show a Type II functional response to prey density—that is, curvilinear with a negatively accelerating rise to an upper limit, which often results in inverse density-independent mortality (Holling 1959, Hassell 1978, Luff 1983, Murdoch 1990, O'Neil 1990, Döbel and Denno 1994). When both prey species are offered simultaneously, the mirid shows a stronger preference for aphids (see also Hansen et al. 1999). Although *Dicyphus hesperus* can complete its life cycle on a diet of twospotted spider mites, this omnivorous mirid develops faster and adults are larger when nymphs are reared on whiteflies (Gillespie et al. 1998, McGregor et al. 1999). *Tetranychus turkestani* is consumed by, but is not a preferred prey of, *M. melanotoma* (Fauvel et al. 1987).

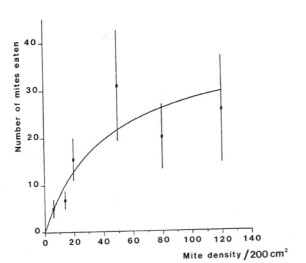

Fig. 14.4. Functional response *of Macrolophus melanotoma* to increasing densities of twospotted spider mite (*Tetranychus urticae*) adults, under standardized conditions of 22°C, 80% relative humidity, and 16-hour photoperiod. Data are recalculated for a 24-hour period; means ± 95% confidence intervals. (Reprinted, by permission of J. C. Malausa and the International Organization for Biological Control [IOBC], from Foglar et al. 1990.)

ERIOPHYOID MITES AS PREY

Mirids appear to be infrequent predators of eriophyoids, although the life histories and natural enemies of few Eriophyoidea are thoroughly studied. Observing predation on these mites is difficult because of their small size and the galls that protect many species.

Glen (1977b) concluded that predatory mirids on apple trees in England do not feed on the apple rust mite (*Aculus schlechtendali*), probably because individuals are too small to be suitable prey. In addition, Lindquist (1986) remarked that mites such as eriophyoids and tarsonemids probably are too small to be used as prey by most arthropodan predators. But Herbert and Sanford (1969) reported that *Campylomma verbasci*, *Compsidolon salicellum*, and *Hyaliodes harti* feed voraciously on the apple rust mite in laboratory trials, and suggested that this eriophyid provides alternative food for various mirids on apple trees in Nova Scotia. *Hyaliodes vitripennis* will also feed on *A. schlechtendali* (Horsburgh 1969), and *Campylomma verbasci* feeds on the pearleaf blister mite (*Eriophyes pyri*) during its active stages in British Columbia (McMullen and Jong 1970, Thistlewood and Smith 1996). *Deraeocoris ruber* is a possible predator of the currant bud mite (*Cecidophyopsis ribis*) on black currant in England (Mumford 1931). In French studies on predators of the eriophyid that induces ash spangle gall, mirids were collected in galls inhabited by predatory species (Fauvel et al. 1975); the mirids probably also prey on the eriophyid.

Mirids will attack predacious mites that feed on tetranychids. Horsburgh (1969) reported predation by *Hyaliodes vitripennis* on the stigmaeid *Zetzellia mali*. *Blepharidopterus provancheri* feeds on the phytoseiid *Typhlodromus pyri* in Nova Scotian apple orchards (Herbert 1962). Phytoseiids do not occur in high numbers on apple trees in Nova Scotia, probably because they are fed on by other predators, mainly mirids (Hussey and Huffaker 1976). *Blepharidopterus angulatus*, *Deraeocoris lutescens*, *Phytocoris longipennis*, and *Psallus ambiguus* are predators of phytoseiids on apple trees in Germany (Kramer 1961). Kramer also listed *Capsus ater* as a phytoseiid predator, but this record seems doubtful and might be based on a misidentification of *Atractotomus mali*. Although *C. ater* sometimes migrates to fruit trees (see Chapter 12), it most likely does not attack phytoseiids. *Campylomma liebknechti* attacks phytoseiids and stigmaeids in Australian apple orchards (Readshaw 1975). Collyer (1952) said that predatory laelaptid mites occasionally fall prey to *B. angulatus* in England.

Although in England *Typhlodromus* species serve as alternative prey of plant bugs when European red mite populations are low (Muir 1965a, Dicker 1967), preliminary field trials indicate that predacious mirids do not feed extensively on *T. pyri* and seem unlikely to disrupt phytoseiid–European red mite interactions (Cranham et al. 1981). Gilliatt (1935) reported that a mirid's stylets are incapable of penetrating the dorsum of certain adult phytoseiids, and *Psallus ambiguus* readily attacks *T. pyri* only when the phytoseiid is stuck to an apple leaf with water (Morris 1965). In this case, the immobilized prey is presumably more easily penetrated. Oribatids, a nonpredatory group, appear unsuitable as mirid prey (Morris 1965).

Thrips

Entomologists often are unaware of the extent to which mirids prey on thrips, but species of *Termatophylidea* deserve attention from biological control workers: members of this Neotropical genus might be obligate thrips predators. Specialized predation in insects has generally been neglected (e.g., Thompson 1951, Tauber and Tauber 1987, Bristow 1988, Gilbert 1990). Other plant bugs are facultative predators of Thysanoptera. Nearly all references to facultative thrips predation pertain to mirids occurring outside North America, but the extent of thrips feeding by Nearctic plant bugs almost certainly is greater than the literature would suggest.

SPECIALIZED PREDATION ON THRIPS

Three species of *Termatophylidea* are known to feed on Thysanoptera. These deraeocorines are specialized for thrips predation, thus occupying a niche filled elsewhere by anthocorids (Callan 1975).

Myers (1931a) associated an undetermined predatory mirid with the redbanded or cocoa thrips (*Selenothrips rubrocinctus*) in Jamaica; the mirid was later identified as *T. pilosa* (Myers 1935). The nymphs and adults of another West Indian species, *T. maculata* (as *T. "maculosa"* in Callan [1943]), feed on redbanded thrips larvae on the lower surfaces of cocoa and cashew leaves, and on the larvae of a grass thrips, *Caliothrips insularis* (Callan 1943, 1975). Callan (1943) reared nymphs of *T. maculata* in the laboratory on a diet of redbanded thrips larvae. *Termatophylidea opaca* is the third member of the genus known to prey on redbanded thrips. In Suriname, van Doesburg (1964) found that the larvae are attacked on the lower leaf surfaces and that a fine webbing over the thrips, perhaps a shelter once occupied by spiders, affords protection and makes the mirids difficult to detect. Callan (1975) thought these apparently obligate predators might be too rare and locally distributed to reduce pest populations effectively. Each *T. opaca* individual, however, consumes numerous thrips each day, and a single bug can eliminate a small thrips colony. This predator once occurred regularly on cocoa trees sprayed with dieldrin, suggesting it is somewhat insecticide tolerant (van Doesburg 1964).

Basic biological studies on *Termatophylidea* species are lacking (Cassis 1995), and their capacity for suppressing populations of pest thrips under greenhouse conditions is unknown. *Termatophylidea* species apparently have not been evaluated for their biological control potential against important pests such as the western flower thrips (*Frankliniella occidentalis*). This thrips, one of the most important vectors of the tomato spotted wilt virus family, has spread throughout the United States, become established in greenhouses in other parts of the world, and generally plagues the floral industry (De Angelis 1988, Zitter et al. 1989). Acceptable pest densities are generally much lower on flower than on food crops because the former are important for their aesthetic value (Tauber and Helgesen 1981). Although an almost zero tolerance for pests hinders the implementation of IPM for ornamental greenhouse crops (van Lenteren et al. 1980, Parrella and Jones 1987, van Lenteren and Woets 1988, Ravensberg 1994), predatory mirids should be considered potential natural enemies of thrips in greenhouses. Any use of *Termatophylidea* species in the greenhouse would depend on their ability to feed and reproduce on novel prey, *Frankliniella* species, and their generally herbaceous host plants, and to search areas of the plant that harbor thrips.

FACULTATIVE PREDATION ON THRIPS

Although *Nesidiocoris tenuis* and related dicyphines injure the foliage, stems, and reproductive structures of tobacco and other solanaceous crops, China (1931) suggested that injury to tobacco is actually inflicted by *Thrips tabaci*, which might serve as prey of *N. tenuis*. This dicyphine preys on western flower thrips larvae in the laboratory (Riudavets and Castañé 1998). The dicyphine *Macrolophus costalis* is the most abundant of 28 predator species occurring in Bulgarian tobacco fields. Females destroy an average of 32.5 *T. tabaci* per day; males, an average of 27.4 thrips. An individual mirid (sex not stated) can consume 450–500 thrips during its life span. The predator complex so reduced populations of *T. tabaci* and aphids that chemical treatments were unnecessary for three years. Early-season releases of *M. costalis* in tobacco fields—one bug to 30–35 thrips—might provide useful biological control (Dimitrov 1975, Dirimanov and Dimitrov 1975). Waterhouse and Norris (1989) listed additional dicyphine predators (also *Deraeocoris* spp.) of *T. tabaci* in Bulgaria and the former USSR.

Beginning around the mid-1980s, European workers began to evaluate the use of facultatively predacious mirids for suppressing populations of thrips in greenhouses. The possibilities for using mirids to help control pests of protected crops were enhanced by the widespread adoption of IPM, which led to decreased pesticide use. Use of the native predators *Dicyphus tamaninii* and *M. melanotoma* (as *M. caliginosus*) was suggested to help manage populations of the western flower thrips in Spain. Both dicyphine species can complete their nymphal development on a diet of thrips larvae and show a higher consumption rate than a phytoseiid mite sold commercially for thrips control (Riudavets et al. 1993, 1995; Riudavets and Castañé 1998). Nymphs of *M. melanotoma* develop faster on a diet of western flower thrips larvae than on a diet of greenhouse whitefly nymphs (Riudavets and Castañé 1998).

Cage trials with *D. tamaninii* demonstrated that this predator can maintain populations of the western flower thrips at low densities on cucumber. This thrips feeds mainly on cucumber leaves rather than on flowers, which could serve as a potential refuge for the prey because the mirid does not search the flowers of this plant. Even at a relatively low predator-prey ratio (1 : 10), *D. tamaninii*, when released as late instars, is an efficient natural enemy when thrips populations are low. With a higher ratio, 3 : 10, the bug is able to keep the thrips from exceeding the economic injury level. This plant bug does not injure cucumber fruit. Preliminary results thus suggest the mirid's potential for managing thrips populations on cucumber in commercial greenhouses (Fig. 14.5; Gabarra et al. 1995, Castañé et al. 1996). In laboratory feeding trials, *D. tamaninii* shows a high consumption rate of western flower thrips on beans (Riudavets et al. 1993).

Less is known about dicyphines as thrips predators in North America than in the Palearctic region, although *Tupiocoris rhododendri* adults and nymphs prey on *Heterothrips azaleae* on deciduous azaleas in Georgia. In the laboratory, nymphs will capture and consume *H. azaleae* adults (Braman and Beshear

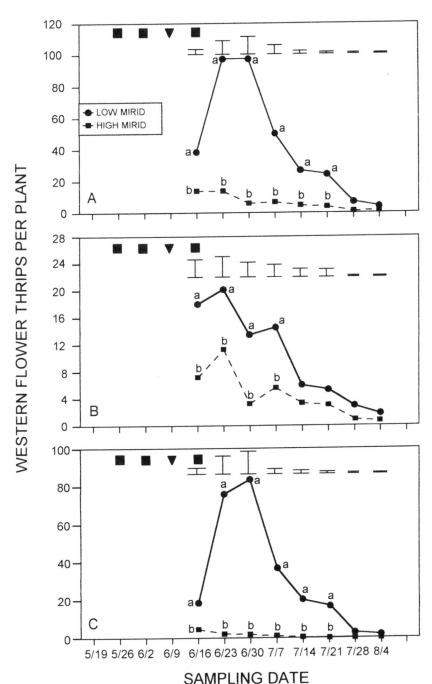

Fig. 14.5. Mean numbers of western flower thrips (*Frankliniella occidentalis*) per plant in field cages at two release rates of *Dicyphus tamaninii* in Spain. A. Numbers of thrips (adults + larvae). B. Numbers of adult thrips. C. Numbers of larval thrips. Each value is the average of 12 cages in which three cucumber plants in each were sampled. ■ and ▼ indicate the date of thrips and mirid introductions, respectively. Means followed by no letter are not significantly (*P* > 0.05) different. Data were transformed by log (*x* + 1) before analysis. Error bars represent weekly pooled standard errors. (Modified, by permission of R. Gabarra and Carfax Publishing, from Gabarra et al. 1995.)

1994). *Dicyphus hesperus*, an omnivore evaluated as a predator of spider mites and whiteflies, also feeds on western flower thrips in the laboratory (Gillespie et al. 1998).

Mirids of other subfamilies are also natural enemies of thrips. These include the phylines *Rhinocapsus vanduzeei* on *H. azaleae* in the southeastern United States (Braman and Beshear 1994) and *Campylomma lividum* on *Thrips palmi* in the Philippines (Calilung et al. 1994; Table 14.3). In Taiwan, *C. chinense* is common on *T. palmi*–infested eggplant, and in the laboratory fifth instars consume about 20 thrips larvae per day. A greater percentage of *C. chinense* nymphs can develop on thrips than on a diet of spider mites or lepidopteran eggs (Wang 1995).

Leafhoppers

Specialized predation on the adults and nymphs of cicadellids is unknown in the Miridae, although certain deraeocorines (e.g., *Hyaliodes* spp.) and orthotylines

Table 14.3. Examples of mirids as facultative predators of thrips

Mirid	Prey	Locality	Reference
Deraeocorinae			
Deraeocoris punctulatus	Thrips	Russia, Ukraine	Putschkov 1961a, 1961b; Zavodchikova 1974; see also Waterhouse and Norris 1989
Deraeocoris serenus	*Frankliniella occidentalis*	Spain	Riudavets and Castañé 1998
Deraeocoris spp.	*Frankliniella occidentalis*	Spain	Riudavets et al. 1993
Fingulus porrectus	*Leeuwenia karnyiana*	India	Stonedahl and Cassis 1991
Paracarnus sp.	*Selenothrips rubrocinctus*	West Indies	Ballou 1919, Reyne 1921
Termatophylum hikosanum	*Thrips* spp.	Japan	Takeno 1998
Termatophylum insigne	*Gynaikothrips ficorum*	Egypt	Lewis 1973
Orthotylinae			
"*Reuteria*" sp.[a]	*Scirtothrips aurantii*	South Africa	Bedford 1976
Phylinae			
Campylomma verbasci	*Frankliniella occidentalis*	British Columbia, Canada	Thistlewood and McMullen 1989
	Haplothrips verbasci	Canada	Thistlewood and Smith 1996
	Thrips	England	Thomas 1943, Groves 1969
Campylomma sp.	*Thrips palmi*	Thailand	Hirose 1990, Hirose et al. 1993
Moissonia flavomaculata	Thrips on cotton	India	Ballard 1921, Nair 1975
Orthonotus rossicus	Thrips	Former USSR	Putshkov 1961a, 1961b
Psallus sp.	*Taeniothrips nigricornis*	India	Rajasekhara et al. 1964
Rhinacloa forticornis	*Thrips palmi*	Hawaii, USA	Johnson and Nafus, 1995

Note: See text for additional examples and discussion.

[a] Another orthotyline genus (or that of another mirid subfamily) likely is involved because *Reuteria* species are not known to occur in Africa (Schuh 1974, 1995; Henry 1976a).

(e.g., *Blepharidopterus, Zanchius* spp.) often co-occur with leafhoppers on trees and shrubs. Leafhopper nymphs are included in the diet of plant bugs that specialize on planthopper or leafhopper eggs and of those that are facultative predators on various soft-bodied arthropods (Table 14.4).

Riley (1871) reported predation by *Hyaliodes vitripennis* on grape leafhoppers in the United States: "Leaves were actually covered on the underside with the dead carcasses [and] on a single leaf not so large as a man's hand a half hundred of these skeleton leafhoppers could be counted." Riley even suggested that living bugs be sent through the mail and then released in vineyards to help control leafhoppers. Prominent mid-nineteenth century entomologists such as Asa Fitch and B. D. Walsh had suggested importing natural enemies to help control exotic pests (Caltagirone and Doutt 1989), but Riley (1871) might have been the first to mention the possible use of mirids in biological control. He was also one of the first to advocate the augmentative release of a native predator to help suppress populations of native pests.

Summers (1891) reported that *H. vitripennis* often appears in increasing numbers as leafhoppers on grape vines are beginning to cause serious injury, and in such cases, he suggested withholding control measures to conserve this predator. These early references to *H. vitripennis* might refer to the morphologically similar *H. harti*, which Knight (1941) described as a new species.

Three mirid species—*B. provancheri, H. vitripennis,* and *Plagiognathus politus*—prey on the rose leafhopper (*Edwardsiana rosae*) on apple trees in Pennsylvania. *Blepharidopterus provancheri* and *H. vitripennis* were abundant on foliage during the summer of 1924, and nymphs fed on rose leafhopper nymphs in orchards and in cages. These mirids might have been primarily responsible for reducing leafhopper numbers in the second generation (Stear 1925). Ackerman and Isely (1931) noted that *H. vitripennis* feeds on several leafhopper species found in apple orchards in Arkansas, especially the white apple leafhopper (*Typhlocyba pomaria*). Steiner (1938) regarded *H. vitripennis* as an incidental predator of *T. pomaria* in New York and reported that older nymphs and adults of *B. provancheri* feed preferentially on this leafhopper rather than on European red mites. The identity of apple-infesting leafhoppers discussed as mirid prey—*E. rosae* and *T. pomaria*—was often confused in the early literature (Elsner and Beers 1988).

Blepharidopterus angulatus helps suppress populations of *T. pomaria* in Nova Scotia by attacking nymphs of the second generation; in feeding tests, nymphs show a functional response to prey density, consuming about half of the prey at densities of 2, 5, and 10 per cage (MacPhee 1979). The phyline *Campylomma verbasci* and orthotyline *Paraproba capitata* feed on immature leafhoppers in Quebec apple orchards (Braimah et al. 1982). In England, the black-kneed capsid (*B. angulatus*) feeds on leafhopper inhabitants of apple trees, such as

Table 14.4. Examples of mirids as facultative predators of leafhoppers

Mirid	Prey	Locality	Reference
Bryocorinae			
Macrolophus melanotoma	Leafhoppers	France	Carayon 1986
Nesidiocoris tenuis	*Empoasca* sp.	Australia	Atherton 1933
Tupiocoris cucurbitaceus	Leafhoppers	Argentina	Carpintero and Carvalho 1993
Tupiocoris rhododendri	Leafhoppers	Georgia, USA	Braman and Beshear 1994
Deraeocorinae			
Deraeocoris brevis	*Erythroneura ziczac*	Utah, USA	Knowlton 1946
Deraeocoris histrio	*Dikraneura* sp.	Iowa, USA	Hendrickson 1930
Deraeocoris oculatus	*Empoasca facialis*	Democratic Republic of the Congo	Leroy 1936
Mirinae			
Adelphocoris lineolatus	*Evacanthus interruptus*	England	Jennings 1903
Lygus lineolaris	*Empoasca fabae*	New York, USA	Wheeler 1974b
Pithanus maerkelii	Leafhoppers	England	Dolling 1991
Orthotylinae			
Blepharidopterus provancheri	*Erythroneura comes*	Pennsylvania, USA	Johnson 1914
Hyalochloria denticornis	*Empoasca* sp.	Peru	Hsiao 1945
Malacocoris chlorizans	*Empoasca vitis, Zygina rhamni*	Italy	Arzone et al. 1988
Zanchius mosaicus, Z. tarasovi	*Limesolla diospyri*	China	Zheng and Liang 1991, Zheng and Liu 1993
Phylinae			
Plagiognathus politus	*Empoasca fabae*	New York, USA	Wheeler 1974b
Rhinocapsus vanduzeei	Leafhoppers	Georgia, USA	Braman and Beshear 1994

Note: See text for additional examples and discussion.

Erythroneura and *Typhlocyba* species (Muir 1965a, 1966b), and also attacks *Alnetoidea alneti* on European linden. The bugs' tarsi are especially important for capturing and restraining leafhoppers (Glen 1975b), the foretarsi being used to hold the prey's body or appendages. On alder trees in England, *Psallus ambiguus* nymphs and adults feed on immature *Typhlocyba* species (Morris 1965). In Japan, at least three species of the orthotyline genus *Zanchius* apparently prey on typhlocybine leafhoppers (Yasunaga 1999a; see Table 14.4).

Psyllids

Mirids are common predators of psyllids, but whether any feed preferentially on these homopterans or might be specialized for psyllid predation is not known. Much of the literature pertains to attacks on psyllid pests of fruit trees, mainly the pear psylla (*Cacopsylla pyricola*) and the apple sucker (*C. mali*), but records are also available for predation on psyllids that develop on other host plants.

PSYLLIDS OF APPLE AND PEAR

Deraeocoris brevis (*Deraeocoris* sp. in early literature [Westigard et al. 1968]) is an important natural enemy of the pear psylla in the Pacific Northwest (Westigard

et al. 1979, Fye 1981a). It is generally more abundant on pear trees than are anthocorids (Riedl 1991, Booth and Riedl 1996). Nymphal development requires about 25 days at 21°C in the laboratory, and each nymph can consume about 400 psyllid eggs and nymphs during its development (Westigard 1973).

In the first generation, large populations of *D. brevis*, a multivoltine generalist, are sometimes attained on pear trees. At other times an increase in numbers sufficient to suppress psyllid populations does not occur until the prey's second generation, or not until pears have been marred by psyllid honeydew (Westigard 1973). In unsprayed orchards of southern Oregon, *D. brevis* begins to appear in mid to late May, reaches its highest densities from late June to July, and with other predators, maintains psyllid densities below economic levels. Suppression may not occur if cool, early-season temperatures delay the influx and buildup of predators (Westigard et al. 1979). A late arrival in pear orchards, compared to early-season predators such as certain neuropterans and Cantharidae, might actually typify this mirid (Gut et al. 1982b). The composition and timing of the predator complex colonizing pear trees depend mainly on the surrounding or background vegetation and habitats (Gut et al. 1982a, 1982b, 1991).

Because organophosphates are toxic to *D. brevis* and other predatory mirids (Westigard 1973), pyrethroids and other synthetic chemicals are recommended in Oregon's IPM program for pear orchards (Westigard et al. 1979). But synthetic pyrethroids such as permethrin can also adversely affect predacious mirids (Hagley and Simpson 1983). Life-history and ecological factors, rather than detoxification enzyme activities such as higher esterase activity, might explain the different effects that azinphos-methyl and fenvalerate have on *D. brevis* (lack of resistance) compared to their effects on the pear psylla (development of rapid resistance). The mirid perhaps is susceptible to these insecticides because of its lower fecundity and higher immigration rate relative to the psyllid (van de Baan and Croft 1990). A pear pest-management program using mating disruption rather than organophosphates to control the codling moth and based on an insect growth regulator, diflubenzuron, helps conserve *D. brevis* populations (Westigard and Moffitt 1984, Booth and Riedl 1996).

Following the eruption of Mount St. Helens, populations of *D. brevis* in an eastern Washington orchard declined initially because adults were affected by the volcanic ash. Their numbers, however, recovered within a few months because the eggs laid in pear foliage before the eruption remained unaffected (Fye 1983a).

McMullen and Jong (1967) considered the phyline *Campylomma verbasci* a significant early-season predator of the pear psylla in British Columbia, and the planting of mullein, an important alternative host, might enhance the benefits from this psyllid predator in pear orchards (Thistlewood and Smith 1996). McMullen and Jong (1967) also recorded *D. fasciolus* and *Blepharidopterus provancheri* as pear psylla enemies. Kelton (1983) mentioned other *Deraeocoris* species as psyllid predators on Canadian fruit crops and unspecified psyllids as prey of *B. angulatus*, *Hyaliodes harti*, *Phytocoris neglectus*, and *Plagiognathus ribesi*. Westigard et al. (1986) noted that mirids, including a *Phytocoris* species and *Pilophorus* species, are abundant predators of the pear psylla in Oregon. A *Phytocoris* species, near *conspurcatus*, is one of the most common natural enemies of pear psylla nymphs in southwestern Oregon (Stonedahl 1988). Nymphs of a *Phytocoris* species and *C. verbasci* hatching from excised pear shoots containing their eggs can decimate laboratory cultures of the pear psylla (Fye 1981b).

Deraeocoris nebulosus is a polyphagous predator that sometimes feeds on pear psylla in eastern North America (Wheeler et al. 1975; pers. observ.). *Orthotylus nassatus* and *O. viridinervis*, European orthotylines now established in the Nearctic region (Wheeler and Henry 1992), are similarly associated with heavy infestations of pear psylla in the eastern United States (Henry 1977; Kelton 1982b, 1983).

The phyline *C. verbasci* and a *Deraeocoris* species occur on psyllid-infested pear trees in Nova Scotia, and the latter feeds on *Cacopsylla pyricola* eggs and nymphs in the laboratory (Rasmy and MacPhee 1970). *Campylomma verbasci* also preys on the apple sucker (*Cacopsylla mali*) in Canada (Thistlewood and Smith 1996).

Another phyline, *Atractotomus mali*, fed on apple sucker nymphs during laboratory studies in Nova Scotia (Sanford 1964b).

Atractotomus mali, as well as *O. marginalis* and *Psallus ambiguus*, preys on the apple sucker in Norwegian apple orchards (Jonsson 1983a). Jonsson (1987) reared *A. mali* in the laboratory on the apple sucker and aphids. Based on comparative growth rates on these classes of prey, she categorized *A. mali* as a psyllophagous rather than an aphidophagous predator. The plant bug and apple sucker develop synchronously in the blossom clusters of apple trees, with the psyllid providing a predictable food source. This mirid, however, is a less efficient predator compared to certain other heteropterans occurring on Norwegian apple trees (Jonsson 1987). Santas (1987) recorded *A. mali* as a predator of *Cacopsylla* species on pear trees in Greece and reared it on psyllids in the laboratory. *Atractotomus mali* also preys on *Cacopsylla* species that develop on hawthorn, the bug's densities in Germany being positively correlated with those of its psyllid prey (Novak and Achtziger 1995).

In England, apple sucker nymphs are attacked by *P. ambiguus*, a rather general predator already discussed in this chapter as a mite feeder. Psyllid eggs are not accepted, but the mirid feeds on newly hatched first instars and all other nymphal stages. This phyline develops normally in the laboratory—that is, females produce eggs—on a diet of psyllid nymphs (Morris 1965). Although the apple sucker represents essential food for *P. ambiguus*, the bug does not effectively suppress psyllid populations in orchards (Niemczyk 1968).

The orthotylines *Heterocordylus genistae* and *O. marginalis* and the phyline *Plagiognathus arbustorum* were reported to prey on apple sucker nymphs in Poland (Minkiewicz 1927), but the occurrence of both *H. genistae* and *P. arbustorum* on apple is doubtful and misidentifications seem likely. *Psallus ambiguus* and *D. olivaceus* feed on this pest in Belarus (Byazdenka et al. 1975).

Myers (1927a) observed *Pilophorus perplexus* feeding on a late-stage nymph of *Cacopsylla pyricola* in France. *Orthotylus nassatus*, *Campyloneura virgula*, and *Heterotoma merioptera* (perhaps a misidentification of *H. planicornis* [see Wheeler and Henry 1992]) are rather abundant predators of *Cacopsylla pyri* in France (Herard 1985). Populations of the last-named mirid are reduced by regularly pruning pear trees in the winter, a practice that eliminates many of its overwintering eggs (Herard 1986). Herard (1986) gave additional information on, and references to, mirids as natural enemies of *C. pyri* in Europe.

PSYLLIDS OF BROOM, COCOA, AND OTHER HOSTS

Mirids living on Scotch broom in England attack the third through fifth instars and adults of broom psyllids (Dempster 1964). Overwintered eggs of *Heterocordylus* tibialis are the first to hatch in the spring, and the occurrence of this plant bug overlaps that of the univoltine psyllid *Arytainilla spartiophila* and the first generation of the bivoltine *Arytaina genistae*. *Heterocorylus tibialis* probably is more important than other broom-associated mirids in reducing psyllid numbers in the spring. Because mirids are the most numerous predators found on broom and feed extensively on psyllids, they are considered the most important group of psyllid predators associated with this host (Watmough 1968).

In Ghana, *Deraeocoris crigi*, the "cocoa deraeocorine," preys on *Tyora tessmanni* within the leaf buds of cocoa, both fan and chupon tips (Leston and Gibbs 1968), and within the psyllid galls on a euphorbiaceous tree (Leston 1980a). A "*Psallus*-like" phyline mirid collected within the chupon tips of cocoa may also feed on *T. tessmanni* (Leston and Gibbs 1968). During its nymphal development, *D. delagrandei* kills an average of 103 nymphs of *Euphyllura olivina* on olive trees, whereas adults destroy an average of 236 nymphs of this injurious psyllid in Turkey (Yayla 1986). A search for natural enemies of the leucaena psyllid (*Heteropsylla cubana*) in its native tropical America and in areas where it has become established revealed several mirid predators of this pest, including the phylines *Campylomma* species and *Rhinacloa forticornis* in Hawaii, Mexico, Trinidad, and Thailand (Funasaki et al. [1990], Napompeth et al. [1990], Waage [1990b]). Additional examples of mirids as facultative predators of psyllids are given in Table 14.5.

Whiteflies

Predators generally have received less emphasis as natural enemies of whiteflies than have parasitoids (Gerling 1992), and the biological control potential of predatory heteropterans has been neglected (Cohen and Byrne 1992). Mirids are not included in Gerling's (1986) review of natural enemies of the sweetpotato whitefly (*Bemisia tabaci*); the only "mirid" listed (*Orius* sp.) is actually an anthocorid. The sweetpotato whitefly's increasing importance as a major worldwide pest of field-grown and greenhouse crops and as a vector of gemini viruses (Brown 1994) intensified the search for effective biological control agents, including generalist predators that can be inundatively released (Nordlund and Legaspi 1994). It might be necessary to use several natural enemy species to suppress this whitefly at both low and high densities (Osborne et al. 1994). Mirids are now among the complementary natural enemies used to help manage this pest. *Bemisia tabaci* "strain B" was recently described as a new species, *B. argentifolii* (Bellows et al. 1994), and many earlier references to the sweetpotato whitefly as a pest of field and greenhouse crops might refer to the silverleaf whitefly (*B. argentifolii*) or other members of a species complex (Brown et al. 1995).

Until the mid to late 1980s (preliminary work was conducted in the early 1980s by B. Kaspar; cited in Malausa and Trottin-Caudal [1996]), information on mirids as natural enemies of aleyrodids and on the bionomics of these predators was scant (but see Kajita

Table 14.5. Examples of mirids as facultative predators of psyllids

Mirid	Prey	Locality	Reference
Deraeocorinae			
Deraeocoris brevis	*Paratrioza cockerelli*	Utah, USA	Knowlton and Allen 1936
Deraeocoris limbatus	Psyllids	Agalega Islands	Miller 1956
?*Deraeocoris* sp.	*Trioza erytraea*	South Africa	Van den Berg et al. 1987
Mirinae			
Megacoelum infusum	Psyllids	Poland, Sweden, former USSR	Kullenberg 1944, Putshkov 1961a, Strawiński 1964a
Phytocoris intricatus	*Psylla alni*	Finland	Kanervo 1946
Orthotylinae			
Dryophilocoris flavoquadrimaculatus	Psyllids	Poland	Strawiński 1964a
Orthotylus tenellus	Psyllids	Poland	Strawiński 1964a
Phylinae			
Campylomma lividum	*Psylla insitis*	India	Grove and Ghosh 1914
Chlamydatus pullus	*Psylla* sp.	Greenland	Böcher 1971
Opuna sharpiana	*Psylla uncatoides*	Hawaii, USA	Gagné 1975b
Psallus flavellus	*Psyllopsis fraxini*	England	Hodkinson and Flint 1971
Sejanus albisignatus	*Psyllopsis fraxini, P. fraxinicola*	New Zealand	Dumbleton 1964, Valentine 1967
	Psyllids	New Zealand	Collyer 1964a

Note: See text for additional examples and discussion.

[1978]). Whitefly predation often was inferred from the co-occurrence of a particular mirid species and a whitefly (e.g., Bruner 1934), or information was based on casual field observations or limited experimentation in the laboratory. European workers led research on the use of mirids to suppress greenhouse pests, and in North America, mirids are being considered for the biological control of whiteflies (e.g., Hoelmer et al. 1994, Gillespie et al. 1998, McGregor et al. 1999). *Macrolophus melanotoma* (as *M. caliginosus*), field tested in France and the Netherlands (Koppert 1994a, 1994b), is available from commercial suppliers of beneficial organisms in Europe (Ravensberg 1994, van Lenteren et al. 1997, Barnadas et al. 1998). Although this predatory mirid is listed as being available in North America (Hunter 1997), the U.S. Department of Agriculture does not permit its introduction (R. V. Flanders, pers. comm.). Agriculture and Agri-Food Canada similarly prohibits the importation of *Macrolophus* species (D. J. Parker, pers. comm.). With North American importation denied for the dicyphine mirids used widely in European greenhouses, attention has turned to native, omnivorous dicyphines, for example, *Dicyphus hesperus*, that might be used to control greenhouse pests in North America (Gillespie et al. 1998, 1999; McGregor et al. 2000).

With small prey such as whiteflies, the gains or energy rewards for acquiring prey will exceed the losses in time or energy devoted to capturing, handling, and digesting prey only at high prey densities (e.g., Schoener 1971, Luck 1984). Whiteflies often occur in dense, localized populations, and laboratory determination of energy budgets for the predatory lygaeoid *Geocoris punctipes* indicates that adult whiteflies represent energetically suitable prey for heteropterans (Cohen and Byrne 1992).

DICYPHINES AS WHITEFLY PREDATORS

The recent emphasis on several dicyphine mirids as predators of whiteflies attacking field- and greenhouse-grown vegetables in Europe (e.g., Gabarra et al. 1995, Alomar and Albajes 1996) has included biological studies to assess their role in the integrated management of pests of protected crops. Goula and Alomar (1994) provided descriptions, keys, and color photographs of dicyphines—*Dicyphus*, *Macrolophus*, and *Nesidiocoris* species—found in tomato fields in Spain. They noted that *M. caliginosus* and *M. pygmaeus* probably have been confused in the biological control literature (see also Kerzhner and Josifov [1999] for nomenclatural clarification). *Macrolophus caliginosus* has since been synonymized with *M. melanotoma* (Carapezza 1995).

Nesidiocoris tenuis, sometimes an important pest of tomato and other solanaceous crops, attacks all stages of the greenhouse whitefly (*Trialeurodes vaporariorum*) on tobacco plants under laboratory conditions. Kajita (1978) showed that adults feed on an average of 12.1 fourth-instar whitefly larvae per day. This predator also feeds facultatively on larval and adult whiteflies infesting tomato and gerbera in Sicily (Nucifora and Calabretta 1986, Nucifora and Vacante 1987). It preys on *T. vaporariorum* infesting tomato and zucchini plants in Italian greenhouses (Arzone et al. 1990) and has been accidentally introduced into greenhouses in France (Malausa and Trottin-Caudal 1996). *Dicyphus errans* sometimes occurs in considerable numbers on greenhouse-grown tomatoes in Italy, where it attacks whiteflies and aphids (Petacchi and Rossi 1991, Quaglia et al. 1993). Khristova et al. (1975) noted that the predatory *D. eckerleini* occurs in Bulgarian greenhouses.

Dicyphus tamaninii shows a functional response to the numbers of greenhouse whiteflies in unsprayed tomato plots in northeastern Spain; whitefly densities are consistently negatively correlated with mirid numbers during the preceding two to five weeks. Injury to fruit occurs only when the prey are scarce—that is, fall below a threshold density (Salamero et al. 1987, Gabarra et al. 1988). This bug is used successfully in an IPM program for the greenhouse whitefly in outdoor tomato crops in Spain (Alomar and Albajes 1996). Under field conditions, predator-prey ratios of 1:5 provide good whitefly control. The development of sooty mold on whitefly honeydew is generally retarded at predator-prey ratios of 1:10. At ratios above 1:5, the mirid can potentially injure tomato fruit. Using a working hypothesis that injury is related to a shortage of prey (some plant feeding by dicyphines may be necessary even when prey are abundant [Gillespie et al. 1999]), O. Alomar, R. Albajes, and colleagues developed a decision chart that employs variable thresholds (Fig. 14.6). Use of the chart allows crop advisors or growers to consider both mirid and whitefly densities and to spray in such a way as to minimize fruit injury from either insect. The result has been a 60–75% reduction in insecticide applications against whiteflies compared to usage in 1983 (Alomar et al. 1988, 1990; Alomar and Albajes 1996). A recent decline in the numbers of *D. tamaninii*

		Whitefly adults per plant	
		< 20	>20
Dicyphus adults & nymphs	> 4	Spray for *Dicyphus*	No action
	< 4	No action	No action

Fig. 14.6. Decision chart showing densities of the facultative predator *Dicyphus tamaninii* and the greenhouse whitefly (*Trialeurodes vaporariorum*) used in making pest-management decisions for fresh-market tomatoes in Spain. Insecticides are applied only when needed to limit potential injury from the mirid or damage by the whitefly. (Reprinted, by permission of O. Alomar and the Entomological Society of America, from Alomar and Albajes 1996.)

(and *M. melanotoma*) in tomato fields might be the result of residual toxicity from insecticides used to control populations of the western flower thrips on tomato seedlings (Figuls et al. 1999).

Dicyphus tamaninii can also reproduce on cucumber plants and reduce greenhouse whitefly densities in field cages. Because the mirid appears to need prey to reproduce on cucumber, its early establishment on this crop outdoors might be hindered (Gabarra et al. 1995). IPM specialists involved in using this mirid to help suppress whitefly densities on cucumber plants might take into account the bug's apparent preferences for lower leaf surfaces of cucumber in contrast to the upper surfaces of tomato leaves (Gesse Solé 1992).

Dicyphus tamaninii and other predatory mirids also colonize greenhouse-grown crops, preying on whitefly eggs, larvae, and adults (Albajes et al. 1988). Their role in complementing, or even replacing, classical IPM techniques in the greenhouse (e.g., use of parasitic wasps such as *Encarsia formosa*) has been assessed (Alomar et al. 1990). Further evaluation is needed to determine the potential of *D. tamaninii* for injuring tomato and other crops and its risk as a vector of plant pathogens (e.g., Salamero et al. 1987, Gabarra et al. 1988; see also Chapter 8).

Another dicyphine, *Macrolophus pygmaeus*, has been evaluated as an enemy of whiteflies injurious to vegetables grown outdoors or under glass in Spain (Nucifora and Vacante 1987), Russia (Slobodyanyuk et al. 1993), and Greece (Perdikis and Lykouressis 2000). In Ukraine, this mirid develops rapidly on a whitefly diet, completing development in 22 days, or about half the time needed without access to aleyrodids (Tsybul'skaya and Kryzhanovskaya 1980). Perdikis and Lykouressis (2000) studied the nymphal development and survival rates of *M. pygmaeus* on different host plants, with and without prey, at six temperatures. Nymphs were reared to adulthood on all host plants tested, but development was significantly shorter when prey were provided.

Macrolophus costalis nymphs (adults tend to be phytophagous) destroy numerous greenhouse whitefly larvae and eggs on tobacco and tomato plants in Bulgarian greenhouses. More prey are consumed at 22°C than at 24.8° or 25.5°C. This mirid, which breeds continuously in greenhouses, is considered a reliable whitefly predator, but possible injury to solanaceous crops by adult *M. costalis* has not been assessed (Khristova et al. 1975). *Macrolophus costalis* has also been studied as a whitefly predator in Poland (references cited by Salamero et al. 1987, Malausa et al. 1987).

Macrolophus melanotoma is another dicyphine occurring in untreated greenhouses of the Mediterranean region (Llorens and Garrido 1992). It has been evaluated for the possible biological control of whiteflies, mainly *T. vaporariorum*, in France (Fauvel et al. 1987; Malausa 1987, 1989; Malausa et al. 1987), Italy (Arzone et al. 1990), the Netherlands (Fransen 1994, Schelt 1994, Schelt et al. 1996), and Spain (Barnadas et al. 1998). This polyphagous predator develops not only on whiteflies of all immature stages (Barnadas et al. 1998), but also on alternative prey such as eggs of the pyralid *Ephestia kuehniella* (Fauvel et al. 1987), which can be used to mass-produce the bugs (Grenier et al. 1989, Ferran et al. 1996). Grenier et al. (1989) also reared it on an artificial medium and on the medium with a geranium leaf added. The ability of *M. melanotoma* to be reared on alternative prey and artificial media and to survive at low temperatures (10–15°C [Carayon 1986]) enhances its potential as a biological control agent against *Bemisia tabaci*, *T. vaporariorum*, and other insect pests of greenhouses. Barnadas et al. (1998) evaluated the potential of *M. melanotoma* (and *D. tamaninii*) to control the sweetpotato whitefly when it co-occurs with the greenhouse whitefly. Both mirid species were reared on *T. vaporariorum*, and Barnadas et al. (1998) suggested that the bugs' previous experience might have affected their preference for that whitefly species over *B. tabaci*.

Macrolophus melanotoma can be used in conjunction with the hymenopteran parasitoid *E. formosa* (Malézieux et al. 1995) and when temperatures are too low for the parasitoid to be effective (Malausa 1987, Malausa and Trottin-Caudal 1996). The release of two adult mirids per plant in early season—that is, when whiteflies are detected—followed by a second release at the same rate when second-generation whiteflies begin to emerge in tomatoes will give good control. Injury to the leaves, petioles, stems, or fruit occurs only when large numbers of *M. melanotoma* (30 adults/15-cm-high plant) are caged on young tomato plants; feeding symptoms are not observed in commercial greenhouses in France (Malausa and Trottin-Caudal 1996; cf. Sampson and Jacobson 1999). Malausa and Trottin-Caudal (1996) noted that when releases are made late in the whitefly cycle, the predator's functional and numerical responses cannot catch up with the prey's biotic potential.

Dicyphus hesperus, a native North American plant bug, has been evaluated for possible use as an omnivorous predator of pests attacking greenhouse-grown tomatoes in British Columbia. This mirid can be reared on a diet of greenhouse whiteflies, with adult bugs consuming as many as 20 immature prey in 24 hours. When released in a greenhouse containing whitefly-infested tomato plants, *D. hesperus* oriented to and oviposited on infested hosts (Gillespie et al. 1998, McGregor et al. 1999). In contrast to *D. tamaninii*, which injures tomato fruits when prey are scarce (Alomar and Albajes 1996), *D. hesperus* under experimental conditions injures the fruits when tomato leaves are not provided; the presence of prey does not significantly affect the degree of fruit feeding (McGregor et al. 2000).

OTHER MIRIDS AS WHITEFLY PREDATORS

An example of whitefly predation by a North American mirid is Butler's (1965) observation that the nymphs and adults of *Spanagonicus albofasciatus* significantly reduce populations of the bandedwinged whitefly (*Trialeurodes abutiloneus*) at all stages when caged on infested cotton seedlings. Kelton (1983) recorded *Ceratocapsus modestus* as a predator of whiteflies on Canadian fruit trees. A South American record of whitefly predation is that by the phyline *Rhinacloa* species in Peru (Vergara and Raven 1988). *Campyloneura virgula* is one of the principal predators of the viburnum whitefly (*Aleurotrachelus jelinekii*) in England; although the mirid feeds on prey of all stages, the effect on host population dynamics is slight (Southwood and Reader 1988).

Kajita (1984) studied predation on the greenhouse whitefly by a Japanese species of *Campylomma*. The bugs insert their stylets dorsally in young larvae but laterally in the older stages. In the laboratory, a *Campylomma* species is 70–100% efficient in attacking its prey except when early-instar bugs attempt to feed on fourth-stage larvae or pupae. The attack efficiencies of the fourth and fifth instars are at least 90% against all whitefly immature stages.

Deraeocoris pallens feeds on immature stages of *Bemisia tabaci* in Israel; its development in summer can be completed in 17 days on a whitefly diet. Under experimental conditions, handling time increases as the ratio of predator size to prey size decreases (Susman 1988; see also Gerling [1990], Ghavami et al. [1998]). The polyphagous predator *Deraeocoris nebulosus*, which feeds outdoors on the greenhouse whitefly (*T. vaporariorum*) on scarlet runner bean in Pennsylvania (Wheeler et al. 1975, pers. observ.), can complete its development on a diet of immature silverleaf whiteflies (*B. argentifolii*) (Jones and Snodgrass 1998). Other mirids that prey on the sweetpotato whitefly include *Deraeocoris punctulatus* and *Campylomma diversicorne* in Syria (Stam and Elmosa 1990), a *Deraeocoris* species and *C. verbasci* (as *C. nicolasi*) in India (Kapadia and Puri 1991), *Engytatus varians* in Cuba (Castineiras 1995), and *Lygus hesperus* in the western United States (Hagler and Naranjo 1994). Additional mirid-whitefly associations are *Deraeocoris serenus* and *T. vaporariorum* in Italy (Arzone 1976), *Campylomma diversicorne* and *T. rara* in Iraq (Georgis 1977), and *D. pallens* and *Dialeurodes citri* in Turkey (Soylu 1980).

Aphids, Adelgids, and Phylloxerans

Plant bugs—both carnivorous and mainly phytophagous taxa—feed on aphids. Even though many mirids have been identified as aphid predators, little information is available on their effectiveness in keeping pest species below economic thresholds.

Mirids can be among the most diverse groups of entomophagous arthropods associated with an aphid species or complex on a particular host. Bouchard et al. (1986) found that plant bugs contribute the largest number of species (21) predacious on the apple aphid (*Aphis pomi*) in southwestern Quebec, although the bugs do not

prevent prey populations from increasing. Bouchard et al. (1988) reported voracities (see Table 7.2)—numbers of *A. pomi* killed by second to fifth instars and adults—for four species: *Deraeocoris fasciolus* (Deraeocorinae), *Phytocoris canadensis* (Mirinae), *Blepharidopterus provancheri* (Orthotylinae), and *Phoenicocoris minusculus* (Phylinae). In the laboratory, *D. fasciolus* and *P. minusculus* show the highest daily consumption rates (7–9 *A. pomi*/day) among eight mirid species occurring on apple trees. Mirids accounted for 64% of all aphidophages collected in apple orchards in Poland (Niemczyk 1966b), and eight predatory mirid species are associated with *A. pomi* on apple trees in Germany (Asgari 1966). Smith (1966) reported feeding rates of five broom-inhabiting mirids on *Acyrthosiphon spartii* in an outdoor insectary in England and assessed their role as natural enemies on the basis of field counts when aphids were most abundant. Mirids and other predatory arthropods were considered unimportant in checking the population rise of this broom aphid.

Examples of prey to be discussed here (Table 14.6) include not only the Aphididae but also other groups of the superfamily Aphidoidea (drepanosiphids, lachnids, and pemphigids), as well as adelgids and phylloxerids in the Phylloxeroidea. Presented first is predation by members of three mainly predacious genera: *Deraeocoris* (Deraeocorinae), *Phytocoris* (Mirinae), and *Pilophorus* (Phylinae). Additional discussion includes information on aphid predation by *Blepharidopterus angulatus*, an orthotyline already discussed as a mite feeder; species of several dicyphine genera; principally phytophagous Mirinae, including species of *Adelphocoris*, *Closterotomus*, *Lygocoris*, and *Lygus*; and plant bugs of other taxa.

DERAEOCORIS SPECIES AS APHID PREDATORS

More records of aphid predation probably are available for this genus than for any other in the Miridae. Its members range from multivoltine, polyphagous predators such as *Deraeocoris nebulosus* and *D. brevis*, which track prey on numerous hosts and have many attributes of *r*-selected natural enemies (e.g., Ehler and Miller 1978), to univoltine specialists that are restricted to prey associated with a single plant genus, often either hardwood or coniferous trees.

Knight (1921), the first to emphasize the predacious tendencies of Nearctic *Deraeocoris*, found *D. fasciolus* closely associated with the rosy apple aphid (*Dysaphis plantaginea*). The nymphs often live within the aphid-curled foliage of apple trees. On hawthorn, *D. fasciolus* feeds on *Eriosoma crataegi*, and the variety *D. f. castus* preys on *Phyllaphis fagi* on the leaves of American beech. Knight (1921) also observed *D. aphidiphagus* in elm leaves curled by the woolly elm aphid (*Eriosoma americanum*), *D. nitenatus* in association with a pemphigid forming rosette galls on elm, and *D. pinicola* preying on the pine bark adelgid (*Pineus strobi*) on white pine. The nymphs and adults of *D. nubilus* are

Table 14.6. Examples of mirids as predators of adelgids, aphids, and phylloxerans

Mirid	Prey	Host plant	Locality	Reference
Bryocorinae				
Tupiocoris rhododendri	Aphids	Azalea	Georgia, USA	Braman and Beshear 1994
Deraeocorinae				
Deraeocapsus ingens	Cinara ponderosae	Ponderosa pine	California, USA	Voegtlin and Dahlsten 1982
Hyaliodes vittaticornis	Aphids	Unspecified	Cuba	Bruner 1934
Mirinae				
Barberiella formicoides	Aphis pomi	Apple	Pennsylvania, USA	Wheeler and Henry 1980a
Orthotylinae				
Globiceps fulvicollis	Aphids	Clover, trefoil	England	Woodroffe 1958a
Heterocordylus tibialis	Acyrthosiphon spartii	Scotch broom	England	Smith 1966
Lopidea cuneata	Aphids	Balsam poplar	New York, USA	Van Duzee 1910
Orthotylus adenocarpi	Acyrthosiphon spartii	Scotch broom	England	Smith 1966
Orthotylus marginalis	Aphids, including Aphis pomi	Apple	Switzerland	Schaub and Baumgärtner 1989
Orthotylus ramus	Monellia caryella, Monelliopsis pecanis	Pecan	Alabama, USA	Edelson and Estes 1987
Orthotylus virescens	Acyrthosiphon spartii	Scotch broom	England	Smith 1966
Orthotylus viridinervis	Eriosoma ulmi	Elm	Netherlands	Cobben 1958
Paraproba nigrinervis	Myzocallis coryli	Filbert	British Columbia, Canada	Messing and AliNiazee 1985, 1986
Paraproba pendula	Aphid complex	European white birch	California, USA	Hajek and Dahlsten 1988
Saileria irrorata	Eriosoma americanum[a]	Slippery elm	Indiana, USA	Henry 1976b
Phylinae				
Atractotomus mali	"Green apple aphid"	Apple	Nova Scotia, Canada	Knight 1924
	Hyalopterus pruni	Peach	France	Remaudière and Leclant 1971
Compsidolon salicellum	Aphids	Raspberry	Eastern Canada	Kelton 1982a
	Myzocallis coryli	Filbert	British Columbia, Canada	Messing and AliNiazee 1985
Cremnocephalus albolineatus	Cinara pinicola	Spruce	Poland	Strawiński 1964a
Lopus decolor	Myzocallis coryli	Filbert	British Columbia, Canada	Messing and AliNiazee 1985
Microphylellus sp. [prob. M. tumidifrons]	Cinara pilicornis	Conifers	New Jersey, USA	Leonard 1971
Phoenicocoris dissimilis	Mindarus abietinus	White fir	Pennsylvania, USA	D. J. Shetlar, pers. comm.; pers. observ.
Phoenicocoris minusculus	Aphids	Fruit crops	Canada	Kelton 1983
Phylus coryli	Aphids [prob. Myzocallis coryli]	Filbert	British Columbia, Canada	Kelton 1982c
Plagiognathus negundinis	Periphilus negundinis	Boxelder	Iowa, USA	Webster 1917, Knight 1929b
Plagiognathus politus	Brevicoryne brassicae	Cole crops	New York, USA	Pimentel 1961
Plagiognathus sp.	Aphids	Apple	Nova Scotia, Canada	Hey 1933b
Psallovius piceicola (as Psallus piceicola)	Adelges piceae	Balsam fir	Canadian Maritime Provinces	Brown and Clark 1956
	Adelgid	Tamarack (larch)	Canadian Maritime Provinces	Brown and Clark 1956
Psallus ambiguus	Aphids	Apple	Finland	Kanervo 1961
Psallus betuleti	Aphids	Cascades azalea, birch	Eastern Canada	Schwartz and Kelton 1990
Psallus pseudoplatani	Drepanosiphum platanoides	Sycamore maple	Netherlands	Aukema 1986
Rhinocapsus vanduzeei	Aphids	Fruit crops	Canada	Kelton 1983
	Aphids	Azalea	Georgia, USA	Braman and Beshear 1994
Sthenaridea carmelitana	Aphids	Corn	Peru	Marín and Sarmiento 1979

Note: See text for additional examples and discussion.

[a] Predation suspected.

sometimes common in Pennsylvania on Scotch pines infested with pine bark adelgids, which they appear to attack in Christmas tree plantations; a predatory relationship to the adelgid was confirmed in the laboratory (pers. observ.). *Deraeocoris piceicola* was found in Colorado only on spruce trees heavily infested with adelgid galls (Knight 1927c), a pattern that also holds for this plant bug in the northeastern United States (Wheeler and Hoebeke 1990; pers. observ.).

Deraeocoris nebulosus, widespread in eastern North America, feeds on the woolly apple aphid (*Eriosoma lanigerum*) (Gillette 1908), hop aphid (*Phorodon humuli*) (Parker 1913), clover aphid (*Nearctaphis bakeri*) (Smith 1923), cotton aphid (*Aphis gossypii*) (Whitcomb and Bell 1964, Snodgrass 1991), *Phyllaphis fagi* on American beech (Wheeler et al. 1975), and pecan aphids (Edelson and Estes 1987). In Virginia apple orchards, aphids are considered important in attracting *D. nebulosus* before it transfers its feeding to other pests (Parrella et al. 1981). Mizell and Schiffhauer (1987) mentioned *D. nebulosus* as an important predator of pecan aphids that apparently uses the crapemyrtle aphid (*Tinocallis kahawaluokalani*) as alternative prey.

Several prey records are available for the common western species *D. brevis*. Essig (1926) said it preys on *Chromaphis juglandicola* in California, and it will feed on *Prociphilus fraxinifolii* on ash and on a winged form of the woolly elm aphid in Utah (Knowlton 1946). The apple grain aphid (*Rhopalosiphum fitchii*) serves as one of several early-season prey of *D. brevis* in Oregon (Westigard 1973). This mirid is important in limiting the numbers of the filbert aphid (*Myzocallis coryli*) before this pest undergoes exponential growth (Messing and AliNiazee 1985). Cage-exclusion methods and laboratory experiments on feeding potential have shown that *D. brevis* and other predacious mirids on filbert help regulate aphid numbers. An integrated management program is recommended in filbert orchards of the Pacific Northwest to conserve this predator complex (Messing and AliNiazee 1986). In central Washington, *D. brevis* is an important predator of the apple aphid (*Aphis pomi*) (Carroll and Hoyt 1984). Although the numbers of *D. brevis* and *Campylomma verbasci* remain at low to moderate levels from late April to early August on apple trees in British Columbia, the buildup of populations in August, coupled with their response to increases in apple aphid numbers, suggests that these mirids play an important role in the predator complex associated with the aphid (Haley and Hogue 1990).

Several other western species of *Deraeocoris* feed on aphids in the field or in the laboratory (Razafimahatratra 1980, Cooper 1981). In Indiana, nymphs of a *Deraeocoris* species (and also *Plagiognathus* sp.) possibly feed on *Phylloxera caryaecaulis* when galls begin to open on shagbark hickory (Caldwell and Schuder 1979).

Deraeocoris annulipes feeds on the Cooley spruce gall adelgid in Europe (Strawiński 1964a). *Deraeocoris ruber*, long known as an aphid predator (Verhoeff 1891),

preys on the woolly apple aphid in France (Marchal 1929); this mirid (and *P. arbustorum*) also feeds on *Chaetosiphon fragaefolii* on strawberry plants in England (Dicker 1952) and is a predator of *Corylobium avellanae* and *Myzocallis coryli* on European filbert in Italy (Viggiani 1971b, 1983). A species sometimes considered mainly phytophagous, *D. lutescens*, feeds on *C. avellanae* in England (Massee 1949).

Deraeocoris punctulatus, although reportedly injurious to alfalfa, cotton, and other crops, was one of the most abundant natural enemies of the green peach aphid (*Myzus persicae*) on tobacco plants in Armenia (Mardzhanyan and Ust'yan 1965). Zavodchikova (1974) established a predatory relationship between *D. punctulatus* and aphids in Russia. Some predation on mites and thrips occurs, but mortality is complete when the diet consists of these arthropods or plant material; *D. punctulatus* can develop successfully on a diet of green peach aphids. Triggiani (1973) found that *D. flavilinea*, which preys on several aphid species associated with almond and plum trees in Italy, requires some plant material to complete its development. *Deraeocoris signatus*, previously thought to cause early-season injury to cotton in eastern Australia, can develop on an aphid diet. This mirid's occasional imbibing of plant sap does not produce injury but might enhance its survival when prey densities decline (Chinajariyawong and Harris 1987).

Additional examples of predation by *Deraeocoris* include *D. aphidicidus* on the cotton aphid in India (Ballard 1927), *D. punctulatus* on this aphid in China (Zhang 1992), *D. oculatus* on cotton aphids in Africa (Leroy 1936), and an unidentified species of the genus on *Pemphigus mordvilkoi* in India (Ghosh et al. 1988).

PHYTOCORIS SPECIES AS APHID PREDATORS

Members of this large genus are mostly general predators that often feed on aphids. LeRoux (1960), for example, cited *Phytocoris conspurcatus* as an important predator of the apple aphid in Quebec. At least eight species of the genus prey on aphids in Quebec apple orchards (Braimah et al. 1982), and Kelton (1983) mentioned 12 species as predators of aphids occurring on fruit trees in Canada. Martin (1966) suggested that a *Phytocoris* species (cited as *P. eximius*, but probably *P. canadensis* [Henry and Stonedahl 1984, Stonedahl 1988]) is an important natural enemy of aphids on red pine in Ontario; an increase in the numbers of the mirid coincided with a decline of aphid populations. Several Palearctic species of *Phytocoris* are also aphid predators (e.g., Gulde 1921; Kullenberg 1944; Southwood and Leston 1959; Putshkov 1961a, 1961b; Strawiński 1964a, 1964b).

PILOPHORUS SPECIES AS APHID PREDATORS

Aphid predation by Palearctic species of *Pilophorus* has been known since the nineteenth century (e.g., Breddin 1896). *Pilophorus perplexus*, a European bug unintentionally introduced into North America (Wheeler and Henry 1992), feeds on aphids within curled foliage of apple trees. In an abandoned New York orchard, Fulton (1918) found that aphids provided the principal food source for nymphs and adults. Aphids being attacked secreted droplets from their cornicles, which sometimes contacted a bug's stylets, causing it to withdraw and clean the labium with its foretarsi (Fulton 1918). Hagley (1975) referred to *P. perplexus* as an effective aphid predator in Ontario apple orchards but gave no supporting data or observations; in later studies, it was found only in low densities on apple trees (Hagley and Allen 1990). This predator once exterminated an aphid colony on hawthorn trees in England (Groves 1969). Cranham et al. (1982) observed that on apple trees in England, *P. perplexus* was one of the few predators that successfully preyed on ant-attended aphids. *Pilophorus perplexus*, *P. confusus*, and *P. clavatus* are natural enemies of aphids occurring on fruit trees and berry shrubs in Lithuania (Rakauskas 1985).

Bradley and Hinks (1968) studied ant-aphid-mirid interactions on jack pine in Manitoba. The ants obtain honeydew from the aphids and protect their colonies from predation. Five *Pilophorus* species prey on aphids that stray from ant colonies; sometimes one or two mirid nymphs can be seen near every colony. Their predation is considered important in preventing aphids from colonizing new jack pine trees.

BLEPHARIDOPTERUS ANGULATUS AS AN APHID PREDATOR

The black-kneed capsid (*Blepharidopterus angulatus*) feeds extensively on the lime aphid (*Eucallipterus tiliae*) on European linden trees in England (Glen 1975b, Glen and Barlow 1980), its overwintered eggs hatching with the production of first-generation aphids. Retrorse barbs at the tip of the mandibular stylets make it difficult for prey to escape once they are pierced. First instars of this orthotyline, which are able to capture only young aphids, are inefficient in capturing prey compared to later stages—they require a prey density nine times greater than that for third to fifth instars (Glen 1975b). About half the total weight of lime aphids consumed during nymphal development is attributable to fifth instars. Food consumption by each instar tends to decrease toward the end, particularly in the fifth instar (Fig. 14.7; Glen 1973). The first and third instars are more efficient than the fifth instar in converting food into body weight when aphids are scarce. The fifth instar does not show an increase in efficiency at low feeding levels.

Glen (1975b) constructed a model allowing the prediction of the aphid densities required for optimum development of *B. angulatus* in each instar and the relationship between prey consumption and prey density. When aphid numbers are high, the mirid's rate of increase on European linden trees is hindered on some trees by a parasitism rate that can reach 80% and by a high rate of emigration of prereproductive females. Because of high emigration, an increase in reproductive rate will not increase this plant bug's effectiveness as an aphid predator (Luff 1983). Simulation also indicates that even increased numbers of *B. angulatus* fail to stabilize predator-prey populations: larger predator numbers lead to overexploitation of prey (Glen 1977a, Glen and Barlow 1980). This predator was one of the main components used to model the population dynamics of the lime aphid (Dixon and Barlow 1979, Barlow and Dixon 1980).

Blepharidopterus angulatus also attacks aphids living on alder trees—for example, *Pterocallis alni* (Gange and Llewellyn 1989). Nymphs from alder feed readily on aphids but will only gradually accept European red mites as prey (Cranham et al. 1974). As previously noted in this chapter, the European red mite represents this mirid's principal prey on apple trees. The idea that populations of mainly carnivorous mirids that develop on unrelated host plants show different prey preferences, depending on the host, warrants further study.

DICYPHINAE AS APHID PREDATORS

Engytatus modestus preys on tomato-infesting aphids in Hawaii (Plate 23; Illingworth 1937b, Holdaway and Look 1940) and is sometimes a common predator of the green peach aphid (*Myzus persicae*) in Florida. This mirid nearly eliminated an aphid infestation in a 0.8-hectare field of shade tobacco that had been allowed to produce suckers, and it so reduced populations in a tobacco seed bed that insecticides could no longer be evaluated for aphid control (Wilson 1948). Other dicyphines—*Dicyphus* species, *Macrolophus tenuicornis*, and *Usingerella bakeri*—prey on aphids (unspecified species) in Canada (Kelton 1983).

Dicyphus errans attacks aphids on zonal geranium in Germany. This species is also phytophagous, although its developmental period is lengthened with a vegetable diet. This mainly predacious mirid requires some plant food because it will not develop on a diet consisting exclusively of arthropods (Schewket Bey 1930). In France, *D. errans* sometimes enters greenhouses from populations developing on weeds outside, and it can be used with hymenopteran parasitoids of aphids to avoid the need for insecticides on tomato crops (Lyon 1986). This species also feeds on *M. persicae* on greenhouse-grown tomatoes in Italy (Petacchi and Rossi 1991, Quaglia et al. 1993). British populations of *Tupiocoris rhododendri* are consistently associated with colonies of *Masonaphis* species on rhododendron (Dolling 1991).

Macrolophus pygmaeus feeds on the green peach aphid in Armenia (Mardzhanyan and Ust'yan 1965) and Greece (Perdikis et al. 1999), and *M. costalis* has been evaluated for inclusion in integrated management programs for controlling aphids in Polish greenhouses. When tested against *Macrosiphoniella sanborni*

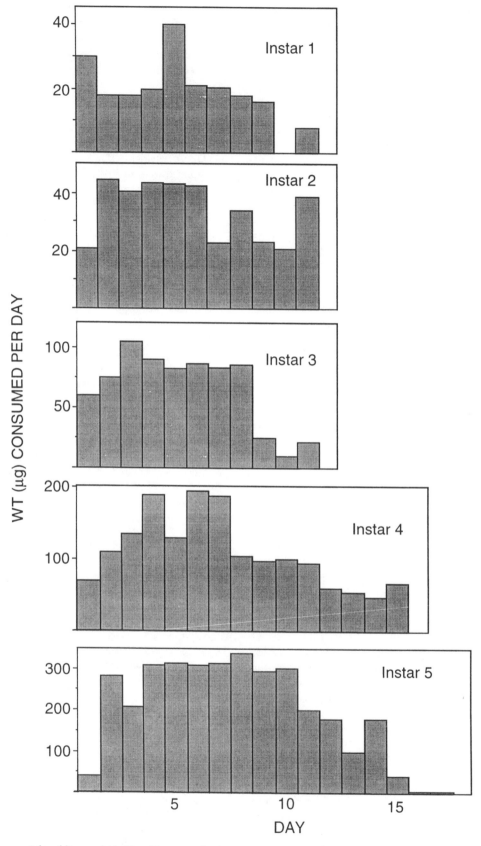

Fig. 14.7. Mean weight of lime aphids (*Eucallipterus tiliae*) consumed by *Blepharidopterus angulatus* nymphs on each day for each instar with excess food at 14°C. Food consumption is not uniform in later instars; in each instar, consumption increases to a peak before declining toward the end, particularly in the fifth instar. (Modified, by permission of D. M. Glen and Kluwer Academic Publishers Group, from Glen 1973.)

infesting chrysanthemums, *M. costalis* reduced aphid populations at predator-prey ratios of 1:2 or 1:3 and was more effective at higher prey densities (Brzeziński 1988). The functional response of *M. melanotoma* (=*M. caliginosus*) females to the green peach aphid has been determined in France (Fig. 14.8; Foglar et al. 1990). Early releases of this mirid are sufficient to maintain the density of green peach aphids below economic levels, but the predator does not suppress cotton aphid populations effectively because of the aphid's higher reproductive rate (Malausa and Trottin-Caudal 1996). Alvarado et al. (1997) assessed the functional response of *D. tamaninii* females to varying densities of the cotton aphid on cucumber and also evaluated *M. melanotoma* as a predator of *Aphis gossypii* on cucumber and *Macrosiphum euphorbiae* on tomato plants.

MAINLY PHYTOPHAGOUS MIRINAE
AS APHID PREDATORS

Phytophagous plant bugs sometimes exploit aphids as a food source. Although species such as the western tarnished plant bug (*Lygus hesperus*) can be reared on a plant diet alone, the addition of prey generally enhances the bug's life-history traits (Musa and Butler 1967, Naranjo and Gibson 1996). The tarnished plant bug (*L. lineolaris*), for instance, will feed on the pea aphid (*Acyrthosiphon pisum*) and spotted alfalfa aphid (*Therioaphis maculata*). Lindquist and Sorensen (1970) found that aphid-free alfalfa varieties are equally attrac-

tive to the bugs, but their populations increase in the laboratory more rapidly on aphid-susceptible varieties than on resistant varieties (Table 14.7). Mirid mortality is lower on the susceptible variety. Latson et al. (1977) observed two instances of predation by *L. lineolaris* on cotton aphids in field cages, Gupta et al. (1980) reported predation by lygus bugs on aphids associated with alfalfa in Washington, and Scott (1983) gave circumstantial evidence for predation by *L. hesperus* on pea aphids in western North America. Bańkowska et al. (1975) classified mirids as omnivores and polyphagous predators, whose exact ecological role in Polish alfalfa fields is uncertain.

Several mirines are facultative aphid predators in England: *Capsodes flavomarginatus* inhabiting hogweed flowers (Morley 1905) and *Adelphocoris ticinensis* and *C. gothicus* in captivity when placed on leguminous hosts (Woodroffe 1955b, 1959b). Goddard (1935) reported aphid predation by other British mirines under laboratory conditions. Circumstantial evidence for aphidophagous habits in *Lygocoris pabulinus* is available (Petherbridge and Thorpe 1928b, Goddard 1935), and Wightman (1969b) reported predation by this species on *Myzus persicae* in the laboratory. Many individuals of *Closterotomus norwegicus* and *Apolygus spinolae* tested positive for predation on cereal aphids using enzyme-linked immunosorbent assay (ELISA), and they also fed on living aphids in the laboratory (Sunderland et al. 1987). *Apolygus spinolae* might also prey on *Hyalopterus pruni*, the mealy plum aphid (Remaudière and Leclant 1971). *Closterotomus biclavatus* is a frequent aphid predator in Germany (Gäbler 1937). But mirines generally do not show as strong a tendency toward predation as do phylines and many orthotylines, and species such as *Creontiades dilutus* do not attack healthy aphids even when caged with potential prey (Hori and Miles 1993).

Additional references to aphid feeding by North

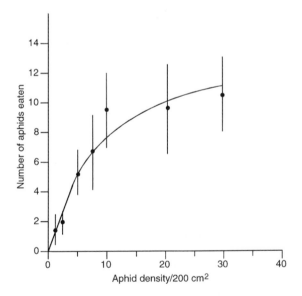

Fig. 14.8. Functional response (Type II) of *Macrolophus melanotoma* to increasing densities of fourth-instar green peach aphids (*Myzus persicae*) under standardized conditions of 22°C, 80% relative humidity, and 16-hour photoperiod. Data are recalculated for a 24-hour period; means ± 95% confidence intervals. (Modified, by permission of J. C. Malausa and the International Organization for Biological Control [IOBC], from Foglar et al. 1990.)

Table 14.7. Numbers of *Lygus lineolaris* on aphid-resistant and aphid-susceptible alfalfa varieties uninfested and infested with spotted alfalfa aphids (*Therioaphis maculata*)

Variety	No. of *L. lineolaris*[a]			
	Uninfested	Infested	Difference	% Increase
Kanza	59.5	70.0	10.5 NS	17.6
Cody	42.2	62.8	20.6 NS	48.8
Alfa[b]	64.5	144.0	79.5[c]	123.2
Cherokee[b]	55.2	93.2	38.0[c]	68.8
LSD 0.10	34.2	34.2		

Source: Lindquist and Sorensen (1970).
Note: NS, nonsignificant.
[a] Means of 4 replications; bugs were introduced 1 week after aphid infestation and were counted daily for 9 days.
[b] Aphid-susceptible varieties.
[c] Significant at 0.10.

American and Old World mirines, including the tarnished plant bug and other *Lygus* species, can be found in papers by Wheeler (1974b, 1976b), Latson et al. (1977), Gupta et al. (1980), Batulla and Robinson (1983), and Ghosh et al. (1988).

APHID PREDATION BY OTHER MIRIDS

Additional records of aphid predation are discussed below by mirid subfamily (see also Table 14.6). Mainly phytophagous species feed facultatively on aphids, as noted earlier, but the frequency of this behavior in nature generally is unknown. In the case of *Lygus lineolaris*, Latson et al. (1977) recorded two instances of aphid predation on cotton, lasting a total of 6.5 minutes, during 45.5 hours of observation (8 A.M. to dark over 8 days) of nine individuals. McIver and Stonedahl's (1987b) behavioral observations of the phyline *Orectoderus obliquus* revealed only two instances of predation—both involving aphids—in nearly 15 hours of field observation.

In addition to *Deraeocoris* species (discussed earlier in this chapter), members of other deraeocorine genera feed on aphids. An early North American record of predation on phylloxerans is Lugger's (1900) observation of *Hyaliodes vitripennis* as an enemy of the grape phylloxera (*Daktulosphaira vitifoliae*). He noted, "Wherever the leaves of the wild grape are covered with the peculiar leaf-galls of the Phylloxera . . . we can be certain to find the bug in large numbers, actively engaged in sucking the lice. For this purpose the young bugs even enter the inside of the galls, and are thus surrounded by an abundance of food."

The phyline *Campylomma verbasci*, sometimes an apple pest (see Chapter 12) that also preys on mites and psyllids, is considered an important predator of aphids such as *Aphis pomi* in Ontario apple orchards (Hagley 1975). The application of food sprays—sucrose and a nutritional yeast material—to apple trees fails to increase significantly the numbers of this aphidophage (Hagley and Simpson 1981). More recent studies in Ontario and Quebec showed that its rate of prey consumption is significantly lower than that for many other predators found on apple trees (Hagley and Allen 1990, Arnoldi et al. 1992; see also Thistlewood and Smith [1996]). This Holarctic bug is one of several early-season generalist predators that suppress small aphid colonies in Poland (Niemczyk and Pruska 1986). Its biological control potential, however, is diminished by a low feeding rate and occasional low densities (Niemczyk 1978), as well as by a tendency to injure its host plants.

The occurrence of *Atractotomus mali*, another phyline apple pest in the Old and New World, is closely synchronized with aphid abundance in Hungarian orchards and is considered important in reducing their numbers (Rácz 1988). During nymphal development, each individual can consume 290–360 *Aphis pomi* nymphs (Harizanova 1989).

Five species of the orthotyline genus *Ceratocapsus* are aphid predators on Canadian fruit crops (Kelton 1983). The orthotyline *Schaffneria davisi*, whose antlike nymphs occur in or near ant-attended aphid colonies on scrub oak in eastern North America, feeds on oak-associated aphids in the laboratory (Wheeler 1991b). In Peruvian cotton fields, the cotton aphid (*Aphis gossypii*) provides the most important winter prey of the orthotyline *Hyalochloria denticornis* (Beingolea 1959b, 1960), but lepidopteran eggs are needed for this bug's egg production.

Scale Insects and Mealybugs

Mirids, often overlooked as predators of scale insects, are more important natural enemies of these homopterans than is generally assumed. They are mostly facultative predators of scale insects and mealybugs, but one mirid subfamily, the Isometopinae, contains species that specialize on armored scales.

ISOMETOPINAE AS PREDATORS OF ARMORED SCALES

Once placed as a separate family, the Isometopinae are now recognized as a mirid subfamily (Carayon 1958, Schuh 1995, Schuh and Slater 1995). Nearly all observations and circumstantial evidence point to scale predation by isometopines, although a Chinese species, *Isometopus shaowuensis*, apparently feeds on spider mites (Ren 1987). At least a few isometopine species can be considered specialized predators of armored scales (Diaspididae).

Palomiella bedfordi nymphs and adults feed on the California red scale (*Aonidiella aurantii*) in South Africa. Although relatively scarce on citrus trees, the bugs can control the scale on heavily infested trees (Hesse 1947). Before Hesse's work, only circumstantial evidence for predatory habits of isometopines had been available (reviewed by Wheeler and Henry [1978a]). Another South African isometopine, *I. transvaalensis*, occurs on citrus heavily infested with the Florida red scale or "circular plum scale" (*Chrysomphalus aonidum*), on which it presumably feeds (Slater and Schuh 1969, Jacobs 1985).

Isometopus intrusus is associated with several diaspidids in the Czech Republic (Štusák and Štys 1958), and in the Netherlands it feeds on the oystershell scale (*Lepidosaphes ulmi*) on apple trees (observations of R. H. Cobben cited by Wheeler and Henry [1978a]). In China, Ren (1987) observed *I. citri* on citrus trees heavily infested with a *Unaspis* species. Ghauri and Ghauri (1983), in describing the new species *Totta zaherii* from India as a predator of tea scale (*Fiorinia theae*), suggested the use of isometopines in the integrated management of scale insects.

Habits of New World isometopines were first documented for *Corticoris signatus* and *Myiomma cixi-*

iforme (Wheeler and Henry 1978a). Both species occur on the trunks of pin oak infested with obscure scale (*Melanaspis obscura*) in Pennsylvania (Plate 24), and *C. signatus* can be associated with walnut scale (*Quadraspidiotus juglansregiae*). The bugs overwinter as eggs deposited under old covers of female scales, with egg hatch beginning in mid to late April. *Corticoris signatus* is bivoltine, whereas *M. cixiiforme* is univoltine. The nymphs and adults of both species feed on female scales by curving the rostrum under a cover and penetrating the body; with the cover removed, a gradual deflating of the scale is seen as feeding progresses (Plate 24; Wheeler and Henry 1978a).

Myiomma cixiiforme occurs in association with mixed populations of obscure scale and *Diaspidiotus osborni* in Alabama (H. J. Hendricks, pers. comm.). *Corticoris signatus* inhabits branches of pecan trees infested with obscure scale in Texas, and *Lidopus heidemanni* occurs on obscure scale-infested branches of black oak in Arkansas (pers. observ.). Miller and Williams (1985) associated populations of *C. signatus* and *L. heidemanni* with gloomy scale (*Melanaspis tenebricosa*) on silver maple in Alabama. Nymphs of *C. signatus* can also be collected from gloomy scale–infested branches of silver maple in Tennessee, and in Arkansas, *Diphleps unica* can be abundant on large branches of osage-orange that are covered with this scale (pers. observ.). A conifer-associated isometopid, *C. libertus*, attacks scale insects on piñon pine in Colorado (Polhemus 1994).

FACULTATIVE PREDATION ON DIASPIDIDS

North America

One of the first North American observations of a mirid preying on scale insects took place in Florida: "These little Capsids [mirids] are very instrumental in destroying scale insects, as I have detected them destroying various species of *Aspidioti . . .* on my Orange trees" (Ashmead 1887b). This predacious mirid was described as a new species, *Rhinacloa citri*, which is now considered a synonym of *Halticus bractatus*, the garden fleahopper. Ashmead's observations, however, probably included several similar-appearing, black species, and the scale predator he reported was most likely an arboreal *Rhinacloa* species (or another dark phyline mirid) rather than *H. bractatus*. Even pestiferous species such as the garden fleahopper sometimes feed on other insects or their eggs, but it would be unusual for this mirid of mainly herbaceous, low-growing plants to occur on trees.

Fulton (1918) observed *Pilophorus perplexus* feeding on San Jose scale (*Quadraspidiotus perniciosus*) on apple trees in New York, and *P. juniperi* is consistently associated with scale insects (*Carulaspis* spp.) on ornamental juniper and feeds on them under laboratory conditions (Wheeler and Henry 1977). *Phytocoris breviusculus*, which feeds on San Jose scale in Delaware

and Ohio, can be reared on balsam fir infested with *Aspidiotus cryptomeriae* in Pennsylvania (Wheeler and Henry 1977). They also found *Deraeocoris nebulosus* and *D. nubilus* apparently feeding on crawlers of *A. cryptomeriae*. *Deraeocoris brevis* attacks the black pineleaf scale (*Nuculaspis californica*) in Oregon (Westigard 1973), and *D. nubilus* appears to prey on pine needle scale (*Chionaspis pinifoliae*) on Scotch pine in Pennsylvania (Stinner 1975; pers. observ.).

Eurychilopterella luridula feeds on *Parlatoreopsis chinensis* in Missouri (Froeschner 1949), scale insects of apple in Nova Scotia (Kelton 1983), *Hemiberlesia diffinis* on pecan in Georgia (Tedders et al. 1990), and on obscure scale in Washington, D.C. (pers. observ.) and in Alabama (H. J. Hendricks, pers. comm.). The powdery-white nymphs of *E. luridula*, which resemble mealybugs, hide in bark crevices (Heidemann 1910). Stonedahl et al. (1997) suggested that the powdery coating of *Eurychilopterella* and certain other deraeocorine nymphs is a waste product from feeding on waxy prey such as some sternorrhynchans. In addition, *Atractotomus magnicornis* preys on the elongate hemlock scale (*Fiorinia externa*) and *Tsugaspidiotus tsugae* in Connecticut (McClure 1979, 1981). Although *A. magnicornis* nymphs and adults can be common on hemlock during June, this phyline appears not to respond numerically to prey densities. It (and a *Phytocoris* sp.) does not exert substantial regulatory pressure on either species of hemlock scale (McClure 1981). A *Phytocoris* species, near *conspurcatus*, is a scale predator in Oregon (Stonedahl 1983b), and *Phytocoris* species might feed on pear-infesting scale insects in Oregon orchards (Gut et al. 1991).

Old World

In Europe, *Pilophorus clavatus* has long been suspected of attacking scale insects (Breddin 1896, Butler 1923). In addition, *P. perplexus* is sometimes abundant in English apple orchards, its distribution on trees coinciding with that of the scale *Quadraspidiotus pyri*. It can consume 10 adult females a day in laboratory tests and might inflict considerable mortality in the field (Solomon 1976). Laboratory tests show daily consumption to be about four adult females, which is sufficient to account for the decline in scale numbers observed in the field (Solomon and Langeslag 1977). *Pilophorus perplexus* apparently also feeds on oystershell scale in English orchards (Easterbrook et al. 1985). Predation on several European diaspidids by *P. cinnamopterus*, *P. clavatus*, *P. perplexus*, *Deraeocoris lutescens*, *D. ruber*, and *Closterotomus biclavatus* also is known (Kosztarab and Kozár 1988), and Kullenberg (1944) observed scale predation by *Miris striatus*. *Alloeotomus chinensis*, a conifer-inhabiting deraeocorine recorded from China and Korea, feeds on "scale insects" (Lee 1971), and *D. pilipes* is considered a natural enemy of the olive scale (*Parlatoria oleae*) on peach trees in Pakistan (Ahmad and Ghani 1972).

Several mirids are natural enemies of soft scales or coccids (Coccidae). Murtfeldt (1894) observed *Deraeocoris nebulosus* preying on crawlers of the terrapin scale (*Mesolecanium nigrofasciatum*) in Missouri, and Howard (1895) might have had her observations in mind when he referred to *D. nebulosus* as one of the most important hemipteran predators of scale insects. In Pennsylvania, *Pilophorus perplexus* is associated with infestations of the globose scale (*Sphaerolecanium prunastri*) on purpleleaved plum (Stinner 1975). The bugs feed on female scales, remaining with their stylets inserted for 5–10 minutes, and also pierce the crawlers. Mirids (taxa unspecified) prey on *Lecanium coryli* in Nova Scotian apple orchards (MacPhee and MacLellan 1971, MacPhee 1975).

FACULTATIVE PREDATION ON MEALYBUGS

Ashmead (1887b) mentioned mirid predation on "Dactylopii" of orange trees in Florida (this predator's probable identity was noted earlier in a discussion of facultative predation on diaspidids), and *Hyaliodes vitripennis* once fed on a "*Dactylopius*" on sycamore in the Washington, D.C., area (Knight and McAtee 1929). *Ranzovius clavicornis* (Henry [1984] clarified the nomenclature) feeds on an undetermined pseudococcid in Washington, D.C. (Knight 1927b). Upholt (1941) reported that *Eurychilopterella luridula* substantially controlled a heavy infestation of the Comstock mealybug (*Pseudococcus comstocki*) on apple trees in South Carolina; the mirid also feeds on this mealybug in West Virginia (Wheeler et al. 1983) and once was abundant in mealybug-infested apple orchards in Virginia (Haeussler and Clancy 1944). Several mirids on apple trees in Nova Scotia were evaluated as predators of the apple mealybug (*Phenacoccus aceris*), and *Deraeocoris fasciolus*, *Blepharidopterus provancheri*, *Hyaliodes harti*, and *Campylomma verbasci* showed promise as important natural enemies (Chachoria 1967). In Washington State, an increasing importance of the grape mealybug (*Pseudococcus maritimus*) in apple and pear orchards prompted surveys for its native natural enemies. The mirids *C. verbasci* and *Deraeocoris brevis*, although not observed feeding on the mealybug, were among potential predators that deserve further study (Grasswitz and Burts 1995).

Nymphs and adults of the clivinematine *Hemicerocoris bicolor* feed on soft scales on guava and orange crops in Mexico (Ferreira 1998). *Engytatus modestus* attacks all stages of Hawaiian "pineapple mealybugs" in rearing cages (Illingworth 1937b; see also Holdaway and Look [1940]). In Kenyan coffee plantations, *Deraeocoris oculatus* might be a major predator of *Planococcus lilacinus* (Anderson 1931), and the phyline *Trichophthalmocapsus jamesi* also feeds on this mealybug of coffee in Kenya (China 1932). *Deraeocoris pallens* feeds on *P. citri* in Israel, but aphids and whiteflies are preferred as prey (Susman 1988). The phyline *Campylomma* species is a probable predator of the cassava mealybug (*Phenacoccus manihoti*) in Nigeria (Neuenschwander et al. 1987).

PREDATION ON ENSIGN SCALES

Records of predation by mirids on Ortheziidae are restricted to plant bugs of the deraeocorine tribe Clivinematini. *Clivinema sericea* was observed "preying on Orthezia" in New Mexico (Knight 1928), and *C. coalinga* feeds on *Orthezia annae* in California (Miller and Schuh 1994). The original description of *Ofellus guaranianus* from Brazil mentioned an association with an *Orthezia* species (Carvalho 1984b). Although Ferreira (1998) stated that Carvalho and Sailer (1953) reported predation on ortheziids by *Ofellus mexicanus*, they actually noted only a host-plant association for this mirid. Another clivinematine, *Ambracius dufouri*, feeds on Ortheziidae in Brazil (Ferreira 1998).

Lace Bugs

Certain mirids are specialized predators of other insects (see previous discussion of *Termatophylidea* spp. as thrips predators). Here, I focus on the genus *Stethoconus*, whose species specialize on lace bugs. Examples of facultative predation on tingids are also included.

STETHOCONUS SPECIES AS LACE BUG PREDATORS

The first references to an association of these mirids with lace bugs were those by Rey (1881) and Nawa (1910), as discussed by Henry et al. (1986). In describing the new species *Stethoconus frappai* as a natural enemy of *Dulinius unicolor* on coffee trees in Madagascar, Carayon (1960) also reviewed most of the other known *Stethoconus*-tingid associations. These include *S. japonicus* and *Stephanitis ambigua* in Japan, *S. cyrtopeltis* and the pear lace bug (*Stephanitis pyri*) in Europe, and *S. scutellaris* and *Habrochila* species on coffee in Africa. Kerzhner (1970) recognized the presence of two *Stethoconus* species in Europe, stating that *S. cyrtopeltis* is associated with *Stephanitis oberti* and that *S. pyri* (previously considered a synonym of *S. cyrtopeltis*) feeds on *Stephanitis pyri*.

In addition, Hoffmann (1935) recorded an undetermined plant bug species attacking the nymphs of *Stephanitis typicus* on bananas in China. This predator might be *S. praefectus* or a related species (Mathen et al. 1967). Cheng (1967) made additional observations on apparently the same mirid species as a predator of *Stephanitis typicus* on banana in Taiwan, noting its developmental time is 8–18 days shorter than that of its prey. Mathen and Kurian (1972) found that *Stethoconus praefectus* is an "ideal predator" that reproduces on coconut trees throughout the year and in India has a life cycle shorter than that of its prey. Fewer than 10 days are required for nymphs to develop, with a mean of 62 tingid nymphs consumed during development. They recommended minimal use of insecticides against lace bugs when *S. praefectus* is present on coconut.

The mirid, cited as an *Appolodotus* species, preying on an adult *Tingis buddleiae* in India (Livingstone 1968) probably also was *S. praefectus*. According to Livingstone, this predator occurs most often with the lace bugs *Stephanitis typicus*, *Corythauma ayyari*, and *Naochila sufflata*. The tingid *Stephanitis subfasciata* apparently is preyed on by *S. praefectus* in southern Japan (Yasunaga et al. 1997).

In India, another lace bug attacked by *S. praefectus* is *Teleonemia scrupulosa* (Ganga Visalakshy and Jayanath 1994), which was imported from Australia to help suppress infestations of lantana. The plant was introduced into India as an ornamental, but it escaped from cultivation and became a pest of pastures and forests (Muniappan and Viraktamath 1986). Whether the mirid's predation on *T. scrupulosa* might hinder biocontrol efforts against lantana is unknown.

Outbreaks of *Habrochila placida* on coffee, which followed DDT spraying west of the Rift Valley in Kenya, might have been triggered by the destruction of its mirid predator, a species of *Stethoconus* (LePelley 1957). With DDT no longer used, *Stethoconus* populations are able to keep the lace bugs under partial control. Organophosphorous insecticides should be used only if mirids are not suppressing the pest (de Pury 1968).

Although *Stethoconus* species help suppress lace bug populations (Schumacher 1917a), their biological control potential against immigrant, injurious tingids is not fully appreciated. However, the discovery of an adventive species now established in the United States has emphasized the role that these specialized predators might play in the integrated management of tingids, including the azalea lace bug (*Stephanitis pyrioides*) and *S. takeyai* on Japanese pieris (Henry et al. 1986, Yasunaga et al. 1997). Henry et al. (1986) reported the

Old World *Stethoconus japonicus* from Maryland, where nymphs and adults feed on the azalea lace bug. They transferred the genus *Stethoconus* from the tribe Clivinematini of the subfamily Deraeocorinae to the deraeocorine tribe Hyaliodini. Their paper includes additional references to *Stethoconus* species as lace bug predators.

Further research on the biological control potential of *S. japonicus* (Neal et al. 1991) showed that nymphs develop at a rate similar to that of their azalea lace bug prey and that, on average, 17.8 fifth-instar tingids are consumed during nymphal development. First instars can kill fifth-instar lace bugs (Fig. 14.9). *Stethoconus japonicus* nymphs approach prey from the side or rear with their antennae recurved. Azalea lace bug nymphs do not respond to the predator's approach. The apparent ease with which the mirid approaches its prey might reflect an environmental insensitivity to predation that is associated with tingid defense mechanisms (Oliver et al. 1990). Stylet insertion by the mirid results in almost immediate paralysis or death of the prey, presumably from injection of potent venoms or salivary enzymes. The predator feeds on nymphal lace bugs in situ (see Fig. 14.9), but *S. japonicus* adults use their forelegs to invert adult prey (Fig. 14.10; Neal et al. 1991). Although adult females will accept the hawthorn lace bug (*Corythucha cydoniae*) as prey, azalea lace bugs are clearly preferred in tests of prey preference (Table 14.8; Neal et al. 1991).

Overwintered eggs of *S. japonicus* do not hatch until June, or not until the azalea lace bug produces a second generation. If diapausing eggs of predator and prey were to hatch synchronously, the mirid would quickly deplete its sole food source (Neal et al. 1991, Neal and Haldemann 1992).

Stethoconus species, like members of certain other mainly carnivorous heteropterans such as phymatine reduviids (Balduf 1943), nabids (Cobben 1978), and the pentatomid *Podisus maculiventris* (Ruberson et al. 1986), occasionally imbibe plant sap (also reviewed by

Fig. 14.9. *Stethoconus japonicus*, first instar, with stylets inserted in a fifth-instar azalea lace bug (*Stephanitis pyrioides*) (×50). (Reprinted, by permission of J. W. Neal Jr. and the Entomological Society of America, from Neal et al. 1991.)

Fig. 14.10. Male azalea lace bug (*Stephanitis pyrioides*) that the predacious mirid *Stethoconus japonicus* immobilized and inverted with its forelegs before beginning to feed (×50). (Reprinted, by permission of J. W. Neal Jr. and the Entomological Society of America, from Neal et al. 1991.)

Table 14.8. Numbers and frequency of azalea lace bug (*Stephanitis pyrioides*) and hawthorn lace bug (*Corythucha cydoniae*) females consumed by *Stethoconus japonicus* females in a preference test

Female no.	Test[a]		
	1	2	3
1	96: 44 (8)	100: 26 (9)	91: 10 (5)
2	115: 7 (5)	75: 39 (7)	74: 3 (2)
3	112: 22 (7)	75: 29 (9)	101: 14 (5)
4	94: 9 (6)	80: 49 (9)	119: 11 (4)
5	106: 14 (5)	82: 45 (9)	89: 26 (8)
6	101: 41 (8)	107: 7 (5)	98: 19 (8)

Source: Neal et al. (1991).

[a] Number at left of colon is azalea lace bug; at right is hawthorn lace bug. In parentheses are the number of times dead hawthorn lace bugs were counted in 10 observations over 21 days.

Naranjo and Gibson [1996]). Carayon (1960) referred to Putshkov and Putshkova's observations in the former USSR of *S. pyri* (as *S. cyrtopeltis*) feeding on the leaf petioles and buds of linden trees.

FACULTATIVE PREDATION ON LACE BUGS

Despite the feeding aggregations characteristic of many lace bug species, the family has relatively few predatory enemies (e.g., Sheeley and Yonke 1977, Oliver et al. 1985; cf. Horn et al. 1983a, Trumbule et al. 1995). The first North American reference to plant bugs as possible lace bug predators was that of P. R. Uhler, who

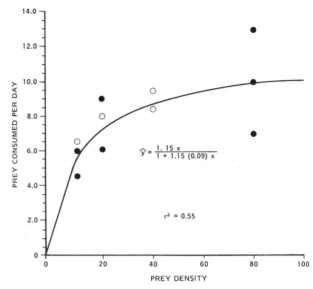

Fig. 14.11. Functional response (Type II) of *Deraeocoris nebulosus* adults to increasing densities of oak lace bug (*Corythucha arcuata*) nymphs at 20–21°C and 30–40% relative humidity. Curve represents predicted feeding rate at any prey density; the points, the observed feeding rates at prey densities of 10, 20, 40, and 80. The open circles represent two data values at the same position. (Reprinted, by permission of the Entomological Society of America, from Wheeler et al. 1975.)

informed Riley (1871) that mirids sometimes live among immature tingids. Another early observation was that of *Hyaliodes vitripennis* as "common on mulberry among *Corythucha pallida*" in the Washington, D.C., area (Knight and McAtee 1929). In the eastern United States, the generalist predator *Deraeocoris nebulosus* helps reduce populations of certain tingids, particularly the oak lace bug (*Corythucha arcuata*). A large second generation of *D. nebulosus* can develop on oak, and in the laboratory, consumption of lace bug nymphs increases with increasing prey density (Fig. 14.11; Wheeler et al. 1975).

Deraeocoris nebulosus, which tracks prey populations on various plants, also feeds on the hawthorn lace bug (*C. cydoniae*) in the field and laboratory (Stinner 1975, Wheeler et al. 1975) and on the sycamore lace bug (*C. ciliata*) (Horn et al. 1983a). Following the detection and spread of the Nearctic sycamore lace bug in Europe, surveys of its natural enemies were conducted in the United States, and a *Deraeocoris* species (probably *D. nebulosus*) and *Reuteria* species (probably *R. bifurcata*) were found on tingid-infested leaves in 1979 and 1981 (Balarin and Maceljski 1986). *Deraeocoris flavilinea* and *D. lutescens* are native European mirids that now include the adventive *C. ciliata* among their prey (Arzone 1986, Tavella and Arzone 1987).

A Nearctic phyline, the azalea plant bug (*Rhinocapsus vanduzeei*), preys on the azalea lace bug (*S. pyrioides*) on evergreen azaleas in Georgia. In the laboratory, fifth-instar mirids kill fifth-instar lace bugs (Braman and Beshear 1994).

Other Arthropods as Prey

In addition to the arthropods already discussed as mirid prey, other groups—Collembola, Psocoptera, Cercopidae, Heteroptera, and adult Diptera—can serve as a food source. The discussion of heteropterans as prey includes mirids, but interspecific predation only; cannibalism, or intraspecific predation, is covered at the end of this chapter.

COLLEMBOLA

Mirids do not often feed on collembolans, or at least few records of such predation appear in the literature. Reuter (1875a), the first to mention plant bugs as possible collembolan predators, said that *Chlamydatus pulicarius* might prey on "small Podurids." Maclagan (1932) noted that *Lygus pratensis* (probably *L. rugulipennis*) "sucked the juices of its *Smynthurus* [sic] victims fairly well," referring to predation on the lucerne flea (*Sminthurus viridis*) under confined conditions in Scotland. Kullenberg (1944) recorded collembolans as prey of *Psallus variabilis* in Sweden. B. D. Turner (1984) studied predation on epiphytic herbivores grazing on bark and leaf surfaces of European larch trees in southern England, noting that an individual of *Phytocoris dimidiatus* reacted positively to collembolan antiserum. *Schaffneria davisi* fed on an unidentified collembolan on post oak in Georgia (Henry 1994), and a fourth-instar *Psallus ambiguus* fed on an unidentified collembolan in laboratory feeding trials (Morris 1965).

PSOCOPTERA

Shelford (1913) suggested a predatory relationship between *Hyaliodes vitripennis* and psocids, noting this mirid usually occurs with booklice ("*Psocus*") on the bark of oaks in Illinois. The nymphs and adults of *Deraeocoris* species feed on barklice or psocids in western North America (Razafimahatratra 1980). Several predacious, bark-inhabiting species of *Phytocoris* also feed on Psocoptera, including *P. neglectus* on psocids living on apple trees in the eastern United States (Knight 1923). *Phytocoris populi*, *P. reuteri*, and the dicyphine *Campyloneura virgula* feed on psocid nymphs and eggs in England (Southwood and Leston 1959, Groves 1976). Schumacher (1917c) discussed psocid predation by *C. virgula* and other European dicyphines. Immature psocids are accepted, though not readily, as prey of *Psallus ambiguus* in laboratory feeding trials (Morris 1965).

China (1931) suggested that the nymph of a remarkable Cuban mirid, possessing a pronotal "anchor" and possibly belonging to the deraeocorine genus *Paracarnus*, feeds on web-building psocids. Whether the prono-

tal process is an adaptation for life in psocid webs has not been determined. The nymphs and adults of the orthotyline *Loulucoris kidoi* occur on a fan palm in Hawaii, specifically in the deep folds of dead, still-attached leaves inhabited by psocids (Asquith 1995a). Although predation was not observed, this mirid might prey on psocids, as do certain anthocorids that consistently occur in dead leaves on trees (Lattin 1999; pers. observ.). Perhaps the only other mirid recorded from a similar microhabitat is the orthotyline *Pseudoloxops* (= *Aretas*) *nigribasicornis*, which in Tahiti was collected in dead (presumably attached) leaves of fehi banana (Knight 1937).

CERCOPIDAE

Cercopid nymphs are protected from many natural enemies by their frothy anal secretions or spittle (Whittaker 1970). The only record of mirid predation on spittlebugs in nature might be that of *Deraeocoris borealis*. In New York, Drake (1922) observed this plant bug with its proboscis inserted in a spittle-covered nymph of *Clastoptera obtusa*. Tedders (1995) found that *D. nebulosus* was the most common predatory insect associated with *C. achatina* on pecan trees in Georgia. When nymphs and adults of this mirid were caged on pecan limbs harboring spittlemasses, the bugs fed on cercopid nymphs of all stages. Under laboratory conditions, the Asian mirine *Castanopsides potanini* fed on the nymphs of *Aphrophora costalis* (Yasunaga 1998).

LYGAEOIDEA, MIRIDAE, AND OTHER HETEROPTERA

Lygaeoid bugs (see Henry [1997] for discussion of taxonomic status of current lygaeoid families traditionally placed as subfamilies of the Lygaeidae) are included among arthropods sometimes attacked by mirids. *Phytocoris pini* occasionally feeds on the eggs and nymphs of a pine cone bug (*Gastrodes grossipes*) in the former USSR (Putshkov 1961a). Duda (1886) suggested that this lygaeoid also serves as prey for *Deraeocoris annulipes*.

In the New World fauna, Champlain and Sholdt (1967) suggested a mirid-lygaeoid relationship between *Lygus hesperus* and *Geocoris punctipes*. Although one might assume that only the geocorid is predacious in this relationship (e.g., Gonzalez and Wilson 1982), they actually prey on one another, but the lygus bug's more rapid development at lower temperatures favors its increase early in the season. On the basis of laboratory observations, Champlain and Sholdt (1967) speculated that early-season predation by lygus nymphs could significantly lower nymphal populations of the geocorid in Arizona (see also Cohen [1982]). In assessing the suitability of various diets for rearing *G. punctipes*, Dunbar and Bacon (1972) reported mortality when the geocorids are provided lygus bug nymphs and green beans. Because they previously observed lygus nymphs attacking first-instar *Geocoris* species, they thought some of the mortality resulted from mirid predation rather than from the low nutritive value of the lygus bug nymphs and beans.

The anthocorid *Orius insidiosus* furnishes occasional food for *Deraeocoris nebulosus* in Virginia apple orchards (McCaffrey and Horsburgh 1986), and *D. pallens* nymphs feed on *Orius* nymphs in cotton fields in Israel (Susman 1988). In laboratory experiments in India, *Nesidiocoris caesar* fed on nymphs of the berytid *Metacanthus pulchellus* (Chatterjee 1986).

Mirids also attack species of their own family (Kullenberg 1944, Collyer 1952, Wheeler 1974b). A *Deraeocoris ruber* adult fed on a *Lygocoris pabulinus* nymph under artificial conditions (Petherbridge and Thorpe 1928b), and *Deraeocoris* species attacked newly molted nymphs of *Platylygus luridus* in the laboratory (Rauf et al. 1984b). The tomato bug (*Engytatus modestus*) feeds on the nymphs and adults of another dicyphine, the suckfly (*Tupiocoris notatus*), on tobacco plants in North Carolina, its predation perhaps limiting the buildup of large *Tupiocoris* populations (Thomas 1945). *Lygus* species were among radiolabeled predators of the cotton fleahopper (*Pseudatomoscelis seriata*) in Texas (Breene et al. 1989a). Under laboratory conditions, adults of *Rhinacloa forticornis*, the western plant bug, can substantially reduce the numbers of second-instar *Lygus* nymphs; a smaller reduction in the numbers of first-instar lygus bugs is obtained with the facultative predator *Spanagonicus albofasciatus* (Stoner and Bottger 1965). *Deraeocoris bakeri* preys on the nymphs of the phyline *Chlamydatus* species in Oregon (Razafimahatratra 1980), and in Quebec, the adults and nymphs of *Neurocolpus nubilus* feed on the adults of another mirine, *Lygocoris communis* (Boivin and Stewart 1983c). Because *N. nubilus* belongs to a mostly phytophagous genus of mirids, Henry and Kim (1984) suggested that Boivin and Stewart (1983c) observed predation on weakened adults. Their suggestion is strengthened by the failure of *N. nubilus* to prey on *L. communis* nymphs or adults under confined conditions (Arnoldi et al. 1991).

The Palearctic mirine *Rhabdomiris striatellus* and orthotyline *Dryophilocoris flavoquadrimaculatus* sometimes feed on nymphs of other mirid species (Kullenberg 1944, Southwood and Leston 1959). Cooper (1981) speculated that *Deraeocoris piceicola* feeds on the early instars of *Phytocoris neglectus*, another predacious mirid found on noble fir in western North America. *Deraeocoris oculatus* and *D. ostentans*, which feed on the nymphs and adults of mirids injurious to cotton in southeastern Africa, are considered valuable biological control agents (da Silva Barbosa 1959; see also Delattre [1947]). In the absence of food, mirid species that co-occur on sorghum in India will prey on each other (Sharma and Lopez 1990a).

Predation might allow the coexistence of five British broom-inhabiting mirid species that use plant and animal food resources sequentially through the season. The nymphs of early-appearing mirids on broom attack the early instars of plant bug species hatching later (Dempster 1960, 1964, 1966; Southwood 1978a). The last species to hatch, *Orthotylus concolor*, is usually the least abundant of the complex. On Scotch broom in California, where this plant has been accidentally introduced, the adventive *O. concolor* flourishes in the absence of other Palearctic mirids that specialize on broom (Waloff 1966, 1968).

ADULT DIPTERA

Nowicki (1871) reported that *Leptopterna dolabrata* attacked an adult of the chloropid *Chlorops pumilionis* in Poland. This likely is the record questioned by Kirkaldy (1906), who apparently did not realize the extent of predatory tendencies among plant bugs. Another early record is a nymph of an unspecified plant bug species feeding on an adult chironomid, *Cricotopus* species, in England (Poulton 1906).

Species of *Rhinacloa* feed on cecidomyiid flies in the laboratory (Herrera 1965). In India, Hiremath et al. (1984) reported that the nymphs and adults of the sorghum earhead bug (*Calocoris angustatus*) attack ovipositing *Contarinia sorghicola*, a cecidomyiid pest of sorghum. Predation also occurs in the laboratory when the bugs and midges are released in the same container. Other mirid pests of sorghum feed opportunistically on the sorghum midge (Sharma and Lopez 1990a). In North America, the dicyphine *Tupiocoris rhododendri* and the phyline *Rhinocapsus vanduzeei* prey on small dipterans occurring on deciduous azaleas in Georgia (Braman and Beshear 1994), and *Taedia scrupea* adults occasionally prey on midges found on grape vines (Martinson et al. 1998).

Cannibalism

Cannibalism, or intraspecific predation, is generally defined as the killing and eating of an individual of the same species. Cannibalism often is the outcome of interactions between individuals in an asymmetrical contest (e.g., Polis 1981, Elgar and Crespi 1992). Conspecific feeding can involve attacks on siblings or parents and include mating, courtship, and competitive interactions (Fox 1975, Polis 1981, Elgar and Crespi 1992). Some authors broaden the definition to include scavenging on members of the same species, or even the same family (Lockwood 1989), but the traditional, more restricted definition—that is, intraspecific predation—is used in the following discussion.

Cannibalism often has a genetic basis (e.g., Fox 1975, Polis 1981, Elgar and Crespi 1992) but is generally induced by environmental cues. It can be triggered by nutritional deficiency, crowding, or other stress factors (Fox 1975, Duelli 1981, Polis 1981, Al-Zubaidi and Capinera 1983, Joyner and Gould 1987). However, cannibalism is not restricted to stressed populations; it occurs in the bollworm or corn earworm (*Helicoverpa zea*) when food resources are not limiting (Joyner and

Gould 1985, 1987). Size, age, developmental stage, and sex influence the frequency of cannibalism (Polis 1981, Dial and Adler 1990).

Once considered abnormal and rare, or merely an artifact of laboratory rearing, cannibalism is an adaptive behavior that is common in many groups of organisms. Such behavior has associated costs as well as benefits. It might be important as a major mortality factor that limits the growth and size of populations and affects age structure and spatial distribution (Polis 1981, Elgar and Crespi 1992). Intraspecific predation is a self-regulating, homeostatic process that can function more efficiently than certain other agents of population regulation (Polis 1981). Individuals can benefit through increased food quality and quantity and decreased intraspecific competition. Increased survivorship, fecundity, developmental rate, and body size are associated with cannibalism in various insects (Quiring and McNeil 1985, Joyner and Gould 1987). Because the extent of cannibalism can vary among members of a genus, any generalizations about its influence as a selective mechanism should be made only after thorough research on a particular system (Spence and Cárcamo 1991).

Cannibalism in the Heteroptera has been studied most extensively in anthocorids (Arbogast 1979, Nasser and Abdurahiman 1993), gerrids (Spence et al. 1980, Spence 1986, Spence and Cárcamo 1991), and notonectids (Fox 1975, Orr et al. 1990, Streams 1992). On the basis of laboratory data, cannibalism in gerrids might influence behavioral patterns such as habitat selection and territoriality (Nummelin 1989).

EXTENT OF CANNIBALISM IN MIRIDS

Cannibalism might be expected to occur frequently in mirids, a group characterized by diet plasticity, numerous omnivores, and species using ephemeral food sources such as inflorescences or other limited resources. Such behavior frequently occurs in laboratory cultures of mirids (e.g., Schewket Bey 1930; Woodroffe 1958b, 1959a; Curtis and McCoy 1964; Stewart 1969a; Readshaw 1975; Bentur and Kalode 1987; Libutan and Bernardo 1995; van Dam and Hare 1998). Fryer (1916) remarked that in biological studies on Lygocoris rugicollis, a pest in English apple orchards (see Chapter 12), "it is difficult . . . to keep more than one alive in each cage." Collyer (1952) commented similarly on the cannibalistic behavior of Blepharidopterus angulatus under laboratory conditions. Often the victim is undergoing ecdysis (e.g., Reed 1959, Taksdal 1961, Morris 1965, Vanderzant 1967, Wightman 1968, Stewart 1969a, Henry and Wheeler 1988) or is moribund or injured (e.g., Ballard 1921, Roberts 1930, Beards and Leigh 1960, Horsburgh 1969, El-Dessouki et al. 1976, Hori and Miles 1993). Lygocoris communis nymphs and adults sometimes feed in aggregations on conspecific individuals infected with an entomophthoraceous fungus (see Chapter 6); the diseased bugs show a ruptured dorsal wall but remain active (Dustan 1924). In the laboratory, cannibalism can occur between newly emerged first instars when they are deprived of food (Smith and Franklin 1961), or between nymphs when their host plants deteriorate (e.g., Khattat and Stewart 1977). When rearing Lygus lineolaris, Stevenson and Roberts (1973) observed greater cannibalism on a green bean than on a lettuce diet. They suggested that the latter food source reduces cannibalism by providing greater shelter for the nymphs. In the largely predacious Macrolophus melanotoma, cannibalism increases in the laboratory on a host plant on which typical prey are difficult to find (Constant et al. 1996a). Apparently because of their random search for prey (see Chapter 7), some predatory mirids, under crowded conditions, will cannibalize even when other prey are abundant (Morris 1965).

Although intraspecific predation in plant bugs most often involves nymphs as victims, eggs and adults also are attacked by conspecific individuals. The mirine Polymerus cognatus sometimes feeds on its eggs (Strawiński 1964a). Cyrtorhinus fulvus and C. lividipennis behave similarly when their preferred prey, planthopper eggs, are scarce (O'Connor 1952, Hinckley 1963a, Matsumoto and Nishida 1966; see also Döbel and Denno [1994]), which might allow these predators to persist during periods of prey scarcity and to stabilize mirid-planthopper interactions. Overwintered adults of three European species of Lygus occasionally cannibalize (Boness 1963), and Taedia scrupea adults exhibit similar behavior (Martinson et al. 1998). Miller (1971) referred to an unusual record of cannibalism: a male of the myrmecomorphic mirine Myrmecoris gracilis, in attempting to copulate, inserted his stylets in the female's thorax and emptied her body fluids. After copulation, the female of Nesidiocoris caesar sometimes attacks the male (Chatterjee 1984a), behavior that might reflect a lack of animal prey in the diet (e.g., Torreno and Magallona 1994).

Most workers have noted cannibalism in mirids without considering its effects on population structure and dynamics. Among the numerous species that cannibalize in nature are the Nearctic Blepharidopterus provancheri (Steiner 1938), Deraeocoris species (Razafimahatratra 1980), Lygus species (Wheeler 1976b), Ranzovius clavicornis (Wheeler and McCaffrey 1984), and Taedia hawleyi (Hawley 1917). In Europe, cannibalism occurs in numerous mirids, including Atractotomus mali, Deraeocoris ruber, Orthotylus marginalis (Collyer 1953b), Cyrtorhinus caricis (Rothschild 1963), Plagiognathus arbustorum (Taksdal and Sørum 1971), and Miris striatus (Couturier 1972). Cannibalism can also be inferred from studies on the English mirids living on Scotch broom (Dempster 1964, 1966; Southwood 1978a), and Waloff (1968) stated that broom mirids indulge in "fratricidal predation."

Cannibalistic tendencies are also apparent among Deraeocoris brachialis in Japan (Yasunaga 1990), Nesidiocoris tenuis in Egypt and the Philippines (El-Dessouki et al. 1976, Torreno and Magallona 1994), N. volucer in Zimbabwe (Roberts 1930), and N. caesar (Chatterjee

1983b, 1986) and *Calocoris angustatus* in India (Hiremath et al. 1984, Natarajan and Sundara Babu 1988c, Sharma and Lopez 1990a). Adults of the last-named species, an important sorghum pest, cannibalize in the field and laboratory. Both sexes will feed in the sternum of their prey for 5–10 minutes. Cannibalism is less frequent in the laboratory when milky earheads of sorghum are provided. No cannibalism occurs at densities fewer than 50 bugs per earhead, but it ranges from 2% to 5% when populations exceed 50 (Hiremath et al. 1984).

CANNIBALISM AND MIRID POPULATION DYNAMICS

Only one study was designed to assess cannibalism's effects on mirid population dynamics: that by Collyer (1965) on *Blepharidopterus angulatus* in England. Using plots of apple trees, she removed nymphs of *B. angulatus* (and anthocorids and other predacious mirids, which occurred only in small numbers) from the central trees. The peripheral trees in each plot served as controls. The numbers of *B. angulatus* were counted on experimental and control trees after all eggs had hatched and before many adults were present. Even though the number of nymphs removed from the experimental trees was approximately equal to the final counts on the control trees, the reduction in numbers was estimated at only 28% in 1959. Collyer (1965) suggested that cannibalism was largely responsible for the surprisingly small population found on the control trees, and she advised that this possibility "be considered when interpreting the results of field experiments in which mortality due to other factors is under study."

CANNIBALISM AND INTERPRETATION OF EXPERIMENTS

Cannibalism's potential for affecting the interpretation of experimental results is shown by the group rearing of *Lygus hesperus* on a meridic diet (see Chapter 13). Strong and Kruitwagen (1969) developed a diet apparently superior to green beans; although survival was poor, adults reared on the diet matured faster, were more fecund, and lived longer. Because cannibalism was observed when bugs were reared on the diet or on beans, its effects were discounted when evaluating the diet's success. They later found that all *L. hesperus* reared in isolation on the diet died as nymphs, but that 90–95% of the bugs survived when they were isolated on green beans. The occurrence of cannibalism among the group-reared bugs actually was responsible for the "survival" reported on the meridic diet (Strong and Kruitwagen 1970).

Bentur and Kalode (1987) stated that prolonged survival of the nearly obligate predator *Cyrtorhinus lividipennis* on plant material in the absence of insect prey might be explained by cannibalism among group-reared individuals (or the inadvertent presence of mites or aphids). Bentur and Kalode (1987) observed cannibalism while rearing this mirid.

Studies on the economic importance of mirids on birdsfoot trefoil provide another example of how cannibalism can influence experimental results. When assessing injury to buds, Wipfli et al. (1990a) found that higher infestation levels generally resulted in greater bud death. They suggested that when the opposite trend is observed, cannibalism within experimental cages could explain it.

UNDERSTANDING CANNIBALISM IN MIRIDS

Although cannibalism is common among mirids, occurring in both phytophagous and predacious groups, many fundamental questions about this behavior remain unanswered. Does it represent opportunistic predation in most species while having evolutionary significance in others? Is intraspecific predation more common among phytophagous or predacious species, or more common in females than in males? Is it more frequent in polyphagous mirids or in host-restricted specialists that develop on ephemeral resources such as oak catkins? Does cannibalism occur among gregarious plant bugs, for example, species of *Caulotops*, *Halticotoma*, and *Mertila*, in which various nymphal instars aggregate on host leaves?

Few examples of cannibalism's effects on population dynamics have been documented for plant bugs, but substantial effects probably are more common than the literature would suggest. Considering the potential economic importance of cannibalism, as well as its ecological and evolutionary implications, studies are needed to determine the frequency of intraspecific predation in mirids, to assess its effects on populations, and to clarify the factors that trigger this behavior.

15/ Scavenging and Use of Other Animal-Associated Foods

The distinction between saprophagy, phytophagy and predatism is not always easily drawn.

—C. T. Brues 1946

Scavenging includes the use of dead and decaying animals or plants and animal wastes as food sources. Plant bugs exploit dead animals as a convenient source of nitrogen and perhaps also trace elements and critical salts. Mirids also exploit other alternative or aphytophagous foods such as insect hemolymph and blood. They appear not to feed on decaying plant matter.

Scavenging

The use of invertebrate carrion has received scant attention from ecologists (Seastedt et al. 1981). An exception is the well-studied ants, which tend to monopolize invertebrate carrion on the forest floor, quickly locating this food source and removing the smaller carcasses to their colonies (Fellers and Fellers 1982). Mirids are not obligate users of invertebrate carrion, but they often use this nutrient source on their host plants.

Mirids commonly feed on dead individuals of their species under laboratory conditions (e.g., Petherbridge and Husain 1918, Reinhard 1926, Guppy 1963, Wheeler et al. 1979, Araya and Haws 1988, Wheeler 1989, Wipfli et al. 1990a, Asquith 1993b). The scavenging or necrophagous tendencies of mirids can be used to advantage in rearing European and North American species of *Lygus* (Wheeler 1976b; see Chapter 13). This tendency toward scavenging is potentially a cause of false-positive results from certain serological techniques used to assess predation in the field. Under experimental conditions, access to arthropod cadavers can affect the results of studies on life-history attributes such as fecundity and longevity (see Chapter 6).

The literature contains numerous references to scavenging on aphids and other small arthropods (e.g., Goddard 1935, Rothschild 1963, Stinner 1975, Braman and Beshear 1994), as well as occasional mention of feeding on larger cadavers such as a butterfly larva (Reuter 1881) and a bumble bee, *Bombus* species (Stephens 1982). Similar casual references to scavenging in the laboratory or field are omitted from this chapter, except for an association with fungus-killed anthomyiid flies and several other notable examples discussed later.

Examples of scavenging in the Miridae also include *Ranzovius* species that live in spider webs and feed mainly on dead insects, and plant bugs that scavenge on arthropods that die after becoming entrapped on the bugs' glandular host plants.

RANZOVIUS SPECIES IN SPIDER WEBS

Species of the phyline genus *Ranzovius* exhibit a habit unique among mirids—an obligate relationship with web-building spiders. They are associated with funnel weavers (Agelenidae) and combfooted spiders (Theridiidae). The only other heteropterans that live as commensals in spider webs are certain plokiophilids (Plokiophilinae), some emesine reduviids, and members of the nabid genus *Arachnocoris* (Eberhard et al. 1993, Schuh and Slater 1995).

Ranzovius fennahi (Henry [1984] provided additional references and clarified nomenclature) sometimes attacks egg sacs of the subsocial theridiid *Anelosimus eximius* in Trinidad (Carvalho 1954a), but other species of the genus feed mainly as scavengers in webs of their "hosts." *Ranzovius californicus*, which inhabits the nonadhesive sheet webs of the agelenid *Hololena curta*, feeds on entrapped insects or ones tied to the web by the spider. When small, stunned flies are thrown on the agelenid web, *R. californicus* attacks, its struggles with the prey eliciting a response from the spider, which rushes out to capture the struggling fly from the mirid (Davis and Russell 1969). Davis and Russell (1969) considered the mirid-spider relationship commensal, but if the mirid feeds on insects representing potential food for the spider, the relationship is one of kleptoparasitism (e.g., Thornhill 1975; see also Vollrath [1984], Eberhard et al. [1993]). The mirids could be classified as "pilfering kleptobionts" (*sensu* Vollrath 1984)—that is, nonaggressive competitors that take a resource from its owner by stealth, often without interacting with the owner. These small plant bugs, according to Vollrath, seem unable to challenge their much larger spider hosts for prey.

In the eastern United States, the multivoltine *R. clavicornis* occurs most frequently in the webs of another subsocial theridiid, *Anelosimus studiosus*, but also inhabits the webs of agelenid spiders. The association with agelenids, observed mainly when their webs are built on shrubs harboring webs of the theridiid, is considered secondary (Wheeler and McCaffrey 1984). Nymphs and adults are active on the webs by day, walking upright or upside down along the bottom of the webs (Plate 24). Their movement usually does not elicit a response from the spiders (cf. Vollrath 1984). Wheeler and McCaffrey (1984) observed *R. clavicornis* scavenging on insects too small to trigger feeding responses or those ignored by the theridiids.

Food sources available to *Ranzovius* species fluctuate throughout the season and depend partly on the plant species colonized by host spiders. When boxwood serves as the host plant, for example, large numbers of the boxwood psyllid (*Cacopsylla buxi*) become entrapped in the webs. Wheeler and McCaffrey (1984) considered the relationship of *R. clavicornis* to its spider hosts one of "benign commensalism"—that is, not lowering host fitness—rather than one of kleptoparasitism or predation. If, however, small insects should become important prey items for the spider when larger insects are scarce, as might be the case for certain tropical araneids (Nentwig 1985), then the mirids might be partly kleptoparasitic (Vollrath 1984). In addition, Wheeler and McCaffrey (1984) observed an example of predation on molting spiderlings of *A. studiosus*.

Ranzovius agelenopsis, known only from the type locality (Knoxville, Tenn.), lives in the webs of the agelenid *Agelenopsis pennsylvanica*. On boxwood, this mirid coexists with *R. clavicornis* and might feed mainly as a scavenger on invertebrates caught in agelenid webs (Wheeler and McCaffrey 1984). Wheeler and McCaffrey (1984) gave additional details on the life history and behavior of *Ranzovius* species, raised questions regarding the sympatry of *R. clavicornis* and *R. agelenopsis*, and speculated on the origin of the feeding habits in the genus.

The four *Ranzovius* species that have been studied move across webs without difficulty. The webs of agelenids are not sticky (e.g., Turnbull 1965), and those of theridiids are only selectively so; the loose, irregular webs have special trapping threads that contain sticky droplets (Foelix 1982). The claws of *Ranzovius* are held straight down parallel to the tarsus when the bugs walk on top of the web, or the claws turn in, almost perpendicular to the tarsus, when the bugs hang under the web (Davis and Russell 1969).

Further studies are needed to clarify the relationships of different *Ranzovius* species with their spider hosts. Some species might be facultative predators, at times attacking spider egg sacs or spiderlings, whereas others might be mostly kleptoparasites or merely commensals. Given the plasticity of mirid trophic habits, a particular species (or any individual) could interact variously with its host—as a predator, kleptoparasite, or commensal.

INHABITANTS OF GLANDULAR AND INSECTIVOROUS PLANTS

Trichomes (plant hairs) help deter insect herbivory. Glandular trichomes of solanaceous plants such as potato, tobacco, and tomato, which impede or immobilize small insects, are well known (e.g., Tingey 1985, Duffey 1986, Gregory et al. 1986). These plants are viscid because of mucilage, but the stickiness of South African roridulas is due to resin (Lloyd 1934, Dolling and Palmer 1991). Mirids are able to live on both types of sticky plants.

Dicyphine mirids are mostly restricted to glandular plants, especially those of the families Geraniaceae, Lamiaceae, Scrophulariaceae, and Solanaceae (Cassis 1984). These bugs are specialized for traversing the surfaces of viscid-hairy plants. In contrast, *Lygus lineolaris* nymphs and adults adhere to the substances exuding from their feeding punctures on a factitious host in laboratory studies (Khattat and Stewart 1977), and the anthocorid *Orius insidiosus*, when released on tomato plants for biological control purposes, dies quickly when it adheres to the host's glandular hairs (Koppert 1993; see also Coll and Ridgway [1995]). Dicyphines often feed as scavengers on arthropods entrapped on their hosts (predation was discussed in the previous chapter).

Reuter (1881) first called attention to the ease with which dicyphine and certain other mirids move over plants bearing adhesive secretions. He also discussed the phyline *Macrotylus quadrilineatus*, which feeds on small insects entrapped on a sticky salvia known as Jupiter's distaff. He suggested that several plant bug species occurring on viscid plants live mainly on animal food. The tarsi of some Dicyphini bear short claws, perhaps enabling these bugs to move over glandular plants without becoming entrapped (Reuter 1881). Seidenstücker (1967) believed that the pretarsal structure in dicyphines inhabiting glandular plants is an adaptation for a way of life on hosts possessing sticky hairs. Schuh (1976), however, pointed out that some mirids living on sticky plants have pretarsal structures similar to those of species associated with nonglandular hosts, and that some species having a pretarsal structure similar to those of glandular-plant inhabitants live on hosts that are not sticky.

Southwood (1986) provided additional information on the pretarsal structure and function in dicyphines. He examined the pretarsus of *Dicyphus errans*, which is common in England on restharrow, and noted that the tooth at the claw base and the expanded pulvilli (pseudoarolia) are adaptations for walking on plants having glandular hairs. During resting or moving, the tibiae are held almost vertically so that the bug is on tiptoe, the angle between the tibia and tarsus being greater than 145°; the pretarsus is sometimes used to grasp the stem of a trichome (Fig. 15.1). If a leg becomes stuck to the plant surface, the bug moves its body forward and upward to free it. Southwood (1986) also presented scanning electron micrographs of the pretarsal structure in *D. errans* and a sundew-inhabiting species of *Setocoris* from Queensland, as well as a mathematical analysis of the pattern of movement for *Dicyphus*.

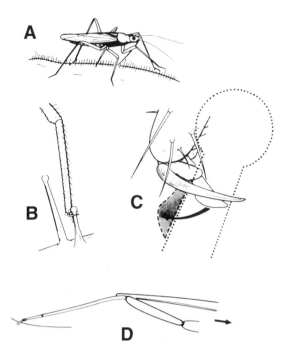

Fig. 15.1. *Dicyphus errans* on restharrow (*Ononis* sp.). The illustrations were drawn from films of the bugs walking on the host plant. A. Typical posture. B. Pretarsus of front leg grasping a trichome below glandular head. C. Detail of pretarsal structures (from electronmicrograph) with position of trichome and displacement of pulvilli in dotted outlines. D. Angle of hind leg when "trapped," showing bending of trichome. (Reprinted, by permission of Sir Richard Southwood and Cambridge University Press, from Southwood 1986.)

Several dicyphines in Western Australia are specialized for living on insectivorous plants; these bugs are well-camouflaged opportunists that scavenge or prey on arthropods entrapped on their hosts (Norris 1991, Falkingham 1995). The brachypterous, antlike *S. byblliphilus* walks freely over sticky leaves of a sundewlike byblidaceous plant, preferring the less glandular upper surfaces. When disturbed, the bugs retreat in any direction without becoming entangled in the plant's mucilage. Their food might consist of small flies trapped by the hosts (Lloyd 1942, China and Carvalho 1951a). *Setocoris droserae* and *S. russellii* live on several species of sundews. When walking, they usually avoid entrapment by placing only one or two legs on the sticky glands, although their movements are sometimes impeded, especially when they are alarmed (Russell 1953). Russell did not observe their feeding habits (except for an adult that appeared to insert its proboscis in a sepal) but referred to China's (1953) assertion that sundew mirids are carnivorous. In describing these two new species from Western Australia, China (1953) had actually stated, "It is probable that the Dicyphini are all carnivorous."

Watson et al. (1982) categorized *Setocoris* species on red-ink sundew in Western Australia as "opportunistic predator-scavengers." Yet the large populations J. A. Slater observed on Australian sundews led him to suggest that the bugs are phytophagous (cited in Cobben [1978:217]; see also concurring opinion of Cassis [1984]), and a *Setocoris* species feeds on its host in addition to trapped prey (Matthews 1976). *Setocoris* species might exploit the nutrient-rich plant sap resulting from absorption of the nitrogenous compounds from entrapped insects. Watson et al. (1982) determined that the nitrogen content of rosette leaves of red-ink sundew is greater than the content in other organs and that plant-incorporated nitrogen is derived principally from arthropod prey (a collembolan) rather than from the nutrient-deficient, sandy soil.

Lamont (1994) noted that *Setocoris* species feed on prey captured by their sundew hosts. Like the arthropods that live in the digestive fluid of the plant, the mirids compete with their host plants for the same resource. Because this feeding has no apparent substantial effect on plant fitness, Lamont (1994) referred to the mirid-sundew relationship as commensal, while acknowledging a mutualism is possible if the bugs also prey on larval herbivores on sundew. Further research might eventually elucidate the interactions of different *Setocoris* species with their insectivorous host plants.

In South Africa, the dicyphines *Pameridea marlothi* and *P. roridulae* have an obligate and potentially kleptoparasitic association with two species of the genus *Roridula* and might help pollinate these viscid shrubs (see Chapter 11). Lloyd (1934) reported that a *Pameridea* species can move swiftly over the adhesive leaves of its host to suck juices from a recently entrapped fly. He regarded species of the genus as commensals that live principally on insects trapped in host secretions. Dolling and Palmer (1991) also reported feeding by *Pameridea* on insects trapped by the resinous hairs of roridulas. The bugs find trapped insects efficiently: 17 of 20 experimentally placed flies were located within 15 minutes (Ellis and Midgley 1996).

The nutritional dynamics of roridulas and the effect of *Pameridea* species on their host plants have been debated. Recent studies by Ellis and Midgley (1996) showed that the plants are only indirectly rather than directly or functionally carnivorous; they do not produce digestive enzymes. The plants do not benefit indirectly from the leaching of nutrients from trapped insects after leaf fall, as has been suggested, but their leaves apparently absorb nitrogen from the bugs' excrement that is deposited mainly on the lower surface. The plant-mirid relationship is thus considered mutualistic.

Dicyphines also feed on dead (and still living) insects adhering to the glandular leaves and stems of tobacco and probably other solanaceous plants. On tobacco and tomato plants, for instance, small insects such as aphids become stuck in exudate and soon die on the leaf surfaces (Roberts 1930, McKinney 1938, Johnson 1956). In

Puerto Rico, Cotton (1917) found nymphs of *Engytatus varians* feeding on the body juices of insects stuck to tobacco foliage. *Nesidiocoris tenuis* and *N. volucer* feed on entrapped insects in Zimbabwe and the Philippines, including ones that are still alive (Roberts 1930, Torreno and Magallona 1994).

The mucilaginous secretions of certain ericaceous plants trap large numbers of insects on buds and flowers (e.g., Eisner and Aneshansley 1983). In England, the Holarctic *Tupiocoris rhododendri* feeds on aphids, a *Masonaphis* species, on rhododendrons possessing glandular hairs (Dolling 1972, McGavin 1982). Similar aphid-feeding habits of this dicyphine are exhibited on azaleas and rhododendrons in the eastern United States (Wheeler and Henry 1992, Braman and Beshear 1994). Gagné (1976) observed *Engytatus modestus* feeding on insects stuck to the viscid hairs of an ornamental rhododendron in Hawaii. In addition, an orthotyline, *Orthotylus gotohi*, scavenges on insects entrapped on the glandular hairs of a wild azalea in Japan (Yasunaga 1999a). Southwood (1973) regarded arthropods entrapped by glandular hairs "as a small but vital secondary source of nutrition" for many dicyphine mirids.

FEEDING ON FUNGUS-KILLED INSECTS

Mirids feed opportunistically on diseased insects. *Lygus lineolaris*, for example, fed on an adult meadow spittle bug (*Philaenus spumarius*) killed by an entomophthoralean fungus (Wheeler 1974b).

A similar example of scavenging by plant bugs involves anthomyiid flies (*Delia* spp.) that succumb to infections by *Entomophthora muscae*. The flies are usually attached by their mouthparts to plant terminals (e.g., Miller and McClanahan 1959). Epizootics occur in certain years, perhaps under conditions of cool temperatures and high humidity. In New York alfalfa fields, cadavers of *D. platura* and *D. florilega* provide supplemental nutrition for various mirids. *Plagiognathus politus* was a frequent scavenger (78 observations); during a 30-minute period in July 1967, 40 instances of feeding were observed (Wheeler 1971). Other species that fed on dead flies were *L. lineolaris* (n = 25), *Adelphocoris lineolatus* (n = 3), *Plagiognathus chrysanthemi* (n = 2), and *Neurocolpus nubilus* (n = 1). The mirids consistently pierce the eyes, and even though the flies soon desiccate, the bugs' salivary secretions are used to moisten the cadavers.

During periods of host-plant deterioration, anthomyiid flies might provide an important food source for mirids that develop on alfalfa, but experimental studies are needed to establish the beneficial effects of scavenging. In fact, the role of scavengers in agroecosystems and the importance of invertebrate carcasses as food sources of insects have yet to be clarified (e.g., Seastedt et al. 1981). O. P. Young (1984) emphasized the presence of a scavenger guild in row crops, although only species associated with the soil surface were considered.

USE OF OTHER INSECT CADAVERS

Scavenging, or necrophagy, is a strategy that helps mirids minimize the effects of low nitrogen levels in host plants and might be an important means allowing late instars of some inflorescence-feeding species to reach adulthood after the flowers of their hosts fall. For example, nymphs of the azalea plant bug (*Rhinocapsus vanduzeei*), an inflorescence feeder as well as predator of small arthropods on azaleas (Braman and Beshear 1994; see Chapters 10, 14), feed extensively on dead alate aphids after host flowers become dry. They have also fed on a dead whitefly and the cadaver of a larval gypsy moth (*Lymantria dispar*). A nymph probed and apparently fed on a cast skin of the gypsy moth (pers. observ.). This observation, with others of similar behavior (see discussion below), suggests that mirid salivary secretions enable nutrients to be obtained even from exuviae.

An adult *Lygus hesperus* flew into the web of an araneid spider, *Araneus* species, in Idaho but was not entrapped. The mirid inserted its stylets into a conspecific adult that minutes before had flown into the web and had been wrapped with silk (Plate 24; D. J. Schotzko, pers. comm.). Somewhat similar behavior was observed in Britain: *Deraeocoris ruber* attacked a pierid butterfly that had been caught and killed by a web-building spider (Hamm 1916). In Japan, Yasunaga (1990) observed the mirine *Mermitelocerus annulipes* feeding on a zygaenid moth that had been caught in a spider web. During emergence of the periodical cicada (*Magicicada septendecim*) in Maryland, a fourth-instar *Deraeocoris quercicola* was observed for more than 90 minutes, probing and feeding on a cicada that had died during eclosion (pers. observ.). Two other nymphs probed and appeared to feed on cicada exuviae on the trunk of an oak tree (Plate 24). Five nymphs of *R. vanduzeei* fed concurrently on a periodical cicada carcass on azalea, and a fifth instar probed or fed on a cast skin of a cicada (pers. observ.).

USE OF VERTEBRATE CARRION AND EXCREMENT

The only heteropterans known to feed to any extent on dead remains of plant or animal origin or on animal dung are seed feeders (Putshkov 1956). Examples include the Lygaeoidea, a group containing many species that specialize on mature seeds, as well as some pentatomids, coreids, and pyrrhocorids. Some coreoids will feed on bird and mammal feces that lack seeds and are not infested by arthropods (Steinbauer 1996). Carrion- and excrement-feeding habits are rather poorly developed in the Miridae. The importance of such food sources would seem minimal, although the dead flesh of animals could supply protein, in addition to amino acids, lipids, carbohydrates, vitamins, and minerals (e.g., Roubik 1989). The term *excrement*, as used in the following discussion, is limited mainly to vertebrate dung. Mirids occasionally feed on insect excrement—

for example, *Atractotomus mali* on caterpillar frass (Kullenberg 1944).

Adler and Wheeler (1984) compiled published records of heteropterans observed on carrion or dung, noting whether actual feeding was reported and speculating on the nutritional significance. The only mirid records cited were of an unidentified plant bug feeding on bird droppings in India (Maxwell-Lefroy and Howlett 1909), *Chlamydatus pulicarius* on excrement in Greenland (China 1934), *Fulvius imbecilis* on pig carrion in South Carolina (Payne et al. 1968), and *Rhinacloa forticornis* on a cow pat in Wyoming (Kumar et al. 1976). Subsequently, Scudder (1985) reported *Macrotylus multipunctatus* beneath old cattle dung in Oregon.

Mirids also occur in birds' nests (Hicks 1959, 1971). Dobroscky (1924, 1925) believed that because the mirid adults she collected in robins' nests in New York belonged to a family assumed to be mainly phytophagous, the bugs were using nests merely for hibernation. Hindwood (1951) thought the plant bugs found in the nests of finches in Australia are casual visitors, acknowledging, however, that such species might be seeking food, shelter, or both. Nymphal specimens represented the only mirids Judd (1962) recorded from the nests of cardinals in Ontario, but he did not comment on the nature of the relationship (a misidentification of the heteropteran family involved is possible). Because mirids exploit various nitrogen-rich food sources, the possibility that the bugs collected from birds' nests are actually feeding on bird droppings cannot be dismissed.

Other Animal-Associated Foods

The use of animal-associated foods not already covered in this chapter, however infrequent its occurrence, illustrates the dietary breadth or trophic plasticity of the family. This behavior might enable the bugs to obtain supplemental nutrition and buffer them from a shortage or decline in quality of their usual food sources.

HEMOLYMPH OF MELOID BEETLES

One of the most biologically interesting, yet enigmatic and paradoxical, aspects of mirid feeding behavior involves their relationship with the beetle family Meloidae (Plate 24) and cantharidin, a defensive, blister-causing, sesquiterpenoid-derived compound. This chemical typically acts as a feeding deterrent to many insects (Carrel and Eisner 1974, D. K. Young 1984a). Certain plant bugs, however, orient to cantharidin.

Only a few references to mirid-meloid relationships were available before Pinto's (1978) study. This earlier literature included notes on the orthotyline *Hadronema militare* "attacking" *Lytta nuttalli* in Saskatchewan (Fox 1943) and the same mirid species feeding through intersegmental membranes on the hemolymph of *Lytta*

species in the Canadian prairie provinces (Church and Gerber 1977). Working in Arizona and California, Pinto (1978) reported the additional associations of *Aoplonema uhleri* with meloids of three genera (*Cordylospata*, *Lytta*, and *Tegrodera*), *H. bispinosum* with *Epicauta* species, and the bryocorine *Halticotoma nicholi* with *Megreta cancellata* and *Meloe laevis*. He described the feeding behavior of *A. uhleri* females on *Lytta moerens*: "The mirids, with rostrum directed forward, periodically advanced slowly toward a beetle ... and, upon reaching it, inserted the mouthparts into a membranous area of the beetle's body. Areas commonly probed included the membrane between the tarsal claws, between the various leg segments, between the coxae and venter, and between the abdominal terga and sterna." Meloids so attacked reacted by "kicking, scraping with the legs, and/or decamping." Pinto (1978) observed that when the beetles move to another part of the same host plant, a bug follows by flying or walking. He also reported the attraction of *A. uhleri* and a bryocorine, *Sixeonotus* sp., to traps baited with cantharidin.

D. K. Young (1984a), intrigued by the attraction of various insects (six families in four orders) to meloids, cantharidin, or both, summarized the literature on insect-meloid and insect-cantharidin associations and discussed the nature of the relationship. On the basis of fieldwork in the United States and Mexico, Young (1984b), using cantharidin-baited filter paper to attract mirids, gave additional records of species that orient to the chemical: the orthotylines *Aoplonema princeps*, *A. uniforme*, and *Hadronema breviatum*, and the bryocorines *Caulotops* species, *Eurychilella* species near *pallida*, *Halticotoma valida*, *Pycnoderes quadrimaculatus*, *Sixeonotus tenebrosus*, and *Sysinas linearis*.

Young (1984b) reviewed the various hypotheses proposed to account for cantharidin orientation in the Miridae. The bugs were once thought to prey on the beetles (Fox 1943, Selander 1960), or the association was considered "ancillary to their primary food sources" and was termed *parasitization* (Pinto 1978). As an alternative hypothesis, Young (1984b) suggested that cantharidin receptors might have evolved in an ancestor that fed on meloids, and although modern species no longer rely on the beetles for food in an obligate way, the sensory apparatus persisted. He noted that species richness for meloids and cantharidin-orienting plant bugs is greatest in the tropics and that temperate species could have lost their dependence on meloids as the bugs radiated northward. Alternatively, organisms unrelated to the Meloidae might produce cantharidin. Mirids might use cantharidin (or a functionally related compound) as a pheromone, defensive mechanism, or both; or cantharidin might be present in host plants with which the bugs evolved. Young (1984b) pointed out that several Ranunculaceae produce a cantharidin-like blistering of epithelial tissue. Cantharidin or a related compound possibly mimics some chemical in the mirids' host plants. Unfortunately, the hosts of temperate and tropical Bryocorinae in the New World are largely

Table 15.1. Examples of mirids known to bite humans

Species	Locality	Reference
Bryocorinae		
Campyloneura virgula	Oregon, USA	Lattin and Stonedahl 1984
Haematocapsus bipunctatus	Nigeria	Poppius 1914
Deraeocorinae		
Deraeocoris manitou	Kansas, USA	Pers. observ.
Deraeocoris nebulosus	South Carolina, USA	D. W. Boyd, pers. comm.
Deraeocoris quercicola	Maryland, USA	Pers. observ.
Deraeocoris ruber	England	Morley 1914
Hyaliodes harti	Kentucky, USA	University of Louisville Insect Collection
Hyaliodes vitripennis	Washington, D.C., area, USA	Knight and McAtee 1929
	New York, USA	Torre-Bueno 1931
Mirinae		
Creontiades pacificus	Malaysia	Miller 1971
Creontiades pallidus	Sudan	Lewis 1958
Irbisia brachycera	California, USA	Usinger 1934
Lygocoris communis	Nova Scotia, Canada	Brittain 1916a
Lygocoris omnivagus	South Carolina, USA	Pers. observ.
Lygus lineolaris	New York, USA	Crosby and Leonard 1914
	Quebec, Canada	Khattat and Stewart 1977
Lygus pratensis	England	Myers 1929
Lygus sp.	[USA]	Horn 1988
Taedia scrupea	Pennsylvania, USA	A. J. Muza, G. L. Jubb Jr., pers. comm.
Trigonotylus brevipes	East Africa	Reuter 1913; see also Strong et al. 1926
Orthotylinae		
Blepharidopterus angulatus	England	Smith 1990
Blepharidopterus provancheri	New York?, USA	Van Duzee 1912
Brachynotocoris puncticornis	Algeria	Bergevin 1924; see also Strong et al. 1926
Cyrtorhinus lividipennis	Japan or Korea	Esaki 1934
Lopidea chandleri	North Carolina, USA	Asquith 1991
Lopidea marginata	California, USA	Usinger 1934
Lopidea media	Minnesota, USA	Lugger 1900
Lopidea robiniae	Pennsylvania, USA	Pers. observ.
Lopidea robusta[a]	British Columbia, Canada	Downes 1927
Orthotylus flavosparsus	New York, USA	Riley and Johannsen 1915
Orthotylus ramus	Texas, USA	Chung et al. 1991
Reuteria irrorata	Arkansas, USA	Pers. observ.
Phylinae		
Atractotomus miniatus	Maryland, USA	Pers. observ.
Camptotylus yersini	Sudan	Lewis 1958
Campylomma verbasci	Nova Scotia, Canada	Gilliatt 1935
Chlamydatus associatus	New York, USA	Crosby and Leonard 1914
Maurodactylus albidus	N. Africa	Bergevin 1926
Microphylellus tsugae	Pennsylvania, USA	Pers. observ.
Phylus melanocephalus	Sweden?	Reuter 1879
Pilophorus crassipes	Arkansas, USA	Pers. observ.
Plagiognathus caryae	South Carolina, USA	D. W. Boyd, pers. comm.
Plagiognathus obscurus	Washington, D.C., area, USA	Caudell 1901; see also Strong et al. 1926
Plagiognathus politus	New York, USA	Culliney et al. 1986
Plagiognathus repetitus	Massachusetts, USA	Franklin 1950
Psallus sp.	Sudan	Lewis 1958
Pseudatomoscelis seriata	Texas, USA	Tucker 1911
Rhinocapsus vanduzeei	Pennsylvania, USA	Wheeler and Herring 1979
	Washington, D.C., area, USA	Miller 1993
Sejanus albisignatus	New Zealand	Dumbleton 1938

Note: See text for discussion.

[a] Although listed as a valid species in *Lopidea* by Schuh (1995), *L. robusta* was not included in Asquith's (1991) revision of North American *Lopidea*.

unknown. These species, as well as those of the Old World, in which cantharidin orientation apparently has not been demonstrated, deserve attention from researchers. Young's (1984b) hypotheses to explain cantharidin orientation in mirids challenge the ingenuity of heteropterists, ecologists, and evolutionary biologists to elucidate the nature and evolutionary origin of the relationships.

HUMAN BLOOD

Heteropterans (and some homopterans) pierce human skin or "bite," especially under unnatural conditions. Usinger (1934) remarked that he had been bitten so often by heteropterans that such behavior seemed a rather common occurrence. As Myers (1929) and Usinger (1934) pointed out, the altered behavior of hemipterans can be triggered by changes in chemical, tactile, thermal, or visual stimuli. Biting by hemipterans might not involve attempts to ingest blood as a nutrient source (e.g., Lawson 1926). Ryckman (1979) and Ryckman and Bentley (1979) compiled an annotated bibliography of most of the scattered references to biting by Hemiptera. Because they did not include a list of taxa, a reader must consult all references cited for a family to determine the hemipteran species involved. Therefore, mirid references from their papers are repeated in Table 15.1 using current names; also included are records they omitted and literature published subsequently. I have attempted to cite original sources rather than secondary literature such as textbooks of medical entomology and review articles; in one case, a textbook (Riley and Johannsen 1915) contains original observations. Mirids known to bite include more or less strict phytophages—bryocorines and stenodemines—and strict or near obligate predators such as deraeocorines (Table 15.1).

Whereas bites from bed bugs, fleas, mosquitoes, and other specialized hematophagous arthropods are painless (e.g., Usinger 1966), those inflicted by mirids can be quite painful. The effects of mirid bites, though, are usually transitory (itching sometimes lasts several days [Myers 1929]) compared to those that hematophagous and certain predacious heteropterans can inflict (Ryckman 1979, Ryckman and Bentley 1979). Consequently, mirids rarely attract attention because of their bites, and they are omitted from most reference books on arthropods of medical importance (e.g., Goddard 1993). Persons bitten by mirids are unlikely to seek medical help, and even if they do, the bite almost certainly would not be attributed to a plant bug.

Mirids that insert their stylets into human skin produce sensations ranging from mild irritation (Culliney et al. 1986) or intense local irritation (Myers 1929) to considerable pain (Lugger 1900; pers. observ.).

The bites and their effects are likened to those of mosquitoes (Tucker 1911, Crosby and Leonard 1914, Torre-Bueno 1931) or a "gnat" (Reuter 1879, 1903), or to irritation from contact with stinging nettle (Morley 1914). Aftereffects such as itching, red spots, and swelling can occur (Culliney et al. 1986). In other cases, mirids cause a small lump (Plate 24; Torre-Bueno 1931), chiggerlike welts (Wheeler and Herring 1979), or hard welts (Lattin and Stonedahl 1984).

Campers in Minnesota were once frequently bitten by *Lopidea media*, large numbers of this bug having invaded the tents (Lugger 1900). Homeowners and gardeners in the eastern United States are sometimes bitten so often by the azalea plant bug (*Rhinocapsus vanduzeei*) that they request chemical controls for this mirid (Wheeler and Herring 1979, Miller 1993). Workers in West Texas pecan orchards have been forced to turn up their shirt collars to keep adult *Orthotylus ramus* from biting their necks and upper bodies (Chung et al. 1991).

For several species listed in Table 15.1, authors refer to blood sucking without elaborating on the extent of imbibition. Lugger (1900), however, observed that *L. media* imbibed so much blood that it could scarcely fly; Brittain (1916a) said *Lygocoris communis* fed on his hand or neck until engorged with blood. Myers (1929) stated that after a fifth-instar *Lygus pratensis* (perhaps *L. rugulipennis*) fed for an hour, its abdomen became slightly distended and reddish; the bug later excreted two drops of reddish fluid. Similarly, Culliney et al. (1986) observed that a *Plagiognathus politus* adult, after feeding for 10 minutes, reinserted its stylets several times to probe more deeply; the abdomen became "visibly distended, indicating that a substantial quantity of fluid had been ingested."

Mirids will pierce human skin when it is either moist with perspiration (Van Duzee 1912, Usinger 1934) or dry (Torre-Bueno 1931). Lewis (1958) stated that hemipterans occasionally bite when they are attracted to sweat or are hungry, and Al-Houty (1990) suggested that swarms of the lygaeid *Nysius* species alighting on and biting humans in Kuwait were under water stress. Southwood (1973) remarked that phytophagous heteropterans "will probe many parts of the plant (and humans!) for water." The reasons for this behavior are unclear. Usinger (1934) emphasized the importance of perspiration in causing heteropterans to bite and suggested that the bugs might be seeking nourishment. Does sodium perhaps trigger the probing of moist skin, as it does for puddling behavior in Lepidoptera (Arms et al. 1974, Adler 1982)? Do plant bugs ever imbibe blood while seeking supplemental food sources following host deterioration or, in the case of predacious mirids, a decline in prey numbers? This is another aspect of mirid behavior that deserves attention from researchers.

16/ Ancestral Feeding Habits of the Heteroptera and Miridae

In plant bugs the entomophagous habit is not too widely separated from the phytophagous.

—C. E. Lilly 1958

The above assertion is scarcely equivocal. But what are the ancestral feeding habits of mirids—is zoophagy or phytophagy primitive? The evolution of feeding habits, not only in mirids but also in the Heteroptera, has been much discussed and debated (Cobben 1978, 1979; Sweet 1979; Schuh 1986a).

Heteroptera

Several early workers (e.g., Myers and China 1929, China 1933) believed that primitive heteropterans were homopteroid and arose from an ancestral phytophagous stock, yet presented little evidence for this conclusion. Most authors assume that the Homoptera (i.e., auchenorrhynchans + sternorrhynchans; paraphyly of the Homoptera is noted in Chapter 2), a phytophagous group, had an earlier origin than the Heteroptera, but the data establishing homopterans as more primitive or ancestral are equivocal (Cobben 1978, Polhemus 1985; Sorensen et al. [1995] reviewed the paleontological evidence). Drake and Davis (1960) envisioned that heteropteran mouthparts eventually became modified for carnivory, and that a gula (see Chapter 7) developed later. Goodchild (1966), Miles (1972), and others (see Cobben [1978]) also considered the Heteroptera to be primitively phytophagous.

Cobben (1978) argued on the basis of mouthpart structure that carnivory rather than phytophagy was the original feeding habit of heteropterans and apparently also homopterans, a conclusion he had reached earlier based on a study of egg structure (Cobben 1968). In Cobben's (1978) view, the barbed maxillary stylets presented a major structural barrier to ancestral phytophagy in heteropterans; plant-feeding habits could develop only after the dentition was lost. With a gradual reduction in the maxillary barbs, there was an increasing capacity for mandibular protrusion. An increased role of the mandibles during probing and feeding allowed progressively deeper stylet penetration (Cobben 1978).

Publication of Cobben's (1978) analysis of evolutionary trends in heteropteran mouthparts led Sweet (1979) to challenge Cobben's ideas and to marshal evidence supporting ancestral phytophagy in Heteroptera and, implicitly, in Homoptera. Schuh (1986a) summarized Sweet's (1979) principal arguments for considering the Heteroptera originally phytophagous and terrestrial: (1) homopterans are phytophagous and terrestrial; (2) carnivorous lineages develop from phytophagous taxa in the Heteroptera, but the reverse situation is unusual; (3) a salivary sheath (see Chapter 7) is present in both Homoptera and Heteroptera (some Pentatomomorpha) and is a primitive trait in the latter; and (4) evolution through carnivory of complex piercing-sucking mouthparts in the Heteroptera would be difficult. In Sweet's (1979) view, the Pentatomomorpha are the most primitive heteropteran group.

Cobben (1979) rebutted Sweet's arguments, noting that he (Cobben 1978) had not merely argued for ancestral carnivory in Heteroptera based on the widespread occurrence of this habit among extant infraorders, as Sweet (1979) said he had done. A common-equals-primitive assumption should not be used as a line of evidence (e.g., Schuh 1986a). Instead, Cobben (1979) explained that he first had developed the hypothesis that evolution of mouthpart structure in heteropterans supports ancestral carnivory. Cobben further pointed out that Sweet (1979) referred to protohemipterans as originally phytophagous and terrestrial, rather than carnivorous in a "semi-aquatic shoreline type habitat" as supposedly envisioned by Cobben (1978). Cobben (1979), however, emphasized that he considered the ancestral heteropteran habitat to be the "damp litter-zone," not a semiaquatic habitat along the shore, and that the heteropteran archetype might have had a small gula that facilitated a predatory habit. Cobben (1979) also commented on Sweet's other points, mentioning inconsistencies in his reasoning and stating that no new data were offered. Ultimately, Cobben concluded that his earlier studies on heteropteran egg systems (Cobben 1968) and mouthparts (Cobben 1978) established unequivocally that heteropterans are ancestrally carnivorous. Neither Cobben nor Sweet used rigorous cladistic methodology for their arguments (e.g., Schuh 1979), but Cobben's conclusions are supported by W. C. Wheeler et al. (1993), who used both morphological and molecular characters in their analysis.

Carnivory as the ancestral heteropteran feeding type now seems reasonably well established and is the most parsimonious explanation of the group's trophic habits (Cobben 1968, 1978, 1979; Kerzhner 1981; Schaefer 1981, 1997; Schuh 1986a; Mitter et al. 1988; Stonedahl and Dolling 1991; W. C. Wheeler et al. 1993; Schuh and Slater 1995). Originally heteropterans might have been "obligatory" carnivores (Schaefer 1997). That carnivory is primitive holds regardless of whether the Gerromorpha, Dipsocoromorpha, or Enicocephalomorpha are considered the probable sister group of all other Heteroptera (China 1955, Cobben 1968, Štys 1970a, Polhemus 1985, Schuh 1986a, W. C. Wheeler et al. 1993). Members of all three infraorders are predatory (Schuh and Slater 1995) or, in the case of the Dipsocoromorpha, are at least presumed to be generalized predators (Štys 1995). Southwood (1973) discussed the evolutionary hurdles of moving from a life on the ground (see Schaefer [1981]) to one on green plants, including that of nutrition on an often suboptimal resource, a problem not faced by predators. In Southwood's (1973) opinion, the step from scavenging and feeding on microorganisms to phytophagy is more difficult nutritionally than the step to predation.

Kerzhner (1981; pers. comm.) suggested that aggressive heteropteran predators do not evolve into phytophagous lineages. Instead, he hypothesized the sequence of preying on litter inhabitants, feeding on inactive prey such as eggs and scale insects, and phytophagy.

A preference of litter-inhabiting predators for certain plants as "hunting grounds" might have influenced speciation and the evolution of phytophagy in the Heteroptera. Relatively few heteropteran groups, though, have become phytophagous. Yet once phytophagy evolved, it could have permitted adaptive radiation and diversification that contribute to the biological success of the Heteroptera (Cobben 1978). Within the Insecta, phytophagous lineages are consistently more diverse than their sister groups (Mitter et al. 1988).

Miridae

The Miridae, like the suborder to which they belong, are often assumed to be ancestrally phytophagous. Adding to a plant-feeding bias might be the common name "plant bugs"; that phytophagy is dominant in the family can lead to a common-equals-primitive assumption (e.g., Myers 1927b; see also Whitman et al. [1994]). Some workers stress the ease with which primitive mirids could make a transition to carnivory: "In the evolution of these bugs it would be easy to pass from piercing plant tissue and sucking sap to piercing smaller insects such as Aphides and sucking the plant sap ingested by the prey" (China 1953). Miller (1971) similarly noted how a primitively phytophagous mirid could have alternated sap feeding with attacks on small insects living on the same host plant.

Goodchild (1952, 1966), however, suggested that mirids evolved from predatory ancestors. He emphasized their carnivore-like feeding habits, noting that they flushed out cell contents with a stream of saliva. He further pointed out that possession of a vesicular accessory salivary gland and structure of the intestine allied the Miridae more closely to predatory than to phytophagous heteropterans. The lacerate (or macerate) and flush feeding mode (see Chapter 7) resembles that of predacious heteropterans (Miles 1972, Schaefer 1981). Cimicomorphans, including mirids, tend to have more potent salivary enzymes and venoms, as well as shorter handling times, than do pentatomomorphans, including the asopine pentatomids that have been studied. The mainly phytophagous *Lygus hesperus*, for instance, can incapacitate aphid prey within a few seconds after attack (Cohen 1996). According to Cobben (1978), unaggressive predators (sometimes termed "cautious" or "timid") seem behaviorally more preadapted for phytophagy than do aggressive predators, which he felt helped explain why plant-feeding heteropterans all belong to taxa—Cimicomorpha and Pentatomomorpha—that contain relatively unaggressive predators (Cobben 1978). Cobben's Cimicomorpha excluded the Reduvioidea, a group he considered aggressive predators (in the current classification, reduvioids are placed in the Cimicomorpha [Schuh and Slater 1995]).

Supporting ancestral carnivory in plant bugs is the predatory habit that characterizes the most primitive mirid subfamily, Isometopinae (Leston 1961b; Schuh 1974, 1976; Wheeler and Henry 1978a; Schuh and Štys 1991; Schuh and Slater 1995). Even Sweet (1979), who hypothesized that carnivorous taxa evolve from phytophagous ones in the Heteroptera, suggested that the mirid-tingid lineage was derived from carnivorous ancestors. Schaefer (1997) suggested that if the Thaumasticoridae represent the sister clade to Miridae + Tingidae (Schuh and Štys 1991), then carnivory might be secondary rather than primary in Isometopinae; Schaefer, however, regarded as weak the evidence for such a placement of thaumastocorids. I concur with Schuh (1974), Cobben (1978), Schuh and Slater (1995), and others who consider the Miridae to be ancestrally predacious.

This concurrence with R. T. Schuh, R. H. Cobben, and other authors is based on recognition of the predatory isometopines as the most primitive mirids and the need to consider the evolution of feeding habits, as with other ecological characters (Miller and Wenzel 1995), only within a phylogenetic context. Arguments for phytophagy as primitive within Miridae, as noted earlier, are not rooted in rigorous methodology. Ecological traits such as trophic behavior are potentially labile and convergent (e.g., Lipscomb 1991); their change over evolutionary time should be evaluated against an existing cladistic analysis of the group. Schuh's (1976) work provides testable hypotheses of heteropteran phylogeny. He considered the Isometopinae the most primitive mirid subfamily; available evidence (see Chapter 14) has established this group, sister to all other mirids, as predatory. Phytophagy has evolved at least twice in the Heteroptera: in the sometimes stylet-sheath feeding Pentatomomorpha, in which this habit is ancestral, and in the mainly lacerate-flush feeding Cimicomorpha, which are sister to the pentatomomorphans. Plant feeding is a derived trait among cimicomorphs; within this group, the presumably predacious Microphysidae are considered sister to the Joppeicidae, Thaumastocoridae, and Miridae + Tingidae (Schuh and Slater 1995). On the basis of a rigorous cladistic hypothesis (Schuh 1976), ancestral predation is the most parsimonious explanation of feeding habits both in the Heteroptera and in the Miridae.

Schaefer (1981) speculated that actively predacious cimicomorphs moved from dry ground onto flowering plants. Some retained an active predatory lifestyle, whereas others became unaggressive predators of sessile or sluggish prey. Mirids arose from such unaggressive predators and became mostly phytophagous. Southwood (1985) referred to mirids as showing an "intermediate condition along the route to phytophagy via plant parts that are high in protein." In addition, as noted in Chapter 7, maxillary barbs in the Miridae—roughened only on the inner surfaces—are intermediate between those of carnivorous heteropterans such as the Gerromorpha, in which maxillary stylets are strongly serrated, and strictly phytophagous taxa such as the tingids, which have smooth maxillae (Cobben 1978).

In Chapter 17, mirid feeding habits are mapped onto a cladogram to infer the evolutionary history of this ecological character. An increased knowledge of phylogenetic relationships within the Heteroptera and the Miridae, as well as additional information on nutritional requirements, biochemical analysis of digestive enzymes, comparative biochemistry of venoms, and feeding behavior, should improve the interpretations of ancestral feeding habits and their subsequent evolution.

17/   Feeding Trends among Mirid Higher Taxa

Without the unifying perspective of evolution our growing knowledge of natural history is in danger of becoming a mere catalogue of curious facts.
—R. H. L. Disney 1994

The preceding chapters of this book emphasized the diversity of feeding habits in the Miridae. In some quite small families of insects, trophic behavior can be readily characterized—all species more or less do the same thing. Their morphological and behavioral adaptations are associated with a particular feeding strategy. In a family as diverse as the Miridae it is unrealistic to assume that a single feeding mode can be recognized, or even that many clear-cut trends will be apparent above the generic level. To a certain extent, trophic habits in the Miridae, as in other diverse arthropod groups, must be considered on a species-by-species basis rather than characteristic of a higher taxon.

The unevenness of biological information available for the different plant bug groups makes it difficult to generalize about their food habits—or at least renders definitive statements about trophic behavior premature or misleading. For some higher taxa, knowledge of what their members eat is unknown or is based on only a few species. Even when multiple observations are available, it is uncertain if the reported behavior is typical or was affected by artificial study conditions.

Phylogenetic relationships involve shared constraints on morphology, physiology, and behavior. What can be said about feeding habits of mirids above the generic level? Are trends evident among tribes or subfamilies? How well does mirid trophic behavior reflect current ideas about their phylogenetic relationships?

Leston (1961b) was the first to attempt to categorize mirid feeding behavior at the subfamily level. His analysis (Table 17.1) revealed marked differences among the subfamilies. Three subfamilies—Cylapinae, Deraeocorinae, and Isometopinae—were considered exclusively predatory. The Bryocorinae were thought to be strict phytophages. Leston (1961b) recognized the dicyphines as a separate subfamily rather than as members of the Bryocorinae. In Leston's analysis, dicyphines were regarded as phytophagous and predacious. The taxon now equivalent to the subfamily Dicyphinae of Leston (1961b) is the Dicyphina, one of three subtribes that make up the tribe Dicyphini, subfamily Bryocorinae

(Schuh 1995, Schuh and Slater 1995). Among the other subfamilies, the Mirinae were considered mainly phytophagous, and mixed habits were said to characterize the Orthotylinae and Phylinae (see Table 17.1). Since Leston's (1961b) analysis of mirid feeding habits, Schuh (1974) categorized their habits similarly, and Schuh and Slater (1995) briefly commented on the trophic patterns apparent in several tribes and subfamilies.

In the following treatment, the feeding habits of mirids are summarized alphabetically by subfamily and tribe (see Table 3.1), and trends are considered in relation to an existing cladogram. This diagnosis is not intended to be comprehensive and does not mention all the types of feeding behavior that are covered in the other chapters. Nor within a tribe does it mention all genera that feed in a similar manner. Details and documentation of feeding behaviors, as well as most references, are available in Parts III and IV on phytophagy and zoophagy, respectively. Kullenberg's (1944) study of Swedish Miridae remains invaluable for its detailed information on feeding habits.

Mirid Feeding Habits: Trends among Higher Taxa

BRYOCORINAE

The subfamily Bryocorinae contains many important plant pests, especially in the tropics (Leston 1961b). It is a diverse group morphologically and cytologically, as well as in feeding habits. Monophyly of the Bryocorinae perhaps has not been fully established; Leston (1970, 1978a) considered the subfamily "at least diphyletic."

Members of the small tribe Bryocorini (Schuh 1995) are associated with ferns; some *Bryocoris* and *Monalocoris* species feed on the sporangia. Three subtribes of the Dicyphini, a tribe once accorded subfamilial status, are recognized (Schuh 1995): Dicyphina, Monaloniina, and Odoniellina. Within the Dicyphina, species of several genera, including *Dicyphus*, *Engytatus*, *Macrolophus*, and *Nesidiocoris*, feed on solanaceous plants, such as tobacco and tomato, and on other glandular-hairy hosts belonging to additional families. They cause stem lesions and cankers (see Chapter 7) and feed on the inflorescences, fruits, leaves, and glandular hairs of their hosts. Extensive vascular feeding by *Engy-*

Table 17.1. Mirid feeding habits according to Leston (1961b)

Subfamily	Feeding habits
Bryocorinae	Exclusively phytophagous
Cylapinae	Exclusively predatory
Deraeocorinae	Exclusively predatory
Dicyphinae	Predatory and phytophagous
Isometopinae	Exclusively predatory
Mirinae	Mainly phytophagous
Orthotylinae	Mixed habits; some exclusively predatory
Phylinae	Mostly mixed habits

Note: Dicyphines are now placed in the Bryocorinae as subtribe Dicyphina of the Dicyphini (Schuh 1995).

tatus and *Nesidiocoris* species is unusual among phytophagous mirids, mesophyll feeding being the more typical behavior. Predatory tendencies, though, are well developed, with some species preying on aphids, psyllids, and whiteflies as well as lepidopteran larvae and arthropod eggs. Although *Dicyphus pallicornis* might be strictly phytophagous (Cobben 1978), most Dicyphina are omnivorous, and *Campyloneura virgula* seems mainly predacious. *Pameridea* and *Setocoris* species apparently scavenge on arthropods entrapped on the viscid surfaces of their hosts. Monaloniina are noted mainly for the lesions and cankers they cause on leaves, stems, and fruits, as well as symptoms such as shot holing and witches'-brooming. This subtribe contains important cocoa pests of the genera *Helopeltis* and *Monalonion*. Habits of the Odoniellina are similar, the species of *Distantiella* and *Sahlbergella* having been intensively studied as pests of cocoa. In contrast to the Dicyphina, zoophagy apparently is unknown among monaloniines and odoniellines, taxa that frequently display gregarious behavior. On the basis of feeding habits, modifications of the alimentary canal (Goodchild 1952, 1966), and morphological characters such as trichobothrial and pretarsal structure (Schuh 1975, 1976) and structure of the dorsal abdominal scent glands (Aryeetey and Kumar 1973), the taxonomic status of these three subtribes might deserve reevaluation.

Most species of the Eccritotarsini, subtribe Eccritotarsina, cause foliar chlorosis on plants such as agaves, cacti, orchids, and yuccas. Some *Neoneella* species are associated with the inflorescences of philodendron. As in the Monaloniina and Odoniellina, there is little if any tendency toward predation or scavenging. The habits of species in the small subtribe Palaucorina are unknown. Their tarsal claws are similar to those of certain emesine reduviids that live in spider webs, suggesting that the Palaucorina might also be associated with spider webs (Schuh 1976).

CYLAPINAE

Whereas three (or even four) tribes are sometimes recognized in this mainly tropical subfamily (Gorczyca and Chérot 1998), only the Cylapini are listed in the most recent catalog; the tribes formerly recognized are considered not to be monophyletic (Schuh 1995). The fulviines, though, have often been considered a tribe, Fulviini, and have even been accorded subfamily status (Schmitz and Štys 1973).

Many cylapines (Cylapini *sensu stricto*) are consistently associated with fungi, especially pyrenomycetes. Speculation surrounding the trophic behavior of these bugs has focused on mycophagy and predation, with scant evidence available to support either possibility. But the recent observation of fungal spores in the guts of a temperate and a tropical species of *Cylapus* suggests that they are at least partially mycophagous. The extent of mycophagy among litter- and bark-inhabiting cylapines remains undetermined. Some *Fulvius* species prey on arthropod eggs, and it is likely that the *Peritropis* species living on tree trunks and branches will also prove to be at least facultatively predacious.

DERAEOCORINAE

Six tribes are recognized in this subfamily of worldwide distribution (Schuh 1995). All deraeocorines are considered predacious (Schuh and Slater 1995), with many showing specialized feeding habits (Schuh 1974). The habits of members of the Saturniomirini, a tribe restricted to Australia and New Guinea, are unknown. Information is similarly lacking for the Surinamellini. Species of the only North American genus of this tribe, *Eustictus*, are mostly nocturnal (Knight 1927a) and are assumed to be predatory (e.g., Van Duzee 1912, Knight 1941, Wheeler 1991b), but documentation of their habits is needed.

Of the remaining tribes, the Clivinematini prey on scale insects. Members of the Deraeocorini feed on the eggs, larvae, and pupae of various arthropod groups. This tribe includes *Eurychilopterella* species, which feed on scale insects. *Deraeocoris brevis* and *D. nebulosus* are generalist predators that track prey populations on a diversity of hosts. The Hyaliodini include mite-feeding species of *Hyaliodes* and obligate lace bug predators of the genus *Stethoconus*. Termatophylines are specialized predators of thrips (*Termatophylidea* spp.) or are reported to attack lepidopterans in their larval galleries on woody hosts.

Some plant feeding occurs in this subfamily, but its extent and importance need clarification. Kullenberg (1944) suggested that members of the genus *Alloeotomus*, the habits of which are little known, are phytophagous, but the former placement of this genus in the Mirinae might have influenced his thinking on food habits. At least one species, *A. chinensis*, preys on scale insects. *Deraeocoris* adults occasionally disperse to flowers, where they might exploit nectar, pollen, or both. Several species of this genus feed on succulent growth under laboratory conditions; *D. fasciolus* can be reared from the first instar to the adult stage on hawthorn foliage (Knight 1921). *Deraeocoris olivaceus* feeds partly on hawthorn fruits (Southwood and Leston 1959), *D. brevis* feeds occasionally on the leaves and young fruit of pear trees (McMullen and Jong 1967), and *D. flavilinea* seems to require both a plant and an animal diet (Triggiani 1973). Having been reared on clean currant shoots, *D. ruber* was considered both phytophagous and predacious (Petherbridge and Thorpe 1928a), but the work of Viggiani (1971b) established *D. ruber* as almost exclusively predatory. Strawiński (1964a), however, considered all *Deraeocoris* species occurring in Poland, including *D. ruber*, to be zoophytophagous. Plant feeding might enhance the survival and persistence of deraeocorines during periods when prey are scarce (McMullen and Jong 1967, Chinajariyawong and Harris 1987), and some species may yet prove to be substantial phytophages, as suggested by Knight (1921).

ISOMETOPINAE

The habits of relatively few members of this chiefly tropical, bark-inhabiting group are known. Nearly all species that have been studied prey on diaspidid scale insects—several are specialized predators of armored scales. A Chinese isometopine reportedly feeds on spider mites. Observations documenting predation in this most primitive subfamily of the Miridae are particularly crucial in establishing the family as ancestrally predacious rather than phytophagous (see Chapter 16).

MIRINAE

Of worldwide distribution, mirines are considered mainly phytophagous, even though the subfamily includes the chiefly predatory genus *Phytocoris*, the most speciose of all plant bug genera. Approximately 300 mirine genera are placed in six tribes (Schuh 1995).

The habits of the Herdoniini, a predominantly New World group of myrmecomorphs (Schuh and Slater 1995), are poorly known. Species of *Barberiella* and *Dacerla* prey on aphids, and this habit may be common in the tribe. The hyalopepline *Hyalopeplus pellucidus* feeds on the flower buds of guava; Carvalho and Gross's (1979) doubts about predatory behavior in this mirid (Kirkaldy 1904) are justified. *Guianerius typicus* has been observed "hovering over flowers in jungle" (Distant 1903). Otherwise, little has been recorded about the habits of this tribe of the Old World tropics. The Mecistoscelidini are a small Oriental group whose species develop on grasses (Schuh and Slater 1995). Feeding by *Mecistoscelis scirtetoides* causes white, tetragonal spots on bamboo foliage.

The widely distributed Mirini, the largest tribe of the subfamily, contain numerous crop pests, including those in the well-studied genus *Lygus*. Many mirines feed on meristematic tissue, new foliage, buds, various flower parts, fruits, and developing seeds, causing growth and differentiation disorders such as leaf crin-

kling, excessive branching, shortened internodes, thickened leaves and stems, and stunting. Species of *Adelphocoris, Calocoris, Closterotomus, Creontiades, Dagbertus, Eurystylus, Lygocoris, Neurocolpus, Orthops,* and *Taedia* are often associated with the inflorescences of host plants, though some mirines feed on foliage and cause tattering and shot holing (e.g., *Closterotomus, Lygocoris, Lygus,* and *Taylorilygus* spp.). In the genus *Proba, P. sallei* feeds on the growing tips of guayule rather than on its seeds (Romney et al. 1945). *Irbisia* and *Tropidosteptes* species cause foliar chlorosis, and *Poecilocapsus lineatus*, a strict phytophage, is characterized by the prominent water-soaked lesions it produces on host leaves and by its apparent specialization on cells of the palisade layer. *Capsus ater* is mainly a stem feeder on grasses. *Phytocoris* species are well-known predators of various arthropods. Some species feed on the developing fruits, buds, and young leaves of their hosts (Kullenberg 1944), and others are considered mainly phytophagous (Knight 1941). The mostly predatory *P. tiliae* can be reared from the second stage to maturity on currant plants (Petherbridge and Thorpe 1928b). Careful studies are needed to assess the extent and importance of phytophagy in this genus. *Miris striatus*, although categorized as a strict predator (e.g., Strawiński 1964a), can injure pear fruit and feeds on the developing seeds of various willows (Kullenberg 1944). Facultative predation, including cannibalism, is relatively common in genera such as *Capsodes, Closterotomus, Lygus,* and *Megacoelum*, but predation seems less common in *Creontiades, Eurystylus, Liocoris, Polymerus, Tropidosteptes,* and certain other mirine genera. Even some *Creontiades* and *Eurystylus* species will cannibalize in the absence of other food sources (Sharma and Lopez 1990a). At least one species of the east Asian genus *Castanopsides, C. potanini*, appears to be mainly predacious (Yasunaga 1998).

The restheniine *Opistheurista clandestina* causes foliar chlorosis on its host plants, and *Prepops* species are common on host inflorescences. The stem-feeding *Platytylus bicolor* causes split lesions and gum-filled blisters on citrus. The feeding habits of members of this New World tribe generally are poorly known.

Stenodemines cause foliar chlorosis and feed on the heads of host grasses and sedges. Several species feed on the mesophyll of grass leaves before switching to host inflorescences to complete their development. Some stenodemines also feed on grass stems. This widely distributed tribe is sometimes stated to consist entirely of phytophagous species (e.g., Southwood and Leston 1959, McNeill and Southwood 1978, Gibson 1980, Dolling 1991). Goddard (1935), for example, did not observe predation by stenodemines on aphids (cf. Parfitt 1884) or other arthropods under confined conditions, even though members of other tribes accepted animal food; Kullenberg (1944) determined that several *Stenodema* species are strict phytophages. Facultative predation, however, occurs in *Leptopterna, Myrmecoris, Notostira, Teratocoris,* and *Trigonotylus*. Possible myrmecophagy by the myrmecomorphic and partially phytophagous *Myrmecoris gracilis* (Gulde 1921) is doubtful and needs confirmation.

ORTHOTYLINAE

Orthotylines perhaps exhibit less homogeneity in cytological, morphological, and other characters than do plant bugs of other subfamilies (Leston 1961b, Thomas 1987). Classification of this diverse taxon has remained somewhat unstable, and monophyly of the tribes has yet to be resolved (Henry 1995); further changes at the tribal level seem probable. Schuh (1995) recognized three tribes: Halticini, Nichomachini, and Orthotylini.

Halticines are phytophagous, typically causing chlorosis on host foliage. This feeding mode, occurring in genera such as *Anapus, Euryopicoris, Halticus, Labops,* and *Orthocephalus,* might generally prevail in the tribe. The Halticini sometimes have been considered a separate subfamily, the Halticinae (e.g., Wagner 1973). The habits of the Nichomachini, a small, mainly ground-living Afrotropical tribe, are unknown. This tribe might be most closely related to the Halticini (Schuh 1974).

Mixed feeding habits characterize the large and widely distributed Orthotylini. The presence in some genera of species that are chiefly phytophagous and others that are mainly predacious makes it difficult to generalize about feeding habits at the generic level. As in certain other mirid groups, trophic behavior is largely a species-level phenomenon, as exemplified by the *Orthotylus* species associated with broom in England. Of the broom inhabitants, predation is common in some species but seldom observed in others (Dempster 1960). The species of several genera—*Brooksetta, Hadronema, Ilnacora, Labopidea, Lopidea, Nesiomiris, Pseudopsallus,* and *Slaterocoris*—are plant feeders that cause foliar chlorosis; zoophagous habits in these bugs seem poorly developed. Orthotylines are infrequently associated with foliar symptoms such as shot holing or lesions and cankers, and they generally do not cause the various plant-growth disorders so commonly encountered in the host plants of many mirines. Species of *Lopidea, Orthotylus, Parthenicus, Pseudoxenetus,* and undoubtedly other genera feed on inflorescences, but this habit is not as prevalent in orthotylines as it is in mirines and phylines.

Other orthotylines are mostly or even exclusively predacious. Specialized homopteran egg predators of the genus *Cyrtorhinus* are used successfully to control rice planthoppers. *Blepharidopterus angulatus* is a thoroughly studied predator of spider mites, aphids, and other arthropods. Predation is also common in genera such as *Ceratocapsus, Cyllecoris, Dryophilocoris, Globiceps, Heterotoma, Hyalochloria, Malacocoris, Paraproba,* and *Pseudoloxops.* Species of the coleopteroid genus *Pseudoclerada* are apparently predacious (Asquith 1997).

This large subfamily, currently composed of five tribes (Schuh 1995), typifies mirids as omnivores; most species probably are both plant feeders and facultative predators. Many phylines feed on inflorescences and, as noted in Chapter 11, pollen feeding and carnivory are evolutionarily connected. Dispersal to the inflorescences of plants that do not serve as breeding hosts is common.

The habits of species in the small Oriental tribe Auricillocorini remain unknown. The myrmecomorphic, often ground-inhabiting Hallodapini have radiated extensively in relatively dry areas of the Old World, such as parts of Africa and the Mediterranean (Schuh 1974). Among hallodapines, bud and flower feeding is known in *Coquillettia, Orectoderus,* and *Teleorhinus;* facultative predation occurs in *Cremnocephalus* and *Systellonotus* but is uncommon in at least the mainly phytophagous *O. obliquus* (McIver and Stonedahl 1987b). Most myrmecomorphic mirids are assumed to be at least partially predacious (Douglas 1895, Breddin 1896, Bergroth 1903, Mjöberg 1906), and some might be myrmecophilous (e.g., Linnavuori 1961), but studies are needed to determine the extent of myrmecophagy in *Hallodapus* and *Systellonotus* species. In many cases, reference to myrmecophagous habits in plant bugs seems based on speculation involving the mere co-occurrence of bugs and ants. Leucophoropterines attain their greatest diversity in the Orient, and in addition to the wide-ranging genus *Tytthus,* they occur in Australia and southern Africa (Schuh and Slater 1995). Predation is common in *Sejanus albisignatus,* and *Tytthus* species are specialized predators of homopteran eggs. Suppression of the sugarcane delphacid by *T. mundulus* in Hawaii is one of the greatest successes of classical biological control. When prey are scarce, even these specialized predators can sustain themselves for a time by feeding on their host plants.

Members of the largest tribe—Phylini—tend to be omnivorous. Species in genera such as *Atractotomus, Campylomma, Glaucopterum, Harpocera, Orthonotus, Phoenicocoris, Phylus, Plagiognathus, Psallus, Rhinacloa,* and *Rhinocapsus* feed on the flower buds, flowers (including catkins of deciduous trees), and fruits of their hosts and also prey and scavenge on small arthropods and their eggs. *Atractotomus mali* and *C. verbasci* can be important pests as well as beneficial predators of plant pests occurring on the same hosts. The foliar chlorosis caused by *Macrotylus, Moissonia,* and *Rhinacloa* species often consists of a yellowing rather than the white spotting or stippling that characterizes mesophyll-feeding mirines and orthotylines. Some species of *Parapsallus* and *Phoenicocoris* develop on the new growth of conifers without inducing chlorosis or other obvious external symptoms on the needles. *Pseudatomoscelis seriata* is noted for causing stem lesions and various growth disorders of cotton, in addition to feeding on the flower buds and flowers of various hosts. *Lopus decolor* develops on the inflorescences of host grasses (especially *Agrostis* spp.). Inflorescence feeding seems particularly well developed in *Cariniocoris*, *Hoplomachus*, *Keltonia*, *Megalocoleus*, *Oncotylus*, *Placochilus*, and *Plesiodema*. As in the Mirinae, pollen and nectar feeding appear to be common. Species of *Decomia* are possibly important pollinators of dipterocarps in the Old World tropics.

Plant bugs belonging to the widely distributed Pilophorini are frequently myrmecomorphic, including those of the largest genus, *Pilophorus*. Its species are chiefly predacious, feeding on mites, aphids, scale insects, and other arthropods. Some phytophagy, though, occurs in this tribe. At least one species of *Pilophorus* (*P. clavatus*) is an occasional fruit pest, and *Sthenaridea carmelitana* feeds on the inflorescences of grasses as well as on aphids and lepidopteran eggs and larvae. *Aloea australis* causes a foliar yellowing on aloes.

PSALLOPINAE

Almost nothing is known about the behavior of this small subfamily that occurs in Africa, Australia, the neotropics, Japan, and the tropical Pacific; several new species are yet to be described. Nearly all specimens have been collected at lights, and the feeding habits of psallopines remain unknown (Schuh and Slater 1995). Two Japanese species of *Psallops* have been collected on trees (Yasunaga 1999b).

Mirid Feeding Habits: Evolutionary Perspective

It is desirable—many would say imperative—to place ecological characters within a historical framework. For most insect groups, a cladistic analysis is unavailable, but the evolutionary history of trophic habits in the Miridae can be inferred from the cladistic hypothesis of Schuh (1976).

In Figure 17.1, mirid feeding habits—classified as mycophagy, phytophagy, and zoophagy—are mapped onto Schuh's (1976) cladogram. Mixed habits prevail in nearly all subfamilies, and the proportion of different habits is only roughly estimated in the rectangle appearing above each clade. In subfamilies characterized by omnivory, zoophagy is shown below phytophagy to suggest that this habit is ancestral.

Schuh (1976) relied on morphological characters and did not use feeding habits per se to construct his cladogram. Here, I assume that his characters are independent of trophic behavior, which may not necessarily be the case. There might exist a functional relationship between feeding habits and some of the characters he used in his cladistic analysis. Given the potential for circularity of the relationship, and the possibility of other problems with a phylogenetic interpretation of mirid feeding habits, some tentative conclusions can be reached.

The first is that zoophagy is the ancestral feeding habit of plant bugs, as discussed in Chapter 16. Another

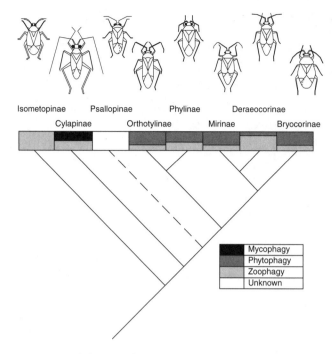

Fig. 17.1. Cladogram of proposed subfamily relationships in the Miridae (Schuh 1976) showing trophic habits, and their estimated relative proportion, for each subfamily. Illustrations represent typical members of the subfamilies.

is that their trophic habits appear to be evolutionarily labile—phytophagy has arisen at least five times (Fig. 17.1). When the largest subfamilies are reexamined cladistically, phytophagy might be shown to have evolved more than once in each subfamily. Whether the most basal clades will prove to be zoophagous rather than phytophagous is problematical.

Schuh's (1976) cladogram hypothesizes the isometopines as the sister group to all remaining subfamilies, and also the sister group relationship of Phylinae + Orthotylinae and of Mirinae + Deraeocorinae (Schuh and Slater 1995). Other relationships await greater resolution. Refinements in phylogenetic reconstruction and availability of information on trophic habits of additional higher taxa—Psallopinae and tribes of several other subfamilies—might help elucidate the extent of convergence, parallel evolution, and evolutionary reversal that is reflected in their feeding. As Lipscomb (1991) and others have noted, characters relating to nutritional mode can be particularly labile and subject to acquisition by unrelated lineages, though tissue preferences in the Hemiptera—xylem, phloem, or mesophyll—tend to be conservative (e.g., Schaefer and Mitchell 1983, Tonkyn and Whitcomb 1987). Mapping mirid feeding habits onto a cladogram of the family provides hypotheses to be tested and may enable one to predict the habits for taxa not yet studied (e.g., Platnick 1977).

Although trophic habits of mirids appear to be labile in the sense that phytophagy has presumably arisen multiple times in the family, and perhaps more than once in some of the larger subfamilies, the basic feeding modes are likely much more conservative characters. In the Bryocorinae, for example, eccritotarsines are consistent in causing foliar chlorosis. The habits of the Dicyphini differ substantially from those of the Eccritotarsini and reflect a different evolutionary history. Within Dicyphini (*sensu* Schuh 1995), the different trophic behavior of the Dicyphina versus Monaloniina + Odoniellina, coupled with this tribe's morphological diversity, suggests that the Dicyphini as currently conceived are not a monophyletic group. It seems likely that the subtribe Dicyphina, members of which are generally predators or scavengers on glandular-pubescent plants, will be reinstated as a tribe, as might both the canker- and lesion-causing Monaloniina and Odoniellina. Another group in which feeding habits tend not to corroborate the current classification (Schuh 1995) are the Cylapinae. Instead of acknowledging validity of the Bothriomirini and Cylapini + Fulviini, which appear not to be monophyletic, Schuh (1995) recognized only an inclusive Cylapini. Future classifications, based on a reassessment of morphological characters, might recognize the largely pyrenomycete-associated and perhaps mostly mycophagous Cylapini *sensu stricto* as distinct from presumably more predatory taxa such as *Fulvius* and *Peritropis* species that were formerly placed in the Fulviini. Among the Orthotylinae, members of the Halticini are relatively uniform in their trophic behavior—they usually cause foliar chlorosis—but similar patterns are mostly absent in the diverse Orthotylini as treated by Schuh (1995). The orthotylines sometimes considered to make up the tribe Ceratocapsini appear to be mostly predacious. Although Schuh (1995) was unable to recognize this group as a valid tribe, it might be recognized as such with additional morphological study (see Hernández and Henry 1999).

Further studies and additional data on feeding behavior of various higher taxa will likely alter my tentative conclusions about the evolution of feeding habits in the Miridae. Yet the relevance of a phylogenetic perspective to both basic and applied work should be apparent. Knowing more about the evolutionary history of mirid feeding habits should enhance both agricultural and basic ecological research.

18/ Future Research

Within the family [Miridae] there are many fascinating specialist species which variously feed on carrion, honeydew, blood and mummified aphids, and even rob spiders' webs and carnivorous plants of their prey.
—G. C. McGavin 1993

In a single sentence McGavin managed to capture the trophic plasticity inherent in this group—to portray mirids as opportunists and as omnivores. I might quibble over his use of "specialist species," for it implies that within the Miridae there is at least one specialist species for each of the resources mentioned when most of the resources are used merely as alternative foods.

Continuing with a synopsis or "diagnosis" of mirid trophic habits, I would add that they are principally cytoplasm and sap suckers on mesophyll, lacerating many cells at a time and ingesting liquefied solids. As tissue specialists, they do relatively little vascular feeding, or at least do not tap into the sieve tubes of the phloem. Mirid feeding combines the use of flexible stylets, a macerate and flush strategy, and extraoral digestion. Selective in their feeding sites, mirids generally exploit high-nitrogen resources such as meristems, buds, young leaves, pollen, ovules, and developing (and sometimes mature) seeds while often avoiding older tissues that might be more resistant to stylet penetration. Such trophic behavior tends to stimulate vigorous regrowth of host plants and to maintain the bugs' preferred food resources. The sites chosen for feeding tend to reflect the bugs' complement of salivary enzymes. Plant bugs are important crop pests that cause diverse symptoms on their hosts, including those that are easily mistaken for symptoms induced by plant pathogens. The presence of salivary pectinase contributes to the severity of host injury.

Some members of the family, though, are beneficial: predators that warrant conservation in pest-management programs and deserve attention from biological control workers. With an abundance of facultative predators and scavengers, the family to a large degree is omnivorous—an example of a group that is successful in feeding on both plants and animals (see Yodzis [1984], Walter [1987]). This trophic plasticity, which includes cannibalism, allows plant bugs to compensate for suboptimal resources and to persist under adverse condi-

tions. Yet it is trophic switching in mirids that complicates our attempts to assess the role of phytophagy versus zoophagy in meeting the essential dietary needs of these bugs and to understand how different plant and animal foods affect their reproductive fitness.

It could be inferred from the foregoing sketch of mirid feeding that most of the critical questions have been asked and answered—that relatively little remains to be discovered about their behavioral, biochemical, morphological, and physiological adaptations in relation to trophic habits. A wealth of biological literature has indeed accumulated on mirids as pests of crops such as alfalfa, apple, cocoa, and cotton. The information we have at hand suggests that availability of plant nitrogen is often crucial in interpreting the trophic behavior of mirids. In addition to using facultative predation and scavenging, some plant bugs can minimize changes in host nitrogen by increasing their ingestion rate and decreasing assimilation efficiencies, as demonstrated by research on *Leptopterna dolabrata*. This species also exhibits host switching, moving from the foliage of *Holcus mollis*, as the leaf nitrogen level drops, to the flower heads of a congener, *H. lanatus*. Other plant bugs develop rapidly on ephemeral resources such as catkins of deciduous trees or new growth, whereas others vary their feeding sites, using different parts of the same host individual as the season progresses.

But the overall knowledge of the family is fragmentary, involving relatively few species that make up this diverse group. Habits remain undiscovered for several tribes of large subfamilies, as well as for the small subfamily Psallopinae. Even for economically important species whose feeding habits have been scrutinized, key questions remain unanswered (e.g., Cohen and Wheeler 1998). For noneconomic plant bugs, information is available mostly for temperate-region species; observations on tropical mirids and those active mainly at night are scant. Moreover, some fundamental questions regarding sensory and digestive physiology, phytopathogenicity, and foraging strategies are only beginning to be addressed for the Miridae. Little is known about their feeding efficiency, growth rates, and respiration. Comparative studies are needed so that our understanding of trophic habits is not biased by studies on species of *Lygus* and a few other genera of agricultural importance.

An obvious need for the majority of mirid species is simply to determine what they eat. Studies on food habits not only will benefit many types of applied research but also will contribute to an understanding of the relationship between phytophagy and diversification of mirid lineages, and of the evolution of trophic behavior in the family. More detailed information might help us address questions such as whether tropical mirid species are more polyphagous than those in temperate regions, whether species of temporary or ephemeral habitats show wider host ranges than those occupying permanent habitats, whether specialized predators are more common in the tropics, whether prey specificity in the Miridae represents a derived condition, and whether colonization has been more important than cospeciation in the evolution of mirid-host relationships.

Myrmecomorphs

The habits of a diverse myrmecomorphic fauna associated with ant nests and the litter layer are little known in the Miridae, with the notable exception of Kullenberg's (1944) observations. Studies of myrmecomorphs in general—those occurring on the ground and on plants—need to be conducted. Research involving myrmecomorphic species that live mainly on plants might be patterned after that designed by J. D. McIver and colleagues. As McIver and Stonedahl (1993) pointed out, myrmecomorphic arthropods, including mirids, "offer abundant opportunities for exploring both ecological and evolutionary aspects of mimicry and can lead to a deeper understanding of related fields, such as ant social organization, systematics, and predator-prey relationships." The feeding habits of coleopteroid and other mainly ground-dwelling mirids also need to be determined.

I mentioned in Chapter 17 the need for information on the trophic habits of *Hallodapus* and *Systellonotus* species, but numerous other myrmecomorphic mirids offer research possibilities. Leston (1980b) called attention to *Xenetomorpha carpenteri*, an African herdoniine referred to as the "sewing-ant mirid." Living in intimate association with *Oecophylla* ants, the bugs run about the leaf nests of the ants, remaining unmolested by major workers of their presumed model. Whether the mirid is phytophagous or predacious is unknown, as is the adaptive significance of the myrmecomorphy. Ant-mirid associations of undetermined adaptive significance continue to be discovered, such as the hallodapines *Laemocoris orphanus*, described from Iraq and observed to live in colonies of a small red ant (Linnavuori 1984), and *Vitsikamiris madecassus* described from Madagascar (Polhemus and Razafimahatratra 1990).

Mirids of Cryptic Microhabitats

Cylapines that live on and under bark, in fallen fruit, among dead leaves, or in rotten logs might be predacious or mycophagous, but observations of their trophic behavior are almost totally lacking. Information is similarly needed for plant bugs of other groups that live in cryptic microhabitats—for example, the orthotyline *Pseudoclerada morai* that can be found in hollow plant stems and beneath bark (Swezey 1954).

Phytophages

In addition to the need for information on the food habits of particular mirid taxa, some fundamental questions remain to be considered, such as how phytophagous mirids orient to their hosts. The role of plant volatiles in mirids such as *Campylomma verbasci* and *Lygocoris pabulinus* (Groot et al. 1999), which alternate between woody and herbaceous hosts, warrants investigation. The chemosensory system allowing mirids to discriminate host from nonhost plants remains to be elucidated. Research on how they perceive secondary plant compounds—attractants and stimulants as well as deterrents and inhibitors—is particularly needed. Researchers might pursue the suggestion (see Chapter 6) that a female's rostral probing of the oviposition site and moistening of the substrate serve to deter subsequent oviposition at that site by conspecific females, as well as the suggestion that in other species the female, after laying eggs, deposits a pheromone that attracts additional females to the site.

I hope that studies on mirid probing and feeding behavior will make greater use of electronic monitoring, which has been used successfully in homopterans such as aphids, leafhoppers, and whiteflies (and only preliminarily in mirids), and will emphasize transmission electron and confocal microscopy and histochemistry to trace stylet pathways, determine the importance of ingestion from vascular tissues, and assess the reaction to various nutrients and secondary metabolites. Research similar to that of K. Hori on *Lygus rugulipennis* is needed on other species in an attempt to unravel the complexity of mirid injury to host plants, including the relationship between the bugs' salivary secretions and plant responses, and the role of auxins, inhibitors of indoleacetic acid-oxidase, and amino acids. Eventually we will have a better understanding of not only the peroxidase systems in mirids, but also the interactions between the phenol-polyphenol oxidase system in mirid saliva and that in host plants, and how such interactions translate into various types of plant injury. How microorganisms introduced during feeding might affect plant growth regulators also needs study.

Our understanding of mirid feeding would be advanced by learning how various foliage feeders discriminate between specific areas of leaves or leaves of different ages, and how nutritional requirements differ among the various nymphal stages and adults. The feeding intensity on different plant parts can vary even between adult males and females (e.g., Tan 1974b), but how such feeding disparity might relate to different physiological properties of the plant parts and nutritional requirements of the bugs is unknown. On dioecious plants, do mirids show a preference for male over female leaves? Do mesophyll feeders distinguish between "sun" and "shade" leaves, which would generally differ in their photosynthetic rate, water content, protein concentration, and tannin content? How are development, reproduction, and longevity in a particular species affected by feeding on different parts of the same host, and are feeding rates higher on more nutritious hosts? Does competition occur among plant bug species living on the same host, and do they compete with homopterans or other phytophagous heteropterans? In addition, the effects on mirids of varying nutrient levels of flowering versus nonflowering trees in species that flower irregularly—for example, some fruit trees (Sutton 1984)—need clarification. As omnivores characterized by rather plastic feeding habits, mirids would seem particularly able to cope with the absence of flowers and other fluctuations in food sources by feeding on growing shoots or using alternative behaviors such as scavenging and predation.

Research on phytophagous species might also include changes in plants that are induced by water stress and the effects of these changes on mirid fecundity and survival. Is host suitability for mirids enhanced by nitrogen availability in water-stressed plants? Somewhat similarly, do changes in plants under stress from pathogens affect the feeding of resident Miridae? Disease-induced changes in plant metabolism (e.g., Hammond and Hardy 1988) and their potential beneficial or detrimental effects on mirid performance offer an especially profitable area for research. Do nitrogen fertilization of crop plants and decreases in phosphorus and potassium routinely trigger increases in mirid fecundity or favor outbreaks of pestiferous species? In general, the relative importance of vigorous plants and plant parts versus stressed plants in understanding the patterns of resource use by phytophagous mirids requires clarification.

Plant nitrogen generally is crucial in interpreting mirid trophic behavior. These bugs use the various strategies that McNeill and Southwood (1978) recognized for buffering or minimizing changes in host nitrogen levels. Most of what is known about how mirids cope with low or fluctuating nitrogen levels in their hosts is based on S. McNeill's work with *Leptopterna dolabrata* or on studies of pestiferous *Lygus* species. This type of research should be extended to other mirid species. For example, numerous mirids disperse from breeding hosts to other, usually flowering, plants, but the significance of these strictly adult hosts in enhancing fecundity, fertility, and survival in plant bugs deserves more attention from researchers.

Zoophages

Predatory mirids offer other kinds of research opportunities. Researchers can study generalists that track abundant prey groups from one plant species to another, or they can focus on specialists that feed on only one class of prey. They can assess the benefits of facultative predation to species that are mainly phytophagous, or at least assumed so. Like certain thrips that can be either important predators of crop pests or pests themselves (Wilson et al. 1996), a particular facultatively predacious mirid might either warrant conservation as a natural enemy or require control measures to limit its plant feeding, thus challenging pest-management specialists to clarify or redefine the role of these bugs in agroecosystems. Work on predacious mirids can contribute to our basic understanding of optimal foraging strategy, dietary mixing that involves nutrient balance and energy gain, and the structure of ecological communities. Or research can be mainly applied, so that it focuses on the role of predatory plant bugs in helping suppress populations of crop pests and their use in pest-management programs. As Döbel and Denno (1994) emphasized, "We know very little about the actual predation rates of most predators in the field, their life history traits and interactions among factors which influence their dispersal and aggregation behavior."

Predation by mirids is often overlooked by researchers—its extent in both natural and managed systems should be identified more accurately. As predators of arthropod eggs and neonate larvae, plant bugs can affect pest survivorship and keep populations of pests below damaging levels. During periods of low prey availability (or regardless of whether prey are present [Schaub and Baumgärtner 1989]), mirids seem generally capable of switching to alternative prey, cannibalizing, or feeding on host plants—all of which should help stabilize predator-prey interactions (cf. Carayon 1961). The use of extraoral digestion by mirids increases the size of prey they can use—that is, their "predatory scope" (Cohen 1998a). This feeding process might be advantageous when prey are scarce (Cohen 1995, 1998a). In addition, a mixed diet typically enhances life-history traits such as fecundity and age to first reproduction.

Predatory mirids, therefore, should be conserved in agroecosystems and, in some cases, might warrant augmentative releases despite the constraints of possible injury from plant feeding (but some phytophagy in predacious mirids can be advantageous in biological control [e.g., Alomar and Wiedenmann 1996, Schelt et al. 1996]), lower reproductive potential compared to many of their prey, inability of some species to aggregate in areas of high prey density or to respond numerically, and difficulties in their mass production. Yet a better understanding of their contribution to pest regulation is requisite to any increased use of mirids in integrated pest management (IPM), as is a greater appreciation of the use of natural enemies in IPM programs having different objectives and strategies (Wiedenmann et al. 1996). In a group as diverse as the Miridae, one would expect the existence of species that could prevent or delay crop pests from reaching economic thresholds and others that could suppress such pests after they reach economic levels. The exploitation of mirids in IPM programs is hampered by a lack of studies designed to elucidate the adaptive life-history traits of potentially useful species, as well as a frequent lack of appreciation that some predacious mirids might be "major investors" (Cohen and Tang 1997). Such predators, risking injury in attacking larger prey and investing substantially in digestive enzymes and venoms, would not necessarily display Type II functional responses (Cohen and Tang 1997, Cohen 1998a).

The recently published review of the ecology of predatory heteropterans in agroecosystems and their use in biological control (Coll and Ruberson 1998) summarizes the information available on anthocorids, lygaeoids (geocorids), pentatomids, and reduviids. Mirids are scarcely mentioned. That in itself is useful, for it serves to contrast how little is known about the Miridae relative to other heteropteran families of agricultural importance. Most of what is known about predatory mirids also has not been made readily available to biological control and pest-management specialists. An exception is the useful collection of papers on zoophytophagous heteropterans in the publication by Alomar and Wiedenmann (1996).

The research needed on predatory mirids includes studies (1) to determine the life-history traits of various species, comparing and contrasting generalists and specialists; (2) to determine the traits and factors that affect their ability to disperse and aggregate in relation to prey populations and to examine potential trade-offs between survival and fecundity; (3) to test the hypothesis that certain mirids show dietary mixing which optimizes the proportion of essential amino acids or other nutrients in the diet; (4) to learn whether prey consumption is generally a simple function of relative abundance in the habitat or whether the most numerous prey are attacked supraproportionally; (5) to determine differences in strict predators versus omnivores in their capacity to stabilize prey populations; (6) to examine the relationship between relative weights of predatory mirids and their prey, and handling times, in studies of functional response; (7) to assess the role of age structure in determining diet, including size of prey taken; (8) to assess the benefits of plant feeding in species sometimes regarded as obligate predators and to determine if facultative phytophagy influences the age structure of populations; (9) to determine if facultative plant feeding by chiefly predacious species renders them susceptible to systemic insecticides; (10) to discover trade-offs between reproduction and longevity that allow populations to survive periods of prey scarcity or adverse environmental conditions; (11) to determine dispersal and colonization rates into new habitats; (12) to determine the relative importance of prey versus the host plant in facilitating colonization and stimulating reproduction by predacious mirids; (13) to identify plant volatiles and prey semiochemicals that mirids might use to locate prey and to elucidate potential interaction between plant- and prey-produced chemicals; (14) to

understand how plant chemistry and morphology might hinder the effectiveness of predatory mirids; and (15) to clarify multiple-predator interactions—additive, synergistic, or antagonistic (e.g., Losey and Denno 1998)—in natural and managed systems. Alomar and Wiedenmann (1996) and Wiedenmann and Wilson (1996) pose many other important questions concerning the feeding strategies and nutrition of predatory mirids, particularly in relation to the exploitation of zoophytophagous species in pest-management programs.

Expectations

Some of our ideas about the feeding behavior of phytophagous and predacious mirids will soon be outmoded. Yet, as Price (1980:vii) pointed out, "Early endeavors . . . may turn out to be far from the truth in many ways, but they are justifiable, even when wrong, if further study is stimulated." What progress toward understanding the trophic habits and other biological aspects of plant bugs can be expected in the next 25 or 50 years?

It may become possible to predict the pest potential or the predatory tendency of a particular mirid species solely from an analysis of its salivary constituents—for example, the presence of amylase, pectinase, and phenoloxidase versus phospholipase, proteinase, and venoms—and a better understanding of its digestive physiology. A zoophage, for example, would be characterized by its nonuse of plant sterols and plant sugars, as well as by a need for large amounts of nitrogen. Perhaps to a lesser extent, stylet structure might prove helpful in predicting trophic habits—for example, strict phytophages might have blunter stylet tips than more predatory species.

We eventually will know much more about how different free amino acids affect mirid feeding behavior, as well as have a better resolution of the interactions among factors such as previous host, mirid, host age, and physiological condition and the influence of these factors on the pathogenicity of plant bugs. We will know more about the mechanisms (perhaps involving altered levels of auxin) responsible for phytostimulation of bean and carrot seed. How mirids balance their diet between carbohydrate- and protein-rich food sources and how they might modify host physiology to create more nutrient-rich feeding sites will be much better understood, as will any effects of diet on chemical composition of the metathoracic glands. Detailed information on feeding behavior and injury, similar to that of Varis (1972) and Handley and Pollard (1993) for lygus bugs on sugar beet and strawberry, respectively, will be available. Ecological studies comparable to those on the mirid fauna of Scotch broom in England (Waloff and Southwood 1960; Waloff and Bakker 1963; Dempster 1964, 1966; Waloff 1967, 1968; Waloff and Richards 1977) will have been conducted, as will research on mirid energetics and population dynamics similar to that by McNeill (1971, 1973) on *Leptopterna dolabrata*. Studies similar to those by Braune (1983) on *L. dolabrata* will clarify the environmental and hereditary factors affecting wing polymorphism in this species and in other wing-dimorphic mirids.

Crop varieties might be developed that use cations to inhibit mirid polygalacturonase, as proposed by Strong (1970) and discussed by Tingey (1976) and Tingey and Pillemer (1977). We will know more about how volatile phytochemicals and other characteristics of pest-resistant crop varieties affect discovery and colonization by predacious mirids. We will better understand how facultative predation by pest mirids and plant feeding by chiefly predacious species affects their fecundity, longevity, and population dynamics. Researchers will routinely consider the availability of animal food under experimental conditions—via cannibalism, hard-to-detect populations of small arthropods such as mites and their eggs, and scavenging—and how such feeding potentially affects the measurement of life-history parameters in mostly phytophagous mirids. Studies similar to those by Eubanks and Denno (1999) on *Geocoris* (Lygaeoidea: Geocoridae) will help clarify the effects of variations in plant and prey quality on omnivorous mirids. With greater appreciation of mirids as solid-to-liquid feeders and emphasis on size relationships between predator and prey (Cohen and Tang 1997, Cohen 1998a), the Miridae will be used more effectively in pest-management programs—that is, with predictable results (e.g., Gross 1987). For predacious mirids, especially omnivores, to be used predictably and confidently in pest management, the factors that affect the bugs' switching between plant and animal diets, the benefits of a mixed diet, and the factors that influence their tendency to injure host plants must be better understood (e.g., Eubanks and Denno 1999, Gillespie et al. 1999, McGregor et al. 2000). Predacious mirids will become even more important components of thrips- and whitefly-management programs for protected crops and probably those grown outdoors.

A subsequent reviewer of mirid biology, with access to data from studies using electronic monitoring devices, will be able to discuss probing and feeding behavior in greater detail and will be able to draw on substantially more compelling evidence for mirids as fungus feeders, vascular tissue feeders, significant pollinators, and effective vectors of plant pathogens. With the recognition of mirids as suitable subjects for diverse types of biological studies, these and other advances seem likely.

Appendix 1/ Valid Names, Authors (Authorities), and Subfamilies of Mirids Mentioned in the Text

Acetropis gimmerthalii (Flor); Mirinae
Adelphocoris lineolatus (Goeze); Mirinae
Adelphocoris quadripunctatus (Fabricius); Mirinae
Adelphocoris rapidus (Say); Mirinae
Adelphocoris seticornis (Fabricius); Mirinae
Adelphocoris superbus (Uhler); Mirinae
Adelphocoris suturalis (Jakovlev); Mirinae
Adelphocoris taeniophorus Reuter; Mirinae
Adelphocoris ticinensis (Meyer-Dür); Mirinae
Adelphocoris vandalicus (Rossi); Mirinae
Adelphocoris variabilis (Uhler); Mirinae
Agnocoris pulverulentus (Uhler); Mirinae
Agnocoris rubicundus (Fallén); Mirinae
Agraptocoris margaretae (Hutchinson); Phylinae
Alloeotomus chinensis Reuter; Deraeocorinae
Aloea australis Schuh; Phylinae
Amblytylus nasutus (Kirschbaum); Phylinae
Ambracius dufouri Stål; Deraeocorinae
Americodema nigrolineatum (Knight); Phylinae
Anapus spp.; Orthotylinae
Aoplonema princeps (Uhler); Orthotylinae
Aoplonema uhleri (Van Duzee); Orthotylinae
Aoplonema uniforme (Knight); Orthotylinae
Apachemiris vigilax (Van Duzee); Orthotylinae
Apolygus lucorum (Meyer-Dür); Mirinae
Apolygus nigritulus (Linnavuori); Mirinae
Apolygus spinolae (Meyer-Dür); Mirinae
Asciodema obsoleta (Fieber); Phylinae
Atomoscelis modesta (Van Duzee); Phylinae
Atractotomus magnicornis (Fallén); Phylinae
Atractotomus mali (Meyer-Dür); Phylinae
Atractotomus miniatus (Knight); Phylinae
Atractotomus prosopidis (Knight); Phylinae
Austropeplus sp.; Mirinae

Barberiella formicoides Poppius; Mirinae
Blepharidopterus angulatus (Fallén); Orthotylinae
Blepharidopterus chlorionis (Say); Orthotylinae
Blepharidopterus provancheri (Burque); Orthotylinae
Bolteria juniperi Knight; Mirinae
Bolteria luteifrons Knight; Mirinae
Bothynotus johnstoni Knight; Deraeocorinae
Bothynotus pilosus (Boheman); Deraeocorinae
Boxiopsis madagascariensis Lavabre; Bryocorinae
Brachynotocoris puncticornis Reuter; Orthotylinae

Bromeliaemiris bicolor Schumacher; Bryocorinae
Brooksetta althaeae (Hussey); Orthotylinae
Bryocoris pteridis (Fallén); Bryocorinae
Bryocoropsis laticollis Schumacher; Bryocorinae
Bryophilocapsus tosamontanus Yasunaga; Bryocorinae

Calocoris angustatus Lethierry; Mirinae
Calocoris barberi Henry & Wheeler; Mirinae
Calocoris roseomaculatus (De Geer); Mirinae
Camptotylus yersini (Mulsant & Rey); Phylinae
Campylomma agalegae Miller; Phylinae
Campylomma angustius Poppius; Phylinae
Campylomma austrinum Malipatil; Phylinae
Campylomma chinense Schuh; Phylinae
Campylomma diversicorne Reuter; Phylinae
Campylomma liebknechti (Girault); Phylinae
Campylomma lividum Reuter; Phylinae
Campylomma plantarum Lindberg; Phylinae
Campylomma verbasci (Meyer-Dür); Phylinae
Campyloneura virgula (Herrich-Schaeffer);
   Bryocorinae
Campyloneuropsis cincticornis (Stål); Bryocorinae
Capsodes flavomarginatus (Donovan); Mirinae
Capsodes gothicus (Linnaeus); Mirinae
Capsodes sulcatus (Fieber); Mirinae
Capsus ater (Linnaeus); Mirinae
Capsus cinctus (Kolenati); Mirinae
Cariniocoris geminatus (Knight); Phylinae
Cariniocoris ilicis (Knight); Phylinae
Cariniocoris nyssae Henry; Phylinae
Carvalhofulvius gigantochloae Stonedahl & Kovac;
   Cylapinae
Carvalhoisca jacquiniae Schaffner & Ferreira;
   Orthotylinae
Carvalhoisca michoacana Schaffner & Ferreira;
   Orthotylinae
Carvalhoma taplini Slater & Gross; Cylapinae
Castanopsides falkovitshi (Kerzhner); Mirinae
Castanopsides hasegawai Yasunaga; Mirinae
Castanopsides potanini (Reuter); Mirinae
Caulotops distanti (Reuter); Bryocorinae
Ceratocapsus bahiensis Carvalho & Fontes;
   Orthotylinae
Ceratocapsus dispersus Carvalho & Fontes;
   Orthotylinae

Ceratocapsus guaratibanus Carvalho & Fontes;
  Orthotylinae
Ceratocapsus mariliensis Carvalho & Fontes;
  Orthotylinae
Ceratocapsus modestus Uhler; Orthotylinae
Ceratocapsus punctulatus Reuter; Orthotylinae
Chamopsis conradti Reuter & Poppius; Bryocorinae
Chamopsis tuberculata (Distant); Bryocorinae
Charagochilus sp.; Mirinae
Chlamydatus associatus (Uhler); Phylinae
Chlamydatus evanescens (Boheman); Phylinae
Chlamydatus pulicarius (Fallén); Phylinae
Chlamydatus pullus (Reuter); Phylinae
Chorosomella spp.; Orthotylinae
Clivinema coalinga Bliven; Deraeocorinae
Clivinema sericea Knight; Deraeocorinae
Closterotomus biclavatus (Herrich-Schaeffer); Mirinae
Closterotomus fulvomaculatus (De Geer); Mirinae
Closterotomus norwegicus (Gmelin); Mirinae
Closterotomus trivialis (Costa); Mirinae
Coccobaphes frontifer (Walker); Mirinae
Collaria columbiensis Carvalho; Mirinae
Collaria meilleurii Provancher; Mirinae
Collaria oleosa (Distant); Mirinae
Collaria scenica (Stål); Mirinae
Compsidolon salicellum (Herrich-Schaeffer); Phylinae
Conostethus americanus Knight; Phylinae
Coquillettia insignis Uhler; Phylinae
Corticoris libertus (Gibson); Isometopinae
Corticoris signatus (Heidemann); Isometopinae
Cremnocephalus albolineatus Reuter; Phylinae
Creontiades debilis (Van Duzee); Mirinae
Creontiades dilutus (Stål); Mirinae
Creontiades pacificus (Stål); Mirinae
Creontiades pallidus (Rambur); Mirinae
Creontiades rubrinervis (Stål); Mirinae
Criocoris saliens (Reuter); Phylinae
Cylapocoris pilosus Carvalho; Cylapinae
Cylapocoris tiquiensis Carvalho; Cylapinae
Cylapus citus Bergroth; Cylapinae
Cylapus ruficeps Bergroth; Cylapinae
Cylapus tenuicornis (Say); Cylapinae
Cyllecoris histrionius (Linnaeus); Orthotylinae
Cyrtocapsus caligineus (Stål); Bryocorinae
Cyrtopeltis kahakai Asquith; Bryocorinae
Cyrtorhinus caricis (Fallén); Orthotylinae
Cyrtorhinus fulvus Knight; Orthotylinae
Cyrtorhinus lividipennis Reuter; Orthotylinae

Dacerla mediospinosa Signoret; Mirinae
Dagbertus fasciatus (Reuter); Mirinae
Dagbertus olivaceus (Reuter); Mirinae
Decomia spp.; Phylinae
Deraeocapsus ingens (Van Duzee); Deraeocorinae
Deraeocoris albigulus Knight; Deraeocorinae
Deraeocoris annulipes (Herrich-Schaeffer);
  Deraeocorinae
Deraeocoris aphidicidus Ballard; Deraeocorinae

Deraeocoris aphidiphagus Knight; Deraeocorinae
Deraeocoris bakeri Knight; Deraeocorinae
Deraeocoris barberi Knight; Deraeocorinae
Deraeocoris borealis (Van Duzee); Deraeocorinae
Deraeocoris brachialis Stål; Deraeocorinae
Deraeocoris brevis (Uhler); Deraeocorinae
Deraeocoris crigi Leston & Gibbs; Deraeocorinae
Deraeocoris delagrangei (Puton); Deraeocorinae
Deraeocoris diveni Knight; Deraeocorinae
Deraeocoris erythromelas Yasunaga & Nakatani;
  Deraeocorinae
Deraeocoris fasciolus Knight; Deraeocorinae
Deraeocoris flavilinea (Costa); Deraeocorinae
Deraeocoris histrio (Reuter); Deraeocorinae
Deraeocoris incertus Knight; Deraeocorinae
Deraeocoris laricicola Knight; Deraeocorinae
Deraeocoris limbatus Miller; Deraeocorinae
Deraeocoris lutescens (Schilling); Deraeocorinae
Deraeocoris manitou (Van Duzee); Deraeocorinae
Deraeocoris maoricus Westwood; Deraeocorinae
Deraeocoris nebulosus (Uhler); Deraeocorinae
Deraeocoris nigrifrons Knight; Deraeocorinae
Deraeocoris nitenatus (Knight); Deraeocorinae
Deraeocoris nubilus Knight; Deraeocorinae
Deraeocoris oculatus (Reuter); Deraeocorinae
Deraeocoris olivaceus (Fabricius); Deraeocorinae
Deraeocoris ostentans (Stål); Deraeocorinae
Deraeocoris pallens (Reuter); Deraeocorinae
Deraeocoris piceicola Knight; Deraeocorinae
Deraeocoris pilipes (Reuter); Deraeocorinae
Deraeocoris pinicola Knight; Deraeocorinae
Deraeocoris punctulatus (Fallén); Deraeocorinae
Deraeocoris quercicola Knight; Deraeocorinae
Deraeocoris ruber (Linnaeus); Deraeocorinae
Deraeocoris schwarzii (Uhler); Deraeocorinae
Deraeocoris serenus (Douglas & Scott); Deraeocorinae
Deraeocoris signatus (Distant); Deraeocorinae
Deraeocoris trifasciatus (Linnaeus); Deraeocorinae
Dichrooscytus elegans Heidemann; Mirinae
Dichrooscytus repletus (Heidemann); Mirinae
Dichrooscytus suspectus Reuter; Mirinae
Dicyphus constrictus (Boheman); Bryocorinae
Dicyphus eckerleini Wagner; Bryocorinae
Dicyphus errans (Wolff); Bryocorinae
Dicyphus globulifer (Fallén); Bryocorinae
Dicyphus gracilentus Parshley; Bryocorinae
Dicyphus hesperus Knight; Bryocorinae
Dicyphus pallicornis (Fieber); Bryocorinae
Dicyphus tamaninii Wagner; Bryocorinae
Dionconotus neglectus (Fabricius); Mirinae
Diphleps sp.; Isometopinae
Diplozona sp.; Deraeocorinae
Distantiella collarti (Schouteden); Bryocorinae
Distantiella theobroma (Distant); Bryocorinae
Dolichomiris linearis Reuter; Mirinae
Dryophilocoris flavoquadrimaculatus (De Geer);
  Orthotylinae
Dryophilocoris miyamotoi Yasunaga; Orthotylinae

Eccritotarsus catarinensis (Carvalho); Bryocorinae
Engytatus modestus (Distant); Bryocorinae
Engytatus nicotianae (Koningsberger); Bryocorinae
Engytatus varians (Distant); Bryocorinae
Eocalocoris albicerus Yasunaga & Takai; Mirinae
Ephedrodoma multilineata Polhemus & Polhemus;
 Orthotylinae
Eumecotarsus breviceps (Reuter); Phylinae
Eumecotarsus kiritshenkoi Kerzhner; Phylinae
Europiella decolor (Uhler); Phylinae
Eurychilella pallida Reuter; Bryocorinae
Eurychilopterella luridula Reuter; Deraeocorinae
Euryopicoris nitidus (Meyer-Dür); Orthotylinae
Eurystylus bellevoyei (Reuter); Mirinae
Eurystylus marginatus Odhiambo; Mirinae
Eurystylus oldi Poppius; Mirinae
Eustictus spp.; Deraeocorinae

Falconia intermedia (Distant); Orthotylinae
Felisacus elegantulus (Reuter); Bryocorinae
Fieberocapsus flaveolus (Reuter); Orthotylinae
Fingulus porrectus (Bergroth); Deraeocorinae
Fulvius angustatus Usinger; Cylapinae
Fulvius brevicornis Reuter; Cylapinae
Fulvius imbecilus (Say); Cylapinae
Fulvius nigricornis Poppius; Cylapinae
Fulvius quadristillatus (Stål); Cylapinae
Fulvius slateri Wheeler; Cylapinae
Fulvius variegatus Poppius; Cylapinae

Garganus albidivittis (Stål); Mirinae
Garganus fusiformis (Say); Mirinae
Glaucopterum atraphaxius Putshkov; Phylinae
Globiceps flavomaculatus (Fabricius); Orthotylinae
Globiceps fulvicollis Jakovlev; Orthotylinae
Gracilomiris sp.; Mirinae
Grypocoris sexguttatus (Fabricius); Mirinae
Guianerius typicus Distant; Mirinae

Hadronema bispinosum Knight; Orthotylinae
Hadronema breviatum Knight; Orthotylinae
Hadronema militare Uhler; Orthotylinae
Haematocapsus bipunctatus Poppius; Bryocorinae
Hallodapus albofasciatus (Motschulsky); Phylinae
Hallodapus montandoni (Reuter); Phylinae
Halticotoma nicholi Knight; Bryocorinae
Halticotoma valida Townsend; Bryocorinae
Halticus apterus (Linnaeus); Orthotylinae
Halticus bractatus (Say); Orthotylinae
Halticus chrysolepis Kirkaldy; Orthotylinae
Halticus intermedius Uhler; Orthotylinae
Halticus minutus Reuter; Orthotylinae
Halticus saltator (Geoffroy); Orthotylinae
Hambletoniola sp.; Phylinae
Harpocera thoracica (Fallén); Phylinae
Helopeltis anacardii Miller; Bryocorinae
Helopeltis antonii Signoret; Bryocorinae
Helopeltis bergrothi Reuter; Bryocorinae

Helopeltis bradyi Waterhouse; Bryocorinae
Helopeltis clavifer (Walker); Bryocorinae
Helopeltis corbisieri Schmitz; Bryocorinae
Helopeltis fasciaticollis Poppius; Bryocorinae
Helopeltis orophila Ghesquiere; Bryocorinae
Helopeltis pernicialis Stonedahl, Malipatil &
 Houston; Bryocorinae
Helopeltis schoutedeni Reuter; Bryocorinae
Helopeltis theivora Waterhouse; Bryocorinae
Helopeltis westwoodi (White); Bryocorinae
Hemicerocoris bicolor Carvalho; Deraeocorinae
Hesperolabops gelastops Kirkaldy; Bryocorinae
Hesperophylum arizonae Knight; Deraeocorinae
Hesperophylum heidemanni Reuter & Poppius;
 Deraeocorinae
Heterocordylus genistae (Scopoli); Orthotylinae
Heterocordylus malinus Slingerland; Orthotylinae
Heterocordylus tibialis Hahn; Orthotylinae
Heterotoma merioptera (Scopoli); Orthotylinae
Heterotoma planicornis (Pallas); Orthotylinae
Hoplomachus affiguratus (Uhler); Phylinae
Hoplomachus thunbergii (Fallén); Phylinae
Horcias nobilellus (Berg); Mirinae
Horciasinus signoreti (Stål); Mirinae
Horistus infuscatus (Brullé); Mirinae
Horistus orientalis (Gmelin); Mirinae
Hyaliodes beckeri Carvalho; Deraeocorinae
Hyaliodes harti Knight; Deraeocorinae
Hyaliodes vitripennis (Say); Deraeocorinae
Hyaliodes vittaticornis Bruner; Deraeocorinae
Hyalochloria denticornis Hsiao; Orthotylinae
Hyalochloria longicornis Henry; Orthotylinae
Hyalopeplus pellucidus (Stål); Mirinae
Hyalopeplus similis Poppius; Mirinae
Hyalopeplus smaragdina Roepke; Mirinae

Ilnacora malina (Uhler); Orthotylinae
Irbisia bliveni Schwartz; Mirinae
Irbisia brachycera (Uhler); Mirinae
Irbisia californica Van Duzee; Mirinae
Irbisia cascadia Schwartz; Mirinae
Irbisia pacifica (Uhler); Mirinae
Irbisia sericans (Stål); Mirinae
Irbisia solani (Heidemann); Mirinae
Isometopus citri Ren; Isometopinae
Isometopus intrusus (Herrich-Schaeffer); Isometopinae
Isometopus shaowuensis Ren; Isometopinae
Isometopus transvaalensis (Slater & Schuh);
 Isometopinae

Kamehameha lunalilo Kirkaldy; Orthotylinae
Keltonia pallida Henry; Phylinae
Keltonia tuckeri (Poppius); Phylinae
Kundakimuka queenslandica Cassis; Deraeocorinae

Labopidea allii Knight; Orthotylinae
Labopidea simplex Uhler; Orthotylinae
Labops burmeisteri Stål; Orthotylinae

Labops hesperius Uhler; Orthotylinae
Labops hirtus Knight; Orthotylinae
Labops utahensis Slater; Orthotylinae
Laemocoris orphanus Linnavuori; Phylinae
Lampethusa anatina Distant; Mirinae
Lampethusa collaris Reuter; Mirinae
Largidea davisi Knight; Deraeocorinae
Larinocerus balius Froeschner; Phylinae
Lepidargyrus ancorifer (Fieber); Phylinae
Leptopterna dolabrata (Linnaeus); Mirinae
Leptopterna ferrugata (Fallén); Mirinae
Lidopus heidemanni Gibson; Isometopinae
Lincolnia lucernina Eyles & Carvalho; Mirinae
Lineatopsallus biguttulatus (Uhler); Phylinae
Liocoris tripustulatus (Fabricius); Mirinae
Litomiris debilis (Uhler); Mirinae
Lopidea chandleri Moore; Orthotylinae
Lopidea cuneata Van Duzee; Orthotylinae
Lopidea dakota Knight; Orthotylinae
Lopidea davisi Knight; Orthotylinae
Lopidea heidemanni Knight; Orthotylinae
Lopidea hesperus (Kirkaldy); Orthotylinae
Lopidea instabilis (Reuter); Orthotylinae
Lopidea marginata Uhler; Orthotylinae
Lopidea media (Say); Orthotylinae
Lopidea minor Knight; Orthotylinae
Lopidea nigridia Uhler; Orthotylinae
Lopidea robiniae Uhler; Orthotylinae
Lopidea robusta (Uhler); Orthotylinae
Lopidea staphyleae Knight; Orthotylinae
Lopidea teton Knight; Orthotylinae
Lopus decolor (Fallén); Phylinae
Loulucoris kidoi Asquith; Orthotylinae
Lycidocoris mimeticus Reuter & Poppius; Bryocorinae
Lygidea mendax Reuter; Mirinae
Lygidolon laevigatum Reuter; Mirinae
Lygocorides rubronasutus (Linnavuori); Mirinae
Lygocoris aesculi (Knight); Mirinae
Lygocoris atrinotatus (Knight); Mirinae
Lygocoris belfragii (Reuter); Mirinae
Lygocoris caryae (Knight); Mirinae
Lygocoris communis (Knight); Mirinae
Lygocoris fagi (Knight); Mirinae
Lygocoris geneseensis (Knight); Mirinae
Lygocoris hirticulus (Van Duzee); Mirinae
Lygocoris inconspicuus (Knight); Mirinae
Lygocoris knighti Kelton; Mirinae
Lygocoris kyushuensis Yasunaga; Mirinae
Lygocoris laureae (Knight); Mirinae
Lygocoris neglectus (Knight); Mirinae
Lygocoris nyssae (Knight); Mirinae
Lygocoris omnivagus (Knight); Mirinae
Lygocoris ostryae (Knight); Mirinae
Lygocoris pabulinus (Linnaeus); Mirinae
Lygocoris quercalbae (Knight); Mirinae
Lygocoris rugicollis (Fallén); Mirinae
Lygocoris semivittatus (Knight); Mirinae
Lygocoris tiliae (Knight); Mirinae
Lygocoris tinctus (Knight); Mirinae

Lygocoris viburni (Knight); Mirinae
Lygocoris viridanus (Motschulsky); Mirinae
Lygocoris viridus (Fallén); Mirinae
Lygocoris vitticollis (Reuter); Mirinae
Lygus abroniae Van Duzee; Mirinae
Lygus atriflavus Knight; Mirinae
Lygus borealis (Kelton); Mirinae
Lygus bradleyi Knight; Mirinae
Lygus elisus Van Duzee; Mirinae
Lygus gemellatus (Herrich-Schaeffer); Mirinae
Lygus hesperus Knight; Mirinae
Lygus keltoni Schwartz; Mirinae
Lygus lineolaris (Palisot de Beauvois); Mirinae
Lygus mexicanus Kelton; Mirinae
"Lygus" muiri Poppius; Mirinae
Lygus oregonae Knight; Mirinae
Lygus pratensis (Linnaeus); Mirinae
Lygus robustus (Uhler); Mirinae
Lygus rugulipennis Poppius; Mirinae
Lygus shulli Knight; Mirinae
Lygus unctuosus (Kelton); Mirinae
Lygus vanduzeei Knight; Mirinae

Macrolophus costalis Fieber; Bryocorinae
Macrolophus melanotoma (Costa); Bryocorinae
Macrolophus praeclarus (Distant); Bryocorinae
Macrolophus pygmaeus (Rambur); Bryocorinae
Macrolophus separatus (Uhler); Bryocorinae
Macrolophus tenuicornis Blatchley; Bryocorinae
Macrotylus amoenus Reuter; Phylinae
Macrotylus multipunctatus Van Duzee; Phylinae
Macrotylus nigricornis Fieber; Phylinae
Macrotylus quadrilineatus (Schrank); Phylinae
Macrotylus sexguttatus (Provancher); Phylinae
Malacocoris chlorizans (Panzer); Orthotylinae
Mansoniella nitida Poppius; Bryocorinae
Maurodactylus albidus (Kolenati); Phylinae
Mecistoscelis scirtetoides Reuter; Mirinae
Mecomma angustatum (Uhler); Orthotylinae
Mecomma dispar (Boheman); Orthotylinae
Megacoelum apicale Reuter; Mirinae
Megacoelum infusum (Herrich-Schaeffer); Mirinae
Megaloceroea recticornis (Geoffroy); Mirinae
Megalocoleus molliculus (Fallén); Phylinae
Megalocoleus tanaceti (Fallén); Phylinae
Megalopsallus atriplicis Knight; Phylinae
Megalopsallus latifrons Knight; Phylinae
Mermitelocerus annulipes Reuter; Mirinae
Mertila malayensis Distant; Bryocorinae
Metriorrhynchomiris dislocatus (Say); Mirinae
Metriorrhynchomiris fallax (Reuter); Mirinae
Microphylellus flavipes (Provancher); Phylinae
Microphylellus modestus Reuter; Phylinae
Microphylellus tsugae Knight; Phylinae
Microphylellus tumidifrons Knight; Phylinae
Mircarvalhoia arecae (Miller & China); Bryocorinae
Miris striatus (Linnaeus); Mirinae
Moissonia flavomaculata (Ballard); Phylinae
Moissonia importunitas (Distant); Phylinae

Moissonia schefflerae (Schuh); Phylinae
Monalocoris filicis (Linnaeus); Bryocorinae
Monalonion annulipes Signoret; Bryocorinae
Monalonion bondari Costa Lima; Bryocorinae
Monalonion dissimulatum Distant; Bryocorinae
Monalonion schaefferi Stål; Bryocorinae
Monopharsus annulatus Eyles & Carvalho; Mirinae
Monosynamma bohemanni (Fallén); Phylinae
Myiomma cixiiforme (Uhler); Isometopinae
Myrmecophyes oregonensis Schuh & Lattin; Orthotylinae
Myrmecoris gracilis (Sahlberg); Mirinae

Nanopsallus sp.; Phylinae
Neocapsus leviscutatus Knight; Mirinae
Neoneella bosqi Carvalho; Bryocorinae
Nesidiocoris caesar (Ballard); Bryocorinae
Nesidiocoris callani (Odhiambo); Bryocorinae
Nesidiocoris tenuis (Reuter); Bryocorinae
Nesidiocoris volucer Kirkaldy; Bryocorinae
Nesidiorchestes hawaiiensis Kirkaldy; Orthotylinae
Nesiomiris pallasatus Gagné; Orthotylinae
Nesiomiris sinuatus Gagné; Orthotylinae
Neurocolpus arizonae Knight; Mirinae
Neurocolpus flavescens Blatchley; Mirinae
Neurocolpus jessiae Knight; Mirinae
Neurocolpus longirostris Knight; Mirinae
Neurocolpus mexicanus Distant; Mirinae
Neurocolpus nubilus (Say); Mirinae
Neurocolpus pumilus Henry; Mirinae
Neurocolpus tiliae Knight; Mirinae
Niastama punctaticollis Reuter; Mirinae
Notostira elongata (Geoffroy); Mirinae
Notostira erratica (Linnaeus); Mirinae

Occidentodema polhemusi Henry; Phylinae
Ofellus mexicanus Carvalho & Sailer; Deraeocorinae
Oncotylus punctipes Reuter; Phylinae
Oncotylus viridiflavus (Goeze); Phylinae
Opistheurista clandestina (Van Duzee); Mirinae
Opuna sharpiana (Kirkaldy); Phylinae
Orectoderus obliquus Uhler; Phylinae
Orthocephalus coriaceus (Fabricius); Orthotylinae
Orthocephalus funestus Jakovlev; Orthotylinae
Orthocephalus saltator (Hahn); Orthotylinae
Orthonotus rossicus (Reuter); Phylinae
Orthonotus rufifrons (Fallén) Phylinae
Orthops basalis (Costa); Mirinae
Orthops campestris (Linnaeus); Mirinae
Orthops kalmii (Linnaeus); Mirinae
Orthops montanus (Schilling); Mirinae
Orthops palus (Taylor); Mirinae
Orthops scutellatus Uhler; Mirinae
Orthotylus adenocarpi (Perris); Orthotylinae
Orthotylus aesculicola Blinn; Orthotylinae
Orthotylus australianus Carvalho; Orthotylinae
Orthotylus catulus Van Duzee; Orthotylinae
Orthotylus concolor (Kirschbaum); Orthotylinae
Orthotylus dorsalis (Provancher); Orthotylinae

Orthotylus ericetorum (Fallén); Orthotylinae
Orthotylus flavosparsus (Sahlberg); Orthotylinae
Orthotylus gotohi Yasunaga; Orthotylinae
Orthotylus iolani Kirkaldy; Orthotylinae
Orthotylus juglandis Henry; Orthotylinae
Orthotylus leviculus (Knight); Orthotylinae
Orthotylus marginalis Reuter; Orthotylinae
Orthotylus mimus (Knight); Orthotylinae
Orthotylus modestus Van Duzee; Orthotylinae
Orthotylus nassatus (Fabricius); Orthotylinae
Orthotylus ramus Knight; Orthotylinae
Orthotylus tenellus (Fallén); Orthotylinae
Orthotylus virescens (Douglas & Scott); Orthotylinae
Orthotylus viridinervis (Kirschbaum); Orthotylinae

Pachymerocista pilosa (Carvalho); Bryocorinae
Pachypeltis maesarum (Kirkaldy); Bryocorinae
Pachypeltis polita (Walker); Bryocorinae
Pachypeltis vittiscutis (Bergroth); Bryocorinae
Pachytomella sp.; Orthotylinae
Palomiella bedfordi (Hesse); Isometopinae
Pameridea marlothi Poppius; Bryocorinae
Pameridea roridulae Reuter; Bryocorinae
Pantilius tunicatus (Fabricius); Mirinae
Paracarnus sp.; Deraeocorinae
Paradacerla formicina (Parshley); Mirinae
Paraproba capitata (Van Duzee); Orthotylinae
Paraproba nigrinervis Van Duzee; Orthotylinae
Paraproba pendula Van Duzee; Orthotylinae
Parapsallus vitellinus (Scholtz); Phylinae
Pararculanus piperis Poppius; Bryocorinae
Parthenicus aureosquamis Knight; Orthotylinae
Peritropis spp.; Cylapinae
Phoenicocoris australis (Blatchley); Phylinae
Phoenicocoris claricornis (Knight); Phylinae
Phoenicocoris dissimilis (Reuter); Phylinae
Phoenicocoris minusculus (Knight); Phylinae
Phylus coryli (Linnaeus); Phylinae
Phylus melanocephalus (Linnaeus); Phylinae
Phytocoris americanus Carvalho; Mirinae
Phytocoris breviusculus Reuter; Mirinae
Phytocoris californicus Knight; Mirinae
Phytocoris calli Knight; Mirinae
Phytocoris canadensis Van Duzee; Mirinae
Phytocoris conspurcatus Knight; Mirinae
Phytocoris dimidiatus Kirschbaum; Mirinae
Phytocoris eximius Reuter; Mirinae
Phytocoris intricatus Flor; Mirinae
Phytocoris longipennis Flor; Mirinae
Phytocoris michiganae Knight; Mirinae
Phytocoris neglectus Knight; Mirinae
Phytocoris nigrifrons Van Duzee; Mirinae
Phytocoris obscuratus Carvalho; Mirinae
Phytocoris olseni Knight; Mirinae
Phytocoris pini Kirschbaum; Mirinae
Phytocoris populi (Linnaeus); Mirinae
Phytocoris relativus Knight; Mirinae
Phytocoris reuteri Saunders; Mirinae
Phytocoris tiliae (Fabricius); Mirinae

Phytocoris tillandsiae Johnston; Mirinae
Phytocoris ulmi (Linnaeus); Mirinae
Phytocoris vanduzeei Reuter; Mirinae
Phytocoris varipes Boheman; Mirinae
Pilophorus amoenus Uhler; Phylinae
Pilophorus cinnamopterus (Kirschbaum); Phylinae
Pilophorus clavatus (Linnaeus); Phylinae
Pilophorus confusus (Kirschbaum); Phylinae
Pilophorus crassipes Heidemann; Phylinae
Pilophorus furvus Knight; Phylinae
Pilophorus juniperi Knight; Phylinae
Pilophorus perplexus Douglas & Scott; Phylinae
Pinalitus approximatus (Stål); Mirinae
Pinalitus cervinus (Herrich-Schaeffer); Mirinae
Pinalitus coccineus (Horvath); Mirinae
Pinalitus conspurcatus (Reuter); Mirinae
Pinalitus viscicola (Puton); Mirinae
Pinophylus carneolus (Knight); Phylinae
Pithanus maerkelii (Herrich-Schaeffer); Mirinae
Placochilus seladonicus (Fallén); Phylinae
Plagiognathus albatus (Van Duzee); Phylinae
Plagiognathus albifacies Knight; Phylinae
Plagiognathus arbustorum (Fabricius); Phylinae
Plagiognathus brevirostris Knight; Phylinae
Plagiognathus caryae Knight; Phylinae
Plagiognathus chrysanthemi (Wolff); Phylinae
Plagiognathus cornicola Knight; Phylinae
Plagiognathus cuneatus Knight; Phylinae
Plagiognathus delicatus (Uhler); Phylinae
Plagiognathus gleditsiae Knight; Phylinae
Plagiognathus guttulosus (Reuter); Phylinae
Plagiognathus laricicola Knight; Phylinae
Plagiognathus luteus Knight; Phylinae
Plagiognathus medicagus Arrand; Phylinae
Plagiognathus negundinis Knight; Phylinae
Plagiognathus nigronitens Knight; Phylinae
Plagiognathus obscurus Uhler; Phylinae
Plagiognathus politus Uhler; Phylinae
Plagiognathus punctatipes Knight; Phylinae
Plagiognathus repetitus Knight; Phylinae
Plagiognathus repletus Knight; Phylinae
Plagiognathus ribesi Kelton; Phylinae
Platycapsus acaciae Reuter; Deraeocorinae
Platylygus luridus (Reuter); Mirinae
Platyngomiriodes apiformis Ghauri; Bryocorinae
Platytylus bicinctus (Walker); Mirinae
Platytylus bicolor (Le Peletier & Serville); Mirinae
Plesiodema sericea (Heidemann); Phylinae
Poecilocapsus lineatus (Fabricius); Mirinae
Polymerus basalis (Reuter); Mirinae
Polymerus cognatus (Fieber); Mirinae
Polymerus cuneatus (Distant); Mirinae
Polymerus fulvipes Knight; Mirinae
Polymerus testaceipes (Stål); Mirinae
Polymerus tinctipes Knight; Mirinae
Polymerus venaticus (Uhler); Mirinae
Polymerus venustus Knight; Mirinae
Polymerus vulneratus (Panzer); Mirinae

Polymerus wheeleri Henry; Mirinae
Poppiusia leroyi (Schouteden); Bryocorinae
Prepops eremicola (Knight); Mirinae
Prepops insignis (Say); Mirinae
Prepops insitivus (Say); Mirinae
Prepops latipennis (Stål); Mirinae
Proba fraudulenta (Stål); Mirinae
Proba sallei (Stål); Mirinae
Proba vittiscutis (Stål); Mirinae
Proboscidotylus carvalhoi Henry; Orthotylinae
Prodromus thaliae China; Bryocorinae
Psallovius piceicola (Knight); Phylinae
Psallus aethiops (Zetterstedt); Phylinae
Psallus albicinctus (Kirschbaum); Phylinae
Psallus ambiguus (Fallén); Phylinae
Psallus betuleti (Fallén); Phylinae
Psallus flavellus Stichel; Phylinae
Psallus lepidus (Fieber); Phylinae
Psallus perrisi (Mulsant & Rey); Phylinae
Psallus physocarpi Henry; Phylinae
Psallus pictipes (Van Duzee); Phylinae
Psallus pseudoplatani Reichling; Phylinae
Psallus vaccinicola Knight; Phylinae
Psallus variabilis (Fallén); Phylinae
Pseudatomoscelis flora (Van Duzee); Phylinae
Pseudatomoscelis seriata (Reuter); Phylinae
Pseudoclerada morai Kirkaldy; Orthotylinae
Pseudodoniella chinensis Zheng; Bryocorinae
Pseudodoniella pacifica China & Carvalho; Bryocorinae
Pseudodoniella typica (China & Carvalho); Bryocorinae
Pseudoloxops nigribasicornis (Knight); Orthotylinae
Pseudoloxops takaii Yasunaga; Orthotylinae
Pseudophylus stundjuki Kulik; Phylinae
Pseudopsallus puberus (Uhler); Orthotylinae
Pseudopsallus viridicans (Knight); Orthotylinae
Pseudoxenetus regalis (Uhler); Orthotylinae
Punctifulvius kerzhneri Schmitz; Cylapinae
Pycnoderes dilatatus Reuter; Bryocorinae
Pycnoderes medius Knight; Bryocorinae
Pycnoderes monticulifer Reuter; Bryocorinae
Pycnoderes quadrimaculatus Guérin-Méneville; Bryocorinae

Ragwelellus festivus (Miller); Bryocorinae
Ragwelellus horvathi (Poppius); Bryocorinae
Ragwelellus suspectus (Distant); Bryocorinae
Ranzovius agelenopsis Henry; Phylinae
Ranzovius californicus (Van Duzee); Phylinae
Ranzovius clavicornis (Knight); Phylinae
Ranzovius fennahi Carvalho; Phylinae
Rayieria sp.; Bryocorinae
Reuteria bifurcata Knight; Orthotylinae
Reuteria irrorata (Say); Orthotylinae
Reuteroscopus ornatus (Reuter); Phylinae
Rhabdomiris striatellus (Fabricius); Mirinae
Rhinacloa basalis (Reuter); Phylinae

Rhinacloa callicrates Herring; Phylinae
Rhinacloa cardini (Barber & Bruner); Phylinae
Rhinacloa clavicornis (Reuter); Phylinae
Rhinacloa forticornis Reuter; Phylinae
Rhinacloa luridipennis (Reuter); Phylinae
Rhinocapsus vanduzeei Uhler; Phylinae
Rhinomiridius ogoouensis Odhiambo; Cylapinae
Romna capsoides (Buchanan-White); Deraeocorinae
Ruspoliella coffeae (China); Mirinae

Sahlbergella singularis Haglund; Bryocorinae
Saileria irrorata Henry; Orthotylinae
Salicarus roseri (Herrich-Schaeffer); Phylinae
Salignus duplicatus (Reuter); Mirinae
Salignus tahoensis (Knight); Mirinae
Sarona mokihana Asquith; Orthotylinae
Saundersiella moerens (Reuter); Mirinae
Scalponotatus albibasis (Knight); Orthotylinae
Schaffneria davisi (Knight); Orthotylinae
Schaffneria pilophoroides (Knight); Orthotylinae
Schizopteromiris carayoni Schuh; Cylapinae
Sejanus albisignatus (Knight); Phylinae
Semium hirtum Reuter; Phylinae
Sericophanes obscuricornis Poppius; Orthotylinae
Setocoris bybliphilus China & Carvalho;
  Bryocorinae
Setocoris droserae (China); Bryocorinae
Setocoris russellii (China); Bryocorinae
Sidnia kinbergi (Stål); Mirinae
Sixeonotus albicornis Blatchley; Bryocorinae
Sixeonotus areolatus Knight; Bryocorinae
Sixeonotus tenebrosus (Distant); Bryocorinae
Sixeonotus unicolor Knight; Bryocorinae
Slaterocoris atritibialis (Knight); Orthotylinae
Slaterocoris hirtus (Knight); Orthotylinae
Slaterocoris pallipes (Knight); Orthotylinae
Slaterocoris stygicus (Say); Orthotylinae
Spanagonicus albofasciatus (Reuter); Phylinae
Stenodema calcarata (Fallén); Mirinae
Stenodema holsata (Fabricius); Mirinae
Stenodema laevigata (Linnaeus); Mirinae
Stenodema trispinosa Reuter; Mirinae
Stenodema vicina (Provancher); Mirinae
Stenodema virens (Linnaeus); Mirinae
Stenotus binotatus (Fabricius); Mirinae
Stenotus gestroi Poppius; Mirinae
Stenotus rubrovittatus (Matsumura); Mirinae
Stenotus transvaalensis (Distant); Mirinae
Stethoconus cyrtopeltis (Flor); Deraeocorinae
Stethoconus frappai Carayon; Deraeocorinae
Stethoconus japonicus Schumacher; Deraeocorinae
Stethoconus praefectus (Distant); Deraeocorinae
Stethoconus pyri (Mella); Deraeocorinae
Stethoconus scutellaris (Schouteden); Deraeocorinae
Sthenaridea carmelitana (Carvalho); Phylinae
Sthenaridea suturalis (Reuter); Phylinae
Sthenarus mcateei Knight; Phylinae
Sthenarus rotermundi (Scholtz); Phylinae

Sthenarus viticola Johnston; Phylinae
Sysinas linearis Distant; Bryocorinae
Systellonotus triguttatus (Linnaeus); Phylinae

Taedia casta (McAtee); Mirinae
Taedia colon (Say); Mirinae
Taedia deletica (Reuter); Mirinae
Taedia evonymi (Knight); Mirinae
Taedia externa (Herrich-Schaeffer); Mirinae
Taedia floridana (Knight); Mirinae
Taedia gleditsiae (Knight); Mirinae
Taedia hawleyi (Knight); Mirinae
Taedia heidemanni (Reuter); Mirinae
Taedia johnstoni (Knight); Mirinae
Taedia multisignata (Reuter); Mirinae
Taedia pallidula (McAtee); Mirinae
Taedia scrupea (Say); Mirinae
Taedia virgulata (Knight); Mirinae
Taylorilygus apicalis (Fieber); Mirinae
Taylorilygus ghesquierei (Schouteden); Mirinae
Taylorilygus ricini (Taylor); Mirinae
Taylorilygus virens (Taylor); Mirinae
Taylorilygus vosseleri (Poppius); Mirinae
Teleorhinus tephrosicola Knight; Phylinae
Tenthecoris bicolor Scott; Bryocorinae
Tenthecoris colombiensis Hsiao & Sailer;
  Bryocorinae
Tenthecoris orchidearum (Reuter); Bryocorinae
Teratocoris sp.; Mirinae
Termatophylidea maculata Usinger; Deraeocorinae
Termatophylidea opaca Carvalho; Deraeocorinae
Termatophylidea pilosa Reuter & Poppius;
  Deraeocorinae
Termatophylum hikosanum Miyamoto; Deraeocorinae
Termatophylum insigne Reuter; Deraeocorinae
Tinginotum sp.; Mirinae
Totta zaherii Ghauri & Ghauri; Isometopinae
Trichophthalmocapsus jamesi China; Phylinae
Trigonotylus brevipes Jakovlev; Mirinae
Trigonotylus caelestialium (Kirkaldy); Mirinae
Trigonotylus ruficornis (Geoffroy); Mirinae
Trigonotylus tenuis Reuter; Mirinae
Tropidosteptes adustus (Knight); Mirinae
Tropidosteptes amoenus Reuter; Mirinae
Tropidosteptes cardinalis Uhler; Mirinae
Tropidosteptes chapingoensis Carvalho & Rosas;
  Mirinae
Tropidosteptes illitus (Van Duzee); Mirinae
Tropidosteptes pacificus (Van Duzee); Mirinae
Tropidosteptes plagifer Reuter; Mirinae
Tropidosteptes pubescens (Knight); Mirinae
Tropidosteptes vittifrons (Knight); Mirinae
Trynocoris lawrencei Herring; Cylapinae
Tupiocoris cucurbitaceus (Spinola); Bryocorinae
Tupiocoris notatus (Distant); Bryocorinae
Tupiocoris rhododendri (Dolling); Bryocorinae
Tytthus chinensis (Stål); Phylinae
Tytthus mundulus (Breddin); Phylinae

Tytthus parviceps (Reuter); Phylinae
Tytthus pygmaeus (Zetterstedt); Phylinae

Usingerella bakeri (Knight); Bryocorinae

Valdasus sp.; Cylapinae
Vitsikamiris madecassus Polhemus &
  Razafimahatratra; Phylinae
Volumnus obscurus Poppius; Mirinae

Xenetomorpha carpenteri Poppius; Mirinae
Xenocylapus nervosus Bergroth; Cylapinae

Zanchius alatanus Hoberlandt; Orthotylinae
Zanchius buddleiae Schuh; Orthotylinae
Zanchius mosaicus Zheng & Liang; Orthotylinae
Zanchius tarasovi Kerzhner; Orthotylinae

Sources: Mainly Schuh (1995) but also Carvalho (1957–1960),
Henry and Wheeler (1988), Cassis and Gross (1995), and I. M.
Kerzhner (pers. comm.).
Note: For synonyms of selected economically important
Miridae, see Table 2.1; a generic name alone indicates that no
member of the genus is specifically cited in the text.

Appendix 2/ Equivalent Common and Scientific (Latin) Names of Mirid Species Mentioned in the Text

alfalfa plant bug, *Adelphocoris lineolatus*
"apple brown bug," *Atractotomus mali*
apple capsid (mirid), *Lygocoris rugicollis*
apple dimpling bug, *Campylomma liebknechti*
apple red bug, *Lygidea mendax*
ash plant bug, *Tropidosteptes amoenus*
azalea plant bug, *Rhinocapsus vanduzeei*

"banded grape bug," *Taedia scrupea*
bean capsid (mirid), *Pycnoderes quadrimaculatus*
bee bug, *Platyngomiriodes apiformis*
black-kneed capsid (mirid), *Blepharidopterus angulatus*
boll shedder bug, *Creontiades pallidus*
bracken bug, *Monalocoris filicis*
brown smudge bug, *Deraeocoris signatus*
brown wattle mirid, *Lygidolon laevigatum*

caragana plant bug, *Lopidea dakota*
catkin bug, *Pantilius tunicatus*
clouded plant bug, *Neurocolpus nubilus*
"cocoa deraeocorine," *Deraeocoris crigi*
common green capsid (mirid), *Lygocoris pabulinus*
cotton fleahopper, *Pseudatomoscelis seriata*
"cotton leaf-bug"   see rapid plant bug
crop mirid, *Sidnia kinbergi*

dark-green apple capsid (mirid), *Orthotylus marginalis*
delicate apple capsid (mirid), *Malacocoris chlorizans*
dimpling bug, *Niastama punctaticollis*

European tarnished plant bug, *Lygus rugulipennis*

fern bug, *Bryocoris pteridis*
fourlined plant bug, *Poecilocapsus lineatus*

garden fleahopper, *Halticus bractatus*
grape-vine flower bug   see hop capsid (mirid)
grass fleahopper, *Halticus chrysolepis*
green mirid, *Creontiades dilutus*

hickory plant bug, *Lygocoris caryae*
hollyhock plant bug, *Brooksetta althaeae*
honeylocust plant bug, *Blepharidopterus chlorionis*
hop capsid (mirid), *Closterotomus fulvomaculatus*
hop plant bug, *Taedia hawleyi*
"humped-back melon bug"   see bean capsid

meadow plant bug, *Leptopterna dolabrata*
mosquito bugs, *Helopeltis* spp., *Monalonion* spp.
"mullein bug," *Campylomma verbasci*

onion plant bug, *Labopidea allii*
orange blossom bug, *Dionconotus neglectus*

pale legume bug, *Lygus elisus*
pear plant bug, *Lygocoris communis*
phlox plant bug, *Lopidea davisi*
pine conelet bug, *Platylygus luridus*
potato mirid, *Closterotomus norwegicus*

ragweed plant bug, *Chlamydatus associatus*
rapid plant bug, *Adelphocoris rapidus*
rice leaf bug (Japan), *Trigonotylus caelestialium*

"sewing-ant capsid (mirid)," *Xenetomorpha carpenteri*
sorghum earhead bug, *Calocoris angustatus*
suckfly, *Tupiocoris notatus*
sunnhemp mirid, *Moissonia importunitas*
superb plant bug, *Adelphocoris superbus*
"sycamore plant bug," *Plagiognathus albatus*

tarnished plant bug, *Lygus lineolaris*
tomato bug, *Engytatus modestus*
"tomato girdler"   see tomato bug
transparentwinged plant bug, *Hyalopeplus pellucidus*
trefoil plant bug, *Plagiognathus chrysanthemi*
twospotted grass bug, *Stenotus binotatus*

western plant bug, *Rhinacloa forticornis*
western tarnished plant bug, *Lygus hesperus*
whitemarked fleahopper, *Spanagonicus albofasciatus*

yucca plant bug, *Halticotoma valida*

Sources: Mainly Southwood and Leston (1959), Naumann (1993), and Bosik (1997).
Note: Quotation marks are used for unofficial names of species occurring in the United States and for informal names of species occurring elsewhere. See Appendix 1 for authors' names and subfamilies and Table 2.1 for synonymy and misspellings involving some plant bugs of economic importance.

Appendix 3/ Common Names of Plants Mentioned in the Text with Their Latin Name, Author or Authority, and Family

acacia, *Acacia* spp.; Fabaceae

agave, *Agave* spp.; Agavaceae

albizia, *Albizia* spp.; Fabaceae

alder, *Alnus* spp., especially *A. glutinosa* (L.) Gaertn.; Betulaceae

alfalfa, *Medicago sativa* L.; Fabaceae

alkali weed, *Bassia* sp.; Chenopodiaceae

allspice, *Pimenta dioica* (L.) Merr.; Myrtaceae

almond, *Prunus dulcis* (Mill.) D. A. Webb; Rosaceae

aloe, *Aloe* spp.; Aloeaceae

amaranth, *Amaranthus cruentus* L.; Amaranthaceae

American basswood, *Tilia americana* L.; Tiliaceae

American beech, *Fagus grandifolia* Ehrh.; Fagaceae

American elm, *Ulmus americana* L.; Ulmaceae

American fewerfew, *Parthenium integrifolium* L.; Asteraceae

American holly, *Ilex opaca* Aiton; Aquifoliaceae

American hydrangea, *Hydrangea arborescens* L.; Hydrangeaceae

American lotus, *Nelumbo lutea* (Willd.) Pers. = *N. pentapetala* (Walt.) Fern.; Nelumbonaceae

angel's trumpet   see downy thorn apple

annatto, *Bixa orellana* L.; Bixaceae

annual bluegrass, *Poa annua* L.; Poaceae

annual fleabane, *Erigeron annuus* (L.) Pers.; Asteraceae

antelope bitterbrush, *Purshia tridentata* (Pursh) DC; Rosaceae

Appalachian groundsel, *Senecio anonymus* Wood = *S. smallii* Britt.; Asteraceae

apple, *Malus domestica* Borkh., *M. sylvestris* Mill.; Rosaceae

araliaceous tree, *Reynoldsia sandwicensis* Gray; Araliaceae

Areca palm, *Areca catechu* L.; Arecaceae

Arizona sycamore, *Platanus wrightii* S. Watson; Platanaceae

aromatic aster, *Aster oblongifolius* Nutt.; Asteraceae

arrow-wood viburnum, *Viburnum dentatum* L.; Caprifoliaceae

artichoke, *Cynara scolymus* L.; Asteraceae

arum lily, *Arum* sp.; Araceae

ash, *Fraxinus* spp.; Oleaceae

asparagus, *Asparagus officinalis* L.; Liliaceae

aspen, *Populus* spp.; Salicaceae

asphodel, *Asphodelus aestivus* Brot. = *A. microcarpa* Salzm.; Liliaceae

aster, *Aster* spp.; Asteraceae

astragalus, *Astragalus* spp.; Fabaceae

astrotricha, *Astrotricha floccosa* DC; Araliaceae

atraphaxis, *Atraphaxis badghysi* Kult.; Polygonaceae

avocado, *Persea americana* Mill.; Lauraceae

azalea, *Rhododendron* spp.; Ericaceae

azuki bean, *Vigna angularis* (Willd.) Ohwi & H. Ohashi; Fabaceae

bald cypress, *Taxodium distichum* (L.) Rich.; Taxodiaceae

balsam fir, *Abies balsamea* (L.) Mill.; Pinaceae

balsam poplar, *Populus balsamifera* L.; Salicaceae

bamboo, mainly *Bambusa* spp., *Dendrocalamus* spp.; Poaceae

banana, *Musa* spp.; Musaceae

baneberry, *Actaea spicata* L.; Ranunculaceae

barley, *Hordeum vulgare* L.; Poaceae

basswood, *Tilia* spp.; Tiliaceae

bayberry, *Myrica pensylvanica* Loisel.; Myricaceae

beach ragweed, *Ambrosia chamissonis* (Less.) Greene; Asteraceae

bean, *Phaseolus* spp.; Fabaceae

bear oak   see scrub oak

bedstraw, *Galium* spp.; Rubiaceae

beet, *Beta vulgaris* L.; Chenopodiaceae

bell pepper, *Capsicum annuum* L.; Solanaceae

bentgrass, *Agrostis* sp.; Poaceae

bermudagrass, *Cynodon dactylon* (L.) Pers.; Poaceae

betel nut   see Areca palm

betel pepper, *Piper betle* L.; Piperaceae

biennial gaura, *Gaura biennis* L.; Onagraceae

bigflower cinquefoil, *Potentilla fissa* Nutt. = *Drymocallis fissa* (Nutt.) Rydb.; Rosaceae

bigtooth aspen, *Populus grandidentata* Michx.; Salicaceae

bindweed, *Calystegia sepium* (L.) R. Br., *Convolvulus arvensis* L.; Convolvulaceae

birch, *Betula* spp.; Betulaceae

birdsfoot trefoil, *Lotus corniculatus* L.; Fabaceae

bittersweet nightshade, *Solanum dulcamara* L.; Solanaceae

blackberry, *Rubus* spp.; Rosaceae

black currant, *Ribes nigrum* L.; Grossulariaceae

black-eyed pea   see cowpea

black gram, *Vigna mungo* (L.) Hepper; Fabaceae

black gum, *Nyssa sylvatica* Marsh.; Cornaceae

black knapweed, *Centaurea nigra* L.; Asteraceae

black locust, *Robinia pseudoacacia* L.; Fabaceae

black pepper, *Piper nigrum* L.; Piperaceae

black sanicle, *Sanicula marilandica* L.; Apiaceae

black walnut, *Juglans nigra* L.; Juglandaceae

black wattle, *Acacia mearnsii* DeWild.; Fabaceae

blackwood, *Acacia melanoxylon* R. Br.; Fabaceae

bladdernut, *Staphylea trifolia* L.; Staphyleaceae

bladder sage, *Salazaria mexicana* Torr.; Lamiaceae

bluebeard, *Caryopteris* x *clandonensis* Rehder;
Verbenaceae

bluebunch wheatgrass, *Pseudoroegneria spicata*
(Pursh) A. Löve; Poaceae

blue buttons   see field scabious

bluegrass, *Poa pratensis* L.; Poaceae

bluejoint reedgrass, *Calamagrostis canadensis*
(Michx.) P. Beauv.; Poaceae

boneset, *Eupatorium perfoliatum* L.; Asteraceae

bottle gourd, *Lagenaria siceraria* (Molina) Standl.;
Cucurbitaceae

boxelder, *Acer negundo* L.; Aceraceae

boxwood, *Buxus* sp.; Buxaceae

bramble, *Rubus* spp.; Rosaceae

branching draba, *Draba ramosissima* Desv.;
Brassicaceae

Brazil nut, *Bertholletia excelsa* Bonpl.; Lecythidaceae

broadleaf meadowsweet, *Spiraea alba* Du Roi var.
latifolia (Aiton) Dippel; Rosaceae

broom, *Cytisus scoparius* (L.) Link; Fabaceae

buddleja, *Buddleja* sp.; Loganiaceae

Bumald spirea, *Spiraea* x *bumalda* Burv.; Rosaceae

bushy seaoxeye, *Borrichia frutescens* (L.) DC.;
Asteraceae

buttercup, *Ranunculus lyallii* Hook.f.; Ranunculaceae

buttonbush, *Cephalanthus occidentalis* L.; Rubiaceae

cabbage, *Brassica oleracea* L.; Brassicaceae

cactus, *Opuntia* spp.; Cactaceae

Canada goldenrod, *Solidago canadensis* L.; Asteraceae

Canada thistle, *Cirsium arvense* (L.) Scop.; Asteraceae

Canadian anemone, *Anemone canadensis* L.;
Ranunculaceae

Canadian wild lettuce, *Lactuca canadensis* L.;
Asteraceae

canola, *Brassica campestris* L., *B. napus* L.;
Brassicaceae

cantaloupe, *Cucumis melo* var. *cantalupensis* Naudin;
Cucurbitaceae

caragana, *Caragana arborescens* Lam.; Fabaceae

cardamom, *Elettaria cardamomum* (L.) Maton;
Zingiberaceae

Carolina vetch, *Vicia caroliniana* Walter; Fabaceae

carrot, *Daucus carota* L. subsp. *sativus* (Hoffm.)
Arcang.; Apiaceae

Cascades azalea, *Rhododendron albiflorum* Hook.;
Ericaceae

cashew, *Anacardium occidentale* L.; Anacardiaceae

cassava, *Manihot esculenta* Crantz; Euphorbiaceae

castanopsis, *Castanopsis* spp.; Fagaceae

castor, *Ricinus communis* L.; Euphorbiaceae

cattleya, *Cattleya* spp.; Orchidaceae

cauliflower, *Brassica oleracea* L. var. *botrytis* L.;
Brassicaceae

ceanothus   see hoaryleaf ceanothus

celery, *Apium graveolens* L.; Apiaceae

chard, *Beta vulgaris* L. subsp. *cicla* (L.) W. Koch;
Chenopodiaceae

chayote   see christophine

cherry, *Prunus* spp.; Rosaceae

chicory, *Cichorium intybus* L.; Asteraceae

Chinese cabbage, *Brassica pekinensis* (Lour.) Rupr.;
Brassicaceae

Chinese cinnamon, *Cinnamomum cassia* Blume;
Lauraceae

christophine, *Sechium edule* (Jacq.) Sw.; Cucurbitaceae

chrysanthemum, *Chrysanthemum* spp., especially *C.*
x *morifolium* Ramat.; Asteraceae

cinquefoil, *Potentilla* sp.; Rosaceae

citrus, *Citrus* spp.; Rutaceae

clasping heart-leaf aster, *Aster undulatus* L.;
Asteraceae

clematis, *Clematis* spp.; Ranunculaceae

cliff-bush, *Jamesia americana* Torr. & A. Gray;
Hydrangeaceae

clover, *Trifolium* spp.; Fabaceae

cockspur hawthorn, *Crataegus crus-galli* L.; Rosaceae

cocoa, *Theobroma cacao* L.; Sterculiaceae

coconut, *Cocos nucifera* L.; Arecaceae

coffee, *Coffea* spp., mainly *C. canephora* Pierre ex A.
Froehner; Rubiaceae

cole, *Brassica oleracea* L. var. *acephala* DC.;
Brassicaceae

combretum, *Combretum racemosum* P. Beauv.;
Combretaceae

common hoptree, *Ptelea trifoliata* L.; Rutaceae

common knotweed, *Polygonum aviculare* L.;
Polygonaceae

common privet, *Ligustrum vulgare* L.; Oleaceae

common ragweed, *Ambrosia artemisiifolia* L.;
Asteraceae

common snowberry, *Symphoricarpos albus* (L.) S. F.
Blake; Caprifoliaceae

continental lady's-tresses, *Spiranthes romanzoffiana*
Cham. = *Ibidium strictum* (Rydb.) House;
Orchidaceae

corn, *Zea mays* L.; Poaceae

corn chrysanthemum, *Chrysanthemum segetum* L.;
Asteraceae

cotton, *Gossypium hirsutum* L.; Malvaceae

cottonwood, *Populus deltoides* Marsh.; Salicaceae

cowparsnip, *Heracleum spondylium* L.; Apiaceae

cowpea, *Vigna unguiculata* (L.) Walp.; Fabaceae

crabapple, *Malus* spp.; Rosaceae

crape myrtle, *Lagerstroemia indica* L.; Lythraceae
creeping softgrass   see German velvetgrass
creosotebush, *Larrea tridentata* (Sessé & Moç ex DC.);
   Zygophyllaceae
crested wheatgrass, *Agropyron cristatum* (L.) Gaertn.;
   Poaceae
crotalaria, *Crotalaria* spp.; Fabaceae
croton, *Croton* spp.; Euphorbiaceae
crownvetch, *Coronilla varia* L.; Fabaceae
cucumber, *Cucumis sativus* L.; Cucurbitaceae
cucumber (wild), *Ecballium elaterium* (L.) A. Rich.;
   Cucurbitaceae
cuphea (ornamental), *Cuphea jorullensis* HBK.;
   Lythraceae
cup rosinweed, *Silphium perfoliatum* L.; Asteraceae
currant, *Ribes* spp.; Grossulariaceae
cutleaf geranium, *Geranium dissectum* L.; Geraniaceae
cutleaf primrose, *Oenothera laciniata* Hill;
   Onagraceae
cyperus, *Cyperus difformis* L.; Cyperaceae

daffodil, *Narcissus pseudonarcissus* L.; Liliaceae
dahlia, *Dahlia* spp.; Asteraceae
daimyo oak, *Quercus dentata* Thunb.; Fagaceae
daisy fleabane, *Erigeron strigosus* Muhl. ex Willd.;
   Asteraceae
dandelion, *Taraxacum officinale* F. H. Wigg; Asteraceae
datil yucca, *Yucca baccata* Torr.; Agavaceae
dayflower, *Commelina* spp.; Commelinaceae
daylily, *Hemerocallis* sp.; Liliaceae
delphinium   see larkspur
desert evening-primrose, *Oenothera californica*
   (S. Wats.) S. Wats.; Onagraceae
desert hackberry, *Celtis pallida* Torr.; Ulmaceae
desert tobacco, *Nicotiana trigonophylla* Dunal;
   Solanaceae
devilsbit, *Succisa pratensis* Moench; Dipsacaceae
digitalis, *Digitalis* spp.; Scrophulariaceae
dill, *Anethum graveolens* L.; Apiaceae
dillwynia, *Dillwynia retorta* Druce; Fabaceae
dipterocarps, *Shorea* spp.; Dipterocarpaceae
discoic beggarticks, *Bidens discoidea* (Torr. & A.
   Gray) Britt.; Asteraceae
dogwood, *Cornus* spp.; Cornaceae
Douglas-fir, *Pseudotsuga menziesii* (Mirb.) Franco;
   Pinaceae
downy serviceberry, *Amelanchier arborea* (F. Michx.)
   Fernald; Rosaceae
downy thorn apple, *Datura innoxia* Mill. = *D. wrightii*
   Regel; Solanaceae
downy viburnum, *Viburnum rafinesquianum*
   Schultes; Caprifoliaceae
downy wild rye, *Elymus villosus* Muhl. ex Willd.;
   Poaceae
dwarf mistletoe, *Arceuthobium* spp.; Visaceae
dyer's greenwood, *Genista tinctoria* L.; Fabaceae

earpod   see elephant's ear
eggplant, *Solanum melongena* L.; Solanaceae

elderberry, *Sambucus canadensis* L.; Caprifoliaceae
elephant's ear, *Enterolobium cyclocarpum* Griesb.;
   Fabaceae
elm, *Ulmus* sp.; Ulmaceae
English ivy, *Hedera helix* L.; Araliaceae
eucalypts, *Eucalyptus* spp.; Myrtaceae
euphorbiaceous tree, *Ricinodendron heudelotii*
   (Baillon) Pax; Euphorbiaceae
European ash, *Fraxinus excelsior* L.; Oleaceae
European cranberrybush, *Viburnum opulus* L.;
   Caprifoliaceae
European filbert, *Corylus avellana* L.; Betulaceae
European larch, *Larix decidua* Mill.; Pinaceae
European linden, *Tilia cordata* Mill.; Tiliaceae
European meadowsweet, *Filipendula ulmaria* (L.)
   Maxim.; Rosaceae
European white birch, *Betula pendula* Roth; Betulaceae
evening primrose, *Oenothera biennis* L.; Onagraceae

fava bean, *Vicia faba* L.; Fabaceae
false foxglove, *Aureolaria* spp.; Scrophulariaceae
false indigo, *Amorpha fruticosa* L.; Fabaceae
false tamarisk, *Myricaria* spp.; Tamaricaceae
fan palm, *Pritchardia beccariana* Rock; Arecaceae
fehi (fe'i) banana, *Musa fehi* Bertero ex Vieill.;
   Musaceae
fennel, *Foeniculum vulgare* Mill.; Apiaceae
ferns, Pteridophyta (various genera)
fewbracted beggarticks   see discoic beggarticks
field bindweed, *Convolvulus arvensis* L.;
   Convolvulaceae
field scabious, *Knautia arvensis* (L.) Coult.;
   Dipsacaceae
fig, *Ficus* spp.; Moraceae
fine fescue, *Festuca rubra* L.; Poaceae
fir, *Abies* spp.; Pinaceae
flax, *Linum usitatissimum* L.; Linaceae
fleabane   see daisy fleabane
Formosan sweetgum, *Liquidambar formosana* Hance;
   Hamamelidaceae
forsythia, *Forsythia* spp.; Oleaceae
fox grape, *Vitis labrusca* L.; Vitaceae
Fremont's grape holly, *Berberis fremontii* Torr.;
   Berberidaceae
fuchsia, *Fuchsia* spp.; Onagraceae

gaillardia, *Gaillardia* sp.; Asteraceae
garden beet   see beet
garden coreopsis, *Coreopsis lanceolata* L.;
   Asteraceae
geranium, *Geranium sanguineum* L.; Geraniaceae
gerbera, *Gerbera jamesonii* Bolus ex Hook. f.;
   Asteraceae
German velvetgrass, *Holcus mollis* L.; Poaceae
giant bamboo, *Gigantochloa scortechinii* Gamble;
   Poaceae
giant hogweed, *Heracleum mantegazzianum* Somm.
   & Lev.; Apiaceae
giant ragweed, *Ambrosia trifida* L.; Asteraceae

giant wildrye see Great Basin wildrye
gladiolus, *Gladiolus* x *hortulanus* L. H. Bailey;
   Iridaceae
goldenrod, *Solidago nemoralis* Aiton; Asteraceae
goodeniaceous strand plant, *Scaevola sericea* Vahl.;
   Goodeniaceae
gooseberry, *Ribes* spp.; Grossulariaceae
gourds, *Cucumis* sp.; Cucurbitaceae
grain amaranth see amaranth
grape, *Vitis* spp.; Vitaceae
grape (cultivated), *Vitis vinifera* L.; Vitaceae
grapefruit, *Citrus* x *paradisi* Macfad.; Rutaceae
grassleaf goldenrod, *Euthamia graminifolia* (L.) Nutt.
   (often included in *Solidago*); Asteraceae
Great Basin wildrye, *Leymus cinereus* (Scribn. &
   Merr.) A. Löve; Poaceae
green ash, *Fraxinus pennsylvanica* Marsh.; Oleaceae
green bean, *Phaseolus vulgaris* L.; Fabaceae
green-gentian, *Frasera speciosa* Douglas ex Griseb.;
   Gentianaceae
green pepper, *Capsicum annuum* L.; Solanaceae
groundnut see peanut
groundsel, *Senecio vulgaris* L.; Asteraceae
groundsel baccaris, *Baccharis halimifolia* L.;
   Asteraceae
guava, *Psidium guajava* L.; Myrtaceae
guayule, *Parthenium argentatum* A. Gray; Asteraceae

hairy beggarticks, *Bidens pilosa* L.; Asteraceae
hairy golden aster, *Heterotheca viscida* (Gray) Harms
   = *Chrysopsis viscida*; Asteraceae
hairy vetch, *Vicia villosa* Roth; Fabaceae
hawkweed, *Hieracium speluncarum* Arv.-Touv.;
   Asteraceae
hawthorn, *Crataegus* spp.; Rosaceae
hayscented fern, *Dennstaedtia punctilobula* (Michx.)
   T. Moore; Pterodophyta (Dennstaedtiaceae)
hazel see European filbert
head lettuce, *Lactuca sativa* L. var. *capitata* L.;
   Asteraceae
heath, *Erica* spp.; Ericaceae
heather, *Calluna vulgaris* (L.) Hull; Ericaceae
hebe, *Hebe anomala* Cockayne, *H. hulkeana* (F. J.
   Muell.) Cockayne & Allan; Scrophulariaceae
helleborine, *Epipactis palustris* (L.) Crantz;
   Orchidaceae
hemlock, *Tsuga canadensis* (L.) Carrière; Pinaceae
hempnettle, *Galeopsis tetrahit* L.; Lamiaceae
herculesclub pricklyash, *Zanthoxylem clava-herculis*
   L.; Rutaceae
hickory, *Carya* spp.; Juglandaceae
hoaryleaf ceanothus, *Ceanothus crassifolius* Torr.;
   Rhamnaceae
hogweed see cowparsnip
holly, *Ilex* spp.; Aquifoliaceae
hollyhock, *Alcea rosea* L.; Malvaceae
honeylocust, *Gleditsia triacanthos* L.; Fabaceae
honey mesquite, *Prosopis glandulosa* Torr.; Fabaceae
hop, *Humulus lupulus* L.; Cannabaceae

horsemint, *Monarda* spp.; Lamiaceae
horseradish tree, *Moringa pterygosperma* C. F. Gaertn.
   = *M. oleifera* Lam.; Moringaceae
horseweed, *Conyza canadensis* (L.) Cronquist;
   Asteraceae
hosta, *Hosta* spp.; Liliaceae
hyacinth-bean, *Lablab purpureus* (L.) Sweet; Fabaceae
hybrid poplar see poplar

Indian balsam, *Impatiens glandulifera* Royle;
   Balsaminaceae
Indian long pepper, *Piper longum* L.; Piperaceae
inkberry, *Ilex glabra* (L.) A. Gray; Aquifoliaceae
intermediate wheatgrass, *Thinopyrum intermedium*
   (Host) Barkw. & D. R. Dewey; Poaceae

jack pine, *Pinus banksiana* Lamb.; Pinaceae
jacquinias, *Jacquinia* spp.; Theophrastaceae
Japanese clethra, *Clethra barbinervis* Siebold & Zucc.;
   Clethraceae
Japanese pagoda-tree, *Sophora japonica* L.; Fabaceae
juniper, *Juniperus* spp.; Cupressaceae
Jupiter's distaff, *Salvia glutinosa* L.; Lamiaceae

kenaf, *Hibiscus cannabinus* L.; Malvaceae
Kentucky bluegrass, *Poa pratensis* L.; Poaceae
knapweed, *Centaurea scabiosa* L.; Asteraceae
kola, *Cola* spp.; Sterculiaceae
kudzu, *Pueraria lobata* (Willd.) Ohwi; Fabaceae

ladino clover see white clover
lambsquarters, *Chenopodium album* L.;
   Chenopodiaceae
lantana, *Lantana* spp.; Verbenaceae
laportea, *Laportea ovalifolia* (Schumach.) Chew =
   Fleurya ovalifolia (Schumach.) Dandy; Urticaceae
larch, *Larix decidua* Mill.; Pinaceae
largeleaved goldenrod, *Solidago macrophylla* Pursh;
   Asteraceae
larkspur, *Delphinium* spp.; Ranunculaceae
lemon, *Citrus limon* (L.) Burm. f.; Rutaceae
lentil, *Lens culinaris* Medik.; Fabaceae
lettuce, *Lactuca sativa* L.; Asteraceae
leucaena, *Leucaena leucocephala* (Lam.) de Wit;
   Fabaceae
lima bean, *Phaseolus lunatus* L.; Fabaceae
lime see basswood
lime pricklyash, *Zanthoxylem fagara* (L.) Sarg.;
   Rutaceae
linden, *Tilia* spp.; Tiliaceae
linseed see flax
littleleaf linden, *Tilia cordata* Mill.; Tiliaceae
live oak, *Quercus virginiana* Mill.; Fagaceae
lizard's-tail, *Saururus cernuus* L.; Saururaceae
loblolly pine, *Pinus taeda* L.; Pinaceae
loganberry, *Rubus loganobaccus* L. H. Bailey;
   Rosaceae
London plane, *Platanus* x *acerifolia* (Aiton) Willd.;
   Platanaceae

Londonrocket hedgemustard, *Sisymbrium irio* L.;
   Brassicaceae
lotus, *Lotus pedunculatus* Cav.; Fabaceae
lucerne   see alfalfa
lupine, *Lupinus* spp.; Fabaceae

magnolia, *Magnolia grandiflora* L.; Magnoliaceae
maize   see corn
malacothrix, *Malacothrix incana* (Nutt.) Torr. & A.
   Gray; Asteraceae
mango, *Mangifera indica* L.; Anacardiaceae
marsh-marigold, *Caltha leptosepala* DC.;
   Ranunculaceae
maypop passion-flower, *Passiflora incarnata* L.;
   Passifloraceae
meadowsweet, *Spiraea alba* DuRoi; Rosaceae
Mediterranean sumac   see sumac (Mediterranean)
melon, *Cucumis melo* L.; Cucurbitaceae
Merion bluegrass   see Kentucky bluegrass
mesquite, *Prosopis* spp.; Fabaceae
metrosideros, *Metrosideros polymorpha* Gaudich.;
   Myrtaceae
Mexican palo-verde, *Parkinsonia aculeata* L.;
   Fabaceae
mild water-pepper, *Polygonum hydropiperoides*
   Michx.; Polygonaceae
millet, *Panicum miliaceum* L.; Poaceae
mistletoe, *Loranthus europaeus* Jacq.; Loranthaceae,
   and *Viscum album* L.; Viscaceae
Monterey cypress, *Cupressus macrocarpa* Hartw.;
   Cupressaceae
monument plant   see green-gentian
moor matgrass, *Nardus stricta* L.; Poaceae
morning-glory, *Ipomoea* sp.; Convolvulaceae
Morrow's honeysuckle, *Lonicera morrowii* A. Gray;
   Caprifoliaceae
moss phlox, *Phlox subulata* L.; Polemoniaceae
moth orchid, *Phalaenopsis amabilis* Blume;
   Orchidaceae
mountain laurel, *Kalmia latifolia* L.; Ericaceae
mountain maple, *Acer spicatum* Lam.; Aceraceae
mugwort, *Artemisia vulgaris* L.; Asteraceae
mulberry, *Morus* sp.; Moraceae
mullein, *Verbascum thapsus* L.; Scrophulariaceae
mullein (cultivated), *Verbascum* sp. (Scrophulariaceae)
muscadine grape, *Vitis rotundifolia* Michx.; Vitaceae
musk mallow, *Malva moschata* L.; Malvaceae
muskmelon, *Cucumis melo* L. var. *reticulatus*
   Naudin; Cucurbitaceae
mustang grape, *Vitis mustangensis* Buckley; Vitaceae
mustard, *Brassica juncea* L.; Brassicaceae

nannyberry, *Viburnum lentago* L.; Caprifoliaceae
nectarine, *Prunus persica* (L.) Batsch var. *nucipersica*
   (Suckow) C. K. Schneid.; Rosaceae
neem, *Azadirachta indica* A. Juss.; Meliaceae
nettle, *Urtica* spp., mainly *U. dioica* L.; Urticaceae
nettleleaf goosefoot, *Chenopodium murale* L.;
   Chenopodiaceae

New Jersey tea, *Ceanothus americanus* L.;
   Rhamnaceae
ninebark, *Physocarpus opulifolius* (L.) Maxim.;
   Rosaceae
noble fir, *Abies procera* Rehder; Pinaceae
nodding silene, *Silene nutans* L.; Caryophyllaceae
Nordmann fir, *Abies nordmanniana* (Steven) Spach.;
   Pinaceae
northern catalpa, *Catalpa speciosa* (Warder ex Barney)
   Warder ex Engelm.; Bignoniaceae
Norway spruce, *Picea abies* (L.) Karst.; Pinaceae

oak, *Quercus* spp.; Fagaceae
oakleaf hydrangea, *Hydrangea quercifolia* Bartram;
   Hydrangeaceae
oats, *Avena sativa* L.; Poaceae
Ohio buckeye, *Aesculus glabra* Willd.;
   Hippocastanaceae
oil palm, *Elaeis guineensis* Jacq.; Arecaceae
oilseed rape   see canola
olive, *Olea europaea* L.; Oleaceae
onion, *Allium* spp., including *A. cepa* L.; Liliaceae
opuntia, *Opuntia* spp.; Cactaceae
orange, *Citrus sinensis* (L.) Osbeck; Rutaceae
orchardgrass, *Dactylis glomerata* L.; Poaceae
orchids, Orchidaceae (various genera)
oriental spruce, *Picea orientalis* (L.) Link; Pinaceae
osage-orange, *Maclura pomifera* (Raf.) C. K. Schneid.;
   Moraceae
oxeye daisy, *Leucanthemum vulgare* Lam. =
   Chrysanthemum leucanthemum L.; Asteraceae

pale-flowered leafcup, *Polymnia canadensis* L.;
   Asteraceae
panic grasses, *Panicum* spp.; Poaceae
panicled hydrangea, *Hydrangea paniculata* Siebold;
   Hydrangeaceae
paper-bark tree, *Melaleuca quinquenervia* (Cav.) S. T.
   Blake; Myrtaceae
paper mulberry, *Broussonetia papyrifera* (L.) Vent.;
   Moraceae
parsley, *Petroselinum crispum* (Mill.) Nyman ex A. W.
   Hill; Apiaceae
parsnip, *Pastinaca sativa* L.; Apiaceae
passionfruit, *Passiflora edulis* Sims; Passifloraceae
paulownia, *Paulownia* spp.; Scrophulariaceae
pea, *Pisum sativum* L.; Fabaceae
peach, *Prunus persica* (L.) Batsch; Rosaceae
peanut, *Arachis hypogaea* L.; Fabaceae
pear, *Pyrus communis* L.; Rosaceae
pecan, *Carya illinoensis* (Wangenh.) K. Koch;
   Juglandaceae
pepper, *Piper* spp.; Piperaceae
philodendron, *Philodendron* sp.; Araceae
phlox, *Phlox* spp.; Polemoniaceae
photinia, *Photinia glabra* (Thunb.) Maxim.; Rosaceae
pigeon pea, *Cajanus cajan* (L.) Millsp.; Fabaceae
pigweed, *Amaranthus retroflexus* L.; Amaranthaceae
pine, *Pinus* spp.; Pinaceae

pineapple, *Ananas comosus* (L.) Merr.; Bromeliaceae

pineywoods geranium, *Geranium caespitosum* James; Geraniaceae

pin oak, *Quercus palustris* Münchh.; Fagaceae

pistachio, *Pistacia vera* L.; Anacardiaceae

pitch pine, *Pinus rigida* Mill.; Pinaceae

plum, *Prunus domestica* L.; Rosaceae

poison hemlock, *Conium maculatum* L.; Apiaceae

poison ivy, *Toxicodendron radicans* (L.) Kuntze; Anacardiaceae

pole bean   see green bean

ponderosa pine, *Pinus ponderosa* Douglas ex Lawson & C. Lawson; Pinaceae

poplar, *Populus* spp.; Salicaceae

possumhaw, *Ilex decidua* Walter; Aquifoliaceae

post oak, *Quercus stellata* Wangenh.; Fagaceae

potato, *Solanum tuberosum* L.; Solanaceae

pot-marigold, *Calendula officinalis* L.; Asteraceae

prostrate kochia, *Kochia prostrata* (L.) Schrad.; Chenopodiaceae

prunus (ornamental), *Prunus* sp.; Rosaceae

pseudarthria, *Pseudarthria hookeri* Wight & Arn.; Fabaceae

pumpkin, *Cucurbita* sp.; Cucurbitaceae

purpleleaved plum, *Prunus cerasifera* J. F. Ehrh. "Atropurpurea"; Rosaceae

pussy-toes, *Antennaria plantaginifolia* (L.) Richardson; Asteraceae

quackgrass, *Elytrigia repens* (L.) Nevski = *Agropyron repens* (L.) Beauv.; Poaceae

quaking aspen, *Populus tremuloides* Michx.; Salicaceae

quinine, *Cinchona* spp.; Rubiaceae

quinoa, *Chenopodium quinoa* Willd.; Chenopodiaceae

radish, *Raphanus sativus* L.; Brassicaceae

ragweed, *Ambrosia artemisiifolia* L.; Asteraceae

ragweed parthenium, *Parthenium hysterophorus* L.; Asteraceae

rape, *Brassica napus* L.; Brassicaceae

raspberry, *Rubus* spp.; Rosaceae

rattlesnake master, *Eryngium yuccifolium* Michx.; Apiaceae

redbay, *Persea borbonia* (L.) Spreng.; Lauraceae

red chokeberry, *Aronia arbutifolia* (L.) Elliott; Rosaceae

red clover, *Trifolium pratense* L.; Fabaceae

red currant, *Ribes rubrum* L.; Grossulariaceae

red-ink sundew, *Drosera erythrorhiza* Lindl.; Droseraceae

red maple, *Acer rubrum* L.; Aceraceae

red pine, *Pinus resinosa* Aiton; Pinaceae

redroot pigweed   see pigweed

red sorrel, *Rumex acetosella* L.; Polygonaceae

reed canarygrass, *Phalaris arundinacea* L.; Poaceae

restharrow, *Ononis repens* L.; Fabaceae

revolute meadowrue, *Thalictrum revolutum* DC.; Ranunculaceae

rhododendron, *Rhododendron* spp.; Ericaceae

rhubarb, *Rheum rhabarbarum* L.; Polygonaceae

rice, *Oryza sativa* L.; Poaceae

Rocky Mountain juniper, *Juniperus scopulorum* Sarg.; Cupressaceae

roridula, *Roridula dentata* Planchon, *R. gorgonias* L.; Byblidaceae

rose, *Rosa* spp.; Rosaceae

rough fleabane, *Erigeron strigosus* Muhl. ex Willd.; Asteraceae

rough pigweed   see pigweed

rubber-tree, *Hevea brasiliensis* (Willd. ex A. Juss.) Müll. Arg.; Euphorbiaceae

rush, *Juncus* spp.; Juncaceae

rutabaga, *Brassica napus* L. var. *napobrassica* (L.) Rchb.; Brassicaceae

rye, *Secale cereale* L.; Poaceae

safflower, *Carthamus tinctorius* L.; Asteraceae

sagebrush, *Artemisia tridentata* Nutt.; Asteraceae

sainfoin, *Onobrychis viciifolia* Scop.; Fabaceae

salad burnet, *Sanguisorba minor* Scop. = *Poterium sanguisorba* sensu auctt. non L.; Rosaceae

saltwort, *Batis maritima* L.; Bataceae

sandbar willow, *Salix exigua* Nutt. = *S. interior* Rowlee = *S. longifolia* Muhl.; Salicaceae

sand pine, *Pinus clausa* (Chapm. ex Engelm.) Vasey ex Sarg.; Pinaceae

sand sagebrush, *Artemisia filifolia* Torr.; Asteraceae

sapodilla, *Manilkara zapota* (L.) van Royen; Sapotaceae

scarlet globemallow, *Sphaeralcea coccinea* (Nutt.) Rydb. subsp. *coccinea* = *Malvastrum coccineum* (Nutt.) Gray; Malvaceae

scarlet runner bean, *Phaseolus coccineus* L.; Fabaceae

schefflera, *Schefflera* sp.; Araliaceae

Scotch broom   see broom

Scotch pine, *Pinus sylvestris* L.; Pinaceae

scrub oak, *Quercus ilicifolia* Wangenh.; Fagaceae

Sea Island cotton, *Gossypium barbadense* L.; Malvaceae

sea myrtle   see groundsel baccharis

sedge, *Carex* spp.; Cyperaceae

senecio   see Appalachian groundsel

shagbark hickory, *Carya ovata* (Mill.) K. Koch; Juglandaceae

sheep fescue, *Festuca ovina* L.; Poaceae

shrubby cinquefoil, *Potentilla fruticosa* L.; Rosaceae

shrubby St. John's-wort, *Hypericum densiflorum* Pursh; Clusiaceae

Siberian larch, *Larix sibirica* Ledeb.; Pinaceae

Siberian peashrub   see caragana

silky dogwood, *Cornus obliqua* Raf.; Cornaceae

silky loco, *Astragalus crassicarpus* Nutt.; Fabaceae

silver maple, *Acer saccharinum* L.; Aceraceae

slash pine, *Pinus elliottii* Engelm.; Pinaceae

slender deutzia, *Deutzia gracilis* Siebold & Zucc.; Saxifragaceae

slippery elm, *Ulmus rubra* Muhl.; Ulmaceae

smooth brome, *Bromus inermis* Leyss.; Poaceae

smooth hydrangea   see American hydrangea

smooth sumac, *Rhus glabra* L.; Anacardiaceae

sorghum, *Sorghum bicolor* (L.) Moench; Poaceae

southern catalpa, *Catalpa bignonioides* Walter; Bignoniaceae

sowbane, *Chenopodium* spp.; Chenopodiaceae

soybean, *Glycine max* (L.) Merr.; Fabaceae

Spanish bayonet   see datil yucca

Spanish moss, *Tillandsia usneoides* (L.) L.; Bromeliaceae

spartina, *Spartina* spp.; Poaceae

spinach, *Spinacia oleracea* L.; Chenopodiaceae

spiny hackberry   see desert hackberry

spotted beebalm, *Monarda punctata* L.; Lamiaceae

spotted spurge, *Euphorbia maculata* L.; Euphorbiaceae

spruce, *Picea* spp.; Pinaceae

spurge, *Euphorbia* spp.; Euphorbiaceae

squash, *Cucurbita* spp.; Cucurbitaceae

staghorn sumac, *Rhus hirta* (L.) Sudw. = *R. typhina* L.; Anacardiaceae

stemodia, *Stemodia tomentosa* (P. Mill.) Greenm. & Thompson; Scrophulariaceae

sterculiaceous liana, *Byttneria aculeata* Jacq.; Sterculiaceae

stinging nettle, *Urtica dioica* L.; Urticaceae

strawberry, *Fragaria* x *ananassa* Duchesne; Rosaceae

sugar beet   see beet

sugarcane, *Saccharum officinarum* L.; Poaceae

sumac (Mediterranean), *Rhus tripartita* (Bernard da Ucria) Grande; Anacardiaceae

sundew, *Drosera* spp.; Droseraceae

sundewlike byblidaceous plant, *Byblis gigantea* Lindl.; Byblidaceae

sunflower, *Helianthus* spp.; Asteraceae

sunnhemp, *Crotalaria juncea* L.; Fabaceae

sweetpotato, *Ipomoea batatas* (L.) Lam.; Convolvulaceae

sycamore, *Platanus occidentalis* L.; Platanaceae

sycamore maple, *Acer pseudoplatanus* L.; Aceraceae

table mountain pine, *Pinus pungens* Lamb.; Pinaceae

tall buttercup, *Ranunculus acris* L.; Ranunculaceae

tall goldenrod, *Solidago canadensis* L. = *S. altissima* L.; Asteraceae

tall larkspur, *Delphinium barbeyi* (Huth) Huth; Ranunculaceae

tall meadowrue, *Thalictrum pubescens* Pursh; Ranunculaceae

tamarack, *Larix laricina* (Du Roi) K. Koch; Pinaceae

tangle mealybean   see wild bean

tansy, *Tanacetum vulgare* L.; Asteraceae

tansy ragwort, *Senecio jacobaea* L.; Asteraceae

taro, *Colocasia esculenta* (L.) Schott; Araceae

tea, *Camellia sinensis* (L.) Kuntze; Theaceae

teasel, *Dipsacus fullonum* L. = *D. sylvestris* Huds.; Dipsacaceae

thelesperma, *Thelesperma* sp.; Asteraceae

thinleaf sunflower, *Helianthus decapetalus* L.; Asteraceae

thistle, *Carduus* spp., *Cirsium* spp.; Asteraceae

timothy, *Phleum pratense* L.; Poaceae

tobacco, *Nicotiana tabacum* L.; Solanaceae

tomato, *Lycopersicon esculentum* Mill.; Solanaceae

toothpick ammi, *Ammi visnaga* L.; Apiaceae

trefoil, *Lotus* spp.; Fabaceae

tuliptree, *Liriodendron tulipifera* L.; Magnoliaceae

turnip, *Brassica rapa* L.; Brassicaceae

upland cotton   see cotton

velvet ash, *Fraxinus velutina* Torr.; Oleaceae

velvet tobacco, *Nicotiana velutina* Wheeler; Solanaceae

vetch, *Vicia* spp.; Fabaceae

viburnum, *Viburnum* spp.; Caprifoliaceae

villebrunea, *Villebrunea scabra* Wedd. = *V. rubescens* Blume; Urticaceae

Virginia pine, *Pinus virginiana* Mill.; Pinaceae

Virginia sweetspire, *Itea virginica* L.; Grossulariaceae

walnut, *Juglans* sp.; Juglandaceae

water hemlock, *Cicuta maculata* L.; Apiaceae

water-hyacinth, *Eichhornia crassipes* (Mart.) Solms; Pontederiaceae

water-lily, *Nymphaea* sp.; Nymphaeaceae

wattle   see acacia

western wheatgrass, *Elymus smithii* (Rydb.) Gould = *Agropyron smithii* Rydb.; Poaceae

wheat, *Triticum aestivum* L.; Poaceae

wheatgrass, *Agropyron* spp.; Poaceae

white ash, *Fraxinus americana* L.; Oleaceae

white clover, *Trifolium repens* L.; Fabaceae

white fir, *Abies concolor* (Gordon & Glend.) Lindl.; Pinaceae

white heath aster, *Aster pilosus* Willd.; Asteraceae

white mulberry, *Morus alba* L.; Moraceae

white oak, *Quercus alba* L.; Fagaceae

white pine, *Pinus strobus* L.; Pinaceae

white poplar, *Populus alba* L.; Salicaceae

white rubber rabbitbrush, *Chrysothamnus nauseosus* (Pallas) Britt. subsp. *albicaulis* (Nutt.) Hall & Clements; Asteraceae

white spruce, *Picea glauca* (Moench) Voss; Pinaceae

whorled loosestrife, *Lysimachia quadrifolia* L.; Primulaceae

wide-leaved spiderwort, *Tradescantia subaspera* Ker (-Gawler); Commelinaceae

wild bean, *Strophostyles helvola* (L.) Elliott; Fabaceae

wild indigo, *Baptisia tinctoria* (L.) Vent.; Fabaceae

wild lupine, *Lupinus perennis* L.; Fabaceae

wild parsnip   see parsnip

willow, *Salix* spp.; Salicaceae

wilwilli tree, *Erythrina tahitensis* Nadeau = *E. sandwicensis* Degener; Fabaceae

winterberry, *Ilex verticillata* (L.) A. Gray; Aquifoliaceae

wislizenia, *Wislizenia refracta* Engelm.; Capparidaceae
witchgrass, *Panicum capillare* L.; Poaceae
wood woundwort, *Stachys sylvatica* L.; Lamiaceae

yams, *Dioscorea* spp.; Dioscoreaceae
yarrow, *Achillea millefolium* L.; Asteraceae
yaupon, *Ilex vomitoria* Aiton; Aquifoliaceae
yellowbells, *Tecoma stans* (L.) Kunth; Bignoniaceae

yellow bush lupine, *Lupinus arboreus* Sims; Fabaceae
yellow elder   see yellowbells
Yorkshire fog, *Holcus lanatus* L.; Poaceae
yucca, *Yucca filamentosa* L.; Agavaceae

zonal geranium, *Pelargonium* x *hortorum* L. H. Bailey;
   Geraniaceae
zucchini, *Cucurbita pepo* L.; Cucurbitaceae

Glossary

Most definitions are taken or modified from *The Torre-Bueno Glossary of Entomology* (Nichols 1989); secondary sources of definitions include Agrios (1988), Allaby (1991), Begon et al. (1990), Borror et al. (1989), Coombs (1992), Dolling (1991), Evans (1984), Fernald (1950), Futuyma (1986), Gerber and Wise et al. (1995), Gullan and Cranston (1994), Hanson (1962), Lincoln et al. (1982), Pedigo (1989), Raven et al. (1976), Ricklefs (1979), Rieger et al. (1991), Schuh (2000), Schuh and Slater (1995), and Stenesh (1975). Abbreviations used are adj., adjective; n., noun; pl., plural; sing., singular; and v., verb. Cross-referenced entries within a definition are boldfaced.

abaxial   on the side away from the axis; dorsal.

abiotic   not **biotic** or infectious, e.g., in referring to the cause of plant diseases.

abscise   to shed by **abscission**.

abscission   the shedding of leaves or other plant parts.

achene   a small dry and hard one-locular, one-seeded dehiscent fruit.

achiasmatic meiosis   **meiosis** in which chiasma (point of contact between homologous chromosomes) is lacking.

acromania   the condition known as *crazy top* in cotton.

action threshold   see **economic threshold**.

adaptation   the condition of showing fitness for a particular environment, as applied to characteristics of a structure, function, or entire organism.

adaptive radiation   evolutionary divergence of members of a single phyletic line into a series of rather diverse niches or **adaptive zones**.

adaptive zone   a set of similar ecological **niches** occupied by a group of (usually) related species, often constituting a higher **taxon**.

adaxial   toward the axis; ventral.

adventitious   buds, shoots, or other plant parts that develop in an irregular or unusual position.

adventive   not native; nonindigenous.

aedeagus   in insects, the male copulatory organ; that portion of the **phallus** distal to the phallobase, including the proximal phallotheca and distal endosoma.

aeropyle   fine pores of an insect egg connected to air spaces in the outer and inner meshworks of the **chorion**.

aggregation   the coming together of organisms into a group; a group of individuals.

aggressive mimicry   mimicry in which a predator resembles a potential food source or potential mate of prey.

agmatoploidy   increase in chromosome number by fragmentation of chromosomes.

agroecosystem   an **ecosystem** largely created and maintained to satisfy a human want or need, e.g., a cotton field, its associated biota, and physical environment.

alimentary canal   the food tube traversing the body from mouth to anus.

aliphatic   pertaining to an organic compound that has an open chain structure rather than rings.

allele   allelomorph; one of two or more forms of a gene that arise by mutation and occupy the same locus on **homologous** chromosomes.

allelochemical   a chemical functioning in interspecific communication; see also **allomone, kairomone, pheromone**.

allomone   **allelochemical** of adaptive advantage to the organism sending it.

allozyme   a form of **protein**, detectable by electrophoresis, that is produced by a particular **allele** at a single gene locus.

amino acid   building blocks of **protein**; any organic compound containing an amino group (NH_2) and a carboxylic acid group (COOH).

amylase   an **enzyme** that hydrolyzes starch or glycogen.

anachoresis   the phenomenon of living in holes or crevices.

anaerobic   able to live without oxygen.

analogous   similar in function, but differing in origin and structure; see also **homologous**.

anamorph   that part of the life cycle of a fungus characterized by the production of asexual spores borne on conidiomata, or specialized **conidia**-bearing structures.

ancestral   primitive; inherited from an earlier form or ancestor.

anemophilous   referring to plants in which pollen is scattered almost exclusively by wind.

aneuploidy   the condition of chromosome evolution characterized by more or fewer than an exact multiple of the haploid number, i.e., one set of chromosomes.

angiosperms   flowering plants.

anomalous   unusual; departing widely from the usual type.

anther   the polleniferous (pollen-bearing) parts of a **stamen**.

anthesis   the expansion or time of expansion of a flower.

anthophilous   flower frequenting or flower loving.

antibiosis   plant characteristics that affect insects in a negative manner.

antibody   a **protein** produced in a warm-blooded animal in reaction to an injected foreign **antigen** and capable of reacting specifically with that antigen.

antigen   foreign **proteins** (and occasionally complex lipids, carbohydrates, and some nucleic acids) which upon injection into a warm-blooded animal, induce the production of **antibodies**.

antiserum   blood serum containing specific **antibodies**.

aperture   a hole or other opening.

apetalous   having no petals; in the present work, used to include tree species with only inconspicuous petals.

aphidophage   an organism that feeds or preys on aphids.

apical   at, near, or pertaining to the apex (end) of any structure.

apomorphic   relatively derived or specialized, when comparing two or more **homologous character states**; see also **synapomorphic**.

aposematic   having warning coloration, indicating that an animal is unpalatable or distasteful.

apterous   without wings.

arboreal   living in, on, or among trees.

ascospore   a fungal spore produced within an **ascus**.

ascus (pl. asci)   a saclike cell of the sexual state of a fungus belonging to Ascomycotina, in which **ascospores** form.

assimilation   the process by which nourishment is changed into living tissue.

asymptomatic   lacking symptoms.

augmentation   a biological control practice designed to increase the number or effectiveness of existing natural enemies; see also **conservation**.

auricular   of or pertaining to the auricle, an appendage resembling a little ear.

autapomorphy   a **derived character** or **character state** unique to a terminal **taxon** on a **cladogram**.

autosome   one of the chromosomes other than a sex chromosome.

autotomy   loss of appendages of arthropods by reflex sloughing or shedding; self-amputation.

auxin   a substance that controls the growth of plants.

Batesian mimicry   **mimicry** in which an edible species (**mimic**) obtains security by counterfeiting the appearance of an inedible species (**model**); see also **Müllerian mimicry**.

biological control   the human use of selected living organisms to suppress **populations** of **pest** species.

bionomics   the habits, breeding, and **adaptations** of living forms.

biotic   pertaining to life; referring to infectious causes of plant disease.

blasting   the shriveling or withering of fruiting or tender growing portions of **host** plants resulting from feeding by mirids and certain other sucking insects.

boreal   northern.

brachypterous   with short or abbreviated wings.

bract   a more or less modified leaf subtending (underlying and enclosing) a flower or belonging to an **inflorescence**.

callosity   a flattened elevation not necessarily harder than the surrounding **tissues**.

callous   having the texture of a **callus**.

callus (pl. calli)   in Heteroptera, the paired or fused impression or elevation in anterior part of pronotum (dorsal part of prothorax) behind **collar**; a hard lump or swelling of the cuticle.

calyx   outer perianth (floral envelope) of a flower.

cambium   a **meristem** that gives rise to parallel rows of cells; commonly applied to the **vascular** cambium and cork cambium.

camouflage   coloration that blends with the background.

campaniform sensillum   a sense organ consisting of a dome-shaped portion of the cuticle with associated sensory neuron.

canker   a necrotic, often sunken **lesion** on a stem, branch, or twig of a plant.

cannibalism   the act of preying on other members of the same species.

canopy   the uppermost layer (overarching branches) of a shrub or tree; the upper, dense layer of a row crop such as cotton.

cantharidin   an irritating, sesquiterpenoid-derived compound produced by blister beetles.

capsid   a mirid, from Capsidae, a synonym of the family-group name Miridae.

carnivory   in insects, the preying or feeding on arthropods or their flesh.

carpel   a simple **pistil**, or one member of a compound pistil.

carrion   dead and decaying flesh.

cataleptic   the state of immobilization in which an insect is insensible (unresponsive) to stimulation.

catfacing   uneven growth and deformation of a developing fruit resulting from the feeding of a sucking insect.

catkin   an ament or dry scaly spikelike **inflorescence** of unisexual flowers.

cecidogenic   **gall** forming.

cellulase   a complex **enzyme** system that hydrolyzes cellulose to sugars of lower molecular weight.

center of origin   the hypothesized geographical area from which any group or particular species has spread.

chalky referring to the whitened appearance of plant parts, e.g., the **cotyledons** of lentil on which lygus bugs feed.

chaparral low, often dense scrub vegetation characterized by shrubs or dwarf trees with mostly evergreen, often hard leaves.

character (states) characteristics or attributes used to recognize, describe, define, or differentiate **taxa**.

chemoreception perception through chemical stimuli.

cherelle young fruit (pod) of the cocoa tree.

chitin a major polysaccharide constituent of arthropod cuticle (epidermal secretion covering the insect body).

chlorophyll the green pigment of plant cells that is necessary for photosynthesis.

chloroplast a membrane-bounded organelle in algal or green plant cells in which **chlorophylls** are contained; site of photosynthesis.

chlorosis a fading or yellowing of the color of plant leaves caused by loss of **chlorophyll**; see also **stippling**.

chorion the outer shell or covering of an insect egg.

chorionated a stage in mirid egg development characterized by **oocytes** with **chorions** in ovaries and oviducts; see also **previtellogenic**, **vitellogenic**.

chupon the upright stem or shoots of the cocoa tree.

cibarium preoral cavity or food pouch between base of the hypopharynx and undersurface of clypeus, often with a muscular pump.

circadian (rhythm) an endogenous (originating within) oscillation with a natural **photoperiod** of approximately the duration of a solar day or 24 hours.

circulative transmission pertaining to plant viruses that are acquired by their **vectors** through their mouthparts, accumulated internally, passed through their tissues, and introduced into plants again via mouthparts of the vectors; see also **propagative transmission**.

cladistic pertaining to **phylogenetic** analysis in which **taxa** are grouped on the basis of relative recency of common ancestry.

cladogram a branching diagram based on the distribution of synapomorphies, or the sharing of one or more derived **character states**.

claspers see **parameres**.

classical biological control the importation and establishment of a natural enemy to help suppress an **adventive** (or sometimes native) **pest**.

clavate clublike; thickened gradually toward the tip.

clavus usually parallel-sided, sharply pointed anal area of **hemelytron**.

coleopteroid beetlelike.

collar rounded or flattened anterior margin of **prothorax**.

collenchyma in plants, the supporting tissue of elongated living cells with irregularly thickened primary cell wall.

colonization the evolution of host associations in **phytophagous** insects in which they adopt new **host** plants or undergo a host "shift" or transfer; see also **cospeciation**.

commensal a species that benefits through commensalism, i.e., a **symbiosis** in which members of one species are benefited while those of the other species are neither benefited nor harmed.

community an association of interacting **populations**, usually defined by the nature of their interaction of the place in which they live; the plants or animals of a given **habitat**.

conelet in pines and other conifers, a first-year cone or **strobilus**.

congener a species belonging to the same **genus** as another.

congeneric belonging to the same **genus**.

conidia (sing. conidium) asexual fungal spores formed from the end of a conidiophore, a specialized **hypha**.

conservation a biological control practice that includes any activity designed to protect and maintain existing populations of natural enemies; see also **augmentation**.

conservation tillage a cropping practice in which conventional tillage is reduced or eliminated to minimize soil erosion, to conserve moisture and lower soil temperatures in summer, and to yield savings in time and energy.

conspecific belonging to the same species.

contagious distribution nonrandom occurrence of individuals of a species.

continuum with respect to mirid feeding habits, the idea that **trophic** behavior shows plasticity and is often difficult to classify, most species falling somewhere along a series with strict **zoophagy** at one extreme and strict **phytophagy** at the other.

convergence resemblance between two forms derived from widely distant ancestries or origins, either by adoption of similar habits or through reduction or elimination of original differences; see also **parallelism**.

convergent becoming closer distally, i.e., near or toward the free end.

coriaceous leatherlike in texture.

corium **proximal** coriaceous (leatherlike) or otherwise differentiated part of a heteropteran forewing exclusive of **clavus** and distinct from the **membrane**.

cornicle in aphids, a peglike or tubular structure on the abdomen that secretes alarm **pheromones**.

corolla the inner perianth; the petals, collectively.

cortex in plants, the rind or bark; ground-tissue region of a stem or root.

cortical in insects, pertaining to the outer skin or layer.

cosmopolitan occurring throughout most of the world.

cospeciation pertaining to a strict, prolonged, pairwise coevolution or parallel diversification of insect and plant species; see also **colonization**.

costal fracture in mirids, a short, usually transverse line of weakness or break in costal margin of forewing separating a well-differentiated **cuneus** from rest of **corium**.

cotyledon   seed leaf; foliar portion or first leaves of the embryo as found in the seed.

crawler   newly emerged, active immature stage of a scale insect.

crepuscular   active or flying at dusk or twilight (before sunrise and after sunset).

cross-pollination   the transfer of pollen from the **anther** of one plant to the **stigma** of a flower of another plant.

cryptic   hidden, concealed, or camouflaged; protectively colored.

culm   the peculiar stem of grasses and sedges.

cultivar   an artificially bred, cultivated variety.

cuneus   in mirids, the usually triangular posterolateral area of the **corium** bounded proximally by **costal fracture** and distally by **membrane**.

cyathium (pl. cyathia)   in *Euphorbia*, the ultimate **inflorescence**.

cymose   bearing cymes or cymelike, i.e., with a usually broad and flattened determinate **inflorescence**.

damage   **injury** resulting in loss of value, or a measurable loss of host utility; see also **injury**.

day-neutral   referring to a plant that blooms when day length is either short or long.

DDT   a synthetic chlorinated hydrocarbon insecticide (dichlorodiphenyltrichloroethane).

declivent   sloping downward.

defaunate   to remove all animals, e.g., from a small island.

degree-day   a measure of physiological time, the product of time and temperature above a threshold, used to monitor growth; sometimes referred to as *day-degree*.

degree-hour   see **growing degree-hour**

dehiscence   method of opening of **anthers**, fruits, or other structures at maturity.

density   the number of individuals per unit of measure.

density-dependent   having influence on individuals in a **population** and varying with the degree of crowding or density of that population.

derived   pertaining to a **character** that is modified (advanced) relative to the **ancestral** or primitive condition.

desiccation   drying.

deutonymph   the third **instar** of a mite.

diagnosis   the determination of the causal agent responsible for insect **injury**.

diapause   a delay in development that is not the direct result of prevailing environmental conditions; see also **dormancy, quiescence**.

dicot   a plant whose embryo has two **cotyledons**; a dicotyledon.

dieback   progressive death of shoots, branches, or roots, usually starting at tip.

diel   referring to the 24-hour period of day and night.

differentiation   a process by which a relatively unspecialized cell undergoes a progressive change to a more specialized cell.

diffusion   the movement of suspended or dissolved particles from a more concentrated to a less concentrated region as a result of a random movement of individual molecules.

dimorphic   occurring in two distinct forms.

dioecious   having male and female elements on different individuals of the same plant species; unisexual.

diphyletic   describing a **taxon** whose members are derived from two **ancestral** species.

diploid   having a double set of chromosomes, i.e., the full complement of maternal and paternal chromosomes.

direct pest   an insect that damages a harvested or marketed part of a plant, such as a fruit.

disc flower   in Asteraceae or composites, the tubular flowers of the head, as distinct from the ray (straplike, marginal) flowers.

disease   any malfunctioning of **host** cells and **tissues** that results from continuous irritation by a **pathogenic** agent or environmental factor and leads to development of **symptoms**.

dispersal   the spreading of individuals away from each other.

dispersion   pattern of spacing of individuals of a population.

distal   farthest from the point of attachment or origin; see also **proximal**.

diurnal   pertaining to the day; active during the day.

dormancy   a seasonally recurring period in an insect life cycle when growth, development, and reproduction are suppressed; **quiescence** or **diapause**.

dormant oil   a petroleum oil applied to trees only when foliage is not present.

dorsum   the upper surface.

drupe   a simple, fleshy fruit, derived from a single **carpel**, usually one seeded, in which the inner fruit coat adheres to the seed, e.g., a peach.

drupelet   a diminutive **drupe**, as in raspberry or blackberry.

ecdysis   the final stage of molting; the process of casting the skin.

eclosion   hatching of the egg; also the escape of the adult insect at the terminal molt.

economic injury level   amount of insect injury causing losses in yield equal to the costs of control; see also **economic threshold**.

economic threshold   pest density at which control or management action should be applied to prevent the **economic injury level** from being reached.

ecosystem   all interacting parts of the biological and physical worlds; a **community** and its **abiotic** environment.

ectoparasitoid   a **parasitoid** that develops externally on and kills its **host**.

edaphic   relating to the soil.

egestion   evacuation; pertaining to excretion from the body.

egg burster   a projecting point, hard spines, or ridges on the head or other part of the embryo used to break the shell when hatching.

electroantennogram   record of the summed receptor potentials of a number of olfactory receptors responding to a stimulus.

electrophoretic   pertaining to electrophoresis, a process of separating molecules, particularly polypeptides, owing to their differential rates of migration in an electric field.

ELISA   a serological test in which one **antibody** carries with it an **enzyme** that releases a colored compound; enzyme-linked immunosorbent assay.

embolium   in the forewing of mirids, a broadened submarginal part of the **corium proximal** to the **costal fracture**.

embryogenesis   development and growth of an embryo.

embryolessness   in umbellifers (Apiaceae), the occurrence of seeds with normal **endosperm** but lacking embryos, often the result of mirid feeding.

emigration   movement out of an area.

encapsulate   to enclose or surround a **parasitoid** larva within the blood of the **host** by a layer of **hemocytes**.

endemic   sometimes used to refer to the restriction of a **taxon** to a given geographic region; indigenous or native.

endocarp   the innermost layer of the mature plant ovary wall, or **pericarp**.

endoparasitoid   a **parasitoid** developing internally in and killing its **host**.

endophytic oviposition   insertion of eggs into plant tissue; see also **exophytic oviposition**.

endosperm   in **angiosperms**, the reserve food stored around the embryo.

endosymbiont   a partner in a **symbiosis** that lives inside the cells of the other.

entomopathogenic   referring to an attack of an insect by a **pathogen** or **disease**-causing organism.

entomophagous   feeding or preying on insects.

enzyme   a **protein** that regulates the rate of chemical reactions.

ephemeral   lasting for a day or less; short-lived.

epicarp   outer layer of the **pericarp** or matured plant ovary.

epidemiology   the study of factors affecting the outbreak and spread of infectious diseases.

epidermis   the outermost layer of cells of the leaf, roots, and young stems.

epiphytic   referring to an organism that grows on another plant but is not parasitic on it.

epithelium   the layer of cells that covers a surface or lines a cavity.

epizootic   in invertebrate pathology, an outbreak of a **disease** involving an unusually large number of cases.

esterase   an **enzyme** that hydrolyzes an ester (compound formed as the condensation product of an acid and an alcohol) into an alcohol and an acid.

etiology   the study of the causes of **diseases**.

eukaryote   an organism with membrane-bound nuclei.

evagination   an outpocketing, or a saclike structure on the outside.

evaporatory area   in many heteropterans, an area of specialized cuticle on the metathoracic pleuron associated with, and usually surrounding, an orifice and auricle of metathoracic **scent glands**.

excrement   waste products or fecal matter eliminated by an insect, mainly after digestion.

exine   the outer wall layer of a pollen grain or spore.

exogenous   originating from outside the organism.

exophytic oviposition   deposition of eggs on the outside of plant **tissue**; see also **endophytic oviposition**.

exopterygote   referring to insects in which wings form progressively in sheaths lying externally on the body surface.

exoskeleton   the external skeleton of insects and certain other invertebrates, consisting of hard cuticle, to the inner side of which muscles are attached.

extrafloral nectary   nectar-producing organ on some part of a plant other than the flower.

extragenital insemination   in certain heteropterans, the puncturing of the body wall or wall of inner genitalia by the **phallus** during mating and deposition of sperm outside the usual reproductive tract; **traumatic insemination**.

extraoral digestion   external digestion in which saliva containing digestive **enzymes** is placed on or injected into food, followed by a sucking up of soluble products.

exudate   any exuded substance, such as plant sap.

exude   to ooze or flow slowly through minute openings.

exuviae (pl. n.)   cast skin of larvae or **nymphs** at **ecdysis**.

facultative   not compulsory; optional behavior, such as facultative predation.

fat body   a loose or compact aggregation of cells, mostly trophocytes, suspended in the **hemocoel**, responsible for storage and excretion.

fecundity   the average number of eggs laid by an insect; rate at which a female produces offspring.

feeding rings   feeding punctures or encircling lesions created when certain hemipterans, such as the three-cornered alfalfa hopper (*Spissistilus festinus*), **girdle** plant stems.

fertility   rate at which fertilized eggs are produced.

filter chamber   a part of the **alimentary canal** in some homopterans in which the two ends of the midgut and the beginning of the hindgut are bound together in a membranous and muscular sheath.

flitting   in mirids, **trivial flight** consisting of short movements and often associated with mating.

floret   one of the small flowers that make up the composite **inflorescence** or spike of grasses.

food canal   canal anterior to the **cibarium** in sucking insects, through which liquefied food is ingested.

food chain   transfer of energy from the primary producers (green plants) through a series of organisms that eat and are eaten, assuming each organism feeds only on one other type of organism; see also **food web**.

food web a diagram that represents the feeding relationships of organisms within an **ecosystem**; see also **food chain**.

foretarsus (pl. foretarsi) tarsus (leg segment attached to the apex of the tibia) of one of the prothoracic (fore) legs.

functional response change in the rate of exploitation of **prey** by an individual predator as a result of change in prey **density**.

funiculus the stalk of the **ovule**.

fuscous dark brown, approaching black.

gall an aberrant plant growth produced in response to the activities of another organism, often an insect.

gametophyte the plant or generation that produces gametes (sex cells) and contains the haploid number of chromosomes.

gastric caecum (pl. caeca) midgut caecum or blind-ending sac or tube.

gel chromatography a form of chromatography (technique for separating or analyzing mixtures of liquids, gases, compounds in solution, or particles) used to separate molecules on the basis of molecular weight.

genotype the genetic constitution of an individual or **taxon**.

genus an assemblage of one or more species united by one or more **derived** features and, therefore, believed to have a single evolutionary origin.

geophilous living on the ground.

geostatistics use of statistical procedures to analyze and model spatial relationships.

germarium structure within an **ovariole** in which the **oogonia** give rise to **oocytes**.

germ band area of thickened cells on the ventral side of the blastoderm (continuous, peripheral cell layer surrounding the yolk of an insect egg following cleavage) that becomes the embryo.

girdle (v.) to puncture a plant stem with **stylets** and produce encircling lesions or **feeding rings**; often used to describe feeding by the mirid *Engytatus modestus* or the threecornered alfalfa hopper (*Spissistilus festinus*).

glabrous not hairy or pubescent; smooth.

glume a chafflike **bract**.

gluten **proteins** found in cereal grains that are used as an adhesive and a flour substitute.

Gondwanan referring to the Mesozoic southern supercontinent that was composed of continental blocks of South America, Africa, Madagascar, India, Antarctica, and Australia.

gravid full of ripe eggs.

growing degree-hour accumulated temperature (in hours) above a certain threshold, used to monitor growth; see also **degree-day**.

guild in a broad sense, groups of species that use or exploit a resource in a similar manner.

gula the throat area behind the **rostrum** in the head of Heteroptera.

gustatory relating to the sense of taste.

habitat the place in which an animal or plant normally lives, or **community** characterized by its physical or **biotic** properties.

habitus general form and appearance.

haustellate formed for sucking; applied chiefly to mouthparts.

hematophagous feeding on blood.

hemelytron (pl. hemelytra) heteropteran forewing with a thickened **proximal** portion and membranous **distal** portion.

hemimetabolous having incomplete development or **metamorphosis**; see also **holometabolous, paurometabolous**.

hemocoel the main body cavity of insects in which the blood flows.

hemocyte blood cells suspended in fluid plasma of the **hemolymph**.

hemolymph the lymphlike fluid (blood) filling the **hemocoel**.

herbaceous referring to any herb or nonwoody plant.

herbivore an organism that eats plants; sometimes restricted to the use of nonwoody plant **tissue**.

hirsute hairy; clothed with long dense **setae**.

histological referring to histology, the study of **tissues** of organisms.

histolysis a breaking down, degeneration, and dissolution of organic **tissue**.

holidic pertaining to a medium whose intended constituents have an exactly known chemical structure before the medium is compounded; see also **meridic**.

holometabolous having complete **metamorphosis;** see also **hemimetabolous, paurometabolous**.

homeostasis maintenance of a steady state by means of self-regulation through internal feedback responses.

homologous pertaining to a pair of features occurring in different organisms, one being **derived** from another, i.e., being states of a single **character**; see also **analogous**.

honeydew watery, sugar-containing fluid excreted from the anus of certain hemipterans.

hopperburn foliar **necrosis** resulting from the feeding of certain leafhoppers; often used to refer to injury to alfalfa or potato by the potato leafhopper (*Empoasca fabae*).

host a plant on which an insect feeds; often restricted to a plant on which development ("breeding") occurs; see also **nonhost plant**; an organism in or on which a **pathogen**, parasite, or **parasitoid** lives.

hydrolysis the catalytic breakdown of macromolecules or polymers into constituent building blocks with incorporation of the elements of water.

hyperparasitoid a secondary **parasitoid** developing on another parasitoid.

hyperplasia an abnormal increase in the number of cells or **tissues** of an organism.

hypertrophy plant overgrowth from abnormal cell enlargement.

hypha (pl. hyphae) a single branch of a mycelium (mass of hyphae composing the body of a fungus).

imago the adult insect.

immigrate move into an area.

impervious resistant to penetration.

importation see **introduction.**

incertae sedis of uncertain taxonomic position.

incrassate thickened; rather suddenly swollen, especially near tip.

indeterminate pertaining to growth or flowering that is unrestricted or continues indefinitely.

indigenous native to an area.

inflorescence a flower cluster, with a definite arrangement of flowers; any **aggregation** of flowers.

infraorder an optional taxonomic category below the suborder.

ingestion the taking in of food.

injury physical or physiological effects, including visible **symptoms**, on a **host** plant; see also **damage.**

inoculate to bring a **pathogen** into contact with a **host** plant or organ.

inoculum any stage or part of a **pathogen**, such as spores or virus particles, that can infect a **host.**

in situ in its natural place or normal position.

instar the growth stage between two successive molts; often used to refer to individuals in a particular stage.

insular pertaining to an island.

integrated pest management (IPM) integration of chemical means of insect control, selectively used, with other methods such as **biological control** and **habitat** manipulation.

integument outer layer of the insect, i.e., epidermis and cuticle.

intercellular between and among cells.

internode region of a stem between two successive **nodes**, i.e., area where one or more leaves are attached.

intine inner wall layer of a pollen grain or spore.

intracellular occurring within a cell.

intraguild predation the killing and eating of species that use similar, often limiting resources and thus are potential competitors.

introduction a **biological control** practice involving identification of natural enemies that regulate a **pest** in its original location and introduction of these into the pest's new location.

intumescence condition or process of being enlarged or swollen.

invertase an **enzyme** that catalyzes the **hydrolysis** of the terminal nonreducing fructose residue from fructose polymers to sucrose.

in vitro in an artificial environment outside a living organism.

in vivo in the living organism.

involucre a rosette or whorl of **bracts** surrounding an **inflorescence**.

isoelectric focusing an analytical or separation procedure similar to gel electrophoresis, used to separate **proteins** and other charged molecules on the basis of their isoelectric point.

isogenic differing genetically only at one locus (position of gene along a chromosome).

isoline a group of individuals from a common ancestry (line) that are genetically similar except for one gene, i.e., **isogenic.**

iteroparous in insects, referring to reproductive effort distributed over time and space, i.e., females that do not deposit a single clutch of eggs in one place during their lifetime.

kairomone a communication chemical that benefits the receiver and is disadvantageous to the producer.

karyotype chromosome set.

key-factor analysis a statistical treatment of population data designed to identify factors most responsible for changes in **population density** in successive generations.

key pest a perennial, severe **pest** that causes serious and difficult crop-production problems.

kinematics the branch of mechanics (study of interactions between matter and the forces acting on it) concerned with the motions of objects without being concerned with the forces that cause the motion.

kleptoparasitism the stealing of food stored by another insect.

K-strategist *K*-selected species, in relatively stable habitats, without high reproductive potential but with relatively high survival of young; see also r-**strategist.**

k-value a mortality factor in an analysis of insect population dynamics; the loss of individuals from a given stage of a life cycle. A **dispersion** parameter used in measuring the clumping in the spatial distribution of an animal population and whose values increase as the distribution becomes less aggregated; see **negative binomial.**

labile unstable; readily changing.

labium (pl. labia) the lower lip in insect mouthparts; in mirids and other heteropterans, the segmented **sheath** that encloses the **stylets.**

labrum the upper lip in insect mouthparts; in mirids and other heteropterans, the usually elongate triangular structure covering the base of the labial groove.

laccase see **polyphenol oxidase**.

lacerate-flush referring to the feeding of **phytophagous** heteropterans in which the **stylets** lacerate and **macerate** cells and the liquefied plant material is imbibed or ingested.

laciniate ovipositor an **ovipositor** with elongate, often laterally compressed blades or **valvulae.**

lamellate sheetlike or leaflike; composed of or covered with **laminae** or thin sheets.

lamina (pl. laminae) the blade, or broad expanded part of a leaf.

leader the main shoot of a shrub or tree.

leafminer an insect that lives in and feeds on **mesophyll** between the upper and lower surfaces of a leaf.

lenticel   a loose-structured opening in the periderm (protective layer) beneath the **stomata** in the stem of many woody plants that facilitates gas transport.

lesion   a localized area of discolored, diseased **tissue**.

life form   the characteristic structure of a plant.

life table   a tabulation of the life stages of an insect with a cumulative record of mortality (age-specific deaths) and survival.

lipase   an **enzyme** that catalyzes the **hydrolysis** of fats to glycerol and fatty acids.

longevity   length of life.

lumen   the enclosed space or cavity of any hollow or vesicular (sactlike or bladderlike) organ or structure.

macerate   to make soft by soaking in a liquid.

macropterous   with fully developed, functional forewings and hindwings.

Malaise trap   tentlike structure of fine netting that insects fly into and eventually move upward into a collection apparatus.

mandible   the first pair of jaws in insects, which are typically fitted for biting, but in heteropterans are modified into an outer pair of needlelike **stylets**.

mandibulate   possessing **mandibles** or chewing mouthparts.

maxilla   the second pair of jaws in insects, which are often fitted for shredding, but in heteropterans are modified into an inner pair of needlelike **stylets**.

m-chromosome   a supernumerary (additional) **autosome**.

mechanoreceptor   a **sensillum** (or group of sensilla) functioning in mechanoreception, i.e., perception of a mechanical distortion of the body.

meiosis   the two successive nuclear divisions in which the chromosome number is reduced from **diploid** (2n) to haploid (1n).

membrane   in heteropterans, the apical, thin, flexible, and usually transparent part of the **hemelytron**.

membranous   thin, more or less transparent; like a membrane.

meridic   pertaining to a medium in which the chemical identity of certain, but not all, of the absolutely essential molecules has been established; see also **holidic**.

meristem   the undifferentiated plant **tissue** from which new cells arise.

mesocarp   the middle layer of the mature ovary wall, or **pericarp**, between exocarp and **endocarp**.

mesofemur   midfemur, or femur of the midleg.

mesophyll   the photosynthetic ground **tissue** (**parenchyma**) of a leaf, located between the layers of **epidermis**, i.e., nonepidermal and nonvascular tissue.

metafemur   hind femur.

metamorphosis   series of changes through which an insect passes in its growth from egg to adult.

microgametogenesis   development of microgametes (male reproductive cells).

microhabitat   the immediate **habitat** in which an organism lives, comprising its environment.

micropyle   opening in the **chorion** of an egg through which sperm pass during the process of fertilization.

microsporangiate strobilus (pl. strobili)   the male cone of gymnosperms (especially conifers), bearing pollen sacs.

middle lamella   in plants, the cementing layer between adjacent cell walls.

midrib   the central or main rib (vein) of a leaf.

mimetic   imitative; pertaining to **mimicry**.

mimic (n.)   an individual, population, or species that resembles a **model**; see also **mimicry**.

mimicry   the resemblance of a **mimic** to a **model**, by which the mimic derives protection from predation provided to the model.

mixed feeder   in mirids, a species that feeds on both plant and animal matter; see also **omnivore**.

model   the protected (distasteful) animal copied by a **mimic**; see also **mimicry**.

monoclonal antibodies   identical **antibodies** produced by a single clone of lymphocytes (phagocytic **hemocytes**).

monocot   a plant whose embryo has one **cotyledon**; a monocotyledon.

monoculture   cultivation of a single crop plant in successive years to the exclusion of other crops, or exclusive production of a crop plant; see also **polyculture**.

monoecious   having the **anthers** and **carpels** produced in separate flowers but borne on the same individual.

monograph   a special essay or treatise dealing in detail or exhaustively with a single subject, such as a taxonomic treatment of a particular insect group.

monooxygenase   an **enzyme** that catalyzes a reaction with molecular oxygen in which only one of the oxygen atoms is introduced into a compound.

monophagy   restriction of feeding to one kind or a few kinds of food; in a strict sense, feeding on only one plant species; see also **oligophagy**, **polyphagy**.

monophyletic   derived from a single **ancestral** form; referring to a group that includes an ancestor and all of its descendants.

monotypic   containing only one immediately subordinate **taxon**, as a genus with only one species.

monsoon   a wind system, mostly in Southeast Asia, that reverses its direction with the season and produces dry and wet seasons.

moribund   about to die.

mottled   showing a variegated or irregular pattern of indistinct light and dark areas.

mucilaginous   **viscid** or like mucilage.

Müllerian mimicry   similarity of several species that are distasteful, poisonous, or otherwise harmful and gain protection from predation by resembling each other; see also **Batesian mimicry.**

multivoltine   having more than one generation in a year; see also **univoltine**.

mummified   converted into a **mummy**.

mummy   in sternorrhynchans, the empty **exoskeleton** containing the mature larva or pupa of the **parasitoid** that has consumed its body contents.

mutualism   **symbiosis** that benefits the members of both participating species, usually being obligatory.

mycetome   a structure housing **intracellular** symbionts such as bacteria.

mycophagous   feeding on fungi.

myrmecomorphic   resembling an ant.

myrmecophagous   feeding on ants.

myrmecophilous   ant loving; typically applied to insects that live in ant nests.

natality   birth rate.

natural enemies   living organisms found in nature that kill insects, weaken them, or reduce their reproductive potential.

Nearctic   referring to the region that includes temperate North America from southern Mexico northward.

necrophagous   feeding on dead or decaying animals.

necrosis   decay; localized death of living **tissues**.

nectariless   lacking leaf and **extrafloral nectaries**.

nectary   in **angiosperms**, a gland that secretes a sugary fluid that is used as food by potential **pollinators**.

negative binomial   a mathematical model used to describe a clumped **dispersion** pattern.

neonate   a newly born or developed animal.

Neotropical   referring to the region that includes South and Central America and tropical North America.

niche   the functional position of an organism in a **community**, i.e., its precise **habitat** plus its behavior in that habitat.

nocturnal   pertaining to the night; active during the night.

node   the part of a stem where one or more leaves are attached.

nomenclatorial   see **nomenclatural**.

nomenclatural   pertaining to the science of scientific names or standardized names assigned to organisms; nomenclatorial.

nominate (adj.)   in taxonomy, referring to a subordinate **taxon** (subspecies, subgenus) containing the type of the higher taxon and bearing the same name; nominotypical.

noncirculative transmission   pertaining to a virus-vector relationship in which the virus is borne on the **stylets** of the **vector** and does not migrate to its salivary glands.

nonhost plant   in mirids, a plant species used only for adult feeding rather than actual development; see also **host**.

nonpersistent transmission   pertaining to a virus-vector relationship in which the retention time or half-life of the virus is measured in minutes.

nonsuberized   referring to a cell wall not impregnated with suberin, a fatty substance found in cell walls of cork tissue.

novel   new; something not formerly used.

nucellar   pertaining to the nucellus (**tissue** composing the main part of the young **ovule**).

numerical response   change in the population size of a predator as a result of a change in **density** of its **prey**.

nutrient sink   nutrient-rich site created by certain hemipterans when they **girdle** a plant stem, which blocks food transport in the **vascular** system.

nymph   the immature stage of **hemimetabolous** insects; often termed *larva*.

obligate   able to develop or survive in a single environment; unavoidable or without alternatives.

occluded   to become closed or covered, as a **lesion**.

ochreous   pale yellow; also ocherous, ochraceous.

oligophagy   accepting a limited range of foods, such as members of related plant families; restricted to feeding on **congeneric** or confamilial hosts; see also **monophagy, polyphagy**.

omnivore   an organism that uses both plant and animal foods, or feeds at more than one **trophic** level; see also **mixed feeder**.

ontogenetic   relating to the developmental history of an individual.

oocyte   the immature egg cell within the **ovariole** differentiated from the **oogonium**.

oogenesis flight syndrome   a distinct behavioral and physiological **syndrome** closely intertwined with reproductive timing and strategy in which migration takes place before egg development.

oogonium (pl. oogonia)   first stage in development in the **germarium** of an insect egg from a female germ cell.

operculum (pl. opercula)   an egg cap; a lid or cover.

original description   the formal description of a nominal **taxon** when it is established.

osmoeffector   a substance that brings about osmosis, a **diffusion** of water across a semipermeable membrane or diffusion that takes place between two miscible liquids through a permeable membrane.

ostiole   external opening of the heteropteran metathoracic **scent gland**.

ovariole   one of several ovarian tubes that form the **ovary**.

ovary   paired structures in the female insect, each consisting of a number of **ovarioles**; in plants, an enlarged basal portion of a **carpel**, which becomes the fruit.

ovicide   a material used to kill eggs of a **pest** insect.

oviposition   the act of egg laying.

ovipositor   the organ used for laying eggs.

ovule   the structure in seed plants that after fertilization becomes the seed.

oxidase   an **enzyme** that catalyzes an oxidation; see also **polyphenol oxidase**.

paedogenesis   reproduction by immature insects.

Palearctic   referring to the region that includes Europe, Africa north of the Sahara, and Asia north of the Himalayas (nontropical Asia).

palisade parenchyma   leaf tissue composed of **chloroplast**-bearing **parenchyma** cells; upper layer of a leaf's photosynthetic cells.

panicle   a loose, irregularly compound **inflorescence** with pedicellate (**pedicel**-borne) flowers.

parallelism   independent acquisition of similar **characters** (attributes used to recognize or differentiate **taxa**) in related evolutionary lines; see also **convergence**.

parameres   paired male genital structures independent of the **phallus**; claspers.

paraphyletic   describing a taxonomic group that does not include all the descendants of a common ancestor; see also **monophyletic.**

paraphyly   the condition of being **paraphyletic.**

parasitoid   an insect that lives in its immature stages in or on another insect, which it kills after completing its own feeding.

parempodia   paired setiform (having the shape of a **seta**) or **lamellate** processes arising distally from the **unguitractor plate**, between claw bases.

parenchyma   soft plant **tissue** of cells with unthickened walls.

parsimony   in **cladistics**, a criterion used to select **phylogenetic** hypotheses most in accord with available observation; simplicity of explanation.

parthenogenesis   egg development without fertilization.

patchily   pertaining to an irregular or nonuniform distribution.

pathogen   any entity that can incite or cause **disease**.

pathogenic   capable of causing **disease**.

pathovar   a category in bacterial classification below the level of species that is characterized by **pathogenic** reaction in one or more **hosts**, generally only a certain genus or species.

paurometabolous   having an incomplete **metamorphosis** in which changes in form are gradual or inconspicuous; see also **hemimetabolous, holometabolous.**

pectin   a methylated polymer of galacturonic acid found in the **middle lamella** and the primary cell wall of plants.

pectinase   an **enzyme** capable of degrading pectic substances, such as the hydrolase polygalacturonase.

pedicel   the stem of an individual flower; see also **peduncle**. In insects the duct connecting an **ovariole** with the oviduct (also spelled "pedicle").

peduncle   the stem of an **inflorescence**; see also **pedicel**.

pericarp   the wall of the matured plant **ovary**.

pericycle   plant **tissue**, generally of root, bounded externally by the **epidermis** and internally by the **phloem**.

peristalsis   wave motion of the intestines that moves the gut contents toward the anal extremity; waves of contraction.

perithecium   the globular or flask-shaped ascocarp (ascomycete fruiting body) of the Pyrenomycetes (fungi), having an opening or pore.

peritreme   the sclerotic plate about any body opening, such as that of the metathoracic **scent gland** in heteropterans.

peroxidase   an iron-porphyrin–containing **enzyme** that catalyzes a reaction in which hydrogen peroxide is an electron acceptor.

pest   any living organism that is undesirable and of economic or aesthetic concern.

petiole   the stalk of a leaf.

phagostimulant   a natural plant substance that induces feeding by an insect.

phallus   in insects, the unpaired median intromittent organ; see also **aedeagus**.

phenol   an aromatic compound that bears one or more hydroxyl groups.

phenolase   a copper-containing **enzyme** that promotes the oxidation of **phenols**.

phenology   periodicity of biological phenomena; the study of the effect of climate on the seasonal occurrence of plants and animals.

phenotypic   referring to the outward appearance of an organism, or to the totality of characteristics of an individual as a result of interaction between **genotype** and environment.

pheromone   a chemical used in communication between individuals of the same species, releasing a specific behavior or development in the receiver; see also **allelochemical, allomone, kairomone**.

phloem   food-conducting **tissue** of plants, composed of **sieve elements**, various kinds of **parenchyma** cells, fibers, and sclereids.

phoretic   pertaining to phoresy, a relationship in which one organism is carried on the body of a larger organism but does not feed on the latter.

phosphatase   an **enzyme** that catalyzes the **hydrolysis** of monophosphate esters.

photoperiod   the period of light in the daily cycle, measured in hours; day length.

photophase   **photoperiod**, day length; see also **scotophase**.

photophobic   shunning or showing an intolerance of light.

phyllode   a flat, expanded **petiole** replacing the blade of a leaf in photosynthetic function.

phylogenetic   relating to phylogeny or evolutionary history of a group of organisms.

phytophagous (adj.)   feeding on plants.

phytophagy (n.)   the eating of plants.

phytostimulation   increase in plant growth as a result of insect feeding, e.g., plants grown from seed on which lygus bugs feed.

phytotoxemia   a diseaselike plant condition produced by the injection of **toxic** substances by insects.

phytotoxic   **toxic** or poisonous to plants.

phytozoophage   a mainly plant-feeding insect that also feeds to some extent on animal matter; see also **zoophytophage**.

pilose   hairy; covered with fine, long **setae** or hairs.

pinhead square   in cotton, a **square** about 3 mm long or less.

pink stage   stage in apple bud development in which all blossom buds in a cluster are pink and stems are fully extended; see also **tight cluster stage**.

pistil   the seed-bearing organ of the flower, consisting of the **ovary**, **stigma**, and **style**.

pistillate pertaining to a flower with one or more **carpels** but lacking functional **stamens**; referring to the **pistil**; see also **staminate**.

pith ground **tissue** of a plant, usually **parenchyma**, occupying the center of a stem or root within the **vascular** cylinder.

plant bug member of the family Miridae; a mirid.

plasticity capacity of an organism to adapt to various or changing environmental conditions, as a plasticity in feeding habits.

plastron a bed of very dense and very fine hairs used to hold an air bubble close to the body and across which gas exchange takes place.

plectrum ordinarily movable portion of a stridulatory mechanism; see also **stridulation, stridulitrum**.

plesiomorphic relatively primitive in comparing two or more **homologous character states**.

podomere a podite or limb segment of an arthropod that has independent musculature.

poikilothermy the state of being cold-blooded; body temperature rising or falling with ambient temperatures.

Poisson distribution a discrete probability distribution useful to approximate a binomial distribution and the probability of an event that occurs randomly over time.

pollenivorous pollen feeding.

pollen sac a cavity in the **anther** that contains the pollen grains.

pollen tube a tube, formed after germination of the pollen grain, that carries male gametes into the **ovule**.

pollination the transfer of pollen from where it was formed to a receptive surface.

pollinator any insect or other agent that is able to effect **pollination**.

polyculture a mixed stand of crop plants; see also **monoculture**.

polymorphism the presence of two or more distinct, structurally different types of individuals within the same stage of a species.

polyphagy the condition of feeding on a broad array of plant or animal species, often with decided preferences; see also **monophagy, oligophagy**.

polyphenol oxidase a copper-containing **enzyme** such as laccase that catalyzes the oxidation of diphenols and polyphenols to **quinones**.

polyploidy the condition of having more than two entire chromosome complements.

population a group of individuals of the same species within given space and time constraints.

porrect extending forward horizontally.

precibarial sensillum anterior and posterior groups of **sensilla** located in the epipharynx (epipharyngeal organ), appearing to have a sensory function related to feeding.

precipitin test a technique used in immunological studies.

predacious living by feeding (preying) on other living organisms; often spelled *predaceous*.

predator an organism that obtains energy by consuming, usually killing, another, the **prey**.

predatory see **predacious**.

pretarsal pertaining to the pretarsus or last segment of an insect leg; in mirids, referring to structures such as the **pulvillus, pseudopulvilli, parempodia**, claws, and **unguitractor plate**.

previtellogenic a stage in mirid egg development in which **oocytes** at the base of **ovarioles** are small and white and contain no yolk; see also **vitellogenic, chorionated**.

prey an organism consumed or killed by a **predator**.

primordium a cell or organ in its earliest stage of **differentiation**.

proboscis the heteropteran **rostrum**, i.e., combined **labium** and mandibular and maxillary **stylets**.

procrypsis an organism's behavior or coloration that affords protection against its enemies.

prokaryote a unicellular microorganism lacking an organized nucleus and organelles.

propagative transmission pertaining to a circulative plant virus that multiplies in its insect vector; see also **circulative transmission**.

propagule any part of a plant that, when separated, will give rise to a new individual; any reproductive structure.

propleural (adj.) referring to the propleuron, or pleuron (lateral region) of the **prothorax**.

protease an **enzyme** that catalyzes the **hydrolysis** of **proteins** to **amino acids**.

protein a chain of **amino acids** that makes up cell structure and controls cell function.

proteinase a proteolytic **enzyme** that partially hydrolyzes **protein** to form small peptides.

protelean parasite an **entomophagous** arthropod that attacks its **prey** only when the attacking species is immature, the adult being free living.

prothorax first thoracic segment of an insect, bearing the anterior legs but no wings.

protocormic of or pertaining to the trunk of an insect embryo.

protoplasm the living substance of all cells.

protoplast the entire contents, both protoplasmic and nonprotoplasmic, of a cell exclusive of the cell wall.

provenance the place of origin of seeds or other **propagules**.

proximal near the point of attachment or origin; see also **distal**.

pseudoarolium in mirids, the **pulvillus**; see also **pretarsal**.

pseudoperculum an insect egg cap without a distinct sealing bar and in which **eclosion** is not the result of fluid pressure.

pseudopulvilli in mirids, paired **pretarsal** structures arising laterally from the **unguitractor plate**, distinct from **parempodia** and often superficially resembling **pulvilli**.

pteridophyte a fern or fern ally.

pubescence the condition of being covered or clothed with short, soft, fine hairs or **setae**.

pulvillus in mirids, the bladderlike **pretarsal** structures arising from ventral or mesal surfaces of the claws.

puparium in certain Diptera, the hardened skin of the final-**instar** larva in which the pupa forms.

pustule a small blisterlike elevation of **epidermis** created as spores form underneath and push outward; a blisterlike elevation or swelling.

pygophore in heteropterans, abdominal segment 9 in males, enclosing the **phallus** or intromittent organ; the genital capsule.

pyrethroid an organic synthetic insecticide with a structure based on pyrethrum, a botanical insecticide derived from *Chrysanthemum* flowers.

quiescence a slowing down of metabolism and development in response to adverse environmental conditions; see also **diapause, dormancy**.

quinone a benzene derivative in which two hydrogen atoms are replaced by two oxygen atoms.

raceme an **inflorescence** in which the main axis is elongated but the flowers are borne on **pedicels** that are about equal in length.

rachis main axis of an **inflorescence** or axis of a compound leaf.

radiolabel (v.) to label an insect radioactively for ecological studies, e.g., **dispersion**.

receptacle the part of the axis of a flower stalk that bears the floral organs.

rectal organ in mirids, the protruded or exserted rectum or hind portion of the hindgut that, upon disturbance, facilitates reestablishment of contact with the **host** plant.

recurved curved upward or backward.

reproductive isolation a condition in which interbreeding between two or more **populations** is prevented by intrinsic factors.

repugnatorial repellent; so offensive as to drive away.

resilin a rubberlike, proteinaceous constituent of the insect procuticle.

resource something that an organism may consume, such as food.

respiratory horn protuberances with **plastrons** on the surface of eggs of some heteropterans.

resurgence a situation in which a **pest population**, after having been suppressed, rebounds to numbers higher than presuppression levels.

retrorse turned or bent backward.

ritualistic referring to behavior resulting from ritualization, the evolutionary process by which a behavior changes to become a display or signal used in communication.

rosette a circular cluster of leaves or other organs.

rostrum in heteropterans, the combined **labium** and mandibular and maxillary **stylets**; see also **proboscis**.

r-strategist a species characterized by having rapid development, high motility (power of movement),

and a high reproductive rate compared to a ***K*-strategist**.

ruderal (n.) a plant inhabiting fields, waste places, or other sites of human disturbance.

ruderal (adj.) growing in waste places or among rubbish.

russetting referring to brown areas on the skin of fruit as a result of cork formation.

sac cell see **teratocyte**.

salivary canal in heteropterans, the posterior of the two canals formed by the maxillary **stylets**, through which salivary secretions are ejected by the salivary pump.

salivary sheath lipoprotein sheath left in plant **tissue**, formed from hardened salivary secretions, encasing the **stylets** as they penetrate plant tissue.

saltatorial adapted for jumping, usually describing the hind legs.

samara an indehiscent winged fruit.

saprophytic pertaining to the use of dead organic material as food; see also **scavenging**.

scavenging pertaining to the feeding on dead plants or animals and on animal wastes; see also **saprophytic**.

scent gland exocrine gland (i.e., secretions are discharged outside the body) on the abdomen of heteropteran **nymphs**, and referring to several types of glands (especially metathoracic) in adults, producing various **allelochemicals**.

scion a piece of twig or shoot inserted on another in grafting.

sclerenchyma a supporting plant **tissue** composed of sclerenchyma cells (i.e., with thick, often lignified, secondary walls), including fibers and sclereids.

sclerite a plate on the insect body wall surrounded by membrane or **sutures**.

scotophase the dark period or night time of a **diel** cycle; see also **photophase**.

scutellar (adj.) pertaining to the **scutellum**.

scutellum in heteropterans, the triangular part of the mesothorax, generally between the bases of the **hemelytra**.

secondary metabolite a substance produced by a plant that plays no role in the basic metabolism but may help defend against herbivory; a secondary plant compound.

sedentary referring to insects with limited movement.

self-pollination the transfer of pollen from an **anther** to the **stigma** of flowers on the same plant.

seminal depository in mirids and certain other heteropterans, paired or unpaired saclike **spermathecae**.

semiochemical any chemical used in intraspecific and interspecific communication.

semipersistent virus pertaining to foregut-borne plant viruses in which retention in their insect **vectors** is intermediate between nonpersistent and persistent, i.e., referring to the length of time (half-life in hours) that viruses, once acquired, remain inoculative in the vector.

sensillum (pl. sensilla) a sense organ, either simple and isolated, or part of a more complex organ; see also **sensillum basiconicum, sensillum trichodeum**.

sensillum basiconicum basiconic peg, i.e., a thin-walled, peg-shaped **sensillum** with minute pores, functioning in **chemoreception**; see also **sensillum trichodeum**.

sensillum trichodeum a hairlike (trichoid) mechanoreceptor or **sensillum**; see also **sensillum basiconicum**.

sensu stricto (Latin) in the strict sense.

sepal a unit of the **calyx**; one of the outermost flower structures that usually enclose the other flower parts in the bud.

septicemia in invertebrate pathology, a morbid condition caused by the multiplication of microorganisms in the blood.

sequential sampling a sampling program in which the number of samples is not fixed in advance and which is based on insect **dispersion** patterns and economic decision levels, allowing a **population** to be placed in one of two or more categories, e.g., economic or noneconomic.

sequester to store **secondary metabolites** or plant compounds for defense.

sericeous silky, as leaves covered with long, soft, mostly appressed hairs.

serological pertaining to serology, the study of the nature and interactions of **antigens** and **antibodies**.

serosal cuticle in an insect egg, a cuticle produced by the serosa (membrane covering the embryo) inside the **chorion**.

serosal plug see **yolk plug**.

serpentine winding like a serpent; often used to describe the curved or coiled mine of a **leafminer**.

serrate sawlike; with notched edges like the teeth of a saw.

seta (pl. setae) a cuticular extension or sclerotized hairlike projection; a hair or bristle.

setiform bristle- or **seta**-shaped.

shale barrens discontinuous shale outcrops and talus (sloping mass of debris at base of cliff) of steep southern exposure, characterized by high insolation temperatures and low moisture availability at the surface, and often undercut by a stream.

sheath the base of a leaf that wraps around the stem, as in grasses; a **tissue** layer surrounding another tissue, as a bundle sheath.

shot holing pertaining to a condition in which small leaf fragments drop out, as a result of **disease** (including feeding by certain insects), leaving small holes; see also **tattering**.

sibling species cryptic species, i.e., closely related and difficult to distinguish.

sieve element the cell of the **phloem** concerned with long-distance transport of food substances in the plant.

sign any manifestation of insect feeding (e.g., cast skins or **honeydew**) or a pathogen observed on a **host** plant; see also **symptom**.

sister group the most closely related **monophyletic** group of another monophyletic group.

spadix a spike (simple **inflorescence**) with a fleshy axis, as in aroids.

spathe a large **bract** enclosing an **inflorescence**.

species richness the number of species in a **community**.

speciose species rich; referring to a **taxon** rich in numbers of species.

spermatheca (pl. spermathecae) median, dorsal, unpaired sclerotized diverticulum (offshoot) of the vagina serving as a sperm-storage receptacle in female Heteroptera.

spongy parenchyma irregularly shaped plant cells on lower side of a leaf that make up the **parenchyma**; see also **palisade parenchyma**.

sporangium a hollow unicellular or multicellular structure in which spores are produced.

sporopollenin a complex polymer of carotenoids and carotenoid esters that comprises the outer layer, or **exine**, of a pollen grain.

square flower bud of the cotton plant.

stalked with a stalk or stem, such as a chrysopid egg.

stamen the part of the flower that produces pollen, usually consisting of **anther** and filament (the stalk of an insect **stamen**).

staminate pertaining to a flower having **stamens** but not functional **carpels**; see also **pistillate**.

stenophagous restricted to eating a few foods.

steppe treeless plains of southeastern Europe and Siberia.

sternite a subdivision of a sternum (ventral division of an insect segment).

stigma region of a **carpel** serving as a receptive surface for pollen grains and on which they germinate.

stippling a localized **phytotoxemia** consisting of numerous points or dots, as in **injury** to foliage on which certain sucking arthropods feed; see also **chlorosis**.

stomate (stoma; pl. stomata) a minute pore and two surrounding guard cells in the **epidermis** of leaves.

stridulation the production of sound by rubbing two rough or ridged surfaces together; see also **plectrum, stridulitrum**.

stridulitrum ordinarily stationary portion of a stridulatory mechanism; see also **plectrum, stridulation**.

strip cropping the growing of crops in narrow strips to reduce wind and water erosion.

strobilus (pl. strobili) a cone; a number of modified leaves (sporophylls) or **ovule**-bearing scales grouped terminally on a stem.

stunting an abnormal reduction in plant size.

style in plants, the slender column of **tissue** arising from the top of the **ovary**, through which the **pollen tube** grows.

stylets needlelike mouthparts of heteropterans, consisting of the paired mandibular and maxillary **stylets** enclosed by the **labium**.

stylet sheath in certain plant-feeding hemipterans, a more or less permanent duct, surrounding the mouthparts between the plant surface and the **phloem**, composed of tanned lipoprotein derived from the saliva; see also **salivary sheath**.

stylet track path taken by heteropteran **stylets** as revealed by histological study, often characterized by collapsed or disorganized cells, brown streaks, or other types of cellular reaction to stylet penetration.

subsocial applied to the condition, or to the group showing it, in which adults care for immature individuals for some period of time.

substrate chemical substance acted on, often by an **enzyme**; an underlying layer or any object or material on which an organism grows or to which it is attached.

suture groove marking the line of fusion of two formerly distinct plates; in heteropterans, the line of juncture of the **hemelytra**.

switching the tendency of a **predator** to switch between **prey** categories according to their relative abundance in the environment.

symbiosis a long-lasting, close and dependent relationship between organisms of two different species; see also **commensal(ism)**, **mutualism**.

sympatric speciation speciation (multiplication of species) without geographic isolation.

sympatry occurrence of two or more populations in the same area; the existence of a breeding population within the cruising range of individuals of another population.

symptom any external or internal expression of **disease**, including insect feeding, in a plant.

symptomatology the complex of **disease symptoms**; use of symptoms to help diagnose plant disease.

synapomorphic the sharing of relatively **derived** or specialized **character states** of two or more **homologous** characters; see also **apomorphic**.

syncytium **tissue** containing many nuclei, which is not divided into separate compartments by cell membranes.

syndrome the complex of **signs** and **symptoms** that indicate or characterize a plant **disease**.

synergist a chemical substance whose use with another agent will result in greater total effect than the sum of their individual effects.

systemic insecticide an insect-killing agent capable of being absorbed into plant sap and translocated (moved) within the plant through the **vascular** system to act against a **pest**.

tattering the condition of leaves that have become torn or ragged; see also **shot holing**.

taxon (pl. taxa) any taxonomic unit, such as species, genus, tribe, or subfamily.

teleotrophic ovariole an **ovariole** in which the nurse cells occur only within the **germarium**.

teneral the condition of a newly eclosed adult insect, which is unsclerotized and unpigmented.

teratocyte a cell that has originated from an unenclosed **parasitoid** and that is liberated into the **host**'s body cavity when the parasitoid hatches, often being a remnant of the **trophamnion**.

teratological pertaining to teratology, the study of structural abnormalities.

tergite a dorsal **sclerite** or part of a segment, especially when it consists of a single sclerite.

tergum (pl. terga) the upper (dorsal) surface of any body segment of an insect, whether it consists of one or more than one **sclerite**.

terminalia the terminal abdominal segments modified to form the genital segments.

territoriality broadly, any space-associated intolerance of others; more narrowly, an intolerance based on real-estate holdings.

testis follicle tubular structure in a testis within which spermatogenesis (sperm formation) occurs.

tetrad a group of four spores formed from a spore mother cell by **meiosis**, as in pollen formation.

thanatosis feigning death, i.e., remaining motionless for a period of time when disturbed.

tight cluster stage stage in apple bud development in which blossom buds are exposed but tightly appressed and stems are short; see also **pink stage**.

timid predator one that does not subdue its **prey** but feeds only on sluggish or stationary prey.

tissue a group of cells of similar structure and function.

tomentose densely pubescent, covered with tomentum or short, matted, woolly hair.

toxic destructive, harmful; pertaining to a toxin or poisonous substance.

toxicity the capacity of a compound to produce **injury**; the inherent poisonous potency of a material.

toxicogenic producing **toxic** substances.

trade-off (n.) in life-history theory, the process of giving up part of one trait for part of one that is more advantageous, e.g., the **phenotypic** trade-off between **fecundity** and **longevity**; benefits from one process that are bought at the expense of another.

transformation mimicry **mimicry** in which different **instars** imitate different **models**.

transgenic pertaining to the genetic manipulation and improvement involving the insertion of genetic material of one species into the genome of another, e.g., to reduce the reproductive capacity of **pests** or to enhance a natural enemy used in **biological control**.

transmission the spread of an infective agent from one **host** to another.

transovarial transmission the conveying of microorganisms from one generation to the next by way of the egg.

transpiration the loss of water vapor by plant parts, most often through **stomata**.

trap crop a small area of crop used to divert **pests** from a larger area of the same or another crop.

traumatic insemination copulation in which the body

wall of the female is pierced and sperm enters the body cavity; see also **extragenital insemination**.

trehalase an **enzyme** that catalyzes the **hydrolysis** of trehalose to two molecules of glucose.

trichobothrium (pl. trichobothria) specialized, slender, hairlike sensory **setae** arising from and including tubercles or pits (bothria) on many body regions and appendages of heteropterans; more strictly, the receptacle in which such setae are inserted.

trichome any hairlike outgrowth of the plant **epidermis**; trichomelike hairs or **setae** on the antennae of certain mirids.

trivial flight nonmigratory flight within or near the breeding **habitat**; see also **flitting**.

trophamnion in **parasitoids**, the enveloping membrane surrounding the polyembryonically derived multiple individuals that arise from a single egg, derived from the host's **hemolymph**.

trophic of or pertaining to food or eating.

truncate cut off squarely at the tip.

turgor pressure pressure within the cell resulting from the movement of water into the cell.

umbel an **inflorescence**, the individual **pedicels** of which all arise from the apex of the **peduncle**.

unguitractor plate in mirids, the **sclerite** lying between the bases of claws, with which bases of claws articulate distally, to which the retractor tendon is attached proximally, and from which **parempodia** arise distally.

univoltine having only 1 generation in a year; see also **multivoltine**.

vacuole a sac or cavity within the cytoplasm filled with a watery fluid, the cell sap.

vagility capacity of or inherent power of individuals to disperse.

valid name the available name correctly applied to a **taxon**.

valvulae small valves or valvelike processes; in female heteropterans, the **ovipositor** blades that in two pairs form the egg-laying apparatus and that proximally attach to the body wall via one or two pairs of corresponding rami or branches.

vascular pertaining to fluid-conducting vessels or ducts, which in plants is the **xylem** and **phloem**.

vector (n.) an animal able to transmit a **pathogen**.

vegetative growth the growth of roots, stems, and leaves, as distinguished from the development of flowers and fruit.

venation arrangement of veins in a leaf blade.

venom a **toxic** fluid injected into **prey** or enemies that causes death, paralysis, or pain.

venter lower surface or undersurface of the body.

vertigo sensation of dizziness or disorientation.

vesica in some heteropterans, the **apical**, sclerotized part of **aedeagus** beyond the conjunctiva (**membranous,** usually eversible **distal** portion of aedeagus).

vesicle a little sac, bladder, or air cavity.

vesicular pertaining to or consisting of **vesicles**.

vicariance the division of the range of a species by events in the Earth's history, such as ocean or mountain formation.

viscid sticky, glutinous; see also **mucilaginous**.

viscous thick, resistant to flow.

vitellogenesis yolk formation in the developing egg.

vitellogenic a stage in mirid egg development in which **oocytes** at the base of **ovarioles** are yellow and contain yolk but have not developed **chorions**; see also **previtellogenic**, **chorionated**.

volatile (adj.) readily vaporized; sometimes used as a noun in referring to a plant chemical that is readily volatilized.

voltinism the state of having a specified number of generations per year; see also **multivoltine**, **univoltine**.

water-soaked a **disease symptom** in which **host tissue** appears wet and dark or somewhat translucent (allowing light to pass through).

waveform during electronic monitoring of aphids, leafhoppers, and other hemipterans, the electrical signal that passes from the insect-plant interface, leaving a record (tracing) that reflects the duration and types of feeding behaviors.

whip (n.) a budding and grafted tree at the end of its first season of growth, consisting of a single, unbranched shoot.

witches'-broom broomlike growth or massed proliferation caused by the dense clustering of branches of woody plants.

X-chromosome the chromosome that at least partly determines the sex of an individual.

xeric pertaining to a dry **habitat**.

xylem a complex **vascular tissue** through which most of the water and minerals of a plant are conducted.

yolk plug pertaining to the serosal cap in heteropteran embryogenesis; a serosal plug, which in mirids, persists until **eclosion**.

zoophagous (adj.) feeding on animals.

zoophagy (n.) the eating of animals.

zoophytophage a mainly predacious insect that also feeds to some extent on plants; see also **phytozoophage**.

References

Note: Multiauthor works—that is, references with the same first author and two or more coauthors—appear chronologically rather than alphabetically and follow two-author works by the same first author.

Aamodt, O. S. and J. Carlson. 1938. Grimm alfalfa flowers in spite of lygus bug injury. Univ. Wis. Agric. Exp. Stn. Bull. 440:67.

Abeles, F. B., P. W. Morgan, and M. E. Saltveit Jr. 1992. Ethylene in plant biology, 2nd ed. Academic, San Diego. 414 pp.

Ables, J. R., J. L. Goodenough, A. W. Harstack, and R. L. Ridgway. 1983. Entomophagous arthropods. Pages 103–127 in R. L. Ridgway, E. P. Lloyd, and W. H. Cross, eds., Cotton insect management with special reference to the boll weevil. U.S. Dep. Agric. Handb. 589.

Abou-Donia, M. B. 1976. Physiological effects and metabolism of gossypol. Residue Rev. 61:125–160.

Abraham, C. C. 1991. Occurrence of *Helopeltis theivora* Waterhouse (Miridae: Hemiptera) as a pest of Indian long pepper *Piper longum* Linn. Entomon 16:245–246.

Abraham, E. V. 1958. Pests of cashew *(Anacardium occidentale)* in South India. Indian J. Agric. Sci. 28:531–543.

Abraham, R. 1935. Wanzen (Heteroptera) an Obstbäumen. (III. Mitteilung). Die anatomische Untersuchung geschädigter Früchte. Z. Pflanzenkr. 45:463–474.

——. 1936. Wanzen (Heteroptera) an Obstbäumen. (IV. Mitteilungen). *Orthotylus marginalis* Reut. (Hemiptera-Heteroptera) an der Niederelbe. Z. Pflanzenkr. 46:225–240.

——. 1937a. *Halticus saltator* Geoffr. als Schädling der Ringelblume (*Calendula officinalis* L.). Arb. Physiol. Angew. Entomol. Berl. 4:244–246.

——. 1937b. Beobachtungen über die Eiablage einiger Capsiden. Arb. Physiol. Angew. Entomol. Berl. 4:321–324.

Abreu, J. M. 1977. Mirideos neotropicais associados ao cacaueiro. Pages 85–106 in E. M. Lavabre, ed., Les mirides du cacaoyer. Institut français du Cafe et du Cacao, Paris.

Abreu, J. M. [et al.]. 1989. Manejo de pragas do cacaueiro. CEPLAC/CEPAC, Ilhéus, Brazil. 30 pp.

Ackerman, A. J. and D. Isely. 1931. The leaf hoppers attacking apples in the Ozarks. U.S. Dep. Agric. Tech. Bull. 263:1–40.

Ackerman, J. K. and R. D. Shenefelt. 1973. Organisms, especially insects, associated with wood rotting higher fungi (Basidiomycetes) in Wisconsin forests. Wis. Acad. Sci. Arts Lett. 61:185–206.

Adair, E. W. 1918. Preliminary list of insects associated with cotton in Egypt. J. Agric. Egypt 8:80–88.

Adams, C. C. 1915. An ecological study of prairie and forest invertebrates. Bull. Ill. State Lab. Nat. Hist. 11(2):33–280.

Adams, J. B. and J. W. McAllan. 1958. Pectinase in certain insects. Can. J. Zool. 36:305–308.

Adams, J. R. and J. R. Bonami, eds. 1991. Atlas of invertebrate viruses. CRC, Boca Raton, Fla. 684 pp.

Addicott, F. T. and V. E. Romney. 1950. Anatomical effects of lygus injury to guayule. Bot. Gaz. 112:133–134.

Adjei-Maafo, I. K. and L. T. Wilson. 1983. Factors affecting the relative abundance of arthropods on nectaried and nectariless cotton. Environ. Entomol. 12:349–352.

Adkisson, P. L. 1957. Influence of irrigation and fertilizer on populations of three species of mirids attacking cotton. FAO Plant Prot. Bull. 6:33–36.

——. 1971. Objective use of insecticides in agriculture. Pages 43–51 in J. E. Swift, ed., Agricultural chemicals—harmony or discord for food, people, environment. Univ. Calif. Div. Agric. Sci., Berkeley.

——. 1973a. The integrated control of the insect pests of cotton. Proc. Tall Timbers Conf. Ecol. Anim. Control Habitat Manage. 4:175–188.

——. 1973b. The principles, strategies and tactics of pest control in cotton. Pages 274–283 in P. W. Geier, L. R. Clark, D. J. Anderson, and H. A. Nix, eds., Insects: Studies in population management. Ecological Society of Australia (Mem. 1), Canberra.

Adkisson, P. L., G. A. Niles, J. K. Walker, L. S. Bird, and H. B. Scott. 1982. Controlling cotton's insect pests: A new system. Science (Wash., D.C.) 216:19–22.

Adler, P. H. 1982. Soil- and puddle-visiting habits of moths. J. Lepid. Soc. 36:161–173.

Adler, P. H. and A. G. Wheeler Jr. 1984. Extra-phytophagous food sources of Hemiptera-Heteroptera: Bird droppings, dung, and carrion. J. Kans. Entomol. Soc. 57:21–25.

Afscharpour, F. 1960. Ökologische Untersuchungen über Wanzen und Zikaden auf Kulturfeldern in Schleswig-Holstein (ein Beitrag zur Agrarökologie). Z. Angew. Zool. 47:257–301.

Agarwal, R. A. and G. P. Gupta. 1983. Insect pests of fibre crops. Pages 147–164 in P. D. Srivastva et al., eds., Agricultural entomology. Vol. II. All India Scientific Writers' Society, New Delhi.

Agblor, A. 1992. Characterization of alpha-amylase and polygalacturonase from *Lygus* spp. (Heteroptera: Miridae). M.S. thesis, University of Manitoba, Winnipeg. 123 pp.

Agblor, A., H. M. Henderson, and F. J. Madrid. 1994. Characterisation of alpha-amylase and polygalacturonase from *Lygus* spp. (Heteroptera: Miridae). Food Res. Int. 27:321–326.

Agnello, A. M., W. H. Reissig, J. P. Nyrop, J. Kovach, and R. A. Morse. 1993. Biology and management of apple arthropods. Cornell Coop. Ext. Inf. Bull. 231:1–32.

Agnew, C. W., W. L. Sterling, and D. A. Dean. 1982. Influence of cotton nectar on red imported fire ants and other predators. Environ. Entomol. 11:629–634.

Agrios, G. N. 1980. Insect involvement in the transmission of fungal pathogens. Pages 293–324 in K. F. Harris and K. Maramorosch, eds., Vectors of plant pathogens. Academic, New York.

——. 1988. Plant pathology, 3rd ed. Academic, San Diego. 803 pp.

Agustí, N. and A. C. Cohen. 2000. *Lygus hesperus* and *L. lineolaris* (Hemiptera: Miridae), phytophages, zoophages, or omnivores: Evidence of feeding adaptations suggested by the salivary and midgut digestive enzymes. J. Entomol. Sci. 35:176–186.

Agustí, N., J. Aramburu, and R. Gabarra. 1999a. Immunological detection of *Helicoverpa armigera* (Lepidoptera: Noctuidae) ingested by heteropteran predators: Time-related decay and effect of meal size on detection period. Ann. Entomol. Soc. Am. 92:56–62.

Agustí, N., M. C. De Vicente, and R. Gabarra. 1999b. Development of sequence amplified characterized region (SCAR) markers of *Helicoverpa armigera*: A new polymerase chain reaction–based technique for predator gut analysis. Mol. Ecol. 8:1467–1474.

Ahmad, I. and C. W. Schaefer. 1987. Food plants and feeding biology of the Pyrrhocoroidea (Hemiptera). Phytophaga 1:75–92.

Ahmad, R. and M. A. Ghani. 1972. Coccoidea and their natural enemy complexes in Pakistan. Commonw. Inst. Biol. Control Tech. Bull. 15:59–104.

Akingbohungbe, A. E. 1969. Some effects of mirid feeding on developing cocoa pods. Niger. Entomol. Mag. 2:4–8.

——. 1974a. Chromosome numbers of some North American mirids (Heteroptera: Miridae). Can. J. Genet. Cytol. 16:251–256.

——. 1974b. Nymphal characters and higher classification analysis in the Miridae (Hemiptera: Heteroptera) with a subfamily key based on the nymphs. Can. Entomol. 106:687–694.

——. 1979. A new genus and four new species of Hyaliodinae (Heteroptera: Miridae) from Africa with comments on the status of the subfamily. Rev. Zool. Afr. 93:500–522.

——. 1983. Variation in testis follicle number in the Miridae (Hemiptera: Heteroptera) and its relationship to the higher classification of the family. Ann. Entomol. Soc. Am. 76:37–43.

Akingbohungbe, A. E., J. L. Libby, and R. D. Shenefelt. 1972. Miridae of Wisconsin (Hemiptera: Heteroptera). Univ. Wis.-Madison Coll. Agric. Res. Div. R2396:1–24.

——. 1973. Nymphs of Wisconsin Miridae. Hemiptera: Heteroptera. Univ. Wis.-Madison Coll. Agric. Res. Div. R2561:1–25.

Alayo, D. P. 1974. Los Hemipteros de Cuba. Parte XIII. Familia Miridae. Torreia (Havana) 32:1–38.

Albajes, R., R. Gabarra, C. Castañé, E. Bordas, O. Alomar, and A. Carnero. 1988. Pest problems in field tomato crops in Spain. Pages 197–207 in R. Cavalloro and C. Pelerents, eds., Progress on pest management in field vegetables. Balkema, Rotterdam.

Albajes, R., O. Alomar, J. Riudavets, C. Castañé, J. Arno, and R. Gabarra. 1996. The mirid bug *Dicyphus tamaninii*: An effective predator for vegetable crops. Int. Organ. Biol. Control/West. Palearctic Reg. Sect. Bull. 19(1):1–4.

Alcock, J. 1994. Postinsemination associations between males and females in insects: The mate-guarding hypothesis. Annu. Rev. Entomol. 39:1–21.

Aldrich, J. R. 1988. Chemical ecology of the Heteroptera. Annu. Rev. Entomol. 33:211–238.

——. 1995. Chemical communication in the true bugs and parasitoid exploitation. Pages 318–363 in R. T. Cardé and W. J. Bell, eds., Chemical ecology of insects II. Chapman & Hall, New York.

——. 1996. Sex pheromones in Homoptera and Heteroptera. Pages 199–233 in C. W. Schaefer, ed., Studies on hemipteran phylogeny. Thomas Say Publ. Entomol.: Proceedings. Entomological Society of America, Lanham, Md.

Aldrich, J. R., M. S. Blum, and S. S. Duffey. 1976. Male specific natural products in the bug, *Leptoglossus phyllopus*: Chemistry and possible function. J. Insect Physiol. 22:1201–1206.

Aldrich, J. R., W. R. Lusby, J. P. Kochansky, M. P. Hoffmann, L. T. Wilson, and F. G. Zalom. 1988. Lygus bug pheromones vis-a-vis stink bugs. *In* Proc. Beltwide Cotton Prod. Res. Conf., Jan. 3–8, 1988, New Orleans, La., pp. 213–216. National Cotton Council, Memphis.

Alford, D. V. 1995. A color atlas of pests of ornamental trees, shrubs, and flowers. Halstead, New York. 448 pp.

Alford, D. V. and D. C. Gwynne. 1983. Pests and diseases of fruit and hops. Pages 295–374 in N. Scopes and M. Ledieu, eds., Pest and disease control handbook, 2nd ed. British Crop Protection Council, Croydon, UK.

Al-Ghamdi, K. M., R. K. Stewart, and G. Boivin. 1995. Synchrony between populations of the tarnished plant bug, *Lygus lineolaris* (Palisot de Beauvois) (Hemiptera: Miridae), and its egg parasitoids in southwestern Quebec. Can. Entomol. 127:457–472.

Al-Houty, W. 1990. *Nysius* (Hem., Lygaeidae) sucking human blood in Kuwait. Entomol. Mon. Mag. 126:95–96.

Alias, A. see Awang, A.

Allaby, M., ed. 1991. The concise Oxford dictionary of zoology. Oxford University Press, Oxford. 508 pp.

Allen, D. C., F. B. Knight, and J. L. Foltz. 1970. Invertebrate predators of the jack-pine budworm, *Choristoneura pinus*, in Michigan. Ann. Entomol. Soc. Am. 63:59–64.

Allen, T. C. 1947. Suppression of insect damage by means of plant hormones. J. Econ. Entomol. 40:814–817.

——. 1951. Deformities caused by insects. Pages 411–415 in F. Skoog, ed., Plant growth substances. University of Wisconsin Press, Madison.

Allen, W. W. 1959. Strawberry pests in California: A guide for commercial growers. Calif. Agric. Exp. Stn. Circ. 484:1–39.

Allen, W. W. and S. E. Gaede. 1963. The relationship of lygus bugs and thrips to fruit deformity in strawberries. J. Econ. Entomol. 56:823–825.

Allsopp, P. G. and R. M. Bull. 1990. Sampling distributions and sequential sampling plans for *Perkinsiella saccharicida* Kirkaldy (Hemiptera: Delphacidae) and *Tytthus* spp. (Hemiptera: Miridae) on sugarcane. J. Econ. Entomol. 83:2284–2289.

Alma, A. 1995. Ricerche bio-etologiche ed epidemiologiche su *Holocacista rivillei* Stainton (Lepidoptera Heliozelidae). Redia 78:373–378.

Almand, L. K., W. L. Sterling, and C. L. Green. 1976. Seasonal abundance and dispersal of the cotton fleahopper as related to host plant phenology. Tex. Agric. Exp. Stn. B-1170:1–15.

Almeida, A. A. 1980. Influence of the glandless and glanded cottonseeds on development, fecundity and fertility of *Dysdercus fasciatus* Signoret (Hemiptera, Pyrrhocoridae). Rev. Bras. Biol. 40:475–483.

Al-Munshi, D. M., D. R. Scott, and H. W. Smith. 1982. Some host plant effects on *Lygus hesperus* (Hemiptera: Miridae). J. Econ. Entomol. 75:813–815.

Alomar, O. and R. Albajes. 1996. Greenhouse whitefly (Homoptera: Aleyrodidae) predation and tomato fruit injury by the zoophytophagous predator *Dicyphus tamaninii* (Heteroptera: Miridae). Pages 155–177 in O. Alomar and R. N. Wiedenmann, eds., Zoophytophagous Heteroptera: Implications for life history and integrated pest management. Thomas Say Publ. Entomol.: Proceedings. Entomological Society of America, Lanham, Md.

Alomar, O. and R. N. Wiedenmann, eds. 1996. Zoophytophagous Heteroptera: Implications for life history and integrated pest management. Thomas Say Publ. Entomol.: Proceedings. Entomological Society of America, Lanham, Md. 202 pp.

Alomar, O., C. Castañé, R. Gabarra, E. Bordas, J. Adillón, and R. Albajes. 1988. IPM in tomato crops in Catalonia (Spain). *In* Proc. 18th Int. Congr. Entomol., Vancouver, p. 385.

Alomar, O., C. Castañé, R. Gabarra, and R. Albajes. 1990. Mirid bugs—another strategy for IPM on Mediterranean vegetable crops? Proc. Working Group, "Integrated Control in Glasshouses," Copenhagen, Denmark, 5–8 June 1990. Int. Organ. Biol. Control/West. Palearctic Reg. Sect. Bull. 13(5):6–9.

Al-Rawy, M. A., I. K. Kaddou, and P. Starý. 1969. Predation of *Chrysopa carnea* Steph. on mummified aphids and its possible significance in population regulation (Neuroptera, Hymenoptera, Homoptera). Bull. Biol. Res. Cent. (Baghdad)4:30–40.

Alvarado, P., O. Baltà, and O. Alomar. 1997. Efficiency of four Heteroptera as predators of *Aphis gossypii* and *Macrosiphum euphorbiae* (Hom.: Aphididae). Entomophaga 42:215–226.

Alvarado-Rodriguez [Rodriquez], B., T. F. Leigh, and K. W. Foster. 1986a. Oviposition site preference of *Lygus hesperus* (Hemiptera: Miridae) on common bean in relation to bean age and genotype. J. Econ. Entomol. 79:1069–1072.

Alvarado-Rodriguez [Rodriquez], B., T. F. Leigh, K. W. Foster, and S. S. Duffey. 1986b. Resistance in common bean (*Phaseolus vulgaris*) to *Lygus hesperus* (Heteroptera: Miridae). J. Econ. Entomol. 79:484–489.

Alvarado-Rodriguez, B., T. F. Leigh, K. W. Foster, and S. S. Duffey. 1987. Life tables for *Lygus hesperus* (Heteroptera: Miridae) on susceptible and resistant common bean cultivars. Environ. Entomol. 16:45–49.

Al-Zubaidi, F. S. and J. L. Capinera. 1983. Application of different nitrogen levels to the host plant and cannibalistic behavior of beet armyworm, *Spodoptera exigua* (Hübner) (Lepidoptera: Noctuidae). Environ. Entomol. 12:1687–1689.

Ambika, B. and C. C. Abraham. 1979. Bio-ecology of *Helopeltis antonii* Sign. (Miridae: Hemiptera) infesting cashew trees. Entomon 4:335–342.

——. 1983. New record of *Helopeltis theivora* Waterhouse and an undetermined species of *Helopeltis* (Miridae: Hemiptera) as potential pests of cashews, *Anacardium occidentale* Linn. Indian J. Entomol. 45:183–184.

Ammar, E. D. 1994. Propagative transmission of plant and animal viruses by insects: Factors affecting vector specificity and competence. Adv. Dis. Vector Res. 10:289–331.

Amyot, C. J. B. and J. G. A. Serville. 1843. Histoire naturelle des insectes Hémiptères. Fain et Thunot, Paris. 675 pp.

Anasiewicz, A. and A. Winiarska. 1995. Insects occurring on celery grown for seeds. Folia Hortic. 7:49–57.

Anderson, G. J. 1976. The pollination biology of *Tilia*. Am. J. Bot. 63:1203–1212.

Anderson, R. A. and M. F. Schuster. 1983. Phenology of the tarnished plant bug on natural host plants in relation to populations on cotton. Southwest. Entomol. 8:131–134.

Anderson, R. C., T. Leahy, and S. Dhillion. 1989. Numbers and biomass of selected insect groups on burned and unburned sand prairie. Am. Midl. Nat. 122:151–162.

Anderson, R. F. 1960. Forest and shade tree entomology. Wiley, New York. 428 pp.

Anderson, R. S. 1993. Weevils and plants: Phylogenetic versus ecological mediation of evolution of host plant associations in Curculioninae (Coleoptera: Curculionidae). Mem. Entomol. Soc. Can. 165:197–232.

——. 1995. An evolutionary perspective on diversity in Curculionoidea. Mem. Entomol. Soc. Wash. 14:103–114.

Anderson, T. J. 1931. Annual report of the Senior Entomologist, 1930. Dep. Agric. Kenya Rep. 1930:190–205.

Andison, H. 1956. Common strawberry insects and their control. Can. Dep. Agric. Publ. 990:1–21.

Ando, Y. 1972. Egg diapause and water absorption in the false melon beetle, *Atrachya menetriesi* Faldermann (Coleoptera: Chrysomelidae). Appl. Entomol. Zool. 7:142–154.

Andow, D. 1982. Miridae and Coleoptera associated with tulip tree flowers at Ithaca, New York. J. N.Y. Entomol. Soc. 90:119–124.

André, M. 1928. Une nouvelle forme larvaire de Thrombidion: *Parathrombium teres* n.sp. Bull. Soc. Zool. Fr. 53:514–519.

——. 1929. Note complémentaire sur *Parathrombium teres* M. André. Bull. Soc. Zool. Fr. 54:644–645.

Andres, L. A., V. E. Burton, R. F. Smith, and J. E. Swift. 1955. DDT tolerance by lygus bugs on seed alfalfa. J. Econ. Entomol. 48:509–513.

Andrews, E. A. [1923]. Factors affecting the control of the tea mosquito bug (*Helopeltis theivora*-Waterh.). Worrall & Robey, London. 260 pp.

Anitha, N. and L. Rajamony. 1991. Occurrence of *Prodromus clypealis* Distant (Heteroptera miridae) [*sic*] on banana (*Musa* sp.) in India: A new record. Trop. Pest Manage. 37:439.

Anonymous. 1847. The potato bug. Gard. Chron. (Lond.) July, 17. p. 468.

——. 1940. Bug attacking fruit and vegetables. (*Megacoelum modestum*.). Agric. Gaz. N.S.W. 51:151–153.

——. 1969. Pests of cane. Exp. Stn. S. Afr. Sugar Assoc. Rep. 1968–9:46–48.

——. 1970. Forage legumes. Garden flea-hopper (*Halticus bractatus*). U.S. Dep. Agric. Coop. Econ. Insect Rep. 20(29): 494.

Ansley, R. J. and C. M. McKell. 1982. Crested wheatgrass vigor as affected by black grass bug and cattle grazing. J. Range Manage. 35:586–590.

Appanah, S. 1981. Pollination in Malaysian primary forests. Malay. For. 44:37–40.

——. 1987. Insect pollinators and the diversity of dipterocarps. Pages 277–291 *in* A. J. G. H. Kostermans, ed., Proceedings of the Third Round Table Conference on Dipterocarps. UNESCO, Jakarta, Java.

——. 1990. Plant-pollinator interactions in Malaysian rain forests. Pages 85–101 *in* K. S. Bawa and M. Hadley, eds., Reproductive ecology of tropical forest plants (Man and the biosphere series, Vol. 7). UNESCO, Paris; Parthenon, Casterton Hall, UK.

Appanah, S. and H. T. Chan. 1981. Thrips: The pollinators of some dipterocarps. Malay. For. 44:234–252.

Araya, J. E. and B. A. Haws. 1988. Arthropod predation of black grass bugs (Hemiptera: Miridae) in Utah ranges. J. Range Manage. 41:100–103.

——. 1991. Arthropod populations associated with a grassland infested by black grass bugs, *Labops hesperius* and *Irbisia brachycera* (Hemiptera: Miridae), in Utah, USA. FAO Plant Prot. Bull. 39:75–81.

Arbogast, R. T. 1979. Cannibalism in *Xylocoris flavipes* (Hemiptera: Anthocoridae), a predator of stored-product insects. Entomol. Exp. Appl. 25:128–135.

Arčanin, B. and I. Balarin. 1972. Predatorske vrste Heteroptera zastupljene u fauni jabučnih nasada hrvatske. Acta Entomol. Jugoslav. 8:1–2.

Ark, P. A. 1944. Studies on bacterial

canker of tomato. Phytopathology 34: 394–400.

Arms, K., P. Feeny, and R. C. Lederhouse. 1974. Sodium: Stimulus for puddling behavior by tiger swallowtail butterflies, *Papilio glaucus*. Science (Wash., D.C.) 185:372–374.

Armstrong, J. A. 1979. Biotic pollination mechanisms in the Australian flora—a review. N.Z. J. Bot. 17:467–508.

Armstrong, J. E., W. H. Kearby, and E. A. McGinnes Jr. 1979. Anatomical response and recovery of twigs of *Juglans nigra* following oviposition injury inflicted by the two-spotted treehopper, *Enchenopa biontata* [sic]. Wood Fiber 11:29–37.

Arnason, A. P., K. M. King, R. Glen, and L. C. Paul. 1939. Insects of the season 1938 in Saskatchewan. Can. Insect Pest Rev. 17:63–73.

Arnaud, P. H. Jr. 1978. A host-parasite catalog of North American Tachnidae (Diptera). U.S. Dep. Agric. Misc. Publ. 1319:1–860.

Arnett, R. H. Jr. 1961. [Review of] *A manual of common beetles of eastern North America* by E.S. and L. S. Dillon. Coleopt. Bull. 15:15–16.

Arnold, G. 1913. The tarnished plant-bug on the aster. Florists' Exch. 36(11):576.

Arnoldi, D., R. K. Stewart, and G. Boivin. 1991. Field survey and laboratory evaluation of the predator complex of *Lygus lineolaris* and *Lygocoris communis* (Hemiptera: Miridae) in apple orchards. J. Econ. Entomol. 84:830–836.

——. 1992. Predatory mirids of the green apple aphid *Aphis pomi*, the two-spotted spider mite *Tetranychus urticae* and the European red mite *Panonychus ulmi* in apple orchards in Québec. Entomophaga 37:283–292.

Arnott, D. A. 1956. Some factors reducing carrot seed yields in British Columbia. Proc. Entomol. Soc. B.C. 52:27–30.

Arnott, D. A. and I. Bergis. 1967. Causal agents of silver top and other types of damage to grass seed crops. Can. Entomol. 99:660–670.

Arnqvist, G. and M. Mäki. 1990. Infection rates and pathogenicity of trypanosomatid gut parasites in the water strider *Gerris odontogaster* (Zett.) (Heteroptera: Gerridae). Oecologia (Berl.) 84:194–198.

Aryeetey, E. A. and R. Kumar. 1973. Structure and function of the dorsal abdominal gland and defence mechanism in cocoa-capsids (Miridae: Heteroptera). J. Entomol. (A) 47:181–189.

Arzone, A. 1976. Indagini su *Trialeurodes vaporariorum* ed *Encarsia tricolor* in pien'aria. Inf. Fitopatol. 26:5–10.

——. 1983. Due fitomizi dannosi al Nocciuolo: L'Acaro delle gemme e il Miride degli amenti. *In* Atti Convegno Int. Nocciuolo, Avellino 22–24 Sept. 1983, pp. 199–204.

——. 1986. Preliminary reports on natural enemies of *Corythuca* [sic] *ciliata* (Say) in Italy. Working Group "Integrated Control of *Corythuca* [sic] *ciliata*." Int. Organ. Biol Control/West Palearctic Reg. Sect. Bull. 9(1):34–36.

Arzone, A., C. Vidano, and C. Arno. 1988. Predators and parasitoids of *Empoasca vitis* and *Zygina rhamni* (Rhynchota Auchenorrhyncha). *In* Proc. 6th Auchenorrhyncha Meet., Turin, Italy, 7–11 Sept. 1987, pp. 623–629. Inst. Agric. Apicul., University of Turin.

Arzone, A., A. Alma, and L. Tavella. 1990. Ruolo dei Miridi (Rhynchota Heteroptera) nella limitazione di *Trialeurodes vaporariorum* Westw. (Rhynchota Aleyrodidae): Nota preliminare. Boll. Zool. Agrar. Bachic. 22:43–52.

Asche, M. and M. R. Wilson. 1989. The three taro planthoppers: Species recognition in *Tarophagus* (Hemiptera: Delphacidae). Bull. Entomol. Res. 79:285–298.

Asgari, A. 1966. Untersuchungen über die im Raum Stuttgart-Hohenheim als wichtigste Prädatoren der grünen Apfelblattlaus (*Aphidula pomi* Deg.) auftretenden Arthropoden. Z. Angew. Zool. 53:35–93.

Ashe, J. S. 1993. Mouthpart modifications correlated with fungivory among aleocharine staphylinids (Coleoptera: Staphylinidae: Aleocharinae). Pages 105–130 *in* C. W. Schaefer and R. A. B. Leschen, eds., Functional morphology of insect feeding. Thomas Say Publ. Entomol.: Proceedings. Entomological Society of America, Lanham, Md.

Ashmead, W. H. 1887a. Report on insects injurious to garden crops in Florida. U.S. Dep. Agric. Div. Entomol. Bull. 14:9–29.

——. 1887b. Hemipterological contributions (No. 1.). Entomol. Am. 3:157–158.

——. 1895. Notes on cotton insects found in Mississippi. Insect Life (Wash., D.C.) 7:320–326.

Ashton, P. S. 1988. Dipterocarp biology as a window to the understanding of tropical forest structure. Annu. Rev. Ecol. Syst. 19:347–370.

Ashton, P. S., T. J. Givnish, and S. Appanah. 1988. Staggered flowering in the Dipterocarpaceae: New insights into floral induction and the evolution of mast fruiting in the aseasonal tropics. Am. Nat. 132:44–66.

Asquith, A. 1990. Taxonomy and variation of the *Lopidea nigridia* complex of western North America (Heteroptera: Miridae: Orthotylinae). Great Basin Nat. 50:135–154.

——. 1991. Revision of the genus *Lopidea* in America north of Mexico (Heteroptera: Miridae: Orthotylinae). Theses Zoologicae Vol. 16. Koeltz, Koenigstein, Germany. 280 pp.

——. 1993a. Patterns of speciation in the genus *Lopidea* (Heteroptera: Miridae: Orthotylinae). Syst. Entomol. 18:169–180.

Asquith, A. 1993b. A new species of *Cyrtopeltis* from coastal vegetation in the Hawaiian Islands (Heteroptera: Miridae: Dicyphinae). Pac. Sci. 47:17–20.

——. 1994. Revision of the endemic Hawaiian genus *Sarona* Kirkaldy (Heteroptera: Miridae: Orthotylinae). Bishop Mus. Occas. Pap. No. 40. 81 pp.

——. 1995a. *Loulucoris*, a new genus, and two new species of endemic Hawaiian plant bug (Heteroptera: Miridae: Orthotylinae). Proc. Entomol. Soc. Wash. 97:241–249.

——. 1995b. Evolution of *Sarona* (Heteroptera, Miridae): Speciation on geographic and ecological islands. Pages 90–120 *in* W. L. Wagner and V. A. Funk, eds., Hawaiian biogeography: Evolution on a hot spot archipelago. Smithsonian Institution Press, Washington, D.C.

——. 1997. Hawaiian Miridae (Hemiptera: Heteroptera): The evolution of bugs and thought. Pac. Sci. 51:356–365.

Asquith, D. and R. L. Horsburgh. 1969. Integrated versus chemical control of orchard mites. Pa. Fruit News 48(3):38, 40–42, 44.

Atherton, D. O. 1933. The tomato "green fly" association. Queensl. Agric. J. 40:291–298.

Atwal, A. S. 1976. Agricultural pests of India and South-East Asia. Kalyani, Delhi. 502 pp.

Auclair, J. L. 1969. Nutrition of plant-sucking insects on chemically defined diets. Entomol. Exp. Appl. 12:623–641.

Auclair, J. L. and J. R. Raulston. 1966. Feeding of *Lygus hesperus* (Hemiptera: Miridae) on a chemically defined diet. Ann. Entomol. Soc. Am. 59:1016–1017.

Aukema, B. 1986. *Psallus (Hylopsallus) assimilis* Stichel, 1956 en *P. (H.) pseudoplatani* Reichling, 1984, twee Miriden nieuw voor de Nederlandse fauna (Heteroptera: Miridae, Phylinae). Entomol. Ber. (Amst.) 46:117–119.

——. 1994. Zeldzame terrestrische wantsen en natuurontwikkeling (Heteroptera). Entomol. Ber. (Amst.) 54:95–102.

Austin, M. D. 1929. Observations on the eggs of the apple capsid (*Plesiocoris rugicollis* Fall.) and the common green capsid (*Lygus pabulinus* Linn.). J. South-East. Agric. Coll. Wye, Kent 26:136–144.

——. 1930a. Capsid damage to the fruits of red currant. Gard. Chron. (Lond.) (3)88:94.

——. 1930b. Oviposition of *Lygus pabulinus*. Gard. Chron. (Lond.) (3)87: 191.

——. 1931a. A contribution to the biology of the apple capsid (*Plesiocoris rugicollis* Fall.) and the common green capsid (*Lygus pabulinus* Linn.). J. South-East. Agric. Coll. Wye, Kent 28: 153–168.

——. 1931b. Observations on the hibernation and spring oviposition of *Lygus pratensis* Linn. Entomol. Mon. Mag. 67:149–152.

——. 1932. A preliminary note on the tarnished plant bug (*Lygus pratensis* Linn.). J. R. Hortic. Soc. 57:312–320.

——. 1933. A note on *Lygus pabulinus* L. J. South-East. Agric. Coll. Wye, Kent 32:168–170.

Austin, M. D. and A. M. Massee. 1947. Investigations on the control of the fruit tree red spider mite (*Metatetranychus ulmi* Koch) during the dormant season. J. Pomol. 23:227–253.

Austin, M. D., S. G. Jary, and H. Martin. 1932. Studies on the ovicidal action of winter washes, 1931 trials. J. South-East. Agric. Coll. Wye, Kent 30:63–86.

Austreng, M. P. and L. Sømme. 1980. The fauna of predatory bugs (Heteroptera, Miridae and Anthocoridae) in Norwegian apple orchards. Fauna Norv. Ser. B 27:3–8.

Avé, D., J. L. Frazier, and L. D. Hatfield. 1978. Contact chemoreception in the tarnished plant bug *Lygus lineolaris*. Entomol. Exp. Appl. 24:217–227.

Avidov, Z. and I. Harpaz. 1969. Plant pests of Israel. Israel University Press, Jerusalem. 549 pp.

Awang, A., R. Muhamad, and K. C. Khoo. 1988. Comparative merits of cocoa pod and shoot as food sources of the mirid, *Helopeltis theobromae* Miller. Planter (Kuala Lumpur) 64:100–104.

Awati, P. R. 1914. The mechanism of suction in the potato capsid bug, *Lygus pabulinus* Linn. Proc. Zool. Soc. Lond. 1914:685–733.

Ayal, Y. and I. Izhaki. 1993. The effect of the mirid bug *Capsodes infuscatus* on fruit production of the geophyte *Asphodelus ramosus* in a desert habitat. Oecologia (Berl.) 93:518–523.

Ayala Sifontes, J. L., H. Grillo Rabelo, and E. R. Vera Catalá. 1982. Enemigos naturales de *Heliothis virescens* (Fabricius) (Lepidoptera, Noctuidae)

en las provincias centrales de Cuba. Cent. Agric. 9(3):3–14.

Azhar, I. 1986. Seasonal and daily activity patterns of *Helopeltis theobromae* Mill (Hemiptera: Miridae) and their implications in field control. Pages 305–316 *in* E. Pushparajah and Chew Poh Soon, eds., Cocoa and coconuts: Progress and outlook (Proc. Int. Conf. on Cocoa and Coconuts, Kuala Lumpur, 15–17 Oct. 1984). Incorporated Society of Planters, Kuala Lumpur.

Bacheler, J. S. and R. M. Baranowski. 1975. *Paratriphleps laeviusculus*, a phytophagous anthocorid new to the United States (Hemiptera: Anthocoridae). Fla. Entomol. 58:157–163.

Bacheler, J. S., J. R. Bradley Jr., and C. S. Eckel. 1990. Plant bugs in North Carolina: Dilemma or delusion? *In* Proc. Beltwide Cotton Prod. Res. Conf., Jan. 9–14, 1990, Las Vegas, Nev., pp. 203–205. National Cotton Council, Memphis.

Back, E. A. and W. J. Price Jr. 1912. Stop-back of peach. J. Econ. Entomol. 5:329–334.

Backus, E. A. 1985. Anatomical and sensory mechanisms of leafhopper and planthopper feeding behavior. Pages 163–194 *in* L. R. Nault and J. G. Rodriguez, eds., The leafhoppers and planthoppers. Wiley, New York.

——. 1988a. Sensory systems and behaviours which mediate hemipteran plant-feeding: A taxonomic overview. J. Insect Physiol. 34:151–165.

——. 1988b. Observations on the feeding behavior of *Empoasca fabae* (Harris) (Cicadellidae: Typhlocybinae) and the cause of hopperburn. *In* Proc. 6th Auchenorrhyncha Meet., Turin, Italy, 7–11 Sept. 1987, pp. 493–500. Inst. Agric. Entomol. Apicul., University of Turin.

——. 1989. Host acceptance and feeding behavior. Pages 10–17 *in* E. J. Armbrust and W. O. Lamp, eds., Proceedings of a symposium: History and perspectives of potato leafhopper (Homoptera: Cicadellidae) research. Entomological Society of America, Lanham, Md.

——. 1994. History, development, and applications of the AC electronic monitoring system for insect feeding. Pages 1–51 *in* M. M. Ellsbury, E. A. Backus, and D. L. Ullman, eds., History, development, and application of AC electronic insect feeding monitors. Thomas Say Publ. Entomol.: Proceedings. Entomological Society of America, Lanham, Md.

——. 2000. Our own Jabberwocky: Clarifying the terminology of piercing-sucking behaviors of homopterans.

Pages 1–13 *in* G. P. Walker and E. A. Backus, eds., Principles and applications of electronic monitoring and other techniques in the study of homopteran feeding behavior. Thomas Say Publ. Entomol.: Proceedings. Entomological Society of America, Lanham, Md.

Backus, E. A. and W. B. Hunter. 1989. Comparison of feeding behavior of the potato leafhopper *Empoasca fabae* (Homoptera: Cicadellidae) on alfalfa and broad bean leaves. Environ. Entomol. 18:473–480.

Backus, E. A., W. B. Hunter, and C. N. Arne. 1988. Techniques for staining leafhopper (Homoptera: Cicadellidae) salivary sheaths and eggs within unsectioned plant tissue. J. Econ. Entomol. 81:1819–1823.

Bacon, O. G., W. D. Riley, and G. Zweig. 1964. The influence of certain biological and environmental factors on insecticide tolerance of the lygus bug, *Lygus hesperus*. J. Econ. Entomol. 57:225–230.

Bae, S. H. and M. D. Pathak. 1966. A mirid bug, *Cyrtorhinus lividipennis* Reuter, predator of the eggs and nymphs of the brown planthopper. Int. Rice Comm. Newsl. 15:33–36.

Bagga, H. S. and M. L. Laster. 1968. Relation of insects to the initiation and development of boll rot of cotton. J. Econ. Entomol. 61:1141–1142.

Bagnoli, B. 1975. Contributo alla conoscenza della entomofauna pronuba del girasole. Redia 56:135–145.

Bahana, J. W. 1976. *Helopeltis bergrothi* Reut (Hemiptera: Miridae) an important pest of cocoa (*Theobroma cacao* L.) in Uganda. Pages 53–58 *in* Veme conference des Entomologistes du Cacaoyer de l'Ouest-Africain. ONAREST (l'Office National de la Recherche Scientifique et Technique), Yaounde, Cameroon.

Bailey, J. C. and G. W. Cathey. 1985. Effect of chlordimeform on tarnished plant bug (Heteroptera: Miridae) nymph emergence. J. Econ. Entomol. 78:1485–1487.

Bailey, J. C., A. L. Scales, and W. R. Meredith Jr. 1984. Tarnished plant bug (Heteroptera: Miridae) nymph numbers decreased on caged nectariless cottons. J. Econ. Entomol. 77:68–69.

Bailey, N. S. 1951. The Tingoidea of New England and their biology. Entomol. Am. 31:1–140.

Bajan, C. and T. Bilewicz-Pawińska. 1971. Preliminary studies on the role of *Beauveria bassiana* (Bals.) Vuill. in reduction of *Lygus rugulipennis* Popp. Ekol. Pol. 19(A):35–46.

Baker, D. N., V. R. Reddy, J. M. McKinion, and F. D. Whisler. 1993. An analysis of the impact of lygus on cotton. Comput. Electron. Agric. 9:147–161

Baker, H. G. 1961. The adaptation of flowering plants to nocturnal and crepuscular pollinators. Q. Rev. Biol. 36:64–73.

——. 1983. An outline of the history of anthecology, or pollination biology. Pages 7–28 in L. Real, ed., Pollination biology. Academic, Orlando, Fla.

Baker, H. G. and I. Baker. 1983a. A brief historical review of the chemistry of floral nectar. Pages 126–152 in B. Bentley and T. Elias, eds., The biology of nectaries. Columbia University Press, New York.

——. 1983b. Floral nectar sugar constituents in relation to pollinator type. Pages 117–141 in C. E. Jones and R. J. Little, eds., Handbook of experimental pollination biology. Van Nostrand Reinhold, New York.

——. 1986. The occurrence and significance of amino acids in floral nectar. Plant Syst. Evol. 151:175–186.

Baker, H. G. and P. D. Hurd Jr. 1968. Intrafloral ecology. Annu. Rev. Entomol. 13:385–414.

Baker, H. G., R. W. Cruden, and I. Baker. 1971. Minor parasitism in pollination biology and its community function: The case of Ceiba acuminata. BioScience 21:1127–1129.

Baker, H. G., P. A. Opler, and I. Baker. 1978. A comparison of the amino acid complements of floral and extrafloral nectars. Bot. Gaz. 139:322–332.

Baker, J. E. and W. A. Connell. 1963. The morphology of the mouthparts of Tetranychus atlanticus and observations on feeding by this mite on soybeans. Ann. Entomol. Soc. Am. 56:733–736.

Baker, K. F., W. C. Snyder, and A. H. Holland. 1946. Lygus bug injury of lima bean in California. Phytopathology 36:493–503.

Baker, R. T. 1978. Damage to passionfruit and strawberry caused by the Australian crop mirid. In Proc. 31st N.Z. Weed Pest Control Conf., pp. 151–153.

Balarin, I. and M. Maceljski. 1986. The results of investigations done on Corythuca [sic] ciliata in Yugoslavia from 1970 on. Working Group "Integrated Control of Corythuca [sic] ciliata." Int. Organ. Biol. Control/West Palearctic Reg. Sect. Bull. 9(2):11–19.

Balasubramanian, C. and R. Janakiraman. 1966. Weather in the incidence of earhead bug (Calocoris Angustatus [sic] D) on Co. 12 irrigated cholam at Coimbatore. Indian J. Agron. 11:167–172.

Balasubramanian, S., S. N. Rawat, and M. C. Diwakar. 1988. Population fluctuations of rice hopper pests and the predatory mirid bug. Indian J. Plant Prot. 16:63–65.

Balciunas, J. K., D. W. Burrows, and M. F. Purcell. 1994. Insects to control melaleuca I: Status of research in Australia. Aquatics 16(4):10–13.

Balduf, W. V. 1943. New food records of entomophagous insects (Hym., Dip., Col., Orth., Hemip.). Entomol. News 54:12–15.

Balduf, W. V. and J. A. Slater. 1943. Additions to the bionomics of Sinea diadema (Fabr.) (Reduviidae, Hemiptera). Proc. Entomol. Soc. Wash. 45:11–18.

Balkwill, W. H. 1846. [Potato disease]. Gard. Chron. (Lond.) Aug. 15, p. 557.

Ball, E. D. 1920. The life cycle in Hemiptera (excl. aphids and coccids). Ann. Entomol. Soc. Am. 13:142–155.

Ballard, E. 1916. Calocoris angustatus, Leth. Bull. Agric. Res. Inst. (Pusa) 58:1–8.

——. 1921. Two new species of Ragmus from South India. Rec. Indian Mus. 22:509–510.

——. 1927. Some new Indian Miridae (Capsidae). Mem. Dep. Agric. India Entomol. Ser. 10(4):61–68.

Ballou, C. H. 1933. Insect conditions in Costa Rica during 1932 and early 1933. U.S. Dep. Agric. Insect Pest Surv. Bull. 13:50–58.

Ballou, H. A. 1919. Insect notes. Miscellaneous insects. Agric. News (Barbados) 18:74.

Balsbaugh, E. U. Jr. and K. L. Hays. 1972. The leaf beetles of Alabama (Coleoptera: Chrysomelidae). Ala. Agric. Exp. Stn. Bull. 441:1–223.

Baltazar, C. R. 1981. Biological control attempts in the Philippines. Philipp. Entomol. 4:505–523.

Bamber, M. K. 1893. A text book on the chemistry and agriculture of tea, including the growth and manufacture. Law-Publishing, Calcutta. 258 pp.

Banerjee, B. 1983a. Arthropod accumulation on tea in young and old habitats. Ecol. Entomol. 8:117–123.

——. 1983b. Pests of tea. Pages 261–276 in P. D. Srivastva et al., eds., Agricultural entomology. Vol. II. All India Scientific Writers' Society, New Delhi.

Banerjee, B. and N. N. Kakoti. 1969. Biology, population cycle and control of Ragmus importunitas Distant. Indian J. Entomol. 30:257–262.

Bańkowska, R., E. Kierych, W. Mikolajczyk, J. Palmowska, and P. Trojan. 1975. Aphid-aphidophage community in alfalfa cultures (Medicago sativa L.) in Poland. Part I. Structure and phenology of the community. Ann. Zool. Pol. Akad. Nauk Inst. Zool. 32:299–345.

Banks, N. 1893. Two uncommon insects. Entomol. News 4:268.

——. 1912. At the Ceanothus in Virginia. Entomol. News 23:102–110.

Baptist, B. A. 1941. The morphology and physiology of the salivary glands of Hemiptera-Heteroptera. Q. J. Microsc. Sci. 83:91–139.

Barbagallo, S. 1969. Appunti morfo-biologici su Tetrastichus miridivorus Domenichini (Hymenoptera Eulophidae) parassita oofago di Eterotteri Miridi. Boll. Zool. Agrar. Bachic. (2) 9:115–122.

——. 1970. Contributo alla conoscenza del Calocoris (Closterotomus) trivialis (Costa) (Rhynchota-Heteroptera, Miridae). Morfologia dell'adulto e biologia. Entomologica (Bari) 6:1–104.

Barber, H. G. and S. C. Bruner. 1946. Records and descriptions of miscellaneous Cuban Hemiptera. Bull. Brooklyn Entomol. Soc. 41:52–61.

Bardner, R. 1983. Pests of Vicia faba L. other than aphids and nematodes. Pages 371–390 in P. D. Hebblethwaite, ed., The faba bean (Vicia faba L.): A basis for improvement. Butterworths, London.

Bardner, R. and K. E. Fletcher. 1974. Insect infestations and their effects on the growth and yield of field crops: A review. Bull. Entomol. Res. 64:141–160.

Bariola, L. A. 1969. The biology of the tarnished plant bug, Lygus lineolaris (Beauvois), and its nature of damage and control on cotton. Ph.D. dissertation, Texas A & M University, College Station. 102 pp.

Barker, R. J. and Y. Lehner. 1972. The resistance of pollen grains and their degradation by bees. Bee World 53:173–177.

Barlow, N. D. and A. F. G. Dixon. 1980. Simulation of lime aphid population dynamics. Centre for Agricultural Publishing & Documentation, Wageningen. 165 pp.

Barlow, V. M., L. D. Godfrey, and R. F. Norris. 1999. Population dynamics of Lygus hesperus (Heteroptera: Miridae) on selected weeds in comparison with alfalfa. J. Econ. Entomol. 92: 846–852.

Barnadas, I., R. Gabarra, and R. Albajes. 1998. Predatory capacity of two mirid bugs preying on Bemisia tabaci. Entomol. Exp. Appl. 86:215–219.

Barnes, H. F. 1948. Gall midges of economic importance. Vol. III: Gall midges of fruit. Crosby Lockwood, London. 184 pp.

Barnes, J. K. 1985. Insects in the new

nation: A cultural context for the emergence of American entomology. Bull. Entomol. Soc. Am. 31:21–30.

Barnett, W. W. and R. E. Rice. 1989. Insect and mite pests. Pages 94–117 in J. H. La Rue and R. S. Johnson, tech. eds., Peaches, plums, and nectarines: Growing and handling for fresh market. Univ. Calif. Div. Agric. Nat. Resour. Publ. 3331.

Barnett, W. W., B. E. Bearden, A. Berlowitz, C. S. Davis, J. L. Joos, and G. W. Morehead. 1976. True bugs cause severe pear damage. Calif. Agric. 30(10):20–23.

Barnett, W. W., W. J. Bentley, R. E. Rice, and C. V. Weakley. 1990. Insects and mites. Pages 1–22 in Peach and nectarine pest management guidelines (UCPMG Publ. 10). Univ. Calif. Div. Agric. Nat. Resour. Publ. 3339.

Barreto, T. N. and E. Martinez G. 1996. La chinche de los pastos Collaria columbiensis, en la Sabana de Bogotá. Carta Fedegan (Colombia) 37:42–49.

Barrows, E. M. 1979. Flower biology and arthropod associates of Lilium philadelphicum. Mich. Bot. 18:109–115.

Barteneva, R. V. 1986. Diseases and pests of castor and their control; pests. Pages 284–286 in V. A. Moshkin, ed., Castor. (Russian Translations Ser. 43). Balkema, Rotterdam.

Barth, F. G. 1985. Insects and flowers: The biology of a partnership. (Transl. from German by M. A. Biederman-Thorson). Princeton University Press, Princeton. 297 pp.

Bartlett, D. 1996. Feeding and egg laying behaviour in Campylomma verbasci Meyer (Hemiptera: Miridae). M.S. thesis, Simon Fraser University, Burnaby, British Columbia. 100 pp.

Basset, Y. 1991. The taxonomic composition of the arthropod fauna associated with an Australian rainforest tree. Aust. J. Zool. 39:171–190.

Basset, Y., G. A. Samuelson, and S. E. Miller. 1996. Similarities and contrasts in the local insect faunas associated with ten forest tree species of New Guinea. Pac. Sci. 50:157–183.

Bateman, D. F. and H. G. Basham. 1976. Degradation of plant cell walls and membranes by microbial enzymes. Pages 316–335 in R. Heitefuss and P. H. Williams, eds., Physiological plant pathology. Springer, New York.

Bateman, D. F. and R. L. Millar. 1966. Pectic enzymes in tissue degradation. Annu. Rev. Phytopathol. 4:119–146.

Batra, L. R. 1973. Nematosporaceae (Hemiascomycetidae): Taxonomy, pathogenicity, distribution, and vector

relations. U.S. Dep. Agric. Agric. Res. Serv. Tech. Bull. 1469:1–71.

Batulla, B. A. and A. G. Robinson. 1983. A list of predators of aphids (Homoptera: Aphididae) collected in Manitoba, 1980–1981. Proc. Entomol. Soc. Manit. 39:25–45.

Baumann, P., N. A. Moran, and L. Baumann. 1997. The evolution and genetics of aphid endosymbionts. BioScience 47:12–20.

Bawa, K. S. and P. A. Opler. 1975. Dioecism in tropical forest trees. Evolution 29:167–179.

Bawa, K. S., S. H. Bullock, D. R. Perry, R. E. Coville, and M. H. Grayum. 1985. Reproductive biology of tropical lowland rain forest trees. II. Pollination systems. Am. J. Bot. 72:346–356.

Bawden, F. C. 1943. Plant viruses and virus diseases, 2nd rev. ed. Chronica Botanica, Waltham, Mass. 294 pp.

———. 1950. Plant viruses and virus diseases, 3rd ed. Chronica Botanica, Waltham, Mass. 335 pp.

Baxendale, R. W. and W. T. Johnson. 1990. Efficacy of summer oil spray on thirteen commonly occurring insect pests. J. Arboric. 16:89–94.

Beards, G. W. and T. F. Leigh. 1960. A laboratory rearing method for Lygus hesperus Knight. J. Econ. Entomol. 53:327–328.

Beards, G. W. and F. E. Strong. 1966. Photoperiod in relation to diapause in Lygus hesperus Knight. Hilgardia 37(10):345–362.

Beardsley, J. W. 1958. Notes and exhibitions. Orthotylus iolani Kirkaldy. Proc. Hawaii. Entomol. Soc. 16:323.

Beattie, A. J., D. E. Breedlove, and P. R. Ehrlich. 1973. The ecology of the pollinators and predators of Frasera speciosa. Ecology 54:83–91.

Bech, R. 1964. Schädliche Wanzen an Zierpflanzen. Gartenbau (Berl.) 11:218–220.

———. 1965. Licht- und Farbreaktionen der Lygus-Arten. Biol. Zentbl. 84:635–640.

———. 1967. Zur Bedeutung der Lygus-Arten als Pflanzenschädlinge. Biol. Zentbl. 86:205–232.

———. 1969. Untersuchungen zur Systematik, Biologie und Ökologie wirtschaftlich wichtiger Lygus-Arten (Hemiptera: Miridae). Beitr. Entomol. 19:63–103.

Beck, S. D., chairman et al. 1975. Pest control: An assessment of present and alternative technologies. Vol. III. Cotton pest control. National Academy of Science, Washington, D.C. 139 pp.

Becker, G. G. 1918. Lopidea media, a persistent pest of phlox. J. Econ. Entomol. 11:431.

Becker, M. and F. A. Frieiro-Costa. 1988.

Natality and mortality in the egg stage in Gratiana spadicea (Klug, 1829) (Coleoptera: Chrysomelidae), a monophagous cassidine beetle of an early successional Solanaceae. Rev. Bras. Biol. 48:467–475.

Becker, P[eter]. 1975. Island colonization by carnivorous and herbivorous Coleoptera. J. Anim. Ecol. 44:893–906.

———. 1992. Colonization of islands by carnivorous and herbivorous Heteroptera and Coleoptera: Effects of island area, plant species richness, and "extinction" rates. J. Biogeogr. 19:163–171.

Becker, P[eterson]. 1974. Pests of ornamental plants. Minist. Agric. Fish. Food Bull. 97. Her Majesty's Stationery Office, London. 175 pp.

Beckham, C. M., W. S. Hough, and C. H. Hill. 1950. Biology and control of the spotted tentiform leaf miner on apple trees. Va. Agric. Exp. Stn. Tech. Bull. 114:1–19.

Bedford, E. C. G. 1976. Citrus pest management in South Africa. Proc. Tall Timbers Conf. Ecol. Anim. Control Habitat Manage. 6:19–42.

Bedford, H. W. 1938. Entomological Section, Agricultural Research Service. Pages 50–67 in Annu. Rep. Part II. Sudan Dep. Agric. For. Agric. Res. Serv.

———. 1940. Entomological Section, Agricultural Research Service. Pages 50–71 in Annu. Rep. Part II. Sudan Dep. Agric. For. Agric. Res. Serv.

Beemster, A. B. R. and J. A. de Bokx. 1987. Survey of properties and symptoms. Pages 84–113 in J. A. de Bokx and J. P. H. van der Want, eds., Viruses of potatoes and seed-potato production, 2nd ed. Pudoc, Wageningen.

Beers, E. H., L. A. Hull, and V. P. Jones. 1994. Sampling pest and beneficial arthropods of apple. Pages 383–416 in L. P. Pedigo and G. D. Buntin, eds., Handbook of sampling methods for arthropods in agriculture. CRC, Boca Raton, Fla.

Beetle, A. A. 1970. Recommended plant names. Wyo. Agric. Exp. Stn. Res. J. 31:1–124.

Begon, M., J. L. Harper, and C. R. Townsend. 1990. Ecology: Individuals, populations and communities, 2nd ed. Blackwell, Boston. 945 pp.

Beingolea, G. O. 1959a. Notas sobre la bionómica de arañas e insectos benéficos que ocurren en el cultivo de algodón. Rev. Peru. Entomol. Agric. 2:36–44.

———. 1959b. Notas sobre Hyalochloria denticornis Tsai Yu-Hsiao (Hemip.: Miridae), predator de los huevos de Anomis texana Riley (Lepidop.:

Noctuidae). Rev. Peru. Entomol. Agric. 2:51–59.

——. 1960. Notas adicionales sobre *Hyalochloria denticornis* Tsai Yu-Hsiao (Hemip: Miridae), predator de los huevos de *Anomis texana* Riley (Lep: Noctuidae). Rev. Peru. Entomol. Agric. 3:1–5.

Beirne, B. P. 1970. Effects of precipitation on crop insects. Can. Entomol. 102: 1360–1361.

——. 1972. Pest insects of annual crop plants in Canada IV. Hemiptera-Homoptera V. Orthoptera VI. Other groups. Mem. Entomol. Soc. Can. 85. 73 pp.

Bekker-Migdisova, E. E. 1962. Order Heteroptera. True bugs. Pages 208–226 *in* B. B. Rohdendorf, ed., Fundamentals of paleontology. Vol. 9. Arthropoda, Tracheata, Chelicerata. Akademia Nauk SSSR, Moscow. [in Russian, English translation, 1991, Smithsonian Institution Libraries and National Science Foundation, Washington, D.C.]

Bell, A. A. and R. D. Stipanovic. 1977. The chemical composition, biological activity, and genetics of pigment glands in cotton. *In* Proc. Beltwide Cotton Prod. Res. Conf., Jan. 10–12, 1977, Atlanta, Ga., pp. 244–258. National Cotton Council, Memphis.

Bell, C. R. 1971. Breeding systems and floral biology of the Umbelliferae or evidence for specialization in unspecialized flowers. Pages 93–107 *in* V. H. Heywood, ed., The biology and chemistry of the Umbelliferae. Academic, New York.

Bell, E. A. 1987. Secondary compounds and insect herbivores. Pages 19–23 *in* V. Labeyrie, G. Fabres, and D. Lachaise, eds., Insects-plants. Proc. 6th Int. Symp. on Insect-Plant Relationships, Pau, France, 1986. Junk, Dordrecht.

Bell, R. R., E. J. Thornber, J. L. L. Seet, M. T. Groves, N. P. Ho, and D. T. Bell. 1983. Composition and protein quality of honeybee-collected pollen of *Eucalyptus marginata* and *Eucalyptus calophylla*. J. Nutr. 113:2479–2484.

Bellows, T. S. Jr. and E. F. Legner. 1993. Foreign exploration. Pages 25–41 *in* R. G. Van Driesche and T. S. Bellows, eds., Steps in classical arthropod biological control. Thomas Say Publ. Entomol.: Proceedings. Entomological Society of America, Lanham, Md.

Bellows, T. S. Jr., T. M. Perring, R. J. Gill, and D. H. Headrick. 1994. Description of a species of *Bemisia* (Homoptera: Aleyrodidae). Ann. Entomol. Soc. Am. 87:195–206.

Belsky, A. J. 1986. Does herbivory benefit plants? A review of the evidence. Am. Nat. 127:870–892.

Benedek, P. and V. E. Jászai. [Mrs. E. Virág] 1968. Lucernat karosito mezei poloskak (Heteroptera, Miridae) rajzasvizgalatanak novenyvedelmi tanulsagai. Növényvédelem 4:257–260.

Benedek, P., Cs. Erdélyi, and V. E. Jászai. 1970. Seasonal activity of heteropterous species injurious to lucerne and its relations to the integrated pest control of lucerne grown for seed. Acta Phytopathol. Acad. Sci. Hung. 5:81–93.

Benedict, J. H. [1986?] Biological control of pests in cotton. Pages 409–423 *in* R. E. Frisbie and P. L. Adkisson, eds., Integrated pest management on major agricultural systems. Tex. Agric. Exp. Stn. MP–1616.

Benedict, J. H., T. F. Leigh, W. Tingey, and A. H. Hyer. 1977. Glandless Acala cotton: More susceptible to insects. Calif. Agric. 31(4):14–15.

Benedict, J. H., T. F. Leigh, A. H. Hyer, and P. F. Wynholds. 1981. Nectariless cotton: Effect on growth, survival, and fecundity of lygus bugs. Crop Sci. 21:28–30.

Benedict, J. H., A. H. Hyer, T. F. Leigh, and W. M. Tingey. 1982. Evaluations of various cottons (*Gossypium* spp.) for resistance to *Lygus hesperus* Knight. U.S. Dep. Agric. Agric. Res. Serv. ARM-W-33:1–29.

Benedict, J. H., T. F. Leigh, and A. H. Hyer. 1983. *Lygus hesperus* (Heteroptera: Miridae) oviposition behavior, growth, and survival in relation to cotton trichome density. Environ. Entomol. 12:331–335.

Benjamin, D. M. 1968. Economically important insects and mites on tea in East Africa. East Afr. Agric. For. J. 34: 1–16.

Bennett, C. W. and A. S. Costa. 1961. Sowbane mosaic caused by a seed-transmitted virus. Phytopathology 51: 546–550.

Bennett, F. A., D. Rosen, P. Cochereau, and B. J. Wood. 1976. Biological control of pests of tropical fruits and nuts. Pages 359–395 *in* C. B. Huffaker and P. S. Messenger, eds., Theory and practice of biological control. Academic, New York.

Bennett, F. D. and H. Zwölfer. 1968. Exploration for natural enemies of the water hyacinth in northern South America and Trinidad. Hyacinth Control J. 7:44–52.

Bennett, F. D., J. W. Smith Jr., and H. W. Browning. 1990. Pests of sugarcane. Pages 81–86 *in* D. H. Habeck, F. D. Bennett, and J. H. Frank, eds., Classical biological control in the southern United States. South. Coop. Ser. Bull. 355.

Benrey, B. and W. O. Lamp. 1994. Biological control in the management of planthopper populations. Pages 519–550 *in* R. F. Denno and T. J. Perfect, eds., Planthoppers: Their ecology and management. Chapman & Hall, New York.

Ben Saad, A. A. and G. W. Bishop. 1976. Attraction of insects to potato plants through use of artificial honeydews and aphid juice. Entomophaga 21:49–57.

Benson, R. B. 1962. Holarctic sawflies (Hymenoptera: Symphyta). Bull. Brit. Mus. (Nat. Hist.) Entomol. 12:379–409.

Bentur, J. S. and M. B. Kalode. 1985a. Technique for rearing the predatory mirid bug (*Cyrtorhinus lividipennis* Reut) on *Corcyra* eggs. Curr. Sci. (Bangalore) 54:513–514.

——. 1985b. Natural enemies of rice leaf- and plant-hoppers in Andhra Pradesh. Entomon 10:271–274.

——. 1987. Off-season survival of the predatory mirid bug, *Cyrtorhinus lividipennis* (Reuter). Curr. Sci. (Bangalore) 56:956–957.

Ben-Ze'ev, I. S., S. Keller, and A. B. Ewen. 1985. *Entomophthora erupta* and *Entomophthora helvetica* sp. nov. (Zygomycetes: Entomophthorales), two pathogens of Miridae (Heteroptera) distinguished by pathobiological and nuclear features. Can. J. Bot. 63:1469–1475.

Berberet, R. C. and W. D. Hutchison. 1994. Sampling methods for insect management in alfalfa. Pages 357–381 *in* L. P. Pedigo and G. D. Buntin, eds., Handbook of sampling methods for arthropods in agriculture. CRC, Boca Raton, Fla.

Berenbaum, M. [R.]. 1983. Coumarins and caterpillars: A case for coevolution. Evolution 37:163–179.

Berenbaum, M. R. 1990. Evolution of specialization in insect-umbellifer associations. Annu. Rev. Entomol. 35:319–343.

Berenbaum, M. R. and M. B. Isman. 1989. Herbivory in holometabolous and hemimetabolous insects: Contrasts between Orthoptera and Lepidoptera. Experientia (Basel) 45:229–236.

Berenbaum, M. R. and A. R. Zangerl. 1992. Quantification of chemical coevolution. Pages 69–87 *in* R. S. Fritz and E. L. Simms, eds., Plant resistance to herbivores and pathogens: Ecology, evolution, and genetics. University of Chicago Press, Chicago.

Berg, C. O., B. A. Foote, L. Knutson, J. K. Barnes, S. L. Arnold, and K. Valley. 1982. Adaptive differences in phenology in sciomyzid flies. Mem. Entomol. Soc. Wash. 10:15–36.

Bergevin, E. de. 1924. Nouvelles observations sur les Hémiptères suceurs de sang humain. Bull. Soc. Hist. Nat. Afr. Nord 15:259–262.

——. 1926. Note à propos d'un nouvel Hémiptère Capsidae se rélévant suceur de sang humain. Bull. Soc. Hist. Nat. Afr. Nord 17:173–174.

Bergroth, E. 1903. Neue myrmecophile Hemipteren. Wien. Entomol. Ztg. 22: 253–256.

Berker, J. 1958. Die natürlichen Feinde der Tetranychiden. Z. Angew. Entomol. 43:115–172.

Bernays, E. A. 1982. The insect on the plant—a closer look. Pages 3–17 in J. H. Visser and A. K. Minks, eds., Proc. 5th Int. Symp. Insect-Plant Relationships, Wageningen, the Netherlands, 1–4 March 1982. Centre for Agricultural Publishing & Documentation, Wageningen.

Bernays, E. A. and N. Graham. 1988. On the evolution of host specificity in phytophagous arthropods. Ecology 69:886–892.

Berry, J. A. and J. T. S. Walker. 1989. Dasineura pyri (Bouché), pear leafcurling midge and Dasineura mali (Kieffer), apple leafcurling midge (Diptera: Cecidomyiidae). Pages 171–175 in P. J. Cameron, R. L. Hill, J. Bain, and W. P. Thomas, eds., A review of biological control of invertebrate pests and weeds in New Zealand 1874 to 1987. Tech. Commun. No. 10. CAB Int. Inst. Biol. Control, Wallingford, UK.

Bertin, R. I. 1989. Pollination biology. Pages 23–86 in W. G. Abrahamson, ed., Plant-animal interactions. McGraw-Hill, New York.

Bethell, R. S. and W. W. Barnett, compilers. 1978. Insect and mite pests. Pages 9–132 in R. S. Bethell, tech. ed., Pear pest management. Univ. Calif. Div. Agric. Sci., Berkeley.

Betrem, J. G. 1953. Het optreden van plaatselijke rassen bij Helopeltis antonii Sign. op Java. Tijdschr. Plziekt. 59:174–177.

Betsch, W. D. 1978. A biological study of three hemipterous insects on honeylocust in Ohio. M.S. thesis, Ohio State University, Columbus. 72 pp.

Betts, C. R. 1986a. The comparative morphology of the wings and axillae of selected Heteroptera. J. Zool. (Lond.) (B) 1:225–282.

——. 1986b. Functioning of the wings and axillary sclerites of Heteroptera during flight. J. Zool. (Lond.) (B) 1:283–301.

——. 1986c. The kinematics of Heteroptera in free flight. J. Zool. (Lond.) (B) 1:303–315.

Beutler, R. 1953. Nectar. Bee World 34:106–116, 128–136, 156–162.

Beyer, A. H. 1921. Garden flea-hopper in alfalfa and its control. U.S. Dep. Agric. Bur. Entomol. Bull. 964:1–27.

Bianchi, F. A. 1966. Nesiomiris sp. Proc. Hawaii. Entomol. Soc. 19:132.

——. 1977. Cyril Eugene Pemberton, 1886–1975: A biographical sketch. Proc. Hawaii. Entomol. Soc. 22:417–441.

Bibby, F. F. 1946. Neurocolpus nubilus, a cotton pest. J. Econ. Entomol. 39:815.

——. 1961. Notes on miscellaneous insects of Arizona. J. Econ. Entomol. 54:324–333.

Bigger, M. 1981. Observations on the insect fauna of shaded and unshaded Amelonado cocoa. Bull. Entomol. Res. 71:107–119.

——. 1993. Time series analysis of variations in abundance of selected cocoa insects and fitting of simple linear predictive models. Bull. Entomol. Res. 83:153–169.

Biggs, A. R. and E. A. C. Hagley. 1988. Effects of two sterol-inhibiting fungicides on populations of pest and beneficial arthropods on apple. Agric. Ecosyst. Environ. 20:235–244.

Bilewicz-Pawińska, T. 1967. From studies on the heteropterofauna of the sugar beet. Ekol. Pol. (A) 15:373–384.

——. 1970. Z badań nad naturalną redukcją niektórych zmieników występujących w agrocenozach. Rocz. Nauk Roln. Ser. E 1:193–204. [English summary.]

——. 1973. Uwagi o trzech gatunkach z rodzaju Peristenus Foerster (Hym., Braconidae) i ich pasożytach Mesochorus spp. (Hym., Ichneumonidae). Pol. Pismo Entomol. 43:841–845. [English summary.]

——. 1976. Wrogowie naturalni niektórych polnych pluskwiaków w Polsce. Rocz. Nauk Roln. Ser. E 6:125–135. [English summary.]

——. 1982. Plant bugs (Heteroptera, Miridae) and their parasitoids (Hymenoptera, Braconidae) on cereal crops. Pol. Ecol. Stud. 8:113–191.

Bilewicz-Pawińska, T. and M. Kamionek. 1973. Lygus rugulipennis Popp. (Het., Miridae) nżywicielem nicienia (Mermitidae) [sic]. Pol. Pismo Entomol. 43:847–849. [English summary.]

Bilewicz-Pawińska, T. and A.-L. Varis. 1985. Structure of mirid communities (Heteroptera) and the parasitism of the main bug populations on wheat in eastern parts of North and Central Europe. Ann. Entomol. Fenn. 51:19–23.

Bilsing, S. W. 1920. Quantitative studies on the food of spiders. Ohio J. Sci. 20:215–260.

Bin, F. 1970. Lygus viscicola Put. (Miridae) e Psylla visci Curt. (Psyllidae), Rincoti del Vischio nuovi per la fauna Italiana. Boll. Zool. Agrar. Bachic. (2) 10:133–143.

Bird, L. S., C. Liverman, R. G. Percy, and D. L. Bush. 1979. The mechanism of multi-adversity resistance in cotton: Theory and results. In Proc. Beltwide Cotton Prod. Res. Conf., Jan. 7–11, 1979, Phoenix, Ariz., pp. 226–228. National Cotton Council, Memphis.

Bisabri-Ershadi, B. and L. E. Ehler. 1981. Natural biological control of western yellow-striped armyworm, Spodoptera praefica (Grote), in hay alfalfa in northern California. Hilgardia 49(5):1–23.

Bishop, A. L. 1980. The potential of Campylomma livida Reuter, and Megacoelum modestum Distant (Hemiptera: Miridae) to damage cotton in Queensland. Aust. J. Exp. Agric. Anim. Husb. 20:229–233.

Black, L. M. 1937. A study of potato yellow dwarf in New York. Cornell Univ. Agric. Exp. Stn. Mem. 209:1–23.

——. 1954. Arthropod transmission of plant viruses. Exp. Parasitol. 3:72–104.

Blackburn, T. 1888. Notes on the Hemiptera of the Hawaiian Islands. Proc. Linn. Soc. N.S.W. 3:343–354.

Blacklock, J. S. 1954. A short study of pepper culture with special reference to Sarawac. Trop. Agric. 31:40–56.

Blackman, M. W. 1918. On the insect visitors to the blossoms of wild blackberry and wild Spiraea—a study in seasonal distribution. N.Y. Coll. For. Tech. Publ. 18(10):119–144.

Blanchard, E. 1840. Histoire naturelle des insectes Orthoptères, Nevroptères, Hémiptères, Hyménoptères, Lépidoptères et Diptères. Vol. 3. Duménil, Paris. 672 pp.

Blatchley, W. S. 1926. Heteroptera or true bugs of eastern North America with especial reference to the faunas of Indiana and Florida. Nature, Indianapolis. 1116 pp.

——. 1928. Notes on the Heteroptera of eastern North America with descriptions of new species, I. J. N.Y. Entomol. Soc. 36:1–23.

——. 1934. Notes on a collection of Heteroptera taken in winter in the vicinity of Los Angeles, California. Trans. Am. Entomol. Soc. 60:1–16.

Blathwayt, P. 1889. Remarks on some Hemiptera-Heteroptera taken in the neighbourhood of Bath. Proc. Bath Nat. Hist. Antiq. Field Club 6:315–327.

Blattný, C. and B. Starý. 1942. Poškozeni bramborových listů od ploštic. Ochr. Rostl. 18:201.

Blattný, C., A. Kac, and A. Hoffer. 1948. Pozorování a pokusy s pěstováním vojtěšky na semeno, zejména s ohledem na boj proti plodomorce vojteškové a j. škodlivým činitelům vojtěšky. Ochr. Rostl. 19–20:40–46.

Blinn, R. L. 1988. *Pseudoxenetus regalis* (Heteroptera: Miridae: Orthotylinae): Seasonal history and description of fifth instar. J. N.Y. Entomol. Soc. 96: 310–313.

——. 1992. Seasonal occurrence of the Miridae (Heteroptera) associated with Ohio buckeye, *Aesculus glabra* Willd., in Missouri. J. N.Y. Entomol. Soc. 100:480–487.

Blinn, R. L. and T. R. Yonke. 1985. An annotated list of the Miridae of Missouri (Hemiptera: Heteroptera). Trans. Mo. Acad. Sci. 19:73–98.

——. 1986. Laboratory life history of *Trigonotylus coelestialium* (Kirkaldy) (Heteroptera: Miridae). J. Kans. Entomol. Soc. 59:735–737.

Blommers, L. H. M. 1994. Integrated pest management in European apple orchards. Annu. Rev. Entomol. 39:213–241.

Blommers, L. [H. M.], V. Bus, E. de Jongh, and G. Lentjes. 1988. Attraction of males by virgin females of the green capsid bug *Lygocoris pabulinus* (Heteroptera: Miridae). Entomol. Ber. (Amst.) 48:175–179.

Blommers, L. H. M., F. W. N. M. Vaal, and H. H. M. Helsen. 1997. Life history, seasonal adaptations and monitoring of common green capsid *Lygocoris pabulinus* (L.) (Hem., Miridae). J. Appl. Entomol. 121:389–398.

Blum, M. S. 1996. Semiochemical parsimony in the Arthropoda. Annu. Rev. Entomol. 41:353–374.

Blumberg, A. J. Y., P. F. Hendrix, and D. A. Crossley Jr. 1997. Effects of nitrogen source on arthropod biomass in no-tillage and conventional tillage grain sorghum agroecosystems. Environ. Entomol. 26:31–37.

Blumenthal, M. A. 1978. The metathoracic gland system of *Lygus lineolaris* (Heteroptera: Miridae). M.S. thesis, Cornell University, Ithaca. 142 pp.

Blümke [no initials]. 1937. Wie lässt sich der Kartoffelabbau bekämpfen? Mitt. Landw. (Berl.) 52:1048–1050.

Bobb, M. L. 1970. Reduction of cat-facing injury to peaches. J. Econ. Entomol. 63:1026–1027.

Böcher, J. 1971. Preliminary studies on the biology and ecology of *Chlamydatus pullus* (Reuter) (Heteroptera: Miridae) in Greenland. Medd. Grønl. 191(3):1–29.

Bochkareva, Z. A. and S. M. Vdovichenko. 1974. The protection of lucerne seed crops. Zashch. Rast. (Mosc.) 7:19. [in Russian.]

Bockwinkel, G. 1990. Food resource utilization and population growth of the grassbug *Notostira elongata* (Heteroptera: Miridae: Stenodemini). Entomol. Gen. 15:51–60.

Bodenheimer, F. 1921. Zur Kenntnis der Chrysanthemum-Wanzen, sowie der durch sie hervorgerufenen Gallbildung. Z. Pflanzenkr. 31:97–100.

——. 1951. Citrus entomology in the Middle East with special references to Egypt, Iran, Iraq, Palestine, Syria, Turkey. Junk, The Hague. 663 pp.

Bodnaruk, K. P. 1992. Daily activity patterns of adult *Creontiades dilutus* (Stål) and *Campylomma liebknechti* (Girault) (Hemiptera: Miridae) in early-flowering cotton. J. Aust. Entomol. Soc. 31:331–332.

Bodnaryk, R. P. 1996. Physical and chemical defences of pods and seeds of white mustard (*Sinapis alba* L.) against tarnished plant bugs, *Lygus lineolaris* (Palisot de Beauvois) (Heteroptera: Miridae). Can. J. Plant Sci. 76:33–36.

Boenisch, A., G. Petersen, and U. Wyss. 1997. Influence of the hyperparasitoid *Dendrocerus carpenteri* on the reproduction of the grain aphid *Sitobion avenae*. Ecol. Entomol. 22:1–6.

Boggs, C. L. 1987. Ecology of nectar and pollen feeding in Lepidoptera. Pages 369–391 *in* F. Slansky Jr. and J. G. Rodriguez, eds., Nutritional ecology of insects, mites, spiders, and related invertebrates. Wiley, New York.

Bohart, G. E. 1966. The need for organized information on crop pollination and pollinators. *In* E. Åkerberg and E. Crane, eds., Proc. 2nd Int. Symp. on Pollination, London, July 1964. Bee World 47(1, Suppl.): 209–211.

Bohart, G. E. and T. W. Koerber. 1972. Insects and seed production. Pages 1–5 *in* T. T. Kozlowski, ed., Seed biology. Vol. III. Insects, and seed collection, storage, testing, and certification. Academic, New York.

Bohart, G. E. and W. P. Nye. 1960. Insect pollinators of carrots in Utah. Utah Agric. Exp. Stn. Bull. 419:1–16.

Bohart, G. E., W. P. Nye, and L. R. Hawthorn. 1970. Onion pollination as affected by different levels of pollinator activity. Utah Agric. Exp. Stn. Bull. 482:1–57.

Bohart, R. M. and B. Villegas. 1976. Nesting behavior of *Encopognathus rufiventris* Timberlake (Hymenoptera: Sphecidae). Pan-Pac. Entomol. 52:331–334.

Bohlen, E. 1973. Crop pests in Tanzania and their control. Parey, Berlin. 142 pp.

Bohmfalk, G. T. 1982. Key pests. Pages 4–8 *in* G. T. Bohmfalk, R. E. Frisbie, W. L. Sterling, R. B. Metzer, and A. E. Knutson, eds., Identification, biology and sampling of cotton insects. Tex. Agric. Ext. Serv. B-933.

Bohning, J. W. and W. F. Currier. 1967. Does your range have wheatgrass bugs? J. Range Manage. 20:265–267.

Boivin, G. and R. K. Stewart. 1982a. Attraction of male green apple bugs, *Lygocoris communis* (Hemiptera: Miridae), to caged females. Can. Entomol. 114:765–766.

——. 1982b. Identification and evaluation of damage to McIntosh apples by phytophagous mirids (Hemiptera: Miridae) in southwestern Quebec. Can. Entomol. 114:1037–1045.

——. 1983a. Sampling technique and seasonal development of phytophagous mirids (Hemiptera: Miridae) on apple in southwestern Quebec. Ann. Entomol. Soc. Am. 76:359–364.

——. 1983b. Spatial dispersion of phytophagous mirids (Hemiptera: Miridae) on apple trees. J. Econ. Entomol. 76: 1242–1247.

——. 1983c. Seasonal development and interplant movements of phytophagous mirids (Hemiptera: Miridae) on alternate host plants in and around an apple orchard. Ann. Entomol. Soc. Am. 76:776–780.

——. 1984. Effect of height and orientation of flight traps for monitoring phytophagous mirids (Hemiptera: Miridae) in an orchard. Rev. Entomol. Qué. 29:17–21.

Boivin, G., G. Mailloux, R. O. Paradis, and J.-G. Pilon. 1981. La punaise terne, *Lygus lineolaris* (P. de B.) (Hemiptera: Miridae), dans le sud-ouest du Québec 1—Information additionelle sur son comportement dans les fraisières et framboisières. Ann. Soc. Entomol. Qué. 26:131–141.

Boivin, G., R. K. Stewart, and I. Rivard. 1982. Sticky traps for monitoring phytophagous mirids (Hemiptera: Miridae) in an apple orchard in southwestern Quebec. Environ. Entomol. 11:1067–1070.

Boivin, G., J.-P. R. Le Blanc, and J. A. Adams. 1991. Spatial dispersion and sequential sampling plan for the tarnished plant bug (Hemiptera: Miridae) on celery. J. Econ. Entomol. 84:158–164.

Bolton, B. 1973. A bacterial pathogen of cocoa capsids. Cocoa Res. Inst. Ghana Rep. 1970–71:155.

Bolton, J. L. and O. Peck. 1946. Alfalfa seed production in northern Saskatchewan as affected by lygus bugs, with a report on their control by burning. Sci. Agric. (Ottawa) 26:130–137.

Bondar, G. 1937. Cancro dos fructos do cacáo, causado por *Monalonion xan-*

thophyllum, Walk, "chupança do cacáu." Rodriguesia 10:179–186.

——. 1939. Notas entomológicas da Bahia, IV. Rev. Entomol. (Rio J.) 10:1–14.

Bonde, R. and E. S. Schultz. 1953. Purpletop wilt and similar diseases of the potato. Maine Agric. Exp. Stn. Bull. 511:1–30.

Boness, M. 1963. Biologisch-ökologische Untersuchungen an Exolygus Wagner (Heteroptera, Miridae) (ein Beitrag zur Agraökologie). Z. Wiss. Zool. 168:376–420.

Booker, R. H. 1969. Resistance of Sahlbergella singularis Hagl. (Hemiptera, Miridae) to the cyclodiene insecticides in Nigeria. Bull. Entomol. Soc. Niger. 2:39–44.

Booth, C. and J. M. Waterston. 1964. Calonectria rigidiuscula. C. M. I. descriptions of pathogenic fungi and bacteria No. 21. Commonwealth Mycological Institute, Surrey, UK. 2 pp.

Booth, C. L. 1990. Biology of Largus californicus (Hemiptera: Largidae). Southwest. Nat. 35:15–22.

Booth, S. R. and H. Riedl. 1996. Diflubenzuron-based management of the pear pest complex in commercial orchards of the Hood River Valley in Oregon. J. Econ. Entomol. 89:621–630.

Borror, D. J., C. A. Triplehorn, and N. F. Johnson. 1989. An introduction to the study of insects, 6th ed. Saunders, Philadelphia. 875 pp.

Bos, L. and J. E. Parlevliet. 1995. Concepts and terminology on plant/pest relationships: Toward concensus in plant pathology and crop protection. Annu. Rev. Phytopathol. 33:69–102.

Boselli, F. B. 1932. Studio biologico degli emitteri che attaccano le nocciuole in Sicilia. Boll. Lab. Zool. Portici 26:142–309.

Bosik, J. J. 1997. Common names of insects & related organisms 1997. Entomological Society of America, Lanham, Md. 232 pp.

Bosque-Perez, N. A., K. W. Foster, and T. F. Leigh. 1987. Heritability of resistance in cowpea to the western plant bug. Crop Sci. 27:1133–1136.

Bostanian, N. J. and L. J. Coulombe. 1986. An integrated pest management program for apple orchards in southwestern Quebec. Can. Entomol. 118:1131–1142.

Bostanian, N. J. and G. Mailloux. 1990. Threshold levels and sequential sampling plans for tarnished plant bug in strawberries. Pages 81–101 in N. J. Bostanian, L. T. Wilson, and T. J. Dennehy, eds., Monitoring and integrated management of arthropod pests of small fruit crops. Intercept, Andover, UK.

Bostanian, N. J., C. Vincent, D. Pitre, and L. G. Simard. 1989. Chemical control of key and secondary arthropod pests of Quebec apple orchards. Appl. Agric. Res. 4:179–184.

Bostanian, N. J., G. Mailloux, M. R. Binns, and P. O. Thibodeau. 1990. Seasonal fluctuations of Lygus lineolaris (Palisot de Beauvois) (Hemiptera: Miridae) nymphal populations in strawberry fields. Agric. Ecosyst. Environ. 30:327–336.

Bostanian, N. J., N. Larocque, C. Vincent, G. Chouinard, and Y. Morin. 2000. Effects of five insecticides used in apple orchards on Hyaliodes vitripennis (Say) (Hemiptera: Miridae). J. Environ. Sci. Health 35B:143–155.

Bostock, R. M. and B. A. Stermer. 1989. Perspectives on wound healing in resistance to pathogens. Annu. Rev. Phytopathol. 27:343–371.

Bostock, R. M., C. S. Thomas, J. M. Ogawa, R. E. Rice, and J. K. Uyemoto. 1987. Relationship of wound-induced peroxidase activity to epicarp lesion development in maturing pistachio fruit. Phytopathology 77:275–282.

Bottger, G. T. 1966. Lygus bugs. Pages 425–427 in C. N. Smith, ed., Insect colonization and mass production. Academic, New York.

Bottger, G. T., E. T. Sheehan, and M. J. Lukefahr. 1964. Relation of gossypol content of cotton plants to insect resistance. J. Econ. Entomol. 57:283–285.

Bottrell, D. G. and P. L. Adkisson. 1977. Cotton insect pest management. Annu. Rev. Entomol. 22:451–481.

Bouchard, D., J.-C. Tourneur, and R. O. Paradis. 1982. Le complexe entomophage limitant les populations d'Aphis pomi de Geer (Homoptera: Aphididae) dans le sud-ouest du Québec. Donnees preliminaires. Ann. Soc. Entomol. Qué. 27:80–93.

Bouchard, D., J. G. Pilon, and J. C. Tourneur. 1986. Role of entomophagous insects in controlling the apple aphid, Aphis pomi, in southwestern Quebec. Pages 369–374 in I. Hodek, ed., Ecology of aphidophaga: Proceedings of the 2nd symposium held at Zvikovské Podhradi, September 2–8, 1984. Junk, Dordrecht.

——. 1988. Voracity of mirid, syrphid, and cecidomyiid predators under laboratory conditions. Pages 231–234 in E. Niemczyk and A. F. G. Dixon, eds., Ecology and effectiveness of aphidophaga. Proceedings of an international symposium, Teresin, Poland, Aug. 31–Sept. 5, 1987. SPB, The Hague.

Bourne, A. I. 1931. Tarnished plant bug (Lygus pratensis L.). U.S. Dep. Agric. Insect Pest Surv. Bull. 11:480.

Bournoville, R. 1975. Relations entre les Hétéroptères Mirides nuisibles à la luzerne porte-graines et la phénologie de la plante. Ann. Zool. Ecol. Anim. 7:197–210.

Bovien, P. and O. Wagn. 1951. Skadedyr på landbrugsplanter. Månedsovers. Plantesyg. No. 319:67–77.

Bowden, J. 1965. Sorghum midge, Contarinia sorghicola (Coq.), and other causes of grain-sorghum loss in Ghana. Bull. Entomol. Res. 56:169–189.

——. 1970. Cotton pests. Pages 178–185, 187 in J. D. Jameson, ed., Agriculture in Uganda, 2nd ed. Oxford University Press, Oxford.

Bowden, J. and W. R. Ingram. 1958. A revised interpretation of the causes of loss of crops of cotton in the drier regions of Uganda. Nature (Lond.) 182:1750.

Boyd, M. L. and G. L. Lentz. 1999. Seasonal occurrence and abundance of the tarnished plant bug (Hemiptera: Miridae) and thrips (Thysanoptera: Thripidae) on rapeseed in West Tennessee. J. Agric. Urban Entomol. 16:171–178.

Boyd, M. L. and J. D. Thomas. 1994. Plant bugs. Pages 74–75 in L. G. Higley and D. J. Boethel, eds., Handbook of soybean insect pests. Entomological Society of America, Lanham, Md.

Boyes, D. G. 1964. The bionomics and control of mirids attacking castor. Pages 12–15 in Symposium on entomological problems, Pretoria, 17–21 July 1961. Dep. Agric. Tech. Serv., Pretoria. Tech. Commun. No. 12.

Bradbury, J. F. 1986. Guide to plant pathogenic bacteria. CAB International, Farnham Royal, Slough, UK. 332 pp.

Bradley, G. A. and J. D. Hinks. 1968. Ants, aphids, and jack pine in Manitoba. Can. Entomol. 100:40–50.

Braimah, S. A., L. A. Kelton, and R. K. Stewart. 1982. The predaceous and phytophagous plant bugs (Heteroptera: Miridae) found on apple trees in Quebec. Nat. Can. (Que.) 109:153–180.

Brako, L., A. Y. Rossman, and D. F. Farr. 1995. Scientific and common names of 7,000 vascular plants in the United States. APS, St. Paul, Minn. 294 pp.

Braman, S. K. and R. J. Beshear. 1994. Seasonality of predacious plant bugs (Heteroptera: Miridae) and phytophagous thrips (Thysanoptera: Thripidae) as influenced by host plant phenology of native azaleas (Ericales: Ericaceae). Environ. Entomol. 23:712–718.

Braman, S. K. and K. V. Yeargan. 1988. Comparison of developmental and reproductive rates of *Nabis americoferus*, *N. roseipennis*, and *N. rufusculus* (Hemiptera: Nabidae). Ann. Entomol. Soc. Am. 81:923–930.

Branson, T. F. and J. L. Krysan. 1981. Feeding and oviposition behavior and life cycle strategies of *Diabrotica*: An evolutionary view with implications for pest management. Environ. Entomol. 10:826–831.

Braudeau, J. 1974. The cocoa tree: Agronomic aspects. Pages 1–12 *in* P. H. Gregory, ed., Phytophthora disease of cocoa. Longman, London.

Braune, H. J. 1971. Der Einfluss der Temperature auf Eidiapause und Entwicklung von Weichwanzen (Heteroptera, Miridae). Oecologia (Berl.) 8:223–266.

———. 1976. Effects of temperature on the rates of oxygen consumption during morphogenesis and diapause in the egg stage of *Leptopterna dolabrata* (Heteroptera, Miridae). Oecologia (Berl.) 25:77–87.

———. 1980. Ökophysiologische Untersuchungen über die Steuerung der Eidiapause bei *Leptopterna dolabrata* (Heteroptera, Miridae). Zool. Jahrb. Abt. Syst. Oekol. Geogr. Tiere 107: 32–112.

———. 1983. The influence of environmental factors on wing polymorphism in females of *Leptopterna dolobrata* [sic] (Heteroptera, Miridae). Oecologia (Berl.) 60:340–347.

Breddin, G. 1896. Nachahmungserscheinungen bei Rhynchoten. Z. Naturwiss. 69:17–45.

Breene, R. G. and W. L. Sterling. 1988. Quantitative phosphorus-32 labeling method for analysis of predators of the cotton fleahopper (Hemiptera: Miridae). J. Econ. Entomol. 81:1494–1498.

Breene, R. G., W. L. Sterling, and D. A. Dean. 1988. Spider and ant predators of the cotton fleahopper on woolly croton. Southwest. Entomol. 13:177–183.

———. 1989a. Predators of the cotton fleahopper on cotton. Southwest. Entomol. 14:159–166.

Breene, R. G., W. R. Martin Jr., D. A. Dean, and W. L. Sterling. 1989b. Rearing methods for the cotton fleahopper. Southwest. Entomol. 14:249–253.

Breene, R. G., A. W. Hartstack, W. L. Sterling, and M. Nyffeler. 1989c. Natural control of the cotton fleahopper, *Pseudatomoscelis seriatus* (Reuter) (Hemiptera, Miridae), in Texas. J. Appl. Entomol. 108:298–305.

Breene, R. G., W. L. Sterling, and M. Nyffeler. 1990. Efficacy of spider and ant predators on the cotton fleahopper

[Hemiptera: Miridae]. Entomophaga 35:393–401.

Brett, C. H., R. R. Walton, and E. E. Ivy. 1946. The cotton flea hopper, *Psallus seriatus* (Reut.), in Oklahoma. Okla. Agric. Exp. Stn. Tech. Bull. T-24:1–31.

Brew, A. H. 1992. Towards the use of pathogenic micro-organisms in the control of some cocoa insect pests in Ghana. Cocoa Grow. Bull. 45:26–30.

Brewer, P. S. and W. F. Campbell. 1977. Anatomical studies of *Labops hesperius* damage to crested wheatgrass leaves. Agron. Abstr. 1977:50.

Brewer, P. S., W. F. Campbell, and B. A. Haws. 1979. How black grass bugs operate. Utah Sci. 40:21–23.

Brice, A. T., K. H. Dahl, and C. R. Grau. 1989. Pollen digestibility by hummingbirds and psittacines. Condor 91:681–688.

Brierley, P. 1933. Dahlia mosaic and its relation to stunt. Boyce Thompson Inst. Plant Res. Prof. Pap. 1(25):240–246.

Brindley, M. D. H. 1930. On the metasternal scent-glands of certain Heteroptera. Trans. Entomol. Soc. Lond. 78:199–207.

Brindley, M. D. [H.] 1935. The means by which the resemblance of the British capsid bug, *Pilophorus cinnamopterus* Kb., to the "wood ant," *Formica rufa* Linn., is produced. Proc. R. Entomol. Soc. Lond. 9:91–92.

Brinkhurst, R. O. 1963. Observations on wing-polymorphism in the Heteroptera. Proc. R. Entomol. Soc. Lond. (A) 38:15–22.

Bristow, C. M. 1988. What makes a predator specialize? Trends Ecol. Evol. 3:1–2.

Bristowe, W. S. 1941. The comity of spiders. Vol. II. Ray Society, London. pp. 229–560.

Brittain, W. H. 1916a. The green apple bug (*Lygus invitus* Say.) in Nova Scotia. 46th Annu. Rep. Entomol. Soc. Ont. 1915:65–78.

———. 1916b. [Discussion of mirid papers by Brittain and Crawford]. 46th Annu. Rep. Entomol. Soc. Ont. 1915:88.

———. 1919. Notes on *Lygus campestris* Linn. in Nova Scotia. Proc. Entomol. Soc. N.S. 1918:76–81.

———. 1929. Insects of the season 1928 in Nova Scotia. 59th Annu. Rep. Entomol. Soc. Ont. 1928:8–10.

Brittain, W. H. and L. G. Saunders. 1918. Notes on the biology of *Lygus pratensis* Linn., in Nova Scotia. Proc. Entomol. Soc. N.S. 1917:85.

Brodbeck, B. and D. Strong. 1987. Amino acid nutrition of herbivorous insects and stress to host plants. Pages 347–364 *in* P. Barbosa and J. C. Schultz,

eds., Insect outbreaks. Academic, New York.

Broersma, D. B. and W. H. Luckmann. 1970. Effects of tarnished plant bug feeding on soybean. J. Econ. Entomol. 63:253–256.

Brooks, F. E. 1921. Walnut husk-maggot. U.S. Dep. Agric. Bur. Entomol. Bull. 992:1–8.

Brown, H. B. and J. O. Ware. 1958. Cotton, 3rd ed. McGraw-Hill, New York. 566 pp.

Brown, J. K. 1994. Current status of *Bemisia tabaci* as a plant pest and virus vector in agroecosystems worldwide. FAO Plant Prot. Bull. 42:3–32.

Brown, J. K., D. R. Frolich, and R. S. Rosell. 1995. The sweetpotato or silverleaf whiteflies: Biotypes of *Bemisia tabaci* or a species complex? Annu. Rev. Entomol. 40:511–534.

Brown, J. M. 1924. A contribution to our knowledge of the life-history of *Heterocordylus genistae* Scop. Entomol. Mon. Mag. 60:249–251.

Brown, J. Z., D. C. Steinkraus, N. P. Tugwell, and T. G. Teague. 1997. The effects and persistence of the fungus *Beauveria bassiana* (Mycotrol) and imidacloprid (Provado) on tarnished plant bug mortality and feeding. *In* Proc. Beltwide Cotton Prod. Res. Conf., Jan. 6–10, 1997, New Orleans, La., pp. 1302–1305. National Cotton Council, Memphis.

Brown, K. S. Jr. 1991. Conservation of neotropical environments: Insects as indicators. Pages 349–404 *in* N. M. Collins and J. A. Thomas, eds., The conservation of insects and their habitats. Academic, San Diego.

Brown, N. R. and R. C. Clark. 1956. Studies on predators of the balsam woolly aphid, *Adelges piceae* (Ratz.) (Homoptera: Adelgidae) II. An annotated list of the predators associated with the balsam woolly aphid in eastern Canada. Can. Entomol. 88:678–683.

Brown, V. K. 1982a. Size and shape as ecological discriminants in successional communities of Heteroptera. Biol. J. Linn. Soc. 18:279–290.

———. 1982b. The phytophagous insect community and its impact on early successional habitats. Pages 205–213 *in* J. H. Visser and A. K. Minks, eds., Proc. 5th Int. Symp. Insect-Plant Relationships, Wageningen, The Netherlands, 1–4 March 1982. Centre for Agricultural Publishing & Documentation, Wageningen.

———. 1990. Insect herbivores, herbivory and plant succession. Pages 183–196 *in* F. Gilbert, ed., Insect life cycles:

Genetics, evolution and co-ordination. Springer, London.

Brown, V. K. and T. R. E. Southwood. 1983. Trophic diversity, niche breadth and generation times of exopterogote insects in a secondary succession. Oecologia (Berl.) 56:220–225.

Brues, C. T. 1946. Insect dietary. Harvard University Press, Cambridge. 466 pp.

Bruin, J., M. Dicke, and M. W. Sabelis. 1992. Plants are better protected against spider-mites after exposure to volatiles from infested conspecifics. Experientia (Basel) 48:525–529.

Bruneau de Miré, P. 1969. Une fourmi utilisée au Cameroun dans la lutte contre les mirides du cacaoyer *Wasmannia auropuncta* Roger. Café Cacao Thé 13:209–212.

——. 1970. Observations sur les fluctuations saisonnières d'une population de *Sahlbergella singularis* au Cameroun. Café Cacao Thé 14:202–208.

——. 1977. La dynamique des populations de Mirides et ses implications. Pages 171–186 in E. M. Lavabre, ed., Les Mirides du cacaoyer. Institut français du Café et du Cacao, Paris.

——. 1985. Enquête sur la tolérance des mirides du cacaoyer aux insecticides au Cameroun. Café Cacao Thé 29:183–196.

Bruner, L. 1895. Insect enemies of the grape-vine. Nebr. State Hortic. Soc. Annu. Rep. 1895:68–162.

Bruner, S. C. 1934. Notes on Cuban Dicyphinae (Hemiptera, Miridae). Mem. Soc. Cub. Hist. Nat. 'Felipe Poey' 8:35–49.

Bryan, D. E., C. G. Jackson, R. L. Carranza, and E. G. Neemann. 1976. *Lygus hesperus*: Production and development in the laboratory. J. Econ. Entomol. 69:127–129.

Bryson, H. R. 1937. A plant bug (*Labopidea allii* Knight). U.S. Dep. Agric. Insect Pest Surv. Bull. 17:186.

Brzeziński, K. 1988. Prospects for using *Macrolophus costalis* Fieb. (Heteroptera: Miridae) in integrated programs for controlling aphids on glasshouse crops. Pages 285–288 in E. Niemczyk and A. F. G. Dixon, eds., Ecology and effectiveness of aphidophaga. Proceedings of an international symposium, Teresin, Poland, Aug. 31–Sept. 5, 1987. SPB, The Hague.

Buchner, P. 1965. Endosymbiosis of animals with plant microorganisms; revised English version. Interscience, New York. 909 pp.

Buckell, E. R. 1939. Insects of the season 1938 in British Columbia. Can. Insect Pest Rev. 17:82–86.

Buczek, D. 1956. Obserwacje nad biologią i morfologią stadiów larwalnych

pluskwiaków (Hem. Heter.) z podrodziny Mirinae (Miridae) stwierdzonych na łąkach w okolicach Lublina. Ann. Univ. Marie Curie-Sklowdowska (C) 10:269–314. [English summary.]

Budgen, L. M. [Acheta Domestica, pseud.]. 1851. Episodes of insect life. Vol. 3. Reeve, Benham & Reeve, London. 434 pp.

Bugg, R. L. and C. Waddington. 1994. Using cover crops to manage arthropod pests of orchards: A review. Agric. Ecosyst. Environ. 50:11–28.

Bugg, R. L. and L. T. Wilson. 1989. *Ammi visnaga* (L.) Lamarck (Apiaceae): Associated beneficial insects and implications for biological control, with emphasis on the bell-pepper agroecosystem. Biol. Agric. Hortic. 6:241–268.

Bugg, R. L., L. E. Ehler, and L. T. Wilson. 1987. Effect of common knotweed (*Polygonum aviculare*) on abundance and efficiency of insect predators of crop pests. Hilgardia 55(7):1–52.

Bugg, R. L., S. C. Phatak, and J. D. Dutcher. 1990a. Insects associated with cool-season cover crops in southern Georgia: Implications for pest control in truck-farm and pecan agroecosystems. Biol. Agric. Hortic. 7:17–45.

Bugg, R. L., F. L. Wäckers, K. E. Brunson, S. C. Phatak, and J. D. Dutcher. 1990b. Tarnished plant bug (Hemiptera: Miridae) on selected cool-season leguminous cover crops. J. Entomol. Sci. 25:463–474.

Bugg, R. L., F. L. Wäckers, K. E. Brunson, J. D. Dutcher, and S. C. Phatak. 1991. Cool-season cover crops relay intercropped with cantaloupe: Influence on a generalist predator, *Geocoris punctipes* (Hemiptera: Lygaeidae). J. Econ. Entomol. 84:408–416.

Bull, R. M. 1981. Population studies on the sugar cane leafhopper (*Perkinsiella saccharicida* Kirk.) in the Bundaberg District. Proc. Aust. Soc. Sugar Cane Technol. 1981:293–303.

Bulyginskaya, M. A. and V. M. Kalinkin. 1985. Effects of chemical treatments against the codling moth *Laspeyresia pomonella* L. (Lepidoptera, Tortricidae) on associated lepidopterous species and their natural enemies. Entomol. Obozr. 64(3):441–449. [in Russian, English translation in Entomol. Rev. 65(3):14–23, 1986.]

Büning, J. 1994. The insect ovary: Ultrastructure, previtellogenic growth and evolution. Chapman & Hall, London. 400 pp.

Buntin, G. D. 1988. *Trigonotylus doddi* (Distant) as a pest of bermudagrass: Damage potential, population dynam-

ics, and management by cutting. J. Agric. Entomol. 5:217–224.

Buntin, G. D., S. K. Braman, D. A. Gilbertz, and D. V. Phillips. 1996. Chlorosis, photosynthesis, and transpiration of azalea leaves after azalea lace bug (Heteroptera: Tingidae) feeding injury. J. Econ. Entomol. 89:990–995.

Burden, B. J., P. W. Morgan, and W. L. Sterling. 1989. Indole-acetic acid and the ethylene precursor, ACC, in the cotton fleahopper (Hemiptera: Miridae) and their role in cotton square abscission. Ann. Entomol. Soc. Am. 82:476–480.

Burel, F. 1996. Hedgerows and their role in agricultural landscapes. Crit. Rev. Plant Sci. 15:169–190.

Burgess, L., J. Duck, and D. L. McKenzie. 1983. Insect vectors of the yeast *Nematospora coryli* in mustard, *Brassica juncea*, crops in southern Saskatchewan. Can. Entomol. 115:25–30.

Burghardt, G., W. Riess, and E. M. Wolfram. 1975. Zur Bedeutung der Wanzen als Aufzuchtnahrung für die Nestlinge einheimischer in Hecken brütender Vogelarten (Insecta: Heteroptera; Aves: Passiformes). Waldhygiene 11:21–25.

Burke, H. E. 1923. Black plant-bug (*Irbisia brachycerus* Uhler). U.S. Dep. Agric. Insect Pest Surv. Bull. 3:95.

Burks, B. D. 1979. Family Trichogrammatidae. Pages 1033–1043 in K. V. Krombein et al., eds., Catalog of Hymenoptera in America north of Mexico. Vol. 1. Symphyta and Apocrita (Parasitica). Smithsonian Institution Press, Washington, D.C.

Burmeister, H. 1835. Handbuch der Entomologie. Zweiter Band. Erste Abtheilung. Schnabelkerfe. Rhynchota. Theod. Ehr. Friedr. Enslin, Berlin. 400 pp.

Burton, V. E. 1978. A beltwide review of the impact of the nectariless character of cotton on pest management programs. In Proc. Beltwide Cotton Prod. Res. Conf., Jan. 9–11, Dallas, Tex., pp. 125–126. National Cotton Council, Memphis.

——. 1991. Insects. Pages 1–17 in Dry bean pest management guidelines (UCPMG Publ. 19). Univ. Calif. Div. Agric. Nat. Resour. Publ. 3339.

Bus, V. G. M., P. J. M. Mols, and L. H. M. Blommers. 1985. Monitoring of the green capsid bug *Lygocoris pabulinus* (L.) (Hemiptera: Miridae) in apple orchards. Meded. Fac. Landbouwwet. Rijksuniv. Gent 50(2b):505–510.

Buschman, L. L., W. H. Whitcomb, R. C. Hemenway, D. L. Mays, N. Ru, N. C. Leppla, and B. J. Smittle. 1977. Predators of velvetbean caterpillar eggs in Florida soybeans. Environ. Entomol. 6:403–407.

Bushing, R. W., V. E. Burton, and C. L. Tucker. 1974. Dry large lima beans benefit from lygus bug control. Calif. Agric. 28(5):14–15.

Butani, D. K. 1979. Insects and fruits. Periodical Expert Book Agency, Delhi. 415 pp.

Butler, E. A. 1918. On the association between the Hemiptera-Heteroptera and vegetation. Entomol. Mon. Mag. 54:132–136.

——. 1922. A contribution to the life-history of *Deraeocoris ruber* L. Entomol. Mon. Mag. 58:200–204.

——. 1923. A biology of the British Hemiptera-Heteroptera. Witherby, London. 682 pp.

——. 1924. Notes on the early stages of British Heteroptera made during 1924. Entomol. Mon. Mag. 60:265–268.

——. 1925. An invasion of *Halticus saltator* Geoffr. (Hemiptera). Entomol. Mon. Mag. 61:276–279.

Butler, E. J. and S. G. Jones. 1949. Plant pathology. Macmillan, London. 979 pp.

Butler, G. D. Jr. 1965. *Spanogonicus* [sic] *albofasciatus* as an insect and mite predator (Hemiptera: Miridae). J. Kans. Entomol. Soc. 38:70–75.

——. 1968. Sugar for the survival of *Lygus hesperus* on alfalfa. J. Econ. Entomol. 61:854–855.

——. 1970. Temperature and the development of *Spanogonicus* [sic] *albofasciatus* and *Rhinacloa forticornis*. J. Econ. Entomol. 63:669–670.

——. 1972. Flight times of *Lygus hesperus*. J. Econ. Entomol. 65:1299–1300.

Butler, G. D. Jr. and T. J. Henneberry. 1976. Temperature-dependent development rate tables for insects associated with cotton in the Southwest. U.S. Dep. Agric. Agric. Res. Serv. ARS W-38:1–36.

Butler, G. D. Jr. and P. L. Ritchie Jr. 1971. Feed Wheast and the abundance and fecundity of *Chrysopa carnea*. J. Econ. Entomol. 64:933–934.

Butler, G.D. Jr. and A. Stoner, 1965. The biology of *Spanogonicus* [sic] *albofasciatus*. J. Econ. Entomol. 58: 664–665.

Butler, G. D. Jr. and A. L. Wardecker. 1970. Fluctuations of populations of *Lygus hesperus* in alfalfa in Arizona. J. Econ. Entomol. 63:1111–1114.

——. 1971. Temperature and the development of eggs and nymphs of *Lygus hesperus*. Ann. Entomol. Soc. Am. 64: 144–145.

Butler, G. D. Jr. and F. L. Watson. 1974. A technique for determining the rate of development of *Lygus hesperus* in fluctuating temperatures. Fla. Entomol. 57: 225–230.

Butler, G. D. Jr., M. H. Schonhorst, and F. Watson. 1971. Cutting alfalfa for hay timed to reduce buildup of lygus bug populations. Progr. Agric. Ariz. 23(6): 12–13.

Butler, G. D. Jr., G. M. Loper, S. E. McGregor, J. L. Webster, and H. Margolis. 1972. Amounts and kinds of sugars in the nectars of cotton (*Gossypium* spp.) and the time of their secretion. Agron. J. 64:364–368.

Butler, L., G. A. Chrislip, V. A. Kondo, and E. C. Townsend. 1997. Effect of diflubenzuron on nontarget canopy arthropods in closed, deciduous watersheds in a central Appalachian forest. J. Econ. Entomol. 90:784–794.

Buttery, R. G., J. A. Kamm, and L. C. Ling. 1984. Volatile components of red clover leaves, flowers, and seed pods: Possible insect attractants. J. Agric. Food Chem. 32:254–256.

Butts, R. A. 1984. Factors influencing lygus bug damage in alfalfa seed grown in the Peace River region of Alberta. Proc. Entomol. Soc. Manit. 40:18.

Butts, R. A. and R. J. Lamb. 1990a. Injury to oilseed rape caused by mirid bugs (*Lygus*) (Heteroptera: Miridae) and its effect on seed production. Ann. Appl. Biol. 117:253–266.

——. 1990b. Comparison of oilseed *Brassica* crops with high or low levels of glucosinolates and alfalfa as hosts for three species of *Lygus* (Hemiptera: Heteroptera: Miridae). J. Econ. Entomol. 83:2258–2262.

——. 1991a. Seasonal abundance of three *Lygus* species (Heteroptera: Miridae) in oilseed rape and alfalfa in Alberta. J. Econ. Entomol. 84:450–456.

——. 1991b. Pest status of lygus bugs (Hemiptera: Miridae) in oilseed *Brassica* crops. J. Econ. Entomol. 84:1591–1596.

Buyckx, E. J. E., ed. 1962. Précis des maladies et des insectes nuisibles rencontrés sur les plantes cultivées au Congo, au Rwanda et au Burundi. Publ. Inst. Nat. Etude Agron. Congo, Brussels. Hors sér. 708 pp.

Byazdenka, T. T., V. G. Osipaw, and T. Ya. Palyakova. 1975. Role of entomophages in reducing the numbers of the apple sucker. Vyestsi Akad. Navuk BSSR Syer. Syel'skhaspad Navuk 1:65–69. [in Belorussian.]

Byerly, K. F., A. P. Gutierrez, R. E. Jones, and R. F. Luck. 1978. A comparison of sampling methods for some arthropod populations in cotton. Hilgardia 46(8):257–282.

Byrne, D. N. and W. B. Miller. 1990. Carbohydrate and amino acid composition of phloem sap and honeydew produced by *Bemisia tabaci*. J. Insect Physiol. 36:433–439.

Byrne, D. N., A. C. Cohen, and E. A. Draeger. 1990. Water uptake from plant tissues by the egg pedicel of the greenhouse whitefly, *Trialeurodes vaporariorum* (Westwood) (Homoptera: Aleyrodidae). Can. J. Zool. 68:1193–1195.

Caccia, R., M. Baillod, and G. Mauri. 1980. Dégâts de la punaise verte de la vigne dans les vignobles de la Suisse italienne. Rev. Suisse Vitic. Arboric. Hortic. 12:275–279.

Cadou, J. 1993. Les Miridae du cotonnier en Afrique et à Madagascar. CIRAD-CA, Paris. 74 pp.

Caesar, L. [I.] 1913. Some new or unrecorded Ontario insect pests. 43rd Annu. Rep. Entomol. Soc. Ont. 1912: 100–105.

Caesar, L. I. 1919. Insects as agents in the dissemination of plant diseases. 49th Annu. Rep. Entomol. Soc. Ont. 1918: 60–66.

Caesar, L. [I.] 1921. Notes on leaf bugs (Miridae) attacking fruit trees in Ontario. 51st Annu. Rep. Entomol. Soc. Ont. 1920:14–16.

Cagampang, G. B., M. D. Pathak, and B. O. Juliano. 1974. Metabolic changes in the rice plant during infestation by the brown planthopper, *Nilaparvata lugens* Stål (Hemiptera: Delphacidae). Appl. Entomol. Zool. 9:174–184.

Cagle, L. R. and H. W. Jackson. 1947. Life history of the garden fleahopper. Va. Agric. Exp. Stn. Tech. Bull. 107:1–27.

Caldwell, D. L. 1981. The control and impact of *Lygus lineolaris* (P. de B.) on first-year peach scion stock in Missouri. Ph.D. dissertation, University of Missouri, Columbia. 141 pp.

Caldwell, D. L. and D. L. Schuder. 1979. The life history and description of *Phylloxera caryaecaulis* on shagbark hickory. Ann. Entomol. Soc. Am. 72:384–390.

Calilung, V. J., E. Mituda-Sabado, and M. Malabayabas. 1994. Management of thrips and mites in lowland potato: I. Identity, biology and natural enemies of the pests. Philipp. Entomol. 9:435–442.

Callan, E. McC. 1943. Natural enemies of the cacao thrips. Bull. Entomol. Res. 34:313–321.

——. 1975. Miridae of the genus *Termatophylidea* [Hemiptera] as predators of cacao thrips. Entomophaga 20:389–391.

Calnaido, D. 1959. Notes on the distribution and biology of the lygus bug—*Lygus viridanus* Motsch—(Heteroptera-Miridae), a pest of tea in Ceylon. Tea Q. 30:108–112.

Caltagirone, L. E. and R. L. Doutt. 1989. The history of the vedalia beetle importation to California and its impact on the development of biological control. Annu. Rev. Entomol. 34:1–16.

Camargo, E. P. and F. G. Wallace. 1994. Vectors of plant parasites of the genus *Phytomonas* (Protozoa, Zoomastigophorea, Kinetoplastida). Adv. Dis. Vector Res. 10:333–359.

Campbell, B. C. and P. J. Shea. 1990. A simple staining technique for assessing feeding damage by *Leptoglossus occidentalis* Heidemann (Hemiptera: Coreidae) on cones. Can. Entomol. 122: 963–968.

Campbell, B. C., J. D. Steffen-Campbell, and R. J. Gill. 1994. Evolutionary origin of whiteflies (Hemiptera: Sternorrhyncha: Aleyrodidae) inferred from 18S rDNA sequences. Insect Mol. Biol. 3: 73–88.

Campbell, J. M. 1982. A revision of the genus *Lordithon* Thomson of North and Central America (Coleoptera: Staphylinidae). Mem. Entomol. Soc. Can. 119:1–116.

Campbell, W. F. and P. S. Brewer. 1978. Physical characteristics of grasses related to insect damage and resistance. Pages 27–28 *in* B. A. Haws, compiler, Economic impacts of *Labops hesperius* on the production of high quality range grasses. Final Rep. Utah Agric. Exp. Stn. to Four Corners Comm. Utah State University, Logan.

Campbell, W. F., B. A. Haws, K. H. Asay, and J. D. Hansen. 1984. A review of black grass bug resistance in forage grasses. J. Range Manage. 37:365–369.

Čamprag, D., R. Sekulić, T. Kereši, R. Almaši, R. Thalji, and I. Balarin. 1986a. Višegodišnja proučavanja pojave stenica iz roda *Lygus* (Heteroptera, Miridae) na suncokretu u Vojvodini. Zašt. Bilja 37:21–30.

Čamprag, D., R. Sekulić, T. Kereši, R. Almaši, and D. Stojanovìc. 1986b. Uticaj napada stenice *Lygus rugulipennis* Popp. (Heteroptera, Miridae) na kvalitet semena suncokretu. Zašt. Bilja 37:101–110.

Čamprag, D., R. Sekulić, T. Kereši, R. Almaši, R. Thalji, and D. Stojanović. 1986c. Proučavanje štetnosti stenice *Lygus rugulipennis* Popp. na suncokretu u Vojvodini. Savrem. Poljopr. 34:41–51.

Cancelado, R. and T. R. Yonke. 1970. Effect of prairie burning on insect populations. J. Kans. Entomol. Soc. 43:274–281.

Cantelo, W. W. and M. Jacobson. 1979. Corn silk volatiles attract many pest species of moths. J. Environ. Sci. Health A14:695–707.

Cao, R. 1986. Field survey of the predators of cotton plant-bug and their effects. Chinese J. Biol. Control 2(4):182. [in Chinese.]

Capco, S. R. 1941. Notes on the orchid bug, *Mertila malayensis* Distant, on white mariposa (*Phalaenopsis amabilis* Blume). Philipp. J. Sci. 75:185–193.

Capinera, J. L. and M. R. Walmsley. 1978. Visual responses of some sugarbeet insects to sticky traps and water pan traps of various colors. J. Econ. Entomol. 71:926–927.

Capizzi [J. Jr.] and [R. L.] Penrose. 1978. A plant bug (*Psallus ancorifer*). U.S. Dep. Agric. Coop. Plant Pest Rep. 3(29):368.

Carapezza, A. 1995. The specific identities of *Macrolophus melanotoma* (A. Costa, 1853) and *Stenodema curticolle* (A. Costa, 1853) (Insecta Heteroptera, Miridae). Nat. Sicil. 19:295–298.

Carayon, J. 1949. *Helopeltis* (Hem. Miridae) nouveaux nuisibles aux quinquinas en Afrique française. Bull. Mus. Natl. Hist. Nat. 21(2):558–565.

———. 1954. Organes assumant les fontions de la spermathèque chez divers Hétéroptères. Bull. Soc. Zool. Fr. 9: 189–197.

———. 1958. Études sur les Hémiptères Cimicoidea. Mém. Mus. Natl. Hist. Nat. Ser. A. Zool. 16:141–172.

———. 1960. *Stethoconus frappai* n. sp., miridé prédateur du tingidé du caféier, *Dulinius unicolor* (Sign.), a Madagascar. J. Agric. Trop. Bot. Appl. 7:110–120.

———. 1961. Quelques remarques sur les Hémiptères-Hétéroptères: Leur importance comme insectes auxiliaires et les possibilités de leur utilisation dans la lutte biologique. Entomophaga 6:133–141.

———. 1966. Traumatic insemination and the paragenital system. Pages 81–166 *in* R. L. Usinger. Monograph of Cimicidae (Hemiptera-Heteroptera). Thomas Say Found. Vol. 7. Entomological Society of America, College Park, Md.

———. 1971. Notes et documents sur l'appareil odorant métathoracique des Hémiptères. Ann. Soc. Entomol. Fr. 7:737–770.

———. 1977. Caractères généraux des Hémiptères Bryocorinae. Pages 13–34 *in* E. M. Lavabre, ed., Les Mirides du cacaoyer. Institut français du Cafe et du Cacao (I.F.C.C.), Paris.

———. 1984. Faits remarquables accompagnant l'insémination chez certains Hétéroptères Miridae. Bull. Soc. Entomol. Fr. 89:982–998.

———. 1986. *Macrolophus caliginosus*, Hémiptère Miridae, à reproduction hivernale. Entomologiste (Paris) 42: 257–262.

———. 1989. Parthénogénèse constante prouvée chez deux Hétéroptères: Le miride *Campyloneura virgula* et l'anthocoride *Calliodis maculipennis*. Ann. Soc. Entomol. Fr. 25:387–391.

Carayon, J. and R. Delattre. 1948. Les *Helopeltis* (Hem. Heteroptera) nuisibles de Côte d'Ivoire. Rev. Pathol. Vég. Entomol. Agric. Fr. 27:185–194.

Carayon, J. and J.-R. Steffan. 1959. Observations sur le régime alimentaire des *Orius* et particulièrement d'*Orius pallidicornis* (Reuter) (Heteroptera Anthocoridae). Cah. Nat. 15 (n.s.):53–63.

Carl, K. [P.] 1965. Mirids on cultivated plants in Europe and their natural enemies: A literature review. Rep. Commonw. Inst. Biol. Control Eur. Stn., Delémont, Switzerland. 16 pp.

Carl, K. P. 1982. Alfalfa plant bug (*Adelphocoris lineolatus*): Work in Europe in 1982. Rep. Commonw. Inst. Biol. Control Eur. Stn., Delémont, Switzerland. 6 pp.

Carlson, E. C. 1956. Lygus bug injury and control on carrot seed in northern California. J. Econ. Entomol. 49:689–696.

———. 1959. Evaluation of insecticides for lygus bug control and their effect on predators and pollinators. J. Econ. Entomol. 52:461–466.

———. 1961. Investigations of lygus bug damage to table beet seed plants. Calif. Agric. 15(6):12–14.

———. 1964. Damage to safflower plants by thrips and lygus bugs and a study of their control. J. Econ. Entomol. 57:140–145.

———. 1966. Further studies of damage to safflower plants by thrips and lygus bugs. J. Econ. Entomol. 59:138–141.

Carlson, J. W. 1940. Lygus bug damage to alfalfa in relation to seed production. J. Agric. Res. (Wash., D.C.) 61:791–815.

Carlson, R. W. 1979. Family Ichneumonidae. Pages 315–740 *in* K. V. Krombein et al., eds., Catalog of Hymenoptera in America north of Mexico. Vol. 1. Symphyta and Apocrita (Parasitica). Smithsonian Institution Press, Washington, D.C.

Caron, D. M. 1978. Other insects. Pages 158–185 *in* R. A. Morse, ed., Honey bee pests, predators, and diseases. Cornell University Press, Ithaca.

Caron, D. M., R. C. Lederhouse, and R. A. Morse. 1974. Insect pollinators of onion in New York State. Univ. Md. Coop. Ext. Serv. Entomol. Leafl. 89:1–5.

Carpenter, G. H. 1912. Injurious insects and other animals observed during the year 1911. Econ. Proc. R. Dublin Soc. 2:53–78.

——. 1920. Injurious insects and other animals observed in Ireland during the years 1916, 1917, and 1918. Econ. Proc. R. Dublin Soc. 2:259–272.

Carpintero, D. L. and J. C. M. Carvalho. 1993. An annotated list of the Miridae of the Argentine Republic (Hemiptera). Rev. Bras. Biol. 53:397–420.

Carrel, J. E. and T. Eisner. 1974. Cantharidin: Potent feeding deterrent to insects. Science (Wash., D.C.) 183:755–757.

Carroll, C. R. and C. A. Hoffman. 1980. Chemical feeding deterrent mobilized in response to insect herbivory and counteradaptation by Epilachna tredecimnotata. Science (Wash., D.C.) 209: 414–416.

Carroll, D. P. and S. C. Hoyt. 1984. Natural enemies and their effects on apple aphid, Aphis pomi DeGeer (Homoptera: Aphididae), colonies on young apple trees in central Washington. Environ. Entomol. 13:469–481.

Carroll, S. P. and J. E. Loye. 1990. Male-biased sex ratios, female promiscuity, and copulatory mate guarding in an aggregating tropical bug, Dysdercus bimaculatus. J. Insect Behav. 3:33–48.

Carter, W. 1952. Injuries to plants caused by insect toxins. II. Bot. Rev. 18:680–721.

——. 1973. Insects in relation to plant disease. Wiley, New York. 759 pp.

Caruthers, R. I. and J. A. Onsager. 1993. Perspective on the use of exotic natural enemies for biological control of pest grasshoppers (Orthoptera: Acrididae). Environ. Entomol. 22:885–903.

Carvalho, J. C. M. 1951. Neotropical Miridae, XLI: Tenthecoris orchidearum (Reuter, 1902) in Britain, and a key to the species of the genus (Hemiptera). Ann. Mag. Nat. Hist. (12)4:294–304.

——. 1952. On the major classification of the Miridae (Hemiptera) (with keys to subfamilies and tribes and a catalogue of the world genera). An. Acad. Bras. Cienc. 24:31–110.

——. 1954a. XIV. Neotropical Miridae, LXVII: Genus Ranzovius Distant, predacious on eggs of Theridion (Araneida) in Trinidad (Hemiptera). Ann. Mag. Nat. Hist. (12)7:92–96.

——. 1954b. Neotropical Miridae, LXXIV: Two new genera of Cylapinae from Brazil (Hemiptera). Proc. Iowa Acad. Sci. 61:504–510.

——. 1955. Keys to the genera of Miridae of the world (Hemiptera). Bol. Mus. Para Emilio Goeldi Nova Ser. Zool. 11(2):1–151.

——. 1956. Insects of Micronesia-

Heteroptera: Miridae. Pages 1–100 in Insects of Micronesia. Vol. 7, No. 1. Bernice P. Bishop Museum, Honolulu.

——. 1957–1960. Catalogue of the Miridae of the world. Arq. Mus. Nac. Rio J. Part I. Cylapinae, Deraeocorinae, Bryocorinae 44(1):1–158 (1957); Part II. Phylinae 45(2):1–216 (1958); Part III. Orthotylinae 47(3):1–161 (1958); Part IV. Mirinae 48(4):1–384 (1959); Part V. Bibliography and general index 51(5): 1–194 (1960).

——. 1976. On Ragmus srilankensis n.sp. found in wasp nest in Sri Lanka (Hemiptera, Miridae). Rev. Bras. Biol. 36:317–319.

——. 1984a. On the subfamily Paulocorinae [sic] Carvalho (Hemiptera, Miridae). Rev. Bras. Biol. 44:81–86.

——. 1984b. Mirideos neotropicais, CCXLVIII: Dois generos e quartorze espécies novas da tribo Clivinemini Reuter (Hemiptera). Rev. Bras. Biol. 44:313–327.

Carvalho, J. C. M. and R. C. Froeschner. 1987. Taxonomic names proposed in the insect order Heteroptera by José Candido de Melo Carvalho from 1943 to January 1985, with type depositories. J. N.Y. Entomol. Soc. 95:121–224.

——. 1990. Taxonomic names proposed in the insect order Heteroptera by José Candido de Melo Carvalho from January 1985 to January 1989, with type depositories. J. N.Y. Entomol. Soc. 98:310–346.

——. 1994. Taxonomic names proposed in the insect order Heteroptera by José Candido de Melo Carvalho from January 1989 to January 1993. J. N.Y. Entomol. Soc. 102:481–508.

Carvalho, J. C. M. and G. F. Gross. 1979. The tribe Hyalopeplini of the world (Hemiptera: Miridae). Rec. S. Aust. Mus. (Adel.) 17:429–531.

Carvalho, J. C. M. and R. F. Hussey. 1954. On a collection of Miridae (Hemiptera) from Paraguay, with descriptions of three new species. Occas. Pap. Mus. Zool. Univ. Mich. 552:1–11.

Carvalho, J. C. M. and D. Leston. 1952. The classification of the British Miridae (Hem.), with keys to genera. Entomol. Mon. Mag. 88:231–251.

Carvalho, J. C. M. and L. M. Lorenzato. 1978. The Cylapinae of Papua New Guinea (Hemiptera, Miridae). Rev. Bras. Biol. 38:121–149.

Carvalho, J. C. M. and Y. A. Popov. 1984. A new genus and species of mirid bug from the Baltic amber (Hemiptera, Miridae). An. Acad. Bras. Cienc. 56: 203–205.

Carvalho, J. C. M. and R. I. Sailer. 1953. Neotropical Miridae, XLVII—The genus Ofellus Distant, 1883, with

descriptions of three new species. Proc. Entomol. Soc. Wash. 55:234–238.

Carvalho, J. C. M. and T. R. E. Southwood. 1955. Revisão do complexo Cyrtorhinus Fieber-Mecomma Fieber (Heteroptera, Miridae). Bol. Mus. Para. Emilio Goeldi 11:1–72.

Carvalho, J. C. M. and R. L. Usinger. 1957. A new genus and two new species of myrmecomorphic Miridae from North America (Hemiptera). Wasmann J. Biol. 15:1–13.

Carvalho, J. C. M., A. V. Fontes, and T. J. Henry. 1983. Taxonomy of the South American species of Ceratocapsus, with descriptions of 45 new species (Hemiptera: Miridae). U.S. Dep. Agric. Tech. Bull. 1676:1–58.

Carver, M., G. F. Gross, and T. E. Woodward. 1991. Hemiptera (Bugs, leafhoppers, cicadas, aphids, scale insects, etc.). Pages 429–448 in CSIRO. The insects of Australia: A textbook for students and research workers, 2nd ed. Vol. I. Cornell University Press, Ithaca; Melbourne University Press, Melbourne.

Casper, R. 1988. Luteoviruses. Pages 235–258 in R. Koenig, ed., The plant viruses. Vol. 3. Polyhedral virions with monopartite RNA genomes. Plenum, New York.

Cassidy, T. P. 1937. A mirid (Melanotrichus mimus Knight (?). U.S. Dep. Agric. Insect Pest Surv. Bull. 17:197.

Cassidy, T. P. and T. C. Barber. 1938. Hemipterous cotton insects of Arizona and their economic importance and control. U.S. Dep. Agric. Bur. Entomol. Plant Quar. E-439:1–14.

——. 1940. Investigations in control of hemipterous cotton insects in Arizona by the use of insecticides. U.S. Dep. Agric. Bur. Entomol. Plant Quar. E-506:1–5.

Cassis, G. 1984. A systematic study of the subfamily Dicyphinae (Heteroptera: Miridae). Ph.D. dissertation, Oregon State University, Corvallis. 389 pp.

——. 1995. A reclassification and phylogeny of the Termatophylini (Heteroptera: Miridae: Deraeocorinae), with a taxonomic revision of the Australian species, and a review of the tribal classification of the Deraeocorinae. Proc. Entomol. Soc. Wash. 97:258–330.

Cassis, G. and G. F. Gross. 1995. Hemiptera: Heteroptera (Coleorrhyncha to Cimicomorpha). in W. W. K. Houston and G. V. Maynard, eds., Zoological catalogue of Australia. Vol. 27.3A. CSIRO Australia, Melbourne. 506 pp.

Casson, D. S. and I. D. Hodkinson. 1991. The Hemiptera (Insecta) communities of tropical rain forest in Sulawesi. Zool. J. Linn. Soc. 102:253–275.

Castañé, C., O. Alomar, and J. Riudavets. 1996. Management of western flower thrips on cucumber with *Dicyphus tamaninii* (Heteroptera: Miridae). Biol. Control 7:114–120.

Castineiras, A. 1995. Natural enemies of *Bemisia tabaci* (Homoptera: Aleyrodidae) in Cuba. Fla. Entomol. 78:538–540.

Caswell, G. H. 1962. Agricultural entomology in the tropics. Edward Arnold, London. 152 pp.

Cate, J. R. 1985. Cotton: Status and current limitations to biological control in Texas and Arkansas. Pages 537–556 *in* M. A. Hoy and D. C. Herzog, eds., Biological control in agricultural IPM systems. Academic, Orlando, Fla.

Cate, J. R., P. C. Krauter, and K. E. Godfrey. 1990. Pests of cotton. Pages 17–29 *in* D. H. Habeck, F. D. Bennett, and J. H. Frank, eds., Classical biological control in the southern United States. South. Coop. Ser. Bull. 355.

Caudell, A. N. 1901. [Short notes and exhibition of specimens.] Proc. Entomol. Soc. Wash. 4:485.

Cave, R. D. and A. P. Gutierrez. 1983. *Lygus hesperus* field life table studies in cotton and alfalfa (Heteroptera: Miridae). Can. Entomol. 115:649–654.

Cech, T. 1989. Beeinflussung der Trieb- und Blattentwicklung von Eichen durch Gelege von Weichwanzen (Het., Miridae) und Zwergzikaden. Anz. Schädlingskd. Pflanzenschutz Umweltschutz 62:81–84.

Cecil, R. 1940. A mirid (*Engytatus geniculatus* Reut.). U.S. Dep. Agric. Insect Pest Surv. Bull. 20:12.

Chachoria, H. S. 1967. Mortality in apple mealybug, *Phenacoccus aceris* (Homoptera: Coccidae), populations in Nova Scotia. Can. Entomol. 99:728–730.

Champlain, R. A. and G. D. Butler Jr. 1967. Temperature effects on development of the egg and nymphal stages of *Lygus hesperus* (Hemiptera: Miridae). Ann. Entomol. Soc. Am. 60:519–521.

Champlain, R. A. and L. L. Sholdt. 1967. Temperature range for development of immature stages of *Geocoris punctipes* (Hemiptera: Lygaeidae). Ann. Entomol. Soc. Am. 60:883–885.

Chandler, S. C. 1955. Biological studies of peach catfacing insects in Illinois. J. Econ. Entomol. 48:473–475.

Chandra, G. 1978. Natural enemies of rice leafhoppers and planthoppers in the Philippines. Int. Rice Res. Newsl. 3(5):20–21.

Chang, K. P. and A. J. Musgrave. 1970. Ultrastructure of rickettsia-like microorganisms in the midgut of a plant bug, *Stenotus binotatus* Jak. (Heteroptera: Miridae). Can. J. Microbiol. 16:621–622.

Chang, S. J. and H. I. Oka. 1984. Attributes of a hopper-predator community in a rice field. Agric. Ecosyst. Environ. 12:73–78.

Chang, Y. C. 1981. Morphology, damage and control of the bamboo mirid *Mecistoscelis scirtetoides* Reuter (Miridae, Hemiptera). Plant Prot. Bull. (Taichung) 23:15–23. [in Chinese, English summary.]

——. 1982. The life history, population density and control of the bamboo mirid *Mecistoscelis scirtetoides* Reuter (Miridae, Hemiptera). Natl. Sci. Counc. Mon. 10:335–347. [in Chinese, English summary.]

Chant, D. A. 1956. Predacious spiders in orchards in south-eastern England. J. Hortic. Sci. 31:35–46.

——. 1959. Phytoseiid mites (Acarina: Phytoseiidae). Part I. Bionomics of seven species in southeastern England. Part II. A taxonomic review of the family Phytoseiidae, with descriptions of 38 new species. Can. Entomol. Suppl. 12:1–164.

Chant, D. A. and C. A. Fleschner. 1960. Some observations on the ecology of phytoseiid mites (Acarina: Phytoseiidae) in California. Entomophaga 5:131–139.

Chapman, P. J., G. W. Pearce, and A. W. Avens. 1941. The use of petroleum oils as insecticides III. Oil deposit and the control of fruit tree leafroller and other apple pests. J. Econ. Entomol. 34:639–647.

Chapman, R. B. 1976. Vegetable crop pests. Pages 78–104 *in* D. N. Ferro, ed., New Zealand insect pests. Lincoln University College of Agriculture, Canterbury, New Zealand.

——. 1984. Seed crop pests. Pages 143–152 *in* R. R. Scott, ed., New Zealand pest and beneficial insects. Lincoln University College of Agriculture, Canterbury, New Zealand.

Chapman, R. F. 1991. General anatomy and function. Pages 33–67 *in* CSIRO. The insects of Australia: A textbook for students and research workers, 2nd ed. Cornell University Press, Ithaca.

Chatt, E. M. 1953. Cocoa: Cultivation, processing, analysis. Interscience, New York. 302 pp.

Chatterjee, V. C. 1983a. *Nesidiocoris caesar* (Ballard) (Heteroptera: Miridae), a new pest of bottle gourd and tobacco in western Uttar Pradesh. Uttar Pradesh J. Zool. 3:149–153.

——. 1983b. Effect of some climatic factors on the seasonal cycle of *Nesid-iocoris caesar* (Ballard) (Heteroptera: Miridae). Pages 38–44 *in* S. C. Goel, ed., Insect ecology and resource management. Sanatan Dharm College, Muzaffarnagar, India.

——. 1984a. Copulation and oviposition behaviour of *Nesidiocoris caesar* (Ballard) (Heteroptera: Miridae). Entomon 9:35–37.

——. 1984b. Gustatory receptor organ of *Nesidiocoris caesar* (Ballard) (Heteroptera: Miridae). Uttar Pradesh J. Zool. 4:220–222.

——. 1986. Biological control of the hemipteran pests of *Lagenaria vulgaris* Ser. (Cucurbitaceae). Pages 223–227 *in* S. C. Goel, ed., Insect and environment. Vol. 2. Pesticide residues and environmental pollution. Symposium, Muzaffarnagar, India, 2–4 Oct. 1985. Sanatan Dharm College, Muzaffarnagar, India.

Chazeau, J. 1985. Predaceous insects. Pages 211–246 *in* W. Helle and M. W. Sabelis, eds., Spider mites: Their biology, natural enemies and control. Vol. IB. Elsevier, Amsterdam. 458 pp.

Chelliah, S. and E. A. Heinrichs. 1980. Factors affecting insecticide-induced resurgence of the brown planthopper, *Nilaparvata lugens* on rice. Environ. Entomol. 9:773–777.

Cheng, C. H. 1967. An observation on ecology of *Stephanitis typica* Distant (Hemiptera, Tingidae) on banana. J. Taiwan Agric. Res. 16:54–69.

——. 1985. Studies on integrated control of brown planthoppers, *Nilaparvata lugens* (Stal) in Taiwan. Pages 149–167 *in* Y. I. Chu, ed., Proceedings of ROC-Japan Seminar on the ecology and control of the brown planthopper. National Science Council, Taipei, Taiwan.

Cheng, L. 1965. The feeding mechanism of a heteropteran bug. Malay. Med. J. 20:176–177.

Cheng, S. A., J. C. Chen, H. Si, L. M. Tan, T. L. Chu, C. T. Wu, J. K. Chien, and C. S. Yan. 1979. Studies on the migration of brown planthopper *Nilaparvata lugens* Stål. Acta Entomol. Sin. 22:1–21.

Cherian, M. C., M. S. Kylasam, and P. S. Krishnamurti. 1941. Further studies on *Calocoris angustatus* Leth. Madras Agric. J. 29:66–69.

Cherrill, A., S. Rushton, R. Sanderson, and J. Byrne. 1997. Comparison of TWINSPAN classifications based on plant bugs, leaf hoppers, ground-beetles, spiders and plants. Entomologist 116:73–83.

Chessin, M. and A. E. Zipf. 1990. Alarm system in higher plants. Bot. Rev. 56:193–235.

Cheung, W. W. K. and A. T. Marshall. 1973. Water and ion regulation in cicadas in relation to xylem feeding. J. Insect Physiol. 19:1801–1816.

Chiang, H. C. 1977. Pest management in the People's Republic of China—monitoring and forecasting insect populations in rice, wheat, cotton and maize. FAO Plant Prot. Bull. 25:1–8.

Childers, C. C. and W. R. Enns. 1975. Predaceous arthropods associated with spider mites in Missouri apple orchards. J. Kans. Entomol. Soc. 48:453–471.

Chin, P. K., A. B. Sipat, and K. C. Khoo. 1988. Studies on the predator-prey relationship between *Oecophylla smaragdina* and *Helopeltis theobromae* using the radiotracer technique. Pages 427–435 *in* Modern insect control: Nuclear techniques and biotechnology. International Atomic Energy Agency, Vienna.

China, W. E. 1925a. Notes on the life-history and habits of *Notostira* (*Megaloceraea*) [*sic*] *erratica* L. Entomol. Mon. Mag. 61:28–33.

——. 1925b. *Notostira erratica* L. bred from *Notostira tricostata* Costa, a further note on the life-history of *N. erratica* L. Entomol. Mon. Mag. 61:279–280.

——. 1931. A remarkable mirid larva from Cuba, apparently belonging to a new species of the genus *Paracarnus*, Dist. (Hemiptera, Miridae). Ann. Mag. Nat. Hist. (10)8:283–288.

——. 1932. A new species of *Trichophthalmocapsus* Popp. (Hemiptera Heteroptera, Capsidae) from Kenya. Ann. Mag. Nat. Hist. (10)10:594–597.

——. 1933. A new family of Hemiptera-Heteroptera with notes on the phylogeny of the suborder. Ann. Mag. Nat. Hist. (10)12:180–196.

——. 1934. Hemiptera collected by the Oxford University Expedition to West Greenland, 1928. Ann. Mag. Nat. Hist. (10)13:330–333.

——. 1935. Hemipterous predators of the weevils *Cosmopolites* and *Odoiporus*. Bull. Entomol. Res. 26:497–498.

——. 1944. New and little known West African Miridae (Capsidae) (Hemiptera Heteroptera). Bull. Entomol. Res. 35:171–191.

——. 1953. Two new species of the genus *Cyrtopeltis* (Hemiptera) associated with sundews in Western Australia. West. Aust. Nat. 4:1–8.

——. 1955. A reconsideration of the systematic position of the family Joppeicidae Reuter (Hemiptera-Heteroptera), with notes on the phylogeny of the suborder. Ann. Mag. Nat. Hist. (12)8:353–370.

China, W. E. and J. C. M. Carvalho. 1951a. A new ant-like mirid from Western Australia (Hemiptera, Miridae). Ann. Mag. Nat. Hist. (12)4:221–225.

——. 1951b. A remarkable genus and species of Cylapinae from British Guiana (Hemiptera, Miridae). Ann. Mag. Nat. Hist. (12)4:289–292.

China, W. E., H. Henson, B. M. Hobby, H. E. Hinton, T. T. Macan, O. W. Richards, and V. B. Wigglesworth. 1958. The terms "larva" and "nymph" in entomology. Trans. Soc. Brit. Entomol. 13:17–24.

Chinajariyawong, A. 1988. The sap-sucking bugs attacking cotton: Biological aspects and economic damage. Ph.D. dissertation, University of Queensland, Brisbane, Australia. 141 pp.

Chinajariyawong, A. and V. E. Harris. 1987. Inability of *Deraeocoris signatus* (Distant) (Hemiptera: Miridae) to survive and reproduce on cotton without prey. J. Aust. Entomol. Soc. 26:37–40.

Chinajariyawong, A. and G. H. Walter. 1990. Feeding biology of *Campylomma livida* Reuter (Hemiptera: Miridae) on cotton, and some host plant records. J. Aust. Entomol. Soc. 29:177–181.

Chinery, M. 1993. Insects of Britain & northern Europe, 3rd ed. Harper Collins, London. 320 pp.

Chinta, S., J. C. Dickens, and J. R. Aldrich. 1994. Olfactory reception of potential pheromones and plant odors by tarnished plant bug, *Lygus lineolaris* (Hemiptera: Miridae). J. Chem. Ecol. 20:3251–3267.

Chinta, S., J. C. Dickens, and G. T. Baker. 1997. Morphology and distribution of antennal sensilla of the tarnished plant bug, *Lygus lineolaris* (Palisot de Beauvois) (Hemiptera: Miridae). Int. J. Insect Morphol. Embryol. 26:21–26.

Chippendale, G. M. 1978. The functions of carbohydrates in insect life processes. Pages 1–55 *in* M. Rockstein, ed., Biochemistry of insects. Academic, New York.

Chittenden, F. H. 1899. Notes on the garden flea-hopper. Pages 57–62 *in* Some insects injurious to garden and orchard crops. U.S. Dep. Agric. Div. Entomol. Bull. 19 (n.s.).

Chiu, S. C. 1979. Biological control of the brown planthopper. Pages 335–355 *in* Brown planthopper: Threat to rice production in Asia. International Rice Research Institute, Los Baños, Philippines.

——. 1984. Recent advances in the integrated control of rice insects in China. Bull. Entomol. Soc. Am. 30:41–46.

Chiykowski, L. N. 1987. Vector relationships of xylem- and phloem-limited fastidious prokaryotes. Pages 313–320 *in* E. L. Civerolo, A. Collmer, R. E. Davis, and A. G. Gillaspie, eds., Plant pathogenic bacteria. Proceedings of the 6th International Conference on Plant Pathogenic Bacteria, University of Maryland, June 2–7, 1985. Nijhoff, Dordrecht.

Christenson, L. D. and F. F. Smith. 1952. Insects and plant viruses. Pages 179–180 *in* F. C. Bishopp, chairman, Insects: The yearbook of agriculture, 1952. U.S. Dep. Agric., Washington, D.C.

Chu, H. F. and L. K. Cutkomp. 1992. How to know the immature insects, 2nd ed. Brown, Dubuque, Iowa. 346 pp.

Chu, H. F. and H. L. Meng. 1958. Studies on three species of cotton plant-bugs, *Adelphocoris taeniophorus* Reuter, *A. lineolatus* (Goeze), and *Lygus lucorum* Meyer-Dur (Hemiptera, Miridae). Acta Entomol. Sin. 8:97–118. [in Chinese, English summary.]

Chua, T. H. and E. Mikil. 1986. Effects of prey quantity and stage on *Cyrtorhinus lividipennis* (Reuter) (Miridae: Hemiptera), a predator of rice brown planthopper. J. Singapore Natl. Acad. Sci. 15:18–20.

——. 1989. Effects of prey number and stage on the biology of *Cyrtorhinus lividipennis* (Hemiptera: Miridae): A predator of *Nilaparvata lugens* (Homoptera: Delphacidae). Environ. Entomol. 18:251–255.

Chung, C. S., M. K. Harris, T. Li, P. Glogoza, S. G. Helmers, and J. White. 1991. *Orthotylus ramus*, a potential mirid pest of pecan in Texas. Southwest. Entomol. 16:243–249.

Chung, G. F. and B. J. Wood. 1989. Chemical control of *Helopeltis theobromae* Mill. and crop loss assessment in cocoa. J. Plant Prot. Trop. 6:35–48.

Church, N. S. and G. H. Gerber. 1977. Observations on the ontogeny and habits of *Lytta nuttalli*, *L. viridana*, and *L. cyanipennis* (Coleoptera: Meloidae): The adults and eggs. Can. Entomol. 109:565–573.

Ciampolini, M. and A. Servadei. 1973. Il *Pantilius tunicatus* F. (Rhynchota, Heteroptera, Miridae) fitofago su *Corylus avellana* L., in Piemonte. Boll. Zool. Agrar. Bachic. 11:217–221.

Cibrián Tovar, D., J. T. Méndez Montiel, R. Campos Bolaños, H. O. Yates III, and J. E. Flores Lara. 1995. Forest insects of Mexico. Universidad Autónoma Chapingo, Chapingo, Mexico. 453 pp.

Clancy, D. W. 1968. Distribution and parasitization of some *Lygus* spp. in western United States and Mexico. J. Econ. Entomol. 61:443–445.

Clancy, D. W. and H. J. McAlister. 1956. Selective pesticides as aids to biological control of apple pests. J. Econ. Entomol. 49:106–202.

Clancy, D. W. and H. D. Pierce. 1966. Natural enemies of lygus bugs. J. Econ. Entomol. 59:853–858.

Clancy, D. W. and H. N. Pollard. 1952. The effect of DDT on mite and predator populations in apple orchards. J. Econ. Entomol. 45:108–114.

Claridge, M. F. 1959. A new species of trichogrammatid (Hymenoptera, Chalcidoidea) parasitic in mirid eggs (Hemiptera-Heteroptera). Proc. R. Entomol. Soc. Lond. (B) 28:128–131.

——. 1989. Electrophoresis in agricultural pest research—a technique of evolutionary biology. Pages 1–6 in H. D. Loxdale and J. den Hollander, eds., Electrophoretic studies on agricultural pests. Systematics Assoc. Spec. Vol. 39. Clarendon, Oxford.

Clark, E. W. and M. J. Lukefahr. 1956. A partial analysis of cotton extrafloral nectar and its approximation as a nutritional medium for adult pink bollworms. J. Econ. Entomol. 49:875–876.

Clark, K. M., W. C. Bailey, and R. L. Myers. 1995. Alfalfa as a companion crop for control of Lygus lineolaris (Hemiptera: Miridae) in amaranth. J. Kans. Entomol. Soc. 68:143–148.

Clark, L. R., P. W. Geier, R. D. Hughes, and R. F. Morris. 1967. The ecology of insect populations in theory and practice. Methuen, London. 232 pp.

Clark, T. B. 1982. Spiroplasmas: Diversity of arthropod reservoirs and host-parasite relationships. Science (Wash., D.C.) 217:57–59.

Clausen, C. P. 1940. Entomophagous insects. McGraw-Hill, New York. 688 pp.

——. 1978. Delphacidae. Pages 74–77 in C. P. Clausen, ed., Introduced parasites and predators of arthropod pests and weeds: A world review. U.S. Dep. Agric. Agric. Handb. 480.

Clayton, R. A. 1982. A phylogenetic analysis of Lygocoris Reuter (Heteroptera: Miridae) with notes on life histories and zoogeography. M.S. thesis, University of Connecticut, Storrs. 78 pp.

——. 1989. Preparation of phalluses of Miridae (Heteroptera) for scanning electron microscopy and a redescription of the vesica of Mirinae. Trans. Am. Microsc. Soc. 108:419–423.

Clements, F. E. and F. L. Long. 1923. Experimental pollination; an outline of the ecology of flowers and insects. Carnegie Institution of Washington, Washington, D.C. 274 pp.

Cleveland, C. R. 1925. Meadow plant bug (Miris dolabratus L.). U.S. Dep. Agric. Insect Pest Surv. Bull. 5:163.

Cleveland, T. C. 1982. Hibernation and host plant sequence studies of tarnished plant bugs, Lygus lineolaris, in the Mississippi delta. Environ. Entomol. 11:1049–1052.

——. 1985. Toxicity of several insecticides applied topically to tarnished plant bugs. J. Entomol. Sci. 20:95–97.

——. 1987. Predation by tarnished plant bugs (Heteroptera: Miridae) of Heliothis (Heteroptera: Noctuidae) eggs and larvae. Environ. Entomol. 16:37–40.

Clifford, P. T. P., J. A. Wightman, and D. N. J. Whitford. 1983. Mirids in 'Grasslands Maku' lotus seed crops: Friends or foes? Proc. N.Z. Grassland Assoc. 44:42–46.

Cloutier, C. and M. Mackauer. 1977. The effect of parasitism on aphid feeding. Eucarpia/OILB Working Group Breeding for Resistance to Insects and Mites. Rep. 1st Meet. Wageningen, The Netherlands, 7 to 9 Dec. 1976. Int. Organ. Biol. Control/West. Palearctic Reg. Sect. Bull. 7:137–142.

Cmoluchowa, A. 1982. Morfologia i bionomia stadiów rozwojowych Macrolophus rubi Woodroffe, 1957 (Heteroptera, Miridae). Ann. Univ. Mariae Curie-Sklodowska 37C:95–103.

Coad, B. R. 1929. Cotton insect problems in the United States. In Trans. 4th Int. Congr. Entomol., Ithaca, N.Y., pp. 241–247.

——. 1931. Insects captured by airplane are found at surprising heights. Pages 320–323 in U.S. Department of Agriculture Yearbook of Agriculture, 1931. Government Printing Office, Washington, D.C.

Coaker, T. H. 1957. Studies of crop loss following insect attack on cotton in East Africa. II.—Further experiments in Uganda. Bull. Entomol. Res. 48:851–866.

Cobben, R. H. 1953. Bemerkungen zur Lebensweise einiger höllandischen Wanzen (Hemiptera-Heteroptera). Tijdschr. Entomol. 96:169–198.

——. 1958. Biotaxonomische einzelkeiten über niederländische Wanzen (Hemiptera, Heteroptera). Tijdschr. Entomol. 101:1–46.

——. 1960. De eerste vondsten in Nederland van een met Sedum en een met Carex geassocieerde wants (Heteroptera: Miridae). Entomol. Ber. (Amst.) 20:195–208.

——. 1968. Evolutionary trends in Heteroptera. Part I. Eggs, architecture of the shell, gross embryology and eclosion. Centre for Agricultural Publishing & Documentation, Wageningen. 475 pp.

——. 1978. Evolutionary trends in Heteroptera. Part II. Mouthpart-structures and feeding strategies. Meded. Landbouwhogesch. Wageningen 78-5:1–407.

——. 1979. On the original feeding habits of the Hemiptera (Insecta): A reply to Merrill Sweet. Ann. Entomol. Soc. Am. 72:711–715.

——. 1986. A most strikingly myrmecomorphic mirid from Africa, with some notes on ant-mimicry and chromosomes in hallodapines (Miridae, Heteroptera). J. N.Y. Entomol. Soc. 94:194–204.

Cochran, D. G. 1975. Excretion in insects. Pages 177–281 in D. J. Candy and B. A. Kilby, eds., Insect biochemistry and function. Chapman & Hall, London.

Cochrane, T. W. and P. F. Entwistle. 1964. Preliminary world bibliography of mirids (= capsids) and other Heteroptera associated with cocoa (Theobroma cacao L.). In Proceedings of the Conf. on Mirids and other Pests of Cocoa, Ibadan, Nigeria, 1964, pp. 123–131. West African Cocoa Research Institute, Ibadan, Nigeria.

Cockerell, T. D. A. 1899. Some insect pests of Salt River Valley and the remedies for them. Ariz. Agric. Exp. Stn. Bull. 32:273–295.

Cockfield, S. D. 1988. Relative availability of nitrogen in host plants of invertebrate herbivores: Three possible nutritional and physiological definitions. Oecologia (Berl.) 77:91–94.

Cockfield, S. D., D. A. Potter, and R. L. Houtz. 1987. Chlorosis and reduced photosynthetic CO_2 assimilation of Euonymus fortunei infested with euonymus scale (Homoptera: Diaspididae). Environ. Entomol. 16:1314–1318.

Cohen, A. C. 1982. Water and temperature relations of two hemipteran members of a predator-prey complex. Environ. Entomol. 11:715–719.

——. 1984. Food consumption, food utilization, and metabolic rates of Geocoris punctipes fed Heliothis virescens eggs. Entomophaga 29:361–367.

——. 1989. Ingestion efficiency and protein consumption by a heteropteran predator. Ann. Entomol. Soc. Am. 82:495–499.

——. 1990. Feeding adaptations of some predaceous Hemiptera. Ann. Entomol. Soc. Am. 83:1215–1223.

——. 1993. Organization of digestion and preliminary characterization of salivary trypsin-like enzymes in a predaceous heteropteran, Zelus renardii. J. Insect Physiol. 39:823–829.

——. 1995. Extra-oral digestion in predaceous terrestrial Arthropoda. Annu. Rev. Entomol. 40:85–103.

——. 1996. Plant feeding by predatory Heteroptera: Evolutionary and adaptational aspects of trophic switching. Pages 1–17 in O. Alomar and R. N. Wiedenmann, eds., Zoophytophagous Heteroptera: Implications for life history and integrated pest management. Thomas Say Publ. Entomol.: Proceedings. Entomological Society of America, Lanham, Md.

——. 1998a. Solid-to-liquid feeding. The inside(s) story of extra-oral digestion in predaceous Arthropoda. Am. Entomol. 44:103–116.

——. 1998b. Biochemical and morphological dynamics and predatory feeding habits in terrestrial Heteroptera. Pages 21–32 in M. Coll and J. R. Ruberson, eds., Predatory Heteroptera: Their ecology and use in biological control. Thomas Say Publ. Entomol.: Proceedings. Entomological Society of America, Lanham, Md.

——. 2000. New oligidic production diet for Lygus hesperus Knight and L. lineolaris (Palisot de Beauvois). J. Entomol. Sci. 35:301–310.

Cohen, A. C. and D. N. Byrne. 1992. Geocoris punctipes as a predator of Bemisia tabaci: A laboratory evaluation. Entomol. Exp. Appl. 64:195–202.

Cohen, A. C. and J. W. Debolt. 1983. Rearing Geocoris punctipes on insect eggs. Southwest. Entomol. 8:61–64.

——. 1984. Fatty acid and amino acid composition of teratocytes from Lygus hesperus (Miridae: Hemiptera) parasitized by two species of parasites, Leiophron uniformis (Braconidae: Hymenoptera) and Peristenus stygicus (Braconidae: Hymenoptera). Comp. Biochem. Physiol. 79B:335–337.

Cohen, A. C. and R. Tang. 1997. Relative prey weight influences handling time and biomass extraction in Sinea confusa and Zelus renardii (Heteroptera: Reduviidae). Environ. Entomol. 26:559–565.

Cohen, A. C. and A. G. Wheeler Jr. 1998. Role of saliva in the destructive fourlined plant bug (Hemiptera: Miridae: Mirinae). Ann. Entomol. Soc. Am. 91:94–100.

Cohen, A. C, J. W. Debolt, and H. A. Schreiber. 1985. Profiles of trace and major elements in whole carcasses of Lygus hesperus adults. Southwest. Entomol. 10:239–243.

Cohen Stuart, C. P. 1922. Iets over den steek van Helopeltis. Meded. Proefstn. Thee 81:24–25.

Cole, B. J. 1980. Growth ratios in holometabolous and hemimetabolous insects. Ann. Entomol. Soc. Am. 73:489–491.

Coley, P. D. 1983. Herbivory and defensive characteristics of tree species in a lowland tropical forest. Ecol. Monogr. 53:209–233.

Coley, P. D., J. P. Bryant, and F. S. Chapin. 1985. Resource availability and plant antiherbivore defense. Science (Wash., D.C.) 230:895–899.

Coll, M. and R. L. Ridgway. 1995. Functional and numerical responses of Orius insidiosus (Heteroptera: Anthocoridae) to its prey in different vegetable crops. Ann. Entomol. Soc. Am. 88:732–738.

Coll, M. and J. R. Ruberson, eds. 1998. Predatory Heteroptera: Their ecology and use in biological control. Thomas Say Publ. Entomol.: Proceedings. Entomological Society of America, Lanham, Md. 233 pp.

Collinge, W. E. 1912. Remarks upon an apparently new apple pest, Lygus pratensis, Linn. J. Econ. Biol. 7:64–65.

Collingwood, C. A. 1972. Cocoa in West Africa: The economics of pest control. SPAN (Shell Public Health Agric. News) 15:74–77.

——. 1977a. African mirids. Pages 71–76 in E. M. Lavabre, ed., Les Mirides du cacaoyer. Institut français du Cafe et du Cacao (I.F.C.C.), Paris.

——. 1977b. Biological control and relations with other insects. Pages 237–255 in E. M. Lavabre, ed., Les Mirides du cacaoyer. Institut français du Cafe et du Cacao (I.F.C.C.), Paris.

——. 1977c. Insecticide resistance in West Africa. Pages 279–284 in E. M. Lavabre, ed., Les Mirides du cacaoyer. Institut français du Cafe et du Cacao (I.F.C.C.), Paris.

Collins, R. P. and T. H. Drake. 1965. Carbonyl compounds produced by the meadow plant bug, Leptopterna dolabrata (Hemiptera: Miridae). Ann. Entomol. Soc. Am. 58:764–765.

Collyer, E. 1952. Biology of some predatory insects and mites associated with the fruit tree red spider mite (Metatetranychus ulmi (Koch)) in south-eastern England I. The biology of Blepharidopterus angulatus (Fall.) (Hemiptera-Heteroptera, Miridae). J. Hortic. Sci. 27:117–129.

——. 1953a. Biology of some predatory insects and mites associated with the fruit tree red spider mite (Metatetranychus ulmi (Koch)) in south-eastern England II. Some important predators of the mite. J. Hortic. Sci. 28:85–97.

——. 1953b. Biology of some predatory insects and mites associated with the fruit tree red spider mite (Metatetranychus ulmi (Koch)) in south-eastern England III. Further predators of the mite. J. Hortic. Sci. 28:98–113.

——. 1953c. Biology of some predatory insects and mites associated with the fruit tree red spider mite (Metatetranychus ulmi (Koch)) in south-eastern England IV. The predator-mite relationship. J. Hortic. Sci. 28:246–259.

——. 1953d. The effect of spraying materials on some predatory insects. East Malling Res. Stn. Rep. 1952:141–145.

——. 1964a. Phytophagous mites and their predators in New Zealand orchards. N.Z. J. Agric. Res. 7:551–568.

——. 1964b. A summary of experiments to demonstrate the role of Typhlodromus pyri Scheut. in the control of Panonychus ulmi (Koch) in England. In Proc. 1st Int. Congr. Acarology, Fort Collins (Colorado, U.S.A.), 2–7 Sept. 1963, pp. 363–371. F. Paillart, Abbeville, France.

——. 1965. Cannibalism as a factor affecting mortality of Blepharidopterus angulatus (Fall.) (Heteroptera: Miridae). East Malling Res. Stn. Rep. 1964:177–179.

——. 1976. Integrated control of apple pests in New Zealand 6. Incidence of European red mite, Panonychus ulmi (Koch), and its predators. N.Z. J. Zool. 3:39–50.

Collyer, E. and A. H. M. Kirby. 1959. Further studies on the influence of fungicide sprays on the balance of phytophagous and predacious mites on apple in south-east England. J. Hortic. Sci. 34:39–50.

Collyer, E. and A. M. Massee. 1958. Some predators of phytophagous mites, and their occurrence, in southeastern England. Proc. 10th Int. Congr. Entomol., Montreal (1956) 4:623–626.

Collyer, E. and M. van Geldermalsen. 1975. Integrated control of apple pests in New Zealand 1. Outline of experiment and general results. N.Z. J. Zool. 2:101–134.

Commonwealth Institute of Biological Control. 1983. Possibilities for the use of natural enemies in the control of Helopeltis spp. (Miridae). Biocontrol News Inf. 4(1):7–11.

Condit, B. P. and J. R. Cate. 1982. Determination of host range in relation to systematics for Peristenus stygicus [Hym.: Braconidae], a parasitoid of Miridae. Entomophaga 27:203–210.

Connell, A. D. 1970a. Aspects of the morphology and bionomics of Batrachomorphus cedaranus Naudé and Lygidolon laevigatum Reut. on black wattle (Acacia mearnsii de Wild.). Ph.D. dissertation, University of Natal, Pietermaritzburg, South Africa.

——. 1970b. The integrated control of the brown wattle mirid, Lygidolon laevigatum Reut. Wattle Res. Inst. Pietermaritzburg Rep. 1969–70:53–56.

———. 1974. The biology and pest status of the wattle leafhopper *Iassomorphus cedaranus* (Naudé) (Homoptera: Cicadellidae). J. Entomol. Soc. S. Afr. 37:15–21.

Connin, R. V. and R. Staples. 1957. Role of various insects and mites in the transmission of wheat streak-mosaic virus. J. Econ. Entomol. 50:168–170.

Constant, B., S. Grenier, and G. Bonnot. 1994. Analysis of some morphological and biochemical characteristics of the egg of the predaceous bug *Macrolophus caliginosus* (Het.: Miridae) during embryogenesis. Entomophaga 39:189–198.

———. 1996a. Artificial substrate for egg laying and embryonic development by the predatory bug *Macrolophus caliginosus* (Heteroptera: Miridae). Biol. Control 7:140–147.

Constant, B., S. Grenier, G. Febvay, and G. Bonnot. 1996b. Host plant hardness in oviposition of *Macrolophus caliginosus* (Hemiptera: Miridae). J. Econ. Entomol. 89:1446–1452.

Conti, E., W. A. Jones, F. Bin, and S. B. Vinson. 1997. Oviposition behavior of *Anaphes iole*, an egg parasitoid of *Lygus hesperus* (Hymenoptera: Mymaridae; Heteroptera: Miridae). Ann. Entomol. Soc. Am. 90:91–101.

Conway, G. R. 1969. Pests follow the chemicals in the cocoa of Malaysia. Nat. Hist. 78(2):46–51.

———. 1971. Pests of cocoa in Sabah and their control (with a list of the cocoa fauna). Ministry of Agriculture and Fisheries, Sabah, Malaysia. 125 pp.

Cook, A. G. and T. J. Perfect. 1985. The influence of immigration on population development of *Nilaparvata lugens* and *Sogatella furcifer* and its interaction with immigration by predators. Crop Prot. 4:423–433.

———. 1989. The population characteristics of the brown planthopper, *Nilaparvata lugens*, in the Philippines. Ecol. Entomol. 14:1–9.

Cook, A. J. 1876. A new insect enemy. Cultiv. Ctry. Gentleman (Phila.) 41:535.

———. 1891. Kerosene emulsion. Mich. Agric. Exp. Stn. Bull. 76:1–16.

Cook, M. T. 1935. Index of the vectors of virus diseases of plants. J. Agric. Res. P.R. 19:407–420.

Cook, O. F. 1924. Acromania, or "crazy-top," a growth disorder of cotton. J. Agric. Res. (Wash., D.C.) 28:803–828.

Cooke, D. A. 1992. Pests of sugar beet in the UK. Agric. Zool. Rev. 5:97–127.

Cooley, D. R., W. F. Wilcox, J. Kovach, and S. G. Schloemann. 1996. Integrated pest management programs for strawberries in the northeastern United States. Plant Dis. 80:228–237.

Cooley, R. A. 1900. Injurious fruit insects; insecticides; insecticide apparatus. Mont. Agric. Exp. Stn. Bull. 23:64–114.

Coombs, E. M. 1985. Growth and development of the black grass bug (*Labops hesperius* Uhler) in the state of Utah. M.S. thesis, Utah State University, Logan. 163 pp.

Coombs, J. 1992. Dictionary of biotechnology, 2nd ed. Stockton, New York. 364 pp.

Cooper, D. J. 1981. The role of predatory Hemiptera in disseminating a nuclear polyhedrosis virus of *Heliothis punctipes*. J. Aust. Entomol. Soc. 20:145–150.

Cooper, G. M. 1981. The Miridae (Hemiptera: Heteroptera) associated with noble fir, *Abies procera* Rehd. M.S. thesis, Oregon State University, Corvallis. 135 pp.

Copeland, L. O., R. H. Leep, R. F. Ruppel, and M. B. Tesar. 1984. Birdsfoot trefoil seed production in Upper Michigan. Mich. State Univ. Ext. Bull. E–1745:1–8.

Cory, E. N. and P. A. McConnell. 1927. The phlox plant bug. Md. Agric. Exp. Stn. Bull. 292:15–22.

Costa, A. S. and A. M. B. Carvalho. 1961. Studies on Brazilian tobacco streak. Phytopathol. Z. 42:113–138.

Costa Lima, A. M. da. 1940. Insetos do Brasil. Vol. 2. Hemípteros. Escola Nacional de Agronomia, Rio de Janeiro. 351 pp.

Coste, R. 1992. Coffee: The plant and the product. (English translation by J. N. Wolf.) Macmillan, London. 328 pp.

Cotterell, G. S. 1926. A preliminary study of the life-history and habits of *Sahlbergella singularis* Hagl. and *Sahlbergella theobroma* Dist., attacking cocoa on the Gold Coast, with suggested control measures. Dep. Agric. Gold Coast Bull. 3:1–26.

———. 1928. Cotton pests of southern British Togoland and Trans-Volta District. Dep. Agric. Gold Coast Bull. 12:1–43.

Cottier, W. 1956. Part 5. Insect pests. Pages 209–481 in J. D. Atkinson, E. E. Chamberlain, J. M. Dingley, W. D. Reid, R. M. Brien, W. Cottier, H. Jacks, and G. G. Taylor. Plant protection in New Zealand. R. E. Owen, Govt. Printer, Wellington, Australia.

Cotton, R. T. 1917. The large tobacco suck-fly. *Dicyphus luridus* Gibson; the small tobacco suck-fly. *Dicyphus prasinus* Gibson. Pages 113–119 in Report of the Assistant Entomologist. Annu. Rep. Insular Exp. Stn. P.R., July 16, 1916 to June 30, 1917. Rio Piedras, Porto Rico.

———. 1918. Insects attacking vegetables in Porto Rico. J. Dep. Agric. P.R. 2:265–317.

Cottrell, C. B. 1984. Aphytophagy in butterflies: Its relationship to myrmecophily. Zool. J. Linn. Soc. 80:1–57.

Coulson, J. R. 1987. Studies on the biological control of plant bugs (Heteroptera: Miridae): An introduction and history, 1961–83. Pages 1–12 in R. C. Hedlund and H. M. Graham, eds., Economic importance and biological control of *Lygus* and *Adelphocoris* in North America. U.S. Dep. Agric. Agric. Res. Serv. ARS–64.

Coulson, R. N. and J. A. Witter. 1984. Forest entomology: Ecology and management. Wiley, New York. 669 pp.

Coutin, R., H.-G. Milaire, Y. Monnet, and J. Robin. 1984. Cultures fruitières. Punaises et poires pierreuses. Phytoma 355:31–34.

Couturier, G. 1972. Contribution à l'étude du peuplement en Hétéroptères dans un "verger naturel" de la région parisienne. Bull. Soc. Entomol. Fr. 77:201–207.

Cowland, J. W. 1934. Gezira Entomological Section, G.A.R.S. Final report on experimental work, 1932–33. Gezira Agric. Res. Serv. Sudan Gov. Rep. 1933:107–125.

Craig, C., R. G. Luttrell, S. D. Stewart, and G. L. Snodgrass. 1997. Host plant preferences of tarnished plant bug: A foundation for trap crops in cotton. Proc. Beltwide Cotton Prod. Res. Conf., Jan. 6–10, 1997, New Orleans, La., pp. 1176–1181. National Cotton Council, Memphis.

Craig, C. H. 1963. The alfalfa plant bug, *Adelphocoris lineolatus* (Goeze) in northern Saskatchewan. Can. Entomol. 95:6–13.

———. 1983. Seasonal occurrence of *Lygus* spp. (Heteroptera: Miridae) on alfalfa in Saskatchewan. Can. Entomol. 115:329–331.

Craig, C. H. and C. C. Loan. 1981. *Lygus* spp., plant bugs (Heteroptera: Miridae). Pages 45–47 in J. S. Kelleher and M. A. Hulme, eds., Biological control programmes against insects and weeds in Canada 1969–1980. Commonwealth Agricultural Bureaux, Slough, UK.

Crane, E. and P. Walker. 1984. Pollination directory for world crops. International Bee Research Association, London. 183 pp.

Crane, P. R. 1989. Patterns of evolution and extinction in vascular plants. Pages 153–187 in K. C. Allen and D. E. G. Briggs, eds., Evolution and the fossil record. Smithsonian Institution Press, Washington, D.C.

Cranham, J. E. 1966. Tea pests and their control. Annu. Rev. Entomol. 11:491–514.

Cranham, J. E., M. G. Solomon, and N. S. Sengupta. 1974. Toxicity of pesticides to predators. East Malling Res. Stn. Rep. 1973:162.

Cranham, J. E., M. G. Solomon, M. A. Easterbrook, E. F. Souter, G. M. Tardivel, E. Kapetanakis, S. I. Firth, M. W. Richards, and R. N. Skinner. 1981. The role of predacious insects. East Malling Res. Stn. Rep. 1980:100.

Cranham, J. E., M. G. Solomon, M. A. Easterbrook, E. F. Souter, G. M. Tardivel, E. Kapetanakis, M. W. Richards, and R. N. Skinner. 1982. Orchard studies on apple aphids. East Malling Res. Stn. Rep. 1981:102.

Cranshaw, W. [S.] 1992. Pests of the West: Prevention and control for today's garden and small farm. Fulcrum, Golden, Col. 275 pp.

Cranshaw, W. S., B. C. Kondratieff, and T. Qian. 1990. Insects associated with quinoa, Chenopodium quinoa, in Colorado. J. Kans. Entomol. Soc. 63:195–199.

Cranston, P. S. and I. D. Naumann. 1991. Biogeography. Pages 180–197 in CSIRO. The insects of Australia: A textbook for students and research workers, 2nd ed. Vol. 1. Cornell University Press, Ithaca; Melbourne University Press, Melbourne.

Cravedi, P. and G. Carli. 1988. Mirides nuisibles au pecher. In H. Audemard, ed., IOBC/WPRS Working Group "Integrated Protection in Fruit Orchards," sub-group "Peach Orchards." Proc. Workshop held at Valence (France), 31 Aug. to 2 Sept. 1988. Int. Organ. Biol. Control/West. Palearctic Reg. Sect. 11(7):22–23.

Crawford, H. G. 1916. A capsid attacking apples (Neurocolpus nubilus Say.). 46th Annu. Rep. Entomol. Soc. Ont. 1915: 79–87.

Crawley, M. J. 1987. Beneficial herbivores? Trends Ecol. Evol. 2:167–168.

Crawley, M. J. and M. Akhteruzzaman. 1988. Individual variation in the phenology of oak trees and its consequences for herbivorous insects. Funct. Ecol. 2:409–415.

Crepet, W. L. 1979. Insect pollination: A paleontological perspective. BioScience 29:102–108.

——. 1983. The role of insect pollination in the evolution of the angiosperms. Pages 29–50 in L. Real, ed., Pollination biology. Academic, Orlando, Fla.

Cressey, P. J., J. A. K. Farrell, and M. W. Stufkens. 1987. Identification of an insect species causing bug damage in New Zealand wheats. N.Z. J. Agric. Res. 30:209–212.

Crocker, R. L. and W. H. Whitcomb. 1980. Feeding niches of the big-eyed bugs Geocoris bullatus, G. punctipes, and G. uliginosus (Hemiptera: Lygaeidae: Geocorinae). Environ. Entomol. 9: 508–513.

Croft, B. A. 1982. Arthropod resistance to insecticides: A key to pest control failures and successes in North American apple orchards. Entomol. Exp. Appl. 31: 88–110.

——. 1990. Arthropod biological control agents and pesticides. Wiley, New York. 723 pp.

Croft, B. A. and A. W. A. Brown. 1975. Responses of arthropod natural enemies to insecticides. Annu. Rev. Entomol. 20:285–335.

Croft, B. A. and L. A. Hull. 1983. The orchard as an ecosystem. Pages 19–42 in B. A. Croft and S. C. Hoyt, eds., Integrated management of insect pests of pome and stone fruits. Wiley, New York.

Croft, B. A. and M. E. Whalon. 1982. Selective toxicity of pyrethroid insecticides to arthropod natural enemies and pests of agricultural crops. Entomophaga 27:3–21.

Crosby, C. R. 1911. Notes on the life-history of two species of Capsidae. Can. Entomol. 43:17–20.

——. 1915. The apple redbugs. Cornell Univ. Agric. Exp. Stn. Bull. 291:212–230.

Crosby, C. R. and C. H. Hadley Jr. 1915. The rhododendron lace-bug, Leptobyrsa explanata Heidemann (Tingitidae, Hemiptera). J. Econ. Entomol. 8:409–414.

Crosby, C. R. and M. D. Leonard. 1914. The tarnished plant-bug. Cornell Univ. Agric. Exp. Stn. Bull. 346:463–526.

Crosby, C. R. and R. Matheson. 1915. An insect enemy of the four-lined leaf-bug (Poecilocapsus lineatus Fabr.). Can. Entomol. 47:181–183.

Cross, D. J. 1970. Feeding stimulus and artificial diets in D. theobroma. Cocoa Res. Inst. Ghana Rep. 1968–69:52–53.

——. 1971. Water stress in cocoa and its effects on Distantiella theobroma in the laboratory. In Proc. III Int. Cocoa Res. Conf., Accra, Ghana, 23–29 Nov. 1969, pp. 252–256. Cocoa Research Institute, Tafo, Ghana.

——. 1972. Laboratory rearing of Distantiella theobroma. Cocoa Res. Inst. (Ghana) Rep. 1969–70:94–96.

Cross, D. J. and A. B. S. King. 1973. Observations on the diel feeding activity of cocoa capsids. Cocoa Res. Inst. (Ghana) Rep. 1970–71:98–102.

Crowdy, S. H. 1947. Observations on the pathogenicity of Calonectria rigidiuscula (Berk. & Br.) Sacc. on Theobroma cacao L. Ann. Appl. Biol. 34:45–59.

Crowe, T. J. 1977. Helopeltis spp. Pages 289–292 in J. Kranz, H. Schmutterer, and W. Koch, eds., Diseases, pests and weeds in tropical crops. Parey, Berlin.

Crowson, R. A. 1981. The biology of the Coleoptera. Academic, New York. 802 pp.

Culliney, T. W., D. Pimentel, O. S. Namuco, and B. A. Capwell. 1986. New observations of predation by plant bugs (Hemiptera: Miridae). Can. Entomol. 118:729–730.

Culver, J. N., A. G. C. Lindbeck, and W. O. Dawson. 1991. Virus-host interactions: Induction of chlorotic and necrotic responses in plants by tobamoviruses. Annu. Rev. Phytopathol. 29:193–217.

Cunningham, G. H. 1950. Plant disease and pest investigations. 24th Annu. Rep. N.Z. Dep. Sci. Ind. Res. 1950:63–70.

Cunningham, H. S. 1928. A study of the histologic changes induced in leaves by certain leaf-spotting fungi. Phytopathology 18:717–751.

Cuong, N. L., P. T. Ben, L. T. Phuong, L. M. Chau, and M. B. Cohen. 1997. Effect of host plant resistance and insecticide on brown planthopper Nilaparvata lugens (Stål) and predator population development in the Mekong Delta, Vietnam. Crop Prot. 16:707–715.

Curtis, C. E. and C. E. McCoy. 1964. Some host-plant preferences shown by Lygus lineolaris (Hemiptera: Miridae) in the laboratory. Ann. Entomol. Soc. Am. 57:511–513.

Curtis, J. 1849. Observations on the natural history and economy of various insects affecting the potato-crops, including plant-lice, plant-bugs, frog-flies, caterpillars, crane-flies, wireworms, millipedes, mites, beetles, flies, &c. J. R. Agric. Soc. 10:70–118.

——. 1860. Farm insects: Being the natural history and economy of the insects injurious to the field crops of Great Britain and Ireland, and also those which infest barns and granaries. With suggestions for their destruction. Blackie, Glasgow. 528 pp.

Curtis, W. E. 1942. A method of locating insect eggs in plant tissues. J. Econ. Entomol. 35:286.

Custer, P. K. 1981. Insects causing catfacing and associated injuries and their control on peach in West Virginia. M.S. thesis, West Virginia University, Morgantown. 99 pp.

Cutright, C. R. 1963. Insect and mite

pests of Ohio apples. Ohio Agric. Exp. Stn. Res. Bull. 930:1–78.

Dale, D. 1994. Insect pests of the rice plant—their biology and ecology. Pages 363–485 in E. A. Heinrichs, ed., Biology and management of rice insects. Wiley, New York.

Dale, J. E. and T. H. Coaker. 1958. Some effects of feeding by *Lygus vosseleri* Popp. (Heteroptera, Miridae) on the stem apex of the cotton plant. Ann. Appl. Biol. 46:423–429.

Daly, H. V., J. T. Doyen, and P. R. Ehrlich. 1978. Introduction to insect biology and diversity. McGraw-Hill, New York. 564 pp.

Dammerman, K. W. 1929. The agricultural zoology of the Malay Archipelago. J. H. de Bussy, Amsterdam. 473 pp.

Danks, H. V. 1979. Terrestrial habitats and distribution of Canadian insects. *In* H. V. Danks, ed., Canada and its insect fauna. Mem. Entomol. Soc. Can. 108:195–210.

——. 1981. Arctic arthropods: A review of systematics and ecology with particular reference to the North American fauna. Entomological Society of Canada, Ottawa. 608 pp.

——. 1986. Insect plant interactions in arctic regions. Rev. Entomol. Qué. 31: 52–75.

——. 1987. Insect dormancy: An ecological perspective. Biol. Surv. Can. Monogr. Ser. No. 1. Biological Survey of Canada, Ottawa. 439 pp.

Danthanarayana, W. 1983. Population ecology of the light brown apple moth, *Epiphyas postvittana* (Lepidoptera: Tortricidae). J. Anim. Ecol. 52:1–33.

Darrow, G. M. 1966. The strawberry: History, breeding, and physiology. Holt, Rinehart & Winston, New York. 447 pp.

Das, G. M. 1963. Some important pests of tea. Two Bud (Assam) 10(2):4–8.

Das, S. C. 1984. Resurgence of tea mosquito bug, *Helopeltis theivora* Waterh., a serious pest of tea. Two Bud (Assam) 31(2):36–39.

Dasch, G. A., E. Weiss, and K.-P. Chang. 1984. Endosymbionts of insects. Pages 811–833 in N. R. Krieg, ed., Bergey's manual of systematic bacteriology. Vol. 1. Williams & Wilkins, Baltimore.

da Silva Barbosa, A. J. 1959. The capsid complex of cotton in Mocambique. S. Afr. J. Sci. 55:147–153.

Daugherty, D. M. 1967. Pentatomidae as vectors of yeast-spot diseases of soybeans. J. Econ. Entomol. 60:147–152.

Daughtrey, M. L. and M. Semel. 1987. Herbaceous perennials: Diseases and insect pests. Cornell Coop. Ext. Inf. Bull. 207:1–25.

Davies, E. 1987. Plant responses to wounding. Pages 243–264 in D. D. Davies, ed., The biochemistry of plants, a comprehensive treatise. Vol. 12. Physiology of metabolism. Academic, New York.

Davies, J. C. and F. K. Kasule. 1964. A note on the relative importance of Heteroptera and bollworms as pests of cotton in eastern Uganda. East Afr. Agric. For. J. 30:69–73.

Davies, R. G. 1958. The terminology of the juvenile phases of insects. Trans. Soc. Brit. Entomol. 13:25–36.

——. 1988. Outlines of entomology, 7th ed. Chapman & Hall, London. 408 pp.

Davis, A. C., F. L. McEwen, and R. W. Robinson. 1963. Preliminary studies on the effect of lygus bugs on the set and yield of tomatoes. J. Econ. Entomol. 56:532–533.

Davis, B. N. K. 1975. The colonization of isolated patches of nettles (*Urtica dioica* L.) by insects. J. Appl. Ecol. 12:1–14.

Davis, C. J., S. Matayoshi, and E. R. Yoshioka. 1985. *Halticus bractatus* Say. Proc. Hawaii. Entomol. Soc. 25:7.

Davis, D. W., B. A. Haws, and G. F. Knowlton. 1976. Insects injurious to seed production. *In* D. W. Davis, ed., Insects and nematodes associated with alfalfa in Utah. Utah Agric. Exp. Stn. Bull. 494:19–27.

Davis, G. C. 1893. Insects injurious to celery. Mich. Agric. Exp. Stn. Bull. 102: 23–52.

——. 1897. Report of the consulting entomologist. Pages 135–138 in 35th Annu. Rep. Secr. State Board Agric. State of Mich.; 9th Annu. Rep. Agric. Coll. Exp. Stn., July 1, 1895 to June 30, 1896. Robert Smith, Lansing, Mich.

Davis, J. J. 1927. Tarnished plant bug (*Lygus pratensis* L.). U.S. Dep. Agric. Insect Pest Surv. Bull. 7:244.

Davis, N. T. 1955. Morphology of the female organs of reproduction in the Miridae. Ann. Entomol. Soc. Am. 48: 132–150.

Davis, R. M. and M. P. Russell. 1969. Commensalism between *Ranzovius moerens* (Reuter) (Hemiptera: Miridae) and *Hololena curta* (McCook) (Araneida: Agelenidae). Psyche (Camb.) 76:262–269.

Day, M. F. and H. Irzykiewicz. 1954. On the mechanism of transmission of nonpersistent phytopathogenic viruses by aphids. Aust. J. Biol. Soc. 7:251–273.

Day, R. A. 1988. How to write and publish a scientific paper, 3rd ed. Oryx, Phoenix. 211 pp.

Day, W. H. 1987. Biological control efforts against *Lygus* and *Adelphocoris* spp. infesting alfalfa in the United States with notes on other associated mirid species. Pages 20–39 in R. C. Hedlund and H. M. Graham, eds., Economic importance and biological control of *Lygus* and *Adelphocoris* in North America. U.S. Dep. Agric. Agric. Res. Serv. ARS-64.

——. 1991. The peculiar sex ratio and dimorphism of the garden fleahopper, *Halticus bractatus* (Hemiptera: Miridae). Entomol. News 102:113–117.

——. 1994. Estimating mortality caused by parasites and diseases of insects: Comparisons of the dissection and rearing methods. Environ. Entomol. 23:543–550.

——. 1995. Biological observations on *Phasia robertsonii* (Townsend) (Diptera: Tachinidae), a native parasite of adult plant bugs (Hemiptera: Miridae) feeding on alfalfa and grasses. J. N.Y. Entomol. Soc. 103:100–106.

Day, W. H. and L. B. Saunders. 1990. Abundance of the garden fleahopper (Hemiptera: Miridae) on alfalfa and parasitism by *Leiophron uniformis* (Gahan) (Hymenoptera: Braconidae). J. Econ. Entomol. 83:101–106.

Day, W. H., R. C. Hedlund, L. B. Saunders, and D. Coutinot. 1990. Establishment of *Peristenus digoneutis* (Hymenoptera: Braconidae), a parasite of the tarnished plant bug (Hemiptera: Miridae), in the United States. Environ. Entomol. 19:1528–1533.

Deacon, G. E. 1948. The green capsid bug (*Lygus pabulinus*). Rose Annu. 1948: 89.

Dean, D. A., W. L. Sterling, M. Nyffeler, and R. G. Breene. 1987. Foraging by selected spider predators on the cotton fleahopper and other prey. Southwest. Entomol. 12:263–270.

Dean, G. A. and R. C. Smith. 1935. Insects injurious to alfalfa in Kansas. 29th Bien. Rep. Kans. State Board Agric. 1933–34:202–249.

Deang, R. T. 1969. An annotated list of insect pests of vegetables in the Philippines. Philipp. Entomol. 1:313–333.

De Angelis, J. 1988. Greenhouse management of western flower thrips and tomato spotted wilt virus. Ornamentals Northwest Newsl. 12(6):7–11.

DeBach, P. 1974. Biological control by natural enemies. Cambridge University Press, Cambridge. 323 pp.

DeBach, P. and D. Rosen. 1991. Biological control by natural enemies, 2nd ed. Cambridge University Press, Cambridge. 440 pp.

Debolt, J. W. 1981. Laboratory biology and rearing of *Leiophron uniformis* (Gahan) (Hymenoptera: Braconidae), a parasite of *Lygus* spp. (Hemiptera: Miridae). Ann. Entomol. Soc. Am. 74:334–337.

——. 1982. Meridic diet for rearing successive generations of *Lygus hesperus*. Ann. Entomol. Soc. Am. 75:119–122.

——. 1987. Augmentation: Rearing, release, and evaluation of plant bug parasites. Pages 82–87 in R. C. Hedlund and H. M. Graham, eds., Economic importance and biological control of *Lygus* and *Adelphocoris* in North America. U.S. Dep. Agric. Agric. Res. Serv. ARS-64.

——. 1989a. Encapsulation of *Leiophron uniformis* by *Lygus lineolaris* and its relationship to host acceptance behavior. Entomol. Exp. Appl. 50:87–95.

——. 1989b. Resistance of *Leiophron uniformis* to the immune response of *Lygus lineolaris*: Non-recognition or immunosuppression. J. Cell. Biochem. Suppl. 13C:67 (Abstr.).

——. 1991. Behavioral avoidance of encapsulation by *Leiophron uniformis* (Hymenoptera: Braconidae), a parasitoid of *Lygus* spp. (Hymenoptera: Miridae): Relationship between host age, encapsulating ability, and host acceptance. Ann. Entomol. Soc. Am. 84:444–446.

Debolt, J. W. and A. C. Cohen. 1984. Composition of teratocytes from *Lygus hesperus* nymphs parasitized by *Leiophron uniformis*. Southwest. Entomol. 9:69–72.

Debolt, J. W. and R. Patana. 1985. *Lygus hesperus*. Pages 329–338 in P. Singh and R. F. Moore, eds., Handbook of insect rearing. Vol. I. Elsevier, Amsterdam.

Decazy, B. 1974. Seasonal variations of populations of *Boxiopsis madagascariensis* Lavabre, a devastating Madagascar cocoa mirid (preliminary note). Café Cacao Thé 18:255–262.

——. 1977. Les Mirides du cacaoyer à Madagascar: *Boxiopsis madagascariensis* Lavabre. Pages 123–137 in E. M. Lavabre, ed., Les Mirides du cacaoyer. Institut français du Café et du Cacao, Paris.

Decelle, J. 1955. Un nouvel ennemi des *Citrus: Distantiella collarti* Schout. (Hemiptera-Capsidae-Bryocorinae). Bull. Agric. Congo Belge 46:79–86.

DeCoursey, R. M. 1971. Keys to the families and subfamilies of the nymphs of North American Hemiptera-Heteroptera. Proc. Entomol. Soc. Wash. 73:413–428.

Deeming, J. C. 1981. The hemipterous fauna of a northern Nigerian cotton plot. Samaru Agric. J. Res. 1:211–222.

DeGrandi-Hoffman, G., J. Diehl, D. Li, L. Flexner, G. Jackson, W. Jones, and J. Debolt. 1994. BIOCONTROL-PARASITE: Parasitoid-host and crop loss assessment simulation model. Environ. Entomol. 23:1045–1060.

De Jong, J. K. see Jong, J. K. de

Dekhtiarev, N. S. 1927. *Poeciloscytus cognatus*, Fieb. (Hemiptera, Miridae) as a serious pest of sugar-beets. Bull. Entomol. Res. 18:1–3.

Delattre, R. 1947. Insectes du cotonnier nouveaux ou peu connus en Côte d'Ivoire. Coton Fibres Trop. 2:28–33.

——. 1958. Les parasites du cotonnier à Madagascar. Coton Fibres Trop. 13:335–352.

Delucchi, V., J. P. Aeschlimann, and E. Graf. 1975. The regulating action of egg predators on the populations of *Zeiraphera diniana* Guenée. Mitt. Schweiz. Entomol. Ges. 48:37–45.

Demidov, N. I. 1940. The effect of puncturing and sucking insects on the shedding of the fruits of cotton. Izv. Akad. Nauk Uzb. SSR 6:98–100. [in Russian.]

Demirdere, A. 1956. Çukurovada *Dionconotus cruentatus* ve mücadelesi. Tomurcuk 58:14–16.

Dempster, J. P. 1960. A quantitative study of the predators on the eggs and larvae of the broom beetle, *Phytodecta olivacea* Forster, using the precipitin test. J. Anim. Ecol. 29:149–167.

——. 1964. The feeding habits of the Miridae (Heteroptera) living on broom (*Sarothamnus scoparius* (L.) Wimm.). Entomol. Exp. Appl. 7:149–154.

——. 1966. Arthropod predators of the Miridae (Heteroptera) living on broom (*Sarothamnus scoparius*). Entomol. Exp. Appl. 9:405–412.

Denis, F. 1908. An orchid parasite. Gard. Chron. (Lond.) (3) 43:313.

Denlinger, D. L. 1986. Dormancy in tropical insects. Annu. Rev. Entomol. 31:239–264.

Dennill, G. B. and D. Donnelly. 1991. Biological control of *Acacia longifolia* and related weed species (Fabaceae) in South Africa. Agric. Ecosyst. Environ. 37:115–135.

Denning, D. G. 1948. The crested wheat bug. Wyo. Agric. Exp. Stn. Circ. 33:1–2.

Denno, R. F. 1977. Comparison of the two assemblages of sap-feeding insects (Homoptera-Hemiptera) inhabiting two structurally different salt marsh grasses in the genus *Spartina*. Environ. Entomol. 6:359–372.

——. 1985. Fitness, population dynamics and migration in planthoppers: The role of host plants. Pages 623–640 in M. A. Rankin, ed., Migration: Mechanisms and adaptive significance. Contrib. in Marine Sci., Vol. 27. Marine Science Institute, University of Texas at Austin, Port Aransas.

——. 1994a. Life history variation in planthoppers. Pages 153–215 in R. F. Denno and J. T. Perfect, eds., Planthoppers: Their ecology and management. Chapman & Hall, New York.

——. 1994b. The evolution of dispersal polymorphisms in insects: The influence of habitats, host plants and mates. Res. Popul. Ecol. (Kyoto) 36:127–135.

Denno, R. F. and H. Dingle. 1981. Considerations for the development of a more general life history theory. Pages 1–6 in R. F. Denno and H. Dingle, eds., Insect life history patterns: Habitat and geographic variation. Springer, New York.

Denno, R. F. and G. K. Roderick. 1990. Population biology of planthoppers. Annu. Rev. Entomol. 35:489–520.

Denno, R. F., K. L. Olmstead, and E. S. McCloud. 1989. Reproductive cost of flight capability: A comparison of life history traits in wing dimorphic planthoppers. Ecol. Entomol. 14:31–44.

Denno, R. F., G. K. Roderick, K. L. Olmstead, and H. G. Döbel. 1991. Density-related migration in planthoppers (Homoptera: Delphacidae): The role of habitat persistence. Am. Nat. 138:1513–1541.

Dent, D. 1991. Insect pest management. CAB International, Wallingford, UK. 604 pp.

Denton, R. E. 1979. Larch casebearer in western forests. U.S. For. Serv. Intermountain For. Range Exp. Stn. Gen. Tech. Rep. INT–55:1–62.

DePew, L. J. 1967. Field studies on control of lygus bugs and onion thrips infesting safflower. J. Econ. Entomol. 6:1224–1226.

de Pury, J. M. S. 1968. Crop pests of East Africa. Oxford University Press, Nairobi. 227 pp.

De Silva, M. D. 1957. A new species of *Helopeltis* (Hemiptera-Heteroptera, Miridae) found in Ceylon. Bull. Entomol. Res. 48:459–461.

——. 1961. Biology of *Helopeltis ceylonensis* De Silva (Heteroptera-Miridae), a major pest of cacao in Ceylon. Trop. Agric. (Colombo) 117:149–156.

Dethier, V. G. 1947. Chemical insect attractants and repellents. Blackiston, Philadelphia. 289 pp.

——. 1976. The hungry fly: A physiological study of the behavior associated with feeding. Harvard University Press, Cambridge. 489 pp.

Devasahayam, S. 1985. Seasonal biology of tea mosquito bug *Helopeltis antonii* Signoret (Heteroptera: Miridae)—a pest of cashew. J. Plant. Crops 13:145–147.

——. 1988. Mating and oviposition behaviour of tea mosquito bug *Helopeltis antonii* Signoret (Heteroptera: Miridae). J. Bombay Nat. Hist. Soc. 85:212–214.

Devasahayam, S. and C. P. Radhakrishnan Nair. 1986. The tea mosquito bug *Helopeltis antonii* Signoret on cashew in India. J. Plant. Crops 14:1–10.

Devasahayam, S., K. M. Abdulla Koya, and T. Prem Kumar. 1986. Infestation of tea mosquito bug *Helopeltis antonii* Signoret (Heteroptera: Miridae) on black pepper and allspice in Kerala. Entomon 11:239–241.

DeVries, P. J. 1979. Pollen-feeding rainforest *Parides* and *Battus* butterflies in Costa Rica. Biotropica 11:237–238.

Deyrup, M. A. 1988. Pollen-feeding in *Poecilognathus punctipennis* (Diptera: Bombyliidae). Fla. Entomol. 71:597–605.

Dhileepan, K. 1991. Insects associated with oil palm in India. FAO Plant Prot. Bull. 39:94–99.

Dhileepan, K., R. R. Nair, and S. Leena. 1990. *Aspergillus candidus* Link as an entomopathogen of spindle bug *Carvalhoia arecae* M & C (Miridae: Heteroptera). Planter (Kuala Lumpur) 66:519–521.

Dhiman, S. C. 1985. Feeding behaviour and feeding mechanism of *Metacanthus pulchellus* Dall. (Heteroptera: Berytidae). Ann. Zool. (Agra) 23:21–25.

Dial, C. I. and P. H. Adler. 1990. Larval behavior and cannibalism in *Heliothis zea* (Lepidoptera: Noctuidae). Ann. Entomol. Soc. Am. 83:258–263.

Dicke, F. F. and J. L. Jarvis. 1962. The habits and seasonal abundance of *Orius insidiosus* (Say) (Hemiptera-Heteroptera: Anthocoridae) on corn. J. Kans. Entomol. Soc. 35:339–344.

Dickens, J. C. 1997. Neurobiology of pheromonal signal processing in insects. Pages 210–217 *in* R. T. Cardé and A. E. Minks, eds., Insect pheromone research: New directions. Chapman & Hall, New York.

Dickens, J. C. and F. E. Callahan. 1996. Antennal-specific protein in tarnished plant bug, *Lygus lineolaris*: Production and reactivity of antisera. Entomol. Exp. Appl. 80:19–22.

Dickens, J. C., F. E. Callahan, W. P. Wergin, and E. F. Erbe. 1995. Olfaction in a hemimetabolous insect: Antennal-specific protein in adult *Lygus lineolaris* (Heteroptera: Miridae). J. Insect Physiol. 41:857–867.

Dicker, G. H. L. 1952. Studies in population fluctuations of the strawberry aphid, *Pentatrichopus fragaefolii* (Cock.) I. Enemies of the strawberry aphid. East Malling Res. Stn. Rep. 1951:166–168.

———. 1967. Integrated control of apple pests. Proc. 4th Br. Insectic. Fungic. Conf., Brighton 1:1–7.

Dickerson, E. L. and H. B. Weiss. 1916.

The ash leaf bug, *Neoborus amoenus* Reut. (Hem.). J. N.Y. Entomol. Soc. 24:302–306.

Diener, T. O. 1979. Viruses and viroid diseases. Wiley, New York. 252 pp.

Diener, T. O. and W. B. Raymer. 1971. Potato spindle tuber 'virus.' CMI/AAB Descrip. Plant Viruses No. 66. Association of Applied Biologists, Wellesbourne, UK. 4 pp.

Dietz, L. L., J. W. Van Duyn, J. R. Bradley Jr., R. L. Rabb, W. M. Brooks, and R. E. Stinner. 1980. A guide to the identification and biology of soybean arthropods in North Carolina. N.C. Agric. Res. Serv. Tech. Bull. 238:1–264.

Dimitrov, A. 1975. Propagation control of tobacco thrips with *Macrolophus costalis*. Rastit. Zasht. (Sofia) 23(6):34–37. [in Bulgarian.]

Dingle, H. 1966. Some factors affecting flight activity in individual milkweed bugs (*Oncopeltus*). J. Exp. Biol. 44:335–343.

———. 1968. The influence of environment and heredity on flight activity in the milkweed bug *Oncopeltus*. J. Exp. Biol. 48:175–184.

Dirimanov, M. and A. Dimitrov. 1975. Role of useful insects in the control of *Thryps* [sic] *tabaci* Lind. and *Myzodes persicae* Sulz. on tobacco. *In* Proc. VIII Int. Plant Prot. Congr., Moscow, Sect. V, Biological and genetic control, pp. 71–72. USSR Organizing Committee, Moscow.

Dirr, M. A. 1975. Manual of woody landscape plants: Their identification, ornamental characteristics, culture, propagation and uses, 3rd ed. Stipes, Champaign, Ill. 826 pp.

Disney, R. H. L. 1994. Scuttle flies: The Phoridae. Chapman & Hall, London. 467 pp.

Disney, R. H. L., M. E. N. Majerus, and M. J. Walpole. 1994. Phoridae (Diptera) parasitising Coccinellidae (Coleoptera). Entomologist 113:28–42.

Distant, W. L. 1903. Report on the Rhynchota. Part I. Heteroptera. Fascic. Malay. Zool. 1:219–272.

Dixon, A. F. G. 1971a. The role of aphids in wood formation. I. The effect of the sycamore aphid, *Drepanosiphum platanoides* [sic] (Schr.) (Aphididae), on the growth of sycamore, *Acer pseudoplatanus* (L.). J. Appl. Ecol. 8:165–179.

———. 1971b. The role of aphids in wood formation. II. The effect of the lime aphid, *Eucallipterus tiliae* L. (Aphididae), on the growth of lime, *Tilia x vulgaris* Hayne. J. Appl. Ecol. 8:393–399.

———. 1976. Timing of egg hatch and viability of the sycamore aphid, *Drepanosiphum platanoidis* (Schr.), at

bud burst of sycamore, *Acer pseudoplatanus* L. J. Anim. Ecol. 45:593–603.

———. 1983. Insect-induced phytotoxemias: Damage, tumors, and galls. *In* K. F. Harris, ed., Current Top. Vector Res. 1:297–314.

———. 1985. Aphid ecology. Blackie, Glasgow. 157 pp.

———. 1987a. The way of life of aphids: Host specificity, speciation and distribution. Pages 197–207 *in* A. K. Minks and P. Harrewijn, eds., Aphids: Their biology, natural enemies and control. Vol. A. Elsevier, Amsterdam.

———. 1987b. Seasonal development in aphids. Pages 315–320 *in* A. K. Minks and P. Harrewijn, eds., Aphids: Their biology, natural enemies and control. Vol. A. Elsevier, Amsterdam.

———. 1987c. Aphid reproductive tactics. Pages 3–18 *in* J. Holman, J. Pelikán, A. F. G. Dixon, and L. Weismann, eds., Population structure, genetics and taxonomy of aphids and Thysanoptera. SPB, The Hague.

Dixon, A. F. G. and N. D. Barlow. 1979. Population regulation in the lime aphid. Zool. J. Linn. Soc. 67:225–237.

Dixon, A. F. G. and R. J. Russel. 1972. The effectiveness of *Anthocoris nemorum* and *A. confusus* (Hemiptera: Anthocoridae) as predators of the sycamore aphid, *Drepanosiphum platanoides* [sic]. II. Searching behaviour and the incidence of predation in the field. Entomol. Exp. Appl. 15:35–50.

Dixon, W. N. 1989. The tarnished plant bug, *Lygus lineolaris* (Palisot de Beauvois) in conifer nurseries (Heteroptera: Miridae). Fla. Dep. Agric. Consum. Serv. Entomol. Circ. 320:1–2.

Döbel, H. G. and R. F. Denno. 1994. Predator-planthopper interactions. Pages 325–399 *in* R. F. Denno and T. J. Perfect, eds., Planthoppers: Their ecology and management. Chapman & Hall, New York.

Dobroscky, I. D. 1924. A study of the external parasites of nestling robins (*Planesticus migratorius migratorius* L.), and the fauna of the nest throughout the year. M.S. thesis, Cornell University, Ithaca. 32 pp.

———. 1925. External parasites of birds and the fauna of birds' nests. Biol. Bull. (Woods Hole) 48:274–281.

Doesburg, P. H. van Jr. 1964. *Termatophylidea opaca* Carvalho, a predator of thrips (Hem.-Het.). Entomol. Ber. (Amst.) 24:248–253.

———. 1968. A revision of the New World species of *Dysdercus* Guérin Méneville (Heteroptera, Pyrrhocoridae). Zool. Verh. (Leiden) 97:1–213.

Dolling, W. R. 1972. A new species of *Dicyphus* Fieber (Hem., Miridae) from southern England. Entomol. Mon. Mag. 107:244–245.

——. 1973. Photoperiodically determined phase production and diapause termination in *Notostira elongata* (Geoffroy) (Hemiptera: Miridae). Entomol. Gaz. 24:75–79.

——. 1991. The Hemiptera. Oxford University Press, Oxford. 274 pp.

Dolling, W. R. and J. M. Palmer. 1991. *Pameridea* (Hemiptera: Miridae): Predaceous bugs specific to the highly viscid plant genus *Roridula*. Syst. Entomol. 16:319–328.

Domek, J. M. and D. R. Scott. 1985. Species of the genus *Lygus* Hahn and their host plants in the Lewiston-Moscow area of Idaho (Hemiptera: Miridae). Entomography 3:75–105.

Donaldson, J. F. and D. A. Ironside. 1982. Predatory insects and spiders in Queensland crops—part 2. Queensl. Agric. J. 108(3):xvii–xviii.

Donisthorpe, H. St. J. K. 1927. The guests of British ants: Their habits and life-histories. Routledge, London.

Donnelly, D. 1986. *Rayieria* sp. (Heteroptera: Miridae): Host specificity, conflicting interests, and rejection as a biological control agent against the weed *Acacia longifolia* (Andr.) Willd. in South Africa. J. Entomol. Soc. S. Afr. 49:183–191.

Donnelly, G. P. 1995. Host specificity and biology of *Rhinacloa callicrates* (Hemiptera: Miridae) for the biological control of *Parkinsonia aculeata* (Caesalpiniaceae) in Australia. Page 445 *in* E. S. Delfosse and R. R. Scott, eds., Proc. 8th Int. Symp. Biol. Control Weeds, 2–7 Feb. 1992, Lincoln University, Canterbury, New Zealand. DSIR/CSIRO, Melbourne.

——. 2000. Biology and host specificity of *Rhinacloa callicrates* Herring (Hemiptera: Miridae) and its introduction and establishment as a biological control agent of *Parkinsonia aculeata* L. (Caesalpiniaceae) in Australia. Aust. J. Entomol. 39:89–94.

Donovan, B. J. 1990. Selection and importation of new pollinators to New Zealand. N.Z. Entomol. 13:26–32.

Doolittle, S. P. 1920. The mosaic diseases of cucurbits. U.S. Dep. Agric. Bull. 879:1–69.

Douglas, A. E. 1998. Nutritional interactions in insect-microbial symbioses: Aphids and their symbiotic bacteria *Buchnera*. Annu. Rev. Entomol. 43:17–37.

Douglas, J. W. 1895. *Capsus laniarius* feeding. Entomol. Mon. Mag. 31:238–239.

Douglas, J. W. and J. Scott. 1865. The British Hemiptera. Vol. I. Hemiptera-Heteroptera. Hardwicke, London. 627 pp.

Doumbia, Y. O., K. Conare, and G. L. Teetes. 1995. A simple method to assess damage and screen sorghums for resistance to *Eurystylus marginatus*. Pages 183–189 *in* K. F. Nwanze and O. Youm, eds., Panicle insect pests of sorghum and pearl millet: Proceedings of an international consultative workshop, 4–7 Oct. 1993, ICRISAT Sahelian Center, Miamey, Niger. International Crops Research Institute for the Semi-Arid Tropics, Patancheru, A.P., India.

Doutt, R. L. 1959. The biology of parasitic Hymenoptera. Annu. Rev. Entomol. 4:161–182.

Downes, J. A. 1955. The food habits and description of *Atrichopogon pollinivorus* sp. n. (Diptera: Ceratopogonidae). Trans. R. Entomol. Soc. Lond. 106:439–453.

——. 1974. Sugar feeding by the larva of *Chrysopa* (Neuroptera). Can. Entomol. 106:121–125.

Downes, W. 1927. A preliminary list of the Heteroptera and Homoptera of British Columbia. Proc. Entomol. Soc. B.C. 23:5–22.

——. 1957. Notes on some Hemiptera which have been introduced into British Columbia. Proc. Entomol. Soc. B.C. 54:11–13.

Downes, W. L. Jr. and G. A. Dahlem. 1987. Keys to the evolution of Diptera: Role of Homoptera. Environ. Entomol. 16:847–854.

Dozier, H. L. 1937. Descriptions of miscellaneous chalcidoid parasites from Puerto Rico. (Hymenoptera). J. Agric. Univ. P.R. 21:121–135.

Drake, C. J. 1922. Heteroptera in the vicinity of Cranberry Lake. Pages 54–86 *in* H. Osborn and C. J. Drake, An ecological study of the Hemiptera of the Cranberry Lake Region, New York. N.Y. State Coll. For. Syracuse Univ. Tech. Publ. 16 (Vol. 22, No. 5).

——. 1928. Meadow plant bug (*Miris dolabratus* L.). U.S. Dep. Agric. Insect Pest Surv. Bull. 8:149.

Drake, C. J. and N. T. Davis. 1960. The morphology, phylogeny, and higher classification of the family Tingidae, including the description of a new genus and species of the subfamily Vianaidinae (Hemiptera: Heteroptera). Entomol. Am. 39:1–100.

Drake, C. J. and H. M. Harris. 1932. Asparagus insects in Iowa. Iowa Agric. Exp. Stn. Circ. 134:1–12.

Drake, V. A. and R. A. Farrow. 1988. The influence of atmospheric structure and motions on insect migration. Annu. Rev. Entomol. 33:183–210.

Dreistadt, S. H. 1994. Pests of landscape trees and shrubs: An integrated pest management guide. Univ. Calif. Div. Agric. Nat. Resour. Publ. 3359. 327 pp.

Dreyer, D. L. and B. C. Campbell. 1987. Chemical basis of host-plant resistance to aphids. Plant Cell Environ. 10:353–361.

Dreyer, D. L., K. C. Jones, and R. J. Molyneux. 1985. Feeding deterrency of some pyrrolizidine, indolizidine, and quinolizidine alkaloids towards pea aphid (*Acyrthosiphon pisum*) and evidence for phloem transport of indolizidine alkaloid swainsonine. J. Chem. Ecol. 11:1045–1051.

Dreyer, W. 1984. Zur Biologie wichtiger Weissdorninsekten und ihrer Parasiten. Z. Angew. Entomol. 97:286–298.

Drosopoulos, S. 1993. Is the bug *Calocoris trivialis* a real pest of olives? Int. J. Pest Manage. 39:317–320.

Duda, L. 1886. Beiträge zur Kenntniss der Hemipteren-Fauna Böhmens. Wien. Entomol. Ztg. 5:81–86.

Dudgeon, G. C. 1895. Notes on the oviposition of *Helopeltis theivora* (Waterhouse). ("mosquito blight."). Indian Mus. Notes 3(5):33–38.

——. 1910. Notes on two West African Hemiptera injurious to cocoa. Bull. Entomol. Res. 1:59–62.

Duelli, P. 1981. Is larval cannibalism in lacewings adaptive? (Neuroptera: Chrysopidae). Res. Popul. Ecol. (Kyoto) 23:193–209.

Duffey, E., M. G. Morris, J. Sheail, L. K. Ward, D. A. Wells, and T. C. E. Wells. 1974. Grassland ecology and wildlife management. Chapman & Hall, London. 281 pp.

Duffey, J. E. and R. D. Powell. 1979. Microbial induced ethylene synthesis as a possible factor of square abscission and stunting in cotton infested by cotton fleahopper. Ann. Entomol. Soc. Am. 72:599–601.

Duffey, S. S. 1986. Plant glandular trichomes: Their partial role in defence against insects. Pages 151–172 *in* B. Juniper and T. R. E. Southwood, eds., Insects and the plant surface. Arnold, London.

Dufour, L. 1833. Recherches anatomiques et physiologiques sur les Hémiptères, accompagnées de considérations relatives à l'histoire naturelle, et à la classification de ces insects. Mém. Savants Etrang. Acad. Sci. 4:129–462.

Dumbleton, L. J. 1938. Notes on a new mirid bug (*Idatiella albisignata* Knight). N.Z. J. Sci. Technol. 20(B):58B–60B.

——. 1964. Notes on insects. N.Z. Entomol. 3(3):24–25.

Dun, G. S. 1954. Notes on cacao capsids

in New Guinea. Papua New Guinea Agric. Gaz. 8:7–11.

Dunbar, D. M. and O. G. Bacon. 1972. Feeding, development, and reproduction of *Geocoris punctipes* (Heteroptera: Lygaeidae) on eight diets. Ann. Entomol. Soc. Am. 65:892–895.

Duncan, C. D. and G. Pickwell. 1939. The world of insects. McGraw-Hill, New York. 409 pp.

Duncan, J. and H. Généreux. 1960. La transmission par les insectes de *Corynbacterium* [*sic*] *sepedonicum* (Spieck. & Koth.) Skaptason et Burkholder. Can. J. Plant Sci. 40:110–116.

Dunegan, J. C. 1932. The bacterial spot disease of the peach and other stone fruits. U.S. Dep. Agric. Tech. Bull. 273:1–53.

Dunlap-Pianka, H., C. L. Boggs, and L. E. Gilbert. 1977. Ovarian dynamics in heliconiine butterflies: Programmed senescence versus eternal youth. Science (Wash., D.C.) 197:487–490.

Dunn, J. A. 1963. Insecticide resistance in the cocoa capsid, *Distantiella theobroma* (Dist.). Nature (Lond.) 199:1207.

Dunn, P. H. and B. J. Mechalas. 1963. The potential of *Beauveria bassiana* (Balsamo) Vuillemin as a microbial insecticide. J. Insect Pathol. 5:451–459.

Dunning, R. A. 1957. Mirid damage to seedling beet. Plant Pathol. (Lond.) 6:19–20.

——. 1975. Arthropod pest damage to sugar beet in England and Wales, 1947–74. Rothamsted Exp. Stn. Rep. 1974:171–185.

Dunning, [R.] A. and W. J. Byford. 1982. Pests, diseases and disorders of the sugar beet. Deleplanque et Cie, Maissons-Laffitte, France. 167 pp.

Dupnik, T. and D. A. Wolfenbarger. 1978. A constant distribution exhibited by the cotton fleahopper (*Pseudatomoscelis seriatus* (Hemiptera: Miridae)) on cotton. Can. Entomol. 110:121–124.

Du Porte, E. M. 1919. Insect carriers of plant disease. Qué. Soc. Prot. Plants Rep. 1918–1919:59–65.

Dupuis, C. 1963. Essai monographique sur les Phasiinae (Diptères Tachinaires parasites d'Hétéroptères). Mem. Mus. Natl. Hist. Nat. Sér. A. Zool. 26:1–461.

Dustan, A. G. 1923. The natural control of the green apple bug (*Lygus communis* var. *novascotiensis* Knight) by a new species of *Empusa*. Qué. Soc. Prot. Plants Rep. 1922–1923:61–66.

——. 1924. Studies on a new species of *Empusa* parasitic on the green apple bug (*Lygus communis* var. *novascotiensis* Knight) in the Annapolis Valley. Proc. Acadian Entomol. Soc. (1923) 9:14–36.

Dutcher, J. D., K. C. McGiffen, and J. N. All. 1988. Entomology and horticulture of muscadine grapes. Pages 73–90 *in* M. K. Harris and C. E. Rogers, eds., The entomology of indigenous and naturalized systems in agriculture. Westview, Boulder, Col.

Dyck, V. A. and G. C. Orlido. 1977. Control of the brown planthopper (*Nilaparvata lugens*) by natural enemies and timely application of narrow-spectrum insecticides. Pages 58–72 *in* Food and fertilizer technology for the Asian and Pacific Region, compil. Taipei, Taiwan.

Dykstra, T. P. and W. C. Whitaker. 1938. Experiments on the transmission of potato viruses by vectors. J. Agric. Res. (Wash., D.C.) 57:319–334.

Dymock, J. J. 1989. *Listronotus bonariensis* (Kuschel), Argentine stem weevil (Coleoptera: Curculionidae). Pages 23–26 *in* P. J. Cameron, R. L. Hill, J. Bain, and W. P. Thomas, eds., A review of biological control of invertebrate pests and weeds in New Zealand 1874 to 1987. Tech. Commun. No. 10. CAB Int. Inst. Biol. Control, Wallingford, UK.

Dzolkhifli, O., K. C. Khoo, R. Muhamad, and C. T. Ho. 1986. Preliminary study of resistance in 4 populations of *Helopeltis theobromae* Miller (Hemiptera: Miridae) to -HCH, propoxur and dioxacarb. Pages 317–323 *in* E. Pushparajah and P. S. Chew, eds., Cocoa and coconuts: Progress and outlook (Proc. Int. Conf. on Cocoa and Coconuts, Kuala Lumpur, 15–17 Oct. 1984). Incorporated Society of Planters, Kuala Lumpur, Malaysia.

Easterbrook, M. A. 1996. Damage to strawberry fruits by the European tarnished plant bug, *Lygus rugulipennis*. Brighton Crop Prot. Conf.: Pests and Diseases—1996. 3:867–872. BCPC Registered Office, Farnham, UK.

——. 1997. The phenology of *Lygus rugulipennis*, the European tarnished plant bug, on late-season strawberries, and control with insecticides. Ann. Appl. Biol. 131:1–10.

——. 2000. Relationships between the occurrence of misshapen fruit on late-season strawberry in the United Kingdom and infestation by insects, particularly the European tarnished plant bug, *Lygus rugulipennis*. Entomol. Exp. Appl. 96:59–67.

Easterbrook, M. A., M. G. Solomon, J. E. Cranham, and E. F. Souter. 1985. Trials of an integrated pest management programme based on selective pesticides in English apple orchards. Crop Prot. 4:215–230.

Eastham, J. W. 1915. The part played by insects in the spread of plant-diseases. Proc. Entomol. Soc. B.C. 7:18–21.

Ebel, B. H. 1963. Insects affecting seed production of slash and longleaf pines. Their identification and biological annotation. U.S. For. Serv. Pap. SE-6: 1–24.

Ebeling, W. 1959. Subtropical fruit pests. University of California, Berkeley. 436 pp.

Eberhard, W. G., N. I. Platnick, and R. T. Schuh. 1993. Natural history and systematics of arthropod symbionts (Araneae; Hemiptera; Diptera) inhabiting webs of the spider *Tengella radiata* (Araneae, Tengellidae). Am. Mus. Novit. No. 3065:1–17.

Ecale, C. L. and E. A. Backus. 1995. Mechanical and salivary aspects of potato leafhopper probing in alfalfa stems. Entomol. Exp. Appl. 77:121–132.

Ecale Zhou, C. L. and E. A. Backus. 1999. Phloem injury and repair following potato leafhopper feeding on alfalfa stems. Can. J. Bot. 77:537–547.

Eddy, C. O. 1927. The cotton flea hopper. S.C. Agric. Exp. Stn. Bull. 235:1–21.

——. 1928. Cotton flea hopper studies of 1927 and 1928. S.C. Agric. Exp. Stn. Bull. 251:1–18.

Edelson, J. V. and P. M. Estes. 1987. Seasonal abundance and distribution of predators and parasites associated with *Monelliopsis pecanis* Bissell and *Monellia caryella* (Fitch) (Homoptera: Aphidae) [*sic*]. J. Entomol. Sci. 22:336–347.

Eden, T. 1953. Gnarled stem canker of tea. Tea Res. Inst. East Afr. Rep. 1953: 30–31.

——. 1976. Tea, 3rd ed. Longman, London. 236 pp.

Edgar, W. D. 1970. Prey of the wolf spider *Lycosa lugubris* (Walck.). Entomol. Mon. Mag. 106:71–73.

Edland, T. 1997. Benefits of minimum pesticide use in insect and mite control in orchards. Pages 197–220 *in* D. Pimentel, ed., Techniques for reducing pesticide use: Economic and environmental benefits. Wiley, Chichester.

Edmunds, M. 1974. Defence in animals: A survey of anti-predator defences. Longman, London. 357 pp.

Edwards, J. S. 1963. Arthropods as predators. Viewp. Biol. 2:85–114.

Edwards, P. J. and S. D. Wratten. 1985. Induced plant defences against insect grazing: Fact or artefact? Oikos 44:70–74.

——. 1987. Ecological significance of wound-induced changes in plant chemistry. Pages 213–218 *in* V. Labeyrie, G. Fabres, and D. Lachaise, eds., Insects-plants. Proc. 6th Int. Symp. on Insect-

Plant Relationships, Pau, France. Junk, Dordrecht.

Eggleton, P. and K. J. Gaston. 1990. "Parasitoid" species and assemblages: Convenient definitions or misleading compromises? Oikos 59:417–421.

Eguagie, W. E. 1977. Studies on biotic potential and sex ratio of the cacao mirid, *Sahlbergella singularis* Haglund (Heteroptera). Niger. J. Plant Prot. 3: 68–75.

Ehanno, B. 1965. Notes écologiques sur les Miridae (Insecta-Heteroptera) observés en Bretagne sur le chêne. Vie Milieu 16:517–533.

——. 1976. Aperçu sur la faune entomologique du bocage breton: Punaises Miridae (Hétéroptères) inféodées à des végétaux des talus. Pages 385–389 in J. Missonnier, coord. Les bocages. Histoire, écologie, économie. Table ronde C.N.R.S.: Aspects physiques, biologiques et humains des écosystèms bocagers des régions tempérées humides. Inst. Natl. de la Recherche Agronomique (I.N.R.A.), Rennes, France.

——. 1983–1987. Les hétéroptères mirides de France. Tome I. Les secteurs biogeographiques. Inventaire Faune Flore 25:1–603 (1983); Tome I bis: Les secteurs biogeographiques (suite) 39:1–96c (1987a); Tome II-A: Inventaire et syntheses ecologiques 40:1–647 (1987b); Tome II-B: Inventaire biogeographique et atlas 42:649–1075 (1987c). Secretariat de la Faune et de la Flore, Paris.

Ehler, L. E. 1977. Natural enemies of cabbage looper on cotton in the San Joaquin Valley. Hilgardia 45(3):73–106.

——. 1990. Introduction strategies in biological control of insects. Pages 111–134 in M. Mackauer, L. E. Ehler, and J. Roland, eds., Critical issues in biological control. Intercept, Andover, UK.

——. 1996. Structure and impact of natural enemy guilds in biological control of insect pests. Pages 337–342 in G. A. Polis and K. O. Winemiller, eds., Food webs: Integration of patterns and dynamics. Chapman & Hall, New York.

——. 1998. Invasion biology and biological control. Biol. Control 13:127–133.

Ehler, L. E. and J. C. Miller. 1978. Biological control in temporary agroecosystems. Entomophaga 23:207–212.

Ehler, L. E., K. G. Eveleens, and R. van den Bosch. 1973. An evaluation of some natural enemies of cabbage loopers on cotton in California. Environ. Entomol. 2:1009–1015.

Ehrenfeld, J. G. 1979. Pollination of three species of *Euphorbia* subgenus *Chamaesyce*, with special reference to bees. Am. Midl. Nat. 101:87–98.

Ehrlich, P. R. 1984. The structure and dynamics of butterfly populations. Pages 25–40 in R. I. Vane-Wright and P. R. Ackery, eds., The biology of butterflies. Academic, London.

Ehrlich, P. R. and P. H. Raven. 1964. Butterflies and plants: A study in coevolution. Evolution 18:586–608.

Eickwort, G. C. 1983. Potential use of mites as biological control agents of leaf-feeding insects. Pages 41–52 in M. A. Hoy, G. L. Cunningham, and L. Knutson, eds., Biological control of pests by mites. Univ. Calif. Agric. Exp. Stn. Spec. Publ. 3304.

Eigenbrode, S. D. and K. E. Espelie. 1995. Effects of plant epicuticular lipids on insect herbivores. Annu. Rev. Entomol. 40:171–194.

Eisner, T. and D. J. Aneshansley. 1983. Adhesive strength of the insect-trapping glue of a plant (*Befaria racemosa*). Ann. Entomol. Soc. Am. 76: 295–298.

El-Dessouki, S. A., A. H. El-Kifl, and H. A. Helal. 1976. Life cycle, host plants and symptoms of damage of the tomato bug, *Nesidiocoris tenuis* Reut. (Hemiptera: Miridae), in Egypt. Z. Pflanzenkr. Pflanzenschutz 83:204–220.

Elgar, M. A. and B. J. Crespi. 1992. Ecology and evolution of cannibalism. Pages 1–12 in M. A. Elgar and B. J. Crespi, eds., Cannibalism: Ecology and evolution among diverse taxa. Oxford University Press, New York.

Elias, T. S. 1983. Extrafloral nectaries: Their structure and distribution. Pages 174–203 in B. Bentley and T. [S.] Elias, eds., The biology of nectaries. Columbia University Press, New York.

Elliott, C. and F. W. Poos. 1940. Seasonal development, insect vectors, and host range of bacterial wilt of sweet corn. J. Agric. Res. (Wash., D.C.) 60:645–686.

Ellis, A. G. and J. J. Midgley. 1996. A new plant-animal mutualism involving a plant with sticky leaves and a resident hemipteran insect. Oecologia (Berl.) 106:478–481.

Ellis, E. A. 1940. The natural history of Wheatfen Broad Surlingham. Trans. Norfolk Norwich Nat. Soc., 1939. 15(1):115–128.

Ellis, P. R. and J. A. Hardman. 1992. Pests of umbelliferous crops. Pages 327–378 in R. G. McKinlay, ed., Vegetable crop pests. CRC, Boca Raton, Fla.

Elmore, J. C. 1955. The nature of lygus bug injury to lima beans. J. Econ. Entomol. 48:148–151.

Elsner, E. A. and E. H. Beers. 1988. Distinguishing characteristics of the principal apple-infesting leafhoppers in central Washington. Melanderia 46:43–47.

Elson, J. A. 1937. A comparative study of Hemiptera. Ann. Entomol. Soc. Am. 30:579–597.

Elton, C. S. 1966. The pattern of animal communities. Methuen, London. 432 pp.

Elze, D. L. 1927. De verspreiding van virusziekten van de aardappel (*Solanum tuberosum* L.) door insekten. Meded. Landbouwhogesch. Wageningen 31(2):1–90.

El-Zik, K. M. and P. M. Thaxton. 1989. Genetic improvement for resistance to pests and stresses in cotton. Pages 191–224 in R. E. Frisbie, K. M. El-Zik, and L. T. Wilson, eds., Integrated pest management systems and cotton production. Wiley, New York.

Emmett, B. J. and L. A. E. Baker. 1971. Insect transmission of fireblight. Plant Pathol. (Lond.) 20:41–45.

Encalada, E. and L. Viñas. 1989. *Ceratocapsus dispersus* (Hemiptera, Miridae) en Piura: Biologia y capacidad predatora en insectario. Rev. Peru. Entomol. 32:1–8.

Enns, W. R. 1947. Tarnished plant bug injury to strawberries. Hortic. News (Mo. State Hortic. Soc.) 7(2):10–11.

Entwistle, P. F. 1965. Cocoa mirids. Part 1. A world review of biology and ecology. Cocoa Grow. Bull. 5:16–20.

——. 1966. Cocoa mirids. Part 2. Their control. Cocoa Grow. Bull. 6:17–22.

——. 1972. Pests of cocoa. Longman, London. 779 pp.

——. 1977. World distribution of mirids. Pages 35–46 in E. M. Lavabre, ed., Les Mirides du cacaoyer. Institut français du Café et du Cacao, Paris.

——. 1985a. Insects and cocoa. Pages 366–443 in G. A. R. Wood and R. A. Lass., eds., Cocoa, 4th ed. Longman, London.

——. 1985b. Cocoa mirids (capsids). Pages 67–72 in R. A. Lass and G. A. R. Wood, eds., Cocoa production: Present constraints and priorities for research. World Bank Tech. Pap. 39. World Bank, Washington, D.C.

Entwistle, P. F. and A. Youdeowei. 1965. A preliminary world review of cacao mirids. Pages 383–391 in Congrès de la protection des cultures tropicales. Compte rendue des travaux... Marseille, 23–27 Mars 1965. Chambre de Commerce et d'Industrie de Marseille [often cited as 1964, Pages 71–79 in Proc. Conf. Mirids and Other Pests Cacao, Ibadan, Nigeria, 1964. West African Cocoa Research Institute, Ibadan].

Erdélyi, C., S. Manninger, K. Manninger,

K. Gergely, L. Hangyel, and I. Bernáth. 1994. Climatic factors affecting population dynamics of the main seed pests of lucerne in Hungary. J. Appl. Entomol. 117:195–209.

Erhardt, A. and I. Baker. 1990. Pollen amino acids—an additional diet for a nectar feeding butterfly? Plant Syst. Evol. 169:111–121.

Erickson, E. H. Jr. 1983. Pollination of entomophilous hybrid seed parents. Pages 493–535 in C. E. Jones and R. J. Little, eds., Handbook of experimental pollination biology. Van Nostrand Reinhold, New York.

Erwin, T. L. 1982. Tropical forests: Their richness in Coleoptera and other arthropod species. Coleopt. Bull. 36: 74–75.

——. 1983. Tropical forest canopies: The last biotic frontier. Bull. Entomol. Soc. Am. 29(1):14–19.

——. 1989. Canopy arthropod biodiversity: A chronology of sampling techniques and results. Rev. Peru. Entomol. 32:71–77.

Esaki, T. 1934. A case of the facultative "blood-sucking" in Cyrtorrhinus [sic] lividipennis Reuter, with notes on the same haibit [sic] in some Typhlocybinae (Hemiptera, Miridae). Mushi 7:97–100. [in Japanese.]

Essig, E. O. 1915. Injurious and beneficial insects of California, 2nd ed. Mon. Bull. Calif. State Comm. Hortic. Suppl. 541 pp.

——. 1926. Insects of western North America. Macmillan, New York. 1035 pp.

Essig, E. O. and R. L. Usinger. 1940. The life and works of Edward Payson Van Duzee. Pan-Pac. Entomol. 16:145–177.

Eubanks, M. D. and R. F. Denno. 1999. The ecological consequences of variation in plants and prey for an omnivorous insect. Ecology 80:1253–1266.

Evans, H. C. 1989. Mycopathogens of insects of epigeal and aerial habitats. Pages 205–238 in N. Wilding, N. M. Collins, P. M. Hammond, and J. F. Webber, eds., Insect-fungus interactions. Academic, London.

Evans, H. E. 1969. Notes on the nesting behavior of Pisonopsis clypeata and Belomicrus forbesii (Hymenoptera, Sphecidae). J. Kans. Entomol. Soc. 42: 117–125.

Evans, H. E. 1984. Insect biology: A textbook of entomology. Addison-Wesley, Reading, Mass. 436 pp.

Evans, H. F. and P. F. Entwistle. 1987. Viral diseases. Pages 257–322 in J. R. Fuxa and Y. Tanada, eds., Epizootiology of insect diseases. Wiley, New York.

Evaristo, F. N. and M. H. Pais. 1970. Revisão do género Helopeltis (Hemiptera-Miridae) para Moçambique. Agron. Moçamb. 4:181–189.

Eveleens, K. G., R. van den Bosch, and L. E. Ehler. 1973. Secondary outbreak induction of beet armyworm by experimental insecticide applications in cotton in California. Environ. Entomol. 2:497–503.

Everly, R. T. 1938. Spiders and insects found associated with sweet corn with notes on the food and habits of some species I. Arachnida and Coleoptera. Ohio J. Sci. 38:136–148.

Every, D., J. A. Farrell, and M. W. Stufkens. 1992. Bug damage in New Zealand wheat grain: The roles of various heteropterous insects. N.Z. J. Crop Hortic. Sci. 20:305–312.

Ewen, A. B. 1966. A possible endocrine mechanism for inducing diapause in the eggs of Adelphocoris lineolatus (Goeze) (Hemiptera: Miridae). Experientia (Basel) 22:470.

Ewing, K. P. 1929. Effects on the cotton plant of the feeding of certain Hemiptera of the family Miridae. J. Econ. Entomol. 22:761–765.

Ewing, K. P. and H. J. Crawford. 1939. Egg parasites of the cotton flea hopper. J. Econ. Entomol. 32:303–305.

Ewing, K. P. and R. L. McGarr. 1933. The effect of certain homopterous insects as compared with three common mirids upon the growth and fruiting of cotton plants. J. Econ. Entomol. 26:943–953.

Eyer, J. R. and J. T. Medler. 1942. Control of hemipterous cotton insects by the use of dusts. J. Econ. Entomol. 35:630–634.

Eyles, A. C. 1999. Introduced Mirinae of New Zealand (Hemiptera: Miridae). N.Z. J. Zool. 26:355–372.

Eyles, A. C. and J. C. M. Carvalho. 1988a. Deraeocorinae of New Zealand (Miridae: Heteroptera). N.Z. J. Zool. 15: 63–80.

——. 1988b. A new genus of Mirini (Heteroptera: Miridae) from lucerne crops in New Zealand. N.Z. J. Zool. 15:339–341.

——. 1995. Further endemic new genera and species of Mirinae (Hemiptera: Miridae) from New Zealand. N.Z. J. Zool. 22:49–90.

Fabellar, L. T. and E. A. Heinrichs. 1984. Toxicity of insecticides to predators of rice brown planthoppers, Nilaparvata lugens (Stål) (Homoptera: Delphacidae). Environ. Entomol. 13:832–837.

——. 1986. Relative toxicity of insecticides to rice planthoppers and leafhoppers and their predators. Crop Prot. 5:254–258.

Faegri, K. 1971. The preservation of sporopollenin membranes under natural conditions. Pages 256–270 in J. Brooks et al., eds., Sporopollenin; proceedings of a symposium held at the Geology Department, Imperial College, London, 23–25 September, 1970. Academic, London.

Faegri, K. and J. Iversen. 1975. Textbook of pollen analysis, 3rd rev. ed. Hafner, New York. 295 pp.

Faegri, K. and L. van der Pijl. 1979. The principles of pollination ecology, 3rd rev. ed. Pergamon, Oxford. 244 pp.

Faeth, S. H. and R. F. Rooney III. 1993. Variable budbreak and insect folivory of Gambel oak (Quercus gambelii: Fagaceae). Southwest. Nat. 38:1–8.

Fagan, W. F. and L. E. Hurd. 1994. Hatch density variation of a generalist arthropod predator: Population consequences and community impact. Ecology 75: 2022–2032.

Fajardo, T. G. 1930. Studies on the mosaic disease of the bean (Phaseolus vulgaris L.). Phytopathology 20:469–494.

Falcon, L. A., R. van den Bosch, C. A. Ferris, L. K. Stromberg, L. K. Etzel, R. E. Stinner, and T. F. Leigh. 1968. A comparison of season-long cotton-pest-control programs in California during 1966. J. Econ. Entomol. 61:633–642.

Falcon, L. A., R. van den Bosch, J. Gallagher, and A. Davidson. 1971. Investigation of the pest status of Lygus hesperus in cotton in central California. J. Econ. Entomol. 64:56–61.

Falkingham, C. 1995. Carnivorous plants —carnivorous bugs. Is there a symbiotic relationship? Vic. Nat. (Blackburn) 112:222–223.

Farrow, R. A. 1984. Detection of transoceanic migration of insects to a remote island in the Coral Sea, Willis Island. Aust. J. Ecol. 9:253–272.

Fatuesi, S., P. Tauili'ili, F. Taotua, and A. Vargo. 1991. Cultural methods of pest control on taro (Colocasia esculenta Schott) in American Samoa. Micronesia Suppl. 3:123–127.

Fauchald, K. and P. A. Jumars. 1979. The diet of worms: A study of polychaete feeding guilds. Oceanogr. Mar. Biol. Annu. Rev. 17:193–284.

Faucheux, M.-J. 1975. Relations entre l'ultrastructure des stylets mandibulaires et maxillaires et la prise de nourriture chez les insectes Hémiptères. C. R. Hebd. Séances Acad. Sci. (Ser. D) 281:41–44.

Faulkner, L. R. 1952. Hemipterous insect pests. Their occurrence and distribution in principal cotton producing areas of New Mexico. N.M. Agric. Exp. Stn. Bull. 372:1–25.

Fauvel, G. 1974. Sur l'alimentation pollinique d'un anthocoride prédateur

Orius (Heterorius) vicinus Rib. (Hémiptère). Ann. Zool. Ecol. Anim. 6:245–258.

——. 1976. Die räuberischen Wanzen in Obstanlagen. *In* Nützlinge in Apfelangen. Einführung in den integrierten Pflanzenschutz. Int. Organ. Biol. Control/West Palearctic Reg. Sect. Bull. 3:125–144.

——. 1999. Diversity of Heteroptera in agroecosystems: Role of sustainability and bioindication. Agric. Ecosyst. Environ. 74:275–303.

Fauvel, G. and P. Atger. 1981. Etude de l'évolution des insectes auxiliaries et de leurs relations avec le psylle du poirier (*Psylla pyri* L.) et l'acarien rouge (*Panonychus ulmi* Koch) dans deux vergers du Sud-Est de la France en 1979. Agronomie (Paris) 1:813–820.

Fauvel, G., A. Rambier, and D. Cotton. 1975. Activité prédatrice et multiplication d'*Orius (Heterorius) vicinus* (Het.: Anthocoridae) dans les galles d'*Eriophyes fraxinivorus* (Acarina: Eriophyidae). Entomophaga 23:261–270.

Fauvel, G., J. C. Malausa, and B. Kaspar. 1987. Etude en laboratoire des principales caractéristiques biologiques de *Macrolophus caliginosus* (Heteroptera, Miridae). Entomophaga 32:529–543.

Fawcett, H. S. 1929. Nematospora on pomegranates, citrus, and cotton in California. Phytopathology 19:479–482.

Federal Ministry of Agriculture and Natural Resources, Nigeria. 1996. A guide to insect pests of Nigerian crops: Identification, biology and control. Natural Resources Institute, Chatham, UK. 253 pp.

Feeny, P. 1975. Biochemical coevolution between plants and their insect herbivores. Pages 3–19 *in* L. E. Gilbert and P. H. Raven, eds., Coevolution of animals and plants. University of Texas Press, Austin.

——. 1976. Plant apparency and chemical defense. Recent Adv. Phytochem. 10:1–40.

Feinsinger, P. 1983. Coevolution and pollination. Pages 282–310 *in* D. J. Futuyma and M. Slatkin, eds., Coevolution. Sinauer, Sunderland, Mass.

Feir, D. and S. D. Beck. 1963. Feeding behavior of the large milkweed bug, *Oncopeltis fasciatus*. Ann. Entomol. Soc. Am. 56:224–229.

Fellers, G. M. and J. H. Fellers. 1982. Scavenging rates of invertebrates in an eastern deciduous forest. Am. Midl. Nat. 107:389–392.

Felt, E. P. 1910. Deformed apples. Ctry. Gentleman (Phila.) 75:82.

——. 1915. Injurious insects. Banded grape bug, *Paracalocoris scrupeus*

Say. Pages 41–44 *in* 29th Rep. State Entomol. on Injurious and Other Insects of the State of New York, 1913. N.Y. State Mus. Bull. 175.

——. 1916. Notes for the year. Banded grape bug. Pages 62–65 *in* 30th Rep. State Entomol. on Injurious and Other Insects of the State of New York, 1914. N.Y. State Mus. Bull. 180.

Felton, G. W. 1996. Nutritive quality of plant protein: Sources of variation and insect herbivore responses. Arch. Insect Biochem. Physiol. 32:107–130.

Fennah, R. G. 1947. The insect pests of food-crops in the Lesser Antilles. Departments of Agriculture for the Windward and Leeward Islands, St. George's, Grenada; St. John's, Antigua. 207 pp.

Fenton, F. A. 1921. Progress report on the season's work on the production of potato tipburn. J. Econ. Entomol. 14:71–79.

Fenton, F. A., F. W. Whitehead, and C. H. Brett. 1945. Studies on the cause and prevention of a peach disease in Oklahoma, known as catfacing. Proc. Okla. Acad. Sci. 25:34–37.

Ferguson, A. W., B. D. L. Fitt, and I. H. Williams. 1997. Insect injury to linseed in south-east England. Crop Prot. 16:643–652.

Fernald, M. L. 1950. Gray's manual of botany, 8th (centennial) ed. American Book, New York. 1632 pp.

Fernando, H. E. and P. Manickavasagar. 1956. Economic damage and control of the cacao capsid, *Helopeltis* sp. (fam. Capsidae, ord. Hemiptera) in Ceylon. Trop. Agric. (Colombo) 112:25–36.

Ferran, A., A. Rortais, J. C. Malausa, J. Gambier, and M. Lambin. 1996. Ovipositional behaviour of *Macrolophus caliginosus* (Heteroptera: Miridae) on tobacco leaves. Bull. Entomol. Res. 86:123–128.

Ferreira, J. 1979. Ethology of host plant feeding preference by the tarnished plant bug, *Lygus lineolaris* (Palisot de Beauvois). M.S. thesis, Mississippi State University, Mississippi State. 104 pp.

Ferreira, P. S. F. 1998. The tribe Clivinematini: Cladistic analysis, geographic distribution and biological considerations (Heteroptera, Miridae). Rev. Bras. Entomol. 42:53–57.

——. 1999. Família Miridae. Pages 93–100 *in* C. R. F. Brandão and E. M. Cancello, eds., Invertebrados terrestres. Vol. V. Biodiversidade do Estado de São Paulo. Fundação de Amparo à Pesquisa do Estado de São Paulo (FAPESP), São Paulo, Brazil.

Feucht, J. R. 1987. Misleading mimics: Some injuries are not due to herbicides. Am. Nurseryman 165(6):148–150.

——. 1988. Herbicides injurious to trees—symptoms and solutions. J. Arboric. 14:215–219.

Fieber, F. X. 1861. Die europaischen Hemiptera. Halbflügler. (Rhynchota Heteroptera). Nach der analytischen Methode bearbeitet. Carl Gerold's Sohn, Vienna. 444 pp.

Fiedler, O. G. H. 1951. Entomologisches aus Afrika (Beobachtungen über Kaffeeschädlinge). 1. Biologische Notizen über Cocciden des Kaffeestrauches in Ostafrika. Z. Angew. Entomol. 32:289–306.

Fife, L. C. 1939. Insects and a mite found on cotton in Puerto Rico, with notes on their economic importance and natural enemies. P.R. Exp. Stn. (Mayaguez) Bull. 39:1–14.

Figuls, M., C. Castañé, and R. Gabarra. 1999. Residual toxicity of some insecticides on the predatory bugs *Dicyphus tamaninii* and *Macrolophus caliginosus*. BioControl 44:89–98.

Findlay, R. M. 1975. Pest of autumn-harvested asparagus. N.Z. J. Agric. 130:56–57.

Finţescu, G. N. 1914. Contributions à la biologie de l'hémiptère "*Capsus Mali*" (Meyer), (syn. *Capsus magnicornis* Fallen) *Plytocoris* [sic] *magnicornis* (Macq), *Atractotomus mali* (Fieber), *Capsus plenicornis* [sic]. Bucarest Bull. Acad. Rom. 3:132–140.

Fisher, E. H., A. J. Riker, and T. C. Allen. 1946. Bud, blossom and pod drop of canning string beans reduced by plant hormones. Phytopathology 36:504–523.

Fitch, A. 1870. Black-lined plant-bug, *Phytocoris lineatus*, Fab. (Hemiptera. Capsidae.). Pages 513–522 *in* 13th Rep. Noxious, Beneficial Insects State of New York. Van Benthuysen, Albany.

Fitt, G. P. 1994. Cotton pest management: Part 3. An Australian perspective. Annu. Rev. Entomol. 39:543–562.

Flechtmann, C. H. W. and J. A. McMurtry. 1992. Studies on how phytoseiid mites feed on spider mites and pollen. Int. J. Acarol. 18:157–162.

Fleischer, S. J. and M. J. Gaylor. 1987. Seasonal abundance of *Lygus lineolaris* (Heteroptera: Miridae) and selected predators in early season uncultivated hosts: Implications for managing movement into cotton. Environ. Entomol. 16:379–389.

——. 1988. *Lygus lineolaris* (Heteroptera: Miridae) population dynamics: Nymphal development, life tables, and Leslie matrices on selected weeds and cotton. Environ. Entomol. 17:246–253.

Fleischer, S. J., M. J. Gaylor, N. V. Hue, and L. C. Graham. 1986. Uptake and elimination of rubidium, a physiological marker, in adult *Lygus lineolaris* (Hemiptera: Miridae). Ann. Entomol. Soc. Am. 79:19–25.

Fleischer, S. J., M. J. Gaylor, and N. V. Hue. 1988. Dispersal of *Lygus lineolaris* (Heteroptera: Miridae) adults through cotton following nursery host destruction. Environ. Entomol. 17:533–541.

Flemion, F. 1958. Penetration and destruction of plant tissues during feeding by *Lygus lineolaris* P. de B. Proc. 10th Int. Congr. Entomol., Montreal (1956) 3:475–478.

Flemion, F. and E. T. Henrickson. 1949. Further studies on the occurrence of embryoless seeds and immature embryos in the Umbelliferae. Contrib. Boyce Thompson Inst. 15:291–297.

Flemion, F. and B. T. MacNear. 1951. Reduction of vegetative growth and seed yield in umbelliferous plants by *Lygus oblineatus*. Contrib. Boyce Thompson Inst. 16:279–283.

Flemion, F., H. Poole, and J. Olson. 1949. Relation of lygus bugs to embryoless seeds in dill. Contrib. Boyce Thompson Inst. 15:299–310.

Flemion, F., R. M. Weed, and L. P. Miller. 1951. Deposition of P^{32} into host tissue through the oral secretions of *Lygus oblineatus*. Contrib. Boyce Thompson Inst. 16:285–294.

Flemion, F., L. P. Miller, and R. M. Weed. 1952. An estimate of the quantity of oral secretion deposited by lygus when feeding on bean tissue. Contrib. Boyce Thompson Inst. 16:429–433.

Flemion, F., M. C. Ledbetter, and E. S. Kelley. 1954. Penetration and damage of plant tissues during feeding by the tarnished plant bug (*Lygus lineolaris*). Contrib. Boyce Thompson Inst. 17: 347–357.

Fletcher, B. S. and T. E. Bellas. 1988. Pheromones of Hemiptera, Blattodea, Orthoptera, Mecoptera, other insects, and Acari. Pages 207–271 *in* E. D. Morgan and N. B. Mandava, eds., CRC Handbook of natural pesticides. Vol. IV. Pheromones. Part B. CRC, Boca Raton, Fla.

Fletcher, J., A. Wayadande, U. Melcher, and F. Ye. 1998. The phytopathogenic mollicute-insect vector interface: A closer look. Phytopathology 88:1351–1358.

Fletcher, M. J. 1985. Plant bugs. Dep. Agric. N. S. W. Agfact AE. 38:1–7.

Fletcher, R. K. 1930. A study of the insect fauna of Brazos County, Texas, with special reference to the Cicadellidae. Ann. Entomol. Soc. Am. 23:33–56.

——. 1940a. A mirid (*Sixeonotus aureolatus* [sic] Knight). U.S. Dep. Agric. Insect Pest Surv. Bull. 20:354.

——. 1940b. Certain host plants of the cotton flea hopper. J. Econ. Entomol. 33:456–459.

Flint, H. M., F. D. Wilson, N. J. Parks, R. Y. Reynoso, B. R. Stapp, and J. L. Szaro. 1992. Suppression of pink bollworm and effect on beneficial insects of a nectariless okra-leaf cotton germplasm line. Bull. Entomol. Res. 81:379–384.

Flint, H. M., S. E. Naranjo, J. E. Leggett, and T. J. Henneberry. 1996. Cotton water stress, arthropod dynamics, and management of *Bemisia tabaci* (Homoptera: Aleyrodidae). J. Econ. Entomol. 89:1288–1300.

Flint, M. L. 1990. Pests of the garden and small farm: A grower's guide to using less pesticide. Univ. Calif. Div. Agric. Nat. Resour. Publ. 3332. Oakland, Calif. 276 pp.

——. tech. ed. 1991. Integrated pest management for apples & pears. Univ. Calif. Div. Agric. Nat. Resour. Publ. 3340. Oakland, Calif. 214 pp.

——. tech. ed. 1992. Integrated pest management for potatoes in the western United States. Univ. Calif. Div. Agric. Nat. Resour. Publ. 3316. Oakland, Calif. 146 pp.

Flint, M. L. and P. A. Roberts. 1988. Using crop diversity to manage pest problems: Some California examples. Am. J. Altern. Agric. 3:163–167.

Flood, B., R. Foster, and B. Hutchison. 1995. Sweet corn. Pages 19–40 *in* R. Foster and B. Flood, eds., Vegetable insect management with emphasis on the Midwest. Meister, Willoughby, Ohio.

Flor, G. 1860. Die Rhynchoten Livlands in systematischer Folge beschrieben. Erster Theil: Rhynchota frontirostria Zett. (Hemiptera heteroptera Aut.). Carl Schulz, Dorpat. 85 pp.

Foelix, R. F. 1982. Biology of spiders. (Transl. from German.) Harvard University Press, Cambridge. 306 pp.

Foglar, H., J. C. Malausa, and E. Wajnberg. 1990. The functional response and preference of *Macrolophus caliginosus* [Heteroptera: Miridae] for two of its prey: *Myzus persicae* and *Tetranychus urticae*. Entomophaga 35:465–474.

Fogle, H. W., H. L. Keil, W. L. Smith, S. M. Mircetich, L. C. Cochran, and H. Baker. 1974. Peach production. U.S. Dep. Agric. Agric. Handb. 463:1–90.

Foley, D. H. and B. A. Pyke. 1985. Developmental time of *Creontiades dilutus* (Stål) (Hemiptera: Miridae) in relation to temperature. J. Aust. Entomol. Soc. 24:125–127.

Folsom, D. 1942. Potato virus disease studies with tuber-line seed plots and insects in Maine 1927 to 1938. Maine Agric. Exp. Stn. Bull. 410:215–250.

Folsom, D., G. W. Simpson, and R. Bonde. 1949. Maine potato diseases, insects, and injuries. Maine Agric. Exp. Stn. Bull. 469:1–49.

——. 1955. Maine potato diseases, insects, and injuries. Maine Agric. Exp. Stn. Bull. 469 (rev.):1–52.

Folsom, J. W. 1932. Insect enemies of the cotton plant. U.S. Dep. Agric. Farmers' Bull. 1688:1–28.

Fomina, K. I. and E. G. Lebedeva. 1975. Potato viruses S and M and their vectors in the Primorsk region. Tr. Biol.-Pochv. Inst. 28(2):132–136. [in Russian; English summary in Rev. Plant Pathol. 60(3):1325, 1981.]

Fomina, K. I., E. G. Lebedeva, V. G. Riefman, and S. A. Fisenko. 1979. The ways of spreading of potato viruses S and M in the Primorsk region. Tr. Biol.-Pochv. Inst. 54(157):55–64. [in Russian; English summary in Rev. Plant Pathol. 61(1):30, 1982.]

Fontes, A. V. 1989. Contribuição ao estudo da genitália da fêmea de algumas espécies de *Prepops* Reuter, 1905 (Hemiptera, Miridae). Bol. Mus. Nac. (Rio J.) (Zool.) No. 330:1–31.

Fontes, E. M. G., D. H. Habeck, and F. Slansky Jr. 1994. Phytophagous insects associated with goldenrods (*Solidago* spp.) in Gainesville, Florida. Fla. Entomol. 77:209–221.

Forbes, S. A. 1883. Miscellaneous notes. *Lygus lineolaris*. Page 98 *in* 12th Rep. State Entomol. Ill., 1882. H. W. Rokker, Springfield.

——. 1884a. Insects injurious to the strawberry. Pages 60–180 *in* 13th Rep. State Entomol. Ill., 1883. H. W. Rokker, Springfield.

——. 1884b. The tarnished plant-bug. Farmers Rev. (Chic.) Feb. 28, p. 150.

——. 1885a. On new and imperfectly known strawberry insects. Pages 77–82 *in* 14th Rep. State Entomol. Ill., 1884. H. W. Rokker, Springfield.

——. 1885b. On some insect enemies of the soft maple (*Acer dasycarpum*). Pages 103–111 *in* 14th Rep. State Entomol. Ill., 1884. H. W. Rokker, Springfield.

——. 1892. Bacteria normal to digestive organs of Hemiptera. Bull. Ill. State Lab. Nat. Hist. 4(1):1–6.

——. 1900. The economic entomology of the sugar beet. Pages 49–186 *in* 21st Rep. State Entomol. Noxious and Beneficial Insects State of Illinois. Hack & Anderson, Chicago.

——. 1905. The more important insect injuries to corn. Pages 1–273 *in* 23rd Rep. State Entomol. Ill. R. R. Donnelly, Chicago.

Ford, J., H. Chitty, and A. D. Middleton. 1938. The food of partridge chicks

(*Perdix perdix*) in Great Britain. J. Anim. Ecol. 7:251–265.

Forsslund, K. H. 1936. Några farliga fiender till barrträdens groddplantor i Norrland. Skogen 5:99–101.

Foschi, S. and G. Carlotti. 1956. *Malacocoris chlorizans* Pz. var. *smaragdina* Fieb. predatore del "ragno rosso." Redia 41:105–111.

Fowler, H. G. 1980. New state record for *Halticotoma valida* (Hemiptera: Miridae) with notes on populations. Southwest. Nat. 25:267–268.

Fowler, S. V. and J. H. Lawton. 1985. Rapidly induced defences and talking trees: The devil's advocate position. Am. Nat. 126:181–195.

Fowler, S. V., M. F. Claridge, J. C. Morgan, I. D. R. Peries, and L. Nugaliyadde. 1991. Egg mortality of the brown planthopper, *Nilaparvata lugens* (Homoptera: Delphacidae) and green leafhoppers, *Nephotettix* spp. (Homoptera: Cicadellidae), on rice in Sri Lanka. Bull. Entomol. Res. 81:161–167.

Fox, L. R. 1975. Cannibalism in natural populations. Annu. Rev. Ecol. Syst. 6:87–106.

Fox, L. R. and P. A. Morrow. 1981. Specialization: Species property or local phenomenon? Science (Wash., D.C.) 211:887–893.

Fox, R. M. and J. W. Fox. 1964. Introduction to comparative entomology. Reinhold, New York. 450 pp.

Fox, R. T. V. 1993. Principles of diagnostic techniques in plant pathology. CAB International, Wallingford, UK. 213 pp.

Fox, W. B. 1943. Some insects infesting the "selenium indicator" vetches in Saskatchewan. Can. Entomol. 75:206–207.

Fox-Wilson, G. 1926. Insect visitors to sap-exudations of trees. Trans. Entomol. Soc. Lond. 74:243–254.

——. 1938. II. The tarnished plant bug or bishop fly, *Lygus pratensis* L.—précis of present knowledge. J. R. Hortic. Soc. 63:392–395.

Fraenkel, G. S. 1959. The raison d'être of secondary plant substances. Science (Wash., D.C.) 129:1466–1470.

Francke-Grosmann, H. 1962. Ungewöhnliche Knospenschäden an Sitkafichten. Proc. XI Int. Congr. Entomol., Vienna (1960) 2:189–191.

Frank, A. 1920. Disease and insect troubles of raspberries and their control. Wash. Agric. Exp. Stn. Mon. Bull. 7: 188–190.

Frank, J. H. 1979. The use of the precipitin technique in predator-prey studies to 1975. Pages 1–15 in M. C. Miller, ed., Serology in insect predator-prey studies. Misc. Publ. Entomol. Soc. Am. 11(4).

Frank, J. H. and E. D. McCoy. 1989. Introduction to attack and defense: Behavioral ecology of parasites and parasitoids and their hosts. Behavioral ecology: From fabulous past to chaotic future. Fla. Entomol. 72:1–6.

——. 1993. Introduction to the behavioral ecology of introduction: The introduction of insects into Florida. Fla. Entomol. 76:1–53.

Franklin, H. J. 1950. Cranberry insects in Massachusetts. Parts II–VII. Mass. Agric. Exp. Stn. Bull. 445:1–88.

Fransen, J. J. 1994. *Bemisia tabaci* in the Netherlands; here to stay? Pestic. Sci. 42:129–134.

Frazier, J. L. 1986. The perception of plant allelochemicals that inhibit feeding. Pages 1–42 in L. B. Brattsten and S. Ahmad, eds., Molecular aspects of insect-plant associations. Plenum, New York.

Free, J. B. 1970. Insect pollination of crops. Academic, London. 544 pp.

——. 1993. Insect pollination of crops, 2nd ed. Academic, San Diego. 684 pp.

Freeman, J. A. 1945. Studies in the distribution of insects by aerial currents. The insect population of the air from ground level to 300 feet. J. Anim. Ecol. 14:128–154.

Freeman, R. J. and A. J. Mueller. 1989. Seasonal occurrence of the tarnished plant bug, *Lygus lineolaris* (Heteroptera: Miridae) on soybean. J. Entomol. Sci. 24:218–223.

Freitag, J. H. 1950. Insect transmission and control of plant viruses. Pages 30–35 in 11th Annual Biology Colloquium, April 29, 1950. Oregon State College, Corvallis.

Frick, K. E. and R. B. Hawkes. 1970. Additional insects that feed upon tansy ragwort, *Senecio jacobaea*, an introduced weedy plant, in western United States. Ann. Entomol. Soc. Am. 63: 1085–1090.

Friend, W. G. and J. J. B. Smith. 1971. Feeding in *Rhodnius prolixus*: Mouthpart activity and salivation, and their correlation with changes of electrical resistance. J. Insect Physiol. 17:233–243.

Frisbie, R. E., ed. 1983. Guidelines for integrated control of cotton pests. FAO Plant Prod. Prot. Pap. 48. FAO, Rome. 187 pp.

Frisbie, R. E., J. R. Phillips, W. R. A. Lambert, and H. B. Jackson. 1983. Opportunities for improving cotton insect management programs and some constraints on beltwide implementation. Pages 521–557 in R. L. Ridgway et al., eds., Cotton insect management with special reference to

the boll weevil. U.S. Dep. Agric. Agric. Handb. 589.

Frisbie, R. E., J. L. Crawford, C. M. Bonner, and F. G. Zalom. 1989. Implementing IPM in cotton. Pages 389–412 in R. E. Frisbie, K. M. El-Zik, and L. T. Wilson, eds., Integrated pest management systems and cotton production. Wiley, New York.

Frisbie, R. E., D. D. Hardee, and L. T. Wilson. 1992. Biologically intensive integrated pest management: Future choices for cotton. Pages 57–81 in F. G. Zalom and W. E. Fry, eds., Food, crop pests, and the environment. APS, St. Paul, Minn.

Frith, D. W. 1979. A list of insects caught in light traps on West Island, Aldabra Atoll, Indian Ocean. Atoll Res. Bull. 225. Smithsonian Institution, Washington, D.C. 12 pp.

Fritz, R. S., N. E. Stamp, and T. G. Halverson. 1982. Iteroparity and semelparity in insects. Am. Nat. 120:264–268.

Froeschner, R. C. 1949. Contributions to a synopsis of the Hemiptera of Missouri. Pt. IV. Hebridae, Mesoveliidae, Cimicidae, Anthocoridae, Cryptostemmatidae, Isometopidae, Meridae [sic]. Am. Midl. Nat. 42:123–188.

——. 1985. Synopsis of the Heteroptera or true bugs of the Galápagos Islands. Smithsonian Contrib. Zool. No. 407:1–84.

Frolov, A. O. and S. O. Skarlato. 1988. Localization and modes of anchoring of the flagellate *Blastocrithidia miridarum* in the intestine of the bug *Adelphocoris quadripunctatus*. Parazitologiya (Leningr.) 22:481–487. [in Russian, English summary.]

——. 1991. Description of *Leptomonas mycophilus* sp.n. (Trypanosomatidae) from the bug *Phytocoris* sp. (Miridae). Parazitologiya (Leningr.) 25:99–103. [in Russian, English summary.]

Fronk, W. D., A. A. Beetle, and D. G. Fullerton. 1964. Dipterous galls on the *Artemisia tridentata* complex and insects associated with them. Ann. Entomol. Soc. Am. 57:575–577.

Frost, S. W. 1922. The false apple red-bug (*Lygidea mendax*) in Pennsylvania. J. Econ. Entomol. 15:102–104.

——. 1925. The delayed dormant oil spray for killing apple red-bug eggs. J. Econ. Entomol. 18:516–519.

——. 1952. Miridae from light traps. J. N.Y. Entomol. Soc. 60:237–240.

——. 1955. Response of insects to ultra violet light. J. Econ. Entomol. 48:155–156.

——. 1958. Insects attracted to light traps placed at different heights. J. Econ. Entomol. 51:550–551.

——. 1979. A preliminary study of North

American insects associated with elderberry flowers. Fla. Entomol. 62: 341–355.

Fryer, J. C. F. 1914. Preliminary notes on damage to apples by capsid bugs. Ann. Appl. Biol. 1:107–112.

———. 1916. Capsid bugs. J. Board Agric. (Lond.) 22:950–958.

———. 1929. The capsid pests of fruit trees in England. Trans. 4th Int. Congr. Entomol., Ithaca, N.Y. (1928)2:229–236.

Fryer, J. C. F. and F. R. Petherbridge. 1917. Report on further investigations on the capsids which attack apples. J. Board Agric. (Lond.) 24:33–44.

Fryxell, P. A. [1978]. The natural history of the cotton tribe (Malvaceae, tribe Gossypieae). Texas A & M University Press, College Station. 245 pp.

Fujisaki, K. 1985. Ecological significance of the wing polymorphism of the oriental chinch bug, *Cavelerius saccharivorus* Okajima (Heteroptera: Lygaeidae). Res. Popul. Ecol. (Kyoto) 27:125–136.

———. 1986a. Reproductive properties of the oriental chinch bug, *Cavelerius saccharivorus* Okajima (Heteroptera: Lygaeidae), in relation to its wing polymorphism. Res. Popul. Ecol. (Kyoto) 28:43–52.

———. 1986b. Genetic variation of density responses in relation to wing polymorphism in the oriental chinch bug, *Cavelerius saccharivorus* Okajima (Heteroptera: Lygaeidae). Res. Popul. Ecol. (Kyoto) 28:219–230.

———. 1992. A male fitness advantage to wing reduction in the oriental chinch bug *Cavelerius saccharivorus* Okajima (Heteroptera: Lygaeidae). Res. Popul. Ecol. (Kyoto) 34:173–183.

———. 1993a. Reproduction and egg diapause of the oriental chinch bug, *Cavelerius saccharivorus* Okajima (Heteroptera: Lygaeidae), in the subtropical winter season in relation to its wing polymorphism. Res. Popul. Ecol. (Kyoto) 35:171–181.

———. 1993b. Genetic correlation of wing polymorphism between females and males in the oriental chinch bug, *Cavelerius saccharivorus* Okajima (Heteroptera: Lygaeidae). Res. Popul. Ecol. (Kyoto) 35:317–324.

Fujita, K. 1977. Wing form composition in the field population of two species of lygaeid bugs, *Dimorphopterus pallipes* and *D. japonicus*, and its relation to environmental conditions. Jpn. J. Ecol. 27:263–267. [in Japanese, English summary.]

Fullaway, D. T. 1940. An account of the reduction of the immigrant taro leafhopper (*Megamelus proserpina*) popula-

tion to insignificant numbers by the introduction and establishment of the egg-sucking bug *Cyrtorhinus fulvus*. *In* Proc. 6th Pac. Sci. Congr. Pac. Sci. Assoc., Univ. Calif., Berkeley, July 24 to Aug. 12, 1939, pp. 345–346. University of California Press, Berkeley.

Fullaway, D. T. and N. L. H. Krauss. 1945. Common insects of Hawaii. Tongg, Honolulu. 228 pp.

Fullerton, D. G. 1961. Host preference of ladybird beetles in Wyoming alfalfa fields. M.S. thesis, University of Wyoming, Laramie. 66 pp.

Fulmek, L. 1930. Die grüne Schildcherwanze (*Lygus spinolae* Mey.) in Steiermark. Z. Angew. Entomol. 17:53–105.

Fulton, B. B. 1918. Observations on the life history and habits of *Pilophorus walshii* Uhler. Ann. Entomol. Soc. Am. 11:93–96.

Funasaki, G. Y., P. Y. Lai, and L. M. Nakahara. [1990]. Status of natural enemies of *Heteropsylla cubana* Crawford (Homoptera: Psyllidae) in Hawaii. Pages 153–158 *in* B. Napompeth and K. G. MacDicken, eds., Leucaena psyllid: Problems and management; proceedings of an international workshop held in Bogor, Indonesia, January 16–21, 1989. [Winrock International Institute for Agricultural Development, Arlington, Va.; International Development Research Centre, Ottawa, Canada; Nitrogen Fixing Tree Association, Waimanalo, Hawaii.]

Furth, D. G. 1985. The natural history of a sumac tree, with an emphasis on the entomofauna. Trans. Conn. Acad. Arts Sci. 46:137–234.

Futuyma, D. J. 1983. Evolutionary interactions among herbivorous insects and plants. Pages 207–231 *in* D. J. Futuyma and M. Slatkin, eds., Coevolution. Sinauer, Sunderland, Mass.

———. 1986. Evolutionary biology, 2nd ed. Sinauer, Sunderland, Mass. 600 pp.

Fuxa, J. R. and J. A. Kamm. 1976a. Effects of temperature and photoperiod on the egg diapause of *Labops hesperius* Uhler. Environ. Entomol. 5:505–507.

———. 1976b. Dispersal of *Labops hesperius* on rangeland. Ann. Entomol. Soc. Am. 69:891–893.

Fye, R. E. 1975. Plant host sequence of major cotton insects in southern Arizona. U.S. Dep. Agric. ARS W-24:1–9.

———. 1981a. An analysis of pear psylla populations, 1977–79. U.S. Dep. Agric. SEA ARM-W-24:1–32.

———. 1981b. Method for rearing the pear psylla. J. Econ. Entomol. 74:490–491.

———. 1982a. Overwintering of lygus bugs in central Washington: Effects of pre-overwintering host plants, moisture,

and temperature. Environ. Entomol. 11:204–206.

———. 1982b. Weed hosts of the lygus (Heteroptera: Miridae) bug complex in central Washington. J. Econ. Entomol. 75:724–727.

———. 1982c. Damage to vegetable and forage seedlings by the pale legume bug (Hemiptera: Miridae). J. Econ. Entomol. 75:994–996.

———. 1983a. Impact of volcanic ash on pear psylla (Homoptera: Psyllidae) and associated predators. Environ. Entomol. 12:222–226.

———. 1983b. Dispersal and winter survival of the pear psylla. J. Econ. Entomol. 76:311–315.

———. 1983c. Cover crop manipulation for building pear psylla (Homoptera: Psyllidae) predator populations in pear orchards. J. Econ. Entomol. 76:306–310.

———. 1984. Damage to vegetable and forage seedlings by overwintering *Lygus hesperus* (Heteroptera: Miridae) adults. J. Econ. Entomol. 77:1141–1143.

Gabarra, R., C. Castañé, E. Bordas, and R. Albajes. 1988. *Dicyphus tamaninii* as a beneficial insect and pest in tomato crops in Catalonia, Spain. Entomophaga 33:219–228.

Gabarra, R., C. Castañé, and R. Albajes. 1995. The mirid bug *Dicyphus tamaninii* as a greenhouse whitefly and western flower thrips predator on cucumber. Biocontrol Sci. Technol. 5:475–488.

Gäbler, H. 1937. Beitrag zur Kenntnis des Eies, der Eiablage und der Larven von *Calocoris biclavatus* H.-Sch. Zool. Anz. 119:299–302.

Gadd, C. H. 1937. A disease of salvias. Trop. Agric. (Colombo) 89:335–338.

Gadoury, D. M., W. E. MacHardy, and D. A. Rosenberger. 1989. Integration of pesticide application schedules for disease and insect control in apple orchards of the northeastern United States. Plant Dis. 73:98–105.

Gagné, R. J. 1989. The plant-feeding gall midges of North America. Cornell University Press, Ithaca. 356 pp.

Gagné, S., C. Richard, and C. Gagnon. 1984. La coulure des graminées: État des connaissances. Phytoprotection 65:45–52.

———. 1985. Présence et causes possibles de la coulure des graminés chez la fléole des près au Québec. Can. Plant Dis. Surv. 65:17–21.

Gagné, W. C. 1975a (1975–1978). Notes and exhibitions. *Halticus chrysolepis* Kirkaldy. Proc. Hawaii. Entomol. Soc. 22:5.

———. 1975b (1975–1978). Notes and exhibitions. *Psallus sharpianus* Kirkaldy. Proc. Hawaii. Entomol. Soc. 22:6.

——. 1976 (1975–1978). Notes and exhibitions. *Cyrtopeltis (Engytatus) modestus* (Distant). Proc. Hawaii. Entomol. Soc. 22:168.

——. 1979. Canopy-associated arthropods in *Acacia koa* and *Metrosideros* tree communities along an altitudinal transect on Hawaii Island. Pac. Insects 21:56–82.

——. 1982. Insular evolution and speciation of the genus *Nesiomiris* in Hawaii (Heteroptera: Miridae). Entomol. Gen. 8:87–88.

——. 1997. Insular evolution, speciation, and revision of the Hawaiian genus *Nesiomiris* (Hemiptera: Miridae). Bishop Mus. Bull. Entomol. 7. Bishop Museum Press, Honolulu. 226 pp.

Gahan, A. B. 1949. A new mymarid parasitic in eggs of *Helopeltis cinchonae* Mann (Hymenoptera, Mymaridae). Proc. Entomol. Soc. Wash. 51:75–76.

Gahukar, R. T. 1991. Recent developments in sorghum entomology research. Agric. Zool. Rev. 4:23–65.

Gahukar, R. T., Y. O. Doumbia, and S. M. Bonzi. 1989. *Eurystylus marginatus* Odh., a new pest of sorghum in the Sahel. Trop. Pest Manage. 35:212–213.

Gaines, J. C. and K. P. Ewing. 1938. The relation of wind currents, as indicated by balloon drifts, to cotton flea hopper dispersal. J. Econ. Entomol. 31:674–677.

Galletti, R., K. D. S. Baldwin, and I. O. Dina. 1972. Nigerian cocoa farmers: An economic survey of Yoruba cocoa farming families. Greenwood, Westport, Conn. 744 pp.

Ganeshaiah, K. N. and T. Veena. 1988. Ant-plant mutualism: Selective forces and adaptive changes. Pages 151–164 *in* T. N. Ananthakrishnan and A. Raman, eds., Dynamics of insect-plant interaction: Recent advances and future trends. Oxford & IBH, New Delhi, India.

Ganga Visalakshy, P. N. and K. P. Jayanth. 1994. *Stethoconus praefectus* (Distant), a predator of *Teleonemia scrupulosa* Stal in Bangalore, India. Entomon 19:177–178.

Gange, A. C. and M. Llewellyn. 1989. Factors affecting orchard colonisation by the black-kneed capsid (*Blepharidopterus angulatus* (Hemiptera: Miridae)) from alder windbreaks. Ann. Appl. Biol. 114:221–230.

Gannaway, J. R. 1994. Breeding for insect resistance. Pages 431–453 *in* G. A. Matthews and J. P. Tunstall, eds., Insect pests of cotton. CAB International, Wallingford, UK.

Gantz, H. 1998. Plant bugs: A move from

secondary status. Page 8 (Special Report: Transgenic Technologies) *in* Delta Farm Press 55(1).

Gapud, V. P., L. R. I. Velasco, I. L. Lit Jr., and E. Baradas-Mora. 1993. On the identity of the capsid bug (Hemiptera: Miridae) attacking cacao (*Theobroma cacao* L.) in Los Baños, Philippines. Philipp. Entomol. 9:152–153.

Garcia, M., C. 1974. Primer catalogo de insectos fitofagos de Mexico. Fitofilo 27(69):1–176.

Garcia-Tejero, F. D. 1993. Plagas y enfermedades de las plantas cultivadas, 9th ed. rev. Ediciones Munda-Prensa, Madrid. 821 pp.

Gardner, E. J. 1947. Insect pollination in guayule, *Parthenium argentatum* Gray. J. Am. Soc. Agron. 39:224–233.

Garman, H. 1926. Two important enemies of bluegrass pastures. Ky. Agric. Exp. Stn. Bull. (Res.) 265:29–47.

Garman, P., W. T. Brigham, and A. DeCaprio. 1953. Control of peach insects. Conn. Agric. Exp. Stn. Bull. 575:1–64.

Gaun, S. 1974. Blomstertaeger. Danmarks fauna. Bd. 81. Dansk Naturhistorisk Forening, Copenhagen. 279 pp.

Gaylor, M. J. and W. L. Sterling. 1975a. Effects of temperature on the development, egg production, and survival of the cotton fleahopper, *Pseudatomoscelis seriatus*. Environ. Entomol. 4:487–490.

——. 1975b. Simulated rainfall and wind as factors dislodging nymphs of the cotton fleahopper, *Pseudatomoscelis seriatus* (Reuter), from cotton plants. Tex. Agric. Exp. Stn. Progr. Rep. 3356: 1–2.

——. 1976a. Development, survival, and fecundity of the cotton fleahopper, *Pseudatomoscelis seriatus* (Reuter), on several host plants. Environ. Entomol. 5:55–58.

——. 1976b. Effects of temperature and host plants on population dynamics of the cotton fleahopper, *Pseudatomoscelis seriatus*. Tex. Agric. Exp. Stn. Bull. 1161:1–8.

——. 1977. Photoperiodic induction and seasonal incidence of embryonic diapause in the cotton fleahopper, *Pseudatomoscelis seriatus*. Ann. Entomol. Soc. Am. 70:893–897.

Gaylor, M. J., G. A. Buchanan, F. R. Gilliland, and R. L. Davis. 1983. Interactions among a herbicide program, nitrogen fertilization, tarnished plant bugs, and planting dates for yield and maturity of cotton. Agron. J. 75:903–907.

Gaylor, M. J., S. J. Fleischer, D. P. Muehleisen, and J. V. Edelson. 1984.

Insect populations in cotton produced under conservation tillage. J. Soil Water Conserv. 39:61–64.

Geering, Q. A. 1953. Studies of *Lygus vosseleri* Popp. (Heteroptera, Miridae). East and Central Africa. I. A method for breeding continuous supplies in the laboratory. Bull. Entomol. Res. 44:351–362.

Geetha, N., M. Gopalan, and M. Mohana Sundarum. 1992. Biology of the predatory mirid, *Cyrtorhinus lividipennis* (Reuter) on the eggs of various insect hosts. J. Entomol. Res. 16:300–304.

Geier, P. and M. Baggiolini. 1952. *Malacocoris chlorizans* Pz. (Hem. Het. Mirid.), prédateur des Acariens phytophages. Mitt. Schweiz. Entomol. Ges. 25:257–259.

Gellatley, J. G. 1967. Citrus blossum [sic] bug. Agric. Gaz. N.S.W. 78:377.

Gentry, J. W. 1965. Crop insects of northeast Africa–southwestern Asia. U.S. Dep. Agric. Agric. Res. Serv. Agric. Handb. 273. 210 pp.

[Georgis, R.] 1977. Two new predators of the white fly in Iraq. PANS (Pest Artic. News Summ.) 23:210.

Gerard, B. M. 1966. Mirid ecology. Cocoa Res. Inst. (Ghana). Rep. 1963–1965:42.

Gerber, G. H. 1995. Fecundity of *Lygus lineolaris* (Heteroptera: Miridae). Can. Entomol. 127:263–264.

——. 1997. Oviposition preferences of *Lygus lineolaris* (Palisot de Beauvois) (Heteroptera: Miridae) on four *Brassica* and two *Sinapis* species (Brassicaceae) in field cages. Can. Entomol. 129:855–858.

Gerber, G. H. and I. L. Wise. 1995. Seasonal occurrence and number of generations of *Lygus lineolaris* and *L. borealis* (Heteroptera: Miridae) in southern Manitoba. Can. Entomol. 127:543–559.

Gerhardson, B. and J. Pettersson. 1974. Transmission of red clover mottle virus by clover shoot weevils, *Apion* spp. Swedish J. Agric. Res. 4:161–165.

Gerin, L. 1956. Les *Helopeltis* (Hemipt. Miridae), nuisible aux Quinqinas du Cameroun Français. J. Agric. Trop. Bot. Appl. 3:512–540.

Gerling, D. 1986. Natural enemies of *Bemisia tabaci*, biological characteristics and potential as biological control agents: A review. Agric. Ecosyst. Environ. 17:99–110.

——. 1990. Natural enemies of whiteflies: Predators and parasitoids. Pages 147–185 *in* D. Gerling, ed., Whiteflies: Their bionomics, pest status and management. Intercept, Andover, UK.

——. 1992. Approaches to the biological control of whiteflies. Fla. Entomol. 75:446–456.

Gertsson, C.-A. 1979. Stinkflyn (Hemiptera-Heteroptera) i jordgubbsodlingar. Växtskyddsnotiser 43:81–86.

———. 1980. Förekomsten av stinkflyn i sydsvenska jordgubbsodlingar. Entomol. Tidskr. 101:71–74.

———. 1982. Fångst av jordgubbsstinkflyet med fönsterfällor. Entomol. Tidskr. 103:18–20.

Gess, S. K. 1996. The pollen wasps: Ecology and natural history of the Masarinae. Harvard University Press, Cambridge. 340 pp.

Gessé Solé, F. 1992. Comportamiento alimenticio de *Dicyphus tamaninii* Wagner (Heteroptera: Miridae). Bol. Sanid. Veg. Plagas 18:685–691.

Getzin, L. W. 1983. Damage to inflorescence of cabbage seed plants by the pale legume bug (Heteroptera: Miridae). J. Econ. Entomol. 76:1083–1085.

Ghauri, M. S. K. and F. Y. K. Ghauri. 1983. A new genus and new species of Isometopidae from North India, with a key to world genera (Heteroptera). Reichenbachia 21(3):19–25.

Ghavami, M. D. 1997. Studies on biology and population dynamic [*sic*] of *Deraeocoris pallens* Reut. (Hemiptera: Miridae) in cotton fields. Ph.D. dissertation, University of Çukurova, Adana, Turkey. 75 pp. [in Turkish, English summary.]

Ghavami, M. D., A. F. Özgür, and U. Kersting. 1998. Prey consumption by the predator *Deraeocoris pallens* Reuther [*sic*] (Hemiptera: Miridae) on six cotton pests. Z. Pflanzenkr. Pflanzenschutz 105:526–531.

Ghesquière, J. 1922. Un Réduvide prédateur du *Sahlbergella singularis* Hgl. Rev. Zool. Afr. 10:329.

———. 1939. Un Capside myrmécoïde nuisible au caféier. Rev. Zool. Bot. Afr. 33:30–32.

Ghosh, D., N. Debnath, and S. Chakrabarti. 1988. Predators and parasites of aphids from north-west and western Himalaya. IV. Twelve species of heteropterans (Heteroptera: Insecta) from Garhwal and Kumaon ranges. Proc. Zool. Soc. Calcutta 39:15–19.

Ghosh, G. C., M. H. Ali, S. V. Fowler, and N. R. Maslen. 1992. The development of practical and appropriate IPM methods for irrigated rice in eastern India. Brighton Crop Prot. Conf.: Pests and Diseases—1992. 3:1021–1026. BCPC Registered Office, Farnham, UK.

Giard, A. 1900. Sur un Hémiptère (*Atractotomus mali* Mey.) parasite des chenilles d'*Hyponomeuta malinellus* Zeller et *H. padellus* L. Bull. Soc. Entomol. Fr. 1900:359–360.

Gibb, K. S. and J. W. Randles. 1988. Studies on the transmission of velvet tobacco mottle virus by the mirid, *Cyrtopeltis nicotianae*. Ann. Appl. Biol. 112:427–437.

———. 1989. Non-propagative translocation of velvet tobacco mottle virus in the mirid, *Cyrtopeltis nicotianae*. Ann. Appl. Biol. 115:11–15.

———. 1990. Distribution of velvet tobacco mottle virus in its mirid vector and its relationship to transmissibility. Ann. Appl. Biol. 116:513–521.

———. 1991. Transmission of velvet tobacco mottle virus and related viruses by the mirid *Cyrtopeltis nicotianae*. Adv. Dis. Vector Res. 7:1–17.

Gibbs, A. 1969. Plant virus classification. Adv. Virus Res. 14:263–328.

Gibbs, A. and B. Harrison. 1976. Plant virology: The principles. Wiley, New York. 292 pp.

Gibbs, D. G. 1969. Effects on *Distantiella* of different N, K, P, levels in cocoa seedings. Cocoa Res. Inst. (Ghana) Rep. 1967–68:60.

———. 1977. *Distantiella theobroma* (Dist.). Pages 292–294 in J. Kranz, H. Schmutterer, and W. Koch, eds., Diseases, pests and weeds in tropical crops. Parey, Berlin.

Gibbs, D. G. and D. Leston. 1970. Insect phenology in a forest cocoa-farm locality in West Africa. J. Appl. Ecol. 7: 519–548.

Gibbs, D. G. and A. D. Pickett. 1966. Feeding by *Distantiella theobroma* (Dist.) (Heteroptera, Miridae) on cocoa. I. The effects of water stress in the plant. Bull. Entomol. Res. 57:159–169.

Gibbs, D. G., A. D. Pickett, and D. Leston. 1968. Seasonal population changes in cocoa capsids (Hemiptera, Miridae) in Ghana. Bull. Entomol. Res. 58:279–293.

Gibson, C. W. D. 1976. The importance of foodplants for the distribution and abundance of some Stenodemini (Heteroptera: Miridae) of limestone grassland. Oecologia (Berl.) 25:55–76.

———. 1980. Niche use patterns among some Stenodemini (Heteroptera: Miridae) of limestone grassland, and an investigation of the possibility of interspecific competition between *Notostira elongata* Geoffroy and *Megaloceraea* [*sic*] *recticornis* Geoffroy. Oecologia (Berl.) 47:352–364.

Gibson, C. [W. D.] and M. Visser. 1982. Interspecific competition between two field populations of grass-feeding bugs. Ecol. Entomol. 7:61–67.

Giesberger, G. [1983]. Biological control of the *Helopeltis* pest of cocoa in Java. Pages 91–180 in H. Toxopeus and P. C. Wessel, eds., Cocoa research in Indonesia 1900–1950. Vol. 2. American Cocoa Research Institute, Washington, D.C.; International Office of Cocoa & Chocolate, Brussels.

Gilbert, F. S. 1981. Foraging ecology of hoverflies: Morphology of the mouthparts in relation to feeding on nectar and pollen in some common urban species. Ecol. Entomol. 6:245–262.

Gilbert, F. [S.] 1990. Size, phylogeny and life-history in the evolution of feeding specialization in insect predators. Pages 101–124 in F. Gilbert, ed., Insect life cycles: Genetics, evolution and coordination. Springer, London.

Gilbert, L. E. 1972. Pollen feeding and reproductive biology of *Heliconius* butterflies. Proc. Natl. Acad. Sci. U.S.A. 69:1403–1407.

———. 1980. Ecological consequences of a coevolved mutualism between butterflies and plants. Pages 210–240 in L. E. Gilbert and P. H. Raven, eds., Coevolution of animals and plants, 2nd ed. University of Texas Press, Austin.

Gildow, F. E. 1983. Influence of barley yellow dwarf virus-infected oats and barley on morphology of aphid vectors. Phytopathology 73:1196–1199.

———. 1987. Virus-membrane interactions involved in circulative transmission of luteoviruses by aphids. Curr. Top. Vector Res. 4:93–120.

———. 1991. Barley yellow dwarf virus transport through aphids. Pages 165–177 in D. C. Peters, J. A. Webster, and C. S. Chlouber, eds., Aphid-plant interactions: Populations to molecules. Oklahoma State University Press, Stillwater.

———. 1993. Evidence for receptor-mediated endocytosis regulating luteovirus acquisition by aphids. Phytopathology 83:270–277.

Gillespie, D., R. McGregor, D. Quiring, and M. Foisy. 1998. *Dicyphus hesperus*—this bug's for you. Agric. Agri-Food Can. Pac. Agri-Food Res. Cent. (Agassiz). Tech. Rep. 148:1–2.

———. 1999. You are what you've eaten—prey versus plant feeding in *Dicyphus hesperus*. Agric. Agri-Food Can. Pac. Agri-Food Res. Cent. (Agassiz). Tech. Rep. 154:1–4.

Gillette, C. P. 1908. Notes and descriptions of some orchard plant lice of the family Aphidae. J. Econ. Entomol. 1:302–310.

Gillette, C. P. and C. F. Baker. 1895. A preliminary list of the Hemiptera of Colorado. Col. Agric. Exp. Stn. Bull. 31. Tech. Ser. 1:1–137.

Gilliatt, F. C. 1935. Some predators of the European red mite, *Paratetranychus pilosus* C. & F., in Nova Scotia. Can. J. Res. 13:19–38.

———. 1937. Natural control of the grey banded leaf roller, *Eulia mariana* Fern., in Nova Scotia orchards. Can. Entomol. 69:145–146.

Gilliland, F. R. Jr. 1972. Influence of simulated early-season insect damage on growth and yield of cotton. Ala. Agric. Exp. Stn. Auburn Univ. Bull. 442:1–12.

——. 1981. The lygus problem—decreasing or increasing. In Proc. Beltwide Cotton Prod. Res. Conf., Jan. 4–8, 1981, New Orleans, La., pp. 147–148. National Cotton Council, Memphis.

Gilmer, R. M. and E. C. Blodgett. 1976. X-disease. Pages 145–155 in T. S. Pine et al., eds., Virus diseases and noninfectious disorders of stone fruits in North America. U.S. Dep. Agric. Agric. Handb. 437.

Gilmer, R. M. and F. L. McEwen. 1958. Chlorotic fleck, an eriophyid mite injury of myrobalan plum. J. Econ. Entomol. 51:335–337.

Gimingham, C. T. 1928. An introduced capsid injurious to orchids. Entomol. Mon. Mag. 64:272–274.

Glasgow, H. 1914. The gastric caeca and the cecal bacteria of the Heteroptera. Biol. Bull. (Woods Hole) 26:101–170.

Glen, D. M. 1973. The food requirements of Blepharidopterus angulatus (Heteroptera: Miridae) as a predator of the lime aphid, Eucallipterus tiliae. Entomol. Exp. Appl. 16:255–267.

——. 1975a. The effects of predators on the eggs of codling moth Cydia pomonella, in a cider-apple orchard in south-west England. Ann. Appl. Biol. 80:115–119.

——. 1975b. Searching behaviour and prey-density requirements of Blepharidopterus angulatus (Fall.) (Heteroptera: Miridae) as a predator of the lime aphid, Eucallipterus tiliae (L.), and leafhopper, Alnetoidea alneti (Dahlbom). J. Anim. Ecol. 44:115–134.

——. 1977a. Ecology of the parasites of a predatory bug, Blepharidopterus angulatus (Fall.). Ecol. Entomol. 2:47–55.

——. 1977b. Predation of codling moth eggs, Cydia pomonella, the predators responsible and their alternative prey. J. Appl. Ecol. 14:445–456.

Glen, D. M. and N. D. Barlow. 1980. Interaction of a population of the black-kneed capsid, Blepharidopterus angulatus, and its prey, the lime aphid. Ecol. Entomol. 5:335–344.

Glen, D. M. and P. Brain. 1978. A model of predation on codling moth eggs (Cydia pomonella). J. Anim. Ecol. 47:711–724.

Glick, P. A. 1939. The distribution of insects, spiders, and mites in the air. U.S. Dep. Agric. Tech. Bull. 673:1–150.

——. 1957. Collecting insects by airplane in southern Texas. U.S. Dep. Agric. Tech. Bull. 1158:1–28.

——. 1983. The influence of cultural practices on arthropod populations in cotton. U.S. Dep. Agric. Agric. Res. Serv. Agric. Rev. Man. ARM-S-32:1–52.

Glover, T. 1856. Insects frequenting the cotton-plant. Pages 64–115 in Rep. Commissioner of Patents, 1855. Cornelius Wendell, Washington, D.C.

——. 1876. Report of the Entomologist. Heteroptera, or plant-bugs. Pages 114–140 in Rep. Commissioner of Agric., 1875. Government Printing Office, Washington, D.C.

——. 1878. Manuscript notes from my journal. Cotton, and the principal insects &c., frequenting or injuring the plant, in the United States. Written and etched by Townend Glover. Transferred to and printed from stone by J. C. Entwistle, Washington, D.C. 97 pp.

Glover, T. F. 1978. Economics and selection of management alternatives (predictive models). Pages 61–68 in B. A. Haws, compiler, Economic impacts of Labops hesperius on the production of high quality range grasses. Final Rep. Utah Agric. Exp. Stn. to Four Corners Reg. Comm. Utah State University, Logan.

Goble, H. W. 1969. Insects of the season 1968 related to fruit, vegetables, field crops and ornamentals. Proc. Entomol. Soc. Ont. 99:5–6.

——. 1970. Insects of the season 1969 related to fruit, vegetables, field crops and ornamentals. Proc. Entomol. Soc. Ont. 100:7–8.

Goddard, J. 1993. Physician's guide to arthropods of medical importance. CRC, Boca Raton, Fla. 332 pp.

Goddard, W. H. 1935. A record of the Hemiptera Heteroptera at the Imperial College Biological Field Station, Slough, Bucks, with notes on their food. Trans. Soc. Brit. Entomol. 2:47–67.

Godfrey, K. E., W. H. Whitcomb, and J. L. Stimac. 1989. Arthropod predators of velvetbean caterpillar, Anticarsia gemmatalis Hübner (Lepidoptera: Noctuidae), eggs and larvae. Environ. Entomol. 18:118–123.

Godfrey, L. D. and T. F. Leigh. 1994. Alfalfa harvest strategy effect on lygus bug (Hemiptera: Miridae) and insect predator population density: Implications for use as a trap crop in cotton. Environ. Entomol. 23:1106–1108.

Godley, E. J. 1979. Flower biology in New Zealand. N.Z. J. Bot. 17:441–466.

Goeden, R. D. and R. L. Kirkland. 1981. Interactions of field populations of indigenous egg predators, imported Microlarinus weevils, and puncturevine in southern California. Pages 515–527 in E. S. Delfosse, ed., Proc. 5th

Symp. Biol. Control Weeds, July 1980, Brisbane. CSIRO, Melbourne.

Goeden, R. D. and S. M. Louda. 1976. Biotic interference with insects imported for weed control. Annu. Rev. Entomol. 21:325–342.

Goeden, R. D. and D. W. Ricker. 1974. The phytophagous insect fauna of the ragweed, Ambrosia acanthicarpa, in southern California. Environ. Entomol. 3:827–834.

——. 1986a. Phytophagous insect faunas of two introduced Cirsium thistles, C. ochrocentrum and C. vulgare, in southern California. Ann. Entomol. Soc. Am. 79:945–952.

——. 1986b. Phytophagous insect faunas of the two most common native Cirsium thistles, C. californicum and C. proteanum, in southern California. Ann. Entomol. Soc. Am. 79:953–962.

Goeden, R. D. and J. A. Teerink. 1993. Phytophagous insect faunas of Dicoria canescens and Iva axillaris, native relatives of ragweeds, Ambrosia spp., in southern California, with analyses of insect associates of Ambrosiinae. Ann. Entomol. Soc. Am. 86:37–50.

Goel, S. C. 1972. A short note on the structure of the trochanter in Miridae (Heteroptera). Dtsch. Entomol. Z. 19: 367–368.

Gogala, M. 1984. Vibration producing structures and songs of terrestrial Heteroptera as systematic character. Biol. Vestn. 32:19–36.

Goidanich, A. 1929. Gli insetti dannosi alla Canapa. Ann. Tec. Agrar. (Rome) 1:423–431.

Göllner-Scheiding, U. 1989. Ergebnisse von Lichtfängen in Berlin aus den Jahren 1981–1986 1. Heteroptera. Teil I: Landwanzen (Cimicomorpha et Pentatomorpha) (Insecta). Faun. Abh. (Dresden) 16:111–123.

Gomez-Menor, G. J. M. 1955. Un mirido que ataca al tomate y al tabaco. Bol. Patol. Veg. Entomol. Agric. 21:193–200.

Gonzalez, D. and L. T. Wilson. 1982. A food-web approach to economic thresholds: A sequence of pests/predaceous arthropods on California cotton. Entomophaga 27 (Spec. Issue):31–43.

Goodchild, A. J. P. 1952. A study of the digestive system of the West African cocoa capsid bugs (Hemiptera, Miridae). Proc. Zool. Soc. Lond. 122: 543–572.

——. 1963a. Some new observations on the intestinal structures concerned with water disposal in sap-sucking Hemiptera. Trans. R. Entomol. Soc. Lond. 115:217–237.

——. 1963b. Studies on the functional anatomy of the intestines of Heteroptera. Proc. Zool. Soc. Lond. 141: 851–910.

——. 1966. Evolution of the alimentary canal in the Hemiptera. Biol. Rev. Camb. Philos. Soc. 41:97–140.

——. 1977. Bionomics, aggregated feeding behaviour, and colour variations in the sap-sucking bug *Mygdonia tuberculosa* Sign. (Hemiptera: Coreidae). Rev. Zool. Afr. 91:1032–1041.

Goodell, P. B., R. E. Plant, T. A. Kerby, J. F. Strand, L. T. Wilson, L. Zelinski, J. A. Young, A. Corbett, R. D. Horrocks, and R. N. Vargas. 1990. CALEX/cotton: An integrated expert system for cotton production and management. Calif. Agric. 44(5):18–21.

Goodman, A. 1953. A predator on *Creontiades pallidus*, Ramb. Nature (Lond.) 171:886.

——. 1955. Observations on the status of certain insect pests of cotton at Tokar, Sudan. Empire Cotton Grow. Rev. 32: 194–203.

Goodman, R. N., Z. Kiraly, and M. Zaitlin. 1967. The biochemistry and physiology of infectious plant diseases. Van Nostrand, Princeton, N.J. 354 pp.

Goot, P. van der. 1917. De zwarte cacaomier, *Dolichoderus bituberculatus*, Mayr, en haar beteekenis voor de cacao-cultuur op Java. Meded. Proefstn. Midden Java No. 25:1–142.

——. 1927. Het *Crotalaria*-wantsje. Korte Meded. Inst. Plantenziekten (Buitenzorg) No. 6:1–17.

——. 1929. Eenige dierlijke vijanden van *Vigna hosei* en *Calopogonium mucunoides*. Korte Meded. Inst. Plantenziekten (Buitenzorg) No. 11:1–16.

Gopalan, M. 1976a. Studies on salivary enzymes of *Ragmus importunitas* Distant (Hemiptera: Miridae). Curr. Sci. (Bangalore) 45:188–189.

——. 1976b. Effect of infestation of *Ragmus importunitas* Distant (Hemiptera: Miridae) on the growth and yield of sannhemp [*sic*]. Indian J. Agric. Sci. 46:588–591.

Gopalan, M. and M. Basheer. 1966. Studies on the biology of *Ragmus importunitas* D. (Miridae-Hemiptera) on sunnhemp. Madras Agric. J. 53:22–33.

Gopalan, M. and R. S. Perumal. 1973. Studies on the incidence of tea-mosquito bug (*Helopeltis antonii* S.) on some varieties of guava. Madras Agric. J. 60:81–85.

Gopalan, M. and T. R. Subramaniam. 1978. Effect of infestation of *Ragmus importunitas* Distant (Hemiptera: Miridae) on respiration, transpiration, moisture content and oxidative enzymes [*sic*] activity in sunn-hemp

plants (*Crotalaria juncea* L.). Curr. Sci. (Bangalore) 47:131–134.

Gorczyca, J. 1994. Mirid communities (Heteroptera: Miridae) of the plant assemblages in Wyżyna Częstochowska. Ann. Upper Siles. Mus. Nat. Hist. 14:33–68.

——. 1997. Revision of the *Vannius*-complex and its subfamily placement (Hemiptera: Heteroptera: Miridae). Genus (Wrocław) 8:517–553.

——. 1998. A revision of *Euchilofulvius* (Heteroptera: Miridae: Cylapinae). Eur. J. Entomol. 95:93–98.

Gorczyca, J. and F. Chérot. 1998. A revision of the *Rhinomiris*-complex (Heteroptera: Miridae: Cylapinae). Pol. Pismo Entomol. 67:23–64.

Gordon, S. C., J. A. T. Woodford, and A. N. E. Birch. 1997. Arthropod pests of *Rubus* in Europe: Pest status, current and future control strategies. J. Hortic. Sci. 72:831–862.

Gorham, R. P. 1938. [Field crop and garden insects.] Can. Insect Pest Rev. 16:267.

Goss, R. W. 1930. Insect transmission of potato-virus diseases. Phytopathology 20:136. (Abstr.)

——. 1931. Infection experiments with spindle tuber unmottled curly dwarf of the potato. Nebr. Agric. Exp. Stn. Res. Bull. 53:1–36.

Gossard, H. A. 1918. The false apple red bug: An insect causing considerable injury to fruit in Ohio orchards. Ohio Agric. Exp. Stn. Mon. Bull. 3:153–155.

Gottsberger, G., J. Schrauwen, and H. F. Linskens. 1984. Amino acids and sugars in nectar, and their putative evolutionary significance. Plant Syst. Evol. 145:55–77.

Gottsberger, G., T. Arnold, and H. F. Linskens. 1989. Intraspecific variation in the amino acid content of floral nectar. Bot. Acta 102:141–144.

Goula, M. 1986. Miridae (Heteroptera) de roures, alzines i faigs (Fagaceae) del Montseny. Sess. Conjuncta Entomol. ICHN (Inst. Catalana Hist. Nat.)-SCL (Soc. Catalana Lepid.) No. 4:165–172.

Goula, M. and O. Alomar. 1994. Miridos (Heteroptera Miridae) de interés en el control integrado de plagas en el tomate. Guía para su identificación. Bol. Sanid. Veg. Plagas 20:131–143.

Goulart, B., E. G. Rajotte, J. W. Travis, and C. H. Collison. 1989. Small fruit production and pest management guide, 1989–90. Pennsylvania State University College of Agriculture, University Park. 62 pp.

Grace, J. and M. Nelson. 1981. Insects and their pollen loads at a hybrid *Heracleum* site. New Phytol. 87:413–423.

Graenicher, S. 1909. Wisconsin flowers

and their pollination. Bull. Wis. Nat. Hist. Soc. 7:19–77.

Graf, E. 1974. Zur Biologie und Gradologie des grauen Lärchenwicklers, von *Zeiraphera diniana* Gn. (Lep., Tortricidae), im schweizerischen Mittelland. Teil. 2. Lebenstafel und Ausbreitung. Z. Angew. Entomol. 76:347–379.

Grafius, E. and E. A. Morrow. 1982. Damage by the tarnished plant bug and alfalfa plant bug (Heteroptera: Miridae) to asparagus. J. Econ. Entomol. 75:882–884.

Graham, H. M. 1987. Attraction of *Lygus* spp. males by conspecific and congeneric females. Southwest. Entomol. 12:147–155.

——. 1988. Sexual attraction of *Lygus hesperus* Knight. Southwest. Entomol. 13:31–37.

Graham, H. M. and J. W. Debolt. 1986. Reproductive isolation between colonies of *Lygus lineolaris*. Southwest. Entomol. 11:125–130.

Graham, H. M. and C. G. Jackson. 1982. Distribution of eggs and parasites of *Lygus* spp. (Hemiptera: Miridae), *Nabis* spp. (Hemiptera: Nabidae), and *Spissistilus festinus* (Say) (Homoptera: Membracidae) on plant stems. Ann. Entomol. Soc. Am. 75:56–60.

Graham, H. M., A. A. Negm, and L. R. Ertle. 1984. Worldwide literature of the *Lygus* complex (Hemiptera: Miridae), 1900–1980. U.S. Dep. Agric. Bibliogr. Lit. Agric. No. 30:1–205.

Graham, H. M., C. G. Jackson, and J. W. Debolt. 1986. *Lygus* spp. (Hemiptera: Miridae) and their parasites in agricultural areas of southern Arizona. Environ. Entomol. 15:132–142.

Graham, H. M., B. J. Schaeffer, and R. L. Carranza. 1987. *Lygus elisus* and *L. desertinus*: Mating characteristics and interactions. Southwest. Entomol. 12: 1–6.

Grant, B. R. 1996. Pollen digestion by Darwin's finches and its importance for early breeding. Ecology 77:489–499.

Grasswitz, T. R. and E. C. Burts. 1995. Effect of native natural enemies on the population dynamics of the grape mealybug, *Pseudococcus maritimus* (Hom.: Pseudococcidae), in apple and pear orchards. Entomophaga 40:105–117.

Gravena, S. and H. F. da Cunha. 1991. Predation of cotton leafworm first instar larvae, *Alabama argillacea* (Lep.: Noctuidae). Entomophaga 36:481–491.

Gravena, S. and J. A. Pazetto. 1987. Predation and parasitism of cotton leafworm eggs, *Alabama argillacea* [Lep.: Noctuidae]. Entomophaga 32:241–248.

Gravena, S. and W. L. Sterling. 1983. Natural predation on the cotton leafworm (Lepidoptera: Noctuidae). J. Econ. Entomol. 76:779–784.

Greathead, D. J. 1983. Natural enemies of *Nilaparvata lugens* and other leaf- and planthoppers in tropical agroecosystems and their impact on pest populations. Pages 371–383 *in* W. J. Knight et al., eds., Proc. 1st Int. Workshop on Biotaxonomy, Classification and Biology of Leafhoppers and Planthoppers (Auchenorryncha) of Economic Importance, London, 4–7 October 1982. Commonwealth Institute of Entomology, London.

Greatorex-Davies, J. N., T. H. Sparks, and M. L. Hall. 1994. The response of Heteroptera and Coleoptera species to shade and aspect in rides of coniferised lowland woods in southern England. Biol. Conserv. 67:255–273.

Greber, R. S. and J. W. Randles. 1986. Solanum nodiflorum mottle virus. AAB Descrip. Plant Viruses No. 318. Association of Applied Biologists, Wellesbourne, UK. 5 pp.

Green, E. E. 1901. Biologic notes on some Ceylonese Rhynchota. No. 1. Entomologist 34:113–116.

Greene, J. K., S. G. Turnipseed, M. J. Sullivan, and G. A. Herzog. 1999. Boll damage by southern green stink bug (Hemiptera: Pentatomidae) and tarnished plant bug (Hemiptera: Miridae) caged on transgenic *Bacillus thuringiensis* cotton. J. Econ. Entomol. 92: 941–944.

Greenstone, M. H. 1989. Foreign exploration for predators: A proposed new methodology. Environ. Entomol. 18: 195–200.

Greenstone, M. H. and C. E. Morgan. 1989. Predation on *Heliothis zea* (Lepidoptera: Noctuidae): An instar-specific ELISA assay for stomach analysis. Ann. Entomol. Soc. Am. 82:45–49.

Greenwood, M. and A. F. Posnette. 1950. The growth flushes of cacao. J. Hortic. Sci. 25:164–174.

Gregory, P., D. A. Avé, P. Y. Bouthyette, and W. M. Tingey. 1986. Insect-defensive chemistry of potato glandular trichomes. Pages 173–183 *in* B. Juniper and T. R. E. Southwood, eds. Insects and the plant surface. Arnold, London.

Grenier, S., J. Guillaud, B. Delobel, and G. Bonnot. 1989. Nutrition et élevage du prédateur polyphage *Macrolophus caliginosus* [Heteroptera, Miridae] sur milieux artificiels. Entomophaga 34: 77–86.

Grensted, L. W. 1946. An assemblage of Diptera on cow-parsnip. Entomol. Mon. Mag. 92:180.

Greve, J. E. van S. and J. W. Ismay. 1983. Crop insect survey of Papua New Guinea from July 1st 1969 to December 31st 1978. Papua New Guinea Agric. J. 32:1–120.

Grimes, D. W. [1986?] Cultural techniques for management of pests in cotton. Pages 365–382 *in* R. E. Frisbie and P. L. Adkisson, eds., Integrated pest management on major agricultural systems. Tex. Agric. Exp. Stn. MP-1616.

Grinfeld, E. K. 1959. The feeding of thrips (Thysanoptera) on pollen of flowers and the origin of asymmetry in their mouthparts. Entomol. Obozr. 38:798–804. [in Russian.]

——. 1975. Anthophily in beetles (Coleoptera) and a critical evaluation of the cantharophilous hypothesis. Entomol. Obozr. 54(3):507–514. [in Russian; English translation in Entomol. Rev. 54(3):18–22.]

Grisham, M. P., W. L. Sterling, R. D. Powell, and P. W. Morgan. 1987. Characterization of the induction of stress ethylene synthesis in cotton caused by the cotton fleahopper (Hemiptera: Miridae) and its microorganisms. Ann. Entomol. Soc. Am. 80:411–416.

Groot, A. T., A. Schuurman, G. J. R. Judd, L. H. M. Blommers, and J. H. Visser. 1996. Sexual behaviour of the green capsid bug *Lygocoris pabulinus* L. (Miridae): An introduction. Proc. Sect. Exp. Appl. Entomol. Neth. Entomol. Soc. (N.E.V., Amst.) 7:249–252.

Groot, A. T., E. van der Wal, A. Schuurman, J. H. Visser, L. H. M. Blommers, and T. A. van Beek. 1998. Copulation behaviour of *Lygocoris pabulinus* under laboratory conditions. Entomol. Exp. Appl. 88:219–228.

Groot, A. T., R. Timmer, G. Gort, G. P. Lelyveld, F. P. Drijfhout, T. A. Van Beek, and J. H. Visser. 1999. Sex-related perception of insect and plant volatiles in *Lygocoris pabulinus*. J. Chem. Ecol. 25:2357–2371.

Gross, G. F. and G. Cassis. 1991. Superfamily Miroidea. Pages 491–493 *in* CSIRO. The insects of Australia: A textbook for students and research workers, 2nd ed. Vol. 1. Cornell University Press, Ithaca; Melbourne University Press, Melbourne.

Gross, H. R. Jr. 1987. Conservation and enhancement of entomophagous insects—a perspective. J. Entomol. Sci. 22:97–105.

Grossman, J. 1989. Update: Strawberry IPM features biological and mechanical controls. IPM Practitioner 11(5):1–4.

Grove, A. J. and C. C. Ghosh. 1914. The life-history of *Psylla isitis* Buckt. (*Psyllopa punctipennis*, Crawford) the "psylla" disease of indigo. Mem. Dep. Agric. India Entomol. Ser. 4:329–357.

Groves, E. W. 1968. Hemiptera-Heteroptera of the London area. Part V. Lond. Nat. 47:50–80.

——. 1969. Hemiptera-Heteroptera of the London area. Part VI. Lond. Nat. 48:86–120.

——. 1976. Hemiptera-Heteroptera of the London area. Part X. Lond. Nat. 55:6–15.

——. 1977. Hemiptera-Heteroptera of the London area. Part XI. Lond. Nat. 56:32–43.

Gruys, P. 1975. Integrated control in orchards in the Netherlands. *In* Proceedings of the 5th Symposium on Integrated Control in Orchards, 3–7 Sept. 1974, Bolzano. Int. Organ. Biol. Control/West. Palearctic Reg. Sect. Bull. 3:59–68.

——. 1982. Hits and misses. The ecological approach to pest control in orchards. Entomol. Exp. Appl. 31: 70–87.

Gubbaiah [no initials], M. C. Devaiah, T. S. Thontadarya, and M. Jayaramaiah. 1976. A new host record of betelvine bug, *Disphinctus politus* Wlk. (Heteroptera: Miridae). Curr. Res. (Bangalore) 5:63–64.

Gueldner, R. C. and W. L. Parrott. 1978. Volatile constituents of the tarnished plant bug. Insect Biochem. 8:389–391.

——. 1981. Constituents of mustard, goldenrod, and croton—three host plants of the tarnished plant bug. J. Agric. Food Chem. 29:418–420.

Guinn, G. 1982. Causes of square and boll shedding in cotton. U.S. Dep. Agric. Tech. Bull. 1672:1–22.

Guinn, G. and M. P. Eidenbock. 1982. Catechin and condensed tannin contents of leaves and bolls of cotton in relation to irrigation and boll load. Crop Sci. 22:614–616.

Gulde, J. 1921. Die Wanzen (Hemiptera-Heteroptera) der Umgebung von Frankfurt a. M. und des Beckens. Abh. Senckenb. Natforsch. Ges. 37:327–503.

Gullan, P. J. and P. S. Cranston. 1994. The insects: An outline of entomology. Chapman & Hall, London. 491 pp.

Günthardt, M. S. and H. Wanner. 1981. The feeding behaviour of two leafhoppers on *Vicia faba*. Ecol. Entomol. 6:17–22.

Guozhong, T. and S. P. Raychaudhuri. 1996. Paulownia witches' broom disease in China: Present status. Pages 227–251 *in* S. P. Raychaudhuri and K. Maramorosch, eds., Forest trees and palms: Diseases and control. Science Publishers, Lebanon, N.H.

Guppy, J. C. 1963. Observations on the biology of *Plagiognathus chrysanthemi* (Hemiptera: Miridae), a pest of birdsfoot trefoil in Ontario. Ann. Entomol. Soc. Am. 56:804–809.

Gupta, A. P. 1961. A critical review of the studies on the so-called stink or repugnatorial glands of Heteroptera with further comments. Can. Entomol. 93:482–486.

Gupta, R. K., G. Tamaki, and C. A. Johansen. 1980. Lygus bug damage, predator-prey interaction, and pest management implications in alfalfa grown for seed. Wash. State Univ. Coll. Agric. Res. Cent. Tech. Bull. 92:1–18.

Gut, L. [J.], P. Westigard, W. [J.] Liss, and M. Willett. 1982a. Biological control of pear psylla: A potential within a potential. Proc. Wash. State Hortic. Assoc. 77:194–198.

Gut, L. J., P. H. Westigard, C. Jochums, and W. J. Liss. 1982b. Variation in pear psylla (Psylla pyricola Foerster) densities in southern Oregon orchards and its implications. Acta Hortic. (Wageningen) 124:101–111.

Gut, L. J., W. J. Liss, and P. H. Westigard. 1991. Arthropod community organization and development in pear. Environ. Manage. 15:83–104.

Gutierrez, A. P. 1995. Integrated pest management in cotton. Pages 280–310 in D. Dent, ed., Integrated pest management. Chapman & Hall, London.

——. 1999. Modeling tritrophic field populations. Pages 647–679 in C. B. Huffaker and A. P. Gutierrez, eds., Ecological entomology, 2nd ed. Wiley, New York.

Gutierrez, A. P., T. F. Leigh, Y. Wang, and R. D. Cave. 1977. An analysis of cotton production in California: Lygus hesperus (Heteroptera: Miridae) injury—an evaluation. Can. Entomol. 109:1375–1386.

Gutierrez, A. P., Y. Wang, and U. Regev. 1979. An optimization model for Lygus hesperus (Heteroptera: Miridae) damage in cotton: The economic threshold revisited. Can. Entomol. 111:41–54.

Hachiya, K. 1985. Control threshold of rice leaf bug (Trigonotylus coelestialium Kirkaldy). Bull. Hokkaido Prefect. Agric. Exp. Stn. 53:43–49. [in Japanese, English summary.]

Hackett, K. J. and T. B. Clark. 1989. Ecology of spiroplasmas. Pages 113–200 in R. F. Whitcomb and J. G. Tully, eds., The mycoplasmas. Vol. V. Spiroplasmas, acholeplasmas, and mycoplasmas of plants and arthropods. Academic, San Diego.

Haeussler, G. J. 1952. Losses caused by insects. Pages 141–146 in F. C. Bishopp, chairman, et al., Insects: The yearbook of agriculture, 1952. U.S. Dep. Agric., Washington, D.C.

Haeussler, G. J. and D. W. Clancy. 1944.

Natural enemies of Comstock mealybug in the eastern states. J. Econ. Entomol. 37:503–509.

Hagel, G. T. 1978. Lygus spp.: Damage to beans by reducing yields, seed pitting, and control by varietal resistance and chemical sprays. J. Econ. Entomol. 71:613–615.

Hagen, A. F. 1982. Labops hesperius (Hemiptera: Miridae) management in crested wheatgrass by haying: An eight-year study. J. Econ. Entomol. 75:706–707.

Hagen, K. S. 1976. Role of nutrition in insect management. Proc. Tall Timbers Conf. Ecol. Anim. Control Habitat Manage. 4:221–261.

——. 1986. Ecosystem analysis: Plant cultivars (HPR), entomophagous species and food supplements. Pages 151–197 in D. J. Boethel and R. D. Eikenbary, eds., Interactions of plant resistance and parasitoids and predators of insects. Ellis Horwood, Chichester, UK.

Hagen, K. S. and J. M. Franz. 1973. A history of biological control. Pages 433–476 in R. F. Smith, T. E. Mittler, and C. N. Smith, eds., History of entomology. Annual Reviews, Palo Alto.

Hagen, K. S., R. L. Tassan, and E. F. Sawall Jr. 1970. Some ecophysiological relationships between certain Chrysopa, honeydews and yeasts. Boll. Lab. Entomol. Agrar. "Filippo Sylvestri" 28:113–134.

Hagen, K. S., R. H. Dadd, and J. Reese. 1984. The food of insects. Pages 79–112 in C. B. Huffaker and R. L. Rabb, eds., Ecological entomology. Wiley, New York.

Hagen, K. S., N. J. Mills, G. Gordh, and J. A. McMurtry. 1999. Terrestrial arthropod predators of insect and mite pests. Pages 383–503 in T. S. Bellows and T. W. Fisher, eds., Handbook of biological control: Principles and applications of biological control. Academic, San Diego.

Hagler, J. R. and S. E. Naranjo. 1994. Determining the frequency of heteropteran predation on sweetpotato whitefly and pink bollworm using multiple ELISAs. Entomol. Exp. Appl. 72:59–66.

——. 1996. Using gut content immunoassays to evaluate predaceous biological control agents: A case study. Pages 383–399 in W. O. C. Symondson and J. E. Liddell, eds., The ecology of agricultural pests: Biochemical approaches. Chapman & Hall, London.

Hagler, J. R., A. C. Cohen, F. J. Enriquez, and D. Bradley-Dunlop. 1991. An egg-specific monoclonal antibody to Lygus hesperus. Biol. Control 1:75–80.

Hagley, E. A. C. 1975. The arthropod fauna in unsprayed apple orchards in Ontario II. Some predacious species. Proc. Entomol. Soc. Ont. 105:28–40.

Hagley, E. A. C. and W. R. Allen. 1990. The green apple aphid, Aphis pomi DeGeer (Homoptera: Aphididae), as prey of polyphagous arthropod predators in Ontario. Can. Entomol. 122:1221–1228.

Hagley, E. A. C. and A. Hikichi. 1973. The arthropod fauna in unsprayed apple orchards in Ontario I. Major pest species. Proc. Entomol. Soc. Ont. 103:60–64.

Hagley, E. A. C. and C. M. Simpson. 1981. Effect of food sprays on numbers of predators in an apple orchard. Can. Entomol. 113:75–77.

——. 1983. Effect of insecticides on predators of the pear psylla, Psylla pyricola (Hemiptera: Psyllidae), in Ontario. Can. Entomol. 115:1409–1414.

Hajek, A. E. and D. L. Dahlsten. 1988. Distribution and dynamics of aphid (Homoptera: Drepanosiphidae) populations on Betula pendula in northern California. Hilgardia 56(1):1–33.

Halbert, S. and D. Voegtlin. 1995. Biology and taxonomy of vectors of barley yellow dwarf viruses. Pages 217–258 in C. J. D'Arcy and P. A. Burnett, eds., Barley yellow dwarf: 40 years of progress. APS, St. Paul, Minn.

Haley, S. and E. J. Hogue. 1990. Ground cover influence on apple aphid, Aphis pomi DeGeer (Homoptera: Aphididae), and its predators in a young apple orchard. Crop Prot. 9:225–230.

Hall, D. G. 1988. Insects and mites associated with sugarcane in Florida. Fla. Entomol. 71:138–150.

Hall, D. G. and F. D. Bennett. 1994. Biological control and IPM of sugarcane pests in Florida. Pages 297–325 in D. Rosen, F. D. Bennett, and J. L. Capinera, eds., Pest management in the subtropics: Biological control—a Florida perspective. Intercept, Andover, UK.

Hall, F. H. and V. H. Lowe. 1900. A few fruit-tree foes. N.Y. Agric. Exp. Stn. (Geneva). Bull. 180 (popular ed.):1–8.

Hall, I. M. 1959. The fungus Entomophthora erupta (Dustan) attacking the black grass bug, Irbisia solani (Heidemann) (Hemiptera, Miridae), in California. J. Insect Pathol. 1:48–51.

Hambleton, E. J. 1938. O percevejo "Horcias nobilellus Berg" como nova prága do algodoeiro em S. Paulo. Observações preliminares. Arq. Inst. Biol. (São Paulo) 9:85–92.

Hameed, S. F., N. P. Kashyap, and D. N. Vaidya. 1975. Host records of insect pests of crops. FAO Plant Prot. Bull. 23:191–192.

Hamm, A. H. 1916. [Exhibitions at Entomological Society of London meeting, 17 Nov. 1915]. Proc. Entomol. Soc. Lond. 1916:cxix.

Hammer, O. H. 1939. The tarnished plant bug as an apple pest. J. Econ. Entomol. 32:259–264.

Hammond, A. M. and T. N. Hardy. 1988. Quality of diseased plants as hosts for insects. Pages 381–432 in E. A. Heinrichs, ed., Plant stress-insect interactions. Wiley, New York.

Hamner, A. L. 1941. Fruiting of cotton in relation to cotton fleahopper and other insects which do similar damage to squares. Miss. Agric. Exp. Stn. Bull. 360:1–11.

Hancock, G. L. R. 1926. Annual report of the assistant entomologist. Uganda Dep. Agric. Rep. 1925:25–28.

——. 1935. Notes on Lygus simonyi, Reut. (Capsidae), a cotton pest in Uganda. Bull. Entomol. Res. 26:429–438.

Handford, R. H. 1949. Lygus campestris (L.): A new pest of carrot seed crops. Can. Entomol. 81:123–126.

Handley, D. T. 1991. Strawberry varieties differ in susceptibility to tarnished plant bug injury. Northeast LISA Small Fruits Newsl. 2(2):4–5.

Handley, D. T. and J. E. Pollard. 1989. The nature of strawberry fruit malformation caused by the tarnished plant bug. HortScience 24:221.

——. 1990. Effects of fruit stage and duration of feeding on strawberry malformation caused by the tarnished plant bug. HortScience 25:624.

——. 1993. Microscopic examination of tarnished plant bug (Heteroptera: Miridae) feeding damage to strawberry. J. Econ. Entomol. 86:505–510.

Hanny, B. W. and T. C. Cleveland. 1976. Effects of tarnished plant bug (Lygus lineolaris, Palisot de Beauvois), feeding on presquaring cotton (Gossypium hirsutum L.). In Proc. Beltwide Cotton Prod. Res. Conf., Jan. 5–7, 1976, Las Vegas, Nev., p. 59 (Abstr.). National Cotton Council, Memphis.

Hanny, B. W. and C. D. Elmore. 1974. Amino acid composition of cotton nectar. J. Agric. Food Chem. 22:476–478.

Hanny, B. W., T. C. Cleveland, and W. R. Meredith Jr. 1977. Effects of tarnished plant bug, (Lygus lineolaris), infestation on presquaring cotton (Gossypium hirsutum). Environ. Entomol. 6:460–462.

Hansen D. L., H. F. Brødsgaard, and A. Enkegaard. 1999. Life table characteristics of Macrolophus caliginosus preying upon Tetranychus urticae. Entomol. Exp. Appl. 93:269–275.

Hansen, J. D. 1986. Differential feeding on range grass seedlings by Irbisia pacifica (Hemiptera: Miridae). J. Kans. Entomol. Soc. 59:199–203.

——. 1987. Feeding site selection by Irbisia pacifica (Hemiptera: Miridae) on four cool-season western range grasses. J. Kans. Entomol. Soc. 60:316–323.

——. 1988. Field observations of Irbisia pacifica (Hemiptera: Miridae): Feeding behavior and effects on host plant growth. Great Basin Nat. 48:68–74.

Hansen, J. D. and R. S. Nowak. 1985. Evaluating damage by grass bug feeding on crested wheatgrass. Southwest. Entomol. 10:89–94.

——. 1988. Feeding damage by Irbisia pacifica (Hemiptera: Miridae): Effects of feeding and drought on host plant growth. Ann. Entomol. Soc. Am. 81:599–604.

Hansen, J. D., K. H. Asay, and D. C. Nielson. 1985a. Screening range grasses for resistance to black grass bugs Labops hesperius and Irbisia pacifica (Hemiptera: Miridae). J. Range Manage. 38:254–257.

——. 1985b. Feeding preference of a black grass bug, Labops hesperius (Hemiptera: Miridae), for 16 range grasses. J. Kans. Entomol. Soc. 58:356–359.

Hanson, H. C. 1962. Dictionary of ecology. Philosophical Library, New York. 382 pp.

——. 1963. Diseases and pests of economic plants of Vietnam, Laos and Cambodia; a study based on field survey data and on pertinent records, material, and reports. American Institute of Crop Ecology, Washington, D.C. 155 pp.

Hanssen, H.-P. and J. Jacob. 1982. Monoterpenes from the true bug Harpocera thoracica (Hemiptera). Z. Naturforsch. 37C:1281–1282.

Harborne, J. B. 1988. Introduction to ecological biochemistry, 3rd ed. Academic, San Diego. 356 pp.

Harcourt, D. G. 1953. Note on injury to cucumber by the tarnished plant bug, Lygus lineolaris P. de B. (Hemiptera: Miridae). Can. Entomol. 85:421.

——. 1964. Population dynamics of Leptinotarsa decemlineata (Say) in eastern Ontario II. Population and mortality estimation during six age intervals. Can. Entomol. 96:1190–1198.

Harcourt, D. G., J. C. Guppy, J. Drolet, and J. N. McNeil. 1987. Population dynamics of alfalfa blotch leafminer, Agromyza frontella (Diptera: Agromyzidae), in eastern Ontario: Analysis of numerical change during the colonization phase. Environ. Entomol. 16:145–153.

Hardee, D. D. and W. W. Bryan. 1997. Influence of Bacillus thuringiensis-transgenic and nectariless cotton on insect populations with emphasis on the tarnished plant bug (Heteroptera: Miridae). J. Econ. Entomol. 90:663–668.

Hardin, M. R., B. Benrey, M. Coll, W. O. Lamp, G. K. Roderick, and P. Barbosa. 1995. Arthropod pest resurgence: An overview of potential mechanisms. Crop Prot. 14:3–18.

Hardison, J. R. 1959. Evidence against Fusarium poae and Siteroptes graminum as causal agents of silver top of grasses. Mycologia 51:712–728.

——. 1976. Fire and flame for plant disease control. Annu. Rev. Phytopathol. 14:355–379.

Hardman, J. M. 1992. Apple pest management in North America: Challenge and response. Proc. Brighton Crop Prot. Conf.: Pests and Diseases—1992. 2:507–516. BCPC Registered Office, Farnham, UK.

Hardman, J. M., R. E. L. Rogers, and C. R. MacLellan. 1988. Advantages and disadvantages of using pyrethroids in Nova Scotia apple orchards. J. Econ. Entomol. 81:1737–1749.

Hardy, A. C. and L. Cheng. 1986. Studies in the distribution of insects by aerial currents. III. Insect drift over the sea. Ecol. Entomol. 11:283–290.

Hare, J. D. 1994. Status and prospects for an integrated approach to the control of rice planthoppers. Pages 615–632 in R. F. Denno and T. J. Perfect, eds., Planthoppers: Their ecology and management. Chapman & Hall, New York.

Hargreaves, [H.]. [1926?]. Lycidocoris mimeticus—a potential pest of coffee. In Proc. South and East African Combined Agric., Cotton, Entomol., and Mycol. Conf., Nairobi, Aug. 1926, pp. 196–198. East African Standard, Nairobi.

Hargreaves, H. 1948. List of recorded cotton insects of the world. Commonwealth Institute of Entomology, London. 50 pp.

Hari Babu, R. S., S. Rath, and C. B. S. Rajput. 1983. Insect pests of cashew in India and their control. Pesticides (Bombay) 17(4):8–16.

Harizanova, V. 1989. Biological peculiarities of the predatory apple bug Atractotomus mali (Heteroptera: Miridae). Rastenieved. Nauki 26(9):98–102. [in Bulgarian, English summary.]

Harper, A. M. and H. C. Huang. 1984. Contamination of insects by the plant pathogen Verticillium albo-atrum in an alfalfa field. Environ. Entomol. 13:117–120.

Harper, J. L. and W. A. Wood. 1957. Biological flora of the British Isles. *Senecio jacobaea* L. J. Ecol. 45:617–637.

Harrap, K. A. 1973. Virus infection in invertebrates. Pages 271–299 *in* A. J. Gibbs, ed., Viruses and invertebrates. North-Holland, Amsterdam.

Harris, K. F. 1981. Arthropod and nematode vectors of plant viruses. Annu. Rev. Phytopathol. 19:391–426.

Harris, K. F. and K. Maramorosch, eds. 1980. Vectors of plant pathogens. Academic, New York. 467 pp.

Harris, K. M. 1995. World review of recent research on panicle insect pests of sorghum and pearl millet. Pages 7–25 *in* K. F. Nwanze and O. Youm, eds., Panicle insect pests of sorghum and pearl millet: Proceedings of an International Consultative Workshop, 4–7 Oct. 1993, ICRISAT Sahelian Center, Niamey, Niger. International Crops Research Institute for the Semi-Arid Tropics, Patancheru, A.P., India.

Harris, P. 1974. A possible explanation of plant yield increases following insect damage. Agro-Ecosystems 1:219–225.

Harris, T. W. 1841. A report on the insects of Massachusetts, injurious to vegetation. Folsom, Wells, & Thurston, Cambridge, Mass. 459 pp.

——. 1851. A new insect depredator. N. Engl. Farmer (Boston) 3(17):268.

Harris, W. V. 1937. *Helopeltis* bug. East Afr. Agric. J. 2:387–390.

Harrison, A. L. 1935. Transmission of bean mosaic. N.Y. Agric. Exp. Stn. (Geneva) Tech. Bull. 236:1–19.

Harrison, M. D., J. W. Brewer, and L. D. Merrill. 1980. Insect involvement in the transmission of bacterial pathogens. Pages 201–292 *in* K. F. Harris and K. Maromorosch, eds., Vectors of plant pathogens. Academic, New York.

Harrison, R. G. 1980. Dispersal polymorphisms in insects. Annu. Rev. Ecol. Syst. 11:95–118.

Hartley, J. C. 1965. The structure and function of the egg-shell of *Deraeocoris ruber* L. (Heteroptera, Miridae). J. Insect Physiol. 11:103–109.

Haseman, L. 1913. Peach "stop back" and tarnished plant bug. J. Econ. Entomol. 6:237–240.

——. 1918. The tarnished plant-bug and its injury to nursery stock. Mo. Agric. Exp. Stn. Res. Bull. 29:1–26.

——. 1928. Tarnished plant bug injury to strawberries. J. Econ. Entomol. 21:191–192.

Haslett, J. R. 1983. A photographic account of pollen digestion by adult hoverflies. Physiol. Entomol. 8:167–171.

Hassan, E. 1977. Major insect and mite pest [*sic*] of Australian crops. Ento,

Gatton, Queensland, Australia. 238 pp.

Hassell, M. P. 1978. The dynamics of arthropod predator-prey systems. Princeton University Press, Princeton. 237 pp.

Hassell, M. P. and T. R. E. Southwood. 1978. Foraging strategies of insects. Annu. Rev. Ecol. Syst. 9:75–98.

Hatchett, J. H., K. J. Starks, and J. A. Webster. 1987. Insects and mite pests of wheat. Pages 625–675 *in* E. G. Heyne, eds., Wheat and wheat improvement, 2nd ed. American Society of Agronomy, Crop Science Society of America, Soil Science Society of America, Madison, Wis.

Hatfield, L. D. and J. L. Frazier. 1980. Ultra-structure of the labial tip sensilla of the tarnished plant bug, *Lygus lineolaris* (P. de Beauvois) (Hemiptera: Miridae). Int. J. Morphol. Embryol. 9:59–66.

Hatfield, L. D., J. L. Frazier, and J. Ferreira. 1982. Gustatory discrimination of sugars, amino acids, and selected allelochemicals by the tarnished plant bug, *Lygus lineolaris*. Physiol. Entomol. 7:15–23.

Hatfield, L. D., J. Ferreira, and J. L. Frazier. 1983. Host selection and feeding behavior by the tarnished plant bug, *Lygus lineolaris* (Hemiptera: Miridae). Ann. Entomol. Soc. Am. 76:688–691.

Haukioja, E. 1990. Induction of defenses in trees. Annu. Rev. Entomol. 46:25–42.

Haviland, E. E. 1945. A pest of yucca, *Halticotoma valida*, Reut. Md. Agric. Exp. Stn. Bull. A37:103–112.

Hawkins, B. A., N. J. Mills, M. A. Jervis, and P. W. Price. 1999. Is the biological control of insects a natural phenomenon? Oikos 86:493–506.

Hawkins, R. D. 1989. A further record of *Placochilus seladonicus* (Fallén) (Hem., Miridae). Entomol. Mon. Mag. 125:205.

Hawley, I. M. 1917. The hop redbug (*Paracalocoris hawleyi* Knight). J. Econ. Entomol. 10:545–552.

——. 1918. Insects injurious to the hop in New York with special reference to the hop grub and the hop redbug. Cornell Agric. Exp. Stn. Mem. 15:143–224.

——. 1920. Injuries to beans in the pod by hemipterous insects. J. Econ. Entomol. 13:415–416.

——. 1922. Insects and other animal pests injurious to field beans in New York. Cornell Agric. Exp. Stn. Mem. 55:945–1037.

Haws, B. A. 1978a. Biological control. Pages 59–60 *in* B. A. Haws, compiler, Economic impacts of *Labops hesperius*

on the production of high quality range grasses. Final Rep. Utah Agric. Exp. Stn. to Four Corners Reg. Comm. Utah State University, Logan.

——., compiler. 1978b. Economic impacts of *Labops hesperius* on the production of high quality range grasses. Final Rep. Utah Agric. Exp. Stn. to Four Corners Reg. Comm. Utah State University, Logan. 269 pp.

——., compiler. 1982. An introduction to beneficial and injurious rangeland insects of the western United States. Utah Agric. Exp. Stn. Spec. Rep. 23:1–64.

Haws, B. A. and G. E. Bohart. 1986. Black grass bugs (*Labops hesperius*) Uhler (Hemiptera: Miridae) and other insects in relation to crested wheatgrass. Pages 123–145 *in* K. L. Johnson, ed., Crested wheatgrass: Its values, problems and myths. Symposium proceedings. Utah State University, Logan.

Haws, B. A. and P. Thompson. 1978. Biology of black grass bugs, *Labops hesperius*. Pages 18–27 *in* B. A. Haws, compiler, Economic impacts of *Labops hesperius* on the production of high quality range grasses. Final Rep. Utah Agric. Exp. Stn. to Four Corners Reg. Comm. Utah State University, Logan.

Haws, B. A., D. D. Dwyer, and M. G. Anderson. 1973. Problems with range grass? Look for black grass bugs! Utah Sci. 34:3–9.

Haws, B. A., G. E. Bohart, C. R. Nelson, and D. L. Nelson. 1990. Insects and shrub dieoff in western states: 1986–89 survey results. Pages 127–151 *in* Proceedings—Symposium on Cheatgrass Invasion, Shrub Die-Off, and Other Aspects of Shrub Biology and Management. U.S. Dep. Agric. For. Serv. Intermountain Res. Stn. Gen. Tech. Rep. INT-276.

Hayashi, H. 1989. Studies on the bionomics and control of the sorghum plant bug, *Stenotus rubrovittatus* Matsumura (Hemiptera: Miridae): 2. Relationship between injury time and symptoms of rice kernels. Bull. Hiroshima Prefect. Agric. Exp. Stn. 52:1–8. [in Japanese, English summary.]

Hayashi, H. and K. Nakazawa. 1988. Studies on the bionomics and control of the sorghum plant bug, *Stenotus rubrovittatus* Matsumura (Hemiptera: Miridae) 1. Habitat and seasonal prevalence in Hiroshima Prefecture. Bull. Hiroshima Prefect. Agric. Exp. Stn. 51:45–53. [in Japanese, English summary.]

Hayward, J. A. 1967. Cotton in western Nigeria 2. Entomological problems. Cotton Grow. Rev. 44:117–135.

——. 1972. Relationship between pest infestation and applied nitrogen on cotton in Nigeria. Cotton Grow. Rev. 49:224–235.

Heads, P. A. 1986. Bracken, ants, and extrafloral nectaries. IV. Do wood ants (*Formica lugubris*) protect the plant against insect herbivores? J. Anim. Ecol. 55:795–809.

Hearst, W. H. 1918. Black heart of celery. Page 61 *in* Rep. Minist. Agric. Prov. Ont., 1917. A. T. Wilgress, Toronto.

Heathcote, G. D. 1976. Insects as vectors of plant viruses. Z. Angew. Entomol. 82:72–80.

Hedlund, R. C. 1987. Foreign exploration for natural enemies of *Lygus* and *Adelphocoris* plant bugs. Pages 76–81 *in* R. C. Hedlund and H. M. Graham, eds., Economic importance and biological control of *Lygus* and *Adelphocoris* in North America. U.S. Dep. Agric. Agric. Res. Serv. ARS-64.

Hedlund, R. C. and H. M. Graham. 1987. Economic importance and biological control of *Lygus* and *Adelphocoris* in North America. U.S. Dep. Agric. Agric. Res. Serv. ARS-64:1–95.

Heidemann, O. 1891. Note on the occurrence of a rare capsid, near Washington, D.C. Proc. Entomol. Soc. Wash. 2:68–69.

——. 1892. Note on the food-plants of some Capsidae from the vicinity of Washington, D.C. Proc. Entomol. Soc. Wash. 2:224–226.

——. 1899. Heteroptera found on ox-eye daisy (*Chrysanthemum leucanthemum*). Proc. Entomol. Soc. Wash. 4:217.

——. 1903. Hemiptera. Pages 80–82 *in* A. N. Caudell, Some insects from the summit of Pike's Peak, found on snow. Proc. Entomol. Soc. Wash. 5:74–82.

——. 1910. [Notes and exhibition of specimens]. Proc. Entomol. Soc. Wash. 12:45–46.

Heikertinger, F. 1922. Sind die Wanzen (Hemiptera heteroptera) durch Ekelgeruch geschutzt? Biol. Zentbl. 42:441–464.

Heine, E. M. 1937. Observations on the pollination of New Zealand flowering plants. Trans. Proc. R. Soc. N.Z. 67:133–148.

Heinrichs, E. A. 1994. Host plant resistance. Pages 517–547 *in* E. A. Heinrichs, ed., Biology and management of rice insects. Wiley, New York.

Heinrichs, E. A., G. B. Aquino, S. Chelliah, S. L. Valencia, and W. H. Reissig. 1982. Resurgence of *Nilaparvata lugens* (Stål) populations as influenced by method and timing of insecticide applications in lowland rice. Environ. Entomol. 11:78–84.

Heinze, K. 1951. Die Überträger pflanzlicher Viruskrankheiten. Mitt. Biol. Zentralanst. Land-Forstwirtsch. Berl.-Dahl. 71:1–126.

——. 1959. Phytopathogene Viren und ihre Überträger. Duncker & Himblot, Berlin. 290 pp.

Heliövaara, K. and R. Väisänen. 1993. Insects and pollution. CRC, Ann Arbor, Mich. 393 pp.

Helton, A. W., J. B. Johnson, and R. D. Dilbeck. 1988. Arthropods as carriers of fungal wood-rotting pathogens in pome and stone fruit orchards. Plant Dis. 72:1077. (Abstr.)

Hely, P. C., G. Pasfield, and J. G. Gellatley. 1982. Insect pests of fruit and vegetables in NSW. Inkata, Melbourne. 312 pp.

Heming, B. S. 1993. Structure, function, ontogeny, and evolution of feeding in thrips (Thysanoptera). Pages 3–41 *in* C. W. Schaefer and R. A. B. Leschen, eds., Functional morphology of insect feeding. Thomas Say Publ. Entomol.: Proceedings. Entomological Society of America, Lanham, Md.

Heming-van Battum, K. E. and B. S. Heming. 1986. Structure, function and evolution of the reproductive system in females of *Hebrus pusillus* and *H. ruficeps* (Hemiptera, Gerromorpha, Hebridae). J. Morphol. 190:121–167.

Henderson, A. 1858. The potato rot—its cause and cure. Sci. Am. 13:408.

Henderson, F. C. 1954. Cacao as a crop for the owner-manager in Papua and New Guinea. Papua New Guinea Agric. J. 9:45–74.

Hendrickson, G. O. 1930. Studies on the insect fauna of Iowa prairies. Iowa State Coll. J. Sci. 4:49–179.

Hendrix, D. L., Y. Wei, and J. E. Leggett. 1992. Homopteran honeydew sugar composition is determined by both the insect and plant species. Comp. Biochem. Physiol. 101B:23–27.

Henneberry, T. J., L. A. Bariola, and D. L. Kittock. 1977. Nectariless cotton: Effect on cotton leafperforator and other cotton insects in Arizona. J. Econ. Entomol. 70:797–799.

Henry, T. J. 1976a. Review of *Reuteria* Puton 1875, with descriptions of two new species (Hemiptera: Miridae). Entomol. News 87:61–74.

——. 1976b. A new *Saileria* from eastern United States (Hemiptera: Miridae). Entomol. News 87:29–31.

——. 1977. *Orthotylus nassatus*, a European plant bug new to North America (Heteroptera: Miridae). U.S. Dep. Agric. Coop. Plant Pest Rep. 2(31):605–608.

——. 1978. Review of the Neotropical genus *Hyalochloria*, with descriptions of ten new species (Hemiptera: Miridae). Trans. Am. Entomol. Soc. 104:69–90.

——. 1979a. Review of the New World species of *Bothynotus* Fieber (Hemiptera: Miridae). Fla. Entomol. 62:232–244.

——. 1979b. Descriptions and notes on five new species of Miridae from North America (Hemiptera). Melsheimer Entomol. Ser. No. 27:1–10.

——. 1979c. Review of the New World species of *Myiomma* with descriptions of eight new species (Hemiptera: Miridae: Isometopinae). Proc. Entomol. Soc. Wash. 81:552–569.

——. 1980. New records for *Saileria irrorata* and *Tropidosteptes adustus* (Hemiptera: Miridae). Fla. Entomol. 63:490–493.

——. 1982. The onion plant bug genus *Labopidicola* (Hemiptera: Miridae): Economic implications, taxonomic review, and description of a new species. Proc. Entomol. Soc. Wash. 84:1–15.

——. 1984. Revision of the spider-commensal genus *Ranzovius* Distant (Hemiptera: Miridae). Proc. Entomol. Soc. Wash. 86:53–67.

——. 1985a. *Caulotops distanti* (Miridae: Heteroptera), a potential yucca pest newly discovered in the United States. Fla. Entomol. 68:320–323.

——. 1985b. What is *Capsus frontifer* Walker, 1873 (Heteroptera: Miridae)? Proc. Entomol. Soc. Wash. 87:679.

——. 1985c. Newly recognized synonyms, homonyms, and combinations in the North American Miridae (Heteroptera). J. N.Y. Entomol. Soc. 93:1121–1136.

——. 1989. *Cariniocoris*, a new phyline plant bug genus from the eastern United States, with a discussion of generic relationships (Heteroptera: Miridae). J. N.Y. Entomol. Soc. 97:87–99.

——. 1991. Revision of *Keltonia* and the cotton fleahopper genus *Pseudatomoscelis*, with the description of a new genus and an analysis of their relationships (Heteroptera: Miridae: Phylinae). J. N.Y. Entomol. Soc. 99:351–404.

——. 1994. Revision of the myrmecomorphic plant bug genus *Schaffneria* Knight (Heteroptera: Miridae: Orthotylinae). Proc. Entomol. Soc. Wash. 96:701–712.

——. 1995. *Proboscidotylus carvalhoi*, a new genus and species of sexually dimorphic plant bug from Mexico (Heteroptera: Miridae: Orthotylinae). Proc. Entomol. Soc. Wash. 97:340–345.

——. 1997. Phylogenetic analysis of family groups within the infraorder Pentatomomorpha (Hemiptera:

Heteroptera), with emphasis on the Lygaeoidea. Ann. Entomol. Soc. Am. 90:275–301.

——. 1999. The spider-commensal plant bug genus *Ranzovius* (Heteroptera: Miridae: Phylinae) revisited: Three new species and a revised key, with the description of a new sister genus and phylogenetic analysis. Acta Soc. Zool. Bohem. 63:93–115.

——. 2000. The predatory Miridae: A glimpse at the other plant bugs. Wings (Portland, Ore.) 23:17–20.

Henry, T. J. and J. C. M. Carvalho. 1987. A peculiar case history: *Hemisphaerodella mirabilis* Reuter is the nymphal stage of *Cyrtocapsus caligineus* (Stål) (Heteroptera: Miridae: Bryocorinae). J. N.Y. Entomol. Soc. 95:290–293.

Henry, T. J. and R. C. Froeschner, eds. 1988. Catalog of the Heteroptera, or true bugs, of Canada and the continental United States. E. J. Brill, Leiden. 958 pp.

Henry, T. J. and L. A. Kelton. 1986. *Orthocephalus saltator* Hahn (Heteroptera: Miridae): Corrections of misidentifications and the first authentic report for North America. J. N.Y. Entomol. Soc. 94:51–55.

Henry, T. J. and K. C. Kim. 1984. Genus *Neurocolpus* Reuter (Heteroptera: Miridae): Taxonomy, economic implications, hosts, and phylogenetic review. Trans. Am. Entomol. Soc. 110:1–75.

Henry, T. J. and J. D. Lattin. 1987. Taxonomic status, biological attributes, and recommendations for future work on the genus *Lygus* (Heteroptera: Miridae). Pages 54–68 *in* R. C. Hedlund and H. M. Graham, eds., Economic importance and biological control of *Lygus* and *Adelphocoris* in North America. U.S. Dep. Agric. Agric. Res. Serv. ARS-64.

Henry, T. J. and R. T. Schuh. 1979. Redescription of *Beamerella* Knight and *Hambletoniola* Carvalho and included species (Hemiptera, Miridae), with a review of their relationships. Am. Mus. Novit. No. 2689:1–13.

Henry, T. J. and G. M. Stonedahl. 1984. Type designations and new synonymies for Nearctic species of *Phytocoris* Fallen (Hemiptera: Miridae). J. N.Y. Entomol. Soc. 91:442–465.

Henry, T. J. and A. G. Wheeler Jr. 1973. *Plagiognathus vitellinus* (Scholtz), a conifer-feeding mirid new to North America (Hemiptera: Miridae). Proc. Entomol. Soc. Wash. 75:480–485.

——. 1979. Palearctic Miridae in North America: Records of newly discovered and little-known species (Hemiptera:

Heteroptera). Proc. Entomol. Soc. Wash. 81:257–268.

——. 1982. New United States records for six neotropical Miridae (Hemiptera) in southern Florida. Fla. Entomol. 65:233–241.

——. 1988. Family Miridae Hahn, 1833 (= Capsidae Burmeister). The plant bugs. Pages 251–507 *in* T. J. Henry and R. C. Froeschner, eds., Catalog of the Heteroptera, or true bugs, of Canada and the continental United States. Brill, Leiden.

Henry, T. J., J. W. Neal Jr., and K. M. Gott. 1986. *Stethoconus japonicus* (Heteroptera: Miridae): A predator of *Stephanitis* lace bugs newly discovered in the United States, promising in the biocontrol of azalea lace bug (Heteroptera: Tingidae). Proc. Entomol. Soc. Wash. 88:722–730.

Heong, K. L., A. T. Barrion, and G. B. Aquino. 1990a. Dynamics of major predator and prey species in ricefields. Int. Rice Res. Newsl. 15(6):22–23.

Heong, K. L., S. Bleih, and A. A. Lazaro. 1990b. Predation of *Cyrtorhinus lividipennis* Reuter on eggs of the green leafhopper and brown planthopper in rice. Res. Popul. Ecol. (Kyoto) 32:255–262.

Heong, K. L., S. Bleih, and E. G. Rubia. 1991. Prey preferences of the wolf spider, *Pardosa pseudoannulata* (Boesenberg et Strand). Res. Popul. Ecol. (Kyoto) 32:179–186.

Herard, F. 1985. Analysis of parasite and predator populations observed in pear orchards infested by *Psylla pyri* (L.) (Hom.:Psyllidae) in France. Agronomie (Paris) 5:773–778.

——. 1986. Annotated list of the entomophagous complex associated with pear psylla, *Psylla pyri* (L.) (Hom.:Psyllidae) in France. Agronomie (Paris) 6:1–34.

Herbert, H. J. 1962. Overwintering females and the number of generations of *Typhlodromus* (*T.*) *pyri* Scheuten (Acarina: Phytoseiidae) in Nova Scotia. Can. Entomol. 94:233–242.

Herbert, H. J. and K. H. Sanford. 1969. The influence of spray programs on the fauna of apple orchards in Nova Scotia XIX. Apple rust mite, *Vasates schlechtendali*, a food source for predators. Can. Entomol. 101:62–67.

Herczek, A. 1993. Systematic position of Isometopinae Fieb. (Miridae, Heteroptera) and their intrarelationships. Pr. Nauk. Uniw. Slask. Katowicach 1357:1–86.

Herczek, A. and J. Gorczyca. 1991. A representative of the genus *Deraeocoris* in Baltic amber (Heteroptera, Miridae). Ann. Naturhist. Mus. Wien 92A:89–92.

Heriot, A. D. 1934. The renewal and replacement of the stylets of sucking insects during each stadium, and the method of penetration. Can. J. Res. 11:602–612.

Herms, D. A. 1986. Pest-free honeylocust is a thing of the past. Am. Nurseryman 163(10):73–74, 76–78.

Herms, D. A. and W. J. Mattson. 1992. The dilemma of plants: To grow or defend. Q. Rev. Biol. 67:283–335.

Herms, D. A., D. G. Nielsen, and T. D. Sydnor. 1987. Impact of honeylocust plant bug (Heteroptera: Miridae) on ornamental honeylocust and associated adult buprestids. Environ. Entomol. 16:996–1000.

Hernández, L. M. and T. J. Henry. 1999. Review of the *Ceratocapsus* of Cuba, with descriptions of three new species and a neotype designation for *C. cubanus* Bergroth (Heteroptera: Miridae: Orthotylinae). Caribb. J. Sci. 35:201–214.

Hernandez, S. A., M. Palma, and A. R. Pedrique. 1953. "Chinche" or "mosquilla del cacao" (*Monalonion dissimulatum*) in Venezuela. Cacao 2(37–39):3.

Herrera, C. M. 1989. Pollinator abundance, morphology, and flower visitation rate: Analysis of the "quantity" component in a plant-pollinator system. Oecologia (Berl.) 80:241–248.

Herrera, J. M. 1965. Investigaciones sobre las chinches del género *Rhinacloa* (Hemiptera: Miridae) controladores importantes del *Heliothis virescens* en el algodón. Rev. Peru. Entomol. 8:44–60.

——. 1987. Importancia del control biológico en el cultivo del algodonero. Rev. Peru. Entomol. 30:25–28.

Herrera, J. M. and F. Alvarez. 1979. El control biológico de *Bucculatrix thurberiella* Busck (Lepidoptera: Lyonettidae [sic] en Piura y Chira. Rev. Peru. Entomol. 22:37–41.

Herrich-Schaefer, G. A. W. 1836. Die Wanzenartigen Insecten. 3:33–114. C. H. Zeh'schen Buchhandlung, Nürnburg.

Herring, J. L. 1976. A new genus and species of Cylapinae from Panama (Hemiptera: Miridae). Proc. Entomol. Soc. Wash. 78:91–94.

Herring, J. L. and P. D. Ashlock. 1971. A key to the nymphs of the families of Hemiptera (Heteroptera) of America north of Mexico. Fla. Entomol. 54:207–212.

Hesjedal, K. 1986. Skadedyrmiddel i ulike konsentrasjonar på blad- og nebbteger i frukthagar. Forsk. Fors. Landbruket 37:213–217.

Hesjedal, K. and E. Vangdal. 1986. Integrerte rådgjerder mot teger som er årsak til stein i paere. Forsk. Fors. Landbruket 37:81–88.

Heslop-Harrison, J., ed. 1971. Pollen: Development and physiology. Butterworths, London. 338 pp.

Hespenheide, H. A. 1985. Insect visitors to extrafloral nectaries of *Byttneria aculeata* (Sterculiaceae): Relative importance and roles. Ecol. Entomol. 10:191–204.

Hesse, A. J. 1947. A remarkable new dimorphic isometopid and two other new species of Hemiptera predaceous upon the red scale of citrus. J. Entomol. Soc. S. Afr. 10:31–45.

Hewitt, G. B. 1980. Tolerance of ten species of *Agropyron* to feeding by *Labops hesperius*. J. Econ. Entomol. 73:779–782.

Hewitt, G. B. and W. H. Burleson. 1976. A preliminary survey of the arthropod fauna of sainfoin in central Montana. Mont. Agric. Exp. Stn. Bull. 693:1–11.

Hey, G. L. 1933a. A hymenopterous parasite of the capsid bug *Plagiognathus arbustorum* Fab. Entomol. Mon. Mag. 69:43.

——. 1933b. A list of British species of capsids (Hemiptera-Heteroptera) taken in U.S.A. and Canada in 1932. Entomol. Mon. Mag. 69:43–44.

——. 1935. Notes on Capsidae. Entomol. Mon. Mag. 71:237–238.

——. 1937. *Lygus pabulinus* L. attacking loganberries and cultivated blackberries. Entomol. Mon. Mag. 73:234.

Hicks, E. A. 1959. Check-list and bibliography on the occurrence of insects in birds' nests. Iowa State College Press, Ames. 681 pp.

——. 1971. Check-list and bibliography on the occurrence of insects in birds' nests. Iowa State J. Sci. 46:123–328.

Higgins, K. M., J. E. Bowns, and B. A. Haws. 1977. The black grass bug (*Labops hesperius* Uhler): Its effect on several native and introduced grasses. J. Range Manage. 30:380–384.

High, M. M. 1924. Tomato suck-fly (*Macrolophus separatus* Uhler). U.S. Dep. Agric. Insect Pest Surv. Bull. 4:129.

Hight, S. D. 1990. Available feeding niches in populations of *Lythrum salicaria* (purple loosestrife) in the northeastern United States. Pages 269–278 in E. S. Delfosse, ed., Proc. 7th Int. Symp. Biol. Control Weeds, 6–11 March 1988, Rome, Italy. Ministro dell'Agriculture e delle Foreste, Rome/ CSIRO, Melbourne.

Higley, L. G., L. P. Pedigo, and K. R. Ostlie. 1986. DEGDAY: A program for calculating degree-days, and assumptions behind the degree-day approach. Environ. Entomol. 15:999–1016.

Hill, A. R. 1952a. A survey of insects associated with cultivated raspberries in the east of Scotland. Entomol. Mon. Mag. 88:51–62.

——. 1952b. Observations on *Lygus pabulinus* (L.), a pest of raspberries in Scotland. East Malling Res. Stn. Rep. 1951:181–182.

——. 1952c. Insect pests of cultivated raspberries in Scotland. Trans. 9th Int. Congr. Entomol., Amsterdam 1:589–592.

Hill, L. L. 1932. Protection of celery from tarnished plant bug injury. J. Econ. Entomol. 25:671–678.

——. 1941. A study of the tarnished plant bug *Lygus pratensis* L. and its injury to celery with special emphasis on control measures. Ph.D. dissertation, Cornell University, Ithaca. 82 pp.

Hill, M. P. and C. J. Cilliers. 1996. Biology and host range of *Eccritotarsus catarinensis* (Heteroptera: Miridae), a new potential biological control agent for water hyacinth (*Eichhornia crassipes*) (Pontederiaceae) in South Africa. Page 229 in V. C. Moran and J. H. Hoffmann, eds., Proc. XI Int. Symp. Biol. Control Weeds, 19–26 Jan. 1996, Stellenbosch, South Africa. University of Cape Town, South Africa.

Hill, M. P., C. J. Cilliers, and S. Neser. 1999. Life history and laboratory host range of *Eccritotarsus catarinensis* (Carvalho) (Heteroptera: Miridae), a new natural enemy released on water hyacinth (*Eichhornia crassipes (Mart.) Solms.-Laub.*) (Pontederiaceae) in South Africa. Biol. Control 14:127–133.

Hillocks, R. J. 1992. Cotton diseases. CAB International, Wallingford, UK. 415 pp.

Hills, O. A. 1941. Isolation-cage studies of certain hemipterous and homopterous insects on sugar beets grown for seed. J. Econ. Entomol. 34:756–760.

Hinckley, A. D. 1963. Ecology and control of rice planthoppers in Fiji. Bull. Entomol. Res. 54:467–481.

——. 1965a. Pest of rice in Fiji. Pages 18–21 in Agricultural science No. 1— 1965: Fiji Dep. Agric. Bull. 44.

——. 1965b. Trophic records of some insects, mites, and ticks in Fiji. Fiji Dep. Agric. Bull. 45:1–116.

Hindwood, K. A. 1951. Bird/insect relationships: With particular reference to a beetle (*Platydema pascoei*) inhabiting the nests of finches. Emu 50:179–183.

Hinton, H. E. 1962. The structure of the shell and respiratory system of the eggs of *Helopeltis* and related genera (Hemiptera, Miridae). Proc. Zool. Soc. Lond. 139:483–488.

——. 1981. Biology of insect eggs. Vol. II. Pergamon, Oxford. pp. 475–778.

Hiremath, I. G. 1986. Host preference of sorghum earhead bug, *Calocoris angustatus* Lethierry (Hemiptera: Miridae). Entomon 11:121–125.

——. 1989. Survey of sorghum earhead bug and its natural enemies in Karnataka. J. Biol. Control 3:13–16.

——. 1995. Biology and population dynamics of sorghum head bug *Calocoris angustatus* in India. Pages 81–90 in K. F. Nwanze and O. Youm, eds., Panicle insect pests of sorghum and pearl millett: Proceedings of an International Consultative Workshop, 4–7 Oct. 1993, ICRISAT Sahelian Center, Niamey, Niger. International Crops Research Institute for the Semi-Arid Tropics, Patancheru, India.

Hiremath, I. G. and T. S. Thontadarya. 1983. Natural enemies of the sorghum earhead bug, *Calocoris angustatus* Lethierry (Hemiptera: Miridae). Curr. Res. (Bangalore) 12:10–11.

——. 1984a. Stage and part of sorghum earhead preferred for oviposition by sorghum earhead bug. Curr. Res. (Bangalore) 13:37–38.

——. 1984b. Seasonal incidence of the sorghum earhead bug, *Calocoris angustatus* Lethierry (Hemiptera: Miridae). Insect Sci. Appl. 6:469–474.

Hiremath, I. G. and C. A. Viraktamath. 1992. Biology of the sorghum earhead bug, *Calocoris angustatus* (Hemiptera: Miridae) with descriptions of various stages. Insect Sci. Appl. 13:447–457.

Hiremath, I. G., T. S. Thontadarya, and A. S. Nalini. 1983. Histochemical changes in sorghum grain due to feeding by sorghum earhead bug, *Calocoris angustatus* (Hemiptera: Miridae). Curr. Res. (Bangalore) 12:15–16.

Hiremath, I. G., T. S. Thontadarya, and K. Jairao. 1984. Report on cannibalistic and predatory behaviour of sorghum earhead bug, *Calocoris angustutus* [sic] (Hemiptera: Miridae). Agric. Sci. Dig. 4:67–68.

Hirose, Y. 1990. Prospective use of natural enemies to control *Thrips palmi* (Thysanop., Thripidae). Pages 135–140 in J. Bay-Petersen, ed., The use of natural enemies to control agricultural pests. Food and Technology Center for the Asian and Pacific Region, Taipei, ROC Taiwan.

Hirose, Y., H. Kajita, M. Takagi, S. Okajima, B. Napompeth, and S. Buranapanichpan. 1993. Natural enemies of *Thrips palmi* and their effectiveness in the native habitat, Thailand. Biol. Control 3:1–5.

Hitchings, E. F. 1908. Insect infesting the dahlia. Pages 4–5 in Third Annu. Rep. State Entomol. on the gipsy and browntail moths and other insect pests of the state of Maine, 1907. Sentinel, Waterville.

Hixson, E. 1941. The host relation of the cotton flea hopper. Iowa State Coll. J. Sci. 16:66–68.

Hlavac, T. F. 1975. Grooming systems of insects: Structure, mechanics. Ann. Entomol. Soc. Am. 68:823–826.

Ho, H. Y., R. S. Tsai, and Y. S. Chow. 1995. Chemical constituents of the defensive secretions of the metasternal gland of the stink bug, *Helopeltis fasciaticollis* Poppius (Hemiptera: Miridae) in Taiwan. Zool. Stud. 34:211–214.

Hoberlandt, L. 1972. Ordnung Heteroptera, Wanzen. Pages 114–125 in W. Schwenke, ed., Die Forstschädlinge Europas: Ein Handbuch in fünf Bänden. Paul Parey, Hamburg.

Hochmut, R. 1964. Populační dynamika obaleče hlohového (*Archips crataegana* (Hb.)) v dubinách ČSSR v letech 1957–61. Pr. Výzk. Úst. Lesn. 28:35–80.

Hocking, B. 1953. The intrinsic range and speed of flight of insects. Trans. R. Entomol. Soc. Lond. 104:222–346.

Hodgkiss, H. E. and S. W. Frost. 1921. The apple red bugs and their control. Pa. Agric. Exp. Stn. Ext. Circ. 88:1–8.

Hodgson, W. A., D. D. Pond, and J. Munro, compilers. 1973. Diseases and pests of potatoes. Agric. Can. Publ. 1492:1–73.

Hodkinson, I. D. and D. Casson. 1991. A lesser predilection for bugs; Hemiptera (Insecta) diversity in tropical rain forests. Biol. J. Linn. Soc. 43:101–109.

Hodkinson, I. D. and P. W. H. Flint. 1971. Some predators from the galls of *Psyllopsis fraxini* L. (Hem., Psyllidae). Entomol. Mon. Mag. 107:11–12.

Hoelmer, K. A., L. S. Osborne, F. D. Bennett, and R. K. Yokomi. 1994. Biological control of sweetpotato whitefly in Florida. Pages 101–113 in D. Rosen, F. D. Bennett, and J. L. Capinera, eds., Pest management in the subtropics: Biological control—a Florida perspective. Intercept, Andover, UK.

Hoffman, R. L. 1992. *Bothynotus johnstoni* Knight in Virginia (Heteroptera: Miridae). Banisteria (Hampden-Sydney, Va.) 1:18–19.

Hoffmann, C. H. 1942. Annotated list of elm insects in the United States. U.S. Dep. Agric. Misc. Publ. 466:1–20.

Hoffmann, W. E. 1935. Observations on a hesperid [sic] leaf-roller and a lace-bug, two pests of banana in Kwantgung. Lingnan Sci. J. (Canton) 14:639–649.

Hofmänner, B. 1925. Beiträge zur Kenntnis der Oekologie und Biologie der schweizerischen Hemipteren (Heteropteren und Cicadinen). Rev. Suisse Zool. 32:181–206.

Hogmire, H. W. Jr. and P. K. Custer. 1982. Catfacing insects and their control on peach. Mountaineer Grow. (W.Va.) No. 435:10, 12, 14, 16, 27.

Hogue, C. L. 1993. Latin American insects and entomology. University of California Press, Berkeley. 536 pp.

Hoke, S. 1926. Preliminary paper on the wing-venation of the Hemiptera (Heteroptera). Ann. Entomol. Soc. Am. 19:13–34.

Hokkanen, H. and D. Pimentel. 1984. New approaches for selecting biological control agents. Can. Entomol. 116:1109–1121.

Holdaway, F. G. 1944. Insects of vegetable crops in Hawaii today. Proc. Hawaii. Entomol. Soc. 12:59–80.

——. 1945. Research on DDT for the control of agricultural insects in Hawaii. Proc. Hawaii. Entomol. Soc. 12:301–308.

Holdaway, [F. G.] and [W. C.] Look. 1940. Ecology of the tomato bug, *Cyrtopeltis varians*. Hawaii. Agric. Exp. Stn. Rep. 1939:36–37.

Holdaway, F. G. and W. C. Look. 1942. Insects of the garden bean in Hawaii. Proc. Hawaii. Entomol. Soc. 11:249–260.

Holder, D. G., P. A. Hedin, W. L. Parrott, F. G. Maxwell, and J. N. Jenkins. 1975. Sugars in the leaves of frego bract strain M64 and Deltapine 16 varieties of cotton. J. Miss. Acad. Sci. 19:178–180.

Holdom, D. G., P. S. Taylor, R. J. Taylor, R. J. Mackay-Wood, M. E. Ramos, and R. S. Soper. 1989. Field studies on rice planthoppers (Hom., Delphacidae) and their natural enemies in Indonesia. J. Appl. Entomol. 107:118–129.

Holdsworth, R. P. Jr. 1968. Integrated control: Effect on European red mite and its more important predators. J. Econ. Entomol. 61:1602–1607.

——. 1972. Major predators of the European red mite on apple in Ohio. Ohio Agric. Res. Dev. Cent. Res. Circ. 192:1–18.

Hölldobler, B. and E. O. Wilson. 1990. The ants. Harvard University Press, Cambridge. 732 pp.

Holliday, P. 1980. Fungous diseases of tropical crops. Cambridge University Press, Cambridge. 607 pp.

Holling, C. S. 1959. Some characteristics of simple types of predation and parasitism. Can. Entomol. 91:385–398.

Hollingsworth, R. G., D. C. Steinkraus, and N. P. Tugwell. 1997. Responses of Arkansas populations of tarnished plant bugs (Heteroptera: Miridae) to insecticides, and tolerance differences between nymphs and adults. J. Econ. Entomol. 90:21–26.

Holloway, B. A. 1976. Pollen-feeding in hover flies (Diptera: Syrphidae). N.Z. J. Zool. 3:339–350.

Holmquist, A. M. 1926. Studies in arthropod hibernation. Ann. Entomol. Soc. Am. 19:395–428.

Holopainen, J. K. 1986. Damage caused by *Lygus rugulipennis* Popp. (Heteroptera, Miridae), to *Pinus sylvestris* L. seedlings. Scand. J. For. Res. 1:343–349.

——. 1989. Host plant preference of the tarnished plant bug *Lygus rugulipennis* Popp. (Het., Miridae). J. Appl. Entomol. 107:78–82.

——. 1990a. The relationship between multiple leaders and mechanical and frost damage to the apical meristem of Scots pine seedlings. Can. J. For. Res. 20:280–284.

Holopainen, J. [K.]. 1990b. The role of summer frost and *Lygus* feeding in the induction of growth disturbance in Scots pine seedlings. Publ. Univ. Kuopio. Nat. Sci. Orig. Rep. 6/1990. 46 pp. (Ph.D. dissertation.)

Holopainen, J. K. and A.-L. Varis. 1991. Host plants of the European tarnished plant bug *Lygus rugulipennis* Poppius (Het., Miridae). J. Appl. Entomol. 111:484–498.

Holopainen, J. K., R. Rikala, P. Kainulainen, and J. Oksanen. 1995. Resource partitioning to growth, storage and defence in nitrogen-fertilized Scots pine and susceptibility of the seedlings to the tarnished plant bug *Lygus rugulipennis*. New Phytol. 131:521–532.

Holtzer, T. O. and W. L. Sterling. 1980. Ovipositional preference of the cotton fleahopper, *Pseudatomoscelis seriatus*, and distribution of eggs among host plant species. Environ. Entomol. 9:236–240.

Holtzer, T. O., T. L. Archer, and J. M. Norman. 1988. Host plant suitability in relation to water stress. Pages 111–137 in E. A. Heinrichs, ed., Plant stress-insect interactions. Wiley, New York.

Hori, K. 1967. Studies on the salivary gland, feeding habits and injury of *Lygus disponsi* Linnavuori (Hemiptera, Miridae). I. Morphology of the salivary gland and the symptoms of host plants. Res. Bull. Obihiro Univ. 5:55–74. [in Japanese, English summary.]

——. 1968a. Feeding behavior of the cabbage bug, *Eurydema rugosa* Motschulsky (Hemiptera: Pentatomidae) on the cruciferous plants. Appl. Entomol. Zool. 3:26–36.

——. 1968b. Histological and histochemical observations on the salivary gland of *Lygus disponsi* Linnavuori (Hemiptera, Miridae). Res. Bull. Obihiro Univ. (1) 5:735–744.

——. 1969a. Some properties of salivary amylases of *Adelphocoris suturalis* (Miridae), *Dolycoris baccarum* (Pentatomidae), and several other heteropteran species. Entomol. Exp. Appl. 12:454–466.

——. 1969b. Effect of various activators on the salivary amylase of the bug *Lygus disponsi*. J. Insect Physiol. 15:2305–2317.

——. 1970a. Some variations on the activities of salivary amylase and protease of *Lygus disponsi* Linnavuori (Hemiptera: Miridae). Appl. Entomol. Zool. 5:51–61.

——. 1970b. Some properties of amylase in the salivary gland of *Lygus disponsi* (Hemiptera). J. Insect Physiol. 16:373–386.

——. 1970c. Some properties of proteases in the gut and in the salivary gland of *Lygus disponsi* Linnavuori (Hemiptera, Miraed [*sic*]). Res. Bull. Obihiro Univ. 6:318–324.

——. 1971a. Studies on the feeding habits of *Lygus disponsi* Linnavuori (Hemiptera: Miridae) and the injury to its host plants. I. Histological observations of the injury. Appl. Entomol. Zool. 6:84–90.

——. 1971b. Studies on the feeding habits of *Lygus disponsi* Linnavuori (Hemiptera: Miridae) and the injury to its host plant. II. Frequency, duration and quantity of the feeding. Appl. Entomol. Zool. 6:119–125.

——. 1971c. Nature of gut invertase of *Lygus disponsi* Linnavuori (Hemiptera, Miridae). Res. Bull. Obihiro Univ. 6:666–671.

——. 1971d. Physiological conditions in the midgut in relation to starch digestion and the salivary amylase of the bug *Lygus disponsi*. J. Insect Physiol. 17:1153–1167.

——. 1972a. Utilization of sucrose and starch by the bug, *Lygus disponsi* Linnavuori (Hemiptera: Miridae). Appl. Entomol. Zool. 7:79–82.

——. 1972b. Comparative study of a property of salivary amylase among various heteropterous insects. Comp. Biochem. Physiol. 42B:501–508.

——. 1972c. The digestibility of insoluble starches by the amylases in the digestive system of the bug *Lygus disponsi* and the effect of Cl^- and NO_3^- on the digestion. Entomol. Exp. Appl. 15:13–22.

——. 1973a. Studies on the feeding habits of *Lygus disponsi* Linnavuori (Hemiptera: Miridae) and the injury to its host plant. III. Phenolic compounds, acid phosphatase and oxidative enzymes in the injured tissue of sugar beet leaf. Appl. Entomol. Zool. 8:103–112.

——. 1973b. Studies on enzymes, especially amylases, in the digestive system of the bug *Lygus disponsi* and starch digestion in the system. Res. Bull. Obihiro Univ. 8:173–260.

——. 1973c. Studies on the feeding habits of *Lygus disponsi* Linnavuori (Hemiptera: Miridae) and the injury to its host plant. IV. Amino acids and sugars in the injured tissue of sugar beet leaf. Appl. Entomol. Zool. 8:138–142.

——. 1974a. Enzymes in the salivary gland of *Lygus disponsi* Linnavuori (Hemiptera: Miridae). Res. Bull. Obihiro Univ. 8:173–260.

——. 1974b. Plant growth-promoting factor in the salivary gland of the bug, *Lygus disponsi*. J. Insect Physiol. 20:1623–1627.

——. 1975a. Digestive carbohydrases in the salivary gland and midgut of several phytophagous bugs. Comp. Biochem. Physiol. 50B:145–151.

——. 1975b. Amino acids in the salivary glands of the bugs, *Lygus disponsi* and *Eurydema rugosum*. Insect Biochem. 5:165–169.

——. 1975c. Plant growth-regulating factor, substances reacting with Salkourski reagent and phenoloxidase activities in vein tissue injured by *Lygus disponsi* Linnavuori (Hemiptera: Miridae) and surrounding mesophyll tissues of sugar beet leaf. Appl. Entomol. Zool. 10:130–135.

——. 1975d. Pectinase and plant growth-promoting factors in the salivary glands of the larva of the bug, *Lygus disponsi*. J. Insect Physiol. 21:1271–1274.

——. 1976. Plant growth-promoting factor in the salivary gland of several heteropterous insects. Comp. Biochem. Physiol. 53B:435–438.

——. 1977. Uselessness of Casamino acids (DIFCO H50320) for the rearing of *Lygus disponsi* Linnavuori (Hemiptera, Miridae). Res. Bull. Obihiro Univ. 10:743–747.

——. 1979a. Metabolism of ingested auxins in the bug *Lygus disponsi*: Conversion of several indole compounds. Appl. Entomol. Zool. 14:56–63.

——. 1979b. Metabolism of ingested indole-3-acetic acid in the gut of various heteropterous insects. Appl. Entomol. Zool. 14:149–158.

——. 1980. Metabolism of ingested auxins in the bug *Lygus disponsi*: Indole compounds appearing in the excreta of bugs fed with host plants and the effect of indole-3-acetic acid on the feeding. Appl. Entomol. Zool. 15:123–128.

——. 1992. Insect secretions and their effect on plant growth, with special reference to hemipterans. Pages 157–170 *in* J. D. Shorthouse and O. Rohfritsch, eds., Biology of insect-induced galls. Oxford University Press, New York.

Hori, K. and R. Atalay. 1980. Biochemical changes in the tissue of Chinese cabbage injured by the bug *Lygus disponsi*. Appl. Entomol. Zool. 15:234–241.

Hori, K. and M. Endo. 1977. Metabolism of ingested auxins in the bug *Lygus disponsi*: Conversion of indole-3-acetic acid and gibberellin. J. Insect Physiol. 23:1075–1080.

Hori, K. and T. Hanada. 1970. Biology of *Lygus disponsi* Linnavuori (Hemiptera, Miridae) in Obihiro. Res. Bull. Obihiro Univ. (2) 6:304–317.

Hori, K. and M. Kishino. 1992. Feeding of *Adelphocoris suturalis* Jakovlev (Heteroptera: Miridae) on alfalfa plants and damage caused by the feeding. Res. Bull. Obihiro Univ. 17:357–365.

Hori, K. and K. Kuramochi. 1984. Effects of food plants of the first generation nymph on the growth and reproduction of *Lygus disponsi* Linnavuori (Hemiptera, Miridae). Res. Bull. Obihiro Univ. 14:89–93.

Hori, K. and P. Miles. 1977. Multiple plant growth-promoting factors in the salivary glands of plant bugs. Marcellia 39:399–400.

Hori, K. and P. W. Miles. 1993. The etiology of damage to lucerne by the green mirid, *Creontiades dilutus* (Stål). Aust. J. Exp. Agric. 33:327–331.

Hori, K., D. R. Singh, and A. Sugitani. 1979. Metabolism of ingested indole compounds in the gut of three species of Heteroptera. Comp. Biochem. Physiol. 64C:217–222.

Hori, K., Y. Hashimoto, and K. Kuramochi. 1985. Feeding behaviour of the timothy plant bug, *Stenotus binotatus* F. (Hemiptera: Miridae) and the effect of its feeding on orchard grass. Appl. Entomol. Zool. 20:13–19.

Hori, K., H. Torikura, and M. Kumagai. 1987. Histological and biochemical changes in the tissue of pumpkin fruit injured by *Lygus disponsi* Linnavuori (Hemiptera: Miridae). Appl. Entomol. Zool. 22:259–265.

Horn, D. J. 1988. Ecological approach to pest management. Guilford, New York. 285 pp.

——. 1989. Secondary parasitism and population dynamics of aphid parasitoids (Hymenoptera: Aphidiidae). J. Kans. Entomol. Soc. 62:203–210.

Horn, K. F., M. H. Farrier, and C. G. Wright. 1983a. Estimating egg and first-instar mortalities of the sycamore lace bug, *Corythucha ciliata* (Say). J. Ga. Entomol. Soc. 18:27–37.

———. 1983b. Some mortality factors affecting eggs of the sycamore lace bug, *Corythucha ciliata* (Say) (Hemiptera: Tingidae). Ann. Entomol. Soc. Am. 76:262–265.

Horne, A. S. and H. M. Lefroy. 1915. Effects produced by sucking insects and red spider upon potato foliage. Ann. Appl. Biol. 1:370–386.

Horsburgh, R. L. 1969. The predaceous mirid *Hyaliodes vitripennis* (Hemiptera) and its role in the control of *Panonychus ulmi* (Acarina: Tetranychidae). Ph.D. dissertation, Pennsylvania State University, University Park. 106 pp.

Horsfield, D. 1978. Evidence for xylem feeding by *Philaenus spumarius* (L.) (Homoptera: Cercopidae). Entomol. Exp. Appl. 24:95–99.

Horton, D. R., T. M. Lewis, T. Hinojosa, and D. A. Broers. 1998. Photoperiod and reproductive diapause in the predatory bugs *Anthocoris tomentosus*, *A. antevolens*, and *Deraeocoris brevis* (Heteroptera: Anthocoridae, Miridae) with information on overwintering sex ratios. Ann. Entomol. Soc. Am. 91:81–86.

Horváth, G. 1906. Sur quelques Hémiptères nuisibles de Cochinchine. Bull. Soc. Entomol. Fr. 1906:295–297.

Hossfeld, R. 1963. Synökologischer Vergleich der Fauna von Winter- und Sommerrapsfeldern. Z. Angew. Entomol. 52:209–254.

Houillier, M. 1964a. Régime alimentaire et disponibilité de ponte des miridés dissimulés du cacaoyer. Rev. Pathol. Vég. Entomol. Agric. 43:195–200.

———. 1964b. Étude expérimentale de la répartition des piqures de miridés (*Sahlbergella singularis* Hagl. et *Distantiella theobromae* [sic] Distant) sur cabosse de cacaoyer. Rev. Pathol. Vég. Entomol. Agric. 43:201–208.

Houk, E. J. 1987. Symbionts. Pages 123–129 in A. K. Minks and P. Harrewijn, eds., Aphids: Their biology, natural enemies and control. Vol. A. Elsevier, Amsterdam.

Houk, E. J. and G. W. Griffiths. 1980. Intracellular symbiotes of the Homoptera. Annu. Rev. Entomol. 25: 161–187.

Howard, L. O. 1892. A new enemy to timothy grass. Insect Life (Wash., D.C.) 5:90–92.

———. 1895. Some scale insects of the orchard. Pages 249–250 in Yearb. U.S. Dep. Agric., 1894. Government Printing Office, Washington, D.C.

———. 1898. Notes from correspondence. The so-called "Cotton Flea." Page 101 in Some miscellaneous results of the work of the Division of Entomology.

III. U.S. Dep. Agric. Div. Entomol. Bull. 18 (n.s.).

———. 1899. The principal insects affecting the tobacco plant. Pages 121–150 in Yearb. U.S. Dep. Agric., 1898. Government Printing Office, Washington, D.C.

———. 1901a. Notes from correspondence. Injury by *Lygus invitus* Say. Page 98 in Some miscellaneous results of the work of the Division of Entomology. V. U.S. Dep. Agric. Div. Entomol. Bull. 30.

———. 1901b. The insect book. A popular account of the bees, wasps, ants, grasshoppers, flies and other North American insects exclusive of the butterflies, moths and beetles, with full life histories, tables and bibliographies. Doubleday, Page, New York. 429 pp.

———. 1916. On the Hawaiian work in introducing beneficial insects. J. Econ. Entomol. 9:172–179.

———. 1930. A history of applied entomology (somewhat anecdotal). Smithsonian Misc. Coll. Vol. 84. Smithsonian Institution, Washington, D.C. 545 pp.

———. 1933. Fighting the insects: The story of an entomologist. Macmillan, New York. 333 pp.

Howard, R. J., J. A. Garland, and W. L. Seaman. 1994. Diseases and pests of vegetable crops in Canada: An illustrated compendium. Canadian Phytopathological Society; Entomological Society of Canada, Ottawa. 554 pp.

Howell, D. J. 1974. Bats and pollen: Physiological aspects of the syndrome of chiropterophily. Comp. Biochem. Physiol. 48A:263–276.

Howitt, A. J. 1993. Common tree fruit pests. North Central Reg. Ext. Publ. 63. Michigan State University, East Lansing. 252 pp.

Howitt, A. J., A. Pshea, and W. S. Carpenter. 1965. Causes of deformity in strawberries evaluated in a plant bug control study. Q. Bull. Mich. Agric. Exp. Stn. 48:161–166.

Hoyt, S. C. 1969. Population studies of five mite species on apple in Washington. In Proc. 2nd Int. Congr. Acarol., Sutton Bonington (England), 19–25 July 1967, pp. 117–133. Akadémiai Kiadó, Budapest.

Hoyt, S. C., J. R. Leeper, G. C. Brown, and B. A. Croft. 1983. Basic biology and management components for insect IPM. Pages 93–151 in B. A. Croft and S. C. Hoyt, eds., Integrated management of insect pests of pome and stone fruits. Wiley, New York.

Hsiao, T. H. 1985. Feeding behavior. Pages 471–512 in G. A. Kerkut and L. I. Gilbert, exec. eds., Comprehensive insect physiology, biochemistry, and

pharmacology. Vol. 9. Behaviour. Pergamon, Oxford.

Hsiao, T. Y. 1942. A list of Chinese Miridae (Hemiptera) with keys to subfamilies, tribes, genera and species. Iowa State Coll. J. Sci. 16:241–269.

———. 1945. A new plant bug from Peru, with note on a new genus from North America (Miridae: Hemiptera). Proc. Entomol. Soc. Wash. 47:24–27.

Hsiao, T. Y. and R. I. Sailer. 1947. The orchid bugs of the genus *Tenthecoris* Scott (Hemiptera: Miridae). J. Wash. Acad. Sci. 37:64–72.

Huber, J. T. and V. K. Rajakulendran. 1988. Redescription of and host-induced antennal variation in *Anaphes iole* Girault (Hymenoptera: Mymaridae), an egg parasite of Miridae (Hemiptera) in North America. Can. Entomol. 120:893–901.

Huber, R. T. and P. P. Burbutis. 1967. Some effects of the tarnished plant bug on sweet peppers. J. Econ. Entomol. 60:1332–1334.

Hudler, G. W. 1984. Wound healing in bark of woody plants. J. Arboric. 10: 241–245.

Huffaker, C. B., M. van de Vrie, and J. A. McMurtry. 1970. Ecology of tetranychid mites and their natural enemies: A review II. Tetranychid populations and their possible control by predators: An evaluation. Hilgardia 40(11):391–458.

Huffaker, C. B., P. S. Messenger, and P. DeBach. 1971. The natural enemy component in natural control and the theory of biological control. Pages 16–67 in C. B. Huffaker, ed., Biological control. Plenum, New York.

Hughes, J. H. 1943. The alfalfa plant bug *Adelphocoris lineolatus* (Goeze) and other Miridae (Hemiptera) in relation to alfalfa-seed production in Minnesota. Minn. Agric. Exp. Stn. Tech. Bull. 161:1–80.

Hull, L. A., G. M. Greene II, D. Asquith, and B. A. Croft. 1983. The orchard as a crop production system. Pages 43–65 in B. A. Croft and S. C. Hoyt, eds., Integrated management of insect pests of pome and stone fruits. Wiley, New York.

Hull, L. A., D. G. Pfeiffer, and D. J. Biddinger. 1995. Apple—direct pests. Pages 5–17 in H. W. Hogmire Jr., ed., Mid-Atlantic orchard monitoring guide (NRAES-75). Northeast Regional Agricultural Engineering Service, Ithaca, N.Y.

Hull, R. 1994. Molecular biology of plant virus-vector interactions. Adv. Dis. Vector Res. 10:361–386.

Hunter, C. D. 1997. Suppliers of beneficial organisms in North America. Calif.

Environ. Prot. Agency, Dep. Pestic. Regul. Environ. Monitoring Pest Manage. Branch, Sacramento. 32 pp.

Hunter, W. B. and E. A. Backus. 1989. Mesophyll-feeding by the potato leafhopper, *Empoasca fabae* (Homoptera: Cicadellidae): Results from electronic monitoring and thin-layer chromatography. Environ. Entomol. 18:465–472.

Hunter, W. D. 1926. The cotton hopper, or so-called "cotton flea." U.S. Dep. Agric. Dep. Circ. 361:1–15.

Huque, H. [1970]. Hand book of agricultural pests of Pakistan and their control. Central Department of Plant Protection, Karachi. 125 pp.

Hurd, P. D. Jr. and E. G. Linsley. 1975. Some insects other than bees associated with *Larrea tridentata* in the southwestern United States. Proc. Entomol. Soc. Wash. 77:100–120.

Hussey, N. W. and C. B. Huffaker. 1976. Spider mites. Pages 179–228 *in* V. L. Delucchi, ed., Studies in biological control. Cambridge University Press, Cambridge.

Hussey, N. W., W. H. Read, and J. J. Hesling. 1969. The pests of protected cultivation: The biology and control of glasshouse and mushroom pests. Arnold, London. 404 pp.

Hussey, R. F. 1922a. Hemiptera from Berrien County, Michigan. Occas. Pap. Mus. Zool. Univ. Mich. No. 118:1–23.

——. 1922b. Hemipterological notes. Psyche (Camb.) 29:229–233.

——. 1954a. Concerning the Floridian species of *Fulvius* (Hemiptera, Miridae). Fla. Entomol. 37:19–22.

——. 1954b. Some new or little-known Miridae from the northeastern United States (Hemiptera). Proc. Entomol. Soc. Wash. 56:196–202.

Hussey, R. S. 1989. Disease-inducing secretions of plant-parasitic nematodes. Annu. Rev. Phytopathol. 27:123–141.

Hutchinson, G. E. 1934. Yale North India Expedition. Report on terrestrial families of Hemiptera-Heteroptera. Mem. Conn. Acad. Sci. Arts 10:119–146.

——. 1965. The ecological theater and the evolutionary play. Yale University Press, New Haven. 139 pp.

Hyslop, J. A. 1938. Losses occasioned by insects, mites, and ticks in the United States. U.S. Dep. Agric. Bur. Entomol. Plant Quar. E-444:1–57.

Ibrahim, W. A. 1989. *Erythmelus helopeltidis*—an egg parasitoid of *Helopeltis theobromae* in cocoa. Planter (Kuala Lumpur) 65:211–215.

Idowu, O. L. 1988. Comparative toxicities of new insecticides to the cocoa mirid, *Sahlbergella singularis* in Nigeria. *In* Proc. 10th Int. Cocoa Res.

Conf., Santo Domingo, Dominican Republic, 17–23 May 1987, pp. 531–534. R. Cocoa Producers' Alliance, Lagos, Nigeria.

Illingworth, J. F. 1929. *Engytatus geniculatus* Reuter—an important pest of tomatoes in Hawaii. Proc. Hawaii. Entomol. Soc. 7:247–248.

——. 1937a. A study of blossom-drop of tomatoes and control measures. Proc. Hawaii. Entomol. Soc. 9:457–458.

——. 1937b. Observations on the predaceous habits of *Cyrtopeltis varians* (Dist.) (Hemip.). Proc. Hawaii. Entomol. Soc. 9:458–459.

Imms, A. D. 1926. The biological control of insect pests and injurious plants in the Hawaiian Islands. Ann. Appl. Biol. 13:402–423.

——. 1971. Insect natural history, 3rd ed. Collins, London. 317 pp.

Ingham, D. S., M. J. Samways, and P. Govender. 1995. The development of an effective monitoring method for the wattle mirid *Lygidolon laevigatum* Reuter (Hemiptera: Miridae). *In* Proc. Tenth Entomol. Congress, Entomol. Soc. South. Africa, Grahamstown, 3–7 July 1995, p. 73. Entomological Society of Southern Africa, Pretoria.

——. 1998. Monitoring the brown wattle mirid, *Lygidolon laevigatum* Reuter (Hemiptera: Miridae). Afr. Entomol. 6:111–116.

Ingram, W. R. 1970a. Pests of cocoa. Pages 209–210 *in* J. D. Jameson, ed., Agriculture in Uganda, 2nd ed. Oxford University Press, Oxford.

——. 1970b. Pests of cereals. Pages 227–228 *in* J. D. Jameson, ed., Agriculture in Uganda, 2nd ed. Oxford University Press, Oxford.

——. 1980. Preliminary indications of the status of *Rhinacloa forticornis* as a pest of sea island cotton in Barbados. Trop. Pest Manage. 26:371–376.

Inouye, D. W. 1980. The terminology of floral larceny. Ecology 61:1251–1253.

——. 1983. The ecology of nectar robbing. Pages 153–173 *in* B. Bentley and T. Elias, eds., The biology of nectaries. Columbia University Press, New York.

International Commission on Zoological Nomenclature. 1970. Opinion 898. Miridae Hahn, 1833 (Hemiptera) and Mirini Ashmead, 1900 (Hymenoptera): Removal of homonymy under the plenary powers. Bull. Zool. Nomencl. 26:203–208.

International Rice Research Institute (IRRI). 1981. IRRI annual report for 1981. International Rice Research Institute, Los Baños, Philippines.

Ironside, D. A. 1970. Biology of macadamia flower caterpillar (*Ho-*

moeosoma vagella Zell.). Queensl. J. Agric. Anim. Sci. 27:301–309.

Irvine, A. K. and J. E. Armstrong. 1990. Beetle pollination in tropical forests of Australia. Pages 135–149 *in* K. S. Bawa and M. Hadley, eds., Reproductive ecology of tropical forest plants (Man and the biosphere series, Vol. 7). UNESCO, Paris; Parthenon, Casterton Hall, UK.

Irwin, M. E., R. W. Gill, and D. Gonzalez. 1974. Field-cage studies of native egg predators of the pink bollworm in southern California cotton. J. Econ. Entomol. 67:193–196.

Isely, D. 1920. Grapevine flea-beetles. U.S. Dep. Agric. Bull. 901:1–27.

——. 1927. The cotton hopper and associated leaf-bugs attacking cotton. Univ. Ark. Coll. Agric. Ext. Circ. 231:1–8.

Ishaaya, I. 1986. Nutritional and allelochemic insect-plant interactions relating to digestion and food intake: Some examples. Pages 191–223 *in* J. R. Miller and T. A. Miller, eds., Insect-plant interactions. Springer, New York.

Ishaaya, I. and M. Sternlicht. 1971. Oxidative enzymes, ribonuclease, and amylase in lemon buds infested with *Aceria sheldoni* (Ewing) (Acarina: Eriophyidae). J. Exp. Bot. 22:146–152.

Ishihara, R. and S. Kawai. 1981. Feeding habits of the azalea lace bug, *Stephanitis pyrioides* Scott (Hemiptera: Tingidae). Jpn. J. Appl. Entomol. Zool. 25:200–202. [in Japanese, English summary.]

Iversen, C. E. and J. G. H. White. 1959. Improvement of *Medicago glutinosa*. Annu. Rev. Canterbury Agric. Coll. N.Z. 1959:95–98.

Ives, W. G. H. 1967. Relations between invertebrate predators and prey associated with larch sawfly eggs and larvae on tamarack. Can. Entomol. 99:607–622.

Ives, W. G. H. and H. R. Wong. 1988. Tree and shrub insects of the prairie provinces. Can. For. Serv. North. For. Serv., Edmonton, Alberta. Inf. Rep. NOR-X-292. 327 pp.

Izhaki, I., N. Maestro, D. Meir, and M. Broza. 1996. Impact of the mirid bug *Capsodes infuscatus* (Hemiptera: Miridae) on fruit production of the geophyte *Asphodelus aestivus*: The effect of plant density. Fla. Entomol. 79:510–520.

Jackson, C. G. 1987. Biology of *Anaphes ovijentatus* (Hymenoptera: Mymaridae) and its host, *Lygus hesperus* (Hemiptera: Miridae), at low and high temperatures. Ann. Entomol. Soc. Am. 80:367–372.

Jackson, C. G., J. W. Debolt, and J. J. Ellington. 1995. Lygus bugs. Pages

87–90 *in* J. R. Nechols, L. A. Andres, J. W. Beardsley, R. D. Goeden, and C. G. Jackson, tech. eds., Biological control in the western United States. Univ. Calif. Div. Agric. Nat. Resour. Publ. 3361. Berkeley, Calif.

Jackson, D. A. 1984. Ant distribution patterns in a Cameroonian cocoa plantation: Investigation of the ant mosaic hypothesis. Oecologia (Berl.) 62:318–324.

Jackson, G. V. H. and F. W. Zettler. 1983. Insect potato witches' broom and legume little-leaf diseases in the Solomon Islands. Plant Dis. 67:1141–1144.

Jackson, J. F. and B. A. Drummond III. 1974. A Batesian ant-mimicry complex from the Mountain Pine Ridge of British Honduras, with an example of transformational mimicry. Am. Midl. Nat. 91:248–251.

Jacobs, D. H. 1985. Suborder Heteroptera. Pages 117–148 *in* C. H. Scholtz and E. Holm, eds., Insects of southern Africa. Butterworths, Durban.

Jacobs, W. 1974. Taschenlexikon zur Biologie der Insekten mit besonderer Berüchsichtigung mitteleuropäischer Arten. Fischer, Stuttgart. 635 pp.

Jacobson, L. A. 1939. [Fruit insects.] Can. Insect Pest Rev. 17:273.

Jagdale, G. B., A. B. Pawar, and D. S. Ajri. 1986. Comparative efficacy of some insecticides against betelvine bug. J. Maharashtra Agric. Univ. 11:110–111.

Jahn, E. 1944. Über das Auftreten von *Tortrix viridana* L. im Gebiet der Pollauer Berge und die Parasiten und Räuber dieses Schädlings. Z. Angew. Entomol. 30:252–262.

Jamieson, B. G. M. 1987. The ultrastructure and phylogeny of insect spermatozoa. Cambridge University Press, Cambridge. 320 pp.

Janzen, D. H. 1985. Coevolution as a process: What parasites of animals and plants do not have in common. Pages 83–99 *in* K. C. Kim, ed., Coevolution of parasitic arthropods and mammals. Wiley, New York.

Japan International Cooperation Agency (JICA). 1981. Aphids, mirids and thrips found in the rice paddies of Thailand. Pages 179–189 *in* Contributions to the development of integrated rice pest control in Thailand. JICA, Tokyo.

Jeevaratnam, K. and R. H. S. Rajapakse. 1981a. Biology of *Helopeltis antonii* Sign. (Heteroptera: Miridae) in Sri Lanka. Entomon 6:247–251.

———. 1981b. Studies on the chemical control of the mirid bug, *Helopeltis antonii* Sign., in the cashew. Insect Sci. Appl. 1:399–402.

Jenkins, J. N. 1986. Host plant resistance: Advances in cotton. *In* Proc. Beltwide Cotton Prod. Res. Conf., Jan. 4–9, 1986, Las Vegas, Nev., pp. 34–40. National Cotton Council, Memphis.

Jenkins, J. N. and F. D. Wilson. 1996. Host plant resistance. Pages 563–597 *in* E. G. King, J. R. Phillips, and R. J. Coleman, eds., Cotton insects and mites: Characterization and management. Cotton Foundation, Memphis.

Jenkins, W. A. 1940. A new virus disease of snap beans. J. Agric. Res. (Wash., D.C.) 60:279–288.

Jennings, F. B. 1903. Miscellaneous notes on British Heteroptera. Entomol. Mon. Mag. 39:69–70.

Jensen, B. M., J. L. Wedberg, and D. B. Hogg. 1991. Assessment of damage caused by tarnished plant bug and alfalfa plant bug (Hemiptera: Miridae) on alfalfa grown for forage in Wisconsin. J. Econ. Entomol. 84:1024–1027.

Jensen, D. D. 1946. Virus diseases of plants and their insect vectors with special reference to Hawaii. Proc. Hawaii. Entomol. Soc. 12:535–610.

———. 1951. The North American species of *Psylla* from willow, with descriptions of new species and notes on biology (Homoptera: Psyllidae). Hilgardia 20(16):299–324.

Jeppson, L. R. and G. F. MacLeod. 1946. Lygus bug injury and its effect on the growth of alfalfa. Hilgardia 17(4):165–188.

Jeppson, L. R., H. H. Kiefer, and E. W. Baker. 1975. Mites injurious to economic plants. University of California Press, Berkeley. 614 pp.

Jewett, H. H. 1929. Potato flea-beetles. Ky. Agric. Exp. Stn. Bull. 297:283–301.

Jewett, H. H. and J. T. Spencer. 1944. The plant bugs, *Miris dolabratus* L., and *Amblytylus nasutus* Kirschbaum, and their injury to Kentucky bluegrass, *Poa pratensis* Linn. J. Am. Soc. Agron. 36:147–151.

Jewett, H. H. and L. H. Townsend. 1947. *Miris dolabratus* (Linn.) and *Amblytylus nasutus* (Kirschbaum). Two destructive insect pests of Kentucky bluegrass. Ky. Agric. Exp. Stn. Bull. 508:1–16.

Jewett, H. H., R. C. Buckner, and E. N. Fergus. 1954. Control of *Miris dolabratus* (Linn.) and *Amblytylus nasutus* (Kirsch.) on Kentucky bluegrass. Ky. Agric. Exp. Stn. Bull. 621:1–14.

Jin, K. X., C. J. Liang, and D. L. Deng. 1981. A study of the insect vectors of witches' broom in *Paulownia* trees. Linye Keji Tongxun 12:23–24. [in Chinese, English summary in Rev. Appl. Entomol. A71:77, 1983.]

Johansen, C. A. 1981. Involvement of bee poisoning in integrated pest management with special reference to alfalfa seed crops. Pages 433–444 *in* D. Pimentel, ed., CRC handbook of pest management in agriculture. Vol. II. CRC, Boca Raton, Fla.

Johansen, C. [A.] and J. Eves. 1972. Acidified sprays, pollinator safety, and integrated pest control on alfalfa grown for seed. J. Econ. Entomol. 65:546–551.

Johansen, C. [A.] and A. H. Retan. 1975. Lygus bugs and their predators. Wash. State Univ. Coop. Ext. Serv. E.M. 3440:1–2.

Johnson, B. 1956. The influence on aphids of the glandular hairs on tomato plants. Plant Pathol. (Lond.) 5:131–132.

Johnson, C. G. 1934. On the eggs of *Notostira erratica* L. (Hemiptera, Capsidae). I. Observations on the structure of the egg and the sub-opercular yolk-plug, swelling of the egg and hatching. Trans. Soc. Brit. Entomol. 1:1–32.

———. 1937. The absorption of water and the associated volume changes occurring in the eggs of *Notostira erratica* L. (Hemiptera, Capsidae) during embryonic development under experimental conditions. J. Exp. Biol. 14:413–421.

———. 1960. A basis for a general system of insect migration and dispersal by flight. Nature (Lond.) 186:348–350.

———. 1962. Capsids: A review of current knowledge. Pages 316–331 *in* J. B. Wills, ed., Agriculture and land use in Ghana. Oxford University Press, London.

———. 1969. Migration and dispersal of insects by flight. Methuen, London. 763 pp.

Johnson, C. G. and T. R. E. Southwood. 1949. Seasonal records in 1947 and 1948 of flying Hemiptera-Heteroptera, particularly *Lygus pratensis* L., caught in nets 50 ft. to 3,000 ft. above the ground. Proc. R. Entomol. Soc. Lond. (A) 24:128–130.

Johnson, D. R., C. D. Klein, H. B. Myers, and L. D. Page. 1996. Pre-bloom square loss, causes and diagnosis. *In* Proc. Beltwide Cotton Conf., Jan. 9–12, 1996, Nashville, Tenn., 1:103–105. National Cotton Council, Memphis.

Johnson, F. 1914. The grape leafhopper in the Lake Erie Valley. U.S. Dep. Agric. Bull. 19:1–47.

Johnson, J. B. and M. P. Stafford. 1985. Adult Noctuidae feeding on aphid honeydew and a discussion of honeydew feeding by adult Lepidoptera. J. Lepid. Soc. 39:321–327.

Johnson, M. W. and D. M. Nafus. 1995. Melon thrips. Pages 79–80 *in* J. R. Nechols, L. A. Andres, J. W. Beardsley, R. D. Goeden, and C. G. Jackson, tech. eds., Biological control in the western United States. Univ. Calif. Div. Agric. Nat. Resour. Publ. 3361. Berkeley, Calif.

Johnson, S. J., H. N. Pitre, J. E. Powell, and W. L. Sterling. 1986. Control of *Heliothis* spp. by conservation and importation of natural enemies. Pages 132–154 *in* S. J. Johnson, E. G. King, and J. R. Bradley Jr., eds., Theory and tactics of *Heliothis* population management: 1—Cultural and biological control. South. Coop. Ser. Bull. 316.

Johnson, W. G. 1898. Preliminary notes upon an important peach tree pest. Entomol. News 9:255.

Johnson, W. T. and H. H. Lyon. 1988. Insects that feed on trees and shrubs, 2nd ed. Cornell University Press, Ithaca. 556 pp.

Johnston, H. G. 1928. Host relationships of the family Miridae in North America (Hemiptera). M.S. thesis, Iowa State University, Ames. 202 pp.

——. 1930. *Dicyphus minimus* Uhler, a pest on tomatoes (Hemiptera, Miridae). J. Econ. Entomol. 23:642.

——. 1940. A melon bug (*Pycnoderes quadrimaculatus* Guer.). U.S. Dep. Agric. Insect Pest Surv. Bull. 20:503.

Jolivet, P. 1950. Les parasites, prédateurs et phorétiques des Chrysomeloidea (Coleoptera) de la fauna Franco-Belge. Bull. Inst. R. Sci. Nat. Belg. 26(34):1–39.

Jones, A. L. 1965. Possible relation of insects to fire blight infection in pear orchards of New York. Phytopathology 55:1063. (Abstr.)

Jones, D. H. 1984. Phenylalanine ammonia-lyase: Regulation of its induction, and its role in plant development. Phytochemistry (Oxf.) 23:1349–1359.

Jones, D. L. and W. R. Elliot. 1986. Pests, diseases and ailments of Australian plants, with suggestions for their control. Lothian, Melbourne. 333 pp.

Jones, F. G. W. and R. A. Dunning. 1972. Sugar beet pests, 3rd ed. Minist. Agric. Fish. Food Bull. 162:1–113.

Jones, F. G. W. and M. G. Jones. 1984. Pests of field crops, 3rd ed. Arnold, London. 392 pp.

Jones, T. H. 1921. *Opisthuria clandestina* var. *dorsalis* Knight injurious to legumes. J. Econ. Entomol. 14:501.

Jones, W. A. and C. G. Jackson. 1990. Mass production of *Anaphes iole* for augmentation against *Lygus hesperus*: Effects of food on fecundity and longevity. Southwest. Entomol. 15:463–468.

Jones, W. A. and G. L. Snodgrass. 1998. Development and fecundity of *Deraeocoris nebulosus* (Heteroptera: Miridae) on *Bemisia argentifolii* (Homoptera: Aleyrodidae). Fla. Entomol. 81:345–350.

Jones, W. A., M. H. Ralphs, and L. F. James. 1998. Use of a native insect to deter grazing and prevent poisoning in livestock. Pages 23–28 *in* T. Garland and A. C. Barr, eds., Toxic plants and other natural toxicants. CAB International, Wallingford, UK.

Jones, W. W. 1932. Mesquite injured by *Orthotylus transluciens* [*sic*] Tucker in Arizona. J. Econ. Entomol. 25:136.

Jong, J. K. de. 1934. Anatomische waarnemingen bij *Helopeltis* (*Helopeltis antonii* Sign.). Archf. Theecult. Ned.-Indie 8:38–57.

——. 1936. Over kweekproeven van *Helopeltis* in het laboratorium. Archf. Theecult. Ned.-Indie 10:50–61.

——. 1938. The influence of the quality of the food on the egg-production in some insects. Treubia 16:445–468.

Jonsson, N. 1983a. The life history of *Psylla mali* Schmidberger (Hom., Psyllidae); and its relationship to the development of the apple blossom. Fauna Norv. Ser. B 30:3–8.

——. 1983b. The bug fauna (Hem., Heteroptera) on apple trees in southeastern Norway. Fauna Norv. Ser. B 30:9–13.

——. 1985. Ecological segregation of sympatric heteropterans on apple trees. Fauna Norv. Ser. B 32:7–11.

——. 1987. Nymphal development and food consumption of *Atractotomus mali* (Meyer-Dür) (Hemiptera: Miridae), reared on *Aphis pomi* (DeGeer) and *Psylla mali* Schmidberger. Fauna Norv. Ser. B 34:22–28.

Jordan, K. H. C. 1951a. Bestimmungstabellen der Familien von Wanzenlarven. Zool. Anz. 147:24–31.

——. 1951b. Zoogeographische Betrachtungen über das östliche Sachsen, dargestellt an deutschen Neufunden von Heteropteren. Zool. Anz. 147:79–84.

——. 1972. Heteroptera (Wanzen) Pages 1–5, 40–46, 88–95 *in* Handb. Zool. (Berl.). Vol. 4(20). [translated by Al Ahram Center for Scientific Translations for U.S. Dep. Agric.; Nat. Sci. Found.]

Josifov, M. 1978. Dendrobionte und dendrophile Halbflügler (Heteroptera) an der Eiche in Bulgarien. Acta Zool. Bulg. 9:3–14. [in Bulgarian, German summary.]

Jotwani, M. G. 1983. Insect pests of sorghum, maize and pearl millet. Pages 109–125 *in* P. D. Srivastva, et al., eds., Agricultural entomology. Vol. II. All India Scientific Writers' Society, New Delhi.

Joyner, K. and F. Gould. 1985. Developmental consequences of cannibalism in *Heliothis zea* (Lepidoptera: Noctuidae). Ann. Entomol. Soc. Am. 78:24–28.

——. 1987. Conspecific tissues and secretions as sources of nutrition. Pages 697–719 *in* F. Slansky Jr. and J. G. Rodriguez, eds., Nutritional ecology of insects, mites, spiders, and related invertebrates. Wiley, New York.

Jubb, G. L. Jr. 1979. Little known grape insect pests. Am. Wine Soc. J. 11:46–47.

Jubb, G. L. Jr. and L. A. Carruth. 1971. Growth and yield of caged cotton plants infested with nymphs and adults of *Lygus hesperus*. J. Econ. Entomol. 64:1229–1236.

Judd, G. J. R. and H. L. McBrien. 1994. Modeling temperature-dependent development and hatch of overwintered eggs of *Campylomma verbasci* (Heteroptera: Miridae). Environ. Entomol. 23:1224–1234.

Judd, G. J. R., H. L. McBrien, and J. H. Borden. 1995. Modification of responses by *Campylomma verbasci* (Heteroptera: Miridae) to pheromone blends in atmospheres permeated with synthetic sex pheromone or individual components. J. Chem. Ecol. 21:1991–2002.

Judd, W. W. 1962. Insects and other invertebrates from nests of the cardinal, *Richmondena cardinalis* (L.), at London, Ontario. Can. Entomol. 94:92–95.

——. 1969. Studies on the Byron Bog in southwestern Ontario XXXVIII. Insects associated with flowering boneset, *Eupatorium perfoliatum* L. Proc. Entomol. Soc. Ont. 99:65–69.

——. 1974. Insects associated with flowering musk-mallow (*Malva moschata* L.) at Owen Sound, Ontario. Ont. Field Biol. 28(2):28–36.

——. 1975. Insects associated with flowering silky dogwood (*Cornus obliqua* Raf.) at Dunnville, Haldimand County, Ontario. Ont. Field Nat. 29:26–35.

——. 1980. Insects associated with flowering basswood, *Tilia americana* L., at Dunnville, Haldimand County, Ontario. Ont. Field Nat. 34:33–39.

——. 1984. Insects associated with flowering teasel, *Dipsacus sylvestris*, at Dunnville, Ontario. Proc. Entomol. Soc. Ont. 114:95–98.

Julien, M. H., ed. 1992. Biological control of weeds: A world catalogue of agents and their target weeds, 3rd ed. CAB International, Wallingford, UK. 186 pp.

Juniper, B. and T. R. E. Southwood. 1986. Insects and the plant surface. Arnold, London. 360 pp.

Junnikkala, E. 1960. Life history and insect enemies of *Hyponomeuta malinellus* Zell. (Lep., Hyponomeutidae) in Finland. Ann. Zool. Soc. Zool.-Bot. Fenn. 'Vanamo' 21:1–44.

Juzwik, J. and M. Hubbes. 1984. Association of bacteria with tarnished plant bug stem lesions on hybrid poplars in Ontario. Phytopathology 74:803. (Abstr.)

———. 1986. Bacteria associated with tarnished plant bug stem lesions on hybrid poplars in Ontario. Eur. J. For. Pathol. 16:390–400.

Kabrick, L. R. and E. A. Backus. 1990. Salivary deposits and plant damage associated with specific probing behaviors of the potato leafhopper, *Empoasca fabae*, on alfalfa stems. Entomol. Exp. Appl. 56:287–304.

Kaczmarek, W. 1955. Les perspectives de la lutte biologique contre la doryphore (*Leptinotarsa decemlineata* Say). Bull. Acad. Pol. Sci. Ser. Sci. Biol. 3:219–224.

Kageyama, M. E. 1974. The tarnished plant bug, *Lygus lineolaris* (P. de B.) (Hemiptera: Miridae): Feeding, injury and control on lettuce with preliminary investigations on its cold tolerance. Ph.D. dissertation, Cornell University, Ithaca. 106 pp.

Kain, D., A. Agnello, J. Kovach, and H. Reissig. 1997. Development of an action threshold and management strategies for mirid bugs on apples 1996—final report. NYS IPM Publ. 213:1–7.

Kajita, H. 1978. The feeding behaviour of *Cyrtopeltis tenuis* Reuter on the greenhouse whitefly, *Trialeurodes vaporariorum* (Westwood). Rostria (Osaka) No. 29:235–238.

———. 1984. Predation of the greenhouse whitefly, *Trialeurodes vaporariorum* (Westwood) (Homoptera: Aleyrodidae), by *Campylomma* sp. (Hemiptera: Miridae). Appl. Entomol. Zool. 19:67–74.

Kakizaki, M. and H. Sugie. 1997. Attraction of males to females in the rice leaf bug, *Trigonotylus caelestialium* (Kirkaldy) (Heteroptera: Miridae). Appl. Entomol. Zool. 32:648–651.

Kalode, M. B. 1983. Leafhopper and planthopper pests of rice in India. Pages 225–245 *in* W. J. Knight, et al., eds., Proceedings of the 1st International Workshop on Biotaxonomy, Classification and Biology of Leafhoppers and Planthoppers (Auchenorrhyncha) of Economic Importance, London, 4–7 October 1982. Commonwealth Institute of Entomology, London.

Kalshoven, L. G. E. 1981. Pests of crops in Indonesia. (Rev. and transl. from Dutch by P. A. van der Laan.) P. T. Ichtiar Baru-Van Hoeve, Jakarta.

Kalvelage, H. 1988. *Collaria scenica* (Stal, 1859) (Hemiptera, Miridae): Praga de gramíneas forrageiras na Região do Planalto Catarinense, Brasil. An. Soc. Entomol. Bras. 17:221–222.

Kamm, J. A. 1979. Plant bugs: Effects of feeding on grass seed development; and cultural control. Environ. Entomol. 8:73–76.

———. 1987. Impact of feeding by *Lygus hesperus* (Heteroptera: Miridae) on red clover grown for seed. J. Econ. Entomol. 80:1018–1021.

Kamm, J. A. and P. O. Ritcher. 1972. Rapid dissection of insects to determine ovarial development. Ann. Entomol. Soc. Am. 65:271–274.

Kanervo, V. 1946. Tutkimuksia lepän lehtikuoriaisen, *Melasoma aenea* L. (Col., Chrysomelidae), luontaisista vihollisista. Ann. Zool. Soc. Zool.-Bot. Fenn. 'Vanamo' 12:1–206.

———. 1962. Einfluss der Bekämpfungsmassnahmen im Apfelbau auf die Populationsentwicklung der Obstbaumspinnmilbe (*Metatetranychus pilosus* C. & F.) und ihre natürlichen Feinde in Finnland. Verh. XI. Int. Kongr. Entomol. (1960)2:64–72.

Kapadia, M. N. and S. N. Puri. 1991. Biology and comparative predation efficacy of three heteropteran species recorded as predators of *Bemisia tabaci* in Maharashtra. Entomophaga 36:555–559.

Karban, R. and J. H. Myers. 1989. Induced plant responses to herbivory. Annu. Rev. Ecol. Syst. 20:331–348.

Karman, M. and F. Akşit. 1961. Ege pamuklarinda tarak ve yeşil koza dökümünde rol oynayan zararlilar üzerinde çalişmalar. Bitki Koruma Bul. (Ankara) 2(7):15–17.

Karpova, A. I. 1945. Insects injurious to alfalfa in Hissar range of Tadzhikistan. Rev. Entomol. URSS 28:1–7. [in Russian, English summary.]

Kartohardjono, A. and E. A. Heinrichs. 1984. Populations of the brown planthopper, *Nilaparvata lugens* (Stål) (Homoptera: Delphacidae), and its predators on rice varieties with different levels of resistance. Environ. Entomol. 13:359–365.

Kassanis, B. 1952. Some factors affecting the transmission of leaf-roll virus by aphids. Ann. Appl. Biol. 39:157–167.

Katayev, O. A., G. I. Golutvin, and A. V. Selikhovkin. 1983. Changes in arthropod communities of forest biocoenoses with atmospheric pollution. Entomol. Obozr. 62(1):33–41. [in Russian; English translation in Entomol. Rev. 62(1):20–29.]

Kathirithamby, J. 1992. Strepsiptera of Panama and Mesoamerica. Pages 421–431 *in* D. Quintero and A. Aiello, eds., Insects of Panama and Mesoamerica: Selected studies. Oxford University Press, Oxford.

Kaufmann, O. 1936. Eine gefährliche Viruskrankheit an Rübsen, Raps und Kohlrüben. Arb. Biol. Reichsanst. Land-Forstwirtsch. (Berl.) 21:605–623.

Kavanagh, M. 1979. Flowering forests. Nature (Lond.) 279:374.

Kawasawa, T. and M. Kawamura. 1975. Pictorial guide to the true bugs in Japan. Natl. Assoc. Agric. Educ., Tokyo. 301 pp. [in Japanese.]

Kay, D. 1961. Die-back of cocoa. West African Cocoa Res. Inst. Tech. Bull. 8:1–20.

Kaya, H. K. and S. P. Stock. 1997. Techniques in insect nematology. Pages 281–324 *in* L. A. Lacey, ed., Manual of techniques in insect pathology. Academic, San Diego.

Kaya, M. 1979. Ege bölgesinin önemli zeytin sahalarinda zeytin ağaçlarinin tali zararlilari, taninmalari, zarar şekilleri ve populasyon yoğunluklari üzerinde incelemeler. Zir. Mücad. Zir. Karan. Gn. Müdür. Arast. Eserl. Ser. 31:1–45.

Kearns, C. A. 1992. Anthophilous fly distribution across an elevation gradient. Am. Midl. Nat. 127:172–182.

Kearns, C. A. and D. W. Inouye. 1993. Pistil-packing flies. Nat. Hist. 102(4):30–37.

Keese, M. C. and T. K. Wood. 1991. Host-plant mediated geographic variation in the life history of *Platycotis vittata* (Homoptera: Membracidae). Ecol. Entomol. 16:63–72.

Keifer, H. H., E. W. Baker, T. Kono, M. Delfinado, and W. E. Styer. 1982. An illustrated guide to plant abnormalities caused by eriophyid mites in North America. U.S. Dep. Agric. Agric. Res. Serv. Agric. Handb. 573:1–178.

Keller, S. 1981. *Entomophthora erupta* (Zygomycetes: Entomophthoraceae) als Pathogen von *Notostira elongata* (Heteroptera: Miridae). Mitt. Schweiz. Entomol. Ges. 54:57–64.

———. 1982. *Zoophthora elateridiphaga* (Zygomycetes, Entomophthoraceae) als Ursache von Massensterben der Wanze *Notostira elongata* (Heteroptera, Miridae). Mitt. Schweiz. Entomol. Ges. 55:289–296.

Kellogg, V. L. 1905. American insects. Holt, New York. 674 pp.

Kelton, L. A. 1955. Species of *Lygus*, *Liocoris*, and their allies in the Prairie Provinces of Canada (Hemiptera: Miridae). Can. Entomol. 87:531–556.

———. 1959. Male genitalia as taxonomic characters in the Miridae (Hemiptera). Can. Entomol. Suppl. 11:1–72.

———. 1972. Species of *Dichrooscytus* found in Canada, with descriptions of four new species (Heteroptera: Miridae). Can. Entomol. 104:1033–1049.

———. 1973. Two new species of *Lygus*

from North America, and a note on the status of *Lygus abroniae* (Heteroptera: Miridae). Can. Entomol. 105:1545–1548.

———. 1975. The lygus bugs (genus *Lygus* Hahn) in North America (Heteroptera: Miridae). Mem. Entomol. Soc. Can. No. 95:1–101.

———. 1980a. Description of a new species of *Parthenicus* Reuter, new records of Holarctic Orthotylini in Canada, and new synonymy for *Diaphnocoris pellucida* (Heteroptera: Miridae). Can. Entomol. 112:341–344.

———. 1980b. The insects and arachnids of Canada. Part 8. The plant bugs of the Prairie Provinces. Heteroptera: Miridae. Agric. Can. Publ. 1703:1–408.

———. 1982a. Description of a new species of *Plagiognathus* Fieber, and additional records of European *Psallus salicellus* in the Nearctic Region (Heteroptera: Miridae). Can. Entomol. 114:169–172.

———. 1982b. New records of European *Pilophorus* and *Orthotylus* in Canada (Heteroptera: Miridae). Can. Entomol. 114:283–287.

———. 1982c. New and additional records of Palearctic *Phylus* Hahn and *Plagiognathus* Fieber in North America (Heteroptera: Miridae). Can. Entomol. 114:1127–1128.

———. 1983. Plant bugs on fruit crops in Canada. Heteroptera: Miridae. Res. Branch Agric. Can. Monogr. No. 24:1–201.

———. 1985. Species of the genus *Fulvius* Stål found in Canada (Heteroptera: Miridae: Cylapinae). Can. Entomol. 117:1071–1073.

Kelton, L. A. and J. L. Herring. 1978. Two new species of *Neoborella* Knight (Heteroptera: Miridae) found on dwarf mistletoe, *Arceuthobium* spp. Can. Entomol. 110:779–780.

Kelton, L. A. and H. H. Knight. 1962. *Mecomma* Fieber in North America (Hemiptera: Miridae). Can. Entomol. 94:1296–1302.

———. 1970. Revision of the genus *Platylygus*, with descriptions of 26 new species (Hemiptera: Miridae). Can. Entomol. 102:1429–1460.

Kendall, D. A. and M. E. Solomon. 1973. Quantities of pollen on the bodies of insects visiting apple blossom. J. Appl. Ecol. 10:627–634.

Kenmore, P. E. 1980. Ecology and outbreaks of a tropical insect pest of the green revolution, the rice brown planthopper, *Nilaparvata lugens* (Stål). Ph.D. dissertation, University of California, Berkeley. 226 pp.

Kennedy, C. E. J. 1986. Attachment may be a basis for specialization in oak aphids. Ecol. Entomol. 11:291–300.

Kennedy, G. G. and D. C. Margolies. 1985. Mobile arthropod pests: Management in diversified agroecosystems. Bull. Entomol. Soc. Am. 31(3):21–27.

Kennedy, J. S. 1958. Physiological condition of the host-plant and susceptibility to aphid attack. Entomol. Exp. Appl. 1:50–65.

Kennedy, J. S. and I. H. M. Fosbrooke. 1973. The plant in the life of an aphid. Pages 129–140 in H. F. van Emden, ed., Insect/plant relationships. Wiley, New York.

Kennett, C. E., D. L. Flaherty, and R. W. Hoffmann. 1979. Effect of wind-borne pollens on the population dynamics of *Amblyseius hibisci* (Acarina: Phytoseiidae). Entomophaga 24:83–98.

Kerner, A. 1878. Flowers and their unbidden guests. (Transl. from German by W. Ogle.) C. Kegan Paul, London. 164 pp.

Kerzhner, I. M. 1962. Materials on the taxonomy of capsid bugs (Heteroptera, Miridae) in the fauna of the USSR. Entomol. Obozr. 41(2):372–387. [in Russian; English translation in Rev. Entomol. 41(2):226–235.]

———. 1970. New and little known mirid bugs (Heteroptera, Miridae) from the USSR and Mongolia. Entomol. Obozr. 49(3):634–635. [in Russian; English translation in Rev. Entomol. 49(3):392–399.]

———. 1981. Fauna of the USSR. Bugs. Vol. 13, no. 2. Heteroptera of the family Nabidae. Fauna SSSR (n.s.) No. 124:1–326. [in Russian.]

Kerzhner, I. M. and M. Josifov. 1999. Cimicomorpha II: Miridae. *In* B. Aukema and C. Rieger, eds., Catalogue of the Heteroptera of the Palaearctic Region. Vol. 3. Netherlands Entomological Society, Amsterdam. 577 pp.

Kester, K. M. and D. M. Jackson. 1996. When good bugs go bad: Intraguild predation by *Jalysus wickhami* on the parasitoid, *Cotesia congregata*. Entomol. Exp. Appl. 81:271–276.

Kettlewell, B. 1973. The evolution of melanism: The study of a recurring necessity; with special reference to Industrial Melanism in the Lepidoptera. Clarendon, Oxford. 423 pp.

Kevan, P. G. 1972. Insect pollination of High Arctic flowers. J. Ecol. 60:831–847.

———. 1973. Parasitoid wasps as flower visitors in the Canadian High Arctic. Anz. Schädlingskd. Pflanzenschutz Umweltschutz 46:3–7.

———. 1978a. Anthophilous springtails (Collembola) from the Alaskan north slope, from Signy Island, Antarctica, and from near Ottawa, Ontario. Rev. Ecol. Biol. Sol. 15:373–378.

———. 1978b. Floral coloration, its colorimetric analysis and significance in anthecology. Pages 51–78 in A. J. Richards, ed., The pollination of flowers by insects. Linn. Soc. Symp. Ser. 6. Academic, New York.

———. 1983. Floral colors through the insect eye: What they are and what they mean. Pages 3–30 in C. E. Jones and R. J. Little, eds., Handbook of experimental biology. Van Nostrand Reinhold, New York.

———. 1987. Alternative pollinators for Ontario's crops: Prefatory remarks to papers presented at a workshop held at the University of Guelph, 12 April, 1986. Proc. Entomol. Soc. Ont. 118:109–110.

Kevan, P. G. and H. G. Baker. 1983. Insects as flower visitors and pollinators. Annu. Rev. Entomol. 28:407–453.

———. 1984. Insects on flowers. Pages 607–631 in C. B. Huffaker and R. L. Rabb, eds., Ecological entomology. Wiley, New York.

Kevan, P. G. and D. K. McE. Kevan. 1970. Collembola as pollen feeders and flower visitors with observations from the High Arctic. Quaest. Entomol. 6:311–326.

Kevan, P. G., E. A. Tikhmenev, and M. Usui. 1993. Insects and plants in the pollination ecology of the boreal zone. Ecol. Res. 8:247–267.

Khan, Z. R. and R. C. Saxena. 1985. Mode of feeding and growth of *Nephotettix virescens* (Homoptera: Cicadellidae) on selected resistant and susceptible rice varieties. J. Econ. Entomol. 78:583–587.

———. 1986. Technique for locating planthopper (Homoptera: Delphacidae) and leafhopper (Homoptera: Cicadellidae) eggs in rice plants. J. Econ. Entomol. 79:271–273.

Khattat, A. R. and R. K. Stewart. 1975. Damage by tarnished plant bug to flowers and setting pods of green beans. J. Econ. Entomol. 68:633–635.

———. 1977. Development and survival of *Lygus lineolaris* exposed to different laboratory rearing conditions. Ann. Entomol. Soc. Am. 70:274–278.

———. 1980. Population fluctuations and interplant movements of *Lygus lineolaris*. Ann. Entomol. Soc. Am. 73:282–287.

Kho, Y. O. and J. P. Braak. 1956. Reduction in the yield and viability of carrot seed in relation to the occurrence of the plant bug *Lygus campestris* L. Euphytica 5:146–156.

Khoo, K. C. 1987. The cocoa mirid in peninsular Malaysia and its management. Planter (Kuala Lumpur) 63:516–520.

Khoo, K. C. and G. F. Chung. 1989. Use

of the black cocoa ant to control mirid damage in cocoa. Planter (Kuala Lumpur) 65:370–383.

Khoo, K. C., Y. Ibrahim, D. A. Maelzer, and T. K. Lim. 1982. Entomofauna of cashew in West Malaysia. Pages 289–294 in K. L. Heong, B. S. Lee, T. M. Lim, C. H. Teoh, and Y. Ibrahim, eds., Proceedings of the International Conference on Plant Protection in the Tropics. Malaysian Plant Protection Society, Kuala Lumpur, Malaysia.

Khoo, K. C., P. A. C. Ooi, and C. T. Ho. 1991. Crop pests and their management in Malaysia. Tropical Press, Kuala Lumpur, Malaysia. 242 pp.

Khristova, E., E. Loginova, and S. Petrakieva. 1975. *Macrolophus costalis* Fieb.—predator of white fly (*Trialeurodes vaporariorum* Wstw.) in greenhouses. *In* Proc. VIII Int. Plant Prot. Congr., Moscow. Sect. V. Biological and genetic control, pp. 124–125. USSR Organizing Committee, Moscow.

Killian, J. C. and J. R. Meyer. 1984. Effect of weed management on catfacing damage to peaches in North Carolina. J. Econ. Entomol. 77:1596–1600.

Kiman, Z. B. and K. V. Yeargan. 1985. Development and reproduction of the predator *Orius insidiosus* (Hemiptera: Anthocoridae) reared on diets of selected plant material and arthropod prey. Ann. Entomol. Soc. Am. 78:464–467.

King, A. B. S. 1971. Parasitism of *Sahlbergella singularis* (Hagl.) and *Distantiella theobroma* (Dist.). *In* Proc. III Int. Cocoa Res. Conf., Accra, Ghana, 23–29 Nov. 1969, pp. 237–241. Cocoa Research Institute, Tafo, Ghana.

——. 1973. Studies of sex attraction in the cocoa capsid, *Distantiella theobroma* (Heteroptera: Miridae). Entomol. Exp. Appl. 16:243–254.

King, A. B. S. and J. L. Saunders. 1984. The invertebrate pests of annual food crops in Central America: A guide to their recognition and control. Overseas Development Administration, London. 166 pp.

King, E. G., J. R. Phillips, and R. B. Head. 1988. 41st annual conference report on cotton insect research and control. *In* Proc. Beltwide Cotton Prod. Res. Conf., Jan. 3–8, New Orleans, La., pp. 188–202. National Cotton Council, Memphis.

King, E. G., J. R. Phillips, and R. J. Coleman, eds. 1996a. Cotton insects and mites: Characterization and management. Cotton Foundation, Memphis. 1008 pp.

King, E. G., R. J. Coleman, J. A. Morales-Ramos, K. R. Summy, M. R. Bell, and G. L. Snodgrass. 1996b. Biological con-

trol. Pages 511–538 in E. G. King, J. R. Phillips, and R. J. Coleman, eds., Cotton insects and mites: Characterization and management. Cotton Foundation, Memphis.

King, W. V. 1929. The cotton flea hopper (*Psallus seriatus*). Trans. 4th Int. Congr. Entomol., Ithaca, N.Y. (1928)2:452–454.

King, W. V. and W. S. Cook. 1932. Feeding punctures of mirids and other plant-sucking insects and their effect on cotton. U.S. Dep. Agric. Tech. Bull. 296:1–11.

Kingsolver, J. G. and T. L. Daniel. 1993. Mechanics of fluid feeding in insects. Pages 149–161 in C. W. Schaefer and R. A. B. Leschen, eds., Functional morphology of insect feeding. Thomas Say Publ. Entomol.: Proceedings. Entomological Society of America, Lanham, Md.

Kinkorová, J. and P. Štys. 1989. Heteroptera associated with trees in parks of Prague (Czechoslovakia). Acta Univ. Carol. Biol. 31:359–371.

Kirby, P. 1991. Unusual host plant records for some Heteroptera (Hem., Miridae, Pentatomidae). Entomol. Mon. Mag. 127:38.

——. 1992. A review of the scarce and threatened Hemiptera of Great Britain. UK Nature Conservation No. 2. Joint Nature Conservation Committee, Peterborough, UK. 267 pp.

Kirby, W. F. 1892. Elementary text-book of entomology; second edition revised and augmented. Swan Sonnenschein, London. 281 pp.

Kirchner, T. B. 1977. The effects of resource enrichment on the diversity of plants and arthropods in a shortgrass prairie. Ecology 58:1334–1344.

Kiritani, K. 1979. Pest management in rice. Annu. Rev. Entomol. 24:279–312.

Kiritani, K. and J. P. Dempster. 1973. Different approaches to the quantitative evaluation of natural enemies. J. Appl. Ecol. 10:323–330.

Kiritshenko, A. N. 1951. True bugs of the European USSR. Key and bibliography. Opred. Faune USSR 42:1–423. [in Russian.]

Kirk, W. D. J. 1984. Pollen-feeding in thrips (Insecta: Thysanoptera). J. Zool. (Lond.) 204:107–117.

——. 1985. Pollen-feeding and the host specificity and fecundity of flower thrips (Thysanoptera). Ecol. Entomol. 10:281–289.

——. 1987. How much pollen can thrips destroy? Ecol. Entomol. 12:31–40.

——. 1988. Thrips and pollination biology. Pages 129–135 in T. N. Ananthakrishnan and A. Raman, eds., Dynamics of insect-plant interaction:

Recent advances and future trends. Oxford and IBH, New Delhi, India.

Kirkaldy, G. W. 1904. A preliminary list of the insects of economic importance recorded from the Hawaiian Islands. Hawaii. For. Agric. 1(6):152–159; 1(7): 183–189.

——. 1906. Biological notes on the Hemiptera of the Hawaiian Isles No. 1. Proc. Hawaii. Entomol. Soc. 1:135–161.

——. 1907. A bibliographical note on the food of Miridae (Hemiptera). Entomologist 40:287.

——. 1908. A catalogue of the Hemiptera of Fiji. Proc. Linn. Soc. N.S.W. 33:345–391.

Kirkpatrick, T. W. 1923. Preliminary notes on two minor pests of the Egyptian cotton crop (*Creontiades pallidus*, Ramb., and *Nezara viridula*, L.). Minist. Agric. Egypt. Tech. Sci. Serv. Bull. 33:1–15.

——. 1925. Notes on the fungus, *Rhizopus nigricans*, Eer., in relation to insect pests of cotton in Egypt. Minist. Agric. Egypt Tech. Sci. Serv. Bull. 54:1–28.

——. 1941. *Helopeltis* (Hem., Capsidae) on cinchona. Bull. Entomol. Res. 32: 103–110.

Kisimoto, R. 1979. Brown planthopper migration. Pages 113–124 in Brown planthopper: Threat to rice production in Asia. International Rice Research Institute, Los Baños, Philippines.

——. 1981. Development behaviour, population dynamics and control of the brown planthopper, *Nilaparvata lugens* Stål. Rev. Plant Prot. Res. 14:26–58.

Klausnitzer, B. 1988. Zur Kenntnis der winterlichen Insektenvergesellschaftung unter Platanenborke (Heterop-tera, Coleoptera). Entomol. Nachr. Ber. 32:107–112.

Klein, R. M. 1952. Nitrogen and phosphorus fractions, respiration, and structure of normal and crown gall tissues of tomato. Plant Physiol. (Wash., D.C.) 27:335–354.

Klingauf, F. A. 1987. Host plant finding and acceptance. Pages 209–223 in A.K. Minks and P. Harrewijn, eds., Aphids: Their biology, natural enemies and control. Vol. A. Elsevier, Amsterdam.

Klungness, L. M. and Y. S. Peng. 1984. A histochemical study of pollen digestion in the alimentary canal of honeybees (*Apis mellifera* L.). J. Insect Physiol. 30:511–521.

Knaus, W. 1889. The tarnished plant-bug on pear and apple. Insect Life (Wash., D.C.) 2:49.

Knight, D. W., M. Rossiter, and B. W. Staddon. 1984. Esters from the metathoracic scent gland of two capsid bugs, *Pilophorus perplexus* Douglas and Scott and *Blepharidopterus angulatus* (Fallen) (Heteroptera: Miridae). Comp. Biochem. Physiol. 78B:237–239.

Knight, F. B. and H. J. Heikkenen. 1980. Principles of forest entomology, 5th ed. McGraw-Hill, New York. 461 pp.

Knight, H. H. 1915. Observations on the oviposition of certain capsids. J. Econ. Entomol. 8:293–298.

——. 1917a. A revision of the genus *Lygus* as it occurs in America north of Mexico, with biological data on the species from New York. Cornell Univ. Agric. Exp. Stn. Bull. 391:555–645.

——. 1917b. Notes on species of Miridae inhabiting ash trees *(Fraxinus)* with the description of a new species (Hemip.). Bull. Brooklyn Entomol. Soc. 12:80–82.

——. 1917c. New species of *Lopidea* (Miridae, Hemip.). Entomol. News 28:455–461.

——. 1918a. An investigation of the scarring of fruit caused by apple redbugs. Cornell Univ. Agric. Exp. Stn. Bull. 396:187–208.

——. 1918b. Additional data on the distribution and food plants of *Lygus* with descriptions of a new species and variety (Hemip. Miridae). Bull. Brooklyn Entomol. Soc. 13:42–45.

——. 1918c. Interesting new species of Miridae from the United States, with a note on *Orthocephalus mutabilis* (Fallen) (Hemip. Miridae). Bull. Brooklyn Entomol. Soc. 13:111–116.

——. 1921. Monograph of the North American species of *Deraeocoris* (Heteroptera, Miridae). 18th Rep. State Entomol. Minn. pp. 76–210. Agricultural Experiment Station, St. Paul.

——. 1922. Studies on insects affecting the fruit of the apple with particular reference to the characteristics of the resulting scars. Cornell Univ. Agric. Exp. Stn. Bull. 410:447–498.

——. 1923. Family Miridae (Capsidae). Pages 422–658 in W. E. Britton, ed., The Hemiptera or sucking insects of Connecticut. Conn. Geol. Nat. Hist. Surv. Bull. 34.

——. 1924. *Atractotomus mali* (Meyer) found in Nova Scotia (Heteroptera, Miridae). Bull. Brook. Entomol. Soc. 19:65.

——. 1926a. On the distribution and host plants of the cotton flea-hopper (*Psallus seriatus* Reuter) Hemiptera, Miridae. J. Econ. Entomol. 19:106–107.

——. 1926b. Descriptions of six new Miridae from eastern North America (Hemiptera-Miridae). Can. Entomol. 58:252–256.

——. 1927a. Notes on the distribution and host plants of some North American

Miridae (Hemiptera). Can. Entomol. 59:34–44.

——. 1927b. Descriptions of twelve new species of Miridae from the District of Columbia and vicinity (Hemiptera). Proc. Biol. Soc. Wash. 40:9–18.

——. 1927c. New species and a new genus of Deraeocorinae from North America (Hemiptera, Miridae). Bull. Brooklyn Entomol. Soc. 22:136–143.

——. 1928. Key to the species of *Clivinema* with descriptions of seven new species (Hemiptera, Miridae). Proc. Biol. Soc. Wash. 41:31–36.

——. 1929a. Descriptions of five new species of *Plagiognathus* from North America (Hemip.: Miridae). Entomol. News 40:69–74.

——. 1929b. The fourth paper on new species of *Plagiognathus* (Hemiptera: Miridae). Entomol. News 40:263–268.

——. 1930. An European plant-bug (*Adelphocoris lineolatus* Goeze) found in Iowa (Hemip.:Miridae). Entomol. News 41:4–6.

——. 1934. *Neurocolpus* Reuter: Key with five new species. (Hemiptera, Miridae). Bull. Brooklyn Entomol. Soc. 29:162–167.

——. 1937. Six new species of *Aretas* (Hemiptera: Miridae) from the Society Islands and one from the Philippines. Bernice P. Bishop Mus. Bull. 142:161–167.

——. 1939. *Monalonion* Herrich-Schaeffer; descriptions of cacao species from Brazil (Hemiptera, Miridae). Rev. Entomol. (Rio J.) 10:226–230.

——. 1941. The plant bugs, or Miridae, of Illinois. Ill. Nat. Hist. Surv. Bull. 22:1–234.

——. 1943. Hyaliodinae, new subfamily of Miridae (Hemiptera). Entomol. News 54:119–121.

——. 1958. Forty years of progress on the classification of family Miridae (Hemiptera). *In* Proc. 12th Annu. Meet. No. Cent. Branch Entomol. Soc. Am., p. 27. (Abstr.)

——. 1966. *Schaffneria*, a new genus of ground dwelling plant bugs (Hemiptera, Miridae). Iowa State J. Sci. 41:1–6.

——. 1968. Taxonomic review: Miridae of the Nevada Test Site and the western United States. Brigham Young Univ. Sci. Bull. Biol. Ser. 9:1–282.

Knight, H. H. and W. L. McAtee. 1929. Bugs of the family Miridae of the District of Columbia and vicinity. Proc. U.S. Natl. Mus. 75(13):1–27.

Knipling, E. F. 1979. The basic principles of insect population suppression and management. U.S. Dep. Agric. Agric. Handb. 512:1–659.

Knowlton, G. F. 1942. Range lizards as

insect predators. J. Econ. Entomol. 35:602.

——. 1945. *Labops* damage to range grasses. J. Econ. Entomol. 38:707–708.

——. 1946. *Deraeocoris brevis* feeding observations. Bull. Brooklyn Entomol. Soc. 41:100–101.

——. 1951. Bugs damage grass in Utah. Bull. Brooklyn Entomol. Soc. 46:74–75.

——. 1954. Some Utah insects of 1954. Part 1. Utah State Agric. Coll. Mimeo. Ser. 133:1–18.

——. 1956. Some Hemiptera and Homoptera of Utah—1956. Utah State Agric. Coll. Mimeo. Ser. 158:1–11.

——. 1966. Black grass bug observations in Utah. U.S. Dep. Agric. Coop. Econ. Insect Rep. 16(25):596.

Knowlton, G. F. and M. Allen. 1936. Three hemipterous predators of the potato psyllid. Proc. Utah Acad. Sci. 13:293–294.

Knowlton, G. F. and F. C. Harmston. 1946. Insect food of the mountain bluebird. J. Econ. Entomol. 39:384.

Knowlton, G. F. and D. R. Maddock. 1943. Insect food of the western meadowlark. Great Basin Nat. 4:101–102.

Knowlton, G. F., D. R. Maddock, and S. L. Wood. 1946. Insect food of the sagebrush swift. J. Econ. Entomol. 39:382–383.

Knowlton, G. F., F. V. Lieberman, and C. J. Sorenson. 1951. Lygus bug control for alfalfa seed production. Utah State Agric. Coll. Ext. Bull. 221:1–2.

Knuth, P. 1906. Handbook of flower pollination based upon Hermann Müller's work 'The fertilisation of flowers by insects.' Vol. I. (Transl. by J. R. Ainsworth Davis.) Clarendon, Oxford. 382 pp.

——. 1909. Handbook of flower pollination based upon Hermann Müller's work 'The fertilisation of flowers by insects.' Vol. III. (Transl. by J. R. Ainsworth Davis.) Clarendon, Oxford. 644 pp.

Koch, A. 1967. Insects and their endosymbionts. Pages 1–106 in S. M. Henry, ed., Symbiosis. Vol. II. Associations of invertebrates, birds, ruminants, and other biota. Academic, New York.

Kodys, F. 1971. Možnosti chovu klopušky chlupaté *Lygus rugulipennis* Popp. v laboratorních podmínkách. Ochr. Rostl. 7:149–155.

Koehler, C. S. 1963. *Lygus hesperus* as an economic insect on *Magnolia* nursery stock. J. Econ. Entomol. 56:421–422.

——. 1987. Symptomatology in the instruction of landscape ornamentals entomology. J. Arboric. 13:78–80.

Kogan, M. and D. C. Herzog, eds. 1980. Sampling methods in soybean entomology. Springer, New York. 587 pp.

Koppert. 1993. A new predatory bug for *Bemisia* control in tomato? Koppert Bio-Journal (Berkel en Rodenrijs) No. 6:4.

——. 1994a. First results of *Macrolophus* beyond expectation. Koppert Bio-Journal (Berkel en Rodenrijs) No. 7:3–4.

——. 1994b. Waging war against whitefly. Koppert Bio-Journal (Berkel en Rodenrijs) No. 8:1, 4.

Koptur, S. 1989. Is extrafloral nectar production an inducible defense? Pages 323–339 in J. H. Bock and Y. B. Linhart, eds., The evolutionary ecology of plants. Westview, Boulder, Col.

Korcz, A. 1967. Fauna pluskwiaków drapieżnych (Hemiptera-Heteroptera) na jabłoniach w okolicach Poznania. Pol. Pismo Entomol. 37:581–586.

——. 1977. Biologia, morfologia i występowanie *Lygus campestris* (L.)—zmienika złocieniowca oraz innych gatunków z rodzaju *Lygus* (Heteroptera, Miridae) w Polsce. Pr. Nauk. Inst. Ochr. Rosl. 19:209–240.

Kosztarab, M. and F. Kozár. 1988. Scale insects of Central Europe. Junk, Dordrecht. 455 pp.

Kraft, S. K. and R. F. Denno. 1982. Feeding responses of adapted and non-adapted insects to the defensive properties of *Baccharis halimifolia* L. (Compositae). Oecologia (Berl.) 52:156–163.

Krämer, P. 1961. Untersuchungen über den Einfluss einiger Arthropoden auf Raubmilben (Acari). Z. Angew. Zool. 48:257–311.

Krăsteva, K. T. and I. A. Apostolov. 1990. Quantitative and qualitative studies of Heteroptera insects on wheat. Ekologiya (Sofia) 23:84–92.

Krausse, A. 1923. Über *Camptozygum pinastri maculicollis* Mls. Z. Forst-Jagdwes. 55:174–175.

Krischik, V. A. and R. F. Denno. 1990. Patterns of growth, reproduction, defense, and herbivory in the dioecious shrub *Baccharis halimifolia* (Compositae). Oecologia (Berl.) 83:182–190.

Krishna Prasad, N. K., G. Halappa, S. N. Vajranabhaiah, and P. K. Keshava Prasad. 1979. A note on the incidence of tobacco bug, *Nesidiocoris tenuis* Reuter, on VFC tobacco. Curr. Res. (Bangalore) 8:17.

Krištín, A. 1984. Ernährung und Ernährungsökologie des Feldsperlings *Passer montanus* in der Umgebung von Bratislava. Folia Zool. 33:143–147.

——. 1986. Heteroptera, Coccinea, Coccinellidae a Syrphidae v potrave *Passer montanus* L. a *Pica pica* L. Biológia (Bratisl.) 41:143–150.

Kromer, G. W. 1978. Current status and future market potential for cottonseed.

Pages 1–9 in Glandless cotton: Its significance, status, and prospects. U.S. Dep. Agric. Agric. Res. Serv., Beltsville, Md.

Kroon, G. H., J. P. van Praagh, and H. H. W. Velthuis. 1974. Osmotic shock as a prerequisite to pollen digestion in the alimentary tract of the worker honeybee. J. Apic. Res. 13:177–181.

Kudô, S. and M. Kurihara. 1988. Seasonal occurrence of egg diapause in the rice leaf bug, *Trigonotylus coelestialium* Kirkaldy (Hemiptera: Miridae). Appl. Entomol. Zool. 23:365–366.

——. 1989. Effects of maternal age on induction of egg diapause in the rice leaf bug, *Trigonotylus coelestialium* Kirkaldy (Heteroptera: Miridae). Jpn. J. Entomol. 57:440–447.

Kukalová-Peck, J. 1991. Fossil history and the evolution of hexapod structures. Pages 141–179 in CSIRO. The insects of Australia: A textbook for students and research workers, 2nd ed. Cornell University Press, Ithaca; Melbourne University Press, Melbourne.

Kukor, J. J. and M. M. Martin. 1987. Nutritional ecology of fungus-feeding arthropods. Pages 791–814 in F. Slansky Jr. and J. G. Rodriguez, eds., Nutritional ecology of insects, mites, spiders, and related invertebrates. Wiley, New York.

Kullenberg, B. 1941a. Über Farbenveränderungen unter den Wanzen. Ark. Zool. 33B 7:1–5.

——. 1941b. Zur Kenntnis der Morphologie des männlichen Kopulationsapparates bei den Capsiden (Rhynchota). Zool. Bidr. Upps. 20:415–430.

——. 1942. Die Eier der schwedischen Capsiden (Rhynchota) I. Ark. Zool. 33A(15):1–16.

——. 1943. Die Eier der schwedischen Capsiden (Rhynchota) II. Ark. Zool. 34A(15):1–8.

——. 1944. Studien über die Biologie der Capsiden. Zool. Bidr. Upps. 23:1–522.

——. 1947a. Über Morphologie und Funktion des Kopulationsapparats der Capsiden und Nabiden. Zool. Bidr. Upps. 24:217–418.

——. 1947b. Der Kopulationsapparat der Insekten aus phylogenetischem Gesichtspunkt. Zool. Bidr. Upps. 25: 79–90.

Kumar, R[abinder]., R. J. Lavigne, J. E. Lloyd, and R. E. Pfadt. 1976. Insects of the Central Plains Experiment Range, Pawnee National Grassland. Wyo. Agric. Exp. Stn. Sci. Monogr. 32:1–76.

Kumar, R[aj]. 1970. Occurrence of proteases in the salivary glands of cocoa-capsids (Heteroptera: Miridae). J. N.Y. Entomol. Soc. 78:198–200.

——. 1971a. The natural history of cocoa-capsids. Newsl. Ghana Cocoa Mark. Board 47:21–23.

——. 1971b. Chromosomes of cocoa-capsids (Heteroptera: Miridae). Caryologia 24:229–237.

Kumar, R[aj]. and A. K. Ansari. 1974. Biology, immature stages and rearing of cocoa-capsids (Miridae: Heteroptera). Zool. J. Linn. Soc. 54:1–29.

Kumar, R[aj]. and A. Youdeowei. 1983. Management of cocoa pests. Pages 186–201 in A. Youdeowei and M. W. Service, eds., Pest and vector management in the tropics. Longman, London.

Kunkel, L. O. 1924. Insect transmission of aster yellows. Phytopathology 14:54. (Abstr.)

——. 1926. Studies on aster yellows. Am. J. Bot. 13:646–705.

Kuno, E. 1991. Sampling and analysis of insect populations. Annu. Rev. Entomol. 36:285–304.

Kuno, E. and V. A. Dyck. 1985. Dynamics of Philippine and Japanese populations of the brown planthoppers: Comparison of basic characteristics. Proc. ROC-Japan Seminar on the Ecology and Control of the Brown Planthopper. Nat. Sci. Counc. ROC 4:1–9.

Kuno, E. and N. Hokyo. 1970. Comparative analysis of the population dynamics of rice leafhoppers, *Nephotettix cincticeps* Uhler and *Nilaparvata lugens* Stål, with special reference to natural regulation of their numbers. Res. Popul. Ecol. (Kyoto) 12:154–184.

Kurczewski, F. E. 1968. Nesting behavior of *Plenoculus davisi* (Hymenoptera: Sphecidae, Larrinae). J. Kans. Entomol. Soc. 41:179–207.

Kurczewski, F. E. and D. J. Peckham. 1970. Nesting behavior of *Anacrabro ocellatus ocellatus* (Hymenoptera: Sphecidae). Ann. Entomol. Soc. Am. 63:1419–1424.

Kurian, C. and K. N. Ponnamma. 1983. Pests of coconut and arecanut. Pages 294–309 in P. D. Srivastva, et al., eds., Agricultural entomology. Vol. II. All India Scientific Writers' Society, New Delhi, India.

Kurstak, E., ed. 1991. Viruses of invertebrates. Dekker, New York. 351 pp.

Labandeira, C. C. and T. L. Phillips. 1996. Insect fluid-feeding on Upper Pennsylvanian tree ferns (Palaeodictyoptera, Marattiales) and the early history of the piercing-and-sucking functional feeding group. Ann. Entomol. Soc. Am. 89:157–183.

Labandeira, C. C. and J. J. Sepkoski. 1993. Insect diversity in the fossil record. Science (Wash., D.C.) 61:310–315.

LaBonte, G. A. and L. J. Lipovsky. 1967. Oak-leaf shot-hole caused by

Japanagromyza viridula. J. Econ. Entomol. 60:1266–1270.

Lafferty, H. A., J. G. Rhynehart, and G. H. Pethybridge. 1922. Investigations on flax disease. J. Dep. Agric. Tech. Instr. Irel. 22:103–120.

Lal, R. 1950. Biological observations on *Creontiades pallidifer* Walker (*Megacoelum stramineum* Walker), with notes on various insect pests found on potato at Delhi. Indian J. Entomol. 10:267–278.

Lambert, L. and G. L. Snodgrass. 1989. Tarnished plant bug (Heteroptera: Miridae) populations on a susceptible and resistant soybean. J. Entomol. Sci. 24:378–380.

Lamont, B. B. 1994. Triangular trophic relationships in Mediterranean-climate Western Australia. Pages 83–89 *in* M. Arianoutsou and R. H. Groves, eds., Plant-animal interactions in Mediterranean-type ecosystems. Kluwer, Dordrecht.

La Munyon, C. and T. Eisner. 1990. Effect of mite infestation of the anti-predator defenses of an insect. Psyche (Camb.) 97:31–41.

Landes, D. A. and F. E. Strong. 1965. Feeding and nutrition of *Lygus hesperus* (Hemiptera: Miridae). I. Survival of bugs fed on artificial diets. Ann. Entomol. Soc. Am. 58:306–309.

Landis, B. J. and L. Fox. 1972. Lygus bugs in eastern Washington: Color preferences and winter activity. Environ. Entomol. 1:464–465.

Landolt, P. J. 1997. Cabbage looper (Lepidoptera: Noctuidae) fecundity maximized by a combination of access to water and food, and remating. Ann. Entomol. Soc. Am. 90:783–789.

Lane, C. P. and S. J. Weller. 1994. A review of *Lycaeides* Hübner and Karner blue butterfly taxonomy. Pages 5–21 *in* D. A. Andow, R. J. Baker, and C. P. Lane, eds., Karner blue butterfly: A symbol of a vanishing landscape. Univ. Minn. Agric. Exp. Stn. Misc. Publ. 84–1994.

Lange, W. H. Jr. 1944. Insects affecting guayule with special reference to those associated with nursery plantings in California. J. Econ. Entomol. 37:392–399.

La Rivers, I. 1949. Entomic nematode literature from 1926 to 1946, exclusive of medical and veterinary titles. Wasmann Collect. 7:177–206.

Larsen, O. 1941. Die Autotomie der Capsiden. K. Fysiogr. Sallsk. Lund Forh. 11:241–253.

Larsson, S. 1989. Stressful times for the plant stress-insect performance hypothesis. Oikos 56:277–283.

Lateef, S. S. and W. Reed. 1990. Insect

pests on pigeon pea. Pages 193–242 *in* S. R. Singh, ed., Insect pests of tropical food legumes. Wiley, Chichester, UK.

Latson, L. N., J. N. Jenkins, W. L. Parrott, and F. G. Maxwell. 1977. Behavior of the tarnished plant bug, *Lygus lineolaris* on cotton, *Gossypium hirsutum* L. and horseweed, *Erigeron canadensis.* Miss. Agric. For. Exp. Stn. Tech. Bull. 85:1–5.

Lattin, J. D. 1999. Dead leaf clusters as habitats for adult *Calliodis temnostethoides* and *Cardiastethus luridellus* and other anthocorids (Hemiptera: Heteroptera: Anthocoridae). Gt. Lakes Entomol. 32:33–38.

Lattin, J. D. and N. L. Stanton. 1999. Host records of Braconidae (Hymenoptera) occurring in Miridae (Hemiptera: Heteroptera) found on lodgepole pine (*Pinus contorta*) and associated conifers. Pan-Pac. Entomol. 75:23–31.

Lattin, J. D. and G. M. Stonedahl. 1984. *Campyloneura virgula,* a predacious Miridae not previously recorded from the United States (Hemiptera). Pan-Pac. Entomol. 60:4–7.

Lattin, J. D., T. J. Henry, and M. D. Schwartz. 1992. *Lygus desertus* Knight, 1944, a newly recognized synonym of *Lygus elisus* Van Duzee, 1914 (Heteroptera: Miridae). Proc. Entomol. Soc. Wash. 94:12–25.

Lattin, J. D., A. Christie, and M. D. Schwartz. 1994. The impact of nonindigenous crested wheatgrasses on native black grass bugs in North America: A case for ecosystem management. Nat. Areas J. 14:136–138.

——. 1995. Native black grass bugs (*Irbisia-Labops*) on introduced wheatgrasses: Commentary and annotated bibliography (Hemiptera: Heteroptera: Miridae). Proc. Entomol. Soc. Wash. 97:90–111.

Laubert, R. 1927. Zur 'Krauselkrankheit bei Pelargonien'. Gartenwelt (Hambg.) 31:391.

Laurema, S. and P. Nuorteva. 1961. On the occurrence of pectin polygalacturonase in the salivary glands of Heteroptera and Homoptera Auchenorrhyncha. Ann. Entomol. Fenn. 27:89–93.

Laurema, S. and A.-L. Varis. 1991. Salivary amino acids in *Lygus* species (Heteroptera: Miridae). Insect Biochem. 21:759–765.

Laurema, S., A.-L. Varis, and H. Miettinen. 1985. Studies on enzymes in the salivary glands of *Lygus rugulipennis* (Hemiptera: Miridae). Insect Biochem. 15:211–224.

Lavabre, E. M. 1954. Insectes dangereux aux cultures du cacaoyer au Cameroun.

Agron. Trop. (Nogent-sur-Marne) 9:479–484.

——. 1969. Recent progress in breeding certain pests of African cocoa in the laboratory. FAO Plant Prot. Bull. 17:132–135.

——. 1970. Insectes nuisibles des cultures tropicales (cacaoyer, caféier, colatier, poivrier, théier). Maisonneuve & Larose, Paris. 276 pp.

——. ed. 1977. Les Mirides du cacaoyer. Editions G.-P. Maisonneuve et Larose, Paris. 366 pp.

Lavigne, R., R. Kumar, and J. A. Scott. 1991. Additions to the Pawnee National Grasslands insect checklist. Entomol. News 102:150–164.

Lawrence, J. F. 1989. Mycophagy in the Coleoptera: Feeding strategies and morphological adaptations. Pages 1–23 *in* N. Wilding, et al., eds., Insect-fungus interactions. Academic, London.

Lawson, F. A. and T. R. Yonke. 1991. Key to the families of Hemiptera larvae. Pages 25–30 *in* F. W. Stehr, ed., Immature insects. Vol. 2. Kendall/Hunt, Dubuque, Iowa.

Lawson, P. B. 1926. Some "biting" leafhoppers. Ann. Entomol. Soc. Am. 19:73–74.

Layton, B. 1996. Anticipated changes in Mid-South insect management resulting from adoption of Bt-transgenic cotton. *In* Proc. Beltwide Cotton Conf., Jan. 9–12, 1996, Nashville, Tenn. 1:160–161. National Cotton Council, Memphis.

Leach, J. G. 1940. Insect transmission of plant diseases. McGraw-Hill, New York. 615 pp.

Leach, J. G. and P. Decker. 1938. A potato wilt caused by the tarnished plant bug, *Lygus pratensis* L. Phytopathology 28:13. (Abstr.)

Leach, R. 1935. Insect injury simulating fungal attack on plants. A stem canker, an angular spot, a fruit scab and a fruit rot of mangoes caused by *Helopeltis bergrothi* Reut. (Capsidae). Ann. Appl. Biol. 22:525–537.

Leach, R. and C. Smee. 1933. Gnarled stem canker of tea caused by the capsid bug (*Helopeltis bergrothi* Reut.). Ann. Appl. Biol. 20:691–706.

Lean, O. B. 1926. Observations on the life-history of *Helopeltis* on cotton in southern Nigeria. Bull. Entomol. Res. 16:319–324.

Leather, S. R. 1988. Size, reproductive potential and fecundity in insects: Things aren't as simple as they seem. Oikos 51:386–389.

Leatherdale, D. 1970. The arthropod hosts of entomogenous fungi in Britain. Entomophaga 15:419–435.

Le Baron, W. 1871. Insects injurious to

the potato. Pages 63–78 *in* 1st Annu. Rep. Noxious Insects of the State of Illinois. Illinois Journal Printing Office, Springfield.

Lebedeva, E. G. and K. I. Fomina. 1977. Potato virus S and the means of its spread in the Primorsk region. Tr. Biol.-Pochv. Inst. Dal'nevost. Nauch. Tsentr. AN SSSR 48:70–74. [in Russian, English summary.]

Lebert, C. D. 1935. A plant bug (*Pycnoderes quadrimaculatus* Guer.). U.S. Dep. Agric. Insect Pest Surv. Bull. 15:376.

Ledbetter, M. C. and F. Flemion. 1954. A method for obtaining piercing-sucking mouth parts in host tissue from the tarnished plant bug by high voltage shock. Contrib. Boyce Thompson Inst. 17:343–346.

Lee, C. E. 1971. Heteroptera of Korea. Illustrated encyclopedia of fauna and flora of Korea. Vol. 12 (Insect IV). pp. 99–448, 475–601, 1051–1059. Samhwa, Seoul. [in Korean.]

Lee, L. S., P. E. Lacey, and W. R. Goynes. 1987. Aflatoxin in Arizona cottonseed: A model study of insect-vectored entry of cotton bolls by *Aspergillis flavus*. Plant Dis. 71:997–1001.

Lee, R. F. 1971. A preliminary annotated list of Malawi forest insects. Malawi For. Res. Inst. Res. Rec. 40:1–132.

Lee, Y. I., M. Kogan, and J. R. Larsen Jr. 1986. Attachment of the potato leafhopper to soybean plant surfaces as affected by morphology of the pretarsus. Entomol. Exp. Appl. 42:101–107.

Leefmans, S. 1916. Bijdrage tot het *Helopeltis*-vraagstuk voor thee. Meded. Proefstn. Thee 50:1–213.

——. 1920. Aanteekeningen over voedsterplanten van *Helopeltis*. Thee (Batavia) 1:77–78.

Leferink, J. H. M. and G. H. Gerber. 1997. Development of adult and nymphal populations of *Lygus lineolaris* (Palisot de Beauvois), *L. elisus* Van Duzee, and *L. borealis* (Kelton) (Heteroptera: Miridae) in relation to seeding date and stage of plant development on canola (Brassicaceae) in southern Manitoba. Can. Entomol. 129:777–787.

Lefèvre, P. C. 1942. Introduction à l'étude de *Hélopeltis orophila* Ghesq. Inst. Natl. Étude Agron. Congo Belge. Ser. Sci. 30:1–46.

Leftwich, A. W. 1976. A dictionary of entomology. Crane Russak, N.Y. 360 pp.

Legaspi, B. A. C. Jr., W. L. Sterling, A. W. Hartstack Jr., and D. A. Dean. 1989. Testing the interactions of pest-predator-plant components of the TEXCIM model. Environ. Entomol. 18:157–163.

Lehman, P. S. and J. W. Miller. 1988. Symptoms associated with *Aphelenchoides fragariae* and *Pseudomonas cichorii* infections of Philippine violet. Fla. Dep. Agric. Consum. Serv. Nematol. Circ. 159:1–4.

Lehmann, H. 1932a. Beitrag zur ökologie grasbewohnender Heteropteren Norddeutschlands. Z. Pflanzenkr. Pflanzenschutz 42:1–10.

——. 1932b. Wanzen (Hemiptera-Heteroptera) als Obstbaumschädlinge. Z. Pflanzenkr. Pflanzenschutz 42:440–451.

Lei, J. D., H. J. Su, and T. A. Chen. 1979. Spiroplasmas isolated from green leaf bug, *Trigonotylus ruficornis* Geoffroy. Pages 89–98 *in* H. J. Su and R. E. McCoy, eds., Proc. R.O.C.–United States Coop. Sci. Seminar on Mycoplasma Diseases of Plants. National Science Council, Taipei, Taiwan.

Leiby, R. W. 1927. Tomato suckfly (*Diciphus* [sic] *minimus* Uhler). U.S. Dep. Agric. Insect Pest Surv. Bull. 7:297.

Leigh, T. F. 1963. Life history of *Lygus hesperus* (Hemiptera: Miridae) in the laboratory. Ann. Entomol. Soc. Am. 56:865–867.

——. 1966. A reproductive diapause in *Lygus hesperus* Knight. J. Econ. Entomol. 59:1280–1281.

——. 1976. Detrimental effect of lygus feeding on plants. Pages 1–2 *in* D. R. Scott and L. E. O'Keeffe, eds., Lygus bug: Host plant interactions. University Press of Idaho, Moscow.

——. 1987. The mite, whitefly, and lygus complex on California cotton. Summary Proceedings, Western Cotton Production Conference, Phoenix, Ariz., Aug. 18–20, 1987, pp. 66–67. [rptd. Cotton Gin Oil Mill Press 88(20):10–11, 1987.]

Leigh, T. F. and D. Gonzalez. 1976. Field cage evaluation of predators for control of *Lygus hesperus* Knight on cotton. Environ. Entomol. 5:948–952.

Leigh, T. F. and G. A. Matthews. 1994. *Lygus* (Hemiptera: Miridae) and other Hemiptera. Pages 367–379 *in* G. A. Matthews and J. P. Tunstall, eds., Insect pests of cotton. CAB International, Wallingford, UK.

Leigh, T. F., D. W. Grimes, H. Yamada, J. R. Stockton, and D. Bassett. 1969. Arthropod abundance in cotton in relation to some cultural management variables. Proc. Tall Timbers Conf. Ecol. Anim. Control Habitat Manage. 1:71–83.

Leigh, T. F., D. W. Grimes, H. Yamada, D. Bassett, and J. R. Stockton. 1970. Insects and cotton as affected by irrigation and fertilization practices. Calif. Agric. 24(3):12–14.

Leigh, T. F., A. H. Hyer, and R. E. Rice. 1972. Frego bract condition of cotton in relation to insect populations. Environ. Entomol. 1:390–391.

Leigh, T. F., D. W. Grimes, W. L. Dickens, and C. E. Jackson. 1974. Planting pattern, plant populations, and insect interactions in cotton. Environ. Entomol. 3:492–496.

Leigh, T. F., C. E. Jackson, P. F. Wynholds, and J. A. Cota. 1977. Toxicity of selected insecticides applied topically to *Lygus hesperus*. J. Econ. Entomol. 70:42–44.

Leigh, T. F., A. H. Hyer, J. H. Benedict, and P. F. Wynholds. 1985. Observed population increase, nymphal weight gain, and oviposition nonpreference as indicators of *Lygus hesperus* Knight (Heteroptera: Miridae) resistance in glandless cotton. J. Econ. Entomol. 78:1109–1113.

Leigh, T. F., T. A. Kerby, and P. F. Wynholds. 1988. Cotton square damage by the plant bug, *Lygus hesperus* (Hemiptera: Heteroptera: Miridae), and abscission rates. J. Econ. Entomol. 81:1328–1337.

Leius, K. 1960. Attractiveness of different foods and flowers to the adults of some hymenopterous parasites. Can. Entomol. 92:369–376.

——. 1963. Effects of pollens on fecundity and longevity of adult *Scambus buolianae* (Htg.) (Hymenoptera: Ichneumonidae). Can. Entomol. 95:202–207.

Lenteren, J. C. van and L. P. J. J. Noldus. 1990. Whitefly-plant relationships: Behavioural and ecological aspects. Pages 47–89 *in* D. Gerling, ed., Whiteflies: Their bionomics, pest status and management. Intercept, Andover, UK.

Lenteren, J. C. van and J. Woets. 1988. Biological and integrated pest control in greenhouses. Annu. Rev. Entomol. 33:239–269.

Lenteren, J. C. van, P. M. J. Ramakers, and J. Woets. 1980. World situation of biological control in greenhouses, with special attention to factors limiting application. Meded. Fac. Landbouwwet. Rijksuniv. Gent 45:537–544.

Lenteren, J. C. van, M. M. Roskam, and R. Timmer. 1997. Commercial mass production and pricing of organisms for biological control of pests in Europe. Biol. Control 10:143–149.

Leonard, B. R. 1995. Cotton insect resistance and mid-South management strategies. Pages 5–10 *in* D. M. Oosterhuis, ed., Proceedings of the 1994 Cotton Research Meeting and 1994 summaries of cotton research in progress. Ark. Agric. Exp. Stn. Spec. Rep. 166.

Leonard, M. D. 1915. Further experiments in the control of the tarnished plant-bug. J. Econ. Entomol. 8:361–367.

——. 1916a. The immature stages of *Tropidosteptes cardinalis* Uhler (Capsidae, Hemiptera). Psyche (Camb.) 23:1–3.

——. 1916b. The immature stages of two Hemiptera—*Empoasca obtusa* Walsh (Typhlocybidae) and *Lopidea robiniae* Uhler (Capsidae). Entomol. News 27:49–54.

——. 1916c. A tachinid parasite reared from an adult capsid (Dip., Hom.). Entomol. News 27:236.

——. 1930. An unrecorded food-habit of the large tobacco suck-fly in Porto. Rico. J. Econ. Entomol. 23:640–641.

——. 1971. More records of New Jersey aphids (Homoptera: Aphididae). J. N.Y. Entomol. Soc. 79:62–83.

Leont'eva, Iv. A. 1962. On the identification of some virus diseases of potato. Nauchn. Dokl. Vyssh. Shkol. Biol. Nauki 3:158–162. [in Russian; English summary in Rev. Appl. Mycol. 42:212, 1963.]

Lepage, H. S. 1941. O percevejo das orquídeas. Orquídea (Rio J.) 4:14–19.

——. 1942. O percevejo das orquídeas. Biologico (São Paulo) 8:67–72.

Le Pelley, R. H. 1932. *Lygus simonyi*, Reut. (Hem. Capsid.), a pest of coffee in Kenya Colony. Bull. Entomol. Res. 23:85–100.

——. 1942. The food and feeding habits of *Antestia* in Kenya. Bull. Entomol. Res. 33:71–89.

——. 1957. A survey of coffee entomological problems: Kenya, 1956–1957. Kenya Coffee Board Mon. Bull. 22(256): 92–93.

——. 1968. Pests of coffee. Longmans Green, London. 590 pp.

LeRoux, E. J. 1960. Effects of "modified" and "commercial" spray programs on the fauna of apple orchards in Quebec. Ann. Soc. Entomol. Qué. 6:87–121.

——. 1971. Biological control attempts on pome fruit (apple and pear) in North America, 1860–1970. Can. Entomol. 103:963–974.

Leroy, J. V. 1936. Observations relatives à quelques hémiptères du cotonnier. Publ. Inst. Etude Agron. Congo Belge Ser. Sci. 10:1–20.

LeSar, C. D. and J. D. Unzicker. 1978. Life history, habits, and prey preferences of *Tetragnatha laboriosa* (Araneae: Tetragnathidae). Environ. Entomol. 7: 879–884.

Leston, D. 1951. Notes on the Hemiptera-Heteroptera of Bookham Common. Lond. Nat. 30:49–62.

——. 1952a. Antennal oligomery in Heteroptera. Nature (Lond.) 169:890.

——. 1952b. *Oncotylus viridiflavus* Goeze (Hem., Miridae) and its food-plant, knapweed. Entomologist 85:19–21.

——. 1953. The eggs of Tingitidae (Hem.), especially *Acalypta parvula* (Fallén). Entomol. Mon. Mag. 89:132–134.

——. 1957a. Cyto-taxonomy of Miridae and Nabidae (Hemiptera). Chromosoma (Berl.) 8:609–616.

——. 1957b. Spread potential and the colonisation of islands. Syst. Zool. 6:41–46.

——. 1958. Unisexual dimorphism in a mirid (Hemiptera), *Blepharidopterus angulatus* (Fallén). Proc. S. Lond. Entomol. Nat. Hist. Soc. 1957:100–114.

——. 1959. The mirid (Hem.) hosts of Braconidae (Hym.) in Britain. Entomol. Mon. Mag. 95:97–100.

——. 1961a. The number of testis follicles in Miridae. Nature (Lond.) 191:93.

——. 1961b. Testis follicle number and the higher systematics of Miridae (Hemiptera-Heteroptera). Proc. Zool. Soc. Lond. 137:89–106.

——. 1961c. Observations on the mirid (Hem.) hosts of Braconidae (Hym.) in Britain. Entomol. Mon. Mag. 97:65–71.

——. 1961d. The Miridae (Hemiptera) of Bedfordshire. Proc. Trans. S. Lond. Entomol. Nat. Hist. Soc. 1960:110–123.

——. 1968. Heteroptera survey of Ghana. (ii) *Stenotus affinis* Poppius Miridae. Cocoa Res. Inst. (Ghana) Rep. 1965–66:57.

——. 1970. Entomology of the cocoa farm. Annu. Rev. Entomol. 15:273–294.

——. 1971. Ants, capsids and swollen shoot in Ghana: Interactions and the implications for pest control. *In* Proc. III Int. Cocoa Res. Conf., Accra, Ghana, 23–29 Nov. 1969, pp. 205–221. Cocoa Res. Institute, Tafo, Ghana.

——. 1973a. The flight behaviour of cocoa-capsids (Hemiptera: Miridae). Entomol. Exp. Appl. 16:91–100.

——. 1973b. The ant-mosaic-tropical tree crops and the limiting of pests and diseases. PANS (Pest Artic. News Summ.) 19:311–341.

——. 1973c. The natural history of some West African insects. Entomol. Mon. Mag. 108:110–122.

——. 1975. The ant mosaic: A fundamental property of cocoa farms. *In* Proc. IV Int. Cocoa Res. Conf., St. Augustine, Trinidad, 8–18 Jan. 1972, pp. 570–581. Government of Trinidad and Tobago, W.I.

——. 1978a. Review of *Les Mirides du cacaoyer*, edited by E. M. Lavabre. Bull. Entomol. Soc. Am. 24(1):125–127.

——. 1978b. A Neotropical ant mosaic. Ann. Entomol. Soc. Am. 71:649–653.

——. 1979a. The species of *Dagbertus* (Hemiptera: Miridae) associated with avocado in Florida. Fla. Entomol. 62:376–379.

——. 1979b. The eversible rectal organ of certain Miridae (Hemiptera) and its function. Fla. Entomol. 62:409–411.

——. 1980a. The natural history of some West African insects (Parts 8–13). Entomol. Mon. Mag. 115:35–45.

——. 1980b. The natural history of some West African insects (Parts 17–20). Entomol. Mon. Mag. 116:225–240.

Leston, D. and D. G. Gibbs. 1968. A new deraeocorine (Hemiptera: Miridae) predacious on *Mesohomotoma tessmanni* (Aulmann) (Hemiptera: Psyllidae) on cocoa. Proc. R. Entomol. Soc. Lond. (B) 37:73–79.

——. 1971. Phenology of cocoa and some associated insects in Ghana. *In* Proc. III Int. Cocoa Res. Conf., Accra, Ghana, 23–29 Nov. 1969, pp. 197–204. Cocoa Res. Institute, Tafo, Ghana.

Leston, D. and G. G. E. Scudder. 1956. A key to the larvae of the families of British Hemiptera: Heteroptera. Entomologist 89:223–231.

Leuschner, K., S. L. Taneja, and H. C. Sharma. 1985. The role of host-plant resistance in pest management in sorghum in India. Insect Sci. Appl. 6:453–460.

Lever, R. J. A. W. 1941. Entomological notes. Agric. J. Fiji 12:77–80.

——. 1942. Pests of the vegetable garden and their control. Agric. J. Fiji 13: 109–115.

——. 1949. The tea mosquito bugs (*Helopeltis* spp.) in the Cameron Highlands. Malay. Agric. J. 32:91–109.

Levin, D. A. 1971. Plant phenolics: An ecological perspective. Am. Nat. 105: 157–181.

——. 1976. The chemical defenses of plants to pathogens and herbivores. Annu. Rev. Ecol. Syst. 7:121–159.

Levin, M. D., G. D. Butler Jr., and D. D. Rubis. 1967. Pollination of safflower by insects other than honey bees. J. Econ. Entomol. 60:1481–1482.

Lewis, C. T. and N. Waloff. 1964. The use of radioactive tracers in the study of dispersion of *Orthotylus virescens* (Douglas & Scott) (Miridae, Heteroptera). Entomol. Exp. Appl. 7:15–24.

Lewis, D. J. 1958. Hemiptera of medical interest in the Sudan Republic. Proc. R. Entomol. Soc. Lond. (A) 33:43–47.

Lewis, T. 1969. The diversity of the insect fauna in a hedgerow and neighbouring fields. J. Appl. Ecol. 6:453–458.

——. 1973. Thrips: Their biology, ecology

and economic importance. Academic, London. 349 pp.

——. 1987. Feeding of *Limothrips cerealium* (Hal.): The mouthparts, their action, and effects on plant tissue. Pages 439–447 in J. Holman, J. Pelikän, A. F. G. Dixon, and L. Weismann, eds., Population structure, genetics and taxonomy of aphids and Thysanoptera. SPB, The Hague.

Lewis, T. and L. R. Taylor. 1964. Diurnal periodicity of flight by insects. Trans. R. Entomol. Soc. Lond. 116:393–479.

Libby, J. L. 1968. Insect pests of Nigerian crops. Univ. Wis. Coll. Agric. Res. Div. Bull. 269:1–68.

Libutan, G. M. and E. N. Bernardo. 1995. The host preference of the capsid bug, *Cyrtopeltis tenius* [sic] Reuter (Hemiptera: Miridae). Philipp. Entomol. 9:567–586.

Lidell, M. C., G. A. Niles, and J. K. Walker. 1986. Response of nectariless cotton genotypes to cotton fleahopper (Heteroptera: Miridae) infestation. J. Econ. Entomol. 79:1372–1376.

Light, S. S. 1930. *Helopeltis* in Ceylon. Tea Q. 3:21–26.

Lightfoot, D. C. and W. G. Whitford. 1987. Variation in insect densities on desert creosotebush: Is nitrogen a factor? Ecology 68:547–557.

——. 1989. Interplant variation in creosotebush foliage characteristics and canopy arthropods. Oecologia (Berl.) 81:166–175.

Lilly, C. E. 1958. Observations on predation by the plant bug *Liocoris borealis* Kelton (Hemiptera: Miridae). Can. Entomol. 87:420–421.

Lilly, C. E. and G. A. Hobbs. 1956. Biology of the superb plant bug, *Adelphocoris superbus* (Uhl.) (Hemiptera: Miridae), in southern Alberta. Can. Entomol. 88:118–125.

——. 1962. Effects of spring burning and insecticides on the superb plant bug, *Adelphocoris superbus* Uhl.), and associated fauna in alfalfa seed fields. Can. J. Plant Sci. 42:53–61.

Lim, K. P. and R. K. Stewart. 1976a. Parasitism of the tarnished plant bug, *Lygus lineolaris* (Hemiptera: Miridae), by *Peristenus pallipes* and *P. pseudopallipes* (Hymenoptera: Braconidae). Can. Entomol. 108:601–608.

——. 1976b. Laboratory studies on *Peristenus pallipes* and *P. pseudopallipes* (Hymenoptera: Braconidae), parasitoids of the tarnished plant bug, *Lygus lineolaris* (Hemiptera: Miridae). Can. Entomol. 108:815–821.

Lim, T. K. and K. C. Khoo. 1990. Guava in Malaysia: Production, pests and diseases. Tropical Press, Kuala Lumpur, Malaysia. 260 pp.

Lim, T. K., R. Muhamad, G. F. Chung, and C. L. Chin. 1989. Studies on *Beauveria bassiana* isolated from the cocoa mirid, *Helopeltis theobromae*. Crop Prot. 8:358–362.

Lincoln, C. 1974. Use of economic thresholds and scouting as the basis for using parasites and predators in integrated control programs. Pages 182–189 in F. G. Maxwell and F. A. Harris, eds., Proceedings of the Summer Institute on Biological Control of Plant Insects and Diseases. University of Mississippi Press, Jackson.

Lincoln, C., G. Dean, B. A. Waddle, W. C. Yearian, J. R. Phillips, and L. Roberts. 1971. Resistance of frego-type cotton to boll weevil and bollworm. J. Econ. Entomol. 64:1326–1327.

Lincoln, R. J., G. A. Boxshall, and P. F. Clark. 1982. A dictionary of ecology, evolution and systematics. Cambridge University Press, Cambridge. 298 pp.

Lindig, O. H., P. A. Hedin, and W. E. Poe. 1981. Amino acids in pecan weevil, southwestern corn borer and tarnished plant bug, and at their feeding sites. Comp. Biochem. Physiol. 68A:261–263.

Lindquist, E. E. 1986. The world genera of Tarsonemidae (Acari: Heterostigmata): A morphological, phylogenetic, and systematic revision, with a reclassification of family-group taxa in the Heterostigmata. Mem. Entomol. Soc. Can. 136:1–517.

Lindquist, R. K. and E. L. Sorensen. 1970. Interrelationships among aphids, tarnished plant bugs, and alfalfas. J. Econ. Entomol. 63:192–195.

Lindroth, C. H., H. Andersson, H. Bödvarsson, and S. H. Richter. 1973. Surtsey, Iceland: The development of a new fauna, 1963–1970. Terrestrial invertebrates. Entomol. Scand. 5 (Suppl.):1–280.

Lindsey, A. H. 1984. Reproductive biology of Apiaceae. I. Floral visitors to *Thaspium* and *Zizia* and their importance to pollination. Am. J. Bot. 71:375–387.

Ling, Y. H., W. F. Campbell, B. A. Haws, and K. H. Asay. 1985. Scanning electron microscope (SEM) studies of morphology of range grasses in relation to feeding by *Labops hesperius*. Crop Sci. 25:327–332.

Linnavuori, R. [E.] 1951. Hemipterological observations. Ann. Entomol. Fenn. 17:51–65.

——. 1961. Hemiptera of Israel. II. Ann. Zool. Soc. Zool.-Bot. Fenn. "Vanamo" 22(7):1–51.

——. 1964. Hemiptera of Egypt, with remarks on some species of the adjacent Eremian region. Ann. Zool. Fenn. 1:306–356.

Linnavuori, R. E. 1984. New species of Hemiptera Heteroptera from Iraq and the adjacent countries. Acta Entomol. Fenn. 44:1–59.

Linskens, H. F. and J. M. L. Mulleneers. 1967. Formation of "instant pollen tubes." Acta Bot. Neerl. 16:132–142.

Linskens, H. F. and J. Schrauwen. 1969. The release of free amino acids from germinating pollen. Acta Bot. Neerl. 18:605–614.

Linsley, E. G. and J. W. MacSwain. 1947. Factors influencing the effectiveness of insect pollinators of alfalfa in California. J. Econ. Entomol. 40:349–357.

Lintner, J. A. 1882. Injurious hemipterous insects. *Poecilocapsus lineatus* (Fabr.). Pages 271–281 in First Annu. Rep. Injurious Insects State of N.Y. Weed, Parsons, Albany.

Lipe, J. A. and P. W. Morgan. 1972. Ethylene: Role in fruit abscission and dehiscence processes. Plant Physiol. (Bethesda) 50:759–764.

Lipetz, J. 1970. Wound-healing in higher plants. Int. Rev. Cytol. 27:1–28.

Lipscomb, D. L. 1985. The eukaryotic kingdoms. Cladistics 1:127–140.

——. 1991. Broad classification: The kingdoms and the protozoa. Pages 81–136 in J. P. Kreier and J. R. Baker, eds., Parasitic protozoa, 2nd ed. Vol. 1. Academic, San Diego.

Lipsey, R. L. 1970a. The hosts of *Neurocolpus nubilus* (Say), the clouded plant bug (Hemiptera: Miridae). Entomol. News 81:213–219.

——. 1970b. The life history of *Neurocolpus nubilus* (Say), the clouded plant bug (Hemiptera: Miridae). Entomol. News 81:257–262.

Liquido, N. J. and T. Nishida. 1983. Geographical distribution of *Cyrtorhinus* and *Tytthus* (Heteroptera: Miridae), egg predators of cicadellid and delphacid pests. FAO Plant Prot. Bull. 31:159–162.

——. 1985a. Population parameters of *Cyrtorhinus lividipennis* (Heteroptera: Miridae) reared on eggs of natural and factitious prey. Proc. Hawaii. Entomol. Soc. 25:87–93.

——. 1985b. Observations on some aspects of the biology of *Cyrtorhinus lividipennis* Reuter (Heteroptera: Miridae). Proc. Hawaii. Entomol. Soc. 25:95–101.

——. 1985c. Variation in number of instars, longevity, and fecundity of *Cyrtorhinus lividipennis* Reuter (Hemiptera: Miridae). Ann. Entomol. Soc. Am. 78:459–463.

Lis, J. A. 1992. An influence of industrial pollutions on communities of Heteroptera in selected plant associations in the zincwork "Miasteczko Slaskie" Region (Upper Silesia, Poland). Pages 645–647 in L. Zombori and L. Peregovits, eds., Proceedings of the 4th European Congress of Entomology, Gödöllő, 1991. Vol. 2. Hungarian Natural History Museum, Budapest.

Little, V. A. and D. F. Martin. 1942. Cotton insects of the United States. Burgess, Minneapolis, Minn. 130 pp.

Livingstone, D. 1968. On the morphology and bionomics of Tingis buddleiae Drake (Heteroptera: Tingidae). Part I—Bionomics. Agra Univ. J. Res. 17:1–16.

Llewellyn, M. 1982. The energy economy of fluid-feeding insects. Pages 243–251 in J. H. Visser and A. K. Minks, eds., Proc. 5th Int. Symp. Insect-Plant Relationships, Wageningen, the Netherlands, 1–4 March 1982. Centre for Agricultural Publishing & Documentation, Wageningen.

Llorens, J. M. and A. Garrido. 1992. Homóptera III. Moscas blancas y su control biológico. Pisa Ediciones, Valencia, Spain. 203 pp.

Lloyd, D. C. and R. Ahmad. 1971. Investigations on natural enemies of some pasture and alfalfa insects in Argentina. Commonw. Inst. Biol. Control Mimeo. Rep. 15 pp.

Lloyd, F. E. 1934. Is Roridula a carnivorous plant? Can. J. Res. 10:780–786.

——. 1942. The carnivorous plants. Chronica Botanica, Waltham, Mass. 352 pp.

Lloyd, N. C. 1969. The apple dimpling bug, Campylomma livida Reut. (Hemiptera: Miridae). Agric. Gaz. N.S.W. 80:582–584.

Loan, C. C. 1965. Life cycle and development of Leiophron pallipes Curtis (Hymenoptera: Braconidae, Euphorinae) in five mirid hosts in the Belleville district. Proc. Entomol. Soc. Ont. 95:115–121.

——. 1966. A new species of Leiophron Nees (Hymenoptera: Braconidae, Euphorinae) with observations on its biology and that of its host, Plagiognathus sp. (Heteroptera: Miridae). Ohio J. Sci. 66:89–94.

——. 1974. The North American species of Leiophron Nees, 1818 and Peristenus Foerster, 1862 (Hymenoptera: Braconidae, Euphorinae) including the description of 31 new species. Nat. Can. (Qué.) 101:821–860.

Loan, C. C. and C. H. Craig. 1976. Euphorine parasitism of Lygus spp. in alfalfa in western Canada (Hymenoptera: Braconidae; Heteroptera: Miridae). Nat. Can. (Qué.) 103:497–500.

Loan, C. C. and S. R. Shaw. 1987. Euphorine parasites of Lygus and Adelphocoris (Hymenoptera: Braconidae and Heteroptera: Miridae). Pages 69–75 in R. C. Hedlund and H. M. Graham, eds., Economic importance and biological control of Lygus and Adelphocoris in North America. U.S. Dep. Agric. Agric. Res. Serv. ARS-64.

Lockwood, J. A. 1989. Cannibalism in rangeland grasshoppers (Orthoptera: Acrididae): Attraction to cadavers. J. Kans. Entomol. Soc. 61:379–387.

——. 1993. Environmental issues involved in biological control of rangeland grasshoppers (Orthoptera: Acrididae) with exotic agents. Environ. Entomol. 22:503–518.

Lockwood, S. 1933. Insect and mite scars of California fruits. Calif. Dep. Agric. Mon. Bull. 22:319–345.

Lockwood, S. and E. T. Gammon. 1949. Incidence of insect pests. Pages 190–203 in H. M. Armitage. Bureau of Entomology [Report]. Calif. Dep. Agric. Bull. 38.

Logan, C. and T. H. Coaker. 1960. The transmission of bacterial blight of cotton (Xanthomonas) malvacearum (E. F. Smith) Dowson by the cotton bug, Lygus vosseleri Popp. Emp. Cotton Grow. Rev. 37:26–29.

Lord, F. T. 1956. The influence of spray programs on the fauna of apple orchards in Nova Scotia. IX. Studies on means of altering predator populations. Can. Entomol. 88:129–137.

——. 1971. Laboratory tests to compare the predatory value of six mirid species in each stage of development against the winter eggs of the European red mite, Panonychus ulmi (Acari: Tetranychidae). Can. Entomol. 103:1663–1669.

Lorenz, K. and P. Meredith. 1988. Insect-damaged wheat—effects on starch characteristics. Starch Stärke 40: 136–139.

Losey, J. E. and R. F. Denno. 1998. Positive predator-predator interactions: Enhanced predation rates and synergistic suppression of aphid populations. Ecology 79:2143–2152.

Lotodé, R. 1977. Distribution spatiale des Mirides et étude comparative clonale de l'attractivité. Pages 187–202 in E. M. Lavabre, ed., Les Mirides du cacaoyer. Institut français du Café et du Cacao, Paris.

Loudon, C. 1995. Insect morphology above the molecular level: Biomechanics. Ann. Entomol. Soc. Am. 88:1–4.

Lovett, A. L. and B. B. Fulton. 1920. Fruit grower's handbook of apple and pear insects, rev. 1922. Oreg. Agric. Coll. Exp. Stn. Circ. 22:1–72.

Lucas, H. 1888. Note sur le parasitisme du Myobia pumila, Diptère de la tribu des Tachinaire. Ann. Soc. Entomol. Fr. (6) 8:102–104.

Luck, R. F. 1984. Principles of arthropod predation. Pages 497–529 in C. B. Huffaker and R. L. Rabb, eds., Ecological entomology. Wiley, New York.

Luck, R. F., P. S. Messenger, and J. F. Barbieri. 1981. The influence of hyperparasitism on the performance of biological control agents. Pages 34–42 in D. Rosen, ed., The role of hyperparasitism in biological control: A symposium. Univ. Calif. Div. Agric. Sci., Berkeley.

Luff, M. L. 1983. The potential of predators for pest control. Agric. Ecosyst. Environ. 10:159–181.

Lugger, O. 1900. Bugs (Hemiptera) injurious to our cultivated plants. Minn. Agric. Exp. Stn. Bull. 69:1–259.

Lukefahr, M. J. 1981. [Notes and news, Brazil.] Trop. Grain Legume Bull. 23:29.

Lukefahr, M. J. and J. E. Houghtaling. 1975. High gossypol cottons as a source of resistance to the cotton fleahopper. In Proc. Beltwide Cotton Prod. Res. Conf., Jan. 6–8, 1975, New Orleans, La., pp. 93–94. National Cotton Council, Memphis.

Lukefahr, M. J., L. W. Noble, and J. E. Houghtaling. 1966. Growth and infestation of bollworms and other insects on glanded and glandless strains of cotton. J. Econ. Entomol. 59:817–820.

Lukefahr, M. J., C. B. Cowan Jr., L. A. Bariola, and J. E. Houghtaling. 1968. Cotton strains resistant to the cotton fleahopper. J. Econ. Entomol. 61:661–664.

Lukefahr, M. J., C. B. Cowan Jr., and J. E. Houghtaling. 1970. Field evaluations of improved cotton strains resistant to the cotton fleahopper. J. Econ. Entomol. 63:1101–1103.

Lukefahr, M. J., J. E. Jones, and J. E. Houghtaling. 1976. Fleahopper and leafhopper populations and agronomic evaluations of glabrous cottons from different genetic sources. In Proc. Beltwide Cotton Prod. Res. Conf., Jan. 5–7, 1976, Las Vegas, Nev., pp. 84–86. National Cotton Council, Memphis.

Lupo, V. 1946. Invasion of Calocoris norvegicus (Gml.) in the communes of the Vesuvius region. Int. Bull. Plant Prot. 20:105M–108M.

Luttrell, R. G. 1994. Cotton pest management: Part 2. A US perspective. Annu. Rev. Entomol. 39:527–542.

Lyle, C. 1936. A mirid (Lopidea robiniae Uhler). U.S. Dep. Agric. Insect Pest Surv. Bull. 16:182.

Lyon, J. P. 1986. Use of aphidophagous

and polyphagous beneficial insects for biological control of aphids in greenhouse. Pages 471–474 in I. Hodek, ed., Ecology of aphidophaga. Proc. 2nd Symp. at Zvíkovské Podhradí, Sept. 2–8, 1984. Junk, Dordrecht.

Ma, R., J. C. Reese, W. C. Black IV, and P. Bramel-Cox. 1990. Detection of pectinesterase and polygalacturonase from salivary secretions of living greenbugs, *Schizaphis graminum* (Homoptera: Aphidiidae). J. Insect Physiol. 36:507–512.

Ma, W. K. and S. B. Ramaswamy. 1987. Histological changes during ovarian maturation in the tarnished plant bug, *Lygus lineolaris* (Palisot de Beauvois) (Hemiptera: Miridae). Int. J. Morphol. Embryol. 16:309–322.

———. 1990. Histochemistry of yolk formation in the ovaries of the tarnished plant bug, *Lygus lineolaris* (Palisot de Beauvois) (Hemiptera: Miridae). Zool. Sci. (Tokyo) 7:147–151.

Maas, J. L., ed. 1984. Compendium of strawberry diseases. American Phytopathological Society, St. Paul, Minn. 138 pp.

Mabberley, D. J. 1987. The plant-book: A portable dictionary of the higher plants. Cambridge University Press, Cambridge. 706 pp.

MacCollom, G. B. 1967. Control of Miridae spp. in birdsfoot trefoil seed fields. J. Econ. Entomol. 60:1116–1118.

MacCreary, D. 1965. Flight range observations on *Lygus lineolarius* [sic] and certain other Hemiptera. J. Econ. Entomol. 58:1004–1005.

MacCreary, D. and L. R. Detjen. [1947?]. A southern insect of interest to Delaware tomato growers. Trans. Peninsula Hortic. Soc. 1946:94–95.

MacFarlane, J. 1989. The hemipterous insects and spiders of sorghum panicles in northern Nigeria. Insect Sci. Appl. 10:277–284.

Macfarlane, R. P. and A. M. Ferguson. 1984. Kiwifruit pollination: A survey of the insect pollinators in New Zealand. *In* Proc. 5th Symposium International sur la Pollinisation, Versailles, 27–30 Sept. 1983, pp. 367–373. Colloq. INRA no. 21.

Macfarlane, R. P., J. A. Wightman, R. P. Griffin, and D. N. J. Whitford. 1981. Hemiptera and other insects on South Island lucerne and lotus seed crops 1980–81. *In* Proc. 34th N.Z. Weed Pest Control Conf., pp. 39–42. New Zealand Weed and Pest Control Society, Palmerston North.

MacGowan, J. B., ed. 1988. A plant bug, *Cyrtocapsus caligineus* (Stal). Triology (Fla. Dep. Agric. Consumer Serv. Bur. Plant Ind.) 27(9):5.

Macias, W. and G. I. Mink. 1969. Preference of green peach aphids for virus-infected sugarbeet leaves. J. Econ. Entomol. 62:28–29.

Maclagan, D. S. 1932. An ecological study of the "lucerne flea" (*Smynthurus* [sic] *viridis*, Linn.)—I. Bull. Entomol. Res. 23:101–145.

MacLellan, C. R. 1962. Mortality of codling moth eggs and young larvae in an integrated control orchard. Can. Entomol. 94:655–666.

———. 1963. Predator populations and predation on the codling moth in an integrated control orchard—1961. Mem. Entomol. Soc. Can. 32:41–54.

———. 1972. Codling moth populations under natural, integrated, and chemical control on apple in Nova Scotia (Lepidoptera: Olethreutidae). Can. Entomol. 104:1397–1404.

———. 1973. Natural enemies of the light brown apple moth, *Epiphyas postvittana*, in the Australian Capital Territory. Can. Entomol. 105:681–700.

———. 1979. Pest damage and insect fauna of Nova Scotia apple orchards. Can. Entomol. 111:985–1004.

MacLeod, G. F. 1933. Some examples of varietal resistance of plants to insect attacks. J. Econ. Entomol. 26:62–67.

MacLeod, G. F. and L. R. Jeppson. 1942. Some quantitative studies of *Lygus* injury to alfalfa plants. J. Econ. Entomol. 35:604–605.

MacLeod, N. J. and J. B. Pridham. 1965. Observations on the translocation of phenolic compounds. Phytochemistry 5:777–781.

MacNay, C. G. 1952. Summary of important insect infestations, occurrences, and damage in Canada in 1952. Annu. Rep. Entomol. Soc. Ont. 83:66–94.

MacPhee, A. W. 1975. Integrated control in orchards in Canada. Pages 125–133 *in* Integrated control in orchards [Proceedings of the Fifth Symposium on Integrated Control in Orchards, Bolzano, 3–7 Sept. 1974]. PUDOC, Wageningen.

———. 1976. Predictions of destructive levels of the apple-stinging bugs *Atractotomus mali* and *Campylomma verbasci* (Hemiptera: Miridae). Can. Entomol. 108:423–426.

———. 1979. Observations on the white apple leafhopper, *Typhlocyba pomaria* (Hemiptera: Cicadellidae), and on the mirid predator *Blepharidopterus angulatus*, and measurements of their cold-hardiness. Can. Entomol. 111:487–490.

MacPhee, A. W. and C. R. MacLellan. 1971. Cases of naturally-occurring biological control in Canada. Pages 312–328 *in* C. B. Huffaker, ed., Biological control: Proceedings of an AAAS Symposium on Biological Control, held at Boston, Massachusetts, December 30–31, 1969. Plenum, New York.

———. 1972. Ecology of apple orchard fauna and development of integrated pest control in Nova Scotia. Proc. Tall Timbers Conf. Ecol. Anim. Control Habitat Manage. 3:197–208.

MacPhee, A. W. and K. H. Sanford. 1954. The influence of spray programs on the fauna of apple orchards in Nova Scotia. VII. Effects on some beneficial arthropods. Can. Entomol. 86:128–135.

———. 1956. The influence of spray programs on the fauna of apple orchards in Nova Scotia. X. Supplement to VII. Effects on some beneficial arthropods. Can. Entomol. 88:631–634.

———. 1961. The influence of spray programs on the fauna of apple orchards in Nova Scotia. XII. Second supplement to VII. Effects on beneficial arthropods. Can. Entomol. 93:671–673.

Madden, A. H. and F. S. Chamberlin. 1954. Biology of the tobacco hornworm in the southern cigar-tobacco district. U.S. Dep. Agric. Tech. Bull. 896:1–51.

Maddox, G. D. and R. B. Root. 1987. Resistance to 16 diverse species of herbivorous insects within a population of goldenrod, *Solidago altissima*: Genetic variation and heritability. Oecologia (Berl.) 72:8–14.

Madelin, M. F. 1966. Fungal parasites of insects. Annu. Rev. Entomol. 11:423–448.

Madge, D. S. 1968. The behaviour of the cocoa mirid (*Sahlbergella singularis* Hagl.) to some environmental factors. Bull. Entomol. Soc. Niger. 1:63–70.

Madhusudhan, V. V., G. S. Taylor, and P. W. Miles. 1994. The detection of salivary enzymes of phytophagous Hemiptera: A compilation of methods. Ann. Appl. Biol. 124:405–412.

Madsen, H. F. and B. J. Madsen. 1982. Populations of beneficial and pest arthropods in an organic and a pesticide treated apple orchard in British Columbia. Can. Entomol. 114:1083–1088.

Madsen, H. F., H. F. Peters, and J. M. Vakenti. 1975. Pest management: Experience in six British Columbia apple orchards. Can. Entomol. 107:873–877.

Magnarelli, L. A., J. F. Anderson, and J. H. Thorne. 1979. Diurnal nectar-feeding of salt marsh Tabanidae (Diptera). Environ. Entomol. 8:544–548.

Mailloux, G. and N. J. Bostanian. 1988. Economic injury level model for tarnished plant bug, *Lygus lineolaris* (Palisot de Beauvois) (Hemiptera: Miridae), in strawberry fields. Environ. Entomol. 17:581–586.

———. 1989. Presence-absence sequential decision plans for management of

Lygus lineolaris (Hemiptera: Miridae) on strawberry. Environ. Entomol. 18:829–834.

——. 1991. The phenological development of strawberry plants and its relation to tarnished plant bug seasonal abundance. Adv. Strawberry Prod. 10: 30–36.

Mailloux, G., R. O. Paradis, and J.-G. Pilon. 1979. Développement saissonier de la punaise terne, *Lygus lineolaris* (P. de B.) (Hemiptera: Miridae), sur fraisiers, framboisiers et pommiers dans le sud-ouest du Québec. Ann. Soc. Entomol. Qué. 24:48–64.

Majer, J. D. 1972. The ant mosaic in Ghana cocoa farms. Bull. Entomol. Res. 62:151–160.

——. 1993. Comparison of the arboreal ant mosaic in Ghana, Brazil, Papua New Guinea and Australia—its structure and influence on arthropod diversity. Pages 115–141 in J. LaSalle and I. D. Gault, eds., Hymenoptera and biodiversity. CAB International, Wallingford, UK.

Maki, M. 1918. On the "Kattsau Kua" of the Pescadores. Konch. Sek. [Insect World], Gifu 20:1–8. [in Japanese.]

Malausa, J. C. 1987. Sur l'utilisation de mirides prédateurs (Hétéroptères) dans la lutte biologique contre les ravageurs des cultures maraîchères sous serre. In Proc. CEC/IOBC Experts' Group Meeting, Cabrils, 27–29 May 1987, pp. 63–66. Balkema, Rotterdam.

——. 1989. Sur l'utilisation de mirides prédateurs (Hétéroptères) dans la lutte biologique contre les ravageurs des cultures maraîchères sous serre. Pages 63–66 in R. Cavalloro and C. Pelerents, eds., Integrated pest management in protected vegetable crops. Balkema, Rotterdam.

Malausa, J. C. and Y. Trottin-Caudal. 1996. Advances in the strategy of use of the predaceous bug *Macrolophus caliginosus* (Heteroptera: Miridae) in glasshouse crops. Pages 178–189 in O. Alomar and R. N. Wiedenmann, eds., Zoophytophagous Heteroptera: Implications for life history and integrated pest management. Thomas Say Publ. Entomol.: Proceedings. Entomological Society of America, Lanham, Md.

Malausa, J. C., J. Drescher, and E. Franco. 1987. Perspectives for the use of a predaceous bug *Macrolophus caliginosus* Wagner (Heteroptera, Miridae) on glasshouse crops. In Proc. Working Group, "Integrated Control in Glasshouses," EPRS/WPRS, Budapest, Hungary, 26–30 April 1987. Int. Organ. Biol. Control/West Palearctic Reg. Sect. Bull. 10(2):106–107.

Maldonado Capriles, J. 1969. The Miridae

of Puerto Rico (Insecta, Hemiptera). P.R. Agric. Exp. Stn. Tech. Pap. 45:1–133.

——. 1986. Concerning Cuban Miridae (Insecta: Hemiptera). Caribb. J. Sci. 22:125–136.

Malechek, J. C., A. M. Gray, and B. A. Haws. 1977. Yield and nutritional quality of intermediate wheatgrass infested by black grass bugs at low population densities. J. Range Manage. 30:128–131.

Malézieux, S., C. Giradet, B. Navez, and J.-M. Cheyrias. 1995. Contre l'Aleurode des serres en cultures de tomates sous abris: Utilisation et développement de *Macrolophus caliginosus* associé a *Encarsia formosa*. Phytoma No. 471:29–32.

Malipatil, M. B. 1992. Revision of Australian *Campylomma* Reuter (Hemiptera: Miridae: Phylinae). J. Aust. Entomol. Soc. 31:357–368.

Malipatil, M. B. and G. Cassis. 1997. Taxonomic review of *Creontiades* Distant in Australia (Hemiptera: Miridae: Mirinae). Aust. J. Entomol. 36:1–13.

Mally, F. W. 1893. Report on the boll worm of cotton (*Heliothis armigera* Hübn.). U.S. Dep. Agric. Div. Entomol. Bull. 29:1–73.

Maltais, J. B. 1938. [Field crop and garden insects.] Can. Insect Pest Rev. 16:267.

Mangel, M., S. E. Stefanou, and J. E. Wilen. 1985. Modeling *Lygus hesperus* injury to cotton yields. J. Econ. Entomol. 78:1009–1014.

Manglitz, G. R. and R. H. Ratcliffe. 1988. Insects and mites. Pages 671–704 in A. A. Hanson, D. K. Barnes, and R. R. Hill Jr., eds., Alfalfa and alfalfa improvement. Agron. Monogr. 29.

Mani, M. S. 1962. Introduction to high altitude; insect life above the timberline in the north-west Himalaya. Methuen, London. 302 pp.

——. 1964. Ecology of plant galls. Junk, The Hague. 434 pp.

Manjunath, T. M., P. S. Rai, and G. Gowada. 1978. Natural enemies of brown planthoppers and green leafhopper in India. Int. Rice Res. Newsl. 3:11–12.

Mann, J. 1969. Cactus-feeding insects and mites. U.S. Natl. Mus. Bull. 256:1–158.

Mann, J. E., S. G. Turnipseed, M. J. Sullivan, P. H. Adler, J. A. Durant, and O. L. May. 1997. Effects of early-season loss of flower buds on yield, quality, and maturity of cotton in South Carolina. J. Econ. Entomol. 90:1324–1331.

Manns, T. F. 1942. Peach yellows and little peach. Del. Agric. Exp. Stn. Bull. 236:1–50.

Manti, I. 1989. The role of *Cyrtorhinus*

lividipennis Reuter (Hemiptera, Miridae) as a major predator of the brown planthopper *Nilaparvata lugens* Stål (Homoptera: Delphacidae). Ph.D. dissertation, University of the Philippines, Los Baños. 126 pp.

Maramorosch, K. 1963. Arthropod transmission of plant viruses. Annu. Rev. Entomol. 8:369–414.

March, R. G. del Valle y. 1972. Contribuição para o conhecimento das pragas do ricino em Moçambique. Agron. Moçambicana 6:157–175.

Marchal, P. 1929. Les ennemis du puceron lanigère, conditions biologiques et cosmiques de sa multiplication. Traitements. Ann. Epiphyt. (Paris) 15:125–181.

Marchart, H. 1968. Radiotracer study of the predators on *Distantiella theobroma* (Distant) (Hemiptera: Miridae). Pages 3–14 in Isotopes and radiation in entomology (Proceedings series). International Atomic Energy Agency, Vienna.

——. 1971. Ants, capsids and swollen shoot: A reply to Leston. In Proc. III Int. Cocoa Res. Conf., Accra, Ghana, 23–29 Nov. 1969, pp. 235–236. Cocoa Research Institute, Tafo, Ghana.

——. 1972. Effect of capsid attack on pod development. Cocoa Res. Inst. (Ghana) Rep. 1969–70:99.

Mardzhanyan, G. M. and A. K. Ust'yan. 1965. Integrated control of the green peach aphid, *Myzodes persicae* Sulz., on tobacco. Entomol. Obozr. 44(4):750–761. [in Russian; English translation in Entomol. Rev. 44(4):441–448.]

Maredia, K. M., B. A. Waddle, and N. P. Tugwell. 1993. Evaluation of rolled (frego) bract cottons for tarnished plant bug and boll weevil resistance. Southwest. Entomol. 18:219–227.

Marín, R. and J. Sarmiento. 1979. Biologia y comportamiento de *Orthotylellus carmelitanus* (Carvalho), Hemiptera: Miridae. Rev. Peru. Entomol. 22:43–47.

Markham, P. G. and G. N. Oldfield. 1983. Transmission techniques with vectors of plant and insect mycoplasmas and spiroplasmas. Pages 261–267 in J. G. Tully and S. Razin, eds., Methods in mycoplasmology. Vol. II. Diagnostic mycoplasmology. Academic, San Diego.

Markham, R. and K. M. Smith. 1949. Studies on the virus of turnip yellow mosaic. Parasitology 39:330–342.

Marloth, R. 1903. Some recent observations on the biology of *Roridula*. Ann. Bot. (Lond.) 17:151–157.

——. 1910. Further observations on the biology of *Roridula*. Trans. R. Soc. S. Afr. 2:59–61.

Marples, T. G. 1966. A radionuclide tracer study of arthropod food chains in a *Spartina* salt marsh ecosystem. Ecology 47:270–277.

Martignoni, M. E. and P. J. Iwai. 1977. A catalog of viral diseases of insects and mites, 2nd ed. U.S. Dep. Agric. For. Serv. Gen. Tech. Rep. PNW-40:1–28.

Martin, J. E. H. 1977. The insects and arachnids of Canada. Part I. Collecting, preparing, and preserving insects, mites, and spiders. Agric. Can. Publ. 1643:1–182.

Martin, J. L. 1966. The insect ecology of red pine plantations in central Ontario IV. The crown fauna. Can. Entomol. 98: 10–27.

Martin, M. M. 1979. Biochemical implications of insect mycophagy. Biol. Rev. Camb. Philos. Soc. 54:1–21.

Martin, M. M., J. J. Kukor, J. S. Martin, T. E. O'Toole, and M. W. Johnson. 1981. Digestive enzymes of fungus-feeding beetles. Physiol. Zool. 54:137–145.

Martin, P. J., C. P. Topper, R. A. Bashiru, F. Boma, D. De Waal, H. C. Harries, L. J. Kasuga, N. Katanila, L. P. Kikoka, R. Lamboll, A. C. Maddison, A. E. Majule, P. A. Masawe, K. J. Millanzi, N. Q. Nathaniels, S. H. Shomari, M. E. Sijaona and T. Stathers. 1997. Cashew nut production in Tanzania: Constraints and progress through integrated crop management. Crop Prot. 16:5–14.

Martin, W. R. Jr., M. P. Grisham, C. M. Kenerley, W. L. Sterling, and P. W. Morgan. 1987. Microorganisms associated with cotton fleahopper, *Pseudatomoscelis seriatus* (Heteroptera: Miridae). Ann. Entomol. Soc. Am. 80:251–255.

Martin, W. R. Jr., W. L. Sterling, C. M. Kenerley, and P. W. Morgan. 1988a. Transmission of bacterial blight of cotton, *Xanthomonas campestris* pv. *malvacearum*, by feeding of the cotton fleahopper: Implications for stress ethylene-induced square loss in cotton. J. Entomol. Sci. 23:161–168.

Martin, W. R. Jr., P. W. Morgan, W. L. Sterling, and R. W. Meola. 1988b. Stimulation of ethylene production in cotton by salivary enzymes of the cotton fleahopper (Heteroptera: Miridae). Environ. Entomol. 17:930–935.

Martinson, T., D. Bernard, G. English-Loeb, and T. Taft Jr. 1998. Impact of *Taedia scrupeus* [sic] (Hemiptera: Miridae) feeding on cluster development in Concord grapes. J. Econ. Entomol. 91:507–511.

Maschwitz, U., B. Fiala, and W. R. Dolling. 1987. New trophobiotic symbioses of ants with South East Asian bugs. J. Nat. Hist. 21:1097–1107.

Masner, P. 1965. The structure and function of gonads of alfalfa plant bug—*Adelphocoris lineolatus* (Goeze) (Heteroptera, Miridae) affected by food. *In* Proc. 12th Int. Congr. Entomol., London (1964), p. 175.

Massee, A. M. 1928. Note on apple capsids. Gard. Chron. (Lond.) (3) 83:399–400.

——. 1937. The pests of fruits and hops. Crosby Lockwood, London. 294 pp.

——. 1942a. Some important pests of the hop. Ann. Appl. Biol. 29:324–326.

——. 1942b. Notes on some interesting insects observed in 1941. East Malling Res. Stn. Rep. 1941:47–51.

——. 1944. Notes on some interesting insects observed in 1943. East Malling Res. Stn. Rep. 1943:58–65.

——. 1949. Notes on some interesting insects observed in 1948. East Malling Res. Stn. Rep. 1948:102–107.

——. 1952. Transmission of reversion of black currants. East Malling Res. Stn. Rep. 1951:162–165.

——. 1955. Entomology. East Malling Res. Stn. Rep. 1954:38–40.

——. 1956. Hemiptera-Heteroptera associated with fruits and hops. J. Soc. Brit. Entomol. 5:179–186.

——. 1959. *Campyloneura virgula* (H.-S.) (Hem., Miridae) found in association with the beetle *Malthinus flaveolus* (Payk.) (Col., Cantharidae). Entomol. Mon. Mag. 95:240.

Massee, A. M. and W. Steer. 1928. Capsid bugs. Gard. Chron. (Lond.) (3) 84:154–155.

——. 1929. The hatching of *Calocoris norvegicus* Gmel. Entomol. Mon. Mag. 65:160.

Mathen, K. and C. Kurian. 1972. Description, life-history and habits of *Stethoconus praefectus* (Distant) (Heteroptera: Miridae) predacious on *Stephanitis typicus* Distant (Heteroptera: Tingidae), a pest of coconut palm. Indian J. Agric. Sci. 42:255–262.

Mathen, K., B. Sathimma, and C. Kurian. 1967. Record of *Appolodotus praefectus* Distant (Heteroptera: Miridae), predacious on *Stephanitis typicus* Distant (Heteroptera: Tingidae), a pest of coconut palm. Curr. Sci. (Bangalore) 36:52.

Mathew, A. P. 1935. Transformational deceptive resemblance as seen in the life history of a plant bug (*Riptorus pedestris*), and of a mantis (*Evantissa pulchra*). J. Bombay Nat. Hist. Soc. 37:803–813.

Matrangolo, W. J. R. and J. M. Waquil. 1991. Biologia de *Paramixia carmelitana* (Carvalho, 1948) (Hemiptera: Miridae). An. Soc. Entomol. Bras. 20: 299–306.

Matsuda, R. 1976. Morphology and evolution of the insect abdomen. Pergamon, Oxford. 532 pp.

Matsumoto, B. M. 1964. Predator-prey relationships between the predator, *Cyrtorhinus fulvus* Knight, and the prey, the taro leafhopper, *Tarophagus proserpina* (Kirkaldy). M.S. thesis, University of Hawaii, Honolulu. 70 pp.

Matsumoto, B. M. and T. Nishida. 1966. Predator-prey investigations on the taro leafhopper and its egg predator. Hawaii Agric. Exp. Stn. Tech. Bull. 64:1–32.

Matthews, E. G. 1976. Insect ecology. University of Queensland Press, St. Lucia. 226 pp.

Matthews, G. A. 1989. Cotton insect pests and their management. Longman Scientific & Technical, Essex, UK.

Matthews, R. E. F. 1991. Plant virology, 3rd ed. Academic, San Diego. 835 pp.

Mattson, W. J. [Jr.] 1975. Abundance of insects inhabiting the male strobili of red pine. Gt. Lakes Entomol. 8:237–239.

Mattson, W. J. Jr. 1980. Herbivory in relation to plant nitrogen. Annu. Rev. Ecol. Syst. 11:119–161.

Mattson, W. J. [Jr.] and R. A. Haack. 1987. The role of drought in outbreaks of plant-eating insects. BioScience 37: 110–118.

Mattson, W. J. [Jr.] and J. M. Scriber. 1987. Nutritional ecology of insect folivores of woody plants: Nitrogen, water, fiber, and mineral considerations. Pages 105–146 *in* F. Slansky Jr. and J. G. Rodriguez, eds., Nutritional ecology of insects, mites, spiders, and related invertebrates. Wiley, New York.

Mau, R. F. L. and K. Nishijima. 1989. Development of the transparentwinged plant bug, *Hyalopeplus pellucidus* (Stål), a pest of cultivated guava in Hawaii. Proc. Hawaii. Entomol. Soc. 29:139–147.

Mauney, J. R. and T. J. Henneberry. 1979. Identification of damage symptoms and patterns of feeding of plant bugs in cotton. J. Econ. Entomol. 72:496–501.

——. 1984. Causes of square abscission in cotton in Arizona. Crop Sci. 24:1027–1030.

Maw, M. G. 1976. An annotated list of insects associated with Canada thistle (*Cirsium arvense*) in Canada. Can. Entomol. 108:235–244.

Maxson, A. C. 1920. Principal insect enemies of the sugar beet in the territories served by The Great Western Sugar Company. Agric. Dep. Great Western Sugar, Denver. 157 pp.

Maxwell, F. G. 1977. Plant resistance to cotton insects. Bull. Entomol. Soc. Am. 23:199–203.

Maxwell, F. G., M. F. Schuster, W. R. Meredith, and M. L. Laster. 1976. Influence of the nectariless character in cotton on harmful and beneficial insects. Pages 157–161 *in* T. Jermy, ed., The host-plant in relation to insect behaviour and reproduction. Plenum, New York.

Maxwell-Lefroy, H. and F. M. Howlett. 1909. Indian insect life. A manual of the insects of the plains (tropical India). Thacker, Spink, Calcutta; W. Thacker, London. 786 pp.

McAlpine, J. F. 1965. Observations on anthophilous Diptera at Lake Hazen, Ellesmere Island. Can. Field-Nat. 79: 247–252.

McAtee, W. L. 1915. Psyllidae wintering on conifers about Washington, D.C. Science (Wash., D.C.) 41:940.

——. 1916. Key to the Nearctic species of *Paracalocoris* (Heteroptera, Miridae). Ann. Entomol. Soc. Am. 9:366–390.

——. 1924. Mullen rosettes as winter shelters for insects. J. Econ. Entomol. 17:414–415.

McAtee, W. L. and J. R. Malloch. 1924. Some annectant bugs of the superfamily Cimicoideae (Heteroptera). Bull. Brooklyn Entomol. Soc. 19:69–83.

McBrien, H. L., G. J. R. Judd, and J. H. Borden. 1994. *Campylomma verbasci* (Heteroptera: Miridae): Pheromone-based seasonal flight patterns and prediction of nymphal densities in apple orchards. J. Econ. Entomol. 87:1224–1229.

——. 1996. Potential for pheromone-based mating disruption of the mullein bug, *Campylomma verbasci* (Meyer) (Heteroptera: Miridae). Can. Entomol. 128:1057–1064.

——. 1997. Population suppression of *Campylomma verbasci* (Heteroptera: Miridae) by atmospheric permeation with synthetic sex pheromone. J. Econ. Entomol. 90:801–808.

McCaffrey, J. P. and R. L. Horsburgh. 1980. The egg and oviposition site of *Deraeocoris nebulosus* (Hemiptera: Miridae) on apple trees. Can. Entomol. 112:527–528.

——. 1986. Biology of *Orius insidiosus* (Heteroptera: Anthocoridae): A predator in Virginia apple orchards. Environ. Entomol. 15:984–988.

McClintock, J. A. 1931. Cross-inoculation experiments with Erigeron yellows and peach rosette. Phytopathology 21:373–386.

McClintock, J. A. and L. B. Smith. 1918. True nature of spinach-blight and relation of insects to its transmission. J. Agric. Res. (Wash., D.C.) 14:1–59.

McClure, M. S. 1979. Spatial and seasonal distribution of disseminating stages of *Fiorinia externa* (Homoptera: Diaspididae) and natural enemies in a hemlock forest. Environ. Entomol. 8:869–873.

——. 1980. Foliar nitrogen: A basis for host suitability for elongate hemlock scale, *Fiorinia externa* (Homoptera: Diaspididae). Ecology 61:72–79.

——. 1981. Effects of voltinism, interspecific competition and parasitism on the population dynamics of the hemlock scales *Fiorinia externa* and *Tsugaspidiotus tsugae* (Homoptera: Diaspididae). Ecol. Entomol. 6:47–54.

McCoy, R. E. (chairman), A. Caudwell, C. J. Chang, T. A. Chen, L. N. Chiykowski, M. T. Cousin, J. L. Dale, G. T. N. de Leeuw, D. A. Golino, K. J. Hackett, B. C. Kirkpatrick, R. Marwitz, H. Petzold, R. C. Sinha, M. Sugiura, R. F. Whitcomb, I. L. Yang, B. M. Zhu, and E. Seemüller. 1989. Plant diseases associated with mycoplasma-like organisms. Pages 545–560 *in* R. F. Whitcomb and J. G. Tully, eds., The mycoplasmas. Vol. V. Spiroplasmas, acholeplasmas, and mycoplasmas of plants and arthropods. Academic, San Diego.

McDaniel, E. I. 1931. Insect and allied pests of plants grown under glass. Mich. Agric. Exp. Stn. Spec. Bull. 214:1–117.

McDaniel, S. G. and W. L. Sterling. 1979. Predator determination and efficiency on *Heliothis virescens* eggs in cotton using $^{32}P^2$. Environ. Entomol. 8:1083–1087.

——. 1982. Predation of *Heliothis virescens* (F.) eggs on cotton in east Texas. Environ. Entomol. 11:60–66.

McDermid, E. M. 1956. The insect pests of cotton in Zande district of Equatoria Province, Sudan 2. The insect pests, their status and control. Emp. Cotton Grow. Rev. 33:44–66.

McEwen, F. L. and G. E. R. Hervey. 1960. The effect of lygus bug control on the yield of lima beans. J. Econ. Entomol. 53:513–516.

McGarr, R. L. 1933. Damage to the cotton plant caused by *Megalopsallus atriplicis* Kngt. and other species of Miridae. J. Econ. Entomol. 26:953–956.

McGavin, G. C. 1979. A taxonomic and phylogenetic study of the immature stages of British Miridae (Hemiptera-Heteroptera). Ph.D. dissertation, University of London, London. 471 pp.

——. 1982. A new genus of Miridae (Hem.:Heteroptera). Entomol. Mon. Mag. 118:79–86.

——. 1993. Bugs of the world. Facts on File, New York. 192 pp.

McGiffen, K. C. and H. H. Neunzig. 1985. A guide to the identification and biology of insects feeding on muscadine and bunch grapes in North Carolina. N.C. Agric. Res. Serv. Bull. 470:1–93.

McGregor, E. A. 1927. *Lygus elisus*: A pest of the cotton regions in Arizona and California. U.S. Dep. Agric. Tech. Bull. 4:1–15.

——. 1961. Early cotton insects of the Imperial Valley. Bull. South. Calif. Acad. Sci. 60(1):47–55.

McGregor, E. A. and F. L. McDunough. 1917. The red spider on cotton. U.S. Dep. Agric. Bur. Entomol. Bull. 416: 1–72.

McGregor, M. D. 1970. Biological observations on the life history and habits of *Choristoneura lambertiana* (Lepidoptera: Tortricidae) on lodgepole pine in southeastern Idaho and western Montana. Can. Entomol. 102:1201–1208.

McGregor, R. R., D. R. Gillespie, D. M. J. Quiring, and M. R. J. Foisy. 1999. Potential use of *Dicyphus hesperus* Knight (Heteroptera: Miridae) for biological control of pests of greenhouse tomatoes. Biol. Control 16:104–110.

McGregor, R. R., D. R. Gillespie, C. G. Park, D. M. J. Quiring, and M. R. J. Foisy. 2000. Leaves or fruit? The potential for damage to tomato fruits by the omnivorous predator, *Dicyphus hesperus*. Entomol. Exp. Appl. 95:325–328.

McGregor, S. E. 1976. Insect pollination of cultivated crop plants. U.S. Dep. Agric. Agric. Handb. 496:1–411.

McGregor, W. S. 1942. *Orius insidiosus*, a predator on cotton insects in western Texas. J. Econ. Entomol. 35:454–455.

McIver, J. D. 1987. On the myrmecomorph *Coquillettia insignis* Uhler (Hemiptera: Miridae): Arthropod predators as operators in an ant-mimetic system. Zool. J. Linn. Soc. 90:133–144.

——. 1989. Protective resemblance in a community of lupine arthropods. Natl. Geogr. Res. 5:191–204.

McIver, J. D. and A. Asquith. 1989. Biology of *Lopidea nigridea* Uhler, a possible aposematic plant bug (Heteroptera: Miridae: Orthotylinae). J. N.Y. Entomol. Soc. 97:417–429.

McIver, J. D. and J. D. Lattin. 1990. Evidence for aposematism in the plant bug *Lopidea nigridea* Uhler (Hemiptera: Miridae: Orthotylinae). Biol. J. Linn. Soc. 40:99–112.

McIver, J. D. and G. M. Stonedahl. 1987a. Biology of the myrmecomorphic plant bug *Coquillettia insignis* Uhler (Heteroptera: Miridae: Phylinae). J. N.Y. Entomol. Soc. 95:258–277.

——. 1987b. Biology of the myrmecomorphic plant bug *Orectoderus obliquus* Uhler (Heteroptera: Miridae: Phylinae). J. N.Y. Entomol. Soc. 95: 278–289.

McIver, J. D. and G. [M.]. Stonedahl. 1993. Myrmecophily: Morphological and behavioral mimicry of ants. Annu. Rev. Entomol. 38:351–379.

McIver, J. D. and C. H. Tempelis. 1993. The arthropod predators of ant-mimetic and aposematic prey: A serological analysis. Ecol. Entomol. 18:218–222.

McKendrick, J. D. and D. P. Bleicher. 1980. Observations of a grass bug on bluejoint ranges. Agroborealis 12:15–18.

McKey, D. 1979. The distribution of secondary compounds within plants. Pages 55–133 in G. A. Rosenthal and D. H. Janzen, eds., Herbivores: Their interaction with secondary plant metabolites. Academic, New York.

McKinlay, K. S. and Q. A. Geering. 1957. Studies of crop loss following insect attack on cotton in East Africa. I—Experiments in Uganda and Tanganyika. Bull. Entomol. Res. 48:833–849.

McKinney, K. B. 1938. Physical characteristics on the foliage of beans and tomatoes that tend to control some small insect pests. J. Econ. Entomol. 31:630–631.

McLain, D. K. 1981. Resource partitioning by three species of hemipteran herbivores on the basis of host plant density. Oecologia (Berl.) 48:414–417.

——. 1984. Coevolution: Müllerian mimicry between a plant bug (Miridae) and a seed bug (Lygaeidae) and the relationship between host plant choice and unpalatability. Oikos 43:143–148.

——. 1989. Prolonged copulation as a post-insemination guarding tactic in a natural population of the ragwort seed bug. Anim. Behav. 38:659–664.

McLean, D. L. and M. G. Kinsey. 1964. A technique for electronically recording aphid feeding and salivation. Nature (Lond.) 202:1358–1359.

——. 1968. Probing behavior of the pea aphid, Acyrthosiphon pisum. II. Comparisons of salivation and ingestion in host and non-host plant leaves. Ann. Entomol. Soc. Am. 61:730–739.

McLean, D. M. 1941. Studies on mosaic of cowpeas, Vigna sinensis. Phytopathology 31:420–430.

McLellan, A. R. 1977. Minerals, carbohydrates and amino acids of pollens from some weedy and herbaceous plants. Ann. Bot. (Lond.) 41:1225–1232.

McMichael, S. C. 1954. Glandless boll in upland cotton and its use in the study of natural crossing. Agron. J. 46:527–528.

——. 1960. Combined effects of glandless genes gl2 and gl3 on pigment glands in the cotton plant. Agron. J. 52:385–386.

McMullen, C. K. 1993. Flower-visiting insects of the Galapagos Islands. Pan-Pac. Entomol. 69:95–106.

McMullen, R. D. and C. Jong. 1967. New records and discussion of predators of the pear psylla, Psylla pyricola Forster, in British Columbia. J. Entomol. Soc. B.C. 64:35–40.

——. 1970. The biology and influence of pesticides on Campylomma verbasci (Heteroptera: Miridae). Can. Entomol. 102:1390–1394.

McMurtry, J. A. and H. G. Johnson. 1965. Some factors influencing the abundance of the predaceous mite Amblyseius hibisci in southern California (Acarina: Phytoseiidae). Ann. Entomol. Soc. Am. 58:49–56.

McMurtry, J. A. and J. G. Rodriguez. 1987. Nutritional ecology of phytoseiid mites. Pages 609–644 in F. Slansky Jr. and J. G. Rodriguez, eds., Nutritional ecology of insects, mites, spiders, and related invertebrates. Wiley, New York.

McMurtry, J. A., C. B. Huffaker, and M. van de Vrie. 1970. Ecology of tetranychid mites and their natural enemies: A review I. Tetranychid enemies: Their biological characters and the impact of spray practices. Hilgardia 40(11):331–390.

McNaughton, S. J. 1979. Grazing as an optimization process: Grass-ungulate relationships in the Serengeti. Am. Nat. 113:691–703.

——. 1983. Compensatory plant growth as a response to herbivory. Oikos 40:329–336.

McNeill, S. 1971. The energetics of a population of Leptopterna dolabrata (Heteroptera: Miridae). J. Anim. Ecol. 40:127–140.

——. 1973. The dynamics of a population of Leptopterna dolabrata (Heteroptera: Miridae) in relation to its food resources. J. Anim. Ecol. 42:495–507.

McNeill, S. and R. A. Prestidge. 1982. Plant nutritional strategies and insect herbivore community dynamics. Pages 225–235 in J. H. Visser and A. K. Minks, eds., Proc. 5th Int. Symp. Insect-Plant Relationships, Wageningen, The Netherlands, 1–4 March 1982. Centre for Agricultural Publishing & Documentation, Wageningen.

McNeill, S. and T. R. E. Southwood. 1978. The role of nitrogen in the development of insect/plant relationships. Pages 77–98 in J. B. Harborne, ed., Biochemical aspects of plant and animal coevolution. Academic, London.

McPherson, J. E. 1982. The Pentatomoidea (Hemiptera) of northeastern North America with emphasis on the fauna of Illinois. Southern Illinois University Press, Carbondale. 240 pp.

McPherson, J. E., B. C. Weber, and T. J. Henry. 1983. Seasonal flight patterns of Hemiptera in a North Carolina black walnut plantation. 7. Miridae. Gt. Lakes Entomol. 16:35–42.

Meagher, R. L. Jr. and J. R. Meyer. 1990. Effects of ground cover management on certain abiotic and biotic interactions in peach orchard ecosystems. Crop Prot. 9:65–72.

Medler, J. T. 1941. The nature of injury to alfalfa caused by Empoasca fabae (Harris). Ann. Entomol. Soc. Am. 34:439–450.

——. 1961. A new record of parasitism of Lygus lineolaris (P. de B.) (Hemiptera) by Tachinidae (Diptera). Proc. Entomol. Soc. Wash. 63:101–102.

Medler, J. T. and E. H. Fisher. 1953. Leafhopper control with methoxychlor and parathion to increase alfalfa hay production. J. Econ. Entomol. 46:511–513.

Meeuse, A. D. J. 1978. Entomophily in Salix: Theoretical considerations. Pages 47–50 in A. J. Richards, ed., The pollination of flowers by insects. Linn. Soc. Symp. Ser. 6. Academic, New York.

Meeuse, B. J. D. 1972. [Review of] The Principles of Pollination Ecology (2nd rev. ed.) by K. Faegri and L. van der Pijl, 1971. Ecology 53:984–986.

Meijden, E. van der, M. Wijn, and H. J. Verkaar. 1988. Defence and regrowth, alternative plant strategies in the struggle against herbivores. Oikos 51:355–363.

Melber, A., L. Hölscher, and G. H. Schmidt. 1980. Further studies on the social behaviour and its ecological significance in Elasmucha grisea L. (Hem.-Het.: Acanthosomatidae). Zool. Anz. 205:27–38.

Melton, B., G. Watts, B. Vering, W. Knipe, and R. Ditterline. 1971. Breeding for lygus resistance in alfalfa. N.M. Agric. Exp. Stn. Res. Rep. 197:1–9.

Menken, S. B. J. and L. E. L. Raijmann. 1996. Biochemical systematics: Principles and perspectives for pest management. Pages 7–29 in W. O. C. Symondson and J. E. Liddell, eds., The ecology of agricultural pests: Biochemical approaches. Chapman & Hall, London.

Menken, S. B. J. and S. A. Ulenberg. 1987. Biochemical characters in agricultural entomology. Agric. Zool. Rev. 2:305–360.

Mensah, R. K. and M. Khan. 1997. Use of Medicago sativa (L.) interplantings/trap crops in the management of the green mirid, Creontiades dilutus (Stål) in commercial cotton in Australia. Int. J. Pest Manage. 43:197–202.

Menzel, R. 1922. Over de biologische bestrijding van *Helopeltis*. Meded. Proefstn. Thee 81:21–23.

———. 1928. *Helopeltis* en haar parasieten. Bergcultures (Batavia) 2:1259–1262.

Meredith, P. 1970. "Bug" damage in wheat. N.Z. Wheat Rev. No. 11, 1968–70:49–53.

Meredith, W. R. Jr. 1976. Nectariless cottons. *In* Proc. Beltwide Cotton Prod.-Mech. Conf., Jan. 5, 7–8, 1976, Las Vegas, Nev., pp. 34–37. National Cotton Council, Memphis.

———. 1980. Performance of paired nectaried and nectariless F_3 cotton hybrids. Crop Sci. 20:757–760.

Meredith, W. R. Jr. and M. F. Schuster. 1979. Tolerance of glabrous and pubescent cottons to tarnished plant bug. Crop Sci. 19:484–488.

Meredith, W. R. Jr., C. D. Ranney, M. L. Laster, and R. R. Bridge. 1973. Agronomic potential of nectariless cotton. J. Environ. Qual. 2:141–144.

Meredith, W. R. Jr., B. W. Hanny, and J. C. Bailey. 1979. Genetic variability among glandless cottons for resistance to two insects. Crop Sci. 19:651–653.

Merrifield, F. 1906. Proceedings for 1906: Exhibitions. Trans. Entomol. Soc. Lond. 1906:xc.

Messina, F. J. 1978. Mirid fauna associated with old-field goldenrods (*Solidago*: Compositae) in Ithaca, N.Y. J. N.Y. Entomol. Soc. 86:137–143.

Messing, R. H. and M. T. AliNiazee. 1985. Natural enemies of *Myzocallis coryli* (Hom.:Aphididae) in Oregon hazelnut orchards. J. Entomol. Soc. B.C. 82:14–18.

———. 1986. Impact of predaceous insects on filbert aphid, *Myzocallis coryli* (Homoptera: Aphididae). Environ. Entomol. 15:1037–1041.

Metcalf, C. L., W. P. Flint, and R. L. Metcalf. 1962. Destructive and useful insects: Their habits and control, 4th ed. McGraw-Hill, New York. 1087 pp.

Metzer, R. B. 1975. Glandless cotton and progress that has been made. *In* Proc. Beltwide Cotton Prod. Res. Conf., Jan. 6–8, 1975, New Orleans, La., p. 86. National Cotton Council, Memphis.

Meurer, J. J. 1956. Waarnemingen van wansten (Hem.-Het.) met behulp van een vanglamp. Entomol. Ber. (Amst.) 16:54–63.

Meyer, J. R. and V. G. Meyer. 1961. Origin and inheritance of nectariless cotton. Crop Sci. 1:167–169.

Meyer-Dür, L. R. 1843. Verzeichniss der in der Schweiz einheimischen Rhynchoten. (Hemiptera Linn.). Erstes Heft. Die Familie der Capsini. Jent & Gassmann, Solothurn, Switzerland. 115 pp.

Michailides, T. J. 1989. The "Achilles heel" of pistachio fruit. Calif. Agric. 43(5):10–11.

Michailides, T. J., R. E. Rice, and J. M. Ogawa. 1987. Succession and significance of several hemipterans attacking a pistachio orchard. J. Econ. Entomol. 80:398–406.

Michalk, O. 1933. Einige Bermerkungen über die Genitatasymmetrie der Capsiden. (Hemipt.-Heteropt.). Entomol. Jahrb. 42:153–154.

———. 1935. Zur Morphologie und Ablage der Eier bei den Heteropteren. Dtsch. Entomol. Z. 1935:148–175.

Michaud, O. D. and R. K. Stewart. 1990. Susceptibility of apples to damage by *Lygocoris communis* and *Lygus lineolaris* (Hemiptera: Miridae). Phytoprotection 71:25–30.

Michaud, O. D., G. Boivin, and R. K. Stewart. 1989a. Economic threshold for tarnished plant bug (Hemiptera: Miridae) in apple orchards. J. Econ. Entomol. 82:1722–1728.

Michaud, O. [D.]., R. K. Stewart, and G. Boivin. 1989b. Economic injury levels and economic thresholds for the green apple bug, *Lygocoris communis* (Knight) (Hemiptera: Miridae), in Quebec apple orchards. Can. Entomol. 121:803–808.

Michelbacher, A. E. 1954. Natural control of insect pests. J. Econ. Entomol. 47:192–194.

Middlekauff, W. W. 1956. Relationship of lygus bug populations to blackeye bean necrosis. Bull. Entomol. Soc. Am. 2(3):20.

Middlekauff, W. W. and E. E. Stevenson. 1952. Insect injury to blackeye bean seeds in central California. J. Econ. Entomol. 45:940–946.

Middleton, A. D. and H. Chitty. 1937. The food of adult partridges, *Perdix perdix* and *Alectoris rufa*, in Great Britain. J. Anim. Ecol. 6:322–336.

Mijušković, M. 1966. Prilog poznavanaju *Caliroa varipes* Klug (Hymenoptera, fam. Tenthredinidae). Zb. Rad. Zav. Unapr. Poljopr. Titograd 1:1–99.

Mikhaïlova, N. A. 1979. Factors of dynamics of the numbers of *Trigonotylus coelestialium* (Hemiptera, Miridae). Zool. Zh. 58:838–848. [in Russian, English summary.]

———. 1980. On forecasting the cereal mirid. Zashch. Rast. (Mosc.) 1980 No. 9:47. [in Russian.]

Milam, M. R., J. N. Jenkins, J. C. McCarty Jr., and W. L. Parrott. 1985. Combining tarnished plant bug resistance with frego bract. Miss. Agric. For. Exp. Stn. Bull. 939:1–4.

———. 1989. Breeding upland cotton for resistance to the tarnished plant bug. Field Crops Res. 21:227–238.

Miles, H. W. 1932. The control of fruit pests by winter spraying. J. R. Lancashire Agric. Soc. 1932:7–20.

Miles, P. W. 1958a. The stylet movements of a plant-sucking bug, *Oncopeltis fasciatus* Dall. (Heteroptera: Lygaeidae). Proc. R. Entomol. Soc. Lond. (A) 33:15–20.

———. 1958b. Contact chemoreception in some Heteroptera, including chemoreception internal to the stylet food canal. J. Insect Physiol. 2:338–347.

———. 1964. Studies on the salivary physiology of plant bugs: Oxidase activity in the salivary apparatus and saliva. J. Insect Physiol. 10:121–129.

———. 1968a. Studies on the salivary physiology of plant-bugs: Experimental induction of galls. J. Insect Physiol. 14:97–106.

———. 1968b. Insect secretions in plants. Annu. Rev. Phytopathol. 6:137–164.

———. 1969. Interaction of plant phenols and salivary phenolases in the relationship between plants and Hemiptera. Entomol. Exp. Appl. 12:736–744.

———. 1972. The saliva of Hemiptera. Adv. Insect Physiol. 9:183–255.

———. 1978. Redox reactions of hemipterous saliva in plant tissues. Entomol. Exp. Appl. 24:334–339.

———. 1987a. Plant-sucking bugs can remove the contents of cells without mechanical damage. Experientia (Basel) 43:937–939.

———. 1987b. Feeding process of Aphidoidea in relation to effects on their food plants. Pages 321–339 *in* A. K. Minks and P. Harrewijn, eds., Aphids: Their biology, natural enemies and control. Vol. A. Elsevier, Amsterdam.

———. 1989a. The responses of plants to the feeding of Aphidoidea: Principles. Pages 1–21 *in* A. K. Minks and P. Harrewijn, eds., Aphids: Their biology, natural enemies and control. Vol. C. Elsevier, Amsterdam.

———. 1989b. Specific responses and damage caused by Aphidoidea. Pages 23–47 *in* A. K. Minks and P. Harrewijn, eds., Aphids: Their biology, natural enemies and control. Vol. C. Elsevier, Amsterdam.

Miles, P. W. and K. Hori. 1977. Fate of ingested β-indolyl acetic acid in *Creontiades dilutus*. J. Insect Physiol. 23:221–226.

Miles, P. W. and D. Slowiak. 1976. The accessory salivary gland as the source of water in the saliva of Hemiptera: Heteroptera. Experientia (Basel) 32:1011–1012.

Miles, P. W. and G. S. Taylor. 1994.

'Osmotic pump' feeding by coreids. Entomol. Exp. Appl. 73:163–173.

Miles, P. W., D. Aspinall, and L. Rosenberg. 1982. Performance of the cabbage aphid, *Brevicoryne brassicae* (L.), on water-stressed rape plants, in relation to changes in their chemical composition. Aust. J. Zool. 30:337–345.

Millar, J. G. and R. E. Rice. 1998. Sex pheromone of the plant bug *Phytocoris californicus* (Heteroptera: Miridae). J. Econ. Entomol. 91:132–137.

Millar, J. G., R. E. Rice, and Q. Wang. 1997. Sex pheromone of the mirid bug *Phytocoris relativus*. J. Chem. Ecol. 23:1743–1754.

Miller, D. 1927. Parasites of the pearmidge (*Perrisia pyri*). Report on the 1926 consignments from Europe. N.Z. J. Agric. 35:170–175.

Miller, G. L. and M. L. Williams. 1985. Notes on some little known scale insect predators recently collected in Alabama. J. Ala. Acad. Sci. 56:81. (Abstr.)

Miller, J. R. and K. L. Strickler. 1984. Finding and accepting host plants. Pages 127–157 in W. J. Bell and R. T. Cardé, eds., Chemical ecology of insects. Sinauer, Sunderland, Mass.

Miller, J. S. and J. W. Wenzel. 1995. Ecological characters and phylogeny. Annu. Rev. Entomol. 40:389–415.

Miller, L. A. and R. J. McClanahan. 1959. Note on occurrence of the fungus *Empusa muscae* Cohn on adults of the onion maggot, *Hylemya antiqua* (Meig.) (Diptera: Anthomyiidae). Can. Entomol. 91:525–526.

Miller, N. C. E. 1941. Insects associated with cocoa (*Theobroma cacao*) in Malaya. Bull. Entomol. Res. 32:1–16.

——. 1956. Two new species of Miridae from the Agalega Islands [Hemip. Heteroptera]. Mauritius Inst. Bull. 3:317–320.

——. 1971. The biology of the Heteroptera, 2nd (rev.) ed. Classey, Hampton, UK. 206 pp.

Miller, R. S. and R. T. Schuh. 1994. Predation by *Clivinema coalinga* Bliven (Heteroptera: Miridae: Deraeocorinae: Clivinemini) of *Orthesia annae* Cockerell (Sternorrhyncha: Ortheziidae). J. N.Y. Entomol. Soc. 102:383–384.

Miller, S. E. and W. S. Davis. 1985. Insects associated with the flowers of two species of *Malacothrix* (Asteraceae) on San Miguel Island, California. Psyche (Camb.) 92:547–555.

Miller, W. C. III. 1993. *Rhinocapsus vanduzeei* Uhler, a little known pest of azaleas. Azalean (Bethesda) 15(3):58–59.

Mills, H. B. 1939. Montana insect pests

for 1937 and 1938. Mont. Agric. Exp. Stn. Bull. 366:1–32.

——. 1941. Montana insect pests for 1939 and 1940. Mont. Agric. Exp. Stn. Bull. 384:1–27.

Mink, G. I. 1993. Pollen- and seed-transmitted viruses and viroids. Annu. Rev. Phytopathol. 31:375–402.

Minkiewicz, S. 1927. Studja nad miodówką jabłoniową (*Psylla mali* Schmidberger). Część II. Rozwój i biologja. Mém. Inst. Natl. Pol. Econ. Rurale Pulawy (A) 8:457–528.

Misra, B. C. 1980. The leaf and planthoppers of rice. Central Rice Research Institute, Cuttack, India. 182 pp.

Mitchell, P. L. and L. D. Newsom. 1984. Histological and behavioral studies of threecornered alfalfa hopper (Homoptera: Membracidae) feeding on soybean. Ann. Entomol. Soc. Am. 77:174–181.

Mitchell, W. C. and P. A. Maddison. 1983. Major taro pests in the Pacific Islands. In Proc. Pac. Sci. Assoc. 15th Congr., Dunedin, New Zealand, 1–11 Feb. 1983. Vol. I, p. 167. Royal Society of New Zealand, Wellington.

Mitter, C., B. Farrell, and B. Wiegmann. 1988. The phylogenetic study of adaptive zones: Has phytophagy promoted insect diversification? Am. Nat. 132:107–128.

Mitter, C., B. Farrell, and D. J. Futuyma. 1991. Phylogenetic studies of insect-plant interactions: Insights into the genesis of diversity. Trends Ecol. Evol. 6:290–293.

Mittler, T. E. 1957. Studies on the feeding and nutrition of *Tuberolachnus salignus* (Gmelin) (Homoptera, Aphididae). I. The uptake of phloem sap. J. Exp. Biol. 34:334–341.

——. 1958. Studies on the feeding and nutrition of *Tuberolachnus salignus* (Gmelin) (Homoptera, Aphididae). II. The nitrogen and sugar composition of ingested phloem sap and excreted honeydew. J. Exp. Biol. 35:74–84.

——. 1967. Flow relationships for hemipterous stylets. Ann. Entomol. Soc. Am. 60:1112–1114.

Miyamoto, S. 1957. List of ovariole numbers in Japanese Heteroptera. Sieboldia 2:69–82. [in Japanese.]

——. 1961. Comparative morphology of alimentary organs of Heteroptera, with the phylogenetic consideration. Sieboldia 2:197–259.

Mizell, R. F. III and D. E. Schiffhauer. 1987. Seasonal abundance of the crapemyrtle aphid, *Sarucallis kahawaluokalani*, in relation to the pecan aphids, *Monellia caryella* and *Monelliopsis pecanis* and their common predators. Entomophaga 32:511–520.

Mizell, R. F. III and M. C. Sconyers. 1992. Toxicity of imidacloprid to selected arthropod predators in the laboratory. Fla. Entomol. 75:277–280.

Mjöberg, E. 1906. Über *Systellonotus triguttatus* L. und sein Verhältnis zu *Lasius niger*. Z. Wiss. Insektenbiol. 2:107–109.

Moffett, J. O., L. S. Stith, C. C. Burkhardt, and C. W. Shipman. 1976. Insect visitors to cotton flowers. J. Ariz. Acad. Sci. 11(2):47–48.

Mols, P. J. M. 1990. Forecasting orchard pests for adequate timing of control measures. Proc. Exp. Appl. Entomol. Neth. Entomol. Soc. (N.E.V., Amst.) 1:75–81.

Molz, E. 1917. Die Wiesenwanze, *Lygus pratensis* L., ein geführlicher Kartoffelschädling. Z. Pflanzenkr. 27:337–339.

Monaco, R. 1975. Preoccupante ricomparsa di *Calocoris trivialis* Costa sull'olivo in Puglia. Inf. Fitopatol. 25(9):5–7.

Montealegre, R. J. and D. A. Rodriguez. 1989. Patogenicidad del hongo *Beauveria bassiana* (Bals.) Vuill. sobre la chinche *Monalion* [sic] *dissimulatum* Distant plaga del cacaotero *Theobroma cacao* L. Acta Agron. (Palmira) 39:88–96.

Monteith, J. and E. A. Hollowell. 1929. Pathological symptoms in legumes caused by the potato leafhopper. J. Agric. Res. (Wash., D.C.) 38:649–677.

Montllor, C. B. 1991. The influence of plant chemistry on aphid feeding behavior. Pages 125–173 in E. Bernays, ed., Insect-plant interactions. Vol. III. CRC, Boca Raton, Fla.

Moore, J. B. and C. C. Fox. 1941. *Lygus* injury to peaches in the Pacific Northwest and its prevention. J. Econ. Entomol. 34:99–101.

Moore, T. B., R. Stevens, and E. D. McArthur. 1982. Preliminary study of some insects associated with rangeland shrubs with emphasis on *Kochia prostrata*. J. Range Manage. 35:128–130.

Moran, V. C. and T. R. E. Southwood. 1982. The guild composition of arthropod communities in trees. J. Anim. Ecol. 51:298–306.

Moreby, S. J. 1996. The effects of organic and conventional farming methods on plant bug densities (Hemiptera: Heteroptera) within winter wheat fields. Ann. Appl. Biol. 128:415–421.

Moreby, S. J., N. W. Sotherton, and P. C. Jepson. 1997. The effects of pesticides on species of non-target Heteroptera inhabiting cereal fields in southern England. Pestic. Sci. 51:39–48.

Moreira, C. 1923. Les Capsides du Tabac au Brésil. In Rep. Int. Conf.

Phytopathol. Econ. Entomol. Holland, 1923, Wageningen, pp. 283–286. Veenman, Wageningen.

Moreira, P. H. R., J. J. Soares, A. C. Busoli, V. R. da Cruz, M. H. L. Pimentel, and G. J. B. Pelinson. 1994. Causas do apodrecimento de maçãs do algodoeiro. Pesqui. Agropecu. Bras. 29:1503–1507.

Moreton, B. D. 1958. Beneficial insects, 5th ed. Minist. Agric. Fish. Food Bull. 20:1–49.

———. 1964. Pests, and spraying programmes. Pages 165–188 in A. H. Burgess, ed., Hops: Botany, cultivation, and utilization. Leonard Hill, London; Interscience, New York. 300 pp.

Mori, H. 1986. Water absorption by eggs and serosal specialization as clues to evolutionary trends in Heteroptera. Ann. Entomol. Soc. Am. 79:456–459.

Morley, C. 1905. The Hemiptera of Suffolk. J. H. Keyes, Plymouth, UK. 34 pp.

———. 1914. Garden notes. Entomologist 47:215–218.

———. 1916. Garden notes. Entomologist 49:246–248.

Morrill, A. W. 1910. Plant-bugs injurious to cotton bolls. U.S. Dep. Agric. Bur. Entomol. Bull. 86:1–110.

———. 1917. Cotton plants in the arid and semi-arid Southwest. J. Econ. Entomol. 10:307–317.

———. 1918. Insect pests of interest to Arizona cotton growers. Ariz. Agric. Exp. Stn. Bull. 87:173–205.

———. 1925. Commercial entomology on the west coast of Mexico. J. Econ. Entomol. 18:707–716.

———. 1927a. A plant bug (Pycnoderes incurvatus [sic] Dist.). U.S. Dep. Agric. Insect Pest Surv. Bull. 7:111.

———. 1927b. Pest control problems on the west coast of Mexico. In Proc. 8th Annu. Conf. West. Plant Quar. Board, June 9–11, 1926, Olympia, Wash, pp. 70–82. Calif. Dep. Agric. Spec. Publ. 73.

———. 1928. Sonora cotton square dauber (Creontiades debilis Van D.). J. Econ. Entomol. 21:437.

Morrill, W. L., R. L. Ditterline, and C. Winstead. 1984. Effects of Lygus borealis Kelton (Hemiptera: Miridae) and Adelphocoris lineolatus (Goeze) (Hemiptera: Miridae) feeding on sainfoin seed production. J. Econ. Entomol. 77:966–968.

Morris, M. G. 1965. Some aspects of the biology of Psallus ambiguus (Fall.) (Heteroptera: Miridae) on apple trees in Kent. Entomologist 98:14–31.

———. 1967. Differences between the invertebrate faunas of grazed and ungrazed chalk grassland. I. Responses of some phytophagous insects to cessa-tion of grazing. J. Appl. Ecol. 4: 459–474.

———. 1975. Preliminary observations on the effects of burning on the Hemiptera (Heteroptera and Auchenorhyncha [sic]) of limestone grassland. Biol. Conserv. 7:311–319.

———. 1979. Responses of grassland invertebrates to management by cutting II. Heteroptera. J. Appl. Ecol. 16:417–432.

Morris, R. F. 1972. Predation by insects and spiders inhabiting colonial webs of Hyphantria cunea. Can. Entomol. 104:1197–1207.

Morrison, L. 1938. Surveys of the insect pests of wheat crops in Canterbury and North Otago during the summers of 1936–37 and 1937–38. N.Z. J. Sci. Technol. 20(A):142–155.

Morrow, P. A. 1977. Host specificity of insects in a community of three codominant Eucalyptus species. Aust. J. Ecol. 2:89–106.

Morse, D. H. and R. S. Fritz. 1983. Contributions of diurnal and nocturnal insects to the pollination of common milkweed (Asclepias syriaca L.) in a pollen-limited system. Oecologia (Berl.) 60:190–197.

Moscardi, F. and B. S. Correa Ferreira. 1985. Biological control of soybean caterpillars. Pages 703–711 in R. Shibles, ed., World soybean research conference III: Proceedings. Westview, Boulder, Col.

Mosseler, A. J. and M. Hubbes. 1983. Erwinia spp. and a new canker disease of hybrid poplars in Ontario. Eur. J. For. Pathol. 13:261–278.

Mote, U. N. and S. S. Jadhav. 1990. Incidence and losses caused by sorghum head bug. J. Maharashtra Agric. Univ. 15:121–122.

Motten, A. F., D. R. Campbell, and D. E. Alexander. 1981. Pollination effectiveness of specialist and generalist visitors to a North Carolina population of Claytonia virginica. Ecology 62:1278–1287.

Mound, L. A. 1962. Extra-floral nectaries of cotton and their secretions. Emp. Cotton Grow. Rev. 39:254–261.

Moursi, K. S. and E. M. Hegazi. 1983. Destructive insects of wild plants in the Egyptian Western Desert. J. Arid Environ. 6:119–127.

Moxon, J. E. 1992. Insect pests of cocoa in Papua New Guinea: Importance and control. In P. J. Keane and C. A. J. Putter, eds., Cocoa pest and disease management in Southeast Asia and Australasia. FAO Plant Prod. Prot. Pap. 112:129–143.

Mueller, A. J. and V. M. Stern. 1973a. Lygus flight and dispersal behavior. Environ. Entomol. 2:361–364.

———. 1973b. Effects of temperature on the reproductive rate, maturation, longevity, and survival of Lygus hesperus and L. elisus (Hemiptera: Miridae). Ann. Entomol. Soc. Am. 66:593–597.

———. 1974. Timing of pesticide treatments on safflower to prevent Lygus from dispersing to cotton. J. Econ. Entomol. 67:77–80.

Muenchow, G. and V. A. Delesalle. 1992. Patterns of weevil herbivory on male, monoecious and female inflorescences of Sagittaria latifolia. Am. Midl. Nat. 127:355–367.

Muggeridge, J. 1929. Biological control of pear-midge (Perrisia pyri) in New Zealand. The present position. N.Z. J. Agric. 38:317–320.

Muhamad, R. 1994. Ability of cocoa to tolerate and compensate for pod damage by the mirid, Helopeltis theivora Waterhouse. In Proc. Malaysian Int. Cocoa Conf., 20–21 Oct. 1994, Kuala Lumpur, pp. 64–65. Malaysian Cocoa Board, Kota Kinabalu, Malaysia.

Muhamad, R. and K. C. Khoo. 1983. A technique for rearing the cocoa mirid, Helopeltis theobromae Miller in captivity. Pages 19–21 in Advances in cocoa plant protection in Malaysia. Malaysian Plant Protection Society, Kuala Lumpur, Malaysia.

Muhamad, R. and M. J. Way. 1995a. Damage and crop loss relationships of Helopeltis theivora, Hemiptera, Miridae and cocoa in Malaysia. Crop Prot. 14:117–121.

———. 1995b. Relationships between feeding habits and fecundity of Helopeltis theivora (Hemiptera: Miridae) on cocoa. Bull. Entomol. Res. 85:519–523.

Muir, F. 1920. Report of entomological work in Australia, 1919–1920. Hawaii. Plant. Rec. 23:125–130.

Muir, R. C. 1958. On the application of the capture-recapture method to an orchard population of Blepharidopterus angulatus (Fall.) (Hemiptera-Heteroptera, Miridae). East Malling Res. Stn. Rep. 1957:140–147.

———. 1965a. The effect of sprays on the fauna of apple trees. I. The influence of winter wash, captan, and lime-sulphur on the interaction of populations of Panonychus ulmi (Koch) (Acarina: Tetranychidae) and its predator, Blepharidopterus angulatus (Fall.) (Heteroptera: Miridae). J. Appl. Ecol. 2:31–41.

———. 1965b. The effect of sprays on the fauna of apple trees. II. Some aspects of the interaction between populations of Blepharidopterus angulatus · (Fall.) (Heteroptera: Miridae) and its prey,

Panonychus ulmi (Koch) (Acarina: Tetranychidae). J. Appl. Ecol. 2:43–57.

——. 1966a. The effect of temperature on development and hatching of the egg of *Blepharidopterus angulatus* (Fall.) (Heteroptera, Miridae). Bull. Entomol. Res. 57:61–67.

——. 1966b. The effect of sprays on the fauna of apple trees. IV. The recolonization of orchard plots by the predatory mirid *Blepharidopterus angulatus* and its effect on populations of *Panonychus ulmi*. J. Appl. Ecol. 3:269–276.

Mukerji, M. K. 1973. The development of sampling techniques for populations of the tarnished plant bug, *Lygus lineolaris* (Hemiptera: Miridae). Res. Popul. Ecol. (Kyoto) 15:50–63.

Müller, G. W. and A. S. Costa. 1964. Citrus false exanthema induced by feeding of a myrid [*sic*]. FAO Plant Prot. Bull. 12:97–104.

Müller, H. 1873. Die Befruchtung der Blumen durch Insekten und die gegenseitigen Anpassungen beider. Engelmann, Leipzig. 478 pp.

——. 1881. Alpenblumen, ihre Befruchtung durch Insekten und ihre Anpassungen an dieselben. Engelmann, Leipzig. 611 pp.

——. 1883. The fertilisation of flowers. (Transl. by D'Arcy W. Thompson.) Macmillan, London. 669 pp.

Muller, J. 1979. Form and function in angiosperm pollen. Ann. Mo. Bot. Gard. 66:593–632.

Mullick, D. B. 1977. The non-specific nature of defense in bark and wood during wounding, insect and pathogen attack. *In* F. A. Loewus and V. C. Runeckles, eds., The structure, biosynthesis and degradation of wood. Recent Adv. Phytochem. 11:395–441.

Mullin, C. A. 1986. Adaptive divergence of chewing and sucking arthropods to plant allelochemicals. Pages 175–209 *in* L. B. Brattsten and S. Ahmad, eds., Molecular aspects of insect-plant associations. Plenum, New York.

Mumford, E. P. 1931. On the fauna of the diseased big-bud of the black currant, *Ribes nigrum* L., with a note on some fungous parasites of the gall-mite, *Eriophyes ribis* (Westwood) Nal. Marcellia 27:29–62.

Mundinger, F. G. and G. L. Slate. 1952. Insecticide sprays as a probable control of "sterility" in blackberries. J. Econ. Entomol. 45:135–136.

Muniappan, R. and C. A. Viraktamath. 1986. Status of biological control of the weed, *Lantana camara* in India. Trop. Pest Manage. 32:40–42.

Muraleedharan, N. 1987. Entomological research in tea in southern India. J. Coffee Res. 17:80–83.

——. 1992. Pest control in Asia. Pages 375–412 *in* K. C. Willson and M. N. Clifford, eds., Tea: Cultivation and consumption. Chapman & Hall, London.

Murdoch, W. W. 1990. The relevance of pest-enemy models to biological control. Pages 1–24 *in* M. Mackauer, L. E. Ehler, and J. Roland, eds., Critical issues in biological control. Intercept, Andover, UK.

Murdoch, W. W., J. Chesson, and P. L. Chesson. 1985. Biological control in theory and practice. Am. Nat. 125:344–366.

Murphy, D. D. 1984. Butterflies and their nectar plants: The role of the checkerspot butterfly *Euphydryas editha* as a pollen vector. Oikos 43:113–117.

Murphy, P. A. 1923a. Investigations on the leaf-roll and mosaic diseases of the potato. J. Dep. Agric. Irel. 23:20–34.

——. 1923b. On the use of rolling in potato foliage; and on some further insect carriers of the leaf-roll disease. Sci. Proc. R. Dublin Soc. 17:163–184.

Murphy, P. A. and R. McKay. 1926. Methods for investigating the virus diseases of the potato, and some results obtained by their use. Sci. Proc. R. Dublin Soc. 18:169–184.

——. 1929. The insect vectors of the leaf-roll disease of the potato. Sci. Proc. R. Dublin Soc. 19:341–353.

Murray, D. A. H. and R. J. Lloyd. 1997. The effect of spinosad (Tracer) on arthropod pest and beneficial populations in Australian cotton. *In* Proc. Beltwide Cotton Conf., Jan. 6–10, 1997, New Orleans, La. 2:1087–1091. National Cotton Council, Memphis.

Murray, R. A. and M. G. Solomon. 1978. A rapid technique for analysing diets of invertebrate predators by electrophoresis. Ann. Appl. Biol. 90:7–10.

Murtfeldt, M. E. 1894. Notes on the insects of Missouri for 1893. U.S. Dep. Agric. Div. Entomol. Bull. 32:37–45.

——. 1902. Recent experience with destructive insects. *In* Forty-fifth Annu. Rep. State Hortic. Soc. Missouri, pp. 353–358. Tribune Printing, Jefferson City, Mo.

Musa, M. S. and G. D. Butler. 1967. The stages of *Spanogonicus* [*sic*] *albofasciatus* and their development (Hemiptera: Miridae). J. Kans. Entomol. Soc. 40:596–600.

Mussett, K. S., J. H. Young, R. G. Price, and R. D. Morrison. 1979. Predatory arthropods and their relationship to fleahoppers on *Heliothis*-resistant cotton varieties in southwestern Oklahoma. Southwest. Entomol. 4:35–39.

Myers, J. G. 1922. The order Hemiptera in New Zealand. With special reference to its biological and economic aspects. N.Z. J. Sci. Technol. 5:1–12.

——. 1926. Biological notes on New Zealand Heteroptera. Trans. Proc. N.Z. Inst. 56:449–511.

——. 1927a. Natural enemies of the pear leaf-curling midge, *Perrisia pyri*, Bouché (Dipt., Cecidom.). Bull. Entomol. Res. 18:129–138.

——. 1927b. Ethological observations on some Pyrrhocoridae of Cuba (Hemiptera-Heteroptera). Ann. Entomol. Soc. Am. 20:279–300.

——. 1929. Facultative blood-sucking in phytophagous Hemiptera. Parasitology 21:472–480.

——. 1931a. A preliminary report on an investigation into the biological control of West Indian insect pests. Emp. Mark. Board (E.M.B.) 42:1–173.

——. 1931b. Insect conditions in Haiti during July, 1931. U.S. Dep. Agric. Insect Pest Surv. Bull. 11:645–647.

——. 1935. Notes on cocoa-beetle and cocoa-thrips. Trop. Agric. 12:22.

Myers, J. G. and W. E. China. 1929. The systematic position of the Peloridiidae as elucidated by a further study of the external anatomy of *Hemiodoecus leai*, China (Hemiptera, Peloridiidae). Ann. Mag. Nat. Hist. (10)3:282–294.

Myers, J. G. and G. Salt. 1926. The phenomenon of myrmecoidy, with new examples from Cuba. Trans. Entomol. Soc. Lond. 74:427–436.

Myers, J. H. 1981. Interactions between western tent caterpillars and wild rose: A test of some general plant herbivore hypotheses. J. Anim. Ecol. 50:11–25.

Myers, N. 1988. Threatened biotas: "Hot spots" in tropical forests. Environmentalist 8:187–208.

Nadgauda, D. and H. N. Pitre. 1986. Effects of temperature on feeding, development, fecundity, and longevity of *Nabis roseipennis* (Hemiptera: Nabidae) fed tobacco budworm (Lepidoptera: Noctuidae) larvae and tarnished plant bug (Hemiptera: Miridae) nymphs. Environ. Entomol. 15:536–539.

Nagaraja, K. V., P. S. Bhavanishankara Gowda, V. V. Krishna Kurup, and J. N. John. 1994. Biochemical changes in cashew in relation to infestation by tea mosquito bug. Plant Physiol. Biochem. (New Delhi) 21:91–97.

Nagel, H. G. 1973. Effect of spring prairie burning on herbivorous and non-herbivorous arthropod populations. J. Kans. Entomol. Soc. 46:485–496.

Nair, M. R. G. K. 1975. Insects and mites of crops in India. Indian Council of Agricultural Research, New Delhi, India. 404 pp.

Nair, M. R. G. K. and N. Mohan Das.

1962. On the biology and control of *Carvalhoia arecae* Miller and China (Miridae: Hemiptera), a pest of areca palms in Kerala. Indian J. Entomol. 24:86–93.

Naito, A. 1976. Studies on the feeding habits of some leafhoppers attacking the forage crops. I. Comparison of the feeding habits of the adults. Jpn. J. Appl. Entomol. Zool. 20:1–8. [in Japanese, English summary.]

———. 1977a. Studies on the feeding habits of some leafhoppers attacking the forage crops. III. Relation between the infestation by leafhopper and its feeding habits. Jpn. J. Appl. Entomol. Zool. 21:1–5. [in Japanese, English summary.]

———. 1977b. Feeding habits of leafhoppers. JARQ (Jpn. Agric. Res. Q.) 11: 115–119.

Naito, T. 1983. Genetic dimorphism of thorax color in the rose sawfly. J. Hered. 74:469–472.

Nakabachi, A. and H. Ishikawa. 1999. Provision of riboflavin to the host aphid, *Acyrthosiphon pisum*, by endosymbiotic bacteria, *Buchnera*. J. Insect Physiol. 45:1–6.

Nambiar, K. K. N., Y. R. Sarma, and G. B. Pillai. 1973. Inflorescence blight of cashew (*Anacardium occidentale* L.). J. Plant. Crops 1:44–46.

Napompeth, B. 1973. Ecology and population dynamics of the corn planthopper, *Peregrinus maidis* (Ashmead) (Homoptera: Delphacidae), in Hawaii. Ph.D. dissertation, University of Hawaii, Manoa. 257 pp.

Napompeth, B., A. Winotai, and P. Sommartya. 1990. Utilization of natural enemies for biological control of the leucaena psyllid in Thailand. Pages 175–180 in B. Napompeth and K. G. MacDicken, eds., Leucaena psyllid: Problems and management; proceedings of an international workshop held in Bogor, Indonesia, January 16–21, 1989. [Winrock International Institute for Agricultural Development, Arlington, Va.; International Development Research Centre, Ottawa, Canada; Nitrogen Fixing Tree Association, Waimanalo, Hawaii.]

Naranjo, S. E. and R. L. Gibson. 1996. Phytophagy in predaceous Heteroptera: Effects on life history and population dynamics. Pages 57–93 in O. Alomar and R. N. Wiedenmann, eds., Zoophytophagous Heteroptera: Implications for life history and integrated pest management. Thomas Say Publ. Entomol.: Proceedings. Entomological Society of America, Lanham, Md.

Naranjo, S. E. and J. R. Hagler. 1998. Characterizing and estimating the effect of heteropteran predation. Pages 171–197 in M. Coll and J. R Ruberson, eds., Predatory Heteroptera: Their ecology and use in biological control. Thomas Say Publ. Entomol.: Proceedings. Entomological Society of America, Lanham, Md.

Nasser [Naseer], M. and U. C. Abdurahiman. 1993. Cannibalism in *Cardiasthethus exiguus* Poppius (Hemiptera: Anthocoridae), a predator of the coconut caterpillar *Opisina arenolella* Walker (Lepidoptera: Xylorictidae). J. Adv. Zool. 14:1–6.

Natarajan, N. and P. C. Sundara Babu. 1988a. Culture of the sorghum earhead bug, *Calocoris angustatus* Lethierry in the laboratory. Trop. Pest Manage. 34:356.

———. 1988b. Economic injury level for sorghum earhead bug, *Calocoris angustatus* Lethierry in southern India. Insect Sci. Appl. 9:395–398.

———. 1988c. Damage potential of earhead bug on sorghum panicle. Insect Sci. Appl. 9:411–414.

———. 1990. Importance of the gram pod borer and earhead bug on sorghum. Insect Sci. Appl. 11:851–854.

Natarajan, N., P. C. Sundara Babu, and S. Chelliah. 1988. Influence of weather factors on sorghum earhead bug *Calocoris angustatus* Lethierry incidence. Trop. Pest Manage. 34:413–420.

———. 1989. Seasonal occurrence of sorghum earhead bug, *Calocoris angustatus* Lethierry (Hemiptera: Miridae) in southern India. Trop. Pest Manage. 35:70–77.

Nault, L. R. 1997. Arthropod transmission of plant viruses: A new synthesis. Ann. Entomol. Soc. Am. 90:521–541.

Naumann, I. 1993. CSIRO handbook of Australian insect names: Common and scientific names for insects and allied organisms of economic and environmental importance, 6th ed. CSIRO, East Melbourne. 193 pp.

Nawa, U. 1910. On a new predator of tingid bugs. Konch. Sek. [Insect World], Gifu 14:414–416. [in Japanese.]

Neal, J. W. Jr. 1989. Bionomics of immature stages and ethology of *Neochlamisus platani* (Coleoptera: Chrysomelidae) on American sycamore. Ann. Entomol. Soc. Am. 82:64–72.

Neal, J. W. Jr. and R. H. Haldemann. 1992. Regulation of seasonal egg hatch by plant phenology in *Stethoconus japonicus* (Heteroptera: Miridae), a specialist predator of *Stephanitis pyrioides* (Heteroptera: Tingidae). Environ. Entomol. 21:793–798.

Neal, J. W. Jr., R. H. Haldemann, and T. J. Henry. 1991. Biological control potential of a Japanese plant bug, *Stethoconus japonicus* (Heteroptera: Miridae), an adventive predator of azalea lace bug (Heteroptera: Tingidae). Ann. Entomol. Soc. Am. 84:287–293.

Neal, T. M., G. L. Greene, F. W. Mead, and W. H. Whitcomb. 1972. *Spanogonicus* [sic] *albofasciatus* (Hemiptera: Miridae): A predator in Florida soybeans. Fla. Entomol. 55:247–250.

Needham, J. G. 1903. Button-bush insects. Psyche (Camb.) 10:22–31.

———. 1948. Ecological notes on the insect population of the flower heads of *Bidens pilosa*. Ecol. Monogr. 18:431–446.

Neff, S. E. and C. O. Berg. 1966. Biology and immature stages of malacophagous Diptera of the genus *Sepedon* (Sciomyzidae). Va. Agric. Exp. Stn. Bull. 566:1–113.

Neilson, W. T. A., G. W. Wood, and C. W. Maxwell. 1970. Dimethoate sprays for apple maggot and their effect on predacious insects and mites. J. Econ. Entomol. 63:764–766.

Neiswander, C. R. 1931. The sources of American corn insects. Ohio Agric. Exp. Stn. Bull. 473:1–98.

Neklyudova, E. T. and S. P. Dikii. 1973. Tarnished plant bugs as vectors of stolbur in Solanaceae. Tr. Prikl. Bot. Genet. Sel. 50(2):36–39. [in Russian.]

Nelson, C. R. 1994. Insects of the Great Basin and Colorado Plateau. Pages 211–237 in K. T. Harper, L. L. St. Clair, K. H. Thorne, and W. M. Hess, eds., Natural history of the Colorado Plateau and Great Basin. University Press of Colorado, Niwot.

Nentwig, W. 1985. Prey analysis of four species of tropical orb-weaving spiders (Araneae: Araneidae) and a comparison with araneids of the temperate zone. Oecologia (Berl.) 66:580–594.

Nettles, W. C. 1963. Garden fleahopper (*Halticus bracteatus* [sic]). U.S. Dep. Agric. Coop. Econ. Insect Rep. 13(40): 1175.

Neuenschwander, P., R. D. Hennessey, and H. R. Herren. 1987. Food web of insects associated with the cassava mealybug, *Phenacoccus manihoti* Matile-Ferrero (Hemiptera-Pseudococcidae), and its introduced parasitoid, *Epidinocarsis lopezi* (De Santis) (Hymenoptera: Encyrtidae), in Africa. Bull. Entomol. Res. 77:177–189.

Neumann, P. 1955. Krankheiten der Keimlinge und Jungpflanzen unserer Kohlgewächse. Pflanzenschutz (Vienna) 7:39–44.

Neunzig, H. H. and G. G. Gyrisco. 1955. Some insects injurious to birdsfoot trefoil in New York. J. Econ. Entomol. 48:447–450.

Nevinnykh, V. A. and M. A. Riabov. 1931. On the injury caused to the growing point and on the shedding of buds and ovaries in kenaf (*Hibiscus cannabinus*). Plant Prot. (Leningr.) 8:43–66. [in Russian, English summary.]

Newcomer, E. J. 1932. Tarnished plant bug (*Lygus pratensis* L.). U.S. Dep. Agric. Insect Pest Surv. Bull. 12:93.

Newsom, L. D. 1974. Pest management: History, current status and future progress. Pages 1–18 *in* F. G. Maxwell and F. A. Harris, eds., Proceedings of the Summer Institute on Biological Control of Plant Insects and Diseases. University of Mississippi Press, Jackson.

Newsom, L. D. and J. R. Brazzel. 1968. Pests and their control. Pages 367–405 *in* F. C. Elliot et al., eds., Advances in production and utilization of quality cotton: Principles and practices. Iowa State University Press, Ames.

Newton, A. F. Jr. 1984. Mycophagy in Staphylinoidea (Coleoptera). Pages 302–353 *in* Q. Wheeler and M. Blackwell, eds., Fungus-insect relationships: Perspectives in ecology and evolution. Columbia University Press, New York.

Newton, R. C. and R. R. Hill Jr. 1970. Use of caged adult forage insects to determine their comparative roles in delaying the regrowth of alfalfa. J. Econ. Entomol. 63:1542–1543.

Nichols, S. W., compiler. 1989. The Torre-Bueno glossary of entomology. New York Entomological Society, New York. 840 pp.

Nickle, W. R. 1978. On the biology and life history of some terrestrial mermithids parasitic on agricultural insects. J. Nematol. 10:295. (Abstr.)

Nielson, M. W., H. Don, and J. Zaugg. 1974. Sources of resistance in alfalfa to *Lygus hesperus* Knight. U.S. Dep. Agric. ARS W-21:1–5.

Niemczyk, E. 1963. Heteroptera associated with apple orchards in the district of Nowy Sącz. Ekol. Pol. (A) 11:295–300.

——. 1966a. Food ecology of *Psallus ambiguus* (Fall.) (Heteroptera: Miridae). Page 69 *in* I. Hodek, ed., Ecology of aphidophagous insects. Proc. Symp. Liblice near Prague, Sept. 27–Oct. 1, 1965. Junk, The Hague; Academia, Prague.

——. 1966b. Predators of aphids, associated with apple orchards. Pages 275–276 *in* I. Hodek, ed., Ecology of aphidophagous insects. Proc. Symp. Liblice near Prague, Sept. 27–Oct. 1, 1965. Junk, The Hague; Academia, Prague.

——. 1967. *Psallus ambiguus* (Fall.) (Heteroptera, Miridae) Część I: Morfologia i biologia. Pol. Pismo Entomol. 37: 797–842.

——. 1968. *Psallus ambiguus* (Fall.) (Heteroptera, Miridae). Część II. Odżywianie się i rola w biocenozie sadów. Pol. Pismo Entomol. 38:387–416.

——. 1978. *Campylomma verbasci* Mey-Dur (Heteroptera, Miridae) as a predator of aphids and mites in apple orchards. Pol. Pismo Entomol. 48:221–235.

Niemczyk, E. and M. Pruska. 1986. The occurrence of predators in different types of colonies of apple aphids. Pages 303–310 *in* I. Hodek, ed., Ecology of aphidophaga. Proc. 2nd Symp. at Zvíkovské Podhradí, Sept. 2–8, 1984. Junk, Dordrecht.

Niklas, K. J. 1985. Wind pollination—a study in controlled chaos. Am. Sci. 73:462–470.

Niles, G. A. 1980. Breeding cotton for resistance to insect pests. Pages 337–369 *in* F. G. Maxwell and P. R. Jennings, eds., Breeding plants resistant to insects. Wiley, New York.

Nishida, T. 1958. Extrafloral glandular secretions, a food source for certain insects. Proc. Hawaii. Entomol. Soc. 16:379–386.

Nishijima, Y. and K. Sogawa. 1963. Morphological studies on the salivary glands of Hemiptera I. Heteroptera. Res. Bull. Obihiro Univ. Ser. I. 3:512–521.

Nitobe, I. 1906. On *Heterocordylus flaviceps* Mats. of apple trees in the prefecture Aomori. Konch. Sek. [Insect World], Gifu 10:19–22. [in Japanese.]

Nixon, G. E. J. 1946. Euphorine parasites of capsid and lygaeid bugs in Uganda (Hymenoptera, Braconidae). Bull. Entomol. Res. 37:113–129.

Nizamlioglu, K. 1962. Türkiye pamuklarinda yeni bir haşere *Phytocoris obscurus* Reut. Koruma (Istanbul) 3(24):6–7.

Nokkala, S. 1986. The mechanisms behind the regular segregation of autosomal univalents in *Calocoris quadripunctatus* (Vil.) (Miridae, Hemiptera). Hereditas (Lund) 105:199–204.

Nokkala, S. and C. Nokkala. 1986. Achiasmatic male meiosis of collochore type in the heteropteran family Miridae. Hereditas (Lund) 105:193–197.

Nordlund, D. A. and J. C. Legaspi. 1994. Whitefly predators and their possible use in biological control. International *Bemisia* Workshop, Shoresh, Israel, 3–7 Oct. 1994. *Bemisia* Newsl. 8:25.

Norman, J. W. Jr. and A.[N.] Sparks Jr. 1998. *Creontides* [*sic*] species. Pest Cast (Tex. Agric. Ext. Serv.) 24(8):2.

Norris, D. M. 1979. How insects induce disease. Pages 239–255 *in* J. G. Horsfall and E. B. Cowling, eds., Plant disease: An advanced treatise. Vol. IV. How pathogens induce disease. Academic, New York.

Norris, K. R. 1991. General biology. Pages 68–108 *in* CSIRO. The insects of Australia: A textbook for students and research workers, 2nd ed. Cornell University Press, Ithaca; Melbourne University Press, Melbourne.

Norton, A. P. and S. C. Welter. 1996. Augmentation of the egg parasitoid *Anaphes iole* (Hymenoptera: Mymaridae) for *Lygus hesperus* (Heteroptera: Miridae) management in strawberries. Environ. Entomol. 25:1406–1414.

Norton, A. P., S. C. Welter, J. L. Flexner, C. G. Jackson, J. W. Debolt, and C. Pickel. 1992. Parasitism of *Lygus hesperus* (Miridae) by *Anaphes iole* (Mymaridae) and *Leiophron uniformis* (Braconidae) in California strawberry. Biol. Control 2:131–137.

Norton, B. E. and L. B. Smith. 1975. Plant response to insect herbivory. US/IBP Desert Biome Res. Memo. 75–15. Utah State University, Logan. 6 pp.

Novak, H. and R. Achtziger. 1995. Influence of heteropteran predators (Het., Anthocoridae, Miridae) on larval populations of hawthorn psyllids (Hom., Psyllidae). J. Appl. Entomol. 119:479–486.

Novinenko, A. I. 1928. Insects as carriers of the mosaic disease of sugar beet. Prot. Plants, Ukraine 4:164–168. [in Russian; English summary in Rev. Appl. Mycol. 8:82, 1929.]

Nowicki, M. 1871. Ueber die Weizenverwüsterin *Chlorops taeniopus* Meig. und die Mittel zu ihrer Bekämpfung. K. K. Zoologisch-botanischen Gesellschaft in Wien. 58 pp.

Nucifora, A. and C. Calabretta. 1986. Advances in integrated control of gerbera protected crops. Acta Hortic. (Wageningen) 176:191–197.

Nucifora, A. and V. Vacante. 1987. The state of protected crops in the Mediterranean basin and the present possibilities for a pest integrated control. Proc. Working Group, "Integrated Control in Glasshouses," EPRS/WPRS, Budapest, Hungary, 26–30 April 1987. Int. Organ. Biol. Control/West Palearctic Reg. Sect. 10(2):139–143.

Nuessly, G. S. and W. L. Sterling. 1994. Mortality of *Helicoverpa zea* (Lepidoptera: Noctuidae) eggs in cotton as a function of oviposition sites, predator species, and desiccation. Environ. Entomol. 23:1189–1202.

Nummelin, M. 1989. Cannibalism in waterstriders (Heteroptera: Gerridae):

Is there kin recognition? Oikos 56:87–90.

Nuorteva, P. 1954. Studies on the salivary enzymes of some bugs injuring wheat kernels. Ann. Entomol. Fenn. 20:102–124.

———. 1955. On the nature of the plant injuring salivary toxins of insects. Ann. Entomol. Fenn. 21:33–38.

———. 1956a. Studies on the effect of the salivary secretions of some Heteroptera and Homoptera on plant growth. Ann. Entomol. Fenn. 22:108–117.

———. 1956b. Developmental changes in the occurrence of the salivary proteases in *Miris dolabratus* L. (Hem., Miridae). Ann. Entomol. Fenn. 22:117–119.

———. 1956c. The possibility of distinguishing the symptoms of injury to wheat kernels made by different heteropterous bugs. Ann. Entomol. Fenn. 22:120–121.

———. 1958. Die Rolle der Speichelsekrete im Wechselverhältnis zwischen Tier und Nahrungspflanze bei Homopteren und Heteropteren. Entomol. Exp. Appl. 1:41–49.

Nuorteva, P. and S. Laurema. 1961. Observations on the activity of salivary proteases and amylases in *Dolycoris baccarum* (L.) (Het., Pentatomidae). Ann. Entomol. Fenn. 27:93–97.

Nuorteva, P. and L. Reinius. 1953. Incorporation and spread of C14-labeled oral secretions of wheat bugs in wheat kernels. Ann. Entomol. Fenn. 19:95–104.

Nuorteva, P. and T. Veijola. 1954. Studies on the effect of injury by *Lygus rugulipennis* Popp. (Hem., Capsidae) on the baking quality of wheat. Ann. Entomol. Fenn. 20:65–68.

Nuzzaci, G. 1977. Il *Lygus (Orthops) kalmi* L. dannoso al finocchio. Inf. Fitopatol. 27(3):3–5.

Nwana, I. E. and A. Youdeowei. 1976. The effect of relative humidity on the development and survival of the preimaginal stages of *Bathycoelia thalassina* (H-S) (Pentatomidae) and *Sahlbergella singularis* Hagl. (Miridae) in Nigeria. J. Nat. Hist. 11:445–449.

Nwanze, K. F. 1985. Sorghum insect pests in West Africa. *In* Proc. Int. Sorghum Entomol. Workshop, 15–21 July 1984, Texas A & M University, College Station, pp. 37–43. International Crops Research for the Semi-Arid Tropics, Patancheru, A. P., India.

Nye, W. P. and J. L. Anderson. 1974. Insect pollinators frequenting strawberry blossoms and the effect of honey bees on yield and fruit quality. J. Am. Soc. Hortic. Sci. 99:40–44.

Nyffeler, M., D. A. Dean, and W. L. Sterling. 1987. Predation by green lynx spider, *Peucetia viridans* (Araneae: Oxyopidae), inhabiting cotton and woolly croton plants in East Texas. Environ. Entomol. 16:355–359.

Nyland, G., R. M. Gilmer, and J. D. Moore. 1976. "Prunus" ring spot group. Pages 104–132 *in* T. S. Pine, et al., eds., Virus diseases and noninfectious disorders of stone fruits in North America. U.S. Dep. Agric. Agric. Handb. 437.

Oakman, J. H. 1981. The lygus problem—decreasing or increasing. *In* Proc. Beltwide Cotton Prod. Res. Conf., Jan. 4–8, 1981, New Orleans, La., p. 148. National Cotton Council, Memphis.

O'Brien, L. B. and S. W. Wilson. 1985. Planthopper systematics and external morphology. Pages 61–102 *in* L. R. Nault and J. G. Rodriguez, eds., The leafhoppers and planthoppers. Wiley, New York.

O'Brien, M. H. 1980. The pollination biology of a pavement plain: Pollination visitation patterns. Oecologia (Berl.) 47:213–218.

O'Brien, M. J. and A. E. Rich. 1979. Potato diseases. U.S. Dep. Agric. Agric. Handb. 474 (rev.):1–79.

Occhioni, P. 1944. Da raiz de *Tephrosia toxicaria* Pers. e do seu aproveitamento no combate ao *Tenthecoris bicolor* Scott. Rodriguésia 16:55–61.

O'Connor, B. A. 1952. The rice leaf hopper, *Sogata furcifera kolophon* Kirkaldy and "rice yellows." Fiji Agric. J. 23:97–104.

Odhiambo, T. R. 1958. Notes on the East African Miridae (Hemiptera).—V: New species of *Eurystylus* Stål from Uganda. Ann. Mag. Nat. Hist. (13)1:257–281.

———. 1959. Notes on the East African Miridae (Hemiptera).—XII: New species of the genera *Campylomma* Reuter and *Sthenarus* Fieber. Ann. Mag. Nat. Hist. (13)2:421–438.

———. 1961. A study of some African species of the *Cyrtopeltis* complex (Hemiptera: Miridae). Rev. Entomol. Moçamb. 4:1–35.

———. 1962. Review of some genera of the subfamily Bryocorinae (Hemiptera: Miridae). Bull. Brit. Mus. (Nat. Hist.) Entomol. 11:245–331.

O'Donohue, J. B. 1992. Practical integrated control of pests and diseases of cocoa in Papua New Guinea. Pages 69–74 *in* P. J. Keane and C. A. J. Putter, eds., Cocoa pest and disease management in Southeast Asia and Australasia. FAO Plant Prod. Prot. Pap. 112.

Oetting, R. D. and T. R. Yonke. 1971. Immature stages and biology of *Podisus placidus* and *Stiretrus fimbriatus* (Hemiptera: Pentatomidae). Can. Entomol. 103:1505–1516.

———. 1975. Immature stages and notes on the biology of *Euthyrhynchus floridanus* (L.) (Hemiptera: Pentatomidae). Ann. Entomol. Soc. Am. 68:659–662.

Ogle, W. 1878. Editor's preface. Pages vii–xiv *in* A. Kerner, Flowers and their unbidden guests. (Transl. from German and edited by W. Ogle.) C. Kegan Paul, London.

Ohler, J. G. 1979. Cashew. Koninklijk Instituut voor de Tropen, Amsterdam. 260 pp.

Ojo, A. 1981. Assessment of damage to cacao by *Sahlbergella singularis* Hagl. (Miridae) and the effects of pods and beans. *In* Proc. 6th Int. Cocoa Res. Conf. Caracas, Venezuela 1977, pp. 391–395.

———. 1985. A note on the qualitative damage caused to cocoa pods by *Sahlbergella singularis* (Hagl.) (Hemiptera: Miridae). Turrialba 35:87–88.

Olalquiaga, F. G. 1955. Insect problems in Chile. FAO Plant Prot. Bull. 3:65–70.

Oliveira, P. S. 1985. On the mimetic association between nymphs of *Hyalymenus* spp. (Hemiptera: Alydidae) and ants. Zool. J. Linn. Soc. 83:371–384.

Oliver, J. E., J. W. Neal Jr., W. R. Lusby, J. R. Aldrich, and J. P. Kochansky. 1985. Novel components from secretory hairs of azalea lace bug *Stephanitis pyrioides* (Hemiptera: Tingidae). J. Chem. Ecol. 11:1223–1228.

Oliver, J. E., W. R. Lusby, and J. W. Neal Jr. 1990. Exocrine secretions of the andromeda lace bug *Stephanitis takeyai* (Hemiptera: Tingidae). J. Chem. Ecol. 16:2243–2252.

Olson, D. L. and R. L. Wilson. 1990. Tarnished plant bug (Hemiptera: Miridae): Effect on seed weight of grain amaranth. J. Econ. Entomol. 83:2443–2447.

Önder, F. 1972. İzmir ili çevresinde bitki zararlisi Mirinae (Miridae: Hemiptera) türlerinin taninmalari, konukçulari, yayilişlari ve kisa biyolojileri üzerinde araştirmalar. Ege Univ. Ziraat Fak. Derg. (A) 9:221–241.

Önder, F. and Y. Karsavuran. 1986. İzmir çevresinde çiriş otu (*Asphodelus microcarpus* Salz. Viv.) 'na karşi uygulanacak biyolojik savaşta *Capsodes infuscatus* (Brul.) (Heteroptera: Miridae) 'un etkinliği üzerinde araştirmalar. Page 26 *in* Türkiye I. Biyolojik Mücadele Kongresi (12–14 Subat 1986), Adana. Çukurova Üniversitesi, Adana.

Önder, F., R. Atalay, and Y. Karsavuran. 1983. İzmir ili çevresinde kişi ergin halde geçiren Heteroptera türleri ve kişlak yerleri üzerinde araştirmalar I. Türkiye Bitki Kor. Derg. 7:65–77.

O'Neal, L. H. and A. G. Peterson. 1971. A population study of *Lygus lineolaris* on alfalfa grown for forage and an evaluation of its damage. Proc. North Cent. Branch Entomol. Soc. Am. 26:84–85.

O'Neil, R. J. 1990. Functional response of arthropod predators and its role in the biological control of insect pests in agricultural systems. Pages 83–96 in R. R. Baker and P. E. Dunn, eds., New directions in biological control: Alternatives for suppressing agricultural pests and diseases. Liss, New York.

Ooi, P. A. C. 1979. Flight activities of brown planthopper, whitebacked planthopper, and their predator C. lividipennis in Malaysia. Int. Rice Res. Newsl. 4(6):12.

——. 1988. Ecology and surveillance of Nilaparvata lugens (Stal)—implications for its management in Malaysia. Ph.D. dissertation, University of Malaya, Kuala Lumpur, Malaysia. 275 pp.

——. 1992. Prospects for biological control of cocoa insect pests. Pages 101–107 in P. J. Keane and C. A. J. Putter, eds., Cocoa pest and disease management in Southeast Asia and Australasia. FAO Plant Prod. Prot. Pap. 112.

Ooi, P. A. C. and B. M. Shepard. 1994. Predators and parasitoids of rice insect pests. Pages 585–612 in E. A. Heinrichs, ed., Biology and management of rice insects. Wiley, New York.

Ooi, P. A. C. and J. K. Waage. 1994. Biological control in rice: Applications and research needs. Pages 209–216 in P. S. Teng, K. L. Heong, and K. Moody, eds., Rice pest science and management: Selected papers from the International Rice Research Conference. International Rice Research Institute, Manila.

Opler, P. A. 1983. Nectar production in a tropical ecosystem. Pages 30–79 in B. Bentley and T. Elias, eds., The biology of nectaries. Columbia University Press, New York.

Oppong-Mensah, D. and R. Kumar. 1973. Internal reproductive organs of the cocoa-capsids (Heteroptera-Miridae). Entomol. Mon. Mag. 109:148–154.

Orlob, G. B. 1963. Reappraisal of transmission of tobacco mosaic virus by insects. Phytopathology 53:822–830.

Orr, B. K., W. W. Murdoch, and J. R. Bence. 1990. Population regulation, convergence, and cannibalism in Notonecta (Hemiptera). Ecology 71:68–82.

Osakabe, M., K. Inoue, and W. Ashihara. 1986. Feeding, reproduction and development of Amblyseius sojaensis Ehara (Acarina: Phytoseiidae) on two species of spider mites and on tea pollen. Appl. Entomol. Zool. 21:322–327.

Osborn, H. 1888. Observations on certain species of Hemiptera. Proc. Entomol. Soc. Wash. 1:35.

——. 1898. Additions to the list of Hemiptera of Iowa, with descriptions of new species. Proc. Iowa Acad. Sci. 5:232–247.

——. 1918. The meadow plant bug, Miris dolabratus. J. Agric. Res. (Wash., D.C.) 15:175–200.

——. 1919. The meadow plant bug. Maine Agric. Exp. Stn. Bull. 276:1–16.

——. 1931. Early work and workers in American hemipterology. Ann. Entomol. Soc. Am. 24:679–685.

——. 1939. Meadow and pasture insects. Educators' Press, Columbus, Ohio. 288 pp.

Osborne, D. J. 1973. Mutual regulation of growth and development in plants and insects. Pages 33–42 in H. F. van Emden, ed., Insect/plant relationships. Blackwell, Oxford.

Osborne, L. S., F. L. Pettit, Z. Landa, and K. A. Hoelmer. 1994. Biological control of pests attacking crops grown in protected culture: The Florida experience. Pages 327–342 in D. Rosen, F. D. Bennett, and J. L. Capinera, eds., Pest management in the subtropics: Biological control—a Florida perspective. Intercept, Andover, UK.

Osman, D. H. and W. A. Brindley. 1981. Estimating monooxygenase detoxification in field populations: Toxicity and distribution of carbaryl in three species of Labops grassbugs. Environ. Entomol. 10:676–680.

Ossiannilsson, F. 1946. Orkidéstinkflyet—en icke önskvärd utlänning. Växtskyddsnotiser 10(4):55–56.

——. 1966. Insects in the epidemiology of plant viruses. Annu. Rev. Entomol. 11:213–232.

Ostry, M. E., L. F. Wilson, H. S. McNabb Jr., and L. M. Moore. 1988. A guide to insect, disease, and animal pests of poplars. U.S. Dep. Agric. For. Serv. Agric. Handb. 677:1–118.

Otake, A. 1977. Natural enemies of the brown planthopper. Pages 42–56 in Food and Fertilizer Technology Center for the Asian and Pacific Region, compil., The rice brown planthopper. Taipei, Taiwan.

Otanes, F. Q. 1948. Notes on orchid pests and suggestions for their control. Philipp. Orchid Rev. 1(1):12–18.

Otten, E. 1956. Heteroptera, Wanzen, Halbflügler. Pages 1–149 in H. Blunck, Tierische Schädlinge an Nutzpflanzen. 2. Teil. P. Sorauer. Handbuch der Pflanzenkrankheiten. 5 Band, 5 Auflage, 3 Lieferung. Parey, Berlin.

Ottosson, J. G. and J. M. Anderson. 1983. Number, seasonality and feeding habits of insects attacking ferns in Britain: An ecological consideration. J. Anim. Ecol. 52:385–406.

Oudemans, A. C. 1912. Die bis jetzt bekannten Larven von Thrombidiidae und Erythraeidae mit besonderer Berücksichtigung der für den Menschen schädlichen Arten. Zool. Jahrb. Suppl. 14(1):1–230.

Overhulser, D. L. and A. Kanaskie. 1989. Lygus bugs. Pages 146–147 in C. E. Cordell, et al., tech. coords., Forest nursery pests. U.S. Dep. Agric. For. Serv. Agric. Handb. 680.

Owen, D. F. 1990. The language of attack and defence. Oikos 57:133–135.

Owen, H. 1956. Further observations on the pathogenicity of Calonectria rigidiuscula (Berk. & Br.) Sacc. to Theobroma cacao L. Ann. Appl. Biol. 44:307–321.

Owen, J. 1980. Feeding strategy. University of Chicago Press, Chicago. 160 pp.

Owusu-Manu, E. 1985. The evaluation of the synthetic pyrethroids for the control of Distantiella theobroma Dist. (Hemiptera, Miridae) in Ghana. In Proc. 9th Int. Cocoa Res. Conf., Lomé, Togo, 12 Feb. 1984, pp. 535–538. Cocoa Producers' Alliance, Lagos, Nigeria.

Ozols, E. and J. Zirnits. 1927. Insect pests in 1926. Rep. Latvian Inst. Plant Prot. 1926–1927:13–16. [in Lettish.]

Pack, T. M. and [N.] P. Tugwell [Jr.]. 1976. Clouded and tarnished plant bugs on cotton: A comparison of injury symptoms and damage on fruit parts. Ark. Agric. Exp. Stn. Rep. Ser. 226:1–17.

Packham, J. M. 1982. Holcus, Holcaphis and food quality. Pages 429–430 in J. H. Visser and A. K. Minks, eds., Proc. 5th Int. Symp. Insect-Plant Relationships, Wageningen, the Netherlands, 1–4 Mar. 1982. Centre for Agricultural Publishing & Documentation, Wageningen.

Painter, R. H. 1927. Notes on the oviposition habits of the tarnished plant bug, Lygus pratensis Linn, with a list of host plants. 57th Annu. Rep. Entomol. Soc. Ont. 1926:44–46.

——. 1928. Notes on the injury to plant cells by chinch bug feeding. Ann. Entomol. Soc. Am. 21:232–242.

——. 1929a. A brief note on the occurrence of a mermithid parasite genus Hexamermis, in the tarnished plant bug, Lygus pratensis L. Qué. Soc. Prot. Plants. 21st Annu. Rep. 21:53–55.

——. 1929b. The tarnished plant bug Lygus pratensis L.: A progress report. 16th Annu. Rep. Entomol. Soc. Ont. 1929:102–107.

——. 1930. A study of the cotton flea hopper, Psallus seriatus Reut., with especial reference to its effect on cotton plant tissues. J. Agric. Res. (Wash., D.C.) 40:485–516.

———. 1951. Insect resistance in crop plants. Macmillan, New York. 520 pp.

Palmer, L. S. and H. H. Knight. 1924. Anthocyanin and flavone-like pigments as cause of red colorations in the hemipterous families Aphididae, Coreidae, Lygaeidae, Miridae, and Reduviidae. J. Biol. Chem. 59:451–455.

Palmer, W. A. 1987. The phytophagous insect fauna associated with *Baccharis halimifolia* L. and *B. neglecta* Britton in Texas, Louisiana, and northern Mexico. Proc. Entomol. Soc. Wash. 89:185–199.

Palmer, W. A. and K. R. Pullen. 1995. The phytophagous arthropods associated with *Lantana camara, L. hirsuta, L. urticifolia,* and *L. urticoides* (Verbenaceae) in North America. Biol. Control 5:54–73.

———. 1998. The host range of *Falconia intermedia* (Distant) (Hemiptera: Miridae): A potential biological control agent for *Lantana camara* L. (Verbenaceae). Proc. Entomol. Soc. Wash. 100:633–635.

Pang, T. C. 1981. The present status of cocoa bee bug, *Platyngomiriodes apiformis* Ghauri, in Sabah and its life cycle study. Pages 83–90 *in* Int. Symp. Problems of Insect Pest Management in Developing Countries. Trop. Agric. Res. Ser. 14. Trop. Agric. Res. Cent. Min. Agric., For., Fish., Yatabe, Tsukuba, Ibaraki, Japan.

Pang, T. C. and R. A. Syed. 1972. Some important pests of cocoa in Sabah. Pages 45–49 *in* R. L. Wastie and D. A. Earp, eds., Cocoa and coconuts in Malaysia. Proc. Conf. [on cocoa and coconuts], Kuala Lumpur, 25–27 Nov. 1971. Incorporated Society of Planters, Kuala Lumpur, Malaysia.

Pankanin, M. 1972. Wpływ pokarmu na przeżywalność *Lygus rugulipennis* (Popp.) (Heteroptera, Miridae) w warunkach laboratoryjynch. Pol. Pismo Entomol. 42:223–227.

Paoli, G. 1923. Rincote dannoso all vite. Boll. Soc. Entomol. Ital. 56:110–112.

———. 1924. La "rissetta" delle viti. Redia 15:181–189.

Pape, H. 1925. Neue und wenig bekannte Insektenschäden bei der Tomate. Gartenwelt (Hambg.) 29:628–630.

———. 1935. Über eine Mosaikkrankheit der Kohlrübe. Dtsch. Landw. Pr. 62:319–320.

Paraqueima, O. L. 1977. Some effects of different temperatures on the development of the black grass bug (*Labops hesperius* Uhler) from the egg through the adult stage. M.S. thesis, Utah State University, Logan. 84 pp.

Parencia, C. R. Jr. 1978. One hundred twenty years of research on cotton insects in the United States. U.S. Dep. Agric. Agric. Handb. 515:1–75.

Parfitt, E. 1884. The fauna of Devon. Hemiptera Heteroptera; or, plant bugs. Rep. Trans. Devonshire Assoc. Adv. Sci. Lit. Art 16:749–774.

Parker, B. L. and K. I. Hauschild. 1975. A bibliography of the tarnished plant bug, *Lygus lineolaris* (Hemiptera: Miridae), on apple. Bull. Entomol. Soc. Am. 21:119–121.

Parker, W. B. 1913. The hop aphis in the Pacific Region. U.S. Dep. Agric. Bur. Entomol. Bull. 111:1–43.

Parrella, M. P. and J. A. ["F."] Bethke. 1982. Biological studies with *Cyrtopeltis modestus* (Hemiptera: Miridae): A facultative predator of *Liriomyza* spp. (Diptera: Agromyzidae). Pages 180–185 *in* S. L. Poe, ed., Proceedings of the 3rd Annual Industry Conference on the Leafminer, San Diego, Calif. Society of American Florists, Alexandria, Va.

Parrella, M. P. and V. P. Jones. 1987. Development of integrated pest management strategies in floricultural crops. Bull. Entomol. Soc. Am. 33:28–34.

Parrella, M. P., J. P. McCaffrey, and R. L. Horsburgh. 1981. Population trends of selected phytophagous arthropods and predators under different pesticide programs in Virginia apple orchards. J. Econ. Entomol. 74:492–498.

Parrella, M. P., K. L. Robb, G. D. Christie, and J. A. Bethke. 1982. Control of *Liriomyza trifolii* with biological agents and insect growth regulators. Calif. Agric. 36(11–12):17–19.

Parrott, P. J. 1913. New destructive insects in New York. J. Econ. Entomol. 6:61–65.

Parrott, P. J. and H. E. Hodgkiss. 1913. The false tarnished plant-bug as a pear pest. N.Y. Agric. Exp. Stn. (Geneva) Bull. 368:363–384.

Parrott, W. L., J. N. Jenkins, F. G. Maxwell, and M. L. Bostick. 1975. Improved techniques for rearing the tarnished plant bug, *Lygus lineolaris* (Palisot de Beauvois). Miss. Agric. For. Exp. Stn. Tech. Bull. 72:1–8.

Parshley, H. M. 1915. On the external anatomy of *Adelphocoris rapidus* Say, with reference to the taxonomy of the Miridae or Capsidae (Hemip.). Entomol. News 26:208–213.

———. 1919. On the preparation of Hemiptera for the cabinet. Entomol. News 30:223–227.

———. 1923. Two new comprehensive works on the Hemiptera: A review. Bull. Brooklyn Entomol. Soc. 18:166–169.

Partyka, R. E. 1982. The ways we kill a plant. J. Arboric. 8:57–66.

Parvin, D. W. Jr. and J. W. Smith. 1996. Crop phenology and insect management. Pages 815–829 *in* E. G. King, J. R. Phillips, and R. J. Coleman, eds., Cotton insects and mites: Characterization and management. Cotton Foundation, Memphis.

Parvin, D. W. Jr., J. W. Smith, and F. T. Cooke Jr. 1989. The economics of biological control of *Heliothis*. Pages 89–112 *in* E. G. King and R. D. Jackson, eds., Proceedings of the Workshop on Biological Control of *Heliothis*: Increasing the Effectiveness of Natural Enemies, 11–15 Nov. 1985, New Delhi, India. Far Eastern Regional Research Office, U.S. Department of Agriculture, New Delhi, India.

Patana, R. 1969. Rearing cotton insects in the laboratory. U.S. Dep. Agric. Res. Serv. Prod. Res. Rep. 108:1–6.

———. 1982. Disposable diet packet for feeding and oviposition of *Lygus hesperus* (Hemiptera: Miridae). J. Econ. Entomol. 75:668–669.

Patana, R. and J. W. Debolt. 1985. Rearing *Lygus hesperus* in the laboratory. U.S. Dep. Agric. Agric. Res. Serv. ARS-45:1–9.

Patch, E. M. 1906. Insect notes for 1906. Maine Agric. Exp. Stn. Bull. 134:209–228.

———. 1907. Insect notes for 1907. Maine Agric. Exp. Stn. Bull. 148:261–282.

Patel, N. G. 1980. The bionomics and control measures of tobacco bug *Nesidiocoris tenuis* Reuter (Miridae: Hemiptera). Gujarat Agric. Univ. Res. J. 5(2):60.

Paula, A. S. and P. S. F. Ferreira. 1998. Fauna de Heteroptera de la "Mata do Córrego do Paraíso", Viçosa, Minas Gerais, Brasil. I. Riqueza y diversidad específicas. An. Inst. Biol. Univ. Autón Méx. Ser. Zool. 69:39–51.

Pavis, C. 1987. Les sécrétions exocrines des Hétéroptères (allomones et phéromones). Une mise au point bibliographique. Agronomie (Paris) 7:547–561.

Payne, J. A., F. W. Mead, and E. W. King. 1968. Hemiptera associated with pig carrion. Ann. Entomol. Soc. Am. 61:565–567.

Payne, R. M. 1984. Insects on flowers of *Hieracium speluncarum* Arv.-Touv. Entomol. Mon. Mag. 120:118.

Payne, W. W. 1972. Observations of harmomegathy in pollen of Anthophyta. Grana 12:93–98.

Peal, S. E. 1873. The tea-bug of Assam. J. Agric. Hortic. Soc. India 4(1):126–132.

Pearson, E. O. 1949. Problems of insect

pests of cotton in tropical Africa. Emp. Cotton Grow. Rev. 26:85–99.

——. 1958. The insect pests of cotton in tropical Africa. Empire Cotton Growing Corp.; Commonwealth Institute of Entomology, London. 355 pp.

Pearson, W. D. 1991. Effect of meadow spittlebug and Australian crop mirid on white clover seed production in small cages. N.Z. J. Agric. Res. 34:439–444.

Pedigo, L. P. 1989. Entomology and pest management. Macmillan, New York. 646 pp.

Pegazzano, F. 1958. Osservazioni su alcuni emitteri eterotteri (gen. *Calocoris*, fam. Miridae) e sui danni da essi arrecati al pesco. Redia 43:137–143.

Pellitteri, P. and C. F. Koval. 1981. Ash (*Fraxinus*) disorder: Ash plant bug. Univ. Wis.-Ext. A3126:1–2.

Pellmyr, O. 1984. The pollination ecology of *Actaea spicata* (Ranunculaceae). Nord. J. Bot. 4:443–456.

——. 1992. The phylogeny of a mutualism: Evolution and co-adaptation between *Trollius* and its seed-parasitic pollinators. Biol. J. Linn. Soc. 47:337–365.

Pellmyr, O. and L. B. Thien. 1986. Insect reproduction and floral fragrances: Keys to the evolution of the angiosperms? Taxon 35:76–85.

Pemberton, C. E. 1948. History of the Entomology Department Experiment Station, H.S.P.A. 1904–1945. Hawaii. Plant. Rec. 52:53–90.

——. chairman. 1954. Introduction of beneficial parasites and predators into Guam and the Trust Territory. Pages 42–45 *in* Invertebrate Consultants Committee for the Pacific; Rep. 1949–1954. National Research Council Pacific Science Board, Washington, D.C. 56 pp.

Peña, J. E. 1993. Pests of mango in Florida. Acta Hortic. (Wageningen). 341:395–406.

Peña, J. E., R. M. Baranowski, and H. Nadel. 1996. Pest management of tropical fruit trees in Florida. Pages 349–370 *in* D. Rosen, F. D. Bennett, and J. L. Capinera, eds., Pest management in the subtropics: Integrated pest management—a Florida perspective. [Vol. 2.] Intercept, Andover, UK.

Pendergrass, J. 1989. An overview of pest management in cotton in the USA. Pages 11–15 *in* M. B. Green and D. J. de B. Lyon, eds., Pest management in cotton. Ellis Horwood, Chichester, UK.

Peng, R. K., K. Christian, and K. Gibb. 1995. The effect of the green ant, *Oecophylla smaragdina* (Hymenoptera: Formicidae), on insect pests of cashew trees in Australia. Bull. Entomol. Res. 85:279–284.

——. 1997. Control threshold analysis for the tea mosquito bug, *Helopeltis pernicialis* (Hemiptera: Miridae) and preliminary results concerning the efficiency of control by the green ant, *Oecophylla smaragdina* (Hymenoptera: Formicidae) in northern Australia. Int. J. Pest Manage. 43:233–237.

Peng, Y. S., M. E. Nasr, J. M. Marston, and Y. Fang. 1985. The digestion of dandelion pollen by adult worker honeybees. Physiol. Entomol. 10:75–82.

Penna, R. J., W. M. Morgan, and M. S. Ledieu. 1983. Pests and diseases of protected crops. Pages 375–472 *in* N. Scopes and M. Ledieu, eds., Pest and disease control handbook. British Crop Protection Council, Croydon, UK.

Penny, N. D. and J. R. Arias. 1982. Insects of an Amazon forest. Columbia University Press, New York. 269 pp.

Pepper, J. H., N. L. Anderson, G. R. Roemhild, and L. N. Graham. 1956. Montana insect pests, 1955–1956. Mont. Agric. Exp. Stn. Bull. 526:1–27.

Perdikis, D. [C.] and D. [P.] Lykouressis. 2000. Effects of various items, host plants, and temperatures on the development and survival of *Macrolophus pygmaeus* Rambur (Hemiptera: Miridae). Biol. Control 17:55–60.

Perdikis, D. C., D. P. Lykouressis, and L. P. Economou. 1999. The influence of temperature, photoperiod and plant type on the predation rate of *Macrolophus pygmaeus* on *Myzus persicae*. BioControl 44:281–289.

Peregrine, W. T. H. 1970. A serious disease of annato caused by *Glomerella cingulata*. PANS (Pest Artic. News Summ.) 16:331–333.

——. 1991. Anatto—a possible trap crop to assist control of the mosquito bug (*Helopeltis schoutedeni* Reut.) in tea and other crops. Trop. Pest Manage. 37:429–430.

Perkins, P. V. and T. F. Watson. 1972. *Nabis alternatus* as a predator of *Lygus hesperus*. Ann. Entomol. Soc. Am. 65:625–629.

Pescott, R. T. M. 1940. A capsid plant bug attacking stone fruits. J. Aust. Inst. Agric. Sci. 6:101–102.

Petacchi, R. and E. Rossi. 1991. Prime osservazioni su *Dicyphus (Dicyphus) errans* (Wolff)(Heteroptera Miridae) diffuso sul pomodoro in serre della Liguria. Boll. Zool. Agrar. Bachic. 23:77–86.

Petanidou, T. and W. N. Ellis. 1993. Pollinating fauna of a phryganic ecosystem: Composition and diversity. Biodivers. Lett. 1:9–22.

Petch, T. 1948. A revised list of British entomogenous fungi. Trans. Brit. Mycol. Soc. 31:286–304.

Peterson, A. G. and E. V. Vea. 1969. Silvertop, the elusive mystery. Minn. Sci. 25(2):12–14.

——. 1971. Silvertop of bluegrass in Minnesota. J. Econ. Entomol. 64:247–252.

Peterson, S. S., J. L. Wedberg, and D. B. Hogg. 1992. Plant bug (Hemiptera: Miridae) damage to birdsfoot trefoil seed production. J. Econ. Entomol. 85:250–255.

Petherbridge, F. R and M. A. Husain. 1918. A study of the capsid bugs found on apple trees. Ann. Appl. Biol. 4:179–205.

Petherbridge, F. R. and W. G. Kent. 1926. The control of the apple capsid bug. J. Minist. Agric. (Lond.) 33:50–57.

Petherbridge, F. R. and W. H. Thorpe. 1928a. The common green capsid bug (*Lygus pabulinus*). Ann. Appl. Biol. 15:446–472.

——. 1928b. Notes on the capsid bugs found on species of *Ribes*. Entomol. Mon. Mag. 64:109–113.

Pfannenstiel, R. S. and K. V. Yeargan. 1998. Partitioning two- and three-trophic-level effects of resistant plants on the predator, *Nabis roseipennis*. Entomol. Exp. Appl. 88:203–209.

Pfeiffer, D. G., J. C. Killian, and K. S. Yoder. 1999. Clarifying the roles of white apple leafhopper and potato leafhopper (Homoptera: Cicadellidae) in fire blight transmission in apple. J. Entomol. Sci. 34:314–321.

Phillips, J. H. H. 1958. The tarnished plant bug, *Liocoris lineolaris* (Beauv.) (Hemiptera: Miridae), as a pest of peach in Ontario: A progress report. Annu. Rep. Entomol. Soc. Ont. 88:44–48.

Phillips, J. H. H. and J. H. DeRonde. 1966. Relationship between the seasonal development of the tarnished plant bug, *Lygus lineolaris* (Beauv.) (Hemiptera: Miridae) and its injury to peach fruit. Proc. Entomol. Soc. Ont. 96:103–107.

Phillips, R. L., N. M. Randolph, and G. L. Teetes. 1973. Seasonal abundance and nature of damage of insects attacking cultivated sunflowers. Tex. Agric. Exp. Stn. MP-1116:1–7.

Piart, J. 1970. Étude de quelques caractéristiques biologiques du miride du cacaoyer, *Distantiella theobroma* Dist., au moyen d'un élevage au laboratoire. Café Cacao Thé 14:28–38.

Pickel, C. and R. S. Bethell. 1990. Insects and mites. Pages 1–31 *in* Apple pest management guidelines (UCPMG Publ. 12). Univ. Calif. Div. Nat. Resour. Publ. 3339.

Pickel, C., P. Phillips, H. Otto, J. Trumble, and N. Welch. 1991. Insects and mites. Pages 1–21 *in* Strawberry pest management guidelines (UCPMG

Publ. 22). Univ. Calif. Div. Nat. Resour. Publ. 3339.

Pickel, C., F. G. Zalom, D. B. Walsh, and N. C. Welch. 1994. Efficiency of vacuum machines for *Lygus hesperus* (Hemiptera: Miridae) control in coastal California strawberries. J. Econ. Entomol. 87:1636–1640.

Pickett, A. D. 1938. The mullein leaf bug—*Campylomma verbasci*, Meyer, as a pest of apple in Nova Scotia. Sixty-ninth Annu. Rep. Entomol. Soc. Ont. 1938:105–106.

——. 1959. Utilization of native parasites and predators. J. Econ. Entomol. 52:1103–1105.

——. 1968. Influence on the fecundity of *Distantiella theobroma* of feeding in the laboratory on chupons of various textures. Pages 46–47 *in* Cocoa Res. Inst. (Ghana Acad. Sci.). Annu. Rep. 1965–66.

Pickett, A. D., M. E. Neary, and D. MacLeod. 1944. The mirid, *Calocoris norvegicus* Gmelin, a strawberry pest in Nova Scotia. Sci. Agric. (Ottawa) 24:299–303.

Pieters, E. P. and W. L. Sterling. 1974. Aggregation indices of cotton arthropods in Texas. Environ. Entomol. 3:598–600.

Pillai, G. B. 1987. Integrated pest management in plantation crops. J. Coffee Res. 17:150–153.

Pillai, G. B. and V. A. Abraham. 1975. Tea mosquito—a serious menace to cashew. Indian Cashew J. 10(1):5,7.

Pillai, G. B., O. P. Dubey, and V. Singh. 1976. Pests of cashew and their control in India—a review of current status. J. Plant. Crops 4:37–50.

Pimentel, D. 1961. Competition and the species-per-genus structure of communities. Ann. Entomol. Soc. Am. 54:323–333.

Pinckard, J. A., L. J. Ashworth Jr., J. P. Snow, T. E. Russell, R. W. Roncardori, and G. L. Sciumbato. 1981. Boll rots. Pages 20–24 *in* G. M. Watkins, ed., Compendium of cotton diseases. APS, St. Paul, Minn.

Pinto, J. D. 1978. The parasitization of blister beetles by species of Miridae (Coleoptera: Meloidae; Hemiptera: Miridae). Pan-Pac. Entomol. 54:57–60.

——. 1982. The phenology of the plant bugs (Hemiptera: Miridae) associated with *Ceanothus crassifolius* in a chaparral community of southern California. Proc. Entomol. Soc. Wash. 84:102–110.

——. 1990. The occurrence of *Chaetostricha* in North America, with the description of a new species (Hymenoptera: Trichogrammatidae). Proc. Entomol. Soc. Wash. 92:208–213.

Pinto, J. D. and R. K. Velten. 1986. The plant bugs (Hemiptera: Miridae) associated with *Adenostoma* (Rosaceae) in southern California. J. N.Y. Entomol. Soc. 94:542–551.

Pirone, P. P. 1978. Diseases and pests of ornamental plants, 5th ed. Wiley, New York. 566 pp.

Pirone, P. P., J. R. Hartman, M. A. Sall, and T. P. Pirone. 1988. Tree maintenance, 6th ed. Oxford University Press, New York. 514 pp.

Pitblado, R. E. 1994. Tomato, eggplant, pepper: Other insect pests. Pages 288–289 *in* R. J. Howard, J. A. Garland, and W. L. Seaman, eds., Diseases and pests of vegetable crops in Canada: An illustrated compendium. Canadian Phytopathological Society; Entomological Society of Canada, Ottawa.

Platnick, N. I. 1977. Cladograms, phylogenetic trees, and hypothesis testing. Syst. Zool. 26:438–442.

Pless, C. D. and R. D. Miller. 1986. Injury by the tarnished plant bug to selected potyvirus-resistant burley tobacco lines. Tob. Sci. 30:127–129 (Tob. Int. 188(2):29–31).

Ploaie, P. G. 1981. Mycoplasmalike organisms and plant diseases in Europe. Pages 61–104 *in* K. Maramorosch and K. F. Harris, eds., Plant diseases and vectors: Ecology and epidemiology. Academic, New York.

Podlipaev, S. A. and A. O. Frolov. 1987. Description and laboratory cultivation of *Blastocrithidia miridarum* sp. n. (Mastigophora, Trypanosomatidae). Parazitologiya (Leningr.) 21:545–552. [in Russian, English summary.]

Podlipaev, S. A., A. O. Frolov, and A. A. Kolesnikov. 1990. *Proteomonas inconstans* n. gen., n. sp. (Kinetoplastida, Trypanosomatidae), a parasite of the bug *Calocoris sexguttatus* (Hemiptera, Miridae). Parazitologiya (Leningr.) 24:339–346. [in Russian, English summary.]

Podoler, H. and D. Rogers. 1975. A new method for the identification of key factors from life-table data. J. Anim. Ecol. 44:85–114.

Poe, S. L. 1972. Pest populations on floral crops—*Taylorilygus pallidulus*, the green plant bug. Fla. Coop. Ext. Serv. Fla. Flower Grow. 9(5):1–3.

Poinar, G. O. Jr. 1975. Entomogenous nematodes: A manual and host list of insect-nematode associations. Brill, Leiden. 317 pp.

Poinar, G. O. Jr. and G. G. Gyrisco. 1962. Studies on the bionomics of *Hexamermis arvalis* Poinar and Gyrisco, a mermithid parasite of the alfalfa weevil, *Hypera postica* (Gyllenhal). J. Insect Pathol. 4:469–483.

Polhemus, D. A. 1984. Biological and phytochemical factors affecting host choice among phytophagous Miridae on *Juniperus scopulorum*. Ph.D. dissertation, University of Utah, Salt Lake City. 136 pp.

——. 1988. Intersexual variation in densities of plant bugs (Hemiptera: Miridae) on *Juniperus scopulorum*. Ann. Entomol. Soc. Am. 81:742–747.

——. 1994. An annotated checklist of the plant bugs of Colorado (Heteroptera: Miridae). Pan-Pac. Entomol. 70:122–147.

Polhemus, D. A. and J. T. Polhemus. 1984. *Ephedrodoma*, a new genus of orthotyline Miridae (Hemiptera) from western United States. Proc. Entomol. Soc. Wash. 86:550–554.

——. 1985. Myrmecomorphic Miridae (Hemiptera) on mistletoe: *Phoradendrepulus myrmecomorphus*, n. gen., n. sp., and a redescription of *Pilophoropsis brachypterus* Poppius. Pan-Pac. Entomol. 61:26–31.

——. 1988. A new ant mimetic mirid from the Colorado tundra (Hemiptera: Miridae). Pan-Pac. Entomol. 64:23–27.

Polhemus, D. A. and V. Razafimahatratra. 1990. A new genus and species of myrmecomorphic Miridae from Madagascar (Heteroptera). J. N.Y. Entomol. Soc. 98:1–8.

Polhemus, J. T. 1985. Shore bugs (Heteroptera, Hemiptera; Saldidae): A world overview and taxonomy of Middle American forms. Different Drummer, Englewood, Col. 252 pp.

Polis, G. A. 1979. Prey and feeding phenology of the desert sand scorpion *Pauroctonus mesaenis* (Scorpionidae: Vaejovidae). J. Zool. (Lond.) 188:333–346.

——. 1981. The evolution and dynamics of intraspecific predation. Annu. Rev. Ecol. Syst. 12:225–251.

——. 1991. Food webs in desert communities: Complexity via diversity and omnivory. Pages 383–437 *in* G. A. Polis, ed., The ecology of desert communities. University of Arizona Press, Tucson.

Polis, G. A. and R. D. Holt. 1992. Intraguild predation: The dynamics of complex trophic interactions. Trends Ecol. Evol. 7:151–154.

Polis, G. A. and T. Yamashita. 1991. The ecology and importance of predaceous arthropods in desert communities. Pages 180–222 *in* G. A. Polis, ed., The ecology of desert communities. University of Arizona Press, Tucson.

Polis, G. A., C. A. Myers, and R. D. Holt. 1989. The ecology and evolution of intraguild predation: Potential competitors that eat each other. Annu. Rev. Ecol. Syst. 20:297–330.

Polk, D. F., H. W. Hogmire, and C. M. Felland. 1995. Peach and nectarine—Direct pests. Pages 49–56 in H. W. Hogmire Jr., ed., Mid-Atlantic orchard monitoring guide (NRAES-75). Northeast Regional Agricultural Engineering Service, Ithaca, N.Y.

Pollard, D. G. 1955. Feeding habits of the cotton whitefly, Bemisia tabaci Genn. (Homoptera: Aleyrodidae). Ann. Appl. Biol. 43:664–671.

———. 1959. Feeding habits of the lace-bug Urentius aegyptiacus Bergevin (Hemiptera: Tingidae). Ann. Appl. Biol. 47:778–782.

———. 1968. Stylet penetration and feeding damage of Eupteryx melissae (Curtis) on sage. Bull. Entomol. Res. 58:55–71.

———. 1969. Directional control of the stylets in phytophagous Hemiptera. Proc. R. Entomol. Soc. Lond. 44:173–185.

———. 1973. Plant penetration by feeding aphids (Hemiptera, Aphidoidea): A review. Bull. Entomol. Res. 62:631–714.

Pollard, E. 1968. Hedges. II. The effect of removal of the bottom flora of a hawthorn hedgerow on the fauna of the hawthorn. J. Appl. Ecol. 5:109–123.

Pommerol, F. 1900. Un petit hémiptère destructeur des larves de l'Yponomeute du pommier. Rev. Sci. (Paris) (4)14:348–349.

———. 1901. Un hémiptère destructeur de chenilles du pommier. Rev. Sci. Bourbon. Cent. Fr. 14:18–23.

Poos, F. W. and C. Elliott. 1936. Certain insect vectors of Aplanobacter stewarti. J. Agric. Res. (Wash., D.C.) 52:585–608.

Pophaly, D. J., T. Bhasker Rao, and M. B. Kalode. 1978. Biology & predation of the mirid bug, Cyrtorhinus lividipennis Reuter on plant and leafhoppers. Indian J. Plant Prot. 6:7–14.

Popov, Yu. A. 1981. Historical development and some questions on the general classification of Hemiptera. Rostria (Tokyo) 33(Suppl.):85–99.

Popov, Yu. A. and A. Herczek. 1992. The first Isometopinae from Baltic amber (Insecta: Heteroptera, Miridae). Mitt. Geol.-Paläont. Inst. Univ. Hamburg 73:241–258.

———. 1993. New data on Heteroptera in amber resins. Ann. Upper Siles. Mus. Entomol. Suppl. 1:7–12.

Popova, Ye. A. 1959. Pests of maize in the conditions of the Samarkand region. Tr. Uzbek. Gos. Univ. (Tashk.) 87:189–243. [in Russian.]

Poppius, B. 1912. Die Miriden der äthiopischen Region I. Mirina, Cylap-

ina, Bryocorina. Acta Soc. Sci. Fenn. 41(3):1–203.

———. 1914. Die Miriden der äthiopischen Region II. Macrolophinae, Heterotominae, Phylinae. Acta Soc. Sci. Fenn. 44(3):1–138.

———. 1921. Fam. Miridae. Pages 32–65 in B. Poppius and E. Bergroth. Beiträge zur Kenntnis der myrmecoiden Heteropteren. Ann. Mus. Natl. Hung. 18:31–88.

Porsch, O. 1956. Windpollen und Blumeninsekt. Österr. Bot. Z. 103:1–18.

———. 1957. Alte Insektentypen als Blumenausbeuter. Österr. Bot. Z. 104:115–164.

———. 1966. Insekten als Blutenbesücher. Z. Angew. Entomol. 57:1–72.

Porter, B. A. 1926. The tarnished plant bug as a peach fruit pest. J. Econ. Entomol. 19:43–47.

Porter, B. A., S. C. Chandler, and R. F. Sazama. 1928. Some causes of cat-facing in peaches. Ill. Nat. Hist. Surv. Bull. 17(6):261–275.

Posnette, A. F. and R. W. Smith. 1985. Insects and other pests of cocoa: Perspectives. Pages 65–67 in R. A. Lass and G. A. R. Wood, eds., Cocoa production: Present constraints and priorities for research. World Bank Tech. Pap. 39. World Bank, Washington, D.C.

Poston, F. L. and L. P. Pedigo. 1975. Migration of plant bugs and the potato leafhopper in a soybean-alfalfa complex. Environ. Entomol. 4:8–10.

Poteri, M., R. Heikkilä, and L. Yuan-Yi. 1987. Peltoluteen aiheuttaman kasvuhäiriön kehittyminen yksivuotiailla männyntaimilla. Folia For. (Helsinki) 695:1–14.

Pottinger, R. P. and E. J. LeRoux. 1971. The biology and dynamics of Lithocolletis blancardella (Lepidoptera: Gracillariidae) on apple in Quebec. Mem. Entomol. Soc. Can. 77:1–437.

Potts, G. R. 1986. The partridge: Pesticides, predation and conservation. Collins, London. 274 pp.

Poulton, E. B. 1906. Predaceous insects and their prey. Trans. Entomol. Soc. Lond. 1906:323–409.

Powell, R. D. 1974. The effect of flea-hoppers on cotton under controlled conditions at different ages. In Proc. Beltwide Cotton Prod. Res. Conf., Jan. 7–9, Dallas, Texas, p. 39. National Cotton Council, Memphis.

Prabhakar, B., P. Kameswara Rao, and B. H. Krishnamurthy Rao. 1986. Studies on the seasonal prevalence of certain Hemiptera occurring on sorghum. Entomon 11:95–99.

Pradhan, S. 1969. Insect pests of crops. National Book Trust, India, New Delhi. 208 pp.

Pree, D. J. 1985. Control of the tarnished plant bug Lygus lineolaris Palisot de Beauvois on peaches. Can. Entomol. 117:327–331.

Prentice, A. N. 1972. Cotton, with special reference to Africa. Longman, London. 282 pp.

Prestidge, R. A. and S. McNeill. 1983. The role of nitrogen in the ecology of grassland Auchenorryncha. Pages 257–281 in J. A. Lee, S. McNeill, and I. H. Rorison, eds., Nitrogen as an ecological factor. Blackwell, Oxford.

Price, J. M. 1997. Heteroptera from Hartlebury Common, L.N.R., Worcestershire. Entomol. Mon. Mag. 133:17–26.

Price, P. W. 1980. Evolutionary biology of parasites. Princeton University Press, Princeton. 237 pp.

———. 1984. Insect ecology, 2nd ed. Wiley, New York. 607 pp.

———. 1991. The plant vigor hypothesis and herbivore attack. Oikos 62:244–251.

———. 1997. Insect ecology, 3rd ed. Wiley, New York. 874 pp.

Price, P. W., C. E. Bouton, P. Gross, B. A. McPheron, J. N. Thompson, and A. E. Weis. 1980. Interactions among three trophic levels: Influence of plants on interactions between insect herbivores and natural enemies. Annu. Rev. Ecol. Syst. 11:41–65.

Primack, R. B. 1978. Variability in New Zealand montane and alpine pollinator assemblages. N.Z. J. Ecol. 1:66–73.

———. 1983. Insect pollination in the New Zealand mountain flora. N.Z. J. Bot. 21:317–333.

Primack, R. B. and J. A. Silander Jr. 1975. Measuring the relative importance of different pollinators to plants. Nature (Lond.) 255:143–144.

Prins, G. 1965. A laboratory rearing method for the cocoa mirid Distantiella theobroma (Dist.) (Hemiptera, Miridae). Bull. Entomol. Res. 55:615–616.

Pritts, M. and D. Handley, eds. 1989. Bramble production guide. Northeast Regional Agricultural Engineering Service, Cooperative Extension, Ithaca, N.Y. 189 pp.

Proctor, M. and P. Yeo. 1972. The pollination of flowers. Taplinger, New York. 418 pp.

Proeseler, G. 1964. Injektionsversuche mit dem Rübenkräuselvirus. Z. Angew. Entomol. 54:325–333.

Prokopy, R. J. 1994. Integration in orchard pest and habitat management: A review. Agric. Ecosyst. Environ. 50:1–10.

Prokopy, R. J. and G. L. Hubbell. 1981. Susceptibility of apple to injury by tarnished plant bug adults. Environ. Entomol. 10:977–979.

Prokopy, R. J. and E. D. Owens. 1978. Visual generalist—["with"] visual specialist phytophagous insects: Host selection behaviour and application to management. Entomol. Exp. Appl. 24:609–620.

Prokopy, R. J., R. G. Adams, and K. I. Hauschild. 1979. Visual responses of tarnished plant bug adults on apple. Environ. Entomol. 8:202–205.

Prokopy, R. J., W. M. Coli, R. G. Hislop, and K. I. Hauschild. 1980. Integrated management of insect and mite pests in commercial apple orchards in Massachusetts. J. Econ. Entomol. 73:529–535.

Prokopy, R. J., G. L. Hubbell, R. G. Adams, and K. I. Hauschild. 1982. Visual monitoring trap for tarnished plant bug adults on apple. Environ. Entomol. 11:200–203.

Prokopy, R. J., M. Christie, S. A. Johnson, and M. T. O'Brien. 1990. Transitional step toward second-stage integrated management of arthropod pests of apple in Massachusetts orchards. J. Econ. Entomol. 83:2405–2410.

Pruess, K. P. 1974. Tarnished and alfalfa plant bugs in alfalfa: Population suppression with ULV malathion. J. Econ. Entomol. 67:525–528.

Pruess, K. P. and N. C. Pruess. 1966. Note on a Malaise trap for determining flight direction of insects. J. Kans. Entomol. Soc. 39:98–102.

Purcell, M. and S. C. Welter. 1990a. Degree-day model for development of *Calocoris norvegicus* (Hemiptera: Miridae) and timing of management strategies. Environ. Entomol. 19:848–853.

——. 1990b. Seasonal phenology and biology of *Calocoris norvegicus* (Hemiptera: Miridae) in pistachios and associated host plants. J. Econ. Entomol. 83:1841–1846.

——. 1991. Effect of *Calocoris norvegicus* (Hemiptera: Miridae) on pistachio yields. J. Econ. Entomol. 84:114–119.

Putman, W. L. 1941. The feeding habits of certain leafhoppers. Can. Entomol. 73:39–53.

——. 1962. Life-history and behaviour of the predacious mite *Typhlodromus* (*T.*) *caudiglans* Schuster (Acarina: Phytoseiidae) in Ontario, with notes on the prey of related species. Can. Entomol. 94:163–177.

——. 1965. The predacious thrips *Haplothrips faurei* Hood (Thysanoptera: Phloeothripidae) in Ontario peach orchards. Can. Entomol. 97:1208–1221.

Putshkov, V. G. 1956. Basic trophic groups of phytophagous hemipterous insects and changes in the character of their feeding during the process of development. Zool. Zh. 35:32–44. [in Russian, English summary.]

——. 1960. Ecology of some little-known species of Hemiptera-Heteroptera. I. Entomol. Obozr. 39(2):300–312. [in Russian; English translation in Rev. Entomol. 39(2):192–199.]

——. 1961a. Carnivorous Hemiptera of the USSR beneficial to agriculture and forestry. Pr. Akad. Nauk URSR Inst. Zool. 17:7–18. [in Ukrainian.]

——. 1961b. Plant bugs (Heteroptera, Miridae) of the Poltava region. Pr. Akad. Nauk URSR Inst. Zool. 17:71–85. [in Ukrainian.]

——. 1966. The main bugs—plant bugs—as pests of agricultural crops. Naukova Dumka, Kiev. 171 pp. [in Russian.]

——. 1975. Heteroptera. Plant bugs or mirids—pests of agricultural crops. Zashch. Rast. (Mosc.) No. 12:30–33. [in Russian.]

——. 1977. New and little-known mirid bugs (Heteroptera, Miridae) from Mongolia and Soviet Central Asia. Entomol. Obozr. 56(2):360–374. [in Russian; English translation in Entomol. Rev. 56(2):91–100.]

Putshkov, V. G. and L. V. Putshkova. 1956. Eggs and larvae of Heteroptera—agricultural pests. Trud. Vses. Entomol. Obshch. (Leningr.) 45:218–342. [in Russian.]

Putshkova, L. V. 1971. The functions of the wings in the Hemiptera and trends in their specialization. Entomol. Obozr. 50(3):537–549. [in Russian; English translation in Entomol. Rev. 50(3):303–309.]

Puttarudriah ["Puttarudraiah"], M. 1952. Blister disease "Kajji" of guava fruits (*Psidium guava*). Mysore Agric. J. 28:8–13.

Puttarudriah, M. and M. Appana. 1955. Two new hosts of *Helopeltis antonii* Signoret in Mysore. Indian J. Entomol. 17:391–392.

Qadri, M. A. H. 1959. Mechanism of feeding in Hemiptera. Pages 237–245 in J. F. G. Clarke, chairman, Studies in invertebrate morphology. Smithsonian Misc. Coll. Vol. 137. Smithsonian Institution, Washington, D.C.

Qingcai, W. and M. A. Jervis. 1988. Foraging for patchily distributed preys by *Cyrtorhinus lividipennis* (Reuter) J. Southwest. Agric. Univ. (China) 10:245–252. [in Chinese, English summary.]

Quaglia, F., E. Rossi, R. Petacchi, and C. E. Taylor. 1993. Observations on an infestation by green peach aphids (Homoptera: Aphididae) on greenhouse tomatoes in Italy. J. Econ. Entomol. 86:1019–1025.

Quaintance, A. L. 1896. Insect enemies of truck and garden crops. Fla. Agric. Exp. Stn. Bull. 34:242–327.

——. 1897. Some strawberry insects. Fla. Agric. Exp. Stn. Bull. 42:551–600.

——. 1898. A preliminary report upon the insect enemies of tobacco in Florida. Fla. Agric. Exp. Stn. Bull. 48:154–188.

——. 1912. The peach bud mite. (*Tarsonemus waitei* Banks, MSS.). U.S. Dep. Agric. Bur. Entomol. Bull. 97(6):103–114.

——. 1913. Remarks on some of the injurious insects of other countries. Proc. Entomol. Soc. Wash. 15:54–83.

Quayle, H. J. 1938. Insects of citrus and other subtropical fruits. Comstock Publishing, Ithaca, N.Y. 583 pp.

Quicke, D. L. J. 1997. Parasitic wasps. Chapman & Hall, London. 470 pp.

Quiring, D. T. and M. L. McKinnon. 1999. Why does early-season herbivory affect subsequent budburst? Ecology 80:1724–1735.

Quiring, D. T. and J. N. McNeil. 1985. Effect of larval cannibalism on the development and reproductive performance of *Agromyza frontella* (Rondani) (Diptera: Agromyzidae). Ann. Entomol. Soc. Am. 78:429–432.

Quisenberry, S. and T. R. Yonke. 1981. Effects of *Forcipata loca* feeding on tissue of Kentucky-31 tall fescue. Ann. Entomol. Soc. Am. 74:521–524.

Rabie, A. L., J. D. Wells, and L. K. Dent. 1983. The nitrogen content of pollen protein. J. Apic. Res. 22:119–123.

Rácz, V. 1986. Composition of heteropteran populations in Hungary in apple orchards belonging to different management types and the influence of insecticide treatments on the population densities. Acta Phytopathol. Entomol. Hung. 21:355–361.

——. 1988. The association of the predatory bug *Atractotomus mali* Mey.-D. (Heteroptera: Miridae) with aphids on apple in Hungary. Pages 43–46 in E. Niemczyk and A. F. G. Dixon, eds., Ecology and effectiveness of aphidophaga. Proc. Int. Symp., Teresin, Poland, Aug. 31–Sept. 5, 1987. SPB, The Hague.

Rácz, V. and I. Bernáth. 1993. Dominance conditions and population dynamics of *Lygus* (Het., Miridae) species in Hungarian maize stands (1976–1985), as functions of climatic conditions. J. Appl. Entomol. 115:511–518.

Radcliffe, E. B. and D. K. Barnes. 1970. Alfalfa plant bug injury and evidence of plant resistance in alfalfa. J. Econ. Entomol. 63:1995–1996.

Radcliffe, E. B., R. W. Weires, R. E.

Stucker, and D. K. Barnes. 1976. Influence of cultivars and pesticides on pea aphid, spotted alfalfa aphid, and associated arthropod taxa in a Minnesota alfalfa ecosystem. Environ. Entomol. 5:1195–1207.

Rajakulendran, S. V. and J. R. Cate. 1986. Field studies on the parasitoids of cotton fleahopper in its wild habitat. *In* Proc. Beltwide Cotton Prod. Res. Conf., Jan. 4–9, 1986, Las Vegas, Nev., pp. 241–244. National Cotton Council, Memphis.

Rajasekhara, K., S. Chatterji, and M. G. Ramdas Menon. 1964. Biological notes on *Psallus* species (Miridae: Hemiptera), a predator of *Taeniothrips nigricornis* Schmutz (Thripidae: Thysanoptera). Indian J. Entomol. 26:62–66.

Raju, B. C. and J. M. Wells. 1986. Diseases caused by fastidious xylem-limited bacteria and strategies for management. Plant Dis. 70:182–186.

Rakauskas, R. 1985. Natural enemies of aphids that feed on fruit-trees and berry-shrubs in the Lithuanian SSR. Acta Entomol. Lituanica 8:58–69. [in Lithuanian, English summary.]

Rakickas, R. J. and T. F. Watson. 1974. Population trends of *Lygus* spp. and selected predators in strip-cut alfalfa. Environ. Entomol. 3:781–784.

Ralphs, M. H., W. A. Jones, and J. A. Pfister. 1997. Damage from the larkspur mirid deters cattle grazing of larkspur. J. Range Manage. 50:371–373.

Ralphs, M. H., D. R. Gardner, W. A. Jones, and G. D. Manners. 1998. Diterpenoid alkaloid concentration in tall larkspur plants damaged by larkspur mirid. J. Chem. Ecol. 24:829–840.

Ramachandra Rao, Y. 1915. *Helopeltis antonii* as a pest on nim trees. Agric. J. India 10:412–416.

Ramakrishna Ayyar, T. V. 1940. Handbook of economic entomology for South India. Government Press, Madras, India. 528 pp.

——. 1942. Insect enemies of the cashewnut plant (*Anacardium occidentale*) in South India. Madras Agric. J. 30:223–226.

Ramalho, F. S. 1994. Cotton pest management: Part 4. A Brazilian perspective. Annu. Rev. Entomol. 39:563–578.

Raman, K. and K. P. Sanjayan. 1984a. Host plant relationships and population dynamics of the mirid, *Cyrtopeltis tenuis* Reut. (Hemiptera: Miridae). Proc. Indian Natl. Sci. Acad. B50:355–361.

——. 1984b. Histology and histopathology of the feeding lesions by *Cyrtopeltis tenuis* Reut. (Hemiptera: Miridae) on *Lycopersicon esculentum* Mill. (Solanaceae). Proc. Indian Acad. Sci. 93:543–547.

Raman, K., K. P. Sanjayan, and G. Suresh. 1984. Impact of feeding injury of *Cyrtopeltis tenuis* Reut. (Hemiptera: Miridae) on some biochemical changes in *Lycopersicon esculentum* Mill. (Solanaceae). Curr. Sci. (Bangalore) 53:1092–1093.

Ramesh, P. 1994. Olfactory responses of sorghum earhead bug, *Calocoris angustatus* Leth. (Hemiptera-Miridae) to certain less and more susceptible sorghum germplasm lines. Indian J. Entomol. 56:392–398.

Rammner, W. 1942. Nektar als Nahrung einheimischer Wanzen. Zool. Anz. 140:133–137.

Ramsay, M. J. 1973. Beneficial insect or plant pest? The regulatory dilemma. Pages 40–46 *in* P. H Dunn, ed., Proc. Second Int. Symp. Biol. Control Weeds. Misc. Publ. 6 Commonw. Inst. Biol. Control. Commonwealth Agricultural Bureaux, Farnham Royal, UK.

Rand, F. V. and W. D. Pierce. 1920. A coördination of our knowledge of insect transmission in plant and animal diseases. Phytopathology 10: 189–231.

Randles, J. W., C. Davies, T. Hatta, A. R. Gould, and R. I. B. Francki. 1981. Studies on encapsidated viroid-like RNA. 1. Characterisation of velvet tobacco mottle virus. Virology 108: 111–122.

Rankin, M. A., M. L. McAnelly, and J. E. Bodenhamer. 1986. The oogenesis-flight syndrome revisited. Pages 27–48 *in* W. Danthanarayana, ed., Insect flight: Dispersal and migration. Springer, Berlin.

Rao, G. N. 1970. Tea pests in southern India and their control. PANS (Pest Artic. News Summ.) 16:667–672.

Rapusas, H. R., D. G. Bottrell, and M. Coll. 1996. Intraspecific variation in chemical attraction of rice to insect predators. Biol. Control 6:394–400.

Rasmy, A. H. and A. W. MacPhee. 1970. Studies on pear psylla in Nova Scotia. Can. Entomol. 102:586–591.

Rasweiler, J. J. 1977. The care and management of bats as laboratory animals. Pages 519–617 *in* W. A. Wimsatt, ed., Biology of bats. Vol. III. Academic, New York.

Rathore, Y. K. 1961. Studies on the mouth-parts and feeding mechanism in *Dysdercus cingulatus* Fabr. (Pyrrhocoridae: Heteroptera). Indian J. Entomol. 23:163–185.

Ratnadass, A., B. Cissé, and K. Mallé. 1994. Notes on the biology and immature stages of West African sorghum head bugs *Eurystylus immaculatus* and *Creontiades pallidus* (Heteroptera: Miridae). Bull. Entomol. Res. 84:383–388.

Ratnadass, A., Y. O. Doumbia, and O. Ajayi. 1995a. Bioecology of sorghum head bug *Eurystylus immaculatus*, and crop losses in West Africa. Pages 91–102 *in* K. F. Nwanze and O. Youm, eds., Panicle insect pests of sorghum and pearl millet: Proceedings of an international consultative workshop, 4–7 Oct. 1993, ICRISAT Sahelian Center, Niamey, Niger. International Crops Research Institute for the Semi-Arid Tropics, Patancheru, A.P., India.

Ratnadass, A., O. Ajayi, G. Fliedel, and K. V. Ramaiah. 1995b. Host-plant resistance in sorghum to *Eurystylus immaculatus* in West Africa. Pages 191–199 *in* K .F. Nwanze and O. Youm, eds., Panicle insect pests of sorghum and pearl millet: Proceedings of an international consultative workshop, 4–7 Oct. 1993, ICRISAT Sahelian Center, Niamey, Niger. International Crops Research Institute for the Semi-Arid Tropics, Patancheru, A.P., India.

Ratnadass, A., B. Cissé, D. Diarra, and M. L. Sangaré. 1997. Indigenous host plants of sorghum head-bugs (Heteroptera: Miridae) in Mali. Afr. Entomol. 5:158–160.

Rattan, P. S. 1992. Pest and disease control in Africa. Pages 331–352 *in* K. C. Willson and M. N. Clifford, eds., Tea: Cultivation to consumption. Chapman & Hall, London.

Rauf, A., R. A. Cecich, and D. M. Benjamin. 1984a. Conelet abortion in jack pine caused by *Platylygus luridus* (Hemiptera: Miridae). Can. Entomol. 116:1213–1218.

Rauf, A., D. M. Benjamin, and R. A. Cecich. 1984b. Bionomics of *Platylygus luridus* (Hemiptera: Miridae) in Wisconsin jack pine seed orchards. Can. Entomol. 116:1219–1225.

——. 1985. Insects affecting seed production of jack pine, and life tables of conelet and cone mortality in Wisconsin. For. Sci. 31:271–281.

Raulston, J. R. and J. L. Auclair. 1968. Responses of *Lygus hesperus* to chemically defined diets. Ann. Entomol. Soc. Am. 61:1495–1500.

Raulston, J. R., T. J. Henneberry, J. E. Leggett, D. N. Byrne, E. Grafton-Cardwell, and T. F. Leigh. 1996. Short- and long-range movement of insects and mites. Pages 143–162 *in* E. G. King, J. R. Phillips, and R. J. Coleman, eds., Cotton insects and mites: Characterization and management. Cotton Foundation, Memphis.

Raupp, M. J. and R. F. Denno. 1983. Leaf age as a predictor of herbivore distribution and abundance. Pages 91–124 *in* R. F. Denno and M. S. McClure, eds., Variable plants and herbivores in natural and managed systems. Academic, New York.

Rautapää, J. 1969. Effect of *Lygus rugulipennis* Popp. (Hem., Capsidae) on the yield and quality of wheat. Ann. Entomol. Fenn. 35:168–175.

——. 1970. Effect of the meadow capsid bug, *Leptopterna dolabrata* (L.) (Het., Capsidae), on the yield and quality of wheat. Ann. Entomol. Fenn. 36:145–152.

Raven, J. A. 1983. Phytophages of xylem and phloem: A comparison of animal and plant sap-feeders. Adv. Ecol. Res. 13:135–234.

Raven, P. H. 1983. The challenge of tropical biology. Bull. Entomol. Soc. Am. 29(1):4–12.

Raven, P. H., R. F. Evert, and H. Curtis. 1976. Biology of plants, 2nd ed. Worth, New York. 685 pp.

Ravensberg, W. J. 1994. Biological control of pests: Current trends and future prospects. Brighton Crop Prot. Conf.: Pests and Diseases—1994. 2:591–600. BCPC Registered Office, Farnham, UK.

Raw, F. 1959a. An insectary method for rearing cacao mirids, *Distantiella theobroma* (Dist.) and *Helopeltis singularis* Hagl. Bull. Entomol. Res. 50:11–12.

——. 1959b. Studies on the chemical control of cacao mirids, *Distantiella theobroma* (Dist.) and *Sahlbergella singularis* Hagl. Bull. Entomol. Res. 50:13–23.

Rawlins, J. E. 1984. Mycophagy in Lepidoptera. Pages 382–423 *in* Q. Wheeler and M. Blackwell, eds., Fungus-insect relationships: Perspectives in ecology and evolution. Columbia University Press, New York.

Razafimahatratra, V. P. 1980. A revision of the genus *Deraeocoris* Kirschbaum (Heteroptera: Miridae) from western America north of Mexico. Ph.D. dissertation, Oregon State University, Corvallis. 235 pp.

Readio, P. A. 1924. Notes on the life history of a beneficial reduviid, *Sinea diadema* (Fabr.), Heteroptera. J. Econ. Entomol. 17:80–86.

Readshaw, J. L. 1971. An ecological approach to the control of mites in Australian orchards. J. Aust. Inst. Agric. Sci. 37:226–230.

——. 1975. The ecology of tetranychid mites in Australian orchards. J. Appl. Ecol. 12:473–495.

Reddy, D. B. 1958. Sannhemp [sic] and its insect fauna. Proc. 10th Int. Congr. Entomol., Montreal (1956) 3:439–440.

Reding, M. E. and E. H. Beers. 1996. Influence of prey availability on survival of

Campylomma verbasci (Hemiptera: Miridae) and factors influencing efficacy of chemical control on apples. Pages 141–154 *in* O. Alomar and R. Wiedenmann, eds., Zoophytophagous Heteroptera: Implications for life history and integrated pest management. Thomas Say Publ. Entomol.: Proceedings. Entomological Society of America, Lanham, Md.

Reed, W. 1959. A preliminary study of the toxicity of lime-sulphur and other fungicides to *Psallus ambiguus* (Fall.) (Hemiptera-Heteroptera: Miridae). East Malling Res. Stn. Rep. 1958:127–130.

——. 1974. Selection of cotton varieties for resistance to insect pests in Uganda. Cotton Grow. Rev. 51:106–123.

Reh, L. 1929. Pflanzenschädliche Wanzen. Z. Wiss. Insektenbiol. 24:43–49.

Reichle, D. E., R. A. Goldstein, R. I. Van Hook Jr., and G. J. Dodson. 1973. Analysis of insect consumption in a forest canopy. Ecology 54:1076–1084.

Reid, D. G. 1974. New records of *Hexamermis* (Nematoda: Mermithidae) parasitizing three species of *Slaterocoris* (Hemiptera: Miridae). Can. Entomol. 106:239.

Reid, D. G., C. C. Loan, and R. Harmsen. 1976. The mirid (Hemiptera) fauna of *Solidago canadensis* (Asteracea) in south-eastern Ontario. Can. Entomol. 108:561–567.

Reid, M. R. 1965. Studies on the salivary physiology of the tarnished plant bug, *Lygus lineolaris* (P. de B.) (Hemiptera: Miridae). M.S. thesis, Louisiana State University, Baton Rouge. 41 pp.

——. 1968. Influence of previous host upon the ability of the tarnished plant bug, *Lygus lineolaris* (P. de B.) to injure cotton tissue. Ph.D. dissertation, Louisiana State University, Baton Rouge. 60 pp.

Reinhard, H. J. 1926. The cotton flea hopper. Tex. Agric. Exp. Stn. Bull. 339:1–39.

——. 1928. Hibernation of the cotton flea hopper. Tex. Agric. Exp. Stn. Bull. 377:1–26.

Reissig, W. H., E. A. Heinrichs, and S. L. Valencia. 1982a. Insecticide-induced resurgence of the brown planthopper, *Nilaparvata lugens*, on rice varieties with different levels of resistance. Environ. Entomol. 11:165–168.

——. 1982b. Effects of insecticides on *Nilaparvata lugens* and its predators: Spiders, *Microvelia atrolineata*, and *Cyrtorhinus lividipennis*. Environ. Entomol. 11:193–199.

Remamony, K. S. and C. C. Abraham. 1977. New reccord [sic] of *Pachypeltis maesarum* (Kirkaldy) (Miridae, He-

miptera) as a pest of cashew in Kerala. Sci. Cult. 43:553.

Remaudière, G. and F. Leclant. 1971. Le complexe des ennemis naturels des aphides du pêcher dans le Moyenne Vallée du Rhona. Entomophaga 16:255–267.

Remold, H. 1962. Über die biologische Bedeutung der Duftdrüsen bei den Landwanzen (Geocorisae). Z. Vgl. Physiol. 45:636–694.

——. 1963. Scent-glands of land-bugs, their physiology and biological function. Nature (Lond.) 198(4882):764–768.

Ren, S. Z. 1987. New species and a newly recorded genus of Isometopidae from China (Hemiptera: Heteroptera). Acta Zootaxonomica Sin. 12:402–403.

Renquist, A. R., H. G. Hughes, and M. K. Rogoyski. 1987. Solar injury of raspberry fruit. HortScience 22:396–397.

Rethwisch, M. D., E. T. Natwick, B. R. Tickes, M. Meadows, and D. Wright. 1995. Impact of insect feeding and economics of selected insecticides on early summer bermudagrass seed production in the desert southwest. Southwest. Entomol. 20:187–201.

Reuter, O. M. 1875. Revisio critica Capsinarum, praecipue Scandinaviae et Fenniae. J. C. Frenckell & Son, Helsinki. 190 pp.

——. 1879. Till kännedomen om mimiska Hemiptera och deras lefnads historia. Ofv. F. Vet-Soc. Forh. 21:141–198.

——. 1881. Analecta Hemipterologica. Zur Artenkenntnis, Synonymie und geographischen Verbreitung der palaearktischer Heteropteren. Berl. Entomol. Z. 25:155–196.

——. 1903. The food of capsids. Entomol. Mon. Mag. 39:121–123.

——. 1905. Hemipterologische Spekulationen. I. Die Klassifikation der Capsiden. Festschr. Palmén (Helsinki) No. 1:1–58.

——. 1907a. Eine neotropische Capside als Orchideenschädling in europäischen Warmhäusern. Z. Wiss. Insektenbiol. 3:251–254.

——. 1907b. *Pameridea* nov. gen., eine Capside, die in Sudafrika die Bestaubung von *Roridula gorgonias* besorgt. Bull. Soc. Entomol. Fr. 1907:723–726.

——. 1909. Charackteristik und Entwicklungsgeschichte der Hemipteren-Fauna (Heteroptera, Auchenorrhynchia und Psyllidae) der palearktischen Coniferen. Acta Soc. Sci. Fenn. 36(1):1–129.

——. 1910. Neue Beiträge zur Phylogenie und Systematik der Miriden nebst einleitenden Bermerkungen über die Phylogenie der Heteropteren-Familien. Acta Soc. Sci. Fenn. 37(3):1–171.

——. 1913. Die Familie der Bett- oder Hauswanzen (Cimicidae), ihre Phylogenie, Systematik, Oekologie und Verbreitung. Z. Wiss. Insektenbiol. 9:251–255, 303–306, 325–329, 360–364.

Rey, C. 1881. Note sur le *Stethoconus mamillosus*. Ann. Soc. Linn. Lyon 29:385–386.

Rey, J. R. 1981. Ecological biogeography of arthropods of *Spartina* islands in northwest Florida. Ecol. Monogr. 51: 237–265.

Reyes, T. M. and B. P. Gabriel. 1975. The life history and consumption habits of *Cyrtorhinus lividipennis* Reuter (Hemiptera: Miridae). Philipp. Entomol. 3:79–88.

Reynard, G. B. 1943. "Red ring" of tomato stems caused by an insect, *Cyrtopeltis varians* (Dist.), at Charleston, S.C. Phytopathology 33:613–615.

Reyne, A. 1921. De cacaothrips (*Heliothrips rubrocinctus* Giard). Dep. Landbouw Suriname Bull. 44:1–214.

Reynolds, D. R. and M. R. Wilson. 1989. Aerial samples of macro-insects migrating at night over central India. J. Plant Prot. Trop. 6:89–101.

Reynolds, H. T., P. L. Adkisson, and R. F. Smith. 1975. Cotton insect pest management. Pages 379–443 in R. L. Metcalf and W. H. Luckmann, eds., Introduction to past management. Wiley, New York.

Rhoades, D. F. 1979. Evolution of plant chemical defense against herbivores. Pages 3–54 in G. A. Rosenthal and D. Janzen, eds., Herbivores: Their interaction with secondary plant metabolites. Academic, New York.

——. 1983. Herbivore population dynamics and plant chemistry. Pages 155–220 in R. F. Denno and M. S. McClure, eds., Variable plants and herbivores in natural and managed systems. Academic, New York.

Rhodes, J. M. and L. S. C. Wooltorton. 1978. The biosynthesis of phenolic compounds in wounded plant storage tissues. Pages 243–286 in G. Kahl, ed., Biochemistry of wounded plant tissues. Gruyter, Berlin.

Rhyne, C. L. 1966. Inheritance of extrafloral nectaries in cotton. Adv. Front. Plant Sci. 13:121–137.

Riba, G., T. Poprawski, and J. Maniania. 1986. Isoesterase variability among geographical populations of *Beauveria bassiana* (Fungi Imperfecti) isolated from Miridae. Pages 205–209 in R. A. Samson, J. M. Vlak, and D. Peters, eds., Fundamental and applied aspects of invertebrate pathology. Foundation of

the Fourth International Colloquium of Invertebrate Pathology, Wageningen.

Rice, P. L. [1938?]. Cat-facing of peaches by the tarnished plant bug, *Lygus pratensis* (L.). 51st Trans. Peninsula Hortic. Soc. 1937:131–136.

Rice, R. E., J. K. Uyemoto, J. M. Ogawa, and W. M. Pemberton. 1985. New findings on pistachio problems. Calif. Agric. 39(1–2):15–18.

Rice, R. E., W. J. Bentley, and R. H. Beede. 1988. Insect and mite pests of pistachios in California. Univ. Calif. Div. Agric. Nat. Resour. Publ. 21452. 26 pp.

Richards, O. W. and N. Waloff. 1961. A study of a natural population of *Phytodecta olivacea* (Forster) (Coleoptera, Chrysomeloidea). Philos. Trans. R. Soc. Lond. 244B:205–257.

Richardson, J. K. 1938. Studies on blackheart, soft-rot, and tarnished plant bug injury of celery. Can. J. Res. (C) 16:182–193.

Richardson, K. C. and R. D. Wooller. 1990. Adaptations of the alimentary tracts of some Australian lorikeets to a diet of pollen and nectar. Aust. J. Zool. 38:581–586.

Richardson, K. C., R. D. Wooller, and B. G. Collins. 1986. Adaptations to a diet of nectar and pollen in the marsupial *Tarsipes rostratus* (Marsupialia: Tarsipedidae). J. Zool. (Lond.) (A) 208:285–297.

Ricklefs, R. E. 1979. Ecology, 2nd ed. Chiron, New York. 966 pp.

Ridgway, R. L. and J. C. Bailey. 1978. Assessment of pest management for glandless cotton: Entomological viewpoint. Pages 118–121 in Glandless cotton: Its significance, status, and prospects. U.S. Dep. Agric. Agric. Res. Serv., Beltsville, Md.

Ridgway, R. L. and G. G. Gyrisco. 1960a. Effect of temperature on the rate of development of *Lygus lineolaris* (Hemiptera: Miridae). Ann. Entomol. Soc. Am. 53:691–694.

——. 1960b. Studies of the biology of the tarnished plant bug, *Lygus lineolaris*. J. Econ. Entomol. 53:1063–1065.

Ridley, M. 1988. Mating frequency and fecundity in insects. Biol. Rev. Camb. Philos. Soc. 63:509–549.

Riedl, H. 1991. Beneficial arthropods for pear pest management. Pages 101–118 in K. Williams, gen. ed., New directions in tree fruit pest management. Good Fruit Grower, Yakima, Wash.

Rieger, R., A. Michaelis, and M. M. Green. 1991. Glossary of genetics: Classical and molecular, 5th ed. (English). Springer, Berlin. 553 pp.

Riggs, D. I. 1990. Greenhouse studies of the effect of lygus bug feeding on

'Tristar' strawberry. Adv. Strawberry Prod. 9:40–43.

Riley, C. V. 1870. The tarnished plant-bug—*Capsus oblineatus*, Say. [Heteroptera Capsidae]. Pages 113–115 in Second Annu. Rep. Noxious, Beneficial and Other Insects of the State of Missouri. Jefferson City, Mo.

——. 1871. Beneficial insects. The glassy-winged soldier-bug—*Campyloneura vitripennis*, Say. Pages 137–139 in Third Annu. Rep. Noxious, Beneficial and Other Insects of the State of Missouri. Jefferson City, Mo.

——. 1885. The tarnished plant-bug. (*Lygus lineolaris*, Beauv). Order Heteroptera; family Capsidae. Pages 312–315 in Report of the Commissioner of Agriculture for 1884. Government Printing Office, Washington.

——. 1893a. Report of the Entomologist. New observations on the elm leaf-beetle. Pages 166–167 in Report of the Secretary of Agriculture, 1892. Government Printing Office, Washington.

——. 1893b. New injurious insects of a year. Insect Life (Wash., D.C.) 5:16–19.

Riley, J. R., D. R. Reynolds, and R. A. Farrow. 1987. The migration of *Nilaparvata lugens* (Stål) (Delphacidae) and other Hemiptera associated with rice during the dry season in the Philippines: A study using radar, visual observations, aerial netting and ground trapping. Bull. Entomol. Res. 77:145–169.

Riley, J. R., D. R. Reynolds, S. Mukhopadhyay, M. R. Ghosh, and T. K. Sarkar. 1995. Long-distance migration of aphids and other small insects in northeast India. Eur. J. Entomol. 92: 639–653.

Riley, N. D. [1931?]. Class Insecta (Earwigs, grasshoppers, butterflies, moths, etc.). Pages 194–291 in W. P. Pycraft, ed., The standard natural history from amoeba to man. Warne, London.

Riley, W. A. and O. A. Johannsen. 1915. Handbook of medical entomology. Comstock Publishing, Ithaca, N.Y. 348 pp.

Rincker, C. M. and H. H. Rampton. 1985. Seed production. Pages 417–443 in N. L. Taylor, ed., Clover science and technology. Agron. Monogr. 25. Am. Soc. Agron., Crop Sci. Soc. Am., and Soil Sci. Soc. Am., Madison, Wis.

Ring, D. R., J. H. Benedict, M. L. Walmsley, and M. F. Treacy. 1993. Cotton yield response to cotton fleahopper (Hemiptera: Miridae) infestations on the Lower Gulf Coast of Texas. J. Econ. Entomol. 86:1811–1819.

Rings, R. W. 1958. Types and seasonal

incidence of plant bug injury to peaches. J. Econ. Entomol. 51:27–32.

——. 1959. Oak and hickory plant bugs. Am. Fruit Grow. 79(5):36.

Ripley, L. B. 1927. "Froghopper" in wattles. Farming S. Afr. 1:423.

——. 1929. "Froghopper" in wattles. Farming S. Afr. Reprint 51:1–4.

Ripper, W. E. and L. George. 1965. Cotton pests of the Sudan: Their habits and control. Blackwell, Oxford. 363 pp.

Rita, M. see Muhamad, R.

Riudavets, J. and C. Castañé. 1998. Identification and evaluation of native predators of *Frankliniella occidentalis* (Thysanoptera: Thripidae) in the Mediterranean. Environ. Entomol. 27:86–93.

Riudavets, J., R. Gabarra, and C. Castañé. 1993. *Frankliniella occidentalis* predation by native natural enemies. Int. Organ. Biol. Control/West Palearctic Reg. Sect. Bull. 16:137–140.

Riudavets, J., C. Castañé, and R. Gabarra. 1995. Native predators of western flower thrips in horticultural crops. Pages 255–258 *in* B. L. Parker, M. Skinner, and T. Lewis, eds., Thrips biology and management. Plenum, New York.

Robaux, P. 1974. Recherches sur le développement et al biologie des Acariens "Thrombidiidae." Mem. Mus. Natl. Hist. Nat., Paris. Sér. A Zool. 85:1–186.

Robel, R. J., B. M. Press, B. L. Henning, and K. W. Johnson. 1995. Nutrient and energetic characteristics of sweepnet-collected invertebrates. J. Field Ornithol. 66:44–53.

Roberts, D. A. and C. W. Boothroyd. 1984. Fundamentals of plant pathology, 2nd ed. Freeman, New York. 432 pp.

Roberts, J. I. 1930. The tobacco capsid (*Engytatus volucer*, Kirk.) in Rhodesia. Bull. Entomol. Res. 21:169–183.

Roberts, W. P. and D. J. Pree. 1983. Pest management program for peach series. Tarnished plant bug. Ont. Minist. Agric. Food. 83–27:1–3.

Robertson, C. 1928. Flowers and insects; lists of visitors of four hundred and fifty-three flowers. Carlinville, Ill. 221 pp.

Robinson, A. G. 1952. Annotated list of predators of tetranychid mites in Manitoba. 82nd Annu. Rep. Entomol. Soc. Ont. 1951:33–37.

Robinson, R. W. 1954. Seed germination problems in the Umbelliferae. Bot. Rev. 20:531–550.

Rochow, W. F. and H. W. Israel. 1977. Luteovirus (barley yellow dwarf virus) group. Pages 363–369 *in* K. Maramorosch, ed., The atlas of insect and plant viruses including mycoplasmaviruses and viroids. Academic, New York.

Rodriguez, E. and D. A. Levin. 1976. Biochemical parallelisms of repellents and attractants in higher plants and arthropods. *In* J. W. Wallace and R. L. Mansell, eds., Biochemical interaction between plants and insects. Recent Adv. Phytochem. 10:214–270.

Roepke, W. 1916. Het *Helopeltis*-vraagstuk, in het bijzonder met betrekking tot cacao. Med. Proefstn. Midden Java. No. 21:1–40.

——. 1918. *Mertila malayensis* Dist., een "bloemwants" (Capside), schadelijk voor orchideën. Teysmannia (Batavia) 29:201–212.

——. 1919. *Hyalopeplus smaragdinus* n. sp., eine neue Thee-Capside aus Java (Rhynch.: Hem. Heteropt.). Treubia 1:73–81.

Roff, D. A. 1986. The evolution of wing dimorphism in insects. Evolution 40:1009–1020.

——. 1990. The evolution of flightlessness in insects. Ecol. Monogr. 60:389–421.

——. 1992. The evolution of life histories: Theory and analysis. Chapman & Hall, New York. 535 pp.

Roff, D. A. and D. J. Fairbairn. 1991. Wing dimorphisms and the evolution of migratory polymorphisms among the Insecta. Am. Zool. 31:243–251.

Rogers, C. E. 1985. Extrafloral nectar: Entomological implications. Bull. Entomol. Soc. Am. 31(3):15–20.

Roitberg, B. D. and N. P. N. Angerilli. 1986. Management of temperate-zone deciduous fruit pests: Applied behavioural ecology. Agric. Zool. Rev. 1:137–165.

Roland, G. 1936. Onderzoek van de vergelingsziekte van de biet, met enkele opmerkingen over de mozaiekziekte. Tijdschr. Plziekt. 42(3):54–70.

Rolfs, A. R. [1932?]. The tarnished plant bug. *In* Proc. 27th Annu. Mtg. Wash. State Hortic. Assoc. (Yakima), pp. 13–16.

Romankow, W. 1959. Wyniki badań nad biologią ozdobnika lucernowca—*Adelphocoris lineolatus* Goeze (Heteroptera, Miridae), z uwzględnieniem niektórych momentów jego ekologii. Pol. Pismo Entomol. 29:55–105.

Romney, V. E. and T. P. Cassidy. 1945. *Anaphes ovijentatus*, an egg-parasite of *Lygus hesperus*. J. Econ. Entomol. 38:497–498.

Romney, V. E., G. T. York, and T. P. Cassidy. 1945. Effect of *Lygus* spp. on seed production and growth of guayule in California. J. Econ. Entomol. 38:45–50.

Room, P. M. 1979. Parasites and predators of *Heliothis* spp. (Lepidoptera: Noctuidae) in cotton in the Namoi Valley, New South Wales. J. Aust. Entomol. Soc. 18:223–228.

Room, P. M. and E. S. C. Smith. 1975. Relative abundance and distribution of insect pests, ants and other components of the cocoa ecosystem in Papua New Guinea. J. Appl. Ecol. 12:31–46.

Rose, R. I. 1996. Management alternatives for *Lygus lineolaris* Palisot de Beauvois (Heteroptera: Miridae), the tarnished plant bug, in Iowa strawberries. M.S. thesis, Iowa State University, Ames. 94 pp.

Rosen, D. 1985. Biological control. Pages 413–464 *in* G. A. Kerkut and L. I. Gilbert, eds., Comprehensive insect physiology, biochemistry and pharmacology. Vol. 12. Insect control. Pergamon, Oxford.

Rosen, D. and P. De Bach. 1992. Foreign exploration: The key to classical biological control. Fla. Entomol. 75:409–413.

Rosenheim, J. A., L. R. Wilhoit, and C. A. Armer. 1993. Influence of intraguild predation among generalist insect predators on the suppression of an herbivore population. Oecologia (Berl.) 96:439–449.

Rosenheim, J. A., H. K. Kaya, L. E. Ehler, J. J. Marois, and B. A. Jaffee. 1995. Intraguild predation among biological-control agents: Theory and evidence. Biol. Control 5:302–335.

Rosenheim, J. A., M. W. Johnson, R. F. L. Mau, S. C. Welter, and B. E. Tabashnik. 1996. Biochemical preadaptations, founder events, and the evolution of resistance in arthropods. J. Econ. Entomol. 89:263–273.

Rosenheim, J. A., L. R. Wilhoit, P. B. Goodell, E. E. Grafton-Cardwell, and T. F. Leigh. 1997. Plant compensation, natural biological control, and herbivory by *Aphis gossypii* on pre-reproductive cotton: The anatomy of a non-pest. Entomol. Exp. Appl. 85:45–63.

Rosenthal, G. A. and D. H. Janzen. 1979. Herbivores: Their interaction with secondary plant metabolites. Academic, New York. 718 pp.

Rosenthal, S. S. and K. L. Lipps. 1973. Development and survival of *Lygus hesperus* nymphs feeding on cotton flower parts. Folia Entomol. Mex. 25–26:34.

Rosenzweig, V. Ye. 1997. Revised classification of the *Calocoris* complex and related genera (Heteroptera: Miridae). Zoosyst. Ross. 6:139–169.

Rosewall, O. W. and C. E. Smith. 1930. The predaceous habit of *Cyrtopeltis* varians Dist. J. Econ. Entomol. 23: 464.

Roshko, G. M. 1976. Miridae (Heteroptera) of the Ukrainian Carpathians, Transcarpathia and Ciscarpathia. Entomol. Obozr. 55(4):814–819. [in Russian; English translation in Entomol. Rev. 55(4):51–55.]

Ross, W. A. 1939. Fruit insects of the season 1938 in Ontario. Can. Insect Pest Rev. 17:44–52.

Ross, W. A. and L. Caesar. 1922. Insects of the season in Ontario. 52nd Annu. Rep. Entomol. Soc. Ont. 1921:42–50.

Ross, W. A. and W. Putman. 1934. The economic insect fauna of Niagara peach orchards. 64th Annu. Rep. Entomol. Soc. Ont. 1933:36–41.

Rostrup, S. and M. Thomsen. 1923. Bekaempelse af Taeger paa Aebletraeer samt Bidrag til disse Taegers Biologi. Tidsskr. Planteavl 29:395–461.

Rothschild, G. H. L. 1963. The immature stages and biology of some mirid predators of Delphacidae, with notes on other predatory Heteroptera occurring in *Juncus* areas. Entomol. Mon. Mag. 99:157–161.

——. 1966. A study of a natural population of *Conomelus anceps* (Germar) (Homoptera: Delphacidae) including observations on predation using the precipitin test. J. Anim. Ecol. 35:413–434.

Rotrekl, J. 1973. Poznatky z bionomie klopušky světlé (*Adelphocoris lineolatus* Goeze) a její škodlivost na luscích vojtěšky seté. Sb. Ved. Pr. VSUP Troubsko 3:161–173.

Rotrekl, J., J. Klumpar, B. Cagaš, and J. Bumerl. 1985. Ploštice a totální beloklasost trav. Ochr. Rostl. 21:267–274.

Roubik, D. W. 1989. Ecology and natural history of tropical bees. Cambridge University Press, Cambridge. 514 pp.

Rowley, J. R. 1971. Implications on the nature of sporopollenin based upon pollen development. Pages 174–218 in J. Brooks, ed., Sporopollenin; proceedings of a symposium held at the Geology Department, Imperial College, London, 23–25 September, 1970. Academic, London.

Royce, L. A. and G. W. Krantz. 1989. Observations on pollen processing by *Pneumolaelaps longanalis* (Acari: Laelapidae), a mite associate of bumblebees. Exp. Appl. Acarol. 7:161–165.

Rozhkov, A. S., ed. 1970. Pests of Siberian larch. (Transl. from Russian by S. Nemchonok.) Israel Program for Scientific Translations, Jerusalem. 393 pp.

Ruberson, J. R. and J. T. Kring. 1991. Predation of *Trichogramma pretiosum* by the anthocorid *Orius insidiosus*. Pages 41–43 in E. Wajnberg and S. B. Vinson, eds., *Trichogramma* and other egg parasitoids: 3rd International Symposium, San Antonio (TX, USA), September 23–27, 1990. INRA, Paris.

Ruberson, J. R., M. J. Tauber, and C. A. Tauber. 1986. Plant feeding by *Podisus maculipennis* (Heteroptera: Pentatomidae): Effect on survival, development, and preoviposition period. Environ. Entomol. 15:894–897.

Rubin, B. A. and E. V. Artsikhovskaya. 1964. Biochemistry of pathological darkening of plant tissues. Annu. Rev. Phytopathol. 2:157–178.

Russ, K. 1959. Schäden durch Wanzen an Reben. Pflanzenarzt 12(2):24–25.

Russell, J. E., T. J. Dennehy, L. Antilla, M. Whitlow, R. Webb, and J. Pacheco. 1997. Lygus bugs in Arizona regain susceptibility to key insecticides. In Proc. Beltwide Cotton Conf., Jan. 6–10, 1997, New Orleans. La. 2:1232–1239. National Cotton Council, Memphis.

Russell, J. L. 1947. The tarnished plant bug. Horticulture (Boston) 25:206–207.

Russell, M. C. 1953. Notes on insects associated with sundews (*Drosera*) at Lesmurdie. West. Aust. Nat. 4:9–12.

Rutter, J. F., A. P. Mills, and L. J. Rosenberg. 1998. Weather associated with autumn and winter migrations of rice pests and other insects in southeastern and eastern Asia. Bull. Entomol. Res. 88:189–197.

Ryan, C. A. 1983. Insect-induced chemical signals regulating natural plant protection responses. Pages 43–60 in R. F. Denno and M. S. McClure, eds., Variable plants and herbivores in natural and managed systems. Academic, New York.

Ryan, R. B. 1985. Mortality of eggs of the larch casebearer (Lepidoptera: Coleophoridae) in Oregon. Can. Entomol. 117:991–994.

Ryckman, R. E. 1979. Host reactions to bug bites (Hemiptera, Homoptera): A literature review and annotated bibliography. Part I. Calif. Vector Views 26:1–24.

Ryckman, R. E. and D. G. Bentley. 1979. Host reactions to bug bites (Hemiptera, Homoptera): A literature review and annotated bibliography. Part II. Calif. Vector Views 26:25–49.

Ryder, W. D., M. Neyra, and O. Chong. 1968. Grain sorghum insects in Cuba, and some effects of DDT high-volume emulsion sprays on their abundance and on yield. Rev. Cubana Cienc. Agric. (English ed.) 2:245–252.

Saahlan, R. M., S. G. Tan, R. Muhamad, and Y. Y. Gan. 1986. Peptidase polymorphism in natural populations of the cocoa pest, *Helopeltis theobromae* Miller. Pertanika 9:65–68.

Sabelis, M. W. 1992. Predatory arthropods. Pages 225–264 in M. J. Crawley, ed., Natural enemies: The population biology of predators, parasites and diseases. Blackwell, Oxford.

Sadof, C. S. and J. J. Neal. 1993. Use of host plant resources by the euonymus scale, *Unaspis euonymi* (Heteroptera: Diaspididae). Ann. Entomol. Soc. Am. 86:614–620.

Sadras, V. O. and G. P. Fitt. 1997. Apical dominance-variability among cotton genotypes and its association with resistance to insect herbivory. Environ. Exp. Bot. 38:145–153.

Saharia, D. 1982. The effect of different plant parts on development, reproduction and longevity in *Lipaphis erysimi* (Kltb.) (Homoptera: Aphididae). J. Res. Assam Agric. Univ. 3:67–71.

Sahlberg, R. F. 1848. Monographia Geocorisarum Fenniae. Officina Typographica Frenckelliana, Helsinki. 154 pp.

Sailer, R. I. 1952. A technique for rearing certain Hemiptera. U.S. Dep. Agric. Bur. Entomol. Plant Quar. ET-303:1–5.

Salamero, A., R. Gabarra, and R. Albajes. 1987. Observations on the predatory and phytophagous habits of *Dicyphus tamaninii* Wagner (Heteroptera; Miridae). Proc. Working Group, "Integrated Control in Glasshouses," EPRS/WPRS, Budapest, Hungary, 26–30 April 1987. Int. Organ. Biol. Control/West. Palearctic Reg. Sect. Bull. 10(2):165–169.

Salim, M. and E. A. Heinrichs. 1986. Impact of varietal resistance in rice and predation on the mortality of *Sogatella furcifera* (Horvath) (Homoptera: Delphacidae). Crop Prot. 5:395–399.

Salt, R. W. 1945. Number of generations of *Lygus hesperus* Knt. and *L. elisus* Van D. in Alberta. Sci. Agric. (Ottawa) 25:573–576.

Samal, P. and B. C. Misra. 1977. Notes on the life history of *Cyrtorhinus lividipennis* Reuter, a predatory mirid bug of rice brown plant hopper *Nilaparvata lugens* (Stal) in Orissa. Oryza 14:47–50.

Sampson, C. and R. J. Jacobson. 1999. *Macrolophus caliginosus* Wagner (Heteroptera: Miridae): A predator causing damage to UK tomatoes. Int. Organ. Biol. Control/West Palearctic Reg. Sect. Bull. 22(1):213–216.

Samson, J. A. 1980. Tropical fruits. Longman, London. 250 pp.

Samuel, G. 1927. On the shot-hole disease caused by *Clasterosporium carpophilum* and on the 'shot-hole' effect. Ann. Bot. 41:375–404.

Samways, M. J. 1979. Immigration, population growth and mortality of insects and mites on cassava in Brazil. Bull. Entomol. Res. 69:491–505.

——. 1993. A spatial and process sub-regional framework for insect and biodiversity conservation research and management. Pages 1–27 in K. J. Gaston, T. R. New, and M. J. Samways, eds., Perspectives on insect conservation. Intercept, Andover, UK.

——. 1994. Insect conservation biology. Chapman & Hall, London. 358 pp.

Samy, O. 1958. Green cotton boll shedding in relation to insect infestation of flowers. Agric. Res. Rev. 36:51–58.

Sanders, G. E. and W. H. Brittain. 1916. Spraying for insects affecting apple orchards in Nova Scotia. Can. Dep. Agric. Entomol. Branch Circ. 8:1–11.

Sanderson, E. D. 1906. Report on miscellaneous cotton insects in Texas. U.S. Dep. Agric. Bur. Entomol. Bull. 57:1–63.

Sanford, K. H. 1964a. Eggs and oviposition sites of some predacious mirids on apple trees (Miridae: Hemiptera). Can. Entomol. 96:1185–1189.

——. 1964b. Life history and control of Atractotomus mali, a new pest of apple in Nova Scotia (Miridae: Hemiptera). J. Econ. Entomol. 57:921–925.

Sanford, K. H. and H. J. Herbert. 1970. The influence of spray programs on the fauna of apple orchards in Nova Scotia. XX. Trends after altering levels of phytophagous mites or predators. Can. Entomol. 102:592–601.

Sanford, K. H. and F. T. Lord. 1962. The influence of spray programs on the fauna of apple orchards in Nova Scotia. XIII. Effects of Perthane on predators. Can. Entomol. 94:928–934.

Sannikova, M. F. and L. I. Garbar. 1981. Cereal bugs in the Tyumen region. Zashch. Rast. (Mosc.) No. 4:29. [in Russian.]

Santas, L. A. 1987. The predators' complex of pear-feeding psyllids in unsprayed wild pear trees in Greece. Entomophaga 32:291–297.

Santiago-Blay, J. A. and G. O. Poinar Jr. 1993. Classification of Diphleps (Heteroptera: Miridae: Isometopinae), with the description of D. yenli, a new species from Dominican amber (lower Oligocene-upper Eocene). Proc. Entomol. Soc. Wash. 95:70–73.

Santoro, R. 1960. Notas de entomologia agricola Dominicana. Secretaria de Estado de Agricultura y Comercio, República Dominicana. Editorial "La Nacion," Ciudad Trujillo, República Dominicana. 474 pp.

Sapio, F. J., L. F. Wilson, and M. E. Ostry. 1982. A split-stem lesion on hybrid Populus trees caused by the tarnished plant bug, Lygus lineolaris (Hemiptera [Heteroptera]: Miridae. Gt. Lakes Entomol. 15:237–246.

Sappington, T. W. and W. B. Showers. 1992. Reproductive maturity, mating status, and long-duration flight behavior of Agrotis ipsilon (Lepidoptera: Noctuidae) and the conceptual misuse of the oogenesis-flight syndrome by entomologists. Environ. Entomol. 21:677–688.

Sargent, T. D., C. D. Millar, and D. M. Lambert. 1998. The "classical" explanation of industrial melanism: Assessing the evidence. Evol. Biol. (N.Y.) 30:299–322.

Sasidharan Pillai, K., K. Saradamma, and M. R. G. K. Nair. 1980. Helopeltis antonii Sign. as a pest of Moringa oleifera. Curr. Sci. (Bangalore) 49:288–289.

Satchell, J. E. and T. R. E. Southwood. 1963. The Heteroptera of some woodlands in the English Lake District. Trans. Soc. Brit. Entomol. 15:117–134.

Sathiamma, B. 1977. Nature and extent of damage by Helopeltis antonii S., the tea mosquito on cashew. J. Plant. Crops 5:58–62.

Satterthwait, A. F. 1944. Leucopoecila albofasciata, a pest of golf greens. J. Econ. Entomol. 37:562.

Sauer, H. F. G. 1942. "Horcius nobilellus (Berg)" (Hem. Mir.) praga dos algodoais do Estado de S. Paulo. Arq. Inst. Biol. 13:29–66.

Saulich, A. Kh. and D. L. Musolin. 1996. Univoltinism and its regulation in some temperate true bugs (Heteroptera). Eur. J. Entomol. 93:507–518.

Saunders, E. 1892. The Hemiptera Heteroptera of the British Islands. A descriptive account of the families, genera, and species indigenous to Great Britain and Ireland, with notes as to localities, habitats, etc. L. Reeve, London. 350 pp.

Saunders, J. L. 1981. Cacao pests in Central America. Pages 429–432 in J. De Lafforest, eds., Proc. 7th Int. Cocoa Res. Conf., 4–12 Nov. 1979, Douala, Cameroun. Transla-Inter, London.

Saunders, W. 1883. Insects injurious to fruits. Lippincott, Philadelphia. 436 pp.

Saxena, K. N. 1954. Feeding habits and physiology of digestion of certain leafhoppers Homoptera: Jassidae. Experientia (Basel) 10:383–384.

——. 1963. Mode of ingestion of a heteropterous insect Dysdercus koenigii (F.) (Pyrrhocoridae). J. Insect Physiol. 9:47–71.

Saxena, P. N. and H. J. Chada. 1971. The greenbug, Schizaphis graminum. I. Mouth parts and feeding habits. Ann. Entomol. Soc. Am. 64:897–904.

Say, T. 1832. Descriptions of new species of heteropterous Hemiptera of North America. New Harmony, Ind. 39 pp.

Scales, A. L. 1968. Female tarnished plant bugs attract males. J. Econ. Entomol. 61:1466–1467.

——. 1973. Parasites of the tarnished plant bug in the Mississippi Delta. Environ. Entomol. 2:304–306.

Scales, A. L. and R. E. Furr. 1968. Relationship between the tarnished plant bug and deformed cotton plants. J. Econ. Entomol. 61:114–118.

Scales, A. L. and J. Hacskaylo. 1974. Interaction of three cotton cultivars to infestations of the tarnished plant bug. J. Econ. Entomol. 67:602–604.

Scarbrough, A. G. 1981. Ethology of Eudiotria tibialis Banks (Diptera: Asilidae) in Maryland: Prey, predatory behavior, and enemies. Proc. Entomol. Soc. Wash. 83:258–268.

Scarbrough, A. G. and B. E. Sraver. 1979. Predatory behavior and prey of Atomosia puella (Diptera: Asilidae). Proc. Entomol. Soc. Wash. 81:630–639.

Schaarschmidt, H. 1982. Zwei Plagiognathus-Arten (Heteroptera, Miridae) auf Geranium-Blüten in Leipzig. Entomol. Nachr. Ber. 26:280–281.

——. 1983. Exolygus rugulipennis Popp. (Heteroptera, Miridae) an Caryopteris × clandonensis Simmonds ex Rehd. (Verbenaceae) im Zentrum Leipzigs. Entomol. Nachr. Ber. 27:41.

Schaber, B. D. and T. Entz. 1988. Effect of spring burning on insects in seed alfalfa fields. J. Econ. Entomol. 81:668–672.

——. 1994. Effect of annual and biennial burning of seed alfalfa (lucerne) stubble on populations of lygus (Lygus spp.), and alfalfa plant bug (Adelphocoris lineolatus (Goeze)) and their predators. Ann. Appl. Biol. 124:1–9.

Schaber, B. D., A. M. Harper, and T. Entz. 1990. Effect of swathing alfalfa for hay on insect dispersal. J. Econ. Entomol. 83:2427–2433.

Schaefer, C. W. 1974. Rise and fall of the apple redbugs. Mem. Conn. Entomol. Soc. 1974: 101–116.

——. 1975. Heteropteran trichobothria (Hemiptera: Heteroptera). Int. J. Insect Morphol. Embryol. 4:193–264.

——. 1981. The land bugs (Hemiptera: Heteroptera) and their adaptive zones. Rostria (Tokyo) 33(Suppl.): 67–83.

——. 1990. The Hemiptera of North America: What we do and do not know. Pages 105–117 in M. Kosztarab and C. W. Schaefer, eds., Systematics of the North American insects and arachnids: Status and needs. Va. Agric. Exp. Stn. Info. Ser. 90-1. Virginia Polytechnic Institute & State University, Blacksburg.

——. 1996. Introduction. Pages 1–8 in C. W. Schaefer, ed., Studies on hemipteran phylogeny. Thomas Say Publ. Entomol.: Proceedings. Entomological Society of America, Lanham, Md.

——. 1997. The origin of secondary carnivory from herbivory in Heteroptera (Hemiptera). Pages 229–239 *in* A. Raman, ed., Ecology and evolution of plant-feeding insects in natural and man-made environments. International Scientific Publications, New Delhi, India.

Schaefer, C. W. and P. L. Mitchell. 1983. Food plants of the Coreoidea (Hemiptera: Heteroptera). Ann. Entomol. Soc. Am. 76:591–615.

Schaefers, G. A. 1966. The reduction of insect-caused apical seediness in strawberries. J. Econ. Entomol. 59:698–706.

——. 1972. Insecticidal evaluations for reduction of tarnished plant injury in strawberries. J. Econ. Entomol. 65: 1156–1160.

——. 1979. The tarnished plant bug and other causes of strawberry deformities. Pa. Fruit News 58(4):95–97.

——. 1980. Yield effects of tarnished plant bug feeding on June-bearing strawberry varieties in New York State. J. Econ. Entomol. 73:721–725.

——. 1981. Pest management systems for strawberry insects. Pages 377–393 *in* D. Pimentel, ed., CRC handbook of pest management in agriculture. Vol. III. CRC, Boca Raton, Fla.

——. 1987. Miscellaneous arthropod damage in strawberry. Pages 76–78 *in* R. H. Converse, ed., Virus diseases of small fruits. U.S. Dep. Agric. Agric. Handb. 631.

——. 1991. Lygus bugs. Pages 66–67 *in* M. A. Ellis, R. H. Converse, R. N. Williams, and B. Williamson, eds., Compendium of raspberry and blackberry diseases and insects. APS, St. Paul, Minn.

Schaffner, J. C. and P. S. F. Ferreira. 1995a. *Carvalhoisca*, a new genus of Orthotylini from Mexico (Miridae, Heteroptera). Proc. Entomol. Soc. Wash. 97:373–378.

——. 1995b. *Rolstonocoris*, a new genus of Neotropical Miridae (Heteroptera: Orthotylinae). J. N.Y. Entomol. Soc. 103:374–385.

Schaub, G. A. and B. Breger. 1988. Pathological effects of *Blastocrithidia triatomae* (Trypanosomatidae) on the reduviid bugs *Triatoma sordida*, *T. pallidipennis* and *Dipetalogaster maxima* after coprophagic infection. Med. Vet. Entomol. 2:309–318.

Schaub, L. P. and J. U. Baumgärtner. 1989. Significance of mortality and temperature on the phenology of *Orthotylus marginalis* (Heteroptera: Miridae).

Mitt. Schweiz. Entomol. Ges. 62:235–245.

Schaub, L. [P.], J. Baumgärtner, and V. Delucchi. 1987. Anwendung multivariater Verfahren auf die Darstellung der Heteropterenfauna unter dem Einfluss der Intensivierung des Apfelanbaus. Mitt. Schweiz. Entomol. Ges. 60:15–24.

Schaub, L. [P.], W. A. Stahel, J. Baumgärtner, and V. Delucchi. 1988. Elements for assessing mirid (Heteroptera: Miridae) damage threshold on apple fruits. Crop Prot. 7:118–124.

Schedl, K. 1931. Der Hemlockspanner *Ellopia fiscellaria* Hb. und seine natürlichen Feinde. Z. Angew. Entomol. 18:219–275.

Scheffer, R. P. 1983. Toxins as chemical determinants of plant disease. Pages 1–40 *in* J. M. Daly and B. J. Deverall, eds., Toxins and plant pathogenesis. Academic, New York.

Schelt, J. van. 1994. The use of *Macrolophus caliginosus* as a whitefly predator in protected crops. International *Bemisia* workshop, Shoresh, Israel, 3–7 Oct. 1994. *Bemisia* Newsl. 8:26.

Schelt, J. van., J. Klapwijk, M. Letard, and C. Aucouturier. 1996. The use of *Macrolophus caliginosus* as a whitefly predator in protected crops. Pages 515–521 *in* D. Gerling and R. T. Mayer, eds., *Bemisia*: 1995: Taxonomy, biology, damage, control and management. Intercept, Andover, UK.

Schemske, D. W. and C. C. Horvitz. 1984. Variation among floral visitors in pollination ability: A precondition for mutualism specialization. Science (Wash., D.C.) 225:519–521.

Schewket Bey, N. 1930. Zur Biologie der phytophagen Wanze *Dicyphus errans* Wolff (Capsidae). Z. Wiss. Insektenbiol. 25:179–183.

Schlee, M. A. 1986. Avian predation on Heteroptera: Experiments on the European blackbird *Turdus m. merula* L. Ethology 73:1–18.

——. 1992. La prédation des Hétéroptères par les Oiseaux. EPHE (Ecole Pratique Hautes Etudes). Biol. Evol. Insectes 5:87–96.

Schmidt, A. 1932. Kräuselkrankheit bei Pelargonien. Gartenflora (Berl.) 81:40.

Schmitz, G. 1994. Zum Blütenbesuchsspektrum indigener und neophytischer *Impatiens*-Arten. Entomol. Nachr. Ber. 38:17–23.

Schmitz, G[uy]. 1958. *Helopeltis* du cotonnier en Afrique centrale. Publ. Inst. Natl. Agron. Congo Belge Sér. Sci. No. 71:1–178.

——. 1968. Monographie des espèces africaines du genre *Helopeltis* Signoret (Heteroptera, Miridae) avec un exposé

des problèmes relatifs aux structures génitales. Ann. Mus. R. Afr. Cent. 168:1–247.

Schmitz, G[uy]. and P. Štys. 1973. *Howefulvius elytratus* gen. n., sp. n. (Heteroptera, Miridae, Fulviinae) from Lord Howe Island in the Tasman Sea. Acta Entomol. Bohemoslov. 70:400–407.

Schmutterer, H. 1969. Pests of crops in Northeast and Central Africa with particular reference to the Sudan. Fischer, Stuttgart. 296 pp.

——. 1977. Other injurious Heteroptera. Pages 300–301 *in* J. Kranz, H. Schmutterer, and W. Koch, eds., Diseases, pests and weeds in tropical crops. Parey, Berlin.

——. 1990a. Beobachtungen an Schädlingen von *Azadirachta indica* (Niembaum) und von verschiedenen *Melia*-Arten. J. Appl. Entomol. 109:390–400.

——. 1990b. Crop pests in the Caribbean with particular reference to the Dominican Republic. (GTZ) Deutsche Gesellschaft für Technische Zusammenarbeit, Eschborn. 640 pp.

Schneider, I. 1962. Die Nadelholzspinnmilbe, *Paratetranychus unungius* [*sic*] Jac. (Acari, Trombidiformes), auf den Aufforstingsflächen Schlewsig-Holsteins. Allg. Forst. Jagdztg. 133:144–148.

Schoen, C. 1932. Bestrijding van wantsen in eitoestand. Tijdschr. Plziekt. 38:41–59.

Schoener, T. W. 1971. Theory of feeding strategies. Annu. Rev. Ecol. Syst. 2: 369–404.

Schoenly, K. G., J. E. Cohen, K. L. Heong, G. S. Arida, A. T. Barrion, and J. A. Litsinger. 1996. Quantifying the impact of insecticides on food web structure of rice-arthropod populations in a Philippine farmer's irrigated field: A case study. Pages 343–351 *in* G. A. Polis and K. O. Winemiller, eds., Food webs: Integration of patterns & dynamics. Chapman & Hall, New York.

Schoevers, T. A. C. 1930. Appelwantsen en hunne bestrijding. Tijdschr. Plziekt. 36:75–83.

Scholl, J. M. and J. T. Medler. 1947. Trap strips to control insects affecting alfalfa seed production. J. Econ. Entomol. 40:448–450.

Schoonhoven, L. M. and I. Derksen-Koppers. 1973. Effects of secondary plant substances on drinking behaviour in some Heteroptera. Entomol. Exp. Appl. 16:141–145.

Schotzko, D. J. and L. E. O'Keeffe. 1989a. *Lygus hesperus* distribution and sampling procedures in lentils. Environ. Entomol. 18:308–314.

——. 1989b. Geostatistical description of

the spatial distribution of *Lygus hesperus* (Heteroptera: Miridae) in lentils. J. Econ. Entomol. 82:1277–1288.

——. 1990. Effect of sample placement on the geostatistical analysis of the spatial distribution of *Lygus hesperus* (Heteroptera: Miridae) in lentils. J. Econ. Entomol. 83:1888–1900.

Schowalter, T. D. 1987. Abundance and distribution of *Lygus hesperus* (Heteroptera: Miridae) in two conifer nurseries in western Oregon. Environ. Entomol. 16:687–690.

Schowalter, T. D. and J. D. Stein. 1987. Influence of Douglas-fir seedling provenance and proximity to insect population sources on susceptibility to *Lygus hesperus* (Heteroptera: Miridae) in a forest nursery in western Oregon. Environ. Entomol. 16:984–986.

Schowalter, T. D., J. W. Webb, and D. A. Crossley Jr. 1981. Community structure and nutrient content of canopy arthropods in clearcut and uncut forest ecosystems. Ecology 62:1010–1019.

Schowalter, T. D., D. L. Overhulser, A. Kanaskie, J. D. Stein, and J. Sexton. 1986. *Lygus hesperus* as an agent of apical bud abortion in Douglas-fir nurseries in western Oregon. New For. 1:5–15.

Schøyen, T. H. 1916. Beretning om skadeinsekter og plantesygdommer i land og pavebruket. Pages 37–92 *in* Arsberetn. off. Foranst. Landbr. Frem. 1915.

Schreier, O. and W. Faber. 1954. Schädigungen durch Wanzensaugstiche. Pflanzenarzt 7(10):1–2.

Schroeder, N. C., R. B. Chapman, and P. T. P. Clifford. 1998. Effect of potato mirid (*Calocoris norvegicus*) on white clover seed production in small cages. N.Z. J. Agric. Res. 41:111–116.

Schuh, R. T. 1974. The Orthotylinae and Phylinae (Hemiptera: Miridae) of South Africa with a phylogenetic analysis of the ant-mimetic tribes of the two subfamilies for the world. Entomol. Am. 47:1–332.

——. 1975. The structure, distribution, and taxonomic importance of trichobothria in the Miridae (Hemiptera). Am. Mus. Novit. No. 2585:1–26.

——. 1976. Pretarsal structure in the Miridae (Hemiptera) with a cladistic analysis of relationships within the family. Am. Mus. Novit. No. 2601: 1–39.

——. 1979. [Review of] *Evolutionary Trends in Heteroptera. Part II. Mouthpart-structures and Feeding Strategies* by R. H. Cobben, 1978. Syst. Zool. 28:653–656.

——. 1984. Revision of the Phylinae (Hemiptera, Miridae) of the Indo-Pacific. Bull. Am. Mus. Nat. Hist. 177:1–476.

——. 1986a. The influence of cladistics on heteropteran classification. Annu. Rev. Entomol. 31:67–93.

——. 1986b. *Schizopteromiris*, a new genus and four new species of coleopteroid cylapine Miridae from the Australian region (Heteroptera). Ann. Soc. Entomol. Fr. 22:241–246.

——. 1991. Phylogenetic, host and biogeographic analyses of the Pilophorini (Heteroptera: Miridae: Phylinae). Cladistics 7:157–189.

——. 1995. Plant bugs of the world (Insecta: Heteroptera: Miridae): Systematic catalog, distributions, host list, and bibliography. New York Entomological Society, New York. 1329 pp.

——. 1996. [Review of] *Studies on Hemipteran Phylogeny* edited by C. W. Schaefer. J. N.Y. Entomol. Soc. 104: 231–235.

——. 2000. Biological systematics: Principles and applications. Cornell University Press, Ithaca. 236 pp.

Schuh, R. T. and J. D. Lattin. 1980. *Myrmecophyes oregonensis*, a new species of Halticini (Hemiptera, Miridae) from the western United States. Am. Mus. Novit. No. 2697:1–11.

Schuh, R. T. and M. D. Schwartz. 1985. Revision of the plant bug genus *Rhinacloa* Reuter with a phylogenetic analysis (Hemiptera, Miridae). Bull. Am. Mus. Nat. Hist. 179:379–470.

——. 1988. Revision of the New World Pilophorini (Heteroptera: Miridae: Phylinae). Bull. Am. Mus. Nat. Hist. 187:101–201.

Schuh, R. T. and J. A. Slater. 1995. True bugs of the world (Hemiptera: Heteroptera): Classification and natural history. Cornell University Press, Ithaca. 336 pp.

Schuh, R. T. and G. M. Stonedahl. 1986. Historical biogeography in the Indo-Pacific: A cladistic approach. Cladistics 2:337–355.

Schuh, R. T. and P. Štys. 1991. Phylogenetic analysis of cimicomorphan family relationships (Heteroptera). J. N.Y. Entomol. Soc. 99:298–350.

Schuh, R. T., P. Lindskog, and I. M. Kerzhner. 1995. *Europiella* Reuter (Heteroptera: Miridae): Recognition as a Holarctic group, notes on synonymy, and description of a new species, *Europiella carvalhoi*, from North America. Proc. Entomol. Soc. Wash. 97:379–395.

Schuler, K. 1982. Blütenbesuch durch Insekten an *Solidago canadensis* und *S. virgaurea*, eine vergleichende Studie. Ber. Natwiss.-Med. Ver. Innsbr. 69:127–144.

Schumacher, F. 1917a. Über die Gattung *Stethoconus* Flor. (Hem. Het. Caps.). Sitzungsber. Gesell. Naturf. Freunde Berl. 1916:344–346.

——. 1917b. Die Bedeutung der Hemipteren als Blütenbestäuber. Sitzungsber. Ges. Naturf. Freunde Berl. 1917:444–446.

——. 1917c. Ueber Psociden-Feinde aus der Ordnung der Hemipteren. Z. Wiss. Insektenbiol. 13:217–218.

——. 1917d. [Über *Sthenarus rotermundi* Sz., eine an Silberpappeln Missbildungen erzeugende Wanze]. Dtsch. Entomol. Z. 1917:331.

——. 1919a. Die *Roridula*-Arten und ihre Bewohner. Z. Wiss. Insektenbiol. 14: 218–221.

——. 1919b. Einige schädliche Hemipteren von der Insel Java. Z. Wiss. Insektenbiol. 14:221–224.

Schumann, G. L., W. M. Tingey, and H. D. Thurston. 1980. Evaluation of six insect pests for transmission of potato spindle tuber viroid. Am. Potato J. 57:205–211.

Schuster, M. F. 1977. Plant bugs—key pests on cotton. *In* Proc. Beltwide Cotton Prod. Res. Conf., Jan. 10–12, 1977, Atlanta, Ga., pp. 156–157. National Cotton Council, Memphis.

——. 1980. Insect resistance in cotton. Pages 101–112 *in* M. K. Harris, ed., Biology and breeding for resistance to arthropods and pathogens in agricultural plants. Tex. Agric. Exp. Stn. MP-1451.

Schuster, M. F. and J. C. Boling. 1974. Phenology of early and mid-season predaceous and phytophagous insects in cotton in the Lower Rio Grande Valley of Texas. Tex. Agric. Exp. Stn. MP-1133. 31 pp.

Schuster, M. F. and J. L. Frazier. 1977. Mechanisms of resistance to *Lygus* spp. in *Gossypium hirsutum* L. Eucaria/OLIB Working Group Breeding for Resistance to Insects and Mites. Rep. 1st Meet. Wageningen, The Netherlands, 7 to 9 Dec. 1976. Int. Organ. Biol. Control/West. Palearctic Reg. Sect. Bull. 3:129–135.

Schuster, M. F. and F. G. Maxwell. 1974. The impact of nectariless cotton on plant bugs, bollworms and beneficial insects. *In* Proc. Beltwide Cotton Prod. Res. Conf., Jan. 7–9, 1974, Dallas, Texas, pp. 86–87. National Cotton Council, Memphis.

Schuster, M. F., C. A. Richmond, J. C. Boling, and H. M. Graham. 1969. Host plants of the cotton fleahopper in the Rio Grande Valley: Phenology and hibernating quarters. J. Econ. Entomol. 62:1126–1129.

Schuster, M. F., D. G. Holder, E. T. Cherry, and F. G. Maxwell. 1976a. Plant bugs and natural enemy insect populations on frego bract and smooth-leaf cottons. Miss. Agric. For. Exp. Stn. Tech. Bull. 75:1–11.

Schuster, M. F., M. J. Lukefahr, and F. G. Maxwell. 1976b. Impact of nectariless cotton on plant bugs and natural enemies. J. Econ. Entomol. 69:400–402.

Schwartz, M. D. 1984. A revision of the black grass bug genus *Irbisia* Reuter (Heteroptera: Miridae). J. N.Y. Entomol. Soc. 92:193–306.

———. 1994. Review of the genus *Salignus* Kelton and a character discussion of related genera (Heteroptera: Miridae: Mirinae). Can. Entomol. 126:971–993.

Schwartz, M. D. and R. G. Foottit. 1992a. *Lygus* species on oilseed rape, mustard, and weeds: A survey across the Prairie Provinces of Canada. Can. Entomol. 124:151–158.

———. 1992b. Lygus bugs on the prairies: Biology, systematics, and distribution. Agric. Can. Res. Branch Tech. Bull. 1992–4E:1–44.

———. 1998. Revision of the Nearctic species of the genus *Lygus* Hahn, with a review of the Palaearctic species (Heteroptera: Miridae). Mem. Entomol. Int. Vol. 10. Associated Publishers, Gainesville, Fla. 428 pp.

Schwartz, M. D. and L. A. Kelton. 1990. *Psallus salicicola*, a new species, with additional records of recently discovered Palearctic *Psallus* Fieber from Canada (Heteroptera: Miridae: Phylinae). Can. Entomol. 122:941–947.

Schwartz, P. H. 1983. Losses in yield of cotton due to insects. Pages 329–358 *in* R. L. Ridgway, E. P. Lloyd, and W. H. Cross, eds., Cotton insect management with special reference to the boll weevil. U.S. Dep. Agric. Agric. Res. Serv. Agric. Handb. 589.

Schwartz, P. H. and W. Klassen. 1981. Estimate of losses caused by insects and mites to agricultural crops. Pages 15–77 *in* D. Pimentel, ed., CRC handbook of pest management in agriculture. Vol. 1. CRC, Boca Raton, Fla.

Schwarz, E. A., O. Heidemann, and N. Banks. 1914. Philip Reese Uhler. Proc. Entomol. Soc. Wash. 16:1–7.

Schweizer, J. 1939. Jaarverslag tabak over Juli 1938 t/m Juni 1939. Meded. Besoekisch Proefstn. 64:1–64.

Scott, D. R. 1969. Lygus-bug feeding on developing carrot seed: Effect on plants growing from that seed. J. Econ. Entomol. 62:504–505.

———. 1970. Feeding of *Lygus* bugs (Hemiptera: Miridae) on developing carrot and bean seed: Increased growth and yields of plants grown from that seed. Ann. Entomol. Soc. Am. 63:1604–1608.

———. 1972. Lygus bugs feeding on developing carrot seed: Apparent transmission of a viruslike foliage disorder. J. Econ. Entomol. 65:297–298.

———. 1976a. Phytostimulation by lygus bugs feeding on developing seeds. Pages 17–18 *in* D. R. Scott and L. E. O'Keeffe, eds., Lygus bug: Host plant interactions. University Press of Idaho, Moscow.

———. 1976b. Pollination by *Lygus* spp. Page 36 *in* D. R. Scott and L. E. O'Keeffe, eds., Lygus bug: Host plant interactions. University Press of Idaho, Moscow.

———. 1977a. An annotated listing of host plants of *Lygus hesperus* Knight. Bull. Entomol. Soc. Am. 23(1):19–22.

———. 1977b. Selection for lygus bug resistance in carrot. HortScience 12:452.

———. 1980. A bibliography of *Lygus* Hahn (Hemiptera: Miridae). Idaho Agric. Exp. Stn. Misc. Ser. 58:1–71.

———. 1983. *Lygus hesperus* Knight (Hemiptera: Miridae) and *Daucus carota* L. (Umbelliflorae: Umbelliferae): An example of relationships between a polyphagous insect and one of its plant hosts. Environ. Entomol. 12:6–9.

———. 1987. Biological control of lygus bugs on vegetable and fruit crops. Pages 40–47 *in* R. C. Hedlund and H. M. Graham, eds., Economic importance and biological control of *Lygus* and *Adelphocoris* in North America. U.S. Dep. Agric. Agric. Res. Serv. ARS-64.

Scott, D. R., A. J. Walz, and H. C. Manis. 1966. The effect of *Lygus* spp. on carrot seed production in Idaho (Hemiptera: Miridae). Idaho Agric. Exp. Stn. Res. Bull. 69:1–12.

Scott, H. G. and C. J. Stojanovich. 1963. Digestion of juniper pollen by Collembola. Fla. Entomol. 46:189–191.

Scott, W. P., J. W. Smith, and G. L. Snodgrass. 1986. A two year large-scale field study on differences in arthropod populations in nectaried and nectariless varieties of cotton. *In* Proc. Beltwide Cotton Prod. Res. Conf., Jan. 4–9, 1986, Las Vegas, Nev., pp. 212–214. National Cotton Council, Memphis.

Scott, W. P., G. L. Snodgrass, and J. W. Smith. 1988. Tarnished plant bug (Hemiptera: Miridae) and predaceous arthropod populations in commercially produced selected nectaried and nectariless cultivars of cotton. J. Entomol. Sci. 23:280–286.

Scudder, G. G. E. 1960. *Dictyonota fuliginosa* Costa (Hemiptera: Tingidae) in the Nearctic. Proc. Entomol. Soc. B.C. 57:22.

———. 1963. Heteroptera stranded at high altitudes in the Pacific Northwest. Proc. Entomol. Soc. B.C. 60:41–44.

———. 1979. Hemiptera. Pages 329–348 *in* H. V. Danks, ed., Canada and its insect fauna. Mem. Entomol. Soc. Can. 108.

———. 1985. Heteroptera new to Canada. J. Entomol. Soc. B.C. 82:66–71.

———. 1995. The first record for *Bothynotus pilosus* (Boheman) (Hemiptera; Miridae) in the Nearctic Region. Proc. Entomol. Soc. Wash. 97:396–400.

———. 1997. True bugs (Heteroptera) of the Yukon. Pages 241–336 *in* H. V. Danks and J. A. Downes, eds., Insects of the Yukon. Monogr. Ser. 2. Biological Survey of Canada (Terrestrial Arthropods), Ottawa.

Seastedt, T. R., L. Mameli, and K. Gridley. 1981. Arthropod use of invertebrate carrion. Am. Midl. Nat. 105:124–129.

Šedivý, J. 1972. Die Schädlichkeit der Wanze *Lygus rugulipennis* Popp an Samenluzerne. Z. Pflanzenkr. Pflanzenschutz 79:407–412.

Šedivý, J. and V. Fric. 1999. Harmfulness of mirid bugs (Heteroptera, Miridae) on hop plants. Rostl. Výroba 45:255–257.

Šedivý, J. and A. Honěk. 1983. Flight of *Lygus rugulipennis* Popp. (Heteroptera, Miridae) to a light trap. Z. Pflanzenkr. Pflanzenschutz 90:238–243.

Seidenstücker, G. 1967. Eine Phyline mit Dicyphus-Kralle (Heteroptera, Miridae). Reichenbachia 8(27):215–220.

Selander, R. B. 1960. Bionomics, systematics, and phylogeny of *Lytta*, a genus of blister beetles (Coleoptera, Meloidae). Ill. Biol. Monogr. 28:1–295.

Senguttuvan, T. and M. Gopalan. 1990. Predatory efficiency of mirid bug (*Cyrtorhinus lividipennis*) on eggs and nymphs of brown planthopper (*Nilaparvata lugens*) in resistant and susceptible varieties of rice (*Oryza sativa*). Indian J. Agric. Sci. 60:285–287.

Seshu Reddy, K. V. 1988. Assessment of on-farm yield losses in sorghum due to insect pests. Insect Sci. Appl. 9:679–685.

———. 1991. Insect pests of sorghum in Africa. Insect Sci. Appl. 12:653–657.

Sevacherian, V. 1975. Activity and probing behavior of *Lygus hesperus* in the laboratory. Ann. Entomol. Soc. Am. 68:557–558.

———. 1976. Spatial distribution of *Lygus* in host plants. Pages 3–7 *in* D. R. Scott and L. E. O'Keeffe, eds., Lygus bug: Host plant interactions. University of Idaho Press, Moscow.

Sevacherian, V. and V. M. Stern. 1972a. Spatial distribution patterns of lygus bugs in California cotton fields. Environ. Entomol. 1:695–704.

——. 1972b. Sequential sampling plans for lygus bugs in California cotton fields. Environ. Entomol. 1:704–710.

——. 1974. Host plant preferences of lygus bugs in alfalfa-interplanted cotton fields. Environ. Entomol. 3:761–766.

——. 1975. Movements of lygus bugs between alfalfa and cotton. Environ. Entomol. 4:163–165.

Severin, H. H. P. and J. H. Freitag. 1938. Western celery mosaic. Hilgardia 11(9):493–558.

Seymour, P. R., compiler. 1979. Invertebrates of economic importance in Britain; common and scientific names. H. M. Stationery Office, London. 132 pp.

Shabbir, S. G. and U. D. Choudhry. 1984. Pests of important fibre crops in Pakistan. Pages 93–162 in M. K. Ahmed, ed., Insect pests of important crops in Pakistan. Department of Plant Protection, Karachi.

Shahjahan, M. 1974. *Erigeron* flowers as a food and attractive odor source for *Peristenus pseudopallipes*, a braconid parasitoid of the tarnished plant bug. Environ. Entomol. 3:69–72.

Shahjahan, M. and F. A. Streams. 1973. Plant effects on host-finding by *Leiophron pseudopallipes* (Hymenoptera: Braconidae), a parasitoid of the tarnished plant bug. Environ. Entomol. 2:921–925.

Shapiro, A. M. 1975. The temporal component of butterfly species diversity. Pages 181–195 in M. L. Cody and J. M. Diamond, eds., Ecology and evolution of communities. Harvard University Press, Cambridge.

Shapiro, I. D. 1956. On the destruction of crops in Leningrad District in 1952. Entomol. Obozr. 35(1):139–141. [in Russian.]

Shaposhnikov, G. Ch. 1987. Evolution of aphids in relation to evolution of plants. Pages 409–414 in A. K. Minks and P. Harrewijn, eds., Aphids: Their biology, natural enemies and control. Vol. A. Elsevier, Amsterdam.

Sharma, H. C. 1985a. Screening for host-plant resistance to mirid head bugs in sorghum. In Proc. Int. Sorghum Entomol. Workshop, 15–21 July 1984, Texas A&M Univ., College Station, pp. 317–335. International Crops Research Institute for the Semi-Arid Tropics, Patancheru, A.P., India.

——. 1985b. Strategies for pest control in sorghum in India. Trop. Pest Manage. 31:167–185.

——. 1985c. Oviposition behaviour and host-plant resistance to sorghum head

bug, *Eurystylus marginatus* Odh. Report (limited distribution) of co-operative work carried out at Station de Recherches sur les Cultures Vivriers et Oleagineuses (SRCVO), Sotuba, Mali. International Crops Research Institute for the Semi-Arid Tropics, Patancheru, A.P., India. 53 pp.

Sharma, H. C. and K. Leuschner. 1987. Chemical control of sorghum head bugs (Hemiptera: Miridae). Crop Prot. 6:334–340.

Sharma, H. C. and V. F. Lopez. 1989. Assessment of avoidable losses and economic injury levels for the sorghum head bug, *Calocoris angustatus* Leth. (Hemiptera: Miridae) in India. Crop Prot. 8:429–435.

——. 1990a. Biology and population dynamics of sorghum head bugs (Hemiptera: Miridae). Crop Prot. 9:164–173.

——. 1990b. Mechanisms of resistance in sorghum to head bug, *Calocoris angustatus*. Entomol. Exp. Appl. 57:285–294.

——. 1991. Stability of resistance in sorghum to *Calocoris angustatus* (Hemiptera: Miridae). J. Econ. Entomol. 84:1088–1094.

——. 1992. Genotypic resistance in sorghum to head bug, *Calocoris angustatus* Lethiery [sic]. Euphytica 58:193–200.

Sharma, H. C., Y. O. Doumbia, and N. Y. Diorisso. 1992. A headcage technique to screen sorghum for resistance to mirid head bug, *Eurystylus immaculatus* Odh. in West Africa. Insect Sci. Appl. 13:417–427.

Sharma, H. C., J. W. Stenhouse, and K. F. Nwanze. 1995. Mechanisms and inheritance of resistance to panicle-feeding insects in *Sorghum bicolor*. Pages 171–182 in K. F. Nwanze and O. Youm, eds., Panicle insect pests of sorghum and pearl millet: Proceedings of an International Consultative Workshop, 4–7 Oct. 1993, ICRISAT Sahelian Center, Niamey, Niger. International Crops Research Institute for the Semi-Arid Tropics, Patancheru, A.P., India.

Shaw, D. E. 1984. Microorganisms in Papua New Guinea. Papua New Guinea Dep. Primary Ind. Res. Bull. 33:1–344.

Sheeley, R. D. and T. R. Yonke. 1977. Biological notes on seven species of Missouri tingids (Hemiptera: Tingidae). J. Kans. Entomol. Soc. 50:342–356.

Shek, G. Kh. and H. Ya. Evdokimov. 1981. Pests of grain in Kazakhstan. Zashch. Rast. (Mosc.) No. 8:26–29. [in Russian.]

Sheldon, J. K. and E. G. MacLeod. 1971. Studies on the biology of the Chrysopidae II. The feeding behavior of the

adult of *Chrysopa carnea* (Neuroptera). Psyche (Camb.) 78:107–121.

Shelford, V. E. 1913. Animal communities in temperate America as illustrated in the Chicago region. University of Chicago Press, Chicago. 368 pp.

Shepard, M. and G. S. Arida. 1986. Parasitism and predation of yellow stem borer, *Scirpophaga incertulas* (Walker) (Lepidoptera: Pyralidae) eggs in transplanted and direct seeded rice. J. Entomol. Sci. 21:26–32.

Shepard, M., R. J. Lawn, and M. A. Schneider. 1983. Insects on grain legumes in northern Australia. University of Queensland Press, St. Lucia. 81 pp.

Shetlar, D. J., J. A. Chatfield, K. D. Cochran, and C. W. Ellett. 1992. Evaluation of junipers for mite, disease and insect incidence: Secrest Arboretum–1991. Pages 35–43 in Ornamental plants: A summary of research 1992. Ohio Agric. Res. Dev. Cent., Wooster. Spec. Circ. 140.

Shimizu, J. T. and K. S. Hagen. 1967. An artificial ovipositional site for some Heteroptera that insert their eggs into plant tissue. Ann. Entomol. Soc. Am. 60:1115–1116.

Shindrova, P. 1979. Influence of injuries produced by some pentatomid bugs on the seeding and biochemical characters of sunflower seeds. Rastenieved Nauki 16(9–10):143–149. [in Bulgarian, English summary.]

Shindrova, P. and P. Ivanov. 1982. The effect of injuries produced by *Dolycoris baccarum* L. and *Lygus rugulipennis* Pop. on some biochemical indicators of sunflower seeds. Plant Sci. (Sofia) 19(8):78–84. [in Bulgarian, English summary.]

Sholes, O. D. V. 1980. Response of arthropods to the phenology of host-plant inflorescences, concentrating on the host genus *Solidago*. Ph.D. dissertation, Cornell University, Ithaca. 296 pp.

——. 1984. Responses of arthropods to the development of goldenrod inflorescences (*Solidago*: Asteraceae). Am. Midl. Nat. 112:1–14.

Shorey, H. H., A. S. Deal, and M. J. Snyder. 1965. Insecticidal control of lygus bugs and effect on yield and grade of lima beans. J. Econ. Entomol. 58:124–126.

Shrimpton, G. 1985. Four insect pests of conifer nurseries in British Columbia. In Proc. Western Forest Nursery Council Intermountain Nurseryman's Association, Coeur d'Alene, Idaho, Aug. 14–16, 1984, pp. 119–121. USDA For. Serv. Gen. Tech. Rep. INT-185.

Shrivastava, U. 1991. Insect pollination in some cucurbits. Pages 445–451 in C. van Heemert and A. de Ruijter, eds., Sixth Int. Symp. Pollination, Tilburg, the Netherlands, 27–31 Aug. 1990. Acta Hortic. (Wageningen) No. 288.

Shteynberg, D. M. 1962. The use of entomorphages [sic] to protect apple orchards in eastern Canada. Entomol. Obozr. 41(2):300–305. [in Russian; English translation in Entomol. Rev. 41(2):185–187.]

Shull, W. E. 1933. An investigation of the Lygus species which are pests of beans (Hemiptera, Miridae). Univ. Idaho Agric. Exp. Stn. Res. Bull. 11:1–42.

———. 1941. A mirid (Thyrillus pacificus Uhler). U.S. Dep. Agric. Insect Pest Surv. Bull. 21:241.

Shull, W. E. and C. Wakeland. 1931. Tarnished plant bug injury to beans. J. Econ. Entomol. 24:326–327.

Shull, W. E., P. L. Rice, and H. F. Cline. 1934. Lygus hesperus Knight (Hemiptera: Miridae) in relation to plant growth, blossom drop, and seed set in alfalfa. J. Econ. Entomol. 27:265–269.

Shure, D. J. 1970. Limitations in radiotracer determination of consumer trophic positions. Ecology 51:899–901.

———. 1973. Radionuclide tracer analysis of trophic relationships in an old-field ecosystem. Ecol. Monogr. 43:1–19.

Shurtleff, M. C. 1966. How to control plant diseases in home and garden, 2nd ed. Iowa State University Press, Ames. 649 pp.

Siddique, A. B. and R. B. Chapman. 1987. Effect of prey type and quantity on the reproduction, development, and survival of Pacific damsel bug, Nabis kinbergii Reuter (Hemiptera: Nabidae). N.Z. J. Zool. 14:343–349.

Sidlyarevitsch, V. I. 1965. The importance of predacious mites and bugs in reducing the numbers of Metatetranychus ulmi Koch in the Byelorussian SSR. Tr. Vses. Inst. Zashch. Rast. (Mosc.) 24:240–247. [in Russian; English summary.]

Sidlyarevitsch, V. [I.] 1982. Predators of tetranychous mites in the fruit orchards of Byelorussia. Acta Entomol. Fenn. 40:30–32.

Siemann, E., J. Haarstad, and D. Tilman. 1997. Short-term and long-term effects of burning on oak savanna arthropods. Am. Midl. Nat. 137:349–361.

Sillén-Tullberg, B. 1981. Prolonged copulation: A male 'postcopulatory' strategy in a promiscuous species, Lygaeus equestris (Heteroptera: Lygaeidae). Behav. Ecol. Sociobiol. 9:283–289.

Sillén-Tullberg, B., C. Wiklund, and T. Järvi. 1982. Aposematic coloration in adults and larvae of Lygaeus equestris and its bearing on müllerian mimicry: An experimental study on predation on living bugs by the great tit Parus major. Oikos 39:131–136.

Silva, D. B. da, R. T. Alves, P. S. F. Ferreira, and A. J. A. Camargo. 1994. Collaria oleosa (Distant, 1883) (Heteroptera: Miridae), uma praga potencial na cultura do trigo na região dos cerrados. Pesqui. Agropecu. Bras. 29:2007–2012.

Silverman, J. and R. D. Goeden. 1979. Life history of the lacebug, Corythucha morrilli Osborn and Drake, on the ragweed, Ambrosia dumosa (Gray) Payne, in southern California (Hemiptera-Heteroptera: Tingidae). Pan-Pac. Entomol. 55:305–308.

Silvestri, F. 1932. Contribuzione alla conoscenza del Lopus lineolatus (Brullé) e di un suo parassita [Hemiptera Heteroptera Miridae]. Soc. Entomol. Fr. Livre Centen., pp. 551–565.

Simão, S. and Z. C. Maranhão. 1959. Os insetos como agentes polinizadores da mangueira. An. Esc. Super. Agric. "Luiz Queiroz" Univ. São Paulo 16:299–304.

Simberloff, D. 1981. What makes a good island colonist? Pages 195–205 in R. F. Denno and H. Dingle, eds., Insect life history patterns: Habitat and geographic variation. Springer, New York.

Simko, B. and D. J. Kreigh. 1998. Using degree-day accumulations to predict Lygus hatches in alfalfa seed. Malheur Experiment Station Annual Report, 1997. Ore. State Agric. Exp. Stn. Spec. Rep. 988:13–17.

Simmonds, F. J. 1969. Biological control of sugar cane pests: A general survey. Pages 461–479 in J. R. Williams et al., eds., Pests of sugar cane. Elsevier, Amsterdam.

Simmonds, N. W., ed. 1976. Evolution of crop plants. Longman, London. 339 pp.

Simmons, A. M. and K. V. Yeargan. 1988. Feeding frequency and feeding duration of the green stink bug (Hemiptera: Pentatomidae) on soybean. J. Econ. Entomol. 81:812–815.

Simonet, J. 1955. Note relative aux Hémiptères capturés sur les Champignons. Mitt. Schweiz. Entomol. Ges. 28:111–114.

Simonse, M. P. 1985. Gewasbescherming. Tuinderij (Doetinchem) 65(23):55.

Simpson, B. B. and J. L. Neff. 1983. Evolution and diversity of floral rewards. Pages 142–159 in C. E. Jones and R. J. Little, eds., Handbook of experimental pollination biology. Van Nostrand Reinhold, New York.

Simpson, K. W. 1975. Biology and immature stages of three species of Nearctic Ochthera (Diptera: Ephydridae). Proc. Entomol. Soc. Wash. 77:129–155.

Simpson, S. J. and C. L. Simpson. 1990. The mechanisms of nutritional compensation by phytophagous insects. Pages 111–160 in E. A. Bernays, ed., Insect-plant interactions. Vol. II. CRC, Boca Raton, Fla.

Sinclair, W. A., H. H. Lyon, and W. T. Johnson. 1987. Diseases of trees and shrubs. Cornell University Press, Ithaca. 574 pp.

Singh, D. 1976. Castor, Ricinus communis (Euphorbiaceae). Pages 84–86 in N. W. Simmonds, ed., Evolution of crop plants. Longman, London.

Singh, P. 1977. Artificial diets for insects, mites, and spiders. IFI/Plenum, New York. 594 pp.

Singh, S. R., L. E. N. Jackai, J. H. R. dos Santos, and C. B. Adalla. 1990. Insect pests of cowpea. Pages 43–89 in S. R. Singh, ed., Insect pests of tropical food legumes. Wiley, Chichester, UK.

Singh-Pruthi, H. 1925. The morphology of the male genitalia in Rhynchota. Trans. Entomol. Soc. Lond. 1925:127–267.

Sinha, S. N., A. K. Chakrabarti, and U. Ramakrishnan. 1981. Field evaluation of some insecticides for the control of hemipteran bugs on bottlegourd. Indian J. Agric. Sci. 51:906–910.

Sivapragasam, A. and A. Asma. 1985. Development and reproduction of the mirid bug, Cyrtorhinus lividipennis (Heteroptera: Miridae) and its functional response to the brown planthopper. Appl. Entomol. Zool. 20:373–379.

Škapec, L. and P. Štys. 1980. Asymmetry in the forewing position in Heteroptera. Acta Entomol. Bohemoslov. 77:353–374.

Slansky, F. Jr. 1974. Relationship of larval food-plants and voltinism patterns in temperate butterflies. Psyche (Camb.) 81:243–253.

———. 1982. Insect nutrition: An adaptationist's perspective. Fla. Entomol. 65:45–71.

Slansky, F. Jr. and A. R. Panizzi. 1987. Nutritional ecology of seed-sucking insects. Pages 283–320 in F. Slansky Jr. and J. G. Rodriguez, eds., Nutritional ecology of insects, mites, spiders, and related invertebrates. Wiley, New York.

Slansky, F. Jr. and J. M. Scriber. 1985. Food consumption and utilization. Pages 87–163 in G. A. Kerkut and L. I. Gilbert, eds., Comprehensive insect physiology, biochemistry and pharmacology. Vol. 4. Regulation: Digestion, nutrition, excretion. Pergamon, Oxford.

Slater, J. A. 1950. An investigation of the female genitalia as taxonomic characters in the Miridae (Hemiptera). Iowa State Coll. J. Sci. 25:1–81.

——. 1954. Notes on the genus *Labops*, Burmeister in North America with the descriptions of three new species (Hemiptera: Miridae). Bull. Brooklyn Entomol. Soc. 49:57–65, 89–94.

——. 1964. A catalogue of the Lygaeidae of the world. 2 vols. University of Connecticut, Storrs. 1668 pp.

——. 1974. Class Insecta, order Hemiptera, suborder Heteroptera. Pages 66–74 in W. G. H. Coaton, ed., Status of the taxonomy of the Hexapoda of southern Africa. Entomol. Mem. 38.

——. 1975. On the biology and zoogeography of Australian Lygaeidae (Hemiptera: Heteroptera) with special reference to the southwest fauna. J. Aust. Entomol. Soc. 14:47–64.

——. 1977. The incidence and evolutionary significance of wing polymorphism in lygaeid bugs with particular reference to those of South Africa. Biotropica 9:217–229.

——. 1978. Harry H. Knight: An appreciation and remembrance. Melsheimer Entomol. Ser. No. 24:1–8.

——. 1982. Hemiptera. Pages 417–447 in S. P. Parker, ed., Synopsis and classification of living organisms. McGraw-Hill, New York.

——. 1993. [Review of] *A synthesis of the Holarctic Miridae (Heteroptera): Distribution, biology and origin, with emphasis on North America* by A. G. Wheeler, Jr. and T. J. Henry. Proc. Entomol. Soc. Wash. 95:513–516.

Slater, J. A. and R. M. Baranowski. 1978. How to know the true bugs (Hemiptera-Heteroptera). Brown, Dubuque, Iowa. 256 pp.

Slater, J. A. and N. T. Davis. 1952. The scientific name of the tarnished plant bug (Hemiptera, Miridae). Proc. Entomol. Soc. Wash. 54:194–198.

Slater, J. A. and G. F. Gross. 1977. A remarkable new genus of coleopteroid Miridae from southern Australia (Hemiptera: Heteroptera). J. Aust. Entomol. Soc. 16:135–140.

Slater, J. A. and J. E. O'Donnell. 1995. A catalogue of the Lygaeidae of the world (1960–1994). New York Entomological Society, New York. 410 pp.

Slater, J. A. and [R.] T. Schuh. 1969. New species of Isometopinae from South Africa (Hemiptera: Miridae). J. Entomol. Soc. S. Afr. 32:351–366.

Slaymaker, P. H. and N. P. Tugwell. 1982. Low-labor method for rearing the tarnished plant bug (Hemiptera: Miridae). J. Econ. Entomol. 75:487–488.

——. 1984. Inexpensive female-baited trap for the tarnished plant bug (Hemiptera: Miridae). J. Econ. Entomol. 77:1062–1063.

Slingerland, M. V. 1893. The four-lined leaf-bug. Cornell Univ. Agric. Exp. Stn. Bull. 58:207–239.

——. 1895. A bad fruit bug. Rural New-Yorker (N.Y.) 54:328 (May 11).

Slobodyanyuk, G. A., T. N. Ignat'eva, and O. N. Andreenko. 1993. A system of protection of cucumbers against greenhouse whitefly. Zashch. Rast. (Mosc.) No. 4:45. [in Russian.]

Slosser, J. E., D. G. Bordovsky, and S. J. Bevers. 1994. Damage and costs associated with insect management options in irrigated cotton. J. Econ. Entomol. 87:436–445.

Sluss, T. P., H. M. Graham, and E. S. Sluss. 1982. Morphometric, allozyme, and hybridization comparisons of four *Lygus* species (Hemiptera: Miridae). Ann. Entomol. Soc. Am. 75:448–456.

Smee, C. 1928a. Report of the Government Entomologist. Nyasaland Dep. Agric. Rep. 1927:19–22.

——. 1928b. Tea mosquito bug in Nyasaland (*Helopeltis bergrothi*, Reut.) and notes on two potential pests of tea, (1) the tea leaf weevil (*Dicasticus mlanjensis*, Mshl.) (2) the bean flower capsid (*Callicratides rama*, Kirby). Nyasaland Dep. Agric. Entomol. Ser. Bull. 4:1–10.

Smee, C. and R. Leach. 1932. Mosquito bug the cause of stem canker of tea. Nyasaland Dep. Agric. Bull. 5 (n.s.):1–7.

Smee, L. 1963. Insect pests of *Theobroma cacao* in the territory of Papua and New Guinea: Their habits and control. Papua New Guinea Agric. J. 16:1–19.

Smit, B. 1964. Insects in southern Africa: How to control them. A handbook for students, health officers, gardeners, farmers. Oxford University Press, Cape Town. 399 pp.

Smith, B. D. 1962. Experiments in the transfer of the black currant gall mite (*Phytoptus ribis* Nal.) and of reversion. Long Ashton Res. Stn. Rep. 1961:170–172.

——. 1966. Effects of parasites and predators on a natural population of the aphid, *Acyrthosiphon spartii* (Koch) on broom (*Sarothamnus scoparius* L.). J. Anim. Ecol. 35:255–267.

Smith, C. C. 1940. The effect of overgrazing and erosion upon the biota of the mixed-grass prairie of Oklahoma. Ecology 21:381–397.

Smith, C. M. 1994. Integration of rice insect control strategies and tactics. Pages 681–692 in E. A. Heinrichs, ed., Biology and management of rice insects. Wiley, New York.

Smith, C. W. 1992. History and status of host plant resistance in cotton to insects in the United States. Adv. Agron. 48:251–296.

Smith, E. H. 1940. The biology and control of the hickory capsid, *Lygus caryae* Knight (Hemiptera: Miridae) as a peach pest. M.S. thesis, Cornell University, Ithaca. 28 pp.

——. 1950. The problem of controlling peach pests in New York State. Proc. N.Y. State Hortic. Soc. 1950:213–220.

Smith, E. S. C. 1972. Population fluctuations of the cocoa mirid in the Northern District of Papua New Guinea. In Abstr. 14 Int. Congr. Entomol., Canberra, p. 330.

——. 1973. A laboratory rearing method for the cacao mirid *Helopeltis clavifer* Walker (Hemiptera: Miridae). Papua New Guinea Agric. J. 24:52–53.

——. 1977a. Presence of a sex attractant pheromone in *Helopeltis clavifer* (Walker) (Heteroptera: Miridae). J. Aust. Entomol. Soc. 16:113–116.

——. 1977b. The cardamom mirid (*Ragwelellus horvathi*) Poppius (Heteroptera: Miridae) in Papua New Guinea. Papua New Guinea Agric. J. 28:97–101.

——. 1978. Host and distribution records of *Helopeltis clavifer* (Walker) (Heteroptera: Miridae) in Papua New Guinea. Papua New Guinea Agric. J. 29:1–4.

——. 1979. Descriptions of the immature and adult stages of the cocoa mirid *Helopeltis clavifer* (Heteroptera: Miridae). Pac. Insects 20:354–361.

——. 1985. A review of relationships between shade types and cocoa pest and disease problems in Papua New Guinea. P.N.G. J. Agric. For. Fish. 33:79–88.

Smith, E. S. C., B. M. Thistleton, and J. R. Pippet. 1985. Assessment of damage and control of *Helopeltis clavifer* (Heteroptera: Miridae) on tea in Papua New Guinea. P.N.G. J. Agric. For. Fish. 33:123–131.

Smith, F. F. 1940. Certain sucking insects causing injury to rose. J. Econ. Entomol. 33:658–662.

Smith, F. F. and P. Brierley. 1956. Insect transmission of plant viruses. Annu. Rev. Entomol. 1:299–322.

Smith, F. F. and F. W. Poos. 1931. The feeding habits of some leaf hoppers of the genus *Empoasca*. J. Agric. Res. (Wash., D.C.) 43:267–285.

Smith, G. L. 1942. California cotton insects. Calif. Agric. Exp. Stn. Bull. 660:1–50.

Smith, J. B. 1896. Economic entomology for the farmer and fruit-grower, and for use as a text-book in agricultural schools and colleges. Lippincott, Philadelphia. 481 pp.

Smith, J. J. B. 1985. Feeding mechanisms. Pages 33–85 in G. A. Kerkut and L. I. Gilbert, eds., Comprehensive insect physiology, biochemistry and pharmacology. Vol. 4. Regulation: Digestion, nutrition, excretion. Pergamon, Oxford.

Smith, K. G. V. 1990. Hemiptera (Anthocoridae and Miridae) biting man. Entomol. Mon. Mag. 126:96.

Smith, K. M. 1920a. Investigation of the nature and cause of the damage to plant tissue resulting from the feeding of capsid bugs. Ann. Appl. Biol. 7:40–55.

——. 1920b. The injurious apple capsid (*Plesiocoris rugicollis*, Fall.). J. Minist. Agric. (Lond.) 27:379–381.

——. 1925. Note on the egg-laying of *Calocoris bipunctatus* Fab. Entomol. Mon. Mag. 61:91–92.

——. 1926. A comparative study of the feeding methods of certain Hemiptera and of the resulting effects upon the plant tissue, with special reference to the potato plant. Ann. Appl. Biol. 13:109–139.

——. 1927. Observations on the insect carriers of mosaic disease of the potato. Ann. Appl. Biol. 14:113–131.

——. 1929. Studies on potato virus diseases V. Insect transmission of potato leaf-roll. Ann. Appl. Biol. 16:209–229.

——. 1931a. Virus diseases of plants and their relationship with insect vectors. Biol. Rev. Camb. Philos. Soc. 6:302–344.

——. 1931b. A textbook of agricultural entomology. University Press, Cambridge, UK. 285 pp.

——. 1934. The mosaic disease of sugarbeet and related plants. J. Minist. Agric. (Lond.) 41:269–274.

——. 1951. Recent advances in the study of plant viruses. Blakiston, Philadelphia. 300 pp.

——. 1958. Transmission of plant viruses by arthropods. Annu. Rev. Entomol. 3:469–482.

——. 1965. Plant virus–vector relationships. Adv. Virus Res. 11:61–96.

——. 1977. Plant viruses, 6th ed. Chapman & Hall, London. 241 pp.

Smith, P. M., A. V. Brooks, E. J. Evans, and A. J. Halstead. 1983. Pests and diseases of hardy nursery stock, bedding plants and turf. Pages 473–557 in N. Scopes and M. Ledieu, eds., Pest and disease control handbook. British Crop Protection Council, Croydon, UK.

Smith, R. B. and T. P. Mommsen. 1984. Pollen feeding in an orb-weaving spider. Science (Wash., D.C.) 226:1330–1332.

Smith, R. C. and W. W. Franklin. 1961. Research notes on certain species of

alfalfa insects at Manhattan (1904–1956) and at Ft. Hays, Kansas (1948–1953). Kans. Agric. Exp. Stn. Rep. Progr. 54:1–121.

Smith, R[ay] F. and L. A. Falcon. 1973. Insect control for cotton in California. Cotton Grow. Rev. 50:15–27.

Smith, R[ay] F. and A. E. Michelbacher. 1946. Control of lygus bugs in alfalfa seed fields. J. Econ. Entomol. 39:638–648.

Smith, R[obert] F. 1991. The mullein bug, *Campylomma verbasci*. Pages 199–214 in K. Williams, eds., New directions in tree fruit pest management. Good Fruit Grower, Yakima, Wash.

Smith, R[obert] F. and J. H. Borden. 1990. Relationship between fall catches of *Campylomma verbasci* (Heteroptera: Miridae) in traps baited with females and density of nymphs in the spring. J. Econ. Entomol. 83:1506–1509.

——. 1991. Fecundity and development of the mullein bug, *Campylomma verbasci* (Meyer) (Heteroptera: Miridae). Can. Entomol. 123:595–600.

Smith, R[obert] F., H. D. Pierce Jr., and J. H. Borden. 1991. Sex pheromones of the mullein bug, *Campylomma verbasci* (Meyer) (Heteroptera: Miridae). J. Chem. Ecol. 17:1437–1447.

Smith, R[obert] F., S. O. Gaul, J. H. Borden, and H. D. Pierce Jr. 1994. Evidence for a sex pheromone in the apple brown bug, *Atractotomus mali* (Meyer) (Heteroptera: Miridae). Can. Entomol. 126:445–446.

Smith, R. H. 1923. The clover aphis: Biology, economic relationships and control. Idaho Agric. Exp. Stn. Res. Bull. 3:1–75.

Smith, R. I. 1907. Some Georgia insects during 1906. Proc. 19th Annu. Mtg. Assoc. Econ. Entomol. U.S. Dep. Agric. Bur. Entomol. Bull. 67:101–106.

Smith, S. M. and C. R. Ellis. 1983. Economic importance of insects on regrowths of established alfalfa fields in Ontario. Can. Entomol. 115:859–868.

Snapp, O. I. 1947a. Experiments in 1946 on the control of bugs that cause deformed peaches. J. Econ. Entomol. 40:135–136.

——. [1947b]. Sucking bugs and plum curculio on peaches. Indiana Hortic. Soc. Trans. 1946:98–104.

Snodgrass, G. L. 1991. *Deraecoris* [sic] *nebulosus* (Heteroptera: Miridae): Little-known predator in cotton in the Mississippi Delta. Fla. Entomol. 74:340–344.

——. 1993. Estimating absolute density of nymphs of *Lygus lineolaris* (Heteroptera: Miridae) in cotton using drop cloth and sweep-net sampling

methods. J. Econ. Entomol. 86:1116–1123.

——. 1996. Insecticide resistance in field populations of the tarnished plant bug (Heteroptera: Miridae) in cotton in the Mississippi Delta. J. Econ. Entomol. 89:783–790.

——. 1998. Distribution of the tarnished plant bug (Heteroptera: Miridae) within cotton plants. Environ. Entomol. 27:1089–1093.

Snodgrass, G. L. and G. W. Elzen. 1995. Insecticide resistance in a tarnished plant bug population in cotton in the Mississippi Delta. Southwest. Entomol. 20:317–323.

Snodgrass, G. L. and Y. H. Fayad. 1991. Euphorine (Hymenoptera: Braconidae) parasitism of the tarnished plant bug (Heteroptera: Miridae) in areas of Washington County, Mississippi disturbed and undisturbed by agricultural production. J. Entomol. Sci. 26:350–356.

Snodgrass, G. L. and J. M. McWilliams. 1992. Rearing the tarnished plant bug (Heteroptera: Miridae) using a tissue paper oviposition site. J. Econ. Entomol. 85:1162–1166.

Snodgrass, G. L. and W. P. Scott. 1988. Tolerance of the tarnished plant bug to dimethoate and acephate in different areas of the Mississippi Delta. In Proc. Beltwide Cotton Prod. Res. Conf., Jan. 3–8, 1988, New Orleans, La., pp. 294–296. National Cotton Council, Memphis.

——. 1997. A conversion factor for correcting numbers of adult tarnished plant bugs (Heteroptera: Miridae) captured with a sweep net in cotton. Southwest. Entomol. 22:189–193.

Snodgrass, G. L. and E. A. Stadelbacher. 1994. Population levels of tarnished plant bugs (Heteroptera: Miridae) and beneficial arthropods following early-season treatments of *Geranium dissectum* for control of bollworms and tobacco budworms (Lepidoptera: Noctuidae). Environ. Entomol. 23:1091–1096.

Snodgrass, G. L., W. P. Scott, and J. W. Smith. 1984a. A survey of the host plants and seasonal distribution of the cotton fleahopper (Hemiptera: Miridae) in the delta of Arkansas, Louisiana, and Mississippi. J. Ga. Entomol. Soc. 19:34–41.

——. 1984b. An annotated list of the host plants of *Lygus lineolaris* (Hemiptera: Miridae) in the Arkansas, Louisiana, and Mississippi Delta. J. Ga. Entomol. Soc. 19:93–101.

——. 1984c. Host plants of *Taylorilygus pallidulus* and *Polymerus basalis* (Hemiptera: Miridae) in the delta of

Arkansas, Louisiana, and Mississippi. Fla. Entomol. 67:402–408.

Snodgrass, G. L., C. C. Loan, and H. M. Graham. 1990. New species of *Leiophron* (Hymenoptera: Braconidae, Euphorinae) from Kenya. Fla. Entomol. 73:492–496.

Snodgrass, R. E. 1935. Principles of insect morphology. McGraw-Hill, New York. 667 pp.

——. 1944. The feeding apparatus of biting and sucking insects affecting man and animals. Smithsonian Misc. Coll. 104(7):1–113.

Soderstrom, T. R. and C. E. Calderón. 1971. Insect pollination in tropical rain forest grasses. Biotropica 3:1–16.

Sohati, P. H., R. K. Stewart, and G. Boivin. 1989. Egg parasitoids of the tarnished plant bug, *Lygus lineolaris* (P. de B.) (Hemiptera: Miridae), in Quebec. Can. Entomol. 121:1127–1128.

Sohmer, S. H. and D. F. Sefton. 1978. The reproductive biology of *Nelumbo pentapetala* (Nelumbonaceae) on the Upper Mississippi River. II. The insects associated with the transfer of pollen. Brittonia 30:355–364.

Solbreck, C. 1986. Wing and flight muscle polymorphism in a lygaeid bug, *Horvathiolus gibbicollis*: Determinants and life history consequences. Ecol. Entomol. 11:435–444.

Solbreck, C. and I. Pehrson. 1979. Relations between environment, migration and reproduction in a seed bug, *Neacoryphus bicrucis* (Say) (Heteroptera: Lygaeidae). Oecologia (Berl.) 43:51–62.

Solbreck, C., D. B. Anderson, and J. Förare. 1990. Migration and the coordination of life-cycles as exemplified by Lygaeinae bugs. Pages 197–214 *in* F. Gilbert, ed., Insect life cycles: Genetics, evolution and co-ordination. Springer, London.

Solomon, M. G. 1975. The colonization of an apple orchard by predators of the fruit tree red spider mite. Ann. Appl. Biol. 80:119–122.

——. 1976. Natural enemies of the oystershell scale, *Quadraspidiotus pyri*. East Malling Res. Stn. Rep. 1975:130.

——. 1981. Windbreaks as a source of orchard pests and predators. Pages 273–283 *in* J. M. Thresh, ed., Pests, pathogens and vegetation. Pitman Advanced Publishing Program, London.

——. 1982. Phytophagous mites and their predators in apple orchards. Ann. Appl. Biol. 101:201–203.

——. 1987. Fruit and hops. Pages 329–360 *in* A. J. Burn, T. H. Coaker, and P. C. Jepson, eds., Integrated pest management. Academic, London.

Solomon, M. G. and S. Langeslag. 1977. Predators of the oystershell scale,

Quadraspidiotus pyri (Licht.). East Malling Res. Stn. Rep. 1976:157.

Sömermaa, K. 1961. Untersuchungen über die "Bollnäser Krankheit." III. Studien über die "Trübe Feldwanze" *Lygus rugulipennis*. Medd. Statens Växtskyddsanst. 12(86):79–93.

Song, Y. H. and K. L. Heong. 1997. Changes in searching responses with temperature of *Cyrtorhinus lividipennis* Reuter (Hemiptera: Miridae) on the eggs of the brown planthopper, *Nilaparvata lugens* (Stål) (Homoptera: Delphacidae). Res. Popul. Ecol. (Kyoto) 39:201–206.

Sorensen, C. 1988. The rise of government sponsored applied entomology, 1840–1870. Agric. Hist. 62:98–115.

Sorensen, E. L., R. A. Byers, and E. K. Horber. 1988. Breeding for insect resistance. Pages 859–902 *in* A. A. Hanson, D. K. Barnes, and R. R. Hill Jr., eds., Alfalfa and alfalfa improvement. Agron. Monogr. 29.

Sorensen, J. T., B. C. Campbell, R. J. Gill, and J. D. Steffen-Campbell. 1995. Non-monophyly of Auchenorrhyncha ("Homoptera"), based upon 185 rDNA phylogeny: Eco-evolutionary and cladistic implications within pre-Heteropterodea Hemiptera (s.l.) and a proposal for new monophyletic suborders. Pan-Pac. Entomol. 71:31–60.

Sorenson, C. J. 1932a. The tarnished plant bug, *Lygus pratensis* (Linn.) and the superb plant bug, *Adelphocoris superbus* (Uhler), in relation to flower drop in alfalfa. Proc. Utah Acad. Sci. Arts Lett. 9:67–70.

——. 1932b. Insects in relation to alfalfa-seed production. Utah Agric. Exp. Stn. Circ. 98:1–28.

——. 1939. *Lygus hesperus* Knight and *Lygus elisus* Van Duzee in relation to alfalfa-seed production. Utah Agric. Exp. Stn. Bull. 284:1–61.

Sorenson, C. J. and L. Cutler. 1954. The superb plant bug *Adelphocoris superbus* (Uhler). Its life history and its relation to seed development in alfalfa. Utah Agric. Exp. Stn. Bull. 370:1–20.

Sorenson, C. J. and F. H. Gunnell. 1936. Type of injury caused by lygus bugs to maturing peach fruits. Proc. Utah Acad. Sci. Arts Lett. 13:225–227.

Soroka, J. J. 1991. Insect pests of legume and grass crops in western Canada. Agric. Can. Publ. 1435/E rev.:1–39.

Soroka, J. J. and D. C. Murrell. 1993. The effects of alfalfa plant bug (Hemiptera: Miridae) feeding late in the season on alfalfa seed yield in northern Saskatchewan. Can. Entomol. 125:815–824.

Sørum, O. 1977. Teger som skadedyr på eple og paere. Gartneryrket (Oslo) 67:436, 438–439, 442, 444.

Sørum, O. and G. Taksdal. 1970. Teger som årsak til knartbaer i jordbaer. Gartneryrket (Oslo) 60:223–229.

Sosa, O. Jr. 1985. The sugarcane delphacid, *Perkinsiella saccharicida* (Homoptera: Delphacidae), a sugarcane pest new to North America detected in Florida. Fla. Entomol. 68:357–360.

Sotherton, M. W. 1991. Conservation headlands: A practical combination of intensive cereal farming and conservation. Pages 373–397 *in* L. G. Firbank, N. Carter, J. F. Darbyshire, and G. R. Potts, eds., The ecology of temperate cereal fields. Blackwell, Oxford.

South, D. [B.] 1986. The "tarnished plant bug" can cause loblolly pine seedlings to be "bushy-topped." Auburn Univ. So. For. Nurs. Manage. Coop. No. 27:1–4.

——. 1991a. Tiny tarnished plant bugs cause big problems for pine seedlings. Highlights Agric. Res. (Ala. Agric. Exp. Stn.) 38(3):8.

——. 1991b. *Lygus* bugs: A worldwide problem in conifer nurseries. Pages 215–222 *in* J. R. Sutherland and S. G. Glover, eds., Proc. First Meet. IUFRO Working Party S2.07–09 (Diseases and Insects in Forest Nurseries). For. Can. Pacific Yukon Reg. Inf. Rep. BC-X-331.

South, D. B., J. B. Zwolinski, and H. W. Bryan. 1993. *Taylorilygus pallidulus* (Blanchard): A potential pest of pine seedlings. Tree Plant. Notes 44:63–67.

Southier, F. E. 1974. Integrated pest management in diversified California crops. Proc. Tall Timbers Conf. Ecol. Anim. Control Habitat Manage. 5:81–88.

Southwood, T. R. E. 1955. The morphology of the salivary glands of terrestrial Heteroptera (Geocoridae) and its bearing on classification. Tijdschr. Entomol. 98:77–84.

——. 1956a. A key to determine the instar of an heteropterous larva. Entomologist 89:220–222.

——. 1956b. The structure of the eggs of terrestrial Heteroptera and its relationship to the classification of the group. Trans. R. Entomol. Soc. Lond. 108:163–221.

——. 1956c. The nomenclature and life cycle of the European Tarnished Plant Bug, *Lygus rugulipennis* Poppius (Hem., Miridae). Bull. Entomol. Res. 46:845–848.

——. 1960a. The abundance of the Hawaiian trees and the number of their associated insect species. Proc. Hawaii. Entomol. Soc. 17:299–303.

——. 1960b. The flight activity of Heteroptera. Trans. R. Entomol. Soc. Lond. 112:173–220.

——. 1961a. A hormonal theory of the mechanism of wing polymorphism in Heteroptera. Proc. R. Entomol. Soc. Lond. (A) 36:63–66.

——. 1961b. Notes on light trap catches of Heteroptera made in the tropics. Entomol. Mon. Mag. 96:114–117.

——. 1962. Migration of terrestrial arthropods in relation to habitat. Biol. Rev. Camb. Philos. Soc. 37:171–214.

——. 1964. Hemiptera Heteroptera (Bugs). Pages 80–82 in O. W. Richards. [Natural history of the garden of Buckingham Palace.] Insects other than Lepidoptera. Proc. S. Lond. Entomol. Nat. Hist. Soc. 1963.

——. 1969. Population studies of insects attacking sugar cane. Pages 427–459 in J. R. Williams, J. R. Metcalfe, R. W. Mungomery, and R. Mathis, eds., Pests of sugar cane. Elsevier, Amsterdam.

——. 1973. The insect/plant relationship—an evolutionary perspective. Pages 3–30 in H. F. van Emden, ed., Insect/plant relationships. Wiley, New York.

——. 1977. Habitat, the templet for ecological strategies? J. Anim. Ecol. 46:337–365.

——. 1978a. The components of diversity. Pages 19–40 in L. A. Mound and N. Waloff, eds., Diversity of insect faunas. Symposium of the Royal Entomological Society of London, No. 9. Blackwell, Oxford.

——. 1978b. Ecological methods: With particular reference to the study of insects, 2nd ed. Chapman & Hall, London. 524 pp.

——. 1981. Bionomic strategies and population parameters. Pages 30–52 in R. M. May, ed., Theoretical ecology: Principles and applications, 2nd ed. Blackwell, Oxford.

——. 1985. Interactions of plants and animals: Patterns and processes. Oikos 44:5–11.

——. 1986. Plant surfaces and insects—an overview. Pages 1–22 in B. Juniper and Sir R. [T. R. E.] Southwood, eds., Insects and the plant surface. Arnold, London.

——. 1996. [Review of] Plant Bugs of the World (Insecta: Heteroptera: Miridae): Systematic Catalog, Distributions, Host List and Bibliography by R. T. Schuh. Syst. Entomol. 21:76–77.

Southwood, T. R. E. and B. E. Blackith. 1960. Inter- and intra-specific variation in the mirid genus Plagiognathus. Proc. R. Entomol. Soc. Lond. (C) 25:9–10. (Abstr.)

Southwood, T. R. E. and C. G. Johnson. 1957. Some records of insect flight activity in May 1954, with particular reference to the massed flights of Coleoptera and Heteroptera from concealing habitats. Entomol. Mon. Mag. 93:121–126.

Southwood, T. R. E. and C. E. J. Kennedy. 1983. Trees as islands. Oikos 41:359–371.

Southwood, T. R. E. and D. Leston. 1957. Notes on the nomenclature and zonal occurrence of the Orthotylus species (Hem., Miridae) of British salt marshes. Entomol. Mon. Mag. 93:166–168.

——. 1959. Land and water bugs of the British Isles. Warne, London. 436 pp.

Southwood, T. R. E. and P. M. Reader. 1988. The impact of predation on the viburnum whitefly, (Aleurotrachelus jelinekii). Oecologia (Berl.) 74:566–570.

Southwood, T. R. E. and G. G. E. Scudder. 1956. The immature stages of the Hemiptera-Heteroptera associated with the stinging nettle (Urtica dioica L.). Entomol. Mon. Mag. 92:313–325.

Southwood, T. R. E., V. C. Moran, and C. E. J. Kennedy. 1982. The assessment of arboreal insect fauna: Comparisons of knockdown sampling and faunal lists. Ecol. Entomol. 7:331–340.

Soyer, D. 1942. Miride du cotonnier, Creontiades pallidulus Ramb. Capsidae (Miridae). Publ. Inst. Natl. Etude Agron. Congo Belge Ser. Sci. No. 29:1–15.

Soylu, O. Z. 1980. Akdeniz Bölgesi turunçgillerinde zararli olan turunçgil beyaz sinegi (Dialeurodes citri Ashmead), nin biyolojisi ve mücadelesi üzerinde ara'tirmalar. Bitki Koruma Bül. (Ankara) 20:36–53.

Spahr, U. 1988. Ergänzungen und Berichtigungen zu R. Keilbachs Bibliographie und Liste der Bernsteinfossilien—Überordnung Hemipteroidea. Stuttg. Beitr. Naturkd. (B) 144:1–60.

Spangler, S. [M.] and A. [M.] Agnello. 1989. Insect and mite scouting and management. Pages 53–64 in M. Pritts and D. Handley, eds., Bramble production guide. Northeast Regional Agricultural Engineering Service, Cooperative Extension, Ithaca, N.Y.

Spangler, S. M. and J. A. MacMahon. 1990. Arthropod faunas of monocultures and polycultures in reseeded rangelands. Environ. Entomol. 19:244–250.

Spangler, S. M., A. M. Agnello, and M. D. Schwartz. 1993. Seasonal densities of tarnished plant bug (Heteroptera: Miridae) and other phytophagous Heteroptera in brambles. J. Econ. Entomol. 86:110–116.

Spence, J. R. 1986. Relative impacts of mortality factors in field populations of the waterstrider Gerris buenoi Kirkaldy (Heteroptera: Gerridae). Oecologia (Berl.) 70:68–76.

Spence, J. R. and H. A. Cárcamo. 1991. Effects of cannibalism and intraguild predation on pondskaters (Gerridae). Oikos 62:333–341.

Spence, J. R., D. H. Spence, and G. G. E. Scudder. 1980. The effects of temperature on growth and development in water-striders (Gerridae) and implications for species packing. Can. J. Zool. 58:1813–1820.

Speyer, W. 1933. Wanzen (Heteroptera) an Obstbäumen. Z. Pflanzenkr. Pflanzenschutz 43:113–138.

——. 1934. Wanzen (Heteroptera) an Obstbäumen. (II. Mitteilung). Z. Pflanzenkr. Pflanzenschutz 44:122–150.

Spooner, C. S. 1938. The phylogeny of the Hemiptera based on a study of the head capsule. Ill. Biol. Monogr. 16(3):1–102.

Squire, F. A. 1947. On the economic importance of the Capsidae in the Guinean region. Rev. Entomol. (Rio J.) 18:219–247.

——. 1972. Entomological problems in Bolivia. PANS (Pest Artic. News Summ.) 18:249–268.

Sreeramulu, C., T. Ayanna, and E. Dharmaraju. 1975. Occurrence of Gallobellicus croccicornis [sic] D. (Hemiptera: Capsidae) on bottlegourd. Indian J. Entomol. 37:202–203.

Srivastava, D. S., J. H. Lawton, and G. S. Robinson. 1997. Spore-feeding: A new, regionally vacant niche for bracken herbivores. Ecol. Entomol. 22:475–478.

Stace-Smith, R. 1954. Chlorotic spotting of black raspberry induced by the feeding of Amphorophora rubitoxica Knowlton. Can. Entomol. 86:232–235.

Staddon, B. W. 1979. The scent glands of Heteroptera. Adv. Insect Physiol. 14:351–418.

——. 1986. Biology of scent glands in the Hemiptera-Heteroptera. Ann. Soc. Entomol. Fr. 22:183–190.

Stadelbacher, E. A. 1987. Dynamics and control of early season populations of the tarnished plant bug, Lygus lineolaris in Geranium dissectum. In Proc. Beltwide Cotton Prod. Res. Conf., Jan. 4–8, 1987, Dallas, Texas, pp. 226–228. National Cotton Council, Memphis.

Städler, E. 1984. Contact chemoreception. Pages 3–35 in W. J. Bell and R. T. Carde, eds., Chemical ecology of insects. Sinauer, Sunderland, Mass.

——. 1992. Behavioral responses of insects to plant secondary compounds. Pages 45–88 in G. A. Rosenthal and M. R. Berenbaum, eds., Herbivores: Their interactions with secondary plant metabolites, 2nd ed. Vol. II. Ecological and evolutionary processes. Academic, New York.

Stahl, F. J. 1976. Transmission of Erwinia

amylovora to pear fruit by *Lygus* species. M.S. thesis, Colorado State University, Ft. Collins. 42 pp.

Stahl, F. J. and N. S. Luepschen. 1977. Transmission of *Erwinia amylovora* to pear fruit by *Lygus* spp. Plant Dis. Rep. 61:936–939.

Stam, P. A. 1987. *Creontiades pallidulus* (Rambur) (Miridae, Hemiptera), a pest on cotton along the Euphrates river and its effect on yield and control action threshold in the Syrian Arab Republic. Trop. Pest Manage. 33:273–276.

Stam, P. A. and H. Elmosa. 1990. The role of predators and parasites in controlling populations of *Earias insulans, Heliothis armigera* and *Bemisia tabaci* on cotton in the Syrian Arab Republic. Entomophaga 35:315–327.

Stam, P. A., D. F. Clower, J. B. Graves, and P. E. Schilling. 1978. Effects of certain herbicides on some insects and spiders found in Louisiana cotton fields. J. Econ. Entomol. 71:477–480.

Stanley, J. N. and M. H. Julien. 1999. The host range of *Eccritotarsus catarinensis* (Heteroptera: Miridae), a potential agent for the biological control of waterhyacinth (*Eichhornia crassipes*). Biol. Control 14:134–140.

Stanley, R. G. and H. F. Linskens. 1965. Protein diffusion from germinating pollen. Physiol. Plant. 18:47–53.

———. 1974. Pollen: Biology, biochemistry, management. Springer, New York. 307 pp.

Stapley, J. H. 1975. The problem of the brown planthopper (*Nilaparvata lugens*) on rice in the Solomon Islands. Rice Entomol. Newsl. 2:37.

———. 1976. The brown planthopper and *Cyrtorhinus* spp. predators in the Solomon Islands. Rice Entomol. Newsl. 4:17.

Starks, K. J. and R. Thurston. 1962. Control of plant bugs and other insects on Kentucky bluegrass grown for seed. J. Econ. Entomol. 55:993–997.

Starý, B., P. Bezděčka, M. Čapek, P. Starý, J. Zelený, and J. Šedivý. 1988. Atlas of insects beneficial to forest trees. Vol. 2. Elsevier, Amsterdam. 100 pp.

Starý, P. 1962. Hymenopterous parasites of the pea aphid, *Acyrthosiphon onobrychis* (Boyer) in Czechoslovakia I. Bionomics and ecology of *Aphidius ervi* Haliday. Zool. Listy 11:265–278.

———. 1970. Biology of aphid parasites (Hymenoptera: Aphidiidae) with respect to integrated control. Junk, The Hague. 643 pp.

———. 1988. Natural enemies. Parasites: Aphidiidae. Pages 171–184 *in* A. K. Minks and P. Harrewijn, eds., Aphids: Their biology, natural enemies and control. Vol. B. Elsevier, Amsterdam.

Stear, J. R. 1923. *Orthocephalus mutabilis* Fall. (Hemip., Miridae). Bull. Brooklyn Entomol. Soc. 18:62.

———. 1925. Three mirids predaceous on the rose leaf-hopper on apple. J. Econ. Entomol. 18:633–636.

Stearns, L. A. 1956. Meadow spittlebug and peach gumosis. J. Econ. Entomol. 49:382–385.

Stearns, S. C. 1976. Life-history tactics: A review of the ideas. Q. Rev. Biol. 51:3–47.

———. 1989. Trade-offs in life-history evolution. Func. Ecol. 3:259–268.

Steck, G. J., G. L. Teetes, and S. D. Maiga. 1989. Species composition and injury to sorghum by panicle feeding bugs in Niger. Insect Sci. Appl. 10:199–217.

Stedman, J. M. 1899. The tarnished plant bug. *Lygus pratensis* Linn. Mo. Agric. Exp. Stn. Bull. 47:1–13.

Steer, W. 1929. The eggs of some Hemiptera-Heteroptera. Entomol. Mon. Mag. 65:34–38.

Stehlík, J. L. 1952. Fauna Heteropter Hrubého Jeseníku. Acta Mus. Morav. 37:132–248.

———. 1995. Heteroptera. Pages 147–164 *in* R. Rozkošný and J. Vaňhara, eds., Terrestrial invertebrates of the Pálava Biosphere Reserve of UNESCO, I. Folia Fac. Sci. Nat. Univ. Marasykianae Brunensis Biol. 92.

———. 1998. The heteropteran fauna of introduced Cupressaceae in the southern part of Moravia (Czech Republic). Acta Mus. Morav. (Sci. Nat.) 82:127–155.

Stehr, F. W. 1987. Immature insects. Kendall/Hunt, Dubuque, Iowa. 754 pp.

Steinbauer, M. J. 1996. Notes on extra-phytophagous food sources of *Gelonus tasmanicus* (LeGuillou) (Hemiptera: Coreidae) and *Dindymus versicolor* (Herrich-Schäffer) (Hemiptera: Pyrrhocoridae). Aust. Entomol. 23:121–124.

Steiner, H. M. 1938. Effects of orchard practices on natural enemies of the white apple leafhopper. J. Econ. Entomol. 31:232–240.

Steinhaus, E. A. 1941. A study of the bacteria associated with thirty species of insects. J. Bacteriol. 42:757–790.

Steinkraus, D. C. and N. P. Tugwell. 1997. *Beauveria bassiana* (Deuteromycotina: Moniliales) effects on *Lygus lineolaris* (Hemiptera: Miridae). J. Entomol. Sci. 32:79–90.

Stellwaag, F. 1928. Die Weinbauinsekten der Kulturländer; Lehr- und Handbuch. Parey, Berlin. 884 pp.

Stenesh, J. 1975. Dictionary of biochemistry. Wiley, New York. 344 pp.

Stephens, G. M. III. 1982. The plant bug fauna (Heteroptera: Miridae) of grasses (Poaceae) of the Medicine Bow Moun-tain Ranger District, Wyoming. Wyo. Agric. Exp. Stn. Sci. Monogr. 43:1–175.

Stephens, G. S. 1975. Transportation and culture of *Tytthus mundulus*. J. Econ. Entomol. 68:753–754.

Stephenson, L. W. and T. E. Russell. 1974. The association of *Aspergillis flavus* with hemipterous and other insects infesting cotton bracts and foliage. Phytopathology 64:1502–1506.

Sterling, W. L. 1982. Predaceous insects and spiders. Pages 25–31 *in* G. T. Bohmfalk, R. E. Frisbie, W. L. Sterling, R. B. Metzer, and A. E. Knutson, eds., Identification, biology and sampling of cotton insects. Tex. Agric. Ext. Serv. B-933.

———. 1984. Action and inaction levels in pest management. Tex. Agric. Exp. Stn. B-1480:1–20.

Sterling, W. L. and D. A. Dean. 1977. A bibliography of the cotton fleahopper *Pseudatomoscelis seriatus* (Reuter). Tex. Agric. Exp. Stn. MP-1342:1–28.

Sterling, W. L., D. Jones, and D. A. Dean. 1979. Failure of the red imported fire ant to reduce entomophagous insect and spider abundance in a cotton agroecosystem. Environ. Entomol. 8:976–981.

Sterling, W. L., K. M. El-Zik, and L. T. Wilson. 1989a. Biological control of pest populations. Pages 155–189 *in* R. E. Frisbie, K. M. El-Zik, and L. T. Wilson, eds., Integrated pest management systems and cotton production. Wiley, New York.

Sterling, W. L., L. T. Wilson, A. P. Gutierrez, D. R. Rummel, J. R. Phillips, N. D. Stone, and J. H. Benedict. 1989b. Strategies and tactics for managing insects and mites. Pages 267–305 *in* R. E. Frisbie, K. M. El-Zik, and L. T. Wilson, eds., Integrated pest management systems and cotton production. Wiley, New York.

Sterling, W. L., [D.] A. Dean, and N. M. A. El-Salam. 1992. Economic benefits of spider (Araneae) and insect (Hemiptera: Miridae) predators of cotton fleahoppers. J. Econ. Entomol. 85:52–57.

Sterling, W. L., A. W. Hartstack, and D. A. Dean. 1996. Toward comprehensive economic thresholds for crop management. Pages 251–282 *in* E. G. King, J. R. Phillips, and R. J. Coleman, eds., Cotton insects and mites: Characterization and management. Cotton Foundation, Memphis.

Stern, V. M. 1966. Significance of the economic threshold in integrated pest control. *In* Proc. FAO Symp. Integrated Pest Control, 11–15 Oct. 1965, Rome, 2:41–56. FAO, Rome.

———. 1969. Interplanting alfalfa in cotton

to control lygus bugs and other insect pests. Proc. Tall Timbers Conf. Ecol. Anim. Control Habitat Manage. 1:55–69.

———. 1973. Economic thresholds. Annu. Rev. Entomol. 18:259–280.

———. 1976. Ecological studies of lygus bugs in developing a pest management program for cotton pests in the San Joaquin Valley, California. Pages 8–13 in D. R. Scott and L. E. O'Keeffe, eds., Lygus bug: Host plant interactions. University of Idaho Press, Moscow.

Stern, V. [M.]. 1981. Environmental control of insects using trap crops, sanitation, prevention, and harvesting. Pages 199–207 in D. Pimentel, ed., CRC handbook of pest management in agriculture. Vol. I. CRC, Boca Raton, Fla.

Stern, V. M. and A. Mueller. 1968. Techniques of marking insects with micronized fluorescent dust with especial emphasis on marking millions of Lygus hesperus for dispersal studies. J. Econ. Entomol. 61:1232–1237.

Stern, V. M., R. van den Bosch, and T. F. Leigh. 1964. Strip cutting alfalfa for lygus bug control. Calif. Agric. 18(4):4–6.

Stevens, R. E. and F. G. Hawksworth. 1970. Insects and mites associated with dwarf mistletoes. U.S. Dep. Agric. For. Serv. Res. Pap. RM-59:1–12.

Stevenson, A. B. 1994. Celery, celeriac: Tarnished plant bug. Pages 90–91 in R. J. Howard, J. A. Garland, and W. L. Seaman, eds., Diseases and pests of vegetable crops in Canada: An illustrated compendium. Canadian Phytopathological Society, Entomological Society of Canada, Ottawa.

Stevenson, A. B. and M.-D. Roberts. 1973. Tarnished plant bug rearing on lettuce. J. Econ. Entomol. 66:1354–1355.

Stewart, A. J. A. and D. R. Lees. 1988. Genetic control of colour/pattern polymorphism in British populations of the spittlebug Philaenus spumarius (L.) (Homoptera: Aphrophoridae). Biol. J. Linn. Soc. 34:57–79.

Stewart, R. K. 1969a. The biology of Lygus rugulipennis Poppius (Hemiptera: Miridae) in Scotland. Trans. R. Entomol. Soc. Lond. 120:437–457.

———. 1969b. Species of Lygus (Hahn) (Hemiptera: Miridae) in Scotland. Proc. R. Entomol. Soc. (B) 38:20–26.

Stewart, R. K. and K. Khoury. 1976. The biology of Lygus lineolaris (Palisot de Beauvois) (Hemiptera: Miridae) in Quebec. Ann. Soc. Entomol. Qué. 21:52–63.

Stewart, S. D. and M. J. Gaylor. 1990.

Age-grading adult tarnished plant bugs (Heteroptera: Miridae). J. Entomol. Sci. 25:216–218.

———. 1991. Age, sex, and reproductive status of the tarnished plant bug (Heteroptera: Miridae) colonizing mustard. Environ. Entomol. 20:1387–1392.

———. 1993. Age-grading eggs of the tarnished plant bug (Heteroptera: Miridae). J. Entomol. Sci. 28:263–266.

———. 1994a. Effects of age, sex, and reproductive status on flight by the tarnished plant bug (Heteroptera: Miridae). Environ. Entomol. 23:80–84.

———. 1994b. Effects of host switching on oviposition by the tarnished plant bug (Heteroptera: Miridae). J. Entomol. Sci. 29:231–238.

Stewart, S. D. and W. L. Sterling. 1988. Dynamics and impact of cotton fruit abscission and survival. Environ. Entomol. 17:629–635.

———. 1989a. Susceptibility of cotton fruiting forms to insects, boll rot, and physical stress. J. Econ. Entomol. 82:593–598.

———. 1989b. Causes and temporal patterns of cotton fruit abscission. J. Econ. Entomol. 82:954–959.

Stewart, S. D., A. G. Appel, and M. J. Gaylor. 1992. Ingestion of ^{14}C-labeled diet by Lygus hesperus Knight. Southwest. Entomol. 17:251–254.

Stewart, V. B. 1913a. The fire blight disease in nursery stock. Cornell Univ. Agric. Exp. Stn. Bull. 329:313–372.

———. 1913b. The importance of the tarnished plant bug in the dissemination of fire blight in nursery stock. Phytopathology 3:273–276.

Stewart, V. B. and M. D. Leonard. 1915. The rôle of sucking insects in the dissemination of fire blight bacteria. Phytopathology 5:117–123.

———. 1916. Further studies in the rôle of insects in the dissemination of fire blight bacteria. Phytopathology 6:152–158.

Steyaert, R. L. 1946. Plant protection in the Belgian Congo. Sci. Mon. (Wash., D.C.) 63:268–280.

Steyaert, R. L. and J. Vrydagh. 1933. Étude sur une maladie grave du cotonnier provoquée par les piqûres d'Helopeltis. Mem. Inst. R. Colon. Belge (Sci. Nat.) 1(7):1–53.

Steyskal, G. C. 1973. The grammar of names in the Catalogue of the Miridae (Heteroptera) of the World by Carvalho, 1957–1960. Stud. Entomol. 16:203–208.

———. 1991. On the meaning of the term 'trichobothrium.' Entomol. News 102:95–96.

Stigter, H. 1996. Campylomma verbasci, a new pest on apple in The Nether-

lands. Pages 140–144 in F. Polesny, W. Müller, and R. W. Olszak, eds., Proceedings of the International Conference on Integrated Fruit Production, 28 Aug.–2. Sept. 1995, Cedzyna, Poland. Bull. IOBC, WPRS, Avignon, France.

Stinner, B. R. 1975. Observations on predacious Miridae. Proc. Pa. Acad. Sci. 49:101–102.

Stitt, L. L. 1940. Three species of the genus Lygus and their relation to alfalfa seed production in southern Arizona and California. U.S. Dep. Agric. Tech. Bull. 741:1–19.

———. 1941. An experimental cooperative community program for the cultural control of bugs of the genus Lygus on alfalfa seed crops in the Mohawk area of Arizona in 1939 and 1940. U.S. Dep. Agric. Bur. Entomol. Plant Quar. E-546:1–14.

———. 1944. Difference in damage by three species of Lygus to alfalfa seed production. J. Econ. Entomol. 37:709.

———. 1948. Reduction of the vegetative growth of alfalfa by insects. J. Econ. Entomol. 41:739–741.

———. 1949. Host plant sources of Lygus spp. infesting the alfalfa seed crop in southern Arizona and southeastern California. J. Econ. Entomol. 42:93–99.

Stoffels, E. H. J. 1941. L'improductivité des caféiers Arabica dans le Kivu Nord. Bull. Agric. Congo Belge 32:59–69.

Stoltz, R. L. and C. D. McNeal Jr. 1982. Assessment of insect emigration from alfalfa hay to bean fields. Environ. Entomol. 11:578–580.

Stone, M. W. and F. B. Foley. 1959. Effect of time of application of DDT on lygus bug populations and yield of lima beans. J. Econ. Entomol. 52:244–257.

Stone, T. B., A. C. Thompson, and H. N. Pitre. 1985. Analysis of lipids in cotton extrafloral nectar. J. Entomol. Sci. 20:422–428.

Stonedahl, G. M. 1983a. New records for Palearctic Phytocoris in western North America (Hemiptera: Miridae). Proc. Entomol. Soc. Wash. 85:463–471.

———. 1983b. A systematic study of the genus Phytocoris Fallén (Heteroptera: Miridae) in western North America. Ph.D. dissertation, Oregon State University, Corvallis. 469 pp.

———. 1988. Revision of the mirine genus Phytocoris Fallén (Heteroptera: Miridae) for western North America. Bull. Am. Mus. Nat. Hist. 188:1–257.

———. 1990. Revision and cladistic analysis of the Holarctic genus Atractotomus Fieber (Heteroptera: Miridae: Phylinae). Bull. Am. Mus. Nat. Hist. 198:1–88.

———. 1991. The Oriental species of Helopeltis (Heteroptera: Miridae): A review of economic literature and guide to identification. Bull. Entomol. Res. 81:465–490.

———. 1995. Taxonomy of African *Eurystylus* (Heteroptera: Miridae), with a review of their status as pests of sorghum. Bull. Entomol. Res. 85:135–156.

Stonedahl, G. M. and G. Cassis. 1991. Revision and cladistic analysis of the plant bug genus *Fingulus* Distant (Heteroptera: Miridae: Deraeocorinae). Am. Mus. Novit. No. 3028:1–55.

Stonedahl, G. M. and W. R. Dolling. 1991. Heteroptera identification: A reference guide, with special emphasis on economic groups. J. Nat. Hist. 25:1027–1066.

Stonedahl, G. M. and T. J. Henry. 1991. A new genus of mirine plant bugs, *Gracilimiris*, with three new species from North America (Heteroptera: Miridae). J. N.Y. Entomol. Soc. 99:224–234.

Stonedahl, G. M. and D. Kovac. 1995. *Carvalhofulvius gigantochloae*, a new genus and species of bamboo-inhabiting Fulviini from West Malaysia (Heteroptera: Miridae: Cylapinae). Proc. Entomol. Soc. Wash. 97:427–434.

Stonedahl, G. M. and M. D. Schwartz. 1986. Revision of the plant bug genus *Pseudopsallus* Van Duzee (Heteroptera: Miridae). Am. Mus. Novit. No. 2842:1–58.

Stonedahl, G. M., M. B. Malipatil, and W. Houston. 1995. A new mirid (Heteroptera) pest of cashew in northern Australia. Bull. Entomol. Res. 85:275–278.

Stonedahl, G. M., J. D. Lattin, and V. Razafimahatratra. 1997. Review of the *Eurychilopterella* complex of genera, including the description of a new genus from Mexico (Heteroptera: Miridae: Deraeocorinae). Am. Mus. Novit. No. 3198:1–33.

Stoner, A. and G. T. Bottger. 1965. *Spanogonicus* [sic] *albofasciatus* and *Rhinacloa forticornis* on cotton in Arizona. J. Econ. Entomol. 58:314–315.

Stoner, A. and D. E. Surber. 1969. Notes on the biology and rearing of *Anaphes ovijentatus*, a new parasite of *Lygus hesperus* in Arizona. J. Econ. Entomol. 62:501–502.

Storey, H. H. 1939. Transmission of plant viruses by insects. Bot. Rev. 5:240–272.

Stork, N. E. 1981. The structure and function of the adhesive organs on the antennae of male *Harpocera thoracica* (Fallen) (Miridae; Hemiptera). J. Nat. Hist. 15:639–644.

———. 1991. The composition of the arthropod fauna of Bornean lowland rain forest trees. J. Trop. Ecol. 7:161–180.

Strawiński, K. 1964a. Zoophagism of terrestrial Hemiptera-Heteroptera occurring in Poland. Ekol. Pol. (A) 12:429–452.

———. 1964b. Drapieżnictwo u roślinożernych pluskwiaków różnoskrzydłych (Hemiptera-Heteroptera). Pol. Pismo Entomol. (B) 1–2 (33–34):129–133.

Streams, F. A. 1992. Intrageneric predation by *Notonecta* (Hemiptera: Notonectidae) in the laboratory and in nature. Ann. Entomol. Soc. Am. 85:265–273.

Streams, F. A., M. Shahjahan, and H. G. LeMasurier. 1968. Influence of plants on the parasitization of the tarnished plant bug by *Leiophron pallipes*. J. Econ. Entomol. 61:996–999.

Street, R. S. 1989. The bug sucker. Agrichem. Age 33(3):38–39.

Strickland, A. H. 1951. The entomology of swollen shoot of cacao. II.—The bionomics and ecology of the species involved. Bull. Entomol. Res. 42:65–103.

Strickland, E. H. 1953. An annotated list of the Hemiptera (s.l.) of Alberta. Can. Entomol. 85:193–214.

Stride, G. O. 1968. On the biology and ecology of *Lygus vosseleri* (Heteroptera: Miridae) with special reference to its hostplant relationships. J. Entomol. Soc. S. Afr. 31:17–55.

Strong, D. R. Jr. 1974. Rapid asymptotic species accumulation in phytophagous insect communities: The pests of cacao. Science (Wash., D.C.) 185:1064–1066.

Strong, D. R. [Jr.], J. H. Lawton, and Sir R. [T. R. E.] Southwood. 1984. Insects on plants: Community patterns and mechanisms. Harvard University Press, Cambridge. 313 pp.

Strong, F. E. 1968. The selective advantage accruing to lygus bugs that cause blasting of floral parts. J. Econ. Entomol. 61:315–316.

———. 1970. Physiology of injury caused by *Lygus hesperus*. J. Econ. Entomol. 63:808–814.

———. 1971. A computer-generated model to simulate mating behavior of lygus bugs. J. Econ. Entomol. 64:46–50.

Strong, F. E. and E. C. Kruitwagen. 1968. Polygalacturonase in the salivary apparatus of *Lygus hesperus* (Hemiptera). J. Insect Physiol. 14:1113–1119.

Strong, F. E. and E. [C.] Kruitwagen. 1969. Feeding and nutrition of *Lygus hesperus*. III. Limited growth and development on a meridic diet. Ann. Entomol. Soc. Am. 62:148–155.

———. 1970. Gustatory discrimination between meridic diets by the bug, *Lygus hesperus*. J. Insect Physiol. 16:521–530.

Strong, F. E. and D. A. Landes. 1965. Feeding and nutrition of *Lygus hesperus* (Hemiptera: Miridae). II. An estimation of normal feeding rates. Ann. Entomol. Soc. Am. 58:309–314.

Strong, F. E. and J. A. Sheldahl. 1970. The influence of temperature on longevity and fecundity in the bug *Lygus hesperus* (Hemiptera: Miridae). Ann. Entomol. Soc. Am. 63:1509–1515.

Strong, F. E. and J. Villanueva-Barradas. 1968. Damage to cotton by *Lygus hesperus* (Knight). Folia Entomol. Mex. 18–19:34–36.

Strong, F. E., J. A. Sheldahl, P. R. Hughes, and E. M. K. Hussein. 1970. Reproductive biology of *Lygus hesperus* Knight. Hilgardia 40(4):105–147.

Strong, R. P., G. C. Shattuck, J. C. Bequaert, and R. E. Wheeler. 1926. Medical report of the Hamilton Rice Seventh Expedition to the Amazon, in conjunction with the Department of Tropical Medicine of Harvard University, 1924–1925. Contrib. Harvard Inst. Trop. Biol. Med. No. IV. Harvard University Press, Cambridge. 313 pp.

Stultz, H. T. 1955. The influence of spray programs on the fauna of apple orchards in Nova Scotia. VIII. Natural enemies of the eye-spotted bud moth, *Spilonota ocellana* (D. & S.) (Lepidoptera: Olethreutidae). Can. Entomol. 87:79–85.

Štusák, J. M. and J. L. Stehlík. 1977. First contribution to the teratology of Tingidae (Heteroptera): Reflexion and variability of paranota. Acta Mus. Morav. (Sci. Nat.) 62:119–122.

———. 1978. Second contribution to the teratology of Tingidae (Heteroptera). Antennal anomalies. Acta Mus. Morav. (Sci. Nat.) 63:89–105.

———. 1979. Third contribution to the teratology of Tingidae (Heteroptera). Anomalies of legs. Acta Mus. Morav. (Sci. Nat.) 64:75–84.

———. 1982. Fifth contribution to the teratology of Tingidae (Heteroptera). Anomalies of fore wings (hemelytra). Acta Mus. Morav. (Sci. Nat.) 67:163–180.

Štusák, J. M. and P. Štys. 1958. První nález čeledi Isometopidae v Čechách (Heteroptera). Mus. Zpr. Prazskeko Kraje 3:126–127.

Štys, P. 1970a. On the morphology and classification of the family Dipsocoridae s.lat., with particular reference to the genus *Hypsipteryx* Drake (Heteroptera). Acta Entomol. Bohemoslav. 67:21–46.

———. 1970b. *Orthops coccineus* (Horv.), stat. n.—an unrecognized species of the

European Miridae (Heteroptera). Acta Entomol. Bohemoslav. 67:100–104.

——. 1974. Population explosion of *Acetropis longirostris* in eastern Slovakia (667–670.

——. 1975. *Hypseloecus visci* (Put.) in Czechoslovakia (Heteroptera: Miridae). Acta Univ. Carol. (Biol.) 1973(3–4):155–158.

——. 1985. A new genus of Palaearctic Bryocorinae related to Afrotropical *Rhodocoris* (Heteroptera, Miridae). Acta Entomol. Bohemoslav. 82:407–425.

——. 1995. Dipsocoromorpha. Pages 74–83 in R. T. Schuh and J. A. Slater. True bugs of the world (Hemiptera: Heteroptera): Classification and natural history. Cornell University Press, Ithaca.

Štys, P. and J. Davidová-Vilímová. 1989. Unusual numbers of instars in Heteroptera: A review. Acta Entomol. Bohemoslav. 86:1–32.

Štys, P. and J. Kinkorová. 1985. *Psallus (Hyalopsallus) wagneri* in Czechoslovakia (Heteroptera, Miridae). Acta Entomol. Bohemoslav. 82:355–359.

Subbaiah, C. C. 1983. Fruiting and abscission patterns in cashew. J. Agric. Sci. (Camb.) 100:423–427.

Subinprasert, S. and B. W. Svensson. 1988. Effects of predation on clutch size and egg dispersion in the codling moth *Laspeyresia pomonella*. Ecol. Entomol. 13:87–94.

Sudhakar, M. A. 1975. Role of *Helopeltis antonii* Signoret (Hemiptera: Miridae), in causing scab on guava fruits, its biology and control. Mysore J. Agric. Sci. 9:205–206.

Sugonyaev, E. S. and K. Kamalov. 1976. On the study of biocenotic connections and their effect on number dynamics of pests and their beneficial arthropods of cotton fields in lowlands of Murgab. Pages 29–63 in A. O. Tashlieva, ed., Ecology and economic significance of the insects of Turkmenia. Akad. Nauk Turkmenskoi SSR Inst. Zool., Ashkabad, Turkmenistan. [English translation by Saad Publications, Karachi, Pakistan, 1982.]

Suguiyama, L. and C. Osteen. 1996. The economic impact of cotton insects and mites. Pages 755–780 in E. G. King, J. R. Phillips, and R. J. Coleman, eds., Cotton insects and mites: Characterization and management. Cotton Foundation, Memphis.

Sullivan, D. J. 1988. Hyperparasites. Pages 189–203 in A. K. Minks and P. Harrewijn, eds., Aphids: Their biology, natural enemies, and control. Vol. B. Elsevier, Amsterdam.

Summerfield, R. J., F. J. Muehlbauer, and R. W. Short. 1982. Lygus bugs and seed quality in lentils (*Lens culinaris* Medik.). U.S. Dep. Agric. Agric. Res. Serv. ARM-W-29:1–43.

Summers, C. G. 1976. Population fluctuations of selected arthropods in alfalfa: Influence of two harvesting practices. Environ. Entomol. 5:103–110.

Summers, H. E. 1891. The glassy-winged soldier-bug. Tenn. Agric. Exp. Stn. Bull. 4:32–33.

Sundararaju, D. 1984. Cashew pests and their natural enemies in Goa. J. Plant. Crops 12:38–46.

Sundararaju, D. and P. C. Sundara Babu. 1998. Life table studies of *Helopeltis antonii* Sign. (Heteroptera: Miridae) on neem, guava and cashew. J. Entomol. Res. (New Delhi) 22:241–244.

Sunderland, K. D., N. E. Crook, D. L. Stacey, and B. J. Fuller. 1987. A study of feeding by polyphagous predators on cereal aphids using ELISA and gut dissection. J. Appl. Ecol. 24:907–933.

Surface, H. A. 1906. Order X. The Hemiptera: The bugs, lice, aphids, scale insects, cicadas, and others. Pa. Dep. Agric. Mon. Bull. Div. Zool. 4:48–72.

Susman, I. 1988. The cotton insects of Israel and aspects of the biology of *Deraeocoris pallens* Reuter (Heteroptera, Miridae). M.S. thesis, Tel Aviv University, Israel. 154 pp. [in Hebrew, English summary.]

Sutherland, J. R., G. M. Shrimpton, and R. N. Sturrock. 1989. Diseases and insects in British Columbia forest seedling nurseries. Canada-British Columbia For. Resour. Dev. Agreement Rep. 65:1–85.

Sutton, R. D. 1984. The effect of host plant flowering on the distribution and growth of hawthorn psyllids (Homoptera: Psylloidea). J. Anim. Ecol. 53:37–50.

Swaine, G. 1959. A preliminary note on *Helopeltis* spp. damaging cashew in Tanganyika Territory. Bull. Entomol. Res. 50:171–181.

——. 1971. Agricultural zoology in Fiji. H.M. Stationery Office, London. 424 pp.

Swaine, G., D. A. Ironside, and R. J. Corcoran. 1991. Insect pests of fruit and vegetables, 2nd ed. Queensland Department of Primary Industries, Brisbane, Australia. 126 pp.

Swallow, W. H. and P. J. Cressey. 1987. Historical overview of wheat-bug damage in New Zealand wheats. N.Z. J. Agric. Res. 30:341–344.

Swan, L. A. 1964. Beneficial insects. Harper & Row, New York. 429 pp.

Sweet, H. E. 1930. An ecological study of the animal life associated with *Artemisia californica* Less, at Claremont, California. J. Entomol. Zool. (Claremont) 22:57–70, 75–103.

Sweet, M. H. 1964. The biology and ecology of the Rhyparochrominae of New England (Heteroptera: Lygaeidae). Part I. Entomol. Am. 43:1–124.

——. 1979. On the original feeding habits of the Hemiptera (Insecta). Ann. Entomol. Soc. Am. 72:575–579.

Sweetman, H. L. 1958. The principles of biological control: Interrelation of hosts and pests and utilization in regulation of animal and plant populations. Brown, Dubuque, Iowa. 560 pp.

Swezey, O. H. 1928. Present status of certain insect pests under biological control in Hawaii. J. Econ. Entomol. 21:669–676.

——. 1929. The present status of certain insect pests under biological control in Hawaii. Trans. 4th Int. Congr. Entomol., Ithaca, N.Y. (1928)2:366–371.

——. 1936. Biological control of the sugar cane leafhopper in Hawaii. Hawaii. Plant. Rec. 40:57–101.

——. 1945. Insects associated with orchids. Proc. Hawaii. Entomol. Soc. 12:343–403.

——. 1954. Forest entomology in Hawaii. An annotated check-list of the insect faunas of the various components of the Hawaiian forests. Bernice P. Bishop Mus. Spec. Publ. 44:1–266.

Szent-Ivany, J. J. H. 1961. Insect pests of *Theobroma cacao* in the territory of Papua and New Guinea. Papua New Guinea Agric. J. 13:127–147.

——. 1965. Factors influencing dispersal of insects in rainforest areas converted to cacao lands. In Proc. 12th Int. Congr. Entomol., London (1964), p. 330.

Takanona, T. and K. Hori. 1974. Digestive enzymes in the salivary gland and midgut of the bug *Stenotus binotatus*. Comp. Biochem. Physiol. 47A:521–528.

Takara, J. and T. Nishida. 1981. Eggs of the Oriental fruit fly for rearing the predacious anthocorid, *Orius insidiosus* (Say). Proc. Hawaii. Entomol. Soc. 23:441–445.

Takeno, K. 1998. Enumeration of the Heteroptera in Mt. Hikosan, western Japan with their hosts and preys I. Esakia 38:29–53.

Taksdal, G. 1959. Angrep av skjermplantetege (*Lygus campestris* L.) i gulrotfrøfelt fører til nedsatt spireprosent og avling. Gartneryrket (Oslo) 42:709–710, 712, 714.

——. 1961. Ecology of plant resistance to the tarnished plant bug, *Lygus lineolaris* (P. de B.). M.S. thesis, Cornell University, Ithaca. 94 pp.

——. 1963. Ecology of plant resistance to the tarnished plant bug, *Lygus lineolaris*. Ann. Entomol. Soc. Am. 56:69–74.

——. 1965. Hemiptera (Heteroptera) collected on ornamental trees and shrubs at the Agricultural College of Norway, Ås. Nor. Entomol. Tidsskr. 13:5–10.

——. 1970. Hagetege og stein i paere. Gartneryrket (Oslo) 60:458–463.

——. 1983. *Orthotylus marginalis* Reuter and *Psallus ambiguus* Fallén (Heteroptera, Miridae) causing stony pits in pears. Acta Agric. Scand. 33:205–208.

Taksdal, G. and O. Sørum. 1971. Capsids (Heteroptera, Miridae) in strawberries, and their influence on fruit malformation. J. Hortic. Sci. 46:43–50.

Talhouk, A. S. 1969. Insects and mites injurious to crops in Middle Eastern countries. Parey, Hamburg. 239 pp.

Tallamy, D. W. 1986. Behavioral adaptations in insects to plant allelochemicals. Pages 273–300 *in* L. B. Brattsten and S. Ahmad, eds., Molecular aspects of insect-plant associations. Plenum, New York.

——. 1994. Nourishment and the evolution of paternal investment in subsocial arthropods. Pages 21–55 *in* J. H. Hunt and C. A. Nalepa, eds., Nourishment and evolution in insect societies. Westview, Boulder, Col.; Oxford & IBH, New Delhi, India.

Tallamy, D. W. and V. A. Krischik. 1989. Variation and function of cucurbitacins in *Cucurbita*: An examination of current hypotheses. Am. Nat. 133:766–786.

Tallamy, D. W. and C. [W.] Schaefer. 1997. Maternal care in the Hemiptera: Ancestry, alternatives, and current adaptive value. Pages 94–115 *in* J. C. Choe and B. J. Crespi, eds., The evolution of social behavior in insects and arachnids. Cambridge University Press, Cambridge.

Tamaki, G. and G. T. Hagel. 1978. Evaluation and projection of *Lygus* damage to sugarbeet seedlings. J. Econ. Entomol. 71:265–268.

Tamaki, G., D. P. Olsen, and R. K. Gupta. 1978. Laboratory evaluation of *Geocoris bullatus* and *Nabis alternatus* as predators of *Lygus*. J. Entomol. Soc. B.C. 75:35–37.

Tan, G. S. 1974a. *Helopeltis theivora theobromae* on cocoa in Malaysia. I. Biology and population fluctuations. Malays. Agric. Res. 3:127–132.

——. 1974b. *Helopeltis theivora theobromae* on cocoa in Malaysia. II. Damage and control. Malays. Agric. Res. 3:204–212.

Tanada, Y. and F. G. Holdaway. 1954. Feeding habits of the tomato bug, *Cyrtopeltis (Engytatus) modestus* (Distant), with special reference to the feeding lesion on tomato. Hawaii Agric. Exp. Stn. Tech. Bull. 24:1–40.

Tanangsnakool, C. 1975. Ecological investigation on the brown planthopper, *Nilaparvata lugens* (Stål) (Homoptera: Delphacidae), and its egg predator, *Cyrtorhinus lividipennis* Reuter (Hemiptera: Miridae). M.S. thesis, Kasetsart University, Thailand. 57 pp.

Tanigoshi, L. K. and J. M. Babcock. 1989. Insecticide efficacy for control of lygus bugs (Heteroptera: Miridae) on white lupin, *Lupinus albus* L. J. Econ. Entomol. 82:281–284.

Tanigoshi, L. K., D. F. Mayer, J. M. Babcock, and J. D. Lunden. 1990. Efficacy of the β-exotoxin of *Bacillus thuringiensis* to *Lygus hesperus* (Heteroptera: Miridae): Laboratory and field responses. J. Econ. Entomol. 83:2200–2206.

Tate, H. D. 1940. Insects as vectors of yellow dwarf, a virus disease of onions. Iowa State Coll. J. Sci. 14:267–294.

Tauber, C. A. and M. J. Tauber. 1981. Insect seasonal cycles: Genetics and evolution. Annu. Rev. Ecol. Syst. 12:281–308.

——. 1987. Food specificity in predacious insects: A comparative ecophysiological and genetic study. Evol. Ecol. 1:175–186.

Tauber, M. J. and R. G. Helgesen. 1981. Development of biological control systems for greenhouse crop production in the U.S.A. Pages 37–40 *in* J. R. Coulson, ed., Proceedings of the Joint American-Soviet Conference on Use of Beneficial Organisms in the Control of Crop Pests, Washington, D.C., Aug. 13–14, 1979. Entomological Society of America, College Park, Md.

Tauber, M. J., C. A. Tauber, and S. Masaki. 1986. Seasonal adaptations of insects. Oxford University Press, New York. 411 pp.

Tavella, L. and A. Arzone. 1987. Indagini sui limitori naturali di *Corythucha ciliata* (Say) (Rhynchota Heteroptera). Redia 70:443–457.

Tavella, L., A. Arzone, A. Alma, and A. Galliano. 1996. IPM application in peach orchards against *Lygus rugulipennis* Poppius. Int. Organ. Biol. Control/West. Palearctic Reg. Sect. Bull. 19(4):160–164.

Taylor, B. 1977. The ant mosaic on cocoa and other tree crops in western Nigeria. Ecol. Entomol. 2:245–255.

Taylor, D. E. 1978. The mosquito bug. Rhod. Agric. J. 75:41.

Taylor, D. J. 1954. A summary of the results of capsid research in the Gold Coast. West Afr. Cacao Res. Inst. Tech. Bull. 1:1–20.

——. 1955. Capsid studies: Population studies. W. Afr. Cocoa Res. Inst. Rep. 1954–55:61–63.

Taylor, E. P. 1908. Dimples in apples from oviposition of *Lygus pratensis* L. J. Econ. Entomol. 1:370–375.

Taylor, F. 1981. Ecology and evolution of physiological time in insects. Am. Nat. 117:1–23.

Taylor, G. S. and P. W. Miles. 1994. Composition and variability of the saliva of coreids in relation to phytoxicoses and other aspects of the salivary physiology of phytophagous Heteroptera. Entomol. Exp. Appl. 73:265–277.

Taylor, L. R. 1974. Insect migration, flight periodicity and the boundary layer. J. Anim. Ecol. 43:225–238.

——. 1984. Assessing and interpreting the spatial distributions of insect populations. Annu. Rev. Entomol. 29:321–357.

Taylor, T. H. C. 1945. *Lygus simonyi*, Reut., as a cotton pest in Uganda. Bull. Entomol. Res. 36:121–148.

——. 1947a. On the identity of the cotton capsid of Uganda. Bull. Entomol. Res. 37:503–505.

——. 1947b. Some East African species of *Lygus*, with notes on their host plants. Bull. Entomol. Res. 38:233–258.

Tedders, W. L. Jr. 1965. The biology and effect of two Miridae on pecan nut drop in southwest Georgia. Proc. Southeast. Pecan Grow. Assoc. 58:34–36.

Tedders, W. L. [Jr.] 1978. Important biological and morphological characteristics of the foliar-feeding aphids of pecan. U.S. Dep. Agric. Tech. Bull. 1579:1–29.

——. 1995. Identity of spittlebug on pecan and life history of *Clastoptera achatina* (Homptera: Cercopidae). J. Econ. Entomol. 88:1641–1649.

Tedders, W. L. [Jr.], C. C. Reilly, B. W. Wood, R. K. Morrison, and C. S. Lofgren. 1990. Behavior of *Solenopsis invicta* (Hymenoptera: Formicidae) in pecan orchards. Environ. Entomol. 19:44–53.

Teetes, G. L. 1985. Head bugs: Methodology for determining economic threshold levels in sorghum. Pages 301–315 *in* K. Leuschner and G. L. Teetes, eds., Proceedings of the International Sorghum Entomology Workshop, 15–21 July 1984, Texas A&M Univ., College Station. International Crops Research Institute for the Semi-Arid Tropics, Patancheru, A.P., India.

Teetes, G. L., W. R. Young, and M. G. Jotwani. 1979. Insect pests of sorghum. Pages 17–40 *in* Elements of integrated control of sorghum pests. FAO Plant Prod. Prot. Pap. 19.

Teetes, G. L., K. V. Seshu Reddy, K. Leuschner, and L. R. House. 1983. Sorghum insect identification handbook. Inf. Bull. 12. International Crops Research Institute for the Semi-Arid Tropics, Patancheru, A.P., India. 124 pp.

Telford, A. D. 1957. Arizona cotton insects. Ariz. Agric. Exp. Stn. Bull. 286:1–60.

Terauds, A. 1970. Evaluation of methods for the control of damage to apples by dimpling bug in Tasmania. Aust. J. Exp. Agric. Anim. Husb. 10:647–650.

——. 1971. Detection and control of the dimpling bug *Rhodolygus milleri* Ghauri, in Tasmania. Tasmanian J. Agric. 42:179–181.

Terrell, E. E., S. R. Hill, J. H. Wiersema, and W. E. Rice. 1986. A checklist of names for 3,000 vascular plants of economic importance. U.S. Dep. Agric. Agric. Res. Serv. Agric. Handb. 505:1–241.

Terry, L. I. and B. B. Barstow. 1988. Susceptibility of early-season cotton floral bud types to thrips (Thysanoptera: Thripidae) damage. J. Econ. Entomol. 81:1785–1791.

Thalenhorst, W. 1962. Auftreten, Massenwechsel und Bekämpfung der Fichtenspinnmilbe, *Paratetranychus (Oligonychus) ununguis* (Jacobi). Forst-Holzw. 15:295–300.

Thankamma Pillai, P. K. and G. B. Pillai. 1975. Note on the shedding of immature fruits in cashew. Indian J. Agric. Sci. 45:233–234.

Thatcher, R. W. 1923. Division of Entomology: Peach deforming capsids. Forty-first annual report, with the Director's report for 1922. N.Y. Agric. Exp. Stn., Geneva. p. 39.

Theiling, K. M. and B. A. Croft. 1988. Pesticide side-effects on arthropod natural enemies: A database summary. Agric. Ecosyst. Environ. 21:191–218.

Theobald, F. V. 1895. Notes on the needle-nosed hop bugs. J. South-East. Agric. Coll. Wye, Kent 2:11–16.

——. 1896. On some hop-pests. Entomol. Mon. Mag. 32:60–62.

——. 1911a. Bugs (Hemiptera-heteroptera) damaging apples. Rep. Econ. Zool. 1911:27–30. South-East. Agric. Coll., Wye.

——. 1911b. Animals injurious to hops. The needle nosed hop bug (*Calocoris fulvomaculatus* De Geer). Rep. Econ. Zool. 1911:82–86. South-East. Agric. Coll., Wye.

——. 1913. Some new and unusual insect attacks on fruit trees and bushes in 1912. J. Board Agric. (Lond.) 20:106–116.

——. 1928. Notes on hop insects in 1927. Entomologist 61:121–122.

——. 1929. Some notes on injurious insects and other animals in 1928. J. South-East. Agric. Coll. Wye, Kent 26:104–116.

Thistlewood, H. M. A. 1989. Spatial dispersion and sampling of *Campylomma verbasci* (Heteroptera: Miridae) on apple. Environ. Entomol. 18:398–402.

Thistlewood, H. M. A. and R. D. McMullen. 1989. Distribution of *Campyloma verbasci* (Heteroptera: Miridae) nymphs on apple and an assessment of two methods of sampling. J. Econ. Entomol. 82:510–515.

Thistlewood, H. M. A. and R. F. Smith. 1996. Management of the mullein bug, *Campyloma verbasci* (Heteroptera: Miridae), in pome fruit orchards of Canada. Pages 119–140 *in* O. Alomar and R. N. Wiedenmann, eds., Zoophytophagous Heteroptera: Implications for life history and integrated pest management. Thomas Say Publ. Entomol.: Proceedings. Entomological Society of America, Lanham, Md.

Thistlewood, H. M. A., J. H. Borden, R. F. Smith, H. D. Pierce Jr., and R. D. McMullen. 1989a. Evidence for a sex pheromone in the mullein bug, *Campyloma verbasci* (Meyer) (Heteroptera: Miridae). Can. Entomol. 121:737–744.

Thistlewood, H. M. A., R. D. McMullen, and J. H. Borden. 1989b. Damage and economic injury levels of the mullein bug, *Campyloma verbasci* (Meyer) (Heteroptera: Miridae), on apple in the Okanagan Valley. Can. Entomol. 121:1–9.

Thistlewood, H. M. A., J. H. Borden, and R. D. McMullen. 1990. Seasonal abundance of the mullein bug, *Campylomma verbasci* (Meyer) (Heteroptera: Miridae), on apple and mullein in the Okanagan Valley. Can. Entomol. 122:1045–1058.

Thomas, D. B. Jr. 1987. Chromosome evolution in the Heteroptera (Hemiptera): Agmatoploidy versus aneuploidy. Ann. Entomol. Soc. Am. 80:720–730.

Thomas, D. B. [Jr.] 1996. Role of polyploidy in the evolution of the Heteroptera. Pages 159–178 *in* C. W. Schaefer, ed., Studies on hemipteran phylogeny. Thomas Say Publ. Entomol.: Proceedings. Entomological Society of America, Lanham, Md.

Thomas, D. C. 1938. Report on the Hemiptera-Heteroptera taken in the light trap at Rothamsted Experimental Station, during the four years 1933–1936. Proc. Entomol. Soc. Lond. (A) 13:19–24.

——. 1943. Hemiptera-Heteroptera in Essex. Entomol. Mon. Mag. 79:199.

Thomas, F. L. and W. L. Owen Jr. 1937. Cotton flea hopper, an ecological problem. J. Econ. Entomol. 30:848–850.

Thomas, H. E. and P. A. Ark. 1934. Fire blight of pears and related plants. Calif. Agric. Exp. Stn. Bull. 586:1–43.

Thomas, W. A. 1945. *Cyrtopeltis varians* in some of the tobacco-growing areas of North Carolina. J. Econ. Entomol. 38:498–499.

Thompson, B. G. 1945. DDT to control *Psallus ancorifer* in onions. J. Econ. Entomol. 38:277.

Thompson, J. N. 1983. Selection of plant parts by *Depressaria multifidae* (Lep., Oecophoridae) on its seasonally-restricted hostplant, *Lomatium grayi* (Umbelliferae). Ecol. Entomol. 8:203–211.

——. 1994. The coevolutionary process. University of Chicago Press, Chicago. 376 pp.

Thompson, J. N. and D. Althoff. 1999. Insect diversity and the trophic complexity of communities. Pages 537–552 *in* C. B. Huffaker and A. P. Gutierrez, eds., Ecological entomology, 2nd ed. Wiley, New York.

Thompson, W. R. 1951. The specificity of host relations in predacious insects. Can. Entomol. 83:262–269.

Thompson, W. R. and F. J. Simmonds. 1965. A catalogue of the parasites and predators of insect pests. Section 4. Host predator catalogue. Commonwealth Agricultural Bureaux. Commonwealth Institute of Biological Control, Farnham Royal, Buckinghamshire, UK.

Thomson, J. D., W. P. Maddison, and R. C. Plowright. 1982. Behavior of bumble bee pollinators of *Aralia hispida* Vent. (Araliaceae). Oecologia (Berl.) 54:326–336.

Thontadarya, T. S. and G. P. Channa Basavanna. 1962. Mode of egg-laying in *Helopeltis antonii* Sign. (Hemiptera, Miridae). Curr. Sci. (Bangalore) 31:339.

Thornhill, R. 1975. Scorpionflies as kleptoparasites of web-building spiders. Nature (Lond.) 258:709–711.

Thornton, I. W. B., T. R. New, D. A. McLaren, H. K. Sudarman, and P. J. Vaughan. 1988. Air-borne arthropod fall-out on Anak Krakatau and a possible pre-vegetation pioneer community. Philos. Trans. R. Soc. Lond. (B) 322:471–479.

Thorold, C. A. 1975. Diseases of cocoa. Clarendon, Oxford. 423 pp.

Thorpe, W. H. 1933. Notes on the natural control of *Coleophora laricella*, the larch case-bearer. Bull. Entomol. Res. 24:271–291.

Thresh, J. M. 1960. Capsids as a factor influencing the effect of swollen-shoot disease on cacao in Nigeria. Emp. J. Exp. Agric. 28:193–200.

Thresh, J. M., G. K. Owusu, A. Boamah, and G. Lockwood. 1988. Ghanaian cocoa varieties and swollen shoot virus. Crop Prot. 7:219–231.

Thygesen, T., P. Esbjerg, and H. Eiberg. 1973. Ildsotoverføring med insekter. Tidsskr. Planteavl 77:324–336.

Tiensuu, L. 1936. Insect life on plants attacked by aphids. Ann. Entomol. Fenn. 2:161–169.

Tillyard, R. J. 1926. The insects of Australia and New Zealand. Angus & Robertson, Sydney. 560 pp.

Timberlake, P. H. 1927. Biological control of insect pests in the Hawaiian Islands. Proc. Hawaii. Entomol. Soc. 6:529–556.

Ting, Y. C. 1963a. Studies on the ecological characteristics of cotton mirids I. Effect of temperature and humidity on the development and distribution of the pests. Acta Phytophyl. Sin. 2:285–296. [in Chinese, English summary.]

———. 1963b. Studies on the ecological characteristics of cotton mirid bugs II. The correlation of the injury caused by mirid bugs with the chemical composition of the cotton plant. Acta Phytophyl. Sin. 2:365–370. [in Chinese, English summary.]

———. 1965. Studies on the ecological characteristics of cotton plant bugs III. The pattern of spatial distribution of the plant bugs in cotton fields with analysis of its effective factors. Acta Entomol. Sin. 14:264–273. [in Chinese, English summary.]

Tingey, W. M. 1976. Survey of crop resistance to lygus bugs. Pages 14–16 in D. R. Scott and L. E. O'Keeffe, eds., Lygus bug: Host plant interactions. University of Idaho Press, Moscow.

———. 1985. Plant defensive mechanisms against leafhoppers. Pages 217–234 in L. R. Nault and J. G. Rodriguez, eds., The leafhoppers and planthoppers. Wiley, New York.

Tingey, W. M. and W. J. Lamont Jr. 1988. Insect abundance in field beans altered by intercropping. Bull. Entomol. Res. 78:527–535.

Tingey, W. M. and E. A. Pillemer. 1977. Lygus bugs: Crop resistance and physiological nature of feeding. Bull. Entomol. Soc. Am. 23:277–287.

Tingey, W. M., T. F. Leigh, and A. H. Hyer. 1975a. Lygus hesperus: Growth, survival, and egg laying resistance of

cotton genotypes. J. Econ. Entomol. 68:28–30.

———. 1975b. Glandless cotton: Susceptibility to Lygus hesperus Knight. Crop Sci. 15:251–253.

Tinsley, T. W. 1964. The ecological approach to pest and disease problems of cacao in West Africa. Trop. Sci. 6:38–46.

Tinsley, T. W. and K. A. Harrap. 1978. Viruses of invertebrates. Pages 1–101 in H. Fraenkel-Conrat and R. R. Wagner, eds., Comprehensive virology. Vol. 12. Newly characterized protist and invertebrate viruses. Plenum, New York.

Tischler, W. 1951. Die Überwinterungverhältnisse der landwirtschaftlichen Schädlinge. Z. Angew. Entomol. 32:184–194.

Tjallingii, W. F. and T. Hogen Esch. 1993. Fine structure of aphid stylet routes in plant tissues in correlation with EPG signals. Physiol. Entomol. 18:317–328.

Todd, F. E. and O. Bretherick. 1942. The composition of pollens. J. Econ. Entomol. 35:312–317.

Todd, F. E. and S. E. McGregor. 1952. Insecticides and bees. Pages 131–135 in F. C. Bishopp, chairman, et al., Insects: The yearbook of agriculture, 1952. U.S. Department of Agriculture, Washington, D.C.

Todd, F. E. and G. H. Vansell. 1942. Pollen grains in nectar and honey. J. Econ. Entomol. 35:728–731.

Todd, J. G. and J. A. Kamm. 1974. Biology and impact of a grass bug Labops hesperius Uhler in Oregon rangeland. J. Range Manage. 27:453–458.

Tolstova, Yu. S. and N. M. Atanov. 1982. Action of chemical substances for plant protection on the arthropod fauna of the orchard. 1. Long-term action of pesticides in agrocoenoses. Entomol. Obozr. 61(3):441–453. [in Russian; English translation in Entomol. Rev. 61(3):1–14.]

Tominić, A. 1951. Stjenica loznog cvijeta u vinogradima Konavlja. Zast. Bilja 5:3–12.

Tong, X. S. and L. S. Wang. 1987. Bionomics and control of the fleahopper Halticus minutus Reuter. Acta Entomol. Sin. 30:113–115. [in Chinese.]

Tonhasca, A. Jr. 1987. Seasonal history of the clouded plant bug, Neurocolpus nubilus (Hemiptera: Miridae), and development of damaging populations on cotton in northern Mississippi. M.S. thesis, Mississippi State University, Mississippi State. 84 pp.

Tonhasca, A. Jr. and R. G. Luttrell. 1991. Seasonal history and damage caused by Neurocolpus nubilus (Hemiptera: Miridae) on cotton in northern Mississippi. Environ. Entomol. 20:742–748.

Tonkyn, D. W. 1986. Predator-mediated mutualism: Theory and tests in the Homoptera. J. Theor. Biol. 118:15–31.

Tonkyn, D. W. and R. F. Whitcomb. 1987. Feeding strategies and the guild concept among vascular feeding insects and microorganisms. Curr. Top. Vector Res. 4:179–199.

Torre-Bueno, J. R. de la. 1925. Methods of collecting, mounting and preserving Hemiptera. Can. Entomol. 57:6–10, 27–32, 53–57.

———. 1929. [Review of] Flowers and insects by Charles Robertson. Bull. Brooklyn Entomol. Soc. 24:335–337.

———. 1931. Biting bugs. Bull. Brooklyn Entomol. Soc. 26:176.

Torreno, H. S. 1994. Predation behavior and efficiency of the bug, Cyrtopeltis tenuis (Hemiptera: Miridae), against the cutworm, Spodoptera litura (F.). Philipp. Entomol. 9:426–434.

Torreno, H. S. and E. D. Magallona. 1994. Biological relationship of the bug, Cyrtopeltis tenuis Reuter (Hemiptera: Miridae) with tobacco. Philipp. Entomol. 9:406–425.

Toscano, N. C., F. G. Zalom, E. R. Oatman, J. Trumble, and K. Kido. 1990. Insects and mites. Pages 1–15 in Tomato pest management guidelines (UCPMG Publ. 14). Univ. Calif. Div. Nat. Resour. Publ. 3339.

Townes, H. 1958. Some biological characteristics of the Ichneumonidae (Hymenoptera) in relation to biological control. J. Econ. Entomol. 51:650–652.

Townsend, R. J. and R. N. Watson. 1982. Biology of potato mirid and Australian crop mirid on asparagus. Pages 332–337 in M. J. Hartley, ed., Proc. 35th N.Z. Weed Pest Control Conf., Aug. 9–12, 1982. N.Z. Weed and Pest Control Society, Palmerston North.

Toxopeus, H. 1985. Botany, types and populations. Pages 11–37 in G. A. R. Wood and R. A. Lass, eds., Cocoa, 4th ed. Longman, New York.

Toxopeus, H. and B. M. Gerard. 1968. A note on mirid damage to mature cacao pods. Niger. Entomol. Mag. 1:59–60.

Trapman, M. and L. Blommers. 1992. An attempt to [sic] pear sucker management in the Netherlands. J. Appl. Entomol. 114:38–51.

Travis, J. 1984. Breeding system, pollination, and pollinator limitation in a perennial herb, Amianthium muscaetoxicum (Liliaceae). Am. J. Bot. 71:941–947.

Treacy, M. F., J. H. Benedict, M. H. Walmsley, J. D. Lopez, and R. K. Morrison. 1987. Parasitism of bollworm (Lepidoptera: Noctuidae) eggs on nectaried and nectariless cotton. Environ. Entomol. 16:420–423.

Trehan, K. N. and V. V. Phatak. 1946. Life-history and control of betelvine bug (*Disphinctus maesarum* Kirk). Poona Agric. Coll. Mag. 37(1):25–32.

Treherne, R. C. 1923. The relation of insects to vegetable seed production. 15th Annu. Rep. Que. Soc. Prot. Plants 1922–1923:47–59.

Trelease, W. 1879. Nectar; what it is, and some of its uses. Pages 319–343 *in* J. H. Comstock, Report upon cotton insects. Government Printing Office, Washington, D.C.

Treshow, M. 1970. Environment and plant response. McGraw-Hill, New York. 422 pp.

Trichilo, P. J., L. T. Wilson, and D. W. Grimes. 1990. Influence of irrigation management on the abundance of leafhoppers (Homoptera: Cicadellidae) on grapes. Environ. Entomol. 19:1803–1809.

Triggiani, O. 1973. Note biologiche sulla *Deraeocoris flavilinea* Costa (Rhynchota-Heteroptera). Entomologica (Bari) 9:137–145.

Triplehorn, B. W., L. R. Nault, and D. J. Horn. 1984. Feeding behavior of *Graminella nigrifrons* (Forbes). Ann. Entomol. Soc. Am. 77:102–107.

Trojan, P. 1968. Egg reduction of the Colorado beetle (*Leptinotarsa decemlineata* Say) as a hunger-dependent reaction. Ekol. Pol. (A) 16:171–183.

———. 1989. Bug (Heteroptera) associations in the agricultural landscape of Great Poland. Ekol. Pol. 37:135–155.

Trouvelot, B. 1926a. L'origine des déformations pierreuses des poires en France et recherches sur les moyens de prévenir le dégat dans les vergers. C.R. Acad. Agric. Fr. 12:1024–1029.

———. 1926b. Sur la biologie de *Calocoris fulvomaculatus* De Geer [Hem. Capsidae]. Bull. Soc. Entomol. Fr. 1926:233–235.

Trumbule, R. B., R. F. Denno, and M. J. Raupp. 1995. Management considerations for the azalea lace bug in landscape habitats. J. Arboric. 21:63–68.

Tsai, J. H. 1979. Vector transmission of mycoplasmal agents of plant diseases. Pages 265–307 *in* R. F. Whitcomb and J. G. Tully, eds., The mycoplasmas. Vol. III. Plant and insect mycoplasmas. Academic, New York.

Tsybulskaya, G. N. and T. V. Kryzhanovskaya. 1980. A promising insect control agent. Zashch. Rast. (Mosc.) No. 10:23. [in Russian.]

Tucker, E. S. 1911. Random notes on entomological field work. Can. Entomol. 43:22–33.

Tugwell, N. P. Jr. 1983. Methods: Evaluating cotton for resistance to plant bugs. Pages 46–53 *in* Host plant resis-tance research methods for insects, diseases, nematodes and spider mites in cotton. South. Coop. Ser. Bull. 280.

Tugwell, [N.] P. [Jr.], E. P. Rouse, and R. G. Thompson. 1973. Insects in soybeans and a weed host (*Desmodium* sp.). Ark. Agric. Exp. Stn. Rep. Ser. 214:3–18.

Tugwell, [N.] P. [Jr.], S. C. Young Jr., B. A. Dumas, and J. R. Phillips. 1976. Plant bugs in cotton: Importance of infestation time, types of cotton injury, and significance of wild hosts near cotton. Ark. Agric. Exp. Stn. Rep. Ser. 227:1–24.

Tullgren, A. 1919. Axsugaren *Miris dolobratus* L. (*Leptopterna dolobrata* L.) ett hittills föga beaktat Skadedjur på Sädesslagen och Gräsen. Centranst. Forsoksv. Jordbruksom. Medd. 182. Entomol. Avd. 33:1–19.

Tunstall, A. C. 1928. Vegetable parasites of the tea plant (continued). Blights on the stem (continued). Q. J. Indian Tea Assoc. 4:220–231.

Turka, I. 1985. Composition and concentration of free amino acids in the leaves of potato varieties attractive to bugs that transmit plant viruses. Tr. Latv. S-Kh. Akad. 22:3–8. [in Russian.]

Turnbull, A. L. 1965. Effects of prey abundance on the development of the spider *Agelenopsis potteri* (Blackwall) (Araneae: Agelenidae). Can. Entomol. 97:141–147.

Turner, B. D. 1984. Predation pressure on the arboreal epiphytic herbivores of larch trees in southern England. Ecol. Entomol. 9:91–100.

Turner, V. 1984. *Banksia* pollen as a source of protein in the diet of two Australian marsupials *Cercartetus nanus* and *Tarsipes rostratus*. Oikos 43:53–61.

Turnock, W. J., G. H. Gerber, B. H. Timlick, and R. J. Lamb. 1995. Losses of canola seeds from feeding by *Lygus* species [Heteroptera: Miridae] in Manitoba. Can. J. Plant Sci. 75:731–736.

Twinn, C. R. 1939. A summary of the insect pest situation in Canada in 1938. Can. Insect Pest Rev. 17:1–21.

Tyler, F. J. 1908. Miscellaneous papers, V. The nectaries of cotton. U.S. Dep. Agric. Bur. Plant Ind. Bull. 131:45–54.

Tyndall, D. R. 1958. Evaluation of insect predators in alfalfa. M.S. thesis, University of Wyoming, Laramie. 171 pp.

Udayagiri, S. and S. C. Welter. 2000. Escape of *Lygus hesperus* (Heteroptera: Miridae) eggs from parasitism by *Anaphes iole* (Hymenoptera: Mymaridae) in strawberries: Plant structure effects. Biol. Control 17:234–242.

Ueshima, N. 1979. Animal cytogenetics. Vol. 3: Insecta 6, Hemiptera II: Het-eroptera. Gebrüder Borntraeger, Berlin. 117 pp.

Uhler, P. R. 1872. Notices of the Hemiptera of the western territories of the United States, chiefly from the surveys of Dr. F. V. Hayden, Pages 392–423 *in* F. V. Hayden, Preliminary report of the United States Geological Survey of Montana and portions of adjacent territories. Vol. 5. Washington, D.C.

———. 1876. List of Hemiptera of the region west of the Mississippi River, including those collected during the Hayden explorations of 1873. Bull. U.S. Geol. Geogr. Surv. Terr. 1:267–361.

———. 1877. Report upon the insects collected by P. R. Uhler during the explorations of 1875, including monographs of the families Cydnidae and Saldae, and the Hemiptera collected by A. S. Packard Jr., M.D. Bull. U.S. Geol. Geogr. Surv. Terr. 3:355–475.

———. 1878. Notices of the Hemiptera Heteroptera in the collection of the late T. W. Harris, M.D. Proc. Boston Soc. Nat. Hist. 19:365–446.

———. 1884. Order VI.—Hemiptera. Sub-Order III.—Heteroptera. Pages 249–293 *in* J. S. Kingsley, ed., The standard natural history. Vol. II. Crustacea and insects. Cassino, Boston.

———. 1887. Observations on some Capsidae with descriptions of a few new species (No. 2). Entomol. Am. 3:29–35.

———. 1891. Observations on some remarkable forms of Capsidae. Proc. Entomol. Soc. Wash. 2:119–123.

———. 1893. Summary of the collection of Hemiptera secured by Mr. E. A. Schwarz in Utah. Proc. Entomol. Soc. Wash. 2:366–385.

———. 1901. [Exhibition of specimens and short notes.] Proc. Entomol. Soc. Wash. 4:406.

Ulenberg, S. A., L. J. W. de Goffau, A. van Frankenhuyzen, and H. C. Burger. 1986. Bijzondere aantastingen door insekten in 1985. Entomol. Ber. 46:163–171.

Ullah, G. 1940. Food plants of *Creontiades pallidifer* Walker. Indian J. Entomol. 2:242.

Ullman, D. E. and D. L. McLean. 1988. The probing behavior of the summer-form pear psylla. Entomol. Exp. Appl. 47:115–125.

Undeen, A. H. and J. Vávra. 1997. Research methods for entomopathogenic protozoa. Pages 117–151 *in* L. A. Lacey, ed., Manual of techniques in insect pathology. Academic, San Diego.

University of California. 1984. Integrated pest management for cotton in the western region of the United States.

Univ. Calif. Div. Agric. Nat. Resour. Publ. 3305:1–144.

——. 1990. Integrated pest management for tomatoes, 3rd ed. Univ. Calif. Div. Agric. Nat. Resour. Publ. 3274:1–104.

Upholt, W. M. 1941. Chemical control of *Pseudococcus comstocki* Kuw. J. Econ. Entomol. 34:859–860.

U.S. Department of Agriculture. 1968. Insects not known to occur in the United States. Orange blossom bug (*Dionconotus cruentatus* (Brullé)). U.S. Dep. Agric. Coop. Econ. Insect Rep. 18(18):377–378.

U.S. Department of Interior Fish and Wildlife Service. 1992. Endangered and threatened wildlife and plants; determination of endangered status for the Karner blue butterfly. Fed. Reg. 57(240):59236–59244.

Usher, G. 1962. Fungi associated with capsid lesions. Pages 331–332 *in* J. B. Wills, ed., Agriculture and land use in Ghana. Oxford University Press, London.

Usinger, R. L. 1934. Blood sucking among phytophagous Hemiptera. Can. Entomol. 66:97–100.

——. 1939. Distribution and host relationships of *Cyrtorhinus* (Hemiptera: Miridae). Proc. Hawaii. Entomol. Soc. 10:271–273.

——. 1945. Biology and control of ash plant bugs in California. J. Econ. Entomol. 38:585–591.

——. 1946. Hemiptera Heteroptera of Guam. Pages 11–103 *in* Insects of Guam—II. Bernice P. Bishop Mus. Bull. 189.

——. 1955. Hemiptera. Pages 534–536 *in* E. L. Kessel, ed., A century of progress in the natural sciences, 1853–1953. California Academy of Sciences, San Francisco.

——. 1960. [Review of] *Land and Water bugs of the British Isles* by T. R. E. Southwood and D. Leston. Ann. Entomol. Soc. Am. 53:442.

——. 1966. Monograph of Cimicidae (Hemiptera-Heteroptera). Thomas Say Foundation Vol. VII. Entomological Society of America, College Park, Md. 585 pp.

Uyemoto, J. K., J. M. Ogawa, R. E. Rice, H. R. Teranishi, R. M. Bostock, and W. M. Pemberton. 1986. Role of several true bugs (Hemiptera) on incidence and seasonal development of pistachio fruit epicarp lesion disorder. J. Econ. Entomol. 79:395–399.

Valentine, E. W. 1967. A list of the hosts of entomophagous insects of New Zealand. N.Z. J. Sci. 10:1100–1210.

Vallentine, J. F. 1989. Range development and improvements, 3rd ed. Academic, San Diego. 524 pp.

van Dam, N. M. and J. D. Hare. 1998. Differences in distribution and performance of two sap-sucking herbivores on glandular and non-glandular *Datura wrightii*. Ecol. Entomol. 23:22–32.

van de Baan, H. E. and B. A. Croft. 1990. Factors influencing insecticide resistance in *Psylla pyricola* (Homoptera: Psyllidae) and susceptibility to the predator *Deraeocoris brevis* (Heteroptera: Miridae). Environ. Entomol. 19:1223–1228.

van den Berg, H., B. M. Shepard, J. A. Litsinger, and P. C. Pantua. 1988. Impact of predators and parasitoids on the eggs of *Rivula atimeta*, *Naranga aenescens* (Lepidoptera: Noctuidae) and *Hydrellia philippina* (Diptera: Ephydridae) in rice. J. Plant Prot. Trop. 5:103–108.

van den Berg, H., J. A. Litsinger, B. M. Shepard, and P. C. Pantua. 1992. Acceptance of eggs of *Rivula atimeta*, *Naranga aenescens* [Lep.: Noctuidae] and *Hydrellia philippina* [Dipt.: Ephydridae] by insect predators on rice. Entomophaga 37:21–28.

Van den Berg, M. A. 1982. Hemiptera attacking *Acacia dealbata* Link., *Acacia decurrens* Willd., *Acacia longifolia* (Andr.) Willd., *Acacia mearnsii* de Wild. and *Acacia melanoxylon* R. Br. in Australia. Phytophylactica 14:47–50.

Van den Berg, M. A., V. E. Deacon, C. J. Fourie, and S. H. Anderson. 1987. Predators of the citrus psylla, *Trioza erytreae* (Hemiptera: Triozidae), in the lowland and Rustenburg areas of Transvaal. Phytophylactica 19:285–289.

van den Bosch, R. 1971a. Experimental field studies on upsets and resurgences of pest populations associated with agricultural chemicals. Pages 104–107 *in* J. E. Swift, ed., Agricultural chemicals—harmony or discord for food, people and the environment. University of California, Berkeley.

——. 1971b. The melancholy addiction of Ol' King Cotton. Nat. Hist. 80(10):86–91.

——. 1978. The pesticide conspiracy. Doubleday, Garden City, N.Y. 226 pp.

van den Bosch, R. and K. S. Hagen. 1966. Predaceous and parasitic arthropods in California cotton fields. Calif. Agric. Exp. Stn. Bull. 820:1–32.

van den Bosch, R. and P. S. Messenger. 1973. Biological control. Intext Educational, New York. 180 pp.

van den Bosch, R. and V. M. Stern. 1969. The effect of harvesting practices on insect populations in alfalfa. Proc. Tall Timbers Conf. Ecol. Anim. Control Habitat Manage. 1:47–54.

van den Bosch, R., T. F. Leigh, L. A.

Falcon, V. M. Stern, D. Gonzales, and K. S. Hagen. 1971. The developing program of integrated control of cotton pests in California. Pages 377–394 *in* C. B. Huffaker, ed., Biological control. Plenum, New York.

van der Goot, P. see Goot, P. van der.

van der Meijden, E. see Meijden, E. van der.

Vanderzant, E. S. 1967. Rearing lygus bugs on artificial diets. J. Econ. Entomol. 60:813–816.

van der Zwet, T. see Zwet, T. van der.

van Doesburg, P. H. Jr. see Doesburg, P. H. van Jr.

Van Driesche, R. G. 1983. Meaning of "percent parasitism" in studies of insect parasitoids. Environ. Entomol. 12:1611–1622.

Van Driesche, R. G. and T. S. Bellows Jr. 1993. Introduction. Pages 1–3 *in* R. G. Van Driesche and T. S. Bellows [Jr.], eds., Steps in classical arthropod biological control. Thomas Say Publ. Entomol.: Proceedings. Entomological Society of America, Lanham, Md.

Van Duzee, E. P. 1887. Partial list of Capsidae taken at Buffalo, N.Y. Can. Entomol. 19:69–73.

——. 1910. Descriptions of some new or unfamiliar North American Hemiptera. Trans. Am. Entomol. Soc. 36:73–88.

——. 1912. Hemipterological gleanings. Bull. Buffalo Soc. Nat. Sci. 10:477–512.

——. 1923. Expedition of the California Academy of Sciences to the Gulf of California in 1921. Proc. Calif. Acad. Sci. 12(4):123–200.

Van Dyke, E. C. 1919. A few observations on the tendency of insects to collect on ridges and mountain snowfields. Entomol. News 30:241–244.

Vane-Wright, R. I. 1976. A unified classification of mimetic resemblances. Biol. J. Linn. Soc. 8:25–56.

Van Halteren, I. R. P. 1969. *Cyrtopeltis tenuis*, a pest of tomato. Ghana Farmer 13:78–79.

van Lenteren, J. C. see Lenteren, J. C. van.

van Turnhout, H. M. T. and P. A. van der Laan. 1958. Control of *Lygus campestris* on carrot seed crops in North Holland. Tijdschr. Plziekt. 64:301–306.

Van Velsen, R. J. 1967. "Little leaf," a virus disease of *Ipomoea batatas* in Papua and New Guinea. Papua New Guinea Agric. J. 18:126–128.

van Vreden, G. and A. L. Ahmadzabidi. 1986. Pests of rice and their natural enemies in Peninsular Malaysia. Centre for Agricultural Publishing & Documentation, Wageningen. 230 pp.

Vanwetswinkel, G. and E. Paternotte.

1968. Biologie, schade en bestrijding van de fruitwants *Lygus pabulinus* L. Meded. Rijksfac. Landbouwwet. Gent 33:869–874.

Van Zwaluwenburg [R. H.] and [J. S.] Rosa. 1951. Notes and exhibitions. *Cyrtorhinus mundulus* (Breddin). Proc. Hawaii. Entomol. Soc. 14:218.

Vappula, N. A. 1965. Pests of cultivated plants in Finland. Acta Entomol. Fenn. 19:1–239.

Varis, A.-L. 1972. The biology of *Lygus rugulipennis* Popp. (Het., Miridae) and the damage caused by this species to sugar beet. Ann. Agric. Fenn. 11:1–56.

——. 1974. Distribution of damage caused by *Lygus rugulipennis* Popp. (Het., Miridae) in cultivated fields. Ann. Agric. Fenn. 13:18–22.

——. 1978. *Lygus rugulipennis* (Heteroptera, Miridae) damaging greenhouse cucumbers. Ann. Entomol. Fenn. 44:72.

——. 1991. Effect of *Lygus* (Heteroptera: Miridae) feeding on wheat grains. J. Econ. Entomol. 84:1037–1040.

——. 1995. Species composition, abundance, and forecasting of *Lygus* bugs (Heteroptera: Miridae) on field crops in Finland. J. Econ. Entomol. 88:855–858.

Varis, A.-L. and T. Bilewicz-Pawińska. 1992. Influence of temperature on survival and emergence of diapausing *Peristenus stenodemae* (Hymenoptera, Braconidae). Entomol. Fenn. 3:117–119.

Varis, A.-L. and J. Rautapää. 1976. Chemical control of sugar-beet pests in Finland: Efficiency and economic return. Ann. Agric. Fenn. 15:137–144.

Varis, A.-L., S. Laurema, and H. Miettinen. 1983. Variation of enzyme activities in the salivary glands of *Lygus rugulipennis* (Hemiptera, Miridae). Ann. Entomol. Fenn. 49:1–10.

Varley, G. C. and G. R. Gradwell. 1960. Key factors in population studies. J. Anim. Ecol. 29:399–401.

Varma, R. V. and M. Balasundaran. 1990. Tea mosquito (*Helopeltis antonii*) feeding as a predisposing factor for entry of wound pathogens in cashew. Entomon 15:249–251.

Varty, I. W. 1963. A survey of the sucking insects of the birches in the Maritime Provinces. Can. Entomol. 95:1097–1106.

Vásárhelyi, T. 1990. Poloska lárvák családhatározója (Heteroptera). Folia Entomol. Hung. 51:149–161.

Venables, E. P. and D. B. Waddell. 1943. The influence of leguminous plants on the abundance of tarnished plant bug. Can. Entomol. 75:78.

Verbeke, C. and Y. Verschueren. 1984. La pollinisation de l'épipactis des marais (*Epipactis palustris*) du Bakkersdam (Pays-Bas). Nat. Belg. 65:175–192.

Verdcourt, B. 1949. Miscellaneous records of Arachnida and Insecta from N.W. Hertfordshire. Entomol. Mon. Mag. 85:249–253.

Vergara, C. and K. Raven. 1988. Mirids (Hemiptera) registrados en el Museo de Entomología de la Universidad Nacional Agraria La Molina. Rev. Peru. Entomol. 31:51–56.

Verhoeff, C. 1891. *Capsus capillaris* F. ein Aphiden-Feind. Entomol. Nachr. 17:26–27.

Verkaar, H. J. 1988. Are defoliators beneficial for their host plants in terrestrial ecosystems—a review? Acta Bot. Neerl. 37:137–152.

Verma, J. P. 1986. Bacterial blight of cotton. CRC, Boca Raton, Fla. 278 pp.

Verma, J. S. 1955a. Biological studies to explain the failure of *Cyrtorhinus mundulus* (Breddin) as an egg-predator of *Peregrinus maidis* (Ashmead) in Hawaii. Proc. Hawaii. Entomol. Soc. 15:623–634.

——. 1955b. Comparative toxicity of some insecticides to *Peregrinus maidis* (Ashm.) and its egg-predator. J. Econ. Entomol. 48:205–206.

Vernigor, S. F. 1928. On the biology of the beet bug. Zakhist Rosl. (Kharkov) 3–4:97–105. [in Russian.]

Vickerman, G. P. and M. O'Bryan. 1979. Partridges and insects. Game Conservancy Annu. Rev. 1978:35–43.

Victorian Plant Research Institute. 1971. Plant-feeding bugs in Victoria: A guide to identification and control. J. Agric. (Vic.) 69:160–165.

Viggiani, G. 1971a. A new species of *Chaetostricha* Walk. from Africa (XXVI—Researches on the Hymenoptera Chalcidoidea). J. Entomol. Soc. S. Afr. 34:33–35.

——. 1971b. Osservazioni biologiche sul Miride predatore *Deraeocoris ruber* (L.) (Rhynchota, Heteroptera). Boll. Lab. Entomol. Agrar. "Filippo Silvestri" 29:270–286.

——. 1983. Natural enemies of the filbert aphids in Italy. Pages 109–113 *in* R. Cavalloro, ed., Aphid antagonists. Balkema, Rotterdam.

——. 1989. *Ufensia minuta* sp. n. (Hymenoptera: Trichogrammatidae), ooparassitoide di *Reuteria marqueti* Puton (Hemiptera: Miridae), con note sulle specie paleartiche del genere *Ufensia* Girault. Boll. Lab. Entomol. Agrar. "Filippo Silvestri" 45:15–21.

Vilkova, N. A. 1976. Factors determining host-plant selection behaviour of insects. Acta Phytopathol. Acad. Sci. Hung. 11:99–103.

Villacorta, A. 1977a. Algunas observa-ciones sobre la biologia de *Monalonion annulipes* Sig. en Costa Rica. An. Soc. Entomol. Bras. 6:173–179.

——. 1977b. Fluctuación annual de las poblaciones de *Monalonion annulipes* Sig. y su relación con la 'muerte descendente de *Theobroma cacao*' en Costa Rica. An. Soc. Entomol. Bras. 6:215–223.

Vince, S. W., I. Valiela, and J. M. Teal. 1981. An experimental study of the structure of herbivorous insect communities in a salt marsh. Ecology 62:1662–1678.

Vincent, C. and J. Hanley. 1997. Measure of agreement between experts on apple damage assessment. Phytoprotection 78:11–16.

Vincent, C. and P. Lachance. 1993. Evaluation of a tractor-propelled vacuum device for management of tarnished plant bug (Heteroptera: Miridae) populations in strawberry plantations. Environ. Entomol. 22: 1103–1107.

Vincent, C., D. de Oliveira, and A. Bélanger. 1990. The management of insect pollinators and pests in Quebec strawberry plantations. Pages 177–192 *in* N. J. Bostanian, L. T. Wilson, and T. J. Dennehy, eds., Monitoring and integrated management of arthropod pests of small fruit crops. Intercept, Andover, UK.

Vinokurov, N. N. 1988. Heteroptera of Yakutia. (Transl. from Russian by B. R. Sharma.) Amerind, New Delhi, India. 328 pp.

Vinson, C. G. 1927. Some nitrogenous constituents of corn pollen. J. Agric. Res. (Wash., D.C.) 35:261–278.

Viswanath, B. N. and C. A. Viraktamath. 1969. A new bug pest (Hemiptera: Miridae) on *Cucumis melo* L. in Mysore State. Mysore J. Agric. Sci. 3:475–476.

Voegtlin, D. J. and D. L. Dahlsten. 1982. Observations on the biology of *Cinara ponderosa* (Williams) (Homoptera: Aphididae) in the westside forests of the Sierra Nevada. Hilgardia 50(5):1–19.

Voelcker, O. J. and J. West. 1940. Cacao die-back. Trop. Agric. 17:27–31.

Vojdani, S. and A. Doftari. 1963. The alfalfa weevil, *Hypera postica* Gyll., a destructive beetle in Karaj. Tehran Univ. Publ. Dep. Plant Prot. Phytopharm., Karaj, Iran. 34 pp. [in Persian, English summary.]

Vollema, J. S. 1951. De landbouwkundige zijde der blister blight bestrijding. Bergcultures (Batavia) 20:235–236.

Vollrath, F. 1984. Kleptobiotic interactions in invertebrates. Pages 61–94 *in* C. J. Barnard, ed., Producers and scroungers: Strategies of exploitation and parasitism. Croom Helm, London; Chapman & Hall, New York.

von Dohlen, C. D. and N. A. Moran. 1995. Molecular phylogeny of the Homoptera: A paraphyletic taxon. J. Mol. Evol. 41:211–223.

Voorhees, F. R. 1969. Seven-legged mirid (Hemiptera: Miridae). Ann. Entomol. Soc. Am. 62:1492–1493.

Vosler, E. J. 1913. A new fruit and truck crop pest (*Irbisia brachycerus* Uhler). Mon. Bull. Calif. State Comm. Hortic. 2:551–553.

Vrijdagh, J. M. 1936. Contribution à l'étude de la maladie des chancres des tiges du cotonnier causée par *Helopeltis bergrothi* Reut. Bull. Agric. Congo Belge 27:3–37.

Waage, J. [K.] 1990a. Ecological theory and the selection of biological control agents. Pages 135–157 *in* M. Mackauer, L. E. Ehler, and J. Roland, eds., Critical issues in biological control. Intercept, Andover, UK.

——. [1990b]. Exploration for biological agents of leucaena psyllid in tropical America. Pages 144–152 *in* B. Napompeth and K. G. MacDicken, eds., Leucaena psyllid: Problems and management; proceedings of an international workshop held in Bogor, Indonesia, January 16–21, 1989. [Winrock International Institute for Agricultural Development, Arlington, Va.; International Development Research Centre, Ottawa, Canada; Nitrogen Fixing Tree Association, Waimanalo, Hawaii.]

Waage, J. K. and D. J. Greathead. 1988. Biological control: Challenges and opportunities. Philos. Trans. R. Soc. Lond. (B) 318:111–128.

Waage, J. K. and N. J. Mills. 1992. Biological control. Pages 412–430 *in* M. J. Crawley, ed., Natural enemies: The population biology of predators, parasites and diseases. Blackwell, Oxford.

Wachmann, E. 1989. Wanzen: Beobachten—Kennenlernen. Neumann-Neudamm, Melsungen. 274 pp.

Waddill, V. and M. Shepard. 1974. Biology of a predaceous stinkbug, *Stiretrus anchorago* (Hemiptera: Pentatomidae). Fla. Entomol. 57:249–253.

Wagn, O. 1954. Bladtaeger (Miridae) og forekomst af frø uden kim hos skaermblomstrede (Umbelliferae). Tidsskr. Planteavl 58:58–90.

Wagner, E. 1955. Bemerkungen zum System der Miridae (Hem. Het.). Dtsch. Entomol. Z. 2:230–242.

——. 1958. Ein Männchen von *Campyloneura virgula* H.S. 1839 (Heteropt. Miridae). Nachrbl. Bayer. Entomol. 7:81–83.

——. 1967. Über *Megacoelum* Fieber, 1858 (Hem. Het. Miridae). Mitt. Dtsch. Entomol. Ges. 26:61–65.

——. 1968. Über *Campyloneura virgula* Herrich-Schäffer (Hem. Het. Miridae). Mitt. Dtsch. Entomol. Ges. 27:46–47.

——. 1970. Die Miridae Hahn, 1831, des Mitelmeerraumes und der Makaronesischen Inseln (Hemiptera, Heteroptera). Teil. 1 Entomol. Abh. (Dresd.) 37 Suppl.:1–484.

——. 1973. Die Miridae Hahn, 1831, des Mittelmeeraumes und der Makaronesischen Inseln (Hemiptera, Heteroptera). Entomol. Abh. (Dresd.) 39(2):1–423.

Wagner, E. and J. A. Slater. 1952. Concerning some Holarctic Miridae. Proc. Entomol. Soc. Wash. 54:273–281.

Wagner, F. 1960. Über Untersuchungen zur Ursache und Bekämpfung der totalen Weissährigkeit an Gräsern. Prakt. Blätt. Pflanzenbau Pflanzenschutz 55:137–147.

Wagner, F. and P. Ehrhardt. 1961. Untersuchungen am Stichkanal der Graswanze *Miris dolobratus* L., der Urheberin der totalen Weissährigkeit des Rotschwingels (*Festuca rubra*). Z. Pflanzenkr. Pflanzenschutz 68:615–620.

Wagner, T. L., R. L. Olson, J. L. Willers, and M. R. Williams. 1996. Modeling and computerized decision aids. Pages 205–249 *in* E. G. King, J. R. Phillips, and R. J. Coleman, eds., Cotton insects and mites: Characterization and management. Cotton Foundation, Memphis.

Wakeland, C. 1931. Tarnished plant bug (*Lygus pratensis* L.). U.S. Dep. Agric. Insect Pest Surv. Bull. 11:369.

Walker, G. P., L. R. Nault, and D. E. Simonet. 1984. Natural mortality factors acting on potato aphid (*Macrosiphum euphorbiae*) populations in processing-tomato fields in Ohio. Environ. Entomol. 13:724–732.

Walker, J. K. and G. A. Niles. 1984. Primordial square formation in cotton and the cotton fleahopper. Southwest. Entomol. 9:104–108.

Walker, J. K., G. A. Niles, J. R. Gannaway, J. V. Robinson, C. B. Cowan, and M. J. Lukefahr. 1974. Cotton fleahopper damage to cotton genotypes. J. Econ. Entomol. 67:537–542.

Walker, J. T. S., C. H. Wearing, and A. J. Hayes. 1989. *Panonychus ulmi* (Koch), European red mite (Acari: Tetranychidae). Pages 217–221 *in* P. J. Cameron, R. L. Hill, J. Bain, and W. P. Thomas, eds., A review of biological control of invertebrate pests and weeds in New Zealand

1874 to 1987. Tech. Commun. No. 10. CAB International Institute of Biological Control, Wallingford, Oxon, UK.

Wallace, F. G. 1979. Biology of the Kinetoplastida of arthropods. Pages 213–240 *in* W. H. R. Lumsden and D. A. Evans, eds., Biology of the Kinetoplastida. Vol. 2. Academic, London.

Wallace, H. R. 1973. Nematode ecology and plant disease. Arnold, London. 228 pp.

Waloff, N. 1965. The coexistence of three species of *Orthotylus* (Heteroptera, Miridae). *In* Proc. 12th Int. Congr. Entomol., London (1964), p. 426.

——. 1966. Scotch broom (*Sarothamnus scoparius* (L.) Wimmer) and its insect fauna introduced into the Pacific Northwest of America. J. Appl. Ecol. 3:293–311.

——. 1967. Biology of three species of *Leiophron* (Hymenoptera: Braconidae, Euphorinae) parasitic on Miridae on broom. Trans. R. Entomol. Soc. Lond. 118:187–213.

——. 1968. A comparison of factors affecting different insect species on the same host plant. Pages 76–87 *in* T. R. E. Southwood, ed., Insect abundance. Blackwell, Oxford.

——. 1983. Absence of wing polymorphism in the arboreal, phytophagous species of some taxa of temperate Hemiptera: An hypothesis. Ecol. Entomol. 8:229–232.

Waloff, N. and K. Bakker. 1963. The flight activity of Miridae (Heteroptera) living on broom, *Sarothamnus scoparius* (L.) Wimm. J. Anim. Ecol. 32:461–480.

Waloff, N. and O. W. Richards. 1977. The effect of insect fauna on growth mortality and natality of broom, *Sarothamnus scoparius*. J. Appl. Ecol. 14:787–798.

Waloff, N. and T. R. E. Southwood. 1960. The immature stages of mirids (Heteroptera) occurring on broom (*Sarothamnus scoparius* (L.) Wimmer) with some remarks on their biology. Proc. R. Entomol. Soc. Lond. (A) 35:39–46.

Walsh, B. D. 1860. Entomological notes. Prairie Farmer (Chic.) 21:308.

Walsh, B. D. and C. V. Riley. 1869. Injured strawberry and grapevines. Am. Entomol. (St. Louis) 1:227.

Walstrom, R. J. 1983. Plant bug (Heteroptera: Miridae) damage to first-crop alfalfa in South Dakota. J. Econ. Entomol. 76:1309–1311.

Walter, D. E. 1987. Trophic behavior of "mycophagous" microarthropods. Ecology 68:226–229.

Walton, C. L. and L. N. Staniland. 1929. The common green capsid bug (*Lygus pabulinus*) as a pest of sugar beet. Agric. Hortic. Res. Stn. Long Ashton, Bristol Rep. 1929:99–100.

Wang, C. L. 1995. Predatory capacity of *Campylomma chinensis* Schuh (Hemiptera: Miridae) and *Orius sauteri* (Poppius) (Hemiptera: Anthocoridae) on *Thrips palmi*. Pages 259–262 *in* B. L. Parker, M. Skinner, and T. Lewis, eds., Thrips biology and management. Plenum, New York.

Wang, Q. and J. G. Millar. 1997. Reproductive behavior of *Thyanta pallidovirens* (Heteroptera: Pentatomidae). Ann. Entomol. Soc. Am. 90:380–388.

Ward, L. K. and D. F. Spalding. 1993. Phytophagous British insects and mites and their food-plant families: Total numbers and polyphagy. Biol. J. Linn. Soc. 49:257–276.

Warren, S. D., C. J. Scifres, and P. D. Teel. 1987. Response of grassland arthropods to burning: A review. Agric. Ecosyst. Environ. 19:105–130.

Waser, N. M., L. Chittka, M. V. Price, N. M. Williams, and J. Ollerton. 1996. Generalization in pollination systems, and why it matters. Ecology 77:1043–1060.

Waterhouse, D. F. and K. R. Norris. 1987. Biological control: Pacific prospects. Inkata, Melbourne. 454 pp.

——. 1989. Biological control: Pacific prospects—supplement 1. Australian Centre for International Agricultural Research, Canberra. 123 pp.

Waterhouse, P. M., F. E. Gildow, and G. R. Johnstone. 1988. Luteovirus group. AAB Descrip. Plant Viruses No. 339. 9 pp. Association of Applied Biologists, Wellesbourne, UK.

Waters, H. A. 1943. Rearing insects that attack plants. Pages 3–28 *in* F. L. Campbell and F. R. Moulton, eds., Laboratory procedures in studies of the chemical control of insects. Am. Assoc. Adv. Sci. Publ. 20. Washington, D.C.

Watmough, R. H. 1968. Population studies on two species of Psyllidae (Homoptera, Sternorhyncha) [*sic*] on broom (*Sarothamnus scoparius* (L.) Wimmer). J. Anim. Ecol. 37:283–314.

Watson, A. P., J. N. Matthiessen, and B. P. Springett. 1982. Arthropod associates and macronutrient status of the red-ink sundew (*Drosera erythrorhiza* Lindl.). Aust. J. Ecol. 7:13–22.

Watson, J. S. 1989. Recent progress in breeding for insect resistance in cotton. Pages 44–52 *in* M. B. Green and D. J. de B. Lyon, eds., Pest management in cotton. Ellis Horwood, Chichester, UK.

Watson, M. A. 1973. Plant viruses. Pages 26–37 *in* A. J. Gibbs, ed., Viruses and invertebrates. North-Holland, Amsterdam; Elsevier, New York.

Watson, R. N. and R. J. Townsend. 1981. Invertebrate pests on asparagus in Waikato. *In* Proc. 34th N.Z. Weed Pest Control Conf., pp. 70–75. N.Z. Weed and Pest Control Society, Palmerston North.

Watson, S. A. 1928. The Miridae of Ohio. Ohio Biol. Surv. Bull. 16:1–44.

Watt, G. and H. H. Mann. 1903. The pests and blights of the tea plant, 2nd ed. Office of the Superintendent, Government Printing, Calcutta. 429 pp.

Watts, J. G., E. W. Huddleston, and J. C. Owens. 1982. Rangeland entomology. Annu. Rev. Entomol. 27:283–311.

Watts, J. G., G. B. Hewitt, E. W. Huddleston, H. G. Kinzer, R. J. Lavigne, and D. N. Ueckert. 1989. Rangeland entomology, 2nd ed. Society for Range Management, Denver. 388 pp.

Way, D. W. 1968. Strawberry fruit malformation I. Pomological aspects. East Malling Res. Stn. Rep. 1967:199–203.

Way, M. J. 1963. Mutualism between ants and honeydew-producing Homoptera. Annu. Rev. Entomol. 8:307–344.

Way, M. J. and K. C. Khoo. 1989. Relationships between *Helopeltis theobromae* damage and ants with special reference to Malaysian cocoa smallholdings. J. Plant Prot. Trop. 6:1–11.

——. 1991. Colony dispersion and nesting habits of the ants, *Dolichoderus thoracicus* and *Oecophylla smaragdina* (Hymenoptera: Formicidae), in relation to their success as biological control agents on cocoa. Bull. Entomol. Res. 81:341–350.

——. 1992. Role of ants in pest management. Annu. Rev. Entomol. 37:479–503.

Weatherston, J. and J. E. Percy. 1978. Venoms in Rhyncota [*sic*] (Hemiptera). Pages 489–509 *in* S. Bettini, ed., Arthropod venoms. Springer, Berlin.

Weaver, C. R. and C. J. Olson. 1952. Chrysanthemum crop production speeded up 21 to 30 days by controlling lygus bug. Ohio Farm Home Res. 27(275):26–27.

Weaver, N. 1978. Chemical control of behavior—interspecific. Pages 391–418 *in* M. Rockstein, ed., Biochemistry of insects. Academic, New York.

Webb, F. E. 1953. An ecological study of the larch casebearer, *Coleophora laricella* Hbn. (Lepidoptera: Coleophoridae). Ph.D. dissertation, University of Wisconsin, Madison. 210 pp.

Webb, G. A. 1989. Insects as potential pollinators of *Micromyrtus ciliata* (Sm.) Druce, Myrtaceae. Vic. Nat. (Blackburn) 106:148–151.

Weber, H. 1930. Biologie der Hemipteren: Eine Naturgeschichte der Schnabelkerfe. Springer, Berlin. 543 pp.

Webster, B. N. 1954. Notes on pathological matters. Tea Q. 25:17–19.

Webster, F. M. 1884. Insects affecting fall wheat. Pages 383–393 *in* Report of the Entomologist 1884. Government Printing Office, Washington, D.C.

——. 1886. Insects affecting fall wheat. Pages 311–319 *in* Report of the Entomologist 1885. Government Printing Office, Washington, D.C.

——. 1897a. *Halticus bractatus* Say. Entomol. News 8:209–210.

——. 1897b. Warning colors, protective mimicry and protective coloration. 27th Annu. Rep. Entomol. Soc. Ont. 1896:80–86.

Webster, F. M. and C. W. Mally. 1899. Insects of the year in Ohio. U.S. Dep. Agric. Div. Entomol. Bull. 20(n.s.):68–73.

Webster, R. L. 1917. The box elder aphid, *Chaitophorus negundinis* Thomas. Iowa Agric. Exp. Stn. Bull. 173:95–121.

Webster, R. L. and A. Spuler. 1931. Tarnished plant bug injury to pears in Washington. J. Econ. Entomol. 24:969–971.

Webster, R. L. and D. Stoner. 1914. The eggs and nymphal stages of the dusky leaf bug *Calocoris rapidus* Say. J. N.Y. Entomol. Soc. 22:229–234.

Wehrle, L. P. 1935. Notes on *Pycnoderes quadrimaculatus* Guerin (Hemiptera, Miridae) in the vicinity of Tucson, Arizona. Bull. Brooklyn Entomol. Soc. 30:27.

Weigel, C. A. and E. R. Sasscer. 1935. Insects injurious to ornamental greenhouse plants. U.S. Dep. Agric. Farmers' Bull. 1362:1–80.

Weimer, J. L. 1937. The possibility of insect transmission of alfalfa dwarf. Phytopathology 27:697–702.

Weires, R. W. 1983. The tarnished plant bug—how it effects [*sic*] your apples and your pocketbook. *In* Proc. 89th Annu. Meet. Mass. Fruit Growers' Assoc., pp. 52–54, 56–59. Massachusetts Fruit Growers' Association, North Amherst.

Weires, R. W. and G. L. Smith. 1979. Mite predators in eastern New York commercial apple orchards. J. N.Y. Entomol. Soc. 87:15–20.

Weires, R. W., J. R. Van Kirk, W. D. Gerling, and F. M. McNicholas. 1985. Economic losses from the tarnished plant bug on apple in eastern New York. J. Agric. Entomol. 2:256–263.

Weis, A. E. and M. R. Berenbaum. 1989. Herbivorous insects and green plants. Pages 123–162 *in* W. G. Abrahamson, ed., Plant-animal interactions. McGraw-Hill, New York.

Weiss, E. A. 1971. Castor, sesame and

safflower. Barnes & Noble, New York. 901 pp.

Weiss, H. B. 1916. Notes from New Jersey. *Halticus citri* Ashm. injuring phlox in New Jersey (Hemip.). Can. Entomol. 48:35–36.

——. 1917. Some unusual orchid insects (Hem., Lep., Dip., Col.). Entomol. News 28:24–29.

——. 1918. Some new insect enemies of greenhouse and ornamental plants in New Jersey. N.J. Agric. Exp. Stn. Circ. 100:1–19.

——. 1920. Miscellaneous nursery insects. N.J. Dep. Agric. Circ. 31:1–21.

——. 1921. A summary of the food habits of North American Hemiptera. Bull. Brooklyn Entomol. Soc. 16:116–118.

Weitschat, W. and W. Wichard. 1998. Atlas der Pflanzen und Tiere im Baltischen Bernstein. Dr. Friedrich Pfeil, Munich. 256 pp.

Welbourn, W. C. 1983. Potential use of trombidioid and erythraeoid mites as biological control agents of insect pests. Pages 103–140 *in* M. A. Hoy, G. L. Cunningham, and L. Knutson, eds., Biological control of pests by mites. Calif. Agric. Exp. Stn. Spec. Publ. 3304.

Welbourn, W. C. and O. P. Young. 1987. New genus and species of Erythraeinae (Acari: Erythraeidae) from Mississippi with a key to the genera of North American Erythraeidae. Ann. Entomol. Soc. Am. 80:230–242.

Welch, N. C., C. Pickel, and D. Walsh. 1990. Effects of vacuum devices on population of *Lygus hesperus*, various beneficial insects and fruit quality of strawberries. HortScience 25:1131. (Abstr.)

Wellhouse, W. H. 1922. The insect fauna of the genus *Crataegus*. Cornell Univ. Agric. Exp. Stn. Mem. 56:1039–1136.

Welter, S. C. 1989. Arthropod impact on plant gas exchange. Pages 135–150 *in* E. A. Bernays, ed., Insect-plant interactions. Vol. 1. CRC, Boca Raton, Fla.

Welty, C. 1995. Survey of predators associated with European red mite (*Panonychus ulmi*; Acari: Tetranychidae) in Ohio apple orchards. Gt. Lakes Entomol. 28:171–184.

Wene, G. P. and L. W. Sheets. 1962. Relationship of predatory and injurious insects in cotton fields in the Salt River Valley area of Arizona. J. Econ. Entomol. 55:395–398.

——. 1964. Lygus bug injury to presquaring cotton. Ariz. Agric. Exp. Stn. Tech. Bull. 166:1–25.

Werner-Solska, J. 1983. Przenoszenie wiroida wrzecionowatości bulw ziemniaka przez owady—świetla literatury. Biul. Inst. Ziemniaka 29:57–62.

Westcott, C. 1973. The gardener's bug book, 4th ed. Doubleday, Garden City, N.Y. 689 pp.

Wester, P. J. 1915. Another nursery pest. Philipp. Agric. Rev. 7:420–421.

Westigard, P. H. 1973. The biology of and effect of pesticides on *Deraeocoris brevis piceatus* (Heteroptera: Miridae). Can. Entomol. 105:1105–1111.

Westigard, P. H. and H. R. Moffitt. 1984. Natural control of the pear psylla (Homoptera: Psyllidae): Impact of mating disruption with the sex pheromone for control of the codling moth (Lepidoptera: Tortricidae). J. Econ. Entomol. 77:1520–1523.

Westigard, P. H., L. G. Gentner, and D. W. Berry. 1968. Present status of biological control of pear psylla in southern Oregon. J. Econ. Entomol. 61:740–743.

Westigard, P. H., P. M. Lombard, and D. W. Berry. 1979. Integrated pest management of insects and mites attacking pears in southern Oregon. Oreg. Agric. Exp. Stn. Bull. 634:1–41.

Westigard, P. H., L. J. Gut, and W. J. Liss. 1986. Selective control program for the pear pest complex in southern Oregon. J. Econ. Entomol. 79:250–257.

Westwood, J. O. 1840. An introduction to the modern classification of insects; founded on the natural habits and corresponding organisation of the different families. Vol. II. Longman, Orme, Brown, Green, & Longmans, London. 587 pp.

[Westwood, J. O.]. 1874. The tea bug. Gard. Chron. 1(n.s.) (Lond.) (14):475.

Westwood, J. O. 1877. [Exhibition of specimen of *Tenthecoris* from orchids.] Entomol. Mon. Mag. 14:71.

Wetter, C. 1971. Potato virus S. CMI/AAB Descrip. Plant Viruses No. 60. 3 pp. Commonwealth Mycological Institute, Surrey, UK; Association of Applied Biologists, Wellesbourne, UK.

——. 1972. Potato virus M. CMI/AAB Descrip. Plant Viruses No. 87. 4 pp. Commonwealth Mycological Institute, Surrey, UK; Association of Applied Biologists, Wellesbourne, UK.

Wetton, M. N. and C. W. D. Gibson. 1987. Grass flowers in the diet of *Megaloceraea* [*sic*] *recticornis* (Heteroptera: Miridae): Plant structural defences and interspecific competition reviewed. Ecol. Entomol. 12:451–457.

Whalon, M. E. and B. A. Croft. 1984. Apple IPM implementation in North America. Annu. Rev. Entomol. 29:435–470.

Whalon, M. E. and B. L. Parker. 1978. Immunological identification of tarnished plant bug predators. Ann. Entomol. Soc. Am. 71:453–456.

Wheatley, P. E. 1961. The insect pests of agriculture in the coast province of Kenya II. Cashew. East Afr. Agric. For. J. 26:178–181.

Wheeler, A. G. Jr. 1971. Studies on arthropod fauna of alfalfa: Insect feeding on *Hylemya* flies (Diptera: Anthomyiidae) killed by a phycomycosis. J. N.Y. Entomol. Soc. 79:225–227.

——. 1972. Studies on the arthropod fauna of alfalfa III. Infection of the alfalfa plant bug, *Adelphocoris lineolatus* (Hemiptera: Miridae) by the fungus *Entomophthora erupta*. Can. Entomol. 104:1763–1766.

——. 1974a. Phytophagous arthropod fauna of crownvetch in Pennsylvania. Can. Entomol. 106:897–908.

——. 1974b. Studies on the arthropod fauna of alfalfa VI. Plant bugs (Miridae). Can. Entomol. 106:1267–1275.

——. 1976a. Yucca plant bug, *Halticotoma valida*: Authorship, distribution, host plants, and notes on biology. Fla. Entomol. 59:71–75.

——. 1976b. Lygus bugs as facultative predators. Pages 28–35 *in* D. R. Scott and L. E. O'Keeffe, eds., Lygus bug: Host plant interactions. University of Idaho Press, Moscow.

——. 1977. Studies on the arthropod fauna of alfalfa VII. Predaceous insects. Can. Entomol. 109:423–427.

——. 1979. A comparison of the plant-bug fauna of the Ithaca, New York area in 1910–1919 with that in 1978. Iowa State J. Res. 54:29–35.

——. 1980a. First United States records of *Lygocoris knighti* (Hemiptera: Miridae). Entomol. News 91:25–26.

——. 1980b. Life history of *Plagiognathus albatus* (Hemiptera: Miridae), with a description of the fifth instar. Ann. Entomol. Soc. Am. 73:354–356.

——. 1980c. The mirid rectal organ: Purging the literature. Fla. Entomol. 63:481–485.

——. 1981a. The distribution and seasonal history of *Slaterocoris pallipes* (Knight) (Hemiptera: Miridae). Proc. Entomol. Soc. Wash. 83:520–523.

——. 1981b. Insect associates of spurges, mainly *Euphorbia maculata* L., in eastern United States. Proc. Entomol. Soc. Wash. 83:631–641.

——. 1981c. The tarnished plant bug: Cause of potato rot?—an episode in mid-nineteenth century entomology and plant pathology. J. Hist. Biol. 14:317–338.

——. 1982a. *Coccobaphes sanguinarius* and *Lygocoris vitticollis* (Hemiptera: Miridae): Seasonal history and description of fifth-instar, with notes on other mirids associated with maple. Proc. Entomol. Soc. Wash. 84:177–183.

——. 1982b. Ash plant bug, *Tropi-*

dosteptes amoenus. Reuter. Regul. Hortic. (Harrisburg) 8(1):21–22.

——. 1983a. The small milkweed bug, Lygaeus kalmii (Hemiptera: Lygaeidae): Milkweed specialist or opportunist? J. N.Y. Entomol. Soc. 91:57–62.

——. 1983b. Outbreaks of the apple red bug: Difficulties in identifying a new pest and emergence of a mirid specialist. J. Wash. Acad. Sci. 73:60–64.

——. 1985. Seasonal history, host plants, and nymphal descriptions of Orthocephalus coriaceus, a plant bug pest of herb garden composites (Hemiptera: Miridae). Proc. Entomol. Soc. Wash. 87:85–93.

——. 1989. Texocoris nigrellus: Distribution and hosts of an enigmatic plant bug (Heteroptera: Miridae: Orthotylinae). J. N.Y. Entomol. Soc. 97:167–172.

——. 1991a. Hesperophylum heidemanni, a rare plant bug: Notes and new records (Heteroptera: Miridae). Proc. Entomol. Soc. Wash. 93:636–640.

——. 1991b. Plant bugs of Quercus ilicifolia: Myriads of mirids (Heteroptera) in pitch pine–scrub oak barrens. J. N.Y. Entomol. Soc. 99:405–440.

——. 1993. A maple plant bug, Lygocoris vitticollis (Reuter). Regul. Hortic. (Harrisburg) 19(2):17–19.

——. 1995. Plant bugs (Heteroptera: Miridae) of Phlox subulata and other narrow-leaved phloxes in eastern United States. Proc. Entomol. Soc. Wash. 97:435–451.

——. 1999. Phoenicocoris claricornis and Pinophylus carneolus (Hemiptera: Miridae): Distribution and seasonality of two specialists on microsporangiate strobili of pines. J. N.Y. Entomol. Soc. 107:238–246.

——. 2000a. Plant bugs (Miridae) as plant pests. Pages 37–83 in C. W. Schaefer and A. R. Panizzi, eds., Heteroptera of economic importance. CRC, Boca Raton, Fla.

——. 2000b. Predacious plant bugs (Miridae). Pages 657–693 in C. W. Schaefer and A. R. Panizzi, eds., Heteroptera of economic importance. CRC, Boca Raton, Fla.

Wheeler, A. G. Jr. and T. J. Henry. 1973. Camptozygum aequale (Villers), a pine-feeding mirid new to North America (Hemiptera: Miridae). Proc. Entomol. Soc. Wash. 75:240–246.

——. 1976. Biology of the honeylocust plant bug, Diaphnocoris chlorionis, and other mirids associated with ornamental honeylocust. Ann. Entomol. Soc. Am. 69:1095–1104.

——. 1977. Miridae associated with Pennsylvania conifers 1. Species on arborvitae, false cypress, and juniper. Trans. Am. Entomol. Soc. 103:623–656.

——. 1978a. Isometopinae (Hemiptera: Miridae) in Pennsylvania: Biology and descriptions of fifth instars, with observations of predation on obscure scale. Ann. Entomol. Soc. Am. 71:607–614.

——. 1978b. Ceratocapsus modestus (Hemiptera: Miridae), a predator of grape phylloxera: Seasonal history and description of fifth instar. Melsheimer Entomol. Ser. No. 25:6–10.

——. 1980a. Seasonal history and host plants of the ant mimic Barberiella formicoides Poppius, with description of the fifth-instar (Hemiptera: Miridae). Proc. Entomol. Soc. Wash. 82:269–275.

——. 1980b. Brachynotocoris heidemanni (Knight), a junior synonym of the Palearctic B. puncticornis Reuter and pest of European ash. Proc. Entomol. Soc. Wash. 82:568–575.

——. 1981. Jalysus spinosus and J. wickhami: Taxonomic clarification, review of host plants and distribution, and keys to adults and 5th instars. Ann. Entomol. Soc. Am. 74:606–615.

——. 1983. Seasonal history and host plants of the plant bug Lygocoris atrinotatus, with description of the fifth-instar nymph (Hemiptera: Miridae). Proc. Entomol. Soc. Wash. 85: 26–31.

——. 1985. Trigonotylus coelestialium (Heteroptera: Miridae), a pest of small grains: Seasonal history, host plants, damage, and descriptions of adult and nymphal stages. Proc. Entomol. Soc. Wash. 87:699–713.

——. 1992. A synthesis of the Holarctic Miridae (Heteroptera): Distribution, biology, and origin, with emphasis on North America. Thomas Say Found. Monogr. Vol. 15. Entomological Society of America, Lanham, Md. 282 pp.

——. 1994. Orthotylus robiniae: A Gleditsia rather than Robinia specialist that resembles the honeylocust plant bug, Diaphnocoris chlorionis (Heteroptera: Miridae). Proc. Entomol. Soc. Wash. 96:63–69.

Wheeler, A. G. Jr. and J. L. Herring. 1979. A potential insect pest of azaleas. Q. Bull. Am. Rhododendron Soc. 33:12–14, 34.

Wheeler, A. G. Jr. and E. R. Hoebeke. 1982. Psallus variabilis (Fallén) and P. albipennis (Fallén), two European plant bugs established in North America, with notes on taxonomic changes (Hemiptera: Heteroptera: Miridae). Proc. Entomol. Soc. Wash. 84:690–703.

——. 1985. The insect fauna of ninebark, Physocarpus opulifolius (Rosaceae). Proc. Entomol. Soc. Wash. 87:356–370.

——. 1990. Psallus lepidus Fieber, Deraeocoris piceicola Knight, and Dichrooscytus latifrons Knight: New records of plant bugs in eastern North America (Heteroptera: Miridae). J. N.Y. Entomol. Soc. 98:357–361.

Wheeler, A. G. Jr. and C. C. Loan. 1984. Peristenus henryi (Hymenoptera: Braconidae, Euphorinae), a new species parasitic on the honeylocust plant bug, Diaphnocoris chlorionis (Hemiptera: Miridae). Proc. Entomol. Soc. Wash. 86:669–672.

Wheeler, A. G. Jr. and J. P. McCaffrey. 1984. Ranzovius contubernalis: Seasonal history, habits, and description of fifth instar, with speculation on the origin of spider commensalism in the genus Ranzovius (Hemiptera: Miridae). Proc. Entomol. Soc. Wash. 86:68–81.

Wheeler, A. G. Jr. and G. L. Miller. 1981. Fourlined plant bug (Hemiptera: Miridae), a reappraisal: Life history, host plants, and plant response to feeding. Gt. Lakes Entomol. 14:23–35.

——. 1990. Leptoglossus fulvicornis (Heteroptera: Coreidae), a specialist on magnolia fruits: Seasonal history, habits, and descriptions of immature stages. Ann. Entomol. Soc. Am. 83: 753–765.

Wheeler, A. G. Jr. and C. W. Schaefer. 1982. Review of stilt bug (Hemiptera: Berytidae) host plants. Ann. Entomol. Soc. Am. 75:498–506.

Wheeler, A. G. Jr. and W. A. Snook II. 1986. Biology of Sumitrosis rosea (Coleoptera: Chrysomelidae), a leafminer of black locust, Robinia pseudoacacia (Leguminosae). Proc. Entomol. Soc. Wash. 88:521–530.

Wheeler, A. G. Jr. and J. F. Stimmel. 1988. Heteroptera overwintering in magnolia leaf litter in Pennsylvania. Entomol. News 99:65–71.

Wheeler, A. G. Jr. and K. Valley. 1981. Ragged and shot-holed leaves: Diagnosing insect injury. J. Arboric. 7: 225–229.

Wheeler, A. G. Jr., J. T. Hayes, and J. L. Stephens. 1968. Insect predators of mummified pea aphids. Can. Entomol. 100:221–222.

Wheeler, A. G. Jr., B. R. Stinner, and T. J. Henry. 1975. Biology and nymphal stages of Deraeocoris nebulosus (Hemiptera: Miridae), a predator of arthropod pests on ornamentals. Ann. Entomol. Soc. Am. 68:1063–1068.

Wheeler, A. G. Jr., G. L. Miller, and T. J. Henry. 1979. Biology and habits of Macrolophus tenuicornis (Hemiptera: Miridae) on hayscentedfern (Pteridophyta: Polypodiaceae). Melsheimer Entomol. Ser. No. 27:11–17.

Wheeler, A. G. Jr., T. J. Henry, and T. L. Mason Jr. 1983. An annotated list of the Miridae of West Virginia

(Hemiptera-Heteroptera). Trans. Am. Entomol. Soc. 109:127–159.

Wheeler, Q. D. 1995. Systematics, the scientific basis for inventories of biodiversity. Biodiversity Conserv. 4:476–489.

Wheeler, Q. D. and A. G. Wheeler Jr. 1994. Mycophagous Miridae? Associations of Cylapinae (Heteroptera) with pyrenomycete fungi (Euascomycetes: Xylariaceae). J. N.Y. Entomol. Soc. 102:114–117.

Wheeler, W. C., R. T. Schuh, and R. Bang. 1993. Cladistic relationships among higher groups of Heteroptera: Congruence between morphological and molecular data sets. Entomol. Scand. 24:121–137.

Whitcomb, W. Jr. and H. F. Wilson. 1929. Mechanics of digestion of pollen by the adult honey bee and the relation of undigested parts to dysentery of bees. Wis. Agric. Exp. Stn. Res. Bull. 92:1–27.

Whitcomb, W. D. 1953. Biology and control of Lygus campestris L. on celery. Mass. Agric. Exp. Stn. Bull. 473: 1–15.

Whitcomb, W. H. and K. Bell. 1964. Predaceous insects, spiders, and mites of Arkansas cotton fields. Ark. Agric. Exp. Stn. Bull. 690:1–84.

White, T. C. R. 1968. Uptake of water by eggs of Cardiaspina densitexta (Homoptera: Psyllidae) from leaf of host plant. J. Insect Physiol. 145:1669–1683.

——. 1969. An index to measure weather-induced stress of trees associated with outbreaks of psyllids in Australia. Ecology 50:905–909.

——. 1974. A hypothesis to explain outbreaks of looper caterpillars, with special reference to populations of Selidosema suavis in a plantation of Pinus radiata in New Zealand. Oecologia (Berl.) 16:279–301.

——. 1978. The importance of a relative shortage of food in animal ecology. Oecologia (Berl.) 33:71–86.

——. 1984. The abundance of invertebrate herbivores in relation to the availability of nitrogen in stressed food plants. Oecologia (Berl.) 63:90–105.

White, W. B. and N. F. Schneeberger. 1981. Socioeconomic impacts. Pages 681–694 in C. C. Doane and M. L. McManus, eds., The gypsy moth: Research toward integrated pest management. U.S. Dep. Agric. For. Serv. Tech. Bull. 1584.

Whitehead, D. R. 1983. Wind pollination: Some ecological and evolutionary perspectives. Pages 97–108 in L. Real, ed., Pollination biology. Academic, New York.

Whitfield, G. H. and C. R. Ellis. 1976.

The pest status of foliar insects on soybeans and white beans in Ontario. Proc. Entomol. Soc. Ont. 107:47–55.

Whitham, T. G., J. Maschinski, K. C. Larson, and K. N. Paige. 1991. Plant responses to herbivory: The continuum from negative to positive and underlying physiological mechanisms. Pages 227–256 in P. W. Price, T. M. Lewinsohn, G. W. Fernandes, and W. W. Benson, eds., Plant-animal interactions: Evolutionary ecology in tropical and temperate regions. Wiley, New York.

Whitman, D. W., M. S. Blum, and F. Slansky Jr. 1994. Carnivory in phytophagous insects. Pages 161–205 in T. N. Ananthakrishnan, ed., Functional dynamics of phytophagous insects. Science Publishers, Lebanon, N.H.

Whitney, E. D. and J. E Duffus, eds. 1986. Compendium of beet diseases and insects. APS Press, St. Paul, Minn. 76 pp.

Whittaker, J. B. 1970. Cercopid spittle as a microhabitat. Oikos 21:59–64.

——. 1984. Responses of sycamore (Acer pseudoplatanus) leaves to damage by a typhlocybine leaf hopper, Ossiannilssonola callosa. J. Ecol. 72:455–462.

——. 1992. Green plants and plant-feeding insects. J. Biol. Educ. 26:257–262.

Whittaker, R. H. 1952. A study of summer foliage insect communities in the Great Smoky Mountains. Ecol. Monogr. 22:1–44.

——. 1969. New concepts of kingdoms of organisms. Science (Wash., D.C.) 163: 150–160.

Whitwell, A. C. 1993. The pest/predator/parasitoid complex on mango inflorescences in Dominica. Acta Hortic. (Wageningen) 341:421–432.

Widstrom, N. W. 1979. The role of insects and other plant pests in aflatoxin contamination of corn, cotton, and peanuts—a review. J. Environ. Qual. 8:5–11.

Wiebe, H. W., D. J. Schimpf, C. M. McKell, and J. Ansley. 1978. Physiological effects of grass bug feeding and resistance. Pages 28–32 in B. A. Haws, compiler, Economic impacts of Labops hesperius on the production of high quality range grasses. Final Rep. Utah Agric. Exp. Stn. to Four Corners Comm. Utah State University, Logan.

Wiedenmann, R. N. and L. T. Wilson. 1996. Zoophytophagous Heteroptera: Summary and future research needs. Pages 190–202 in O. Alomar and R. N. Wiedenmann, eds., Zoophytophagous Heteroptera: Implications for life history and integrated pest management. Thomas Say Publ. Entomol.: Proceed-

ings. Entomological Society of America, Lanham, Md.

Wiedenmann, R. N., J. C. Legaspi, and R. J. O'Neil. 1996. Impact of prey density and facultative plant feeding on the life history of the predator Podisus maculiventris (Heteroptera: Pentatomidae). Pages 94–118 in O. Alomar and R. N. Wiedenmann, eds., Zoophytophagous Heteroptera: Implications for life history and integrated pest management. Thomas Say Publ. Entomol.: Proceedings. Entomological Society of America, Lanham, Md.

Wiegert, R. G. and C. E. Petersen. 1983. Energy transfer in insects. Annu. Rev. Entomol. 28:455–486.

Wier, D. B. 1875. The fruit grower and the bugs. Trans. Ill. State Hortic. Soc. 8(1874):29–33.

Wightman, J. A. 1968. A study of oviposition site, mortality and migration in the first (overwintering) generation of Lygocoris pabulinus (Fallen) (Heteroptera: Miridae) on blackcurrant shoots (1966–7). Entomologist 101:269–275.

——. 1969a. Termination of egg diapause in Lygocoris pabulinus (Heteroptera: Miridae). Long Ashton Res. Stn. Rep. 1968:154–156.

——. 1969b. Rearing and feeding Lygocoris pabulinus (Heteroptera: Miridae). Long Ashton Res. Stn. Rep. 1968:157–161.

——. 1972. The egg of Lygocoris pabulinus (Heteroptera: Miridae). N.Z. J. Sci. 15:88–89.

——. 1973. Ovariole microstructure and vitellogenesis in Lygocoris pabulinus (L.) and other mirids (Hemiptera: Miridae). J. Entomol. (A) 48:103–115.

——. 1974. Heteroptera beaten from potato haulms at Long Ashton Research Station. Entomol. Mon. Mag. 109:132–139.

Wightman, J. A. and R. P. Macfarlane. 1982. The integrated control of pests of legume seed crops: 2. Summation and strategy of the 1980–81 season. In Proc. 3rd Australasian Conf. Grassland Invert. Ecol., Adelaide, 30 Nov.–4 Dec. 1981, pp. 377–384. S. A. Government Printer, Adelaide.

Wightman, J. A. and G. V. Ranga Rao. 1994. Groundnut pests. Pages 395–479 in J. Smartt, ed., The groundnut crop: A scientific basis for improvement. Chapman & Hall, London.

Wightman, J. A. and D. N. J. Whitford. 1982. Integrated control of pests of legume seed crops. 1. Insecticides for mirid and aphid control. N.Z. J. Exp. Agric. 10:209–215.

Wiklund, C., T. Eriksson, and H. Lundberg. 1979. The wood white butterfly

Leptides sinapis and its nectar plants: A case of mutualism or parasitism? Oikos 33:358–362.

Wilborn, R. and J. Ellington. 1984. The effect of temperature and photoperiod on the coloration of the *Lygus hesperus, desertinus* and *lineolaris*. Southwest. Entomol. 9:187–197.

Wildbolz, T. and A. Henauer. 1957. Die Bekämpfung der Fruchtwanzen auf Glockenapfel. Schweiz. Z. Obst-Weinbau 66:81–84.

Wildbolz, T., W. Vogel, and A. Henauer. 1955. Wanzenschäden an Glockenäpfeln. Schweiz. Z. Obst-Weinbau 64:531–534.

Wildermuth, V. L. 1915. Three-cornered alfalfa hopper. J. Agric. Res. (Wash., D.C.) 3:343–364.

Wilkinson, D. S. 1927. On two new parasites from West Africa bred from the cacao barksapper (*Sahlbergella*). Bull. Entomol. Res. 17:309–311.

Willcocks, F. C. 1922. A survey of the more important economic insects and mites of Egypt. Sultanic Agric. Soc. Bull. 1:1–482.

——. 1925. The insect and related pests of Egypt. Vol. II. Insects and mites feeding on gramineous crops and products in the field, granary, and mill. Sultanic Agricultural Society, Cairo. 418 pp.

Wille, J. E. 1944. Insect pests of cacao in Peru. Trop. Agric. 21:143.

——. 1951. Biological control of certain cotton insects and the application of new organic insecticides in Perú. J. Econ. Entomol. 44:13–18.

——. 1952. Entomologia agricola del Peru, 2nd ed. rev. and amplified. Ministerio de Agricultura, Lima. 543 pp.

——. 1958. El control biológico de los insectos agrícolas en el Perú. Proc. 10th Int. Congr. Entomol., Montreal (1956) 4:519–523.

Willemstein, S. C. 1987. An evolutionary basis for pollination ecology. Brill, Leiden, The Netherlands. 425 pp.

Williams, F. X. 1931. Handbook of the insects and other invertebrates of Hawaiian sugar cane fields. Experiment Station of the Hawaiian Sugar Planters' Association, Honolulu. 400 pp.

Williams, G. 1953. Field observations on the cacao mirids, *Sahlbergella singularis* Hagl. and *Distantiella theobroma* (Dist.), in the Gold Coast. Part I. Mirid damage. Bull. Entomol. Res. 44:101–119.

——. 1954. Field observations on the cacao mirids, *Sahlbergella singularis* Hagl. and *Distantiella theobroma* (Dist.), in the Gold Coast. Part III. Population fluctuations. Bull. Entomol. Res. 45:723–744.

Williams, I. H. 1977. Behaviour of insects foraging on pigeon pea (*Cajanus cajan* (L.) Millsp.) in India. Trop. Agric. 54:353–363.

Williams, L. III and N. P. Tugwell. 2000. Histological description of tarnished plant bug (Heteroptera: Miridae) feeding on small cotton floral buds. J. Entomol. Sci. 35:187–195.

Williams, L. III, J. R. Phillips, and N. P. Tugwell. 1987. Field technique for identifying causes of pinhead square shed in cotton. J. Econ. Entomol. 80:527–531.

Williamson, D. L., J. G. Tully, and R. F. Whitcomb. 1989. The genus *Spiroplasma*. Pages 71–111 in R. F. Whitcomb and J. G. Tully, eds., The mycoplasmas. Vol. V. Spiroplasmas, acholeplasmas, and mycoplasmas of plants and arthropods. Academic, San Diego.

Willis, J. C. and I. H. Burkill. 1895. Flowers and insects in Great Britain. Part I. Ann. Bot. 9:227–273.

——. 1903. Flowers and insects in Great Britain. Part II. Ann. Bot. 17:313–349.

——. 1908. Flowers and insects in Great Britain. Part IV. Ann. Bot. 22:603–649.

Wills, G. A. 1986. Use of an early warning system for the control of *Helopeltis theivora theobromae* in cocoa. Pages 241–253 in E. Pushparajah and Chew Poh Soon, eds., Cocoa and coconuts: Progress and outlook (Proc. Int. Conf. on Cocoa and Coconuts, Kuala Lumpur, 15–17 Oct. 1984). Incorporated Society of Planters, Kuala Lumpur, Malaysia.

Willson, M. F. 1991. Sexual selection, sexual dimorphism and plant phylogeny. Evol. Ecol. 5:69–87.

Wilson, C. L. and C. W. Ellett. 1980. The diagnosis of urban tree disorders. J. Arboric. 6:141–145.

Wilson, E. O. 1987. Causes of ecological success: The case of the ants. J. Anim. Ecol. 56:1–9.

——. 1992. The diversity of life. Harvard University Press, Cambridge. 424 pp.

Wilson, F. D. and B. W. George. 1986. Smoothleaf and hirsute cottons: Response to insect pests and yield in Arizona. J. Econ. Entomol. 79:229–232.

Wilson, G. F. see Fox-Wilson, G.

Wilson, J. W. 1948. A note on the predacious habit of the mirid *Cyrtopeltis varians* (Dist). Fla. Entomol. 21:20.

Wilson, L. F. and L. M. Moore. 1985. Vulnerability of hybrid *Populus* nursery stock to injury by the tarnished plant bug, *Lygus lineolaris* (Hemiptera: Miridae). Gt. Lakes Entomol. 18:19–23.

Wilson, L. J., L. R. Bauer, and G. H. Walter. 1996. 'Phytophagous' thrips are facultative predators of twospotted spider mites (Acari: Tetranychidae) on cotton in Australia. Bull. Entomol. Res. 86:297–305.

Wilson, L. T. [1986?]. Developing economic thresholds in cotton. Pages 308–344 in R. E. Frisbie and P. L. Adkisson, eds., Integrated pest management on major agricultural systems. Tex. Agric. Exp. Stn. MP-1616.

Wilson, L. T., T. F. Leigh, D. Gonzalez, and C. Foristiere. 1984. Distribution of *Lygus hesperus* (Knight) (Miridae: Hemiptera) on cotton. J. Econ. Entomol. 77:1313–1319.

Wilson, N. S. and L. C. Cochran. 1952. Yellow spot, an eriophyid mite injury on peach. Phytopathology 42:443–447.

Wilson, R. L. and D. L. Olson. 1990. Tarnished plant bug, *Lygus lineolaris* (Palisot de Beauvois) (Hemiptera: Miridae) oviposition site preference on three growth stages of a grain amaranth, *Amaranthus cruentus* L. J. Kans. Entomol. Soc. 63:88–91.

——. 1992. Tarnished plant bug, *Lygus lineolaris* (Palisot de Beauvois) (Hemiptera: Miridae): Effect on yield of grain amaranth, *Amaranthus cruentus* L., in field cages. J. Kans. Entomol. Soc. 65:450–452.

Wilson, R. L. and F. D. Wilson. 1978. A review of natural resistance in cotton to insects in Arizona. J. Ariz.-Nev. Acad. Sci. 13(2):44–46.

Windig, W., H. L. C. Meuzelaar, B. A. Haws, W. F. Campbell, and K. H. Asay. 1983. Biochemical differences observed in pyrolysis mass spectra of range grasses with different resistance to *Labops hesperius* Uhler attack. J. Anal. Appl. Pyrolysis 5:183–198.

Wink, M. 1988. Plant breeding: Importance of plant secondary metabolites for protection against pathogens and herbivores. Theor. Appl. Genet. 75:225–233.

Wink, M., T. Hartmann, L. Witte, and J. Rheinheimer. 1982. Interrelationship between quinolizidine alkaloid producing legumes and infesting insects: Exploitation of the alkaloid-containing phloem sap of *Cytisus scoparius* by the broom aphid *Aphis cytisporum*. Z. Naturforsch. 37(C):1081–1086.

Winokur, J. 1986. Writers on writing. Running Press, Philadelphia. 372 pp.

Wipfli, M. S., J. L. Wedberg, D. B. Hogg, and T. D. Syverud. 1989. Insect pests associated with birdsfoot trefoil, *Lotus corniculatus*, in Wisconsin. Gt. Lakes Entomol. 22:25–33.

Wipfli, M. S., J. L. Wedberg, and D. B. Hogg. 1990a. Damage potentials of three plant bug (Hemiptera: Heteroptera: Miridae) species to birdsfoot trefoil grown for seed in Wisconsin. J. Econ. Entomol. 83:580–584.

——. 1990b. Cultural and chemical control strategies for three plant bug (Heteroptera: Miridae) pests of birds-foot trefoil in northern Wisconsin. J. Econ. Entomol. 83:2086–2091.

——. 1991. Screen barriers for reducing interplot movement of three adult plant bug (Hemiptera: Miridae) species in small plot experiments. Gt. Lakes Entomol. 24:169–172.

Wise, I. L., J. R. Tucker, and R. J. Lamb. 2000. Damage to wheat seeds caused by a plant bug, *Lygus lineolaris* L. Can. J. Plant. Sci. 80:459–461.

Wiygul, G. and G. McKibben. 1997. Tarnished plant bug pheromone-olfactometer studies. *In* Proc. Beltwide Cotton Conf., Jan. 6–10, 1997, New Orleans, La., pp. 1333–1334. National Cotton Council, Memphis.

Wodehouse, R. P. 1935. Pollen grains; their structure, identification and significance in science and medicine. McGraw-Hill, New York. 574 pp.

Wolcott, G. N. 1933. An economic entomology of the West Indies. Entomological Society of Puerto Rico, San Juan. 688 pp.

Wolda, H. 1978. Seasonal fluctuations in rainfall, food and abundance of tropical insects. J. Anim. Ecol. 47:369–381.

——. 1988. Insect seasonality: Why? Annu. Rev. Ecol. Syst. 19:1–18.

Wolfe, H. S., L. R. Toy, and A. L. Stahl. 1946. Avocado production in Florida. Fla. Agric. Ext. Serv. Bull. 129:1–107.

Wolfenbarger, D. O. 1963. Insect pests of avocado and their control. Fla. Agric. Exp. Stn. Bull. 605A:1–52.

Womack, C. L. and M. F. Schuster. 1987. Host plants of the tarnished plant bug (Heteroptera: Miridae) in the northern Blackland Prairies of Texas. Environ. Entomol. 16:1266–1272.

Wood, B. J. and G. F. Chung. 1989. Integrated management of insect pests of cocoa in Malaysia. Planter (Kuala Lumpur) 65:389–418.

——. 1992. Integrated management of insect pests of cocoa in Malaysia. Pages 45–62 *in* P. J. Keane and C. A. J. Putter, eds., Cocoa pest and disease management in Southeast Asia and Australasia. FAO Plant Prod. Prot. Pap. 112.

Wood, G. A. 1991. Three graft-transmissible diseases and a variegation disorder of small fruit in New Zealand. N.Z. Crop Hortic. Sci. 19:313–323.

Wood, G. A. R. 1986. The cocoa tree. Biologist (Lond.) 33:99–104.

Wood, G. A. R. and R. A. Lass. 1985. Cocoa, 4th ed. Longman, London. 620 pp.

Wood, T. K. 1987. Host plant shifts and speciation in the *Enchenopa binotata* Say complex. Pages 361–368 *in* M. R. Wilson and L. R. Nault, eds., Proceedings of the 2nd International Workshop on Leafhoppers and Planthoppers of Economic Importance, Provo, Utah, 28 July–1 Aug. 1986. Commonwealth Institute of Entomology, London.

——. 1993. Diversity in the New World Membracidae. Annu. Rev. Entomol. 38:409–435.

Wood, T. K. and M. C. Keese. 1990. Host-plant-induced assortative mating in *Enchenopa* treehoppers. Evolution 44:619–628.

Wood, T. K., K. L. Olmstead, and S. I. Guttman. 1990. Insect phenology mediated by host plant water relations. Evolution 44:629–636.

Woodburn, T. L. and E. E. Lewis. 1973. A comparative histological study of the effects of feeding by nymphs of four psyllid species on the leaves of eucalypts. J. Aust. Entomol. Soc. 12:134–138.

Woodell, S. R. J. 1979. The role of unspecialized pollinators in the reproductive success of Aldabran plants. Philos. Trans. R. Soc. Lond. (B) 286:99–108.

Woodroffe, G. E. 1954. *Lygus campestris* (L.) (Hem., Miridae) damaging dahlias in Buckinghamshire. Entomol. Mon. Mag. 90:40.

——. 1955a. The Hemiptera-Heteroptera of some cinder-covered waste land at Slough, Buckinghamshire. Entomologist 88:10–17.

——. 1955b. A note on some Hemiptera-Heteroptera from Egham, Surrey. Entomol. Mon. Mag. 91:54.

——. 1958a. A note on *Pilophorus confusus* (Kb.), *Globiceps cruciatus* Reut. and *Adelphocoris ticinensis* (Mey-Duer) (Hem., Miridae) at Virginia Water and *Acompus rufipes* Wolff (Hem., Lygaeidae) at Chobham, Surrey. Entomol. Mon. Mag. 94:64.

——. 1958b. The food-plants of *Capsodes sulcatus* (Fieb.) (Hem., Miridae). Entomol. Mon. Mag. 94:240.

——. 1959a. *Globiceps salicicola* Reuter (Hem., Miridae) new to Britain. Entomologist 92:61–65.

——. 1959b. The host-plants of *Capsodes flavomarginatus* (Don.) (Hem., Miridae). Entomol. Mon. Mag. 95:207.

——. 1966. The *Lygus pratensis* complex (Hem., Miridae) in Britain. Entomologist 99:201–206.

——. 1969. Notes on some Hemiptera-Heteroptera from Aviemore, Inverness-shire. Entomol. Mon. Mag. 105:165–166.

——. 1977. *Notostira erratica* (L.) and *N. elongata* (Geoffroy) (Hem., Miridae) in the British Isles. Entomol. Gaz. 28:123–126.

Woodroffe, G. E. and D. G. H. Halstead. 1959. *Fulvius brevicornis* Reut. (Hem., Miridae) and other insects breeding on stored Brazil nuts in Britain. Entomol. Mon. Mag. 95:130–133.

Woods, W. 1992. Phytophagous insects collected from *Parkinsonia aculeata* [Leguminosae: Caesalpiniaceae] in the Sonoran Desert region of the southwestern United States and Mexico. Entomophaga 37:465–474.

Woodside, A. M. 1947. Weed hosts of bugs which cause cat-facing of peaches in Virginia. J. Econ. Entomol. 40:231–233.

——. 1950. Cat-facing and dimpling of peaches. Va. Agric. Exp. Stn. Bull. 435:1–18.

Woodward, T. E. 1949. Notes on the biology of some Hemiptera-Heteroptera. Entomol. Mon. Mag. 85:193–206.

——. 1952. Studies on the reproductive cycle of three species of British Heteroptera, with special reference to the overwintering stages. Trans. R. Entomol. Soc. Lond. 103:171–218.

Wooller, R. D., K. C. Richardson, and C. M. Pagendham. 1988. The digestion of pollen by some Australian birds. Aust. J. Zool. 36:357–362.

Wootton, R. J. and C. R. Betts. 1986. Homology and function in the wings of Heteroptera. Syst. Entomol. 11:389–400.

Wright, J. E. and L. D. Chandler. 1991. Field evaluation of Naturalis® against the boll weevil: A biorational mycoinsecticide. *In* Proc. Beltwide Cotton Conf., Jan. 8–12, 1991, San Antonio, Texas, pp. 677–679. National Cotton Council, Memphis.

Wukasch, R. T. and M. K. Sears. 1982. Damage to asparagus by tarnished plant bugs, *Lygus lineolaris*, and alfalfa plant bugs, *Adelphocoris lineolatus* (Heteroptera: Miridae). Proc. Entomol. Soc. Ont. 112:49–51.

Wyatt, R. 1983. Pollinator-plant interactions and the evolution of breeding systems. Pages 51–95 *in* L. Real, ed., Pollination biology. Academic, Orlando, Fla.

Wyniger, R. 1962. Pests of crops in warm climates and their control. Recht und Gesellschaft, Basel. 555 pp. (Acta Trop. Suppl. 7.)

Yamada, T. and K. Watanabe. 1952. Studies on the sterility in Sudan grass I. Sterility caused by the insect (*Calocoris rubrovittatus* Matsumura) and its mechanism. Bull. Natl. Inst. Agric. Sci. Ser. G (Anim. Husb.) 4:209–216. [in Japanese, English summary.]

Yamamuro, R. and K. Hoshino. 1940. Observations in [sic] Lygus viridis Fall. Rep. Dep. Seric. Agric. Exp. Stn. Korea 4:37–57. [in Japanese.]

Yang, S. F. and H. K. Pratt. 1978. The physiology of ethylene in wounded plant tissues. Pages 595–622 in G. Kahl, ed., Biochemistry of wounded plant tissues. de Gruyter, Berlin.

Yasuda, K. 1993. The damage by infestation of Taylorilygus pallidulus (Blanchard) (Heteroptera: Miridae). Bull. Okinawa Agric. Exp. Stn. 14:59–63. [in Japanese, English summary.]

Yasunaga, T. 1990. Abnormal food habits observed on [sic] two mirids (Heteroptera: Miridae). Rostria (Tokyo) 40:665–667. [in Japanese, English summary.]

——. 1991a. A revision of the plant bug, genus Lygocoris Reuter from Japan, part II (Heteroptera: Miridae, Lyguscomplex). Jpn. J. Entomol. 59:593–609.

——. 1991b. A revision of the plant bug, genus Lygocoris Reuter from Japan, part III (Heteroptera: Miridae, Lyguscomplex). Jpn. J. Entomol. 59:717–733.

——. 1992a. A revision of the tribe Mirini Hahn from Japan (Heteroptera, Miridae). Ph.D. dissertation, Kyushu University, Fukuoka, Japan. 146 pp.

——. 1992b. A revision of the plant bug, genus Lygocoris Reuter from Japan, part IV (Heteroptera, Miridae, Lyguscomplex). Jpn. J. Entomol. 60:10–25.

——. 1992c. A revision of the plant bug genus Lygocoris Reuter from Japan, part V (Heteroptera, Miridae, Lyguscomplex). Jpn. J. Entomol. 60:291–304.

——. 1992d. New genera and species of the Miridae of Japan (Heteroptera). Proc. Jpn. Soc. Syst. Zool. 47:45–51.

——. 1993. Descriptions of the last-instar nymphs of four mirid species (Heteroptera, Miridae) found in the southern Primorskij Kraj, Russia. Jpn. J. Entomol. 61:285–292.

——. 1995. A new genus of mirine plant bug, Carvalhopantilius, with two new species from Taiwan (Heteroptera, Miridae). Proc. Entomol. Soc. Wash. 97:452–457.

——. 1996a. An aberrant male specimen of Lygocoris spinolae (Heteroptera: Miridae), having "a pair of right parameres." Entomologist 115:59–62.

——. 1996b. Review of Lygocorides Yasunaga (Heteroptera: Miridae). Tijdschr. Entomol. 139:267–275.

——. 1997. Species of the orthotyline genus Pseudoloxops Kirkaldy from Japan (Heteroptera: Miridae). Bull. Biogeogr. Soc. Jpn. 52:11–18.

——. 1998. Revision of the mirine genus Castanopsides Yasunaga from the eastern Asia (Heteroptera: Miridae). Entomol. Scand. 29:99–119.

——. 1999a. The plant bug tribe Orthotylini in Japan (Heteroptera: Miridae: Orthotylinae). Tijdschr. Entomol. 142: 143–183.

——. 1999b. First record of the plant bug subfamily Psallopinae (Heteroptera: Miridae) from Japan, with descriptions of three new species of the genus Psallops Usinger. Proc. Entomol. Soc. Wash. 101:737–741.

——. 2000. An annotated list and descriptions of new taxa of the plant bug subfamily Bryocorinae in Japan (Heteroptera: Miridae). Biogeography (Tokyo) 2:93–102.

Yasunaga, T. and Y. Nakatani. 1998. The eastern Palearctic relatives of European Deraeocoris olivaceus (Fabricius) (Heteroptera: Miridae). Tijdschr. Entomol. 140:237–247.

Yasunaga, T. and M. Takai. 1994. Review of the genus Eocalocoris Miyamoto et Yasunaga (Heteroptera, Miridae) of Japan, with description of a new species from the mountains of Shikoku and southwestern Honshu. Proc. Jpn. Soc. Syst. Zool. 52:75–80.

Yasunaga, T., M. Takai, I. Yamashita, M. Kawamura, and T. Kawasawa. 1993. A field guide to Japanese bugs—terrestrial heteropterans (M. Tomokuni, ed.). Zenkoku Noson Kyoiku Kyokai Publishing, Tokyo. 380 pp. [in Japanese.]

Yasunaga, T., S. Miyamoto, and I. M. Kerzhner. 1996. Type specimens and identity of the mirid species described by Japanese authors in 1906–1917 (Heteroptera: Miridae). Zoosyst. Ross. 5:91–94.

Yasunaga, T., M. Takai, and Y. Nakatani. 1997. Species of the genus Stethoconus of Japan (Heteroptera, Miridae): Predaceous deraeocorine plant bugs associated with lace bugs (Tingidae). Appl. Entomol. Zool. 32:261–264.

Yasunaga, T., N. N. Vinokurov, and M. Takai. 1999. New records of the Heteroptera from Japan. Rostria (Tokyo) 48:1–9.

Yayla, A. 1986. A new beneficial heteroptera [sic] (Miridae, Deraeocorinae) in olive groves in Turkey Deraeocoris delagrangei (Puton). Olivae (Madrid) 14:12–13.

Yeargan, K. V. 1985. Alfalfa: Status and current limits to biological control in the eastern U.S. Pages 521–536 in M. A. Hoy and D. C. Herzog, eds., Biological control in agricultural IPM systems. Academic, Orlando, Fla.

——. 1998. Predatory Heteroptera in North American agroecosystems: An overview. Pages 7–19 in M. Coll and J. R. Ruberson, eds., Predatory Heteroptera: Their ecology and use in biological control. Thomas Say Publ. Entomol.: Proceedings. Entomological Society of America, Lanham, Md.

Yeates, G. W. 1971. Feeding types and feeding groups in plant and soil nematodes. Pedobiologia 11:173–179.

Yodzis, P. 1984. How rare is omnivory? Ecology 65:321–323.

Yokoyama, V. Y. 1978. Relation of seasonal changes in extrafloral nectar and foliar protein and arthropod populations in cotton. Environ. Entomol. 7:799–802.

Yonke, T. R. 1991. Order Hemiptera. Pages 22–65 in F. W. Stehr, ed., Immature insects. Vol. 2. Kendall-Hunt, Dubuque, Iowa.

Youdeowei, A. 1965. A note on the spatial distribution of the cocoa mirid Sahlbergella singularis Hagl. in a cocoa farm in western Nigeria. Niger. Agric. J. 2:66–67.

——. 1970. Physiology and behaviour of cacao mirids. Cocoa Res. Inst. Niger. Annu. Rep. 1968–69:32–33.

——. 1972. The internal reproductive organs of four species of mirids associated with cocoa (Theobroma cacao) in West Africa. Rev. Zool. Bot. 86:93–100.

——. 1973. The life cycles of the cocoa mirids Sahlbergella singularis Hagl. and Distantiella theobroma Dist. in Nigeria. J. Nat. Hist. 7:217–223.

——. 1977. Behaviour and activity. Pages 223–236 in E. M. Lavabre, ed., Les Mirides du Cacaoyer. Institut français du Café du Cacao, Paris.

Young, A. M. 1994. The chocolate tree: A natural history of cacao. Smithsonian Institution Press, Washington, D.C. 200 pp.

Young, D. K. 1984a. Cantharidin and insects: An historical review. Gt. Lakes Entomol. 17:187–194.

——. 1984b. Field records and observations of insects associated with cantharidin. Gt. Lakes Entomol. 17:195–199.

Young, G. R. 1984. A checklist of mite and insect pests of vegetable, grain and forage legumes in Papua New Guinea. P.N.G. J. Agric. For. Fish. 33:13–38.

Young, H. J. 1986. Beetle pollination of Dieffenbachia longispatha (Araceae). Am. J. Bot. 73:931–944.

——. 1988. Differential importance of beetle species pollinating Dieffenbachia longispatha (Araceae). Ecology 69:832–844.

——. 1990. Pollination and reproductive biology of an understory neotropical aroid. Pages 151–164 in K. S. Bawa and M. Hadley, eds., Reproductive ecology of tropical forest plants (Man and the biosphere series, Vol. 7). UNESCO, Paris; Parthenon, Casterton Hall, UK.

Young, J. H. and L. J. Willson. 1987. Use of Bose-Einstein statistics in population dynamics models of arthropods. Ecol. Model. 36:87–99.

Young, L. J. (Willson) and J. H. Young. 1988. The effects of insecticides on the spatial distribution of cotton insects. *In* Proc. Beltwide Cotton Prod. Res. Conf., Jan. 3–8, 1988, New Orleans, La., pp. 323–324. National Cotton Council, Memphis.

Young, O. P. 1984. Utilization of dead insects on the soil surface in row crop situations. Environ. Entomol. 13:1346–1351.

——. 1985. Temporal patterns of tarnished plant bug (*Lygus lineolaris*) population structure and density on three species of *Erigeron* and cotton. Bull. Ecol. Soc. Am. 66:298. (Abstr.)

——. 1986. Host plants of the tarnished plant bug, *Lygus lineolaris* (Heteroptera: Miridae). Ann. Entomol. Soc. Am. 79:747–762.

——. 1989. Relationships between *Aster pilosus* (Compositae), *Misumenops* spp. (Araneae: Thomisidae), and *Lygus lineolaris* (Heteroptera: Miridae). J. Entomol. Sci. 24:252–257.

Young, O. P. and T. C. Lockley. 1985. The striped lynx spider, *Oxyopes salticus* [Araneae: Oxyopidae], in agroecosystems. Entomophaga 30:329–346.

——. 1986. Predation of striped lynx spider, *Oxyopes salticus* (Araneae: Oxyopidae), on tarnished plant bugs, *Lygus lineolaris* (Heteroptera: Miridae). Ann. Entomol. Soc. Am. 79:879–883.

——. 1990. Autumnal populations of arthropods on aster and goldenrod in the Delta of Mississippi. J. Entomol. Sci. 25:185–195.

Young, O. P. and W. C. Welbourn. 1987. Biology of *Lasioerythraeus johnstoni* (Acari: Erythraeidae), ectoparasitic and predaceous on the tarnished plant bug, *Lygus lineolaris* (Hemiptera: Miridae), and other arthropods. Ann. Entomol. Soc. Am. 80:243–250.

——. 1988. Parasitism of *Trigonotylus doddi* (Heteroptera: Miridae) by *Lasioerythraeus johnstoni* (Acari: Erythraeidae), with notes on additional hosts and distribution. J. Entomol. Sci. 23:269–273.

Young, S. C. Jr. and P. Tugwell. 1975. Different methods of sampling for clouded and tarnished plant bugs in Arkansas cotton fields. Ark. Agric. Exp. Stn. Rep. Ser. 219:1–12.

Young, S. Y. and W. C. Yearian. 1987. *Nabis roseipennis* adults (Hemiptera: Nabidae) as disseminators of nuclear polyhedrosis virus to *Anticarsia gemmatalis* (Lepidoptera: Noctuidae) larvae. Environ. Entomol. 16:1330–1333.

Young, W. R. and G. L. Teetes. 1977. Sorghum entomology. Annu. Rev. Entomol. 22:193–218.

Youtie, B. A., M. Stafford, and J. B. Johnson. 1987. Herbivorous and parasitic insect guilds associated with Great Basin wildrye (*Elymus cinereus*) in southern Idaho. Great Basin Nat. 47:644–651.

Yun, Y. M. 1986. Lygus bugs. Page 47 *in* E. D. Whitney and J. E. Duffus, eds., Compendium of beet diseases and insects. APS Press, St. Paul, Minn.

Zak, B. 1986. Florida critters. Common household and garden pests of the Sunshine State. Taylor, Dallas. 294 pp.

Zalom, F. G., C. Pickel, and N. C. Welch. 1990. Recent trends in strawberry arthropod management for coastal areas of the western United States. Pages 239–259 *in* N. J. Bostanian, L. T. Wilson, and T. J. Dennehy, eds., Monitoring and integrated management of arthropod pests of small fruit crops. Intercept, Andover, UK.

Zalom, F. G., C. Pickel, D. B. Walsh, and N. C. Welch. 1993. Sampling for *Lygus hesperus* (Hemiptera: Miridae) in strawberries. J. Econ. Entomol. 86:1191–1195.

Zalucki, M. P., G. Daglish, S. Firempong, and P. Twine. 1986. The biology and ecology of *Heliothis armigera* (Hübner) and *H. punctigera* Wallengren (Lepidoptera: Noctuidae) in Australia: What do we know? Aust. J. Zool. 34:779–814.

Zaugg, J. L. and M. W. Nielson. 1974. *Lygus hesperus*: Comparison of olfactory response between laboratory-reared and field-collected adults. J. Econ. Entomol. 67:133–134.

Zavodchikova, V. V. 1974. The feeding, development and fecundity of *Deraeocoris (Camptobrochis) punctulatus* Fall. (Heteroptera, Miridae) on different diets. Entomol. Obozr. 53(4):861–864. [in Russian; English translation in Entomol. Rev. 53(4):94–97.]

Zeh, D. W., J. A. Zeh, and R. L. Smith. 1989. Oviposition, amnions and eggshell architecture in the diversification of terrestrial arthropods. Q. Rev. Biol. 64:147–168.

Zeletzki, C. and G. Rinnhofer. 1966. Über Vorkommen und Wirksamkeit von Praedatoren in Obstanlagen I. Eine Mitteilung über Ergebnisse zweijähriger Klopffänge an Apfelbäumen. Beitr. Entomol. 16:713–720.

Zera, A. J. and R. F. Denno. 1997. Physiology and ecology of dispersal polymorphism in insects. Annu. Rev. Entomol. 42:207–230.

Zhang, Z. Q. 1992. The natural enemies of *Aphis gossypii* Glover (Hom., Aphididae), in China. J. Appl. Entomol. 114:251–262.

Zheng, L. Y. 1992. A new species of genus *Pseudodoniella* China & Carvalho from China (Insecta, Hemiptera, Miridae). Reichenbachia 29(21):119–122.

Zheng, L. Y. and L. J. Liang. 1991. Mirid bugs preying on persimmon leafhoppers from China (Hemiptera: Miridae). Acta Sci. Nat. Univ. Nankaiensis 3:84–87.

Zheng, L. Y. and G. Q. Liu. 1993. Two new species of genus *Zanchius* Dist. (Insecta: Hemiptera: Heteroptera: Miridae). Reichenbachia 30(5):17–20.

Zhu, K. Y. and W. A. Brindley. 1990a. Acetylcholinesterase and its reduced sensitivity to inhibition by paraoxon in organophosphate-resistant *Lygus hesperus* Knight (Hemiptera: Miridae). Pestic. Biochem. Physiol. 36:22–28.

——. 1990b. Properties of esterases from *Lygus hesperus* Knight (Hemiptera: Miridae) and the roles of the esterases in insecticide resistance. J. Econ. Entomol. 83:725–732.

Zia-ud-Din, [K.]. 1951. Tests of lindane and other insecticides for control of *Lygus oblineatus*. J. Econ. Entomol. 44:773–779.

Zimmerman, E. C. 1948a. Insects of Hawaii. Vol. 3. Heteroptera. University of Hawaii Press, Honolulu. 255 pp.

——. 1948b. Insects of Hawaii. Vol. 4. Homoptera: Auchenorrhyncha. University of Hawaii Press, Honolulu. 268 pp.

Zitter, T. A., M. L. Daughtrey, and J. P. Sanderson. 1989. Tomato spotted wilt virus. Cornell Coop. Ext. Fact Sheet. 6 pp.

Znamenskaya, M. K. 1962. A review of the crop pests of the Murmansk region. Entomol. Obozr. 41(2):310–321. [in Russian; English translation in Entomol. Rev. 41(2):190–194.]

Zoebelein, G. 1956. Der Honigtau als Nahrung der Insekten.Teil I, II. Z. Angew. Entomol. 38:369–416; 39:129–167.

——. 1957. Die Rolle des Waldhonigtaus im Nahrungshaushalt forstlich nützlicher Insekten. Forstwiss. Cbl. 76:24–34.

Zschokke, T. 1922. Ueber das Steinigwerden der Birnen und über Missbildungen an Obstfrüchten. Mit biologischen Notizen und Abbildungen über Capsiden, welche als Schädlinge an den Obstbäumen beobachtet und gesammelt wurden. Landwirtsch. Jahrb. Schweiz 36:575–593.

Zwet, T. van der and S. V. Beer. 1991. Fire blight—its nature, prevention, and control: A practical guide to integrated disease management. U.S. Dep. Agric. Agric. Inf. Bull. 631:1–83.

Zwet, T. van der and H. L. Keil. 1979. Fire blight, a bacterial disease of rosaceous plants. U.S. Dep. Agric. Agric. Handb. 510:1–200.

Zwick, F. B. and R. T. Huber. 1983. Early season population dynamics of lygus bugs in drip vs. furrow-irrigated cotton. In Proc. Beltwide Cotton Prod. Res. Conf., Jan. 2–6, 1983, San Antonio, Texas, pp. 197–200. National Cotton Council, Memphis.

Index to Scientific Names of Animals

Note: Page numbers with an *f* indicate figures; those with a *t* indicate tables.

Epicauta, 321
Epiphyas postvittana, 287
Erinnyis ello, 284t
Eriophyes pyri, 293
Eriosoma americanum, 303t
Eriosoma crataegi, 302
Eriosoma lanigerum, 304
Eriosoma ulmi, 303t
Erythmelus, 93
Erythroneura, 296–97
Erythroneura comes, 297t
Erythroneura ziczac, 297t
Eucallipterus tiliae, 289, 305, 306f
Eucerocoris suspectus. See Ragwelellus suspectus
Eumecotarsus, 223t
Eumecotarsus breviceps, 223t
Eumecotarsus kiritshenkoi, 223t
Euphydryas gilletti, 287
Euphyllura olivina, 299
Europiella decolor, 153t, 159
Eurychilella, 321
Eurychilella pallida, 321
Eurychilopterella, 12t, 40, 329
Eurychilopterella luridula, 309, 310
Eurydema rugosa, 112, 120, 148
Eurygaster, 200
Eurygaster integriceps, 128, 201
Euryopicoris, 331
Euryopicoris nitidus, 153t
Eurystylus, 12t, 330
  as inflorescence feeders, 330
  as legume pests, 204t
  nymphal coloration of, 32
  as sorghum pests, 197
Eurystylus bellevoyei, 198t
Eurystylus immaculatus. See Eurystylus oldi
Eurystylus marginatus, 197
Eurystylus oldi, 8t, 66t
  as castor pest, 218
  color variation in, 29f
  nymphal development of, 76
  as sorghum pest, 197
Euschistus conspersus, 258
Euseius, 232t
Eustictus, 22

Falconia intermedia, 50, 153t
Felisacus, 38
Felisacus elegantulus, 178t
Fenusa dohrnii, 290
Fieberocapsus flaveolus, 98
Fingulus porrectus, 296t
Fiorinia externa, 309
Fiorinia theae, 308
Formica subsericea, 102f
Frankliniella occidentalis, 294, 295f, 296t

Fulvius, 329, 333
  mycophagy and, 269t
  predation and, 270, 284t
Fulvius angustatus, 269t
Fulvius brevicornis, 269
Fulvius imbecilis, 321
Fulvius nigricornis, 284t
Fulvius quadristillatus, 269
Fulvius slateri, 269t
Fulvius variegatus, 284t

Galerucella lineola, 284t
Garganus albidivittis, 196
Garganus fusiformis, 289
Gastrodes grossipes, 313
Gelis, 101
Geocoris, 92, 93
  as mirid prey, 313–14
Geocoris punctipes, 300, 313
Geospiza, 236
Gerris odontogaster, 99
Glaucopterum, 331
Glaucopterum atraphaxius, 223t
Globiceps, 12t, 331
Globiceps flavomaculatus, 231
Globiceps fulvicollis, 121f, 303t
Gonioctena japonica, 286
Gracilomiris, 42
Gratiana spadicea, 284t
Grypocoris sexguttatus, 99
  inflorescence feeding by, 191, 194
  pollination and, 238t
Guianerius typicus, 330
Gynaikothrips ficorum, 296t

Habrochila, 310
Habrochila placida, 311
Hadronema, 331
Hadronema bispinosum, 321
Hadronema breviatum, 321
Hadronema militare, 41, 153t, 321
Haematocapsus bipunctatus, 322t
Hallodapus, 331, 335
Hallodapus albofasciatus, 49f
Hallodapus montandoni, 290
Halticotoma, 12t
  cannibalism and, 316
  defenses of, 103
  feeding aggregations and, 88
  host of, 38
Halticotoma nicholi, 321
Halticotoma valida, 321
  chlorosis from, 152
  feeding aggregations and, 134
  nymphal dispersal of, 75
  oviposition of, 59
  population densities of, 87t
  reproduction in, 56
Halticus, 12t

chlorosis from, 153t, 157, 331
defenses of, 103
hatching of, 82
Halticus apterus, 33f, 59f, 142t
Halticus bractatus, 44f
  annual generations of, 85
  chlorosis from, 154
  collecting of, 88
  hatching of, 74
  longevity of, 69
  nymphal development of, 77
  overwintering of, 79
  oviposition of, 59, 60, 63
  parasitic mites of, 97t
  population densities of, 87t
  as predator, 283, 309
  premating period of, 50
  preoviposition period of, 57
  viral transmission and, 141t
  wing dimorphism of, 43
Halticus chrysolepis, 153t, 158
Halticus intermedius, 153t, 158
Halticus minutus, 8t, 140
Halticus saltator, 103, 153t
Halticus tibialis. See Halticus minutus
Hambletoniola, 22
Haplothrips verbasci, 296t
Harpocera, 12t, 331
Harpocera thoracica
  antennae of, 22, 51f
  deciduous trees and, 227
  egg of, 33f
  hatching of, 82
  longevity of, 69
  mating of, 51, 52f
  pheromones of, 48
  scent glands of, 104
Heliconius, 233t, 235, 236
Helicoverpa armigera, 196, 283, 287
Helicoverpa zea, 214, 283
Heliothis, 284t
Heliothis virescens, 166, 285, 286f, 287
Helopeltis, 6, 9, 12t, 14, 177, 178t
  as avocado pests, 249
  as cashew pests, 184, 250
  as cocoa pests, 176–77, 251–52
  collecting of, 88
  control of, 91
  as cotton pests, 182
  defenses of, 100–101, 104
  eggs of, 61, 64f
  fecundity of, 68, 69
  feeding habits of, 75, 112, 120–22, 329
  fungi of, 98, 99
  as guava pests, 255
  leaf and stem injury from, 163t, 167t, 169

as mango pests, 255
mating of, 55
mimicry and, 101
nymphal development of, 76
parasitic mites of, 97t
parasitic nematodes of, 98
parasitoids of, 94, 95
pheromones of, 49
phytotoxic saliva and, 129
plant diseases and, 137, 138t, 139, 145
premating period of, 50
scutellar spine of, 31f
seasonality of, 86
as tree pests, 184–85
Helopeltis ancardii, 174
Helopeltis antonii
  as apple pest, 248
  as cashew pest, 250
  fecundity of, 68
  genitalia of, 25
  as guava pest, 255
  mating of, 51, 52
  molting of, 75
  nymphal development of, 78
  oviposition of, 59
  plant injury from, 180t
  population fluctuations of, 61
  sex ratio for, 70
Helopeltis bergrothi, 95, 99
Helopeltis bradyi, 176, 251
Helopeltis clavifer
  adult coloration of, 28
  as cocoa pest, 176, 177, 252, 253
  feeding habits of, 110, 111
  nymphal development of, 76
  pheromones of, 48
  population densities of, 87
  salivary enzymes and, 127
  stylets of, 108–9, 109f
Helopeltis corbisieri, 127
Helopeltis fasciaticollis, 104
Helopeltis orophila, 90, 96
Helopeltis pernicialis, 250
Helopeltis schoutedeni
  as cotton pest, 170–71, 213t, 217
  egg of, 71f
  fecundity and longevity of, 69
  fungal diseases and, 137
  incubation period of, 73
  molting of, 75
  parasitic nematodes of, 97
  as tea pest, 173f, 173–74
Helopeltis theivora
  as cashew pest, 174
  as cocoa pest, 251–52
  as guava pest, 255
  insecticides and, 71, 176
  oviposition of, 60, 66t
  parasitoids of, 94
  sibling species and, 28
  as tea pest, 174

Helopeltis westwoodi, 243
Hemiberlesia diffinis, 309
Hemicerocoris bicolor, 310
"Hemisphaerodella
    mirabilis," 56
Hesperolabops, 12t, 38
Hesperolabops gelastops, 152
Hesperophylum, 40, 56
Hesperophylum arizonae, 56
Hesperophylum heidemanni,
    41, 56
Heterocordylus, 12t
Heterocordylus flavipes. See
    Pseudophylus stundjuki
Heterocordylus genistae, 59,
    223t, 298
Heterocordylus malinus
    as apple pest, 245
    bacterial diseases and,
        138t, 139
    collecting of, 87
    defenses of, 103
    hatching of, 83–84
    host finding and, 114
    leaf injury from, 163t
    pretarsal structures of, 23f
Heterocordylus tibialis
    as aphid predator, 303t
    eggs of, 65
    parasitoid of, 95
    as psyllid predator, 298–99
    umbellifer feeding by, 194
Heteropsylla cubana, 299
Heterothrips azaleae, 294–95
Heterotoma, 12t, 331
Heterotoma merioptera
    eggs of, 33f, 91
    mating of, 54f
    predatory behavior of, 123
    as psyllid predator, 298
Heterotoma planicornis
    as egg predator, 284t
    habitat of, 39
    as psyllid predator, 298
Hexamermis arvalis, 97
Hippodamia, 286
Holocacista rivillei, 284t,
    288t
Hololena curta, 317
Homoeosoma vagella, 288t
Hoplomachus, 12t, 332
Hoplomachus affiguratus,
    148, 153t
    nymphal dispersal of, 75
    population densities of, 87t
Hoplomachus thunbergii,
    192–93, 240
Horciasinus signoreti, 204t
Horcias nobilellus
    as cotton pest, 182, 213t
    longevity of, 69–70
    fungal diseases and, 137
    premating period of, 50
Horistus, 12t
    chlorosis from, 156
    overwintering of, 79
Horistus infuscatus
    hatching of, 82
    oviposition of, 57

plant injury from, 122
Horistus orientalis, 34f, 93f,
    94
Horvathiolus gibbicollis, 44
Hyaliodes, 12t, 329
    as egg predators, 284t
    as leafhopper predator,
        295–96
Hyaliodes beckeri, 284t
Hyaliodes harti, 18f
    human bites from, 322t
    as lepidopteran predator,
        287
    as mealybug predator, 310
    as mite predator, 291, 293
    as psyllid predator, 298
Hyaliodes vitripennis, 13,
    312
    as aphid predator, 308
    human bites from, 322t
    as leafhopper predator, 296
    as lepidopteran predator,
        287
    as mealybug predator, 310
    as mite predator, 291, 293
    molting of, 75
    pretarsal structures of, 23f
    as psocid predator, 313
Hyaliodes vittaticornis, 303t
Hyalochloria, 12t
    antennae of, 22, 23f
    as arthropod predators, 331
    mating of, 51
Hyalochloria denticornis
    as aphid predator, 308
    as leafhopper predator,
        297t
    as lepidopteran predator,
        283, 287
    preoviposition period of, 57
Hyalochloria longicornis, 23f
Hyalopeplus, 12t
Hyalopeplus pellucidus, 19f,
    254–55, 330
Hyalopeplus similis, 204t
Hyalopeplus smaragdinus,
    32
Hyalopterus pruni, 289, 303t,
    307
Hypena humuli, 288t
Hypera postica, 286
Hyphantria cunea, 287

Iassomorphus cedarnus, 185
Ilnacora, 331
Ilnacora malina, 153t
Irbisia, 12t, 330
    chlorosis from, 155–56
    collection of, 41
    dispersal to herbs by, 194f
    feeding density of, 156f
    hatching of, 84
    as honeydew feeders, 271
    as peach pests, 257
    pollen feeding by, 233–34
Irbisia bliveni, 228
Irbisia brachycera, 159, 238t,
    322t
Irbisia californica, 155

Irbisia cascadia, 41
Irbisia pacifica, 155
Irbisia sericans, 41, 155
Irbisia solani, 98, 257
Isometopus, 12t, 308
Isometopus citri, 308
Isometopus intrusus, 308
Isometopus shaowuensis,
    308
Isometopus transvaalensis,
    308

Kamehameha lunalilo, 38
Keltonia, 332
Keltonia pallida, 150
Keltonia tuckeri, 217
Kundakimuka
    queenslandica, 288t

Labopidea, 12t, 142t
    chlorosis from, 331
Labopidea allii, 159
Labopidea simplex, 254
Labops, 12t, 159, 181t, 331
Labops burmeisteri, 159
Labops hesperius, 43, 44, 111
    control of, 91
    diapause in, 80
    eggs of, 74
    fecundity of, 68
    flight of, 45, 46
    hatching of, 84
    nematode infections of,
        97
    nymphal development of,
        77
    oviposition of, 60
    plant injury from, 122,
        155, 157–58, 166
    pollination and, 238t
    population densities of, 87t
    premating period of, 50
    sex ratio for, 70
Labops hirtus, 19f
Labops utahensis, 133
Laemocoris orphanus, 335
Lampethusa, 223t
Lampethusa anatina, 223t
Lampethusa collaris, 223t
Lamprocapsidea coffeae. See
    Ruspoliella coffeae
Largidea davisi, 41
Larinocerus, 22
Larinocerus balius, 150
Lasioerythraeus johnstoni,
    96–97, 97t, 98f
Lecanium coryli, 310
Leeuwenia karnyiana, 296t
Leiophron, 94–96
Leiophron uniformis, 94–96
Lema bilineata, 283
Lepidargyrus, 12t
Lepidargyrus ancorifer, 8t
    as inflorescence feeder,
        192t
    as legume pest, 204t
    as onion pest, 222
    as pistachio pest, 260
    setae of, 21f

Lepidopsallus minusculus.
    See Phoenicocoris
    minusculus
Lepidopsallus nyssae. See
    Atractotomus miniatus
Lepidosaphes ulmi, 309
Leptinotarsa decemlineata,
    285, 286
Leptoglossus occidentalis,
    127
Leptopterna, 12t, 169t, 330
Leptopterna dolabrata, 19f,
    43–44, 48, 99, 121, 122f,
    334
    chlorosis from, 157
    control of, 91
    as corn pest, 195
    defenses of, 104
    digestive efficiency and,
        199f
    as dipteran predator, 314
    dispersal to herbs by, 193
    egg of, 33f
    egg diapause of, 84
    fecundity of, 68
    flight of, 45
    fungi of, 98
    as grass feeder, 198–99
    as inflorescence feeder,
        195
    mortality of, 90
    nitrogen and, 133, 135
    oviposition of, 61, 66t
    population densities of, 87t
    research needed on, 338
    salivary enzymes of, 127
    scent glands of, 104
    as umbellifer feeder, 194
    as wheat pest, 200
    wing dimorphism of, 42–44
Leptopterna ferrugata, 43,
    198t
Leptus, 96, 97t
Letaba bedfordi. See
    Palomiella bedfordi
Lidopus heidemanni, 309
Limesolla diospyri, 297t
Lincolnia lucernina, 204
Lineatopsallus biguttulatus,
    170
Liocoris, 12t, 79, 330
Liocoris tripustulatus, 150
    dispersal to herbs by, 193
    as green pepper pest, 220
    as weed feeder, 191
Liriomyza sativae, 288
Liriomyza trifoliearum,
    287–88
Liriomyza trifolii, 288–89
Listronotus bonariensis, 284t
Lithocolletis, 287
Litomiris debilis, 153t, 169t
Lopidea, 12t, 36, 225
    chlorosis from, 153t, 331
    chromosomes of, 25
    dispersal to herbs by, 194f
    as egg predators, 284t
    hosts of, 38
    human bites from, 322t

as inflorescence feeders, 193t, 225, 331
Lopidea chandleri, 322t
Lopidea cuneata, 303t
Lopidea dakota, 249–50, 266
Lopidea davisi, 159
Lopidea heidemanni, 153t, 194t, 225
Lopidea hesperus, 38, 153t
Lopidea instabilis, 101, 289
Lopidea marginata, 322t
Lopidea media, 20f
  human bites from, 322t, 323
  as predator, 276
Lopidea minor, 194t
Lopidea nigridia
  color variation in, 28
  defenses of, 101
  egg hatch in, 73
  male genitalia of, 24
  spatial distribution of, 89
Lopidea robiniae
  hosts of, 38
  human bites from, 322t
  as predator, 286
  leaf injury from, 163t
  mimicry and, 101
  as peach pest, 257
Lopidea robusta, 322t
Lopidea staphyleae, 223
Lopidea teton, 192t
Lopus decolor, 40, 303t, 331–32
Lordithon, 270
Loulucoris kidoi, 313
Lycaeides melissa samuelis, 41
Lycidocoris, 12t
Lycidocoris mimeticus, 178t, 184
Lygidea mendax
  as apple pest, 245
  bacterial diseases and, 138t
  collecting of, 87
  defenses of, 103
  hatching of, 84
  host finding and, 114
  insecticides and, 71
  leaf injury from, 163t, 164
  oviposition of, 63–64
Lygidolon laevigatum, 185
Lygocorides rubronasutus, 226t
Lygocoris, 12t, 163t, 167–68t, 225, 238t
  as aphid predators, 302
  chlorosis from, 154, 156
  deciduous trees and, 226t, 227
  flowering trees and, 228–29
  as foliage feeders, 330
  as fruit pests, 256–59
  as inflorescence feeders, 193t, 225, 330
  plant diseases and, 138t, 139, 145
  plant injury from, 65, 162, 181t, 243

salivary secretions and, 129
  as shrub feeders, 222, 223–24t, 225
Lygocoris aesculi, 149t
Lygocoris atrinotatus, 222
Lygocoris belfragii, 193t, 223t, 224t, 225
Lygocoris caryae, 224t, 225–28, 238, 257
Lygocoris communis, 148, 234
  as apple pest, 245
  cannibalism of, 315
  collecting of, 87
  fungi of, 98
  human bites from, 322t, 323
  as lepidopteran predator, 288t
  as pear pest, 258–59f
  pheromones of, 48
  predation on, 314
  spatial distribution of, 89
Lygocoris fagi, 224t, 227, 228
Lygocoris geneseensis, 229
Lygocoris hirticulus, 228
Lygocoris inconspicuus, 254
Lygocoris knighti, 223t, 225
Lygocoris kyushuensis, 226t
Lygocoris laureae, 222
Lygocoris lucorum. See Apolygus lucorum
Lygocoris neglectus, 223t
Lygocoris nyssae, 226t
Lygocoris omnivagus, 322t
Lygocoris ostryae, 224t
Lygocoris pabulinus, 85, 150, 184, 250
  as aphid predator, 307
  as apple pest, 184f, 248
  as conifer pest, 185
  dispersal to herbs by, 194t
  eggs of, 33f, 64, 70
  fecundity of, 68
  fungi of, 98
  hatching of, 83, 84
  as inflorescence feeder, 192
  as legume feeder, 204t
  mating of, 50, 51, 55
  nymphal development of, 76, 77
  ovarioles of, 27
  overwintering of, 79
  as pear pest, 259
  pheromones of, 49
  plant injury from, 164f, 164–65, 167t, 180t
  predation on, 314
  research needed on, 335
  sex ratio for, 70
  as sugar beet pest, 183
  viral transmission and, 141–42t
Lygocoris quercalbae, 227
Lygocoris rugicollis, 8t
  as apple pest, 169, 247–48, 248
  cannibalism of, 315

as egg predator, 284t
  hatching of, 82, 83
  leaf injury from, 164f
  nymphal dispersal of, 74–75
  oviposition of, 63
Lygocoris semivittatus, 225
Lygocoris spinolae. See Apolygus spinolae
Lygocoris tiliae, 224t, 225, 227
Lygocoris tinctus, 28, 226
Lygocoris viburni, 156, 168t
Lygocoris viridanus, 88, 185
Lygocoris viridus, 243
Lygocoris vitticollis, 38, 166
Lygus, 12t, 150, 276–77, 330
  adaptability of, 14
  as alfalfa pests, 201–3, 207–8, 208f
  annual generations of, 85
  as apple pests, 243–45
  artificial diets for, 272–73
  cannibalism of, 315
  as carrot pests, 220, 221
  collecting of, 88
  color variation of, 28, 188
  control of, 91
  as corn pests, 195
  as cotton pests, 179–82, 205t, 205–13
  defenses of, 103, 104
  distribution of, 37
  eggs of, 56–58
    deposition of, 58–61
    incubation period of, 73
  fecundity and longevity of, 67–69
  feeding behavior of, 118–20
  flight of, 45–46
  as foliage feeders, 330
  as fruit pests, 249
  fungi of, 98–99
  as grain feeders, 200
  hatching of, 84
  as honeydew feeders, 271
  host selection of, 115f
  hosts of, 38, 114–15, 189–90
  human bites from, 322t
  as inflorescence feeders, 187, 188, 192t, 193t
  ingestion by, 120–22
  as legume pests, 204t, 219–20
  as predators, 283, 302, 308
Leptinotarsa decemlineata and, 285
  morphology of, 21f
  nematode infections of, 97–98
  nymphs of, 31, 76, 77
  overwintering of, 79
  parasitic mites of, 97t
  parasitoids of, 94, 96
  as parasitoid predators, 290
  as peach pests, 255–56
  as pear pests, 257–58
  pheromones of, 47–49

phytoplasmas and, 140
plant diseases and, 138t, 136–40, 141–42t, 141–43, 145
plant injury and, 66t, 149t, 154–55, 161t, 162–66, 168t, 180t
phytotoxic saliva and, 129
pollination and, 238t
population densities of, 87t
as predators, 283, 302, 308
predators of, 92, 93, 314
as rapeseed pests, 218
reproduction in, 50, 55, 80, 81f
salivary enzymes of, 127–31
as shrub feeders, 224t, 225
spatial distribution of, 88
as strawberry pests, 261f, 261–66, 263–66f
as sugar beet pests, 217–18
as sunflower pests, 218
as vegetable pests, 222
Lygus abroniae, 28, 192t
Lygus atriflavus, 188
Lygus borealis, 276
Lygus bradleyi, 154, 155
Lygus desertinus. See Lygus elisus
Lygus disponsi. See Lygus rugulipennis
Lygus elisus, 8t
  as apple pest, 244–45
  as cotton pest, 207
  hosts of, 119
  as legume pest, 219–20
  as peach pest, 256
Lygus gemellatus, 9
Lygus hesperus, 3, 285, 286
  as apple pest, 244–45
  artificial diets for, 272–73
  cannibalism of, 316
  as carrot pest, 220
  as conifer seedling pest, 185–86, 186f
  control of, 266
  as cotton pest, 208–13, 209
  defenses of, 104
  as honeydew feeder, 271
  incubation period of, 73t
  leaf injury from, 168t
  as legume pest, 219–20
  mating of, 52, 55
  molting of, 76
  nymphal development of, 76
  as peach pest, 256
  photoresponse curve for, 81f
  as pistachio pest, 260
  pollination and, 238–39
  as predator, 307, 313
  predators of, 93
  salivary glands of, 166
  as scavenger, 320
  spatial distribution of, 89

Lygus hesperus (continued)
    as strawberry pest, 261–66,
        266f
    as vegetable pest, 183
Lygus keltoni, 85
Lygus lineolaris, 1, 9, 13,
        184, 228
    as alfalfa pest, 183
    antennal sensilla of, 116
    as apple pest, 243–44
    artificial diets for, 272–73
    as blackberry pest, 249
    cannibalism of, 315
    color variation of, 28
    as conifer seedling pest,
        185f, 185–86
    as cotton pest, 170, 189f,
        190f, 207–13
    deciduous trees and, 227
    defenses of, 103, 104
    eggs of, 189–90, 190f
        development of, 74
        insecticides and, 72
    endosymbionts and, 113
    feeding cessation by, 122
    fire and, 91
    fungi of, 98
    glandular plants and, 318
    as grape pest, 254
    grooming of, 120
    as herb feeder, 188–91,
        189–90f
    as honeydew feeder, 271
    host selection of, 115f, 116
    hosts of, 114–15, 119, 189f
    human bites from, 322t
    as legume pest, 205, 219
    management of, 244
    as nectar and pollen feeder,
        192, 231
    nematode infections of, 97,
        98
    nymphs of, 29, 191–92
    ovariole structure of, 27
    overwintering of, 79
    parasitic mites of, 97t, 98f
    parasitoids of, 96, 190–91
    as peach pest, 256–57
    phytotoxic saliva and, 129
    plant diseases and, 136,
        138t, 140
    plant injury from, 65, 152,
        165, 166, 179f
    as poplar pest, 169, 184
    as predator, 285, 286f,
        286–90, 307t
    reproductive rate of,
        189–90, 190f
    as scavenger, 320
    seasonal abundance of,
        189f
    sex ratio for, 70
    as strawberry pest, 261f,
        261–66, 263–65f
    stylets of, 109f, 110, 116,
        118f
    as sugar beet pest, 183
    as vegetable pest, 220, 222
    as wheat pest, 199

Lygus mexicanus, 195, 200
"Lygus" muiri, 168, 180
Lygus oblineatus, 9
Lygus oregonae, 192t
Lygus pratensis, 9
    as collembolan predator,
        313
    as herb feeder, 191
    human bites from, 322t,
        323
    as kenaf pest, 217
    as olive pest, 255
    pollination and, 240
    as umbellifer feeder, 194,
        221
Lygus ravus. See Lygus shulli
Lygus robustus, 188
Lygus rugulipennis, 8t, 9, 243
    as collembolan predator,
        313
    as conifer seedling pest,
        186
    dispersion of, 88
    eggs of, 32
    fecundity and longevity of,
        68, 69, 190
    feeding of, 110–11, 111f,
        147
    as flax pest, 218
    flight of, 45
    hosts of, 119
    as herb feeder, 191
    human bites and, 323
    nematode infections of, 97
    overwintering of, 79t
    parasitoids of, 95
    as peach pest, 257
    as pear pest, 259
    phytoplasmas and, 140
    plant injury from, 148,
        149t, 162, 166, 181t
    plant-wound responses
        and, 131
    research needed on, 335
    salivary glands of, 125, 130
    as strawberry pest, 266
    stylets of, 122t
    as sugar beet pest, 183
    survival rate of, 190
    as vegetable pest, 183–84
    as wheat pest, 200
Lygus shulli, 8t, 266
Lygus unctuosus, 188
Lygus vanduzeei, 23f
Lygus varius. See Lygus
        shulli
Lymantria dispar, 14, 320
Lytta, 321
Lytta moerens, 321
Lytta nuttalli, 321

Macrolophus, 12t, 328
    as predators, 276, 300
Macrolophus caliginosus,
        300. See Macrolophus
        melanotoma
Macrolophus costalis, 8t, 124
    as aphid predator, 305–7
    leaf injury from, 167t

    as thrips predator, 294
    as whitefly predator, 301
Macrolophus melanotoma,
        8t
    as aphid predator, 307f
    cannibalism of, 315
    as egg predator, 283
    eggs of, 65, 72f
    fecundity of, 69
    hatching of, 84
    as leafhopper predator,
        297t
    as mite predator, 292, 293f
    overwintering and, 79
    oviposition of, 58, 60, 61
    as thrips predator, 294
    as whitefly predator,
        300–301
Macrolophus nubilus. See
        Macrolophus pygmaeus
Macrolophus praeclarus, 276
Macrolophus pygmaeus, 8t
    as aphid predator, 305
    nymphal development of,
        77
    overwintering of, 78
    as tomato feeder, 220
    as whitefly predator, 300,
        301
Macrolophus rubi. See
        Macrolophus costalis
Macrolophus separatus, 23f,
        152t
Macrolophus tenuicornis, 51,
        149t, 305
Macromischoides, 177
Macromischoides aculeatus,
        177
Macrosiphoniella sanborni,
        305
Macrosiphum euphorbiae,
        307
Macrotylus, 153t, 331
Macrotylus amoenus, 153t
Macrotylus multipunctatus,
        321
Macrotylus nigricornis, 168t
Macrotylus quadrilineatus,
        318
Macrotylus sexguttatus, 153t
Magicicada septendecim,
        320
Malacocoris, 12t, 331
Malacocoris chlorizans
    as egg predator, 284t
    hatching of, 84
    insecticides and, 72
    as leafhopper predator,
        297t
    as mite predator, 292
    oviposition of, 58, 63
Malacosoma americanum,
        288t
Malthinus, 101
Malthodes, 101
Mansoniella nitida, 178t
Mantis religiosa, 92
Masonaphis, 305, 320
Maurodactylus albidus, 322t

Mecistoscelis, 12t
Mecistoscelis scirtetoides,
        157, 330
Mecomma, 28, 101
Mecomma angustatum, 101
Mecomma dispar, 46
Megachile rotundata, 202
Megacoelum, 12t, 79, 330
Megacoelum apicale, 8t, 217
Megacoelum infusum, 33f,
        299t
Megacoelum modestum. See
        Creontiades dilutus
Megaloceraea reticornis. See
        Megaloceroea recticornis
Megaloceroea, 12t
Megaloceroea recticornis, 8t
    chlorosis from, 157
    grass feeding by, 198t
    inflorescence feeding by,
        195
    interspecific competition
        and, 41
    nymphal development of,
        76
    silvertop from, 169t
Megalocoleus, 12t
    as inflorescence feeders,
        332
    as pollen feeders, 193
    pollination and, 240
Megalocoleus molliculus,
        192t
Megalocoleus tanaceti, 191
Megalopsallus, 170
Megalopsallus atriplicis, 217
Megalopsallus latifrons, 170
Megaselia impariseta, 289
Megreta cancellata, 321
Melanaspis obscura, 309
Meloe laevis, 321
Mermitelocerus annulipes,
        193t, 320
Mertila, 12t, 316
Mertila malayensis, 134
    feeding aggregations of, 88
    oviposition of, 59–60
    plant injury from, 66t, 154
Mesochorus, 96
Mesolecanium
        nigrofasciatum, 310
Metacanthus pulchellus, 120,
        314
Metriorrhynchomiris
        dislocatus
    color variation of, 28
    dispersal to herbs by, 194f
    nectar feeding by, 231, 239
Metriorrhynchomiris fallax,
        193t, 227
Microphylellus, 12t, 224t,
        238t, 303t
Microphylellus flavipes, 227
Microphylellus modestus,
        271, 285t
Microphylellus tsugae, 322t
Microphylellus tumidifrons,
        303t
Mindarus abietinus, 303t

Mircarvalhoia, 12t
Mircarvalhoia arecae, 8t, 92, 185
  leaf injury from, 166, 167t
Miris striatus, 330
  cannibalism of, 315
  as egg predator, 283
  as honeydew feeder, 271
  as larval predator, 285–86
  as lepidopteran predator, 288t
  as pear pest, 259
  as scale predator, 309
Moissonia, 12t, 331
Moissonia flavomaculata, 296t
Moissonia importunitas, 8t, 133
  chlorosis from, 159–60
  fecundity of, 68
  mating of, 51–52
  nymphs of, 75, 76
  plant injury from, 181t
  preoviposition period of, 57
  salivary enzymes of, 127
Moissonia schefflerae, 223t
Monalocoris, 12t, 38, 328
Monalocoris filicis
  ants and, 92
  egg of, 33f
  fern sporangia and, 193
  hosts of, 39
  overwintering of, 79
Monalonion, 12t, 329
  as cocoa pests, 176, 253
Monalonion annulipes, 61, 70
Monalonion bondari, 253
Monalonion dissimulatum, 253
Monalonion schaefferi, 253
Monellia caryella, 303t
Monopharsus annulatus, 38
Monosynamma bohemanni, 228
Myiomma, 12t
Myiomma cixiiforme, 18f
  nymph of, 30f
  as scale predator, 308–9
Myrmecophyes oregonensis, 32f, 103
Myrmecoris, 330
Myrmecoris gracilis, 331
  ant predation and, 290
  cannibalism of, 315
  as honeydew feeder, 271
Myzocallis coryli, 303t, 304
Myzus persicae, 292, 304, 305, 307f

Nanopsallus, 79
Naochila sufflata, 311
Neacoryphus bicrucis, 101
Nearctaphis bakeri, 304
Nematocampa filamentaria, 288t
Neoborella, 38
Neocapsus leviscutatus, 227
Neochlamisus platani, 165

Neoneella, 12t, 329
  as inflorescence feeder, 222–23
  pollination and, 237
Neoneella bosqi, 222
Nephotettix virescens, 67t, 112, 280
Nesidiocoris, 12t, 328–29
  as egg predators, 283
  leaf injury from, 163t, 167t
  mating of, 50–51, 55
  as tobacco pests, 217
  as whitefly predators, 300
Nesidiocoris caesar, 8t
  as berytid predator, 314
  cannibalism of, 315–16
  gustatory sensilla of, 116
  mortality of, 90
  population densities of, 87
Nesidiocoris callani, 283
Nesidiocoris tenuis, 8t, 133
  annual generations of, 85
  as dipteran predator, 288–89
  fecundity of, 69
  feeding habits of, 112, 132, 134, 147
  as leafhopper predator, 297t
  as lepidopteran predator, 287, 288t
  nymphal development of, 76, 77
  overwintering and, 79
  oviposition of, 56, 57, 60
  phytoplasmas and, 140
  pollination and, 241
  reproductive behavior of, 50, 189–90
  as scavenger, 320
  as thrips predator, 294
  tomato lesions from, 172
  viral transmission and, 142t
  as whitefly predator, 300
Nesidiocoris volucer, 118
  cannibalism of, 315–16
  hatching of, 84
  as scavenger, 320
Nesidiorchestes hawaiiensis, 103f
  defenses of, 103
  habitat of, 40
Nesiomiris, 36
  chlorosis from, 153t, 331
  colonizing ability of, 47
  host of, 38
  interspecific competition and, 41
  mating of, 51
  predators of, 92
  seasonal activity of, 79
Nesiomiris pallasatus, 100
Nesiomiris sinuatus, 75
Neurocolpus, 12t, 225
  female genitalia of, 26f
  as inflorescence feeders, 228, 330
Neurocolpus arizonae, 223t

Neurocolpus flavescens, 223t
Neurocolpus jessiae, 223t, 225
Neurocolpus longirostris, 260
Neurocolpus mexicanus, 219t
Neurocolpus nubilus
  as apple pest, 245
  bacterial diseases and, 138t
  as cotton pest, 217
  deciduous trees and, 226t
  mating of, 52, 55
  pheromones of, 47
  as predator, 286, 314
  preoviposition period of, 57
  as scavenger, 320
  as shrub feeder, 222, 224t, 225
Neurocolpus pumilus, 193t
Neurocolpus tiliae, 228
Nezara viridula, 145
Niastama punctaticollis, 8t, 249
Nilaparvata lugens, 67t, 280, 282
Nomia melanderi, 202
Notostira, 12t
  chlorosis from, 157
  overwintering of, 79, 80
  as predators, 330
  varieties of, 28
Notostira elongata
  diapause in, 80
  fecundity of, 68
  feeding habits of, 157f
  fungi of, 98
  grazing animals and, 91
  hatching of, 74, 83–84
  as hymenopteran predator, 290
  interspecific competition and, 41
  stylets of, 110
  varieties of, 28
Notostira erratica, 168
  adaptations of, 90f
  eggs of, 33f
  inflorescence feeding by, 195
  mouthparts of, 106–7f
  overwintering of, 79
  parasitic mites of, 97t
Nuculaspis californica, 309
Numicia viridis, 279
Nysius huttoni, 201

Occidentodema polhemusi, 166
Ochthera mantis, 92
Odontota dorsalis, 101
Oecophylla, 335
Oecophylla longinoda, 177
Oecophylla smaragdina, 252
Ofellus mexicanus, 310
Oligonychus ununguis, 292
Oncopeltis fasciatus, 120, 122
Oncotylus, 332

Oncotylus punctipes, 193, 240
Oncotylus viridiflavus, 191
Operophtera brumata, 287
Ophraella sexvittata, 284t
Opistheurista, 12t
Opistheurista clandestina, 156, 330
Opuna sharpiana, 299t
Orectoderus, 12t, 331
Orectoderus obliquus, 20f, 331
  as aphid predator, 308
  feeding time of, 120
  mimicry and, 102
Orgyia pseudotsugata, 287
Orius, 232t, 299
Orius insidiosus, 195, 314, 318
Orthezia annae, 310
Orthocephalus, 12t, 42
  chlorosis from, 153t, 331
Orthocephalus coriaceus, 60f
Orthocephalus funestus, 153t
Orthocephalus saltator, 42, 153t
Orthonotus, 331
Orthonotus rossicus, 296t
Orthonotus rufifrons, 39, 191
Orthops, 12t, 66t
  as carrot pests, 221
  chlorosis from, 154
  flight of, 46
  as inflorescence feeders, 330
  overwintering of, 79
  pollination and, 238t
Orthops basalis, 154, 221
Orthops campestris
  fungal diseases and, 138
  mating of, 53f
  as pear pest, 259
  plant injury and, 129, 180t
  weed feeding by, 191
Orthops kalmii, 33f
Orthops montanus, 97t
Orthops palus, 250, 255
Orthops scutellatus, 76
Orthotylus, 12t, 331
  bacterial diseases and, 138t, 139
  chlorosis from, 153t
  deciduous trees and, 226
  as inflorescence feeders, 331
  parasitoids of, 95
  as predators, 284f, 298, 303t
  seasonal activity of, 41
Orthotylus adenocarpi, 303t
Orthotylus aesculicola, 149t
Orthotylus australianus, 287
Orthotylus catulus, 192t
Orthotylus concolor, 314
Orthotylus dorsalis, 226
Orthotylus ericetorum, 223t

Orthotylus flavosparsus
  camouflage of, 100
  human bites from, 322t
  viral transmission and,
    141t
Orthotylus gotohi, 320
Orthotylus iolani, 153t
Orthotylus juglandis, 226
Orthotylus leviculus, 170
Orthotylus marginalis
  as apple pest, 247, 248
  cannibalism of, 315
  eggs of, 33f
  hatching of, 82
  nymphal development of,
    77
  as pear pest, 259
  as predator, 298, 303t
Orthotylus marginatus, 150
Orthotylus mimus, 160
Orthotylus nassatus, 292,
  298
Orthotylus ramus, 65, 87t
  as aphid predator, 303t
  human bites from, 322t,
    323
  as inflorescence feeder, 187
Orthotylus tenellus, 299t
Orthotylus virescens
  as aphid predator, 303t
  chlorosis from, 159
  flight of, 45
Orthotylus viridinervis, 298,
  303t

Pachymerocista pilosa, 23f
Pachypeltis, 152t
Pachypeltis maesarum, 55,
  167t
Pachypeltis polita, 161t
Pachypeltis vittiscutis, 152t
Pachytomella, 79
Paleacrita vernata, 288t
Palomena prasina, 284t
Palomiella, 12t
Palomiella bedfordi, 8t, 308
Pameridea, 12t, 319, 329
  host of, 38
  pollination and, 240
Pameridea marlothi, 319
Pameridea roridulae, 240f,
  319
Panonychus ulmi, 275, 290,
  292
Pantilius, 82
Pantilius tunicatus, 98, 227
Paracarnus, 12t, 32, 313
Paracentrobia, 93
Paradacerla formicina, plate
  3
Paramermis, 97, 98
Paramixia carmelitana. See
  Sthenaridea carmelitana
Paramixia suturalis. See
  Sthenaridea suturalis
Paraproba, 12t, 331
Paraproba capitata, 291, 296
Paraproba nigrinervis, 303t
Paraproba pendula, 303t

Parapsallus, 331
Parapsallus vitellinus, 149
Pararculanus piperis, 178t
Paraswammerdamia lutarea,
  287
Parathrombium
    megalochirum, 96, 97t
Paratrioza cockerelli, 299t
Paratriphleps laevisculus,
  232t
Parides, 233t
Parlatoreopsis chinensis, 309
Parlatoria oleae, 309
Parthenicus, 12t, 331
Parthenicus aureosquamis,
  217
Pectinophora gossypiella,
  283
Peregrinus maidis, 67t, 278,
  279
Periphilus negundinis, 303t
Peristenus, 94, 95f, 290
Peristenus malatus, 124
Peristenus pallipes, 99
Peristenus pseudopallipes,
  190–91
Peritropis, 333
Perkinsiella saccharicida,
  278
Pestalotiopsis psidii, 255
Phaedon cochleariae, 284t
Phalacrotophora
    berolinensis, 289
Phalacrotophora fasciata,
  289
Phasia, 96
Phenacoccus aceris, 310
Phenacoccus manihoti, 310
Philaenus spumarius, 28,
  320
Phoenicocoris, 331
Phoenicocoris australis, 149,
  227
Phoenicocoris claricornis, 82,
  227
Phoenicocoris dissimilis,
  303t
Phoenicocoris minusculus, 8t
  as aphid predator, 124t,
    302, 303t
Phorodon humuli, 304
Phyllaphis fagi, 302, 304
Phylloxera caryaecaulis, 304
Phylus, 331
Phylus coryli, 303t
Phylus melanocephalus, 322t
Phytocoris, 2t, 12t, 35, 99,
  330
  activity of, 41
  as aphid predators, 302,
    304
  defenses of, 104
  as egg predators, 277,
    282–83
  flight of, 45
  habitat of, 39, 40
  hosts of, 39
  as inflorescence feeders,
    193t

  as lepidopteran predators,
    287
  pheromones of, 48, 49
  as psocid predator, 313
  as psyllid predators, 298
  as scale predators, 309
Phytocoris americanus, 288t
Phytocoris breviusculus, 30f,
  309
Phytocoris californicus,
  47–49
Phytocoris calli, 287
Phytocoris canadensis, 124t,
  302, 304
Phytocoris conspurcatus, 304
Phytocoris dimidiatus, 313
Phytocoris eximius, 304
Phytocoris intricatus, 299t
Phytocoris longipennis
  dispersal to herbs by, 194f
  as mite predator, 293
  molting of, 76f
Phytocoris michiganae,
  227–28
Phytocoris neglectus, 314
  as mite predator, 292
  as psocid predator, 313
  as psyllid predator, 298
Phytocoris nigrifrons, 287
Phytocoris obscuratus, 213t
Phytocoris olseni, 226
Phytocoris pini, 313
Phytocoris populi, 313
Phytocoris relativus, 47, 260
Phytocoris reuteri, 313
Phytocoris salicis, 19f
Phytocoris tiliae, 286, 287,
  330
Phytocoris tillandsiae, 150
Phytocoris ulmi, 39, 138t,
  139
Phytocoris vanduzeei, 133
Phytocoris varipes, 194, 199
Phytodecta olivacea, 284t,
  286
Pieris rapae, 289
Piesma quadrata, 152
Pilophorus, 12t, 332
  as aphid predators, 302,
    304–5
  hosts of, 40
  mimicry and, 102
  as psyllid predators, 298
Pilophorus amoenus, 20f,
  101, 227
Pilophorus cinnamopterus,
  309
Pilophorus clavatus, 259,
  309, 332
Pilophorus confusus, 305
Pilophorus crassipes, 322t
Pilophorus furvus, 41
Pilophorus juniperi, 30f, 309
Pilophorus perplexus
  as aphid predator, 304–5
  as dipteran predator, 289
  egg of, 33f
  as honeydew feeder, 271
  as mite predator, 292

  as psyllid predator, 298
  as scale predator, 309, 310
Pinalitus, 38, 228
Pinalitus approximatus, 194,
  224t
Pinalitus cervinus, 228
Pinalitus coccineus, 38
Pinalitus conspurcatus, 226t
Pinalitus viscicola, 38, 228
Pineus strobi, 302
Pinophylus carneolus, 38,
  227, 234t
Pithanus maerkelii, 55f, 297t
Placochilus, 332
Placochilus seladonicus, 191
Plagiognathus, 12t, 191, 331
  as alfalfa pests, 203
  as aphid predators, 303t,
    304
  dispersal to herbs by, 194f
  as inflorescence feeders,
    192t, 193t, 225, 226t
  leaf injury from, 164
  as legume feeders, 204t
  as nectar feeders, 231
  as shrub feeders, 223–24t,
    225
  as strawberry pests, 266–67
Plagiognathus albatus, 165,
  166
Plagiognathus albifacies,
  192t
Plagiognathus arbustorum
  as aphid predator, 304
  cannibalism of, 315
  mating of, 53f
  pollination and, 238t
  as psyllid predator, 298
  spirea flowers and, 225
Plagiognathus brevirostris,
  192t
Plagiognathus caryae, 322t
Plagiognathus chrysanthemi
  leaf injury from, 163t
  nymphal development of,
    75, 76
  as pear pest, 259
  preoviposition period of, 57
  as scavenger, 320
  as trefoil pest, 205
Plagiognathus cornicola,
  193t, 223–24t
Plagiognathus cuneatus,
  191
Plagiognathus delicatus, 226
Plagiognathus gleditsiae, 226
Plagiognathus guttulosus,
  227
Plagiognathus laricicola,
  285t
Plagiognathus luteus, 223t
Plagiognathus medicagus,
  203
Plagiognathus negundinis,
  303t
Plagiognathus nigronitens,
  192t
Plagiognathus obscurus, 195,
  322t

Subject Index

Note: Page numbers with an *f* indicate figures; those with a *t* indicate tables.

About the Author

Alfred G. ("Al") Wheeler Jr., a native of Peru, Nebraska, received a B.A. in biology from Grinnell College and a Ph.D. in entomology from Cornell University. He spent 25 years as a survey entomologist for the Pennsylvania Department of Agriculture and is currently adjunct professor in the Department of Entomology at Clemson University and adjunct professor in the Department of Entomology at Pennsylvania State University. He is coauthor (with Thomas J. Henry) of *A Synthesis of the Holarctic Miridae (Heteroptera)*. His scientific honors include the Entomological Society of America's (ESA) Distinguished Achievement Award in Regulatory Entomology and the ESA Eastern Branch's L. O. Howard Distinguished Achievement Award.